International Symposium on Food and Agro-biodiversity (ISFA 2017)

Developing Sustainable Agriculture and Food Production

IOP Conference Series: Earth and Environmental Science
Volume 102

Semarang, Indonesia
26 - 27 September 2017

Editor:

Ahmad N. Albaarri

ISBN: 978-1-5108-5771-1
ISSN: 1755-1307

Printed from e-media with permission by:

Curran Associates, Inc.
57 Morehouse Lane
Red Hook, NY 12571

proceedings.com

Some format issues inherent in the e-media version may also appear in this print version.

Printed by Curran Associates, Inc. (2018)

For permission requests, please contact the Institute of Physics
at the address below.

Institute of Physics
Dirac House, Temple Back
Bristol BS1 6BE UK

Phone: 44 1 17 929 7481
Fax: 44 1 17 920 0979

techtracking@iop.org

Additional copies of this publication are available from:

Curran Associates, Inc.
57 Morehouse Lane
Red Hook, NY 12571 USA
Phone: 845-758-0400
Fax: 845-758-2633
Email: curran@proceedings.com
Web: www.proceedings.com

Table of contents

Volume 102

International Symposium on Food and Agro-biodiversity (ISFA) 2017

26–27 September 2017, Semarang, Indonesia

Accepted papers received: 13 December 2017
Published online: 31 January 2018

Preface

Papers

IOP Conf. Series: Earth and Environmental Science **102** (2017) 011001 doi:10.1088/1755-1315/102/1/011001

Preface

Thanks for the supports given to **The 2nd International Symposium on Food and Agro-biodiversity**. The symposium has been held successfully on 26-27 September 2017 at Grand Candi Hotel, Semarang, Indonesia. It is the second symposium after the first one in 2014, and we plan to continue the ISFA as 3-yearly program in the future.

This year's theme is **'Developing Sustainable Agriculture and Food Production'**. The symposium has been honored by the attendance of 4 keynote speakers: Prof. Hayakawa from Kagawa University, Japan; Prof. Cheng Neng-chan from Guangdong Institute of Eco-environmental Science and Technology (GIEST), China; Dr. Bessie M. Burgos from The Southeast Asian Regional Center for Graduate Study and Research in Agriculture (SEARCA), The Philippines; and lastly Dr. Agung P. Murdanoto from PT. Rajawali Nusantara Indonesia (RNI).

The committee has seen a very big interest to the symposium and finally accepted 125 abstracts after careful selection, in which 105 papers have been presented in parallel session and 20 papers in the format of poster. There are 6 international participants, from China and Philippine, while the rest are from many universities, research agencies and government institutions across Indonesia.

Selected papers from this symposium have been through rigorous review by the scientific committee and resulted in 101 papers submitted to **IOP Conference Series: Earth and Environmental Sciences (EES)** and form this proceeding.

On behalf of the symposium organizers, we would like to express our sincerest gratitude to all editors, and all reviewers for their contributions and dedications for the development of this proceeding. We also thank all authors for the trust and support given to the symposium. Last but not least, we thank IOP publisher for their help on publishing this special work.

Below are some interesting pictures from the symposium:

IOP Conf. Series: Earth and Environmental Science **102** (2017) 011001 doi:10.1088/1755-1315/102/1/011001

IOP Conf. Series: Earth and Environmental Science **102** (2017) 011001 doi:10.1088/1755-1315/102/1/011001

Editor

Ahmad N Albaarri, PhD
Diponegoro University
Indonesia

IOP Conf. Series: Earth and Environmental Science **102** (2017) 011002 doi:10.1088/1755-1315/102/1/011002

International Symposium on Food and Agro-biodiversity (ISFA) 2017

Organizing Committee

- Prof. Dr. Anang Mohamad Legowo (Chairperson)
- Yoga Pratama, M.Sc (Secretary)
- Ahmad Nimatullah Al-Baarri, PhD.
- Dr. Yoyok Budi Pramono
- Dr. Heni Rizqiati
- Bhakti Etza Setiyani, MSc.
- Siti Susanti, PhD
- Dr. Siwi Gayatri
- Tutik Dalmiyatun, M.Sc
- Bagus Herwibawa, M.Sc.

Technical committee

- Ana Wijayanti
- Arif Fauzi
- Danang
- Desi W.
- Desita Kencana Sari
- Diah R.
- Digna Rahma P.
- Endra Fajar Pratiwi
- Fatimatuz Zahra
- Hidayatul Munawaroh
- Indri Iriana Putri
- Khairunnisa Febriani
- Maria W.
- Muhammad Aidi Helyanda
- Muhammad Ilham P.
- Nur Wahyuningsih Apriliana
- Oktavia Winda Dewi
- Puspa Irzalifa
- Rani Oktaviani Saraswati
- Reeno Perfecta Gennio
- Remaylina Nurani
- Reza Siskana Lia
- Sekar Anggita Cahyani
- Tania Adinda
- Wita Friyanti Harianja
- Yolanda Yunivia

IOP Conf. Series: Earth and Environmental Science **102** (2017) 011003 doi:10.1088/1755-1315/102/1/011003

Peer review statement

All papers published in this volume of *IOP Conference Series: Earth and Environmental Science* have been peer reviewed through processes administered by the proceedings Editors. Reviews were conducted by expert referees to the professional and scientific standards expected of a proceedings journal published by IOP Publishing.

Soil problems in China and its lessons for other developing countries

N Chen[1], D W Widjajanto[2*], Y Zheng[1]

[1]Guangdong Institute of Eco-environmental Science and Technology, Guangzhou, China
[2]Department of Agriculture, Faculty of Animal and Agricultural Sciences, Diponegoro University, Semarang, Indonesia

Email: dwwidjajanto@gmail.com

Abstract. With the rapid growth of economy in China for more than 35 years since 1980, the soils in China experienced severe chances featured in high input of chemical fertilizers indeed of organic fertilizers to pursue high yield for the ever-growing population and the increase of life style and in high input of a large amount of metals especially cadmium from manure and atmospheric deposition. The shift of fertilizer application pattern and the high-yield output greatly change the soil quality, of which soil acidification is one of the main problems. Soil acidification and high cadmium input not only caused pollution on the soil but also contaminated the food. Cadmium with high percentage based on the strict soil quality standard (cadmium 0.3 mg/kg for soil, 0.2 mg/kg for rice grain). This paper will elucidate the soil pollution process in China during this 35 years and evaluate the soil and food problem properly and it may give a lesson for other developing countries when they are pave the way of modernization.

Keywords *: soil, cadmium pollution, acidification, food safety, China*

1. Introduction

China, a country with big population and limited land resources is highly dependent on soil for food production. Thanks to "green revolution", food production in China has increased several folds now when compared in 1980s. However, large amount of fertilizer and pesticide application have greatly changed the soil properties; what is more, large amount of pollutants mainly from industrial emission have made the situation to worse, resulting in high rate of occurrence of cadmium and other metal tainted rice and pose a health threat to human in some polluted areas.

2. Soil acidification due to fertilization during the 40 years

Farmland soil in China has been acidified due to large amount of fertilizer (especially N-fertilizer) input. Soil acidification is now a major problem in soils of intensive Chinese agricultural systems, It has been estimated that Soil pH declined in the range of 0.13-0.80 units from the 1980s to the 2000s in the major Chinese crop-production areas [1], which means the soil acidity has increased 6 times in farmland. Another study in Guangdong Province indicated all the soil has decline from an average of 5.70 to 5.44 (n=24671) in Pearl River Area [2].

Acid mine drainage (AMD) is another source of acids to deteriorate soil in some mining areas. A case in point is Dabaoshan mine, where the long-term irrigation of AMD has made the soil pH be 4.0 in a village 16 km away from the mining area [3].

Figure 1. Tailing of Dabashan mine with AMD

3. Large amount of metals enter farmland during these 40 years

Farmland soil in China featured in low cadmium with an average concentration of 0.097 mg/kg based on a national soil investigation in 1980s [4]. However large amount of metals has entered the farmland by ways of air deposition and livestock manure application. Study showed more than five, two and nine provinces emitted 100-500t of cadmium in 2000, 2005, 2010, respectively. The sector of cadmium source was mainly zinc smelting, followed by coal combustion, with the former to be 1681t, the later 303t, and total in 2086t in China in 2010 [5]. It was calculated that the amount of cadmium, arsenic and chromium that enter the farmland were1417t, 5925t and 17010t, respectively. In the case of cadmium, major pollution source comes from livestock manure application (55%), followed by air deposition (35%), then by chemical fertilizer (8%, mainly from compound fertilizer 6%), and wastewater irrigation (2%). Only about 13% of entered cadmium was removed by harvest, run-off and other ways and 87% kept left in the farmland. A net input of cadmium was 1239t with a concentration increase of 0.004 mg/kg [6]. Further calculation indicated after 50 years the cadmium concentration in farmland will exceed the current soil quality standard form cadmium (0.3mg/kg) if the pollution rate keeps unchanging. Fortunately, the pollution speed has been slowing down quickly since the 12th Five Year Plan (2011-2015) on Environmental Protection was taken.

4. High rate of occurrence of cadmium tainted rice in some polluted area

Rice is a plant easily contaminated by cadmium during its wet-dry production environment. During the flooded period, CdS is formed in highly reduced condition and Cd^{2+} activity is very low, however, CdS will be oxidized to sulfate and Cd^{2+} in aerobic condition and cadmium can be uptake quickly into rice plant and into grains especially in the reproductive stage.

Based on the aboved statement, it may be concluded that farmland soil in China is characterized with acidic condition and high anthropogenic source of metals. This combination makes metals have high activities for the "foreign" source of cadmium cannot be aging into forms with low activity.

As a country with rice is a staple food of the population, it can be understood if in some polluted areas such as in Hunan Province rice was contaminated with cadmium. Therefore, potential health risk is posed for the time being. Based on the current hygiene standard for cadmium (0.2mg/kg), it was estimated that about 10% of rice has been contaminated with cadmium in China, and about 50% of rice tainted in some heavily polluted areas as in Hunan province.

5. Countermeasures is now being taken to control the pollution and to remediate the polluted soils

After the release of the Bulletin of National Soil Pollution Survey in 2014 [7]. On May 31, 2015 the Action Plan on Prevention and Control of Soil Pollution [8]—10 chapters with 231 detailed measures dealing with monitoring, pollution prevention and remediation to be used in drafting laws and regulations was released by the State Council. The modification of out-of-dated soil heavy metal standards and the establishment of New Chinese Soil Pollution Law are under way. It is expected that 90 percent of contaminated farmland be made safe by 2020, with an increase to 95 percent by 2030.

6. Lessons for Other Developing Countries

The 40 years' economic reform did make China a big success in economics, however, the soil and the total environment was degraded to some extent. China failed to take the lesson from Japan where cadmium pollution in paddy soil results in *itaiitai* disease and paid more attention to the protection of soil environment. Even though the pollution in soil in China have not brought obvious health threat, but the story above did provide a good lesson for other developing countries when they pave the way of modernization.

References

[1] Guo J H, Liu1X J, Zhang Y, Shen J L, Han W X, Zhang W F, Christie P, Goulding K W T, Vitousek P M and Zhang F S 2010 Significant acidification in major Chinese

croplands *Science* **327**1008-1

[2] Zhi-xing G, Jing W and Min C 2011 Spatiotemporal variation of soil pH in Guangdong Province of China in past 30 years *Chinese J. of Applied Ecology* **22** (2) 425—430 (in Chinese with English abstract)

[3] Yan-ling X, Neng-chang C, Sheng-guang X 2009 Breeding rice cultivars with low accumulation of cadmium: cultivars versus types. *J. of Agro-Environment Science* **28** (7) 1346-1352 (in Chinese with English abstract)

[4] China National Environmental Monitoring Center. Element background values of soils in China (In Chinese) Beijing: China Environmental Science Press 1990

[5] Shao X, Cheng H, Li Q and Lin C2013 Anthropogenic atmospheric emissions of cadmium in China *Atmospheric Environment* **79** 155-160

[6] Luo L, Y Ma, S Zhang and D Wei 2009 An inventory of trace element inputs to agricultural soils in China *J. of Environmental Management* **90** 2524–2530

[7] Bulletin of National Soil Pollution Survey http://www.zhb.gov.cn/gkml/hbb/qt/201404/t20140417_270670.htm

[8] The Action Plan on Prevention and Control of Soil Pollution http://www.gov.cn/zhengce/content/2016-05/31/content_5078377.htm

IOP Conf. Series: Earth and Environmental Science **102** (2017) 012002 doi:10.1088/1755-1315/102/1/012002

Effect of various solvent on the specific amino acids of black soybean (*Glycine soja*) sprout

B Kanetro[*], A Slamet, and A Wazyka

Food Technology Department, Mercu Buana University of Yogyakarta, Indonesia

Email: bayu_kanetro@yahoo.co.id

Abstract. The objective of this research was to study the effect of various solvent extractions on the specific amino acids as small peptide or free amino acids that was contained in the extract after removal of the macromolecule protein of black soybean sprouts. The experimental design of this research was randomized complete design with one factor, which was the three various solvent, *i.e.* hexane, ethanol and water. The black soybean seed was germinated for 36 h. The small peptide and free amino acids of black soybean sprout were isolated at 3 various of solvents extraction, and then the macromolecule proteins in the extracts were precipitated at the pH 4. The extracts of black soybean sprout after removal of the macromolecule protein were analysed by HPLC to determine the profile of amino acids for stimulation of insulin secretion. The result of this research showed that the extracts contained the small peptide and free amino acid for stimulation of insulin secretion. The best solvent extraction was water that was due to the content of Leu, Arg, Ala, Phe, Ile, and Lys of water extract was higher than hexane and ethanol extracts.

***Keywords**: solvents, amino acids, insulin, black soybean, sprout*

1. Introduction

Soybean seed has popularly known as functional food for diabetes, it is due to its ability to reduce blood glucose. There were two hypoglycaemic components of soybean seed. The first component is trypsin inhibitor, however the consumptions of food that contain trypsin inhibitor may cause pancreatic hypertrophy. The second component is specific amino acids that are known as insulin stimulation. Amino acids may influence insulin secretion via a number of possible mechanisms, including generation of metabolic coupling factor, depolarization of the plasma membrane, or enhancement of mitochondrial function [11,3]. However the availability of amino acids in soybean seed naturally combined with each other to make long chains in the form of protein-macromolecule complex. The acceleration of amino acids in the form of macromolecule to be utilized by pancreatic β-cell for stimulation of insulin secretion may be lower than small peptide and free amino acids.

Seed germination increased protease activity that could degrade protein, hence the small peptide and free amino acid in soybean sprout may be higher than the original seed [1]. Based on in vivo and in vitro bio-assay, the specific amino acids were known as stimulation of insulin secretion such as Leucine/Leu, Lisoleucine/Ile [13,16,], Arginine/Arg [8,16], Alanine/Ala, Phenylalanine/Phe, Lysine/Lys [14,4,15]. Therefore the consumption of food that contains the specific amino acids as insulin stimulation may accelerate an increase in plasma insulin [12] which was useful for diabetes therapy. The research had showed that soybean sprouts protein could increase stimulation of insulin secretion due to the specific amino acids, such as Leu and Arg were higher than the seeds [6]. The profile of amino acid for stimulation of insulin secretion in the water extract of local legumes sprout (winged bean/*Psophocarpus tetragonolobus*, velvet bean/*Mucuna pruriens*, cowpea/*Vigna unguiculata*, and soybean/*Glycine max*) had been determined in the previous research [7]. This research was conducted to know what the black soybean sprouts contain the small peptide and free amino acids for stimulation of insulin secretion, such as amino acids Leu, Arg, Ala, Phe, Ile, and Lys.

Amino acids are carboxylic acids which have amino group and functional or R group or side chain in the structure. Based on the side chain of amino acids are classified into amino acids with non polar or hydrophobic side chains, polar but neutral side chains, polar but negative or acidic side chains, and polar but positive or basic side chains [9]. There are amino acids for stimulation of insulin secretion that are classified into non polar side chain amino acids, such as Leu, Ile, Ala, and Phe. While the others are classified into basic amino acids, such as Arg and Lys. The objective of this research was to study the effect of various solvent extraction on the amino acids Leu, Arg, Ala, Phe, Ile, and Lys in the extract after removal of the macromolecule protein of black soybean sprout.

2. Material and methods

The black soybean was obtained from Research Center of Legumes, Malang, Indonesia. Chemical agents such as amino acids standard, OPA (Ortho Phentaldialdehyde) were purchased from Sigma, Fluka and Merck Chemical Co.

2.1. Preparation of the extracts of black soybean sprout in various solvent

Black soybean (*Glicine soja*) seeds were soaked for 8 h, and then germinated for 36 h. Germination of black soybean for 36 h was selected for producing soybean sprout. The black soybean sprout was dried at 50°C and then was milled in order to be flour. The extracts of black soybean sprout were prepared by mixing the flour and 3 various solvents, that were hexane, ethanol or water at 40°C, 15 minute and the ratio of the flour and solvent was 1:10. After centrifugation, the macromolecule protein in the extracts were precipitated at pH 4 and centrifugation. Then the extracts of black soybean sprout after removal of the macromolecul protein were analysed by HPLC to determine the profile of amino acids [6]

2.2. Preparation of sample and amino acids standar solution for HPLC analysis

Amino acids profile was analyzed by OPA (ortho phentaldialdehyde) method [2] with modification using HPLC (High Performance Liquid Chromatography). Ethanol and hexane in sample extracts were heated to evaporate the organic solvent. 5ml of water extract was added 5ml 12N HCl, while 5 ml of hexane and ethanol extracts were added 5ml 6N HCl, and then were hydrolyzed by autoclafe at 110°C, 12 h. After that, the samples were cooled and neutralized by adding 6N NaOH. Add 2.5 ml 40 % Pb-acetate and 1 ml 15 % oxalate acid and then was fixed 50 ml with aquabides. Take 3 ml and filtered by millex 0,45 μm, and then mixed with OPA solution for 3 minute. The 30 μL sample solution was injected to HPLC.

Take 50 μL 2.5 ppm amino acid standard solution (Leu, Arg, Ala, Phe, Ile, and Lys) and then was mixed with 950 μL OPA solution for 3 minute. The 30 μL standar solution was injected to HPLC. The conditions of HPLC of the analysis were Column: Eurospere 100-5 C18, 250x4,6mm with precolumn P/N: 1115Y535; Eluen: A = 0.01M Acetic buffer pH 5.9; Eluen B = MeOH: 0.01M Acetic buffer pH 5.9:THF 80:15:5; λ Fluorescence: Ext = 340 nm; Em = 450 nm.

2.3. *Experimental design and data analysis*

The experimental design of this research was randomized complete design with 1 factor, which was the three various solvent, *i.e.* hexane, ethanol and water. The experiment was done with 2 replication. Statistical analysis by Analysis of Varian and Duncan Multiple Range Test with 0.05 significant level

3. Result and discussion

The analysis of amino acids profile of the sample of black soybean sprout extracts were preceded by preparation of sample to hydrolyse small peptide and liberate amino acids. After that the sample could be injected to HPLC. The amount of free amino acids in samples could be known by comparison of retention time of samples chromatogram with standard chromatogram of amino acids. Based on the chromatograms could be determined the amount of amino acids in the extract of black soybean sprout with the various of solvent after removal of the protein, that was seen at Table 1. The chromatograms of this analysis could be seen at Fig. 1a for chromatogram of amino acids standard and Fig. 1b, 1c, and 1d for chromatogram of sample that was extracted by solvent of hexane, ethanol, and water respectively.

Table 1. The amount of amino acids for stimulation of insulin secretion in the extract of black soybean sprout with the various solvent (hexane, ethanol, and aquadest) after removal of the protein.

Amino acids	The amount of amino acid (µg/ml of extract) *)		
	Hexane	Ethanol	Water
Alanine (Ala)	4.40ᵃ	30.80ᵇ	360.00ᶜ
Arginine (Arg)	1.40ᵃ	5.80ᵇ	614.40ᶜ
Phenylalanine (Phe)	1.80ᵃ	7.40ᵇ	192.00ᶜ
Isoleucine (Ile)	1.40ᵃ	8.00ᵇ	216.00ᶜ
Leucine (Leu)	2.20ᵃ	10.40ᵇ	312.00ᶜ
Lycine (Lys)	4.80ᵃ	15.20ᵇ	480.00ᶜ

*) The different alphabetic notation after the data in the table showed significant differences at the same row and the different column.

Data in Table 1 showed that the extracts of black soybean sprout after removal of the protein still contained amino acids for stimulation of insulin secretion, indicating that the extract that was the waste of protein isolation had the potency as the source of amino acids. Protein isolate is raw material of meat analog. Usually, the extract after removal of the protein in the processing of protein isolate as raw material of meat analog is not to be utilized. The extract is the waste of this processing. This research also indicated if the protein isolate as raw material of meat analog was made of germinated legumes, the extract would contain small peptide and free amino acids. That was due to germination could increase protease activity and then proteins are hydrolysed by proteases, resulting in several small peptides and free amino acids [1]. Hence, amino acids for stimulation of insulin secretion of the extract could be recovered and processed as the side product of protein isolation.

This research also showed that the best solvent extraction was water it was due to the content of Leu, Arg, Ala, Phe, Ile, and Lys of water extract was higher than hexane and ethanol extracts (Table 1). This phenomena indicated that the solubility of small peptides and free amino acids in the water solvent was higher than the hexane and ethanol solvent, although some amino acids (Leu and Ile) are known as non polar amino acids. That may be due to the mechanism of free amino acids in the extracts was trapped in the small peptide that was soluble in the water solvent. The research had been showed model the amino acid solubility in water with very good precision, including L-tyrosine, L-leucine, L-aspartic acid, L-tryptophan, L-glutamic acid, L-alanine, DL-alanine, DL-valine, L-phenylalanine, DL-serine, L-proline, L-serine, and glycine [5]. Hydrophilic peptides containing more than 25% charged residues of amino acids and 25% hydrophobic amino acids are usually soluble in water or aqueous buffers. Peptides containing 50% and more hydrophobic residues might be insoluble or only partly soluble in aqueous solutions. Researchers should always attempt to dissolve peptides in sterile water first, especially when the peptide contains less than five residues of amino acids. Peptide solubility is also effected by electron acceptor and electron donor solvent properties [10].

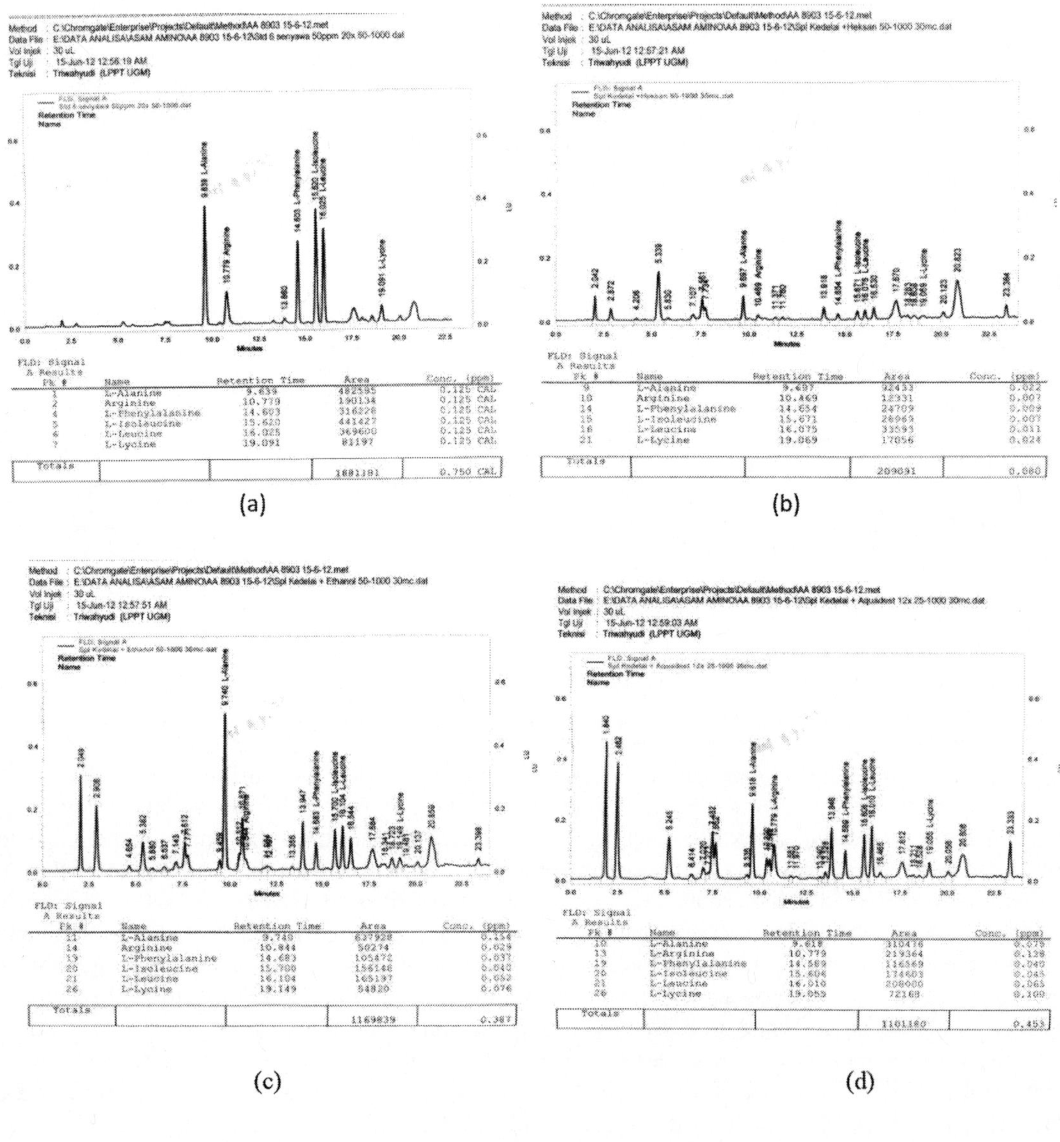

Figure. 1. Chromatogram of amino acid for stimulation of insulin secretion: (a) standard amino acid Ala, Arg, Phe, Ile, Leu, and Lys; Extract of black soybean sprout after removal of the protein in (b) hexane (c) ethanol and (d) water solvent

stimulation of insulin secretion was water that was due to the content of Leu, Arg, Ala, Phe, Ile, and Lys of water extract was higher than hexane and ethanol extracts. Hence, the extract could be processed as the side product of protein isolation which was useful for diabetes therapy.

References

[1] Ahmed F A D, Rahim E A M A, Fatah O M A, Erdmann V A and Lippmann C 1995 The changes of protein pattern during one week of germination of some legumes seeds and roots *Food Chemistry* **52** 433-437

[2] Antoine F R, Wei C I, Little R C and Marshall M R 1999 HPLC method for analysis of free amino acids in fish using o-ptaldehyde precollumn derivatization. Journal of Agriculture *Food Chemistry* **47** 5100-5107

[3] Argmann C and Auwerx J 2006. Insulin secretion: SIRT 4 gets in on the act. *Cell* **126** 837-839.

[4] Calbet J A L and Maclean D A 2002 Plasma glucagons and insulin responses depend on the rate of apperance of amino acids after ingestion of different protein solutions in human. *Journal of Nutrition* **132** 2174-2182

[5] Ji P and Fang W 2008 Solubility of amino acids in water and aqueous solutions by the statistical associating fluid theory. *Industrial and Engineering Chemistry Reserach* **47** 6275–6279

[6] Kanetro B, Noor Z, Sutardi and Indrati R 2008The Potential of germinated soybean protein to stimulate insulin secretion in the pancreas of normal and diabetic rats *Agritech* **28** 50-57

[7] Kanetro B and Setyowati A 2013 Profile of amino acid for stimulation of insulin secretion in the extract after protein removal of local legumes sprout. *Agritech* **33** 258-264

[8] Kim S W, Mcpherson R L and Wu G 2004 Dietary arginine supplementation enhances the growth of milk-fed young pigs. *Journal of Nutrition* **134** 625-630.

[9] Lieberman M, Allan D M, Colleen M S, and Dawn B M 2007. *Marks' essential of medical biochemistry a clinical approach.* (Philadelphia: Lippincott company) p 49-56

[10] Malayolta L, Pinto M R S, Cuvero J H, and Nakaie, C R 2006 Interpretation of the dissolution of insoluble peptide sequences based on the acid-base properties of the solvent *Protein Science* **15**1476-1488

[11] Newsholme P, Brennan L and Bender K 2006 Amino acid metabolism, β-cell function, and diabetes *Diabetes* **55** S39-S47

[12] Newsholme P, Brennan L and Bender K 2007 Amino acid metabolism, insulin secretion, and diabetes *Biochemical Society Transactions* **35** 1180-1186.

[13] Sans M D, Tashiro M and Vogel N L 2006 Leucine activates pancreatic translational machinery in rats and mice through mTOR independently of CCK and insulin 1-3 *Journal of Nutrition* **136** 1792-1799.

[14] Vanloon L J, Kruijshoop M, Verhagen H, Sarris W H M and Wagenmakers A J M 2000 Ingestion of protein hydrolysate and amino acid-carbohydrate mixture increases postexcercise plasma insulin response in men *Journal of Nutrition* **130** 2508-2513

[15] Vanloon L J, Kruijshoop M, Menheere P P, Sarris W H M, Wagenmakers A J M and Keizer, H A 2003 Amino acids ingestion strongly enhances insulin secretion in patients with long -term type 2 diabetes. *Diabetes Care* **26** 625-630

[16] Yang J, Wong R K, Park M J and Wu J 2006 Leucine regulation of glukokinase and ATP synthase sensitizes glucose-induced insulin secretion in pancreatic β-cells *Diabetes* **55** 193-201

The browning value changes and spectral analysis on the Maillard reaction product from glucose and methionine model system

A.N. Al-Baarri[1*], A.M. Legowo[1], Widayat[2]

[1]Food Technology Department, Faculty of Animal and Agricultural Sciences, Diponegoro University, Indonesia
[2]Chemical Engineering Department, Faculty of Engineering, Diponegoro University, Indonesia

*email: albari@undip.ac.id

Abstract. D-glucose has been understood to provide the various effect on the reactivity in Maillard reaction resulting in the changes in physical performance of food product. Therefore this research was done to analyse physical appearance of Maillard reaction product made of D-glucose and methionine as a model system. The changes in browning value and spectral analysis model system were determined. The glucose-methionine model system was produced through the heating treatment at 50°C and RH 70% for 24 hours. The data were collected for every three hour using spectrophotometer. As result, browning value was elevated with the increase of heating time and remarkably high if compare to the D-glucose only. Furthermore, the spectral analysis showed that methionine turned the pattern of peak appearance. As conclusion, methionine raised the browning value and changed the pattern of spectral analysis in Maillard reaction model system.

Keywords: D-glucose, methionine, browning value, spectral analysis, Maillard reaction

1. Introduction

Maillard reaction has been widely known as a three step of chemical reaction involving both carbonyl groups of reducing sugars and free amino groups of proteins that contributes to a very important role in food industry that applies thermal process. These three steps are early, intermediate and final step [1]. Maillard reaction has been recognized as critical effects on food properties, such as colour, texture, and flavour that might provide attractive odour such as caramel-like and burning [2]. The generation of these senses are rely on the applied amino acid and sugar. The reactivity of Maillard reaction was reported as sugar variety dependent [3, 4] and applied of protein [5].

Besides their generation in specific odour, Maillard reaction provides beneficial functions as antioxidant [6]. The utilization of glucose might generate the production of 3-deoxy-glucosone, as one of the intermediate products of the Maillard reaction. This compound acts as a precursor of 5-hydroxymethylfurfural generation, resulting in the high antioxidant activity and brown colour of Maillard reaction product [2].

Since glucose has been widely known as ingredient in Maillard reaction, this study aimed to determine the browning value and spectral analysis in Maillard reaction between D-glucose and methionine.

2. Material and method

D-glucose was obtained from Sigma Aldrich US. Methionine was purchased from Cheil Jedang Indonesia, Co. Ltd with the purity index of 99%. Unless otherwise specified, all other chemicals were reagent grade.

2.1 Preparation of D-glucose-methionine model system

The preparation method of D-glucose-methionine model system has been adopted from other researcher with some modifications [7]. The D-glucose as a control and 1:1 D-glucose-methionine were dissolved in 400 mL of 10 mM carbonate buffer (pH 9) solution. The solution was then heated at 50°C and RH 70% for 24 hours.

2.2 Spectral analysis of Maillard reaction product

The heated samples were diluted with phosphate buffer until 200 μL in microtube to measure the emission spectrum starting from 190 to 620 nm using spectrophotometer (UV-1280; Shimadzu, Kyoto, Japan). This spectroscopic measurement method has been adopted from previous researcher [7].

2.3 Browning value measurement

The browning value of the MRPs was measured according to the method of earlier researcher [7] without any modification using spectrophotometer at 420 nm.

2.4 Data analysis

The results were reported in figures and the description upon figure was provided. The data were obtained from at least three time of replication.

3. Result and discussion

Figure 1 shows the browning value through the determination of absorbance at 420 nm on the Maillard reaction product using D-glucose and methionine model system when was heated at 50°C for 24 hour. It can be described that the increase of both sample was able to be detected clearly. D-glucose–methionine model system showed remarkably higher (about two times higher) in the absorbance value than D-glucose only. As seen on this figure, the Maillard reaction model system generating absorbance at 1.218 while sugar alone only generated the absorbance at 0.644. The higher browning value might able to be recognized starting at heating time of 9 hour that was resulting the value of 0.660 while sugar–only–sample provided about a half of absorbance (0.338).

In this research, methionine was chosen to generate the Maillard reaction in model system, due to its high reactivity [5]. D-glucose was applied as the control because it widely used in Maillard reaction process. Hexoses–amino acid system model generated it specific reactivity based on the sugar and amino acid that was used [8]. Methionine was known as the high reactivity of amino acid and it reduced at about 60% when Maillard processing was applied [9]. Spectral analysis showed similar patterns or peak appearance on the D-glucose model system when the heating time of 15 and 18 hour was applied (Fig. 2a). However, the remarkable differences in the pattern or peak appearance was occurred when the methionine was applied (Fig. 2b). This could be seen at the appearance of peak at 18-h heating time. As can be seen, no peak was detected at the D-glucose-methionine model system on 18-h heating time. Therefore, the presence of amino acid in the model system turned the spectral appearance remarkably.

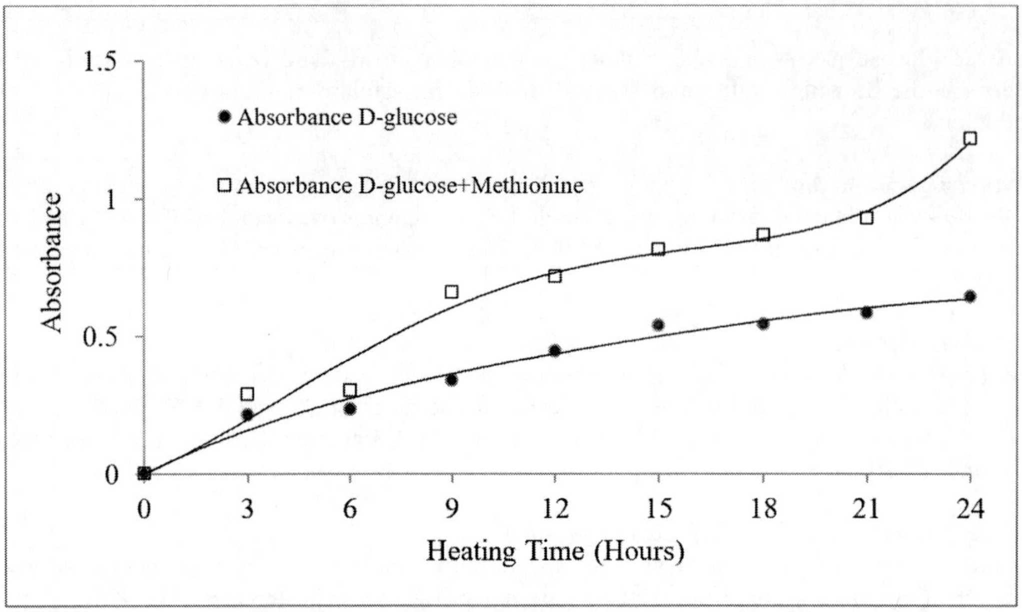

Figure 1. Browning development through the determination of absorbance at 420 nm on Maillard reaction product of D-glucose and D-glucose–methionine model system that was heated at 50°C for 24 hour.

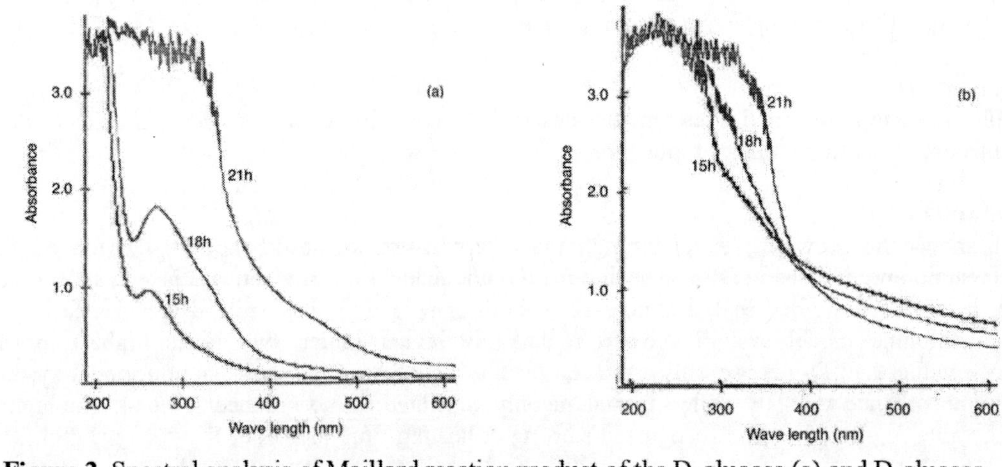

Figure 2. Spectral analysis of Maillard reaction product of the D-glucose (a) and D-glucose-methionine model system (b) that was heated at 50°C.

It was recognized that the heating process might reduce remarkably the luminosity (L*) in Maillard reaction system [10]. This reduction in the L* value was due to the formation of brown pigments from Maillard reaction that might interference the spectral analysis. Brown pigments was easily detected at 420 nm [11], therefore based on the obtained spectral analysis at 420 nm, the change on the absorbance at those wavelength might be detected with ease.

4. Conclusion

As conclusion, the browning value of D-glucose–methionine model system were higher than D-glucose model system and it could be seen starting at the 9-h of heating time. The heating time at 18 h showed the change in peak appearance and pattern of spectral analysis. This result may provide the beneficial information about Maillard reaction process during 24 hour.

5. Reference

[1]. Jiang W, He X, Yang H, Xiang X, Hu S, Li S and Liu Y 2017 Histamine reduction by Maillard reaction with glucose, Food Control 82(Supplement C) 136-144.

[2]. Asikin Y, Kamiya A, Mizu M, Takara K, Tamaki H and Wada K 2014 Changes in the physicochemical characteristics, including flavour components and Maillard reaction products, of non-centrifugal cane brown sugar during storage, Food Chemistry **149** 170–177.

[3]. Sun Y, Hayakawa S, Puangmanee and K. Izumori 2006 Chemical properties and antioxidative activity of glycated α-lactalbumin with a rare sugar, D-allose, by maillard reaction, Food Chemistry **95** 509-517.

[4]. Sun Y, Hayakawa S, Chuamanochan M , Fujimoto M, Innun A and Izumori K 2006 Antioxidant effect of maillard reaction products obtained from ovalbumin and different D-Aldohexoses, Bioscience, Biotechnology, and Biochemistry **70**(3) 598-605.

[5]. Suminar AT, Al-Baarri AN and Legowo AM 2017 Demonstration of physical phenomenas and scavenging activity from d-psicose and methionine maillard reaction products, Potravinarstvo Slovak *Journal of Food Sciences* **11**(1) 417–424.

[6]. Chen X.-M and Kitts D D 2011 Antioxidant and anti-inflammatory activities of Maillard reaction products isolated from sugar–amino acid model systems *Journal of Agricultural and Food Chemistry* **59** 11294–11303.

[7]. Arum SK, Legowo AM, Al-Baarri AN, Hayakawa S and Ogawa M 2017 Prolongation the Shelf Life of Indonesian Soft Cheese (Dangke) Derived from Bovine Milk using Lactoperoxidase System and Lysozyme, Czech *Journal of Food Science Article* in Press.

[8]. Tan T.-C, AlKarkhi AFM and Easa AM 2012 Assessment of the ribose-induced Maillard reaction as a means of gelatine powder identification and quality control, Food Chemistry **134** 2430-2436

[9]. Hou C, Wu S, Xia Y, Phillips G O and Cui SW 2017 A novel emulsifier prepared from Acacia seyal polysaccharide through Maillard reaction with casein peptides, Food Hydrocolloids **69** 236–241.

[10].Cai L, Li D, Dong Z, Cao A, Lin and Li J. 2016 Change regularity of the characteristics of Maillard reaction products derived from xylose and Chinese shrimp waste hydrolysates, LWT - Food Science and Technology **65** 908–916.

[11].Delgano-Andrade C, Seiquer I, Nieto R and Navarro MP 2004 Effects of heated glucose–lysine and glucose–methionine model-systems on mineral solubility, Food Chemistry **87** 329–337.

Quantitative risk assessment of *E. coli* in street-vended cassava-based delicacies in the Philippines

I C P Mesias

Department of Food Science and Technology
Visayas State University, Baybay City, Leyte,Philippines

Email: inishchris.mesias@vsu.edu.ph

Abstract. In the Philippines, rootcrop-based food products are gaining popularity in street food trade. However, a number of street-vended food products in the country are reported to be contaminated with *E. coli* posing possible risk among consumers. In this study, information on quantitative risk assessment of *E. coli* in street-vended cassava-based delicacies was generated. The assessment started with the prevalence and concentration of *E. coli* at post production in packages of the cassava-based delicacies. *Combase* growth predictor was used to trace the microbial population of *E. coli in* each step of the food chain. The @Risk software package, version 6 (Palisade USA) was used to run the simulations. Scenarios in the post-production to consumption pathway were simulated. The effect was then assessed in relation to exposure to the defined infective dose. In the worst case scenario, a minimum and most likely concentration of 6.3 and 7.8 log CFU of *E. coli* per serving respectively were observed. The simulation revealed that lowering the temperature in the chain considerably decreased the *E. coli* concentration prior to consumption and subsequently decreased the percentage of exposure to the infective dose. Exposure to infective dose however was increased with longer lag time from postproduction to consumption.

Keywords: *quantitative risk assessment, street-vended cassava-based delicacies, E. coli*

1. Introduction

The Food and Agricultural Organization defines street food as a ready-to eat food and beverage prepared and/or sold by vendors on the street from pushcarts, buckets, balance pole, stalls or shops having fewer than four permanent walls. It encompasses a wide range of ready-to-eat foods and beverages prepared and/or sold by mobile or stationary vendors and hawkers especially on streets and around public places [1-5]. Such foods feed millions of people daily with a wide variety of foods that are relatively cheap and easily accessible. In selling snacks, complete meals, and refreshments at relatively low prices, vendors provide an essential service to students, workers, shoppers, travellers, and people on low income [6]. People who depend on such food are often more interested in its convenience than in questions of its safety, quality and hygiene. However, to date, it is well known that majority of the reported contamination of food that contributes to food-borne disease in countries is associated with street-vended foods [7-8]. In developing countries, since infrastructure development was relatively limited, with restricted access to potable water, toilets, refrigeration and washing and waste disposal facilities [9], food-borne illnesses are occurring and recurring.

In the Philippines, increasing pace of urbanization has resulted in proliferation of street food vendors and hawkers as the movement of people from rural to urban areas has led to the need to feed large numbers of working people away from their place of residence. Street food vending became a popular type and distinctive part of a large informal sector providing the vendors with a means to sustain their livelihood. While the importance of street-vended food is not to be negated as an important source of nourishment as well as income and work opportunity offered for the informal sector, the safety of the food is a particular problem, especially from a microbiological point of view. However, in the country, studies investigating the epidemiological link between street foods and food borne diseases, profile of street food vendors, hygiene practices as well as the microbial quality of street-vended foods were considerably limited. In the Philippines, as much as 70% of diarrhea diseases are believed to be of foodborne origin. Although the government authority is making initiatives to ensure public health and consumers safety in street-vended foods, nevertheless, strategies for efficiently improving food safety can only be developed after obtaining science-based studies on local foods, conditions and practices along each area from the production to consumption. Quantitative microbial risk assessment (QMRA) for instance is considered as a very indispensable tool in estimating the risk of a certain pathogen along the food chain. With QMRA, scientifically-supported information can be provided to help risk managers set priorities, decisions and strategies for the most effective intervention. In this paper therefore, risk assessment of *E. coli* in the street-vended cassava-based delicacies in the Philippines is presented. These delicacies are contributing a large portion in the wide range of the most commonly marketed street foods in the country.

2. Quantitative risk assessment methodology

2.1 Statement of purpose

A number of street-vended food products in the Philippines are reported to be contaminated with *E. coli* [10]. This has caused concern among regulatory agencies, particularly because a lot have been patronizing such products across population. This assessment aimed to estimate the risk to a general consumer of the street vended cassava delicacies. A simplified retail to consumption pathway was constructed as series of unit operations. The assessment started with the prevalence and concentration of *E. coli* at postproduction/retail in packages of these cassava-based delicacies and traced growth of the microbial population in contaminated product through to the point where the consumer eats a portion. Cross-contamination was not considered. This was to simulate the prevalence and levels of *E. coli* in those portions that, along with serving size, determine the size of the dose of *E. coli* that a consumer might ingest. To characterize the risk to consumers, the variables that were considered in this exposure assessment include: *E. coli* prevalence and concentration in the cassava-based delicacies, growth rate, pH, water activity and time and temperature during transport and storage.

2.2 Approach taken/ model design

A simplified post-production to consumption pathway was constructed as series of unit operations along with their associated variables for microbial events (Figure 1). The pathway comprised of initial contamination after production (node 1), growth during transport from production area to market (node 2), growth during storage at market (node 3), growth during transport from market to home (node 4), growth during storage at home prior to consumption (node 5) and dose response after consumption (node 6).

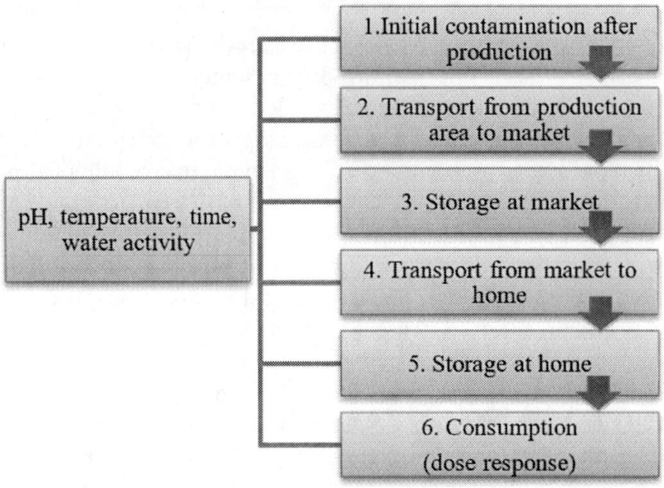

Figure 1. Simplified post-production to consumption pathway of the cassava-based delicacies alongside with the variables influencing *E. coli* growth in different node.

The input settings for initial contamination at retail in node 1 are contributed by the level of *E. coli* contamination and the prevalence of *E. coli* in cassava-based delicacies at post production. Primary data of the initial contamination of the products were used: minimum and maximum contamination of 0.48 and 1 log CFU/g respectively. In node 2 through 6, microbial growth was expected due to pH, water activity (a_w) and transport and storage variables such as temperature and time. Combase growth predictor was used to predict the growth of *E. coli* in each step of the chain. The pH value of 5.97 (primary data) and a_w of 0.97 were made constant. The time required to transport from the production area to market were presumably between 1-2.25 hours based on a typical production to retail set up in the country. A maximum time required for storage at market of 8 hours, maximum time for transport from market to home of 2.25 hours and maximum storage time at home prior to consumption of 12.25 hours were used in modelling the growth of the *E. coli*. The surrogate temperatures used were the recorded minimum and maximum ambient temperature in the Philippines, 26.9°C and 31°C respectively (*Philippine Atmospheric Geophysical and Astronomical Services Administration*). Since the available data was for generic *E. coli* and only a portion of them can be pathogenic, 4.5 and 5% of pathogenic *E. coli* specifically for ready to eat products were used to determine the number of pathogenic *E. coli* from the total concentration obtained. The fraction of pathogenic *E. coli* in street-vended cassava delicacies was obtained by multiplying the fraction of pathogenic *E. coli* by that of the total *E. coli*.

For the dose response, infective doses of 10^6 to 10^8 CFU for enterotoxigenic (ETEC), enteroinvasive (EIEC) and enteropathogenic *E. coli* (EPEC) were used. ETEC is recognized as the causative agent of travelers' diarrhea and illness is characterized by watery diarrhea with little or no fever and occurs commonly in under-developed countries. EIEC closely resemble *Shigella* and causes an invasive, dysenteric form of diarrhea in humans while EPEC causes a profuse watery diarrheal disease and it is a leading cause of infantile diarrhea in developing countries (USFDA). Pert distribution was used for the dose-response assessment with a minimum value of 10^5, most likely value of 10^6 and a maximum of 10^8. The risk on illness per serving was obtained by dividing the exposure per serving by the dose response.

2.3 Model simulation

The @Risk software package version 6 (Palisade USA) was used to run the simulations. Monte Carlo simulations were run three times using 10,000 iterations.

2.4 Sensitivity analysis

Various scenarios in the post-production to consumption pathway were simulated. The effect was then assessed in relation to exposure to the defined infective dose. First scenario evaluated was the impact of the temperature all throughout the chain. Three temperature conditions were evaluated: Philippines' average maximum temperature (29°C), average minimum temperature (22°C) and refrigeration temperature (10°C) was used since this is the lowest possible temperature in *Combase*). The second scenario dealt with the effect of lag time after production prior to consumption: after 13, 19, 25 and 37 hours.

3. Results and discussion

3.1 Frequency distribution of E. coli concentration at the time of consumption

As depicted in figure 2, in 100% of the cases, the consumers would be exposed to the infective dose of 6-8 log CFU of *E. coli* per serving. A minimum and most likely concentration of 6.3 and 7.8 log CFU of *E. coli* per serving were observed. The risk on illness per serving was 63.61. The result was anticipated in this case since aside from the initial concentration of *E. coli* contamination used, the surrogate storage and transport time/temperature combination was at its maximum or in the worst case scenario.

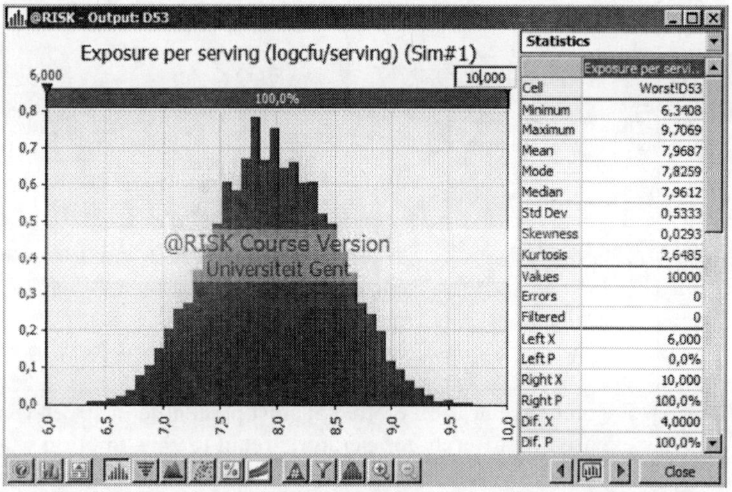

Figure 2. Frequency distribution of *E. coli* concentration in the street-vended cassava delicacies at the time of consumption. Values in x axis are in log CFU/serving

3.2 Effect of temperature on the E. coli concentration: postproduction to consumption

The effect of temperature on the growth rate and concentration of *E. coli* in the street-vended cassava delicacies was observed distinctly in the simulation. Using the three different temperature conditions, results revealed that lowering the temperature in the chain considerably decreased the *E. coli* concentration prior to consumption (Table 1 and Figure 3). Subsequently, a decrease in percentage of exposure to the infective dose of *E. coli* was attained with lower temperature (Figure 4). It should be noted that at the average minimum temperature of 22°C and at 10°C, percentage exposure to 10^8 infective dose is zero. However, for the infective dose of 10^6, 0% exposure can be attained as the temperature approaches < 5°C. This implies for the importance of storing the cassava delicacies at refrigeration temperature at any point of the chain.

Using the average maximum temperature condition in the Philippines, the risk on probability of illness was 63.61 while 1.92 for average minimum temperature.

Table 1. Effect of temperature on the exposure per serving of *E. coli* (log cfu/serving) in the street-vended cassava delicacies.

Philippine Temperature Condition	Min*	Max*	Mean*	SD	Percentile (%)*				
					25	50	75	90	95
Average Maximum Temperature (29^{0}Celcius)	6.22	9.68	7.97	0.53	7.60	7.96	8.34	8.67	8.84
Average Minimum Temperature (22^{0}Celcius)	5.30	7.52	6.45	0.34	6.21	6.45	6.68	6.89	7.01
At 10^{0}Celcius	4.69	6.94	5.83	0.34	5.60	5.83	6.06	6.26	6.38

*Min- minimum; Max- maximum; SD- Standard Deviation; * Values are in log cfu/serving*

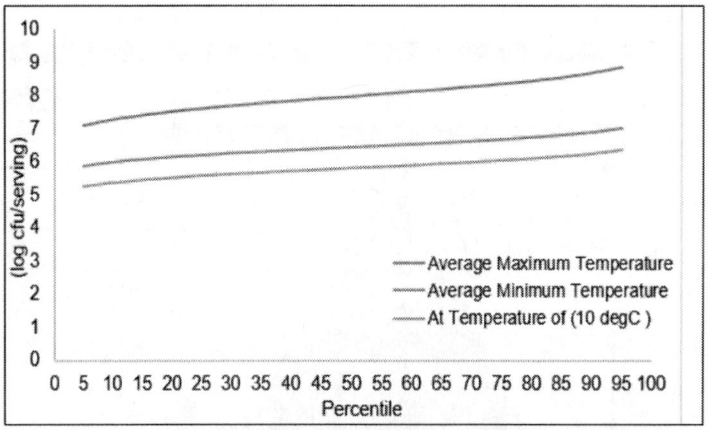

Figure 3. E. coli concentration plotted against percentile in street-vended cassava-based delicacies exposed to different temperature: retail to consumption

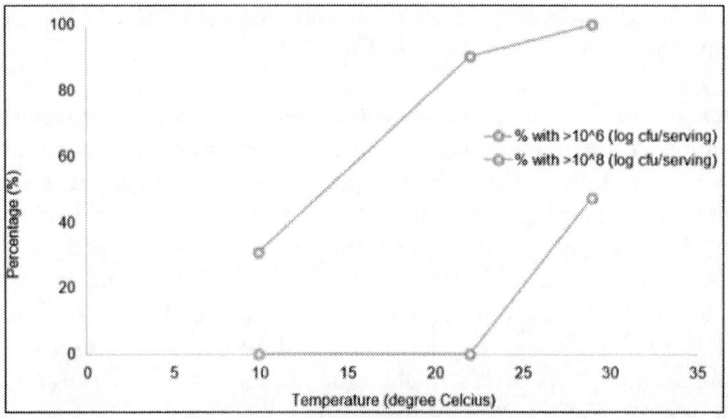

Figure 4. Effect of temperature in the exposure to the infective dose of *E. coli*

3.3 Effect of time on the E. coli concentration: postproduction to consumption

The impact of lag time from postproduction to consumption (13, 19, 25 and 37 hours) on the concentration of *E. coli* was also evaluated. Typically, cassava delicacies can last 2-7 days depending on the handling and storage condition. Based on the simulation, not much appreciable increase in the *E. coli* concentration was indicated within 24 hours and values lower than the minimum infective dose of 6 log cfu/serving were noted (Table 2 and Figure 5). On the other hand, exposure to infective dose was increased with longer lag time from postproduction to consumption (Figure 6).

Table 2. Effect of time on the exposure per serving of *E. coli* (log cfu/serving) in the street-vended cassava delicacies: retail to consumption period.

Postproduction to Consumption Period	Min*	Max*	Mean*	SD	Percentile (%)*				
					25	50	75	90	95
After 13hours	5.08	7.24	6.17	0.33	5.94	6.18	6.40	6.59	6.71
After 19hours	5.12	7.39	6.24	0.33	6.02	6.24	6.47	6.67	6.79
After 25hours	5.30	7.52	6.45	0.34	6.21	6.45	6.68	6.89	7.01
After 37hours	6.14	8.51	7.31	0.35	7.06	7.31	7.56	7.77	7.89

Min- minimum; Max- maximum; SD- Standard Deviation; * Values are in log cfu/serving

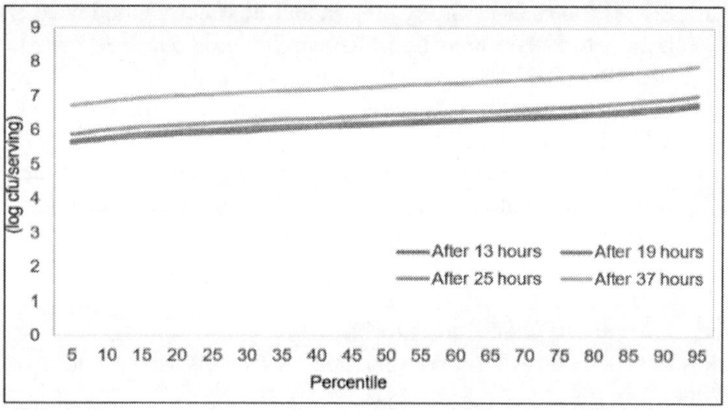

Figure 5. E. coli concentration plotted against percentile in street-vended cassava-based delicacies exposed to different time: retail to consumption.

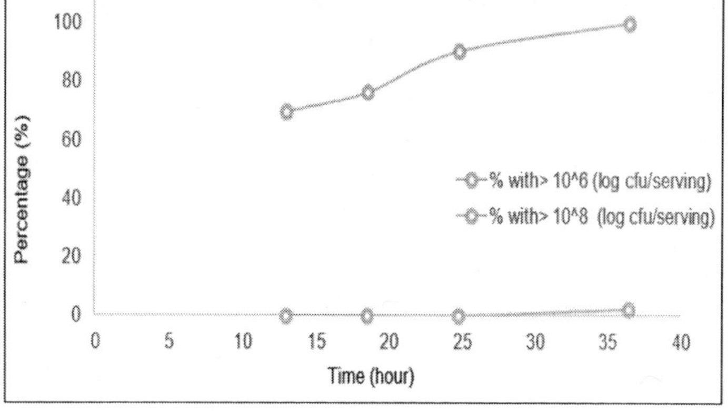

Figure 6. Effect of time in the exposure to the infective dose of *E. coli.*

4. Conclusion and recommendations

The risk assessment study revealed high exposure to the infective dose of E. coli with temperature and time abuse. Thus, both temperature and time should be controlled to reduce the risk of food borne illness from street-vended cassava delicacies. There is a need for stricter implementation of the food sanitation code. It is recommended that every food handler has to undergo proper training with regard to basic food hygiene, transport and storage of food and lag time between preparation and dispensing of food items to the consumers. Consumer awareness as to how to handle and store such products must also be promoted.

References

[1] Dardano C 2003 Caribbean regional working group on street food vendors. Report of FAO, PAHO and BNSI. Retrieved from ftp://ftp.fao.org/es/esn/food/carribean_report.pdf

[2] Nirathon N 2006 Fighting poverty from the street: a survey of street food vendors in Bangkok. international labour organization, informal economy, poverty and employment Thailand series 1. p 14-15.

[3] Faruque Q, Haque Q.F, Shekhar H.U and Begum S 2010 Institutionalization of healthy street food system in Bangladesh: A pilot study with three wards of Dhaka city corporation as model. National food policy capacity strengthening programme. Final report PR#7/07

[4] Cho K, Park S. and Joo N 2011 Study on a model of street vended food choices by Korean high school students. Nutrition Research and Practice, **5**(5), 481–488. doi: 10.4162/nrp.2011.5.5.481

[5] Monterjc 2013 Street Food-Term Paper. Retrieved from www.studymode.com/ essays/Street-Foods-1802701.html

[6] Rheinländer T, Olsen M, Bakang J.A, Takyi H, Konradsen F and Samuelsen H 2008 Keeping up appearances: perceptions of street food safety in urban Kumasi, Ghana. *Journal of Urban Health* **6** 952-64. doi: 10.1007/s11524-008-9318-3.

[7] Donkor E. S 2009 Application of the WHO five keys of food safety to improve food handling practices of food vendors in a poor resource community in Ghana. *East African Journal of Public Health*, **6**(2), 148-5

[8] Bhowmik S.K (2005) Street Vendors in Asia. A Review Economic and Political Weekly 2257-64.

[9] Muzaffar A.T 2009 Entrepreneurs of the streets: an analytical work on the street food vendors of Dhaka City. *International Journal of Business and Management*, **4**(2), 1-9.

[10] Canini N.D, Bala J.J.O, Maraginot E.N and Median B.C.B. 2013 Evaluation of Street Food Vending in Ozamiz City. *J Multidisciplinary Studies* **1**(1) ISSN: 2350-7020 doi:http://dx.doi.org/10.7828/jmds.v2i1.391

Acknowledgement

The author would like to thank the Faculty of Bioscience Engineering, Department of Food Safety and Food Quality, Ghent University, Belgium and Flemish Interuniversity Council (VLIR—UOS) for organizing the training programme on Food Safety, Quality Assurance Systems and Risk Analysis. Above all, interminable thanks are due to the Philippine Council for Industry, Energy and Emerging Technology Research and Development (PCIEERD) Human Resource and Institution Development Division for the travel grant awarded to the author.

Prospect for the development of salted egg agro industry: an analysis on marketing distribution aspect

W Sumekar[1*], A N Al-Baarri[1] and E Kurnianto[2]

[1]Department of Agriculture, Faculty of Animal and Agricultural Sciences Diponegoro University
[2]Department of Animal Science, Faculty of Animal and Agricultural Sciences Diponegoro University
Tembalang Campus, Semarang 50275 – Indonesia

*E-mail : wulan_sumekar@yahoo.co.id

Abstract. Marketing distribution is an important of the strategy in business development in agroindustries. The aim of the research was to introduce marketing (distribution pattern, margin and marketing efficiency) at the salted egg agro industries in Brebes Regency. Survey method had been conducted on 52 salted egg agro industries which had active PIRT certificate. The data collection was conducted by means of interview and observation. Descriptive analysis was used to determine the marketing distribution of salted eggs. Marketing efficiency was obtained by calculating marketing margin and farmer share. The results show that the salted egg agro industries implemented two marketing distribution patterns; direct marketing pattern (consumer→producers) and indirect marketing pattern (producer→retailer→consumer). The number of the salted egg agro industries which apply indirect marketing pattern is 57.69%. The implementation of direct and indirect marketing patterns was classified as efficient according to the farmer's share values of 87.13% and 78.21%. It can be recommended the direct marketing.

Keywords: *efficiency, marketing margin, marketing, pattern, salted egg*

1. Introduction

Competitiveness and market orientation are the factors that continue to receive attention in sustainable agricultural development. In the early 1990s, a strategy of agricultural industrialization was launched to reduce the imbalance between the growth of agricultural and industrial sector. Industrialization of agriculture is an effort to increase the added value of agricultural goods or services so as to increase the income of farmers, to be able to absorb labor, and to create new jobs [1]. One of the three principles of agricultural industrialization is rural development stimulated by agro-industry. In this principle, the importance of forward and backward linkages of industrialization process is emphasized that will accumulate generating substantial added value. The problem is that agricultural commodities are perishable, so it needs immediate handling, such as storage and processing. Producers' decision to get closer to consumers is common in primary agricultural commodities. One of livestock agro industry commodities is salted egg.

In Indonesia, Brebes Regency is known as the center of salted egg small industries because in 2004 it became the champion of sustainable development competition in the category of awareness in the management of small industry environment center. The object of the industrial center is Brebes salted egg. The driving factor is the potential availability of duck eggs in Brebes Regency which is the highest compared to other districts in duck egg production reaching 5,341,711 kg [7]. The availability of duck eggs in Brebes District varies throughout the year due to seasonal influences. In the dry season (August to October) the production of duck eggs is indicated by the price of eggs which is more expensive than that in the rainy season (November to February). The existence of local institutions has not shown optimal role so that many small entrepreneurs of salted eggs (family) get loss and/ or bankrupt. It was indicated by business uncertainty (risk) faced by craftsmen as result of long selling time (product circulation), fluctuating raw material price, and standardization of salted egg products. Long selling time allows the growth of *Staphylococcus aureus* and *Clostridium perfringens,* [11] that endanger consumers. Similarly, the marketing constraints are related to the perception of the public on the danger from excessive salt consumption. The salt content of various salted egg products may be avoided by consumers who want to reduce their salt intake (SNI-SNI Quality Standard, SNI-01-4277-1996). The tendency of reduced extensive maintenance systems in rice fields is opposed to consumer demand on salted eggs from the eggs with grazed ducks. Consumers' perception is that the price of salted egg products from the eggs with ducked ducks and grazing patterns is higher because it is preferred by consumers [8].

Salted egg agro industry in Brebes Regency is in small scale. The application of price per item varies from IDR 2,500 to IDR 3,500. 40% of the producers markets salted egg products in stores around their homes and 25% of them have no place to market [9]. The components of the curing material used depend on the scale of the business in salted egg agro industry [10]. Based on the problems, it is necessary to conduct a research with the aim of analyzing the prospects for the development of salted egg agro industry by reviewing the aspects of distribution and marketing efficiency. The benefit of this research are as a recommendation to the agroindustry of salted egg and the holder of the livestock product policy in determining the efficient marketing pattern

2. Research methods

The survey for the research on salted egg agro industry with PIRT certificate had been conducted in Brebes Regency known as a small industry center of salted eggs. The respondents consisted of producers, intermediary traders, and consumers. The number of respondents as the salted egg producers was 52 people with active PIRT certificates from 69 salted egg agro industries with PIRT certificates [3].Interview method with questionnaires and observation was used for data collection, and then they were a nalyzed. Descriptive analysis was used to determine the marketing distribution of salted eggs. The analysis of marketing efficiency was conducted by calculating marketing margin and farmer's share. If farmer' s share > 50%, it can be concluded as an efficient marketing; and if farmer's share < 50%, the marketing is not efficient [4].

3. Results and discussion

3.1. Characteristics of salted egg agroindustry

The existence of salted egg agro industry in Brebes Regency has contributed to the regional economy with a value of IDR 12,044,700,00 from 6,882,383 eggs [3].The business management is still implemented traditionally and only 23.08% of producers used business bookkeeping. It is in line with the research results of Sumekar *et al* [9] that 65% of agro-industries had relatively small capital related to the purchasing ability of duck eggs of 5,000 eggs per week on average. 69.23% of salted egg agro industries had conducted their business activity based on product orientation, that is, typical salted egg product with bright yellow color in yolk.

The salted egg products were classified as having low durability. 57.70% of the manufacturers stated that the duration of preservation is < 2 weeks. Salted eggs were produced with a saltiness diversity associated with 65.38% of manufacturers who did not have information about the salinity limit for health. Arthur *et al* [2] states that consumers buy salted eggs by considering brand name, quality, preference, health and traditionality. The thing that did not become consumer consideration when buying salted egg products is the salt content.

3.2. Marketing pattern of salted eggs

The marketing pattern of salted egg agro industry in Brebes district shows 57.69% of the respondents relied on intermediaries with retail status (Table 1) It is possible because of the short duration of durability. The constraints faced by the producers in the marketing of salted eggs through intermediaries (retailers) were the indirect payment systems (waiting after salted eggs are sold) and the risk insurance was charged to the producers.

Retailer as an intermediary in marketing of salted eggs has a status as a regular customer. 36.37% of them were from out of Brebes District, 60.61% of them had low educational level (less than 9 years) which was contradictive with 100% of consumers with the educational level over 9 years. In accordance with Arthur *et al* [2], the typical and traditional products are consumer consideration in the purchase of food

Table 1. Number of Respondents By Marketing Pattern in Salted Egg Agro Industry.

No.	Marketing Pattern	Number of Respondents	
		N	%
1.	Producer → Consumer	22	42.31
2.	Producer → Retailer → Retailer	30	57.69

3.3. Marketing efficiency in salted egg agro industry

The marketing of salted eggs in direct marketing pattern shows higher margin value (IDR 409.90 / egg) than the indirect marketing pattern (IDR 706.67 / egg) (Table 2). In accordance with Soekartawi's opinion [6], the amount of marketing margin is determined by the costs incurred by the marketing agency (intermediaries) and the expected profit.

Table 2. Marketing Margin in Salted Egg Agro Industry.

No.	Aspects in Marketing Pattern	Marketing Margin	
		Price	Margin
		---------------- IDR/egg----------------	
1.	Producer → Consumer (direct marketing pattern)		
	Average Price in Producer Level	2775.00	
	Average Price in Consumer Level	3184.00	409.90
2.	Producer → Retailer → Retailer (indirect marketing pattern)		
	Average Price in Producer Level	2536.67	
	Average Price in Retailer Level	2893.33	356.67
	Average Price in Consumer Level	3243.33	706.67

The value of efficiency in direct marketing pattern (producer → consumer) is higher than that in indirect marketing pattern (producer →retailer → consumer) as indicated from the shares received by the producers (farmer's share); 87.13% and 78.21% (Table 3). Along with the opinion of Santoso [5], the low farmer's share in the marketing pattern does not indicate the weak bargaining power of producers especially for the fragile food commodities.

Salt eggs agro industry in Brebes Regency has implemented an efficient marketing pattern, which is the share received by producers > 50%. Therefore, salted egg agro industry can be further developed. According Soekartawi [6], marketing efficiency indicates the factors of buying and selling transactions, market competition, and competition among market participants running well

Table 3. Marketing Efficiency in Salted Egg Agro Industry.

Efficiency Aspect	Efficiency in Marketing Pattern (%)	
	Producer → Consumer	Producer → Retailer → Consumer
Marketing		
Margin	12.87	21.79
Farmer's share	87.13	78.21

4. Conclusion

It is concluded that 57.69% of salted egg agro industries in Brebes district implemented indirect marketing pattern (producer → retailer → consumer). The implementation of direct and indirect marketing pattern was classified as efficient that can be seen from the farmer's shares of 87.13% and 78.21% respectively.

References

[1] Arifin, B. Agricultural development : The policy paradigm and the revilalization. Jakarta : Grasindo. 2005.

[2] Arthur J, Wisemann K and Cheng K M. Salted and preserved duck egg : A consumer market segmentation analysis. Poult. Sci. 2015. **94** (8) : 1942-1956

[3] The Animal Husbandry Office of Brebes Regency. Data of livestock products that already have 'PIRT' in Brebes Regency. 2016

[4] Ramadhani, D K , Rahayu E S and Setyowati. 2013. The analysis of marketing efficiency of maize in Grobogan Regency (case study in Geyer Sub-district). *J. Agrib.* **4** (2) : 15 - 26

[5] Santoso T I. 2014. Marketing efficiency of some vegetable commodities in Indramayu Regency. *J. Agri. Wiraloda* . **6** (2) : 9-18

[6] Soekartawi. The basic principles of agricultural economic, theory and application. Raja Grafindo Persada. Bogor. 1993.

[7] Animal Husbandry and Veterinary Office of Central Java Province. Livestock statistic of Central Java Province. 'Tarubudaya' Ungaran, Indonesia. 2014

[8] Sumekar W , Setiadi A and Rossali W. Duck farmer perception on raising pattern in Brebes regency, Central Java Indonesia. Proceedings of the 4th International Conference on Sustainable Animal Agriculture for Developing Countries, Lanzhou University, Lanzhou, China, 2013. July 27 – 31, p 442 – 443.

[9] Sumekar W , Al-Baari A N and Kurnianto E. 2016. Business Characteristic of Salted Egg In the Agro Industrial Center, Brebes, Central Java. Proceeding of the 3rd Animal Production International Seminar (3rd APIS) & The 3rd ASEAN Regional Conference on Animal Production (3rd ARCAP), Malang, 2016 19 – 21 October. p. 496-498

[10] Sumekar, W, Al-Baari A N and Kurnianto E. 2017. Application of salted egg technology based on local environment in the agroindustrial center of Brebes, Central Java. Adv. Sci. Letter. **23** (3) : 2627-2628

[11] Wongvilairat, R. Quality and control of Staphylococcus aureus and Clostridium perfingens in salted egg production. *J. NU Science* 2007. **4** (1):31-41

The potential of avocado paste (*Persea americana*) as fat substitute in non-dairy ice cream

Ervina[1*], I Surjawan[1] and E Abdillah[1,2]

[1]Food Technology Department, Indonesia International Institute for Life Sciences, Jakarta, Indonesia 13210
[2]Department of Food Technology, HAS University of Applied Sciences, 5223 DE-Hertogenbosch, Netherlands

*Email: e.ervina@i3l.ac.id

Abstract. Consumer preferences towards plant-based food have shifted significantly due to sustainable and healthy reasons. Dairy products consist of high Saturated Fatty Acid (SFA) and overconsumption of SFA could lead to cardiovascular diseases. Avocado contains high levels of fat dominated by Monounsaturated Fatty Acid (MUFA) and phytosterol that have the potential as a plant-based fat source to substitute dairy-fat in ice cream. The objective of this study was to analyze the physicochemical, rheological and sensorial properties of ice cream substituted with different concentrations of avocado paste ranging from 0%, 25%, 50%, 75% and 100% respectively against dairy fat to produce non-dairy fat ice cream. The psychochemical properties and total fat were determined. Sensorial quality and hedonic attributes of ice cream were investigated using 60 semi-trained panelists. There were significant differences ($p<0.05$) for overrun, melting rate, and viscosity of the ice cream substituted with avocado paste. The addition of avocado paste lead to the increase in viscosity and hardness of the ice cream significantly ($p<0.05$) while the sensorial properties for airiness and creaminess were perceived the same ($p>0.05$). The addition of 50% avocado paste was the most preferred among the panelists. Avocado could provide a potential substitution for dairy-fat in ice cream.

Keywords: avocado, ice cream, plant-based food, non-dairy, MUFA

1. Introduction

Consumers prefer plant-based diets [1], which are more sustainable [2] and able to provide better health benefit. The preference towards plant-based diets has significantly increased [3], with the rise of millennial generation who are increasingly adopting vegan and free-lactose diets [4]. It also has been reported that dairy milk sales have decreased globally by 7% in 2015 due to the negative health perception of dairy-based product consumption. Plant-based food diets however have become more inclusive within personalized nutrition plans for more than 70% of consumers [5]. Several factors such as lactose intolerance, gastrointestinal discomfort, and allergies have become the principal reasons for people avoiding dairy-based products [6]. Moreover, the clean label trend has also become one of the major concerns that affect consumer preference towards dairy consumption [7]. Furthermore, traces of antibiotics and hormones potentially carried in dairy and dairy-based products can also contribute to the harmful effects to the human health if consumed at certain levels [8]. From the healthy aspect, dairy and dairy-based products such as cream and butter provide high SFA (10 g/100 g) and high cholesterol (20 mg/100 g) levels. The overconsumption of SFA and cholesterol increased the risk of cardiovascular diseases [9], [10], [11]. Therefore plant-based alternative ingredients are deemed suitable substitutes in dairy-based products. An investigation is required to answer this challenge.

Fruits and vegetables have been reported as the largest category that can promote plant-based diets [12]. Avocado fruits contain high total fat (15.4 g/100 g) dominated by MUFA (9.8 g/100 g), high in phytosterol (57 mg/100 g) and contain zero cholesterol [13]. It has been reported that people who consumed avocado regularly tended to have a sufficient nutrient supply of dietary fibre, vitamin K, vitamin E, potassium, magnesium, as well as vitamin B complex [14]. These nutrients provide a wide range of potential health benefits such as preventing hypercholesterolemia [15], vascular damage [16], and reduce the risk of atherosclerosis [17]. The consumption of avocado has been investigated to have positive effects on lipid profiles [18] by increasing HDL cholesterol levels, reducing lipid serum peroxidation, and promoting cardiovascular health [19]. Furthermore, phytosterol has also been proven to be an effective compound against type-2-diabetes mellitus, lowering the risk of cardiovascular diseases, and preventing cancer [20].

Avocado has high fat content [13] which is a good source for substituting the milk fat in frozen dairy desserts [21]. Ice cream is one of the most widely consumed foods in the frozen dairy dessert category [21]. A series of different types of non-dairy based ice cream have been developed due to the rise of health concerns associated with milk consumption [9], milk-fat consumption that dominated by SFA [9], [10], [11], and clean label issues [8]. A non-dairy fat source derived from avocado could be a potential alternative in order to create an option for free-dairy fat ice cream as well as provide better health benefits to the consumers. Plant-based fat sources have great potentials to be used as a replacement for dairy-based fat sources in the ice cream formulation. Furthermore, plant-based sources are generally less expensive compared to dairy fat [21] and provide advantages towards health [15], [16], [17], [18], [19]. The objective of this study was to analyze the physicochemical, rheological and sensorial properties of ice cream that have been substituted with different concentrations of avocado paste.

2. Material and methods

2.1 Avocado paste

The avocado used in this research originated from Indonesia local horticulture. The avocados were bought in the local market and served at room temperature (25°C) until the desired ripening was achieved. The softness of the avocado fruits indicated the ripeness of the fruits. The fruits that used were soft when lightly touched and does not have a mushy texture. The avocado paste was obtained by separating the flesh of the ripe avocado from its seed. The flesh was then mixed and stored at low temperature (4°C) in a sealed vacuum bag to be used further.

2.2. Ice cream formulation

Preliminary experiments were conducted to determine the optimal formulation of the ice cream. The final ice cream formulation used in this experimentation contained water (48.83%), sugar (12.7%), cream (29.3%), glucose syrups (4.88%), skimmed milk powder (2.93%), egg yolk powder (0.49%), vanilla flavour (0.49%), and stabilizer consisting of carrageenan and xanthan gum (0.20%). All dry ingredients were weighed and mixed with wet ingredients. The mixture was then stirred and pasteurized until 80°C. During pasteurization process the mixture was stirred to homogenize all the ingredients and to distribute the heat evenly. The ice cream mixture was then cooled using a metal basin placed in ice water. The mixture was homogenized again continuously and aging at 4°C for 12 hours then stored at -20°C for hardening process. Five batches of ice cream substituted with avocado paste (AV) concentration with ranges from AV0%, AV25%, AV50%, AV75% and AV100% were produced and stored at -20°C in a sealed vacuum packaging. The samples were measured and evaluated further for their rheological, psychochemical and sensorial properties.

2.3 Psychochemical and rheological properties
The physicochemical and rheological properties of the ice cream mixtures were obtained. Several data such as pH, total solids, overrun, viscosity, moisture, and melting properties were measured. The pH was measured using a pH meter (Benchtop meter Satorius). The total soluble solid (TSS) and moisture content were determined using a moisture analyzer (MB45 Ohaus). The overrun was determined using an exact measurement of volume and weight of the ice cream mixture (g) before and after mixing [22]. The viscosity was measured using Lamyrheology refers to ASTM/ISO 2555 employed ASTM 3 with 40 rpm disc speed [23]. The viscosity measurement was conducted at room temperature (25°C) equipped with the MK-ASTM probe using standard number 3. All measurements were carried out in triplicates. The melting rate was quantified using the dripping method by which a certain amount of ice cream (ml) that was dripped over time (ml/min) at room temperature [24]. A funnel (60 mm) was placed on the top of the measuring cylinder (100 ml) and certain samples were place inside the funnel. The measurement started from -20°C (from hardening temperature) for 60 minutes that carried out at the room temperature (25°C). The drip was observed every five minutes and pictures of the ice cream sample deformation was taken [24]. The chemical properties such as total fat content were determined using the Soxhlet method. The fat extraction samples were hydrolyzed with 25 ml of 3% HCl solution in Erlenmeyer flask and then heated for one hour until all compound evaporated. The solution was then filtered using filter paper and dried at 110°C. The filter continued to be analyzed through fat extraction using petroleum ether 40—60°C for 2 hour.

2.4. Sensory evaluation
The sensory evaluation was conducted using Quantitative Descriptive Analysis (QDA) by 12 trained panelists followed by hedonic testing by semi-trained panelist. The QDA was used to quantify the intensity of ice cream profiles that were substituted with different concentrations of avocado paste. A selection of attributes and terms that were used to determine sensory properties of the ice cream were determined at this section. Nine attributes consist of colour, hardness, melting profile, creaminess, airiness, sweetness, bitterness, taste and overall liking were determined as attributes of investigation. A semantic different scaling was implemented for each attribute. A hedonic test was also conducted in order to evaluate the liking preference of the ice cream. A selection of 60 semi-trained panelist participated in the hedonic evaluation test. They were screened through a two-step process, firstly through a questionnaire, and second through their ability to pass the matching test for the screening evaluation. The potential panelist that had milk allergies, heavy smokers or consumed alcohol frequently, and did not pass the matching test were disqualified.

During the sensory evaluation ice cream samples were evaluated in a 50 ml plastic cup served at room temperature and were stored at -12°C before being served to the panelists. The maximum holding time was set at three minutes in order to avoid the samples melting. The samples were served in monadic with 3-digit random code and evaluated by panelists in a random order. The sensory panelists consisted of 30 males and 30 females adults with age range from 18—54 years old. The degree of liking was measured using a 9-point hedonic scale.

Analysis of variance (ANOVA) was performed in order to evaluate the effect of avocado paste substitution against milk fat in different concentrations at a level of AV0%, AV 25%, AV50%, AV75%, and AV100% respectively. The interaction of these parameters towards dependent variables was also analyzed using Duncan test. Significant means will be determined at $\alpha < 0.05$. Statistical analysis was conducted using XLSTAT ver. 2014.5.03 (Addinsoft).

3. Results and discussion

3.1 Value of pH, total solid, and moisture

The pH showed no significant difference between the formulations. However, the trend showed that the addition of avocado paste lowered the acidity level of the ice cream mixture. The acidity level of avocado was between pH 6.2-6.6 depending on the type of cultivar, ripening stage, and maturity of the fruit. The range of the pH from each formulation of AV0%, AV25%, AV50%, AV75% and AV100% demonstrated that they were on the standard acidity level of the ice cream mixture [21]. The acidity of the ice cream mixture is important because it determines the whipping capability that contributes to the stability of the ice cream [21].

Tabel 1. Value of pH, TSS and moisture

Parameter	AV 0%	AV 25%	AV 50%	AV 75%	AV 100%
pH	6.65 ± 0.03	6.56 ± 0.01	6.50 ± 0.01	6.51 ± 0.02	6.29 ± 0.03
TSS	0.32 ± 0.0004	0.33 ± 0.0006	0.31 ± 0.0016	0.33 ± 0.0017	0.32 ± 0.0014
Moisture (%)	65.86 ± 0.014	65.75 ± 0.006	65.39 ± 0.002	64.37 ± 0.004	63.77 ± 0.007

The total solid measurement demonstrated that there were no significance difference between AV0%, AV25%, AV50%, AV75% and AV100% of each formulation. The moisture content also showed similar values. AV100% had the lowest moisture content (63.77 ± 0.007) followed by AV75% (64.37 ± 0.004), AV50% (65.39 ± 0.002), AV25% (65.75 ± 0.006) and AV0% (65.86 ± 0.014) respectively. Ice cream should contain 55-65% water [25]. Ice cream mixture with high water content will resulted in a low total solid content in general and proportionately higher water that available to freeze when hardened at the same temperature compared to the ice cream mixture with low water content and high total solid. This condition will affect the quality of ice cream particularly for texture and hardness [26].

3.2 Viscosity and melting rate

The viscosity showed significant difference ($p < 0.05$) among the formulations. AV100% had the highest viscosity levels (1900 ± 67 m.Pas) followed by AV75% (1065 ± 89 m.Pas), AV50% (1002 ± 55 m.Pas), AV25% (245 ± 30 m.Pas), and AV0% (167 ± 38 m.Pas). The greater the concentration of avocado paste in the ice cream resulted an increase in overall viscosity. The viscosity in the ice cream mixture was particularly affected by the ingredients used as well as type of stabilizer. Carrageenan and Xanthan gum were used as stabilizers for this experimentation. Sugar has also been reported to play a significant role in determining the viscosity profile in ice cream [21]. The higher concentration of avocado paste in the ice cream mixture also directly contributed to the increase in sugar levels that are naturally contain in avocado fruits. This could have resulted the change in the ice cream mixture consistency into a thicker and more viscous [21]. Moreover, avocado contains about 7 grams of dietary fibre (28%) which has the ability as a water binding [13], [28] resulting in a thick mixture and therefore affects the viscosity profile. The avocado paste itself has a viscosity of 1818 m.Pas. The substitution of milk fat with avocado paste contributed to the change of viscosity in ice cream mixture and showed a positive correlation when higher concentration of avocado paste added.

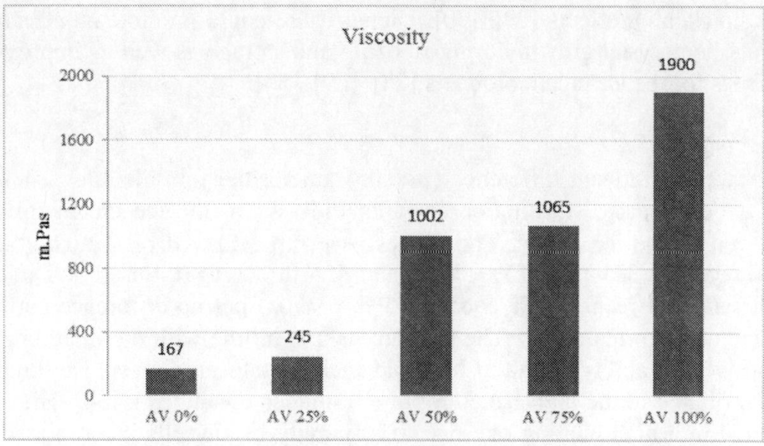

Figure 1. Viscosity (m.Pas) of samples

The melting rate showed significant difference between formulations ($p<0.05$). It is demonstrated from the **figure 2.** that AV0% has the highest melting rate (36 ml), followed by AV25% (18 ml), AV50% (13ml), AV75% (12ml), and AV100% (4 ml) respectively. The higher concentration of avocado paste resulted in the increase of viscosity of the ice cream mixture followed by decrease in the melting rate as shown in **figure 2**. The fibre content in avocado paste could have affected the melting rate of the ice cream. It has been reported that frozen dairy dessert formulated with dietary fibre could improve the melting quality of the ice cream [27] depending on the type of fibre used [27]. Fibre with high soluble fraction capability such as inulin affects the rheology of the ice cream and could provide in potential cryoprotective action [27]. The result showed that fibre in insoluble compound was able to increase significantly the viscosity and the shear thinning properties of the ice cream mixture [28].

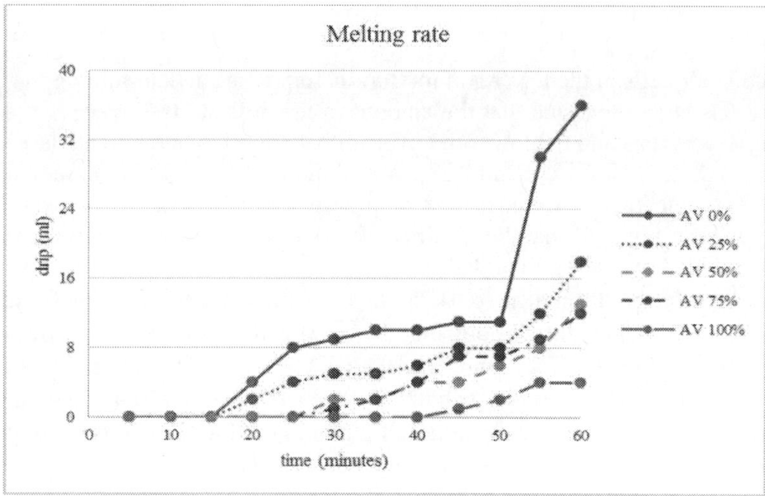

Figure 2. Melting rate (ml/minutes) of samples

Fibre has water binding properties that create a higher viscosity of ice cream mixture due to a reduction in water molecule mobility from the bulk aqueous phase into the ice crystal surface phase [28], [27]. The addition of fibre from fruits such as avocado in the ice cream mixture elevated the viscosity properties of the ice cream [28]. Fibre can contribute to the modification and improvement of food texture and sensory characteristic due to its water-binding capacity, gel-forming ability, texturizing, and thickening effects [29], [30]. Dietary fibre could promote an effective control of ice crystallization and ice crystal growth during freezing and storage as well as improve the melting and viscosity properties for the ice cream products [28], [29], [30].

3.3 Overrun and total fat

The overrun showed significant difference ($p<0.05$) among the formulations. The data showed that increasing the avocado paste substitution towards milk fat in the ice cream mixture significantly reduce the overrun of the ice cream. The highest overrun achieved by AV0% (56%), followed by AV25% (35%), AV50% (32%), AV75% (17%) and AV100% (16%). Based on Duncan test it showed significant difference between AV0% and AV25% ($p<0.05$) but no difference between AV75% and AV100% and the other formulations. The overrun data is in line with the melting rate where AV0% had the highest overrun ability followed by rapid melting rate properties. The higher overrun value significantly contributed to the higher melting rate in the ice cream [31], [34]. High overrun levels in ice cream resulted in a high melting rate due to instability of air cells incorporated in the ice cream mixture [34]. This instability is due to the increase in temperature levels when ice cream is unfrozen. Moreover, the overrun also affected the hardness profile of ice cream [31]. Overrun and air cell distribution became critical parameters in the ice cream industry that continuously need to be controlled and monitored. There are numerous factors that could influence the development of air cells and affect overrun in ice cream products such as mixing, freezing, hardening, the use of stabilizers, homogenization, and ingredients used in the ice cream mixture [21], [31]. Air content in the ice cream mixture contributed to the physical properties of the ice cream such as texture, hardness, melting properties and other sensorial characteristics [31]. The amount of air incorporated in the colloidal system in the ice cream mixture influenced the distribution of the air size that could provide melting profile and directly affect the sensory quality parameters of the ice cream [31].

Table 2. Overrun (%) and total fat (%) analysis in samples

Parameter	AV 0%	AV 25%	AV 50%	AV 75%	AV 100%
Overrun (%)	56 ± 1.37	35 ± 0.87	32 ± 0.29	17 ± 0.94	16 ± 1.45
Total Fat (%)	12.28 ± 0.58	11.53 ± 0.25	10.69 ± 0.63	9.75 ± 0.57	8.68 ± 0.84

The presence of air cells in the ice cream mixture is important in determining the texture quality of the ice cream. It has been suggested that the amount of air cells at 70% overrun is enough to prevent collision among ice crystals and the lower the overrun with subsequently create larger ice crystal, thus lead to increase the hardness of ice cream [32]. Air is one of the important ingredients that need to be incorporated in the ice cream mixture. It makes up between 30%-50% of the total volume of the ice cream [34]. It has been reported that the stability of air cells could slow down the melting rate of the ice cream as determined by adding saturated monoglycerides to the ice cream mixture [33].

There was a significant difference ($p<0.05$) in total fat content for each formulation. The data showed that the highest fat content was obtained by AV0% (12.28%), followed by AV25% (11.53%), AV50% (10.69%), AV75% (9.75%), and AV100% (8.68%). Dairy cream act as one of the base ingredients in ice cream products and contributed up to 35.1% (w/w) of total fat content while avocado fruits contributed around 15.4% (w/w) of total fat including the MUFA and the phytosterol [13]. The differences in fat content of each formulation is due to the reducing fat source from dairy fat particularly from dairy cream that has subsequently substituted with avocado paste. This fat content contributes to the fat crystallization and was able to influence the textural properties of the ice cream [33].

Ice cream contains at least 10% of dairy fat, 20% of total milk solid and able to establish a certain percentage of overrun, which is increasing in volume by whipping air into the ice cream mixture [21], [34], [36]. The percentage of fat, total solid and overrun are important factors in ice cream quality [21]. It has been demonstrated that the increase in dairy fat and total solids lead to higher costs in ice cream products, since dairy fat is considered as an expensive ingredients. However, this ingredient is the main structure in constructing sensorial properties and determines quality of the ice cream [35]. The role of fat, particularly milk fat in ice cream is important to act as an air stabilizer during the freezing process which creates a dispersed phase of air bubbles [21]. During the freezing process the crystalized fat and partially coalesced fat globule stabilize the air bubbles of the ice cream mixture and provide the desired foamy texture which contributes to the hardness quality of ice cream [21]. Milk fat acts as a thermal insulator which affects the melting properties of the ice cream [25], [37]. Ice cream with high milk fat content can retain its shape and take longer time to melt. In this experiment the substitution of milk fat with avocado paste proportionally at 0%, 25%, 50%, 75% and 100% resulted in reducing the amount of total fat content and retain the melting rate as shown in **figure 2.** This substitution of dairy-fat into plant-based fat from avocado was able to lower the total fat content of the ice cream mixture and lowered the melting rate of the ice cream. This might be due to the fibre content in avocado paste that has the ability to lower the water mobility in the ice cream [27], [28], [29], [30].

3.4 Sensorial properties

The QDA showed that there were significant differences between the formulation for colour ($p< 0.01$), hardness ($p< 0.01$), sweetness ($p< 0.01$), bitterness ($p< 0.01$), and taste ($p< 0.01$). It has been shown that the colour of each formulation showed significant difference towards green colour intensity. The higher substitution of avocado paste lead to the higher intensity of green colour. Based on pairwise multiple comparisons the texture of AV0% was perceived as the most hardest compared to the other formulations ($p< 0.05$). The amount of fat in AV0% affected the crystallization of fat and fat globules that contribute to the hardness and texture quality of ice cream [21]. Ice cream with a higher fat content will be able to retain its shape and able to stabilize the fat crystal [33], [35]. This condition creates a positive influence on the textural properties of ice cream [37].

Table 3. QDA[1] for each samples

Properties	AV 0%	AV 25%	AV 50%	AV 75%	AV 100%
Colour	1.43^{a}	2.97^{b}	5.13^{c}	5.90^{c}	7.27^{d}
Hardness	6.17^{a}	4.83^{b}	4.37^{bc}	3.37^{c}	4.73^{b}
Melting profile	5.00	5.60	5.53	5.27	4.73
Creaminess	5.47	6.27	6.07	6.60	5.90
Airiness	4.93	4.63	4.90	4.13	4.27
Sweetness	6.23^{ab}	6.43^{a}	6.27^{ab}	5.33^{b}	3.67^{c}
Bitterness	1.87^{a}	2.23^{b}	3.47^{c}	5.83^{d}	7.73^{d}
Taste	4.77^{d}	5.23^{c}	5.57^{b}	5.73^{a}	6.23^{a}

[a,b,c,d] different alphabets in one row showed significant different ($p< 0.05$) for selected attributes
[1]sensory evaluations scores for intensity: 1 = not to, 9 = very

Melting properties ($p>0.05$), airiness ($p>0.05$), and creaminess ($p>0.05$) were perceived the same. Although the overrun, melting rate, viscosity and total fat content showed significant difference in the rheological properties, however this result did not affect the sensorial properties towards melting profile, airiness, and creaminess of the ice cream. These results showed that avocado paste could be a potential plant-based substitution to milk fat in the ice cream products in terms of organoleptic since the important textural properties of the ice cream such as melting, creaminess, and airiness were perceived the same by the panelist.

It has been reported that the fat replacement in ice cream should be effectively replaced in order to match the texture, mouthfeel, and functional properties of fat. Moreover, it also should convey the desired flavour profile [38]. The amounts of fat will affect the size of fat globules, and dairy fat is an important determinant of the texture and body of ice cream [39]. The textural properties in ice cream such as creaminess are one of the desirable attributes determined by milk fat [39] since there was no significant difference for creaminess between the formulations it indicate that substituting milk fat with avocado paste could maintain the same creaminess properties as ice cream formulated with milk fat. This showcased that avocado paste could replace the fat properties of milk fat and able to maintain the creaminess profile of ice cream. This is also followed by airiness and melting profile that showed the same trends.

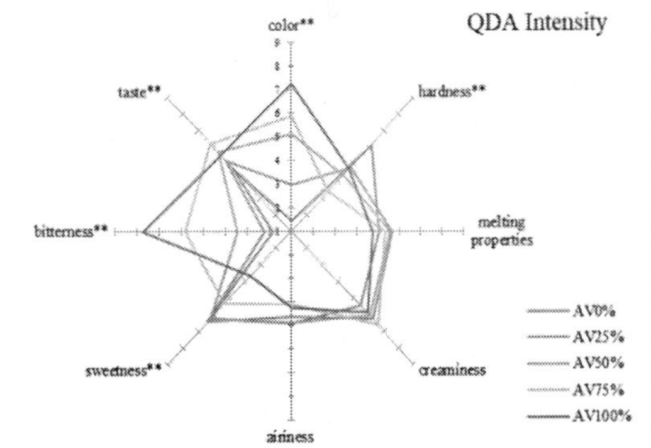

Figure 3. QDA test
*significant difference at p< 0.05
**significant difference at p<0.0

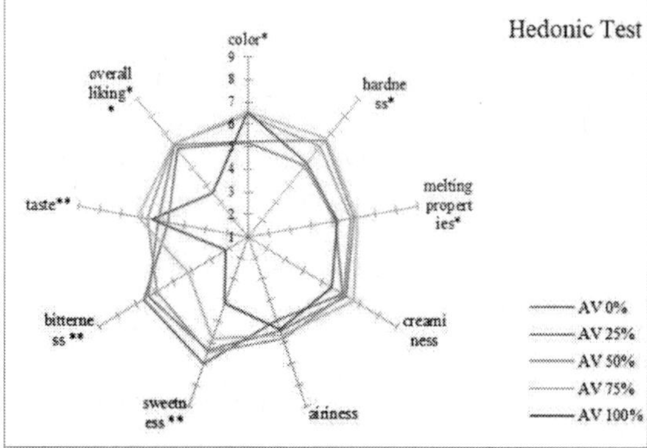

Figure 4. Hedonic testing
*significant difference at p< 0.05
**significant difference at p<0.01

Bitterness and taste were perceived significantly difference ($p < 0.05$). The strong bitterness flavour was perceived at AV100% ($p<0.05$). The higher avocado paste substitution could contribute to the bitterness flavour of ice cream that resulted in the low preferences of consumers as demonstrated in the **figure 4**. Bitterness has been reported as a natural flavour that aversive by consumers. This flavour has been removed from any food because could lead to human rejection [40]. The sweetness perceived as a reverse of the bitterness. The higher the sweetness intensity lead to the lower bitterness perception. The taste also perceived significant difference ($p < 0.05$) due to different intensity of sweetness, bitterness and avocado flavour that has developed in line with the higher substitution of avocado paste.

The bitterness flavour in ice cream could be developed from avocado paste. Avocado flavour compounds derived from 4-keto-2-hydroxyl-1-acetate, 1,2-dihydroxyl-4-acetoxy compound, and 1,4-dihydroxyl-2-acetoxy are present naturally in fruits and vegetables [41] including in avocado, these compounds contribute to the bitterness flavours [41] and commonly found in unripe fruits. Fresh fruit also contains large amounts of degradative enzymes such as lipase, lipoxygenase, polyphenoloxidase, and methyl pectinase that are also able to produce bitterness [42]. Moreover, incorrect preparation and processing from fresh avocado paste could lead to the development of bitterness [42], [43]. For example, the treatment of fresh avocado using antioxidants and citric acid have been reported to suppress the bitter taste in freshly cut avocado [43] whereas the avocado paste used in this experimentation was processed without any addition of antioxidants and used minimal heating process.

The hedonic evaluation showed that there were significant differences for colour ($p < 0.05$), hardness ($p < 0.05$), melting profile ($p < 0.05$), sweetness ($p < 0.01$), bitterness ($p < 0.01$), and overall taste ($p < 0.01$). These results were in line with the profile intensity from QDA except for melting profile ($p<0.05$). Based on pairwise comparison using Duncan test the difference was found between AV0% and AV75% ($p<0.05$) and no significant difference was found between the other formulations for the melting profile. The panelist preference towards colour showed an increase according to the higher percentage of avocado paste substitution as shown in **figure 4.** AV100% was the most preferred due to its green colour intensity based on colour preferences. Sample AV75% was the most preferred for the hardness while the preference of AV100% was dropped for this attribute. This could be happened due to the higher polysaccharides and fibre content in the AV100% that affect the textural characteristic especially the hardness and the melting profile of the ice cream [13], [21], [28]. Creaminess and airiness were perceived the same ($p > 0.05$) among the formulations which in alignment with the QDA as shown in **figure 3**. The liking towards sweetness were showed in opposite with bitterness as AV0% was the most preferred in terms of sweetness and this preference of liking decrease with the incremental substitution of avocado paste which lead to the bitter taste. Bitterness could lead to human rejection due to unpleasant taste [40]. The avocado itself naturally contains natural compounds that contribute to the bitter flavour [41], [42] coupled with incorrect avocado processing techniques which could have resulted in the development of a strong and unpleasant bitter taste [42], [43]. The overall liking showed that AV50% was the most preferred (6.38). There were significant differences ($p < 0.01$) between formulation for the overall liking, pairwise comparison test using Duncan showed that the differences was in AV100% which showed low preference compared to AV0%, AV25%, AV50%, and AV75% respectively. This was due to the strong bitterness that leads to consumers disliking the samples.

Table 4. Hedonic test of samples

Properties	AV0%	AV25%	AV50%	AV75%	AV100%
Colour	5.17^{a}	5.23^{a}	6.40^{b}	6.57^{b}	6.57^{b}
Hardness	5.17^{a}	6.60^{c}	6.23^{bc}	6.70^{c}	5.30^{ab}
Melting profile	5.13^{a}	6.03^{ab}	5.87^{ab}	6.17^{b}	5.23^{a}
Creaminess	6.10^{ab}	6.27^{ab}	6.20^{ab}	6.70^{a}	5.53^{b}
Airiness	4.97^{a}	5.47^{ab}	5.83^{b}	5.63^{ab}	5.37^{ab}
Sweetness	7.00^{a}	$6.47a^{b}$	$6.33a^{b}$	5.83^{b}	4.17^{c}
Bitterness	6.57^{a}	6.33^{a}	5.90^{a}	4.20^{b}	2.20^{c}
Taste	4.77^{a}	$5.23a^{b}$	$5.73a^{b}$	6.23^{b}	5.57^{ab}

^{a,b,c,d} different alphabets in one row showed significant different ($p < 0.05$) for selected attributes
^{1}a-9-point hedonic scale: 1 = dislike extremely to 9= like extremely

4. Conclusion and recommendation

The substitution of milk fat with avocado paste in ice cream formula could decrease the overrun and total fat content in ice cream. However it was demonstrated that the avocado paste could retain the melting rate of the ice cream due to polysaccharides and natural dietary fibre content in the avocado. Higher concentration of avocado paste in the ice cream formula provides more hydrocolloids from the dietary fibre and thus influences the viscosity properties and contributes to the stabilization of air during aeration. The substitution of avocado paste at 0%, 25%, 50%, 75% and 100% against milk fat did not showed significant difference in creaminess, airiness and melting profile based on QDA while the bitter taste strongly developed from the higher concentration of avocado paste that could lead to rejection from consumers. While colour, hardness, bitterness, sweetness and taste were perceived differently, the important parameter of ice cream such as creaminess, airiness and melting profile were perceived the same. This indicated that avocado could be a great potential as a substitute for milk fat. There were no significant differences in overall liking between samples with 0%, 25%, 50% and 75% avocado paste concentration, while the concentration of 100% fat from avocado paste showed a significant difference ($p<0.01$) compared to other formulations and 50% substitution was the most preferred. The processing of avocado paste was not thoroughly investigated in this research. It is recommended to investigate the post-harvest condition and assess the ripening stage of avocado fruits used in the experimentation since avocado is a sensitive climacteric fruits and the ripeness state of avocado could affect the properties and characteristic of avocado paste.

5. References

[1] Ranganathan J 2016 Shifting diets for a sustainable food future. working paper, installment 11 of creating a sustainable food future. Washington, DC: World Resources Institute. [cited 2017 March 18] Available from http://www.worldresourcesreport.org.

[2] Salonen R, Nyyssonen K, Kaikkonen J, Porkkala-Saratabo E, Voutilainen S and Rissanen T 2003 Six-year effect of combined vitamin C and E supplementation on atherosclerotic progression. *Circulation* **107** 947–953.

[3] Granato D, Gabriel F, Filomena N, Andriano G and Jose AF 2010 *Functional Food and Non-Dairy probiotic food development: Trends, concepts and products*. Washington DC p 276.

[4] Flávera C, Pradoa J, Paradaa, Ashok P and Carlos R 2008 Trends in non-dairy probiotic beverages. *Food Research International* **41**(2) 111–123.

[5] Food Marketing Institute (FMI). The power of meat [cited 2017 March 29] Available from http://: www.meatconference.com/POM2015

[6] Deloitte 2015 Capitalizing on the shifting consumer food value equation. Deloitte consulting. Washington DC. [cited 2017 March 29] Available from http:// www.deloitte.com/id/ en.html

[7] Bhandari SD, Schmidt RH and Rodrick GE 2005 Hazards resulting from environmental, industrial, and agricultural contaminants. *Food Safety Handbook*. Hoboken, N.J. John Wiley&Sons, Inc. p 451.

[8] Baars AJ, Bakker MI and Baumann RA 2004 Dioxins, dioxin-like PCBs and nondioxin-like PCBs in foodstuffs: occurrence and dietary intake in the Netherlands. *Toxicol Lett*. **151** 51–61.

[9] Peter JH and Keigan MP 2012 Influence on dairy product and milk fat consumption on cardiovascular diseases risk: A review of evidence. *American Society for Nutrition. Adv. Nutr* **3** 266–285.

[10] Patty WS, Qi Sun, Frank BH and Ronald MK 2010 Saturated Fatty Acid Risk of Coronary heart diseases: modulation by replacement nutrients. *Curr atherosclerosis rep*. **12(6)** 384-390

[11] Sonia SA 2016 Food consumption and its impact on cardiovascular diseases: importance of solutions focused on the globalized food system. *J Am Coll Cardiol* **66**(14) 1590-1614.

[12] Organic Trade Association (OTA) 2015 Quick statistic. Organic Trade Association. Washington DC [cited 2017 April 3] Available from: http://ota.com.

[13] USDA (United States Department of Agriculture) 2011 Avocado Composition. Nutrient Data Laboratory. USDA National Nutrient Database for Standard Reference, Release 24. U.S. Department of Agriculture Washington, DC.

[14] Fulgoni VL, Dreher ML and Davenport AJ 2010 Consumption of avocados in diets of US adults: NHANES 2011–2006. *American Dietetic Association. Abstract* **54**.

[15] Salonen RM, Nyyssonen K, Kaikkonen J, Porkkala-Saratabo E, Voutilainen S and Rissanen TH 2003 Six-year effect of combined vitamin C and E supplementation on atherosclerotic progression. *Circulation*. **67**(107) 947–953.

[16] Hozawa A, Jacobs DR, Steffes MW, Gross MD, Steffen LM and Lee LH 2007 Relationships of circulating carotenoid concentrations with several markers of inflammation, oxidative stress, and endothelial dysfunction: The coronary artery risk development in young adults (CARDIA): young adult longitudinal trends in antioxidants study. *Clin. Chem.* **53**(3) 447–455.

[17] Dwyer JH, Paul-Labrador MJ, Fan J, Shircore AM, Bairey-Merz CN and Dwyer KM 2004 Progression of carotid intima-media thickness and plasma antioxidants: The Los Angeles atherosclerosis study. *Arterioscler. Thromb. Vasc. Biol.* **24** 313–319.

[18] Carranza MJ, Herrera JE, Alvizouri MM, Alvarado MDR and Chavez CF. 1997. Effects of a vegetarian diet vs. a vegetarian diet enriched with avocado in hyper-cholesterolemic patients. *Arch. Med. Res.* **28**(4) 537–41.

[19] Wu X, Gu L, Holden J, Haytowitz DB, Gebhardt SE and Beecher GR 2007 Development of a database for total antioxidant capacity in foods: A preliminary study. *J. Food. Comp. Anal.* **17** 407–422.

[20] Jones PJ and Abumweis SS 2009 Phytosterols as functional food ingredients: linkages to cardiovascular disease and cancer. *Curr Op Clin Nutr Metab Care.* **12**(2) 147-51.

[21] Goff D and Richard WH. 2013. *Ice Cream.* Springer: New York.

[22] Hammink C, Son M, Heijnis W 2014 Practical Analysis. Den Bosch: HAS University of Applied Sciences.

[23] Ismail EA, Al-Saleh AA and Metwali AM 2013 Effect of inulin supplementation on rheological properties on low-fat Ice Cream. *Life Science Journal* **10**(3) 1742-1746.

[24] Muse MR and Hartel RW 2004 Ice Cream Structural Elements that Affect Melting Rate and Hardness. *J. Dairy Science.* **87** 1–10.

[25] Mi JC and Kwang SS 2014 Studies on physical and sensory properties of premium vanilla ice cream distributed in Korean Market. *Korean J Food Sci Anim Resourc* **34** (6) 757-762.

[26] El Owni OA and Zeinab KO 2009 Chemical composition of ice cream produced in Khartoum state, Sudan. *Pakistan J. Nutr.* **8** 158–160.

[27] Stieven D. 2009. Fibre-rich ice-cream gets formulation boost: Study. [cited 2017 August 22] Available from: http://www.foodnavigator.com/Science/Fibre-rich-ice-cream-gets-formulation-boost-study.

[28] Christous S, Dimitra L and Constantina T 2009 Enrichment of ice cream with dietary fibre: effect of rheological properties, ice crystallization, and glass transition phenomena. *Food chemistry* **115** 665-671.

[29] Dello SM, Bertola N, Martino M and Bevilaqcua A 2004 Influence of dietary fibre addition on sensory and rheological properties of yogurt. *International Dairy Journal.* **14** 263–268.

[30] Gelroth, J and Ranhotra GS 2001 Food uses of fibre. In Sungsoo Cho & In MS. Dreher (Eds.), *Handbook of dietary fibre*. New York: Taylor and Francis 117 p.

[31] Rosalina PS and Richard WH 2004 Effects of overrun on structural and physical characteristic of ice cream. *International dairy journal.* **14** 255-262.

[32] Prindiville EA, Marshall RT and Heymann H 1999 Effect of milk fat on the sensory properties of chocolate ice cream. *Journal of Dairy Science* **82** 1425–1432.

[33] Pelan BMC, Watts KM, Campbell IJ and Lips A 1997 The stability of aerated milk protein emulsions in the presence of small molecule surfactants. *Journal of Dairy Science.* **80** 2631–2638.

[34] Brian R 2014 The ice cream and chemistry. Chemmatters. February/march 2014 [cite 2017 August 18]. Available from: https:www.acs.org/chemmatters.

[35] Hui Y, Isabel GL, Miang HL, Murrell and Wai-Kit KD 2004 *Handbook of Frozen Foods*, Marcel Dekker, Inc: New York. p 88.

[36] Karaman S and Kayacier A 2012 Rheology of ice cream mix flavoured with black tea or herbal teas and effect of flavouring on the sensory properties of ice cream. *Food Bioprocess Technol.* **5** 3159-3169.

[37] Li Z, Marshall RH and Fernando HL 1997 Effect of milk fat content on flavour perception of vanilla ice cream. *J. Dairy Sci.* **80** 3133-3141.

[38] Bradley RLJ, Arnold Jr, Barbano DM, Semerad RG, Smith DE and Vines BK 1993 Standard Methods for the Examination of Dairy Products. 16th ed. R. T. Marshall, ed. *Am. Publ. Health Assoc.*, Washington, DC. **8**(16) 171-179.

[39] Ohmes RL, Marshall RT and Heyman H. 1998. Sensory and physical properties of ice cream containing milk fat or fat replacers. *Journal of dairy sciences.* **81** 1222-1228.

[40] Drewnowski A and Carmen-Gomez Ca 2000 Bitter taste, phytonutrients, and the consumer: a review. *American society for clinical nutrition.* **72** (6) 1424-1435.

[41] Brian IB 1972 Isolation of unpleasant flavour compounds in the avocado (*Persea americana*). *Journal of Agriculture Food Chemistry.* **20** (4) 753-757.

[42] Jonathan MG and Coolen KB. 2004. Patent (WO 20014100670 A1). Method of treating avocados and method of preparing guacamole therefrom. Application number PCT/US2003/008746. Patentscope, Espacenet. General Mills, Inc
[43] Bower JP and Dennison MT. 2003. Progress in the development of avocado products. South African avocado growers' association yearbook. **39** (26) 35-37.

Effect of green spinach (*Amaranthus tricolor* L.) and tomato (*Solanum lycopersicum)* addition in physical, chemical, and sensory properties of marshmallow as an alternative prevention of iron deficiency anemia

B Yudhistira, D R Affandi and P N Nusantari*

Food Science and Technology, Universitas Sebelas Maret, Surakarta 57101, Indonesia

*Email : nimasnusantari17@gmail.com

Abstract. Iron deficiency anemia is the most common nutritional disorder in the world. Consuming vegetable which contain iron, including spinach, is an alternative to fulfill iron requirement. Fe will be more easily absorbed in the presence of vitamin C. Tomato is one of vitamin C source that can be used. Spinach can be applied into confectionary products in the form of marshmallow. This research aimed to find out the physical, chemical and sensory properties of green spinach Marshmallow in addition of Tomato, the best formula, and define the category of nutrition contents based on Acuan Label Gizi (ALG). This study used a completely randomized design (CRD) with one factor that was different proportion of spinach:tomato (75%: 25%; 50%: 50%; 25%: 75%). The data were analyzed using One Way Anova with 5% significance level. The result of this study showed that the difference of spinach and tomato proportion affect tensile strength, moisture, ash content, Fe content, crude fiber, vitamin C, color and marshmallow's flavor. Best marshmallow formulation of 25% spinach in addition of 75% tomato had Fe content of 1.159 mg/ 100g and vitamin C of 44 mg/ 100g.

Keywords: *marshmallow, green spinach, tomato*

1. Introduction

Anemia is the most common and widespread nutritional disorder in the world. The most common anemia is iron deficiency anemia which caused by the deficiency of iron and can be suffered by anyone including infants, children, even adults, both men and women [1]. Data in 2013 shown that the proportion of anemia suffered by people aged 15-24 years old in amount of 18,4% from the whole group of 1-75 years old [2]. One effort that can be done to overcome iron deficiency anemia is iron fortification. Iron can be obtained from vegetables, including spinach (*Amaranthus tricolor* L), which has an iron content of 3.9 mg/100g [3]. Spinach is considered as a good source of vitamins (ascorbic acid, riboflavin, niacin, and folic acid), minerals (iron and calcium) and dietary fibers [4].

The nutritional component in spinach is a series of complementary compositions, e.g. Fe, will be more easily absorbed in the presence of vitamin C. Also described by Hallberg et al [5] that vitamin C is a strong promoter of iron absorption from food and can counter the inhibitory effects of phytate and tannin. However, vitamin C content easily decreases during processing, so apart from spinach, iron absorption can be helped by vitamin C from other ingredients such as tomato. The benefits contained in spinach as a source of iron will not support the prevention of anemia if the amount of consumption is low. So it is needed to have vegetable products in other forms that potential to be accepted by the community. Spinach can be applied to confectionery food product, which is one form of food products. One of confectionery product that can be consumed by wide range of age is marshmallow. The innovation of marshmallow-shaped vegetable processing will ease the increase in vegetable consumption. The formation of marshmallow texture is influenced by gelatinizing agent, one of them is pectin [6]. Spinach has 11.58% pectin content and tomato (*Solanum lycopersicum*), which has 4.63% pectin content. This spinach and tomato added can play a role in the formation of marshmallow texture.

Making green spinach marshmallow with the addition of tomato is expected to be one of the processed foods that can meet the needs of iron and can be an alternative food fulfillment of iron in the wide range of community. Furthermore, the addition of tomato in the manufacture of marshmallow is expected to help the maximum absorption of iron levels in the body. This research conducted to know the physical, chemical, and organoleptic characteristics of marshmallow, the best formula, and the nutritional group of marshmallow based on Acuan Label Gizi (ALG).

2. Materials and methods

2.1. Materials

Materials used in the manufacture of marshmallows are green spinach (*Amaranthus tricolor* L.), red tomato (*Solanum lycopersicum*) (Karanganyar, Indonesia), linear gelatine (Edible Bovine Gelatine), liquid sorbitol, citrus essence, icing sugar, cornstarch, as well as ingredients for other analysis of jewawut (Setaria Italica), xylene, aquades, concentrated HCl solution, blue bromophenol indicator, sodium acetate, sterilizing solution, ascorbic acid, acetic acid, thio urea, 2,4-dinitrophenyl hydrazine, sulphuric acid, bromine water, potassium iodide, potassium iodate (KIO3), sodium thiosulphate, starch Solution, NaOH, K_2SO_4 solution, and 95% alcohol.

2.2. Spinach and Tomato Extract

Spinach and tomato steamed for 5 minutes for spinach and 7 minutes for tomato at 100° C and drained. After slicing, smoothing or spinning of spinach and tomato is done using a juicer.

2.3. Marshmallow

The making of marshmallow divided into two stages, the first was mixing the spinach and tomato extract that has been cooled according to the formula (75%: 25%; 50%: 50%; 25%: 75%) with gelatin 43.2 grams until it dissolved and let it for 10 minutes. In stage two, spinach and tomato according to the formula were heated by adding 200 ml sorbitol for 5 minutes to be boiled and foamed. Next, mixed the results of stages 1 and 2 in the basin by mixer until formed foam. If foam started to form, added essence citrus and mixing for 30 minutes. Marshmallow poured on a baking sheet covered with bread paper and icing sugar mixed with cornstarch. After flattening, the top surface of marshmallow sprinkled with icing sugar that mixed with cornstarch. The next stage was curing for 10 to 12 hours [7] with modification.

2.4. Physical, Chemical, and Sensory Analysis

Physical analyzes included tensile strength using the Zwick /Z0.5 Universal Testing Machine (BL-GRS500N), density test [8]. While chemical analyzes include of water content of thermovolumetric method [9], ash oven method [10], Atomic Absorption Spectrophotometer [11], crude fiber acidic method [11], Vitamin C concentration of spectrophotometric methods [12], and organoleptic tests with hedonic Scoring and Ranking tests [13].

2.5. Data Analysis

This study used a completely randomized design (CRD) with one factor that was different proportion of spinach : tomato (75%: 25%; 50%: 50%; 25%: 75%). The data were analyzed using One Way Anova with 5% significance level.

3. Results and discussion

3.1 Physical Characteristics of Marshmallow Green Spinach (Amaranthus tricolor L) in addition of Tomato (Solanum lycopersicum)

Tabel 1. Physical Characteristics of Marshmallow Green Spinach (*Amaranthus tricolor* L) in addition of Tomato (*Solanum lycopersicum*).

Formulation	Parameter	
	Tensile Strength (Fmax) (N)	Density (g/ml)
M1	$0.558^{b} \pm 0.027$	$0.476^{a} \pm 0.008$
M2	$0.428^{a} \pm 0.059$	$0.485^{a} \pm 0.008$
M3	$0.376^{a} \pm 0.007$	$0.538^{a} \pm 0.050$

a = subset a, b = subset b, a and b in one column show a significant difference at $\alpha = 0.05$, M1 : Spinach 75%:Tomato 25%, M2 : Spinach 50%:Tomato 50%, M3 : Spinach 25%:Tomato 75%

Physical characteristic of marshmallow determined by the value of tensile strength and density. The value of tensile strength marshmallow green spinach with the addition of tomato increased with the greater concentration of spinach, but density test shown the different result. The difference of spinach and tomato concentration on marshmallow has no effect on density value. It can be seen in Table 1. that the highest tensile strength value in marshmallow M1 (0.558 N). The high value of tensile strength shown that the quality of marshmallow produced is better.

The formation of marshmallow texture is influenced by the presence of pectin, pectin in spinach and tomato in the form of polysaccharide can be used as a thickener in food, besides gelatin which is a protein and used as an emulsifier. Gelatin is responsible for gel formation, with the help of pectin providing a protective coating so that marshmallows can be stable during storage [14]. Therefore, marshmallow requires conjugation of polysaccharides and protein to obtain froth [15]. The ionic interaction effect between proteins and pectins results in increase of foam stability and no effect indicating that the foam layer is thickened in any sample. Proteins contribute to the foaming capacity and foam formation, while the soluble pectin complex decreases the drainage which ultimately gives foam stability [16]. Foam is a thermodynamically unstable system. As for the foam, it can be stable because of the factors that one of them is drainage. The presence of protein combined with carbohydrates will reduce the drainage to become a stable system [17].

3.2 Chemical Characteristics of Marshmallow Green Spinach (Amaranthus tricolor L) in addition of Tomato (Solanum lycopersicum)

There are 5 parameters observed in chemical characteristics. Table 2 shows that the concentration of spinach and tomatoes used has an effect for all parameters, but the kind of effect is different for each parameter. The highest water content value was found on marshmallow M1 (17.562%). This is due to high water content in foodstuffs influenced by several things including chemical content, such as protein clusters, polysaccharides, or fiber. The fiber (polysaccharide), in food affects the process of water absorption because in the fiber there are enough polar free hydroxyl groups [18], so that the water can form hydrogen bonds with hydroxyl groups (-OH) [19]. So the less fiber content the higher the water content. According to the Badan Standarisasi Nasional [20] for soft jelly candy (Indonesia

National Standard 3547-2-2008), the maximum water content is 20%. It means that the water content of marshmallow is accordance with the Indonesian National Standard.

Table 2. Chemical Characteristics of Marshmallow Green Spinach (*Amaranthus tricolor* L) in Addition of Tomato (*Solanum lycopersicum*).

Formulation	Parameter				
	Water Content (%)*	Ash Content (%db)*	Fe Content (ppm)*	Crude Fiber (%db)*	Vitamin C (%wb)*
M1	17.562^{c}±0.276	0.441^{c}±0.013	45.744^{c}±4.767	0.079^{a}±0.039	0.021^{a}±0.001
M2	15.572^{b}±0.001	0.351^{b}±0.013	31.838^{b}±0.567	0.104^{b}±0.007	0.024^{a}±0.001
M3	11.709^{a}±0.016	0.223^{a}±0.008	11.594^{a}±1.053	0.160^{c}±0.008	0.044^{b}±0.002

a = subset a, b = subset b, a and b in one column show a significant difference at α = 0.05, M1 : Spinach 75% : Tomato 25%, M2 : Spinach 50%:Tomato 50%, M3 : Spinach 25%:Tomato 75%

The ash content ranged from 0.2-0.4 (% db) or 0.16-0.37 (% wb). Marshmallow as one of the jelly soft candy products has values that meet the standards set by SNI, according to Badan Standarisasi Nasional [20] for soft jelly candy (Indonesia National Standard 3547-2-2008), ash content has a maximum limit of 3%. Spinach has a content of ash of 0.6-3.1% [21] and tomato has an ash content of 0.39-0.6% [22] so the higher concentration of spinach added the higher the ash content in marshmallows. The highest ash content was found on marshmallow M1 (0.441%). The lowest ash content was found on marshmallow M3 (0.223%). Marshmallow M3 is the best because the less ash the better is the quality of marshmallow produced [23].

Spinach has a high Fe content of 3.9 mg/100 g [3] compared to tomato that containing only 0.5 mg /100 g [24] to 0.81 mg/100 g (MOH , 1990 in Fitri [25]. So the addition of more green spinach will increase iron levels. Marshmallow with the highest iron content found in marshmallow M1 that is equal to 45.7437 ppm (4.574 mg/ 100 gram). The iron content in marshmallow has the same value as the ingredients used. This is because according to [26], iron content will survive at high temperatures.

According to Kusharto [27], tomato has a crude fiber content of 1 gram per 100 grams higher than spinach with 0.8 grams per 100 grams. Rough fibers comprising cellulose, hemicellulose, and lignin are insoluble fibers. Since this fiber is not water soluble, its presence is still present in the final product [28]. Marshmallow M3 has the highest crude fiber content is 0.160%.

Vitamin C levels in marshmallows were not significantly different for M1 marshmallow and marshmallow M2 but differed significantly with M3 marshmallows. The highest vitamin C values were found in marshmallow M3 (0.044%). The higher the concentration of tomato used, the higher the vitamin C. Even though tomato has vitamin C content of 40 mg/100 grams [29] lower than spinach that has vitamin C content of 80 mg/100 grams to 629 mg/100 grams in fresh condition and 299.7 mg/100 gram with blanching for 5 minutes [30], the pretreatment is different between tomato and spinach. During the processing, tomato is treated to avoid the decreasing of vitamin C level by cooling in the refrigerator, which according to Syafutri et al [31], the cold temperature used (10°C) were causing the metabolic activity in the fruit runs slower so that the decreasing of organic acids (ascorbic acid) contained also slower. While spinach more treated to the heat, in mixing with sorbitol it is heated by boiling. Glucose, sucrose, and sorbitol can protect ascorbic acid from degradation at low temperatures (≤40°C), but at high temperatures (≥70°C) will cause ascorbic acid damage [32].

According to Hallberg et al. [5,40] that vitamin C is a strong promoter of iron absorption from food and can counteract the inhibitory effects of phytate and tannin. Vitamin C can increase iron absorption when taken at the same time because vitamin C will convert iron from ferrite form to ferrous form. Iron in the form of fero is more easily absorbed. Spinach and tomato used in marshmallow making materials were given preliminary treatment to increase the availability of iron. Spinach and tomato are given a blanching treatment first. Heat treatment increases iron availability by 32.5% [33] and according to Kurniawan (2007) in [26], blanching and cooking increase iron availability.

Acuan Label Gizi (ALG) is a reference of information inclusion about nutritional content on food product label. According to the Regulation of the Head of Food and Drug Supervisory Authority of the Republic of Indonesia Number 13 of 2016 [34], on Claim Supervision on Labels and Advertisements of Processed Foods, a vitamin or mineral is said to have nutrients as a source if it contains 15% ALG per 100 grams (in solid form) or 7.5% ALG per 100 ml (in liquid form) and is said to be high/rich if the product contains twice of vitamins or minerals content for "source". On the Regulation of the Head of Food and Drug Supervisor of Republic of Indonesia Number 9 Year 2016 [35], Label of Nutrition shows that ALG value for iron is 22 mg (15% ALG: 3.3 mg for "source" and 6, 6 mg for "high/rich"), and for vitamin C of 90 mg (15% ALG: 13.5 mg for source and 27 mg for "high / rich").

In green spinach marshmallows with the addition of tomatoes, the best formula is found in the M3. Marshmallow M3 has iron content of 1.159 mg / 100 gram and vitamin C at 0.044% or 44 mg / 100 gram. M3 marshmallow iron levels do not include "source" or "high/rich" because they are lower than 15% ALG per 100 grams. While the vitamin C content of marshmallow M3 including "high/rich" because higher than the provision of the number of "source" ALG. Thus, green spinach marshmallow with the addition of tomato allowed as an alternative food to prevent iron deficiency anemia. Although the best formulation of marshmallow has less iron content than source which is only 1.159 mg / 100 gram, the amount is meet the recommended daily Nutrient Adequacy Ratio (AKG) of iron for age 0-9 About 3-10 mg. Men over 9 years of age of 13-17 mg, and women over 9 years of age 14-26 mg (Supariasa et al 2000 in [36]) should consume according to the number of marshmallows required to meet the amount of the iron needed.

3.3 Sensory Characteristics of Marshmallow Green Spinach (Amaranthus tricolor L) in addition of Tomato (Solanum lycopersicum)

Table 3. Level of Preferences Marshmallow Panellist Spinach Green with Addition of Tomatoes.

Formulation	Parameter				
	Color	Aroma	Flavor	Texture	*Overall*[*]
M1	3.23^a	3.23^a	2.63^a	3.55^a	$-0{,}30^a$
M2	3.03^a	3.60^a	3.00^a	3.65^a	$-0{,}15^a$
M3	3.85^b	3.33^a	3.45^b	3.73^a	$0{,}45^b$

a = subset a, b = subset b, a and b in one column show a significant difference at $\alpha = 0.05$, M1 : Spinach 75% : Tomato 25%, M2 : Spinach 50%:Tomato 50%, M3 : Spinach 25%:Tomato 75%
* Ranking Test: The largest number shows the highest level of favorite and followed by the next small number. M1: Spinach 75%:Tomato 25%, M2: Spinach 50%:Tomato 50%, M3: Spinach 25%:Tomato 75%

The determination of the value done by panelist giving score from 1 to 5, the higher value obtained indicate it is more preferred by panelist. Table 3 shows the panelist's preference for the color of marshmallow not significantly different for M1 and M2 but significantly different from M3. Marshmallow M1 and M2 colors is brownish yellow with green spots obtained from green spinach, while on M3 has whitish orange color. Thus panelists prefer the M3 marshmallow with the colors tend to be bright (red). Differences in green spinach concentration with the addition of tomatoes have an effect on the color characteristics of marshmallows. The panelist's favorite level on this color parameter is allegedly influenced by the color of the marshmallow material, which is chlorophyll from spinach [21] to contribute green color, while the tomato is known to have a red color contributed by carotenoid and lycopene. Lycopene is a natural pigment found in plants [37].

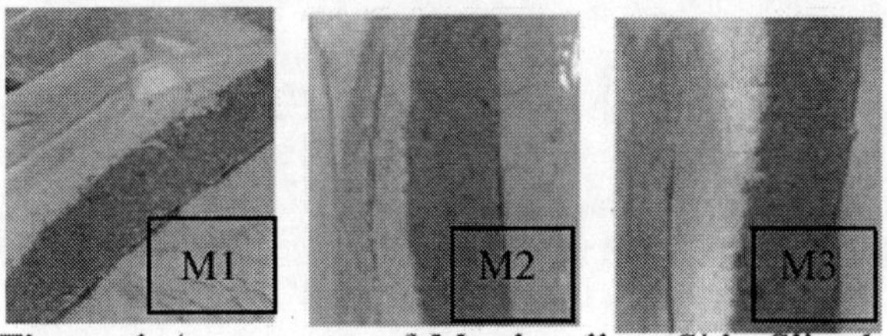

Figure 1 Appearance of Marshmallow Side Sliced (spinach: tomato): A) M1 (75%: 25%); B) M2 (50%: 50%); C) M3 (25%: 75%)

The level of panelist acceptance of the scent is not significantly different between formulas. The absence of this difference is due to the addition of citrus essence with the same concentration for all treatments which aimed to provide a preferred scent to cover the typical odor of gelatin due to cooking [23]. Indonesia National Standard (2008) mentioned that the quality requirement of soft jelly soft candy oil is normal.

The panelist acceptance level for taste is not significantly different for marshmallow M1 and M2 but different from M3. The highest panelist favor value for taste parameter is for marshmallow M1 because tomato has a varied flavor from sour to sweet [38].

The texture of the marshmallow did not differ between the formulas. The process of making is done by the same method to produce a uniform texture. Gelatin is used to improve marshmallow aeration and texture. Gelatin absorbs water to form a thick solution. Thus gelatin used in equal amounts causing the resulting marshmallow texture to be the same.

3.4 Determination of Green Spinach Marshmallow Formulation (Amaranthus tricolor L) with the Best Addition of Tomato (Solanum lycopersicum)

The best formula determination of green spinach marshmallow with the addition of tomato is done by compensatory model or weighted test [39]. Thus, after passing the organoleptic test in determining the panelist's preference for the product, we then conduct a weighted test to select the preferred formula of the organoleptic parameters as well as the physical and chemical characteristics. The best formula is the treatment with the highest value score of the degree of product interest expected by the consumer. The best formula with the highest yield value is M3. Therefore, Formula M3 with 25% green spinach concentration and 75% tomato is the right combination in marshmallow formula of green spinach and tomato.

4. Conclusion

The physical characteristics of green spinach marshmallow (*Amaranthus tricolor* L) with the addition of tomato (*Solanum lycopersicum*) are tensile strength of 0.376-0.558 N and density of 0.476-0.538 g/ml while the chemical characteristics are moisture content of 11.709-17.562% (wb), ash content of 0.223-0.441% (db), Fe content of 11.594-45,744 ppm (db), crude fiber content of 0.079-0.160% (db), and vitamin C levels of 0.021-0.044% (wb). The higher the tomato concentration affected to the sensory characteristic, the color and taste of marshmallows tend to be preferred. Overall, marshmallow green spinach 25% with the addition of 75% tomato is the most preferred formula. Thus proportion of spinach and tomato also the best formula which classified as high / rich in vitamin C that is 44 mg / 100 gram according to the Nutritional Label Reference (ALG).

5. References

[1] World Health Organization 2008 *Worldwide Prevalence of Anaemia 1993-2005 WHO Global Database on Anaemia*. Ed. Benoist et al. (accesed at March 16th 2016)

[2] Kadafi M 2015 Influence of Iron Tablet and Vitamin C to Hemoglobin Levels of Undergraduate Students of Nursing University of Muhammadiyah Surakarta *Published Reasarch* Universitas Muhammadiyah Surakarta

[3] Faridah A and Novita S 2014 *The addition of spinach (Amaranthus tricolor L) in the manufacture of cookies as fortification Fe.* Proceedings FKPT-TPI Universitas Riau

[4] Ankita and Prasad K 2013 Studies on Spinach Powder as Affected by Dehydration Temperature

[5] Hallberg L, Brune M, RossanderHulthén L 1987 Is there a physiological role of vitamin c in iron absorption *Annals of the New York Academy of Sciences* **498** p 324-32

[6] Koswara S 2009 *Confectionary Technology* Ebook Pangan.com

[7] Ginting N A, Herla R and Rona N 2014 The effect of comparison of red guava with lemon and gelatin concentration to the quality of red guava marshmallow. *J. Rekayasa Pangan dan Pertanian* **2** p 16-20

[8] Ramli E 2010 *Effect of gelatin concentration and the ratio of sucrose - glucose syrup to the physicochemical and organoleptic properties of marshmallow rosella.* Undergraduate Thesis Universitas Katolik Widya Mandala Surabaya

[9] Sudarmaji S, Suhardi and Haryono B 1989 *Analysis of Food and Agricultural Materials* Yogyakarta: Liberty Yogyakarta

[10] AOAC (Associaton of Official Analytical Chemistry) 2005 *Official Methods of Analysis.* 18th edition. Marylan: Association of Official Analytical Chemist inc. Gaithersburg

[11] AOAC (Associaton of Official Analytical Chemistry) 2002 *Official Methods of Analysis.* 17th edition. Marylan: Association of Official Analytical Chemist inc. Gaithersburg

[12] Majidi M I H A and Y-AlQubury H 2016 Determination of vitamin c (ascorbic acid) content in various fruit and vegetable by UV-spectrophotometry and titration methods *J Chem. and Pharmaceutical Sci* **9**(4) p 2972

[13] Kemp S E, Tracey H and Joanne H 2009 *Sensory Evaluation a Practical Handbook* Macmillan Publishing Solutions. Singapore

[14] Dickinson E 2003 Hydrocolloids at interfaces and influence on the properties of dispersed system. *food hydrocolloids* **17** p 30-1

[15] Suryani R and Fithri C N 2015 Modified cassava starch (*Manihot esculenta*) with α-amylase enzyme as a foaming agent and its application in the process of making marshmallow *J. Pangan dan Agroindustri* **3** p 731

[16] Schmidt I, B. Novales, F Boue, and M.A.V Axelos 2010 Foaming properties of protein/pectin electrostatic complexes and foam structure at nanoscale *J. Colloid and Interface Sci.* **345** p 316

[17] Damodaran S 2005 Protein stabilization of emulsions and foams *J. Food Sci.* ***70*** R54–R66

[18] Winarti S 2010 *Functional Food* Yogyakarta: Graha Ilmu

[19] Wahyuni S 2015 *Production and evaluation of physical, chemical and sensory properties of fruit leather of apple manalagi (Mallus sylvestris mill) with variation of xanthan gum concentration* Undergraduate Thesis Unversitas Sebelas Maret. Surakarta

[20] BSN (Badan Standardisasi Nasional) 2008 Indonesia National Standard 3547.2:2008. http://sisni.bsn.go.id (accesed at November 9th 2016)

[21] Fatimah S 2009 *Study of chlorophyll and iron (fe) levels on some types of spinach on the number of white mouse erythrocytes (Rattus Norvegicus) Anemia* Undergraduate Thesis Universitas Islam Negeri Maulina Malik Ibrahim. Malang

[22] Suarez M, Hernandez E M. Rodriguez, and C D Romero 2008 chemical composition of tomato (*Lycopersicon esculentum)* from Tenerife, the Canary Islands *Food Chemistry* **106** p 1049

[23] Sartika D 2009 *Development of marshmallow products from red snapper fish gelatin (Lutjanus Sp.)* Undergraduate Thesis Institut Pertanian Bogor.

[24] Arifulloh *2013 Lycopene Extraction from Tomato Fruit (Lycopersicum esculentum Mill.) With Various Solvent Compositions* Undergraduate Thesis Universitas Jember.

[25] Fitri A 2016 Pectin from the skin of cocoa fruit (*theobroma cacao l.*) As an edible coating of tomatoes *Research* Universitas Halu Oleo Kendari
[26] Dewi A A 2014 *Evaluation of sensory and chemical properties of mp-ation based on pumpkin flour (Cucurbita moschata), spinach (Amaranthus sp) and red beans (Phaseolus vulgaris L.) by drying method* of spray drying Undergraduate Thesis Universitas Sebelas Maret. Surakarta
[27] Kusharto C M 2006 Dietary fiber and its role for health *J. Gizi dan Pangan* **1** p 48-9
[28] Chinachotti P, M P Steinberg, and R.Villota 1990 A model for quantitating energy and degree of starch gelatinization based on water, sugar and salt contents *J. Food Sci.* **55**
[29] Firmanto, B.H 2011 *Success of Organic Tomatoes* Bandung: Angkasa
[30] Yadav S K, and Salil S 1995 effect of home processing on ascorbic acid and β-carotene content of spinach (*Spnacia Oleracia*) and amaranth (*Amaranthus tricolor*) leaves. *Plant Foods for Human Nutrition* **47** p 128
[31] Syafutri M I, Filli P and Daniel S 2006 Physical and chemical properties of mango fruit (*Mangifera indica* L.) during storage with various packaging methods *J. Teknologi dan Industri Pangan* **17** p 2-9
[32] Octaviani L F and Rahayuni A 2014 Effect of various concentrations of sugar on antioxidant activity and the level of acceptance of buni juice (*Antidesma bunius*) *J. Nutrition College* **3** p 958-65
[33] Shi J and Marc L M 2000 Lycopene in tomatoes: chemical and physical properties affected by food processing *Critical Reviews in Food Science and Nutrition* **40** p 2
[34] Regulation of the Head of the National Agency for Drug and Food Control of the Republic of Indonesia 2016 *Claim Monitoring on Food Label and Advertisement* Number 13
[35] Regulation of the Head of the Food and Drug Supervisory Agency of the Republic of Indonesia 2016 *Nutrition Status Reference* Number 9
[36] Manampiring, Aaltje E. 2008. *Prevalence of Anemia and Iron Adequacy Level in Primary School Children in Minaesa Village, Wori Sub-district, North Minahasa District* Scientific work. National Education Ministry RI. Medical Faculty. Universitas Sam Ratulangi. Manado.
[37] Shi J and Marc L M 2000 Lycopene in tomatoes: chemical and physical properties affected by food processing *Critical Reviews in Food Science and Nutrition* **40** p 2
[38] Maulida D and Zulkarnaen N 2010 *Extraction of antioxidants (lycopene) from tomatoes using mixed solvents, n-hexane, acetone, and ethanol* Undergraduate Thesis Universitas Diponegoro, Semarang.
[39] Sullivant W G, Elin M W and Patrick K C 2015 *Engineering Economy* Canada: Pearson Higher Education, Inc
[40] Ridwan E 2012 Study of iron interactions with other micronutrients in supplementation. *Panel Gizi* **35** p 49-54

IOP Conf. Series: Earth and Environmental Science **102** (2017) 012008 doi:10.1088/1755-1315/102/1/012008

Improving characteristics of goat milk yogurt drink fortified by mangosteen rind (*Garcinia mangostana Lin.*) extract

J M W Wibawanti*, Rinawidiastuti, H D Arifin and Zulfanita

Animal Science Department, Faculty of Agriculture
Muhammadiyah University of Purworejo, Central Java Indonesia*

E-mail: jekiwibawanti@gmail.com

Abstract. Peranakan Etawah (PE) milk is highly nutritive goaty flavour make it not preferable by consumer. Processing of yogurt drink made PE goat milk enhances palatability. Goat yogurt drink can be supplemented high antioxidantive activity by mangosteen rind extract. Comletely Randomized Design (CRD) was throughout the research with different concentrations of mangosteen rind extract (0, 1, 2, 3, and 4% (v/v). The organoleptic analysis showed the significant enhancement by fortification with mangosteen rind extract. The highest antioxidant activity was shown on the yogurt drink 4% v/v of mangosteen extract. In conclusion, the fortification of mangosteen rind extract was enhanced of characteristics of yogurt drink.

Keywords: goat's milk, mangosteen extract, yogurt drink

1. Introduction

Yogurt is daily product manufactured by fermentation of lactic acid bacteria such as *Streptococcus thermophilus*, *Lactobacillus bulgaricus* and *Lactobacillus acidhophilus* [1]. It provides healthy drink from natural ingredients. Yoghurt enriched with antioxidants would be improves the functionality and acceptability of consumers. Yogurt drink could be made from goat's milk. Goat milk's have high nutrition (vitamins and minerals) and benefits for body health [2]. They have also small fat globules and reduce allergic reactions, which is suitable for lactose intolerance people. The fat composition contains volatile fatty acid such as capric, caprilic and caproic. They cause unique odor and flavor of goat milk [3].

Mangosteen (*Garcinia mangostana Lin.*) is one of the most favorite fruits in Indonesia. Mangosteen characteristics have slightly acidic, sweet taste, an edible soft, brown pericarp and purple exterior [4]. The studies have shown that obtained xanthones from mangosteen have benefits for antibacterial, antioxidant, antiallergy, antiviral, antitumor and antifungal activities [5]. Xanthones compound could be obtained from peel, rind, leves, bark and whole fruit. Yellow latex of mangosteen would make the fruit tasted bitter [6].

Yogurt drinks of goat milk were produced as alternative diversification milk product. Fortification of mangosteen extract with goat milk expected to improve health's benefits. It's also contributing to the produces antioxidant beverages. Therefore, the objective of this research was to investigate the influence of fortification between mangosteen's rind on the characteristics and antioxidant activities of yogurt drink from goat milk.

2. Methods

The research has been done at the Terpadu Laboratory of Muhammadiyah Purworejo University and The Chemistry and Biochemistry Laboratory of Gadjah Mada University, Indonesia

2.1.Bacterial culture, raw goat milk and mangosteren rind

There were materials for this research: mangosteen rind, goat's milk, and yogurt culture (*Lactobacillus bulgaricus, Lactobacillus achidophilus* and *Strepcoccus thermophilus*). The fresh goat milk used in this study was supplied by Peranakan Etawah (PE) from Goat Breeding Farm in Kaligesing, Central Java Province.

2.2.Mangosteen rind extract

Mangosteen rinds were collected from local markets Kaligesing, Purworejo, Indonesia. The selection was performed fresh fruits. They were first cleaned by water. After washing, they were put and weight into sterile glass jars. Water was added and the samples were homogenized for 15 minutes. The solvent was removed. The clear supernatant was collected. Mangosteen rind was showed in Figure1.

(a) mangosteen fruits (b) mangosteen rind

Figure 1. Mangosteen

2.3.Yogurt preparation

Yogurt drinks were manufactured from goat's milk. They were pasteurized by temperature at 71^{0}C for 15 seconds and cooling to 45-43^{0}C for incubation. The yogurt cultures (2%) was added to the milk. They were separated into five lots and fortification with mangosteen rind's extract at different level. Following incubation, yogurt drinks were cooled at room temperature for 24 hour's. After incubation, the yogurt drink samples are refrigerate and analyzed. Completely Randomized Design (CRD) was throughout the research with differences level concentration of mangosteen rind extract (0, 1, 2, 3, and 4% (v/v).

2.4. pH value

Yogurt drink sample were monitored for active acidity (pH) by using a digital pH meter. The pH value The tool was standardized by buffer solution (pH 4 and 7). The electrode was dipped into the yogurt drink and then the pH was determined value.

2.5. Antioxidant activity

The antioxidant activity of yogurt drink with mangosteen rind's extract were carried out by DPPH method. Different concentration (20-100 m) of yogurt drink was taken in test tube. 5 ml of 0.1 Mm methanol solution of DPPH were added into sampels. They were shaken vigorous each of tubes. they were put into 37^{0}C for 20 minutes. The absorbance was measured at 517 nm. The % inhibition was calculated, where A_{DPPH} is the absorbance of the control and A_{sample} is the absorbance of the tested sample. The yogurt drink's were provided without mangosteen rind's extract.

$$\% \ inhibition = \frac{A_{DPPH} - A_{sample}}{A_{DPPH}} \times 100\%$$

2.6. Organoleptics analysis

Organoleptics analysis used was hedonic and test scoring [7]. Organoleptics analysis of the samples was evaluated by 25 semi trained panelist who are conversantly with yogurt and consisting of students of Muhammadiyah University of Purworejo for sensory attributes. Each panelist assessed five samples treatment. The panelists evaluated yogurt drink using 4 point scale. The panelist were responsible to his perception of flavour, color, odor, viscosity, and preference. Hedonic score for degree of preference are 4=like extremely; 3=like; 2=like moderately; 1=dislike. Statistical analysis was performed by Friedman test.

2.6. Statistical analysis
Statistical analysis calculated by SPSS programe (Statistical Packaging for social Science). All result was colected in a completely randomized design using analysis of variance (ANOVA) for each experiment. It means comparison in each experiment was performed using Least Significant Differences (LSD) and Duncan in degrees of freedom (df) 0.05.

3. Result and discussion

3.1.Antioxidant activity
In this experiment, the antioxidant activity of mangosteen rind extract was shown in Figure 2. Yogurt drink with fortification of mangosteen rind's extract was showed significant differences ($p<0,05$). Antioxidant activity of yogurt drink of mangosteen rind extract T0; T1; T2; T3; and T4 was founded value 1.09; 4.42; 5.07; 8.50; 11.83%, respectively. The highest was found in yogurt drink with fortification mangosteen rind extract with level 4%, while the lowest values were earned in control samples. The yogurt drinks with fortification 4% of mangosteen rind extracts were found the highest antioxidant activity among these extract when compared to another level. The antioxidant activity of yogurt drink could be attributed xanthoes compound of mangosteen rind extract. They were considered responsible as antioxidant activity. Differences level of mangosteen rind extract could have contributed to the antioxidant activity. It was possible that because the different in level of mangosteen rind extract

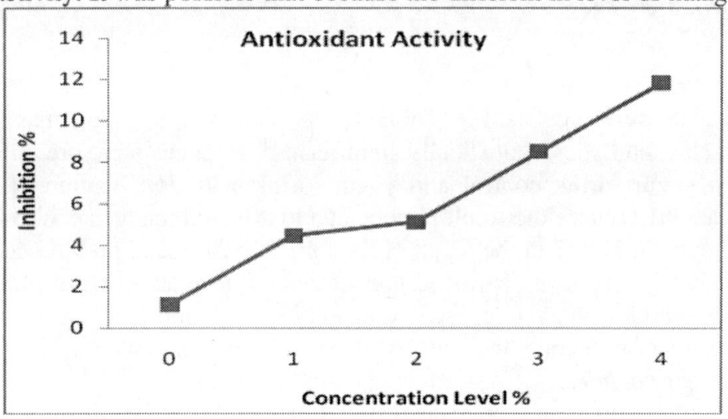

Figure 2. Antioxidant Activity of Yogurt Drink by DPPH

Antioxidant activity was correlated with xanthon content. Among these extract from mangosteen part showed antioxidant activity with different values. [5] reported the antioxidant activities (IC_{50}) with the extract from mangosteen's peel was 0.023μg/ml. These extract from mangosteen part, each part (peel, leaves, and bark) showed antioxidant activity with different IC_{50} values. [8] showed that yellow peels (48.8μg/ml) showed higher antioxidant activity compared with the pulp extract (153.2μg/ml).

3.2. pH result
The pH result of yogurt drink is shown in Figure 3. Based on the statistically analysis, that pH value control were significantly different ($p<0,05$) comparison to yogurt drink of mangosteen rind extract.

The pH value of yogurt drink extract 0, 1, 2, 3, and 4% were obtained 5.41; 5.42; 5.23; 5.24, and 5.22, respectively. Generally, yogurt drinks with mangosteen extract were found value lower than control. They have similarly value each other treatment. The pH values were related viability of microorganism and incubation temperature. It's also may be due to xanthones compounds in mangosteen extract.

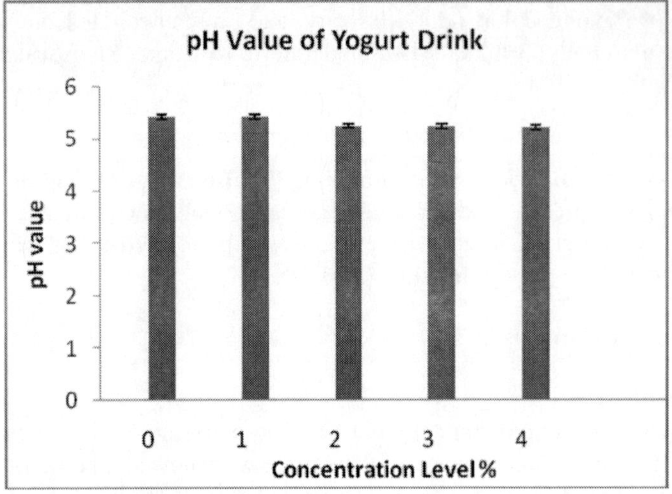

Figure 3. pH value of yogurt drink

The decreasing of pH value was observed through the storage period and this may be because of the growth of bacteria that converts lactose into lactic acid [9]. [2] reported that inoculation and incubation temperature have effects on acidity, pH value and the total bacterium counts in goat yogurt. The longer fermentation time the lower the pH values. Changes of pH value and acidity were caused amoung of bacteria. The lactose was utilized by bacteria such as *Lactobacillus casei* and *L. acidophilus* [10].

3.3. Organoleptics Result

The organoleptic data were presented in Table 1. also demonstrated that mangosteen rinds extract changes of yogurt characteristics. Statistically significant differences were presented analysis for color and odor between yogurt drink control and yogurt drink with fortification of mangosteen extract ($p<0.05$). Significant differences the result may be due to affected mangosteen rink extract. They were attributed level of extract. Hovewer the result of flavour, viscosity, and preference were no statistically significant differences ($p>0.05$) in control sample compared to manggosteen rind's extract treatment. The result of organoleptics analysis caused fermentation of lactic acid bacterium. [11] shown that characteristics composition, texture and sensory affected by bacteria such as *Lactobacillus bulgaricus* and *Streptococcus thermophiles.*

The color score of yogurt with mangosteen rind's extract were significantly higher comparison to the yogurt without extract ($p<0.05$). They were showed yellowness color in Figure 3. Organoleptics analysis for odor were significantly differences ($p<0.05$). The fortification of mongosteen rind extract were exhibited decrease goaty odor. In our opinion, this may be because of the presence of compound of mangosteen rind extract. The differences in odor and flavour were related to lactic acid bacteria's activities with suplementation of mangosteen rind extract.

Table 1. Organoleptics Analysis of Yogurt Drink

Treatment	Color	Odor	Flavors	Viscosity	Preference
0%	1.72[a]	2.12[a]	2.52	2.56	2.82
1%	2.2[b]	2.67[b]	2.08	2.60	3.00
2%	2.4[b]	2.76[b]	2.08	2.84	2.72
3%	3.12[c]	3.28[c]	2.32	2.64	3.26
4%	3.48[c]	3.60[c]	2.56	2.76	3.20

Subscript indicates significant differences ($p < 0.05$)

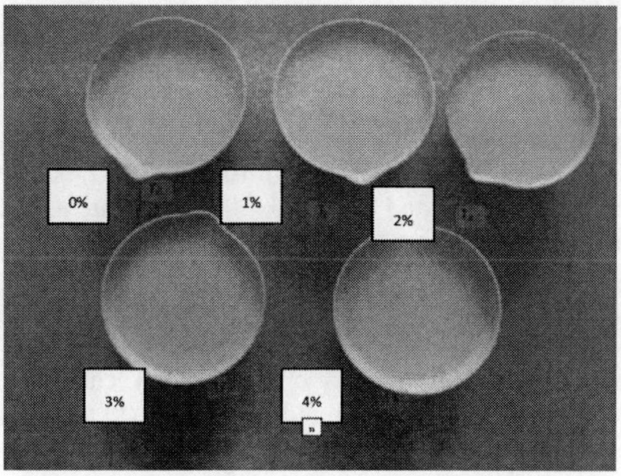

Figure 4. The color of yogurt drink

The fermentation process also reduces the bitter taste of the mangosteen rind's extract. Perhaps, those compounds stimulted viability of lactic acid bacterium. [12] reported that unique flavour of goat milk that caused fatty acids (capric, caproic and caprylic). Based on scoring test of preference, it was selected the highest score yogurt drink with fortification 3% mangosteen rind extract. Different level of mangosteen rind extract affected the formation of flavor compounds.

4. Conclusion

Fortification of mangosteen rind extracts was enhances of characteristics and the consumer preferences of yogurt drink. The results suggest that yogurt drink of mangosteen rind extract tested in this study have the potential to contribute of antioxidant baverage.

5. Refenrences

[1] Cossu M, Juliano C, Pisu R and Alamanni MC. 2009. Effect of enrichment with polyphenolic extract from Sardinian plants on physic-chemical, antioxidant, and microiological properties of yogurt. Ital. *J Food Sci.* 4 **21** 447-459

[2] Shu G, Li C and Wang C. 2014. Effect of inuculum and temperature on the fermentation of goat yogurt. Adv. *J Food Sci Technol.* **6** (1) : 68-71

[3] Boycheva, S., Dimitrov, T., and Naydenova, N. 2011. Quality characteristics of yogurt from goat ' s milk , supplemented with fruit juice. *Biochemistry*, **29** (1) 24–30

[4] Vishnu Priya VV, Jainu M, Mohan SK, Saraswati P and Gopan CS. 2010. Antimicrobial Aktivity of pericarp extract of *Gracinia mangostana Linn. Inter. J. of Pharma Sci and Research.* **18** 278-281

[5] Palakawong C, Sophanodora P, Pisuchpen S and Phongpaichit S. 2010. Antioxidant and antimicrobacterial activities of crude extract from mangosteen (*Garcinia mangostana* Lin.) part and some essential oils. *International Food Research J.* **17** 583-589.

[6] Pereira E, Barros L, Isabel FCR and Fereira. 2013. Relevance of the mention of antioxidant properties in yogurt labels: *in vitro* evaluation and chromatpgrapic analisys. *Antioxidant.* **2** 62-67.

[7] Allger LC, Miller MJ and Lee SY. 2010. Sensory and microbiological quality of yogurt drinks with prebiotics and probitics. *J. Diary. Sci* **93** 4471-4479

[8] Arazo M, Bello A, Rastrelli L, Monterlier M, Delgado L and Panfet C. 2011. Antioxidant properties of pulp and peel of yellow mangosteen fruits. Emir *J. Food Agric.* **23** (6) 517-524.

[9] Damunupola DAPR, Weerathilake WADV and Sumanasekara GSS. 2014. Characteristics of goat milk yogurt incorporate with Beetroot juice. *International J of Scientific and Research Publications.* **4** (10) 1-5

[10] Donkor, Nilmini ON, S. L. I., P. Stolic., T. Vasiljevic, and N. P. Shah. 2006) Survival and Activity of Selected Probiotic Organism in Set Type Yogurt during Cold Storage. *International Dairy Journal* **16** 1181-1189

[11] Seelee W, Tungrajaroenchai W and Natvaratat M. 2009. Development of low fat set -type probiotic yogurt from goat milk. As. *J. Food Ag-Ind.* **2** (4) 771-779.

[12] Sumarmono, J., Sulistyowati, M and Soenarto. 2015. *Fatty acids profiles of fresh milk, yogurt and concentrated yogurt from peranakan etawah goat milk.* Procedia Food Science, **3**, 216–222

Acknowledgment

This research was supported from The Ministry of Research, Technology, and Higher Education of Republic Indonesia (Penelitian Dosen Pemula) and the LPPM (Lembaga Penelitian dan Pengabdian Masyarakat) Muhammadiyah University of Purworejo.

Improvement of broiler meat quality due to dietary inclusion of soybean oligosaccharide derived from soybean meal extract

N Suthama[1*], Y B Pramono[1], and B Sukamto[1]

[1]Faculty of Animal and Agricultural Sciences, Diponegoro University
Semarang 50275 Central Java Indonesia

*E-mail : nsuthama@gmail.com

Abstract. Dietary inclusion of antibiotics as growth promoters (AGPs) in poultry production has been applied for decades worldwide, but recently AGPs have been banned due to the negative consequences for health and food safety. Soybean oligosccharide (SOS) derived from soybean meal extract is one of natural compound without carrying-over the residue to product and is consumer's health friendly. The purpose of the present study was to evaluate dietary inclusion of SOS on broiler meat quality. A total of 120 broilers of 7-day-old were allocated into 3 treatments with 4 replications (10 birds each) in completely randomized design. Treatments applied were D_1: diet without SOS, D_2: D_1 plus 0.15% SOS, and D_3: D_1 plus 0.30% SOS. Intestinal lactic acid bacteria (LAB), protein digestibility, meat protein and fat depositions, and meat cholesterol were the parameters observed. Data were statistically tested using analysis of variance and Duncan test. Dietary SOS inclusion at 0.30% (D_3) significantly ($P<0.05$) increased LAB population (7.21×10^4 cfu/g), protein digestibility (72.80%), and meat protein deposition (90.83 g/bird), but it decreased meat fat (8.27 g/bird) and meat cholesterol (37.28 mg/100 g). In conclusion, dietary SOS inclusion at 0.30% improves meat quality of broiler based on the increase in meat protein deposition with lower fat and cholesterol.

Keywords: soybean oligosccharide, meat protein and fat, meat cholesterol, broiler

1. Introduction

Increasing feed efficiency in meat production is an important key for economical broiler industry. The improvement of feed efficiency for economical meat production recently should be a serious consideration compared to the previous decade. The use of antibiotic growth promoters (AGPs) have dramatically increased for the last few decades with the purposes of improving efficiency of poultry production [1]. It has been reported previously that a large amount of AGPs have long been used by the farmers as growth promoter and desease control [1, 2, 3]. In many countries, dietary inclusion of antibiotic growth promoters or AGPs in poultry has not been reccommended or has been banned due to the concerns in relation to the development of resistant pathogenic bacteria and gain residue in animal product. The ban on AGPs has prompted nutritionists to investigate an alternative substances beneficial for improving the production performance of poultry. Using natural compound is one possible alternative to feeding antibiotic which is able to improve productivity without producing negative effects on either the host or consumen health. Intestinal health and performance improvements by providing feed without antibiotic growth promoters are the possible way to produce chicken's clean product.

It was reported that there are over 600 species of bacteria found in the chicken gastrointestinal and their development affected intestinal morphology and function [4]. Shorter villi growth is the indication of the change in intestinal morphology due to the effect of toxin produced by phatogenic bacteria and leads the reduction in nutrients absorption and decreased productivity. An alternative feed compound known as prebiotic(s) have natural characteristic without carrying-over the residue to product and consumer's health friendly. Prebiotic is a nondigestible feed compound that can be fermented by selective bacteria of the host and stimulates growth as well as activity of beneficial bacteria, and on the other hand, depresses harmfull bacteria counts, thus improves intestinal health and performance of poultry. Prebiotics are commonly derived from plant sources and one of them can be obtained from soybean meal exctract called soybean oligosaccharide (SOS). Some oligosaccharides are known to have function as prebiotic because they are not hydrolyzed by intestinal host enzyme but are able to favorably alter the gut microflora. Soybean oligosaccharide (SOS) could be fermented by selective beneficial bacteria such as *Bifidobacteria* and *Lactobacillus* and resulting short chain fatty acid (SCFA) that modulate the intestinal millieu to be healthy condition with lower pH. For example, greater SCFA production has been found in turkey [5], and in broiler [6] fed oligosaccharides associated-soybean meal. Better fermentability of functional oligosaccharide indentified by the SCFA or butyrate production was found with *in vitro* study [7].

The change in gut bacteria to be stable enhances the animal's resistence to gastrointestinal infections and provides a positive impact on intestinal wall function [8]. This phenomenon leads to the increase in nutrients digestibility and absorption since the main nutrients utilization is started in the gut due to improved villi growth. Intestinal health indicated by higher villi growth and supported by the increased beneficial bacteria counts, such as lactic acid bacteria (LAB), are the two advantages for the host caused by the additional effect of prebiotic. Fisrt, the increased LAB population the higher bile salt hydrolase (BSH) could be produced, and this enzyme has an ability to bind bile salt to be deconyugated form so that diminish the emulsion of lipid fraction and resulted the decresed fat absorption [9], and lowered cholesterol [10]. Second, growth improvement of villi could facilitate the increase in nutrients absorption, especially protein, since these two nutrients are absorbed at low pH condition. It has been described in the previous paragraph that feeding SOS bring about the lower intestinal pH due to the intermedier substances produced by LAB and support absorption. The first phenomenon is supported by the previous study in crossbred local chickens that prebiotic inulin of dahlia tuber at higher level of either powder (1.2%) or extract (1.17%) forms siginificantly decreased meat cholesterol compared to control (5.02 vs. 9.06 mg/dl) and meat fat (1.66 vs. 4.57%) [11]. Similarly, further study was consistence with the second phenomenon that of prebiotic inulin of dahlia tuber (1.2%) combined with single probiotic *Lactobacillus* sp. (1.2 mL equal to 10^8cfu/mL) increased meat protein deposition (74.48 vs. 62.731 g/bird) with the same meat Ca mass [12]. Therefore, according to the background as described previously, the effect of dietary supplementation of soybean oligosaccharide (SOS) derived from extract of soybean meal on broiler meat quality based on meat protein deposition and cholesterol content was evaluated.

2. Materials and Methods

2.1. Experimental Animal and Diet

Two hundred and eighty (280) birds of 7-day old broiler with average body weight of 112.2 ± 6.7 g were used in the present study. Experimental diet was composed of corn, rice bran, soybean meal, fish meal, $CaCO_3$, minerals and vitamins mixture, and soybean meal extract as a source of soybean oligosaccharide (SOS). Extract of soybean meal was insidentally prepared at the laboratory using methanol and water as a dilution with a ratio of 50 and 50. Protein and energy contents of the experimental diets for starter period was 22.8% and 2.933 kcal/kg, respectively, and those for finisher period was 20.9% and 2.950 kcal/kg, respectively (Table 1). The birds were provided dietary treatments for 5 wk starting on day 7 until day 42. Feed and drinking water were available *ad libitum*.

2.2. *Parameter and Experimental Design*

The experimental birds were distributed into 3 (three) treatments with 4 replications (10 birds each) in a completely randomized design. Dietary treatments were D1: ration without SOS extract, D2: D1 plus 0.15% SOS extract, D3: D1 plus 0.30% SOS extract. Intestinal lactic acid bacteria (LAB), protein digestibility, meat protein and fat mass or depositions, and meat cholesterol were the parameters measured. Data were statistically tested based on analysis of variance and continued to Duncan test at 5% probability level [13].

2.3. *Sampel Analysis*

Intestinal LAB counts in the totally collected digesta from all segments were performed using media deman rogosa sharpe (MRS) combined with eosin methylene blue agar (EMBA). Total plate count method [14] was applied to calculate colony of LAB (cfu/g) with the formula as follows : total colony multiplied by 1 per dilution factor per plate. Protein and fat mass were obtained by multiplying protein and fat percentage of meat, respectively, with the weight of all body meat (15). Determination of meat cholesterol contents were performed according to Leibermann and Burchard method using UV-visible spectrophotometer at a wavelength of 680 nm with slight modification.

Table 1. Experimental diet composition and its nutrient content*.

Ingredient	Starter	Finisher
	(%)	
Yellow corn	51.50	56.00
Rice bran	15.50	14.80
Fish meal	9.00	6.50
Soybean meal	23.30	21.50
CaCO3	0.40	0.70
Vitamin-mineral mix	0.30	0.50
Total	100.00	100.00
Nutritional content (%)		
Metabolizable energy (kcal/kg)	2932.97	2950.25
Crude protein	22.85	20.89
Ether extract	6.38	6.02
Crude fiber	6.56	6.43
Mthionine	0.42	0.37
Lysine	1.23	1.07
Arginine	1.38	1.24
Calcium total	1.04	0.96
Phosphorus total	0.65	0.59

*Chemical value of Laboratory analysis results

3. **Results and Discussion**

Lactic acid bacteria (LAB) population, protein digestibility, and meat protein mass or deposition significantly ($P<0.05$) increased due to dietary inclusion of higher level (0.30%/D_3) of soybean meal extract as soybean oligosaccharide (SOS) source (Table 2). However, dietary addition of soybean meal extract as source of SOS at the levels of either 0.15% (D_2) or 0.30% (D_3) significantly ($P<0.05$) decreased meat fat deposition and meat cholesterol content as compared to control (D_1). In case of both protein digestibility as well as meat protein deposition indicated the medium values due to feeding additional soybean meal extract at the level of 0.15% (D_2), and no significantly different compared to D_1 and D_3. It has been known that prebiotic function as "food source" that could be fermented by the beneficial endogenous bacteria of the host such as lactic acid bacteria (LAB).

3.1. Intestinal Lactic Acid Bacteria (LAB) Population

The obvious result was found in D3 treatment (Table 2) that dietary inclusion of higher SOS significantly increased LAB population. Some oligosaccharides including SOS, especially in poultry, could be fermented by LAB to produce short chain fatty acids (SCFAs), namely acetate, butyrate and propionate. These fermented products brought about the intestinal millieu to be lower pH which was favourable for the growth of LAB in particular [16], and lead to the increased their population. In addition, SCFAs as the end products of fermentation of dietary SOS, were known to have beneficial effect on energy metabolism in mammalian [17]. Soybean oligosaccharide (SOS) have been reported to be effectively fermented by intestinal *Lactobacillus* and *Bifidobacteria* to produce SCFAs and predicted to have lowering-effect on intestinal pH which condition facilitates the growth improvement of these beneficial bacteria, in one side, and inhibites the harmful or phatogenic bacteria, in other side.

Table 2. Intestinal lactic acid bacteria (LAB) population, protein digestibility, and meat quality of broiler fed soybean oligosaccharide derived from soybean meal extract.

Parameter	Soybean meal extract inclusion		
	D_1 (none)	D_2 (0.15%)	D_3 (0.30%)
LAB population (10^4 cfu/g)	4.85 b	5.75 b	7.21 a
Protein digestibility (%)	62.92 b	68.59 ab	72.80 a
Meat protein deposition (g/bird)	83.72 b	87.10 ab	90.83 a
Meat fat deposition (g/bird)	11.25 a	8.27 b	8.49 b
Meat cholesterol (mg/100 g)	42.17 a	37.28 b	36.74 b

[a–b]Mean value within the same raw followed by different superscript differ significantly ($P<0.05$).

3.2. Protein Digestibility and Meat Protein Deposition

Phenomenons of either the increased LAB population or the inhibition growth of pahtogenic bacteria lead to the improved intestinal health and activity of digestive enzymes of the host animal. Intestinal health indicated by the improved growth of villi is closely related to the better nutrients digestibility, especially protein. As a comparisson, feeding a combination of oligosaccharide derived from chicory and *Enterococcus faecium* increased villus height (774 vs. 614 μm) and villus height/crypt depth (7.13 vs. 4.86) [18]. Thus, it could be concluded that dietary inclusion of a synbiotic oligosaccharide and *Enterococcus faecium* improved intestinal morphology and nutrient absorption and lead to the increased growth performance. The present study was supported by the previous result [18] that dietary inclusion of SOS at the level of 0.30% (D_3) significantly improved protein digestibility (Table 2). Protein digestibility improvement due to stimulating effect of feeding SOS ensures the higher availability of protein supply as a substrate for protein deposition. Stimulating effect of SOS is similar to that of MOS on intestinal development and health in supporting the improvement of nutrients digestibility for better performance in poultry, especially meat prorein deposition. Intestinal development could be indicated by the growth of villi in term of height and crypt depth [18,19]. Although these two parameters, villus height and crypt depth, were not determined in present study, the results indicated that protein digestibility was consistent with the improvement of protein deposited into meat due to incusion of SOS at 0.30% in the diet (Table 2). The present study agreed with the previous finding [19] that feeding mannan oligosaccharide (MOS) increased growth of intestinal villi, indicated by the villi height, and followed by numerically increasing nutrients digestibility (protein and starch) in broilers. Villi growth would be a factor stimulates the improvement of nutrients digestibility, especially protein, and finally increase protein supply for meat protein deposition.

3.3. Meat Fat Deposition and Cholesterol Content

The decreased meat fat deposition in can be correlated with the activity of gut bacteria as affected by feeding dietary inclusion of SOS at 0.30% (D3). The present study was similar to the previous results [20] that the increased LAB was found in crossbred native chickens fed probiotic inulin derived from dahlia tuber. The present results of either fat meat mass or cholesterol concentration were decreased by dietary addition of SOS was greately possible to be due to the increasing activity of bile salt hydrolase (BSH), an enzyme produced by lactic acid bacteria (LAB) [11]. The biochemical mechanism of reduced cholesterol in particular, will be discussed in the following paragraph. Short chain fatty acids (SCFAs), product of fermentative effect of intestinal bacteria on SOS was assumed to have important contribution to the decrease in meat fat deposition. This is supported by some studies [21, 22] that SCFAs production derived from the fermentation of prebiotic inulin was due to the high activity of endogenous intestinal bacteria. One component of SCFAs, namely propionate, has an important key in controling and lowering hepatic lipogenesis and leads to an inhibition of lipogenesis which is related to the decrease in meat fat mass. The enhanced LAB growth is assumed to be correlated with higher bile salt hydrolase (BSH) that can be produced. BSH enzyme has been reported to exert its effect on the change in bile salt activity to be deconyugated form so that diminish the emulsion of lipid fraction and resulted the decresed fat absorption [9].

Metabolic function and activity of BSH related to either animal or intestinal bacteria could be connected with a given prebiotic such as SOS that was investigated in the present study. It has been described in the previous paragraph that dietary inclusion of SOS at the level of 0.15% (D_2) as well as at 0.30% (D_3) significantly ($P<0.05$) decreased meat cholesterol content (Table 2). The increased intestinal LAB population, due to feeding additional SOS, is assumed to be more BSH production and activity causing the depressed cholesterol absorption and resulted lower deposition in the meat. BSH production derived from the endogenous bacteria (LAB) and its activity were the assemble of probiotic strain which could explain cholesterol-lowering mechanism [10]. Previously discussed that gut BSH produced by LAB could change bile salt into deconyugated form [23, 24] which enzyme can not be completely able to hydrolise dietary lipid. This phenomenon have been reported [25, 26] that the decreased fat meat mass and cholesterol content were found in growing chickens fed a combination of prebiotic inulin and single probiotic *Lactobacillus* sp. The increasing number of lactic acid bacteria (LAB), due to the feeding prebiotic SOS, the higher production and activity of BSH enzyme that could be correlated with meat cholesterol reduction. In relation to the mechanism of fat digestion and absorption, feeding whatever dietary prebiotics including SOS was antagonistic with the usage of antibiotic growth promoters (AGPs). In case of chicken studies demonstrated that inclusion of AGPs in the diet significantly decreased *Lactobacillus* sp. population, the major intestinal BSH-producer [27]. Therefore, in the absence of using AGPs, BSH is a promising microbial enzyme targeting for the improvement of animal production quality with special reference to meat fat deposition and meat cholesterol content.

4. **Conclusion**

Meat quality improvement based on high protein deposition, low fat mass, and cholesterol content, is found in broiler fed dietary inclusion of soybean oligosaccharide (SOS) derived from soybean meal extract at the level of 0.30% (D_3).

References

[1] Zeweil S H, Genedy S G, Bassiouni M 2006 Effect of probiotic and medical plant supplements on the production and egg quality of laying Japanese quail hens. *Proc. 12th Europen Poultry Conference,* Verona. Pp 1-6
[2] Plail, R., 2006. The innovative power of probiotics. *Poult. Int.* **45** 34-36
[3] Akinleye S B, Iyayi E A and Afolabi K D 2008 The performance, haematology and carcass traits of broilers as affected by diets supplemented with or without biomin a natural growth promoter. *World J. Agric. Sci.* **4**(4) 467-470

[4] Apajalahti J, Kettunen A and Graham H 2004 Characterisation of the gastrointestinal microbial communities, with special reference to the chicken. *World's Poult Sci J.* **60** 223–232

[5] Jankowski J, Juskiewicz J, Gulewicz K, Lecewicz A, Slominski B A and Zdunczyk Z 2009 The effect of diets containing soybean meal, soybean protein concentrate, and soybean protein isolate of different oligosaccharide content on growth performance and gut function of young turkeys. *Poult. Sci.* **88** 2132–2140. DOI: 10.3382/ps.2009-00066

[6] Faber T A, Dilger R N, Hopkins A C, Price N P and Fahey Jr G C 2012 Effects of oligosaccharides in a soybean meal-based diet on fermentative and immune responses in broiler chicks challenged with *Eimeria acervulina. Poult. Sci.* **91** 3132–3140. http://dx.doi.org/ 10.3382/ps.2012-02364

[7] Wang H, Shi Y, Zhang S, Gao X, Liu F, Zhang H, *et al.* 2014 The vitro fermentation of six functional oligosaccharides by *Clostridium butyricum* TK2 and *Clostridium butyricum* CB8. *Food Sci. Technol. Res.* **20**(5) 1005–1011

[8] Klasing K C 2007 Nutrition and the immune system. *Br. Poult. Sci.* **48** 525–537

[9] Begley M, Hill C and Gahan C G M 2006 Bile salt hydrolase activity in probiotics. Appl. Environ. Microbiol. **72**(3) 1729–1738. DOI:10.1128/AEM.72.3

[10] Kim G-B and Lee B H 2005 Biochemical and molecular insights into bile salt hydrolase in the gastrointestinal microflora - A Review. *Asian-Australasian J. Anim. Sci.* **18**(10) 1505-1512

[11] Fajrih N, Suthama N and Yunianto V D 2014 Body resistance and productive performances of crossbred local chicken fed inulin of dahlia tubers. *Med. Pet.* **37**(2) 108-114. DOI: 10.5398/medpet.2014.37.2.108

[12] Abdurrahmana Z H, Pramono Y B and Suthama N 2016 Feeding effect of inulin derived from dahlia tuber combined with *lactobacillus* sp. on meat protein mass of cCrossbred kampong chicken. *J. Indonesian Trop. Anim. Agric.* **41**(1) 37–44. DOI: 10.14710/jitaa.41.1.37-44

[13] Steel R G D and J.H. Torrie J H 1991. *Principles and Procedures of Statistics. A Biometrical Approach.* 2nd Edition (McGraw-Hill Boock Co., New York)

[14] Fardiaz S 1993 *Mikrobiologi Pangan I.* (Gramedia Pustaka Utama Jakarta)

[15] Suthama N 2003 Metabolisme protein pada ayam kampung periode pertumbuhan yang diberi ransum memakai dedak padi fermentasi. J. Pengemb. Pet. Trop., Edisi Spesial, Hal.44–48. (In Indonesian with English abstrat)

[16] Pan X-D, Chen F-Q, Wu T-X, Tang H-G and Zhao Z-Y 2009 Prebiotic oligosaccharides change the concentrations of short-chain fatty acids and the microbial population of mouse bowel. *J. Zhejiang Univ. Sci. B* **10**(4) 258–263

[17] den Besten G, van Eunen K , Groen A K, Venema K , Reijngoud D-J and Bakker B M 2013 The role of short-chain fatty acids in the interplay between diet, gut microbiota, and host energy metabolism. *J. Lipid Res.* **54** 2325 – 2340

[18] Awad W, Ghareeb K and Böhm J 2008 Intestinal structure and function of broiler chickens on diets supplemented with a synbiotic containing *Enterococcus faecium* and oligosaccharides. *Int. J. Mol. Sci.* **9** 2205–2216. DOI: 10.3390/ijms9112205

[19] Yang Y, Iji P A and Choct M 2007 Effects of different dietary levels of mannanoligosaccharide on growth performance and gut Development of boiler chickens. *Asian-Austrasian J. Anim. Sci.* **20**(7) 1084–1091

[20] Krismiyanto L, Suthama N and Wahyuni H I 2014 Feeding effect of inulin derived from *Dahlia variabilis* tuber on intestinal microbes in starter period of crossbred native chickens. J. Indonesian Trop. Anim. Agric. 39(4):217–223

[21] Al-Lahham S H, Peppelenbosch M P, Roelofsen H, Vonk R J and Venema K 2010 Biological effects of propionic acid in humans; metabolism, potential applications and underlying mechanisms. *Biochim. Biophys. Acta.* 1801(11) 1175–1183. DOI:10.1016/j.bbalip. 2010.07.007

[22] Weitkunat, K., Schumann S, Petzke K J, Blaut M, Loh G and Klaus. S 2015 Effects of dietary inulin on bacterial growth, short-chain fatty acid production and hepatic lipid metabolism in gnotobiotic mice. *J. Nut. Biochem.* **26** 929-937. DOI: org/10.1016/j.jnutbio.2015.03.010

[23] Sudha, M. R., P. Chauhan, K. Dixit, S. Babu, & K. Jamil. 2009. Probiotics as complementary therapy for hypercholesterolemia. *Biol. Med.* **1** 1–13

[24] Rai, A. K., Debetto P and Sala F D 2013 Molecular regulation of cholesterol metabolism: HDL-based intervention through drugs and diet. *Indian J. Exp. Biol.* **51** 885–894

[25] Ooi L-G and Liong M-T 2010 Cholesterol-lowering effects of probiotics and prebiotics: A Review of *in Vivo* and *in Vitro* findings. *Int. J. Mol. Sci.* **11** 2499-2522. DOI: 10.3390/ijms11062499

[26] Abdurrahmana Z H, Pramono Y B and Suthama N 2016 Meat characteristic of crossbred Local chicken fed inulin of dahlia tuber and *Lactobacillus* sp. *Med. Pet.* **39**(2) 112 – 118. DOI: 10.5398/medpet.2016.39.2.112

[27] Lin J 2014 Antibiotic growth promoters enhance animal production by targeting intestinal bile salt hydrolase and its producers. Frontier Microbiol. **5**(33) 1–4. DOI: 10.3389/fmicb. 2014.00033

Formulation and characterization of bread using coconut-pulp flour and wheat flour composite with addition of xanthan-gum

Erminawati, W Sidik, R Listanti, and H Zulfakar

University of Jenderal Soedirman, Purwokerto, Central-Java, Indonesia

*E-mail: erminawati.w@gmail.com

Abstract. Coconut-pulp flour is coconut flour made from by-product of coconut-milk based food products. The flour contains no gluten and high fibre, which can be considered as functional potential food. Bread made from composite-flour of coconut-pulp flour and wheat flour was studied for its physic-chemical and sensory characteristics. Addition of hydrocolloid, like xanthan-gum, was aimed to provide viscoelasticity for the dough which is essential for baked product. Composite-flour proportion used in this study was; 10CPF/90WF, 15CPF/85WF and 20CPF/80WF; and xanthan gum to total flour of 0,1% and 0,4%. Variable observed were; crumb-texture, crumb-colour, taste of coconut, preference and flavour; moisture, ash, fiber and soluble-protein contents. The research showed that addition of coconut-pulp flour in the composite-flour decreased specific volume value and increased the bread texture produced. It also increased the bread moisture-content, ash-content, fibre-content and soluble protein-content. Moreover, the xanthan-gum addition resulted in decreased specific-volume value and increased texture and fiber-content of the bread produced. Overall, the sensory characteristic of crumb colour, flavour and panellist preference revealed better than control bread made from wheat flour, however its crumb texture harder compare to control bread made from wheat flour. This study showed that coconut-pulp flour potential to be developed for production of functional food.

Keywords *: bread, coconut-pulp flour, fibre, functional-food, and gum-xanthan*

1. Introduction

Coconut-pulp flour (CPF) is coconut flour made from by-product of coconut milk based food products. One method to improve the nutritional and functional value of CPF prior to use is by modification through the fermentation process [3,5]. The flour contains no gluten and high fibre, which can be considered as functional potential food. The flour can provide not only value added income to the country, but also a nutritious and healthy source of dietary fiber [7,9]. The flour can be utilised to produce various food products, such as cokies, biscuit, brownies or bread [10]. Bread is a food product based on wheat flour (WF) which poses flexible use and appropriate to be vehicle of various ingredients like CPF through formulation of composite flour [7,8]. CPF contain no gluten, therefore produce bread with poor charcteristics. It requires the addition of other gluten-like substances to optimize the characteristics of the resulting bread. Hydrocolloids compounds are often added in the manufacture of gluten-free bread, because hydrocolloids have the ability to bind starch grains, create a three-dimensional network and are emulsifiers [6,11]. The inclusion of CPF on WF to form composite flour will cause the decrease gluten content; it will decrease in development of the dough. This limitation can be overcome by the addition of colloid such as xanthan gum. The formulation of composite flour using CPF to bread with the addition of hydrocolloid such as xanthan-gum can be considered as innovation of new product by utilising the by – products from coconut. Nowadays, people were aware about the consumption of healthier food in their daily life. The innovation was in coherent with consumers demand for a healthier choice of food product [7]. Therefore, the present study was aimed to produce bread from composite flour of CPF and wheat flour (WF) with addition of xanthan gum.

2.Materials and methods

2.1 Preparation of CPF and composite flour formulation

The CPF was prepared as described in [2,4] and the formulation of the composite flour and the xanthan gum are presented in Table 1:

Tabel 1. The composite flour and the xanthan gum formulation

No. Formula	% Coconut pulp flour	5 Wheat flour	% Xanthan Gum
CIX1	10	90	0.1
C1X2	10	90	0.4
C2X1	15	85	0.1
C2X2	15	85	0.4
C3X1	20	80	0.1
C3X2	20	80	0.4

2.2 Prearation of the bread

The ingredients formulation of the bread were; *the composite flour*, *the xanthan-gum*, 1 yolk, 10% of Shortening (white margarine), 10% sugar, 0.1% yeast, 0.1% salt, 5% full cream milk and 50 mL of cold water (ice) (modification of [2]).

The bread was prepared straightly through dough preparation method. All the ingredients were mixed using mixer for 4 min. The dough was fermented for 60 min at room temperature (28 ± 2°C), weighed to 100 g, rolled, shaped, and baked in pre-heated oven operating at 250°C for 15 min. The bread loaves produced then allowed to cool then packed in low density PE bags for subsequent analysis. Bread prepared use the same ingredients, but only wheat flour and without xanthan-gum was used as the control.

2.3 Analysis

Sensory evaluation of coded samples bread was carried out by twenty semi-trained panelists of students of Food Science and Technology University of Jenderal Soedirman, Purwokerto. The characteristics evaluated were; crumb-colour, texture, flavour, taste of coconut and panellist preference. Scoring was done in 4 scales. Hedonic scale of 4 represent extremely like and scale of 1 represent extremely dislike. In addition, moisture, ash, fibre and soluble-protein contents were determined according to the standard AOAC [1] methods.

Data obtained was analysed using Analysis of variance and the samples means were separated using Duncan Multiple Range Test at 5% level.

3. Results and discussion

3.1.Sensory characteristics of bread crumb produced

The sensory evaluation of the bread produced showed that overall sensory characteristics better than control in colour and panellists preference Data regarding the organoleptic evaluation is presented in figure 1.

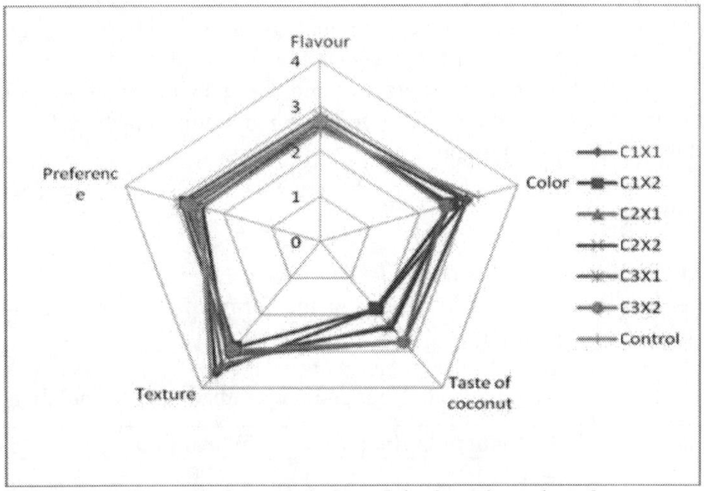

Figure 1. Sensory characteristics of the bread produced

Statistical analysis showed that the treatments combination gave significantly different on the bread crumb colour. The color scale of the bread produced were between 2.53 – 3 (yellowish white). The highest colour was obtained from C3X1 treatment. However, that the treatments combination gave no different on the bread flavor. The flavor scores of the bread of 2.53 (rather delicious) to 2.81(delicious), while for the control of 2,47(rather delicious). Furthermore, in this study the taste of coconut revealed that the scores increased with the increase of CPF concentration on the composite flour; the highest score was obtained from C3X1 and C3X2. These result can be related to panelist preference on the bread; in which showed increased with the CPF concentration on the composite flour. The highest panellists preference score given by C3X1 and C3X2; that were 2.89 and 2.73; respectively, while 2.50 for the control.

3.2. Effect of utilization of coconut-pulp flour and xanthan-gum on physico-chemical characteristics of the bread produced

Table 2. Physico-chemical characteristics of the bread produced

Formulation	Specific volume, mL/g	Texture, N	Moisture-content, %	Ash-content, %	Soluble-protein, %	Crude fiber, %
C1X1	5.57^{a}	0.19^{a}	36.19	0.87	2.89	6.56^{a}
C1X2	5.33^{ab}	0.31^{abc}	38.41	0.89	2.53	7.05^{ab}
C2X1	5.00^{ab}	0.21^{ab}	38.18	0.86	2.73	6.17^{ab}
C2X2	4.73^{ab}	0.41^{bc}	37.57	0.83	2.53	6.62^{a}
C3X1	4.07^{b}	0.29^{abc}	37.54	0.91	2.73	6.46^{ab}
C3X2	4.33^{ab}	0.47^{c}	37.13	1.00	2.76	8.34^{b}

Note : * Mean values with the same *superscript* letters within the same column do not differ significantly at 5% significant level. Values are means of triplicates.

Specific volume produced from the bread are between 4.07 – 5.57 mL/g; which is lower than that of control of 6.8 mL/g. The decrease of its specific volume was relate to the increase of CPF concentration on the composite flour. The more the CPF addition the lower the specific volume; that because of the decrease on gluten availability on the composite flour, which contributed from WF. In addition, the decrease on gluten availability on the composite flour also affect the texture of bread produced. The CPF addition also affect the bread moisture content; that relate to crude fibre content of the CPF. The moisture content of the bread produced were between 36.19% to 38.41%. The values was higher than that of control. The crude fibre content of CPF > that of WF; while xanthan-gum also contributed on the increase of the bread moisture content. In this study the water content of bread produced also influenced by the addition of xanthan-gum at concentrations of 0.1% and 0.4%. Xanthan-gum is a hydrocolloids which capable to form gel when interacting with water [6]. Xanthan-gum also has the ability to retain moisture [11]. The addition of xanthan-gum, resulted in the formation a gel structure in the bread dough which capable to bind and retain water, therefore in the baking not much water that evaporates which resulted the moisture content of bread becomes quite high. As the xanthan-gum addition was mean to overcome the limit on gluten content because of the use CPF; however, the study resulted that the increase xanthan-gum addition , decrease the specific volume of the bread produced.

4. Conclusion

In conclusion, the addition of CPF on the preparation of bread resulted in significant improvement of crude-fibre and soluble protein content. However, its decrease the specific volume, increase in the bread texture .Furthermore, the addition of xanthan-gum on the bread preparation also increase of its crude-fibre, decrease the specific volume and increase the texture of bread produced. Overall, the combination treatments has increase the crude fibre content and soluble protein of the bread produced, and has sensory characteristics better than control in colour and panellist preference.

5. References

[1] AOAC Association of Official Analytical Chemists 2000 Official methods of analysis of the association of official analytical chemists **17** issues 1-2 Gaithersburg MD USA
[2] Dipayani A A 2016 *Thesis Report Departement Food Technology Faculty of Agriculture Jenderal Soedirman University* (Purwokerto: Indonesia/Unsoed).
[3] Erminawati, Wuryatmo S and Riana L 2017 *Proc. Int. Conf. on Health Ingridient South East Asia (Jakarta)* (Jakarta: Indonesia/P3FN-UBM) 39
[4] Erminawati. Coconut-pulp Flour 2017 Fiber rich flour. Appropriate Technology Book ISBN 978 602 1643 174
[5] Igbabul B D, Bello F A and Ani E C 2014 *Sky Journal of Food Science* **3**(5) 34
[6] Manaff W Z W A 2012 Natural hydrokolids for food industries. *MARDI Technology Bulletin* 2: 135-143. Food Technology Research Centre, Serdang
[7] Muthia F and Ninik R 2013 *Research Article, Departement of Science Food Nutrient Faculty of Medical Diponegoro University Semarang* (Semarang: Indonesia/UNDIP)
[8] Olapade A A and Oluwole O B 2013 *Nigerian ood Journal* **31**(1):6
[9] Sujirtha N and Mahendran T 2005 Int. *Journal of Innovative Research in SET* **4**(8) 7344
[10] Trinidad T P, Valdez D H, and Masa D B 2003 *British Journal of Nutrition* **90** (3): 551-556
[11] Waruwu F , Julianti E and Ginting S 2015 Evaluation on physico-chemical and sensory characteristics of rice, casava, potato and soybean composite flours with addition of Xanthan Gum. *Journal of Agriculture and Food Engineering* 3(4) 448-457.

Acknowledgments

The authors wish to express their sincere gratitude to LPPM Unsoed Purwokerto, for providing the fund and facilities for this research through Riset Unggulan BLU Unsoed 2017.

Nutrition content of brisket point end of part Simental Ongole Crossbred meat in boiled various temperature

J Riyanto*, Sudibya, M Cahyadi, and A P Aji

Department of Animal Science, Faculty of Agriculture, Universitas Sebelas Maret, Surakarta, Indonesia 57126

*Email: jokoriyanto@staff.uns.ac.id

Abstract. This aim of this study was to determine the quality of nutritional contents of beef brisket point end of Simental Ongole Crossbred meat in various boiling temperatures. Simental Ongole Crossbred had been fattened for 9 months. Furthermore, they were slaughtered at slaughterhouse and brisket point end part of meat had been prepared to analyse its nutritional contents using Food Scan. These samples were then boiled at 100°C for 0 (TR), 15 (R15), and 30 (R30) minutes, respectively. The data was analysed using Randomized Complete Design (CRD) and Duncan's multiple range test (DMRT) had been conducted to differentiate among three treatments. The results showed that boiling temperatures significantly affected moisture, and cholesterol contents of beef ($P<0.05$) while fat content was not significantly affected by boiling temperatures. The boiling temperature decreased beef water contents from 72.77 to 70.84%, on the other hand, the treatment increased beef protein and cholesterol contents from 20.77 to 25.14% and 47.55 to 50.45 mg/100g samples, respectively. The conclusion of this study was boiling of beef at 100°C for 15 minutes and 30 minutes decreasing water content and increasing protein and cholesterol contents of brisket point end of Simental Ongole Crossbred beef.

Keywords: Nutrition content, brisket point, Simental Ongole, meat, temperature

1.Introduction

Good quality meat is high in protein and low in cholesterol [1]. Factors that may affect meat quality include age, muscle type and cooking method. Handling meat when cooked will affect the chemical composition of the meat so that processing is a major factor to reduce the loss of nutrients in meat [2]. Cooking methods determine the nutritional content of meat because it is influenced by the temperature and duration used. The length of boiling may affect the chemical composition of meat, while the temperature affects the strength of the myofibrillar. The duration of boiling in a water bath varies from 30 minutes to 24 hours.

Boiling at temperatures up to 80ºC, does not affect the tenderness of beef when compared to boiling with lower temperatures [2]. Boiling meat at high temperatures (90°C) will cause a shrinkage of about 30% [3]. The chemical composition of meat may change due to heating [4]. Therefore, the duration of boiling should be in accordance with the characteristics of the meat so that the nutrients in the meat remain awake and not easily lost. Based on the above statement, this research needs to be done to find out the influence of boiling duration toward the chemical composition of brisket point end meat on Simental Ongole Crossbred cattle.

2. Materials and Methods

2.1. Fattening system and slaughtering

This study uses beef that is taken from the brisket point end Simental Ongole Crossbred cattle. Cow fattened with feedlot system for 90 days. The feed to be used in the study consisted of 30% fermented rice straw, 55% basal concentrate and 15% soybean and protected *lemuru* fish oil. The nutrient content of feed is 10.45% crude protein, 6.28% crude fat, crude fiber 19.87%, organic material 88.91%, dry matter 91.10% and TDN 60.92% [1].

Carcass was prepared using Australian Meat and Livestock Corporation (AMLC) method (1993). Bulls were slaughtered in a slaughterhouse. Then, carcass and non-carcass parts were separated by splitting into two halves right along the spine into two halves of the right carcass and the left carcass. Right carcass was used for experiment. The brisket point end part of the front row carcass or "Force Quarter" was collected for nutritional content analysis.

2.2. Nutritional content analysis

Meat samples were weighed at 180-200 grams for each boiling sample. Meat samples were placed in polyethylene plastic bags tightly clipped around the sample. The sample is then cooked by dipping all the meat in a water bath using an electric stove at a time and temperature controlled. Temperature is measured using a thermometer, while time is measured using stopwatch. Boiled samples are square, boiling is done at an external temperature of 100ºC for 15 minutes and 30 minutes.

The nutrient content of meat is known to use Food Scan type 78810 foss electric A / S 69 DK Siangerupgade Denmark [1]. Place of research in Laboratory of Animal Technology and Products Faculty of Animal Husbandry Universitas Gadjah Mada Yogyakarta. Meat nutrients tested include moisture content, fat content and protein content. Considering the already homogeneous sample of 30g. Inserting into a petridish cup (15 cm in diameter) then flattened until the surface is tightly closed. Turn on the computer connected with the foodscan tool. Then turn on the foodscan machine and open the application on the computer. Insert the petridish plate into the scan room. Wait a while until the count is complete. Repeat the calculation for duplicated data to be generated. Then save the result data out of the foodscan machine.

Meat cholesterol levels are known using the Libberman Burchad method. Place of research in Laboratory Biochemistry Faculty of Animal Husbandry Gadjah Mada University of Yogyakarta. Insert 1 g of meat sample into centrifuge tube containing 10 ml of acetone alcohol (1: 1) solution, boil the tube until boiling solution. Centrifuge at 3000 rpm for 15 minutes, take the supernatant and then evaporate on the boiling waterbath until the residue remains. Perform dilution with chloroform. Residue + 2 ml chloroform + 2 ml mixture (concentrated sulfuric acid: anhydride acetate = 1:30) by vortex. 2 ml standard solution (2 mg cholesterol / 1 ml chloroform) + 2 ml mixture of concentrated sulfate and acetic anhydride, vortex. Blanko: 2 ml of chloroform + 2 ml mixture of concentrated sulfate and acetic aldehyde, by vortex. Put it into a dark room.

The reading at 680 nm after the colour turns green (10 min). Calculate with the equation $Y = 0.034190303 + 26.861515\ X$ X in mg / 100 ml (%). Calculation of cholesterol level is regression equation $Y = 0.00565 + 2.355436\ X$, where Y = absorbance, X = rate, Weight of sample = 1.9211 g,% T, = 20.2, Absorbance = 0.6946 and Dilution = 2 X.

2.3. Data analysis

Data obtained in this study was analysed Randomized Complete Design (CRD) and Duncan's multiple range test (DMRT) had been conducted to differentiate among three treatments.

3. Results and Discussion

The content of beef nutritional contents of Simental Ongole Crossbred cattle at various boiling times is presented in Table 1.

Table 1. Nutritional content of breast meat in Simental Ongole Crossbred male at various times of boiling.

Nutrient Content	Treatment			*P-value*
	TR	R15	R30	
Water content (%)	72.77 ± 1.74^{a}	70.88 ± 4.32^{b}	70.84 ± 0.68^{b}	0.05
Protein content (%)	$20, 77 \pm 1.10^{a}$	24.24 ± 0.72^{b}	25.14 ± 0.71^{b}	0.05
Fat content (%)	4.94 ± 1.95	2.19 ± 1.97	$2.33 \pm 0, 99$	NS
Cholesterol content (mg/100g)	47.55 ± 3.89^{a}	50.45 ± 3.64^{b}	50.37 ± 4.57^{b}	0.05

[a,b]Different superscript in the same row indicates significant effect ($P<0.05$). NS is not significant; TR is un-boiled beef; R15 is boiled beef at 100°C for 15 minutes; and R30 is boiled beef at 100°C for 30 minutes

The effect of boiling time on water content showed a significant difference among treatments ($P<0.05$). Decreased moisture content due to boiling process with a temperature of 100°C for 30 minutes of 4.6% compared with meat without boiling. [2] states that the higher the boiling temperature, the greater the level of the lost meat fluid until it reaches a constant level. Added by [5] boiling will affect the water content of meat because the heat will evaporate water. Mean moisture content of meat in this study, which is 70.4% still in the normal range in accordance with the opinion of [1], which states that the meat muscle contains water about 75% (65 - 80%) [6]. Cooking resulted moisture losses, on the other hand, the nutritional effects of microwaves cooking on protein, lipid, and minerals appear minimal [3].

The effect of boiling time on protein content showed a significant difference (P <0.05). This is because long boiling at 100°C for 15 and 30 minutes can increase meat protein levels by 3.5% and 4.4% compared to meat without boiling. Meat protein levels increased at 100°C for 30 minutes, this is suspected because the protein has not been perfectly denatured, especially against the type of protein that is stable to heat. This is in line [4], protein content and water content of meat have a close relationship, where it is explained that muscle protein is hydrophilic means having the ability to bind water. Another factor that affects the increase in protein levels due to shrinkage of meat is due to the shrinkage of muscle fibres due to protein denaturation by heat induction, thereby pushing the meat fluid out of the sample and evaporating [3]. The shrinking flesh has the least water content and high meat protein. The bonds that are affected by the process of protein denaturation due to boiling include hydrogen bonds (glycine), hydrophobic bonds (leucine, valine, phenylalanine, and tryptophan), ionic bonds and intra-molecular bonds such as disulfide groups in the system [2].

The high boiling temperatures in the damaged flesh have different effects on the protein, the proteins undergoing coagulation to the side by denaturation [4]. This is in the opinion [5] that the normal meat protein content of 16 - 22%. According to [1] meat protein levels are relatively fixed and are not affected by age or diet. Heat may even improve nutritive value of processing foods [4].

The results showed no effect of long boiling on fat content. This is presumably because the fat content does not damage at 100°C so that the fat content value is relatively constant. This arena of heating process can damage the cholesterol levels. According to [6] the chemical composition of meat may change due to heating. [2] reported that cooking treatments did not affect fat content in beef-based foods. However, cooking treatments may cause higher cholesterol oxidation in hamburger.

4. Conclusions

The conclusion of this study was boiling for 15 and 30 minutes decreasing water content and increasing protein and cholesterol contents at brisket point end of Simental Ongole Crossbred beef.

References

[1] J Riyanto, S D Widyawati and W Pratitis 2011 *Proc. Int. Conf. On Sustainable Animal Agriculture for Developing Countries (SAADC)* vol 1 (Nakhon Ratchasima, Thailand) p 26-29

[2] Rodriguez-Estrada M T, G Penazzi, M F Caboni, G Bertacco and G Lercker 1997 *Meat Science* **45** 365 – 375.

[3] Gwendolyn A. Cross , Daniel Y C Fung and Robert V Decareau 1982 *Critical Reviews in Food Sci. and Nut.* **16** 355 – 381.

[4] Bender, E 1966 *Int. J. Food Sci. Technol.* **1** 261 – 289.

[5] Lawrie, R A 1979 *Meat Science* (Pergamon Press, Oxford) p 105

[6] Judge M D, Aberle E D, Forrest J C, Hendrick H B and R A Merkel 1989 (Kendall Hunt Publishing Co., Dubuque, Iowa) p 85

Improvement of lipid yield from microalgae *Spirulina platensis* using ultrasound assisted osmotic shock extraction method

NP Adetya[1] and H Hadiyanto[1,2]

[1]Department of Chemical Engineering Faculty of Engineering, Diponegoro University, Jl. Prof Soedarto, SH Tembalang, Semarang, 50275, Indonesia
[2]Master Program of Environmental Science, School of Postgraduate Studies, Diponegoro University, Jl. Imam Bardjo No 3-5, Semarang, Indonesia

E-mail: naispinta26@gmail.com

Abstract. Microalgae *Spirulina sp.* has been identified as potential source of natural food supplement and food colorant. The high water content of microalgae (70-90%) causes an obstacle in biomass dehydration which requires large amounts of energy, eventually damaging the lipid in the microalgae. Therefore, the lipid must be extracted by using a suitable method which complies to wet biomass conditions. One of the methods is applying osmotic shock. This study was aimed to investigate the influence of osmotic agent (NaCl) concentration (10-30%) and extraction time (20-50 min) on yield of lipid and also to determine the optimal conditions in the extraction process through response surface methodology. The extraction was conducted at a temperature of 40°C under ultrasound frequency of 40 kHz. The result showed that the optimum yield lipid obtained was 6.39% in 16.98% NaCl concentration for 36 minutes 10 seconds.

Keywords: extraction; lipid; microalga; osmotic shock; ultrasound.

1. Introduction

Food is not only used as energy and nutrient source, but also can improve the immune system for our body. Functional food can be produced by adding ingredients which have special function for our health into the food product. Most people living in developing countries are in the state of starvation due to limited source and access to minerals, proteins, vitamins, and nutritious food [1]. The lack of functional compound can be overcome by finding alternative compounds from natural sources, one of which is microalgae. Microalgae is the most efficient plant in capturing and harnessing the solar energy for photosynthesis, fast growth and high productivity [2].

Spirulina sp. is becoming one of the most promising microalgae to be cultivated in Indonesia due to considerable potential as food source and natural coloring agent. It is because *Spirulina sp.* is the highest protein source from microbe which is 46-63% of its dry weight and also high potential compared to soy (40% dry weight) [3,4]. This microalgae has high nutritional values which are: 63% protein content, 8-14% carbohydrate and the rest are, vitamin, mineral also pigments that consists of: chlorophyll, carotene, xanthophyll and phycocyanin [4].

Components of microalgae consist of three main parts which are lipid, protein, and carbohydrate. The important stage in producing lipid from microalgae is extraction [4]. The high water content of microalgae (70-90%) requires process of biomass dehydration so that the lipid contained in the microalgae can be extracted optimally. The process of dehydration needs a large amount of energy and potentially damages the lipid contained in the microalgae. This is the reason why the use of direct extraction methods by using wet biomass microalgae as the most interesting to develop [5].

Lipid accumulated in the microalgae has tendency to increase and improve if the organism was faced with stress [6]. Lipid extraction from wet biomass can be done efficiently by doing cell disruption, for example osmotic shock. The osmotic method is chosen considering *Spirulina sp.* is a microscopic multicellular organism which has semipermeable membrane that is vulnerable to the changes of osmotic pressure [7]. Osmotic shock is the sudden change of soluble substance concentration around the cell by adding osmotic agent solution which causes a rapid change of water flow through the cell's membrane. Osmotic agent is a substance which can produce osmotic pressure that passes through the semipermeable membrane of the cell [5]. The amount of osmotic pressure is inversely proportional to the weight of osmotic agent's molecule. The amount of soluble substance in the solution will affect the osmotic pressure on microalgae's cell, as a result, water from inside the cell move to the solution and finally shrinking lysis happen. The cell's lysis will extract more lipid from inside the cell [6].

The osmotic shock method has a weakness which is the requirement of relatively long soaking time [8]. To deal with this problem, the ultrasound assisted extraction method can be used to speed up the cell disruption process [9]. The process of extraction is influenced by many factors such as particle size, type of soluble and solvent substance temperature, stirring and extraction time [5]. Optimizations are needed in order to obtain the optimum variables during the extraction process one of them is by using Response Surface Methodology (RSM) and Central Composite Design (CCD) optimization method. RSM is a technique used to find out the value of the best respond while CCD is the research design used in Respond Surface Methodology [10]. This study was aimed to investigate the effect of NaCl concentration as osmotic agent and extraction time to lipid yield. The aim was also to determine the optimal condition for lipid extraction by using ultrasound assisted osmotic shock method.

2. Materials and Methods

2.1. Materials

Spirulina platensis wet biomass was provided by CV. Neoalgae (Sukoharjo, Central Java, Indonesia), n-hexane (Merck), NaCl (Merck) and aquadest from Membrane Research Centre (MerC) Laboratory Diponegoro University.

2.2. Osmotic shock

Lipid from wet biomass extracted by using osmotic shock method [5, 11]. NaCl solution was used as an osmotic agent with concentration variations 10, 20 and 30%. Twenty grams of *Spirulina sp.* wet biomass were added to 50 ml of NaCl solution. The sample was left for 1 hour and then filtered to get the extract.

2.3. Ultrasound assisted lipid extraction

Lipid were extracted by using n-hexane solvent method with ultrasound assisted extraction (UAE) [11, 12]. n-hexane solvent was added to the sample with variation of solvent to biomass ratio 10:1. Then the sample was agitated for an hour to make it well mixed. The mixture was then sonicated with of 20, 35 and 50 minutes time variation. The mixture between residue and supernatant was separated by using filter paper and vacuum pump. The supernatant was evaporated with rotary vacuum evaporator in 60°C temperature to separate lipid from n-hexane solvent. Lipid yield was determined by using the following formula (Eq.1).

$$Total\ Lipid = \frac{weight\ lipid\ produced}{total\ weight\ of\ biomass} x\ 100\% \quad (1)$$

2.4. Optimization with RSM-CCD

Response surface method was used to observe the effect of NaCl concentration treatment and extraction time. Two independent variables of the experiment, named NaCl solution concentration (X_1) and extraction time (X_2) were used as controlled factors. Lower, upper, and center point from the design where +1 as high level, -1 low level, $\alpha = 2^n/4$ (n=number of variables or factors) as star point, and 0 is the center point. Star point was added into the design to get estimation curve in the model [10]. Based on the type of experimental design used, 12 experiments were required. In this study Design Expert 7.0.0 (Trial Version State-Ease., MN) program was used for statistical data processing. Design of the experiment shown in Table 1 and Table 2.

Table 1. Factor of the experimental formulation

Faktor	Star Point *(Low)*	Low Level (-1)	High Level (+1)	Star Point *(High)*
NaCl concentration (%)	5.86	10	30	34.14
Extraction time (min)	13.79	20	50	56.21

Table 2. Central Composite Design trial table

Run	NaCl (%)	Time (min)	Lipid Yield (Y)	
	X1	X2	Experiment	Prediction
1	10.00	50.00	3.92	3.63
2	5.86	35.00	4.51	4.53
3	34.14	35.00	1.53	1.80
4	20.00	35.00	5.91	6.24
5	20.00	13.79	2.63	2.46
6	30.00	20.00	0.92	0.90
7	10.00	20.00	4.34	4.49
8	20.00	56.21	3.12	3.58
9	20.00	35.00	6.83	6.24
10	30.00	50.00	3.82	3.36
11	20.00	35.00	6.41	6.24
12	20.00	35.00	5.82	6.24

3. Result and Discussion

3.1. The effect of NaCl concentration to lipid yield

The extraction of lipid was conducted through osmotic shock method using wet biomass (water content ≥80%). Osmotic shock is the sudden change of soluble substance concentration around the cell by adding osmotic agent solution which cause rapid change of water that flow through the cell's membrane. Osmotic agent can be either electrolyte or non-electrolyte such as NaCl, CH_3COONa, Na_2SO_4, glucose, sucrose and lactose [8].

Osmotic pressure is inversely proportional to weight of molecule. NaCl is an electrolyte solution with a small molecule weight (58.44 g/mol). Ionic osmotic agent is more effective for lipid extraction compared to non-ionic osmotic agent [5]. This is because electrolyte solution has more particles than non-electrolyte in the same concentration which cause the osmotic pressure of electrolyte solutions is higher than non-electrolyte [13]. The use of NaCl as osmotic agent successfully improved the amount of yield lipid from *Spirulina sp.* wet biomass to almost twice compared to conventional method (3.5%) [5].

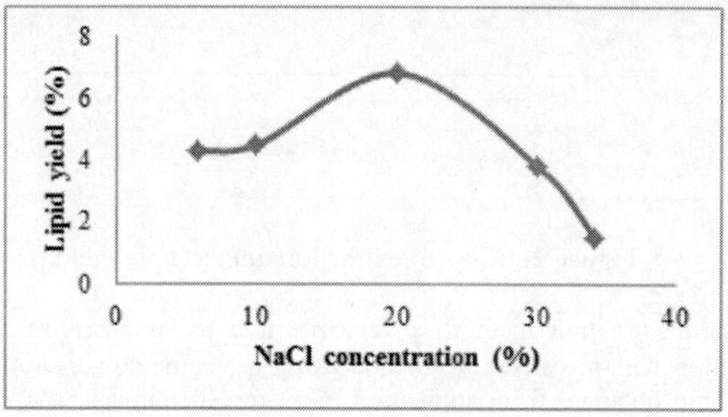

Figure 1. Effect of NaCl concentration to lipid yield.

The effect of NaCl as osmotic agent in the extraction is displayed in Figure 1. The results of the study is consistent with the initial hypotheses that osmotic shock can improve the extraction of lipid yield. In this study three different concentrations of NaCl were tested and the result was the optimum amount of lipid obtained through osmotic shock method was 6.83% in 20% concentration of NaCl solution. Yield lipid decreases as the use NaCl concentration more than 20%, this happened because not all NaCl dissociates. Even some of the Na^+ and Cl^- ions quickly convert back to NaCl and this is more likely to occur at larger concentrations [14]. This NaCl settling will reduce the dissolved ions and the Van't Hoff NaCl factor which further affects the osmotic pressure. In addition, the higher concentration of salt will also increase the consumption of salt and may cause environmental effect when released [5].

3.2. The effect of extraction time to lipid yield

The amount of extraction time shows exponential increase of yield lipid, as the time goes. At 20 minutes of extraction there was gradual increase of yield lipid until 35 minutes later it became relatively constant (Figure 2). This increase of yield lipid shows that ultrasound irradiation affect the numbers of cell disruption during extraction process. The forming of micro-bubble due to the effect cavitation that gave off large amount of energy to the surrounding biomass that cause disruption of cell wall [15]. The disruption of cell wall will release lipid from microalgae's cell. The amount of time for extraction, can increase the transfer rate of the mass but on the other side, continuous ultrasound radiation may cause the solvent become saturated and thus will decrease the lipid yield [16]. The condition occurs because the solvent is not able to dissolve the solute, or the amount of solute has exceeded the capacity of the solvent. In addition, a decrease in extraction yield after 35 minutes occurred due to lipid damage after a balanced condition was achieved with longer extraction time.

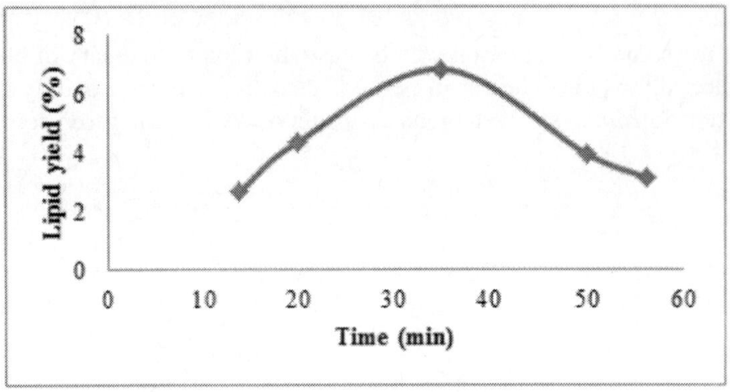

Figure 2. Effect of extraction time to lipid yield.

Ultrasound assisted extraction method is very positive for intensifying the extraction process. Conventional extraction with soxhletation method using n-hexane solvent for 2 hours was done as a comparison. Yield lipid obtained from sonication (6%) for 30 minutes was almost twice more than yield lipid of conventional extraction (3.5%). Ultrasonic wave can facilitate expansion cell and the dissolving of the components which were caused by the cell wall pores expansion. Bigger pores expansion will increase the transfer rate of mass so it will increase efficiency and reduce extraction time [17].

3.3. Optimization with RSM-CCD

The suitable model for respond is quadratic model with 0.0002 of p value which shows that the margin of error is less than 5% so the quadratic model has significant influence on yield lipid respond. Table 3 showed the quadratic selection analysis of the lipid yield.

Table 3. Quadratic selection analysis

Source	Sum of Squares	df	Mean Square	F Value	p-value Prob > F
Model	37.87	5	7.57	34.83	0.0002 (Significant)
A-Concentration	7.48	1	7.48	34.39	0.0011
B- Time	1.26	1	1.26	5.79	0.0529
AB	2.76	1	2.76	12.67	0.0119
A^2	15.09	1	15.09	69.40	0.0002
B^2	16.55	1	16.55	76.11	0.0001
Residual	1.30	6	0.22		
Lack of fit	0.64	3	0.21	0.97	0.5097
Pure error	0.66	3	0.22		
Deviation Standard = 0.47					
R^2 = 0.9667					
R^2 Adjusted = 0.9389					
R^2 Predicted = 0.8533					

Quadratic model has 0.47 standard deviation with $R^2 = 0.9667$ and 0.8533 of predicted R^2 almost close to 1. The value of coefficient X_1^2 and X_2^2 are negative, which indicate the existence of maximum stationery point from the surface respond or open bottomed graphic [18]. The polynomial equation is as follows.

$$Y = (-3.77387) + 0.32390\,X_1 + 0.41608X_2 - 0.015356X_1^2 - 0.00714X_2^2 + 0.00553X_1X_2 \quad (2)$$

where:
Y = yield lipid respond (%)
X_1 = NaCl concentration (%)
X_2 = extraction time (minutes)

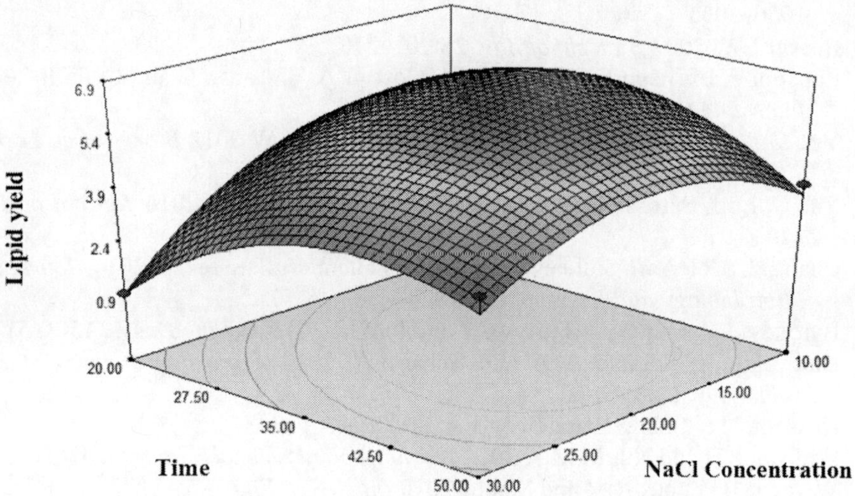

Figure 3. Interaction of NaCl concentration and time to lipid yield.

The Equation 2 is the actual equation needed to find out yield lipid respond which will be obtained if the value of needed variables are different. Interaction graphic between variables over yield lipid is shown in Figure 3. P_{value} regression of 0.000 is smaller than the degree of significance $\alpha = 5\%$. This means that all the variables (X_1 and X_2) have a significant effect in the model, both for linear and quadratic models [18]. The quadratic factor of NaCl concentration (X_1^2) and quadratic time of extraction (X_2^2) turned out to have a negative effect on the lipid yield response. Overall yield of lipid yields ranged from 0.92-6.83%, which varied according to process variables. Increased extraction time gave the most significant effect on the yield of lipid yield. Figure 3 shows that an increase in NaCl concentration from 20-35 min tends to increase lipid yield. However, a longer time leads to a decrease in lipid yield. The longer the contact time, the more Na^+ ions will move to the plasma membrane, thus reducing the osmotic pressure present [14].

From optimization procedure the optimum value for NaCl concentration (X_1) and extraction time (X_2) were obtained. By using this optimal parameters, yield lipid was recalculated with the equation 2. Additional experiment in optimal condition was conducted to confirm the result of prediction. Optimization shows that the optimal condition for extraction is $X_1 = 16.98\%$ and $X_2 = 36$ minutes 10 seconds and produce 6.39% yield lipid. The result show that extraction can be conducted with less amount of time compared to conventional method.

4. Conclusion

Extraction of lipid from microalgae by using ultrasound assisted osmotic shock method was studied for *Spirulina sp.* The result of extraction was then optimized using respond surface methodology. The result of the study shows that NaCl concentration as osmotic agent and extraction time has significant influence on yield lipid. Optimal condition was achieved by using 16.89% concentration NaCl and conducted for 36 minutes 10 seconds. The amount of yield lipid obtained in optimal condition was 6.39%.

References

[1] Saranraj P and Sivasakthi S 2014 *Asian Journal of Pharmaceutical Science & Technology.* **4**(1), 26–33.
[2] Benelhadj S, Gharsallaoui A, Degraeve P, Attia H and Ghorbel D 2016 *Food Chemistry.* **194** 1056–1063.
[3] Becker E W 2007 *Biotechnol Adv.* **25** 207–210.
[4] Lupatini A L, Bispo L O, Colla L M, Costa J A V, Canan C and Colla E 2016 *Food Research International.* 1–8.
[5] Yoo G, Park W K, Kim C W, Choi Y E and Yang J W 2012 *Bioresource Technology.* **123** 717–722.
[6] Yao S, Lu J, Sarossy Z, Baggesen C, Brandt A and An Y 2016 *Journal of Applied Phycology.* **120** 1–9.
[7] Chang J, Coleman S, Lee Y E and Schellenberg-Beaver T 2014 *Journal of Experimental Microbiology and Immunology.* **18** 54–59.
[8] Byreddy A R, Gupta A, Barrow C J and Puri M 2015 *Marine Drugs.* **13**(8) 5111–5127.
[9] Hadiyanto H, Sutanto A A and Suharto Y 2014 *Capartian Journal of Food Science and Tchnology.* **6**(1): 58-65.
[10] Hosseini N S, Shang H and Scott J A 2017 *Energy.* 3–37.
[11] Hadiyanto H and Nur M M A 2014 *World Applied Sciences Journal.* **31**(5) 959-967.
[12] Wiyarno B, Yunus R M and Mel M 2010 *Int. J. Sci. Eng* 1–5.
[13] Dunlap C A, Evans K O Theelen B, Boekhout T and Schisler D A 2007 *FEMS Yeast Research.* **7**(3) 449–458.
[14] Brown, T. L., LeMay, H. E., and Bursten, B. E 2002. Chemistry: The Central Science (10 ed.): Prentice Hall.
[15] Wang M, Yuan W, Jiang X, Jing Y and Wang Z 2014 *Bioresource Technology.* **153** 315–321.
[16] El-Shimi H I, Attia N K, El-Sheltawy S T and El-Diwani G I 2013 *Journal of Sustainable Bioenergy Systems.* **3** 224–233.
[17] Melecchi M A I, Peres V F, Dariva C, Zini C A, Abad F C, Martinez M M and Caramao E B 2006 *Ultrasonics Sonochemistry.* **13**(3) 242–250.
[18] Hadiyanto H and Suttrisnorhadi 2016 *Emirates Journal of Food and Agriculture.* **28**(4) 227–234.

IOP Conf. Series: Earth and Environmental Science **102** (2017) 012013 doi:10.1088/1755-1315/102/1/012013

The application of biotic elicitor on *Artemisia annua* L. to increase artemisinin content

I Darwati*, D Manohara, Rohimatun and H Nurhayati

Indonesian Spices and Medicinal Crops Research Institute
Jl. Tentara Pelajar No. 3. Cimanggu, Bogor, Indonesia

*E-mail: darwatikadarso2011@gmail.com

Abstract. Artemisinin-based Combination Therapy (ACT) has been recommended by WHO as an alternative to treat malaria overcoming drug resistance. The secondary metabolic products in plants, including artemisinin, can be increased by utilizing biotic elicitor from fungi. The research was conducted in Gunung Putri Research Installation, Cipanas, West Java from 2010 to 2011. *Phytophthora sp.* from eggplant and *Colletotrichum* sp. from *Artemisia annua* were applied as biotic elicitor. The types of biotic elicitor applied to the plants were 1) the medium of potato dextrose broth were inoculated with fungi and harvested after 10 days (filtrate), 2) powdery mycelium of both fungi. There were 16 treatments: control negative, control positive (uninoculated medium) 1%, 2%, 3% (v/v)], *Phytophthora* sp. filtrate [1, 2% and 3% (v/v)], *Colletotrichum* sp. filtrate [1, 2% and 3% (v/v)], *Phytophthora* sp. mycelium [1%, 2% and 3% (w/v)], *Colletotrichum* sp mycelium [1%, 2% and 3% (w/v)]. The elicitor application increased plant production by 26.21% and artemisinin yield by 72% compared to control. Furthermore, the artemisinin production of the plants treated with medium inoculated with 2% filtrate of *Phytophthora sp* (FP2) (25.19 kg/ha) and 1% powdery mycelium of *Colletotrichum sp* (MC1) (26.42 kg/ha) were higher than control (K) (11.17 kg/ha).

Keywords: *biotic elicitor, Artemisia annua L., artemisinin content*

1. Introduction

Artemisinin is active compound of *Artemisia annua* L. which is sub-tropical plant. It is introduced from China and spread in Vietnam and Malaysia. Drugs contain artemisinin have been proved particularly effective to treat severe malaria, even for multi drug-resistant malaria [1]. Currently, WHO recommended malaria drug artemisinin combination based therapy (ACT) to treat malaria [2].

Malaria is one of the most important tropical diseases [3], causes great human suffering and loss of life in affected countries. Indonesia is one of the countries with high number of malaria infection, hence requires high demand of artemisinin drug for malaria. However, in Indonesia, artemisia has not been cultivated widely and is planted in limited area only as trial.

Artemisinin, a sesquiterpene lactone containing an endoperoxide bridge isolated from artemisia plants was effective against both drug–resistant and cerebral malaria caused by *Plasmodium falciparum* [4] [5]. Artemisinin had efficacy as antiviral, antiparasitic, antiinflammmation and anticancer [6]. In addition to being used as a drug, artemisinin also has phytotoxic activity as natural herbicide. The artemisinin application was reported could decrease the radicle length and seedling vigor index of *Arabidopsis thaliana* [7].

Artemisinin is obtained by extracting the leaves and stems of artemisia plants. These compounds belong to sesquiterpen lactone groups with endoperoxide bridges that are rarely encountered in nature. Artemisinin is a compound difficult to synthesize. Thus, the easiest and cheapest way to obtain artemisinin were to extract directly from the plant [8]. In Indonesia, the planted accessions had very low artemisinin content, ranged from 0.25-0.41% [9]. Thus, it considered inefficient to develop the plants in industrial scale.

In recent years, several elicitors have been used for stimulating artemisinin biosynthesis to increase artemisinin content. The elicitors can be divided into two groups, biotic and abiotic elicitors. Abiotic elicitors consisted of metal ions and inorganic components. The biotic elicitors included fungi, bacteria, viruses, herbivores, the components of plant cell walls, chemical compounds released by plants parts attacked by pathogens or herbivores, miconazole, chlorocholine chloride, homobrassinolide and crude extracts of fungal mycelia of *Verticillium dahliae, Rhizopus stolonife* and *Colletotrichum dematium* [10]. Elicitors may lead to cellular processes and regulatory principles for the activation of secondary metabolite biosynthesis. This led to the presence of extracellular or intracellular signals captured by receptors on the surface of plasma membrane or endomembrane. The elicitor signal was perceived as the initiation of the transduction signal to form de-novo activation or biosynthesis, or was a transcription factor expressing the biosynthetic gene regulation involved in plant secondary metabolism. Further, it yielded an enzyme for secondary metabolite biosynthesis [11]. The use of elicitors was proved more effective to increase secondary metabolite content because it was more natural, hence easier to enter the plant metabolism system. The types of elicitors commonly used were mycelial extract, filtrate or toxin from the fungi. The use of mycelial extract from fungus *Colletotrichum* sp. increased artemisinin content from 0.8 mg to 1 mg per dry weight [12]. The objective of this research was to obtain the type of elicitor and its concentration to increase artemisinin content.

2. Materials and methods

Phytophthora sp was isolated from eggplant and endophytic fungi (*Colletotrichum* sp) from artemisia. Both fungi were propagated for the examination of their capability as elicitor on artemisia plant in the field. The endophytic fungi was propagated by culturing the fungi in 50 ml dextrose potato broth in the erlenmeyer flask and was incubated at room temperature for 30 days. The mycelium was then separated from the filtrate, washed several times with sterile distillation water, grinded using mortar and centrifuged for 5 minutes. The supernatant was suspended with sterile distillation water following the concentration treatments, then autoclaved for 20 minutes at 120°C. The mycelium and filtrate (the media used to culture *Phytophthora* sp and *Colletotrichum* sp) were used as elicitor.

Artemisia seeds were germinated in the nursery for one month and transplanted into polybag, then planted in the field after 2 months. The basic fertilizer was applied one month after planting (MAP), consisted of 0.5 kg manure, 3.5 g Urea, 1.5 g SP-36 and 1.5 g KCl/plant. Plant maintenance such as weeding was performed once a month and pest control was done if it was necessary.

The elicitor was applied by spraying 2 months-old plants (2 MAP). Water was used as control negative whereas un-inoculated dextrose potato broth as control positive. The elicitor application was performed every week until two weeks before harvesting. The plants were harvested at 6 MAP or when 10% of the population started flowering.

The trial was arranged in randomized block design consisted of 16 treatments: control negative (water); control positive (1%, 2%, 3% (v/v) of un-inoculated medium); *Phytophthora* sp. filtrate (1, 2%, 3% (v/v); *Colletotrichum* sp. filtrate (1, 2%, 3% (v/v)); *Phytophthora* sp. mycelium (1, 2%, 3% (w/v)); *Colletotrichum* sp mycelium (1, 2%, 3% (w/v)). Every plot consisted of 20 plants, repeated three times.

Parameters observed were plant height, leaf dry weight, stem dry weight, total biomass dry weight and artemisinin content. The data were analyzed with ANOVA and contrast orthogonal for further analysis if the treatments were significantly affected the parameters.

3. Result and discussion

3.1. Plant growth

Naturally, plants were able to express their natural defence and triggered the defence signal through intermediary receptor by recognizing microbe structure known as pathogen-associated molecular patterns (PAMPs) or elicitor [13]. Elicitor was considered as signal molecule activated transduction signal wave (Cascade) and caused the activation and expression of gene related to secondary metabolite biosynthesis [14].

The data was analysed using orthogonal contrast and indicated significant differences among the treatments at 5% level of significance (Table 1). The parameters affected by elicitor application showed significant differences compared to both control negative and control positive, except for plant height. However, there was no significant effect between the elicitor from the filtrate of *Phytophthora* sp. and *Colletotrichum* sp on parameters observed (Table 1). The biochemical compound from fungi secretion on growth medium might have low concentration or the fungi secreted no chemical compound into growth medium. The pathogen lived in host plant rich in carbohydrate would arrange the production of polysaccharide-degrading enzyme. A small alteration in composition, structure or accessibility of host-plant carbohydrate was able to change the composition of polysaccharide-degrading enzyme secreted by pathogen, both qualitatively and quantitatively [15]. The phytohormones involved in plant defence, such as gibberellin [16] and salicylic acid [17], also possessed important role in artemisinin accumulation. However in this experiment, chemical compound secreted by fungi on filtrate was unable to give signal to induce secondary metabolite increase in artemisia. Filtrate treatment might be unable to increase gibberellin which functioned to trigger plant growth. Thus, plant growth parameters represented by plant height and total biomass dry weight, were not significantly different (Table 1).

Elicitor from both fungi mycelium gave significant effect on stem and biomass dry weight. The *Phytophthora* mycelium 2% gave higher stem and biomass dry weight than other mycelium treatments (Table 1). Mycelium was not dissolved in water. However, proper handling such autoclave treatment or alkali application could dissolve active elicitor from mycelium cell. The dissolve mycelium-cell comprised of 1,3-glucan or glycoprotein contained mannan. The chemical composition of mycelium linked to plant cell-wall was detected by plant cell during phatogenesis process. The un-dissolved elicitor was released from fungi mycelium wall soon after contacted with plant tissue. This process indicated that the released-elicitor was responsible for initiating the accumulation of phytoalexin and pythohormone in fungi-infected plants [18]. The increase of phytohormone would enhance biomass weight of the plant.

Table 1. The effect of type and concentration of biotic elicitor on artemisia plant growth.

Treatments comparison	Plant Height (cm)	Leaf dry weight /ha (ton)	Stems dry weight /ha (ton)	Biomass dry weight /ha (ton)
Control vs Elicitor				
Control	176.33 a	0.72 a	0.73 a	1.45 a
Elicitor	178.17 a	**0.87 b**	**0.96 b**	**1.83 b**
Control negative vs Control positive				
Control negative	176.33 a	0.72 a	0.73 a	1.45 a
Control positive	182.16 a	0.66 a	0.70 a	1.35 a
Control negative vs Filtrate				
Control negative	176.33 a	0,72 a	0.73 a	1.45 a
Filtrate	182.28 a	**0.84 b**	**0.95 b**	**1.78 b**
Control negative vs Mycelium				
Control negative	176.33 a	0.72 a	0.73 a	1.45 a
Mycelium	172.96 a	**1.00 b**	**1.11 b**	**2.11 b**
Control positive vs Filtrate				
Control positive	182.16 a	0.66 a	0.70 a	1.35 a
Filtrate	182.28 a	**0.84 b**	**0.95 b**	**1.78 b**
Control positive vs Mycelium				
Control positive	182.28 a	0.84 a	0.95 a	1.78 a
Mycelium	172.96 a	**1.00 b**	**1.11 b**	**2.11 b**
Filtrate *Colletotricum* vs Filtrate *Phytophthora*				
Filtrate *Colletotricum*	179.89 a	0.88 a	0.98 a	1.87 a
Filtrate *Phytophthora*	184.67 a	0.87 a	0.95 a	1.82 a
Colletotricum mycelium vs *Phytophthora* mycelium				
Colletotricum mycelium	170.45 a	1.01 a	1.02 a	1.99 a
Phytophthora mycelium	177.00 a	1.03 a	**1.20 b**	**2.23 b**
1% *Colletotricum* filtrate (v/v)	191.33 a	0.86 a	1.12 a	1.99 a
2% *Colletotricum* filtrate (v/v)	172.67 a	0,86 a	0.97 a	1.83 a
3% *Colletotricum* filtrate (v/v)	175.67 a	0.89 a	0.86 a	1.75 a
1% *Phytophthora* filtrate (v/v)	189.67 a	0,85 a	0.93 a	1.78 a
2% *Phytophthora* filtrate (v/v)	181.33 a	0,86 a	1.02 a	1.88 a
3% *Phytophthora* filtrate (v/v)	183.00 a	0,89 a	0.89 a	1.79 a
1% *Colletotricum* mycelium (w/v)	165.67 a	1.09 a	1.13 a	2.22 a
2% *Colletotricum* mycelium (w/v)	181.00 a	1.03 a	1.08 a	2.11 a
3% *Colletotricum* mycelium (w/v)	164.67 a	0.90 a	0.84 b	1.74 b
1% *Phytophthora* mycelium (w/v)	186.33 a	0.99 a	1.13 a	2.12 a
2% *Phytophthora* mycelium (w/v)	174.67 a	1.04 a	**1.28 a**	**2.32 a**
3% *Phytophthora* mycelium (w/v)	170.00 a	1.07 a	1.19 a	2.26 a

Note : Number followed by the same letters in the same column were not significantly different at 5% level of significance

3.2. *Artemisinin production of artemisia*

Gibberelin (GA) was diterpenoid compound and involved in plant defense response and plant growth. GA was synthesized following terpenoid pathway, the same pathway as artemisinin biosynthesis. Artemisinin was synthesized through farnesyl diphosphate (FPP), which was the first intermediary compound. On the other hand, GA was produced from geranyl diphosphate (GGPP), which was condensed product from isopentenyl diphosphate (IPP) and geranyl diphosphate (GPP) [16]. Therefore, GA increase would enhance plant growth, represented by increase in biomass weight. The same pathway in biosynthesis of GA and artemisinin caused the competition in GA and artemisinin production within the plant. The increase in GA biosynthesis, indicated by high biomass production, would be followed by artemisinin content decline. As shown in Figure 1, filtrate treatments generated lower biomass weight (1.44 -1.99 ton/ha) than mycelium treatments (1.64-2.32 ton/ha). On the contrary, the low biomass weight (on filtrate treatments) was followed by higher artemisinin content (1.03%-1.34%) than mycelium treatments (0.95%-1.39%). The exogenous GA_3 application on artemisia increased artemisinin contents [16]. This was linked to the enhancement of key enzyme expression in biosynthesis pathway of artemisinin. Furthermore, exogenous GA continuously increased artemisinin content from vegetative phase until flower initiation and produced higher leaf weight than control plants. Consequently, artemisinin content of plants applied with exogenous GA was much higher than control plants.

Artemisinin yield related to biomass dry weight because it was the multiplication of biomass dry weight and artemisinin content. Generally, artemisinin content tended to be low on plants with high biomass dry weight. However, plants with high biomass dry weight was also able to produce high artemisinin content as indicated by *Colletotrichum* sp mycelium 1% treatment (MC1), resulted in high artemisinin yield (26.42 kg/ha) (Figure 1).

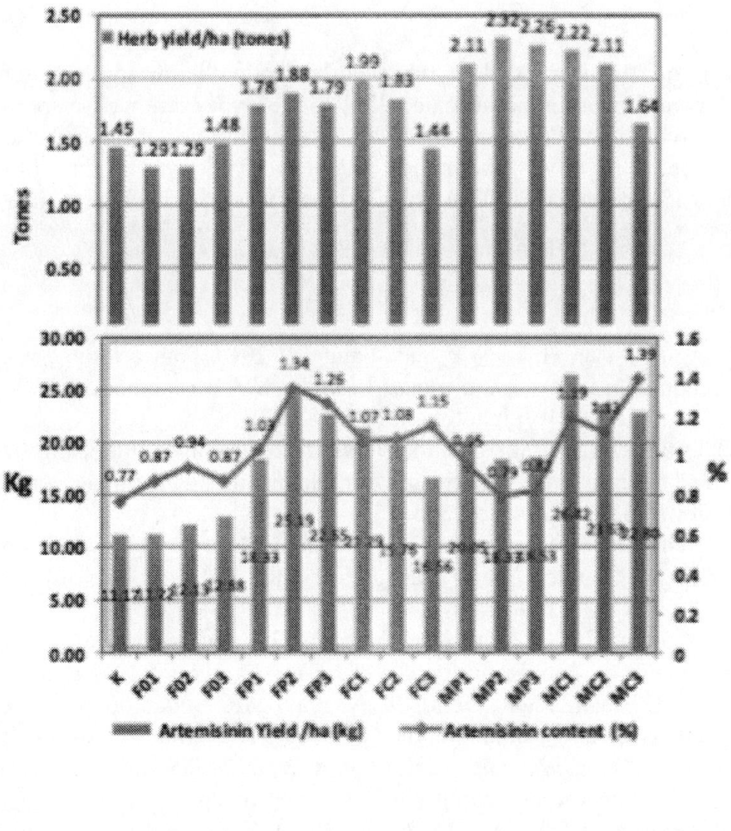

Note :

K		Control negative (water)
F01	:	Filtrate (un-inoculated medium) 1%
F02	:	Filtrate (un-inoculated medium) 2%
F03	:	Filtrate (un-inoculated medium) 3%
FP1	:	*Phytophthora* sp filtrate 1%
FP2	:	*Phytophthora* sp filtrate 2%
FP3	:	*Phytophthora* sp filtrate 3%
FC1	:	*Colletotrichum* sp filtrate 1%
FC2	:	*Colletotrichum* sp filtrate 2%
FC3	:	*Colletotrichum* sp filtrate 3%
MP1	:	*Phytophthora* sp mycelium 1%
MP2	:	*Phytophthora* sp mycelium 2%
MP3	:	*Phytophthora* sp mycelium 3%
MC1	:	*Colletotrichum* sp mycelium 1%
MC2	:	*Colletotrichum* sp mycelium 2%
MC3	:	*Colletotrichum* sp mycelium 3%

Figure 1. The application of *Colletotrichum sp.* and *Phytophthora sp* elicitor on biomass dry weight, artemisinin content and artemisinin yield of artemisia.

4. Conclusion

Artemisinin content of artemisia could be enhanced by biotic elicitor application derived from fungi. The type of elicitor was important because not all of elicitor type were well-responsed by plants. Elicitor derived from mycelium of *Phytophthora* sp at 2% concentration increased biomass dry weight by 60% compared to control negative. The highest artemisinin yield was produced by *Colletotrichum sp* powdery mycelium 1% treatments (MC1) (26.42 kg/ha), was 136.5% higher than control (K) (11.17 kg/ha).

References

[1] Pan WS, Zheng LP, Tian H, Li WY and Wang JW 2014 Transcriptome responses involved in artemisinin production in *Artemisia annua* L. under UV-B radiation. *J Photochem Photobiol B Biol* **40** 292–300. doi:10.1016/j.jphotobiol.2014.08.013.

[2] Ferreira JFS, Luthria DL, Sasaki T and Heyerick A 2010 Flavonoids from *Artemisia annua* L. as antioxidants and their potential synergism with artemisinin against malaria and cancer. *Molecules* **15**:3135–70. doi:10.3390/molecules15053135.

[3] Hasugian AR, Tjitra E 2014 Artesunat-amodiakuin and klorokuin for vivaks malaria treatment in Puskesmas Kopeta, Maumere, Nusa Tenggara Timur, 2007 (in Bahasa) *Media Litbangkes* **24** 161–8.

[4] Devi S and Premkumar R 2012 Physicochemical analysis of groundwater samples near industrial area, Cuddalore District, Tamilnadu, India. *Int J ChemTech Res* **4** 29–34. doi:10.7904/2068.

[5] Zare Mehrjerdi M, Bihamta MR, Omidi M, Naghavi MR, Soltanloo H and Ranjbar M 2013 Effects of exogenous methyl jasmonate and 2-isopentenyladenine on artemisinin production and gene expression in *Artemisia annua*. *Turk J Botany* **37** 499–505. doi:10.3906/bot-1206-12.

[6] Ho WE, Peh HY, Chan TK and Wong WSF 2014 Artemisinins: Pharmacological actions beyond anti-malarial. *Pharmacol Ther* **142** 126–39. doi:10.1016/j.pharmthera.2013.12.001.

[7] Hussain MI, Reigosa MJ 2014 Evaluation of herbicide potential of sesquiterpene lactone and flavonoid: Impact on germination, seedling growth indices and root length in *Arabidopsis thaliana*. *Pakistan J Bot* **46** 995–1000.

[8] Aftab T, Ferreira JFS, Khan MMA and Naeem M 2014 *Artemisia annua* - Pharmacology and biotechnology doi:10.1007/978-3-642-41027-7.

[9] Gusmaini and Nurhayati H 2007 Potential of *Artemisia annua* L cultivation development in Indonesia (in Bahasa). *Perspective* **6** 57–67.

[10] Zhang B, Zou T, Lu YH and Wang JW 2010 Stimulation of artemisinin biosynthesis in *Artemisia annua* hairy roots by oligogalacturonides. *African J Biotechnol* **9** 3437–42.

[11] Vasconsuelo A and Boland R 2007 Molecular aspects of the early stages of elicitation of secondary metabolites in plants. *Plant Sci* **172**:861–75. doi:10.1016/j.plantsci.2007.01.006.

[12] Wang JW, Zhang Z and Tan RX 2001 Stimulation of artemisinin production in *Artemisia annua* hairy roots by the elicitor from the endophytic *Colletotrichum* sp. *Biotechnol Lett* **23** 857–60. doi:10.1023/A:1010535001943.

[13] Mao J, Liu Q, Yang X, Long C, Zhao M, Zeng H, *et al* 2010 Purification and expression of a protein elicitor from *Alternaria tenuissima* and elicitor-mediated defence responses in tobacco. *Ann Appl Biol* **156** 411–20. doi:10.1111/j.1744-7348.2010.00398.x.

[14] Gadzovska-Simic S, Tusevski O, Antevski S, Atanasova-Pancevska N, Petreska J, Stefova M, *et al* 2012 Secondary metabolite production in *Hypericum perforatum* L. cell suspensions upon elicitation with fungal mycelia from *Aspergillus flavus*. *Arch Biol Sci* **64** 113–21. doi:10.2298/ABS1201113G.

[15] Albersheim P, Jones TM and English PD 1969 Biochemistry of the cell wall in relation to infective processes. *Annu Rev Phytopathol* **7**:171–94. doi:10.1146/annurev.py.07.090169.001131.

[16] Banyai W, Mii M and Supaibulwatana K 2011 Enhancement of artemisinin content and biomass in *Artemisia annua* by exogenous GA_3 treatment. *Plant Growth Regul* **63** 45–54. doi:10.1007/s10725-010-9510-9.

[17] Pu G Bin, Ma DM, Chen JL, Ma LQ, Wang H, Li GF, *et al* 2009 Salicylic acid activates artemisinin biosynthesis in *Artemisia annua* L. *Plant Cell Rep* **28** 1127–35. doi:10.1007/s00299-009-0713-3.

[18] Yoshikawa M, Matama M and Masago H 1981 Release of a soluble phytoalexin elicitor from mycelial walls of *Phytophthora megasperma* var. *sojae* by soybean tissues. *Plant Physiol* **67** 1032–5.

Acknowledgment

The authors would like to express the gratitude for Indonesian Agency for Agricultural Research and Development (IAARD) for funding the research; Ir. Rosita SMD, MS for her guidance in the research; laboratory and field technician of Indonesian Spices and Medicinal Crops Research Institute (Iim Rochimat and Susi Noorsyamsiah).

Nutritional composition, glycemic index, glycemic load, and organoleptical quality of glucomannan-enriched soy milk ice cream

S Sa'adah, O M Candra, G Nugrahani, A Pramono, and D N Afifah*

Department of Nutrition Science, Faculty of Medicine, Diponegoro University, Jl. Prof. Soedarto, SH Tembalang, Semarang, Jawa Tengah, Indonesia 50275

*Email : d.nurafifah.dna@gmail.com

Abstract. Over the past decades, the number of childhood obesity cases has increased significantly, which led to an increase in the number of adults suffering from degenerative diseases such as diabetes mellitus (DM). Glucomannan-Enriched Soy Milk Ice Cream (GSMIC) may prevent obesity in children. The aim of the study was to test the level of carbohydrates, protein, fat, dietary fiber, glycemic index, glycemic load, and organoleptic quality of GSMIC. This experiment used a completely randomized design to test three formulations of glucomannan flour and soy milk (0.5%, 1.5%, and 2.5%). The products were tested for nutritional composition, and evaluated on glycemic index, glycemic load, and organoleptic quality. GSMIC 2.5% had higher levels of dietary fiber and high carbohydrate, protein, and fat content compared to ice cream (3.99%, 30.7%, 1.50%, 1.33%, respectively). The glycemic index of ice cream and 2.5% GSMIC were 75.83 (75%) and 51.48 (51%), respectively, while the glycemic load of ice cream and 2.5% GSMIC were 9.04 and 11.61, respectively. Based on the organoleptic analysis, formulation preferred by the panellists was 2.5% glucomannan flour. Glucomannan flour affected the level of carbohydrates, protein, fat, dietary fiber, glycemic index, glycemic load, and organoleptic quality in soy milk ice cream.

Keywords: glucomannan flour, soy milk, glycemic index, glycemic load, organoleptic quality

1. Introduction

Obesity is a nutritional problem caused by positive energy balance [1]. The prevalence of overweight and obese cases in children has increased gradually from 4.2% in 1990 to 6.7% in 2010. This trend is expected to reach up to 9.1% or 60 million in 2020. In Indonesia, according to the Basic Health Research, 10.8% of children aged 5 to 12 years old will be overweight and 8.8% will be obese in 2020 [2]. Childhood obesity increases the risk of other various diseases such as cardiovascular disease and diabetes mellitus (DM) in adulthood.

DM is a metabolic disease characterized by elevated blood sugar levels caused by impaired insulin secretion and/or insulin function [3]. According to the Basic Health Research, the percentage of cases of DM has increased from 1% in 2007 to 2.1% in 2013 [4]. Obesity is a major risk factor of DM [5]. One solution to overcome this problem is by consuming foods containing high fiber with a low glycemic index.

Ice cream is a food product favored by all people in Indonesia especially children and adolescents. The consumption of ice cream in Indonesian increased from 0.3 liters per capita in 1999 to 0.5 liters per capita in 2004. In Indonesia, ice cream is generally made from cow's milk and is high in fat. However, we propose an alternative formulation of ice cream by replacing the cow's milk with soy milk. Soy milk has protein content and amino acid composition similar to cow's milk. In addition, soy milk contains minerals and vitamins in sufficient quantities, contains no lactose, and is low-fat, cholesterol-free, and high in nutrition [6].

Ice cream is usually made using Carboxyl Methyl Cellulose (CMC) as a thickening agent. CMC is one of the most widely used additives in the food industry as a thickening agent. In this study, CMC was replaced with glucomannan flour, obtained from konjac (*Amorphophallus konjac*) and used as a thickener or stabilizer to improve the nutritional value of the soymilk ice cream. Glucomannan flour derived from konjac contains 5% crude fiber is a soluble fiber that is high in hydrocolloid, low in calories, and has a low glycemic index. It is widely used in the food industry both as a functional food or food additive. Glucomannan flour can be used as an alternative stabilizer or thickening agent and reduces blood glucose and lipids in pre-clinical studies of DM [7]-[9]. The use of food products such as ice cream enriched by glucomannan flour is a new innovative method to produce healthier ice cream for patients with obesity and DM. The aim of the study was to test the levels of carbohydrates, protein, fat, dietary fiber, glycemic index, glycemic load, and the organoleptic quality of Glucomannan-Enriched Soy Milk Ice Cream (GSMIC).

2. Materials and Methods

This study was an experimental study with a complete randomized design. The independent variable of this research was glucomannan flour and soy milk ice cream, while the dependent variable was the level of fiber, carbohydrates, protein, fat, glycemic index, glycemic load, and organoleptic quality.

2.1. Food Production

There were 3 treatments and 1 control (t = 4) noted by symbols P0 (0% glucomannan flour, 0.3% CMC; 83.7% soy milk), P1 (0.5% glucomannan flour; 83.5% soy milk), P2 (1.5% glucomannan flour; 82.5% soy milk), P3 (2.5% glucomannan flour; 81.5% soy milk). Soy milk and glucomannan flour with 3 different concentrations were obtained from a preliminary study based on the composition of a stabilizer in ice cream making, whereas the control of the standard stabilizer formula was CMC. Each treatment group was analyzed in repeated testing including analysis of levels of dietary fiber, carbohydrate, protein, and fat, while the glycemic index test on ice cream and 2.5% GSMIC were conducted without repetition.

Soy milk was made using ratio of soybean and water of 1:4. Soy milk was made by sorting soy beans and soaking, washing, grinding, filtering, and cooking by heating at a temperature of 80°C for 10-15 minutes. Ice cream was made from raw materials including soy milk, 10% stevia sugar Tropicana Slim® Alergon, 5% whipped cream, vanilla essence, water, and 1% salt. The ice cream was made by mixing all the ingredients, stiring, and heating at 80°C for 10 minutes. Then the mixture was put into the ice cream maker (Donper, USA) for 30-60 minutes and frozen for 24 hours at a temperature of -45° to (-23)°C.

2.2. Product Analysis

The protein concentration test was carried out with the Kjeldahl method, the fat concentration test was carried out with the Soxhlet method, the carbohydrates was carried out by different method, dietary fiber was carried out with the enzymatic method, and the glycemic index was carried out with the Incremental Area Under the Blood Glucose Response Curve (IAUC) method [10]-[12].

For the glycemic index test, a minimum of 8 panelists were used [12]. To avoid drop out, this study used 9 subjects. One day prior to the experiment, the panelists fasted for 10 hours (drinking water allowed) starting from 10.00 pm to 8:00 am the next day. The criteria for subjects were Body Mass

Index (BMI) between 18.5-22.9kg/m2, fasting blood glucose <110mg/dl, and age 20-24 years. The capillary blood of the subjects was taken to measure fasting glucose levels. Furthermore, panelists were asked to consume pure glucose, ice cream, and GSMIC. Blood samples were taken every 30 minutes (minutes 30, 60, 90, and 120) after testing the food for 2 hours. Each treatment was given within three days to avoid bias. The glycemic index test was carried out using a glucometer (Autocheck GCU 3 in 1, Taiwan). Secondary data was collected using the Indonesian National Standard (SNI) for ice cream quality requirements.

2.3. Statistical Analysis
The results of the analysis of the nutritional composition were analyzed statistically using One Way ANOVA (95%) and tested further by the Tukey test to determine significant difference between treatment and control groups. Organoleptic tests using 5 scale (hedonic scale), i.e. 1 = Dislike very much, 2 = Dislike, 3 = Neutral, 4 = Like, and 5 = Like Very Much. The 25 semi-trained panelists used were students of Nutrition Science Faculty of Medicine, Diponegoro University. The results of organoleptic tests were analyzed using Friedman test and continued using Wilcoxon to determine significant difference between treatment and control groups. Mean value was obtained then categorized into ≤1.4 "Dislike very much", from 1.5 to 2.4 "dislike", from 2.5 to 3.4 "neutral", 3.5 to 4.4, "like" and ≥4.5 "Like Very Much." The glycemic index data collected from the measurement were analyzed using univariate statistical method by calculating the average value of measurement data. The result used to describe the levels of glycemic index and glycemic load of GSMIC.

3. Results and Discussion

3.1. Nutritional Composition
The results of the analysis of nutrient content of Glucomannan-enriched Soy Milk Ice Cream (GSMIC) shows that there were significant differences between carbohydrate, protein, and fat between four products ($p<0.05$). Ice cream with 2.5% glucomannan flour contained the highest carbohydrate and protein, and the lowest fat composition (Table 1).

Table 1. Nutrition composition of ice cream and GSMIC.

Formulation	Carbohydrate (g)	Protein (g)	Fat (g)
P0	24.26 ± 0.78^{d}	7.71 ± 1.04^{a}	3.06 ± 0.12^{a}
P1	27.06 ± 0.54^{c}	4.91 ± 0.48^{b}	2.36 ± 0.16^{b}
P2	28.34 ± 0.22^{b}	3.20 ± 0.40^{c}	1.95 ± 0.16^{b}
P3	30.17 ± 023^{a}	1.50 ± 0.04^{d}	1.33 ± 0.06^{c}
	p=0.000	*p*=0.000	*p*=0.000

P0=ice cream; P1=0.5% glucomannan flour; P2=1.5% glucomannan flour; P3=2.5% glucomannan flour. Figures followed by different superscript letters (a, b, c, d) in the same column indicate significant differences.

Obesity in children and adolescents can be overcome with the proper diet. Provision of a balanced diet in accordance with Requirement Daily Allowances (RDA) is recommended for obese children. Since children are still growing and developing, they need a diet with 3 main meals and 2 snacks [13]. The nutrient content per serving of a snack is generally up to 10% of daily caloric needs.

GSMIC was formulated to meet nutritional needs without causing weight gain if consumed properly – generally, one serving of ice cream that is equal to 90 grams. Recommended carbohydrate intake for obese children is approximately to 55% of daily recommended calories or 27.5 g per single snack [13]. The composition of carbohydrate in one serving of GSMIC in its highest concentration meet the recommendation at approximately 98.5%, as the higher the concentration of glucomannan flour, the higher the carbohydrate content. Glucomannan is a complex carbohydrate and the soluble fiber that is included in polysaccharides increases carbohydrates [14].

The recommended fat intake is maximum 30% of calories which is equal to 6.6 g per serving. The composition of fat in one serving of the highest concentration GSMIC was 1.1 g and 18%, which meets the recommendation. The recommended protein intake is approximately 15% of caloric needs, equivalent to 7.5g per serving of a snack. The protein content in one serving of ice cream with the highest glucomannan flour substitution is 1.35 g, which meets 18% of recommended protein intake. In GSMIC, the protein composition was lower (2.6-2.8 g) than ice cream.

3.2. Dietary Fiber

The analysis of dietary fiber content of GSMIC showed that 2.5% GSMIC had the highest fiber content across other treatment groups (Table 2). Compared to ice cream with CMC stabilizer (ice cream), dietary fiber was increased by 1.69%. The results showed that there was a significant difference ($p<0.05$) of GSMIC on levels of dietary fiber.

Table 2. Dietary fiber composition of ice cream and GSMIC.

Formulation	Water-Soluble Fiber (%)	Water-Insoluble Fiber (%)
P0	1.06 ± 0.08^c	0.56 ± 0.09^b
P1	1.69 ± 0.16^b	0.82 ± 0.55^a
P2	2.07 ± 0.14^b	0.98 ± 0.55^a
P3	2.75 ± 0.09^a	1.24 ± 0.04^a
	p=0.000*	p=0.000*

P0=ice cream; P1=0.5% glucomannan flour; P2=1.5% glucomannan flour; P3=2.5% glucomannan flour. Figures followed by different superscript letters (a, b, c, d) in the same column indicate significant differences.

The results showed increased levels in each treatment group, such that the higher percentage of glucomannan flour, the higher composition of the dietary fiber in the ice cream. There were significant differences in dietary fiber in the GSMICs compared to ice cream ($p<0.05$). The glucomannan flour increased dietary fiber because glucomannan is a water-soluble fiber [7]-[8]. Glucomannan has been shown prolong gastric-emptying, thus increasing satiety and reducing body weight [15]-[16]. The dietary fiber in the ice cream was lower because the stabilizer used was CMC, a polymeric chain that consists of insoluble cellulose molecule which contains less soluble fiber [17].

Adequate consumption of dietary fiber could lower the risk of obesity. High fiber foods need longer time to chew and digest. Foods that contain insoluble fiber are not digested and increase the volume of food, thus reducing the risk of excessive consumption. Soluble fiber turns into a gel-like substance during digestion, prolonging the food in the intestines, and making the body feel satisfied for longer [18]. According to a research, glucomannan is more effective when administered at a smaller dose because it has high viscosity. Consuming glucomannan 3 g per day for 8 weeks reduces weight by 1.4-2.49 kg [19]. Regulation from the Head of National Agency of Drug and Food of the Republic of Indonesia number HK.00.05.52.6291, recommends 25 g/2.000 calories or 30 g/2500 calories for children and adolescents aged 2 to 20 years old [20]. One serving of 2.5% GSMIC (90 g) contains 3.59 g of fiber. Thus, eating one serving of GSMIC meets the 14% requirement of dietary fiber.

3.3. Blood Glucose Response

Variations in the concentration of GSMIC resulted in differences in glucose response in healthy subjects. The blood sugar tests showed that there was an increase 30 minutes the test foods and continue to decreased gradually (Figure 1). An increment in blood glucose lowering corresponded with the addition of glucomannan flour. The highest blood glucose increment occurred after consuming pure glucose intake and the lowest occurred after consuming 2.5% GSMIC. The blood glucose response after ice cream consumption was higher after 30 and 60 minutes compared to 2.5% GSMIC and lower after 90 and 120 minutes. This condition could be due to the dietary fiber composition, the higher fiber consumed the lower blood glucose response [21].

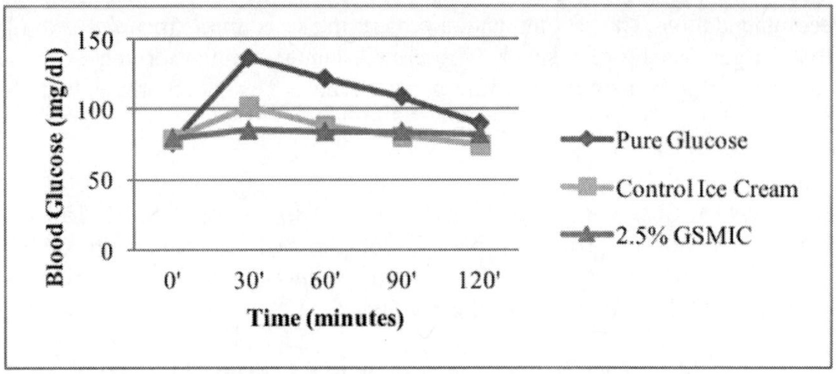

Figure 1. Blood glucose response graph

3.4. Glycemic Index and Glycemic Load

The glycemic index (GI) of GSMIC obtained from the average value of the nine subjects glycemic index is shown in Table 3. The 2.5% GSMIC had lower GI (51.48) compared to the control ice cream (75.83). The results of Glycemic Load (GL) for GSMIC was categorized as low glycemic load for ice cream (9.04) and medium glycemic load for the 2.5% GSMIC (11.61) (Table 4).

The glycemic Index (GI) test is conducted using pure glucose, ice cream, and 2.5% GSMIC. All tested foods were equivalent to 50 g carbohydrates determined by available carbohydrates. Available carbohydrates measures the total available carbohydrates that are easily digest, absorb, and metabolize [22].

Table 3. Glycemic index of Glucomannan-enriched Soy Milk Ice Cream.

Food tested	Glycemic Index	Category*
Control Ice Cream (P0)	75.83	High GI
2.5%glucomannan flour(P3)	51.48	Low GI

*Category: low glycemic index (<55); medium glycemic index (55-70); high glycemic index (>70)[30]

The GI value of foods were divided into three categories as follows: (1) low GI (<55), (2) medium GI (55-70), and (3) high GI (>70) [30]. Based on these categories, 2.5% GSMIC was categorized as a low GI food (51%) and ice cream was categorized as a high GI food, which decreased after addition of 2.5% glucomannan flour. Dietary fiber and production methods can affect GI changes [21].

Table 4. Glycemic load of Glucomannan-enriched Soy Milk Ice Cream.

Formulation	Serving portion (g)	Available Carbohydrate (%)	Available Carbohydrate per serving[a]	Glycemic Load[b]	Category[c]
P0	30	40.18	12.06	9.04	Low GL
P3	30	75.97	22.8	11.61	Medium GL

P0=ice cream; P3=2.5% glucomannan flour
[a] Available carbohydrate/serving =(serving portion/100g)*available carbohydrate
[b] Glycemic load=(glycemic index*available carbohydrate per serving)/100
[c] Categories: low glycemic load (<10), medium glycemic load (11-19), high glycemic load (> 20) [23]

IOP Conf. Series: Earth and Environmental Science **102** (2017) 012014 doi:10.1088/1755-1315/102/1/012014

The low GI value of GSMIC might be caused by glucomannan flour added into production. High fiber glucomannan flour (5.9 g/100g) could affect the GI value due to its role as physical obstacle in the digestion process [24]. Dietary fiber increases viscosity, provides longer satiety, decreases the absorption of macronutrients, influences the change of GI value, and lowers post-prandial blood glucose. Adequate fiber intake could be advantageous in controlling blood glucose and plasma lipid concentration [25]. Fiber addition has anhypoglycemic effect because it can slows down gastric-emptying and diffusion of glucose and glucose absorption so it reduces the increment of blood glucose [26]. The fiber addition process in the GSMIC can affect the GI value. The use of high temperature processing could result in the formation of starch gelatinization that is difficult to digest, therefore reducing the value of the glycemic index [27].

GI provides information on the speed of carbohydrate transformation into glucose but does not provide information about the amount of carbohydrates and the impact of certain food on blood glucose levels. Glycemic Load (GL) may provide more complete information on the effect of actual food consumption on increasing blood glucose levels. Consumption of low GI foods aims to reduce GL. GL is used to assess the impact of carbohydrate consumption considering the GI value of the food. GL is proportional to the carbohydrate composition of food such that the lower the carbohydrate content, the lower GL of the food. A smaller serving of food would trigger an increment raise in blood glucose level [33]. GL was categorized into three categories as follows: low GL (<10), medium GL (11-19), and high GL (> 20) [28].

The GL of the ice cream and 2.5% GSMIC were 9.04 and 11.61, respectively. Due to the carbohydrate composition in the flour, the higher the level of added glucomannan flour, the higher the GL level. 2.5% GSMIC was classified as medium GL food. GSMIC could be used as an alternative snack for diabetics but should be limited to 1-2 servings/day. In addition, the GL diet has been shown to improve glycemic control, serum lipid reduction, and cardiovascular and diabetes risk reduction [29]. A Japanese study concluded that there is a relationship between a diet high in GL and risk of type 2 diabetes in women [30].

3.5. Organoleptic Quality

The results of the organoleptic test for GSMIC is shown in four parameters in Table 5. There were no significant differences in the color, flavor, and taste parameters of the three concentrations of GSMIC (0.5%, 1.5%, 2.5%). However, there were significant differences in color, flavor, and taste of GSMIC (0.5%, 1.5%, 2.5%) compared to the ice cream. There were no significant differences between ice cream and 2.5% GSMIC in texture ($p<0.05$). Ice cream and 2.5% GSMIC shared similar color and texture scores. 2.5% GSMIC had the most preferred flavor and taste (hedonic scale 3.13).

Table 5. Organoleptic quality of Glucomannan-enriched Soy Milk Ice Cream.

Formulation	Categories			
	Color	Flavor	Taste	Texture
P0	4.00 ± 0.86^a (like)	4.52 ± 0.71^a (very like)	2.12 ± 0.78^b (dislike)	$3.36 \pm 1,18^a$ (neutral)
P1	3.08 ± 0.90^b (neutral)	2.84 ± 0.89^b (neutral)	2.92 ± 0.75^a (neutral)	2.68 ± 0.80^b (neutral)
P2	3.00 ± 0.86^b (neutral)	3.12 ± 0.83^b (neutral)	2.84 ± 1.02^a (neutral)	2.72 ± 0.79^b (neutral)
P3	3.44 ± 0.65^b (neutral)	3.20 ± 0.95^b (neutral)	2.96 ± 1.02^a (neutral)	2.92 ± 1.07^a (neutral)
	p=0.001	p=0.000	p=0.005	p=0.041

P0=ice cream; P1=0.5% glucomannan flour; P2=1.5% glucomannan flour; P3=2.5% glucomannan flour. Figures followed by different superscript letters (a, b) in the same column indicate significant differences.

There was no significant difference between GSMIC products (0.5%, 1.5%, 2.5%) in color. The hedonic scale in ice cream texture parameter was high for 2.5% GSMIC and low for 1.5% GSMIC. GSMIC flavor had a neutral hedonic scale and, according to the results of statistical test, the rating showed a significant difference across each GSMIC group. The 2.5% was preferred by the panelists because ice cream production reduced the percentage of soy milk and increased the glucomannan flour added.

The hedonic rating on the texture of ice cream products were neutral across each GSMIC products. However, according to the statistical tests, there was a significant effect of the amount of glucomannan flour added. The panelists preferred the 2.5% GSMIC due to the texture because glucomannan is a water-soluble fiber and contains strong hydrocolloids that are soluble in cold water by a thick gel mass [14]; the more glucomannan flour added, the more viscous ice cream products produced.

Using glucomannan flour in soy milk ice cream as a CMC replacement and as a stabilizer for this product could be accepted by consumers. However, the mean of panelists hedonic scale was higher in the control product which used CMC as the stabilizer.

4. Conclusion

Glucomannan flour addition to soy milk ice cream production affected carbohydrates, protein, fat, dietary fiber, glycemic index, glycemic load, and organoleptic quality, and higher flour glucomannan concentration increased fiber and carbohydrates and lowered fat and protein. GSMIC has a low glycemic index at 2.5% glucomannan concentration (51.48) and a low glycemic load at 0% glucomannan flour concentration (9.04). The addition of glucomannan flour to soy milk ice cream production reduced GI but increased GL. Panelists preferred 2.5% GSMIC due to its organoleptic quality on the parameters of color, flavor, taste, and texture. The 2.5% GSMIC preferred by the panelists contains higher dietary fiber and lower fat.

5. Sugesstion

The concentration of soybean milk protein content in ice cream should be reanalyzed to obtain the best formulations in order to lower the value of glycemic index and glycemic load. In addition, it is important to test the physical properties of ice cream as well as possible metal and microbial contamination.

References

[1] WHO 2000 *Obesity:* Preventing and Managing the Global Epidemic, WHO Technical Report Series Geneva, 894.

[2] Ministry of Health 2013 Basic Health Research (RISKESDAS) 2013. National Research and Development Department of Ministry of Health, Jakarta, Indonesia. p 218.

[3] American Diabetes Association. 2014. Diagnosis and Classification of Diabetes Mellitus. *Diabetes Care* **37**: S81-90.

[4] Ministry of Health 2013 Basic Health Research (RISKESDAS) 2013. National Research and Development Department of Ministry of Health. Jakarta, Indonesia. p 88.

[5] Gibney, M.J 2009 *Nutrition for Community*. EGC, Jakarta, Indonesia. p 311.

[6] Fitrahdini, Sumarwan U, and Nurmalina R 2010 Analysis of consumer perception to the equity of ice cream product trademark. *Family Knowledge and Consumption Journal* **3**: 74 – 81.

[7] Jannah Y R, Thohari I, and Purwadi 2013 The addition of porang flour (*Amorphophallus oncophillus*) in the yogurt ice cream on total plate count, texture, taste, flavor, total solid, and pH. *Bachelor Thesis*. Brawijaya University, Malang, Indonesia.

[8] Widjanarko S B 2011 Assessment of conventional extraction and ultrasonic method in the porang (*Amorphophallus oncophyllus*) glucomannan purification to produce new food additives and functional food. *Thesis*. Brawijawa University, Malang, Indonesia.

[9] Chearskul S, Sangurai S, Nitiyanant W, Kriengsinyos W, Kooptiwut S, Harindhanavudhi T 2007 glycemic and lipid responses to glucomannan in thais with type 2 diabetes mellitus.

Journal of the Medical Association of Thailand **90**: 2150-7.

[10] Apriyantono A, Fardiaz D, Puspitasari N L, Sedarnawati and Budiyanto S 1989 Food analysis. Bogor Agricultural University, Bogor, Indonesia. p 71-84.

[11] Asp NG, Johanson CG, Halmer H and Siljestrom M 1983 Rapid enzymatic assay of insoluble and soluble dietary fiber *Journal of Agriculture Food Chemistry* **31** 476-482

[12] FAO. 1998. *The Role of Glycemic Index in Food Choice. In Carbohydrates in Human Nutrition* (Online). www.fao.org. (Accessed May 2016).

[13] Nutrition Coordination and Metabolic Disease Unit 2014 Diagnosis, treatment, and obesity prevention in children and adolescent Indonesian pediatric doctors association, jakarta, indonesia. p 22 – 23.

[14] Koswara S 2012 *Tuber processing technology part 2 : processing porang*. Bogor Agricultural University, Bogor, Indonesia. p 2–13.

[15] Sood N, Baker W L, Coleman C I. 2008. Effect of Glucomannan On Plasma Lipid and Glucose Concentrations, Body Weight, And Blood Pressure: Systematic Review and Meta-Analysis. *American Journal of Clinical Nutrition* **88** 1167–75.

[16] EFSA 2010 Scientific opinion on the substantiation of health claims related to konjac mannan (glucomannan) and reduction of body weight. *EFSA Journal* 8:1798

[17] Kamal N. 2010. Effect of CMC (Carboxyl Methyl Cellulose) in the Sucrose Solution. *J. Technology* 1: 78-84.

[18] Asian Food Information Centre. *Dietary Fiber – An essential Ally in Weight Management*. www.afic.org (accessed June 2016).

[19] Keithley J and Swanson B 2005 Glucomannan and obesity: a critical review. *Alternative Therapies* 11: 30-34.

[20] Almatsier S 2004 Nutrition science basic principle. Gramedia Pustaka Umum, Jakarta, Indonesia. p 132-133.

[21] Rimbawan and Siagian A 2004 Foods Glycemic Index. Penebar Swadaya, Jakarta, Indonesia. p 1-30

[22] Adha H M 2012 Glycemic index and glycemic load of breadfruit with various processing. Gajah Mada University, Yogyakarta, Indonesia. p 1-35.

[23] Foster-Powell K, Holt S H and Brand-Miller JC 2002 International table of glycemic index and glycemic load values: 2002. *American Journal of Clinical Nutrition* 75: 5-56.

[24] Gallagher ML 2012 The Nutrient and Their Metabolism. In Mahan LK, Stump SE, editors. *Krause's Food and the Nutrition Care Process 13th edition*. WB Saunders Company, Philadelphia, USA. p. 32-41

[25] Nisviyati A 2006 The usage of sweet potato flour (ipomoea batatas l.) klon bb00105.10 as main ingredient of fried products, the nutrition evaluation, and the glycemic indeks *Bachelor Thesis* Bogor Agricultural University, Bogor, Jakarta.

[26] Gropper SS, Smith JL, Groff JL 2009 Carbohydrates. In *Advanced Nutrition and Human Metabolism 5 th edition*. Wadsworth, Canada. p. 69-77.

[27] Riany YE 2006 Effect of Processing on Metroxylon Sagu (Metroxylon sp.)-source foods Glycemic Index. *Bachelor Thesis*. Bogor Agricultural University, Bogor, Jakarta.

[28] Ningrum D R, Nisa F Z, and R Pangestuti. 2011. Glycemic index and glycemic load of bread fruit sponge cake as carbohydrates snacks in non-diabetic individual *Bachelor Thesis*. Gajah Mada University, Yogyakarta, Indonesia.

[29] Fernandez-raudalez D, Diaz-rios LK, Lotton J and Chapman-novakofski K 2012 Effect of beverages with different protein profiles on postprandial blood glucose response in overweight and obese men. *Journal of Diabetes Mellitus* 2:40-6.

[30] Oba S, Nanri A, Kurotani K., Goto A., Kato M., Mizoue T, Tsugane S 2013 Dietary glycemic index, glycemic load and incidence of type 2 diabetes in Japanese men and women: the Japan public health center-based prospective study *J. Nutrion* **12** 165.

Chemical, physical, and sensory characteristics of analog rice developed from the mocaf, arrowroof, and red bean flour

S B Wahjuningsih[1] and S Susanti[2*]

[1]Faculty of Agricultural Technology, Semarang University, Jl. Soekarno Hatta – Semarang

[2]Department of Agriculture, Animal and Agricultural Sciences Faculty, Diponegoro University Semarang

*Email: sitisusanti5678@gmail.com

Abstract. This research was aimed to analyze the chemical, physical, and sensory characteristics of the analog rice developed from a composite formula consisting of mocaf, arrowroot, and red bean flour. Experiment was designed into 5 different composition types i.e B_1 (90%: 0%: 10%), B_2 (80%:10%: 10%), B_3 (70% : 20% : 10%), B_4 (60%: 30%:10%), and B_5 (50%: 40%: 10%) which in each type was repeated in 4 times. Carrageenan was used as a binder in the making process of those analog rice. Investigation procedure was carried out into several stages such as preparation and characterization of raw materials, production of analog rice in composite formula, then the testing of its chemical and sensory properties. Chemical characteristics were evaluated about the level of starch, amylose, dietary fiber, and resistant starch while sensory characteristics were examined about the texture, flavor, and aroma. The result showed that based on the sensory test, the best composite formula of rice analog was B_2 (ratio flour of mocaf: Arrowroot: Red bean = 80:10:10). In addition, B_2 formula possessed the chemical characteristics similar with the truth rice either in water content (12.18%), ash (2.63%), protein (6.17%), fat (1.31%), carbohydrate (89.88%), starch (73.29%), amylose (24.91%), total dietary fiber (7.04%), or resistant starch (6.71%). Furthermore, the higher of arrowroot flour proportion, the greater of amylose, dietary fiber and resistant starch containing in the rice analog. In the opposite, its starch content was getting down.

***Keywords**: analog rice, mocaf, Arrowroot, red bean, flour, composite formula*

1. Introduction

Food diversification in Indonesia as one of efforts to reach the food sovereignity is still inhibited by the high consumption of rice [1]. This is indicated by the high level of community dependence on a single source of carbohydrates only, whereas the potentcy of local carbohydrates sources such as group of tubers and grains in Indonesia are quite diverse. Therefore, the optimization of carbohydrate sources other than rice should be encouraged to reduce rice dependence, which in the last decade rice is an imported food commodity.

Analog rice has become an alternative solution in optimizing the utilization of local carbohydrate sources recently. Analog rice is non-rice ingredients with carbohydrate content approaching or even exceeding than rice such as mocaf, arrowroot, corn, sorghum, and sago and tubers flour are shaped like a rice [2]. It is often regarded as functional rice because of its raw material properties related to the health aspects. Analog rice from mocaf and arrowroot flour have been shown to have a texture like rice in general with the functional properties such as lower in glycemic index that is suitable as a diet for people with Diabetes Mellitus [3]. Previous research by [4] states that red beans combined with mocaf are able to produce rice analogue with a higher nutritional value due to the content of red bean protein amounted to 23.2%. Thus, mocaf, arrowroot and red bean flour are potential to be the local raw material for functional analog rice. As a consecuency, effort to improve its utilization still needs to be done optimally.

This study developed new analog rice through the composite formula of some raw materials such as mocaf, arrowroot, and red bean flour. Mocaf and arrowroot flour which are rich in amylose content mixed with the high protein content in red beans. It is expected to produce the analog rice with a better quality both in texture and functional properties with the result that can be used as determination of consumer acceptance level. The physicochemical properties of raw materials and analog rice products from several combinations of the three raw materials were studied for its optimization. Furthermore, the sensory characteristics of analog rice resulted in this study also were examined to determine which composite formula is most acceptable to consumers. The development of mocaf, arrowroot and red bean flour based analod rice in this study was expected could increase the utilization of local food sources in order to meet the age of food sovereignty.

2. Material and Methods

2.1. Preparation of raw materials

Mocaf flour production was made by using culture method as described earlier [4]. For Arrowroot flour was made by peeling, washing, and slicing arrowroot that continued by blanching it on 80 °C for 5 minutes. Before grinding process in 80 mesh sieve, blanched arrowroot was drained and dried on 50 °C for 12 hours. Meanwhile, red bean flour was purchased directly from the farmer around the the farm land in the Temanggung sub-district, Central Java Province.

2.2. Analog rice production

Analog rice production was initiated by making composite flour consisting of five different ratio on mocaf : arrowroot : red bean i.e. 90%: 0% : 10% (B_1), 80%:10%: 10% (B_2), 70% : 20% : 10% (B_3), 60%: 30%; 10% (B_4) and 50%: 40%: 10% (B_5). Ingredients used for emulsifier were GMS, salt, distilled water, coconut oil, and carrageenan. Those ingredients were mixed with composite flour, homogenized for 5 minutes, and steamed for 15 minutes. Steamed materials were directly inserted into an extruder in order to shape analog rice. After granules were formed, they were dried in a dryer cabinet at a temperature of 50°C for 12 hours.

2.3. Experimental Analysis

The basic materials mocaf flour, arrowroot flour and red bean flour were analyzed according to its gelatinization properties by using Rapid Visco Analyzer (RVA) and water absoption [5]. Analysis of chemical characteristic on analog rice such as proximate content and starch content were conducted by using acid hydrolysis method [6], while dietary fiber by using enzymatic method [7], amylose content by using spectrophotometry method [6] and resistant starch content was performed by using the method that was modified by Goni [8]. The produced analog rice would be tested its sensory characteristics by using multiple comparison test [9] and proximate test. Determination of selected formula was conducted by using exponential comparison method. Exponential comparison method is a technique in decision making by deciding ranking from alternative decision with various criteria [10].

2.4. Statistical analysis

Data resulted from this experiment were analyzed statisticaly by using SAS Software version 9.2 to perform General linier Models (GLM) procedure. Duncan Multiple Range Test (DMRT) was carried out to get more insight about a significant difference among several formulas at the level of 5% [11].

3. Result and discussion

3.1. Physical Characteristics of Raw Material and Composite Flour

The higher amount of arrowroot flour, composite flour would have peak viscosity, hot paste viscosity, decreased viscosity because of heating, final viscosity, and the increase of viscosity because of cooling process was lower (Table 1). This was shown that there was negative slope from composite flour if all viscosity parameters were illustrated in curve. Gelatinization properties of mixed flour were effected by gelatinization properties of its basic materials. Heating in gelatinization caused irreversible swelling of starch granule in the water so that it occurred the increase of viscosity [12]. Gelatinization temperature of material determine the best temperature used during extrusion, because before extrusion process might occur gelatinization on partial material.

Table 1. Physical characteristics of raw material and composite flour

Flour Type	Viscosity (cp)					Peak time (min)	Gelatinization Temperature (°C)
	Peak	Hot Paste	Breakdown	Final	Setback		
Arrowroot	4749	2201	2548	3423	1222	7.47	80.60
Mocaf	6508	1912	4596	3214	1302	6.20	68.40
Red bean	207	207	0	481	274	13	Undetected
B_1	6543	2292	4251	3644	1352	7	67.95
B_2	5663	2008	3655	3394	1386	6.6	68.4
B_3	4930	1744	3186	3061	1317	6.53	68.85
B_4	4504	1663	2841	2767	1104	6.67	69.25
B_5	4030	1500	2530	2493	993	6.8	70
Slope							
Composite Flour	-618.5	-192.9	-425.6	-292.9	-100	0.495	-0.033

B_1, B_2, B_3, B_4, and B_5 are kind of samples ratio (mocaf flour: arrowroot flour: red bean flour) as a composite formula i.e (90:0:10); (80:10:10); (70:20:10); (60:30:10); (50:40:10), respectivelly.

Gelatinization temperature was effected by several factors, such as granule size, amylose, amylopectin, heating instrument condition, and also food material content like fat, dietary fiber, resistant starch and protein. Component like protein and fat can be bound with amylose and form precipitation, so it increases temperature of gelatinization [13]. Amylose content of arrowroot flour was higher than mocaf flour, it might cause the difference of gelation temperature in both materials. The aligned chain amyloses which form double stranded crystallites were resistant to amylases. The intra- and inter-strand hydrogen bonding results hydrophobic structure with low solubility [14].

The amylose content of red bean flour was high, but it was undetected gelatinization temperature, it might be caused by material handling process such as soaking, blanching for 30 minutes, and drying red bean in red bean flour production. Those phase was assumed to cause starch gelatinization in red bean flour had been passed. Mocaf flour had lower gelatinization temperature, breakdown viscosity (stability of starch paste on high temperature), and higher setback viscosity. When starch was heated in sufficient water, the stability of double helix structured hydrogen bonding in crystalline would be damaged and replaced by water hydrogen bonding, amylose would be released from starch granule.

After that amylose and amylopectin would hydrate, therefore viscosity was increasing until it reached the peak and starch suspension became clearer [15]. Peak viscosity showed starch ability to swell freely before it had breakdown. The value of peak viscosity was effected by amylose and amylopectin content in material. The higher amylose content, the lower its peak viscosity. This was caused by amylose binding and fat to form amylose lipid complexes so that swelling of granule was suspended [16].

3.2. Chemical characteristics of basic material

The chemical characteristic of basic material identified in this research was various nutritional components as well as non-nutritional components which was assumed to effect on physical properties of flour. The chemical characteristics identified were water content, ash content, fat content, protein content, carbohydrate content, amylose content, amylopectin content, energy content, and dietary fiber content. Chemical characteristics were only analyzed on its basic component, however the components of composite weren't analyzed. Chemical characteristics of basic material are displayed in Table 2.

Table 2. Chemical characteristics of basic material

Chemical characteristics	Arrowroot flour	Mocaf Flour	Red Bean Flour
Water (%)	7.12	6.43	13.64
Ash (%)	4.14	0.59	2.69
Fat (%)	0.63	0.76	3.20
Protein (%)	6.14	0.74	27.64
Carbohydrate(%)	81.97	91.48	66.45
Energy (Cal/100g)	358	376	405
Amilylose (%)	26.37	22.48	58.28
Amylopectin (%)	73.63	77.52	41.72
Starch (%)	58.70	78.76	43.25
Dietary fiber (%)	6.76	3.78	14.38
Insoluble dietary fiber (%)	4.49	-	8.79
Soluble dietary fiber (%)	2.27	-	5.59

3.3. Organoleptic assessment of analog rice

Organoleptic test was carried out on 20 semi-trained panelists. The results of organoleptic test are shown in Table 3. Determination of selected formula was conducted by using exponential comparison method. Exponential comparison method is a technique in decision making by deciding ranking from alternative decision with various criteria [10]. Aroma, taste, and texture data of analog rice are tabulated by giving high ranking based on certain criteria. The parameters of product are arranged from the lowest ranking to the highest ranking. The lower ranking shows the exacted value. The lowest ranking product is the best product [17].

Table 3. Mode value of texture, taste, and aroma

Organelptic test	Texture	Taste	Aroma
B1	$-0.25^{a} \pm 1.07$	$-0.35^{ab} \pm 1.18$	$-0.25^{a} \pm 1.37$
B2	$-0.10^{a} \pm 1.48$	$0.30^{b} \pm 1.13$	$0.15^{a} \pm 1.49$
B3	$-0.10^{a} \pm 1.48$	$-1.00^{a} \pm 0.79$	$-0.60^{a} \pm 1.50$
B4	$-0.35^{a} \pm 1.57$	$-0.90^{a} \pm 1.33$	$0.20^{a} \pm 1.39$
B5	$0.65^{a} \pm 1.73$	$-0.95^{a} \pm 1.19$	$-0.55^{a} \pm 1.54$

B_1, B_2, B_3, B_4, and B_5 are kind of samples ratio (mocaf flour: arrowroot flour: red bean flour) as a composite formula i.e (90:0:10); (80:10:10); (70:20:10); (60:30:10); (50:40:10), respectivelly.

Based on ranking test conducted, B1, B2, B3, B4, and B5 products had 7, 4, 11, 7, and 12 total ranking, respectively. The sequence of total ranking for B1, B2, B3, B4, and B5 was 2, 1, 3, 2 dan 4. The purpose of exacted value is the lower ranking, the more exacted product will be. The selected formula was B2, it was an analog rice with 80:10:10 mocaf flour, arrowroot flour, and red bean flour ratio. The pictures of mocaf arrowroot red bean flour analog rice from those formulas are displayed in Figure 1.

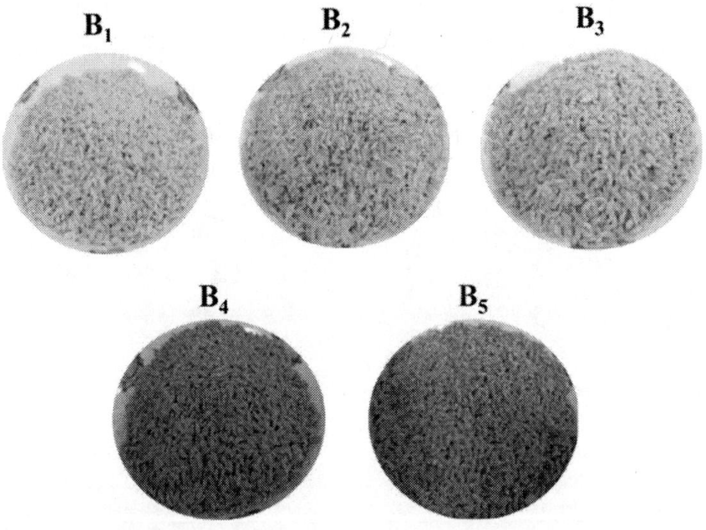

Figure 1. Performance of analog rice developed from mocaf, arrowroot and red bean flour in 5 different composit formulas (mocaf:arrowroot:red bean) such as B_1 (90:0:10), B_2 (80:10:10), B_3 (70:20:10), B_4 (60:30:10), and B_5 (50:40:10).

B2 analog rice had advantage in taste parameter which had positive value. Positive value shows that the product had better taste than standard (rice). The more arrowroot flour proportion caused the taste of analog rice was more disliked. B2 analog rice had disadvantage in texture parameter. This selected formula had -0,10 texture value. Negative value in multiple comparison test shows that selected formula had lower quality than standard (rice). Negative value in multiple comparison test shows the selected formula had lower quality than standard (rice). Amylose content affected on rice texture. Overall B2 analog rice was closest to the most common rice on taste, texture, aroma parameters. Rice with high amylose would tend to be firm. B1 and B4 analog rice had the second lowest total ranking. B1 and B4 analog rice had negative value on aroma, texture, and taste. This was shown that B1 and B4 formula had lower quality than standard (rice).

B3 analog rice had third total ranking. B3 formula rice had negative value on multiple comparison test. Negative value shows that aroma, taste, and texture qualities of product were lower than standard (rice). B3 analog rice had disadvantage in taste, the value was -1,00. The lower its value, the lower quality product would be than rice. B5 analog rice had the highest total ranking from all formulas. The highest total ranking showed the lowest ranking. B5 analog rice had disadvantage in taste parameter. This formula had -0,95 in taste parameter.

3.4. Chemical characteristics of analog rice

Chemical composition of analog rice was influenced by some factors, such as chemical composition of its ingredients (mocaf flour, arrowroot flour and red bean flour), the effect of extrusion process by using high temperature and pressure, steaming before extrusion, other technical factors during processing. The results of the analysis of the chemical characteristic of mocaf analog rice are displayed in Table 4. All of analog rice formulas had a water content below 14%. Food products which had 14% or 15% water content would ease microbes to grow. The water added in the preparation of analog rice significantly affected its water content. This was caused by the more water added during the production of analog rice, the higher the water content of analog rice granules would be after the drying process. In this research, the water added was 50% of the total amount of starch and flour [18]. Analog rice with higher arrowroot percentage tend to have higher water content. Water content in food material will determine the ease of food material to have damage. As the higher water content in food material, then it tends to be easily damaged.

Tabel 4. Chemical characteristics of analog rice

Chemical test (% dry basis)	**Treatments**				
	B_1	**B_2**	**B_3**	**B_4**	**B_5**
Water	12.61^{a}±0.04	12.19^{b}±0.08	12.95^{c}±0.05	13.33^{d}±0.09	12.96^{c}±0.06
Ash	2.44^{a}±0.04	2.64^{b}±0.01	2.83^{c}±0.01	3.05^{d}±0.02	3.12^{d}±0.09
Protein	6.41^{a}±0.06	6.17^{a}±0.07	6.34^{a}±0.10	6.29^{a}±0.01	5.46^{a}±1.59
Fat	1.38^{c}±0.01	1.31^{b}±0.03	1.39^{c}±0.03	1.26^{a}±0.02	1.59^{d}±0.05
Carbohydrate	89.77^{c}±0.03	89.88^{c}±0.09	89.44^{b}±0.15	89.41^{b}±0.04	89.04^{a}±0.04
Starch	74.54^{d}±0.44	73.29^{c}±0.16	72.53^{b}±0.28	72.23^{b}±0.95	70.98^{a}±0.88
Amylose	24.60^{a}±0.07	24.91^{b}±0.07	25.65^{c}±0.03	26.40^{d}±0.003	26.27^{c}±0.09
Amylopectin	49.93^{e}±0.51	48.38^{d}±0.09	46.87^{c}±0.25	45.83^{b}±0.09	44.71^{a}±0.18
Resistant starch	6.65^{a}±0.15	6.71^{ab}±0.16	6.98^{ab}±0.14	7.16^{ab}±0.60	7.25^{b}±0.34
Soluble dietary fiber	2.74^{a}±0.03	2.88^{b}±0.02	3.07^{c}±0.02	3.29^{d}±0.01	3.73^{e}±0.09
Insoluble dietary fiber	3.69^{a}±0.05	4.16^{b}±0.06	4.61^{c}±0.05	4.98^{d}±0.06	5.41^{e}±0.05
Total dietary fiber	6.43^{a}±0.08	7.04^{b}±0.08	7.67^{c}±0.03	8.27^{d}±0.07	9.15^{e}±0.04
Water absorption	214.24^{a}±0.43	230.48^{b}±0.69	239.79^{c}±0.52	249.23^{d}±0.58	256.89^{e}±0.31

Means on the same column with different superscript are significantly different at ($p < 0.05$)

Ash content in analog rice was ranged between 2.4363% and 3.1173%. The more arrowroot flour proportion added to B4 and B5, they had equal ash content to 10 % sago red bean flour analog rice, the ash content was (3.06%) [19]. Based on the test of variance, composite flour ratio displayed significant differences on ash content of mocaf analog rice (p=0,002). The higher arrowroot flour content of mocaf analog rice, the ash content was higher. Ash content in food material can show high mineral content in food material. Ash content showed the amount of mineral and organic substance which were contained in product [20]. The higher ash content resulted the higher mineral content. The results of a research conducted by [10] showed ash content of IR64 rice was 0.56% (dry basis). Mineral content of rice was highly effected by pre harvest condition and variety so that there was possibility of ash content variance.

The protein content of mocaf analog rice was ranged between 5.4572% and 6.4089%. Composite flour ratio do not have significant effects to the protein content of mocaf analog rice ($p > 0.05$). This shows that the amount of arrowroot flour in composite flour has no significant effects in protein content of mocaf analog rice. The protein content of analog rice was from red bean flour, which added constantly. Protein content in food materials can also effect on food glycemic index. The higher protein content, the glycemic index would tend to be higher [21]. All mocaf analog rice had protein content above 5%.

Fat is compound which has non-polar group so that it is insoluble in the water, however it is soluble in organic solvent. Fat content of analog rice reached 0.37%. This fat content was low category if it's compared to research conducted by [22] it ranged from 0.34% to 0.62%. [23] claimed that oxidative damage in high-fat food material was an important problem because it could lower organoleptic quality, i.e rancidity. The fat content of the produced mocaf analog rice was ranged between 1.2589% and 1.5910%. Analog rice with the highest fat content was B2 whereas the lowest one was B4. Fat contents of produced sago analog rice were all higher than granulated rice which made of sweet potato 0.70% - 0.90% [24]. This was caused by high fat content in food materials, particularly on red bean flour which reached up to 3.26%, while other materials were under 1%. High fat content can be used to estimate low glycemic index. The higher fat content in food material would lower its glycemic index [25].

Carbohydrates content of mocaf analog rice was ranged between 89.0382% (B5) and 89.8816% (B2). Composite flour ratio shows significant effect on carbohydrates content of mocaf analog rice ($p > 0.05$). Carbohydrates content of mocaf flour was higher than arrowroot flour, so the higher the amount of arrowroot flour added, the lower its content of carbohydrates would be. According to[26], Carbohydrate gave more than 50% calories with carbohydrate value 4 kcal/g. The slowly absorbed carbohydrate would produce low blood glucose peak and potentially control starch digestibility of rice which was effected by its amylose or amylopectin composition [27].

Composite flour ratio shows significant effect on amylose and amylopectin of mocaf analog rice. Amylose content of analog rice was approximately from 24.6047% (B1) to 26.4032% (B4). Amylopectin content of analog rice was ranged between 44.7073% and 49.9347%. Starch content of composite flour ratio shows significant effects on starch content of mocaf analog rice. Starch content of mocaf analog rice was ranged between from 70.9817% (B5) to 74.5394% (B1). The higher proportion of arrowroot flour in composite flour, the higher amylose content in mocaf analog rice. Arrowroot, mocaf, and red bean flour had high amylose content so that amylose content of produced analog rice was relatively high, it was higher than 20%.

Amylose content determined the texture of produced rice such as the firmness, the softness, and the stickiness of rice [28]. Amylose was polymer which was composed entirely of [1-4]-α-D-Glucopyranose [29]. Amylose affected on texture of rice. Amylose content could be used to estimate the firmness of flour paste gel which was made of rice. Amylopectin component in rice would affect gelatinization temperature and starch paste properties [30]. The higher proportion of arrowroot flour in composite flour would cause analog rice with lower amylopectin content. Amylose and amylopectin contents could be used to estimate glycemic index value of analog rice. The higher amylopectin of rice, the higher glycemic index. Whereas if the amylose content of rice was higher, the glycemic index would tend to be lower. The higher proportion of arrowroot flour in composite flour, the starch content was relatively lower. The value of starch content was calculated from the amount of amylose and amylopectin component.

Resistant starch content mocaf analog rice was ranged between 6.6548% and 7.2492%. Composite flour ratio do not have significant effects on resistant starch content of mocaf analog rice ($p>0,05$). This was caused by several factors, such as amylose content, particle size, and the effect of other compound [31]. The higher amylose content, its resistant starch content will be higher. The result of this research is related to its amylose content. Amylose had linear chain structure on compact matrix so that it lowered accessibility of enzyme [32]. Resistant starch can be used for investigating functional food [33], because resistant starch has characteristic and physiological function like dietary fiber, its slow digestibility can postpone the increase of blood glucose, control glycemic response, give a longer feeling of satiety, as a result it can lower value of glycemic index of that food [32,35].

Total dietary fiber is the total amount of insoluble fiber and soluble fiber content. Composite flour ratio has significant effect on total dietary fiber, soluble dietary fiber, and insoluble dietary fiber content of mocaf analog rice. Analog rice with higher arrowroot flour proportion tended to have higher insoluble dietary fiber content. Insoluble dietary fiber content of analog rice was higher than insoluble dietary fiber of grinded rice which ranged 2.27% to 5.68% [36]. Total dietary fiber content of mocaf analog rice was also higher than total dietary fiber content of grinded rice (4.67% - 7.57%). Soluble fiber content of mocaf analog rice was ranged from 2.7404% to 3.7308%. Grinded rice had lower soluble fiber content than mocaf analog rice (1.57% - 2.95%) [36].

Dietary fiber in human body would be digested become indigestible component and short-chain fatty acid, water, CO_2, methane [37]. Fiber metabolism is related to bacteria activity in colon. The fiber significant effects on metabolism are increasing feces volume, shortening transit time in intestine, softening feces consistency, and producing short chain fatty acid which is useful for health.

Composite flour ratio significantly affected on water absorption of mocaf analog rice. Water absorption content of mocaf analog rice was ranged between 214.2380% (B1) and 256.8897% (B5). The effect of increasing starch content on value of water absorption was related to amylose-amylopectin composition role in starch. [38] claimed that food material with high starch content would be easier to absorb water because of amylopectin molecule availability which was reactive to water molecule, so the amounts of absorbed water in food material.would be higher

Water absorption is affected by some factors, such as fat content, dietary fiber content, amylose and amylopectin content of food material. High amylose and dietary fiber content could increase water absorption [12]. Water absorption characteristic of analog rice was analyzed to find out the effect of flour ratio in composite flour on water absorption and water holding ability in dough. These data were used to estimate water addition during analog rice cooking process. Based on analysis, it shows that the higher proportion of arrowroot flour in composite flour would increase its water absorption. This was assumed because of its high dietary fiber content in arrowroot flour. Dietary fiber influenced on food material ability to absorb and hold the water. Water absorption was inverse compared to water content. The lower water content would increase the water absorption. [39] claimed that the high water absorption is caused by absorbed water in molecules so that it increases water absorption in material and disconnects intermolecular hydrogen bonds, so the water is easier to enter into flour. Absorbing water would be intensive as its increased heating temperature, the granule swells therefore the swelling point of starch granule was irreversible (couldn't return to initial form) [40]. Water absorption ability could be affected by protein component and crude fiber [41, 42].

4. Conclusion

Mocaf flour and red bean utilization on analog rice production can result fine textured rice and similar to the most common rice. According to the sensory test, analog rice made from mocaf flour, arrowroot flour, and red bean flour with ratio of 80:10:10% is the best due to its texture that is almost similar with the common rice, high acceptance by panelists for taste and aroma parameters. Chemical composition of the best analog rice was 12.1890% water content, 2.6376% ash content, 6.1688% protein content, 1.3119% fat content, and 89.8816% carbohydrate content.

References

[1] Buerau of Central Statistic 2014

[2] Samad M Y 2003 The making of artificial rice made from cassava and sago (in Bahasa Indonesia) *Pros. Seminar Teknologi untuk Negeri* 2: 36- 40/ Humas-BPPT/ANY, BPPT Jakarta

[3] Marsono Y 2002 Determination of tubers glycemyc index (in Bahasa Indonesia) *J. Agritech* **22** (1) 13-16

[4] Wahjuningsih S B and Kurnarto B 2013 The making of mocal flour with adding of natural fermentation starter for analog rice (in Bahasa Indonesia) *J.Litbang Provinsi Jawa Tengah* **11** (2) 221-230

[5] Sathe S K and Salunkhe D K 1981 Isolation partical characterization and modification of the great northem bean (*Phaseolus vulgaris* L.) starch *J. Food Science* **46** (2) 617-621

[6] AOAC 2005 Official Methods of Analysis of The Association of Official Agriculture Chemist (sixteenth edition) *Virginia*

[7] Asp NG, Johansson CG, Hallmer H and Siljestroem M 1983 Rapid enzymic assay of insoluble and soluble dietary fiber *J. Agriculture Food Chemical* **31** (3) 476-482

[8] Goni I, Garcia-Diz L, Manas E and Saura-Calixto F 1996 Analysis of resistant starch: a method for food and food products *Food Chemistry* **56** (4) 445-449

[9] Rahayu W P 2001 Guidelines for Organoleptic Practice Assesment (in Bahasa Indonesia) Jurusan Teknologi Pangan dan Gizi Fakultas Teknologi Pertanian Institut Pertanian Bogor

[10] Setyaningsih D, Apriyantono A and Sari MP 2010 Sensory Analysis for Food and Agro Industry (Bogor: IPB Press)

[11] Dawson B and Trap RG 2001 Basic and Clinical Biostatistics 3rd ed (New York: Lange Medical Books/McGrawHill Medical Publishing Division)

[12] Richana N and Sunarti TC 2004 Characterization of tuber flour, elephant foot yam, coconut yam and lesser yam physicochemical properties (in Bahasa Indonesia) *J. Pascapanen* **1** (1) 29-37

[13] Glicksman M 1969 Gum Technology in the Food Industry (*New York: .Academic Press*) pp 214- 224

[14] Chaplin M 2002 Starch http://www.sbu.ac.uk [accessed March 25th 2015]

[15] Schirmer M, Hochstotter A, Jekle M, Arendt E and Becker T 2013 Physicochemical and morphological characterization of different starches with variable amylose/amylopectin ratio *Food Hydrocoloid* **32** (1) 52-63

[16] Sang Y, Bean S, Seib PA, Pedesrsen J, and Sci Y C 2008 Structure and functiona properties of sorghum starches differing in amylase content *J. Agric Food Chem* **56** 6680-6685

[17] Budiyanto A dan Yulianingsih 2008 The influence of temperature and extraction time to character of pectin from siam orange (*Citrus Nobilis* L) waste *J. Pascapanen* **5** (2) Bogor BBP4

[18] Kurachi H 1995 Process for producing artifical rice USA Patent No: 5403606

[19] Wahjuningsih S B, Marsono Y, Praseptiangga D and Haryanto B 2016 Organoleptic, chemical, and physical characteristic of sago (Metroxylon spp.) analogue rice supplemented with red bean (Phaseolus vulgaris) flour as functional food *Oral Presentation in International Food Conference 2016 – Widya Mandala Catholic University Surabaya October 13-14 2016, Surabaya*

[20] Winarno F G 2008 Food Chemistry and Nutrition (in Bahasa Indonesia) (Bogor: Mbrio Press)

[21] Rimbawan A and Siagian A 2004 Food Glicemic Index 'Easy Way to Choose Healthy Food' (in Bahasa Indonesia) (Jakarta: Penebar Swadaya)

[22] Muslikatin 2011 The Making of Extruction Rice with Adding of Seaweed Extract (in Bahasa Indonesia) [Skripsi] IPB: Bogor

[23] Huang X Y, Wang Q, Liu H L, Zhang Y, Xin GR and Shen X 2009 Diastereoisomeric macrocyclic polydisulfi des from the mangrove *Bruguiera gymnorrhiza J. Phytochemistry*, in press

[24] Herawati H and Widowati S 2009 Pearl rice characterization from sweet potato (*Ipomea batatas*) *Buletin Teknologi Pascapanen Pertanian* **5** 37 – 44

[25] Rimbawan and Siagian 2006 Food Glicemic Index (in Bahasa Indonesia) (Jakarta: Penebar Swadaya)

[26] Ginting E, Widodo Y, Rahayuningsih S A and Jusuf M 2005 Starch characterization various sweet potato(in Bahasa Indonesia) *J. Penelitian Tanaman Pangan* **1** (24) 8-17

[27] Willet W, Manson J and Liu S 2002 Glycemic index, glycemic load and risk of type 2 diabetes. *American J. Clinical Nutrition* **76** (1) 274S-280S

[28] Masniawati A, Johannes E, Latunra AI & Paelongan N 2012 Characterization of various red rice physicochemical properties in South Sulawesi (in Bahasa Indonesia) *J. Jurusan Biologi, FMIPA Universitas Hasanudin* 1-10

[29] Brand-Miller J, Kaniz F, Christopher M, Marian B, Vicki L, Fiona A, and Peter P. 2007. Effect of alcoholic beverages on postprandial glycemia and insulinemia in lean, young, healthy adults. *The American Journal of Clinical Nutrition*.

[30] Tran U T, Okadome H, Murata M S, Homma and Ohtsubo K 2001 Comparison of Vietnamese and Japanese rice cultivars in terms of physicochemical croperties *Food Sci. Technol. Res.* **7** 323-330

[31] Raben A, Tagliabue A, Christensen N J, Madsen J, Holst J U and Astrup A 1994 Resistant starch: the effect on postpradial glycemia, hormonal response and satiety *American J. Clinical Nutrient* **60** (4) 544-551

[32] Marsono Y 1998 Changer of resistant starch (RS) and chemical compotition of selected starchy foods puring processing *Food and Nutriton Seminar, Gadjah Mada University, Yogyakarta* Desember 15th 1998

[33] Walter M, Silva L P D and Denardin C C 2005 Rice and resistant starch: different content depending on chosen methodology *J. of Food Comp and Analysis* **18** (4) 279-285

[34] Yamada Y, Hosoya S, Nishimura S, Tanaka T, Kajimoto Y, Nishimura A and Kajimoto O 2005 Effect of bread containing resistant starch on postprandial blood glucose levels in humans *Bioscience, biotechnology, and biochemistry* **69** (3) 559-566

[35] Ashraf S, Anjum FM, Nadeem M and Riaz A 2012 Functional & technological aspects of resistant starch *J. Food Science* **22** (2) 90-95

[36] Widowati S, Astawan M and Santosa B A S 2007 Characterization of quality and influence of parboiling process on glycemic index of various Indonesian rice varieties to improve public health and food security Report

[37] Kusharto C M 2006 Food fiber and its role for health (in Bahasa Indonesia) *J. Gizi dan Pangan* **1**(2) : 45 – 54

[38] Harper J M 1981 Extruction of Food vol. II (Florida: CRC Press Inc) pp 52 – 55

[39] Asgar A and Musaddad D 2006 Optimizing of method, temperature, and time of blanching for processing of dried cabbage *J. Hort.* **16** (3) 245-252

[40] Winarno F G 1992 Food Chemistry and Nutrition (in Bahasa Indonesia) (Jakarta: Gramedia)

[41] Grenus K M, Hsieh F and Huff H E 1993 Extrusion and extrudate properties of rice flour *J. Food Eng.* **18**: 229-245

[42] Kibar Z, Bosoi CM, Kooistra M, Salem S, Finnell RH, De Marco P, Merello E, Bassuk AG, Capra V and Gros P 2009 Novel mutations in VANGL1 in neural tube defects *Hum Mutat.* **30** 706–15

Potential of *Bacillus spp* produces siderophores insuppressing thewilt disease of banana plants

H Kesaulya*,J V Hasinu and G NC Tuhumury

Departement of Agrotechnology, Agriculture Faculty of Pattimura University Ambon-Indonesia

*Email: henry.unpat@gmail.com

Abstract. In nature, different types of siderophore such as hydroxymate, catecholets and carboxylate, are produced by different bacteria. *Bacillus spp*were isolated from potato rhizospheric soil can produce siderophore of both catecholets and salicylate type with different concentrations. Various strains of *Bacillus spp* were tested for pathogen inhibition capability in a dual culture manner. The test results showed the ability of inhibition of pathogen isolated from banana wilt disease. From the result tested were found *Bacillus niabensis* Strain PT-32-1, *Bacillus subtilis* Strain SWI16b, *Bacillus subtilis* Strain HPC21, *Bacillus mojavensis* Strain JCEN3,and *Bacillus subtilis* Strain HPC24 showed different capabilities in suppressing pathogen.

Keywords: *Bacillus spp, wilt disease, banana*

1. Introduction

Current research on microbial siderophore in the rhizosphere is associated with their biocontrol activities due to their competitive effects with plant pathogens. Bacteria produce a large variety of siderophores under iron-limiting conditions: hydroxamates, phenol-catecholates, and carboxylates [2]. In nature, different types of siderophore such as hydroxymate, catecholets and carboxylate, are produced by different bacteria. Hydroxymate siderophore possess N-hydrosylated amide bonds as co-ordination sites, catecholates co-ordinate iron with catecholate hydroxyl group and carboxylates co-ordinate iron with carboxyl and hydroxyl groups [1].Siderophore produced by rhizosphericbacteria improve rhizosphere colonization and play an important role in iron mineralization &supplement to plant.It has been suggested that siderophores are antagonistic by means of sequestering iron from the environment, restricting growth of the pathogen [2-4]. Siderophores facilitate the solubilization and transport of iron into the cell by cognate transport system. Siderophores efficiently deplete iron from the environment making it less available to certain competing microorganisms including plant pathogens.

Bacillus are most predominant colonies in the rhizosphere, potential general due to their spore forming ability [5-7] can elicit plant defenses suppresses the plant pathogens by producing siderophore compounds also induces the Induced systemic resistance (ISR) in host plant and reduces the incidence of disease severity in host plants against pathogens [8]. The objectives of this study were to know the potential of *Bacillus* in producing siderophores to suppress wilt disease in bananas plant.

2. Materials and methods

2.1 Source ofBacillus spp
The *Bacillus* species used are collection of Physiology Laboratories of Agriculture Faculty-Pattimura University which has been identified molecularly based on gen sequence of 16S rRNA and phylogenetic [9].

2.2 Estimated ofsiderophores
The production of bacterial isolate siderephores was tested as follow, The media used is media NB. One ml of bacterial isolate was added to each flask and incubated at 37°C for 7 days. After seven days of incubation, culture of bacterial isolate was centrifuged at 10,000 g for 20 min. Supernatant is used to estimate siderophores salicylate type. Twenty ml of supernatant culture was taken and the pH was set to 2.0 with HCl solution. For 20 ml of supernatant was added 20 ml of ethyl acetate and extraction twice. Five ml of the test solution were added with 5 ml of the Hathway reagent (1 ml of 0.1 M ferric chloride and 1 ml of 0.1 N HCl added to 100 ml of distilled water and 1 ml of 0.1 M potassium ferricyanide added) and the absorbance was measured at 560 nm with sodium salicylate as a standard for salicylate estimation. Standard sodium salicylate is prepared from dilution with a concentration of salicylate nantrium ranging from 0 to 2 mg l-1. To measure the concentration of catechol type siderophores, five ml of the test solution was added with five ml of Hathway reagents and absorbance was determined at 700 nm with 2.3 DHBA as standard. The concentration in the culture filtrate was determined and expressed as mg l-1.

2.3 Confirmation of infected plant samples
Crop samples obtained from farmers' gardens from Central Ceram Manucipal, Maluku Province affected by wilt disease in banana plants were collected for isolation. The infected tissue shows a milky white strand composed of bacterial mass, which seeps out of the infected parts of the crop after being placed in a water-filled test tube.

2.4 Isolation of Pathogenic Bacteria
The source of bacterial inoculum was taken from banana plant from Tial village of Ambon Island, which showed symptoms of bacterial wilt disease allegedly caused by *Ralstonia solanacearum*infection with characteristic of wilting on leaves on old plants and yellowing leaves followed by necrosis. The skin looks normal, some look yellow and black. The inside of the fruit looks reddish brown and slimy mucus. The infected stem part of the plant is removed with a sterile scalpel, weresmall pieces are placed in distilled water for 10 to 15 minutes. The inoculation circle is immersed in the liquid and cultured on the NA medium.

2.5 Biovar determination
Procedure determination of biovar preceded with test uses mineral medium ($NH_4H_2PO_4$ 1.0g, KCl 0.2g, $MgSO_4.7H_2O$ 0.2g, Difcobacto peptone 1.0g, Agar 3.0g and bromothymol blue 80.0 mg per liter) containing 1% sugar. Test biovar based on bacterial abilities utilizing sucrose, lactose, maltose and manitol, sorbitol dan dulcitol were performed following procedure of [10, 11]. A total of 200 µl of liquid media is inserted into microtiter. Suspension of pathogenic bacterial isolate inoculum with a concentration of about 10^8 CFU / ml was prepared and cultured for 24-48 hours in sterile aquades. A total of 20 µl of bacterial suspension was added to microtiter plate wells and incubated at 28-32°C. After 3 days of inoculation was observed discoloration [12].

2.6 Test of inhibition zones in some media
Inhibition zone formation test using several media such as: NA; King`s B Agar; YPDA. 9 ml *Bacillus spp* isolates frommedium NB was added with 1 ml suspension of pathogenic bacteria into a petridish. Filter paper with a diameter 5 mm dipped into *Bacillus spp* bacterial suspension from

various strains is placed in the center of the agar surface. As a control, the piece of filter paper is immersed in a sterile aquadest. Each strain of *Bacillus spp* was tested three times and the inhibition zone diameter was measured .

3. Results and discussion

3.1 Siderophores testing

Microbes capable of producing siderophores may affect the biocontrol, virulence and availability of iron nutrients for plants [13,14,15,16]. Rizosphere bacteria can produce various siderophores.

Generally, siderophores bacteria contain catecholates, and some also contain carboxylates and hydroxamates [17]. The results of this research by testing the production of catechol and salicylate showed that the rhizosphere bacteria from the potato plant produced the species of catechol and salicylate siderophores with different concentration (Table 1).

Table 1. Siderophores Production Capability by Some Strain Bacillus.

Bacteria [18]	Type of Siderophore	
	Catechol (mgl^{-1})	Salicylate ($mg\ l^{-1}$)
Bacillus niabensis Strain PT-32-1	2.26	3.23
Bacillus subtilis Strain SWI16b	3.35	1.71
Bacillus subtilis Strain HPC21	2.87	4.21
Bacillus mojavensis Strain JCEN3	4.21	4.12
Bacillus subtilisStrain HPC24	3.03	2.77

3.2 Isolation of pathogenic bacteria

Result of isolation pathogenic bacteria from banana stem showing symptom of wilt. Isolate pathogenic bacteria grown in milky color and including gram-negative bacteria characterized by round colony, white color and no slime after detection with 3% KOH.

3.3 Biovar determination

The race classification, biovar and phylotype generally aims to divide *R. solanacearum* species. Racial pattern system that classifies *R. solanacearum* strains according to the ability to infect host plants. The results of biovar determination of pathogenic bacterial causes of wilt disease in banana plants including biovar 3. This type of bacteria is able to oxidize carbohydrates (sucrose, lactose, dextrose and cellobiose) and sugar alcohol (mannitol and dulcitol) which is indicated by color changes according to Table 2. and Figure 1.

Table 2. Determination of Bacterial Pathogenic Biovar on Banana Plant Disease.

Usefull	Biovar				
	1	2	3	4	5
Sukrose	-	+	+	-	+
Lactose	-	+	+	-	+
Dextrose	-	+	+	-	+
Cellobiose	-	+	+	-	+
Mannitol	-	-	+	+	+
Dulcitol	-	-	+	+	-

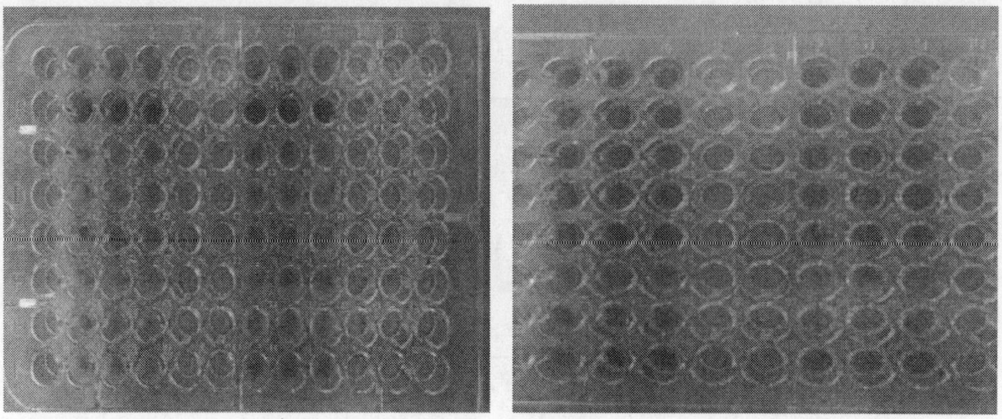

Figure 1. Biovar Determination Model

3.4 In vitro antagonists against bacterial disease on banana plants

In vitro testing to inhibit bacterial wilt diseases from banana plants found that all of *Bacillus spp* strains tried to control wilt disease in banana plants. This is indicated by the formation of halozone around the bacteria tested (Figure 2). The results of antagonistic test showed the presence of inhibitory activity by *Bacillus spp* on the pathogen of wilt disease in banana plants. The difference in inhibition by each strain of *Bacillus spp* is shown in Table 3. *Bacillus spp* is antagonistic to various plant pathogens [19,20,21] including pathogens of wilt banana plants.

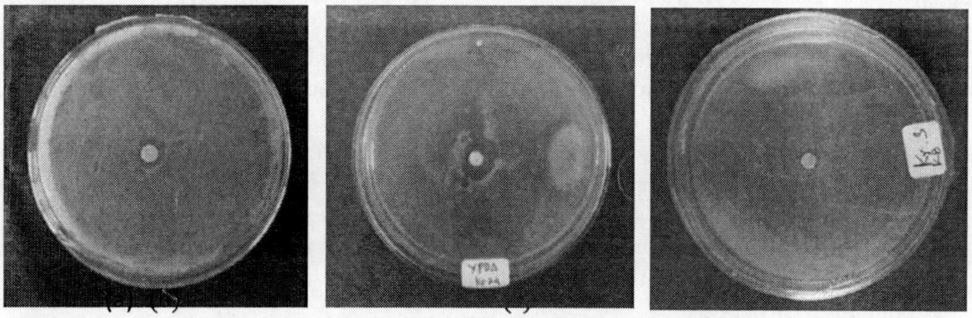

Figure 2. Halozone : a) *Bacillus mojavensis Strain JCEN3* in Nutrien Medium, b)*Bacillus subtilis*Strain HPC2-1 in *Bacillus subtilis* Strain HPC2-1Yest Peptone DextrosAgar (YPDA), and c. *Bacillus niabensis Strain* PT-32-1 in Kings B Medium.

Table 3. Halozone Formation by Bacillus spp on Some Types of Media AgainstBacterial Pathogens Causes of Banana Disease in Banana Plants

Strain Bacillus spp [18]	Halozone Diameter (mm)		
	Nutrien Agar	Yest Pepton Dekstrose Agar	Kings B
Bacillus niabensis Strain PT-32-1	10	8.5	8.5
Bacillus subtilis Strain SWI16b	10.5	6.5	7.5
Bacillus subtilis Strain HPC21	9.5	6.5	7.5
Bacillus mojavensis Strain JCEN3	9.5	6	6
Bacillus subtilis Strain HPC24	8.5	10.6	6

4 Conclusion

Bacteria causing wilt diseases in banana plants belonging to biovar 3 which is indicated by the color change on through the biovar determination test. *Bacillus spp* strains tested in vitro are able to control wilt disease in banana plants, which is shown by the formation of halozone around the bacteria.

5 References

[1] BholayA D, Jadhav P U. Borkhataria B V and Mayuri V Dhalkari 2012 *Fluorescent Pseudomonads* as Plant Growth Promoting Rhizobacteria and Their Siderophoregenesis. *IOSR J. Pharmacy and Biol. Sci., (IOSRJPBS)* **3**pp. 27-32

[2] Bakker P A H M, Lamers J G, Bakker A W , Marugg J D , Weisbeek P Jand Schippers B1986 The role of siderophores in potato tuber yield increase by *Pseudomonas putida* in a short rotation of potato. *Neth J Plant Pathol* **92**249–256

[3] Lemanceau P, Bakker P A H M, De Kogel W J, Alabouvette Cand Schippers B1992 Effect of pseudobactin 358 production by *Pseudomonas putida* WCS358 on suppression of *Fusarium* wilt of carnations by nonpathogenic *Fusarium oxysporum* Fo47. Appl. Environ. Microbiol

[4] Loper J E and Buyer J S1991 Siderophores in microbial interactions on plant surfaces. Mol. Plant-Microbe Interact. **4**5–13

[5]Manikanda R, Saravanakumar D, Rajendran L, Raguchander T and Samiyappan R [2010] Standard ization of liquid formulation of *Pseudomonas fluorescens* Pfl for its efficacy against Fusarium wilt of tomato *Biol control***54**83-89

[6] Joseph B, Patra RR and Lawrence R 2007 Characterization of plant growth promoting rhizobacteria associated with chickpea (*Cicer arietinum* L.). *International Journal of Plant Production***1**(2) 141-151

[7] Liu ZL and Sinclair JB 1993 Colonization of soybean by Bacillus megaterium B153-2-2. *Soil. Biol. Biochem***25** 849-855

[8] Kloepper JW, Ryu CM and Zhang S 2004 Induced systemic resistance and promotion of plant growth by *Bacillius spp*. Phytopathology **94**1259-1266

[9] Kesaulya H 2015 Bioprospek Rizobakteria Asal Kentang (*Solanum tuberosum* L.) Var. Hartapel Sebagai Pemacu Pertumbuhan Tanaman. Desertasi. Unhas.

[10] Hayward A C 1964 Characteristic of *Pseudomonas solanacearumJ. Appl. Bact* **27**265-277

[11] He L Y, Sequeira L and Kelman, A 1983 Characteristic of *Pseudomonas solanacearum* from China. *Plant Dis* **67**1357-1361

[12] Schaad N W, Jones J B and Chun W 2001 Laboratory Guide for Identification of Plant Pathogenic Bacteria.3nd. Ed. APS Press, St. Paul, MN, USA

[13] Loeng J 1952 Siderophores; Their biochemetry and possible role in the biocontrol of plant pathogens. *Annu. Rev. Phytopathol***24**187-209

[14] Leong S A and Expert D 1990 Siderophores in plant-pathogen interaction. Page 62-83 in : Plant Microbe Interaction**3** T. Kosuge and E Nester, eds. Academic Press. New York

[15] Neilands J B and Leong S A 1986 Siderophores in relation to plant growth and disease*Annu. Rev. Plant Physiol* **37** 187-208

[16] Schippers B, Bakker A W and Bakker P A H M 1987 Interactions of deleterious and benefical rhizosphere microorganisms and the effect of cropping practices. *Annu. Rev Phytopathol* **25** 339-358

[17] Matzanke B F 1991 Structures, coordination chemistry and functions of microbial iron chelates. In *CRC Handbook of Microbial Iron Chelates*. Winkelmann, G. (ed.). Boca Raton, FL, USA: CRC Press, pp. 15–64

[18] Kesaulya H, Hasinu J V, and Tuhumury G N C 2017 Potency of *Bacillus spp* from *Potato* Rhizosphere as Active Ingredients for Biostimulant Formulation. *Modern Applied Science***10** 11

[19] Broadbent P, Baker K F, Franks N and Hollad J 1997 Effect of Bacillus spp. on inceased growth of seedlings in stemed and in nontreated soil. *Phythophatology***67**1027-1034

[20] Silo-Suh L A, Lethbridge B J, Raffel S J, He H, Clardy J and Handesman J 1994 Biological Activities of Two Fungistatic Antibiotics Produced by Bacillus cereus UW85. App and Environ. Microbiol **60** 2023-2030

[21] Osburn R M, Miller J L, Oplinger E S, Smith R S and Handelsman J 1995 Effect of *Bacillus cereus* UW85 on the yield of soybean at two field sites in Wisconsin *Plant Dis* **79**551-556

Development of method of optimized flavor production systems design based on nano-emulsification Kawista (*Feronia limonia*) Fruit extraction

A Suyanto[1*], E Noor[2], F Fahma[2], M S Rusli[2] and T Djatna[2]

[1] *PhD student at Agroindustrial Technology, Bogor Agricultural University*
[2]*Department of Agroindustrial Technology, Bogor Agricultural University*

*Email : agus_kh@apps.ipb.ac.id

Abstract. 'Kawista' (*Feronia limonia*) as a tropical fruit has unique flavor that can be applied as a flavor for food products. Flavor as volatile components are unstable by environment factors such as temperature and storage. Flavor nano emulsification form to improve the stability towards environment and increase its use in food products. Research carried out is system development of the nano emulsification Kawista extract flavor with sonication method. The best treatments are selected by Response Surface Methodology (RSM) for independent variable are amplitude (70-100%), time (90-150s) and temperature (5-45°C) controlled by the software of the device. The Flavor Extraction by maceration technique extended highest yield and flavor components. Nano-emulsions made with composition 1% (w/w) flavor extract, 2% (w/w) surfactant (tween 80), 0.25% Gum, and 96.75% (w/w) deionized water. The probe of sonication successfully for preparing stable O/W nano emulsions at amplitude, time and temperature 81.01%, 150s, 45°C, respectively. Characteristic of nano-emulsions i.e energy input (15.489J), viscosity (2.076 mPa.s), droplet size (13.446nm), and Polydispersity index (0.469).

Keywords: *Feronia limonia, flavor, sonication, nano emulsions, droplet size*

1. Introduction

The use of natural food to increase due to awareness of the community to consume safe and nutritious food and have a positive impact on health. New material food excavations including enhancement of existing ingredients have been pursued by improving the cultivation, handling of post harvest, processing, distribution and storage, until of serving to obtain optimal food benefits. The utilization of 'Kawista' fruit (wood apple) especially aroma component that is the provision in the form of food flavor will provide a positive contribution in the needs of natural food through agricultural processing technology.

Kawista (*Feronia limonia*) is a plant that includes the citrus family (*Rutaceae*). As a tropical Asian plant, it spread unevenly and generally grows wildly of the forest and sandy soil like on the beach, and its decreases of population (rare plant). Kawista can be found in Aceh, Rembang, Madura and Bima. It diameter around 10cm, hard peel, and grayish white. As a food, young Kawista can be make for 'Rujak' and ripe fruit to syrup product because the distinctive aroma is very preferred [1].

The perfectly ripe Kawista fruit has a distinctive aroma and can provide a flavor to the food. Flavor component analysis to use Gas Cromatography-Olfactometry (GC-O) and Aroma Extract Dilution Analysis (AEDA) technique, successfully identified and described some character impact odorants that give the feel of aroma to flavor characteristics of Kawista. The dominant aromas of these components include: fruity, sweet, banana like, melon-like, cheese (butter-like), herbaceous (green) and sour. The volatile component of the Kawista flavor that can be identified by GC-MS includes 75 volatile components consisting of 28 ester (37.3%), 11 alcohol (14.7%), 10 aldehyde (13.3%), 1 acetyl (1.3%), 10 ketones (13.3%), 4 lactone (5.3%), 1 heterocyclic aromatic (1.3%), 4 hydrocarbon (5.3%), 1 furan (1.3%) and 5 carboxylic acid (6.7%) [2].

Flavor component extraction aims to obtain flavor components and remove other components. It can be vacuum distillation method, extraction with solvent (maceration), Likens-Nickerson extraction and headspace. The method of extraction with maceration technique to use organic solvent (1:1) got the best of Kawista extract flavor similar to the original fruit [2].

Flavor components of fruits extract as volatile compounds is volatile unstable under normal environmental conditions and during processing. To maintain the stability of food flavors can be done by binding the flavor components in the form of emulsions. Stability of folic acid it can be increased by making triple emulsion, it water-in-oil-in-water (W/O/W) emulsion using Maltodextrin-Whey Protein binder in nano emulsion [3].

Nano technology can be applied in food processing technology in the form of nano emulsion. The definition of nanotechnology in food processing is the manufacture of nano-sized food materials of 20-200 nm. The characteristic of the nano emulsion product are transparent, translucent and uncreaming. Nano emulsion is resistant to creaming because it has sufficient Brownian motion to overcome the separation of gravity forces. It is also resistant to flocculation because of its highly efficient steric stability. Most nano emulsion are stable by synthetic surfactants that tend to have long hydrophilic groups [4].

Technique to make nano emulsions there are two ways i.e. high energy and low energy approach. It is high energy methods to use mechanical equipment such as high pressure homogenizer, micro fluidizer, and sonication methods. The formation of a low energy nano emulsion is produced as a result of a phase transition during an emulsification process in which there is a change in composition with a constant temperature. Frequently used methods are membrane emulsification, spontaneous emulsification, emulsion inversion point, phase inversion point, solvent displacement [5].

Ultrasonic is a cyclical sound pressure with a frequency greater than the upper limit of human hearing in the frequency range 20 kHz-10 MHz. The ultrasonic chemical effect in the liquid comes from several nonlinear acoustic phenomena, where cavitation is the most important. Acoustic cavitation is the formation, growth and breakup of bubbles in a liquid that is irradiated with sound or ultrasonic [6].The intensity of the propagating ultrasonic waves will bring energy to a surface area per unit of time. If the ultrasonic wave energy passes through the tissue, it releases heat energy so that there is a warming which results in increased tissue temperature and then causing the effects of cavitation, namely the formation, growth and bubble burst in a liquid. The rupture of the cavitation bubble can produce a temperature of about 5000 ^{o}C and a pressure of 2000 atm [7].

The objective of the research is to develop a nano emulsion Kawista with ultra-sonication (high energy method). The nano emulsion form of Oil in Water (O/W) to provide flavor of Kawista extract in liquid form. The sonicator probe settings include the intensity of amplitude, time and temperature as independent variables. The research parameters of energy consumption, viscosity, particle size, and polydispersity index (PDI) as dependent variables. The best model form to use Response Surface Methodology (RSM) method, so obtained the mathematical model and its significance level.

Optimization Techniques with RSM

Optimization of nano emulsion process can be done by using one of the process optimization techniques with RSM. It is a statistical study where this technique can be used to analyze and to make the model with one or more treatments in the study [8, 9, 10]. The response of a process as dependent variable can be predicted after to insert the lower limit and upper limit of a value as an independent variables. The formula of the free variables selected through the RSM technique can be used to develop, improve, and optimize the process. The mathematical model of RSM technique can explain chemical or biochemical processes so as to illustrate the best response of a process. Another advantage of RSM method is that it does not require large amount of treatments, so it is efficient in terms of cost and time of research.

Optimization techniques using approaches from statistics and engineering mathematics are facilitated by the Design Expert Program (DX) RSM Box-Behnken Design. The stages in optimization techniques using RSM are 1) creation of formulation and response design, 2) formulation, 3) response analysis, 4) optimization. Verification of results is required to prove the predicted value of the suggested response value.

The design of the formulation and response is done by determining the research variables. At this stage it should establish the lower and upper limits of the variable that provides a response based on the literature review or trial and error with preliminary research. The minimum and maximum limit values entered into the DX 7® RSM Box-Behnken Design program are further processed by the program that will produce the treatment formulas to be analyzed and the measured response.

The formulation stage is the actual stage of the study according to the process conditions i.e. the formula obtained from the DX software. Response analysis phase is to analyze the influence between variables with Analysis of Varian (Anova) then selected the highest significance value of the model obtained whether Linear, Quadratic, Special Cubic, or Cubic. Models that provide significance and non-significance in the lack of fit were selected to analyze the variables. Normality tests are marked by the distribution of data the closer to the normal line, indicating the actual results will be close to the results predicted by the program.

In the optimization stage, the DX program will perform optimization of data and response measurement data with the output in the form of recommendations of some optimum formula according to the program. The best formula is maximum desirability value. The value of desirability is the value of the optimization that indicates the ability of the program to meet the desires based on the criteria set on the final product [10]. The results obtained from the best conditions continued to the stage of verification, which is done processing and analysis of the results in accordance with the best formulation of RSM so that it can be seen suitability in the verification stage.

2. Research Methods

2.1 Preparation of Kawista Flavor Extract Nano emulsion

The necessary ingredients are flavor Kawista extract, Tween 80 (Merck, Germany), Sodium azide as an antimicrobial agent (Sigma Chemicals Company St. Louis, USA) and Deionized water. Procedures for the manufacture of nano emulsion flavor Kawista with ultrasonic adopted from Hashtjin, AM, and Abbasi, S. [11] on the manufacture of nano emulsion from orange peel essential oil which component of flavor extract (1% w/w) as oil phase, added emulsifier Tween 80 (2% w/w), and 0.25% w/w the solvent fraction comprises a mixture of Persian Gum (PG) and Gum Tragacanth (GT) (75:25) and deionized water (96.75% w/w) as an aqueous phase. The first stage of the mixture is placed in a beaker glass homogenized with a magnetic stirrer (700 rpm for 15 minutes) at room temperature. After that, it sonication to the solution (Sonicator 4000, 20 kHz, cylinder sonotrode diameter 19.1 mm, Misonix, Inc, New York) refer to running output the RSM method on amplitude (70-100%), time (90-150 s) and temperature (5-45 °C) (see at the Table 1 and Table 2).

2.2 Nano emulsion measurements

2.2.1. Droplet size and Polydispersity Index (PDI). The droplet size of nano emulsion is determined by photon correlation spectroscopy. Sample (0.1 ml) of nano emulsion was dispersed in 50 ml of water.

Measurements were made using Zetasizer 1000 HS (Zetasizer Nano Zs, Malvern Instrument, Malvern, UK). The scattering light is monitored at 25°C at an angle of 90°. The analysis of the dispersion and the size of the granular nanoparticles were performed by Particulate System-Particle Size Analyzer which can measure the size distribution with the range 2 nm to 7000 nm.

2.2.2. Rheological measurements. Viscosity was measured by Brookfield viscometer (LVDV Pro II, Brookfield Engineering Laboratories, USA) with a spindle S34. The viscosity of the formulation (0.5 g) was determined without dilution using Brookfield DV III ultra-V6.0 RV cone and rheometer plate at 25 ± 0.5°C.

2.3 Research design and Data analysis

Development of Kawista flavor nano emulsion production with ultra-sonication to test the influence of independent variable that is amplitude 70-100% (X_1), time variation 90-150s (X_2) and temperature 5-45°C (X_3) to dependent variable i.e. energy input (Y_1), viscosity (Y_2), droplet size (Y_3) and PDI (Y_4). The experimental design to use 2^3 factorial with central composite design method (Table 2). Data analysis includes descriptive analysis such as tabulation of data, presentation of images, and graphs, as well as statistical data analysis to examine the effect of treatment and choose the best formula. Test RSM optimization technique with RSM using Design Expert Software 7.0 tool.

3. Result and Discussion

3.1 System Development of Kawista Flavor Extract Nano emulsion with RSM

The procedure to make nano emulsion Kawista flavor adopted from Hashtjin, A.M., and Abbasi, S. [11] on the manufacture of nano emulsion orange peel essential oil. Optimization formula of nano emulsions by RSM technique. Nano emulsion formulation with ultra-sonication technique is to set minimum and maximum limit of operating system as independent variable i.e. amplitude value (70-100%), time duration (90-150s) and temperature of sonicator (5-45°C) (Table 1).

Table 1. RSM design of independent variables

Symbols	Independent variable	Min	Max	-alpha	+alpha
X_1	Amplitude (%)	70	100	59.7731	110.2269
X_2	Time (s)	90	150	69.5462	170.4538
X_3	Temperature (°C)	5	45	-8.6359	58.6359

Model of linear regression equation influence between research variables:

$$Y_i = a + a_1X_1 + a_2X_2 + a_3X_3 \quad (1)$$

Where the dependent variable Y_i, a constants, X_1 amplitude, X_2 time and X_3 temperature. Development of production system of nano emulsion Kawista flavor with ultrasonic technique refers to production of nano emulsion orange peel essential [9] obtain dependent variable data that is energy input 7.079-18.889J, viscosity 1.842-2.262 mPa.s, droplet diameter 14.123-33.103 nm, the polydispersity index ranges from 0.43-0.67. Overall result with RSM optimization technique obtained 20 run data presented in Table 2.

Table 2. Influence of different ultra-sonication conditions i.e. amplitude (X_1), time (X_2), and temperature (X_3) to energy input (Y_1), viscosity (Y_2), droplet size (Y_3) and PDI (Y_4) Kawista flavor nano emulsion

Run	Independent variable			Response			
	X_1	X_2	X_3	Y_1	Y_2	Y_3	Y_4
1	85.00	69.55	25.00	7.636	2.314	29.789	0.63
2	100.00	90.00	45.00	12.529	2.202	19.193	0.51
3	70.00	90.00	45.00	7.159	2.382	27.143	0.63
4	100.00	150.00	5.00	18.809	1.842	14.473	0.43
5	100.00	90.00	5.00	12.449	2.082	25.153	0.55
6	110.23	120.00	25.00	17.499	1.961	14.123	0.43
7	70.00	90.00	5.00	7.079	2.262	33.103	0.67
8	70.00	150.00	45.00	13.519	2.142	16.463	0.51
9	85.00	120.00	25.00	12.984	2.112	20.808	0.53
10	59.77	120.00	25.00	8.468	2.263	27.493	0.63
11	85.00	170.45	25.00	18.332	1.910	11.827	0.43
12	100.00	150.00	45.00	18.889	1.962	8.513	0.39
13	85.00	120.00	25.00	12.984	2.112	20.808	0.53
14	85.00	120.00	-8.63	12.917	2.011	25.820	0.56
15	85.00	120.00	25.00	12.984	2.112	20.808	0.53
16	85.00	120.00	25.00	12.984	2.112	20.808	0.53
17	85.00	120.00	25.00	12.984	2.112	20.808	0.53
18	85.00	120.00	58.63	13.051	2.213	15.796	0.50
19	70.00	150.00	5.00	13.439	2.022	22.423	0.55
20	85.00	120.00	25.00	12.984	2.112	20.808	0.53

This sonication process converts electrical signals into physical vibrations with ultrasonic waves with a frequency range 20 kHz-10 MHz. This vibration has a very strong effect on the solution, causing the outbreak of the molecule and the breaking of the cell. The sonication probe sends vibration to the sonication solution causing an effect of cavitation, the process of bubbles bursting in the fluid resulting from a sudden drop in pressure at a constant temperature. The cavitation bubble makes the particles have a diameter in the nano scale.

3.2 Energy Input

The energy input of nano emulsion production are 7.079-18.889J. The result of Anova test obtained p value <0.0001 which means independent variables (amplitude, temperature and operating time sonication) significant effect on dependent variable i.e. energy input. The linear regression equation effect of sonication operation condition to energy input is as follows:

$$Y_1 = 12.98 + 2.68X_1 + 3.18X_2 + 0.040X_3 \quad (2)$$

Based on the mathematical model can be illustrated that all the independent variables positively affect the energy (Y_1) is characterized by positive-positive constants. The largest value of constant is at time (seconds) of sonication operation equal to 3.18 times, amplitude (%) 2.68 times, and temperature (°C) gives the least influence is 0.04 times. The amount of energy required to manufacture nano emulsion droplets to a minimum, the factors contributing to the increase in energy can be reduced. Three-dimensional image of RSM method calculation duration time and amplitude can be seen in Figure 1.

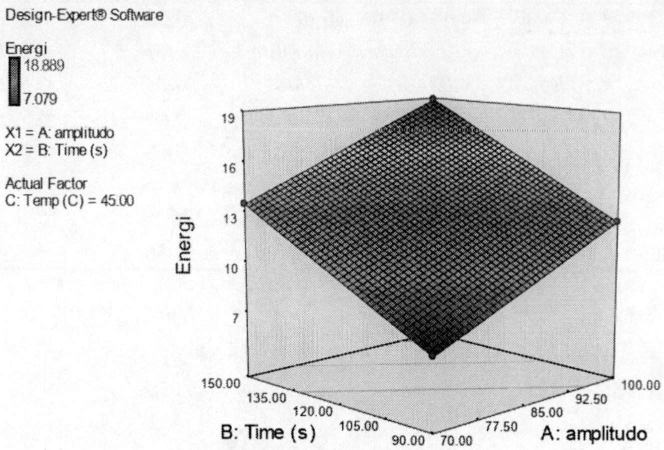

Figure 1. Response surface plot effect of time and amplitude of sonication on energy input

3.3 Rheology

The characteristic of rheology is expressed as the viscosity and flow of liquid food products measured using a viscometer. The widely used viscometer is rotational viscometer. The principle it by measuring the magnitude of the resistance due to the viscous nature of a fluid. Nano emulsion viscosity ranged from 1.842-2.262 mPa.s and Anova test results obtained p value <0.0001 which means independent variable (amplitude, temperature and operating time of sonication) significantly affect the dependent variable (viscosity). The linear regression model of the effect of sonication operation condition on viscosity is as follows :

$$Y_2 = 2.11 - 0.090X_1 - 0.12X_2 + 0.060X_3 \quad (3)$$

Based on the mathematical model can be illustrated that the factor of amplitude and time have negative effect marked with negative constants, on the other hand the temperature factor has positive effect on viscosity (Y_2). The desirable nature of nano emulsion rheology, the negative notation on amplitude and time is desirable to decrease the rheology value, whereas the positive notation at the temperature is undesirable. In other words, to decrease the viscosity of the sonication operation, the amplitude and time value need to be increased while the temperature need to be decreased.

IOP Conf. Series: Earth and Environmental Science **102** (2017) 012017 doi:10.1088/1755-1315/102/1/012017

Table 3: Anova regression on energy input and viscosity

Source	Energy input (Y_1) Sum of Squares	F Value	p-value	Source	Viscosity (Y_2) Sum of Squares	F Value	p-value
Model	236.58	6.37E+07	< 0.0001	Model	0.36	6.37E+07	< 0.0001
A-amplitudo	*98.46*	*6.37E+07*	*< 0.0001*	*A-amplitudo*	*0.11*	*6.37E+07*	*< 0.0001*
B-Time (s)	*138.1*	*6.37E+07*	*< 0.0001*	*B-Time (s)*	*0.2*	*6.37E+07*	*< 0.0001*
C-Temp (C)	*0.022*	*6.37E+07*	*< 0.0001*	*C-Temp (C)*	*0.049*	*6.37E+07*	*< 0.0001*
Residual	0			Residual	0		
Lack of Fit	*0*			*Lack of Fit*	*0*		
Pure Error	*0*			*Pure Error*	*0*		
Cor Total	236.58			Cor Total	0.36		

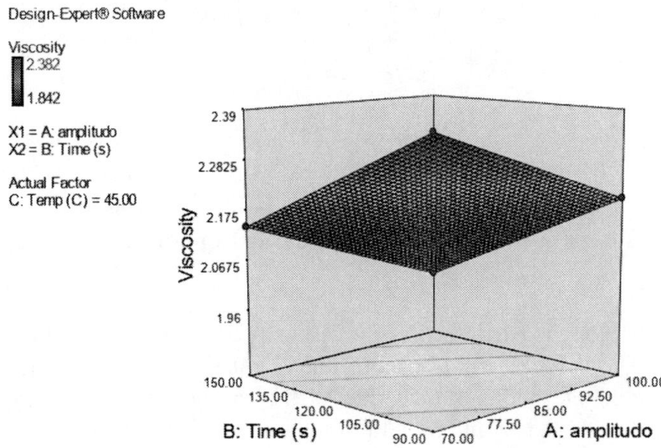

Figure 2. Response surface plot effect of time and amplitude of sonication on viscosity

3.4 Size of droplet diameter

The sonication process per unit time will correlate to the homogeneous properties of the solution and reduce the particle size to a stable nanoscale size and reduce flocculation properties. This condition is caused by shock wave in sonication method can separate agglomeration and there is perfect dispersion with the addition of surfactant as stabilizer.

The size of nanoemulsion droplet diameter ranged 14.123-33.103 nm and anova test results obtained p value <0.0001 which means independent variables (amplitude, temperature and operating time sonicator) significantly the dependent variable (droplet diameter). The linear regression model of the effect of sonication operation condition on droplet size is as follows:

$$Y_3=20.81-3.97X_1-5.34\ X_2-2.98\ X_3 \quad (4)$$

Based on the mathematical model can be illustrated that all free variables negatively affect the droplet size as indicated a negative constants. The smaller droplet size to be more stable the nano emulsions, so the negative notation on the linear regression equation is actually positive. In other words, to decrease the droplet size, the sonication operation is to increase the amplitude, time and operating temperature. The largest value of the constants is at the time of sonication that is 5.34 times, then amplitude 3.97 times, while the temperature gives the smallest effect that is 2.98 times.

According to McClements [12] droplet size is the result of dynamic equilibrium between effects that tend to decrease particle size and the effect that makes the droplet join becomes larger. Droplet size affects the stability of an emulsion. The smaller droplet size makes nano emulsion stable kinetic so as to prevent the occurrence of sedimentation and aggregation during storage.

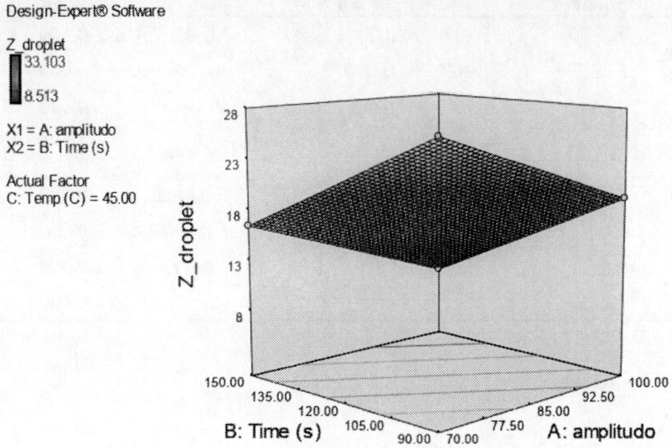

Figure 3. Response surface plot effect of time and amplitude of sonication on sizeof droplet

3.5 Polydispersity Index (PDI)

The PDI of an emulsion is a value that describes the uniformity of the droplet size of an emulsion. The lower the polydispersity value indicates the size of the resulting particles is relatively more uniform. PDI nano emulsion Kawista index ranged from 0.43-0.67 and Anova test results obtained p value <0.0001 which means free variable (amplitude, temperature and operating time of sonication) significantly affect the PDI. The linear regression model of the effect of the sonication operation condition on PDI is as follows:

$$Y_4=0.53-0.060\ X_1-0.060X_2-0.020\ X_3 \quad (5)$$

Based on the mathematical model it can be illustrated that all the independent variables have a negative effect on the PDI. Looking at the linear equations can provide clues that the greater the value of the contest of the independent variables the better the PDI value or the smaller the PDI value representing the more uniform droplet size. The value of the amplitude constant and the sonication operation time are equal to 0.06 times, while the temperature gives that is 0.02 times.

Table 4. Anova droplet size and PDI

Droplet	Source	Sum of Squares	F Value	p-value Prob > F	PDI	Source	Sum of Squares	F Value	p-value Prob > F
	Model	726.50	6.37E+07	< 0.0001		Model	0.100	6.37E+07	< 0.0001
	A-amplitudo	*215.79*	*6.37E+07*	*< 0.0001*		*A-amplitudo*	*0.049*	*6.37E+07*	*< 0.0001*
	B-Time (s)	*389.43*	*6.37E+07*	*< 0.0001*		*B-Time (s)*	*0.049*	*6.37E+07*	*< 0.0001*
	C-Temp (C)	*121.28*	*6.37E+07*	*< 0.0001*		*C-Temp (C)*	*5.46E-03*	*6.37E+07*	*< 0.0001*
	Residual	0.00				Residual	0.000		
	Lack of Fit	*0.00*				*Lack of Fit*	*0.000*		
	Pure Error	*0.00*				*Pure Error*	*0.000*		
	Cor Total	726.50				Cor Total	0.100		

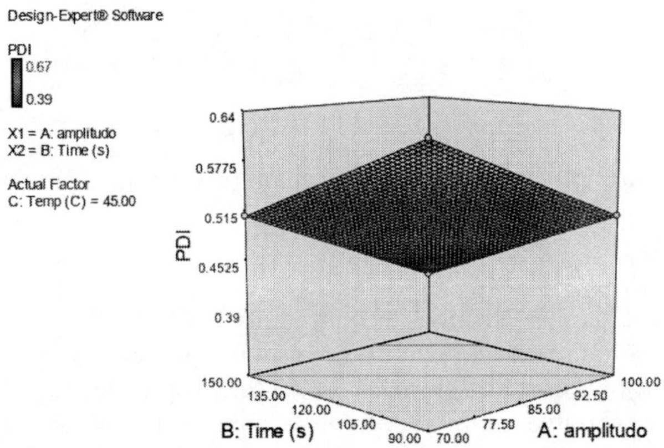

Figure 4. Response surface plot effect of time and amplitude of sonication on PDI

3.6. Optimization of nano emulsion with ultra-sonication

The optimization given to the RSM DX program (Table 5) is based on variable data and response measurement data included with output in the form of recommendations of some optimal new formulas according to the program. The most optimal formula is the formula with maximum desirability value. The range of desirability values is 0 to 1.0, where values closer to 1.0 indicate the ability of the program to produce the desired product more perfectly. The goal of optimization is not to get the value of desirability 1.0, but to find the best conditions that bring together all the purpose functions.

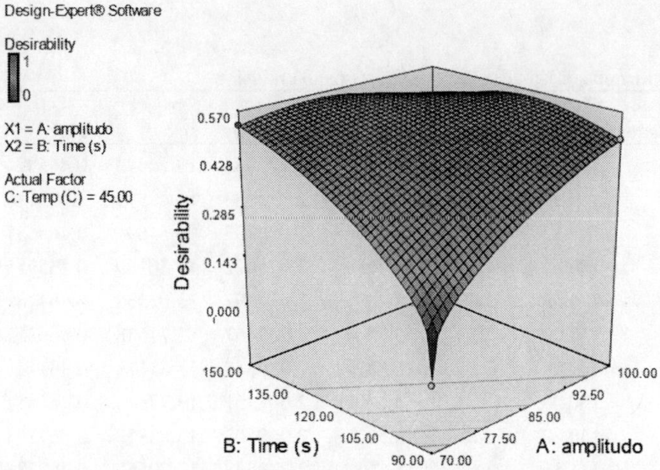

Figure 5. Response surface plot of optimization with time and sonication amplitude

The best formula recommended by DX program it can be seen the Table 6 with desirability value 0.554. It the sonication to use amplitude value, time and temperature are 81.01%, 150s, and 45°C respectively, which will to produce Kawista flavor nano emulsion with energy input, viscosity, droplet size , and the PDI are 15.489J; 2.076 mPa.s; 13.446 nm, and 0.466 respectively. The results obtained from the optimum process conditions are continued to the verification stage, i.e. the processing and analysis of the results in accordance with the best formulation obtained from the optimization of RSM can be seen in the suitability stage verification.

Table 5. Optimization procedure of Kawista flavor extract nano emulsion

Name	Goal	Lower Limit	Upper Limit	Lower Weight	Upper Weight	Importance
amplitudo	is in range	70.000	100.000	1	1	3
Time (s)	is in range	90.00	150.000	1	1	3
Temp (C)	is in range	5.000	45.000	1	1	3
Energi	minimize	7.079	18.889	1	1	3
Viscosity	minimize	1.842	2.382	1	1	3
Z_droplet	minimize	8.513	33.103	1	1	3
PDI	minimize	0.390	0.670	1	1	3

Table 6. Optimum formula of Kawista flavor extracts nano emulsion

Solution	amplitude	Time (s)	Temp (C)	Energy	Viscosity	Droplet size	PDI	Desirability	
1	81.01	150.00	45.00	15.4892	2.075955	13.54643	0.465974	0.554516	Selected
2	81.24	149.97	44.97	15.52757	2.074583	13.49515	0.465136	0.554494	
3	81.33	149.52	45.00	15.49568	2.075957	13.54697	0.465651	0.554467	
4	81.06	150.00	44.80	15.49864	2.075022	13.56199	0.465955	0.554455	
5	83.41	146.03	45.00	15.49919	2.077392	13.61538	0.464289	0.554107	
6	81.28	150.00	43.38	15.53442	2.069467	13.71638	0.466512	0.553992	
7	80.99	149.99	43.26	15.48231	2.070848	13.81118	0.46779	0.553963	
8	81.31	150.00	42.94	15.53999	2.067951	13.7714	0.466802	0.553841	
9	81.09	150.00	42.89	15.50028	2.069105	13.83858	0.467744	0.553838	
10	86.08	141.55	45.00	15.50122	2.079333	13.70666	0.462588	0.553628	
11	96.68	124.33	45.00	15.57384	2.084595	13.96195	0.454617	0.551592	
12	99.70	119.25	45.00	15.57488	2.08683	14.06759	0.45272	0.550954	
13	100.00	117.64	20.63	15.40969	2.018345	17.90491	0.479098	0.537719	

4. Conclusion

The development of the method of making nanoemulsion flavor kawista can be done by sonication technique by using RSM for selection of best formula. The results of ANOVA showed a very significant difference (P<0.0001) of the sonication condition on the characteristic of kawista nanoemulsion. The higher the sonication state produces higher energy inputs, which is not desired in the manufacture of nanoemulsions. The lower of viscosity nanoemulsion will increase its stability, which can be obtanined through sonication by increasing the amplitude and time, and decreasing the temperature. The higher the value of the sonication condition will decrease the droplet size (nm) and the uniformity of droplet size (the smaller PDI value approaching 0). Small droplet and PDI measurements show nanoemulsion stability and particle uniformity. The best formula (desirability value 0.554) is sonication conditions with amplitude, time and temperature are 81.01%, 150s, and 45°C respectively. Output of nanoemulsions flavor kawista that is energy input, viscosity, droplet size, and PDI i.e. 15,489 J, 2,076 mPa.s, 13,446 nm, and 0,466 respectively.

References

[1] Jones DT. 1992. Edible Fruits and Nuts. Di dalam: Verheij EWM, Coronel RE, editor. Plant Resources of South-East Asia **2**. Bogor: Prosea. Hlm 190-191.

[2] Apriyantono Anton dan Bakti Kumara. 2004. Identifikasi Character Impact Odorants Buah Kawista (Feronia limonia). Jurnal.Teknologi dan Industri Pangan, Vol. XV, No.**1** Th. 2004

[3] Assadpour E, Maghsoudlou Y, Jafari SM, Ghorbani M, Aalami M. 2016. Optimization of Folic Acid Nano-emulsification and Encapsulation by Maltodextrin-Whey Protein Double Emulsions. Biological Macromolecul Juornal. http://dx.doi.org/10.1016/j.ijbiomac.2016.01.064

[4] Wooster, T. J., Golding M., Sanguansri P. 2008. Impact of Oil Type on Nanoemulsion Formation and Ostwald Ripening Stability. Langmuir **24**: 12758-12765.

[5] Sanguansri, P. and Augustin M. A. 2006. Nanoscale Materials Development-A Food Industry Perspective. Trends in Food Science & Technology. **17**(10), 547-556.

[6] Cintas, P. and Cravotto, G., (2005), Power Ultrasound in Organic Synthesis: Moving Cavitational Chemistry from Academia to Innovative and Large-Scale Applications, The Royal Society Journal of Chemistry (**35**), pp. 180-196,

[7] Bendicho, C. and Lavilla, I. 2000. Ultrasound Extractions, QuO & mica, Spain, pp. 1448-1453

[8] Montgomery, D.C. 2001. Design and Analysis of Experimental. John Wiley & Sons Inc, New York.

[9] Bas, D. and Boyaci, I.H. 2007. Modelling and optimization I: usability of response surface methodology. J. Food Eng. **78**: 836–845.

[10] Raissi, S., and Farzani, R.E. 2009. Statistical process optimization through multi-response surface methodology. World Academy of Science, Engineering and Technology. pp. 267–271.

[11] Hashtjin AM and Abbasi S. 2014. Nano-emulsification of orange peel essential oil using sonication and native gums. Food hydrocolloid **44** (2015) 40-48. http://dx.doi.org/10.1016/j.foodhyd.2014.08.017

[12] McClements, David Julian. 2005. Food Emulsion: Principles, Practices, and Techniques. CRC Press, New York.

Organoleptic characteristics and nutritive value estimation of baked food products from Manonjaya variety salacca flour

Sumarto[1*], D Aprianty[1], R A Bachtiar[1], L Kristiana[2]

[1] Department of Nutrition, Health Polytechnic of Tasikmalaya, Ministry of Health RI, 35 Cilolohan Street, Tasikmalaya, Indonesia
[2] Research and Development Center of Humaniora and Health Management, National Institute of Health Research and Development, Ministry of Health RI, Indrapura 17, Surabaya, Indonesia

*Email: sumarto@dosen.poltekkestasikmalaya.ac.id

Abstract. Manonjaya salacca (snake fruit) is one of the original varieties of Indonesia which is currently declining due to the quality of taste less favoured than the snake fruit on the market. This variety in the future is feared to be lost, so it is necessary to revitalize the utilization of this snake fruit by diversifying processed products, one of them is baked food products from Manonjaya salacca flour. The purpose of this research was to know the acceptance level of baked food products from Manonjaya salacca flour organoleptically and the nutritional value estimation. This research method was observational with a descriptive explanation. Panellists in this study were consumers with a total of 61 people. Organoleptically, respondents tend to value cake, muffin, cookies, and flakes in every color, flavor, taste, and texture parameters. Nutritional value per 100 g of baked food products from Salacca flour (cake, muffin, cookies, flakes) were energy 287.5-479.0 kcal, water 0.8-3.8 g, protein 6.0-6.7 g, fat 0.8-31.0 g, carbohydrates 45.0-98.8 g, and fiber 1.1-4.6 g. Panellists were accepted the organoleptic characteristics and the estimated nutritional values on baked food products from Manonjaya variety salacca flour were varied.

Keywords*: Manonjaya, salacca, baked, organoleptic, nutritive*

1. Introduction

Manonjaya has long been known as a producer of *salak*/ salacca (snake fruit) that grows naturally both in the garden and in the yard of the population. But at this time, the pamor is decreasing. Even according to data from the Tasikmalaya District Department of Agriculture Food Crops in 2016 decline in a production rate of Manonjaya salacca up to 64% in 2013-2015 [1]. Based on interviews with some residents of Manonjaya obtained information that at this time many farmers who change the allotment of land for planting salacca replaced to plant trees due to the price of it is declining to lose compete with another one, such as Pondoh salacca. Manonjaya salacca is only valued at IDR 300,-/ kg at the farm level far below the price of Pondok and Bali salacca, so it is not able to encourage farmers to plant it. This is very worrying about the loss of these varieties in the future. Even the latest research efforts by looking at the diversity of genes of several varieties of fruits in Indonesia as one of the basic food diversification has not included Manonjaya salacca in it [2]. Whereas Manonjaya salacca has larger pulp quality and distinctive taste. Therefore, it is necessary to revitalize Manonjaya salacca to keep it sustainable and even increase its prestige.

One of the weaknesses owned by Manonjaya salacca is that it is not as sweet as the salacca which is commonly circulating in the market today such as Pondoh salacca. For that reason, one effort that needs to be done to increase public consumption of Manonjaya salacca is innovation into processed products. Research on farming analysis and a prospect of Manonjaya salacca development has been done [3]. Based on the result of the research [3] give a suggestion for the development of quality of Manonjaya salacca for example by processing into the processed product and the existence of processing plant around the development area. However, based on a preliminary survey conducted by current researchers, processed products from Manonjaya salacca were still very limited. Manonjaya salacca only processed into sweets, dodol, and chips. Currently, the product was only made by 1 group of craftsmen and experienced equipment constraints.

Some efforts in diversity products with raw materials Manonjaya salacca need to be done to increase the level of public acceptance of the product. Food products are developed based on the type of food that is currently widely consumed by the community and considered the benefits of nutrients contained in it.

Generally, salacca has nutritional content and health benefits. Nutrient Adequacy Score (NAS) analyses on all fruits consumed by Indonesians based on Susenas 2008 data and Food Composition List has been studied, the result shows that fruit was the fruit with the highest NAS among other fruits for the content Fe (30.8%) and Vitamin A (31.2%) [4]. Salacca contains several nutrients such as per 100 grams of salacca has the composition of energy, protein, and carbohydrates consecutive 77 calories, 0.4 grams, 20.9 grams. Mineral content of calcium, phosphorus, and iron respectively 28 mg, 18 mg, and 4.2 mg per 100 grams [5]. In addition to the content of these nutrients, the consumption of salacca fruit also proved beneficial for health, salacca has high antioxidant activity compared with other tropical fruit species [6]. This is reinforced by the results of research in Malaysia which shows that the fruits have a high phytochemical and antioxidant properties and potentially have properties that are beneficial to health [7]. A healthy diet using salacca prevents the increase in total liver cholesterol, inhibits elevated plasma lipids, inhibits decreased antioxidant status in mice fed a diet containing cholesterol [8].

The Salacca flour has been prepared and analyzed its nutritional content [9]. For further development, the researcher was interested to make some processed snack products from flour, such as cake, muffins, flakes, and cookies. The products were then tested organoleptically and calculated the estimated nutritional value.

2. Materials and Methods

2.1. Materials

Manonjaya salacca used in this study was a variety that has been certified by Center for Plant Variety Protection, Ministry of Agriculture RI No. 62 / PVL / 2008 [10]. Before making the baked food product, Manonjaya salak first made into flour. The stages and yields of the flour were obtained from previous studies [9]. The main basic ingredients used for all products were Manonjaya salacca, added with other ingredients such as wheat flour (substance to be substituted), tapioca, chicken egg, salt, water, butter, and skim milk. The equipment used in the manufacture of such products was with household appliances, such as stoves, pots, pans, and other kitchen utensils.

2.2. Methods

This research has obtained the ethical approval of Health Research Ethics Commission of Balitbangkes RI No. LB.02.01 / 5.2 / KE.301 / 2016. This research was an observational research with a descriptive explanation. Type of processed food was based on baking process technology to minimize the loss of nutrients, especially minerals. Manonjaya salacca first made into flour for substitution with main ingredients such as wheat flour can be done with an equivalent amount [9]. Processed products made from Manonjaya salacca flour were cakes, muffins, cookies, and flakes.

The processed products of Manonjaya salacca were tested through an organoleptic test with consumer panel consisting of teachers, canteens managers, student's parents, and community of Manonjaya salacca center with the number of 61 people [11]. The products were evaluated for taste, aroma, texture, and color. The ratings were on a 5-point hedonic scale ranking from 5 (like extremely) to 1 (dislike extremely) [12]. Mineral water at room temperature was provided to rinse the mouth during evaluations.

Moreover, the nutrients of salacca products were measured so that the nutritional composition contained in it. Nutritional substances of Manonjaya salacca processed products calculated were macro nutrients (energy, carbohydrate, fat, protein, and water) and fiber. The nutritional values of Manonjaya salacca products were calculated based on the nutritional value of each raw material according to the List of Food Composition - DKBM [5] and the result of the research of Manonjaya salacca flour which has been done [9].

This research was conducted in Food Technology Laboratory of Nutrition Department of Tasikmalaya Health Polytechnic, Ministry of Health RI and in State Junior High School 1 Manonjaya which is work area of Manonjaya Health Center of Tasikmalaya Regency. The study was conducted in May to August 2016 (4 months). The data were processed using MS Excel software. For data processing of nutrient estimation used Nutrisurvey software. Then the data were analyzed descriptively.

3. Results and Discussion

3.1. *Acceptance through organoleptic test of Manonjaya Salacca Product*

The processed products that produced from Manonjaya salacca flour can be seen in the following figures. Figure 1. Products made based on baking principle ie cakes, muffins, cookies, and flakes.

Figure 1. (A). Cake Salak Manonjaya (Cak Salman), (B). Muffin Salak Manonjaya (Muf-on), (C). Cookies Salak Manonjaya (Cokiman), (D). Flakes Salak Manonjaya (Flas-on)

The results of an organoleptic test on Manonjaya salacca processed products can be divided into 2 types of data, they were the ordinal and interval scale. If the assumption of data type was ordinal scale, the result can be seen in Figure 2, while the interval data type can be seen in Figure 3.

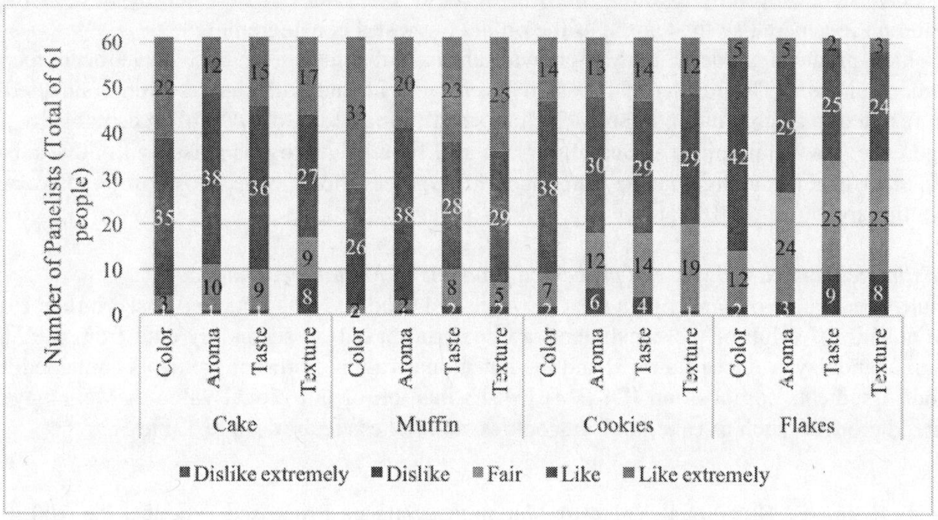

Figure 2. Result of Organoleptic Test on Manonjaya Salacca Processed Products (Ordinal Data)

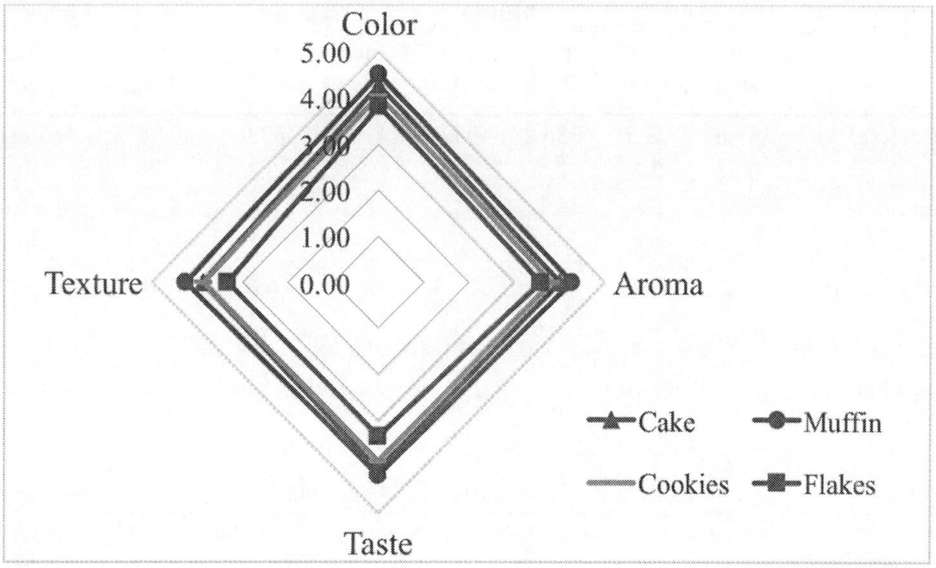

Figure 3. Result of Organoleptic Test on Manonjaya Salacca Processed Products (Interval Data)

Organoleptic testing is a test to assess the quality and safety of a food and drink [11]. Organoleptically, on a scale of 1-5 the panellists/ respondents tend to value cake, muffin, cookies, and flakes in every color, flavor, taste, and texture parameter. When viewed on an ordinal scale, most of the panellists/ respondents like the Manonjaya salacca processed products whether Cake, Muffin, Cookies, or Flakes in any color, flavor, taste, and texture parameters (Figure 2). Whereas, when viewed on an interval scale, overall panellists/ respondents favored Manonjaya salacca processed products on any color, aroma, taste, and texture parameters with average values between 3,33-4,51 on a scale of 1-5 (Figure 3). Among the four processed products of Manonjaya salacca, based on each parameter tested organoleptically, muffin products were the most preferable one. Whereas the other products that was low incomes compared to the other three processed products is Flakes in each parameter tested, especially texture. This was because it has not been brewed by milk as the way of presentation in general, so the texture of the product was still considered hard.

Cookies products made with Manonjaya salacca substitution of 50% have been received by consumer panelists. The results of this study were also in line with the research conducted by [13] which resulted in a substitution of spinach flour substitution of up to 60% still acceptable by panelists. Sponge cake from Manonjaya salacca flour can still be received by panelists with 50% substitution. While other studies using mango pulp and mango peel flour were most preferable one in the substitution treatment of 10% [14].

3.2. *Estimated nutritional value of processed products of Manonjaya Salacca*

The nutritional value of Manonjaya salacca processed products was obtained based on the calculation of the nutritional value of the constituent raw material based on secondary data from the nutritional value of Manonjaya salacca flour [9] and the nutritional value of other ingredients contained in the list of food ingredients composition (DKBM) [5]. Estimation of nutritional value in Manonjaya salacca processed products such as cake, muffin, cookies, and flakes can be seen in Table 1.

Table 1. Estimated Nutritional Value of Manonjaya Salacca Processed Products Per 100 g and Per Portion.

Nutrient	Product							
	Cake		Muffin		Cookies		Flakes	
	per 100 g	per portion	per 100 g	per portion	per 100 g	per portion	per 100 g	per portion
Energy (Kcal)	381.6	93.5	356.2	93.5	479.0	47.9	287.5	69.0
Water (g)	0.8	0.2	0.8	0.2	3.0	0.3	3.8	0.9
Protein (g)	6.5	1.6	6.1	1.6	6.0	0.6	6.7	1.6
Fat (g)	22.4	5.5	21.0	5.5	31.0	3.1	0.8	0.2
Carbohydrate (g)	48.2	11.8	45.0	11.8	66.0	6.6	98.8	23.7
Fiber (g)	1.2	0.3	1.1	0.3	3.0	0.3	4.6	1.1
Number of Servings Per Recipes	40.0		40.0		66,0		4.0	
Weight Per Portion	24.5 (1 piece)		26.3 (1 big cup)		10.0 (1 chip)		24.0 (160 chip)	

Source: The data were based on laboratory test results of Manonjaya Salacca Flour [9] and Nutrisurvey (2007).

Estimated nutritional value of Manonjaya salacca processed products made (cake, muffin, cookies, and flakes) has been calculated based on data from laboratory analysis of Manonjaya salacca flour [9] and data from the 2007 Nutrisurvey Program (Table 1). The nutritional value of Manonjaya salacca products varies for each product. Energy per portion donated from Manonjaya salacca processed products varies from 47.9 to 94.5 kcal. One serving of cake and muffins can almost meet the standard snack of 100 kcal per serving. The fiber content per serving of Manonjaya salacca products varies from 0.3 to 1.11 g. Based on the requirement of fiber per day 25 gram, then per portion of Manonjaya salacca processed product can donate 1.2-4.4% from fiber requirement per day.

In Malaysia, we have developed sponge cake product from a raw material of mango porridge and mango skin flour [14]. The research claims that the products produces high fiber. Based on the research, the content of crude fiber in wheat flour, mango porridge flour, and mango leaf powder were 0.63%, 8.07%, and 12.61% (dry basis), respectively. If calculated based on dry weight, Manonjaya salacca flour has a crude fiber content of 6.42%. The content of ash (total minerals) of Manonjaya salacca flour is also greater than that of wheat flour, mango pulp, and mango leaf powder of the research results [14] ie 1.42%, 1.29%, and 2.50% (respectively based on dry weight). If calculated based on dry weight, Manonjaya salacca flour has the ash content of 2.68%, higher than mango porridge meal and mango leather flour of the research results [9,14]. Thus, the processed products produced from Manonjaya salacca flour have higher fiber and ash content estimation than similar products in the market with wheat flour.

The addition of certain nutrient-rich ingredients to baking-based products is potential to be developed as a source of micronutrient and macronutrients, such as the addition of pomegranate powder to cookies made by researchers from Pakistan [15]. Cookies with pomegranate flour supplementation have an effect on the increase of fiber content, total phenol, and inorganic residue on the resulting product. Although in this research the product of cookies developed from Manonjaya salacca raw material has not been analyzed on fiber content, total phenol, and inorganic residues in a laboratory, the possibility will be the same with cookies product with pomegranate flour supplementation.

In addition, research on bioactive components (total phenolic and vitamin C) correlates to antioxidant capacity in other variety of salacca (Pondoh, Nglumut, and Bali) has been done [16]. The results show that the three salacca cultivars have a capacity as the antioxidants and have a correlation with the total phenolic content and vitamin C contained therein. This is also an indication of the potential of Manonjaya salacca as the same species with all three salacca that have the antioxidant capacity and bioactive components, especially the total phenolic and vitamin C.

In addition to the potential for functional food development from Manonjaya salacca because of the content of the compounds as described above, Manonjaya salacca can also be used as a source of prebiotic and probiotics. A characteristic study of probiotic microcapsules of *Lactobacillus plantarum* in Pondoh salacca has been done [17]. The results indicate that the jam of Pondoh salacca potentially becomes a growth medium of microencapsulated probiotics. The results of this study also become a potential in the development of processed products Manonjaya salacca as a medium for growth of probiotics, especially the fiber content of Manonjaya salacca flour is quite high. In brief, Manonjaya salacca flour and its processed products are expected to become prebiotic which is a food for probiotics, beneficial bacteria.

4. Conclusion and Recommendation

Organoleptic test results show that cake, muffin, cookies, and flakes products from Manonjaya salacca flour can be received by panelists in every color, flavor, taste, and texture parameters. The estimated nutritional values on baked food products from Manonjaya variety salacca flour (cake, muffins, cookies, flakes) were varied. Nutritional values per 100 g of baked food product from Manonjaya salacca flour were energy 287.5-479.0 kcal, water 0.8-3.8 g, protein 6.0-6.7 g, fat 0.8-31.0 g, carbohydrates 45.0-98.8 g, and fiber 1.1-4.6 g. The recommendation of this research is needed nutrition analysis in the laboratory to obtain a more accurate value. Processed products from Manonjaya salacca flour can be developed as a functional food.

Acknowledgement
The authors are thankful to the Research and Development Center of Humaniora and Health Management, National Institute of Health Research and Development, Ministry of Health RI for providing the Health Intervention Research (Riset Intervensi Kesehatan Berbasis Budaya Lokal 2016 – in Bahasa Indonesia) [grant number HK.02.04/V.1/758/2016].

References

[1] Department of Food Crop Agriculture of Tasikmalaya District 2016 Report on Agricultural Crop Production (in Bahasa). Tasikmalaya District
[2] Budiyanti T, Hadiati S, Prihatini R and Sobir 2015 Genetic Diversity of Indonesian Snake Fruits as Food Diversification Resources. *Int J Adv Sci Eng Inf Technol* **5**(3):41–4.
[3] Hadaka B. Analysis of Farming and Development Prospect of Manonjaya Salak (in Bahasa). 2002. Bogor Agricultural University
[4] Jati IRA, Vadivel V, Nöhr D, Biesalski HK. Nutrient density score of typical Indonesian foods and dietary formulation using linear programming *Public Health Nutr*. 2012;**15**(12):1–8.
[5] Ministry of Health. List of Indonesia Food and Nutrition Composition (in Bahasa) 1995 Jakarta: Ministry of Health RI
[6] Lim YY, Lim T and Tee J 2007 Antioxidant properties of several tropical fruits: a comparative study *Food Chem* **103**:1003–8.
[7] Aralas S, Mohamed M, Fadzelly M and Bakar A 2009 Antioxidant properties of selected salak (*Salacca zalacca*) varieties in Sabah, Malaysia *Nutr Food Sci* **39**(3):243–50.
[8] Leontowicz M, Leontowicz H, Drzewiecki J and Jastrzebski Z 2007 Food Chemistry Two exotic fruits positively affect rat ' s plasma composition *Food Chem* **02**:192–200.
[9] Sumarto, Aprianty D, Bachtiar HRA and Kristiana L. Efforts to rescue Manonjaya salacca, Tasikmalaya through the making of flour as raw material of nutritious food products (in Bahasa). In: Setiasih IS, Marsetio, Sunyoto M, Andoyo R, Hanidah I-I, Kurniati D, editors. National Seminar on Building Food Security through Local Commodity Empowerment. Jatinangor, Sumedang, West Java: Unpad Press; November 30th, 2016. p. 16–22.
[10] Tasikmalaya District. Certificate of Registration of Local Variety, Plant Variety Protection Center, Ministry of Agriculture RI. Indonesia; 62/PVL/2008, 2008.
[11] Setyaningsih D, Apriyantono A and Sari MP 2010 Sensory Analysis for Food and Agro Industry (in Bahasa). Bogor: IPB Press
[12] Meilgaard M, Civille GV and Carr BT 1997 Sensory Evaluation Techniques. 3rd ed. Florida: CRC Press; 247-248 p.
[13] Chauhan A, Saxena DC and Singh S 2016Physical , textural , and sensory characteristics of wheat and amaranth flour blend cookies *Cogen t Food Agric* [Internet] **21**(1):1–8. Available from: http://dx.doi.org/10.1080/23311932.2015.1125773
[14] Aziah AAN, Min WLEE and Bhat R 2011 Nutritional and sensory quality evaluation of sponge cake prepared by incorporation of high dietary fiber containing mango (*Mangifera indica var . Chokanan*) pulp and peel flours. *Int J Food Sci Nutr* **62**(September):559–67.
[15] Ismail T, Akhtar S, Riaz M and Ismail A 2014 Effect of pomegranate peel supplementation on nutritional , organoleptic and stability properties of cookies *Int J Food Sci Nutr* **65**(6):661–6.
[16] Ariviani S and Parnanto NHR 2013 Antioxidant capacity of salak fruit (Salacca edulis REINW) cultivar Pondoh, Nglumut and Bali and correlation with total phenolic content and vitamin C. *Agritech* **33**(3):324–33.
[17] Purnasari N 2015 Characteristics of *Lactobacillus plantarum* microcapsules and stability in salak jam during storage (in Bahasa). Bogor Agricultural University

Microbial quality of yellow seasoned "pindang" fish treated with turmeric and tamarind

B R Handayani [1]*, B Dipokusumo[2], W Werdiningsih[1], T I Rahayu[1] and D L Sugita[1]

[1] Faculty of Food Technology and Agroindustry-University of Mataram, Indonesia
[2] Faculty of Agriculture-University of Mataram, Indonesia

*Email: baiqrienhs@yahoo.com

Abstract. The objective of this study was to determine the microbial quality of yellow seasoned pindang fish. The fish was treated using combination of turmeric and tamarind at different ratio. This research used Randomized Block Design with 2 (two) factors ie concentration of turmeric (0%, 2%, and 6%) and concentration of tamarind (0%, 3%, and 6%). Each treatment was replicated 3 times to obtain 27 experimental units. The parameters observed were total microbe, total fungi and some pathogenic bacteria. Some microbial data were analyzed using descriptive method, however, the number of *S. aureus* was analyzed at 5% significance level by using software co-Stat and if there was a real difference then tested further by test Honestly Significant Difference (HSD). The results showed that increasing the use of curcumin and tamarind tended to decrease the total number of microbial from treatment control 5.1 x 10^5 CFU/gram to <1.0 x 10^3 CFU/gram. All the treatment produced yellow seasoned pindang fish with fungi <1.0 x 10^2 CFU/gram. The products contain pathogenic bacteria *E. coli* < 3 MPN/ gram; *S. aureus* <1.0 x 10^3 CFU/gram; *Salmonella* and *V. cholerae* were negative in 25 gram of sample. Based on microbial quality, it is recommended that the use of 3-6% of turmeric and 2-4% of tamarind are the best spices combination to produce safe consumption of yellow seasoned pindang fish.

Keywords: *curcumin, microbial, quality, seasoned, yellow*

1. Introduction

Fishing techniques is one of the fish processing efforts to extend the shelf life and increase the economic value of fish. Fish processing by boiling "pindang" ranks second after the salted fish. This process can be done by heating the fish in a salty condition for a certain time. The use of salt is generally low in concentrations less than 5% all. This causes the pindang fish in terms of the preferred flavor of the salted fish but has a short shelf life. According to [1] pindang fish can only survive for 3-4 days after the cover of the container is opened, even pindang fish processed in one of the centers of yellow seasoned pindang fish in Rumbuk village, East Lombok only able to survive for 1 day. This product is processed by utilizing turmeric, tamarind and salt. So far, the use of such spices has not been standardized, so that the quality of the products varies greatly between producers and the shelf life is short due to very limited sanitation. Pindang fish generally have a relatively high water activity, so it is very suitable for the growth of microorganisms. [2] showed that the water content of the yellow seasoned pindang fish ranged from 60.92 to 68-65%.

Traditionally processed yellow seasoned pindang fish are very susceptible to damage marked by mucus and growth of fungi. Fungi and bacteria are very common in pindang especially from Micrococcus group [3] According to the National Advisory Committee on Microbiological Criteria for Foods [4] and [5] there are several types of pathogenic microbial grown in fishery products such as *Bacillus cereus, Salmonella, Staphylococcus aureus, Vibrio parahaemolyticus.* In addition, *Escherichia coli* is often found as an indicator of poor sanitation. According to [14] to control or suppress the growth of microorganisms there are some hurdle techniques that can be done that are controlling intrinsic and extrinsic factor. Control of extrinsic factors, among others, by providing antimicrobial compounds both natural and synthetic.

Natural antimicrobials such as spices are very commonly used in traditional Asian cooking processing as in processing with the technique of boiling ('pemindangan' –in Bahasa). With a good technique of preservation by regulating the concentration of natural preservatives such as turmeric and tamarind is expected to suppress the growth of some pathogenic bacteria in yellow seasoned "pindang" fish. Turmeric contains various compounds such as curcumin and essential oil [6]. Turmeric is known as a natural dye also and has antibacterial effect. Some studies have shown that turmeric antibacterial effect can inhibit *Clostridium botulinum* (MIC 500 ug / mL) and bactericidal effect against Gram positive bacteria such as *Bacillus subtilis* and *Lactobacillus acidophilus* [7]. In addition to turmeric, tamarind is also known to have an inhibitory effect on a number of bacteria and fungi. [8] showed that the use of 1.5% salt and 3.5% tamarind is the best treatment to produce the best quality and the longest shelf life of dried lemuru fish. The purpose of this study was to determine the microbial quality (total microbial) and some potential pathogenic bacteria in yellow seasoned "pindang" fish processed by using combination of turmeric and tamarind.

2. Materials and Methods

2.1. Materials

The materials used in this research were fresh Cakalang fish and spices such as turmeric, tamarind and salt obtained from Kebon Roek market, Mataram, West Nusa Tenggara (NTB). Materials used for analysis consisted of butterfield's phosphate buffered, Plate Count Agar (PCA) (Merck., Germany), Potato Dextrose Agar (PDA) (Merck., Germany), Lauryl Tryptose Broth (LTB) (Oxoid Ltd., UK), EC Broth (Oxoid Ltd., UK), Eosin Methylen Blue Agar (EMBA) (Oxoid Ltd., UK), Lactose broth (Merck., Germany), Rappaport-Vassilidis (RV) (Oxoid Ltd., UK), Tetreathionate Broth (TTB), Hectoen Enteric (HE) (Oxoid Ltd., UK), Xylose Lysine Desoxycholate (XLD) Merck., Germany), Tripel Sugar Iron (TSI) (Oxoid Ltd., UK), Lysine Iron Agar (LIA) (Merck., Germany), Alkaline Peptone Water (APW), Thiosulfate-Citrate-Bile-Salt-Sucrose Agar (TCBS) (Oxoid Ltd., UK), Trypticase Soy Agar (TSA), NaCl, Baird Parker Agar (BPA) (Oxoid Ltd., UK), and egg yolk telurit (Oxoid Ltd., UK).

2.2. Methods

2.2.1. Processing of yellow seasoned pindang fish. The yellow seasoned pindang fish was made by modifying the processing of the yellow seasoned "pindang fish of Rumbuk's producer at east Lombok by following the process stages: sortation to obtain the fish with fresh condition, weeding to clean the fish from the stomach and gill contents, filleting to obtain slices meat of fish with size ± 4-5 cm (width) x 7-8 cm (long) and 2-3 cm (height) with average fish weight ± 70 gram. Next, fish fillet was washing and sieving, weighing and mixing with turmeric spices (0%, 3% and 6%) and tamarind (0%, 2% and 4%). Fish fillet used ± 750 gram per unit of experiment with combination of treatment in processing as follows: % turmeric: % tamarind (p1 = 0: 0; p2 = 0: 2; p3 = 0: 4; p4= 3: 0; p5 = 3: 2; p6 = 3: 6; p7 = 6: 0; p8 = 6: 2; and p9 = 6: 4). Before the spice used added 2% salt and water as much as 275 mL. Further, cooking for 10 minutes at a temperature of 100°C was done.

2.2.2. Microbial Assay. Determination of total microbial count in yellow pindang was conducted by referring to [9] procedure with modification on the number of samples used. Fungi determination followed modified procedure by [10]. Determination of the amount of *E. coli* in yellow pindang was done with reference to Most Probable Number techniques 3 series tube procedure of [11] with modification on the number of samples used. The positive results were shown from the formation of colonies with black dots, with or without metallic greens (pure culture control of *E. coli* forms metallic green color). Determination of Salmonella in yellow pindang was done by referring to Indonesia National Standard [12] procedure with modification on the number of samples used. Determination of *V. cholerae* was done by referring to the procedure of [13] with modification to the number of samples used. Observed growth by seeing changes in the media. For further confirmation observations were made with gram staining. Determination of the amount of *S. aureus* in yellow pindang fish was done by referring to procedure [14] with modification on the number of samples used.

2.3. Statistical Analysis

The experimental design used in this study was Randomized Block Group with two factors: turmeric concentration (0%, 3% and 6%) and Tamarind (0%, 2% and 4%). Each treatment performed three replications so that 27 units of experiments were obtained. The data of observation parameter of *S. aureus* were analyzed by analysis of variance at 5% real level using Co-stat software and the real difference was tested using 5% of Honestly Different Test [15]. While the total microbial parameters, total fungi, and some pathogenic bacteria (*Escherichia coli, Salmonella, Vibrio cholerae, Staphylococcus aureus*) were analyzed using descriptive method.

3. Results and Discussions

Fish is known to be an excellent growing medium for many types of microorganisms [16] because it contain important nutrients such as water (60-84%), protein (18-30%), fat (0.1-0.2%) and carbohydrates (0-0.1%). Analysis of microbial quality in yellow fish was done to determine total microbial growth and growth of some common pathogenic microbes in fishery products such as *Escherichia coli, Salmonella, Staphylococcus aureus* and *Vibrio cholerae*. According to [13] that *Staphylococcus aureus, Salmonella thypimurium* and *Vibrio* belong to a class of pathogens found in fishery products that are important due to cause disease or poisoning for consumers.

3.1. Total Microbes

Based on [17], pindang fish products had maximum limit of total microbial content of 1.0×10^5 CFU / gram. Table 1 below shows total microbials in yellow seasoned pindang fish with turmeric and turmeric treatment.

Table 1. Total microbe of yellow pindang seasoning fish with turmeric and tamarind treatment.

Treatment	Total Microbe (CFU/g)		
	Tamarind (0%)	Tamarind (2%)	Tamarind (4%)
Turmeric (0%)	5.1×10^5	4.5×10^4	3.4×10^4
Turmeric (3%)	2.6×10^4	$<1.0 \times 10^3$	$<1.0 \times 10^3$
Turmeric (6%)	8.3×10^3	$<1.0 \times 10^3$	$<1.0 \times 10^3$

Table 1 shows that the use of turmeric and tamarind can suppress microbial growth in total. The higher the concentration of turmeric and the acid used, the lower the microbial content of the yellow seasoned pindang fish. The highest total microbial found in pindang fish without the use of turmeric and Tamarind that exceeds the standard of SNI that is equal to 5.0×10^5 CFU / gram, so the opportunity to cause a very short shelf life. While the use of turmeric and tamarind reduced the total amount of microbes to $<1.0 \times 10^3$ CFU / gram. The use of turmeric up to 6% without using tamarind resulted in total microbial of $8.3x10^3$ CFU / gram. The table also shows that the use of turmeric and tamarind together can decrease the total number of microbes up to $<1.0x10^3$ CFU / gram. Results of research conducted by [18] in vitro, proved that the active compound in the turmeric rhizome is able to inhibit the growth of fungi, viruses and bacteria both gram positive and gram negative. In addition, according to [16] and [19], that bioactive components of tamarind such as tannins, flavonoids and saponins are known to have antibacterial effects through reactions with cell membranes, enzyme inactivation and inactivation of genetic material functions.

3.2. Total Fungi
Fungi are one of the most common types of microbes contaminating food products including fishery products. Table 2 shows that the mold content in yellow spiny pindang fish with all treatments <1.0 x 10^2 CFU / gram.

Table 2. Total fungi of yellow pindang seasoning fish with turmeric and tamarind treatment.

Treatment	Total Fungi (CFU/g)		
	Tamarind (0%)	**Tamarind (2%)**	**Tamarind (4%)**
Turmeric (0%)	$<1.0x10^2$	$<1.0x10^2$	$<1.0x10^2$
Turmeric (3%)	$<1.0x10^2$	$<1.0x10^2$	$<1.0x10^2$
Turmeric (6%)	$<1.0x10^2$	$<1.0x10^2$	$<1.0x10^2$

It is suspected that the absence of fungi, without and addition to the antimicrobial effects of turmeric and tamarind is also due to the cooking process used sufficiently to kill the fungi in general. It is known that the processing of yellow seasonings was done by cooking temperature 100 °C for 10 minutes. The absence of fungi in fishery products in accordance with the requirements of fungi free (negative result) as indicated in Indonesia National Standard Quality Requirement for Pindang Indonesia National Standard 2717.1: 2009 [20]

3.3. Total Escherichia coli
Escherichia coli is known to have the optimum temperature for growth is 37°C (with a range of 10-40°C), optimum pH 7.0-7.5 (minimum 4.0 and maximum 8.5). *E. coli* is relatively heat sensitive and immediately destroyed by pasteurization temperature and by cooking. *E. coli* is known as an indicator bacterium whose existence is closely related to poor hygiene (indicator of sanitation) [22].

Table 3. Total *Escherichia coli* of yellow pindang seasoning fish with turmeric and tamarind treatment.

Treatment	Total *E.coli* (Colony/g)		
	Tamarind (0%)	**Tamarind (2%)**	**Tamarind (4%)**
Turmeric (0%)	< 3	< 3	< 3
Turmeric (3%)	< 3	< 3	< 3
Turmeric (6%)	< 3	< 3	< 3

Table 3 shows that the addition of turmeric and acid treatment produced yellow pindang seasoning fish with very low of total *E. coli*. The amount of *E coli* obtained is in the safe position according to Indonesia National Standard 2717.1: 2009 [20], based on the Most Probable Number test method about <3 colony / gram. The low amount of *E.coli* in the yellow spicy pindang fish is suspected to be due to the antimicrobial effect of the tamarind by decreasing the internal pH of the cell, thus interfering with *E.coli* growth also due to the phenolic component of the antibacterial turmeric. It is also suspected that the processing and sanitation processes are appropriate. This is in line with the opinion of [21] that generally the content of pathogenic microbes in fishery products will be low if the process fulfills the requirements of good processing.

3.4. Total number of Salmonella

Salmonella is a microorganism that can cause poisoning if ingested in food consumed. [9] and [16] emphasize that *Salmonella* is one type of pathogenic bacteria whose existence is very harmful to consumer health. In addition, Indonesian National Standard requires that bacteria not be found in the processed fishery products. The following table shows Salmonella content in yellow seasoned pindang fish that is processed with the use of turmeric and tamarind. Based on Table 4 it can be seen that the yellow pindang fish produced does not contain *Salmonella* in all treatments, meaning that no Salmonella contamination in the yellow seasoning pindang fish was produced. The Indonesia National Standard states that *Salmonella* should not be found (negative) for every 25 grams of material weight [20].

Table 4. Total Salmonella of yellow pindang seasoning fish with turmeric and tamarind treatment.

Treatment	Total Salmonella		
	Tamarind (0%)	**Tamarind (2%)**	**Tamarind (4%)**
Turmeric (0%)	-	-	-
Turmeric (3%)	-	-	-
Turmeric (6%)	-	-	-

Description: - = no growth found (Indonesia National Standard requirement for *Salmonella* : negative / 25 g)

3.5. Total number of Staphylococcus aureus

Staphylococcus aureus is a class of pathogenic microbes that cause disease through intoxication. This bacterium has the optimum temperature characteristics for the growth of 35°C-37°C, with a minimum temperature of 6.7°C and a maximum temperature of 45.5°C. These bacteria can grow at pH 4.0 to 9.8 with an optimum pH of 7.0-7.5 [9]. The following figure shows the amount of *S aureus* in the resulting of yellow pindang seasoning fish.

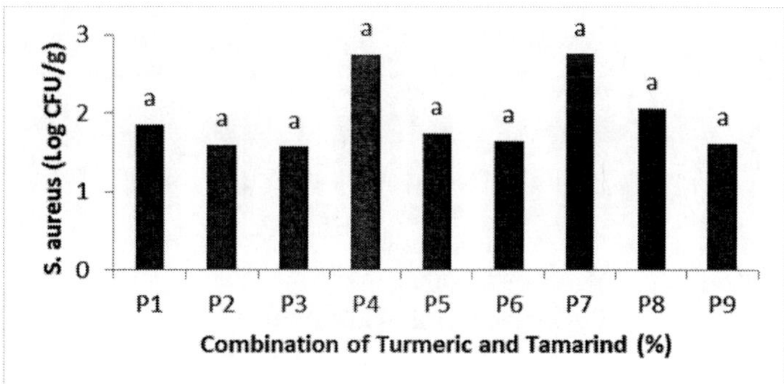

Description : % Turmeric :% Tamarind (p1= 0:0; p2= 0:2; p3 = 0:4; p4= 3:0; p5=3:2; p6=3:6; p7= 6:0; p8=6:2; and p9=6:4)

Figure 1. The content of S*taphylococcus aureus* on the Yellow Pindang Seasoning fish

Figure 1 shows that turmeric and tamarin combination treatments have no effect on the bacteria content of *S. aureus*, with a range <3 log CFU / gram or <of $1.0x10^3$CFU / gram. The maximum limit

of this pathogen content in fishery products according to Indonesian National Standard is $1.0x10^3$ CFU / gram. From Figure 1 it appears that the yellow pindang produced contains *Staphylococcus aureus* below the maximum standard so that it can be said that the resulting product is safe from the pathogen. Although safe, the number of such microorganisms was higher than the content of other microbes tested. This is allegedly due to the growing ability of *S aureus* is much wider range of pH 4.0-9.8.

3.6. Total Vibrio cholerae

Vibrio cholerae is one of the pathogenic microbes that can cause cholera disease. The following table shows that no vibrio was found in yellow seasoned pindang fish either without or usig a combination of turmeric and tamarind. Indonesian National Standard requires no Vibrio to be found in fishery products.

Table 5. Total *Vibrio cholerae* of Yellow Pindang Seasoning fish with turmeric and tamarind treatment.

Treatment	Total Fungi		
	Tamarind (0%)	**Tamarind (2%)**	**Tamarind (4%)**
Turmeric (0%)	-	-	-
Turmeric (3%)	-	-	-
Turmeric (6%)	-	-	-

Description: - = no growth found (SNI requirement: negative / 25 g)

3.7. pH value

The storage capacity of food products in general can be determined by the pH of the product. Figure 2 shows the changes in the pH value of the yellow spiny fish during storage. Figure 2 shows that the pH tends to be stable by the addition of turmeric and tamarind treatment with a pH range of 4.9 to 5.6. The addition of turmeric did not affect the pH value but the increased acid concentration decreased the pH of the yellow spice pindang. This is in line with [22] which says that the addition of acid with a certain portion can increase the acidity causing the decrease of pH of the material. Decreased pH is one of the factors that can inhibit microbial growth even though microbes have been clearly sensitive to low pH. According to [23] tamarind contains a number of organic acids such as malic acid, tartaric acid, citric acid, formic acid, succinic acid, and acetic acid.

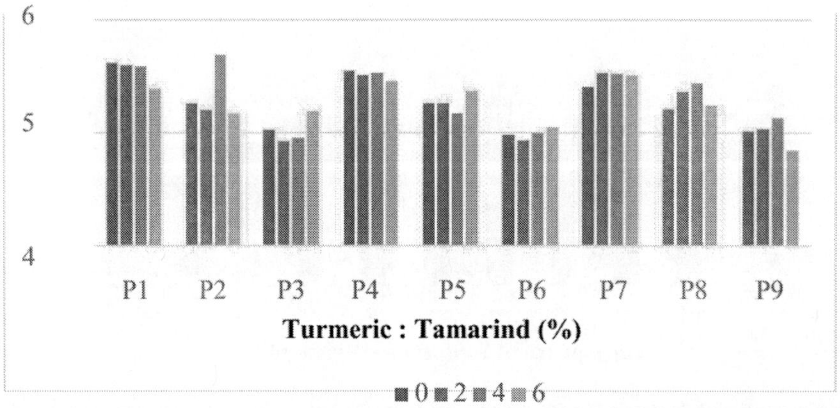

Figure 2. Changes in pH value during storage of yellow pindang seasoning fish.
Description : % Turmeric: % Tamarind (p1= 0:0; p2= 0:2; p3 = 0:4; p4= 3:0; p5=3:2; p6=3:6; p7= 6:0; p8=6:2; and p9=6:4)

4. Conclusions

The use of a combination of turmeric and tamarind can suppress the total growth of microbes. The higher the concentration of turmeric or tamarind, the lower the total microbes up to the safe limit according to SNI. Processing of yellow seasoned pindang fish without the use of turmeric and acid resulted in total microbes exceeding the Indonesia National Standard by the number of days to 0 by 5.0×10^5 CFU / gram. No fungi and pathogenic bacteria of *Salmonella* and *Vibrio cholerae* found in yellow seasoned pindang fish. In addition, pindang fish containing *E coli* and *S. aureus* are under the requirements of Indonesia National Standard 2717.1: 2009.

Acknowledgment

Thanks to the Higher Education Research and Technology Ministry who has funded this research activity through the MP3EI Skim Corridor Fisheries 2017.

References

[1] Saleh 2002 Research Collection on Post Harvest Fishery (in Bahasa Indonesia). The Research Centre of Processed Product and Social Economy of Marine and Fishery Badan Riset Kelautan Dan Perikanan Departemen Kelautan Dan Perikanan. Jakarta.

[2] Sugita DL, Handayani BR and Werdiningsih W 2017 The Effect of Turmeric and Thamarind Concentraton on The Quality Components of Pindang Yellow Spiced Salted Fish (in Bahasa Indonesia). Undergraduate Thesis. Faculty of Food Technology and Agro-industrial Sciences. Universitas Mataram. Mataram

[3] Linggarwati ED 2007 Protein Content and Acceptability of Milkfish by Smoking, Pressure Cooking, Frying and Salting Methods. Undergraduate thesis. Public Health. Universitas Jember. Jember.

[4] Iwamoto M T, Ayers B E, Mahon and Swerdlow D L 2010 Epidemiology of seafood-associated infections in the United States. *Clin. Microbiol. Rev.* **23** *399–411.*

[5] NACMF 2008 Response to the questions posed by the food and drugadministration and the national marine fisheries service regardingdetermination of cooking parameters for safe seafood for consumers.national advisory committee on microbiological criteria for foods. *J. Food Prot.* **71** 1287–1308.

[6] Said A 2001 *Tumeric efficacy and benefits* (in Bahasa) (Jakarta : PT Sinar Wadja Lestari)

[7] Sihombing PA 2007 Application of Turmeric Extract (*Curcuma Domestica*) as Preservative on Wet Noodles. Undergraduate Thesis. Faculty of Agricultural Technology. Institut Pertanian Bogor, Bogor

[8] Yulianti 2016 The Effect of Salt and Thamarind Addition on The Quality and Shelf Life of Lemuru Fish (*Sardinella sp.*) (in Bahasa Indonesia). Undergraduate Thesis. Faculty of Food Technology and Agro-industrial Sciences, Universitas Mataram. Mataram.

[9] National Standardization Institution 2006 How to test microbiology part 3*: Determination of Total Plate Count (TPC) on fishery product*. Indonesia National Standard 01-2332.3-2006

[10] National Standardization Intitution 2015 How to test microbiology part 7 *:* Mold and khamir calculation in fishery products (in Bahasa) Indonesia National Standard 2332.7:2015. Jakarta

[11] National Standardization Institution 2006a How to test microbiology part 1: *Coliform* and *Escherichia coli* in fishery products determination (In Bahasa Indonesia) Indonesia National Standard 01-2332.1-2006. Jakarta

[12] National Standardization Institution 2006b How to test microbiology part 1: *Salmonella* in fishery products determination (In Bahasa) Indonesia National Standard 01-2332.2-2006. Jakarta

[13] National Standardization Institution 200 How to test microbiology part 5: *Vibrio parahaemoliticus* examination in fishery products determinaton (in Bahasa Indonesia) Indonesia National Standard 01-2332.5-2006

[14] National Standardization Institution 2011 How to test microbiology part 9: *Staphylococcus aureus* in fishery products determination (in Bahasa Indonesia) Indonesia National Standard 2332.9:2011. Jakarta

[15] Hanafiah KA 2012 Teory and application trial planning (in Bahasa). Raja Grafindo Permata, Jakarta

[16] Mun'im A, Endang H and Rahmadiah 2009 Characterization on Ethanolic Extract of Tamarind Leaves (*Tamarindus indica L.)* (in Bahasa Indonesia) *Majalah Ilmu Kefarmasian* **4**(1) 38-44

[17] National Standardization Institution 2009 Ikan Pindang (salted fish) part 1 (in Bahasa Indonesia) Spesification Indonesia National Standard. Jakarta.

[18] Hidayati E, Juli N and Marwani E 2002 Isolation of *Enterobacteriaceae* Pathogen from Spiced and Not Spiced Food by Turmeric (*Curcuma longa L.)* and Determination of The Effect of Turmeric Extract (*Curcuma longa L.*) towards The Growth of Isolated Bacteria. Biology Department. Faculty of Mathematics and Natural Sciences. Institut Teknologi Bandung. Bandung.

[19] Ajizah A 2004 Sensitivity of Salmonella typhimurium towards Psidium Guajava L. leave extract. *Bioscientieae* **1** (1): 31-38

[20] National Standardization Institution 2009 Ikan Pindang (salted fish) part 1 (in Bahasa Indonesia) Spesification Indonesia National Standard. Jakarta.

[21] Feldhusen F 2000 The role of seafood in bacterial food borne diseases *DOI:* ***2****(13): 1651-1660. /doi.org/10.1016/S1286-4579(00)01321 -6*

[22] Jay 2006 *Modern Food Microbiology 6th Edition* An Aspen Publication

[23] Bhadoriya SS, Ghanespurkar A, Narwaria J, Rai G, and Jain AP 2011 Tamarindus indica: extend of explored potential Pharmacognosy Reviews, **5**(9):73-81. DOI:10.4103/0973-7847.79102 Pharmacogn

Reproductive performance of peranakan ongole (PO) bull at different altitudes areas in Lampung province

N D Suretno[1*], I Supriyatna[2], B Purwanto[2], R Priyanto[2]

[1]Assessment Institute for Agricultural Technology (AIAT) Lampung
[2]Bogor Agricultural Technology

*Email: nandari.dyah@yahoo.co.id

Abstract. Peranakan Ongole (PO) cattle are famous for their bulls and known for their toughness, rapid growth and natural tolerance to tropical heat and disease resistance. This study was conducted to determine the reproductive performance of PO bull in some altitude in Lampung Province. A total of PO bulls used were 48 heads, with age three to four years. The variables observed were volume, consistency, pH, semen colour, concentration, mass movement, motility, percentage of living spermatozoa and abnormality. Data were collected during the rainy season and dry season, calculated with the average and standard deviation and then analyzed descriptively. Based on the macroscopic examination, Peranakan Ongole bull living in low temperature place had a normal semen volume, the consistency mostly watery to moderate, fresh semen pH is in the normal range and the color of semen are all normal both in the rainy and dry seasons. The results of microscopic examination showed that PO cattle had a good reproduction conditions at all three

***Keywords** : altitude, macroscopic, microscopic, PO cattle*

1. Introduction

Some of the foods we eat come from animal and plant. Animal serves a source of nutrients, especially protein and fat. One source of protein from animal that is increased every year is beef. Provision of a local beef is fulfilled around 65.24% from the total of national demands, and other demand is imported from other country. The effort to increase the meat production has been done through empowerment people to increase population of cattle.

The increasing of cattle population can be attained by various ways such as improving genetic quality and reproductive efficiency through the program Artificial Insemination (AI). The AI program is one of the most successful reproductive technologies and has successfully improved the genetic quality of livestock, so that in a short time it can produce a good quality calf in large quantities by using semen from superior bull [1].

Many factors affect the success of AI programs including: (a) the quality of frozen sperm; (b) the knowledge, understanding and awareness of farmers to heat detection; (c) body condition score (BCS); (d) health and nutritional management; (e) and inseminator skills [2,3]. Feradis [4] stated that every bull has a different quality of sperm that is depended on the age, condition of livestock, libido and races. The most influencing factor on variation in semen quality are breed and environment, whether it is temporary or permanent.

Indonesia has many breed for local beef cattle such as Peranakan Ongole (PO). The PO cattle has a high adaptability to traditional livestock conditions. The biological potential and productivity of PO cattle are widely and highly responsive to environmental change and improvement. In order to improve the productivity of PO cattle in livestock farms, improving the reproduction and feed management can be done immediately. The reproductive performance of PO cattle was more efficient when the cattle reared in the lowland area than those in the highland area according to the days open (DO) and calving interval (CI) periods. The great environmental influence in exposing the genetic potential of the reproduction is justification for this research so the purpose of this study is to evaluate the condition of the reproduction of PO cattle at different altitudes.

2. Methodology

The study was conducted in three districts in Lampung province from December 2013-December 2014, December 2013-February 2014 was the rainy season and July-September 2014 was the dry season. Those three districts were Central Lampung district (lowland about 0-100 m above sea level), Pesawaran districts (medium land about 100-250 m above sea level) and West Lampung (highland > 400 m above sea level). Central Lampung average temperature 29.45 C, humidity 78.60%, THI 80.82, Pesawaran average temperature 27.27 C, humidity 81.43%, THI 78.60, and West Lampung average temperature 23.18 C, humidity 81.18%, THI 71.94.

A total of PO bulls used were 48 heads, with age three to four years. The variables observed were volume, consistency, pH, semen colour, concentration, mass movement, motility, percentage of living spermatozoa and abnormality. Data was collected during the rainy season and dry season, calculated with the average and standard deviation and then analyzed descriptively.

3. Result and Discussion

Examination of semen quality can be performed using macroscopic and microscopic examination. The macroscopic examination includes volume, consistency, pH, colour and smell. While microscopic examination includes concentration, motility, viability, and abnormality. However, [13] said that the aspects that is determined the quality of semen are the motility, concentration and morphology of spermatozoa (abnormality).

3.1 Macroscopic Quality

3.1.1Fresh Semen Volume. The volume of semen is the number of millilitres of semen per ejaculation [5]. The volume of semen volume in the rainy season is varied from 2.17 ± 1.61 ml to 6.00 ± 0.50 ml and in the dry season is varied from 2.70 ± 1.75 ml to 5.20 ± 2.38 ml. The volume of PO cattle at the third altitude in both seasons is in the normal range, according to [6] between 1-15 ml. However, [4] stated that that the range of volume is 5-8 ml and there is an PO cattle in Central Lampung during the rainy season and PO cattle in West Lampung in the dry season.

The highest semen volume is from cattle that are kept in Central Lampung (6.00 ± 0.50 ml). The volume difference is probably caused by the amount of nutrients that is given when the bull is in puberty, although the amount of feed given is does not differ significantly. Number of feeding of PO cattle in Central Lampung is 60,63 kg / bull, Pesawaran is 44,19 kg / bull and West Lampung is 65,35 kg / bull [7]. It is supported by the opinion of Nuryadi [8], he stated thatfirstly spermatogenesis process is taken place when the age of bull around 10-12 months. At this time, the nutrition is used optimally for the formation of sex cells, include the sperm production. Furthermore, at the age of 2-3 years, the bull has an adult body so most of the nutrients that is absorbed in the body is used for the body cells growth. Beside the age and quality of feed, the differences in semen volume are caused by body size, breed, shelter frequency [5], temperature, aged sex of a male [9] and environmental stress [10].

Table1.Macroscopic quality of PO bull's fresh semen in rainy and dry season.

Parameter	Center Lampung	Pesawaran	West Lampung
Rainy season			
Volume (ml)	6.00±0.50	3.77±1.54	2.17±1.61
Consistency	watery, medium	watery, medium, thick	watery, medium
pH	7.00±0.00	7.00±0.00	7.00±0.00
Colour	white milk	white milk	white milk
Dry season			
Volume (ml)	4.83±2.25	2.70±1.75	5.20±2.38
Consistency	watery, medium	watery, medium, thick	medium
pH	6.73±0.46	7.00±0.00	6.67±0.58
Colour	white milk	clear, white milk	white milk

3.1.2 Fresh Semen Consistency. Consistency or degree of viscosity is examined by shaking the tube that contains the semen slowly. The results showed that the consistency of bull's semen in the farms both in the rainy and dry season is mostly watery to medium. On the other hand, the bull's semen in Pesawaran is consistently thick. Consistency during the rainy season in lowland is moderate. In spite of this, the result from [11] research showed that consistency of PO cattle in lowland rainy season is watery. Also, [12] reported that the consistency of PO cattle's semen in both the rainy and dry seasons is watery, with a consistent percentage of 64%, it means the result of this study lower than Aeren's result. This difference could be happened because different management of farming system. The cattle's semen as samples in this study is the cattle that is managed by farmers, while the research of [12] cattle's semen is obtained from the bull that has been selected in Singosari National Artificial Insemination.

3.1.3 The pH of Fresh Semen. The average pH of fresh semen of PO cattle in the dry and rainy seasons at three altitudes is ranged from 6.67-7.00. The fresh semen's pH at all altitudes is still in the normal range according to [14] is around 6.4--7.8.

The pH of PO cattle's fresh semen at livestock farms in Lampung during the rainy season has a high score is about 7.00. While during the dry season, only bulls in Pesawaran have pH 7.00 for fresh semen. The pH of semen is 7.0 or higher due to the number of dead spermatozoa [5].

3.1.4 Fresh Semen Color. The color of fresh semen PO cattle at farms is normal in the rainy season at all three altitudes. While the colour of fresh PO cattle that are all normal in the dry season is in Central Lampung and West Lampung. According to Feradis [4], the normal colour of fresh semen is white milk or whitish and whitish cream. The color is caused by riboflavin pigments that is carried by an autosomal recessive gene but does not affect the fertility. The requirement that is established by Lembang Artificial Insemination Centre, based on SOP [2] about the minimum criterion of semen that could be used in the manufacture of frozen semen is the colour should be normal. Therefore, the bulls in Pesawaran must have rigorous testing if they want to be used as semen source.

3.2 Microscopic Quality

3.2.1 Mass Movement of Fresh Semen. Spermatozoa in a group tend to move together in one direction seem like thick and thin waves. The speed of the movement depends on the viability. Observations of mass movement indicate that during the wet season PO cattle at all altitude have mass movement ≥ 2+. While in the dry season only PO cattle in Central Lampung has mass movement 2+. These results showed only PO cattle in Central Lampung that can be used as a source of fresh semen. According to [15], the requirement for the manufacture of liquid semen is the mass movement of 2+ to 3+. Moreover, based on the SOP of frozen semen manufacture from [2], there is no bull is required for the source of semen because the minimum requirement for the manufacture of frozen semen is 3+.

The movement average of PO cattle is 2.11 (rainy season) and 1.78 (dry season). This result is better than the mass movement of Limousin cross/Simental cross that is 1.33 (rainy season) and 1.33 (dry season) [16]. However, this result is similar with the research of [17] that is conducted on cattle from livestock managed by common people, the mass movement of PO cattle (2.81) higher than the mass movement of limpo cattle (2.00) and simpo cattle (1.90). It indicates that all PO cattle have better adaptation than imported cattle [18,19].

Table2. Microscopic quality of PO bull's fresh semen in rainy and dry season.

Parameter	Center Lampung	Pesawaran	West Lampung
Rainy season			
Mass Movement	2.00±0.00	2.00±0.00	2.33±0.58
Concentration (million/ml)	1528.33±412.94	1314.33±289.79	837.00±170.71
Motility (%)	56.67±11.55	46.67±5.77	56.67±11.55
Living spermatozoa (%)	73.33±5.77	50.00±0.00	70.00±17.32
Abnormality (%)	1.85±1.09	18.34±9.79	5.98±3.31
Dry season			
Mass Movement	2.00±0.00	1.67±1.53	1.67±0.58
Concentration (million/ml)	1491.67±396.32	833.33±544.94	1007.33±325.47
Motility (%)	53.33±5.77	53.33±5.77	43.33±20.82
Living spermatozoa (%)	63.33±5.77	56.67±35.12	53.33±20.82
Abnormality (%)	4.40±1.27	6.04±0.77	7.87±1.67

3.2.2 The concentration of Fresh Semen. The concentration of semen is the amount of spermatozoa contained in one ml of ejaculation. Concentration examination is one of the necessary conditions to know the quantity of fresh semen after shelter.The concentration of PO cattle's fresh semen in this study both in rainy and dry season is required for the criteria as a source of semen. [20] stated that the minimum standard for a semen source is 800 million/ml, whereas according to [15] the concentration could be required for the manufacture of liquid semen if the concentration is bigger than 750 million/ml.

3.2.3 The percentage of Fresh Semen Motility. Motility is the ability of sperm to move properly through the female reproductive tract or through water to reach the egg and fertilize the ovum. The best spermatozoa movement is progressive movement or active movement forward [4] stated that the percentage of motility of bull's spermatozoa below 40% indicates that the semen has a poor quality. The fertile bull has 50% to 80% progressive motile spermatozoa. Based on these criteria, in rainy season, PO cattle in Central Lampung and in West Lampung (56.67 ± 11.55%) is a fertile bull. On the other hand, in dry season, PO cattle in Central Lampung and Pesawaran (53.33 ± 5.77%) are fertile bulls. However The [21] recommends the minimal motility is 30%. Based on these criteria, all PO cattle at all altitude have a good percentage of motility.

The fertile bulls can be further processed into frozen semen if their motility is> 60% [18, 2]; at least 65% [22]; at least 60-65% [14, 23] . Moreover, if it will be produced as liquid semen, the minimum requirement of cattle motility is > 70%. The result of this study indicates that PO bulls, in both seasons could not be a source of frozen semen. Based on the statement from [9], high motility will improve the ability of spermatozoa for fertilization.

One effort that can be done to increase the percentage of semen motility is the provision of traditional supplements such as the research of [24]. Provision of traditional supplements can significantly increase the motility of Bali bull's spermatozoa from 80.3 ± 8.2 to 88.7 ± 5.5. The content of supplement is protein, aphrodisiac, vitamin E and boron. Protein deficiency in young bulls will decrease libido and poor quality of semen[25]. Aphrodisiac is useful for stimulating libido and blood circulation of the genitals [26]. Vitamin E as an antioxidant that can protect spermatozoa against peroxidative that could damage and decreasing the number and motility of spermatozoa [27]. Boron is used to increase testosterone levels, the most responsible hormone in generating sexual desire (libido) [28].

3.2.4 The Percentage of Living Spermatozoa of Fresh Semen. The percentage of living spermatozoa is a good category in rainy season only for PO cattle in West Lampung (70.00 ± 17.32%), while in dry season, PO cattle in Central Lampung (63.33 ± 5.77%) is a good category. Hafez [20] assumed that the percentage of live fresh cattle semen is around 60-80%. The highest percentage of living spermatozoa in rainy season is PO cattle in Lampung Tengah (73.33 ± 5.77%) and the same result in dry season is PO cattle in Central Lampung (63.33 ± 5.77%). Moreover, the percentage of living spermatozoa of PO cattle in the lowland is still higher than the results of [11] that is 53.7 ± 24%, however it is less than the result of[32] that is 78.18 ± 12.25% and [9] that is 93.5 ± 2.1%.

3.2.5 The Abnormality of Fresh Semen. Semen abnormality is the damage that occurs in spermatozoa. Defected spermatozoa have two primary abnormalities that occur during spermatogenesis and secondary abnormalities if they occur after spermatogenesis [29]. Based on the recommendation of The BSE systems of the Society for The riogenology and the WCABP in [21], the minimum normal sperm is 70%, it means all bulls in both seasons of abnormality can still be categorized as a normal sperm. However according to [30], fertility of semen is a good category if the amount of spermatozoa abnormal is not more than 20%. Different opinion is given by [31], he said that if the primary abnormality is more than 10% so it can affect fertility.

[13]assumed that the abnormality is divided into four categories ie low <5%; moderate 5-10%; high 10.1-15% and very high >15%. Based on this, PO cattle in Central Lampung both the rainy season (1.85 ± 1.09%) and the dry season (4.40 ± 1.27%) is in the low abnormality category. Abnormalities during the rainy season are more varied, ie entering the category of low, medium and very high. Abnormalities in the dry season are in the low and medium category.

The determination of environmental compatibility for bulls is based on microscopic examination of concentration, motility and abnormality. Based on these three criteria, PO cattle has a good reproductive conditions at all three elevations.

4. Conclusion

According to the data it was concluded that PO bull living in low temperature place has a normal semen volume, the consistency is mostly watery to moderate, fresh semen pH is in the normal range and the color of semen are all normal both in the rainy and dry seasons. The suitability of the environment for bulls is based on the microscopic examination of concentration, motility and abnormality. Based on these criteria, PO cattle have a good reproduction conditions at all three altitudes.ppp

References

[1] Susilawati T 2013 Guidelines for artificial insemination in livestock Malang Brawijaya University Press
[2] Artificial Insemination Institute 2011 Smart book of Artificial Insemination Lembang Artificial Insemination Institute Jakarta Directorate General of Animal Health and Husbandary
[3] Diwyanto K 2012 Optimization of artificial insemination technology to support agribusiness of dairy cows and beef cattle Jakarta Bunga Rampai. Puslitbangnak Kementerian Pertanian
[4] Feradis 2010 Reproduction biotechnology in livestock Bandung Alfabeta
[5] Toelihere MR 1985 Reproduction physiology in livestock Bandung Angkasa publisher
[6] Toelihere MR 1993 Artificial insemination in livestock Bandung Angkasa publisher
[7] Suretno ND, Purwanto B, Priyanto R and Supriyatna I 2017 Environmental suitability evaluation based on the performance production four breed cattles on some different altitudes in Lampung Province *Jurnal Veteriner* **18** 478

[8] Nuryadi 2000 Livestock Reproduction Basics Malang Faculty of Animal Science of Brawijaya University

[9] Pratiwi WC, Pamungkas D, Affandhy L and Hartati 2006 Evaluation of spermatozoa quality of the sexing results on cold straw packaging stored at 5°C for 7 days In : Mathius IW, Sendow I, Nurhayati, Murdiati TB, Thalib A, Suparyanto A, Prasetyo LH, Darmono and Wina E, editor *Science New Horizon to Support Animal Farm Revitalization* 2006 September 5-6 (Bogor: IAARD Press)

[10] Hafez ESE and Hafez B 2000 Reproduction in farm animal. 7thed Philadelphia USA Lippincott Williams &Wikins

[11] Ratnawati D and Affandhy L 2013 Reproduction performance of bulls with palm oil feed In : Purwantari ND, Saepulloh M, Iskandar S, Anggraeni A, Ginting SP, Priyanti A, Wiedosari E, Yulistiani D, Inounu I, Bahri S, Puastuti W, editor. *Animal Farm Technology and Veterinary Inovation Based on Adaptive and Mitigative Local Resources Against Climate Change* 2013 September 3-5; Bogor (Indonesia.Jakarta:@IAARD Press. page 49-52)

[12] Aerens CDC, Ihsan MN and Isnaini N 2013 Quantitative and qualitative differences of fresh semen on various beef cattles [internet]. Available in :http://fapet.ub.ac.id/wp-content/uploads/2013/04/Perbedaan-Kuantitatif-Dan-Kualitatif-Semen-Segar-Pada-Berbagai-Bangsa-Sapi-Potong.pdf.

[13] Arifiantini RI, Purwantara B and Riyadhi M 2009 PO currence of sperm abnormality of beef cattle at several artificial insemination centers in Indonesia *Anim Prod.* **12** 44

[14] Garner DL and Hafez ESE 2000 Semen evaluation. In: Hafez B and Hafez ESE (Eds) Reproduction in farm animals7th Ed Philadelphia, USA Lippincott Williams and Wilkins

[15] Affandhy L, Pamungkas D, Rasyid A and Situmorang P 2004 Fertility test of liquid and frozen semen on beef cattle Grati :Research report. Seminar of Beef Cattle Research

[16] Suretno ND, Purwanto B, Priyanto R and Supriyatna I 2016 Physiological status of four breed cattles on some different altitudes in lampung province *International Journal of Sciences: Basic and Applied Research (IJSBAR)* **26** 384

[17] Ramsiyati DT, Sriyana and Sudarmadi B 2004 Evaluation of semen quality of beef cattle at various ages in community farms *Proceedings of National Technical Meeting of Agricultural Functional Workers* 2004

[18] Sarastina, Susilawati T and Ciptadi G 2012 Analysis of several parameters of sperm motility in various cattle using computer assisted semen analysis (casa) *J Ternak Tropika* **6** 1

[19] Adhyatma M, Isnaini N and Nuryadi 2013 Effect of body weight on quality and quantity of simmental cattle semen [internet]. Available in :http://fapet.ub.ac.id/wp-content/uploads/2013/04/Pengaruh-Bobot-Badan-terhadap-Kualitas-danKuantitas-Semen-Sapi-Simmental.pdf.

[20] Hafez ESE 2000 Semen evaluation in reproduction in farm animals 7th Edition Maryland, USA : Lippincott Wiliams And Wilkins

[21] Albert D. Barth and Cheryl L. Waldner 2002 Factors affecting breeding soundness classification of beef bulls examined at the Western College of Veterinary Medicine *Can Vet J.* **43** 274

22] Indriani, Susilawati T and Wahyuningsih S 2013 Limousin cattle spermatozoa survival is preserved by *Water Jacket* dan *Free Water Jacket* methods *J Vet.***14** 379

[23] Michael AJ, Alexopoulos C, Pontiki EA, Hadjipavlou-Litina DJ,Saratsis P and Ververidis HN 2010 Effect of N-acetyl-L-cysteine supplementation in semen extenders on semen quality and reactive oxygen species of chilled canine spermatozoa *Reprod Domest Anim.* **45** 201

[24] Ratnawati D, Affandhy L, Pratiwi WC and Prihandini PW 2008 The effect of traditional supplementation on the quality of Balinese cow malesIn :Sani Y, Martindah E, Puastuti W, Satika T, Parede L, Anggraeni A and Natalia L, editor. 2008 Technology innovation promoting the sustainable agricultural development (In Bahasa Indonesia) November 11-12 (Bogor :IAARD Press)

[25] Nix J. 2006. Nutritional Requirement of Bulls at Maintenance and Regaining Body Condition [Internet]. Available in: http://www.sweetlix.com/userfiles/file/articles/CattleDon'tForgetAboutYour Bulls.pdf

[26] Anonimus. 2007 Afrodisiaka (passionate food). [internet]. Available in: [http://sobatbatam.com/forum/showthread.php?t=1284.

[27] Lubis TM, Dasrul, Hamdan and Fauziah 2012 The effects of enervon-c and santa-e supplementation in feeding on the motility of chicken spermatozoa *Agripet* **12** 34

[28] Samiadi LA 2016 Boron [internet]. [diacu 1 Juni 2016]. Available in :https://hellosehat.com/herbal/boron/

[29] Chenoweth PJ 2005 Genetic Sperm Defect *Theriogenology* **64** 457

[30] Dewi AS, Ondho YS and Kurnianto E 2012 Quality of cement based on age in Java bulls *Anim Agric J.* **1** 126

[31] Makhzoomi A, Lundeheim N, Haard M and Rodriguez-Martinwz H 2007 Sperm morphology and fertility of progeny-tested AI Swedish dairy bull *J of Anim And Vet Advances* **8** 975

[32] Costa ND, Susilawati T, Isnaini N and Ihsan MN 2016 Effect of different dilution materials usage on Indonesian ongole bull sperm crossbreed quality during cooling process. *IAJPS* **3** 379

Producing chicken eggs containing isoflavone as functional food due to feeding effect of soy sauce by-product

L D Mahfudz, T A Sarjana, R Muryani and N Suthama

Faculty of Animal and Agricultural Sciences, Diponegoro University, Semarang 50275, Central Java, Indonesia

E-mail: inditik@yahoo.com

Abstract. The present study was aimed to verify the impact of feeding soy sauce by-product in producing isoflavone-enriched chicken eggs as functional food. Experiment used 200 laying hens of 80-week old with average body weight of 1,932.75±189.50 g. Experimental diets were formulated using yellow corn, rice bran, soybean meal, fish meal, meat bone meal, poultry meal, premix, $CaCO_3$, and soy sauce by-product (SSBP). A completely randomized design with 4 treatments and 5 replication (10 birds each), was assigned in this experiment. Inclusion levels of SSBP were the treatments, namely, none (T_0), 10 (T_1), 12.5 (T_2), and 15.0% (T_3). Parameters observed were colour, index, and weight of egg yolk, and isoflavone content. Analysis of variance was applied and continued to Duncan test at 5% probability. Results indicated that yolk colour index and weight were not affected by the treatments, but isoflavone content was significantly ($P<0.05$) increased by feeding SSBP. Egg yolk isoflavone in T2 (0.41 mg/g) and T3 (0.47 mg/g) were higher than those in T0 (0.31 mg/g) and T1 (0.35 mg/g). In conclusion, dietary inclusion of soy sauce by-product at higher level (12.5 and 15.0%) can produce isoflavone-enriched eggs as functional food.

Keywords: *laying hen, soy sauce by-product, isoflavon, egg yolk*

1. Introduction

Eggs are very nutritious foodstuff and function as a source of protein, thus they play an important role in basic nutrition. Also, due to the high nutrients availability and digestibility with achievable price, eggs become a common food ingredient consumed by all levels community all over the world. However, eggs are conventionally associated with negative effects on human health due to their cholesterol content. Although eggs are known to have complete and high nutrients content, but they also contain high cholesterol level, especially in the egg yolk. Cholesterol content in the egg may varied depending on the srtain of the hens and feed ingredients composition provided. Previous researcher [1] reported that one egg contains about 213 mg, but other result [2] indicated higher level was about 410 mg, the quite high amount. Considering cholesterol content in the eggs is strongly influenced by feed ingredients given to laying hens [3], thus feed modification is one of many other ways that can be applied. Exploring the possibilities of the development of functional eggs containing low cholesterol by technological methods have been proposed [2]. In the present study, the effort had been explored with special reference to feeding manipulation in laying poultry through dietary inclusion of isoflavones-containing feedstuff.

Feed ingredients containing active compound such as isoflavones can be greatly possible in lowering egg's cholesterol levels [4]. Isoflavones can be found in high concentrations in soybean and soybean products. Soybean or its by-product such as sauce by- product (SSBP) is one feed ingredient containing high enough isoflavones and also protein of about 13.68 mg/g and 27%, respectively [5]. Isoflavones derived fromsoy productpassed the intestinal lumen and mainly accumulated into egg yolkin the form of daidzein [6]. Isoflavones in soy products function as antioxidant and phytoestrogenic substances (phytoesterogen/phytosterol) which were known to decrease the formation of very low density lipoprotein and to increase high density lipoprotein [5,7].

Isoflavones are plant sterols (phytosterols) that can inhibit dietary cholesterol absorption and also decrease liver fat metabolism. Isoflavones derived from soybean or soy products are structurally and functionally similar to natural estrogens and can bind to estrogen receptors and may exert antiestrogenic effect. It was very logical that isoflavones exert a depressing effect on lipid profiles of the product due to their metabolic function and physiological effects as described above. Isoflavones are also known to play the role of anti-cholesterol which can lowered blood cholesterol, decreased, low density lipoprotein, triglycerides, and increased high density lipoprotein.This phenomenon was supported by the finding that the content of yolk cholesterol reduced from 13.9 to 11.0 g/mg in Isa Brown laying hens fed the increasing levels (100 up to 300 mg/kg) isoflavones daidzein and genistein [8].It was further proved in hens fed alfalfa sprout containing high antioxidants and phytosterolsof 15 to 50 times higher than control diet, produced eggs with much higher isoflavone daidzein (15.7 vs. 5.28 mg/100g) and decreased egg yolk cholesterol from 11.8 to 10.4 mg/g [9].As a comparisson, previously reported that dietary supplementation of genistein, a compound of isoflavones, at a level of 800 mg/kg improved performance and egg quality indicated by the increased egg yolk genistein deposition and decreased egg yolk malondialdehyde in quail [10].

Low level of cholesterol-containing eggs would be categorized as functional food to ensure the believe and confidence of the consumers. In recent decades, there has been an increasing demand for functional foods, especially eggs, which is possibly to be a main attractive foodstuff protein sources in the future. Egg yolk enrichment with isoflavones provided egg to be a functional food that could potentially benefit human health [10]. Whether dietary fortification with isoflavone derived from SSBP in laying chicken can be transformed and deposited into the yolk has not been clearly verified yet. Therefore, the present research was conducted in order to clarify the beneficial effect of isoflavones on egg quality, based on egg yolk characteristics and isoflavone concentration, in 80-week laying hens fed soy sauce by-product (SSBP).

2. Materials and methods

2.1. Experimental animal and feed

The experimental animals of the present study were 200 birds of 80-week old laying chickens of Lohman strain with initial body weight of 2.02 ± 1.17 kg. The experimental rations were formulated using yellow corn, rice bran, soybean meal, fish meal, meat bone meal, poultry meat meal, premix and soy sauce by-product (SSBP), and completed with lysine and methionin. Composition and nutritional content of the experimental ration after the inclusion level of SSBP (Table 1) were created as treatments, namely none (T_0), 10% (T_1), 12.5% (T_2), and 15% (T_3).

2.2. Procedures and parameter

Dietary treatments were started to be fed when the birds were 560 day old and terminated on day 595. The birds were provided free access to feed and drinking water. The SSBP was obtained from the company of Mirama soy sauce producer in Semarang, Central Java. Desalted processing was performed [11] to minimize salt residue of the soy sauce prior to mixing it with other ingredients for feed formulation (Illustration 1). Parameters observed were colour, weight and index of egg yolks, and isoflavone content in egg yolk. Determination of isoflavones content in egg yolk was performed [12] with slight modification.

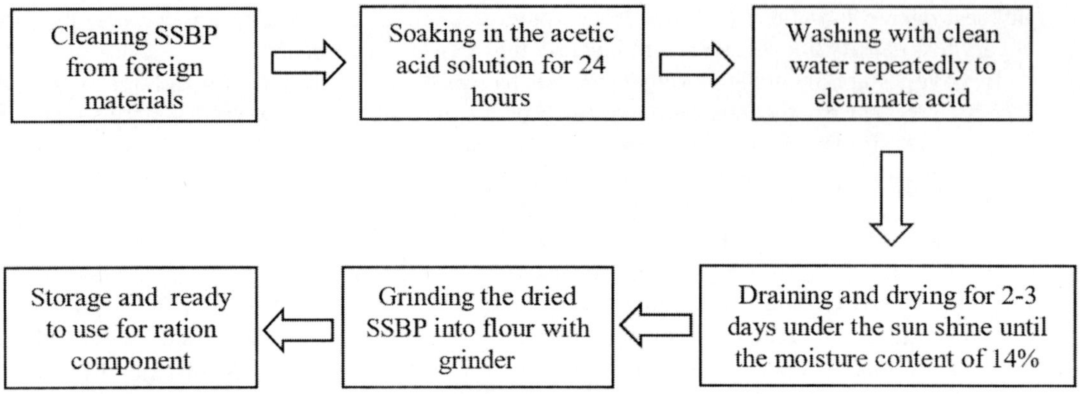

Figure 1. Flow chart of soy sauce by-product (SSBP) meal processing procedure

2.3. Experimental design and statistical analysis

Experiment was conducted using completely randomized design (CRD) with 4 treatments and 5 replications and each replication consisted of 10 birds of laying hens. Dietary treatments tested were as follows:

T_0: ration without soy sauce by-product
T_1: ration with soy sauce by-product 10%
T_2: ration with soy sauce by-product 12.5%
T_3: ration with soy sauce by-product 15%

Data of all parameters were statistically evaluated based on analysis of variance and continued to Duncan test at 5% probability level [13].

3. Results and discussion

Color, weight, and index of egg yolkswere not affected by feeding soy sauce by-product (SSBP) at whatever levels (Table 2). However, dietary inclusion of soy sauce at higher levels either at 12.5 or 15% significantly ($P<0.05$) increased isoflavone content of egg yolk. The extense of egg yolk color is commonly correlated with the ingredients of the diet provided especially those containing carotenoids and/ or xanthophylls.

3.1. Egg yolk color

Feed formulationused in the present study composed of yellow corn in equal amount (Table 1), this means that the contribution containing carotenoids and/or xanthophylls also at the same level. Since feed consumption was not affeted by the dietary inclusion of soy sauce (unpublished data),and soy sauce does not contain xanthophyll, thus the same intake of xanthophyllhad no different effect on egg yolk color. However, the present result was quite different from that of laying hens given additional paprika extraxt increased yolk color by 20.4% [14], and those fed red pepper either normally processed or fine powder produced higher yolk color by 57.7 and 70.5%, respectively [15]. It has been discussed previously that the difference yolk color was due to the content of additional ingredient fed to the chicken with special reference to carotenoids and/or xanthophylls. The intensity of yolk color is influenced by the presence and role of xanthophyll, a pigmentthat could be transfered from the feed.Egg yolk color is a quality indicator that can be altered by manipulating hen's diet. It is well documented that the color of yolk is dependent on the amount of xanthophylls, fat soluble pigments, in hen's diet, but this carrying capacity mechanism was not different in the present study since dietary fat content was the same (Table 1).

Table 1. Composition and nutrient content of experimental ration

Ingredient	Dietary treatment			
	T_0	T_1	T_2	T_3
	%			
Yellow corn	55.00	55.00	55.00	55.00
Rice bran	15.00	12.00	11.00	10.00
Soybean meal	16.00	12.00	10.50	9.00
Soy sauce by-product	0.00	10.00	12.50	15.00
Fish mealn	4.00	2.00	2.00	2.00
Meat bone meal	3.00	2.00	2.00	2.00
Poultry meat meal	3.00	2.00	2.00	2.00
Lysine	0.10	0.10	0.10	0.10
Methionine	0.10	0.10	0.10	0.10
$CaCO_3$	3.80	3.80	3.80	3.80
Premix	1.00	1.00	1.00	1.00
Total	100.00	100.00	100.00	100.00
Nutritional content[a] (%)				
Metabolisable energy (Kcal/kg) [b]	2,837.38	2,895.63	2,909.56	2,923.48
Crude protein	19.17	19.08	19.03	18.98
Crude fiber	5.95	6.16	6.16	6.17
Ether extract	5.26	6.29	6.57	6.86
Calcium	4.15	4.03	4.03	4.04
Phosphorus	1.47	1.16	1.09	1.03
Lysine [c]	1.09	0.96	0.94	0.92
Methionine [c]	0.41	0.50	0.53	0.57
Feed cost (IDR/kg)	5,664.00	5,081.00	4,964.50	4,848.00

[a] Determined value based on proximate analysis of all ingredients used in feed formulation
[b] Calculated value based on Balton fromula (EM (kkal/kg) = 40.81 [0.87 (PK + 2.25 x LK + BETN) + k].[19]
[c] Calculated value based on Berdasarkan Tabel Kandungan Nutrisi Bahan Pakan [20]

3.2. Egg yolk wight and index

Since feed consumption was the same, as it has been described in the previus pharagraph, so that protein intake was also not different. Protein is a basic constituent of yolk, thus, the same protein intakecontributing the impact to produce relatively similar egg yolk weight. This is in accordance with the result reported previously [16] that the yolk index was unchangedwith the same yellow color score.It appears that the egg yolks are predominantly composed of proteins transferred from the feed, so the same feed protein (Table 1) produces the same weight of the egg yolk, which was supported by the fact that fat in egg did not affected by dietary inclusion of isoflavones [8]. Fat in egg yolk varied between 4.25 and 4.30 g/yolk in isoflavones treated groups compared to 4.27 g/yolk in control group. Therefore, it is clear that yolk weight is closely related to the dietary protein supply because isoflavones are much more effective when it bound to protein rather than fat.The size of the egg was strongly influenced by protein and amino acids in the diet, and 50% of egg dry weight is protein [17]. Previously has been reported [9]that yolk weight (between 17.0 and 17.3 g) was not affected by enriched sprout containing daidzein in white Leghorn hens, and this is the same as found in the present study in which yolk weight was between 16.9 and 17.0 g (Table 2). No difference in yolk weight supports the unchangeable in egg yolk index due to feeding soy sauce by-product. When either egg weight in a whole or yolk weight are the same, it can be a certain of providing similar egg yolk index.

Tabel 2. Colour, weight, and index of egg yolks, and isoflavone content in egg yolk due to feeding soy sauce by-product (SSBP)

Parameter	Dietary treatment			
	T_0	T_1	T_2	T_3
Egg yolk colour	7.88	7.13	7.40	7.30
Egg yolk weight (g)	16.90	16.68	16.58	17.00
Egg yolk index	4.23	4.14	4.32	4.30
Isoflavone in egg yolk (mg/g)	0.31^b	0.35^b	0.41^a	0.47^a

$^{a-b}$Mean value in the same raw followed by different superscript differ significantly (P<0.05)

3.3. *Isoflavones in egg yolk*

The interesting phenomenon is that feeding soy sauce by-product at the levels 12.5% (T_2) and 15% (T_3) significantly increased isoflavones deposited in the egg yolk although they were not followed by the change in egg yolk color and weight (Table 2). It is assumed that the higher level of soy sauce was given, the much more amount of isoflavones could be consumed although with similar feed consumption. Therefore, the more isoflavones derived from soy sauce by-product consumed, the greater possibility of being transferred into the egg yolk as found in the present study. The mechanism of isoflavones transportation into the egg yolk have been partly clarified wheter they changed to a conjugated form or not [12]. Isoflavones was known to reduce to 30% during their meabolic fate, suggesting that 70% of the isoflavones existed in the egg yolk in the form of conjugated. Thus, a brief conclusion can be formulated for the treatments of T_3 and T_4that isoflavones changed to a conjugated form, a soluble isoflavones, in order to make it to be easier for the transfer into the egg yolk. Dietary supplementation of isoflavone genestein in quail increased the content of isoflavones in quail yolk [10, 18] because isoflavones from feed can be transferred to the yolk [6]. Biotransfromation of isoflavones indicating the transfer from feed to blood and finally into egg yolk as indicated in the Figure 2 [21].

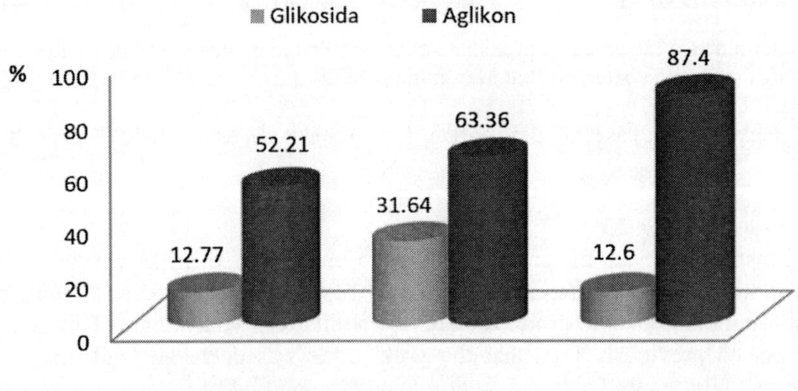

Figure 2 Comparisson of isoflavones in feed, blood and egg yolk (*Japanese PoultryScience, in progress*) [21]

4. Conclusion

Dietary inclusion of soy sauce by-product meal containing isoflavones at higher levels of either 12.5% (T_2) or 15% (T_3) increase the content of isoflavones in egg yolk, and the eggs are possible to become functional food.

5. Implication

Isoflavones-enriched yolk ensures the egg can be a source of high quality food for people, and giving rise to the possibilities for production of functional eggs.

References

[1] Sudaryani, T 2006 Eggs quality (in Bahasa) (*Penebar Swadaya, Jakarta*).

[2] Miranda J M, Anton X, Redondo-Valbuena C, Roca-Saavedra P, Rodriguez J A, Lamas A, *et al* 2015 Egg and egg-derived foods: Effects on human health and use as functional foods. *Nutrients* **7**(1) 706 – 729. doi:10.3390/nu7010706.

[3] Herdiana R M, Marchal Y, Dewanti R and Sudiyono 2014 Pengaruh penggunaan ampas kecap terhadap pertambahan bobot badan harian, konversi pakan, rasio efisiensi protein dan produksi karkas itik lokal jantan umur delapan minggu. *Bul. Peternakan* **38** (3) 157 – 162.

[4] Ariyani E 2006 Penetapan Kandungan Kolesterol dalam Kuning Telur pada Ayam Petelur. *Temu Teknis Nasional Tenaga Fungsional Pertanian, Bogor*. p 12 – 15.

[5] Malik A, Suprijatna E, Yunianto, V D and Mahfudz L D 2015 *Pengaruh isoflavon ampas kecap terhadap antioksidan dan rasio kholesterol LDL/HDL darah ayam petelur. Seminar Nasional. Hasil - Hasil Penelitian dan Pengabdian* LPPM.p 252 – 258.

[6] Saitoh S, Sato T, Harada H and Matsuda T 2004 Bioternsformation of soy isoflaone glycosides in laying hens: intestinal absorption and perential sccumulation into the egg yolk of equal, more estrogenic metabolite of daidzien. *Biochem. Biophys. Acta* **1674**122 – 130.

[7] Amirthaveni S and Vijayalakshmi P 2000 Role of soyflour supplementation on lipid profile among cardiovascular patients.*Proceeding "TSPUC-III"*October 15-20, Tsukuba, Japan. Pp.185-186.

[8] Gjorgovska N, Filev K, Levkov V and Nastova R 2015 Effect of supplemented isoflavones in feed on plasma and egg cholesterol concentration in isa brown layers. *Vet. Med. Zoot.***70** (92) 16 – 21.

[9] Mattioli S, Dal Bosco A, Martino M, Ruggeri S, Marconi O, Sileoni V, *et al* 2016 Alfalfa and flax sprouts supplementation enriches the content of bioactive compounds and lowers the cholesterol in hen egg. *J. Funct. Foods* **22** 454 – 462.

[10] Akdemir F and Sahin K 2009 Genistein supplementation to the quail: Effects on egg production and eggyolk genistein, daidzein, and lipid peroxidation levels. *Poult. Sci.* **88** 2125–2131. doi: 10.3382/ps.2009-00004

[11] Sukarini N E, Mahfudz L D and Legowo A M 2004 Studi penggunaan ampas kecap yang diproses dengan larutan asam asetat untuk pakan terhadap komposisi kimia dan karakteristik fisik daging ayam broiler. *J. Indonesian Trop. Anim. Agri.* **29** (3) 129 – 135 (*In Indonesian with English abstract*).

[12] Saitoh S, Sato T, Harada H and Takita T 2001 Transfer of soy isoflavone into the egg yolk of chickens. *Biosci. Biotechnol. Biochem* **65** (10) 2220 – 2225.

[13] Steel R G D and Torrie J H 1995. The princip and procedure of statistic (in Bahasa). Translated by Bambang Sumantri (*Gramedia, Jakarta*).

[14] Lokaewmanee K, Yamauchi K, Komori T and Saito K 2011 Enhancement of egg yolk color by paprikacombined with a probiotic. *J. Appl. Poult. Res.***20** 90 – 94. doi: 10.3382/japr.2009-00140.

[15] Li H, Jin L, Wu F, Thacker P, Li X, You J,*et al*. 2012 Effect of red pepper (*Capsicum frutescens*) powder or red pepper pigment on the performance and egg yolk color of laying

hens. *Asian-Australas. J. Anim. Sci.***25** (11) 1605 – 1610. http://dx. doi.org/10.5713/ajas.2012.12235.

[16] Yang H M, Wang Z Y and Lu L 2009 Study on the relationship between eggshell colors and egg quality as well as shell ultrastructure in Yangzhou chicken. *Afr. J. Biotechnol.***8** (12) 2898 – 2902.

[17] Zaheer K 2015 An update review on chicken eggs : Production, consumption, management, aspects and nutritional benefits to human health. *Food Nutr. Sci.***6**: 1208 – 1220.

[18] Lin F, Wu J, Abdelnabi M, Ottinger M and Giusti M M 2004 Effects of dose and glycosylation on the transfer of genestin into the eggs of Japanese quail (*Coturnix japonica*)*.J. Agric. Food Chem.* **52** 2397 – 2403.

[19] Murwani R 2010 Modern broiler (in Bahasa). Widya Karya, Semarang.

[20] Amrullah I K 2004 Broiler chicken nutrient (in Bahasa). Lembaga Satu Gunungbudi, Bogor.

[21] Malik A, Suprijatna E, Yunianto, V D and Mahfudz L D 2016 The effect of isoflavones ketchup by-product (IKBP) on digestive and reproductive organs of laying hens. Japanese Poultry Science. (in progress).

IOP Conf. Series: Earth and Environmental Science **102** (2017) 012022 doi:10.1088/1755-1315/102/1/012022

The physicochemical quality and meat microstructure of post laying hen with addition of Biduri (*Calotropis gigantea*) latex extract

A M P Nuhriawangsa, B S Hertanto, L R Kartikasari, W Swastike, M Cahyadi and S Rasid

Laboratory of Industry and Animal Product Technology, Animal Science Departement, Universitas Sebelas Maret, Jl. Ir. Sutami 36A Kentingan, Surakarta, Central Java, Indonesia +6257126, Phone/Fax. +62271637457

Email : adimagna@staff.uns.ac.id

Abstract. The objective of this study was to evaluate the effect of extract level of Biduri latex on the meat quality of laying hens. The materials of this research were Biduri latex and thigh meat from hens strain Lohman. The latex was tapped from a young tissue stem and centrifuged for its supernatant. Meats were smeared with latex, punctured and incubated for 30 minutes. Concentrations of latex were 0, 3, 6 and 9% from the weight of meat (w/w). The variables were water, dissolved protein, crude fat content, tenderness and microstructure of meat. The statistical analysis method using ANOVA and if there was a mean difference, Duncan test was used. Descriptive analysis was used for microstructures of meat by comparing its hydrolysis conditions. The study showed that fat had significant difference ($P < 0.05$), dissolved protein and tenderness had very significance ($P < 0.01$). Descriptive analysis showed that there were different compositions of microstructures on meat structure. The fat content increased with addition of 3% latex. The value of dissolved protein increased but tenderness decreased by addition extract of 6% latex. The addition of Biduri latex extract showed that hydrolysis in the microstructure of meat. The addition of 6% latex was the best meat quality.

Keywords: Physicochemical quality of meat, microstructure of meat, Biduri latex, post laying hen

1. Introduction

In Indonesia, post laying hen is consumed as one source of animal protein, such as a broiler chicken meat. It has good protein and the fatty is low, but the flesh is hard because it 3was produced from older chicken People prefer to choose tender one than hard one, so the post laying hen meat must be made more tender.

We can use physical and chemical method to make meat more tender. Method used were using protease. Protase used among others, bromelain, papain, ficin, actinidin, and calpain in the food industry [1]. Papain can be used to increase tenderness of post laying hen meat [2] and post laying duck meat [3]. Similarly, bromelin can be used to made post laying duck meat [4] and beef meat [5] more tender. Injection of bromelain and papain on beef meat can increased tenderness on preservation vacuum packaging [6]. Protease are mostly taken from fruit, and fruit are human food. This may cause competition in its use. We should be used proteases from other sources e.g. from plants.

Biduri plants (*Calotropis gigantean*) [7] and *Calotropis procera* [8] can produce protease-containing sap. Protease from *Calotropis gigantean* have enzymes those included in exopeptidase class and they can hydrolysis miofibril protein on fish meat with high specific value [9]. It can also hydrolyze systein [10], casein, fibrinogen and fibrin [11]. With the hydrolysis of the protein then the meat can be more tender, so the protease from the plant can be used as the tenderness ingredient [12]. The physic and hedonic quality of post laying hen can increased with adding crude extract of Biduri latex [13]. Crude enzyme extracted from *Calotropis procera* can improved physicochemical properties of meat [12].

Raw extract of biduri latex can be used to raise the quality of meat. The purpose of this research was to understand the effect of raw extract biduri latex and to increase the quality of meat (water, protein, crude lipid, tenderness and microstructure of meat).

2. Materials and methods

2.1. Materials and samples preparation

Material latex Biduri (*Calotropis gigantean*) was collected to product crude enzyme in this study. Post laying hen strain Lohman aged 90 weeks was processed and thigh samples was analyzed for chemical composition, physical quality and meat microsturucture. Young stem tissue of Biduri was tapping for latex collection [14] and the latex was centrifuged to produce crude fluid of biduri latex (supernatant) [15]. Biduri latex was applied to meat sample and punctured using needle. The samples were then incubated for about 30 minutes inside sterile laminar air flow (LAF) in room temperature [16].

2.2. Methods

Moisture content was measured with gravitimetric test [17], dissolved protein content by Biuret [18], crude fat by Soxhlet extraction [19], tenderness by breaking strength with Warner Blatzer method [16]. The microstructure of meat was analyzed by microtom and captured with a microscope at magnification of 40 times [20].

2.3. Experimental designs

The study used experiment design by One Way Randomized Design [21]. The levels of biduri latex concentration used were 0, 3, 6 and 9% from sample weight of meat.

The data of this study was analyzed by ANOVA on meat quality (water, protein, crude lipid, tenderness) and if there was a significant difference then further analyzed with Duncan's New Multiple Range Test [21]. Microstructure of meat was analyzed by descriptive analysis [21] by looking at the damage to the longitudinal and transverse structure of the meat.

3. Results and discussion

The meat quality of post laying hen with treatment of crude extract latex biduri can be seen in Table 1. Results showed that the value of dissolved protein increased by adding the concentration of 6% of crude extract biduri latex and similarly, crude fat content enhanced up to a level of 3%. However, the concentration of extract biduri latex 6% decreased tenderness and water content was not significant.

Table 1. The result of meat quality of post layin hen with the addition of Biduri latex concentration on 0, 3, 6 and 9%

Variables	Levels of addition of Biduri latex (%)				P
	0	3	6	9	
Tenderness	6,12±0,91[a]	6,44±1,33[a]	5,28±1,56[b]	4,60±1,72[b]	0,000
Dissolved protein	4,73+1,97[a]	4,74+2,37[a]	5,17+1,70[b]	4,39+1,89[a]	0,010
Crude fat	2,51+0,11[a]	3,44+0,03[b]	3,61+0,50[b]	3,76+0,20[b]	0,050
Moisture content	74,43+0,42	74,78+0,19	75,74+0,29	75,90+0,07	0,100

Description: [a,b]Different superscript in the same row indicates significantly

The addition of 3% crude extract Biduri latex did not show an increase in meat tenderness, but the addition of 6% extract Biduri latex increased the tenderness that indicated by the decreased of breaking strength meat value. The meat tenderness was not different between the addition of 6 and 9% of extract biduri latex . Several studies have shown that the addition of proteases can improve the tenderness of the meat, but differ in the concentration given to the meat. This data is supported by dissolved proteins which also appear to increase at a concentration of 6% and the occurrence of microstructural the meat broken to also at a concentration of 6%.

Collagen solubility and TCA-soluble peptides were significant increased with the addition 2×10^3 to 6×10^3 activity units/100 g of muscle of protease from *Calotropis procera*, papaya and pineapple to chicken, giant catfish, pork and beef meat. Other than that on the microstructure level of meat, tissue fibers were broken, and the connections between the sarcolemma and the myofibrils were loosened on the meat in the presence of hydrolysis by proteases. Thus the tenderness of meat increases due to protease activity [22]. The same study also showed similar results using 0.05%, 0.1%, 0.2%, 0.3% and 0.5% (w/w) of crude enzyme extract powder from *Calotropis procera latex* in chicken, pork and beef meat [23].

Dissolved protein was increased with addition 6% crude extract latex biduri in this research. Some studies show similar with this results. Water soluble proteins on beef meat increased with the addition extract crude bromelain and commercial bromelain [24]. This is due to the release of hydrophilic groups on the surface of protein molecules by proteases [25].

In this research elevated of fat content appear in the addition of latex biduri with a concentration of 3%. This is different from other studies, utilization unripe latex extract from papaya at levels 0, 0.5, 1.0, 1.5, 2.0, 2.5 and 3.0 mg/100g did not significant on fat content *in beef kababs* [26].

The addition of fresh crude extract of latex biduri until the percentage of 9% has no effect on the water content of meat in this research. The same result, addition of 0.05% and 0.1% (w/w) of crude enzyme extract powder from *Calotropis procera latex* is not significant in chicken and beef meat moisture content [23]. Utilization unripe latex extract from papaya at levels 0, 0.5, 1.0, 1.5, 2.0, 2.5 and 3.0 mg/100g did not significant on moisture content *in beef kababs* [26]. As well as bromelain and papain added to pork and beef do not affect water content [27].

The image of microstructure (40x magnification) of post laying hen with crude extract biduri latex can be seen in Table 2. Microstucture of meat was seen by longitudinal cuts.

Table 2. The picture of microscopic structure (40x longitude magnification) post laying hen with added concentrations of Biduri latex on 0, 3, 6 and 9%

The concentration of Biduri latex 0%	The concentration of Biduri latex 3%	The concentration of Biduri latex 6%	The concentration of Biduri latex 9%
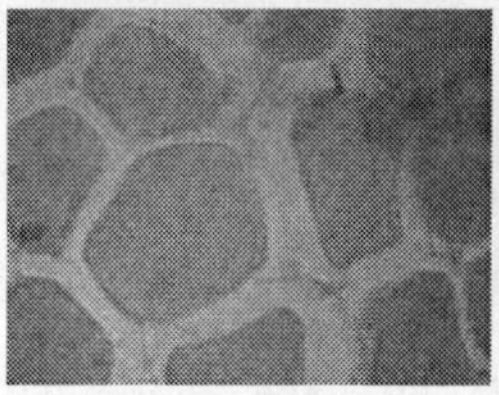	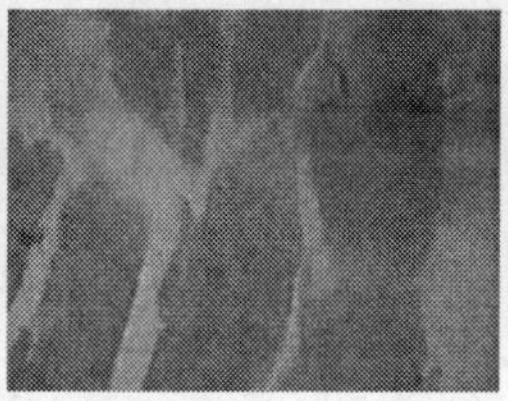	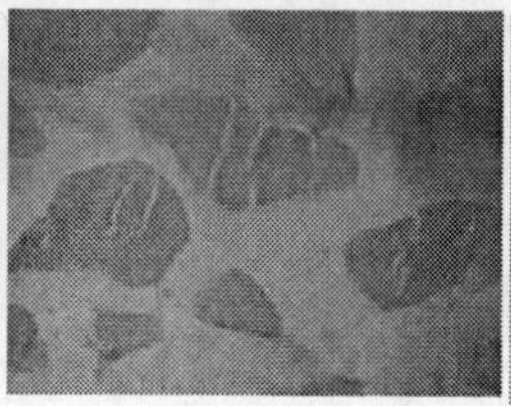	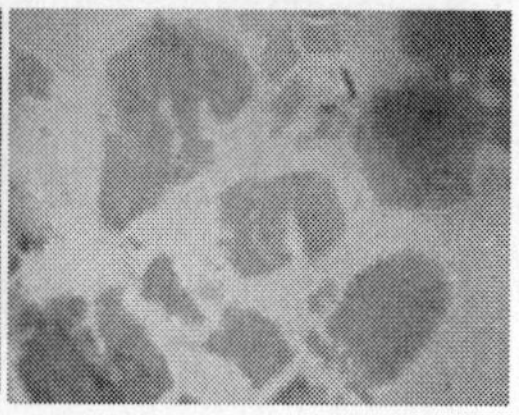

Research shows the increasing concentration of extract of latex biduri resulted in increasingly break down microstructure of meat in post laying hen. This is in accordance with research, microstructure of meat shown was rupturing of tissue on the collagen fiber by adding extract crude bromelain and commercial bromelain on beef meat. So it seems to happen a breaking down of myofibrial tissue [24].

4. Conclusions

The treatment of biduri latex with the concentration of 6% could increase physicochemical quality of meat of post laying hen (water, protein, crude lipid and tenderness of meat) . Hence, the microstructure of meat increased degradation with increasing levels of protease Biduri extract.

References

[1] Arshad M Sajid., Joong H Kwon, Muhammad I, Sohaib M, Aslam A, Nawaz I, Amjad Z, Khan U and Javed M 2016 *Cogent Food and Agriculture* **1**

[2] Widiastuti A, Pudjomartatmo dan Nuhriawangsa A M Patriadi 2012 *Sains Peternakan* **10** 100

[3] Nuhriawangsa A M Patriadi dan Pudjomartatmo 2002 *Kegunaan Enzim Papain dan Pemanggangan untuk Meningkatkan Kualitas Daging Itik Afkir* (Surakarta: Fakultas Pertanian Universitas Sebelas Maret)

[4] Utami D Putri, Pudjomartatmo dan Nuhriawangsa A M Patriadi 2011 *Sains Peternakan* **9** 82

[5] Chaurasiya R Saran, P Z Sakhare, N. Bhaskar and Hebbar H Umesh 2014 *Journal of Food Science and Technology* **52** 3870

[6] Istrati D, Camelia V and Rodica D 2011 *Journal of Agroalimentary Processes and Technologies* **17** 275

[7] Witono Y 2009 Specificity and stability of proteases from biduri plants (Calotropis gigantea) *Proseding Seminar Nasional Peran Ilmu dan Teknologi Pangan dalam Mewujudkan Ketahanan Pangan* (Padang: Fakultas Teknologi Pertanian Universitas Andalas) p 245 (In Bahasa Indonesia)

[8] Rawdkuen S, Punyawatt P, Phanuphong C, Soottawat B 2011 Food and Bioproduct Processing **89** 73

[9] Saputri D S 2011 *Spesifitas Katalik Protease Biduri (Calotropis gigantean)* (Indonesia: Universitas Jember)

[10] Rajesh R, Gowda C D R, Nataraju A, Dhananjaya B L, Kemparaju K and Vishwanath B S 2005 *Toxicon* **46** 84

[11] Joshi H, Gururaja M P and Suares D 2011 *International Journal of Pharmacy and Research* **3** 975

[12] Rawdkuen S, Manon J and Soottawat B 2013 Food Chemistry **136** 909

[13] Nuhriawangsa, A M Patriadi., Swastike W, Bayu S Hertanto, Wahyudi A and Pradisha E Diecky., 2017. The Effect of Biduri (*Calotropis Gigantean*) Latex on Meat Quality of Post Laying Hen. (Surakarta: Sebelas Maret University. OP Conf. Series: Materials Science and Engineering) p 1

[14] Nuhriawangsa A M P, Swastike W, Cahyadi M dan Gunawan D 2013 The application of crude extract of biduri (Calotropis gigantea) as a substitute of rennet to the quality of cow milk cheese*Seminar Nasional Peternakan Berkelanjutan ke-5* (Bandung: Fakultas Peternakan Universitas Padjajaran) p 272 (In Bahasa Indonesia)

[15] Wang D I C, Cooney C L, Demain A L, Dunnill P, Humphrey A E and Lilly MD 1979 *Fermentation and enzyme technology* (New York-Chichester-Brisbane-Toronto-Singapore: A Wiley-Interscience Pub. John Wiley & Sons)

[16] Soeparno 2005 *Ilmu dan teknologi daging* (Indonesia: Gadjah Mada University Press Yogyakarta)

[17] AOAC 2005. *Official Methods of Analysis*. 12th ed. (Washington D.C.: Association of Official Analytical Chemist)

[18] Sudarmadji, S, Haryono B dan Suhardi 1989 Analisa bahan makanan dan pertanian (Yogyakarta: Liberty)

[19] Atkinson T, V R Fowler, Garton G A and Lough A, 1972. *Analyst* **97** 562

[20] Glauert, A. M. (1974). Practical methods in electron microscopy. (Amsterdam: North-Holland Publishing Company)

[21] Steel R G D dan James H Torrie 1995 Principal and Statistical Procedure(Indonesia: PT Gramedia Pustaka Utama Jakarta)

[22] Rawdkuen S and Benjakul S 2012 *African J. Biotech.* **11** 14088

[23] Rawdkuen S, Jaimakreu M and Benjakul S 2013 *Food Chemystry* **136** 909

[24] Chaurasiya R Saran, P Z Sakhare, N. Bhaskar and H. Umesh Hebbar. 2015 *Journal Food Science and Technology* **6** 3870

[25] Doneva M, D. Miteva, S. Dyankova, I. Nacheva, P. Metodieva, K. Dimov 2015 *Biotechnology in Animal Husbandry* **3** 407

[26] Saeed M, Sajjad U Rahman, Shabbir A Muhammad, N Khanand and A Shakeel 2017 *Pakistan Journal of Agriculture Science* **1** 153

[27] Manea I, L Manea and V Marinescu 2016 Annalysis Food Science and Technology **2** 421

Optimum carrageenan concentration improved the physical properties of cabinet-dried yoghurt powder

Y Pratama[1], S B M Abduh, A M Legowo, Y B Pramono, AN Albaarri[1,2]

[1]Department of Food Technology, Faculty of Animal and Agricultural Sciences, Diponegoro University, Semarang, Central Java Indonesia
[2]Integrated Laboratory, Diponegoro University, Semarang, Central Java Indonesia

Email: albari@undip.ac.id

Abstract. Carrageenan is a hydrocolloid which able to improve the characteristic of dried powder. The purpose of this study was to produce and evaluate the characteristics of yoghurt powder which incorporated carrageenan as stabilizer and dried in a cabinet dryer. Carrageenan of 1 %, 2 % and 3 % (w/v) concentration were added to yoghurt prior to cabinet drying process. Physical properties of the yoghurt powder and the reconstituted one were evaluated. The best result was shown in carrageenan concentration of 2 % where it showed the highest bulk density (0.62 g/ml), and best particle size distribution (65.49 % in the range of 250-500 µm). Moisture and water activity (a_w) were 8.02% and 0.37, respectively. The value is lower than what spoilage microorganisms require to grow ($a_w > 0.6$) thus ensuring its long shelf life when combined with proper packaging. Reconstitution at 50°C showed that 2 % carrageenan resulted in a stable yoghurt product with no visible syneresis even after 3 hours. The proposed method shows promising application in the production of long shelf-life yoghurt powder production.

Keywords: cabinet, carrageenan, drying, powder, yoghurt

1. Introduction

Yoghurt is a kind of fermented food which is recognized to poses some health benefits due to a better nutrient utilisation as influenced by more digestible protein, more calcium, certain B group vitamins, and hypocholesterolaemic effect [1]. However the shelf life of yoghurt is relatively short, about 1 day at ambient temperature and 4-5 days at 7 °C refrigeration. Such a short shelf life is a big limitation for archipelago country such as Indonesia to achieve the goodness of yoghurt for people around the country. In order to cover the problem, preservation is necessary to be performed. Dried product is one popular form of preserved food product due to its practical aspect as well as economical.

Convective cabinet drying technique is one of popular drying practices. The technique is relatively available around the country due to the fact that the system does not require sophisticated technology. The technique is based on flowing of dehumidified air through the product surface to carry the water vapour freed from the product. Dehumidification is usually performed by heating the atmospheric ambient air. The heated air is then blown over the products surface. The drying cabinet system is usually consisted of two chamber for heating the atmosphere and drying the freed vapour of the product respectively [2].

Prior to consumption, dried yoghurt is reconstituted by addition of certain amount of water, considering the initial moisture content of the fresh yoghurt. The emulsion stability of the reconstituted yoghurt is critical for consumer preference due to the fact that syneresis is easily observable. The stability can be maintained by using plant origin polysaccharides [3].

Indonesia is the second longest coastal line country over the world, approximately 54,716 km providing abundant number of seaweed that part of them are potential as the source of carrageenan. Carrageenan is popularly used in dairy industry to prevent phase separation [4]. Studies have already carried out on its influence on the yoghurt [5, 6] and on reconstituted yoghurt [4]. In the present paper, its influence on the physical properties of yoghurt powder dried by convective drying cabinet is discussed.

2. Material and Methods

2.1. Material

Bacterial culture containing *L. delbrueckii subsp. bulgaricus* and *S. thermophilus* was obtained from the Diponegoro University culture collection. UHT sterilized full cream milk was obtained from the market. The raw milk was obtained from the Faculty of Animal and Agricultural Sciences' farm. The kappa (κ)-carrageenan was obtained from the collection of Faculty of Fisheries and Marine Sciences, Diponegoro University.

2.2. Inoculum preparation

Freeze dried bacterial culture amounting 1.4 g was inoculated into 50 ml UHT sterilized full cream milk prior to incubation at 38°C for 6 h. The liquid inoculum obtained then was inoculated into similar UHT sterilized full cream milk to achieved 10 % (v/v) and incubated at the same condition. The same procedure was repeated to obtain the working inoculum.

2.3. Yoghurt preparation

Milk containing 3 % of fat was pasteurized at 75°C for 20 s. One percent (v/v) of working culture was inoculated into the pasteurized milk prior to incubation at 38° C for 6 h. The yoghurt resulted then was refrigerated for approximately 15 min prior to addition of carrageenan in varied concentration: 1, 2, and 3 % (w/v) and gently mixed with spoon then homogenized using Ultra Turax Homogenizer (IKA, Germany) at 10,000 rpm for 5 min.

2.4. Yoghurt drying and powdering

The fresh yoghurt was spread on 40 cm x 50 cm trays and kept in convective drying cabinet for 20 h at 50 °C. The dried yoghurt was then kept in water-vapor tight boxes prior to be milled in beater bar miller Maksindo FCT-Z300 for 45 s. The powdered yoghurt was kept in PET bags prior to further analysis.

2.5. Yoghurt powder rehydration

Rehydration was carried out on the dried yoghurt with aquadest at ambient temperature to achieve 10% (w/v).

2.6. Water-activity and moisture content measurement

The water activities of the dried yoghurt were measured with Novasina Aw meter. The moisture content of the dried yoghurt was measured by mean of gravimetry.

2.7. Microscopy analysis

The microscopy analysis was carried out to observe the structure of the powder particle. The analysis was carried out under a software assisted Olympus microscope, coupled with Mshot camera. The device was set on 400x magnification under controlled light to get clear images.

2.8. Particle size distribution
One hundred gram of dried yoghurt was passed through Retsech AS200 Control (Germany) using DIA X siever with 500, 250, 125 μm mesh size.

2.9. Bulk density
The bulk density of the dried yoghurt was determined by putting 10 g of it in measuring glass cup. The measured volume after 10 times knocking was recorded. The bulk density is expressed in g/ml.

2.10. pH and viscosity
pH measurement was carried out using pH meter (Hanna, Italy). An Ostwald viscometer was used for viscosity measurement.

2.11. Sedimentation index
Slight modification from method by [7] was used to measure sedimentation index. Fifteen ml of reconstituted yoghurt was centrifuged in 6000 rpm for 25 min. The centrifuge tube were then put upside-down for 30 min. Sedimentation index was calculated as the weight of sediment left in the tube divided by initial weight of reconstituted yoghurt and expressed in % (w/w).

2.12. Syneresis observation
Only for syneresis observation, two methods were used to reconstitute the yoghurt powder. First was the aforementioned method using aquadest at ambient temperature to get 10% (w/v). The second method was the same procedure, only the aquadest used was at 50°C. Subsequently, 10 ml of reconstituted was poured into flask and visually observed for syneresis/whey layer formation.

2.13. Statistics
All measurements were taken in triplicate and presented as mean value and standard deviation. Experimental data were then analyzed by analysis of variance (ANOVA) which coupled with Duncan test wherever significant result found. All statistical analysis was performed using SAS (Statistical Analysis System, Version 9.13) software.

3. Results and discussions
Yoghurt powder has been successfully created by the proposed method using carrageenan as stabilizer and utilizing cabinet drying technique. Carrageenan concentrations added (1%, 2% and 3% (w/v)) were quite high compared to the normal carrageenan and other hydrocolloid incorporation in yoghurt product which normally 0.5 % (w/v) max [8]. Its high concentration aimed to help drying process by increasing the mixture's dry matter prior to drying step and at the same time to improve the characteristics of the resulted yoghurt powder. An increase in carrageenan concentration will increase the dry matter of the yoghurt's mixture prior to drying step, which is essential. A certain degree of dry matter is needed as a seed for crystallization during drying process. For example, an operation for milk powder which includes spray drying step normally requires the mixture to be brought into 45% - 55% dry matter before drying takes place [9].

3.1. Moisture content and water activity (Aw) of yoghurt powder
The results of the effect of carrageenan addition on yoghurt powder physical characteristics are shown in **Table 1**. The result showed that the moisture content were in the range of 6.44-8.04 %. It is higher than yoghurt powder produced by other procedures such as spray drying and freeze drying technique which range 0.3-0.6% [1]. The possible reasons were the drying temperature was not high enough and high hydrocolloid concentration. The drying temperature was set at 50°C in order to keep lactic acid bacteria survival on the finish product, which resulted in 3 log reduction from the initial load (data not shown). As an addition, carrageenan as hydrocolloid has water binding characteristics, thus it may reduce the drying rate and the finish product's moisture content will be higher. However, the

moistures of the yoghurt powder were not significantly different despite different carrageenan concentration. Authors assumed that after 20 hours of drying, the process has reached equilibrium where the moisture level was kept unchanged.

Table 1. Moisture content, water activity and bulk density of yoghurt powder

Treatment	Moisture (%)	Aw	Bulk Density (g/ml)
1%	7.18±0.05	0.31±0.01	0.50±0.00[a]
2%	8.02±1.09	0.37±0.01	0.62±0.02[b]
3%	6.44±1.21	0.35±0.04	0.52±0.02[a]

Means within a column with different superscript are significantly different (P < 0.05)

Water activity (Aw) signifies the amount of free water in the product which available to microorganism to grow. A low Aw value is an important characteristic for a powder product so that it will prevent microbial growth, especially mould. It was shown that Aw values of yoghurt powder were 0.31, 0.37, and 0.35 for 1%, 2% and 3% carrageenan, respectively. P value > 0.05 showed that the result were not significantly different. It was similar to the moisture result, showing the relation between two parameters. However, it indicated that carrageenan's ability to bind free water did not give a significant difference in the product. Availability of free water would be reduced, thus reducing Aw value, if it is bond to water-binding substances such as sugar and salt. Albeit the insignificant result, the Aw values were lower than the least demanding xerophilic moulds which need around Aw 0.6 minimum to grow [10]. Therefore, the resulted yoghurt powder has the ability to prevent most microbial growth, in the condition that it is properly packaged so that the Aw value remains unchanged.

3.2. Bulk density of yoghurt powder

Bulk density is an important parameter in powder filling and packaging operation. A consistent bulk density will ensure a consistent product weight when volumetric-type filler is used. Bulk density value also determines the amount of packaging material would be needed. When its value is low, it means that the powder is bulky and thus needs more packaging material to pack a certain mass. The highest yoghurt powder's bulk density was observed from 2% carrageenan which was 0.62 g/ml (P<0.05), while 1% and 3% carrageenan were 0.50 g/ml and 0.52 g/ml, respectively. It might be due to the 2% carrageenan had slightly higher moisture level compared to other treatments. Water which has density of approximately 1 g/ml is significantly higher than the yoghurt powder bulk density. Hence, moisture level influences bulk density of yoghurt powder greatly, where higher moisture leads to higher bulk density.

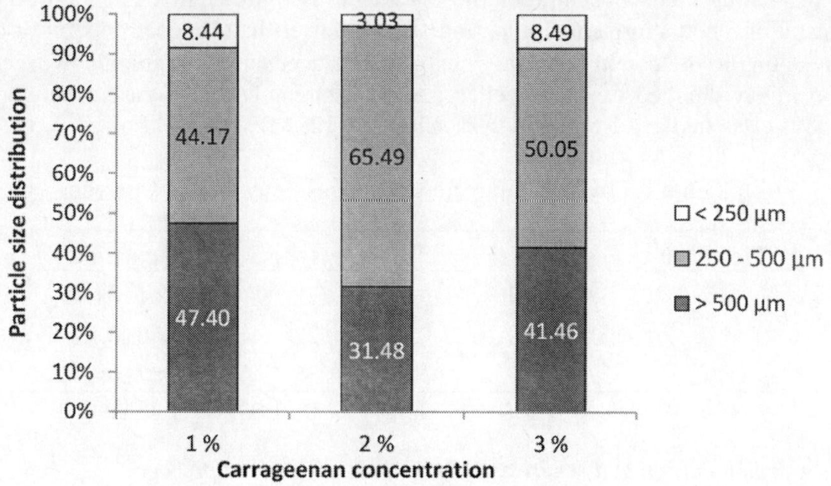

Figure 1. Yoghurt powder particle size distribution

3.3. Particle size distribution of yoghurt powder

Particle size distribution analysis result is shown in **Figure 1**. The carrageenan concentration of 2 % resulted in smaller particle size distribution than the 1 % and 3 % concentration with 65,49 % in the range of 250 – 500 µm. This result was in agreement with the bulk density data, as the 2 % carrageenan has the highest value. Given the composition similarity among the three treatments, smaller particle size means denser and leads to higher bulk density.

Beater bar miller employs impact force as the main size-reducing force. Impacted particle will react based on its brittleness, where brittle substance will break and less brittle one will only deform [11]. Therefore, considering the particle size and bulk density, it can be safely assumed that the 2 % carrageenan yoghurt powder had more brittle characteristic compared with the other two treatments. It considers the fact that the moisture and Aw values were not significantly different. In addition, the appearances of yoghurt powder's optical microstructure were also similar, as can be seen in **Figure 2**. The only notable difference was that the 2 % treatment has smaller size compared to the other treatments which is in agreement with bulk density and particle size distribution data.

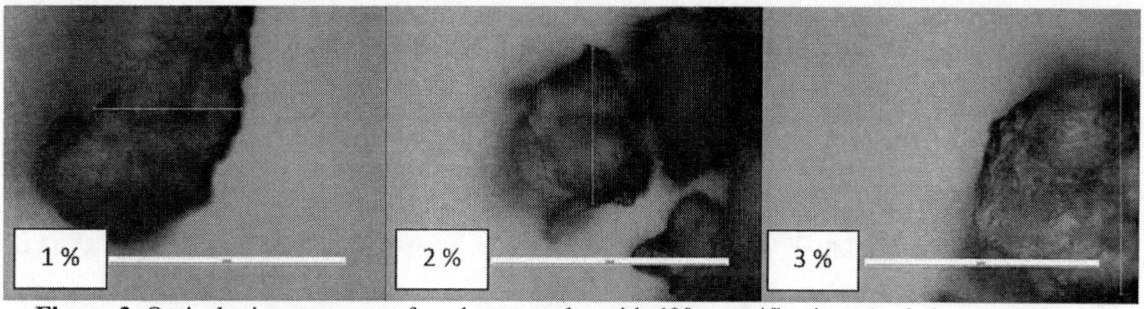

Figure 2. Optical microstructure of yoghurt powder with 400 magnifications (scale bar = 1 mm)

3.4. pH of reconstituted yoghurt powder

Reconstitution of the yoghurt powder was carried out by diluting the powder in both cold water (aquadest in ambient temperature) using spoon to mix it manually. The yoghurt powder was mixed easily, and although some small lumps were formed, it could be immediately broken. The reconstituted yoghurts were then evaluated for its physical characteristics, as can be seen in **Table 2**. It was shown that carrageenan concentration did not affect pH significantly. The mixture pH was 3.93 to 3.96, as a result of 6 hours fermentation period. This is lower than isoelectric point of casein which is at pH 4.6, resulting in the casein positively charged. Carrageenan as an anionic hydrocolloid interacts with this positively charged casein micelles and thus strengthens the casein network and reduces syneresis and is classified as adsorbing polysaccharides [12, 13].

Table 2. Physical parameters of reconstituted yoghurt powder

Treatment	pH	Viscosity (cp)	Sedimentation (%)
1%	3.96±0.01	6.14 ± 0.30^{a}	2.61 ± 0.07^{a}
2%	3.93±0.01	10.81 ± 0.22^{b}	3.40 ± 0.19^{b}
3%	3.95±0.02	17.14 ± 1.35^{c}	3.87 ± 0.44^{b}

Means within a column with different superscript are significantly different ($P < 0.05$)

3.5. Viscosity, sedimentation and syneresis of reconstituted yoghurt powder

Significant increase of viscosity was observed from the increasing concentration of carrageenan. This result is expected, however the value is very low considering the amount of carrageenan added. Furthermore, the sedimentation index result showed increasing trend ($P<0.05$) as the carrageenan concentration increased. Both results indicate that the increasing viscosity was caused by the increasing total solid rather than due to three-dimensional network formed by carrageenan as hydrocolloid. Three-dimensional network was formed into gel during mixing the yoghurt with carrageenan prior to drying, however the network was destroyed during heating and milling process.

Stability of reconstituted yoghurt is one of the most important physical parameters for the consumer acceptance. A visually separated product is deemed unacceptable, thus a stable product is essential. Cold-water reconstitution showed syneresis after only 1 hour. A big layer of whey was observed in both 1% and 2% carrageenan concentration, meanwhile smaller but visible layer was observed in 3% carrageenan. Warm-water reconstitution showed a better performance. Only 1% carrageenan showed small layer of separation, whereas other treatment kept its stability even after 3 hours of observation. Kappa carrageenan is better dissolved in warm water [14] thus give better stabilizing effect than in cool water solution. Therefore it is proposed that the better reconstitution should be done by a portion of warm water to dilute the powder and subsequently added with cool water for organoleptical reason.

4. Conclusions

Yoghurt powder was successfully created using the proposed method. Optimum concentration of carrageenan was 2 % where the resulted yoghurt powder showed the highest bulk density, and smallest particle size distribution. The water activity of the resulted products was low enough to ensure the long shelf life. Better reconstitution was achieved using warm water which resulted in a stable solution with no visible separation.

References

[1] Kumar P, and Mishra H N 2004 Yoghurt Powder — a Review of Process *Food Eng.* **82** 133–142

[2] Singh R P, and Heldman D R 2014 *Introduction to Food Engineering* doi: 10.1016/B978-0-12-398530-9.00012-7

[3] Hussein M M, Hassan F A M, Abdel Daym H H, Salama A, Enab A K, and Abd El-Galil A A 2011 Utilization of some plant polysaccharides for improving yoghurt consistency *Ann. Agric.*

Sci. **56** 97–103

[4] Ji S, Corredig M, and Goff H D 2008 Aggregation of casein micelles and K-carrageenan in reconstituted skim milk *Food Hydrocoll.* **22** 56–64

[5] Michon C, Chapuis C, Langendorff V, Boulenguer P, and Cuvelier G 2005 Structure evolution of carrageenan/milk gels: Effect of shearing, carrageenan concentration and nu fraction on rheological behavior *Food Hydrocoll.* **19** 541–547

[6] Trčková J, Štctina J, and Kánský J 2004 Influence of protein concentration on rheological properties of carrageenan gels in milk *Int. Dairy J.* **14** 337–343

[7] Jensen S, Rolin C, and Ipsen R 2010 Food Hydrocolloids Stabilisation of acidified skimmed milk with HM pectin **24** 291–299

[8] Koksoy A, and Kilic M 2004 Use of hydrocolloids in textural stabilization of a yoghurt drink, ayran *Food Hydrocoll.* **18** 593–600

[9] Bylund G 1995 *Dairy Processing Handbook* (Lund: Tetra Pak Processing Systems AB)

[10] Leong S L L, Pettersson O V, Rice T, Hocking A D, and Schnürer J 2011 The extreme xerophilic mould Xeromyces bisporus - Growth and competition at various water activities *Int. J. Food Microbiol.* **145** 57–63

[11] Varzakas T, and Tzia C (eds) 2015 *Food Engineering Handbook, Food Process Engineering* (Boca Raton, FL: CRC Press)

[12] Phillips G O, and Williams P A (eds) 2009 *Handbook of Hydrocolloids Second Edition* (Boca Raton, FL: CRC Press)

[13] Everett D W, and McLeod R E 2005 Interactions of polysaccharide stabilisers with casein aggregates in stirred skim-milk yoghurt *Int. Dairy J.* **15** 1175–1183

[14] Prajapati V D, Maheriya P M, Jani G K, and Solanki H K 2014 Carrageenan: A natural seaweed polysaccharide and its applications *Carbohydr. Polym.* **105** 97–112

Nutritional comparison of *Spirulina* sp powder by solid-state fermentation using *Aspergillus* sp (FNCL 6088) and *Lactobacillus plantarum* (FNCL 0127)

E N Dewi[1] and U Amalia

Faculty of Fisheries and Marine Science, Diponegoro University, Jl. Prof. Soedarto, SH Tembalang, Semarang, 50275, Indonesia

Email : ulfahamalia0@gmail.com

Abstract. The *Spirulina* sp powder contains high levels of protein and Solid-State Fermentation (SSF) improved protein level. The aims of the study was to find the proximate contents in *Spirulina* sp's powder fermentation. The experiments were conducted by SSF of *Spirulina* sp's powder using fungi *Aspergillus* sp (FNCL 6088) and lactic acid bacteria *Lactobacillus plantarum* (FNCL 0127). SSF was carried out for 10 days at 35% moisture level. The protein contents of *Spirulina* sp's powder fermented by *L. plantarum* were consistently lower ($p < 0.05$) about 43.28% than compare with the other one about 46.12% (SSF by *Aspergillus* sp) until the end of fermentation. The *Spirulina* sp fermented products contained the highest level of protein after 6 days.

Keywords : fermentation, powder, protein, proximate, Spirulina sp

1. Introduction

Dried *Spirulina* powder was containing essential and non-essential amino acids [1]. The amino acid compounds on *Spirulina* are also one potential source of umami such as glutamates and aspartate. The combination of free glutamates and aspartate are high in amount, those commonly found in fermented and aged foods [2]. Solid state fermentation (SSF), an alternative for submerged fermentation for enzyme production, was found to be more favorable, which can be performed under limited financial and labor requirement [3]. SSF have potentially been used for secondary metabolites, for improve the important component in food products. The utilization of *Spirulina* for umami flavor has been made possible by the process of solid-state fermentation (SSF) which involves fermentation in limited moisture levels [4,5]. The aims of this study was to compared the appropriate microorganism (*Aspergillus* sp FNCL 6088 and *Lactobacillus plantarum* FNCL 0127*)* promoting high yield of protein.

2. Material and Methods

2.1 Cultivation for Spore Production and Inoculum Preparation

Selected fungi and LAB strains : *Aspergillus* sp (FNCL 6088) and *L.plantarum* (FNCL 0127) and re-cultured in Potato Dextrose Broth (PDB) (Oxoid) at 30 $\pm$ 1 °C in medium bottles. The spores was harvested after 4-5 days of cultivation with sterile distilled water containing 0.1% (w/v) peptone water.

The spore suspension at 1 x 10^8 spore /ml and volume of 20% (v/w) was mixed with the prior autoclaved growth medium. This medium was used as inoculum for SSF process.

2.2 Substrate Preparation
Solid-state fermentation (SSF) was carried out in 0.5 L conical flasks containing 50 g substrate with the moisture content adjusted to 35% (autoclaved at 121 °C at 15 psi pressure for 1 h) for a period of 10 days. An inoculum size of 2 x 10^7 spores 50 g^{-1} substrate was used for each flask and incubated at 30 $\pm$ 2 °C. Sampling was done at every 48 h interval starting from days 0 to 10 for both experiments [4] . The substrate was mixed for 5 min prior to sampling process. The sample was used to determine the nutrition composition.

2.3. Fatty acid profile
To analyze the fatty acid composition, fat was extracted according to the method of Folch et al. [6], and 2 mL 0.5N NaOH/methanol was added to 20 mg of fat, which was later saponified for 10 minutes at 105 °C. It was examined after applying 2 mL boron trifluoride/methanol, and methylated. Then, 2-3 mL hexane (HPLC grade) and 2 mL saturated NaCl solution were added. The supernatant of the mixture used the separated funnel was analyzed by gas chromatography (Hewlett Packard 6890 series; Palo Alto, CA, USA). The column was set up with an HP-FFAP capillary column (25 m x 0.32 mm internal diameter, 0.5 μm film thickness); initial oven temperature of 130 °C (1 minute), increased at 2.5 °C/min to a final temperature of 230 °C (10 minutes); injector temperature 230 °C, detector temperature 250 °C; helium carrier gas with a spilt ratio of 20 : 1, and flow rate of 1 mL/min.

3. Result and Discussion

3.1. The proximate contents of Spirulina sp with Solid State Fermentation
Proximate contents of *Spirulina* sp with SSF treatment using isolates culture of *Aspergillus* sp FNCL 6088 and *Lactobacillus plantarum FNCL* 0127 shown in Figure 1.

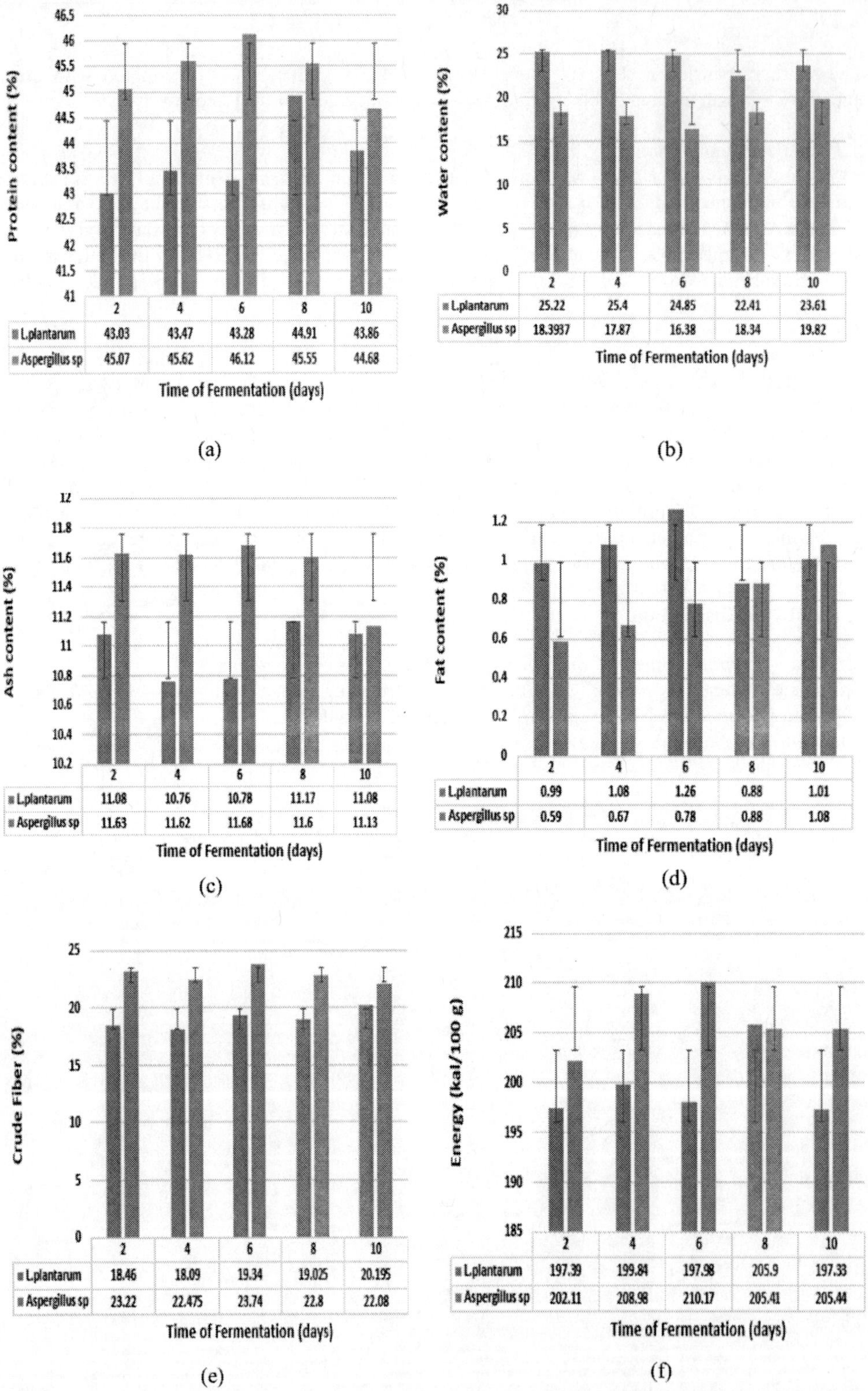

Figure 1. The proximates composition of Dried Powder SSF *Spirulina* sp : (a) protein , (b) water, (c) ash, (d) fat, (e) crude fiber, and (f) energy

The highest protein of 46.12% *Spirulina* sp SSF with *Aspergillus* sp was obtained after 6 days of fermentation process (Figure 1a). Figure 1 (a) showed that the protein content of *Spirulina* sp treated with bacteria increased in significant level. Figure 1 (d) showed that treatment of SSF of *Spirulina* using *Aspergillus* sp generates that fat content more predictable and tends to increase ($p<0.05$) during the SSF as compared to the use of *L. plantarum.* Crude fiber content of *Spirulina* sp on Figure 1 (e), tends to fluctuate. For *Aspergillus* sp optimum value is achieved at day 6 while for *L.plantarum* on the 8th day (Figure 1f). SSF treatment with *Aspergillus* sp tends to produce more higher fat and energy value compared to *L. plantarum* since the occurrence of assimilation between media that *Spirulina* sp growth with microbes, especially *Aspergillus* sp. These microorganism suspected converted fat of *Spirulina* sp into protein. The previous study conducted by [7,8] states that the loss of nutrients of fat in palm oil during SSF is because of the conversion of fats in palm oil into biomass protein.

3.2. Fatty acid composition

The fatty acid composition of *Spirulina* sp treated with SSF are shown in Figure 2.

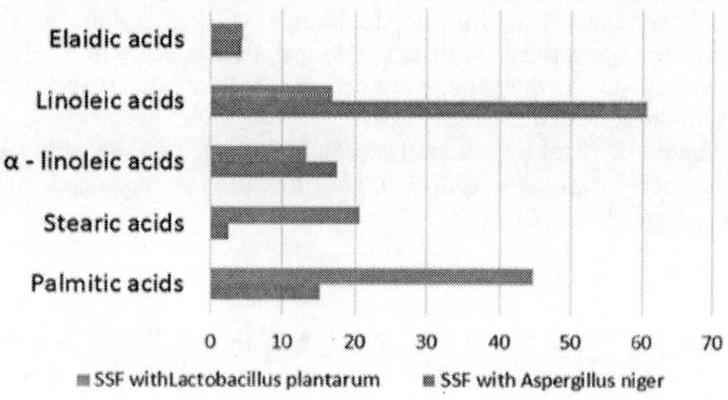

Figure 2. Fatty acids composition of SSF *Spirulina* sp

The *Spirulina* sp SSF with *Aspergillus niger* had the highest unsaturated fatty acids (linoleic acids) concentration at 60.63%, which was significantly higher ($p < 0.05$) than that of SSF with *L. plantarum* (16.93%). On the other hand, the saturated fatty acids composition of *Spirulina* sp powder treated with *L. plantarum* showed highest proportion ($p < 0.05$). This result can be seen as an effect of solid state fermentation. Significant improvement in fatty acid content was detected with reduction of amount of substrate in the fermentation process, because the substrate was used by microbial for growth to produce secondary metabolites. This study was similar with the research studied by Lee et. al [3] stated that time of process in fermentation makes the depth substrate was relatively less than begin.

4. Conclusion

The overall study showed that the 6th days until 10 days of SSF were able to increase the protein content of *Spirulina* sp. *Aspergillus* sp isolate was more potential to improve the *Spirulina* protein value compared to *L. plantarum*. The condition of SSF was suitable for growth of *Aspergillus* sp compared with *L.plantarum.*

5. Acknowledgment

The authors gratefully acknowledge the financial support provided by the Ministry of Research, Technology and Higher Education Republic Indonesia, 2016

6. References

[1] Cepoi, L., Rudi, L., Miscu, V., Cojocari, A., Chiriac, T., Sadovnic, D., 2009. Antioxidative Activity of Ethanol Extracts from *Spirulina platensis* and *Nostoclinckia* Measured by Various Methods. J of Fascicula Biologie 16(2), 43–48.

[2] Mouritsen, O. G.,2015. The Science of Taste. Flavour 4(18), 1–2. DOI :10.1186/si3411-014-0028-3.

[3] Lee, C.K., Darah, I., and Ibrahim, C.O. 2011. Production and Optimization of Cellulase Enzyme Using *Aspergillus niger USM* AI 1 and Comparison with *Trichoderma reesei* via Solid State Fermentation System. Biotechnology Research International ID 658493, 6 pages. Doi: 10.4061/2011/658493.

[4] Andey, A., Radhakrishnan, S., 1992. Packed-bed column bioreactor for production of enzyme. Enzyme and Microbial Technology 14, 486–488.

[5] Miranda, O.A., Salgueiro, A.A., Pimental, N.C.B., Limafilho, J.J., Melo, E.H.M., Duran , N. 1999. Lipase production by a Brazilian strain of *Penicillium citrinium* using industrial residue. *Bioresour. Technol.* 69: 145-149.

[6] Folch J, Lees M and Sloane-Stanley GH. 1957. A simple method for the isolation and purification of total lipid from animal tissues. J Biol Chem ;226:497e509.

[7] Wing Keong, N.G., Lim, H.A., Lim, S., Ibrahim, C.O., 2002. Nutritive evaluation of palm kernel meal pretreated with enzyme or fermented with Trichoderma koningii (Oudemans) as a dietary ingredient for red hybrid tilapia (*Oreochromis* sp.). Aquaculture Research 33, 1199–1207.

[8] Rajesh, N., Imelda, J., Paul-Raj, R. 2010. Value addition of vegetable waste by solid-state fermentation using *Aspergillus niger* for use in aquafeed industry. J. Waste Management. 30, 2223-2227.

Antioxidant activity and total phenolic content in Red Ginger *(Zingiber officinale)* based drinks

Widayat[1], B Cahyono[2], H Satriadi[1] and S Munfarida[3]

[1]Department of Chemical Engineering Diponegoro University
[2]Department of Chemistry Diponegoro University
[3]Magister of Chemical Engineering Department Diponegoro University

Jl. Prof. Soedarto SH Tembalang Semarang 50275 Indonesia

Email : yayat_99@yahoo.com

Abstract. Indonesia is a rich spices country, both as a cooking spice and medicine. One of the most abundant commodities is red ginger, where it still less in application. On the other hand, the level of pollution is higher, so antioxidants are needed to protect the body cells from the bad effects of free radicals. The body can not naturally produce antioxidants as needed, so we need to consume foods with high antioxidant content. The purpose of this study is to know the antioxidant activity and total phenolic content in red ginger (Zingiber officinale) based drinks. Research design with complete randomized design (RAL) with factorial pattern 3 x 3, as the first factor is red ginger extract and water ratio (1: 1, 1: 2 and 1: 3) and second factor is the type of sugar used (cane sugar, palm sugar and mixed sugar). The results of this study indicate that red ginger extract and water ratio of 1: 3 give higher antioxidant. The highest antioxidant obtained in red ginger extract and water ratio of 1: 3 and using mixed sugar. That antioxidants value is 88.56%, it is not significant decreased compared to the antioxidant of pure ginger extract that is 91.46%. For higher phenol total content obtained on syrup that uses palm sugar. The highest phenol total content obtained in red ginger extract and water ratio of 1: 1 and using palm sugar. That total phenol content value is 6299 ppm.

Keywords : Antioxidant activity, phenolic content, red ginger, based drinks

1. Introduction

One of the most abundant commodities in Indonesia is red ginger (*Zingiber officinale)*, and it still less in application. It is an herb plant that has high antioxidant activity. In big cities, the level of pollution is higher, so antioxidants are needed to protect the body cells from the bad effects of free radicals. Free radicals can be obtained naturally and from outside the body. Naturally produced by food digestion. While from outside that is caused by cigarette smoke, polution and radiation. Free radicals that can not be handled by the body, it can trigger chronic diseases, such as cancer and heart disease. The body can not naturally produce antioxidants as needed, so we need to consume foods with high antioxidant content. The benefit of this study is to know the antioxidant activity and total phenolic content in red ginger (Zingeber officinale) based drinks.

In this research was made a drink based from red ginger which rich in antioxidant Antioxidant supplements may not bring about as much as antioxidant benefits in fresh foods. Even some supplements pose a risk. Antioxidant supplements in high doses can bring health risks. More risk will be given if it taken with other drugs. In smokers supplements of high levels beta-carotene can cause lung cancer. High doses of vitamin E may increase the risk of stroke and prostate cancer.

From the previous studies [1], performed supercritical extraction of ginger tubers in which potential compounds were produced. The active compounds are 6-gingerol as much as 238.94 ± 0.79 and 170.50 ± 0.45 mg / g extract, total phenolics of 183.96 ± 1.25 and 0.4 ± 0.72 mg gallic acid / gram extract. Compounds in ginger oil are very large potential in the health / pharmacology field. The pharmacological effects of red ginger are able to strengthen the efficacy of other ingredients that are mixed in the process of drug manufacture [2, 3, 4]. In this study, we were analysis performance and activity of antioxidant in red ginger as a healthy drinks which is the red ginger is mixed with the other spices.

In her study, [5] used ginger extract as an anti-fungal *Candida albicans*. [6] has also conducted a study showing ginger extract as an anti-mutant in mice. [7] conducted the study and looked at the potential of ginger oil as an antioxidant especially gingerol. [8] and [9] used ginger extract as a potential compound as anti-cancer. In this study we would know which is the best condition to extract the antioxidant from the red ginger to get the maximum phenolic compound and to know the effects of adding sugar into the red ginger extract to antioxidant content.

Ginger oil is an essential oil produced from ginger through the distillation process. Characteristics of ginger oil have to fix as the quality standard of EOA (The Essential Oil Association of America).Ginger oil contains two main component groups, namely essential oil and fixed oil. Essential oils in dried ginger root are about 1-3% and these compounds cause a distinctive smelling ginger. The main components in ginger oil are zingiberen and zingiberol, which causes a fragrant odor. The compounds are n-nonilaldehide d-camphene, d-α-phellandrene, metal heptenone, sineol, borneol and geraniol, lineol, acetate, caprylate, citral, chaviol, limonene, and phenol [10, 11, 12]. According to [13] research, red ginger has starch content (52.9%), essential oil (3.9%) and alcohol-soluble extract (9.93%) higher than ginger emprit (41.48 , 3.5 and 7.29%) and elephant ginger (44.25, 2.5 and 5.81%).

2. Methods

2.1. *Material*

In this study used red ginger (*Zingiber officinale*) as the main raw material. In addition, used of side materials such as: sugar, temulawak, lemongrass, cloves, nutmeg, cinnamon and water. As one of the variable, used cane sugar and palm sugar as a sweetener. Water as solvent, cinnamon and lemongrass as a scent maker, while nutmeg, temulawak and cloves as a source of additional antioxidants.Red ginger used is obtained from farmers in the Batang area, central java. While other materials obtained from one of the local traditional market in Semarang.

2.2. *Variables*

The controlled variables in this research are temulawak (1 gr), lemongrass (1 branch), cloves (2 pcs), nutmeg (0.5 gr) and cinnamon (2 cm).The response variable in this study is the antioxidant and phenol total of syrup product. The independent variables in this study were ginger-water ratio (1:1; 1:2; and 1:3) and type of sugar used (cane sugar, palm sugar and mix sugar).

Table 1.Experiment optimization

Red ginger extract (ml)	Water (ml)	Cane Sugar (gr)	Palm Sugar (gr)	Responses
150	150	100		1. Antioxidant
100	200	100		
75	225	100		2. Phenol Total
150	150		100	content
100	200		100	
75	225		100	
150	150	50	50	
100	200	50	50	
75	225	50	50	

2.3. Syrup production

Washed ginger with water, then peeled it. All ginger is smoothed using a scar machine. Separating the ginger juice from the pulp by filtering. Then separate the dissolved starch by depositing it for a moment. Heat the water and add other ingredients such as cinnamon, nutmeg, cloves and lemongrass to boiling. After that, enter water ginger juice until it boil and drain. Separate from the dregs that still carryafter it cold. Then reheat it by adding the sugar until it dissolves and smiles. Take it from the stove and wait until room temperature. Fill in it into the bottle.

2.4. Total Phenolic Determination

2.4.1. Galat Acid Calibration Curves Manufacture. Making galat acid standard curves based on method that Developed by Rungruang and Suwannee (2010). Galat acid solution is made with concentrations of 100, 200, 300, 400 and 500 mg/L. From each concentration is taken 1 mL. Added 10 mL of distilled water and 5 mL Folin-Ciocalteu reagents are whipped up homogeneous. Then the solution is keept For 8 minutes. After hat, 4 mL of solution Na_2CO_3 20% is added, and Shake until homogeneous and keep for 30 minutes at room temperature. The absorbance of each solution is measured at Wavelength 760 nm. Make Calibration curve of the relationship between Concentration of gallic acid with absorbance.

2.4.2. Permination of Total Content Phenolic Compounds. 1 ml of sample is taken and it was added with 10 mL of distilled water And 5 mL of Folin-Ciocalteu reagents. Then the solution is shaken and allowed to stand for 8 minutes. 4 mL of 20% Na_2CO_3 solution added, and shaken Until homogeneous. Keep it for 30 Minutes at room temperature. Determined absorbance at length Wave 760 nm with UV-Vis Spectrophotometry. Gynecology Total phenolate is expressed as mg amount Gallic acid equivalent each gr extract.

2.5.*Antioxidant Activity Test*

2.5.1. Determination of DPPH Maximum Absorption Wavelength. 75 μM DPPH solution was prepared with How to dissolve as much DPPH powder 7.393 mg to 250 mL methanol. Take 5 mL of the solution and Put in vials and bottles Added with 0.2 ml of methanol. Solution Shaken until homogeneous than left for 30 minutes in the dark. The solution Absorption was measured by a UV-Vis spectrophotometer at a wavelength of 400-600 nm. Methanol is used as a blank [14].

2.5.2. Antioxidant Activity Test. Samples were made variations Concentrations of 15, 30, 45, 60, and 75 ppm in Methanol. They has been varied The concentration is taken 0.2 ml into the vial and reacted with 5 ml of 75 μM DPPH solution. The mixture is homogenized and Idle for 30 minutes in dark place. Furthermore, absorbance measurements were read using UV-Vis spectrophotometer on Wavelength 516 nm. Ability To reduce DPPH radicals (% inhibition) Calculated using the equation:

$$\% \, Antioxidant = \frac{Blank\ absorbance - Sample\ absorbance}{Blank\ absorbance} x100\%$$

Then Calculation of IC50 which is Concentration of samples to be able Muffle 50% of DPPH's radical activity. The value of IC50 is obtained from the intersection Line between 50% inhibitory power with Sample concentration [14].

3. Result and discussions

3.1. .Antioxidant

From the diagram can be seen the results that the highest antioxidant found in syrup with the ratio of red ginger and water of 1: 3, then followed by 1: 2 and 1: 1. The results of this study indicate that red ginger extract and water ratio of 1: 3 give higher antioxidant. It obtained in each sugar type variable (cane sugar, palm sugar and mixed sugar). The highest antioxidant obtained in red ginger extract and water ratio of 1: 3 and using mixed sugar. That antioxidants value is 88.56%, it is not significant decreased compared to the antioxidant of pure ginger extract that is 91.46%.

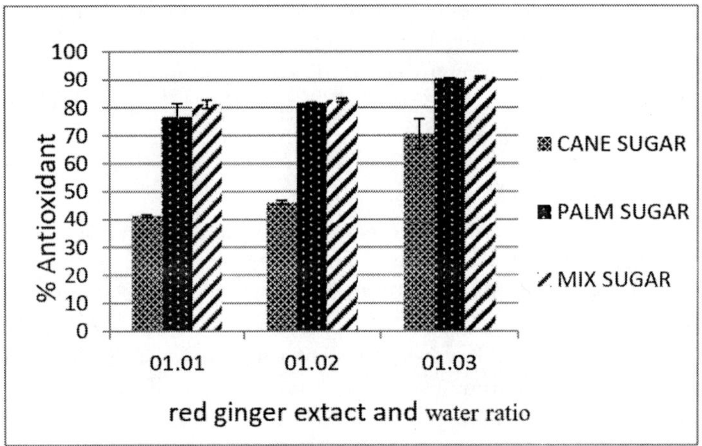

Figure 1. Antioxidant in red ginger syrup product

In the water extract of red ginger, which acts as an antioxidant is a gingerol compound [15]. Gingerol is a compound that is not soluble in water. This is what causes red ginger ratio addition does not increase the antioxidant content. Suspected dissolved antioxidants in the syrup are generated from additional ingredients used. The ingredients include: sugar, temulawak, lemongrass, cloves, nutmeg, and cinnamon. In palm sugar, antioxidants are produced higher than cane sugar because of the presence of phenol compounds contained therein are flavanoid and benzoquinone compounds which are products of Maillard reaction that occur during the sugar-making process [16].

According to [17], the inhibition mechanism of oxidation by antioxidants in two ways, the binding / capture of free radicals (primary antioxidants) and other mechanisms that do not directly capture free radicals (secondary antioxidants), such as binding or chelating metals, oxygen capture, singlet, the conversion of hydrogen peroxide becomes non radical. Meanwhile, according to [18], phenolic compounds have the capacity as primary antioxidants so that the rapid reaction of DPPH radicals occurs with some phenols, therefore the antioxidant capacity of red ginger rhizome water extract, various types of sugar, in red ginger drink in this study measured by analysis of free radical fishing activity (DPPH) and total phenol content.

Antioxidants are also produced from other material addition used. Some researchers claim that these ingredients contain antioxidants. In the water extract of temulawak, which acts as an antioxidant is a water soluble phenol compound [19] that is xanthorrizol [20] which is a primary antioxidant. Research on lemongrass stem was also done by [21], using DPPH method they prove that lemongrass stem extract contains antioxidant. They performed it with mortality tests on shrimp larvae (brine shrimp lethality test). The study was also conducted by [22] and [23] on the content of antioxidants in cloves. According to Dian et al (2013), clove flower oil of 55 μL effectively traps free radicals of up to 60% alkyl, 48.57% hydroxyl, and 35.71% peroxyl. Based on research, the active component in clove oil that has antioxidant activity is eugenol. According to [23] clove leaf extract has activity as a catcher of free radicals or antioxidants and possess phenol content. Along with the antioxidants of nutmeg and cinnamon studied by [24], he claims that the spice contains many antioxidant components, which are known to be rich in phenolic components. High levels of antioxidants obtained in the ratio of ginger and water 1: 3, it contain less ginger than 1:2 and 1:1. Beecause there are more solvent in 1:3 variable,soit allow more dissolved antioxidant compounds.

3.2. *Total Phenolic Content*

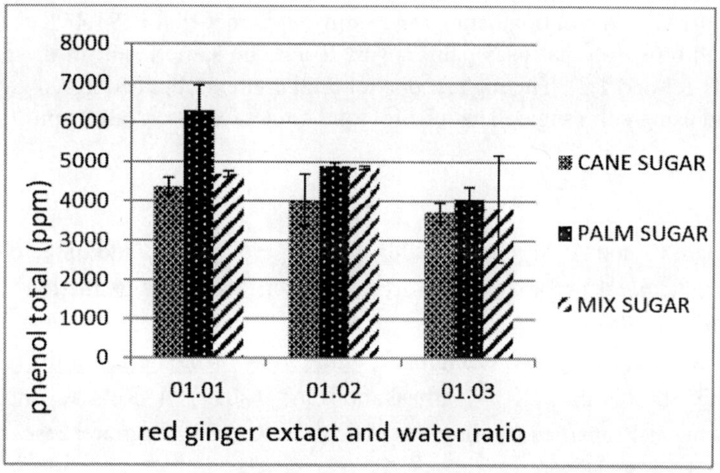

Figure 2. Phenol total content in red ginger syrup product

Based on the research that has been done, from the diagram can be seen that the type of sugar give an effect on the total content of phenols in red ginger syrup. The highest total phenol was obtained on syrup using palm sugar. This occurs in every ginger and water comparison variable (1: 1, 1: 2 and 1: 3). The highest phenol total content obtained in red ginger extract and water ratioof 1: 1 and using palm sugar. That phenol total content value is 6299 ppm. It followed by red ginger and water ratio 1:2 and 1:3 in using palm sugar with the phenol total value 4289 ppm and 4047 ppm. High levels of phenol caused by flavanol and benzoquinon compounds in every sugar contained. The content of flavanols and benzoquinone compounds in palm sugar is higher than sugar. Benzoquinon is a carbonyl compound that plays an active role in the degradation of strikers in the formation of melanoidin in sugar production [25]. Both compounds are the result of maillard reactions that occur during the sugar production.

According to [26], the cane sugar is made from the crystals of sugarcane juice which has the composition of sucrose 8.4-13.4%, inverted sugar 0.2-0.5%, organic acids (carbocyclic, amino) 0.15%, organic substances (Protein, starch, wax, fat, phospholipids) 11-19%. Meanwhile, palm sugar is made from palm sugar which has sucrose composition 13,9-14,9%, carbohydrate 11,28%, and protein 0,02%

[27]. In the manufacture of sugar through evaporation stage with high temperature where can form maillard products in it [28]. Maillard is a reaction between the amino complex (often amino acids, peptides, or proteins) and carbonyl complexes (usually reducing sugars, including glucose, fructose, or lactose). In addition to the influence of the sap composition, the possibility of purification also affects the decrease of flavanols and benzoquinones in sugars. This is due to the decrease in melanoidin levels by the purification process [26].

Processing of sugar through two stages of purification that is when the formation of raw sugar and after dilution back raw sugar. The process of purifying the raw cane sugar solution through two stages, namely the addition of lime and phosphoric acid solution and the activated carbon addition step to remove the color so that the resulting white crystals [29]. In the process of making palm sugar only through the heating or evaporation stage without going through a special purification process as in the process of making sugar. This affects the decreasing levels of brown pigment in sugars which is melanoidin. Therefore, the total phenol of palm sugar is greater than cane sugar.

4. Conclusions

The results of this study indicate that red ginger extract and water ratio of 1: 3 give higher antioxidant. The highest antioxidant obtained on usage mixed sugar that is 88.56%. That value is not significant decreased compared to the antioxidant of pure ginger extract that is 91.46%. For higher phenol total content obtained on syrup that uses palm sugar. It has the same result in the ginger and water ratio variations (1:1; 1:2 and 1:3). The highest phenol total content obtained in red ginger extract and water ratio of 1: 1 and using palm sugar. That phenol total content value is 6299 ppm.

References

[1] Puengphian, C. and A. Sirichote (2008) [6]-gingerol content and bioactive properties of ginger (*Zingiberofficinale*Roscoe) extracts from supercritical CO2 extraction *As. J. Food Ag-Ind. 2008, 1(01), 29-36*

[2] Kemper, Kathi J, MD, 1999, "*Zingiber*", http:/www.mcp.edu/herbal/default.htm

[3] Kumar G., L. Karthik, K. V. Bhaskara Rao (2011) A Review on Pharmacological and Phytochemical Properties of Zingiber officinale Roscoe (Zingiberaceae) *Journal of Pharmacy Research* 4(9),2963-2966

[4] Malhotra, A dan A.P. Sigh (2003), Medicinal Properties of Ginger (Zingiber offinale Roscoe, *Natual Product Radiance*, Vol 2:6

[5] Supreetha S., S. Mannur , S.P., Simon , J.Jain , S. Tikare , dan A. Mahul (2011), Antifungal Activity of Ginger Extract on Candida Albicans:An In-vitro Study, *Journal of Dental Sciences and Research* Vol. 2, Issue 2, Pages 1-5

[6] Nirmala K T, P., Krishna and K. Polasa (2007), In vivo Antimutagenic Potential of Ginger on Formation and Extraction of Urinary Mutagens in Rats, *International Journal of Cancer Research* 3 (3) 134-142

[7] Mishra, A.P., S. Saklani and S. Chandra (2013) Estimation of gingerol content in different brand samples of ginger powder and their anti-oxidant activity: A comparative study, *Recent Research in Science and Technology* 2013, 5(1): 54-59

[8] AL-Sharif, M.M.Z., (2011) Effect of Therapeutic Ginger on Genotoxic of Taxol Drug (Anti-Cancer) in Bone Marrow Cell of Male Mice Journal of Life Sciences 5 (2011) 897-905

[9] Bhanot B., R. Sharma, M. N. Noolvi (2011) Natural sources as potential anti-cancer agents: A review, *International Journal of Phytomedicine* 3 (2011) 09-26

[10] Guenther, E. 1987, "*Minyak Atsiri.*", translated by S. Ketaren, Jilid I, Jakarta: UI Press Hastuti, P., 2007. *Komposisi Bahan Makanan dan Sifatnya*. http://images.gizi kesehatan 07.multiply.multiplycontent.com/. (Diakses tanggal 7 November 2009).

[11] Kamaliroostaa, Z., L. Kamaliroosta b, A. H. Elhamirad (2013)Isolation and Identification of Ginger Essential Oil *Journal of Food Biosciences and Technology,* 3, 73-80, 2013

[12] Kizhakkayil J., and B. Sasikumar (2012), Characterization of ginger (*Zingiberofficinale*Rosc.) germplasm based on volatile and non-volatile components African Journal of Biotechnology Vol. 11(4), pp. 777-786

[13] Hernani and E. Hayani. 2001. Identification of chemical components on red ginger (Zingiber officinale var. Rubrum) by GC-MS. Proc. International Seminar on natural products chemistry and utilization of natural resources. UI-Unesco, Jakarta : 501-505

[14] Molyneux, P. 2003.*Theuse of the stable free radikal diphenylpicrylhydrazyl (DPPH)for estimating antioxidant activity*. Journal Science of Technology. 26(2):211-219.

[15] Yeh, H., Chuang, C., Chen, H., Wan, C., Chen, T., Lin, L. 2014. *Bioactive components analysis of two various gingers (Zingiberofficinale Roscoe) and antioxidant effect of ginger extracts*. LWT - Food Science and Technology 55: 329- 334

[16] Nursten, Harry. 2005. *The Maillard Reaction, Chemistry, Biochemistry and Implications*. Royal Society of Chemistry; Atheneum Press Ltd, Cambridge, UK.

[17] Pokorny, J., Yanishlieva, N,.and Gordon, M. 2001. *Antioxidant in Food*. CRC Press Cambridge. England.

[18] Ningrum, S.S. 2010. Kapasitas Antioksidan Minuman Temulawak (*Curcuma Xanthorrhiza*Roxb.) Menggunakan Gula Kristal Putih, Gula Kristal Merah, Gula Merah, Dan Gula Aren. Sebelas Maret University. Surakarta

[19] Cowan, M. M., 1999. *Plant Products as Antimicrobial Agents*. Clinical Microbiology Reviews Vol.12 No.4.

[20] Anonim, 2010. *Curcumin Tincture (tinctura Curcuma xanthorriza)*. http://pdfsearch engine.com/curcuminoid-compounds-curcuma-xantorrhizapdf. html. (Diakses tanggal 10 Juli 2010).

[21] Hendrik W.G.,Erwin, And Panggabean, A.S., 2013. *Pemanfaatan Tumbuhan Serai Wangi (Cymbopogon Nardus(L.) Rendle) sebagai Antioksidan Alami*. Jurnal Kimia Mulawarman. 10 (2): 74-79.

[22] Nurjannah, D,A,. Retnowati, R. And Juswno, U.P., 2013. AktivitasAntioksidan Dari Minyak Bunga Cengkeh (*Syzygium aromaticum*) Kering Berdasarkan Aktivitas Antiradikal yang Ditentukan Menggunakan *Electron Spin Resonancei*. Kimia.Student journal. 1(2) :283-288

[23] Rorong, J.A. 2008.*Uji Aktivitas Antioksidan Dari Daun Cengkeh (Eugenia carryophyllus) Dengan Metode DPPH*. Chem. Prog. 1 (2):111-116.

[24] Nely, F. 2007. *The Activity of Antioxidant from Spices Market and Powder Spice Factory Using Polyphenolic Method and Active Oxygen Method.* Thesis. Institut Pertanian Bogor.

[25] Sekarini, G.A (2011) *Kajian Penambahan Gula dan Suhu Penyajian Terhadap Kadar Total Feno, Kadar Tanin (Katekin) dan Aktivitas Antioksidan pada Minuman Teh Hijau (Camellia sinensis L.)*. Thesis. Sebelas Maret University

[26] Hugot, E. 1986. *Hand Book Of Cane Sugar Engineering, Third edition.* Elseiver Amsterdam – Oxford – New York

[27] KSU Sukajaya. 2005. *Pengolahan, Produksi, dan Pemasaran Gula Aren*. Bahan Presentasi, Rangkasbitung. Banten

[28] Anonim, 2009. *Sugar Food-Info of Wageningen University Netherlands*. http://www.food-info.net/. (Diakses tanggal 10 November 2009).

[29] Christy, D., 2006. *Invention Edition : Sugar Invention*. PT Elex Media Komputindo Kelompok Gramedia, Jakarta. (In Bahasa Indonesia)

Carcass and Meat Quality Pelung Sentul Kampung Broiler Crossbreed Chicken

S Darwati[1], R Afnan[1], S Prabowo[1] and H Nurcahya[2]

[1]Department of Animal Production and Technology, Faculty of Animal Science IPB
[2]Department of Biology, National University

Email : darwatisri@rocketmail.com

Abstract. Crossbreed chicken of pelung sentul kampung broiler (PSKR) has good growth and ready to slaughter at the age of 10 weeks. So, it has potential as a local chicken for meat producers. Potential of PSKR crossbreed chicken need to know about the percentage of carcass and the physical quality of meat for holistic information. This study aimed to evaluate the carcass and the quality of the physical meat of pelung sentul kampung broiler chicken (PSKR). Material of 12 chickens PSKR 12 weeks unsexing were used and observed for the percentage of carcass in the chest, upper and lower thighs and physical quality of breast meat included pH, water-binding power, cooking impurities, and tenderness. Chickens fed 100% commercial feed for broiler chicken phase starter until age 3 weeks, then gradually added rice bran and age > 5 weeks fed 60% commercial feed plus 40% rice bran. Chicken is slaughter at 12 weeks of age. The data obtained are presented descriptively. Percentage of PSKR carcass was 68%, chest was 27.17%, upper thigh was 17.12%, lower thigh was 16.64% respectively. Physical quality of breast meat has a pH performance of 5.30,% mgH2O of 28.08%, cooking loss of 29.13%, and tenderness of 2.63 respectively. PSKR chicken had potential for meat producers based on carcass percentage with chest meat was very tender because the genetic of broiler in PSKR as much as 25%.

Keywords: carcass, meat quality, PSKR

1. Introduction

The demand for poultry meat increases as the public awareness raises the importance of animal protein for health. Currently poultry meat needs are supplied by broiler products, while local chicken meat production is ranked second after broiler.

Efforts to increase the potential of local chickens as meat producers can be done through selection and crossing. The cross breeding of pelung sentul kampung broiler so that obtained by PSKR chicken is one of effort to increase the productivity of local chicken. PSKR chickens have better growth performance than chicken can reach 1.2 kg at 10 weeks of age (Darwati et al. 2017). PSKR chicken has potential as a producer of meat from local origin. PSKR chicken has a genetic composition of 75% local chicken and 25% broiler chicken.

The advantage of PSKR crossbreed chicken can be slaughter at a young age as a meat producer. The higher carcass obtained from PSKR chickens can therefore be expected to provide information that local chickens can contribute as a producer of poultry meat. Meat is one of the important farm commodities to meet the needs of animal protein community. Poultry is a good source of animal protein because it contains complete and balanced essential amino acids. BPS (2016) reported that in 2016 the production of broiler meat is 1 689 584 tons and local chicken production is 315 538 tons. Until now the needs of chicken meat in Indonesia is much supplied by chicken meat broiler. Besides carcass, the study of the quality of PSKR chicken meat will be better. By the need to continue to develop research to explore the potential of local chicken as a producer of good quality meat. This study aimed to evaluate the carcass and physical quality of PSKR chicken.

2. Materials and method

Chicken samples used in this study were 12 chickens PSKR unsexing age 12 weeks. Chickens were kept from DOC until 12 weeks of age. Chickens fed 100% commercial feed for broiler starter to 3 weeks of age, > 3-4 weeks were given 80% commercial feed broiler phase starter + 20% rice bran, > 4-5 weeks fed 70% commercial feed broiler phase starter + 30% rice bran,> 5-12 weeks fed 60% commercial feed broiler phase starter + 40% rice bran. Chicken is slaughtered to obtain carcass percentage at 12 weeks of age. Drinking water is given ad libitum. Vita chicks is given at DOC until the age of 2 weeks. Furthermore given conditional.

Chickens are cut in accordance with Islamic Shariah in the Field Laboratory of Breeding and Animal Genetics of IPB. Definition of carcass is the remains of plucked and dressed chicken after the legs, breasts and wings have been removed. Commercial pieces of carcasses are divided according to commercial pieces. The commercial cut is weighed. The commercial carcass parts observed include the chest, upper thigh, and lower thigh to obtain commercial carcass percentages of the chest, upper thighs, and lower thighs. Carcass of chest section was skinned and trimed. Furthermore, breast meat analysed physical qualities include meat pH, water holding capacity, cooking loss, and tenderness.

pH measurements of meat based on Van Laack et al. (2000). Calibrated pH meter was pinned into sample of meat, then wait until the pH meter screen shows the pH value of the meat. Subsequently, water holding capacity (WHC) refers to Soeparno (2005). The measurement of water holding capacity of meat was done by carper press and planimeter. 0.3 g of meat samples were placed on filter paper and covered with other filter paper on top then the samples were pressed with carper press maximally (kg cm -2) until the outer circle and inner circle formed, then calculate total H2O content were out of the sample. Afterward, cooking loss is the amount of water lost and nutrients dissolved in water due to cooking. Measurement of cooking loss is based on Bouton et al. (1971). The meat samples were weighted, then boiled until the meat temperature reached 81 ° C. Meat drained to constant weight. Cooking loss is calculated from the gap of weight before and after cooking then compared to the weight before cooking in percentage. Afterward the tenderness level is based on Suryati et al (2008) that the level of meat tenderness is expressed by the amount of strength (kg cm-2) required to cut the meat and shown by the needle pointer of Warner Bratzler meat device that moves above the scale with a measurement sensitivity of 0.1 kg cm^{-2}. The data obtained were analysed descriptively.

3. Results and discussion

3.1. Percentage of carcass

The result of carcass was obtained as shown in Table 1 in this study. Carcass is the remains of plucked and dressed chicken after the legs, breasts and wings have been removed. Carcass performance includes percentage of carcass weight, and percentage of chest, tibia and femur which have high commercial value in chicken carcass. Carcass is a slowly ripe organ, so with increasing age, growth increases and the percentage of live weight increases (Matitaputty et al. 2011). Moreover, genetic composition is influence the carcass crop. PSKR crossbreed chicken has 25 % genetic potential of each Pelung, Sentul, Kampung, and Broiler.

The percentage of chicken carcass of PSKR is higher than research of Pesti and Bakali (1997) that the percentage of broiler carcass of 5 weeks harvest age is 60.52% -69.51%. Means PSKR chicken has the ability as a meat-producing chicken with the genetic composition of 25% broiler and 75% of local chicken in chicken PSKR.

Table 1. Percentage of Carcass, chest part, top thigh, and bottom thigh PSKR crossbreed chicken PSKR

Traits	Mean±Sd (C.V. %)
Carcass (%)	68.06±3.10 (7.86)
Chest part (%)	27.17±1.84 (3.68)
Top thigh (%)	17.12±0.60 (3.51)
Bottom thigh (%)	16.64±0.50 (3.00)

The percentage of chest carcass in this study was higher than the research of Lubis (2017) that the chicken carcasses age of 26 weeks on the chest 25.51%. While in this study amounted to 27.17%. In contrast, upper thigh carcass and lower thighs were lower than Lubis (2017) ie 19.5% and 18.3%. Chest is a place of the largest muscle deposits in chickens, thus PSKR chicken have meat on the carcass which has high economic value.

3.2. Quality of carcass

Result of PSKR meat test for PSKR meat quality indicator. The quality of PSKR chicken is presented in Table 2.

Table 2. Thc quality of carcass crossbreed chicken PSKR

Meat quality	Average	Range
Ph	5.30±0.08	5.17-5.46
%mgH_2O	28.08±2.72	24.00-34.53
Tenderness	2.63±0.34	2.1-2.3
% Cooking loss	29.13±6.11	16.00-36.50

The pH parameters affect the quality of meat sensory, ie color, texture / tenderness, and taste (Min and Ahn 2005). In this study the average pH of PSKR meat was 5.30. Van Laack *et al.* (2000) stated that the pH value of broiler chicken meat ranged from 5.96 to 6.07, Dewi (2013) reported that chicken meat pH ranged from 5.91 to 5.93.

The water holding capacity represents the percentage of the amount of water discharged. Water retained in the muscle is negatively correlated with the water holding capacity. PSKR breast meat has a water holding capacity in this study of 28.02% lower than Winarso (2003) that kampung chicken meat has 29.94% water holding capacity. Suradi (2006) exposed that water holding capacity broiler chicken was in the range 17.8-45.37%, with the result of that PSKR still in the range of broiler chicken. Soeparno (2011) states that water-binding power is the ability of meat proteins to bind or retain water content in response to external force applications such as cutting, cooking, and meat milling. Water-binding power is also associated with meat pH (Alvarado *et al.* 2007). A high pH value can improve the water holding capacity. Pearson and Young (1989) suggest that water will increase if the pH value of the meat increases.

The mean of cooking loss in this research was 29.13%. Winarso (2013) reported the cooking loss of chicken meat in the village by 16.62%. Means cooking loss of chicken breast meat PSKR was in the normal range.

The tenderness of PSKR chicken meat was extremely tender. Suryati *et al.* extremely tender (<3.3 kg cm^{-2}), tender (3.3-5.0 kg cm^{-2}), rather tender (5.0-6.71 kg cm^{-2}), rather tough (6.71-8.42 kg cm^{-2}), tough (8.42-10.12 kg cm^{-2}), and extremely tough (>10.12 kg cm^{-2}). Soeparno (2005) mentioned that the factors that influence meat tenderness are antemortem, genetic, age, sex, and stress level.

4. Conclusion

Carcass crop of PSKR crossbreed chicken in the range of average carcass crop of broiler chicken, so PSKR crossbreed chicken has potential as a local chicken for meat producers. Meat quality of PSKR crossbreed chicken was classified into extremely tender.

5. References

[1] Alvarado C, McKee S. 2007. Marination to improve functional properties and safety of poultry meat. J Appl Poult Res. 16:113-120.

[2] [BPS] Badan Pusat statistik. 2016. *Populasi Unggas Berdasarkan Provinsi dan Jenis Unggas*. Jakarta (ID): BPS RI

[3] Darwati S, Sumatri C, Said Z, Mutopa A, Afnan R, Prabowo S. 2017. Pertumbuhan generasi kedua hasil persilangan ayam lokal dengan ayam ras pedaging (PSKR dan PSRK) umur 1-10 minggu. *Prosiding Seminar Nasional Teknologi dan Agribisnis Peternakan (Seri IV)*. Purwokert (ID): Fakultas Peternkan Universitas Sudirman dan Ikatan Sarjana Peternakan Indonesia.

[4] Dewi SHC. 2013. Kualitas kimia daging ayam kampung dengan ransum berbasis konsentrat broiler. *Jurnal Agri Sains* 6: 1-8.

[5] Lubis NA. 2017. Kualitas karkas dan sifat daging ayam kampung serta persilangannya dengan ayam broiler bibit [skripsi]. Bogor (ID): Fakultas Peternakan IPB.

[6] Mattitaputty PR, Noor RR, Hardjosworo PS, Wijaya CH. 2011. Performa karkas dan nilai heterosis itik alabio, cihateup dan hasil persilangannya pada umur delapan minggu. JITV. 16:90-98.

[7] Min B, Ahn DU. 2005. Mechnism of lipid oxidation in meat and meat products-a review. J Food Sci Biotech. 14:152-163.

[8] Pesti GM, Bakalli RL. 1997. Estimation of composition of broiler carcass from their specific gravity. Poult Sci. 76 (7): 948-951

[9] Soeparno. 2005. *Ilmu dan Teknologi Daging*. Cetakan ke-4. Yogyakarta (ID): Gadjah Mada University Pr.

[10] Soeparno. 2011. *Ilmu dan Teknologi Daging.* Ed ke-4. Yogyakarta (ID): Universitas Gajah Mada Pr.

[11] Suradi, K. 2006. Perubahan Sifat Fisik Daging Ayam Broiler Post Mortem Selama Penyimpanan Temperatur Ruang (Change of Physical Characteristics of Broiler Chicken Meat Post Mortem During Room Temperature Storage). Jurnal Ilmu Ternak. Volume 6. 1: 23-27

[12] Suryati, Arief II, Polii. 2008. Korelasi dan kategori keempukan daging berdasarkan hasil pengujian menggunakan alat dan panelis. *Anim Prod.* **10**:188-193.

[13] Van Laack RLJM, Liu CH, Smith MO, Loveday HD. 2000. Characteristics of pale, soft, exudative broiler breast meat. *Poult Sci.* 79: 1057-1061.

[14] Winarso D. 2003. Perubahan karakteristik fisik akibat perbedaan umur, macam otot, waktu dan temperatur perebusan pada daging ayam kampung. *J Indon Anim Agric*. 28: 119-132

IOP Conf. Series: Earth and Environmental Science **102** (2017) 012027 doi:10.1088/1755-1315/102/1/012027

Combination of ozone and packaging treatments maintained the quality and improved the shelf life of tomato fruit

Zainuri[1*], Jayaputra[2], A Sauqi[2], T Sjah[2] and R Y Desiana[1]

[1] Faculty of Food Technology and Agroindustry, University of Mataram, Lombok, INDONESIA
[2] Faculty of Agriculture, University of Mataram, Lombok, INDONESIA

*Email: nzainuri2@gmail.com

Abstract. Tomato is very important vegetable crop but has short shelf life. The objective of this research was to determine the effect of ozone and packaging combination treatment on the quality and the storage life of tomato fruit. There were six treatments including: control (without ozone and packaging); without ozone and packaged with polyethylene bag; without ozone and polyethylene terephtalate punnet; with ozone but without packaging; with ozone and packaged with polyethylene bag; and with ozone and polyethylene terephtalate punnet. Each treatment was made into 3 replications. Tomato samples were harvested at turning stage. Ozone treatment was applied for 60 seconds. Tomatoes were then treated with and without packaging. The fruit were then stored at room temperature for up to 12 days. The parameters for assessment were water content, color, texture, weight loss and the population of naturally contamination *Escherichia coli.* Each parameter was assessed on day 0, 6 and 12 of storage. The results indicated that combination of ozone and packaging treatments significantly affected the physical and biochemical changes (water content, color, texture and weight loss) of the fruit, suppressed the microbiological contamination on the fruit and maintained fruit freshness or quality after 12 days of storage. The combination of ozone and perforated polyethylene packaging treatment was the best treatment to maintain the quality and prolonged the shelf life of tomato fruit to be 12 days at room temperature.

Keywords: Ozone, Packaging, Tomato, Quality

1. Introduction

Tomato *(Lycopersicum esculentum)* is one of the most widely grown horticultural crops in Indonesia including in West Nusa Tenggara. The production of tomato fruit in the last 5 years has steadily increased. For example, the production in 2012 was 25,045 tons which then increased to 28,839 tons in 2013 [1] and became 36,943 tons in 2015 [2].

Tomato fruit is a very important source of nutrition including vitamins and minerals such as vitamin C, pro-vitamin A and others. Most importantly, tomato fruit is rich of lycopene which is a strong antioxidant that very important for health [3]. The benefit of being rich of nutrition is the reason for high demand of fresh tomatoes. Therefore, sufficient amount of fruit production and continuous supply of fresh tomato fruits are very crucial in order to fulfil the demand.

On the other hand, tomato fruit is a perishable product and has limited shelf life after harvest. As the fruit is categorized into the climacteric vegetable-fruit crop, there is a huge change in the biochemical process in the fruit including high respiration rate, ethylene production and the degradation of some chemical compounds in the fruit. This is in turn not only caused changes in the physical characteristic of the fruit but also losses of nutrients such as vitamins. In addition, the degradation of complex chemical compounds i.e. starch into simple sugar and high water content also provide suitable condition for microbial growth which then lead to fruit decay [4].

Therefore, it is very crucial to develop postharvest technology for fresh fruit to extend fruit shelf life. Creating optimal postharvest condition including low temperature storage, suitable packaging and waxing, are among the potential postharvest intervention methods that can be applied to maintain fruit quality and prolong storage life of the fruit. The use of ozone is another potential method for improving the postharvest life of fruits and vegetables.

Ozone (O3) is a strong oxidant that has potential role as disinfectant agent that can kill microorganisms including bacteria, virus and fungi. Ozone is also an environmentally safe technology that can be applied as postharvest control mean for handling fruits and vegetables [5]. The use of ozone in postharvest technology to improve postharvest storage life for fruits and vegetables has been developed and become more important in the last ten years. Several research have demonstrated that ozone treatment was able to maintain product quality [6]. Results from some research also show that ozone has an important role in reducing microorganism contamination and pesticide residual effect on fruits and vegetables. Other research also reported that treatment using ozone reduced fungal contamination on raspberries, strawberries and blueberries [7]. Ozone also was effective in suppressing *Escherichia coli* growth on chicken [8]. The mechanism of ozone in reducing microbial growth is through modifying cell wall permeability that may cause microbial cell lysis and resulted in microbial inactivation. The inactivation in turn prolong the shelf life of the products [9].

Beside ozone treatment, tomato fruit quality may be further improved with packaging treatment. As mentioned earlier, packaging is very important aspect in postharvest handling technique for fruits and vegetables as it can prevent or reduce damage both physical or mechanical and physiological damage for the product during postharvest handling [10].

Previous research indicated that combination of ozone treatment and packaging can extend product shelf life. Ozone treatment for 60 seconds and combined with polyethylene bag packaging maintained apple fruit quality for 21 days of storage period [11]. Handayani (2016) also reported that ozone treatment for 30 seconds and polietilen packaging treatment maintained chili quality [12]. Other research also shows that treating mango fruit with ozone for 5 minutes maintained fruit quality for 14 days at room temperature [13]. Previous research on tomato indicated that ozone treatment for 45 seconds extended the fruit shelf life for 10 days at 25°C [14]. More research on tomato fruit is needed especially for the room storage condition as a real condition during supply chain in tropical countries like Indonesia. Therefore, this research was carried out to determine the effect of ozone and packaging combination treatment on the quality and the storage life of tomato fruit.

2. Materials and Methods

2.1. Fruits

The tomato fruit cultivar of 'Servo' was harvested at turning stage [15] at the farmer's field in Penedagandor village, East Lombok. The fruits were immediately transported to the Faculty of Food Technology and Agroindustry University of Mataram by road transportation. After arrival at the Quality Control laboratory, the fruits were cleaned and sorted based on the size, color and the present of defects or blemish. Uniform size and maturity level, and free of blemish or physical damage were used for the experiment. The fruits were divided into 18 units as there were 6 treatments (no ozone and no packaging; no ozone and Polietylene (PE) bag; no ozone and Polietylene Terephtalate (PET) punnet; with ozone no packaging; with ozone and PE; and with ozone and PET punnet) and each treatment was made into 3 replications.

2.2. Ozone treatment

Ozone treatment was applied using ozone generator (Tiens *Fruit and Vegetable Cleaner,* model TR-YCA, Output ozone 150 mg/h ± 20%, Tianjin Tianshi Biongfineering Co., Ltd, China). Nine units of the tomato samples were treated with ozone for 60 seconds. After treatment, the fruits were air dried.

2.3. Packaging treatment
Tomato samples (9 units) were either packed according to the treatments i.e. packing with Polietylene 0.05 mm bag (3 units) or Polietylene Terephtalate punnet 0.06 mm (3 units) or left unpacked (3 units). There were two lots of trial regarding the packaging. In the first trial, the packaging both PE bag and PET punnet were without perforation, whereas in the second trial all packaging were perforated (8 holes for each bag or punnet).

2.4. Storage and assessment
The treated and untreated tomato samples were stored at room temperature for up to 12 days for quality and shelf life assessment. The assessment was carried out at day 0, 6, dan day 12 of storage.

2.5. Parameter and statistical analysis
Parameters that were analyzed were water content [16], weight loss [17], texture using *hand penetrometer* [18], color using *colorimeter* (MSEZ *User Manual*) and accessed by panelists and the population of *Escherichia coli* using MPN method. The *E. coli* was assessed as it is one of the indicators for sanitation condition [19]. The data obtained from the trial were analyzed using *Analysis of Variance* at 5% significant level using Co-Stat software. When there was significant different, they were analyzed using honestly significant different test at the same significant level [20].

3. Results and Discussion
3.1. Effect of treatments on water content
Water content is a very important component of fruit and vegetables in general including in tomato fruits as it plays a significant role on product quality. Data obtained from this research indicated that the water content of tomato fruits was not altered by the treatment up to 6 days of storage period. In general, the water content of tomato samples was about 94%. However, as the storage prolonged to 12 days at room temperature, the ozone and packaging treatment had significant effect on water content. As shown in Table 1, tomato fruits treated with ozone and packaging either with PE or PET had higher water content compared to the untreated ozone and without packaging. Combination of ozone treatment and PE packaging performed the best result which maintained the fruit water content of 95.58% compared to the control (92.57%). Low water content of the untreated samples may possibly be associated with high water loss in the samples. Storing the fruit without packaging may create optimal atmosphere condition around the product which trigger high respiration rate [21]. Water produced from the respiration then evaporated through transpiration which lead to high water loss from the sample. With the use of packaging on the other hand, there may be relatively low oxygen concentration around the fruit which associated with the permeability of packaging film to oxygen, and therefore resulted in low respiration rate, which may then lead to low transpiration rate and water loss.

Table 1. Effect of ozone and packaging on the water content (%) of tomato fruits during storage

Treatments	Water content (%) during storage					
	Day 0		Day 6		Day 12	
	NP	P	NP	P	NP	P
No Ozone, no packaging (T1)	94.41	94.49	94.23	93.58	93.84b	92.57b
No Ozone, PE (T2)	94.4	93.85	94.55	94.79	95.05a	94.47ab
No Ozone, PET (T3)	93.60	94.08	94.4a	94.61	94.77ab	94.03ab
Ozone + no packaging (T4)	94.36	93.82	94.35	94.65	94.07ab	94.41ab
Ozone + PE (T5)	94.06	94.31	94.83	94.99	94.84ab	95.58a
Ozone + PET (T6)	94.46	94.00	94.47	94.38	94.69ab	94.68a

Note: Numbers followed by the same letters in the same column were not significant different using HSD at 5% significant level.

3.2. Effect of treatments on weight loss

Weight loss is one of indicators that can be used to monitor changes in fruits and vegetable quality during storage. In general, weight loss occurs in all horticultural products during storage with different rate depending on the product characteristics and postharvest handling technique [21]. As can be seen in Table 2, the treatment of ozone and packaging with either PE or PET reduced weight loss of tomato fruit significantly compared to non-treated ozone and un-packed fruit.

Table 2. Effect of ozone and packaging on the weight loss (%) of tomato fruits during storage

Treatments	Weight loss (%) during storage			
	Day 6		Day 12	
	NP	P	NP	P
No Ozone, no packaging	11.71b	14.12b	18.32a	17.94b
No Ozone, PE	0.96c	1.20c	5.93b	2.28c
No Ozone, PET	1.86c	3.01c	8.89b	5.47c
Ozone + no packaging	15.91a	18.81a	18.89a	25.85a
Ozone + PE	0.78c	1.31c	6.79b	2.86c
Ozone + PET	1.96c	3.78c	7.60b	5.75c

Note: Numbers followed by the same letters in the same column were not significant different using HSD at 5% significant level.

Data in Table 2 also show that there was an increase in the amount of weight loss when the storage period was prolonged. This may be associated with the changes in the cell wall permeability of tomato fruits [22]. As the fruit becomes ripe or even over-ripe which soon turn to the senescence stage, the cell wall is more permeable to oxygen, carbon dioxide and water vapor and cause higher transpiration rate which then lead to higher water loss from the fruit [23, 24]. Suitable postharvest handling technology is very important to suppress or delay water loss from the product. Packaging, for example, plays significant role in atmosphere modification around the products, thus can reduce the rate of respiration and transpiration in the fruit. As Purwadi *et al.* (2007) said that thicker packaging has lower permeability to oxygen, carbon dioxide and water vapor and cause low transpiration rate and lower water loss from the fruit [25].

3.3. Effect of treatments on texture

Texture of fruits and vegetables in general, is closely associated with the metabolism reaction and biochemical degradation of certain chemical compounds (non-soluble pectin into the soluble form) in the fruit after harvest and during ripening process. The texture of mature fruit that is usually firm or hard becomes soft when the fruit start to ripen. Modification of environmental condition around the fruit such as the use of packaging or managing the condition that prevent or reduce the biochemical reaction rate may delay the change in fruit texture. Result from this trial indicated that the combination of ozone and packaging treatment significantly affected tomato fruit texture after 112 days at room temperature storage (Table 3). Ozone treated and packaging with either PE bag or PET punnet had firmer texture of the samples indicated by higher retention value from the penetrometer reading compared to the untreated ozone and without packaging of tomato samples. As mentioned earlier, packaging may provide lower gas environmental condition around the fruit which lead to lower respiration rate and other biochemical degradation in the fruit especially the chemical compounds related to the texture of the fruit. This may then resulted in firmer texture of the ozone treated and packed fruits than the untreated fruit.

Table 3. Effect of ozone and packaging on the texture (kg/cm^2) of tomato fruits during storage

Treatments	Texture (kg/cm^2)					
	Day 0		Day 6		Day 12	
	NP	P	NP	P	NP	P
No Ozone, no packaging	1.45	1.30	0.45	0.73	0.32b	0.28b
No Ozone, PE	1.42	1.70	0.63	1.06	0.55a	0.70a
No Ozone, PET	1.45	0.89	0.60	0.82	0.58a	0.58ab
Ozone + no packaging	1.40	1.53	0.59	1.31	0.57a	0.73a
Ozone + PE	1.55	1.63	0.53	0.76	0.52a	0.90a
Ozone + PET	1.20	1.45	0.60	0.82	0.60a	0.73a

Note: Numbers followed by the same letters in the same column were not significant different using HSD at 5% significant level.

3.4. *Effect of treatments on fruit color*

Color is an important indicator of fruit ripening stage and product quality. Measuring tomato sample color using *Colorimeter* provide information on the L*, a*, and b* which represent the lightness and the changes in the degree of green to red color of the fruit. Results from this research showed that there was no significant different of the lightness and °Hue value (obtained from a* and b* value) of the samples from all treatments. However, after 12 days storage the ozone treated and packaging either PE or PET tend to have lower L value which may related to the lightness degree of red color on the samples. Results from the organoleptic test for the color change (Table 4) also indicated that ozone treated fruit packed with PE packaging had relatively lower score (3.0-3.2 = reddish yellow) compared to the untreated ozone and no packaging treatment (3.4-3.5 = red). In general, there was huge change on color of the fruit during storage from yellowish green (score around 2) to reddish yellow or red color (score of 3.5). This may be associated with the degradation of chlorophyll which represents green color on the fruit and the synthesis of lycopene and other compounds that resulted in red color as the fruit ripen [26].

Table 4. Effect of ozone and packaging on the color (Lightness and Hue) of tomato fruits during storage

Treatments	Lightness (L*)			°Hue			Color (score)		
	Day 0	Day 6	Day 12	Day 6	Day 12	Day 12	Day 0	Day 6	Day 12
T1	64.24	63.97	64.07	174.30	73.27	72.28	2.8	3.4	3.5
T2	64.41	63.87	61.18	143.03	74.12	69.73	2.2	2.9	3.2
T3	65.29	66.18	64.39	111.72	73.81	71.35	2.4	3.0	3.0
T4	62.87	62.09	61.29	115.49	70.98	68.99	2.6	2.8	3.4
T5	64.99	62.94	58.85	174.33	74.31	64.87	2.3	2.7	3.0
T6	59.53	57.19	56.72	113.53	64.12	61.64	2.5	2.4	3.3

3.5. *Effect of treatments on the population of Escherichia coli*

Free of microbial contamination is a very crucial aspect in food quality. Keeping tomato fruits in unsuitable condition such as excessive humidity will trigger microbial growth which lead fruit decay and not safe for the consumers [27]. Fruit may also be contaminated by microorganism through soil in the field or contaminated water used for cleaning the fruits. Some research have demonstrated that ozone treatment has powerful effect in reducing microbial contamination on foods including fruits and vegetables [28, 29, 30]. Results from this research also proved that ozone treatment combined with packaging treatment were effective in reducing *E. coli* contamination on tomato fruit as indicated in Table 5.

Table 5. The effect of ozone and packaging on the freshness (score) of tomato fruits during storage

Treatments	Population of *E. coli* (MPN/gr) during storage					
	Day 0		Day 6		Day 12	
	NP	P	NP	P	NP	P
No Ozone, no packaging	0	0	2	0	18	1
No Ozone, PE	0	0	0	0	0	0
No Ozone, PET	0	0	0	0	0	1
Ozone + no packaging	0	0	0	0	1	0
Ozone + PE	0	0	0	0	0	0
Ozone + PET	0	0	0	0	0	0

4. Conclusion

Ozone and packaging have an important role in postharvest handling of fruit and vegetables including tomatoes. Based on the results from this research the combination of ozone and perforated polyethylene (PE) bag was the best treatment to maintain the quality and prolonged the shelf life of tomato fruit to about 12 days at room temperature. Future research need to be focused on the effect of ozone on pesticide residual level on tomato fruits.

5. Acknowledgments

This research was conducted with funds from DIKTI via the 2017 Master Plan *Percepatan dan Perluasan Pembangunan Ekonomi Indonesia* (MP3EI) research program. The authors gratefully thank DIKTI for this support.

References

[1] Badan Pusat Statistik NTB 2014 *NTB Dalam Angka* (Mataram: Badan Pusat Statistik NTB)

[2] Manurung S 2015 Handling of tomato (Lycopercum escusien Mill) post-harvest to increase the profit in Parahyangan peasent partners Cianjur Regency, West Java (Politeknik Pertanian Negeri Payakumbuh) (In Bahasa Indonesia)

[3] Rahmawati IS *et al* 2011 The effect of concentration Calsium Cloride ($CaCl_2$) treatment and the storage periode of ascorbic acid content of tomato (*Lycopersicum esculentum* Mill.) *J. Buletin Anatomi dan Fisiologi* **29** (1) (In Bahasa Indonesia)

[4] Purwadi A *et al* 2007. The effect of ozonization period on storage period of tomato (*Lycopersicum esculentum* Mill.). *Prosiding PPI-PDIPTN* (Yogyakarta) (In Bahasa Indonesia)

[5] Asgar A *et al* 2015 Ozonization technology to preserve the freshness of Kencana cultivars chilli during storage *J. Penelitian Pascapanen Pertanian* **12**(1) pp 20-26 (In Bahasa Indonesia)

[6] de Candia S *et al* 2014 Antimicrobial efficacy of gaseous ozone on berries and baby leaf vegetables *Proceeding of International Conference on Antimicrobial Research* (Madrid: Spain)

[7] Hecer C *et al* 2007 The Effects of Ozone and Chlorine Applications on Microbiological Quality of Chickens During Processing *J. BIOL. ENVIRON. SCI.* **1**(3) pp 131-138

[8] Asgar A 2014 Ozonization technology to clean vegetables *Iptek Hortikultura* **10** (In Bahasa Indonesia)

[9] Sembiring N 2009 The effect of types of packaging on product quality of red chilli (Capsicum annum L.) (Universitas Sumatera Utara: Medan) (In Bahasa Indonesia)

[10] Isyuniarno and Purwadi A 2007 The effect of using ozone oxidant in polyethylene plastic packaging to store Manalagi Apple (*Malus sylvestris* M.) (*PTAPB-BATAN. Yogyakarta* **X** (1)

[11] Handayani 2016 The effect of ozone combination and packaging on physical characteristics and microbiology of chilli *(Capsicum frutescens L.)* during storing (Universitas Mataram: Mataram)
[12] Yasa WS, Zainuri and Zaini A 2013 Keefektifan Teknologi Ozon Dalam Perbaikan Mutu Buah Mangga Gedong Gincu Lokal Lombok *Laporan Penelitian* (Universitas Mataram: Mataram)
[13] Purwadi A, Usada W and Isyuniarno 2007 Pengaruh Lama Ozonisasi Terhadap Umur Simpan Buah Tomat (*Lycopersicum esculentum* Mill.) *Prosiding PPI-PDIPTN* (Pustek Akselerator dan Proses Badah. Batan. Yogyakarta)
[14] Cantwell M 2008 Impact of Ripening & Storage Conditions
[15] Sudarmadji S, Haryono B and Suhardi 2002 Analysis procedure for food and agriculture (Yogyakarta: Liberty) (In Bahasa Indonesia)
[16] Widyastuti S *et al* 2015 Practicum guide of physiology and post harvest technology. *Fakultas Teknologi Pangan dan Agroindustri* (Universitas Mataram: Mataram)
[17] Muchtadi 1992 Post harvest physiology of fruits and vegetables. (Departemen Pendidikan dan Kebudayaan Direktorat Jenderal Pendidikan Tinggi Pusat Antar Universitas Pangan dan Gizi Institut Pertanian Bogor: Bogor) (In Bahasa Indonesia)
[18] Marriott NG and Gravani RB 2006 Principle of Food Sanitation Fifth Edition (New York, USA: Springer)
[19] Hanafiah KA 2003 (Mataram: PT. Raja Grafindo Permata)The
[20] Lizada C 1993 *Mango The Biochemistry of Fruit Ripening* Ed Seymour GB, Taylor JE and Tucker GA (London: Chapman & Hill)
[21] Wills R *et al* 1998 *Postharvest An Introduction to the Physiology and Handling of Fruit, Vegetables and Ornamentals* (Adelaide, South Australia: UNSW Press)
[22] Hadiwiyoto S and Soehardi 1981 *Penanganan Lepas Panen* (Jakarta: Departemen Pendidikan dan Kebudayaan)
[23] Syarief R and Irawati A 1988 *Pengetahuan Bahan Untuk Industri Pertanian* (Jakarta: Mediyatama Sarana Perkasa)
[24] Purwadi A, Usada W and Isyuniarno 2007 *The effect of ozonization period on storage period of tomato (Lycopersicum esculentum* Mill.)*Prosiding PPI-PDIPTN* (Pustek Akselerator dan Proses Badah. Batan. Yogyakarta)
[25] Pantastico ER 1993 *Fisiologi Pascapanen, Penanganan dan pemanfaatan Buah-buahan dan Sayur-sayuran Tropika dan Subtropika* Translated by Kamariyani (Yogyakarta: Gajah Mada University Press)
[26] Parnell TL 2004 *Tomatoes: Safe methods to store, preserve and enjoy* (Division of Agriculture and Natural Resources University of California)
[27] Kim JG, Yousef AE and Dave S 1999 Application of ozone for enhanching the microbiological safety and quality of foods: A Review *J Plant Prot* **62**(9) pp 1071-1087
[28] Khadre MA, Yousef AE and Kim JG 2001 Microbial aspects of ozone appliations in food: A review *J. Food Sci* **66**(9) pp 1242-1252
[29] Wani S *et al* 2015 Effect of ozone treatment on inactivation of *Escherichia coli* and *Listeria sp.* On spinach *Agriculture* **5** pp 155-169

Detection of aflatoxin M1 in powdered milk and sweetened condensed milk products in several cities in Java with HPLC-fluorescence method

H Wijaya[1], N I Wardayanie[1], R Widjajanti[2] and R F Silitonga[1]

[1]Center for Agro Based Industry, Jl. Ir. H. Juanda No. 11, Paledang, Bogor, West Java, Indonesia
[2]STMI Polytechnic Jakarta, Jl. Letjend. Suprapto No. 26, Cempaka Putih, Central Jakarta, Jakarta, Indonesia

Email : faizawijaya@gmail.com

Abstract. Aflatoxin M1 (AFM1) is a hydroxylated metabolite of aflatoxin B1 (AFB1) produced by lactating animals due to consuming AFB1-contaminated feed. AFM1 can be found in dairy products because it is resistant to heat during processing. This study aimed to detect AFM1 in powdered milk and sweetened condensed milk sold in several cities in Java. The amount of powdered milk sample was 20, while the amount of sweetened condensed milk sample was 16. AFM1 detection in powdered milk and sweetened condensed milk was conducted by HPLC-fluorescence method. The results showed that the concentration of AFM1 in powdered milk ranged from undetectable to 0.549 µg/kg and the highest data (55%) was distributed in concentration range of >0.05 µg/kg - 0.2 µg/kg. On the other hand, AFM1 levels in sweetened condensed milk ranged from undetectable to 0.056 µg/kg and 43.75% data was distributed in concentration range of >0.025 µg/kg - 0.05 µg/kg. All powdered milk and sweetened condensed milk samples have met the maximum level of AFM1 according to Indonesian regulation.

Keywords : Aflatoxin M, milk products, HPLC-fluoresence method

1. Introduction

Milk is one of many foods that consumed a lot by human, because it contains high nutrient that important for health and growth. However, milk also could be an agent for food contaminants, which can migrate into body and harm health. One of the harmful contaminants contained in milk is aflatoxin, which enters the human food chain through aflatoxin M1 (AFM1) residues. AFM1 in milk is stable through pasteurization and storage [1].

Aflatoxin is a secondary metabolite produced by various species of fungi from the genus *Aspergillus,* which, due to its genotoxic, mutagenic and carcinogenic characteristics, has the highest toxicity among all mycotoxins. Chemically, aflatoxin is a derived compound of *difuranocoumarin* (Figure 1) produced by two *Aspergillus* species, *A. flavus* and *A. parasiticus*. *A. flavus* grows in various places, especially on the heading of the plant (leaves, flowers) and produces only aflatoxin B, namely AFB1 and AFB2, while *A. parasiticus* produces aflatoxin B and G, namely AFB1, AFB2, AFG1 and AFG2. Aflatoxin M1 and M2 are the hydroxylated of aflatoxin B1 and B2 metabolites (addition of -OH groups) and can be found in milk and dairy products produced by livestock that consume contaminated feed [2].

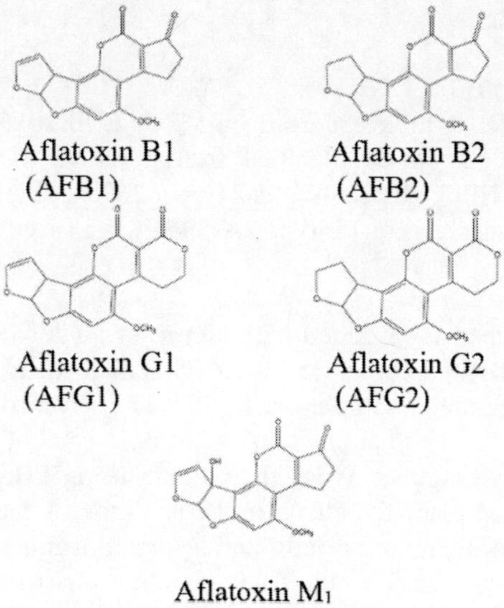

Figure 1. Chemical structure of aflatoxin B, aflatoxin G and aflatoxin M1 [2]

International Agency for Research on Cancer (IARC) has categorized Aflatoxin M1 (AFM1) as a class 2B carcinogenic agent, which is very likely to affect humans, especially infants. The presence of AFM1 in milk, especially cow's milk greatly affects the health condition of humans, where milk is a food that is often consumed, both by adults, and especially children. The existence of AFM1 should not be underestimated and neglected because of the uncertainty of current climatic and environmental conditions and the inability of agricultural systems to deal with and manage mycotoxin contamination [3]. In 2002, IARC has reclassified AFM1 as Group 1 carcinogenic agent, although its carcinogenicity is approximately 2 – 10 % of AFB1 [4].

According to ASEAN Food Reference Laboratories (2015), many countries have determined regulations regarding to the maximum limit of aflatoxin content in food products, especially AFM1 on dairy products. The maximum limit value for each country varies i.e undetectable or 0 μg/kg, max. 0.05 μg/kg, max. 0.2 μg/kg, max. 0.5 μg/kg, max. 5 μg/kg, and max. 15 μg/kg [5,6]. The Indonesian regulation governing the maximum limit of AFM1 in powdered milk can be found in the Regulation of the Head ofIndonesia National Agency of Drug and Food Control No. HK.00.06.1.52.4011 about Determination of the Maximum Limit of Microbial and Chemical Contaminants in Food. Under this regulation, the maximum limit of AFM1 in powdered milk and powdered cream and analog (plain) is 5 μg/kg, while in sweetened condensed milk and its analog is 0.5 μg/kg [7].

Research and study about the content of AFM1 in dairy products has been widely conducted. In 2003, Widiastuti found that 78.38% of fresh cow's milk samples in Darmaga Bogor were detected containing AFM1, with a concentration range of 0.001 to 1.200μg/l [1]. Celik reported his research on AFM1 contamination in pasteurized milk in Ankara, where 88.23% of samples were proven to be contaminated with AFM1, with 64% of it were above 0.05 μg/kg [8]. Jovana Kos has detected AFM1 content in 98.7% of cow's milk sample in Serbia, with concentrations 0.01 to 1.2 μg/kg [9]. Langat has also examined the AFM1 content in milk and dairy products from Bomet County, Kenya, where 84.32% samples were positively contaminated with AFM1 in concentrations above 0.05 μg/l (52% raw milk, and 8,6% processed milk) [10].

This research aims to detect AFM1 content in powdered milk and sweetened condensed milk sold in mini market and traditional market in several cities in Java. These are the most dairy products consumed by the Indonesian population [11]. Data from Statistics Indonesia (BPS) based on Susenas in March 2016, stated that consumption per capita per week for sweetened condensed milk and infant powdered milk is 0.013 kg and 0.542 ounces respectively [12].

2. Materials and methods

2.1 Materials and instruments

Samples were plain powdered milk according to SNI 2970:2015 and sweetened condensed milk according to SNI 2971:2011, which obtained from 4 cities in Java, namely Serang, Bandung, Semarang, and Surabaya. HPLC Brand Waters type e2695 Alliance, with Immunoaffinity column (IAC), coupled with fluorescence detector type 2475 FLR, was used to determine AFM1.

2.2 Methods

The powdered milk is weighed 10 g, added with 50 mL of 50 °C distilled water, stirred, and put in to 50 °C sonicator for 30 min, then cooled to 20-25 °C. The sweetened condensed milk is heated in a water bath until the temperature of sample reach 35 - 37 °C, centrifuged for 15 minutes (1000 rpm), filtered by Whatman paper no. 4 and put in a 50 mL flask. A total of 50 mL for each prepared samples, poured into immunoaffinity columns. Wash the column using PBS, and dissolve aflatoxin from the column using methanol, then place the eluate in the test tube. A total of 0.5 mL of eluate was mixed with 0.5 mL aquabidest, vortex mixed briefly and analyzed using HPLC (Fluorescence detector, 100 C18 chromosome column, and 25% acetonitrile as mobile phase). A total of two replicates for each standard solution and samples injected 50 μL into the HPLC system. The content of AFM1 (mg/kg) is calculated by this following formula:

$$C\,sp = \frac{\frac{A\,sp}{A\,st} \times C\,st \times \frac{Vi\,st}{Vi\,sp} \times VA}{W\,sp} \times fp \tag{1}$$

C sp = concentration of sample (mg/kg)
A st = standard area(mAU)
A sp = sample area (mAU)
C st = concentration of AFM1 (mg/l)
Vi st = standard volume injected (μl)
Vi sp = sample volume injected, (μl)
VA = end volume (ml)
Fp = dilution factor
W sp = weight of sample (gram)

3. Result and Discussion

Sample of sweetened condensed milk and powdered milk were taken randomly without considering any kind of variables, such as packaging material, packaging size, time of production, shelf life, storage area, distribution line, and selling location. Sampling was conducted during February - April 2016. Location, packaging and weather condition data of sampling are seen as Table 1.

Determination of AFM1's content in powdered milk and sweetened condensed milk, focusing on enrichment extraction using immunoafinity chromatography (immunoafinity column) combined with quantification using HPLC. The HPLC system used in this research is kromosil 100 C18 columns, mobile phase acetonitrile 25 %, and Fluorescence detector.

Busman [13] said that HPLC based methods, often require several sample preparation steps, such as extraction, concentration, and isolation, to achieve desired method performance. Sample preparation can involve cleanup steps utilizing liquid-liquid extraction, solid-phase extraction and separation using immunoaffinity column. Enrichment extraction technique like immunoaffinity chromatography provides opportunity for substantial analyte enrichment, while removing matrix components that could potentially attenuate analyte signal.

Table 1. Location, packaging and weather condition data of sampling

Sampling Location	Sample	Packaging	Weather	Remark
Surabaya	powdered milk sweetened condensed milk	sachet 40 g & box 200 g pouch & sachet	sunny	purchased in minimarket
Serang	powdered milk sweetened condensed milk	box 200 g & 400 g can& pouch	sunny	purchased in minimarket
Semarang	powdered milk sweetened condensed milk	sachet 40 g &box 200 g sachet	sunny	purchased in traditional market
Bandung	powdered milk sweetened condensed milk	box 200 g & 400 g pouch & sachet	sunny	purchased in minimarket

Analysis results of AFM1 in powdered milk and in sweetened condensed milk are described on Table 2 and Table 3 respectively. According to Table 2, the concentration of AFM1 in powdered milk was at undetectable to 0.549 µg/kg. All powdered milk samples have met the requirement for maximum limit of AFM1 content in powdered milk, stipulated by Indonesian regulation which is 5 µg/kg [7]. While according to Table 3, the concentration of AFM1 in sweetened condensed milk is ranged from 0 – 0.056 µg/kg. All sweetened condensed milk sample have met the requirement from Indonesian regulatory for maximum limit of AFM1 content in sweetened condensed milk, which is 0.5 µg/kg [7].

Table 2. Analysis results of AFM1 in powdered milk

No.	Packaging	Size (g)	AFM1 (µg/kg)	Sampling location
1	Sachet	40	0.064	Surabaya
2	Sachet	40	0.048	Surabaya
3	Box	200	0.072	Surabaya
4	Box	200	0.117	Surabaya
5	Box	200	0.087	Surabaya
6	Box	200	0.123	Serang
7	Box	400	<0.025	Serang
8	Box	200	0.053	Serang
9	Box	400	0.064	Serang
10	Box	200	0.107	Serang
11	Box	200	<0.025	Semarang
12	Sachet	40	0.151	Semarang
13	Sachet	40	0.549	Semarang
14	Sachet	40	0.203	Semarang
15	Sachet	40	0.204	Semarang
16	Box	200	0.075	Bandung
17	Box	200	0.121	Bandung
18	Box	400	0.026	Bandung
19	Box	200	0.293	Bandung

No.	Packaging	Size (g)	AFM1 (μg/kg)	Sampling location
20	Box	400	<0.025	Bandung

Table 3. Analysis results of AFM1 in sweetened condensed milk

No.	Packaging	Size (g)	AFM1 (μg/kg)	Sampling location
1	Can	385	0.056	Serang
2	Can	385	0.050	Serang
3	Can	385	0.066	Serang
4	Pouch	220	0.035	Serang
5	Pouch	220	0.033	Surabaya
6	Pouch	220	0.042	Surabaya
7	Sachet	40	<0.025	Surabaya
8	Sachet	40	0.027	Surabaya
9	Sachet	40	0.032	Bandung
10	Pouch	220	<0.025	Bandung
11	Sachet	40	0.038	Bandung
12	Sachet	40	0.040	Bandung
13	Sachet	40	0.052	Semarang
14	Sachet	40	<0.025	Semarang
15	Sachet	40	<0.025	Semarang
16	Sachet	40	<0.025	Semarang

Referring to national and international regulatory [6], the data distribution of AFM1 concentration in powdered milk and sweetened condensed milk is shown in Table 4 and 5.

Table 4. Data distribution of AFM1 content in powdered milk

AFM1 (μg/kg)	Amount of Data				Total
	Surabaya	Serang	Semarang	Bandung	
Undetectable	0	1	1	1	3 (15%)
$<0.025< x < 0.05$	1	0	0	1	2 (10%)
$0.05 < x \leq 0.2$	4	4	1	2	11 (55%)
$0.2 < x \leq 0.5$	0	0	2	1	3 (15%)
$x > 0.5$	0	0	1	0	1 (5%)

Table 5. Data distribution of AFM1 content in sweetened condensed milk

AFM1 (μg/kg)	Amount of data				Total
	Surabaya	Serang	Semarang	Bandung	
Undetectable	1	0	3	1	5 (31.25 %)
$0 < x < 0.05$	3	1	0	3	7 (43.75 %)
$0.05 < x \leq 0.2$	0	3	1	0	4 (25%)
$0.2 < x \leq 0.5$	0	0	0	0	0 (0%)
$x > 0.5$	0	0	0	0	0 (0%)

Based on Table 4, the percentage of powdered milk containing AFM1 consist of 15 % undetectable, 10 % ranged from <0.025 to 0.05 μg/kg, 55 % ranged from 0.05 – 0.2 μg/kg, 15 %

ranged from 0.2 – 0.5 μg/kg, and 5 % is above 0.5 μg/kg. These data show that the most AFM1 concentration contained in powdered milk is ranged from 0.05 – 0.2 μg/kg. On the other hand Table 5 show that the percentage of the most AFM1 concentration contained in sweetened condensed milk is ranged from <0.025– 0.05 μg/kg (43.75 %). Concentration AFM1 in sweetened condensed milk for undetectable is 31.25 % samples, ranged from <0.025 – 0.05 μg/kg were 43,75 % samples, and ranged from 0.05 – 0.2 μg/kg were 25 %.

AFM1 that already contaminated milk as raw material for powdered milk and sweetened condensed milk is difficult to control during processing, either by heat treatments, storage at low temperature, concentration and drying of milk, or, chemical and other biological treatment. Some of technologies that have involved in order to decreasing the level of AFM1 in milk processing are by the addition of oxidator agents and adsorber, but unfortunately, this technology has not been used commercially. The most likely effort to decrease the AFM1 levels is by controlling the liquid milk.

The presence of AFM1 in dairy products needs further attention, due to its high consumption, especially for infants and children, and its carcinogenic effect. Therefore it is necessary to apply a technology that can reduce AFM1 contaminants in dairy products. The study about preservative for milk has been conducted. Hypothiocyanite-rich-solution from lactoperoxidase system, obtained from bovine whey, was remarkably inhibited the growth of total bacteria in fresh milk [13,14]. But its activity to decrease the concentration of AFM1 in milk is still unclear. There is a review on prevention efforts to reduce aflatoxin content in animal feed and dairy products. The reduction of aflatoxin content in livestock feed is done indirectly, such as through monitoring of aflatoxin infections on farms, by protecting food from infectious sources, detain the spread of microorganisms, storage of raw materials, and control of humidity. The reduction of aflatoxin content in dairy products can be done through the use of absorbents such as bentonite, verticulate, etc., as well as the use of chemicals, such as hydrogen peroxide that used during the storage. The use of microorganisms such as Flavobacterium aurantiacum can also be used to reduce aflatoxin content in milk [15,16]. Other methods that have been used to reduce AFM1 levels is through fermentation, including the addition of antioxidant compounds such as carotenoids that found in carrots [17].

4. Conclusion

The content of AFM1 in powdered milk is ranged from undetectable to 0.549 μg/kg. According to data distribution, the most AFM1 concentration contained in powdered milk is ranged from 0.05 – 0.2 μg/kg, with the percentage is 55 %. The content of AFM1 in sweetened condensed milk is ranged from 0 – 0.056 μg/kg, and according to data distribution, the most AFM1 concentration contained in sweetened condensed milk is ranged from <0.025 – 0.05 μg/kg, with the percentage is 43.75 %. These results showed that all the samples have met the Indonesian regulatory requirement for maximum level of AFM1 in powdered milk and sweetened condensed milk.

5. Acknowledgment

The authors would like to thank Center for Agro Based Industry for funding this research activity.

6. Bibliography

[1] Widiastuti R, Maryam R, Bahri S and Firmansyah R 2006 *Pros. Semin. Nas. Teknol. Peternak. dan Vet. 2006* 239–43
[2] *EFSA Panel on Contaminants in the Food Chain* 2007 *EFSA J.* **446** 1–127
[3] Prandini A, Tansini G, Sigolo S, Filippi L, Laporta M and Piva G 2009 *Food Chem. Toxicol.* **47** 984–91
[4] Iqbal S Z and Asi M R 2013 *Food Control* **30** 235–9
[5] ASEAN 2015 *Regulatory limits of mycotoxins in ASEAN countries Regulatory limits of*

mycotoxins in ASEAN countries

[6] Mazumder P M and Sasmal D 2001 *Anc. Sci. Life* **20** 1–19

[7] BPOM 2009 *Peraturan Kepala Badan Pengawas Obat dan Makanan Republik Indonesia : Penetapan Batas Maksimum Cemaran Mikroba dan Kimia Dalam Makanan*

[8] Çelik T H, Sarımehmetoğlu B and Küplülü Ö 2005 *Vet. Arh.* **75** 57–65

[9] Kos J, Lević J, Duragić O, Kokić B and Miladinović I 2014 *Food Control* **38** 41–6

[10] Langat G, Tetsuhiro M, Gonoi T, Matiru V and Bii C 2016 *Adv. Microbiol. Kenya. Adv. Microbiol.* **6** 528–36

[11] Ahmad I and Hermiyetti 2008 *Pros. Semiloka Nas. Prospek Ind. Sapi Perah Menuju Perdagang. Bebas-2020* 413–9

[12] Badan Pusat Statistik 2016 *Ringkasan Eksekutif Pengeluaran dan Konsumsi Penduduk Indonesia 2016*

[13] Villa V Y, Legowo A M, Bintoro V P and Al-Baarri A N 2014 *Int. J. Dairy Sci.* **9** 24–31

[14] Nawangsari D N, Al-Baarri A N M, Mulyani S, Legowo A M and Priyo Bintoro V P 2014 *Int. J. Dairy Sci.* **9** 56–62

[15] Hasheminya S M and Dehghannya J 2013 *Int. Res. J. Appl. Basic Sci.* **4** 1506–10

[16] Sumantri I, Murti T W, Poel A F B Van Der, Boehm J and Agus A 2012 *J.Indonesian Trop.Anim.Agric.* **37** 271–277

[17] Al-Delaimy K and Mahmoud I 2015 *Br. J. Appl. Sci. Technol.* **6** 597–605

Characterization of the chemical composition of *Adenostemma lavenia* (L.) Kuntze and *Adenostemma platyphyllum* Cass

A Fauzan[1, 2], D Praseptiangga[2, 3], R Hartanto[2, 3] and B Pujiasmanto[2, 4]

1. Department of Agrotechnology, Faculty of Agriculture, Universitas Pekalongan, Jl. Sriwijaya No. 3 Pekalongan, Indonesia
2. Graduate School Program of Universitas Sebelas Maret, Jl. Ir. Sutami 36 A, Kentingan, Surakarta 57126, Indonesia
3. Department of Food Science and Technology, Faculty of Agriculture, Universitas Sebelas Maret, Jl. Ir. Sutami 36 A, Kentingan, Surakarta 57126, Surakarta, Indonesia
4. Department of Agrotechnology, Faculty of Agriculture, Universitas Sebelas Maret, Jl. Ir. Sutami 36 A, Kentingan, Surakarta 57126, Surakarta, Indonesia

e-mail : zan.anwar@gmail.com

Abstract. The purpose of this study was to characterize the chemical compounds of *Adenostemma lavenia* (L.) Kuntze (Al) and *Adenostemma platyphyllum* Cass (Ap) using Pyrolysis-gas chromatography/mass spectrometry (Py-GCMS) and proximate analysis. Two species of *Adenostemma* samples (roots, stem and leaves) about 1 mg was pyrolyzed directly at the optimum temperature of 600°C. Py-GCMS was relatively fast, easy to use and without samples preparation and identification of the chemical compounds was carried out by comparison of the mass spectra obtained with those stored in Wiley 7th libraries. The data of proximate analysis were statistically analysed using Friedman test followed and hierarchical cluster analysis (HCA) for data of Py-GCMS. The result of proximate analysis showed that *A. lavenia* (L.) Kuntze (Al) and *A. platyphyllum* Cass (Ap) contained 8.27% (*Al*) and 9.18% (*Ap*) of water, 11.52% (*Al*) and 17.84% (*Ap*) of protein, 5.67% (*Al*) and 6.33% (*Ap*) of fat, and 17.32% (*Al*) and 19.94 (*Ap*) of ash. Amines, aldehydes, fatty acids, terpenoids-steroids, alkaloids, aromatic and aliphatic hydrocarbons, phenolic, and oligopeptides as part of 125 chemical compounds of each species are identified by Py-GCMS analysis. Hierarchical cluster analysis of pyrolysis products indicate not similitary of major chemical compounds of two *Adenostemma* species.

Keywords : Adenostemma lavenia, Adenostemma platyphyllum, Py-GCMS, Friedman test, . hierarchical cluster analysis

1. Introduction

The study of *Adenostemma* biology and ecology has been receiving growing attention worldwide [1-7], but study of identification basic ecological relationships and chemical compounds is low. Recent studies have found 11-hydroxilated kauranic acids [8], kaurane [9, 10], *α-cubebene*, *caryophyllene*[10], flavanoids and alkaloids [11]. The chemical composition of *Adenostemma* differs depending on species, growth environment and season [12-16, 18], age and plant parts [17], physiological variations, environmental conditions, geographic variations, genetic factors and evolution [19].

So far, several methods have been developed to analyze the chemical compounds of *Adenostemma*. Phytochemical analysis [11], spectrographic [7, 8], high performance liquid chromatography (HPLC) [8, 9] were used to investigate the chemical compounds of *Adenostemma*. However, troublesome pretreatments are necessary for all the above methods prior to the final analysis such as solvent extraction, purification, and derivatization. Most of the studies were limited to some components of interest, particularly the flavonoids, lacking of a whole chemical composition analysis. Therefore, it is meaningful to develop a simple, rapid and sensitive method for characterization of the whole chemical composition in *Adenostemma*.

Recently, py–GCMS has been used extensively in the characterization of both low and high molecular components in various natural products. This technique yields a pyrogram consisting of the characteristic peaks of constituents in the given natural product without any pretreatment. Py–GCMS has been performed for accurate determination of terpenoids, alcohols in volatile constituents in *Houttuynia cordata* Thunb [16], phenolic compounds in origanum heracleoticum [17], and terpenic acids, aleuritic acid, fatty acids in natural resin shellac [18]. Organic compounds in plants are divided into two main groups, Nitrogen-containing (alkaloids, non-protein amino acids, amines, cyanogenic, glycosides, glucosinolates, alkamides, lectins, peptides, polypeptides) and without nitrogen (terpenoids, flavonoids, tannins, phenylpropanoids, lignin, coumarins, lignans, Polyacetylenes, fatty acids, waxes, Polyketides, Carbohydrates, organic acids) [20].

In this study, Py–GCMS was first applied to analysis of chemical composition in *Adenostemma* without any cumbersome pretreatment. Furthermore, two species samples obtained from The northern slope of Dieng plateau were also compared, based on the chemical compound of *Adenostemma*. The aim of this paper is to analyze and compare the chemical compositions of two *Adenostemma* species using destructive analytical methods.

2. Material and Method

2.1 *Adenostemma sampling*
Fresh *Adenostemma* material from two species was collected in the growing season of 2016 in north slope Dieng plateau (07°3'20.29" - 7°10'25.09" N; 109°40'21.33" – 109°36'20,09" E; 200 – 1,000 m amsl) in tropical rain forest (Java, Indonesia). Only living material with green leaves and fresh flower without signs of decomposition was collected. Samples were identified at Center for Plant Conservation Botanic Garden – Indonesian Institute of Science as *Adenostemma platyphyllum* Cass. and *Adenostemma lavenia* (L.) Kuntze.

2.2 *Sample preparation*
The approximate amount of collected adenostemma material was 50 g dry weight species. The samples was cleaned. The roots, stems and leaves were cut out from some plants; others were kept intact. Afterwards, the roots (Ap1 and Al1), stems (Ap2 and Al2), leaves (Ap3 and Al3) and entire plants (Ap and Al) were oven-dried separately at 40°C until dry weight was constant and were stored at 5°C in air-tight containers before analysis. Samples for proximate analysis and Py-GCMS analysis were dried, ground with a grinder and sieved through a 40 mesh [21].

2.3 *Proximate Analysis*
The proximate analysis (water, proteins, crude fats, ash, carbohydrates and total dietary fiber) of all the samples carried out in triplicate [22]. The water and ash were determined using gravimetric method. The nitrogen value was determined by micro-Kjeldahl method described by [23] involving digestions, distillation and finally titration of the sample. The nitrogen value was converted to protein by multiplying a factor of 6.25. The crude fat was determined by soxhlet extraction method, employing a n-hexane extraction. Carbohydrate was determined by difference method. Carbohydrate was determined when the sum of the percentages of water, ash, crude protein, and crude fat were subtracted from 100. All the proximate values were reported in percentage [22].

2.4 *The Pyrolysis GCMS Analysis*
The chemical compounds in sample were identified by a Pyrolysis GCMS (Shimadzu GCMS-QP2010; Shimadzu Corporation, Kyoto, Japan). One milligram ±0.05 mg of powdered sample was introduced into a pyrolyzer (PY-2020iS). Pyrolysis was performed at a temperature of 600°C with a ramp rate of 20°C/ms with a hold time of 10 s. The pyrolyzer was directly interfaced with a GCMS. A capillary column (RTX-5MS) with 60 m length, 0.25 mm internal diameter and a 0.25 μm stationary phase film was used. The volatiles were trapped on adsorbent trap before being desorbed at 280 °C onto a heated transfer line which was held at 280°C. The purge flow of helium UHP to remove any oxygen from the sample, prior to pyrolysis, was set to 3 ml/min and a split ratio 1:50. The injector temperature was set at 280°C, ion source 200°C. The identification of the chemical compounds was carried out by comparison of the mass spectra obtained with those stored in Wiley 7th libraries [24].

2.5 *Data Analysis*
The data of proximate analysis were statistically analysed using Friedman test followed with Post Hoc test using Wilcoxon Signed Rank test, and hierarchical cluster analysis (HCA) was used to identify relatively homogeneous clusters of samples based on their similarity [25] in chemical compound Adenostemma's. Cluster analysis is technique for grouping object into clusters so that object in same cluster. A cluster are more like one another than they are like object in other cluster [26]. Cluster analysis involves at least two step. The first is the measurement of some form of similarity to determine how many groups really exist in the sample [26, 27]. The second step is to profile the variable to determine their composition [28]. Beginning with N clusters consisting exactly of two entities, the similarity matrix is searched for the most

similar pair of clusters and the number of clusters is reduced by one by merging the most similar pair of clusters with the minimum increase in the total within group error sum of squares [21]. The analysis was performed using IBM SPSS Statistics 23 version software.

Table 1. List of studied *Adenostemma* species with codes and their growth conditions

Species	Code	Growth conditions
A. platyphyllum	Ap	Tropical rainforest with soil pH 6.9 (± 0.93); soil carbon 1.32% (±0.15); total organic matter 2.28% (±0.26); nitrogen 0.25% (± 0.07); available phosphorus 4.38 ppm (± 0.83); K (me) 0.27 (± 0.03); Ca 1.74 mg (±0.49); Mg 1.30 (±0.57); C/N ratio 5.28 (± 0.48); latosol; 591 m amsl; the average max. and min. temperatures are 26.81°C and 18.78°C; the average annual rainfall is 55.91 cm; RH 93%. Ecological indicator [29]: Diversity (b) 3.164; Dominance (Cd) 0.177, Evenness Index (J') 0.8446; Richness Index (al) 23; Shannon-Wiener Index (H') 2.648
roots	Ap1	
stems	Ap2	
leaves	Ap3	
A. lavenia	Al	Tropical rainforest with soil pH 6.8 (± 0.85); soil carbon 1.33% (±0.07); total organic matter 2.30% (± 0.12); nitrogen 0.26% (± 0.04); available phosphorus 4.8 ppm (± 2.16); K (me) 0.31 (± 0.09); Ca 1.20 mg (± 0.69); Mg 0.60 (± 0.03);C/N ratio 5.12 (± 0.55); latosol; 930 m amsl; the average max. and min. temperatures are 22.06°C and 15.5°C; the average annual rainfall is 60.80 cm; RH 95%. Diversity index of herbs [29]: Diversity (b) 3.183; Dominance (Cd) 0.193, Evenness Index (J') 0.85088; Richness Index(al) 24; Shannon-Wiener Index (H') 2.704
roots	Al1	
stems	Al2	
leaves	Al3	

3. Results and discussion

3.1 *Proximate analysis*

The result of proximate analysis shows variant concentration of chemical compounds (water, protein, fat, ash, carbohydrate and total dietary fiber). Based on Friedman test analysis, the water contents of samples are different. The water contents of whole *A platyphyllum* were higher than *A lavenia*. Considering the overall percentage of water composition, it was highest in *A platyphyllum* stems followed by *Aplatyphyllum* roots, *Aplatyphyllum* leaves and *A lavenia* leaves while other had comparatively lesser composition (Table 2). However, the water in dry powder of stems and roots of *A lavenia* are the highest.

Table 2. Proximate analysis of *Adeostemma* sp.

Code	Water		Protein*	Fat*	Ash*	Carbohydrate*	Fiber, total dietary*
	Fresh	Powder					
Ap	89.54 b	9.18 d	17.84 f	6.33 d	19.94	55.89 b	2.41 b
Ap1	90.66 d	7.45 b	8.85 a	3.61 b	16.47	71.07 d	6.27 f
Ap2	91.73 e	9.39 f	12.49 c	7.31 f	21.24	58.96 c	2.36 a
Ap3	89.64 c	7.17 a	18.38 g	10.50 g	21.00	50.12 a	2.62 c
Al	88.56 a	8.27 c	11.52 e	5.67 c	17.32	68.41 c	2.62 c
Al1	89.35 b	9.58 g	11.52 b	6.71 e	15.05	67.64 d	6.75 g
Al2	90.61 c	10.36 h	10.60 b	2.75 a	17.32	68.41 c	3.40 d
Al3	88.82 a	9.34 e	13.37 d	10.93 h	16.92	58.79 d	3.72 e
Friedman test							
Chi-Square	15.667	21.000	18.556	21.000	11.194	19.104	20.556
Df	7	7	7	7	7	7	7
Asymp. Sig	0.028^{s}	0.004^{s}	0.010^{s}	0.004^{s}	0.130^{ns}	0.008^{s}	0.004^{s}

Note: Data are means from three samples. * : dry basis; ns: non-significant, s: significant at P 0.05 Friedman test. Figures followed by the same letter in the same column are not significantly different at 5% level Wilcoxon Signed Rank test.

The protein contents of each species and organs are different. Considering the overall percentage of protein composition, it was highest in *A. platypyllum* leaves by whole of *A. platypyllum*. Whole of *A. lavenia*, and *A. lavenia* leaves while other had comparatively lesser composition (Table 2). Considering the result crude fat. *A. lavenia* leaves and *A. platyphyllum* leaves had prominent levels compared to other plant organs. Considering the resulted achieved from ash analysis. The ash contents of each species are not different. While analyzing the carbohydrate contents in the samples. The results showed that *A platyphyllum roots*. Whole of *A. lavenia*. *A. lavenia* roots and *A.lavenia* stems had highest concentration of carbohydrate to other plant organs (Table 2).

Based on the proximate analysis it is known that two species of *Adenostemma* exhibit different characteristics, except ash content.Different concentrations of chemical compounds are caused by differences in species. plant organs and growing places of each samples [20]. The sample of this study was obtained from wild forests that have different ecology, macro and microclimates (temperature, water, humidity, sunlight intensity)..

3.2 *Pyrolysis GCMS Analysis*

Py-GCMS is a suitable method for the quantitative and qualitative analysis of complex mixtures with high efficiency, precision, and simplicity [30, 31]. In a merbau extractives pyrolysis study, flash pyrolysis-GCMS used in structure analysis provided information at the presence of phenolic forms [24]. Pyrolysis-gas chromatography/mass spectrometry was used to characterize the chemical composition of *A. platyphyllum* and *A lavenia.* The result of Py-GCMS analysis to those sample found 125 chemical compounds. Among all, the 5 compounds dominant were: epoxycyclododecane, 4-allyl-2,6-dimethoxyphenol, Cis,Cis,Cis-8,11,14 eicosatrienoic acid, tetradecahydroanthracene, and levoglucosan. The chemical compounds in the studied two species *Adenostemma* can be seen in Appendix 1. GCMS analysis revealed the presence of terpenoids, phenolic compounds, alkaloids and fatty acids.

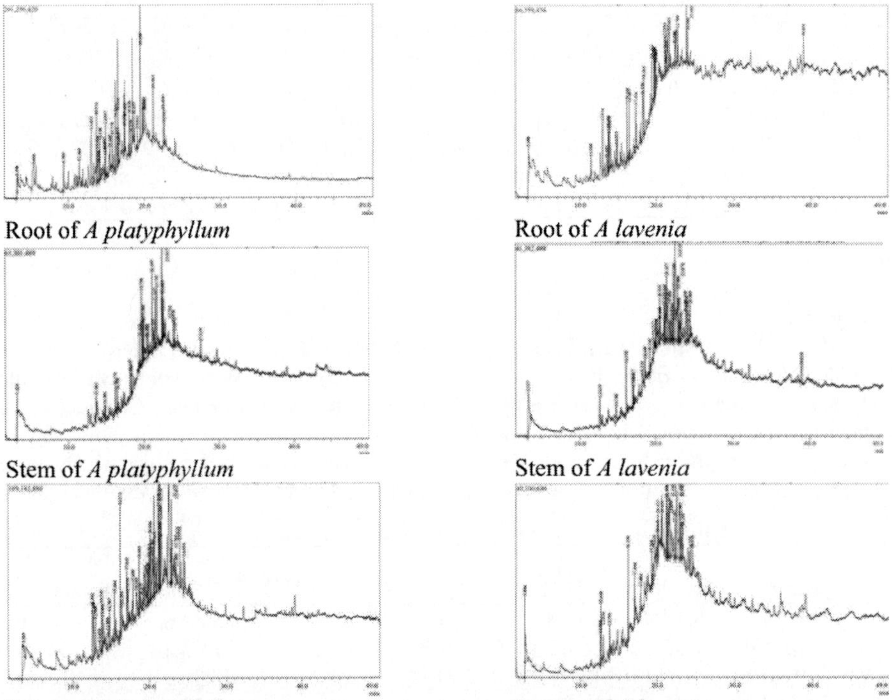

Root of *A platyphyllum* — Root of *A lavenia*
Stem of *A platyphyllum* — Stem of *A lavenia*
Leaves of *A platyphyllum* — Leaves of *A lavenia*

Fig. 1. Chromatogram py-GCMS of *Adenostemma platyphyllum* and *Adenostemma lavenia*

The presence of dominant product groups from each *Adenostemma* species is illustrated in Table 3. The pyrolysates of the studied *Adenostemma* species were dominated by phenolic compounds, lipid orginated compounds and N-bearing compounds. The dominant compounds in *A. platyphyllum* were phenolic compounds and originated lipids (fatty acid). The dominant compounds in *A. lavenia* were N-bearing compounds, alkaloid, aromatic compounds, and terpenoid-steroid. Abundant aromatic compounds products (phenol and derivate) confirm the importance of phenol in the structural make-up of *Adenostemma*. As one of the main products. *A lavenia* species contained more terpenoid-steroid and alkaloid in comparison with *A. platyphyllum* species. *A lavenia* species has a higher alkaloid compounds than *A. platyphyllum*.

4-Allyl-2,6-Dimethoxyphenol ($C_{11}H_{14}O_3$), coniferyl alcohol ($C_{10}H_{12}O_3$), and linoleic acid ($C_{18}H_{32}O_2$) were the dominant phenolics found in *Adenostemma* species. 4-Allyl-2,6-Dimethoxyphenol was phenol detected in the largest quantity (14.2%). 4-O-[3-Acetyl-1-(Trimethylsilyl)-1h-Indolyl]-D-Glucose ($C_{32}H_{62}N_2O_7Si_5$; 2.14%) and 5h-1-Pyrindine (C_8H_7N; 5.07%)were the alkaloid found in the leaf tissue of *A lavenia*. 3-methylindole (C_9H_9N; 0.93%) and 1-cyano-3-methylisoquinoline ($C_{11}H_8N_2$; 1.83%) were the alkaloid found in the stem tissue of *A lavenia*. 6,7-Dihydro-3-Nitro-5h-Cyclopenta[B]Pyridin-2(1h)-One ($C_8H_8N_2O_7$; 2.94%) and 5,10-Diethoxy-2,3,7,8-Tetrahydro-1h,6h-Dipyrrolo[1,2-A;1',2'-D] Pyrazine ($C_{14}H_{22}N_2O_2$; 4.0%) were found in the root tissue of *A lavenia*. The alkaloids were detected in *A platyphyllum* only two substance i.e. 3-Methylindole (1.62%) and 2-O-[3-Acetyl-1-(Trimethylsilyl)-1h-Indolyl]-D-Glucose ($C_{32}H_{62}N_2O_7$; 1.38%).

Table 3. Relative abundance (%) of the main groups of pyrolysis products of *Adenostemma*.

Metabolite Group	N	Alk	MA	C	Al	Alc	Ar	Ph	Lp	TS
Al1	13,39	6,94	6,11	5,44	0	2,61	3,68	44,84	11,9	5,08
Al2	27,97	2,76	0	0	0	3,93	10,71	0	44,22	10,41
Al3	11,88	11,24	3,93	1,51	1,41	0	12,59	4,37	29,62	23,46
Ap1	2,55	0	1,08	4,69	2,01	0	2,37	72,45	14,88	0
Ap2	3,1	0	0	0	5,25	8,66	0	62,74	18,51	1,73
Ap3	12,06	4,27	1,63	1,41	7,21	17,22	13,72	7,42	18,23	16,82

Note: Code of studied species as in Table 1. N, N-bearing compounds (except alkaloids); Alk, Alkaloids; MA, multi-origin aliphatic compounds with C≤6; C, furan originated from carbohydrates. pyran and cyclopentene derivatives; Al. aliphatic compounds with C>6; Alc, Alcohol; Ar, aromatic compounds (except phenolic compounds); Ph, phenolic compounds; Lp, compounds originated from lipids (except terpenoid, steroid); TS, terpenoid-steroid.

Those compounds support the healing of wound or infection, antioxidant and also act as antibacterial agents. Naturally occurring phenolic compounds have been shown to scavenge active oxygen species and to effectively prevent oxidative cell damage [32]. Inhibitory effects of phenolic compounds containing allyl groups were similar to those of flavonoids [33]. 2,6-diphenyl-piperidine, and 5,10-diethoxy-2,3,7,8-tetrahydro-1H,6H-dipyrrolo[1,2- a;1 -d] pyrazine (a member of alkaloids) also acts as antifungal agent [34]. A considerable amount of 3-methylpyridine is used as a starting material for pharmaceuticals and agrochemicals [42] and 4-cyano-3-methylisoquinoline as potential antimalarial agents [35].

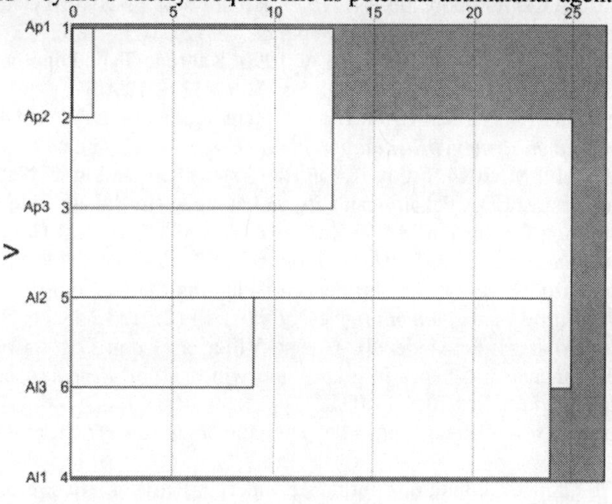

Fig. 2. Cluster analysis of amounts of pyrolysis products of *Adenoetemma*.

Note : Ap1: root of *A platyphyllum*; Ap2: stem of *A platyphyllum*; Ap3: leaves of *A platyphyllum*; Al1: root of *A lavenia*; Ap2: stem of *Alavenia*; Ap3: leaves of *A lavenia*

3.3 *Cluster Analysis*

Cluster analysis of pyrolysis products indicates not similitary of major chemical compounds of the studied *Adenostemma* species (Fig. 2). For example, despited the evident similarity in plants organs composition, the *A. platyphyllum* species were grouped together. For example, despite the evident similarity in plants organs composition, the *A. platyphyllum* species were grouped together. The decisiveness of this classification is reflected in the dendrogram (fig. 2). The initial splitting of the tree forms to two clusters. The top contains *A. platyphyllum*. The bottom contains *A. lavenia*. The *A. platyphyllum* roots and *A. platyphyllum* stems are each more similar than the *A. platyphyllum*leaves. The chemical compounds in chemical taxonomy of *Adenostemma* can be seen in Appendix 1.

4. Conclusions

The chemical compounds *A. lavenia* and *A. platyphyllum* are significantly different. Pyrolysis GCMS showed that the major compound of *A. platyphyllum* is aromatic compounds (phenolic) with a concentration of 26.4% and *A. lavenia* is originated lipids and terpenoid steroid with a concentration of

29.7% and 15.8%. Chemical compounds of *A. platyphyllum* and *A. lavenia* as potential antimicrobial, antioxidant and anti-inflammation agent.

References

[1] Orchard AE. 2011. A review of Australian Adenostemma J.R.Forst. & G.Forst. (Asteraceae: Eupatorieae). *Telopea* 13(1–2): 341–348.

[2] Singh G. Passari AK. Singh BP and Kumar NS. 2017. Traditionally Used Medicinal Plants Belongs to Family Asteraceae for the Treatment of Cancer in Mizoram. Northeast India

[3] Valentin-Silva A. Godinho MAS & Vieira MF. 2016. Life history of *Adenostemma brasilianum* (Eupatorieae. Asteraceae): A Psychophilous Herbaceous Species of the Brazilian Atlantic Forest Understory. Journal of the Torrey Botanical Society 143(1): 87–92. 2016.

[4] Amjad MS. Arshad M. Page S. Qureshi R. dan Mirza SN. 2017. Floristic composition. Biological spectrum and phenological pattern of vegetation in the subtropical forest of Kotli District. AJK. Pakistan. *Pure Appl. Biol.* 6 (2): 426-447. doi. 10.19045/bspab.2017.60043.

[5] Sivaraj N. Pandravada SR. Venkateswaran K. dan Dikshit N. 2017. Ethnic Medicinal Plant Wealth of Eastern Ghats: Status. Knowledge Systems and Conservation Strategies. *Int. J. Curr. Res. Biosci. Plant Biol.* 4 (1). 83-101. DOI 10.20546/ijcrbp.2017.401.010.

[6] Castro. AMO. 2017. Knowledge and Traditional Use of the Phytomedicinal Resource of the Yurumanguí River Community. Buenaventura District. *Thesis.* Facultad de Ciencias Contables Económicas y Administrativas Maestría en Desarrollo Sostenible y Medio Ambiente Manizales. Colombia. (In Spanish)

[7] Godinho MAS. Alvarenga EM. & Vieira MF. 2011. Germination and Seed Quality of Adenostemma brasilianum. Revista Árvore. Viçosa-MG. 35 (6): 1197-1205. (In Portoguese)

[8] Cheng PC. Hufford CD & Doorendos NJ. 1979. Isolation of 11-Hydroxylated Kauranic Acids from *Adenostemma lavenia. Journal of Natural Products.* 42 (2): 183 – 186.

[9] Shimizu S. Miyase T. Umehara K & Ueno A. 1990. Kaurane-Type Diperpenes from *Adenostemma lavenia* O. KUNTZE. Chem. Pham. Bull. 38 5 : 1308 – 1312 1990.

[10] Bardon A. Montanaro S. Catalan CAN. Diaz JG. dan Werner Herz W. 1996. Kauranes and Related Diterpenes from *Adenostemma brasilianum. Phytochemistry.* 42. (2): 479 484. 1996.

[11] Fauzan A and Walid M. 2014. Study of Legetan Warak Utilization as Natural Food Preservative. Research Report. BAPPEDA Pekalongan City. Pekalongan (In Bahasa Indonesia).

[12] Yong-li Y. Shou-jun G. Rui-jun MA & Zui-luan W. 2007. Chemical Composition of the Volatile Oil in *Adenostemma lavenia* L.O.Kutze. *J. Tropical and Subtropical Botany.* 15 (4): 355 – 358.

[13] Fatimah S. Handarto BM. 2008. The Influence of Planting Media Composition Toward Growth and Plant Result of *Andrographis paniculata* Ness. *Embryo* 5 (2): 133-148. (In Bahasa Indonesia)

[14] Gil-Munoz R. Fernández-Fernández JI. Crespo-Villegas O dan Garde-Cerdán T. 2017. Elicitors used as a tool to increase stilbenes in grapes and wines. *Food Research International.* 98 (2017): 34-39. DOI 10.1016/j.foodres.2016.11.035.

[15] Raskin I. Ripoll C. 2004. Can an apple a day keep the doctor away? Current Pharmaceutical Design. 10:1-9

[16] Stanchiu 2017. Climatic conditions influence emerging mycotoxin presence in wheat grown in Romania – A 2-year survey. *Crop Protection.* 100 (2017): 124-133.

[17] Achakzai AKK. Achakzai P. Masood A. Kayanil SA dan Tareen RB. 2009. Response Of Plant Parts And Age On The Distribution Of Secondary Metabolites On Plants Found In Quetta. *Pak. J. Bot.*41 (5): 2129-2135.

[18] Figueiredo AC. Barroso JG. Pedrol LG and Scheffer JJC. 2008. Factors affecting secondary metabolite production in plants: volatile components and essential oils. Flavour And Fragrance J. 23 (2008): 213–226. DOI: 10.1002/ffj.1875.

[19] Figueiredo AC. Barroso JG. Pedrol LG and Scheffer JJC. 2008. Factors affecting secondary metabolite production in plants: volatile components and essential oils. Flavour And Fragrance J. 23 (2008): 213–226. DOI: 10.1002/ffj.1875.

[20] Wink M. 2010. Biochemistry of Plant Secondary Metabolism. Wiley-Blackwell. West Sussex, United Kingdom

[21] Klavina L. Bikovens O. Steinberga I. Maksimova V and Eglite L. 2012. Characterization of chemical composition of some byrophytes common in Latvia. J Enviromental and Experimental Biology (2012 10: 27 – 34.

[22] AOAC. 1990. Official Methods of Analysis. 15th Edn. Association of Official Analytical Chemists Washington, DC, USA.

[23] Husain J., Khan A.L., Rehman N, Hamayun M., Shah T., Nisar M, Bano T, Shinwari Z.K and Lee I.J. 2009. Proximate and nutrient analysis of selected vegetable species: A case study of Karak region, Pakistan. African Journal of Biotechnology. 8 (12): 2725-29.

[24] Malik J., Santoso A., Mulyana Y., and Ozarska O. 2008. Characterization of Merbau Extractives as a Potential Wood-Impregnating Material. BioResources 11(3): 7737-53.

[25] Steinbach M., Ertoz L., Kumar V. 2003. Challenges of clustering in high dimensional data. University of Minnesota Supercomp. Inst. Res. Report 2013: 1-33

[26] Downs G.M., Barnard J.M. 2003. Clustering methods and their uses in computational chemistry. Rev. Comput.Chem. 18: 1–40.

[27] Anderberg, MR. 2014. Cluster Analysis for Applications: Probability and Mathematical Statistics: A Series of Monographs and Textbooks. Academic Press, New York.

[28] Hervada-Sala C. and Jarauta-Bragulat E. 2004. A program to perform Ward`s clustering method on several regionalized variables. Comp. Geosci. 20: 881–886.

[29] Mandal V, Mohan Y & Hemalatha S. 2007. Microwave assisted extractionan innovative and promissing extraction tool for medicinal plant research. Pharmacognosy Reviews 1:7-18

[30] Zheng J.L. 2007. Bio-oil from fast pyrolysis of rice husk: Yields and related properties and improvement of the pyrolysis system. J. of Analytical and Applied Pyrolysis 80(1): 30-35.

[31] Mullen, C.A., and Boateng, A A. 2008. Chemical composition of bio-oils produced by fast pyrolysis of two energy crops. Energy and Fuels 22(3): 2104-2109.

[32] Okuda, T.; Y. Kimura, T. Yoshida, T. Hatano, H. Okuda, and S. Arichi. 1983. Studies on Activities of Tannins and Related Compounds from Medical Plants and Drugs. I. Inhibitory Effects on Lipid Oxidation in Mitochondria and Microsomes of Liver. Chem. Pharm. Bull. 31:1625–31.

[33] Ogata, M.; Hosho, M.; Shimotohno, K.; Urano S.; and Endo, T. 1997. Antioxidant Activity of Magnolol, Honokiol, and Related Phenolic Compounds. JAOCS. 74(5):557-562.

[34] Li, H.; Liu, L.; Zhang, S.; Cui, W.; and Jiaping, LV. 2012. Identification of Antifungal Compounds Produced by Lactobacillus casei AST18. Curr Microbiol (2012) 65:156–161.

[35] Buskes, M.J.; Harvey, K.L.; Richards, B.J.; Kalhor, R.; Christoff, R.M.; Gardhi, C.K.; Littler, D.R.; Cope, E.D.; Prinz, B.; Weiss, G.E.; O'Brien, N.J.; Crabb, B.S.; Deady, L.W., Gilsonb, P.R.; and Abbott, B.M. 2016. Antimalarial activity of novel 4-cyano-3-methylisoquinoline inhibitors against Plasmodium falciparum: design, synthesis and biological evaluation. Organic &Biomolecular Chemistry. 14(20):4617-39.

Appendix 1. Peak assignments and relative abundance (%) of pyrolysis products of *Adenostemma*.

R-Time	Name	MG	WM	Formula	Al1	Al2	Al3	Ap1	Ap2	Ap3
18,643	2-Methyl-8-Propyldodecane	Al		C16H34						1,69
21,225	4,5-Nonadiene (Cas)	Al		C9H16					5,25	
21,333	Cycloundecene	Al	152	C11H20			1,41			
22,758	1-Chloropentadec-11-Yne	Al		C15H27Cl						1,82
22,925	Cyclohexene, 3-Methylene-4-(1,2-Propadienyl)	Al		C10H12						1,54
14,042	N-Heptanal	Al		C7H14O				2,01		
18,265	N-Nonanal	Al		C9H18O						2,16
3,22	(O-D)Ethenol	Alc		C2H3DO					2,77	4,69
14,782	3,5-Heptadien-2-Ol, 2,6-Dimethyl-	Alc	140	C9H16O		1,83				
16,396	Hexa-2,4-Diyne-1,6-Diol	Alc		C6H6O2						1,52
19,438	3-Nonyn-2-Ol (Cas)	Alc		C9H16O						1,93
20,086	Cis-9-Octadecen-1-Ol	Alc		C18H36O						3,86
20,358	Octilin	Alc		C8H18O					5,89	
20,395	Phytol	Alc		C20H40O						3,77
21,229	(Z)6-Pentadecen-1-Ol	Alc	226	C15H30O	2,61					
24,076	N-Eicosanol	Alc	298	C20H42O		2,1				
18,875	Dodecan-7-Ol	Alc		C12H20O						1,45
16,196	5h-1-Pyrindine	Alk	117	C8H7N			5,07			
17,017	3-Methylindole	Alk	131	C9H9N		0,93				1,62
17,064	D-Glucose, 4-O-[3-Acetyl-1-(Trimethylsilyl)-1h-Indolyl]-	Alk	726	C32H62N2O7Si5			2,14			
17,839	D-Glucose, 2-O-[3-Acetyl-1-(Trimethylsilyl)-1h-Indolyl]-	Alk		C32H62N2O7						1,38
18,364	6,7-Dihydro-3-Nitro-5h-Cyclopenta[B]Pyridin-2(1h)-One	Alk	180	C8H8N2O	2,94					
21,523	5,10-Diethoxy-2,3,7,8-Tetrahydro-1h,6h-Dipyrrolo[1,2-A;1',2'-D];	Alk	250	C14H22N2O2	4,00					
22,245	1-Cyano-3methylisoquinoline	Alk	168	C11H8N2		1,83				
23,466	6-Benzyl-3-Methyl-6,7-Dihydro-4h-Isoxazolo[5,4-C]	Alk		C14H14N2O2						1,27
14,55	3,5-Xylenol	Ar		C8H10O						1,68
15,444	Hyacinthin	Ar		C8H8O						1,83
15,465	O-Tolualdehyde	Ar		C8H8O				0,66		
16,413	4-Formylcyclohexene	Ar		C7H10O				1,71		
19,923	Mixture Of 1-Methylidene-4a.Alpha.-Methyl-1,2,3,4,4a,9,10,10a.Alpha.Octahydro	Ar		C16H20						2,08
21,215	Biphenyl, 3,4-Diethyl	Ar	166	C12H22		7,54	3,41			8,13
21,699	5-Beta,8-Beta-Epoxy-3,5,8,8a-Tetrahydro-1h-2-Benzopyr-	Ar	150	C9H10O2	3,68					
22,297	Spiro[Cyclobutane-1,1'(2'h)-Phenanthrene],	Ar	240	C18H24			2,54			
22,396	3,6-Dimethylphenanthrene	Ar	206	C16H14		3,17				
22,451	Anthracene, 9,10-Dimethyl- (Cas)	Ar	206	C16H14			4,95			
23,184	1-Iodo-4-Phenylbicyclo(2	Ar	312	C14H17I			1,69			
9,38	Cyclopentenone	C		C5H6O				0,86		
12,934	Corylon	C	112	C6H8O2	5,44			2,43		1,41
12,936	2,5-Methano-2h-Furo[3,2-B] -8-One, Hexahydro-	C	154	C8H10O3			1,51			

13,326	3-Ethyl-2-Hydroxy-2-Cyclopenten-1-One	C		C7H10O2				1,4		
17,083	2-Methyldecane	Lp	156	C11H24		0,67				
17,884	Carbazic Acid, 3-Pentylidene-, Ethyl Ester	Lp	168	C12H24			3,04			
18,235	Oleic Acid	Lp	282	C18H34O2		1,92				1,55
18,609	N-Dodecane	Lp	170	C12H26		0,81				
19,242	Adacene 12	Lp	168	C12H24		0,53	3,24			
20,048	Z9-Dodecenylacetate	Lp	226	C14H26O2		0,7				
20,123	Trans-2-Dodecenal	Lp	180	C12H20O			2,28			
21,094	Palmitic Acid	Lp	256	C16H32O2	3,92	3,21	4,38	9,83	8,58	5,05
21,234	Cycloundecene, (Z)-	Lp		C11H20						3,15
21,265	1,12-Tridecadiene	Lp	180	C13H24			5,11			
21,316	Ambrettolide	Lp		C16H28O2						1,84
21,477	1a,9b-Dihydro-1h-Cyclopropa[A]Anthracene	Lp	192	C15H12		4,3	5,75		5,97	
21,514	1-Methylanthracene	Lp		C15H12						5,16
21,908	9,12-Octadecadienoyl Chloride, (Z,Z)- $$	Lp	298	C18H31ClO		1,43				
22,078	1-Chlorooctane	Lp	148	C8H17Cl		1,85				
22,525	Epoxycyclododecane	Lp	182	C12H22O		17				
22,583	Stearic Acid	Lp	284	C18H36O2	3,24	1,36		5,05		
22,958	Cis-3-Undecene-1,5-Diyne	Lp	146	C11H14		2,29				
22,982	Cyclododeca-1,5-Dien-9-In	Lp	160	C12H16			3,11			
23,058	2-Octenoic Acid (Cas)	Lp	142	C8H14O2		1,29				
23,228	1-Chlorooctadecane	Lp	220	C17H16		2,8	2,71			
23,333	1-Chlorooctadecane	Lp	288	C18H37Cl		1,32				
23,883	Cis-Octadec-9-Enal	Lp	266	C18H34O	2,73				1,99	1,48
23,899	Trans-Dodec-5-Enal	Lp	182	C12H22O		2,79				
24,062	Cyclohexyleicosane	Lp	364	C26H52	2,01				1,97	
2,804	Cyclopropyl-Cis-1,2,3-D3-Methanol	MA	72	C4H5D3O			3,93			
3,298	2,3-Epoxybutane	MA		C4H8O				0,51		
11,368	6-Oxa-Bicyclo[3,1,0]Hexan-3-One	MA	98	C5H6O2	2,23			0,57		
13,879	N-Pentanal	MA	86	C5H10O	3,88					1,63
19,645	2-Imidazolidinone, 1,3-Diethenyl-	N	138	C7H10N2O			4,03			
3,308	Methanamide	N	45	CH3NO	6,62					
3,317	Ammonium Carbamate	N	61	CH3NO2		7,59				
5,48	Sym-Dimethylhydrazine	N		C2H8N2				2,55		
16,152	7-Cyano(15n)-Cycloheptatriene	N	117	C8H7 15N		1,97			2,07	
16,176	1-Methyl-2-Cyanobenzene	N		C8H7N						3,79
17,175	2-Amino-4,5-Dimethyloxazole	N		C5H8N2O						2,09
19,806	N-Phenyl-N'-Furaldehyde Hydrazone	N	186	C11H10N2O	2,19					
19,819	1-Cyclopentene-1-Carboxamide	N		C12H13NO						1,26
20,158	1,8-Octanediamine	N	144	C8 H20 N2		3,04				
20,608	Pentadecanonitrile	N	223	C15H29N		1,18	1,71			
20,642	Palmitonitrile	N		C16H31N						1,16
21,355	1,4-Diaza-2,5-Dioxo-3-Isobutyl Bicyclo[4,3,0]Nonane	N	210	C11H18N2O2	2,49	5,47	5,22			
22,78	Amide 16	N	255	C16H33NO	2,09		3,41		1,03	
24,46	Oleoamide	N	281	C18H35NO		8,72	1,54			3,76
12,55	Phenol (Cas) Izal	Phenol	94	C6H6O			1,85			2,62
13,458	O-Cresol	Phenol	108	C7H8O	2,83		2,52	1,51		1,12
13,658	Guaiacol	Phenol	124	C7H8O2	2,91				1,97	
13,716	Phenol, 4-Methoxy- (Cas) Hqmme	Phenol		C7H8O2				1,94		
13,746	P-Cresol	Phenol	108	C7H8O	4,47					1,9
14,773	O-Ethylphenol	Phenol	122	C8H10O	2,19					
14,787	M-Ethylphenol	Phenol		C8H10O					2,48	1,78
14,832	2,3-Dimethylphenol	Phenol		C8H10O				0,83		
14,897	2-Methoxy-4-Methylphenol	Phenol		C8H10O2				1,04		
15,778	2,5-Dimethoxytoluene	Phenol		C9H12O2				0,45		
16,161	Carvacrol	Phenol	150	C10H14O	3,52			1,59		
16,483	2,6-Dimethoxyphenol	Phenol	154	C8H10O3	3,29			1,99	1,99	
16,625	5-Propyl-Guaiacol	Phenol		C10H14O2				0,32		
17,326	1,2,3-Trimethoxybenzene	Phenol	168	C9H12O3	2,33					
17,391	2,5-Dimethoxybenzyl Alcohol	Phenol		C9H12O3				5,98		
17,458	Aceteugenol	Phenol		C12H14O3				3,89		
18,028	Toluene, 3,4,5-Trimethoxy-	Phenol		C10H14O3				0,63		
18,22	1,6-Anhydro-Beta-D-Glucopyranose; Levoglucosan	Phenol	162	C6H10O5	8,61				6,27	
18,229	1-(4-Hydroxy-3methoxy	Phenol		C10H12O3				6,73		
18,369	Methyl Meta-Methoxybenzyl Acetate	Phenol		C10H12O3					1,04	
18,657	4-Allyl-2,6-Dimethoxyphenol	Phenol	194	C11H14O3	2,19			14,2	2,01	
19,709	Myristic Acid	Phenol		C14H28O2					6,68	
19,763	Acetosyringone	Phenol	196	C10H12O4	1,84				2,49	
19,866	3,4,5-Trimethoxybenzaldehyde	Phenol		C10H12O4				5,76		
19,905	Coniferyl Alcohol	Phenol	180	C10H12O3	2,88			13,7	6,75	
19,983	2,4-Hexadienedioic Acid, 3,4-Diethyl-, Dimethyl Ester	Phenol	226	C12H18O4	1,93				3,71	
20,583	2-Methyl-4,6-Bis(Ethylamin	Phenol		C8H15N5					2,66	
21,707	1h-2,8a-Methanocyclopenta[A]Cyclopropa[E]Cyclodecen-11-One, 1a,2,5,5a,6,	Phenol		C20H28O6					6,46	
22,4	Linoleic Acid	Phenol	280	C18H32O2	3,32			11,9		
22,417	Tetradecahydroanthracene	Phenol		C26H48					11,8	
22,483	Methyl Linolenate	Phenol		C19H32O2					4,11	
22,525	Mannit, 5-Phenylpent-1-Yl-	Phenol		C17H28O6					2,3	
22,556	N-Hexatriacontane	Phenol	507	C36H74	2,53					
12,668	L-Limonene	TS	136	C10H16		2,31	4,68			2,62
20,195	3A,9B-Dimethyl-1,2,3A,4,5,9B-Hexahydro-Cyclopent	TS		C15H18O						1,65
20,551	Isopropyl Biphenyl	TS	196	C15H16		2,87	1,86			
21,803	N-Pentacosane	TS	352	C25H52		1,8	4,45			
22,551	Cis,Cis,Cis-8,11,14eicosatrienoic Acid	TS	306	C20H34O2			10,6			12,6
24,211	Spiro[Androst-5-Ene-17,1'-Cyclobutan]-2'-One,3-hydroxy	TS	328	C22H32O2		1,35	1,88			
27,529	17-Acetoxy-19 kauranal	TS		C22H34O3					1,73	
38,938	Stigmasta-5,22dien	TS	454	C31H50O2	5,08	2,08				

Note: MG, Metabolite group; WM, weight of molecular; Al_1, root of *Adenostemma lavenia*; Al_1, root of *Adenostemma lavenia*; Al_1, root of *Adenostemma lavenia*; N, N-bearing compounds (except alkaloids); Alk, Alkaloids; MA, multi-origin aliphatic compounds with C≤6; C, furan originated from carbohydrates. pyran and cyclopentene derivatives; Al. aliphatic compounds with C>6; Alc, Alcohol; Ar, aromatic compounds (except phenolic compounds); Ph, phenolic compounds; Lp, compounds originated from lipids (except terpenoid, steroid); TS, terpenoid-steroid.

The Utilization of Blue Swimming Crab *(Portunus pelagicus)* Waste Product, Lemi, as a Food Flavor

[1]**A Y Sasongko, E N Dewi and U Amalia**
[1]Department of Fish Products Technology, Faculty of Fisheries and Marine Sciences, Diponegoro University
Jl. Prof. Soedarto, SH, Tembalang, Semarang, Central Java – 50275

Email: ulfahamalia0@gmail.com

Abstract. *Lemi* is a wasted product that resulted from the meating process of blue swim crab. One of the utilization of blue swim crab *lemi* is processed it into a food flavor. The aim of this research was to know the value of glutamic acid in blue swim crab lemi flavor with the addition of dextrin using different concentration and know the level of consumer preference of lemi flavor by using hedonic test. The research was using a Completely Randomized research Design (CRD) with a factor of 0%, 1%, 2%, and 3% dextrin concentration. The treatment that was tested was the additions of 0%, 1%, 2%, and 3% dextrin. The nonparametric data (panelist hedonic level) was analyzed by Kruskal-Wallis and further analysis using Mann-Whitney. The parametric data (glutamic acid content, protein content, moisture content, and solubility level) were analyzed by analysis of varians and further analysis using Honestly Significant Difference. The results showed that flavor with 1% dextrin addition has the highest hedonic score ($7,07 < \mu < 7,38$) than the others, and its chemical characteristics were as follows; protein content (d/b) was (42,74% ±0,60), glutamic acid content was (66,12% ±0,52), moisture content was (11,78%±0,24) and dissolve level was (80,86%±0,11). The addition of dextrin with different concentration gave effect to the blue swim crab lemi flavor. The flavor resulted from this experiment can be used as an alternative of blue swim crab *lemi* as processing waste so that it can optimalized any further.

Keyword : Blue Swim Crab *Lemi,* Dextrin Concentration, Flavor, Glutamic Acid

1. Introduction

The by product of the processing of blue swim crab consists of solid and liquid waste. The solid waste of the processing of blue swim crab is blue swim crab carapace 55% body reject 5%, while the liquid waste is as much as 25%. Blue swim crab lemi is included in the body reject solid waste group. Lemi has a high amount of protein which is as many as 15.65% that is composed of amino acids. One amino acid that contained in its protein is glutamic amino acid. Glutamic amino acid has a very important role in giving an umami taste in food. Tastes that is formed by food are sweet, sour, salty, bitter, and umami taste [1].

Lemi is an unutilized product in the meating process of blue swim crab, because it could change the color of blue swim crab meat that could lead to the quality and price worsening. Lemi (mustard) is a yellowish product that usually found under the surface of the cooked blue swim crab carapace. This product has a delicious taste like the blue swim crab meat [2]. .

In the production of a powder from a liquid form, filler products that functioned as binding agent (binder) is needed. Dextrin can be used as a volatile compound and oil encapsulation products so that it can protect any compounds that is sensitive to high temperature and oxidation process. It is because dextrin has molecules that stable against high temperature and oxidation process. The use of dextrin is expected to be a good protein binding agent and protect it against high temperature during drying or any further processes [3].

2. Methods

2.1. Materials
The main material that used in this reserach was blue swim crab (*Portunus pelagicus*) lemi. Meanwhile, materialss that is used in the making of blue swim crab flavor powder were dextrin, water, and blue swim crab lemi. The instruments used in the making of the product were analytical scales and spray dryer.

2.2. Research Design
This research was done by making flavor using blue swim crab lemi as main product that added with 3 different dextrin concentrations, with 0% dextrin addition as control. Each concentration of dextrin added is 1%, 2%, and 3%. This research was using Completely Randomized research Design (CRD).

2.3. Research Procedure
Flavor from blue swim crab lemi was made by cooking it with 1:2 comparison of water. Then, the product that already cooked was filtered and then dextrin was added with 0%, 1%, 2%, and 3% concentrations. The homogen mixture then was put into the spray dryer instrument with inlet temperature of 130^{0}C and outlet temperature of 70^{0}C so that flavor powder was formed.

2.4. Testing Procedures
Testing that conducted in this research were hedonic test which refers to National Standardization Institution [4], Protein and moisture content [5], Glutamic acid [6] and Solubility level test [7].

2.5. Data Analysis
The analysis method that used in this research was analysis of variance (ANOVA) with honestly significant difference (HSD) further analysis to determine either the affecting or not affecting point. If the obtained data difference was significant then HSD further analysis should be done.

3. Results and Discussions

3.1. Panelists Hedonic Level of Blue Swim Crab Lemi Flavor
The results of blue swim crab lemi flavor hedonic test can be seen in Table 1. below.

Table 1. The Results of Blue Swim Crab Lemi Flavor Hedonic Test

Specification	Panelists Hedonic Level			
	0% dextrin	1% dextrin	2% dextrin	3% dextrin
Flavor	8.10	7.70	6.20	4.26
Taste	8.20	7.70	5.83	5.00
Color	5.50	6.50	7.20	8.00
Texture	5.30	6.70	7.70	8.20
Averages	6.72±0.36	7.2±0.55	6.73±0.47	6.36±1.33

The results were average point ± standard deviation of 30 panelists

The results of hedonic test above showed that the best concentration of blue swim crab lemi flavor is in the range of 1% and 2% dextrin concentration. The concentration of dextrin that less than 1% will resulted in the flavor and taste of lemi that is too stingy so that it will be less preferred by panelists and the hedonic point results will be less than 7, so that the flavor will be less preferred and less acceptible by panelists. Meanwhile, in the concentration of dextrin that more than 2% will resulted in the less noticeable flavor and taste so it is also less preferred by panelists. The more the concentration of dextrin that being added, the better the color and texture of the product. The addition of dextrin causes the color of the resulting flavor product turn to yellowish white, where that color is more preferable than those flavor product that was made without using dextrin which has a brownish yellow color. The texture of the resulted flavor product also smoother and not clot when it is added with dextrin as the binder.

3.2. Protein Content

The results of blue swim crab lemi flavor protein content with addition of varies dextrin concentrations can be seen in Table 2.

Table 2. The Average Protein Content of Lemi Flavor with the Addition of Dextrin

Dextrin Concentration (%)	Average(s) WW±SD(%)	Average(s) DW±SD(%)
0	31.09±0.61[a]	35.79±0.61[a]
1	43.07±0.54[d]	48.83±0.54[d]
2	38.27±0.29[c]	42.49±0.29[c]
3	34.12±0.52[b]	37.74±0.52[b]

- The data was the average result ± standard deviation with three replication
- Different supperscript letters shown a significant difference among treatments ($p < 0,05$)
- WW: Wet Weight, DW: Dry Weight

The lowest protein content of lemi's flavor was controls; without dextrin addition, thus protein would be denaturate easily during the heating process. While in the 1% addition of dextrin treatment, protein that contained in the flavor will be protected because dextrin has a more stable characteristic against high temperature so that it can protect any volatile compounds and any compounds which are sensitive to high temperature in the drying process using spray dryer so that the protein content will be increased. On the other hand, the other treatments were 2% and 3% dextrin addition causing lower of lemi's flavor moisture content. This was because the amount of solid products that contained in it was increased, while the source of protein was derived from the moisture content of blue swim crab lemi itself. The more the binding agent used the less the protein content because the carbohydrate content of binding agent was acid characterized so that protein will be hydrolized by acid that contains in the dextrin and lead to protein denaturation. This present study in line with previous studies done by Jinap *et al.* [8] stated that the above condiments are added to foods during food preparation in a relatively small amount and they interact with the original food matrix, including the present glutamate and other free amino acids.

The protein denaturation process was occur in the secondary and tertiary protein structure but it does not change the primary structure of protein. The mechanism of protein denaturation caused by acid was; the protein will experience the biggest turbidity upon reaching the isoelectric pH, which is the pH that the protein has the same positive and negative charge, at which time the protein undergoes denaturation that is marked by the increasing turbidity. Acid can also disrupt salt bridges in the presence of ionic charges. A type of double-replacement reaction occurs when positive and negative ions in the salt alternate with positive and negative ions derived from the acid present in dextrin. According to Meiyani *et al.*, [9] the acid characteristic of dextrin will cause starch to undergo chain breaking process by enzyme or acid during heating process to a smaller molecules.

As a comparison, Sutardi research [10] shown that protein content was decreasing as the increasing number of added binder. The higher the binder that added will lead the protein content of sweet corn syrup to decreases, Considering that the binder in the form of dextrin is a carbohydrate group and has no free protein component at all. Thus precisely the addition of dextrin physically lowers the resulting protein powder levels. The highest protein content was found in the addition of 0% dextrin which was 9.20% and the lowest protein content was found in the addition of 7.5% dextrin concentration which was 3.25%.

3.3. Glutamic Acid
The results of blue swim crab lemi flavor glutamic acid value with addition of varies dextrin concentrations can be seen in Table 3.

Table 3. The Average Glutamic Acid Value of Lemi Flavor with the Addition of Dextrin

Dextrin Concentration (%)	Average(s) WW±SD(%)	Average(s) DW±SD(%)
0	48.65±0.27[a]	52.38±0.27[a]
1	66.12±0.63[d]	74.94±0.63[d]
2	60.32±0.49[c]	66.95±0.49[c]
3	54.19±0.19[b]	59.95±0.19[b]

- The data was the average result ± standard deviation with three repeatences
- Different superscript letters shown a significant difference among treatments ($p < 0,05$)
- WW: Wet Weight, DW: Dry Weight

The increased glutamic acid value from 31.09% to 43.07% at 0% and 1% treatment was caused by the characteristics of dextrin as binding agent. Dextrin as binding agent of blue swim crab lemi flavor was functioned to protect the nutrients contained in the blue swim crab lemi flavor along the drying proces period. The increased glutamic acid also affected by the occurence of hydrolysis processes between proteins with acid that contained in dextrin. According to Meiyani *et al.* [9], one of the properties of protein is it can be hydrolized, so that the protein in the shrimp head was hydrolyzed with acid that contained in the dextrin and deamination will occur and form glutamic, those was the process which results in increased glutamic acid. Where glutamic acid consists of 2 carboxyl groups, 1 amino group, hydroxy group and branch chain. The formation of carboxyl groups in glutamic acid was occur from the hydrolysis process between protein and acid. The C = O compound in the protein bids to OH in dextrin of which both are present in dextrin.

The glutamic acid value decreased at 2% and 3% treatments because the higher the concentration of dextrin added the higher the amount of solids it contained. More dextrin added into lemi's flavor formulation causing the content of glutamic acid lower. While the source of glutamic acid comes from the boiling water of blue swim crab lemi itself. Glutamic acid is a part of the main structure of various types of protein molecules that contained naturally in food and low protein content causes the least protein that can decompose into amino acids so that the resulting glutamic acid value was also lower. The resulting glutamic acid value is directly proportional to the protein content. This is because gluamate acid is a forming compound of proteins, so that if the protein content decreases then the glutamic acid value will also decrease. The previous studies Kim *et al.* [11] noted that the most abundant amino acid was glutamic acid in all the pastes, followed by aspartic acid and leucine.

Zuhra and Herlina [12] stated that glutamic acid is one type of amino acid, it is bonded with other amino acids to form a protein structure. Hydrolised protein during the cooking process with high temperature will release free glutamic. This free glutamic is a key component to get the umami food. Glutamic and leucine acids are the largest dominant amino acids with a proportion of 19.90% and 13.01% of all amino acids in the raw product. Glutamic acid is the most important component in the forming of taste and flavor in seafood product.

The average value of glutamic acid in this research's blue swim crab lemi flavor was between 48.65% - 66.12%. The results of glutamic acid in this reearch were higher than the value of glutamic acid flavor of boiled shrimp head in Meiyani *et al.*, [9] research, which was the best glutamic acid value wa found in 2,5% concentration which was 36.85% and the lowest was found in 7.5% concentration which was 31.5%. As regarded to Meiyani *et al.* [9] that he properties of protein is denaturation which is breaking the active structure of proteins and in high temperature conditions or and in a strong and numerous acidic chemicals. So that at 7.5% concentration the glutamic acid value decreases.

3.4. Moisture Content

The results of blue swim crab lemi flavor moisture content with addition of varies dextrin concentrations can be seen in Table 4.

Table 4. The Average Moisture Content of Lemi Flavor with the Addition of Dextrin

Dextrin Concentration (%)	Average(s) ± SD(%)
0	13.14±0.18[c]
1	11.78±0.24[b]
2	9.94±0.38[a]
3	9.61±0.30[a]

- The data was the average result ± standard deviation with three replication
- Different superscript letters shown a significant difference among treatments (p < 0,05)

The moisture content of the sample flavor without binding agent (dextrin) quantitatively shows the highest value (13.14%) than the sample flavor with 3% binding agent (9.61%). The addition of dextrin can increase the product total solids. The higher the total solids that drained, the higher the speed of evaporation, so that the moisture content become lower. Dextrin can reduce the value of moisture content, because dextrin is one of product that has ability to bind other product. According to Meiyani *et al.,* (2014), the addition of dextrin with varies concentrations and the drying process could obtain significant results in lowering the moisture content of the flavor. Purnomo *et al*., (2014), stated that filler product may increase the total solids of the product and the higher the total solids that dried to a certain extent, the higher the evaporation rate so that the moisture content of the products decrease. Sutardi *et al.,* (2010) also stated that, the effect of giving dextrin with varies concentrations was able to reduce moisture content, because the dextrin content has a low molecular weight and simple molecular structure, so that water can easily be evaporated during the drying process either in the form of free water, physically or chemically bound.

The heating process that was done in the production of flavor is by using spray dryer because it was greatly affect the moisture content in the product. Where the higher the heating temperature, the lower the ability of a food product to absorb water from the air, so that the moisture content that contained in the product was decreases. Moisture content is closely related to solubility, this thing is influenced by several factors and one of them is the used binding agent such as dextrin which is hydrophilic. As a comparison, Hastuti and Hidayati [13] research showed that based on the analysis of moisture content in each treatment the highest average value was obtained from the treatment of 0% dextrin concentration which was 4.05%, while the lowest was obtained from the treatment of 2% dextrin concentration which was 3.48%. the high amount of moisture content in the treatment without dextrin was because it does not use any binding agent, while the more the decreasing moisture content of 1% and 2% treatments were because dextrin was able to bind moisture content that contained in the food product.

3.5. Solubility of Blue Swim Crab Lemi Flavor Solids

The results of the solubility blue swim crab lemi flavor solids with addition of varies dextrin concentrations can be seen in Table 5.

Table 5. The Average Solubility Solids of Lemi Flavor with the Addition of Dextrin

Dextrin Concentration (%)	Average(s) WW±SD(%)	Average(s) DW±SD(%)
0	80.00±0.40[a]	80.30±0.40[a]
1	80.86±0.11[b]	81.24±0.11[b]
2	80.93±0.11[b]	81.25±0.11[b]
3	81.26±0.23[b]	81.60±0.23[b]

- The data was the average result ± standard deviation with three replication
- Different superscript letters shown a significant difference among treatments ($p < 0,05$)
- WW: Wet Weight, DW: Dry Weight

The solids solubility value of blue swim crab lemi flavor was increased, this relates to the characteristic of dextrin which is a filler that able to bind flavor component in water. These properties cause the moisture content in the flavor product to be lower. The lower content of moisture content will result in the forming of more hygroscopic powder and it will be easily to absorb water so that the solubility of flavor in water is also getting bigger. Those results was similar with Hardjanti [14] opinion which state that low moisture content will cause the powder to become more hygroscopic so that there is a significant difference in water vapor pressure between solid and liquid. Beside that, the blue swim crab lemi flavor microencapsulation powder with low moisture content are more porous than those with higher water content. As a result the ability to absorb water is greater and the solubility will become larger.

Sutardi and Constantia [10] stated that, the granules structure in dextrin also affecting the solubility. The process of starch repolymerization with acidic help at the time of dextrinization makes the dextrin molecule split in smaller sizes with a more hygroscopic component, so that when it encounter heating process with low pressure, the starch particles will be damaged. As a result the water will easily moves into it while releasing water-soluble component, so that the solubility is high.

As a comparison Mulyadi *et al.* [15] research showed that on the comparison of physical quality for solubility parameters, the average value of solubility test obtained is almost the same, where the best treatment has 75.67% solubility level, while the solubility level of control treatment is 74.67%, this is because for powder product, the higher the solubility the better the product because it is able to dissolve faster when mixed with water. The addition of dextrin causes an increase in solubility.

4. Conclusions

Based on the obtained research results and discussions about the utilization of blue swim crab (*Portunus pelagicus*) waste product, lemi, as a food flavor, we can conclude that:

1. Flavor that made from blue swim crab lemi with addition of 1% dextrin concentration has the best averages in glutamic acid value (main parameter) which was 66.12%, with 43.07% protein content, 11.78% moisture content, and 80.86% solids solubility as supporting parameters.
2. The panelists hedonic level of blue swim crab lemi flavor with addition of 1% dextrin concentration has the highest averages point which was 7,2±0,55, which showed that this blue swim crab lemi flavor was more preferrable by the panelists.

5. References

[1] Multazam. 2002. Prospek Pemanfaatan Cangkang Rajungan (*Portunus sp*) Sebagai Suplemen Pakan Ikan. [Skripsi]. Bogor. Fakultas Perikanan Dan Ilmu Kelautan. Institut Pertanian Bogor.

[2] Flick, G.J dan R. E.Martin. 1990. The Seafood Industry. Published byVan Nostrand Renhold. New York.

[3] Hakim, A. R., dan A. Chamidah. 2013. Arabic Gum and Dextrin Application as A Binding Agent of Head Shrimp Extracted Protein. Journal of Processing and Biotechnology of Marine and Fisheries (In Bahasa Indonesia), 8(1): 45–54.

[4] Badan Standardisasi Nasional. 2006. SNI 01-2346-2006. Petunjuk Pengujian Organoleptik dan atau Sensori.

[5] [AOAC] Association of Official Analytical Chemyst. 2007. Official Method of Analysis of The Association of Official Analytical Chemyst. Arlington

[6] Daniel F., Daniel B., David C. 2011. Defining microbial terroir: The use of native fungi for the study of traditional fermentative processes. *International Journal of Gastronomy and Food Science*. Vol. 1. 64-69.

[7] Kainuma, K, T. Odat, and S. Cuzuki. 1967. Study of Starch Phospate Monoesters. J Technol Soc Starch 14: 24-28.

[8] Jinap, S., Ilya-Nur, AR., Tang, SC., Hajeb, P., Shahrim, K., Khairunnisak, M. 2010. Sensory attributes of dishes containing shrimp paste with different concentrations of glutamate and 5'-nucleotides. *Appetite* (55): 238-244.

[9] Meiyani, D. N. A. T., P. H. Riyadi dan A. D. Anggo. 2014. Pemanfaatan Air Rebusan Kepala Udang Putih (*Penaeus merguensis*) sebagai Flavor dalam Bentuk Bubuk dengan Penambahan Maltodekstrin. Jurnal Pengolahan dan Bioteknologi Hasil Perikanan, 3 (2):67-74. Universitas Diponegoro. Semarang.

[10] Sutardi, Suwedo, H. dan R. Constansia. 2010. Pengaruh Dekstrin dan Gum Arab Terhadap Sifat Kimia dan Fisik Bubuk Sari Jagung Manis (*Zeamays saccharata*). [Jurnal Teknologi dan Industri Pangan, Vol XXI No 2 Th 2010]. Universitas Gajah Mada. Yogyakarta.

[11] Kim, YB., Choi, YS., Ku SK., Jang, DJ., Hasnah, H, Moon, KB. 2014. Comparison of quality characteristics between belacan from Brunei Darussalam and Korean shrimp paste. J Ethn Foods 1; 19-23

[12] Zuhra, S dan C. Herlina. 2012. Pengaruh Kondisi Operasi Alat Pengering Semprot Terhadap Kualitas Susu Bubuk Jagung. J. Rekayasa Kimia Dan Lingkungan. 9(1): 36-44.

[13] Hastuti, A. S dan D. Hidayati. 2012. Pemanfaatan Limbah Cangkang Rajungan (*Portunus pelagicus*) Sebagai Perisa Makanan Alami. Jurusan Teknologi Industri Pertanian. Universitas Trunojoyo Madura.

[14] Hardjanti, S. 2008. Potensi Daun Katuk Sebagai Sumber Zat Pewarna Alami dan Stabilitasnya Selama Pengeringan Bubuk dengan Menggunakan Binder Maltodekstrin. Jurnal Penelitian Saintek, Vol 13, No. 1, April 2008: 1-18.

[15] Mulyadi, A.F., M. M. Jaya, Wignyanto, dan R. Harun. 2013. Karakteristik Organoleptik Serbuk Perisa Alami Dari Cangkang Rajungan (*Portunus pelagicus*): Kajian Konsentrasi Dekstrin Dan Suhu Pengeringan. Jurnal Teknologi Pertanian Vol. 14 No. 3 [Desember 2013] 183-192.

Aflatoxin and nutrient contents of peanut collected from local market and their processed foods

E Ginting, A A Rahmianna, and E Yusnawan

Indonesian Legumes and Tuber Crops Research Institute (ILETRI)
PO Box 66 Malang 65101, East Java, Indonesia

Email : erlianaginting@yahoo.com

Abstract. Peanut is succeptable to aflatoxin contamination and the sources of peanut as well as processing methods considerably affect aflatoxin content of the products. Therefore, the study on aflatoxin and nutrient contents of peanut collected from local market and their processed foods were performed. Good kernels of peanut were prepared into fried peanut, pressed-fried peanut, peanut sauce, peanut press cake, fermented peanut press cake (tempe) and fried tempe, while blended kernels (good and poor kernels) were processed into peanut sauce and tempe and poor kernels were only processed into tempe. The results showed that good and blended kernels which had high number of sound/intact kernels (82,46% and 62,09%), contained 9.8-9.9 ppb of aflatoxin B_1, while slightly higher level was seen in poor kernels (12.1 ppb). However, the moisture, ash, protein, and fat contents of the kernels were similar as well as the products. Peanut tempe and fried tempe showed the highest increase in protein content, while decreased fat contents were seen in all products. The increase in aflatoxin B_1 of peanut tempe prepared from poor kernels > blended kernels > good kernels. However, it averagely decreased by 61.2% after deep-fried. Excluding peanut tempe and fried tempe, aflatoxin B_1 levels in all products derived from good kernels were below the permitted level (15 ppb). This suggests that sorting peanut kernels as ingredients and followed by heat processing would decrease the aflatoxin content in the products.

Keywords: aflatoxin, nutrient, peanut, processing

1. Introduction

Peanut is primarily used for foods in Indonesia with a consumption level of 0.27 kg/capita/year [1]. A number of peanut products is available, such as sauces (*bumbu kacang*) for salads/*satay*, snacks (boiled, deep-fried, roasted, flour-coated, bakery filler, traditional snacks), oil and fermented of defatted peanut (peanut tempe/*oncom*) [2]. As ingredients for such products, peanut is not directly obtained from the farmer. Medium to large-scale food industry normally get supplied from the middle men, while small-scale food processors purchase peanut from the retailers, particularly at local markets [3, 4]. It takes about 40-110 days for marketing of peanut from farmer to consumer [2], suggesting that handling of peanut during such period, particularly storage would ultimately determine the quality of peanut and its safety for human consumption.

The presence of aflatoxins, the secondary metabolites produced by toxigenic fungi of *A. flavus, A. parasiticus,*and *A. nomius* [5, 6] incontaminated peanut is widely highlighted in the literature due to their hazardous effects to humans and animals. Aflatoxins that may belong to B, G, and M are reported to be hepatocarcinogenic, mutagenic, teratogenic, and immunosuppressive [7, 8]. Aflatoxin

B_1 is the most toxic and classified as the first class of carcinogen to humans [9, 10], thus it is normally used to set the permitted limit in foods and feeds. The maximum level of aflatoxin in peanut food products marketed in Indonesia has been established as low as 15 ppb for aflatoxin $B_1$and 20 ppb for total aflatoxins [11], while Codex Alimentarius set a lower level for total aflatoxins (15 ppb) [12].

High *A. flavus* infection and aflatoxin contamination in peanut during distribution/marketing is of concerned, particularly for those samples collected from local/traditional markets [3, 13-15]. The level of aflatoxin B_1in these samples may range from 3.6 up to 1,859 ppb [4], even up to 6,073 ppb in case of Cianjur Regency [14].In fact, most of small-scale peanut food processors and household consumers obtain their ingredients from retailers in local markets. The processors also tend to purchase low price of peanut (meaning low quality peanut) to get more profit since their awareness of aflatoxin hazards to humans is still low [4, 12].This is crucial as processing, such as washing, heat treatment (boiling, drying, roasting, frying), and fermentation cannot eliminate 100% of aflatoxins [7, 16-17]as they are heat resistant and insoluble in water, reflecting the high exposure risk of contaminated peanut.

Selected peanut products, such as peanut sauce for *satay, pecel,* and *gado-gado*, boiled peanut, *oncom*, oil and peanut press cake were reported to contain aflatoxin B_1> 15 ppb, while roasted unshelled peanut and flour coated peanut gave lower levels of aflatoxin B_1 (< 15 pp) [18]. This suggests that initial quality of peanut kernel and processing may affect the final aflatoxin content in peanut products. In addition, nutritional aspects of peanut is also essential as it is rich in fat and protein as well as vitamins and minerals [19], which also may change during processing. Therefore, this project was performed to study the aflatoxin and nutrient contents of selected peanut food products processed from peanut kernels collected from a traditional market as ingredient. This information would be useful for selection of peanut as food ingredient and processing methods in terms of aflatoxin reduction in the products.

2. Materials and Methods

Peanut kernels (Local Ponorogo variety) were obtained from a local market in Ponorogo, East Java. The kernels were then sorted into 3 groups (good, blended, and poor/damaged kernels)[20]. The good kernels were processed into fried peanut, pressed-fried peanut, peanut sauce, peanut press cake, fermented press cake (peanut tempe) and fried peanut tempe. This trial was carried out at the ILETRI Laboratory of Food Chemistry and Technology, Malang using a randomized complete design with three replicates. In addition, the blended kernels (good and poor kernels) were processed into peanut sauce and peanut tempe, while the poor kernels were only processed into peanut tempe since normally peanut tempe in the market was prepared from low quality peanut press cake. Peanut tempeprocessed from blended and poor kernels were also fried.Observations included physical quality of the kernels(intact, shriveled, and damaged kernels) [20], chemical composition of dry kernels and peanut products, including: (a) moisture (gravimetric), ash (muffle furnace) and fat (Soxhlet) contents[21], (b) protein content using Micro Kjeldahl method [22].*A. flavus* infection was observed through growing 100 peanut kernels on 10 petridishes containing *Aspergillus flavus* and *parasiticus* agar (AFPA). The number of mold infected kernels with orange/dark yellow colour was noted on the fourth day and calculated in percentage [2]. The ELISA method was used for analysis of aflatoxin B_1 content both in the kernels and peanut products [23].

3. Results and Discussion

3.1. *Physical quality, A. flavus infection and aflatoxin content of peanut kernels*

The moisture contents of peanut kernels were relatively low and similar for the three kernel categories (Table 1). These levels were slightly higher than the requirement for national quality standard ($\leq$ 6%) [20], however they were already safe for storage (7.5%) [24]. Even though the kernels were stored in a perforated plastic sack when purchased in the market, the moisture content was fairly low due to dry season air conditions.The blended kernels (real peanut quality) obtained at present study had lower moisture content relative to those samples collected from retailers (>8%) in Banjarnegara [2].

Different methods and period of storage as well as temperature and relative humidity of the surrounding air may contribute to such differences in moisture content. In addition to perforated plastic sack, retailers may also store the peanut kernels in an open wooden box, plastic basin, bamboo tray or bamboo basket [2].

Sorting by hand picking gave a high number of intact kernels and low number of shriveled and damaged kernels in good kernels,while poor kernels showed a high number of damaged kernels (Table 1). Normally, these kernels are used for oil extraction purposes and the press cake is further processed into tempe, namely *oncom* [25].The number of damaged kernels in blended peanut kernels (24.12%) was lower compared to those collected from retailers in Banjarnegara which ranged from 26.6% up to 69.9%[2].Peanut maturity and post-harvest handlingconsiderably dictate the physical quality of peanut.Peanut kernels from three categories showed much higher shriveled and damaged kernels than the maximum levels (4% and 2%, respectively) set by the national standard quality [20], reflecting high risk of aflatoxin contamination for further storage/marketed as the samples were collected one month after harvesting season in Ponorogo area.

Table 1. Moisture content, physical quality, *A. flavus* infection and aflatoxin B_1content of peanut Kernels obtained from a local market in Ponorogo, East Java

Category of peanut kernel	Moisture content (%)	Physical quality			*A. flavus* infection (%)	Aflatoxin B_1 (ppb)
		Sound/intact kernels (%)	Shriveled kernels (%)	Damaged kernels (%)		
Good kernels	6.34 ± 0.40	82.46 ± 1.68	5.68 ± 1.81	11.86 ± 0.14	1	9.9 ± 0.9
Blended kernels	6.22 ± 0.16	62.09 ± 1.44	13.79 ± 1.48	24.12 ± 2.66	1	9.8 ± 0.9
Poor kernels	6.62 ± 0.20	12.28 ± 1.61	26.12 ± 2.78	61.6 ± 3.93	2	12.1 ± 0.7

Values are means ± SD from three replicates

The level of aflatoxin B_1 was similar for both good and blended peanut kernels, while it was slightly higher in poor kernels (Table 1). Poor kernels consisted of more damaged and shriveled kernels, which had a higher possibility for aflatoxin contamination [3].However, aflatoxin B_1 levels in three categories of peanut kernels were yet lower relative to the maximum level established in Indonesia (15 ppb). These facts were in agreement with the low levels of *A. flavus* infection which were relatively low(1-2%). Low levels of moisture and relatively dry air storage environment may account for low levels of *A. flavus* infection and aflatoxin contents in peanut kernels as the mold would favourably grow and produce toxins at high moisture (15-30%), temperature (25-30°C), and relative humidity (85%) [26].This is also the reason for relatively low aflatoxin content in poor kernels, even though the number of damaged kernels was quite high. Similar finding was also noted in peanut kernels stored in an opened wooden box for 4 months with an initial moisture content <7%,which contained aflatoxin B_1<10 ppb, even though the damaged kernels was up to 80.9%[27].

The aflatoxin content observed in this study was within the range of aflatoxin levels reported from previous studies in peanut samples collected from traditional markets, which ranged from 0-1,154 ppb [13], 3.6-6.073 ppb [14], 4.4 - 205 ppb [28], and > 5-> 100 ppb [3].A study in Kenya also reported that aflatoxin B_1 content in 37% peanut samples obtained from 1,263 vendors in various markets exceeded the permitted limit (>10 ppb) [10]. This highlights the critical point of storage at market level in aflatoxin development [24]. Moisture content, physical quality, and air environment conditions are the main factors in controlling mold contamination and aflatoxin production [29].

3.2. Nutritient content of peanut kernels and their products

The moisture content of peanut kernels significantly changed through processing into peanut products (Table 2). Deep-fried products, like fried peanut and pressed-fried peanut showed the lowest moisture content due to releasing some moisture during frying. Conversely, peanut tempe gave the highest moisture level as soaking, steaming and fermentation were involved in the preparation method. Fried peanut tempe had similar moisture content to that of peanut tempe as it was covered with wheat flour dough prior to frying, thereby increasing the moisture content and only small amounts of moisture can be released during frying. Peanut sauce had medium moisture content. Even though frying was involved in the preparation method, the addition of water during blending with palm sugar and spices increased its moisture level. Low moisture (<8%) is desired for peanut products to maintain the quality from mold contamination and lipid oxidation (rancidity), thus can be stored longer.

Table 2. Chemical composition and aflatoxin B_1 content of good peanut kernels obtained from a local market and their food products

Peanut products	Moisture (%)	Ash (% dw)	Protein (% dw)	Fat (% dw)	Aflatoxin B_1 (ppbfw)
Good peanut kernels	6.34 ± 0.40^{d}	2.46 ± 0.04^{e}	26.02 ± 0.69^{d}	49.23 ± 1.02^{a}	9.9 ± 0.9^{bcd}
Fried peanut[1]	1.21 ± 0.13^{e}	3.41 ± 0.14^{c}	26.20 ± 1.50^{d}	46.49 ± 1.25^{b}	7.9 ± 1.3^{cd}
Pressed-fried peanut[2]	1.77 ± 0.11^{e}	4.08 ± 0.13^{b}	31.47 ± 1.60^{c}	44.54 ± 0.93^{b}	4.4 ± 1.7^{d}
Peanut sauce[3]	11.11 ± 0.65^{c}	4.52 ± 0.17^{a}	19.18 ± 0.62^{e}	34.48 ± 0.91^{d}	12.4 ± 3.6^{bc}
Peanut press cake [4]	4.31 ± 0.38^{d}	3.33 ± 0.12^{c}	33.31 ± 0.96^{c}	35.87 ± 1.74^{d}	9.2 ± 0.4^{bcd}
Peanut tempe[5]	48.28 ± 2.83^{b}	2.20 ± 0.05^{f}	39.23 ± 1.00^{a}	41.32 ± 2.71^{c}	26.3 ± 7.5^{a}
Fried peanut tempe[6]	51.74 ± 2.00^{a}	2.73 ± 0.20^{d}	37.30 ± 1.02^{b}	41.13 ± 2.43^{c}	16.5 ± 8.9^{b}
LSD 5%	2.28	0.25	1.90	2.29	7.93
CV (%)	7.18	4.20	3.51	3.07	13.04

Values are means ± SD; fw = fresh weight; dw = dry weight
Values in the same coloums followed by different superscript are significantly different at 5% level of LSD test.
[1] Soaking in boiled water (45 min), removing the seed coat, oven drying (80°C, 20 min), soaking (10 min) in garlic and salt solution, deep-frying (120°C, 15 min), centrifuge to remove the excessive oil.
[2] Oven (120°C, 20 min), pressing using hydraulic press, removing the seed coat, soaking (10 min) in garlic and salt solution, deep-frying (120°C, 15 min), centrifuge to remove the excessive oil.
[3] Washing, deep-frying (120°C, 10 min), grinding, mixing with water, salt, palm sugar, spices.
[4] Grinding, oven drying (80°C, 10 min), pressing using a hydraulic press
[5] Grinding, oven drying (80°C, 10 min), pressing using a hydraulic press, soaking (24 h), washing, steaming (1.5 h), cooling (4 h), inoculation of mold (*Rhizopus* spp), fermentation (room temperature, 24 h).
[6]Treatment [5] followed by steeping in wheat flour dough containing salt and garlic, deep-frying (120°C, 10 min).

The ash content of good peanut kernels was about 2.46 % dw (Table 2). Similar values (2.86%-2.89% dw) were also noted for Mahesa and Kancil varieties [27] and slightly higher in Nigerian peanut (3.8%) that primarily consisted of K, Na, P, Ca, and Mg [30]. Preparation into peanut products significantly influenced the ash content. The highest levels were examined for peanut sauce as palm sugar and spices, like salt, chili, garlic, citrus leaves, and tamarind were added. A slight increase in ash content was observed for fried peanut products due to the use of salt, garlic, and edible oil, while peanut tempe showed the lowest ash content due to leaching out of water soluble minerals during overnight soaking.

Good peanut kernels contain 26.02% dw of protein (Table 2) that was slightly lower than those of Mahesa and Kancil varieties (32.06-34.47% dw) [27]and 29 peanut cultivars originated from Peru (26.3-30.9%)[31].The protein content of peanut kernels were significantly different after processing into different products (Table 2). The highest values were seen in peanut tempe, reflecting an increase in protein after fermentation as the mold itself predominantly consists of protein. Peanut tempe in this study contained higher protein compared to *oncom* that made from commercial defatted peanut mixed with cassava starch with a protein content of 26.9% dw[32]. Peanut press cake and pressed-fried peanut also gave considerable increase in protein. Pressing the peanut kernels to remove part of the oil, could reduce 20% of fat and consequently increased the protein content by 16%[33].

Meanwhile, peanut sauce had the lowest value of protein due to a large proportion of palm sugar (40%), hence decreased the proportion of protein derived from peanut kernels.

Higher level of fat was seen in good peanut kernel (Table 2) relative to those of Kancil and Mahesa varieties (45-47% dw) [27]. As a result of processing, fat contents in peanut products were significantly different (Table 2). Peanut sauce gave the lowest value of fat due to the addition of palm sugar, thus decrease the peanut fat proportion in the product. Removing part of the oil through pressing also gave a significant decrease in fat content of peanut press cake. However, it was only a slight decrease in fat values of fried peanut and pressed-fried peanut and were not significant for both products. This was due to a relatively low pressing capacity of the hydraulic press used in this study (15 tons), hence only 8-13% of oil can be removed and subsequently increased after deep-frying. A higher capacity of the hydraulic press (45 tons) can remove oil up to 20% during pressing [33]. An increase of fat was seen in peanut tempe and fried peanut tempe, however the fat contents of both products were similar as the flour dough that covered peanut tempe was excluded prior to analysis. The fat content of peanut tempe in this study was much higher compared to that of *oncom* (6% fw) [32] as commercial defatted peanut press cake was used as ingredient in the latter study which contained very low fat (4.6%).

In terms of blended and poor kernels, the moisture, ash, and protein contents were approximately similar to that of good kernels, while it was slightly lower for fat content of poor kernels (Table 2 and 3). High number of damaged kernels due to discoloured skin, broken, mold and insect attacks in poor kernels may associate with oxidation/degradation of fat in the kernels.

Table 3. Chemical composition and aflatoxin B_1 content of blended and poor peanut kernels obtained from a local market and theirpeanut products

Peanut products	Moisture (%)	Ash (% dw)	Protein (% dw)	Fat (% dw)	Aflatoxin B_1 (ppb fw)
Peanut kernels:					
Blended kernels	6.22 ± 0.16	2.45 ± 0.02	25.49 ± 0.67	49.35 ± 0.68	9.8 ± 0.9
Poor kernels	6.62 ± 0.20	2.58 ± 0.02	25.41 ± 0.52	46.37 ± 0.49	12.1 ± 0.7
Peanut sauce:					
Blended kernels	12.10 ± 1.35	4.49 ± 0.05	17.96 ± 0.84	33.04 ± 0.41	16.8 ± 2.4
Peanut press cake:					
Blended kernels	4.59 ± 0.26	3.26 ± 0.11	34.32 ± 1.33	35.99 ± 1.70	12.4 ± 4.9
Poor kernels	4.91 ± 0.31	3.36 ± 0.10	33.31 ± 1.52	32.88 ± 0.42	15.5 ± 4.5
Peanut					
tempe:Blended	50.55 ± 1.36	2.12 ± 0.09	37.65 ± 0.73	43.09 ± 0.18	65.8 ± 8.3
kernels	55.69 ± 1.72	2.12 ± 0.05	38.75 ± 0.88	40.91 ± 0.76	83.5 ± 9.7
Poor kernels					
Fried tempe *kacang:*	51.71 ± 0.93	2.51 ± 0.28	38.23 ± 0.22	43.00 ± 1.06	17.3 ± 7.0
Blended kernels	56.77 ± 1.44	2.87 ± 0.31	38.80 ± 0.95	41.16 ± 0.94	21.1 ± 5.6
Poor kernels					

Values are means ± SD, dw = dry weight, fw = fresh weight

Peanut sauce prepared from blended and good kernels showed relatively similar moisture, ash, protein and fat contents as their initial levels in peanut kernels were also similar (Table 2 and 3). Both protein and fat contents decreased in peanut sauce processed from blended kernels (Table 3) as also previously discussed in good kernels. Similar moisture, ash, and protein contents was also seen for peanut press cake derived from blended and good kernels, however lower fat was found in poor kernels, following its initial fat content (Table 2 and 3). The moisture, ash, protein and fat contents of peanut tempe and fried peanut tempe were approximately the same for the three kernel categories. The increase in protein and decrease in fat content after fermentation for peanut tempe processed from blended and poor kernels also followed the phenomenon occurred in good kernels. This suggests that peanut samples collected from a local market that stored for about one month did not much affect the nutrient content of the kernels (good and poor kernels) as well as the food products prepared.

3.3. Aflatoxin content of peanut products

Aflatoxin B_1 content was significantly different in peanut products processed from good kernels (Table 2). An increase or decrease in aflatoxin B_1 content in peanut products compared to their initial contents in peanut kernels were shown in Fig. 1. The lowest level of aflatoxin $B_1$was noted for pressed-fried peanut, which was not significantly different with fried peanut and peanut press cake. An average decrease in aflatoxin B_1 content was about 55.6%, 20.2%, and 7.1%for pressed-fried peanut, fried peanut, and peanut press cake, respectively (Fig. 7). In fact, aflatoxins are relatively heat resistant with a melting point about 267-268°C for aflatoxin B_1 [17]. However,it can be reduced up to 80% inboiled peanut [16], 73% in deep-fried peanut at 150°C for 2 min[34], 68.5% in roasted peanut at 150°C for 25 min [35], and 80.2% in oven-dried peanut at 130-150°C [36] due to changes in aflatoxin chemical structure.The use of spices, like garlic and salt in combination with heat treatments would be also effective in degradation of aflatoxins [37-38]. A relatively higher aflatoxin B_1 levels (1.6-20.6 ppb) were obtained previously in fried peanut [39] relative to this study. Both fried peanut and pressed-fried peanuts prepared from good kernels contained aflatoxin B_1<15 ppb, suggesting that they are safe for consumption.

Peanut sauce and peanut press cake showed an increase and a slight decrease in aflatoxin B_1compared to the initial content in good peanut kernels(Fig. 7), however the values were not significant (Table 2). This suggests that the addition of palm sugar, chili, and spices, particularly in peanut sauce may be attributed to a slight increase in aflatoxin B_1 as expectedly it would decrease during deep-frying. However, both peanut sauce and peanut press cake yet contained safe level of aflatoxin B_1 (< 15 ppb). A wide range of aflatoxin levels in peanut sauce samples (0-221 ppb) collected from traditional markets [18], as well as in peanut press cake samples (126 ppb) [18]. Normally, low quality of peanut kernels is used as ingredients for both products, particularly peanut press cake.

The highest aflatoxin B_1 content was observed in peanut tempe (Table 2) with an increase of 165.7% (Fig. 1). This was unexpected as a considerable decrease in aflatoxin B_1level was noted during black and red *oncom* preparation (86.6% and 58.9%, respectively)[25].However, fermentation during *oncom* preparation takes 72 hours, while for peanut tempe performed in this study was only 24 hours, suggesting that longer fermentation might be more effective for *R. oligosporus* to degrade aflatoxin enzymatically as well as to compete with *A. flavus* as both molds are antagonist [40]. Deep-frying (120°C, 10 min) of peanut tempe gave a significant decrease in aflatoxin content (Table 2 and Fig. 1), even though it was slightly higher than the permitted level (<15 ppb).

Peanut products prepared from poor kernels seemed to contain higher aflatoxin B_1 content compared to those derived from blended kernels and good kernels (Table 2 and 3) following its initial aflatoxin in the kernels (Table 1).Peanut sauce, peanut press cake, peanut tempe, and fried-peanut tempe prepared from blended kernels also showed lower aflatoxin relative to those of poor kernels. Physical cleaning and sorting/hand picking may reduce 40-80% aflatoxin level due to removing the mold-damage kernels[41].

Peanut tempe using poor kernels as ingredients showed the highest aflatoxin content (83.5 ppb) compared to other products (Table 3), however it considerably decreased by 74.7% after deep-frying.This final value (21.1 ppb) was only slightly higher compared to those of fried tempe derived from good and blended kernels (16.5 ppb and 17.3 ppb, respectively), while the initial contents of aflatoxin in their raw tempe were much lower (26.3 ppb and 65.8 ppb). This suggests that the higher initial content of aflatoxin in peanut tempe, the higher reduction of aflatoxin occurred during deep-frying. On avarage, the reduction of aflatoxin B_1 in fried peanut tempe was 61.2% for the three kernel categories. *Oncom* that is normally prepared from defatted low quality of peanut contained aflatoxin B_1 about 67 ppb, which was within the range values of peanut tempe derived from blended and poor kernels (Table 2). However, fried *oncom* showed higher content of aflatoxin B_1 (41 ppb)

[18].Differences in initial aflatoxin content as well as methods of fermentation and frying may be attributed to final aflatoxin content in peanut tempe and fried peanut tempe.

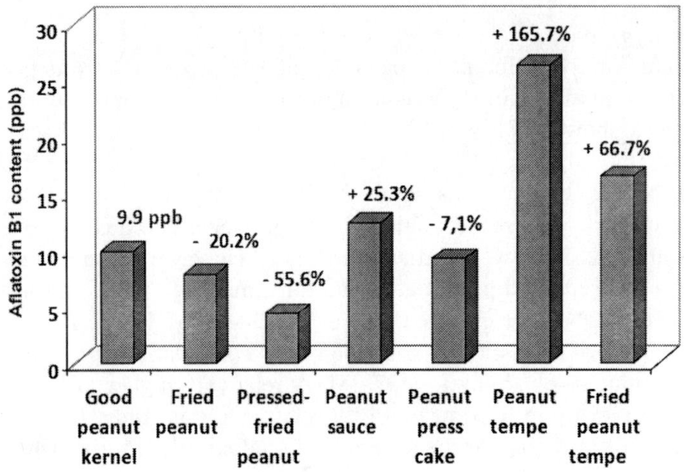

Figure 1. An increase (+) or decrease (-) in aflatoxin B_1 content in peanut products compared to initial aflatoxin B_1 content in good peanut kernels collected from local market.

4. Conclusion

Peanut kernels collected from local market had relatively low moisture and aflatoxin B_1 contents (9.9 ppb), but high number of damaged kernels (24,12%). Higher aflatoxin was seen in particular poor kernels (12.1 ppb). However, the moisture, ash, protein, and fat contents of good, blended, and poor kernels were similar as well as the peanut products. The highest increase in protein content was seen in tempe peanut, while fat content decreased in all products. Peanut tempe prepared from poor kernels showed the highest increase in aflatoxin B_1 that was > blended kernels >good kernels. However, it averagely decreased by 61.2% after deep-fried. Excluding peanut tempe and fried peanut tempe, aflatoxin B_1 levels in all products derived from good kernels were below the permitted level (15 ppb).

5. Acknowledgement

We acknowledge the ACIAR Project #PHT/1997/017 (Reducing Aflatoxin in Peanut Using Agronomic Management and Bio-control Strategies in Indonesia and Australia) for funding this study, Suprapto, SP; Langgeng Sutrisno, Lina Kusumawati, SSi and Maya Indrasari Hariyanto (a student of Agricultural Technology, Faculty of Brawijaya University, Malang) for their technical assistance.

6. References

[1] Data and Information Centre of Agriculture 2014 *Bul. Food Consumption* **4** 16-15 [In Bahasa Indonesia]
[2] Ginting E and Rahmianna AA 2015 *Proc. Food Sci.* **3** 280-88
[3] Rahmianna AA *et al* 2007*J. Pert. Tan. Pangan* **26** 2 137-44 [In Bahasa Indonesia]
[4] Rahmianna AA and Purnomo J 2015 *Iptek. Tan. Pangan* **10** 1 29-37 [In Bahasa Indonesian]
[5] Klich MA 2007 *Molecular Plant Pathol.* **8** 713-22
[6] Reddy KRN *et al* 2010 *Toxin Rev.* **1** 3-26
[7] Mobeen AK*et al* 2011*J. Pharm. Nutr. Sci.* **1** 1 1-3
[8] Abdalla MS *et al* 2014 *Egyp. J. Environ. Res.* **2** 51-56
[9] Bankole SA *et al* 2005. *Food Chem.* **89** 503-06
[10] Mutegi C *et al* 2013 *J. Stored Products Res.* **52** 118-27
[11] Indonesian National Council for Standarization 2009 *National standard of maximum mycotoxin*

content in food (SNI 7385:2009) [In Bahasa Indonesia]
[12] Anukul N *et al* 2013 *J. Food Drug Anal.* **21** 227-41
[13] Dharmaputra OS *et al*1989*Proc. 12th ASEAN Sem. on Grain Postharvest Tech.* (Bangkok:AGGPT) p. 110-23
[14] Dharmaputra OS 2005 *Biotropia* **24** 1-19
[15] Rahmianna AA and Yusnawan E 2015 *J. Exp. Biol. Agri. Sci.* **3** 4 346-52
[16] Siwela et al 2011 *Food Nutr. Sci.* **2** 105-08
[17] Jalili M 2016 *Iran J. Health Safety Environ.* **3** 1 445-59
[18] Mahmud M 1989 *Proc. Int. Workshop on Aflatoxin Contamination of Groundnuts* (Patancheru,India:ICRISAT) p. 215-22
[19] Boli ZA *et al* 2013 *J. App. Biosci.* **72** 5822-29
[20] Indonesian National Council for Standarization 1995. *National standard for peanut (SNI 013921-1995*[In Bahasa Indonesia]
[21] Indonesian National Council for Standarization 1992 *Testing procedure for food and drinks (SNI 01-2891-1992)* [In Bahasa Indonesia]
[22] AOAC 2005 *Microchemical determination of nitrogen using micro Kjeldhal method (12.1.07). OfficialMethods of Analysis of AOAC International.* Vol. I. *Agricultural Chemicals, Contaminants, Drugs*. (Gaithersburs, Maryland, USA:AOAC International)
[23] Lee A 2004. *Workshop on Prevention and Control of Mycotoxin in Food and Feed stuff* (Bogor:SEAMEO BIOTROP) 12 pp.
[24] Waliyar F *et al* 2015 *World Mycotoxin J.* **8** 2 245-52
[25] Fardiaz S 1991*J. Trop.Agr.* **3** 1 27-31.
[26] ICAR 1987 *Aflatoxin in Groundnuts*. (New Delhi: International Centre for Aflatoxin Research)
[27] Ginting E 2006 *J. Agrikultura* **17** 3 165-72 [In Bahasa Indonesia]
[28] Paramawati R 2006 *J. Enjinering Pert.* **4** 1 1-8 [In Bahasa Indonesia]
[29] Hell K and Mutegi C 2011 *African J. Microbiol. Res.* **5** 459-66
[30] Atasie et al 2009 *Pakistan J. Nutr.* **8** 2 194-97
[31] Nelson RG and Carlos AG 1995 *J. Agr. Food. Chem.* **43** 102-05
[32] Sofyan HMI 2003 *Infomatek* **5** 2 74-86 [Indonesian]
[33] Santosa BAS 1993. *Peanut.* Monograph No. 12. (Malang, MARIF) p. 286-303 [In Bahasa Indonesia]
[34] Reddy U 1996*Food Nutr. News Acharya NG RangaAgr. Uni.* **1** 4 1-4
[35] Ogunsanwo BM et al 2004 *African J. Biotechnol.* **3** 9 451-55
[36] Arzandeh S and Jinap S 2011 *Int. J. Food Sci. Technol.* **46** 3 458-91
[37] Farah Z 1983 *Lebensm Wiss Technol.* **16** 122-24
[38] Farag RS 1989. *J. Food Sci.* **54** 74-76
[39] Ansori M2004 *Master Thesis* Faculty of Graduate School, Gadjah Mada University (Yogyakarta) 119 pp
[40] Gourama H and Bullerman B 1995 *J. Food Protec.***58** 12 1395-404
[41] Park DL 2002 *Adv. Exp. Med. Biol.* **504** 173-79.

IOP Conf. Series: Earth and Environmental Science **102** (2017) 012032 doi:10.1088/1755-1315/102/1/012032

Field efficacy of entomopathogenic fungi *Beauveria bassiana* (Balsamo.) for the management of mungbean insect pests

M S Y I Bayu[1] and Y Prayogo[1]

[1]Indonesian Legumes and Tuber Crops Research Institute, Jl. Raya Kendalpayak KM 08, Po. Box 66, Malang 65101, East Java, Indonesia

Email : santi4_nov@yahoo.co.id

Abstract. In order to reduce the use of insecticide, the application of *Beauveria bassiana* may be an alternative control. The objective of this study was to evaluate the efficacy of *B. bassiana* for controlling mungbean pest. The experiment was conducted in Ngale Research Station from February to May 2017, using randomized block design, seven treatments, four replicates. The treatments were frequency of application; P1= six times, P2= five times, P3= four times, P4= three times, P5= once, P6= full protection using chemical insecticide, and P7= no protection. Application of *B. bassiana* four to six times can suppress the population of *Empoasca* sp., *Riptortus linearis*, and *Maruca testulalis*, but did not significantly different with the application of chemical insecticide. Based on the seed weight, application of *B. bassiana* six times (659.7 g/plot) led to significantly high as compare with the application of chemical insecticide (374 g/plot). Application of *B. bassiana* tended to be secure to natural enemies, especially *Coccinella* sp., *Oxyopes javanus*, and *Paederus fuscipes*. Both of those predators were not found on the application of chemical insecticide. Hence, *B. bassiana* can be recommended as a biological agent in integrated pest management component on mungbean because of effective and environmentally friendly.

Keywords : Efficacy, fungi, mungbean insect pests

1. Introduction

Mungbean, *Vigna radiata (L.)* Wilczek, is an important food crop that widely distributed all over the world including tropical and subtropical areas in which south and southeast Asia become the main grown region [1]. Mungbean is easily adopted into multiple cropping systems in the drier and warmer climates because of its short life cycle and most preferred by the farmer due to high protein content [2], [3].

Mungbean is prone to a large number of insect pests that attack mungbean from the early growth up to harvesting time. The infestation by an array of insect pest cause significant yield loss and some of them can destroy seeds in storage. The insect pests that attack mungbean can be classified based on their appearance in the field. They are stem borer, foliage feeders, foliage sucker, pod sucker, pod borer, and pod feeders. According to [4], stem borer, thrips, and pod borer are the most important pest on mungbean. In Indonesia, the major pest on mungbean were dominated by foliage feeder (*Longitarsus suturellus* Duftschmid) and pod borer (*Maruca testulalis* Geyer) that have been considered to be highly responsible for yield losses up to 80% [5].

Recently, more-selective and less-polluting products tend to be used in modern agriculture and farmers have been minimized the use of insecticide to manage the pest population [6].

Entomopathogenic fungi such as *Beauveria bassiana* (Balsamo) Vuillemin is a control agent that infected insect body by contact instead ingestion and widely distributed in nature in wide host range. Moreover, *B. bassiana* known to be safety to non-target organism such as predator and parasitoid and its can be easily cultured on potato dextrose agar (PDA) medium in the laboratory [7]. *B. bassiana* has been evaluated in the laboratory and applied in a field to control numerous of insect pests [8]-[11].

Most study explained the efficacy of *B. bassiana* under laboratory condition. *B. bassiana* is ovisidal for egg of green stinks bug (*Nezara viridula* Linnaeus) which showed the high potential to inhibit the hatchability of egg up to 96% [8]. Moreover, the number of lettuce aphid (*Nasonovia ribisnigri* Mosley) nymph especially fourth instar was significantly affected by the *B. bassiana* in which the mortality increased with increasing the density of conidia [12]. In the two-spotted spider mites, *Tetranychus urticae* Koch, *B. bassiana* known to have a high virulence, but it was shown an opposite trend on the predatory mites in which no pathogenicity of the *B. bassiana* [7]. The laboratory results may be cannot accurately reflect the ecological aspect in the field. Hence, field trials are necessary to confirm the laboratory findings to field conditions, especially for *M. testulalis*, the most damaged of mungbean pest.

In term of integrated pest management (IPM) programs, evaluating of the compatibility of applying *B. bassiana* and natural enemies to control pest of mungbean become a critical issue. In order to improve the utilization of entomopathogenic fungi against insect pest, all factors that contribute to the side effects of entomopathogenic fungi on natural enemies need to be understood [13]. Based on this point of view, the objectives of this study were to evaluate the efficacy of the entomopathogenic fungi *B. bassiana* to control mungbean pest in terms of the level of insect population suppression and seed weight.

2. Materials and Methods

The experiment was conducted in Ngale Experimental Station, Ngawi, East Java, Indonesia, from February to May 2017. The experiment was arranged using randomized block design with seven treatments and four replicates. The treatments were frequency of application of *B. bassiana*; P1= six times, P2= five times, P3= four times, P4= three times, P5= once, P6= full protection using chemical insecticide, and P7= no protection.

2.1 Experimental site

Mungbean Vima-1 was planted on a plot of 4 m x 5 m with the spacing of 40 cm x 15 cm, 2 seeds/hole. Urea 37.5 kg / ha; SP36 73,5 kg / ha; KCl 37.5 kg/ha, and manure 4 kg/ha were applied before planting. Weeding was done manually on 14 and 28 days after planting (DAP). Chemical fungicide was applied to avoid leaf disease and soil borne fungus.

2.2 Entomopathogenic fungi

Isolates of *B. bassiana* was originally collected from Laboratory of Biopesticide of ILETRI then was cultured on PDA. At 21 days after inoculation (DAI), the conidium was taken by using fine brush dipped into the water then was counted by using *haemocytometer* to obtain conidia with 10^7/ml density. The suspension of conidia was added 2 ml/1000 ml of Tween 80 then be shaken for 30 minute before applied into the plant. *B. Bassiana* suspension was applied on the entire surface of plant especially on pod section with the spray volume of 400-600 l/ha.

2.3 Sampling

The observation was made on 10 sample/treatment randomly on pest species and population at 39 and 60 DAP, species and population of natural enemies at 39 dan 60 (DAP), pod damage intensity, the number of pod/plant, the number of empty pod/plant, and seed weight. The diversity of species and population of pest and natural enemies was collected by using sweep net. This kind of trap was used to catch all active insects around the plant. This method was done as many as five times in a single swing. The observation on 39 DAP was aimed to inventory the arthropod on vegetative phase and observation on 60 DAP was aimed to inventory the arthropod on reproductive phase. All arthropod obtained through the sweep net were inserted on the bottle and be brought into the laboratory to be identified based on morphological character then grouped on each family [14], [15].

2.4 Statistical analysis

The analysis was done with one-way ANOVA (SPSS version 22). Means associated with the treatment for each variable was separated using the Tukey's HSD test when significant values were obtained.

3. Result and Discussion

3.1 Pest species and population

Analysis of variance showed that frequency of application of *B. bassiana* did not significantly influence the population of foliage sucker in case of *Empoasca* sp. The number of *Empoasca* sp. found on mungbean field ranged from 1-12 individuals/plot. The population of *Empoasca* sp. was not significantly different between the applications of *B. bassiana* and chemical insecticide. However, the population of *Empoasca* sp. led to significantly higher (12.35 individuals/plant) on plot with no application of *B. bassiana* neither chemical insecticide than that on other treatments (Figure 1).

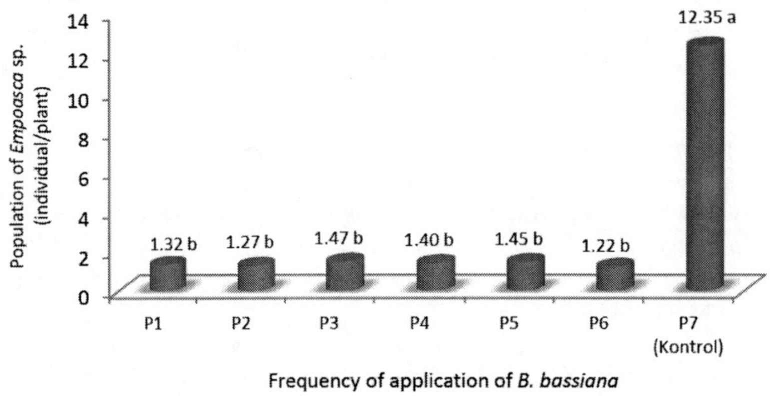

Figure 1. Efficacy of biopesticide *B. bassiana* for the management of foliage sucker, *Empoasca* sp. on mungbean.

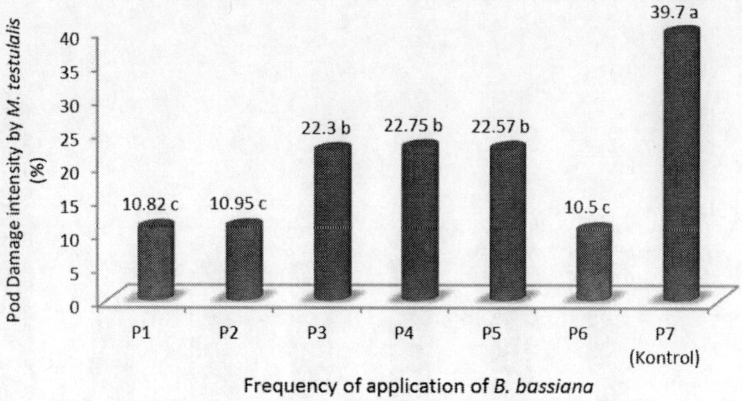

Figure 2. Efficacy of biopesticide *B. bassiana* for the management of pod borer, *M. testulalis* on mungbean.

Figure 2 showed the pod damaged intensity caused by pod borer, *M. testulalis* that was significantly different between each treatment. The highest intensity of pod damaged was found on plot without any application of chemical insecticide or biopesticide, which is 39.7%. An application of chemical insecticide showed the lower intensity of pod damaged (10.5%), did not significantly different as compare with an application of *B. bassiana* six times (10.8%) and five times (10.9%).

In case of pod sucking bugs, *R. linearis*, an application of chemical insecticide led to significantly suppress the population of this major pest as compare with an application of *B. bassiana* (Figure 3). The average number of *R. linearis* found on plot with application of chemical insecticide was 1 individual. However, the population found on plot with no treatment was high (8 individuals/plot). An application of *B. bassiana* with different frequency of application did not significantly different in the population of *R. linearis* which ranged from 2-3 individuals/plot.

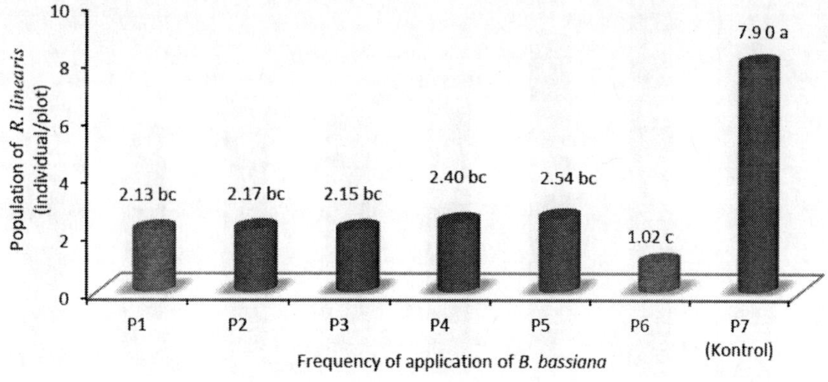

Figure 3. Efficacy of biopesticide *B. bassiana* for the management of brown stink bugs, *R. linearis* on mungbean.

3.2 Yield component

The number of empty pod was observed as an effect of pod sucking bugs attack on mungbean that was significantly different between each treatment (Figure 4). The result showed that an application of *B. bassiana* six times reduced the number of empty pod and it seems no different with an application of chemical insecticide that was only 1 pod/plot. The highest number of empty pod was found on plot with no application of chemical insecticide and biopesticide (2.3 pod/plot).

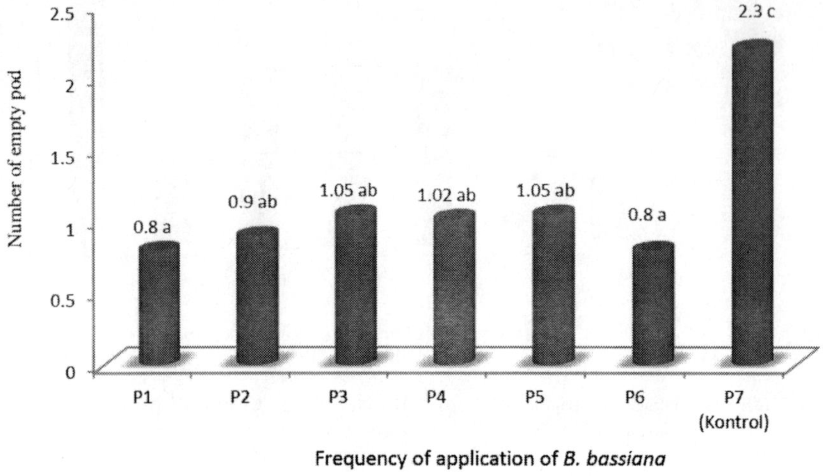

Figure 4. Number of empty pod caused by *R. linearis* on mungben field applied by *B. bassiana*

Analysis of variance showed that seed weight was significantly affected by the treatments, ranged from 338.3 – 659.7 g (Figure 5). The highest seed weight was found on the treatment of six times application of *B. bassiana* (659.7 g), followed by the treatment of five times application of *B. bassiana*. Application of *B. bassiana* four times during the plant growth showed no significantly different in seed weight as compare with three times application and once. A lower seed weight was found on plot with no application of *B. bassiana* neither chemical insecticide (338.3 g) and showed no significantly different with plot in which chemical insecticide applied (374.0 g).

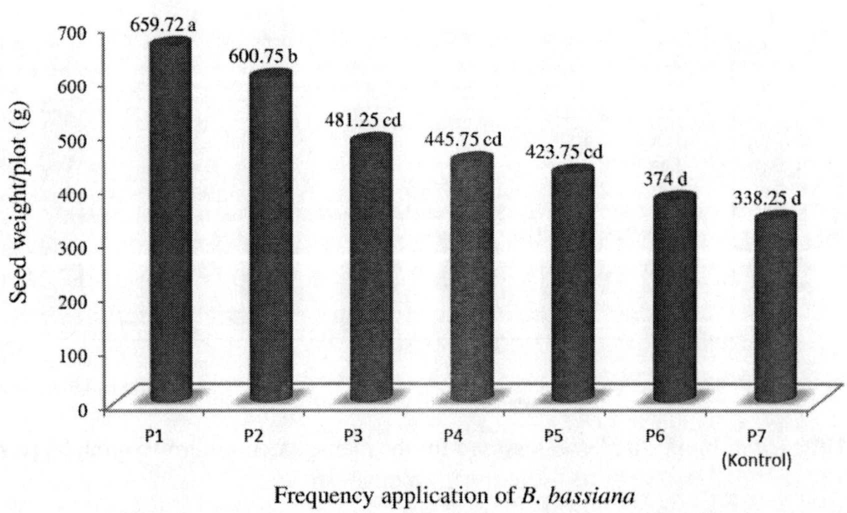

Figure 5. Seed weight of mungbean on each treatment.

3.3 Effect of biopesticide B. bassiana on the survival of predator

The highest population of predator was found on P7 in which no both chemical insecticide and biopesticide applied during the growth season. A total of 17 individuals were found on P7, consists of *Coccinella* sp. *Oxyopes* sp. and *Paederus* sp. However, the application of chemical insecticide resulted in low population and diversity of predator in which only 1 individual of *Paederus* sp. founded (Table 1).

Table 1. Number of predator *Coccinella* sp., *Oxyopes* sp., and *Paederus* sp. on each treatment.

Treatment	Predator population (individuals/plot)		
	Coccinella sp.	*Oxyopes* sp.	*Paederus* sp.
P1 (6x application of *B. bassiana*)	2,50	1,50	1,50
P2 (5x application of *B. bassiana*)	3,25	1,50	0,00
P3 (4x application of *B. bassiana*)	2,75	0,00	4,25
P4 (3x application of *B. bassiana*)	0,00	2,25	3,25
P5 (1x application of *B. bassiana*)	2,50	1,25	5,50
P6 (6x application of chemical insecticide)	0,00	0,00	1,25
P7 (no protection)	5.50	3,50	7,25

Predators found on mungbean field are those who inhabit plant canopy and soil surface. Canopy-dwelling predator was those from the order of Coleoptera, *Coocinella* sp. and *Paederus* sp.. Moreover, the species from order Araneae, *Oxyopes* sp. was also found on mungbean field. These predators usually prey on pest who inhabit plant canopy such as *Aphid* sp., *B. tabaci*, *Empoasca* sp., and pest from the order of Hemiptera.

4. Discussion

Entomopathogenic fungi as microbial pesticide play an important role in sustainable food crop production by providing successful pest management. This present study indicated that *B. bassiana* significantly reduce the population of insect pests and the damaged level of mungbean pod. Moreover, The use of *B. bassiana* is safe to the predators and effectively increased the yield as compared with the application of chemical insecticide and without any treatments.

In case of insect pest existence, the population of *Empoasca* sp. found on mungbean was less and it shows no significantly different between the treatments. The possible reason is that the conidia may not be able to penetrate the cuticle of *Empoasca* sp. even though in abundance colony. This phenomenon occurred maybe due to the presence of protein on their integument that plays as a barrier for conidia penetration [16]. Another reason is that *Empoasca* sp. did not develop their resistance against chemical insecticide yet thus the application of chemical can suppress their population in field.

The major destructive pest on mungbean namely *Maruca* sp., feed on flowers, buds, and pods by webbing them [17]. Tight webbing produced by larva of *Maruca* sp. can protect them selves from pesticide exposure. The application of *B. bassiana* six and five times during the growth stage was more effective than once to four times application, but it shows same effective with the application of chemical insecticide. Based on the number of adult *R. linearis*, the application of chemical insecticide was more effective than the application of *B. bassiana* from once to six times. However, six and five times application of *B. bassiana* are better than other frequencies of application. The abundance colony of *B. bassiana* can be produced when it applied frequently.

According to [18], the distribution of entomopathogenic fungi occurs horizontaly in which the more conidia produced makes the greater distribution of the pathogens and the epizooti on insect body will occur faster than that on less population. Even though the intensity of pod damage was less, the control of *Maruca* sp. by using chemical insecticide is less favorable as compared with the control by using biopesticide. The application of chemical insecticide support an evolution of resistance against chemical and require high cost for traditional farmers in developing country [19], [20]. According to [21], *B. bassiana* endophyte-inoculated common bean plants effective to suppress the population of leaf mining fly larvae, *Liriomyza huidobrensis*, resulting in less pupation and reducing the adult emergence from pupae developing on inoculated plants.

In general, the population of insect pest observed in this study showed no significant different between those applied by *B. bassiana* and chemical insecticide. This may be due to the time of application was not favorable for *B. bassiana* conidia to develop. We applied *B. bassiana* by foliar spray in the morning when the sunlight almost full. It may be influence the persistence of *B. bassiana* on the plant because conidia are not protected from UV radiation. The application of entomopathogenic fungi in the afternoon has many advantages such as low-insolation, low-moderate temperature, favorable moisture conditions in which can enhances the activity of *B. bassiana* conidia [22].

The field efficacy of *B. bassiana* toward various insect pests depends on several factors, some of them are related to the behavior of the insect host in its natural habitat [23]. Soil is known to be a natural habitat of *B. bassiana* and the persistence of this fungal on soil is high [24]. Most of the arthropod dwells in the soil, especially larvae and pupae of lepidopteran and coleopteran. The application of entomopathogenic fungi by foliar spray caused an infection on their body. Although the adults of lepidopteran or other arthropod such as *R. linearis* inhabit plant canopy and feed on plant foliage, they can be seen crawling on the soil surface where the fungal already colonize. It makes them to be possibly contaminated by the fungal conidia [9]. Furthermore, foliar applications of *B. bassiana* against insect pest consistently produce moderate to high level of inoculation and infection, but the unfavorable environmental conditions in the plant canopy may inhibit the development of the conidia thus they cannot penetrate insect body [25].

The application of *B. bassiana* tended to be safe to non-target organism such as predator. This finding is in line with the previous study reported by [26] that the entomopathogenic fungi did not affect the survival rate of *Oxyopes javanus* under laboratory condition. However, some researchers reported that predatory mites namely *Neoseiulus californicus*, *Phytoseiulus persimilis*, and *Amblyseius swirskii* could be infected by *B. bassiana* when exposed to dry residue under laboratory condition [27], [28]. Although *B. bassiana* cause infection on these predatory mites, but the mortality was low (<30%) and showed no different to the control treatment over a 12-day period [28].

Based on the seed weight, the application of *B. bassiana* led to significantly high as compare with other treatment. Although the population of some insect pest showed similar response between application of *B. bassiana* and application of chemical insecticide, the seed weight on the plot with six times application of *B. bassiana* was high. Presumably, insect pest founded on mungbean plot with six times application of *B. bassiana* did not attack the leaf as well as pod but only colonize the plant without any other act.

5. Conclusion

In conclusion, the findings of these field trials are important, especially for mungbean's farmers. *B. bassiana* showed the potential as an alternative to the currently control of insect pest. *B. bassiana* provide safer and more acceptable alternatives than conventional control method due to low-risk pesticide and low cost. Application of *B. bassiana* tended to be secure to natural enemies, especially *Coccinella* sp., *O. javanus*, and *Paederus fuscipes*.

References

[1] Chadha M L 2010 Short duration mungbean: a new success in south Asia. Asian Vegetable

Research and Development Center. Asia-Pacific Association of Agricultural Research Institutions. Bangkok, Thailand pp 1-4

[2] Yaqub M, Mahmood T, Akhtar M, Iqbal M M and Ali S 2010 Induction of mungbean [*Vigna radiata* (L.) Wilczek] as a grain legume in the annual rice-wheat double cropping system *Pakistan Journal of Botani* **42** 3125–3135

[3] Ratnasekera D and Subhashi A P T 2015 Morpho-physiological response of selected mungbean (*Vigna radiata* L.) Sri Lanka genotypes to drought stress *Journal of Agriculture Search* **2**(3) 162-166

[4] Hossain M A, Zaman M S and Alam M J 2011 Management of flower thrips and pod borers in mungbean, *Vigna radiata* L. *Bangladesh Journal of Life Science* **23**(2) 79-86

[5] Indiati S W 2007 The control of pod borer on mungbean *Journal of Agriculture Research and Information* **11**(2) 138-147

[6] Siegwart M, Graillot B, Lopez C B, Besse S, Bardin M, Nicot P C and Lopez-Ferber M 2015 Resistance to bio-insecticides or how to enhance their sustainability: a review. Frontier in Plant Science **6** 381 doi: 10.3389/fpls.2015.00381

[7] Wu S, Gao Y, Zhang Y, Wang E, Xu X and Lei Z 2014 An Entomopathogenic Strain of *Beauveria bassiana* against *Frankliniella occidentalis* with no Detrimental Effect on the Predatory Mite *Neoseiulus barkeri*. Evidence from Laboratory Bioassay and Scanning Electron Microscopic Observation *PLoS ONE* **9** (1): e84732. doi:10.1371/journal.pone.0084732

[8] Prayogo Y 2013 Pathogenicity of *Beauveria bassiana* Bals. Vuill. (Deuteromycotina: Hyphomycetes) on various stages of eggs and nymphs of the green stink bug (*Nezara viridula* L.) *Journal of Tropical Pest and Disease* **13** (1) 75-86

[9] Reddy G V P, Zihua Z and Richard A H 2014 Laboratory and field efficacy of entomopathogenic fungi for the management of the sweetpotato weevil, *Cylas formicarius* (Coleoptera: Brentidae) *Journal of Invertebrate Pathology* **122** 10–15 doi: http://dx.doi.org/10.1016/j.jip.2014.07.009

[10] Lopes R B, Laumann R A, Blassioli-Moraes M C, Borges M and Faria M 2015 The fungistatic and fungicidal effects of volatiles from metathoracic glands of soybean-attacking stink bugs (Heteroptera: Pentatomidae) on the entomopathogen *Beauveria bassiana Journal of Invertebrate Pathology* **132** 77–85

[11] Wakil W and Schmitt T 2015 Field trials on the efficacy of *Beauveria bassiana*, diatomaceous earth and Imidacloprid for the protection of wheat grains from four major stored grain insect pests *Journal of Stored Products Research* **64** 160-167

[12] Shrestha G, Enkegaard E and Steenberg T 2015 Laboratory and semi-field evaluation of *Beauveria bassiana* (Ascomycota: Hypocreales) against the lettuce aphid, *Nasonovia ribisnigri* (Hemiptera: Aphididae) *Biological Control* **85** 37–45

[13] Alavo T B C, Sermann H and Bochow H 2002 Biocontrol of aphids using *Verticillium lecanii* in greenhouse: Factors reducing the effectiveness of the entomopathogenic fungus *Archives of Phytopathology and Plant Protection* **34** 407-424

[14] Norton G, Taylor M, Thiele K and Pickering J 2000 Identification suide to insect orders. The University of Queensland

[15] Schell S and Latchininsky A 2007 Insect identification. University of Wyoming

[16] Wu S, Xie H, Li M, Xu X and Lei Z 2016 Highly virulent *Beauveria bassiana* strains against the two-spotted spider mite, *Tetranychus urticae*, show no pathogenicity against five phytoseiid mite species Experimental Applied Acarology **70** 421–435 DOI 10.1007/s10493-016-0090-x

[17] Grigolli J F J, Laurencila A L F and Avia C J 2015 Field efficacy of chemical pesticides against *Maruca vitrata* Fabricius (Lepidoptera: Crambidae) infesting soybean in Brazil *AJPS* **6**(7) 537-544

[18] Krreutz J, Zimmermann G and Vaupel O 2004 Horizontal transmission of the

entomopathogenic fungus *Beauveria bassiana* among the spruce bark beetle *Ips typographus* (Coleoptera: Scolytidae) in the laboratory and under field conditions *Biocontrol Science and Technology* **14** 837–848 doi: http:// dx.doi.org/10.1080/788222844

[19] Kym A 2006 Transgenic Crops, EU Precaution, and Developing Countries. *International Journal of Technology and Globalization* **2** 65-80

[20] Oparaeke and Mbonu A 2006 The Potential for Controlling *Maruca vitrata* Fab. and *Clavigralla tomentosicollis* Stal. Using Different Concentrations and Spraying Schedules of *Syzigium aromaticum* (L.) Merr and Perr on Cowpea Plants *Journal of Plant Sciences* **1** 132-137 http://dx.doi.org/10.3923/jps.2006.132.137

[21] Akutse K S, Maniania N K, Fiaboe K K M, van den Berg J and Ekesi S 2013 Endophytic colonization of *Vicia faba* and *Phaseo- lus vulgaris* (Fabaceae) by fungal pathogens and their effects on the life-history parameters of *Liriomyza huidobrensis* (Diptera: Agromyzidae) *Fungal Ecol*ogy **6** 293–301

[22] Faria M, Hotchkiss J H, Hajek A E and Wraight S P 2010 Debilitation in conidia of the entomopathogenic fungi *Beauveria bassiana* and *Metarhizium anisopliae* and implications with respect to viability determinations and mycopesticide quality assessments *Journal of Invertebrate and Pathology* **105** 74–83

[23] Gindin G, Levski S, Glazer I and Soroker V 2006 Evaluation of the entomopathogenic fungi *Metarhizium anisopliae* and *Beauveria bassiana* against the red palm weevil *Rhynchophorus ferrugineus Phytoparasitica* **34** 370–379

[24] Ansari M A, Brownbridge M, Shah F A, Butt T M 2008 Efficacy of entomopathogenic fungi against soil-dwelling life stages of western flower thrips, *Frankliniella occidentalis*, in plant-growing media *Entomologia Experimentalis et Applicata* **127** 80–87 doi: http://dx.doi.org/10.1111/j.1570- 7458.2008.00674.x

[25] Wraight S P and Ramos M E 2015 Delayed efficacy of *Beauveria bassiana* foliar spray applications against Colorado potato beetle: Impacts of number and timing of applications on larval and next-generation adult populations *Biological Control* **83** 51-67

[26] Bayu M S Y I and Prayogo Y 2013 The influence of the application of entomopathogenic fungi *Lecanicillium lecanii* on the survival of *Oxyopes javanus* Proceeding of National Conference of Research on Legumes and Tuber Crops pp 673-679

[27] Numa Vergel S J, Bustos R A, Rodriguez C D and Cantor R F 2011 Laboratory and greenhouse evaluation of the entomopathogenic fungi and garlic pepper extract on the predatory mites, *Phytoseiulus persimilis* and *Neoseiulus californicus* and their effect on the spider mite *Tetranychus urticae Biological Control* **57**143–149

[28] Midthassel A, Leather S R, Wright D J and Baxter I H 2016 Compatibility of *Amblyseius swirskii* with *Beauveria bassiana*: two potentially complimentary biocontrol agents *Biological Control* **61** 437–447 DOI 10.1007/s10526-016-9718-3

Response of soybean genotypes against armyworm, *Spodoptera litura* based on no-choice test

M S Y I Bayu[1], A Krisnawati, and M M Adie

[1]Indonesian Legumes and Tuber Crops Research Institute, Jl. Raya Kendalpayak KM. 8, Po Box 66, Malang 65101, East Java, Indonesia.

Email : santi4_nov@yahoo.co.id

Abstract. Armyworm is important polyphagous pest causing economic losses in many agricultural crops including soybean. In Indonesia, there are no soybean varieties which indicated had a resistance against armyworm. The experiment was conducted in Laboratory of Entomology and Green House of Indonesian Legumes and Tuber Crops Research Institute from March to April 2016. The experiment was arranged using randomized block design with a total of 18 soybean genotypes as a treatment in three replicates. The results showed that the difference of soybean genotypes had a significant effect on the leaf damaged intensity. Based on no-choice test and the leaves damaged intensity compared with resistant check genotypes, there was no genotype indicated resistant against *S. litura*. Most of the tested genotype showed moderately resistant and others showed susceptible to highly susceptible. Genotypes that indicated as moderately resistant are G 511 H/Anjs-1-1, G 511 H/Arg//Arg///Arg-30-7, G 511 H/Kaba//Kaba///-4-4, G 511 H/Kaba//Kaba///Kaba////Kaba-16-2, G 511 H/Anjs/Anjs///Anjs-3-3, G 511 H/Anjs/Anjs-1-2, G 511 H/Anjs/Anjs-5-5, G 511 H/Anjs/Anjs///Anjs-6-11, and Argomulyo. In conclusion, those nine genotypes indicated have antixenosis resistance against armyworm and can be considered as a source of gene for improving the soybean resistance to armyworm.

Keywords : Soybean genotypes, resistence, armyworm

1. Introduction

Soybean, *Glycine max* (L.) Merril, is the main staple food of most Indonesian people. In Indonesia, soybean production in 2010 was about 907.031 ton and in 2013 was 779.992 ton. There was a decrease about 127.039 ton during its three years. The decreasing of soybean production is related to the reducing of soybean harvested area about 110.030 ha during 2010 to 2013 [1]. Moreover, the presence of insect pest can be a major factor that adversely influence the reduction of soybean production [2], [3]. According to [3], soybean yield losses caused by insect pest are difficult to be estimated due to the control technology used which generally by applying chemical insecticide intensively, environmental condition, socio-economic status of the farmers, and the genetic variation of soybean.

Among the insect pests that attack soybean from the early growth up to harvesting time, armyworm, *Spodoptera litura* Fabricius (Lepidoptera: Noctuidae) is the major defoliating pest and responsible for huge yield losses [2], [4]. Soybean yield loss caused by *S. litura* in untreated area was about 68% as compare with the yield in treated area with lambda cyhalothrin [5]. Larva is the most destructive stage of *S. litura* in which attacked leaves and pods that plays an important role in the early reproductive phases. The presence of this insects pest should be considered because its known to have many host plants including cultivated plant and wild plant [2], [6], [7].

This pest was known to have ability to develop multiple type of resistance rapidly against chemical insecticide due to its short life cycle and the intensive use of the product [8]-[12]. In term of integrated pest management (IPM) programs and to minimize the residues of chemical, the use of resistant variety for controlling this insect pest is important and become valuable strategy [13]-[15]. Beside its environmentally friendly, the use of resistant variety is low cost and compatible with the utilization of natural enemies such as *Oxyopes javanus* and *Coccinellidae* [16]. In Indonesia, an increasing the resistance of soybean plant against armyworm has been a priority in current breeding program.

Most researchers have searched for and identified armyworm-resistance soybean germ plasm to be used as a resistance donor parents. The mechanisms of resistance against armyworm have been identified through morphological and physiological character of soybean plant [14], [17]-[20]. In Indonesia, there was soybean introduction genotype from Brazil namely IAC 100 were shown to exhibit resistance against armyworm [21]-[23]. IAC 100 was confirmed to have resistance to other herbivorous insects including pod sucking bug and soybean pod borer [24], [25]. In addition, one genotype namely G100 H has confirmed as resistant genotype against armyworm [23], [26]. G100 H known as a result from cross breeding between IAC 100 and Himeshirazu, in which Himeshirazu exhibited resistance against armyworm [27]. Identification of host plant resistant against armyworm is highly required in order to minimize the yield losses. Based on this point of view, the objective of this study was to identify the soybean genotypes resistance to armyworm.

2. Materials and Methods

The experiment was conducted in the Laboratory of Entomology and green house of Indonesian Legumes and Tuber Crops Research Institute, East Java, Indonesia, from March to April 2016. The experiment was arranged using randomized block design with 18 treatments and four replicates. A total of 18 soybean genotypes were assessed: G 511 H/Anjs/Anjs-2-13, G 511 H/Anjs-1-1,G 511 H/Arg//Arg///Arg-30-7, G 511 H/Kaba//Kaba///-4-4, G 511 H/Kaba//Kaba///Kaba////Kaba-16-2, G 511 H/Anjs/Anjs///Anjs-3-3, G 511 H/Anjs/Anjs///-6-13, G 511 H/Anjs/Anjs-1-2, G 511 H/Anjs/Anjs-5-5, G 511 H/Anjs/Anjs///Anjs-6-11, G 511 H/Anjs/Anjs///Anjs-8-1, G 511 H/Anjs/Anjs-1-3, G 511 H/Anjs/Anjs///Anjs-6-12, G 511 H/Anj//Anj///Anj////Anjs-6-8, Anjasmoro and Argomulyo. G100H and IAC 100 plays as resistant check genotypes.

2.1 Experimental site

Each genotype was planted in pot (Φ =18 cm) containing 10 kg of soil that was mixed with manure in a ratio of 4:1, two plants/pot, two pots/genotype/replicate. Planting was arranged such a way that the 2nd larva available when the plant was at 28 days. Each pot was fertilized with 0,4 g urea and 1,2 g NPK in the sowing date. Irrigation was done as required. The control of been fly was done by using sipermetrin at 8 days after sowing (DAS) with a dosage of 1 ml/l and the control of foliage pest was done at 14 and 21 DAS by using sihalotrin with a dosage of 1-2 ml/l. In order to improve quality of plant, then applied by using gandasil D (at 14, 21, and 28 DAS) and gandasil B (at 21 and 28 DAS) with a dosage of 2 g/l.

2.2 S. litura culture

Egg populations of *S. litura* originally collected from soybean field in Kendalpayak Experimental Station, Malang, East Java on January 2016. Eggs were kept inside petri dishes (height (h): 2 cm and Φ: 20 cm), given by fresh soybean leaves in order to provide feed for new emerged larva. It maintained continuously up to pupae. Pupa was then kept in the test tube (h: 20 cm and Φ: 3 cm), given by a piece of yard long been to provide humidity inside the test tube. A new emerged adult were transferred into the cage made from iron frame (h: 50 cm and Φ: 26 cm) then were covered with a white fabric. Some of cotton layers that have been given 10% honey solution were hung on the top of the cage to provide feed for adult. Egg laid by females were collected every day and kept continuously as mention above.

2.3 S. litura infestation

The resistance evaluation was use no-choice test. At 26 DAS, each pot was enclosed in the screen cage (h: 100 cm and Φ: 30 cm). The screen cages were made of nylon fabric that is not translucent to larva *S. litura*. Three individuals of 2nd instar larva were infested into each plant within screen cage at 28 DAS at 14.00 am for one week. Parameter measured on the damaged leaf/plant. Damaged leaf was measured by scoring method with scale from 0–4, in which 0: no damaged, 1: 1-25% leaf damaged, 2: 26-50% leaf damaged, 3: 51-75% leaf damaged, 4: >76% leaf damaged. The leaf damage intensity was calculated as follows:

$$P = \sum \frac{(ni\ x\ vi)}{ZN}\ x\ 100\%$$

in which, P = Leaf damaged intensity, *ni* = Number of damaged leaf with score I, *vi* = Score i (i: 0-4), Z = The highest score (4), N = Total number of leaves.

The criterion of resistance following method by [28]: *x*>*x*+2*SD* = HS (Highly Susceptible), *x*>*x*>*x*+2*SD* = S (Susceptible), *x*>*x*>*x*-1*SD* = MR (Moderately Resistant), *x*-1*SD*>*x*>*x*-2*SD* = R (Resistant), *x*<*x*-2*SD* = HR (Highly Resistant).

2.4 Statistical analysis

Data were subjected to analysis of variance (ANOVA) and mean values were compared using Tukey's test when significant F values were obtained (α = 0.05; SPSS ver. 22).

3. Result and Discussion

The result showed that leaf damaged intensity was significantly influenced by the genotypes (Table 1). The average of leaf damaged intensity at 28 DAS was 21.4%/plant (ranged from 5.36 to 36.2%). The lowest leaf damaged intensity was found on Argomulyo, which was 15.68%, followed by genotype G 511 H/Anjs/Anjs-5-5(15.7%). The highest leaf damaged intensity was found on G 511 H/Anjs/Anjs///Anjs-6-12 (36.2%).

Table 1. Leaf damaged intensity of 18 soybean genotypes at 28 DAS, Malang, Indonesia, 2016.

Genotype	Leaf damaged intensity (%)	Category
G 511 H/Anjs/Anjs-2-13	25.4	S
G 511 H/Anjs-1-1	19.8	MR
G 511 H/Arg//Arg///Arg-30-7	19.3	MR
G 511 H/Kaba//Kaba///-4-4	20.6	MR
G 511 H/Kaba//Kaba///Kaba////Kaba-16-2	20.6	MR
G 511 H/Anjs/Anjs///Anjs-3-3	19.5	MR
G 511 H/Anjs/Anjs///-6-13	23.2	S
G 511 H/Anjs/Anjs-1-2	18.5	MR
G 511 H/Anjs/Anjs-5-5	15.7	MR
G 511 H/Anjs/Anjs///Anjs-6-11	18.7	MR
G 511 H/Anjs/Anjs///Anjs-8-1	23.5	S
G 511 H/Anjs/Anjs-1-3	29.1	HS
G 511 H/Anjs/Anjs///Anjs-6-12	36.2	HS
G 511 H/Anj//Anj///Anj////Anjs-6-8	30.2	HS
Anjasmoro	27.5	S
Argomulyo	15.7	MR

G100 H	17.7	MR
IAC 100	5.4	HR

Based on leaves damaged intensity compared with resistant check genotypes, there was no genotype indicated resistant against *S. litura*. Most of the tested genotype showed moderately resistant and others showed susceptible to highly susceptible. Genotypes that indicated as moderately resistant are G 511 H/Anjs-1-1, G 511 H/Arg//Arg///Arg-30-7, G 511 H/Kaba//Kaba///-4-4, G 511 H/Kaba//Kaba///Kaba////Kaba-16-2, G 511 H/Anjs/Anjs///Anjs-3-3, G 511 H/Anjs/Anjs-1-2, G 511 H/Anjs/Anjs-5-5, G 511 H/Anjs/Anjs///Anjs-6-11, and Argomulyo. The resistant check genotype, IAC 100 showed consistently resistant with the lowest leaf damage intensity (5.4%).

In this research, G100 H showed moderately resistant with the leaves damaged intensity was about 17.7%. In previous study, G100 H exhibited resistance against armyworm based on choice test and preference index [23], [26]. The possible reason to explain this phenomenon is that the different method used on both study. In choice test, larvae have a chance to choose the most preferred genotype to be eaten. Larva will avoid genotypes that potential to inhibit their performance. However, no-choice test does not provide the opportunity for larva to select the most preferred host. It may cause the leaf damaged become high as compared with choice test. In no-choice test, even though larvae were maintained isolated on certain genotypes and have low feeding preference, but larvae should eat for continuing their life. It could generate considerable damaged on soybean [29]. Therefor, no-choice test showed the potential method to obtain resistant genotype against armyworm. This finding is in line with the previous study that the level of leaf consumption on bean genotypes by *Chrysodeixis includes* larva was higher in the no-choice test than that in choice test including in promising resistance genotypes against Lepidoptera, IAC Boreal [30].

Genotypes that indicated as moderately resistant in this study were less preferred by larva of *S. litura* to be eaten. Presumably, leaves of those genotypes are thick so the larva of armyworm could not consume normally. Another reasons maybe due to the presence of organic compound such as methyl salicylate and jasmonic-acid. Methyl salicylate and jasmonic-acid can attract natural enemies to colonize the plant. The existence of natural enemies caused the pest stay away from the plant. Moreover, plant known to have certain chemical compound such as isoflavon that provides detrimental effect on insect behavior especially on their development, mortality, and reproduction. Insect will avoid the plant with such of characteristics.

Generally, the resistance mechanisms of soybean genotype to armyworm are divided into antixenosis or non-preference, antibiosis, and tolerance. Sometimes, it could be a combination of both or various mechanisms of resistance. Antixenosis is the resistance mechanism in which the insect is either repelled by or not attracted to the host plant, as well as for feed, colonize, or laying eggs [13], [14]. It occurred due to the plant morphological structure or secondary metabolites that exhibit from the host plant that play an important role in plant protection against insect pests [31]. Plant morphological structure could include the denser trichomes and the thickness of leaf. Further research to identify the length and density of trichomes as well as the thickness of leaf are required.

4. Conclusion

In conclusion, the findings of these trials are importance for soybean breeders. The genotypes that indicated moderately resistant (G 511 H/Anjs-1-1, G 511 H/Arg//Arg///Arg-30-7, G 511 H/Kaba//Kaba///-4-4, G 511 H/Kaba//Kaba///Kaba////Kaba-16-2, G 511 H/Anjs/Anjs///Anjs-3-3, G 511 H/Anjs/Anjs-1-2, G 511 H/Anjs/Anjs-5-5, G 511 H/Anjs/Anjs///Anjs-6-11, and Argomulyo) can be considered as a source of the gene for improving the soybean resistance against armyworm.

References

[1] FAOSTAT 2017 Agriculture Organization of the United Nations (Statistical database) available: http://www.fao.org/faostat/en/
[2] Souza B H S, Boica-Junior A L, Janini J C, Silva A G and Rodrigues N E L 2012 Feeding of

Spodoptera eridania (Lepidoptera: Noctuidae) on soybean genotypes*Revista Colombiana Entomologia***38** 215-223

[3] Oliveira C M, Auad A M, Mendes S M and Frizzas M R 2014 Crop losses and the economic impact of insect pests on Brazilian Agriculture*Crop Protection***56** 50-54

[4] Punithavalli M, Sharma A N and Balaji R M 2014 Seasonality of the common cutworm *Spodoptera litura* in a soybean ecosystem*Phytoparasitica***42** 213-222

[5] Dudhbale C, Surpam A, Kothikar R and Koche M 2017 Bio-efficacy of chemical insecticides against *Spodoptera litura* infesting soybean *American Journal of Entomology***1**(1) 16-18

[6] Baskar K, Muthu C, Raj G A, Kingsley S, Ignacimuthu S and Duraipandiyan V 2012 Ovicidal activity of *Atalantia monophylla* (L) Correa against *Spodoptera litura* Fab. (Lepidoptera: Noctuidae)*Asian Pacific of Journal Tropical Biomed***2**(12) 987-991

[7] Ahmad M, Ghaffar A and Rafiq M 2013 Host plant of leaf worm, *Spodoptera litura* Fabricius (Lepidoptera: Noctuidae) in Pakistan *Asian Journal of Agriculture and Biology***1**(1) 23-18

[8] Ahmad N, Sarwar M, Khan G Z, Tofique M and Salam A 2011 Efficacy of some plant products and synthetic chemicals to manage the outbreak of mealy bug (*Maconellicoccus hirsutus*) in Cotton*Journal of Agriculture and Biological Sciences***3**(1) 16-21

[9] Muthusamy R, Karthi S, Ramkumar G and Shivakumar M S 2011 Pesticide detoxifying mechanism in field population of *Spodoptera litura* (Lepidoptera: Noctuidae) from South India*Egyptian Acad Journal of Biological Science***3**(1) 51-57

[10] Abbas N, Shad S A and Razaq M 2012 Fitness cost, cross resistance and realized heritability of resistance to imidacloprid in *Spodoptera litura* (Lepidoptera: Noctuidae)*Pesticide Biochemical and Physiology***103** 181-188.

[11] Shad S A, Sayyed A H, Fazal S, Saleem M A, Zaka S M and Ali M 2012 Field evolved resistance to carbamates, organophosphates, pyrethroids, and new chemistry insecticides in *Spodoptera litura* Fab. (Lepidoptera: Noctuidae)*Journal of Pest Science***85** 153-162

[12] Tong H, Su Q, Zhou X, Bai L 2013 Field resistance of *Spodoptera litura* (Lepidoptera: Noctuidae) to organophosphates, pyrethroids, carbamates and four newer chemistry insecticides in Hunan, China*Journal of Pest Science***86** 599-609

[13] Smith C M and Clement S L 2012 Molecular bases of plant resistance to arthropods*Annual Review of Entomology***57** 309-328

[14] Oki N, Komatsu K, Sayama T, Ishimoto M, Takahashi M and Takahashi M 2012 Genetic analysis of antixenosis resistance to the common cutworm (*Spodoptera litura* Fabricius) and its relationship with pubescence characteristics in soybean (*Glycine max* (L.) Merr.)*Breeding Science***61** 608-617

[15] Seifi A, Visser R G F and Bai Y 2013 How to effectively deploy plant resistances to pests and pathogens in crop breeding*Euphytica***190** 321-334

[16] Jesus F G, Boica-Junior A L, Alves G C, Busoli A C and Zanuncio J C 2014 Resistance of cotton varieties to *Spodoptera frugiperdaRevista Colombiana Entomologia***40** 158-163

[17] Glas J J, Schimmel B C J, Alba J M, Escobar-Bravo R, Schuurink R C and Kant M R 2012 Plant glandular trichomes as targets for breeding or engineering of resistance to herbivores*International Journal of Molecular Science***13** 17077- 17103

[18] Zhou Y Y, Luo S H, Yi T S, Li C H, Luo Q, Hua J, Liu Y and Li S H 2011 Secondary metabolites from *Glycine soja* and their growth inhibitory effect against *Spodoptera lituraJournal of Agricultural and Food Chemistry***59** 6004-6010

[19] Souza B H S, Silva A G, Janini J C and Boica-Junior AL 2014 Antibiosis in soybean genotypes and the resistance levels to *Spodoptera eridania* (Cramer) (Lepidoptera: Noctuidae) *Neotropical Entomology*DOI: 10.1007/s13744-014-0241-x

[20] Komatsu K, Takahashi M and Nakazawa Y 2010 Genetic study on resistance to the common cutworm and other leaf-eating insects in soybean *Japan Agriculture Research Quart***44**(2) 117-125

[21] Kraemer M E, Rangappa M and Mohamed A I 1997 Evaluation of vegetable soybean genotypes for resistance to corn earworm (Lepidoptera: Noctuidae) *Journal of Entomological Science*32 25–36

[22] Suharsono and Adie M M 2010 Identifikasi sumber ketahanan aksesi plasma nutfah kedelai untuk ulat grayak *Spodoptera litura* F. *Buletin Plasma Nutfah***16** 29-37

[23] Bayu M S Y I, Tantawizal and Tengkano W 2013 Evaluation of resistance of soybean promising lines tolerant to acid soil and drought against armyworm*Proceedings Seminar of Legumes and Tuber Crops Research***2012**231-241

[24] Silva J P G F,Edson L L, Baldin, Souza E S, Canassa V F, Lourencao A L 2013 Characterization of antibiosis to the redbanded stink bug Piezodorus guildinii (Hemiptera: Pentatomidae) in soybean entries *Journal of Pest Science***86** 649–657

[25] Kuswantoro H, Bayu M S Y I, Baliadi Y and Tengkano W 2017 Resistance of Advanced Soybean Lines to Pod Borrer (*Etiella zinckenella*) *Biosaintifika***9** 317-324

[26] Krisnawati A, Bayu M S Y I and Adie M M 2017 Identification of soybean genotypes based on antixenosis and antibiosis to the armyworm (*Spodoptera litura*) *Nusantara Bioscience***9**164-169

[27] Komatsu K, Okuda S, Takahashi M and Matsunaga R 2004 Antibiotic effect of insect resistant soybean on common cutworm (*Spodoptera litura*) and its inheritance*Breeding Science***54** 27-32

[28] Chiang H S and Talekar N S 1980 Identification of source of resistance and to the bean fly and two other agromyzid flies in soybean and mungbean *Entomologia***7**31-5

[29] Cunningham J P and Zalucki M P 2014 Understanding *Heliothine* (Lepidoptera: Heliothinae) pests: what is a host plant? *Journal of Economic Entomology***107** 881-896

[30] Morando R, Edson L L B, Patricia L C, Andre L L and Alisson F C 2015 Antixenosis of bean genotypes to *Chrysodeixis includens* (Lepidoptera: Noctuidae) *Pesq Agropec Brasilia***50** 450-458

[31] War A R, Paulraj M G, Ahmad T, Buhroo A A, Hussain B, Ignacimuthu S and Sharma H C 2012 Mechanisms of plant defense against insect herbivores *Plant Signal Behavior***7**(10): 1306-1320

Correlation, path analysis and heritability estimation for agronomic traits contribute to yield on soybean

A Sulistyo[1], Purwantoro[1] and K P Sari[1]

[1]Indonesian Legumes and Tuber Crops Research Institute, Jl. Raya Kendalpayak Km 8, Po Box 66, Malang 65101, East Java, Indonesia

E-mail : apri.sulistyo@gmail.com

Abstract. Selection is a routine activity in plant breeding programs that must be done by plant breeders in obtaining superior plant genotypes. The use of appropriate selection criteria will determine the effectiveness of selection activities. The purpose of this study was to analysis the inheritable agronomic traits that contribute to soybean yield. A total of 91 soybean lines were planted in Muneng Experimental Station, Probolinggo District, East Java Province, Indonesia in 2016. All soybean lines were arranged in randomized complete block design with two replicates. Correlation analysis, path analysis and heritability estimation were performed on days to flowering, days to maturing, plant height, number of branches, number of fertile nodes, number of filled pods, weight of 100 seeds, and yield to determine selection criteria on soybean breeding program. The results showed that the heritability value of almost all agronomic traits observed is high except for the number of fertile nodes with low heritability. The result of correlation analysis shows that days to flowering, plant height and number of fertile nodes have positive correlation with seed yield per plot (0.056, 0.444, and 0.100, respectively). In addition, path analysis showed that plant height and number of fertile nodes have highest positive direct effect on soybean yield. Based on this result, plant height can be selected as one of selection criteria in soybean breeding program to obtain high yielding soybean variety.

Keywords : Correlation, path analysis, heritability estimation, soybean

1. Introduction

Processed food from soybean such as tempe and tofu, are still main source of vegetable protein for most of the population in Indonesia. This is evident from the growing demand for soybeans from year to year. According to Sudaryanto and Swastika [1], the average annual growth of demand for soybean consumption in Indonesia reached 2.44%. Most of the domestic demand is met through imports as national soybean production is still low [2]. Therefore, domestic production should be increased to reduce imports [3].

One attempt to improve yields is by genetic improvement of plants through plant breeding program. Germplasm collection, forming the basic population through hybridization and selection are the steps to achieve superior genotypes. Among a series of routine activities in the breeding program, selection is the most important step for obtaining high yielding varieties. The effectiveness of selection is determined by the selection criteria used.

Determining selection criteria can be done in various ways. Information on genetic parameters such as heritability can be used as a consideration to determine selection criteria. In addition, knowledge of correlation is needed as a basis for planning a more efficient selection program [4]. Furthermore, the direct and indirect effects of agronomic traits with seed yield also need to be known in determining selection criteria. The relationship between these properties can be known through path analysis [5]. The aim of this study was to analysis the inheritable agronomic traits and contributes to yield in soybean.

2. Material and method

The study was conducted at Muneng Experimental Station, Probolinggo District, East Java, Indonesia, from August to October in 2016. A total of 91 soybean lines were planted and arranged in a randomized complete block design with two replicates in a plot measuring 1.0 m x 2.5 m for each lines (100 plants each lines per plot). Spacing used was 40 cm between rows, 10 cm in rows, and 20 cm between plots. Fertilizer was given at the planting time with a dose of 50 kg/ha Urea, 100 kg/ha SP36 and 100 kg/ha KCl. Irrigation was done four times, i.e. at planting time, on 3 weeks after planting, during flowering and pods filling. Pest control was done intensively by spraying an insecticide in accordance with the attacking pest.

Agronomic traits observed include days to flowering, days to maturing, plant height, number of branches, number of fertile nodes, number of filled pods, weight of 100 seeds, and yield. The data obtained were analyzed using PKBT software. The estimation of variance components using the tables of analysis of variance follows the methods given by Burton and Devane [6]. Correlation analysis follows the correlation model of Singh and Choudhury [7]. Path analysis were calculated using the Dewey and Lu methods [5].

3. Result and discussion

Table 1 shows the results of the analysis of variance of eight agronomic traits in soybean. Based on the analysis of variance, it can be seen that there are significant differences in almost all agronomic traits observed. Among the eight observed agronomic traits, there is only one character that showing an insignificant difference, namely the number of fertile nodes.

Table 1. Analysis of variance of eight agronomic traits in soybean

Agronomic traits	Mean square	P-value	Significance
Days to flowering	12.66	0.000	**
Days to maturing	19.93	0.000	**
Plant height	112.26	0.000	**
Number of branches	0.90	0.000	**
Number of fertile nodes	88.80	0.115	ns
Number of pods	5.36	0.000	**
Weight of 100 seeds	92,437.74	0.000	**
Yield	23,675.88	0.000	**

** = significant difference, ns = not significant

3.1. Heritability

The estimation value of environmental variance, genotypic variance, phenotypic variance, and broad sense heritability of eight agronomic traits in soybean based on the method of Burton and Devane [6] are showed in Table 2. Based on the results of heritability calculations, it is seen that the days to flowering has the highest heritability (0.88), while the number of fertile nodes has the lowest heritability (0.12). However, except for the character of the number of fertile nodes, all of the observed agronomic traits are classified as characters with high heritability based on heritability groupings proposed by MacWhirter [8]. According to MacWhirter [8], heritability values can divided into three criteria, namely low heritability ($h^2 < 0.2$), medium heritability ($0.2 < h^2 < 0.5$), and high heritability ($h^2 > 0.5$).

Plant breeders usually use heritability values to find out whether the expression of a plant character is genetically inherited or more influenced by the environment. This is understandable because the heritability value is the proportion of the genetic variance (σ^2_g) to phenotypic variance (σ^2_p) [9]. The value of heritability will be close to 0 if the variation of phenotype is more influenced by environmental factors than genetic factors, and will be close to 1 if the expression of phenotype is more controlling by genetic factors than environmental factors.

Table 2. Components of variance and broad sense heritability of eight agronomic traits in soybean

Agronomic traits	σ^2_e	σ^2_g	σ^2_p	h_b^2	Category
Days to flowering	0.83	5.92	6.75	0.88	High
Days to maturing	3.22	8.36	11.58	0.72	High
Plant height	15.25	48.51	63.74	0.76	High
Number of branches	0.26	0.32	0.58	0.55	High
Number of fertile nodes	0.72	0.10	0.82	0.12	Low
Number of pods	22.68	33.06	55.74	0.59	High
Weight of 100 seeds	1.09	2.14	3.23	0.66	High
Yield	23,675.88	34,380.93	58,056.81	0.59	High

σ^2_e = environmental variance, σ^2_g = genotypic variance, σ^2_p = phenotypic variance, h_b^2 = broad sense heritability

In this study, the heritability value of days to flowering, days to maturing, plant height, number of branches, number of pods, weight of 100 seeds, and yield were categorized high (Table 2). These findings reinforce the results of previous studies that find high heritability value in these agronomic traits [10-14]. When compared with the heritability value of other agronomic traits, three traits including days to flowering, days to maturing and plant height, have a heritability value close to 1 (0.88, 0.72 and 0.76, respectively), meaning that the three characters are inherited from the elders to their offspring. This indicates a possible improvement of these three agronomic traits.

In addition, the yield component characteristics (such as number of branches, number of pods, and weight of 100 seeds) and yields have high heritability (0.55, 0.59, 0.66 and 0.59, respectively). In relation to improvement of yield on soybean, there is an opportunity to improve yield and yield components in soybean, because based on the heritability value of each character then number of branches, number of pods, weight of 100 seeds and yields are inherited from parents to offspring.

3.2. Correlation analysis

Table 3 shows the results of correlation analysis among eight agronomic traits on soybean. Based on the results of the correlation analysis, it is seen that only three traits are positively correlated with seed yield, while the others are negatively correlated with seed yield. The three characters are days to flowering, plant height and number of fertile nodes with correlation coefficient values of 0.015, 0.042 and 0.101, respectively. However, of these three characters only plant height is significantly positive correlated with seed yield.

Table 3. Correlation coefficient among eight agronomic traits in soybean

	DF	DM	PH	NB	NF	NP	WO
DM	0.468**						
PH	0.378**	0.008					
NB	0.546**	0.542**	-0.023				
NF	0.279**	0.199	0.466 **	0.274**			
NP	0.431**	0.653**	0.076	0.757**	0.452**		
WO	-0.066	0.255*	-0.242*	0.229*	-0.110	0.212*	
YD	0.015	-0.101	0.442**	-0.348**	0.101	-0.373**	-0.322**

DF = days to flowering, DM = days to maturing, PH = plant height, NB = number of branches, NF = number of fertile nodes, NP = number of pods, WO = weight of 100 seeds, YD = yield, * = significant at $P < 0.05$, ** significant at $P < 0.01$

A significant positive correlation between plant height and seed yield has been widely reported [15-17]. In addition, those researchers also found a significant positive correlation between other agronomic traits with seed yield, among others on the total number of nodes and number of filled pods [18], days to flowering and total number of pods [15]; weight of seed per plant [16]; and days to maturing [17]. Plant height was also reported to be positively correlated with yields on soybean vegetables [19].

A positive correlation between plant height and seed yield means the increase in the plant height will increase the resulting soybean seed. This is understandable because soybean flowers appear along the stem of plant, so the higher the appearance of a plant the more likely it is to produce flowers and pods. Besides the plant height, actually the number of fertile nodes can better describe the resulting soybean seeds. But in this study, the correlation coefficient between the number of fertile nodes and yield is small and positively correlated not significantly. This is probably due to the calculation of the number of fertile nodes involved unfilled pods, so that the addition of the number of fertile nodes can not increase the seed yield. Several previous study reports have also found similar results [16][18][20].

3.3. Path analysis

Path analysis is performed to partition the correlation coefficient between the yields and yield components into direct and indirect effects that are responsible for the increase of yield [21]. The path coefficient analysis of direct and indirect effects of seven agronomic traits on soybean yields are presented in Table 4. Based on the results of path analysis, it is seen that days to flowering, plant height and number of fertile nodes give a positive direct effect on soybean yield. These results are in line with previous research reports [15-17].

Table 4. Path coefficients analysis of agronomic traits on the direct and indirect effects of yield components and yield of soybean

Character	DF	DM	PH	NB	NF	NP	WO	YD
DF	-0.049	0.191	0.151	-0.025	0.035	-0.263	0.017	0.057
DM	-0.025	0.374	0.019	-0.025	0.028	-0.393	-0.041	-0.063
PH	-0.019	0.018	0.392	0.000	0.056	-0.050	0.047	0.444
NB	-0.028	0.212	-0.001	-0.044	0.037	-0.457	-0.037	-0.318
NF	-0.014	0.088	0.184	-0.014	0.119	-0.281	0.019	0.101
NP	-0.022	0.245	0.033	-0.033	0.056	-0.600	-0.042	-0.363
WO	0.004	0.080	-0.097	-0.009	-0.012	-0.131	-0.192	-0.356

Note: DF = days to flowering, DM = days to maturing, PH = plant height, NB = number of branches, NF = number of fertile nodes, NP = number of pods, WO = weight of 100 seeds, YD = yield

The results of the path analysis confirmed the correlation analysis results. Days to flowering, plant height and number of fertile nodes act as the most important of seed yield component determination. According to Hakim [16], this information can be utilized in determining selection criteria in soybean breeding program. However, among these three agronomic traits, plant height has a positive direct effect on soybean yield (Tabel 4). Similar results are also reported by Asadi [15]. The implication is that plant height can be used for selection of soybean seed yields.

4. Conclusion

Based on the results of this study, it can be concluded that almost all agronomic traits in soybean are inherited, indicated by the high heritability value. In relation to the soybean breeding program to obtain high yield soybean varieties, the plant height can be used as the selection criteria. The reason, plant height is not only inherited but also has a positive direct effect on the yield of soybean

5. Acknowledgements

This work was funded by the Ministry of Research, Technology and Higher Education in 2016 through INSINAS scheme with grant number RD-2016-0185.

References

[1] Sudaryanto T and Swastika D K 2007 Soybean economy in Indonesia *Soybean Production Techniques and Development* eds Sumarno *et al* (Jakarta: IAARD Press) pp 1-27 (In Bahasa Indonesia)
[2] Direktorat Akabi 2013 Soybean production only fulfilled 35%. http://tanamanpangan.pertanian.go.id/akabi/berita-142-produksi-kedelai-baru-terpenuhi-35-persen.html
[3] Adisarwanto T 2010 The strategy for improving soybean production as an effort to meet domestic demand and reduce imports *Pengembangan Inovasi Pertanian* **3(4)** 319-33 (In Bahasa Indonesia)
[4] Arsyad M, Aslan M, and Irshad M 2009 Genetic variability and character association among morphological traits of mungbean, *Vigna radiata* L. Wilczek genotypes *J of Agric Res* **47(2)** 121-126
[5] Dewey J R and Lu K H 1959 A correlation and path coefficient analysis of components of crested wheatgrass seed production *Agron J* **51(9)** 515-518
[6] Burton G W and Devane E H 1953 Estimating heritability in Tall Fescue (*Festuca arundinacea*) from replicated clonal material *Agron J* **45(10)** 481-487
[7] Singh R K and Choudhury 1979 *Biometrical methods in quantitative genetic analysis* (Kalyani Publishers, New Delhi)
[8] McWhirter K S 1979 Breeding of cross-pollinated crops: Selection methods for improving cross-pollinating plants species *Plant Breeding* ed Knight R (Academy Press Pty. Ptd, Brisbane)
[9] Roy D 2000 *Plant Breeding, Analysis and Exploitation of Variation* (Narosa Publishing House, New Delhi)
[10] Aditya J P, Bhartiya P and Bhartiya A 2011 Genetic variability, heritability and character association for yield and component characters in soybean (*G. max* (L.) Merrill) *J of Central European Agric* **12(1)** 27-34
[11] Hakim L and Suyamto 2012 Heritability and expected genetic advances of quantitative traits in F4 progenies of soybean crosses *Penelitian Pertanian Tanaman Pangan* **13(1)** 22-26 (In Bahasa Indonesia)
[12] Barmawi M, Yushardi A and Sa'diyah N 2013 Heritability and genetic advance estimate of agronomic traits in the F2 generation of soybean crosses between Yellow Bean and Taichung *J Agrotek Tropika* **1(1)** 20-24 (In Bahasa Indonesia)
[13] Hakim L, Suyamto and Paturohman E 2014 Genetic variability, heritability and expected genetic advances of quantitative characters in F2 progenies of soybean crosses *Indones J Agric Sci* **15(1)** 11-16
[14] Osekita O S and Olorunfemi O 2014 Quantitative genetic variation, heritability and genetic advance in the segregating F3 population in soybean (*Glycine max* (L.) Merrill) *Int J Adv Research* **2(7)** 82-89
[15] Asadi 2012 Path analysis of agronomic characters and resistance to pod sucker bug on yield of soybean germplasm *Buletin Plasma Nutfah* **18(1)** 1-8 (In Bahasa Indonesia)
[16] Hakim L 2012 Yield components and morphological characters determining grain yield of soybean *J Penelitian Pertanian Tanaman Pangan* **31(3)** 173-179 (In Bahasa Indonesia)
[17] Balla M Y and Ibrahim S E 2017 Genotypic correlation and path coefficient analysis of soybean [*Glycine max* (L.) Merr.] for yield and its components *Agric Res Tech* **7(3)** 1-5
[18] Wirnas D, Widodo I, Sobir, Trikoesoemaningtyas and Sopandie D 2006 Selection of agronomic characters to construct selection index on 11 soybean populations F6 generation *Buletin Agronomi* **34(1)** 19-24 (In Bahasa Indonesia)
[19] Li Y S, Du M, Zhang Q Y, Hashemi M,Liu X B and Hebert S J 2013 Correlation and path coefficient analysis for yield components of vegetable soybean in Northeast China *Legume Res* **36(4)** 284-288

[20] Machikowa T and Laosuwan P 2011 Path coefficient analysis for yield of early maturing soybean *Songklanakarin J Sci Technol* **33(4)** 365-368

[21] Kobraee S, Shamsi K and Rasekhi B 2010 Investigation of correlation analysis and relationships between grain yield and other quantitative traits in chikpea (*Cicer arietinum* L.) *Afr J Biotechnol* **9(16)** 2342-2348

IOP Conf. Series: Earth and Environmental Science **102** (2017) 012035 doi:10.1088/1755-1315/102/1/012035

Antibacterial activity of kecombrang flower extract (*Nicolaia speciosa*) microencapsulation with food additive materials formulation

R Naufalin* and H S Rukmini

Departement of Agricultural Technology, Jenderal Soedirman University, Karangwangkal dr Soeparno street, Purwokerto 53123, Indonesia

*E-mail: rnaufalin@yahoo.co.id

Abstract. Kecombrang flower (*Nicolaia speciosa*) contains bioactive components of alkaloids, flavonoids, polyphenols, steroids, saponins, and essential oils as potential antimicrobials. The use of antibacterials in the form of essential oils has constraints; therefore microencapsulation needs to be done to prevent damage to the bioactive components. Microencapsulation can prevent degradation due to radiation or oxygen, easy-mix with foodstuffs and also slow the occurrence of evaporation. This study aimed to determine the effect of types of kecombrang extract, the concentration of microcapsules in food additives (NaCl and sucrose), and concentration of flower extract in the microcapsules. This study used Randomized Block Design (RBD) with 18 treatment combinations and two replications. Factors studied were types of kecombrang flower extract of (semi polar and polar extract), Food Additive types (sucrose and NaCl), the concentration of microcapsules in food additive (0%; 15%; 30% w /v). The results showed that polar and non-polar extract microcapsules produced antibacterial activity of 7.178 mm and 7.145 respectively of *Bacillus cereus* bacteria, while *Escherichia coli* was 7.272 mm and 7.289 mm respectively. A 30 percent microcapsule concentration provides antibacterial activity with inhibiting zone of 7, 818 mm for *B. cereus* and 8,045 for *E.coli.* Food Additive of sucrose concentrations showed that microcapsules produced tend to be more effective in inhibiting the growth of *E.coli* and *B. cereus* bacteria than that of NaCl, with each inhibition zone of 7.499 mm and 7.357 mm

Keywords : Antibacterial activity, kecombrang flower, microencapsulation,

1. Introduction

Most food ingredients are a good for the growth of various microorganisms. The organism will grow and cause food damage or changes in appearance, taste, smell and other properties of the foodstuff. The existence of food additives to be one alternative in improving the quality of food, nutritional value, taste, and appearance when used properly. According to [1], food preservatives are one of the food additives that are used to prevent or reduce chemical and food biological damage. Preservatives to prevent biological damage caused by bacteria are called antibacterials.

[2] have been extracting an antimicrobial compound (additive substance) using non-polar (hexane), semipolar (ethyl acetate) and polar (ethanol) solvents. The results are the extract of ethyl acetate flower kecombrang has large spectrum inhibition include Gram positive, Gram negative and spore-forming bacteria.

The use of antibacterials in the form of essential oils has several constraints. According to [3], antibacterials with high temperatures is easily oxidized; not easily dispersed in dry materials, and it is difficult to handle. According [4], the properties of essential oils of kecombrang is less effective, and it could be overcome by treating essential oils into microcapsules. Microcapsules are small particles containing an active substance or a core material surrounded by a coating. Microencapsulation is a process that makes the material granules wrapped by a coating material to form a microcapsule with a

IOP Conf. Series: Earth and Environmental Science **102** (2017) 012035 doi:10.1088/1755-1315/102/1/012035

diameter of 0.2-5.000 microns [5]. The coating serves as a protective against outside interference (light, temperature, air, water) and extends its shelf life.

The addition of microcapsules of flower extract of kecombrang is very suitable to be applied to process food products from meat, fish and other processed food products, where also use in other ingredients in flavor processing and either sugar or sucrose additives. Sucrose is a sweet taste disaccharide that is often also used as a preservative especially commodities that have been subjected to heat treatment. Sugar, as well as salt, also inhibits the growth and activity of bacteria that cause decay, mold, and yeast. The presence of sucrose in antibacterial food products will affect the antibacterial activity and affect the stability of antibacterial in the food. According to [6] the addition of sucrose in the manufacture of food products serves to provide a sweet taste and can also as a preservative when at high concentrations, which can inhibit the growth of microorganisms by reducing the activity of water from food.

According [7], microcapsules of flower extracts from ethanol fraction have the higher antibacterial activity against Gram-positive and Gram-negative bacteria with better physicochemical properties than microcapsules of ethyl acetate fraction, and the best encapsulation is gelatin-maltodextrin encapsulation with ratio 1: 2 (*b / b*). Antibacterial activity of microcapsules may decrease during food processing. Microencapsulation of kecombrang flower extract is expected to protect antibacterial compounds contained in the flower extract kecombrang from various environmental influences during food processing, therefore this study examines the effectiveness of microcapsule fractions of flowers as an antibacterial kecombrang; the concentration of sucrose and NaCl commonly used in food processing. Thus it is known whether or not there is the influence of antibacterial activity of microcapsule of a fraction of flower kecombrang due to the influence of sucrose and NaCl from some concentration.

This study aimed to determine the effect of types of kecombrang extract, the concentration of microcapsules in food additives (NaCl and sucrose), and concentration of flower extract in the microcapsules.

2. Methodology

2.1. Kecombrang flower powder processing [7].

The flower was cut and spread on trays and dried with a blower dryer at a temperature of 50°C until dry. Kecombrang flower which has been dried crushed in a blender until a homogeneous powder and ready to be extracted.

2.2. The extraction process of kecombrang flower powder [7].

The extraction process is carried out by extraction multilevel, using the two consecutive solvents with ethyl acetate and ethanol as follows: A total of powdered flowers of kecombrang dissolved in ethyl acetate (1:4 w/v), then shaken with a rotation speed of 150 rpm shaker for 2 hours. Then it filtered with filter paper to obtain extract 1 and pulp. Then the extract 1 solvent was evaporated with a rotary evaporator and obtained ethyl acetate fraction. Furthermore, Dregs 1 kecombrang flower extracted again using ethanol solvent and worked the same way as using the ethyl acetate solvent and result in the extract 2 and pulp 2. The kecombrang flower extracts then flowed with N2 gases.

2.3 Microencapsulation of kecombrang flower extract [7].

Microencapsulation process began with making encapsulant in the ratio of distilled water with filler materials (1:1) which is the ratio filler materials are gelatin and maltodextrin (1:1), stirred thoroughly and left at room temperature for 12 hours. The next process is Kecombrang extract was added to the encapsulant and thoroughly stirred for 5 minutes and spread on trays and dried in a cabinet dryer at 40°C for 10 hours.

2.4 Experimental Design

IOP Conf. Series: Earth and Environmental Science **102** (2017) 012035 doi:10.1088/1755-1315/102/1/012035

The experimental design used was Randomized Block Design (RBD) with 18 treatment combinations and 2 replications, so that 36 experimental units were obtained. The treatments include the type of coated fraction consisting of fractions of ethanol (polar) and ethyl acetate fraction (semi-polar); concentration of microcapsules, i.e., 0 percent, 15 percent, 30 percent; sucrose concentration, i.e., 0 percent, 5 percent, 10 percent.

2.5 Microorganism preparation [7]

The tested bacteria were *Bacillus cereus* (FNCC 057), *Pseudomonas aeruginosa* (FNCC 063) obtained from The Center for Food and Nutrients, Gadjah Mada University, and *E. coli* (ATCC 25922). The bacterial stock cultures were maintained on nutrient agar slants and stored at 4°C. The bacterial strains were grown in the Nutrient Broth at 37°C for 24 hr before being used for the antimicrobial activity tests.

2.6 Antibacterial activities analysis [8].

The *N. Speciosa* fruit extract antibacterial activity against *B. cereus, P. aeruginosa* and *E. coli* were carried out using the agar well-diffusion method. The antibacterial activity was expressed as the diameter of the zone of inhibition (mm) formed by the bacteria

2.7 Statistical Analysis

Data obtained from the results of the study were analyzed by using variance analysis (F test). If the results of the analysis show a significant difference, then proceed with Duncans Multiple Range Test (DMRT).

3. Result and Discussion

3.1 Effect of coated fraction type on antibacterial activity of microcapsules in B.cereus and E.coli bacteria

The result of variance analysis showed that fraction type did not have a significant effect on the inhibition zone diameter of *B.cereus* and *E.coli* bacteria. The mean diameter of the inhibitory zone against *B. cereus* on the semi-polar fraction and the polar fractions respectively 7.178 mm; 7.145 mm, while *E. coli* respectively 7.272 mm and 7.289 mm. Microcapsules of ethyl acetate and ethanol fractions are both able to inhibit the growth of *E.coli* and *B.cereus.* For ethyl acetate fraction tends to be more effective in inhibiting the growth of *B. cereus* whereas ethanol fraction tends to more effectively inhibit *E.coli* growth. Figure 1 shows the effect of fraction type on antibacterial activity of microcapsule fraction of flower kecombrang.

B.cereus and *E.coli* bacteria because ethyl acetate or ethanol fractions have similar ability to inhibit bacterial growth. This can be seen from the antibacterial activity of microcapsules in the presence of clear zone formed on media that is not much different. The clear zone is a zone that is not overgrown with bacteria, which means that bacterial growth can be inhibited by antibacterial microcapsules. From these data on *B.cereus* bacteria, there was a tendency that semi-polar fraction resulted in higher inhibition activity than a polar fraction. With the presence of flavonoid compounds cause the ethyl acetate fraction more easily diffuses and able to penetrate the cell wall of *B.cereus* consisting of peptidoglycan and phospholipid. Polar compounds of the ethyl acetate fraction are thought to inhibit bacterial activity by diffusing the protein layer, whereas the non-polar compound diffuses through the lipid layer of phospholipids in the plasma membrane. [9] states that ethnic acetate extract of flower kecombrang can inhibit the growth of Gram-positive and Gram-negative bacteria. [10] reported that hexane-acyl acetate extract could inhibit the growth of Gram-positive bacteria (*B. subtilis*) and Gram-negative (*E.coli*). The dominant antibacterial component of the semi-polar fraction is flavonoids while the polar fraction is phenol [11]. According to [1] flavonoids contained in the flower extract kecombrang have varying solubility according to class and substitution that occurred.

The presence of sugar bound causes the flavonoid to tend to dissolve in water, whereas less polar aglikons such as isoflavones, flavones, flavones, and flavonols tend to be more soluble in semi-polar solvents. In the presence of such compounds, ethyl acetate fractions are easier to diffuse and can easily penetrate the *B.cereus* cell wall consisting of proteins (polar), phospholipids and lipoproteins (nonpolar). According [13], antibacterial substances from flavonoid compounds are lipophilic or easily soluble in lipids. The mechanism of action of flavonoid compounds is not fully known but is thought to be involved in the destruction of cytoplasmic membranes by binding to lipophilic parts [14]. To reach the bacterial cytoplasmic membrane, the flavonoid compound will first pass through the cell wall or the outer membrane to *E.coli*. In the *E.coli* cell wall there are many lipophilic parts that allow the occurrence of bonding with flavonoid compounds so that the flavonoid will be bound to the lipophilic cell wall will likely not reach the cytoplasmic membrane because it is suspected not to bind to the bacterial cell wall and directly bind to the lipophilic portion of the cytoplasmic membrane, so the cytoplasmic membrane is damaged. According to [15], bacterial cell wall serves as a protective component in bacterial cells. Therefore, differences in Gram positive and Gram-negative bacterial cell composition cause differences in both bacterial endurance against various treatments.

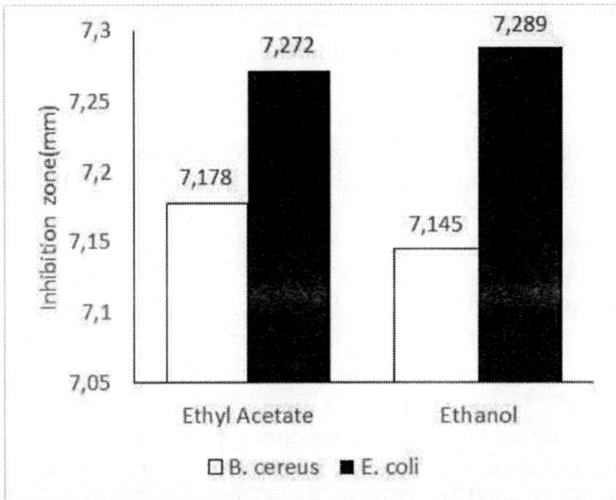

Figure. 1 Effect of coated fraction type on antibacterial activity of microcapsules in *B.cereus* and *E.coli* bacteria

In *E.coli* bacteria, there is a tendency that the polar fraction results in higher inhibition than the semi-polar fraction. It is presumed that the compounds in polar fractions contain a lot of hydroxyl groups which facilitate the water-soluble antibacterial compounds that become bacterial habitats. The polar fraction (ethanol) is thought to have an optimum polarity so that it is easier to diffuse and can inhibit *E.coli* growth. A compound having optimum polarity will have maximum antibacterial activity because for the interaction of an antibacterial compound with bacteria is required a hydrophilic-lipophilic balance [16]. Gram-negative bacteria are more resistant to antibacterial and alkaline dyes; their nutrient requirements are relatively simple but less resistant to physical treatment than Gram-positive [17]. The compounds contained in flowers kecombrang according to [18] include alkaloids, flavonoids, polyphenols, terpenoids, steroids, saponins and essential oils. [19] mentions that the phenolic component found in berry extract can inhibit some Gram-negative bacteria.

3.2. Effect of microcapsule concentration on microbacterial antibacterial activity on B.cereus and E.coli bacteria

The concentration of microcapsules has a very significant effect on the inhibition activity of *B.cereus* and *E.coli* bacteria. The mean values of the diameter of the inhibition zone to *B.cereus* at concentrations of 0 percent, 15 percent, and 30 percent were respectively 6.142 mm; 7,525 mm; 7.818 mm, while *E. coli* respectively 6.141 mm; 7,657 mm; 8,045 mm. This proves that the microencapsulated study can protect extractable bioactive compounds. The effect of microcapsule concentration on the inhibition zone of *B.cereus* and *E.coli* can be seen in Figure 2.

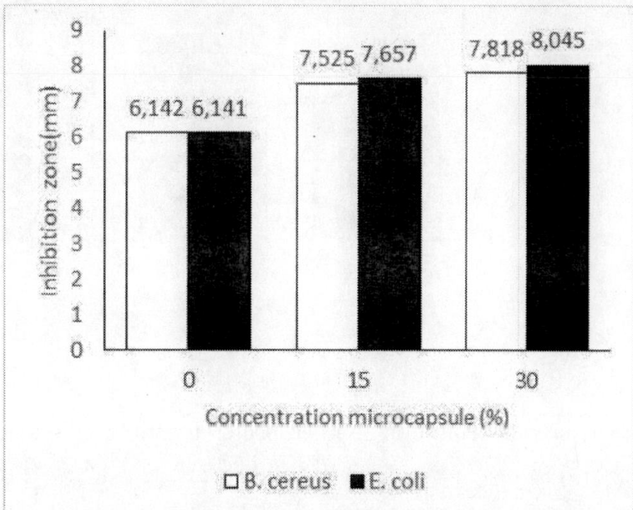

Figure 2. Effect of microcapsule concentration on microbacterial antibacterial activity on *B.cereus* and *E.coli* bacteria

It showed that the higher concentration of microcapsules added, then the higher the antibacterial activity. This is consistent with the research report of [20] that the higher concentration of beluntas leaf extract indicates the greater the inhibition zone formed. [21] adds that the higher concentration of extract kedawung, the number of antibacterial compounds released the greater, thus facilitating the penetration of these compounds into the cell.

3.3. Effect of sucrose concentration on microbacterial antibacterial activity on B. cereus and E.coli bacteria

The result of the analysis of the effect variety of sucrose concentration gave very real effect to the inhibition zone diameter of *B.cereus* and *E.coli* bacteria. The mean diameter of the inhibitory zone against *B. cereus* at concentrations of 0 percent, 5 percent and 10 percent respectively 7.003 mm; 7,124 mm; 7.357 mm, while *E. coli* respectively 7.096 mm; 7,247 mm; 7,499 mm. This proves that the microencapsulated study can protect extracted bioactive compounds in sucrose solution. The effect of sucrose concentration on antibacterial activity of microcapsule fraction of flower kecombrang can be seen in Figure 3.

Based on the results showed that the higher concentration of sucrose added, the higher the antibacterial activity. The increased sugar solution will cause the osmotic pressure outside the high cell to cause water in the bacterial cells out of the membrane so that the bacterial cell will undergo lysis (Hendritomo, 2003)[21]. If the osmotic value in the outer cell medium is increased beyond the osmotic value of the cell contents with the addition of sucrose, the water in the cell is sucked out and the protoplast contracts, causing the cytoplasmic membrane to escape from the cell wall. When occurring in this hypertonic medium are called plasmolysis, the fluid in the cells flows out resulting in dehydration and cell frowning. The mechanism of preservation by sucrose due to sucrose has high osmotic pressure, which can lead to the occurrence of plasmolysis from microbial cells and hygroscopic sucrose that can absorb water from materials and environment so that the activity of food water is low and microbial growth is hampered. According to [22], microbial can only grow in the range of certain water activities therefore to prevent microbial growth of the value of the water activity of the material should be regulated.

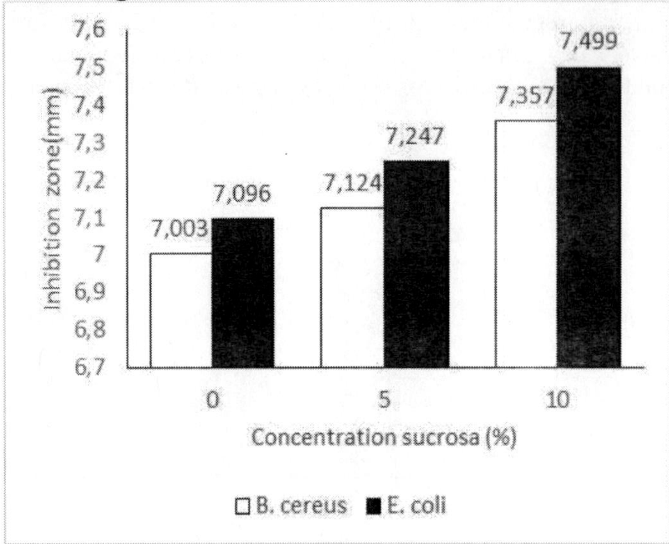

Figure 3. Effect of sucrose concentration on microbacterial antibacterial activity on *B. cereus* and *E.coli* bacteria

3.4. Effect of NaCl Concentration on Antibacterial Activity of Microcapsules

The result showed that NaCl had a significant effect on the inhibition zone diameter of *B. cereus* and *E. coli*. The mean diameter of the inhibition zone against *B. cereus* at a 0 percent NaCl concentration; 5 percent and 10 percent were 8.08 mm, respectively; 8.99 mm; and 9.10 mm. The mean diameter of the inhibitory zone against *E. coli* at a 0 percent NaCl concentration; 2.5 percent and 5 percent were 7.83 mm, respectively; 8.62 mm; and 8.93 mm (Figure 4). The higher the NaCl concentration indicates, the greater the inhibition zone formed.

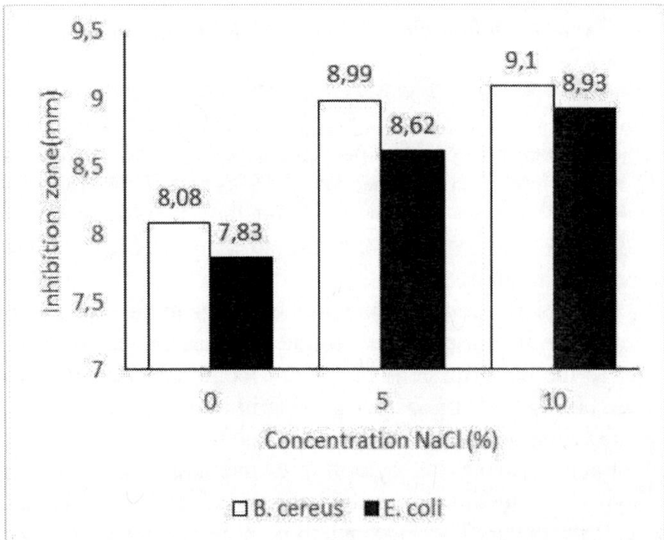

Figure 4. Effect of NaCl Concentration on Antibacterial Activity of Microcapsules

According to [9] the addition of 1 to 4 percent NaCl in ethyl acetate extract and ethanol still showed antibacterial activity. NaCl can inhibit microbial growth. The mechanism of antimicrobial NaCl is by inducing plasmolysis or dehydration of microbial cells. The high concentration of NaCl outside the cell will result in water in the cell out through the membrane to the solution to achieve balance [23]

[23] states although NaCl can not kill all types of microbes, in general most microbes that cause decay can be inhibited its growth. The mechanism of preservation with NaCl is NaCl has high osmotic pressure, so it can cause plasmolysis from microbial cells and hygroscopic NaCl that can absorb water from material and environment, so that the activity of a food, water will be low and microbial growth can be inhibited. First, NaCl will act as a selective inhibitor of certain pollutant microorganisms. Proteolytic microorganisms are the most susceptible, although with a salt content of 6%. Also, NaCl also affects the water activity of the ingredients.

3.5. Interaction Effect Concentration of sucrose, NaCl concentration, and Type of fraction on Antibacterial Activity

The interaction between sucrose concentration, NaCl concentration and coated fraction type (SxMxF) had no significant effect on the antibacterial activity of microcapsules; this was thought to be related to the microencapsulation effectiveness. Microencapsulation may protect the active compounds contained within the fractions of the extracted flowers of the flowers. [3] added that the contents or flavor in the microcapsule could be released at a controlled rate under certain conditions.

The largest inhibitory activity in *B. cereus* was a combination of treatment (10% sucrose concentration, 30% microcapsule concentration and ethyl acetate fraction) of 8.041 mm and for *E. coli* the largest inhibitory activity also in the treatment combination (10% sucrose concentration, 30% microcapsule concentration and ethyl acetate fraction of 8,343 mm. It is predicted that the semi-polar fraction has polar components and semi-polar components. In the presence of such compounds cause the fraction of ethyl acetate more easily diffuses and able to penetrate the cell wall of bacteria *E.coli* and *B.cereus*. [9] stated that ethyl acetate extract has better antibacterial activity than ethanol extract. It is assumed that the semi-polar fraction has an optimum polarity so that it is easier to diffuse and able to inhibit the growth of Gram positive and Gram-negative bacteria. The higher the concentration is expected to be more and more bioactive components against bacteria that can damage the cell wall, thus causing the inhibition zone produced greater. [7] states the higher the concentration of microcapsules, the concentration of extracts coated will be higher so that the antibacterial activity is increasing. Similarly, higher concentrations of sucrose result in large inhibitory zones. The existence of sucrose in the extract will be related each other. This can be seen by the higher concentration of sucrose will result in a larger inhibition zone. Solid sugar solution up to 10% can cause high osmotic pressure so that plasma water of the cell is absorbed by the solution outside the cell resulting in the cell shortage of water and eventually die due to lysis or rupture. All organisms need water for their lives. Water plays a role in the process of cell metabolism in liquid form when the water is bound in a sugar solution; then water can not be used by microbial cells. According to [22] microbes can only grow in the range of certain water activities therefore to prevent the microbial growth of the value of the water activity of the material should be regulated.

4 Conclusions

The fraction type of flower kecombrang has no significant effect on the diameter of inhibition zone of *B.cereus* and *E.coli* bacteria. Micro polar and polar fraction microcapsules resulted in antibacterial activity of 7.178 mm and 7.145 mm in *B.cereus* bacteria, whereas *E.coli* was 7.272 mm and 7.289 mm, respectively. The concentration of microcapsules of flower extract of kecombrang had very significant effect on antibacterial activity against *E.coli* and *B.cereus* bacteria. A 30 percent microcapsule concentration provides antibacterial activity with a drag zone diameter of 7, 818 mm for *B. cereus* and 8,045 mm for *E.coli*. The concentration of sucrose has a very significant effect on the inhibitory zone diameter formed on *B.cereus* and *E.coli* bacteria. A 10 percent effective concentration inhibits the growth of *E.coli* and *B.cereus* bacteria, on *E.coli* inhibit zone diameter of 7.499 mm while *B.cereus* is 7.357 mm. NaCl concentration affects antibacterial activity of microcapsules against *B. cereus* and *E. coli* bacteria. NaCl concentration 5 percent provides antibacterial activity with an inhibitory zone of 9.1 mm in *B. cereus* and 8.93 mm in *E. coli.*

5 Acknowledgement

Thank you to DRPM DIKTI who has provided research fund through Competency Grant Scheme 2017.

6 REFERENCES

[1] Naufalin, R. 2017. Kecombrang Antimikroba dan Pemanfaatannya sebagai Pengawet Pangan. UPT Percetakan Unsoed.
[2] Naufalin, R and S.R. Herastuti. 2012. Pengawet Alami Pada Produk Pangan. UPT Percetakan Unsoed.
[3] Koswara, S. 2007. *Teknologi Enkapsulasi Flavor Rempah-Rempah* (On-line) http://www.ebookpangan.com diakses 06 Maret 2017.
[4] Naufalin, R., and S. R. Herastuti. 2017. Antibacterial activity of *Nicolaia speciosa* fruit extract. *International Food Research Journal* 24.1.
[5] Shahidi, F., dan Han, X.Q. 1993. Encapsulation of food ingredients. CRC Crit. *J. Rev. Food sci. Nutr.* 33:501-547.
[6] Gaman, P. M. dan K. B. Sherington. 1981. Ilmu *Pangan (Pengantar Ilmu pangan, Nutrisi dan Mikrobiologi).* Terjemahan Murdijati Gardjito. Penerbit UGM, yogyakarta. 296 hal.
[7] Naufalin, R., dan S.R. Herastuti. 2013. Microcapsule application of kecombrang flower extract: effect of concentration, type of fraction, pH of medium and NaCl on microbiological properties of minced beef. *Journal Animal Production* 15 (1):8-14.
[8] Carson, C. F. and Riley, T. V. 1995. Antimicrobial Activity of the Major Components of the Essential Oil of *Melaleuca alternifolia.* Journal of Applied Bacteriology 78: 264-269.
[9] Naufalin, R., B.S.L. Jenie., F. Kusnandar., M. Sudarwanto dan H.S. Rukmini. 2006. Pengaruh pH, NaCl dan Pemanasan Terhadap Stabilitas Antibakteri Bunga Kecombrang dan Aplikasinya pada Daging Sapi Giling. *J. Teknol. Dan Industri Pangan.*
[10] Adawiyah, R. A. 1998. Kajian Pengembangan Metode Ekstraksi Komponen Antimikrobia Biji Buah Atung (*Parinarium glaberium hassk*). Tesis. Program Studi Ilmu Pangan. Program Pascasarjana. IPB Bogor.
[11] Houghton, P.J. and A.Raman. 1998. *Laboratory handbook for The Franctination of Natural Extract.* Chapman and Hall, London, UK.199 Pp.
[12] Pramudia, G. 2008. Penggunaan Antioksidan sebagai Penghambat Kerusakan Oksidatif pada Minyak Goreng curah. *Skripsi.* Fakultas Pertanian, Universitas Jendral Soedirman. Purwokerto.
[13] Kumar S, and A.K.Pandey. 2013. Chemistry and biological activities of flavonoids: an overview. The Scientific World Journal.
[14] Cowan MM. Plant products as antimicrobial agents. Clinical microbiology reviews. 1999 Oct 1;12(4):564-82
[15] Schlegel, H.G. 1994. Mikrobiologi Umum Ed Ke-6. Terjemahan Baskoro, R.M.T. Gajah Mada University Press, Yogyakarta.
[16] Kanazawa, A.T., T. Ikeda and T. Endo. 1995. A Novel Approach to Made of Action on Cationic Biocides; Morphological Effect on Antibacterial activity. *J Appl bacterial.*
[17] Fardiaz, S. 1989. *Mikrobiologi Pangan.* Pusat Antar universitas Pangan dan Gizi Institut Pertanian Bogor, Bogor.

[18] Tampubolon, O.T,. S. Suhatsyah,. dan S. Sastra pradja. 1983. Penelitian Pendahuluan Kimia Kecombrang (*Nicolaia speciosa* Horan). *Risalah Simposium Penelitian Tumbuhan Obat III.* Fakultas farmasi, UGM, Yogyakarta.

[19] Puuponen-Pemia R., L. Nohynek., C. Meier., M. Kahkonen. and M Heihoen. 2001. Antimicrobial properties of phenolic compound from barries. *J. Appl Microbial.*

[20] Ardiansyah, Nuraida L, dan N. Andarwulan. 2003. Aktivitas Antimikroba Ekstrak Daun Beluntas (*Plucea indica* L) dan Stabilitas Aktivitasnya Pada Berbagai Konsentrasi Garam dan Tingkat pH. *Jurnal Teknologi dan Industri Pangan* XIV:2 (90-97).

[21] Zuhud, E.A.M.,W.P.Rahayu, C.H. Wijaya and P.P.Sari.2001.Aktivitas Antimikrobia Ekstrak Kedawung (*Parkia roxburgii* G.Don) terhadap Bakteri Pathogen. *J.Teknologi dan Industri Pangan XII*

[21] Hendritomo, H.I. 2003. Perubahan Mutu Kecap Skala Industri Rumah Tangga Selama Tiga Bulan. *Jurnal Teknologi dan Industri Pangan*. XIV No. 3 Tahun 2003.

[22] Winarno, F. G. 1997. Kimia Pangan dan Gizi. Gramedia, Jakarta.

[23] Kuntz, L. A. 1999. Ingredients to Raise The Microbial Bar. Weeks Publishing Company (On-line) http://www.foodproductidesign.com. Diakses 30 April 2017.

Product diversification of banana cv. Mas Kirana off grade by using a double rotating screw extruder

S Setyadjit, E Sukasih, and R Risfaheri

Indonesian Center for Agriculture Postharvest Research and Development
Jl. Tentara Pelajar no 12, Bogor, Kampus Penelitian Pertanian, Cimanggu, Bogor, Indonesia

Email : Pascapanen@yahoo.com

Abstract. Extrusion technology is today's favorite technology since it has a varied, practical and consistent product form. The purpose of this research was to get precise composite flour composition so that the quality of the resulted product has optimum quality for breakfast meals. The experimental design used was Design Expert vs. 7 with response surface box-behnken. The flour composition and level to be inputted to the program were banana flour (10-50g), mung beans (10-30g), egg flour (10-20g). Formula made was based on 200 g with addition of maize flour if the amount is less than 200 g. The extrusion tool used is a Thermo Scientific double rotating screw; with Haake Reomax OS. The best results in terms of carbohydrate content is the Formula 8 with a composition of 60 g of banana flour, 20 g eggs, 20 g of green beans and 100 g maize flour. The proximate chemical content of this formula is carbohydrate 84.04%, protein 8.55%, fat 5.49%, ash content 1.24%. K-calories per 100 g is 419.5 which is higher than the standard of breakfast meals calories.

Keywords : Banana off grade, breakfast meals, double rotating screw, green bean flour, maize flour,

1. Introduction

Banana production in Indonesia in 2014 reached 6,862,558 tonnes, is the highest among fruit commodities [1]. Varieties of bananas grown by Indonesian farmers can be banana table *i.e* Ambon (white), Ambon (yellow), Lampung, Cavendish, Raja Sere, Raja Bulu, Ambon Lumut, Ambon Buai, Mas whilst variety of banana processing and plantain are banana Tanduk, Uli, Nangka, Kapas, Kepok. There are many research result on shelf life on table banana has been done. Requirement of research on banana processing is not only life extension on fresh produce but also research on technique of processing. Usually off grade market of table banana are also considered as raw materials of processing.

Internationally bananas are processed into various types such as the manufacture of ordinary banana flour [2] including the convection drying model [3]; banana flour modified with acid [4]; banana fritters [5]; banana Puree [6]; banana-milk powder, ultrasonic dried banana, cookies, spray dried banana powder, rich fiber powder, high fructose syrup [7]; Banana starch film [8]; composite banana custard [9]; banana starch [10]. One type of banana preparations that has not been studied extensively is the extrusion product for snacks.

Extrusion is a process whereby food is forced to flow during mixing, high temperature heating and shear by means of a mold designed to form a puff dry extrusion pattern in short time [11]. Extrusion technology is not a new technology but has long been found and at this time continues to grow and demand by all technology users. According to SNI No 2886-2015 [12], extruded snacks are ready-to-eat snacks made from food sources of carbohydrates and / or proteins through extrusion processes with or without the addition of other foods and food additives permitted with or without frying process .

There are 2 types of extruder *viz.* single screw and double screw. Single screw extruder has been widely produced by local manufacturers of Indonesia, while twin extruder is still not much to make. One of the modern extruder machines is made by Thermoscientific because it has been equipped with sensors and monitors for a more accurate process [13]. Currently extrusion snack products can be made from various ingredients, one of the foodstuffs often used independently for extrusion materials is corn. This is because corn is cheap and has sufficient expansion, and has good texture [14]. Generally in the manufacture of extrusion products is done by mixing various types of flour and materials to achieve the desired goal.

The purpose of this study was to obtain an appropriate formulation in producing extrudate from banana flour, green beans, and eggs.

2. Materials and Methods

2.1. *Material and equipment*

The research material was obtained in the form of banana (*Musa paradisiaca* L) cv. Mas obtained from 'Pasar Anyar', Bogor with commercial maturity, and not yet ripe. Furthermore, the bananas are brought into the laboratory at ICAPRD for producing banana flour. Other ingredients of corn flour, green bean flour and eggs also come from 'Pasar Anyar'. The raw materials of banana flour, maize flour and green bean flour were , starch (%), amylose (%), amylopectin (%), moisture content (%), ash content (%), fat (%), protein (%), carbohydrates (%). For analyses method see sub chapter analyses.

The extruder used was the Thermoscientific twin barrel extruder (made in Germany), with computer equipment from Haake [13]. In addition, other standard equipment such as, stainless steel pan for preparation of raw materials, glass equipment and laboratory equipment standards for analysis.

2.2. *Determination of Formula composition*

The composition of the ingredients or Formula is the main factor that becomes the basis as well as the treatment factor and becomes the limit determinant on the manufacture of this snack. Formulation using the main ingredient of maize flour, which is mixed with other raw materials with the composition of 10-50% banana flour, 30% and eggs 10-20%. Each Formula was made in a 200 g experimental unit. The amount of maize flour added was the difference of 200 g minus the amount of Formula formed. Each Formula was added 10 g of tapioca starch and 10 ml of cooking oil.

2.3. *Determination of treatment Formula*

Determination of the Formula that was using Design Expert version 7 program with RSM (Response Surface Methodology) design method Box-Behnken Design 9 trial version) [15]. This program automatically performs the Formulation that will be used. The results of the Formula obtained from Design Expert version 7 are 17 Formulas (Table 1). The range or spacing between the initial formulas generated was adjusted automatically with the limits given at the time of the input variables available in the Design Expert program version 7.

Table 1. Formula of extrudate snack from design expert vs. 7

Formula	Banana (g)	Geen bean (g)	Egg (g)	Maize (g)
1	60	20	40	80
2	60	60	40	40
3	60	40	30	70
4	60	60	20	60
5	20	40	20	120
6	100	20	30	50
7	60	40	30	70
8	60	20	20	100
9	20	40	40	100
10	60	40	30	70
11	60	40	30	70
12	60	40	30	70
13	20	60	30	90
14	20	20	30	120
15	100	40	20	400
16	100	40	40	20
17	100	60	30	10

2.4. Preparation of extrudate product
The extruder used for this study was a double screw thread [13]. The initial heating temperature of the extruder was carried out at six points: T1 50°C, T2 50°C, T3 60°C, T4 100°C, T5 140°C, and T6 145°C, threaded rotation of 115 rpm. Injectable water which was added 2.5 ml / min. All ingredients *i.e* banana flour, green bean flour, and eggs were weighed in accordance with the initial formulations made by Design Expert version 7 (Table 1). The extrusion process for 1 Formula was carried out for 1 hour continuously. The extrusion snack that was formed still has high moisture content, so it needs a 3 hour oven with 60°C temperature. Every hour were sampled to see the water content.

2.5. Analyses

Proximate analyses include moisture oven method [17], ash [18], Microkjeldal for protein [22], fat method of soxhlet [17] and total carbohydrate [17] contained in the snack. Total carbohydrate calculation analyses by way of 100 minus the amount of water content, ash, protein and fat. The starch analysis of Luff Schrool [17] method, amylose was determined by spectrophotometry [19]. Amylopectin content can be determined as a difference between the starch content and amylose.

3. Results and Discussion

3.1. Properties of raw material
Starch is an important factor in the manufacture of extrusion snacks, because the starch content in raw materials affects the texture of the snack. Starch consists of amylose and amylopectin, where amylopectin is very influential on the development of extrusion snack so that the amylopectin content must be greater than its amylose content [14]. In the analysis of starch showed that banana flour has the highest starch content among other raw materials that is 70.20%. In the amylose analysis, green bean flour has the highest percentage of 27.33%, and the highest amylopectin content of banana flour with a percentage of 49.55%.

The results of the proximate analysis of raw materials can be seen in Table 2. The green beans have the highest parameter content among other raw materials, the values are ash content, fat content, and protein, the percentage of respective values respectively 3.22%, 1.64% and 10.26% . Green beans are added into Formula of produce snacks for its high protein content. The highest water content is maize with percentage of 10.71%, and highest carbohydrate content is corn with percentage 84.20%. The results of the analysis on these raw materials will be linked to the results of the analysis on extrusion snack, to find out which feedstock has the most influence on water content, ash content, protein content, fat content, carbohydrate, starch, amylose and amylopectin of extrusion snack.

Table 2. Proximate analyses of raw material

Raw material	Moisture content %	Ash content %	Fat content %	Protein Content %	Carbohydrate content %	Starch (%)	Amylose (%)	Amylopectin(%)
Banana	8.72	2.15	0.71	4.44	83.97	70.20	20.65	49.55
Green bean	8.09	3.22	1.64	10.26	76.79	60.29	27.33	32.96
Maize	10.71	0.15	0.68	4.26	84.20	68.26	25.18	43.08

3.2. Extrudate snack

Starch content. Starch is part of the carbohydrate that is polysaccharide composed by two groups of macromolecules, amylose and amylopectin. Both these macromolecules play a significant role in the physical, chemical and functional properties of starch [17]. Generally, starch contains higher amylopectin than amylose [14]. Based on the results of the analysis of variance done with Design Expert version 7 program on the response of starch content to all the formulas, showed that the Formula significantly influence the response of starch content ($p < 0.05$), at 95% confidence interval with $p = 0.0017$; has significantly effect on product starch content. In Table 5 it can be seen that the lowest starch content in Formula 14 is 53.28%, the highest in Formula 6 is 60.28%.

Starch has an important role for extrusion products, in addition to the effect on texture, also affect the durability. The effect of starch is mainly due to ratio of amylose to amylopectin. Amylopectin is known to stimulate the puffing process, so that extrusion products derived from starch with high amylopectin content will be light, porous, crisp, and crunchy. Starch with high amylose tends to produce a hard product due to the limited development process [11]. Another factor affecting the development of snack is the moisture level of the dough that will affect the yield of the dough temperature and the amount of starch that is gelatinised during the extrusion process [22].

Amylose and amylopectin. Amylose and amylopectin are constructed by an α-D-glucose monomer that binds to each other via a glycoside bond. The comparison between amylose and amylopectin varies for different sources of starch. Generally, the amylopectin content is greater than amylose (70-80%) [18]. Based on result of analysis of variance conducted by Design Expert version 7 program on amylose and amylopectin response significantly influenced by Formula made, amylose and amylopectin ($p < 0.05$), at 95% confidence interval with p value of amylose = 0.0017 and p amylopectin = 0.0293 . Snack with the highest amylose content of Formula 6 which is 22.48%. The highest amylopectin level in Formula 17 which is 38.95%.

The content of amylopectin and amylose has a major influence in determining the nature or characteristics of the extrusion snack. Amylopectin can improve the nature of product development and crispy-ness, while amylose can improve the dryness (crunchy-ness) and texture strength of the product [14]. The initial temperature of heating affects the process of starch gelatinization and affects the properties of the ingredients at cooking time and in the development of the extrusion snack [24]. In general the temperature of starch gelatinization at the extrusion will increase as the amylose content increases. Amylose and amylopectin content of banana flour, corn flour, and green bean flour have generally higher amylopectin content than amylose content. The addition of tapioca also has an effect on snack characteristics because tapioca has high enough amylopectin content so that the resulting extrusion product is light, crisp, and crunchy [11].

3.2.3. Proximate analyses. The result of proximate analysis on extrusion snack is moisture content in the range of 0.37-2.95%, ash content 1.03-2.32%, protein content 8.24-14.36%, fat content 5.32-6.88%, and carbohydrate 75.17-84.04%. Proximate analysis of the 17 snack formulas are presented in Table 3.

Table 3. Result of proximate analyses of extrudate snack*

Formula	B:G:E:M[a]	Water content %	Ash content %	Fat content %	Protein content %	Carbohydrate content %	Starch (%)	Amylose (%)	Amylopectin (%)
1	60:20:40:80	2.33	1.03	5.95	10.00	80.69	55.63	19.71	35.92
2	60:60:40:40	2.74	1.98	6.88	13.23	75.17	55.55	19.80	35.75
3	60:40:30:70	2.01	1.64	6.62	10.93	78.80	57.88	20.28	37.60
4	60:60:20:60	0.37	2.32	6.12	11.68	79.51	57.31	20.15	37.16
5	20:40:20:120	0.69	1.59	5.50	11.44	80.78	54.25	18.81	35.44
6	100:20:30:50	1.01	1.31	5.75	8.24	83.69	60.28	22.48	37.80
7	60:40:30:70	2.15	1.58	5.69	10.65	79.93	53.70	19.61	34.09
8	60:20:20:100	0.68	1.24	5.49	8.55	84.04	55.53	20.77	34.76
9	20:40:40:100	2.95	1.68	5.94	12.37	77.06	56.68	19.54	37.14
10	60:40:30:70	2.12	1.39	5.23	10.83	80.43	55.49	21.21	34.28
11	60:40:30:70	1.95	1.62	5.81	11.01	79.62	55.60	21.11	34.49
12	60:40:30:70	1.43	1.86	5.79	11.22	79.70	54.93	21.20	33.73
13	20:60:30:90	0.97	2.27	6.62	14.36	75.78	53.80	19.48	34.32
14	20:20:30:130	2.15	1.35	5.43	11.06	80.01	53.28	19.47	33.81
15	100:40:20:40	0.48	1.94	5.59	9.67	82.32	59.03	22.01	37.02
16	100:40:40:20	2.50	1.80	6.20	11.01	78.50	59.62	21.33	38.29
17	100:60:30:10	0.99	2.22	5.69	12.15	78.95	60.26	21.31	38.95

[a]Remarks: B: Banana, G: Green bean, E:Egg, M: Maize

Water content. Water content is closely related to crispness properties of extruded products. Especially dried snack foods come from cereals. Water content is too high will cause the texture becomes less crisp and crunchy. Analysis of water content is necessary because the water content becomes one of the factors that greatly affect the decline in the quality of food products. Water content of a high material the higher the possibility of the material will be damaged. Based on the results of analysis of variance conducted by the Expert program version 7 shows the Formula made significantly effect on the water content response ($p < 0.05$), at 95% confidence interval with value $p = 0.0001$.

The lowest moisture content in the product is 0.37%, the highest is 2.95% with an average of 1.62% while the water content based on standard [12] is a maximum of 4% (w / w). This indicates that the extrusion snack made meets the SNI water content standard, but the water content of the snack is very low when compared to the SNI, presumably due to drying by using oven done after the finished product. Drying by oven was done because the water content in the extrusion snack was still high so oven drying will lower the water content. In the snack Formula with the addition of 40 g eggs gives the highest water content compared to other treatments (Table 3) because the egg contains moisture content of 73.70%. Other factors that affect the water content is the composition of raw materials. Corn flour has the highest water content when compared with other raw materials that is equal to 10.71%. The addition of corn starch of 100 g also gives the highest water content compared to other treatments.

Water added to the Formula is one of the key factors for controlling the dough behaviour in the extruder and the properties of the final product. High moisture content will result in the starch gelatinization being reduced and the bubble growth will be inhibited resulting in the end product which is not expanding [21].

Ash content. Determination of ash content aims to know the amount of mineral content contained in the resulting product. The principle of the ash analysis is by weighing the residue from the combustion of organic matter components at a temperature of about 550°C [18].

Based on the result of variance analysis conducted by Design Expert version 7 program, it showed that the Formula that made significantly effect ash content ($p < 0.05$), at 95% confidence interval with $p = 0.0001$. The lowest ash content on snack was 1.03% and the highest was 2.32% with an average of 1.69%. Factors that affect the high content of ash in snack is green bean flour, because based on green bean analyst, it has the highest ash content among other raw materials that is equal to 3.22%. The Formula that has the highest ash content is Formula 4 (Table 3).

A similar study conducted [20] on ash content of soy sorghum extrudate products was 2.9% and from the mixture of broken rice and wheat bran showed ash content ranged from 0.7 to 5.1%.

Fat content. Based on result of analysis of variance conducted by Design Expert version 7 programme show that Formula made significantly effect fat content response ($p < 0.05$), at 95% confidence interval with p value = 0.0391.

Table 6 shows the lowest fat content of 5.23% and the highest is 6.88% with an average of 5.90%. The extracted fat content of the extracted snack meets the requirements of standard [12] which is below 30%. Fat content is influenced by the ingredients that exist in the snack. The amount of cooking oil, green beans and eggs in the ingredients affects the fat content of the product. Addition of cooking oil in every Formulas are the same which is 10 ml. An increase in the number of green beans and eggs will increase the fat content of extrusion snacks because the highest levels of green beans fat are compared to corn and bananas. Green bean fat content of 1.64% and eggs also have a high fat content of 11.20%. Formula 2 is the Formula with the highest percentage of fat content with the composition of green beans as much as 60 g and eggs as much as 40 g. In the extrusion process, fat acts as a lubricant because it reduces the friction force between particles in the mix and between the screw and barrel surfaces with molten dough [20].

Protein content. Proteins affect the texture and make the product crunchier. Based on the result of variance analysis conducted by Design Expert version 7 programme showed that the Formula made significantly influence the protein content ($p < 0.05$), at 95% confidence interval with p value = 0.0001.

Formula 2 has the highest protein content of 14.36% and Formula 6 has the lowest protein content of 8.24%. Protein content is strongly influenced by the content of green beans and eggs, because green beans with protein content of 10.26% (Table 2). Eggs also have a high protein content of 12.9% so that eggs and green beans can be a source of protein. Based on 17 Formulations, snacks resulting from the addition of green beans as much as 60 g gave the highest protein content compared with other treatments (Table 3).

Kocherla [16] reported that the extract products of sorghum raw material had a protein of 10.60% while the mixture of sorghum and soybeans increased protein content to 16.60%. Green beans have an effect on the level of protein extrusion products produced.

Carbohydrate content. Carbohydrates play an important role in human life, because carbohydrates become one of the energy sources needed by the body and can provide energy value of 4 Kcal / g. Based on result of analysis of variance conducted by Design Expert version 7 program show that Formula made significant effect to carbohydrate response ($p < 0.05$), at 95% confidence interval with p value = 0.000, meaning Formula made significant effect to product carbohydrate content.

Carbohydrate is present in the dominant number as the composition of nutritional value, in Table 6 it can be seen that the lowest carbohydrate level in Formula 2 is 75.17% and the highest is Formula 8 of 84.04%. The average carbohydrate value of 17 Formula is 79.55%. Raw materials that have the highest carbohydrate *i.e* banana flour so that the addition of banana flour affects the carbohydrate content of the snack. Based on raw material analysis, banana flour containing high carbohydrate that is 83.97% (Table 2). Snack produced by adding 100 g of banana flour gave the highest carbohydrate content compared to other treatments (Table 3).

The results of proximate analysis of extracted snack based on corn and green beans [14] show the percentage of water content, ash content, and higher protein content than the average of formulations made from bananas, green beans, corn and eggs, while the fat content and lower carbohydrate levels. This is because of the difference in composition used in the manufacture.

High levels of fat in formulations made from bananas, green beans, corn and eggs occur due to the addition of oil and eggs. The fat content is higher when compared to extrusion snacks made from corn and green beans, but their fat content meets the standards of [12], where the recommended fat content of SNI if the product without frying process is 30% (w / w) and with the frying process is a maximum of 38% (w / w). Water content in snack made from bananas, green beans, corn and eggs is very low because after the extrusion process finished the finished product in the first because the water content is still high. According to [11] water levels have a close relationship with the properties of "crisp" and the crisp extreme snack. The average carbohydrate content of snack is higher than that of corn extract and nuts. This is because of the difference in composition used, whereas based on analyses, bananas have a high carbohydrate content.

The protein content of corn and green bean extract snacks are higher, due to differences in composition, where green beans were more used in the formulas compared to banana-based snacks, green beans, corn and eggs. High protein content in green beans can affect protein levels in extrusion snacks. The ash content of corn extract snack and green beans is higher than that of banana, green beans, corn and eggs. This shows a consequence that extrudate of maize and green beans have a high mineral content.

4. Conclusion

Addition of green beans on snacks proved to increase protein levels, the addition of banana flour has an effect on the color of snack and starch content. The addition of maize affects moisture content, and carbohydrate content, and starch extrusion snack.

The result of analysis of variance to water content response, ash content, protein content, fat content, carbohydrate content, starch content, amylose content, amylopectin content, and color are significantly different ($p < 0.05$). Carbohydrate levels 75.17-84.04%. The best carbohydrate levels is 84.04% in Formula 8 which is also the best formula.

Acknowledgement
We would like to thank to E N Pangestika for analyzing data.

References

[1] Anonymous 2015 *Horticulture Production Statistics 2014* (Jakarta: Dirjen Hortikultura, Kementrian Pertanian) p 285

[2] Jiang H, Zhang Y, Hong Y, Bi Y, Gu Z, Cheng L, Li Z and Li C 2015 *Food Hydrocoll* **49** 192-9
[3] Zabalaga F Rosa, La Fuente L A Carla and Tadini C Carmen 2016 *J Food Eng.* **187** 62-9
[4] Garcia-Suarez J Francisco, Páramo-Calderón E Delia and Bello-Pérez A Luis 2014 *Lebenson Wiss Technol [dnlm]* **58** 381-6
[5] Daniali G, Jinap S, Hanifah N L and Hajeba P 2013 *Food Control ()* **32** 386-91
[6] Yap M, Fernando M A D B Warnakulasuriya, Brennan S Charles and Coorey J R Ranil 2017 *Lebenson Wiss Technol [dnlm]***80** 10-18
[7] Aurore G, Parfait B and Fahrasmane L 2009 *Trends Food Sci Technol* **20** 78-91
[8] Pelissari M Franciele, Andrade-Mahecha M Margarita, Sobral d J A Paulo and Menegalli C Florencia 2013 *Lebenson Wiss Technol [dnlm]***52** 1-11
[9] Alimi A Buliyaminu, Tilahun S. Workneh S Tilahun and Samson A. Oyeyinka A Samson 2017 *Lebenson Wiss Technol [dnlm]*79 84-91
[10] Mesquitaa M Leonela, Francoa M L Célia, Leonela S, Garciaa L Emerson and dos Santosa P R Thaís 2016 *Int J Biol Macromol* **89** 1-8
[11] Muctadi T R, Purwiyatno, Basuki A 1988 *Extrusion –Cooking Technology* (Bogor: Bogor Agricultural Institute) p 113
[12] Anonymous 2015 *Indonesia National Standard (SNI) 2886-2015: Snack Extrudate* (Jakarta: National Body Standard) p 6 (In Bahasa Indonesia)
[13] Anonymous 2016 *Thermo Scientific Twin-Screw Extrude*r (www:thermoscientific.com) p 17
[14] Atmadja G M 2006 *Development of Food Products from Quality Protein Maize (Zea mays l.) using Extrusion Technology* (Thesis: Bogor Agricultural University) p70
[15] Sukasih E and Setyadjit S 2014 *Proc. of the 2nd International Conference on Sustainable Innovatio*n Eds Taufik T *et al* (Singapore: Springer) p. 29-39
[16] Kocerla P, Aparna K and Laksmi N Dewi 2012 *Agric Eng Int: CIGR Journal* **14(4)** 179-186
[17] Anonymous 1992 *Indonesia National Standard (SNI) 01-2891-1992:Method for analyses food and drink* (Jakarta: National Body Standard) p 25 (In Bahasa Indonesia)
[18] Andarwulan N, Kusnandar F and Herawati D 2011 *Food Analyses* (Jakarta: PT Dian Rakyat) p 328
[19] Jarvis C E and Walker J R L 1993 *J Sci Food Agri* .**63** 53-57
[20] Guy R 2001 E*xtrusion Cooking: Technologies and Application* (Cambrige(UK): CRS Press) p 288
[21] Budi F S, Hariyadi P, Budijanto P and Syah D 2013 *Food* **22(3)** 263-74
[22] Budijanto S, Sitanggan B S, Hasti Wiranti and Koesbiantoro 2012 *J Pascapanen* **9(2)** 63 – 9 (In Bahasa Indonesia)

Physico-chemical characteristics of shallot New-Superior Varieties (NSV) from Indonesia

E Sukasih[1], Setyadjit[1], D Musadad[2]

[1]Indonesian Center for Agriculture Postharvest Research and Development
Jl. Tentara Pelajar no 12, Bogor, Kampus Penelitian Pertanian, Cimanggu, Bogor, Indonesia
[2]Indonesian Research Institute for Vegetable Crops. Jl. Tangkuban Perahu 517 Lembang

Email : ermi_sukasih@yahoo.co.uk

Abstract. Shallot is one of the priority agricultural commodities to be developed in Indonesia to reduce import and to stabilize domestic supply. The efforts include the selection of varieties, seed technology, agronomy, handling and processing to extend the supply and added value. Indonesian Agency for Agricultural Research and Development (IAARD) has purified, cross-pollinated, selected and released new varieties called New Superior Varieties (NSV) to farmers. The purpose of this research was to investigate the characteristic of fresh shallot NSV by understanding its potential for raw material of processed product. A completely randomized design (CRD) of single factor of ten local varieties of shallot such as Cv. Sembrani, Cv. Kuning, Cv. Pancasona, Cv.Bima, Cv. Trisula, Cv. Pikatan, Cv. Katumi, Cv. Kramat-2, Cv. Mentes and Cv. Majalok of three replication was used to arrange the experiment. The results showed that shallot New Superior Varieties (NSV) were significant by effect the physico-chemical parameters, such as diameter, length, weight of both in main bulb and tiller bulb, fat total, carbohydrate, crude fiber, starch content, antioxidant capacity and quercetin. Of the ten varieties of shallot characterized, the largest bulbs are Cv. Sembrani i.e 5.30 ± 0.3g per bulb, the best red color for shallot peeled was Cv. Kuning. Furthermore Cv. Pancasona have the highest protein content of 4.23 ± 0.2%, Cv. Mentes have the highest functional properties of quercetin 1766.4 ± 134 ppm. Shallot varieties such as Cv. Sembrani, Cv. Bima, Cv. Kuning and Cv. Trisula suitable for use as fresh product. Shallot varieties such as Cv. Pikatan, Cv. Pancasona, Cv. Katumi and Cv. Kramat-2 are suitable as raw materials for processed products. Cv. Mentes and Cv. Majalok were potential for raw materials of functional food and pharmaceutical industries.

Keywords: shallot, characteristics, NSV (New-Superior Varieties), Physico-chemical

1. Introduction

The fluctuations of price and scarcity in the market is a problem that occurs in shallot commodity in produced Indonesia. This is caused by crop failures by climate, natural disasters, mafia involvement, and are easily damaged. Shallot price fluctuations occur almost in year-round. The first is associated with the celebration day like Eid Mubarak, Christmas, New Year Eve, long holidays and imports action. Usually shallot is imported when farmers are harvesting, so the price of this commodity would fall and fluctuate.

Shallot is one of strategic horticultural commodities in Indonesia. The average of shallot consumption in Indonesia reaches 4.6 kg per capita per year. Almost all households are known to consume shallot as spices every day. In addition to flavor reasons, some people consume shallot because of the nutrition contained in shallot such as: calcium, phosphorus, vitamin C, and beta-carotene [1]. Shallot production in Indonesia in the last five years tends to increase continuously.

Shallot is one of the priority of agricultural commodities is being developed in Indonesia to reduce import and to stabilize domestic supply. The efforts include the selection of varieties, seed technology, agronomy, handling and processing to extend the supply and added for value. IAARD has purified, cross-pollinated, selected and released new varieties called New Superior Varieties (NSV). These superior varieties are still able to be productive despite being planted out of season. These varieties include Cv. Bima, Cv. Trisula, Cv. Katumi, and Cv. Sembrani [2].

Consumers prefer shallot with characteristics of large tuber size (8-10 g / bulb), have a shiny red skin color and a short harvest age. New varieties will have different characteristics both size and nutrients. Characterization is needed to determine the suitability of shallot varieties for fresh consumption, for processing and which have high functional content as raw materials for the pharmaceutical industry. The purpose of this research was to investigate the physico-chemical characteristic of various shallot varieties from Indonesian Research Institute for Vegetable Crops collections.

2. Materials and Methods

2.1. Material and equipment

Raw materials consisted of ten varieties of local shallot, nine varieties were obtained from Indonesian Research Institute for Vegetable Crops Lembang, were Cv. Pancasona, Cv. Kuning, Cv. Katumi, Cv. Mentes, Cv. Majalok, Cv. Pikatan, Cv. Trisula, Cv. Sembrani, and Cv. Kramat-2 while one variety (Cv.Bima) obtained from farmers in Brebes.

Chemicals for analysis were consisted of DPPH the 2,2-diphenyl-1-picrylhydrazyl, methanol, ascorbic acid, folin-ciocalteau, quercetin standard, acetonitrille, sulfuric acid and sodium hydroxide. Equipment in this research were caliper, ruler, scales, filter paper, soxhlet, kjeldahl, spectrophotometer, chromameter, oven, furnace, HPLC and cuvettes.

2.2. Research methodology

Bulb and tiller shallots were removed from the bunch then sorted and selected and more good quality. Physical characteristic of each variety was taken from about 100 bulbs with 6 replication. These characterization included bulb length (cm), bulb diameter (cm), bulb weight (g), skin color (description), and number of tiller bulbs.

Next, Samples were analysed for proximate such as moisture content, ash content, fat total, protein, carbohydrate, crude fiber and starch [3], functional compound including DPPH [4] and quercetin [5]. This research was used to arrange the experiment was a Completely Randomized Design single factor i.e. varieties with three replication. The result of analysis was calculated of standard deviation and then significance by using Minitab 14 software.

3. Results and Discussion

3.1. Physical Characteristics of Various Shallot

Physical characteristics of various shallot NSV were presented at Table 1. The results showed that varieties of shallot showed were significantly effect on parameters, such as diameter, length, weight of both in main bulb and tiller bulbs. However, varieties did not significantly effect to the number of bulbs.

Differences in the size of the shallot bulbs are influenced by the genetic factors of each variety. Shallots from different varieties that are planted on the same soil will produce different bulb sizes [6].

Table 1. Physical characteristics of various of shallot NSV

No	Varieties	Diameter of bulbs (cm)	Length of bulbs (cm)	Weight of bulbs (g)	Diameter of bulbs (cm)	Length of bulbs (cm)	Weight of bulbs (g)	Number of bulbs	Skin color
		Main Bulbs			Tiller bulbs				
1	Pikatan	1.60±0.2ab	2.50±0.1ab	2.68±0.2a	1.12±0.1ab	1.87±0.1b	1.72±0.1d	3.00±1b	pale red
2	Pancasona	1.80±0.1b	2.45±0.1ab	3.55±0.1b	1.28±0.1bc	1.96±0.2bc	1.92±0.1e	3.00±1b	red
3	Katumi	1.85±0.1b	2.27±0.1a	4.16±0.2c	0.80±0.1a	1.15±0.1a	0.41±0.2a	1.00±0.5a	red
4	Trisula	1.75±0.1ab	2.95±0.1bc	4.22±0.1c	1.53±0.1c	3.03±0.1f	2.84±0.1f	2.00±0.5ab	red
5	Majalok Lembang	1.90±0.1b	2.73±0.2b	3.99±0.1bc	1.20±0.2bc	2.00±0.1c	1.43±0.1c	1.00±0.5a	red
6	Kuning	2.02±0.1b	2.35±0.2a	6.04±0.1d	1.60±0.1c	2.10±0.1c	3.22±0.1g	2.00±1ab	dark red
7	Kramat-2	1.82±0.1b	2.54±0.1ab	4.11±0.1bc	1.26±0.1bc	2.02±0.1c	1.92±0,1e	2.00±0.5ab	red
8	Mentes	1.35±0.1a	2.27±0.1a	2.35±0.1a	1.08±0.1ab	1.94±0.1bc	1.24±0,1b	3.00±0.5b	pale red
9	Sembrani	3.08±0.2c	2.85±0.1bc	12.59±0.1e	2.30±0.1d	2.50±0.1d	5.30±0.2h	2.00±0.5ab	yellowish red
10	Bima Lokal	1.82±0.1b	3.02±0.1c	4.33±0.1c	1.57±0.2c	2.70±0.1e	2.63±0.1f	2.00±0.5ab	red

The largest horizontal diameter of main bulb is on Cv. Sembrani varieties of 3.08 cm while the lowest is Cv. Mentes varieties of 1.35 cm. Indonesian National Standard requires quality of bulb diameter is 1.7 cm and the grade II is 1.3 cm. The varieties which are classified on quality I was Cv. Sembrani of 2.3 cm, while the others variety Cv. Bima, Cv. Kuning and Cv. Trisula were classified in quality II. The diameter bulb of Cv. Bima v is 1.82 cm, this can be categorized as a large tuber size with a diameter range of 1.68 to 2.28 cm. Similarly, for Cv. Majalok varieties of 1.90 cm, the range for large categorized bulbs are in diameter of 1.23 to 2.21 cm [7].

The largest length vertical of shallot main bulb was on Cv. Bima of 3.02 cm and the smallest were Cv. Katumi and Cv. Mentes of 2.27 cm. The largest weight of main bulb was Cv. Sembrani of 12.59 g and the lowest was Cv. Mentes of 2.35 g (Table 1). The irrigation and fertility levels influenced the production of different grades of onion bulb considerably [7].

The highest number of segmen bulb shallot were on Cv. Pikatan, Cv. Pancasona and Cv. Mentes, three segmen of each variety, while the least were Cv. Majalok Lembang and Cv. Katumi, only one tiller of each variety. The largest diameter of segmen bulb was Cv. Sembrani of 2.30 cm and the shortest was Cv. Katumi 0.80 cm. The largest length of segment bulb length was Cv. Trisula of 3.03 cm, and the shortest was Cv. Katumi which is 1.15 cm. The largest weight of bulb weight was Cv. Sembrani varieties of 5.30 g and the lowest was Cv. Katumi of 0.41 g.

The color of shallots were pale red to dark red usually, except Cv. Sembrani has a yellowish red color. Shallots that have a pale red color are varieties of Cv. Mentes and Cv. Pikatan. The highest intensity of shallot color is Cv. Kuning , the other varieties are red color. The red color is caused by the anthocyanin pigment content. Anthocyanin tends to be unstable and easily degraded by various factors. These factors include light, pH, temperature, sulfite, ascorbic acid, enzymes, and others. Increased stability of anthocyanins can be done by binding of o-glycosidic, the presence of acyl groups in sugar molecules, the presence of metal ions, self-association, copigmentation, and encapsulation [9]. Thiosulfinate is a precursor pigment on the red onion [10]. Thiosulfinate can interact with anthocyanin with its own association (Asosiated-self). Such interactions can increase the stability of anthocyanins.

Another factor that makes the anthocyanin unstable is its own association factor and pH. The association itself is the interaction between anthocyanins with different copolymers that can affect the color and stability of anthocyanins. The mechanism of anthocyanin stabilization with its own association is closely related to the pH factor. Anthocyanins are more stable at lower pH (acid) than in high pH (alkaline). The ionic group of anthocyanins activates molecular structure changes depending on the pH state which will result in color differences in different pH conditions.

The color of shallot bulbs is likely to be stable, it will be unstable when the cell of shallot is damaged due to the processing or mechanical damage. Generally consumers prefer the color of shallots, the higher red color intensity will be favored by consumers. However, if shallots are used as raw material, the color can be improved through several treatments such as soaking process or addition with ingredients such as citric acid. The examples of shallot products that have been using this application are shallot paste, dried shallot slices and whole shallot in brine. The visual color of the various shallot new superiorvarieties (NSV) are presented at Figure 1.

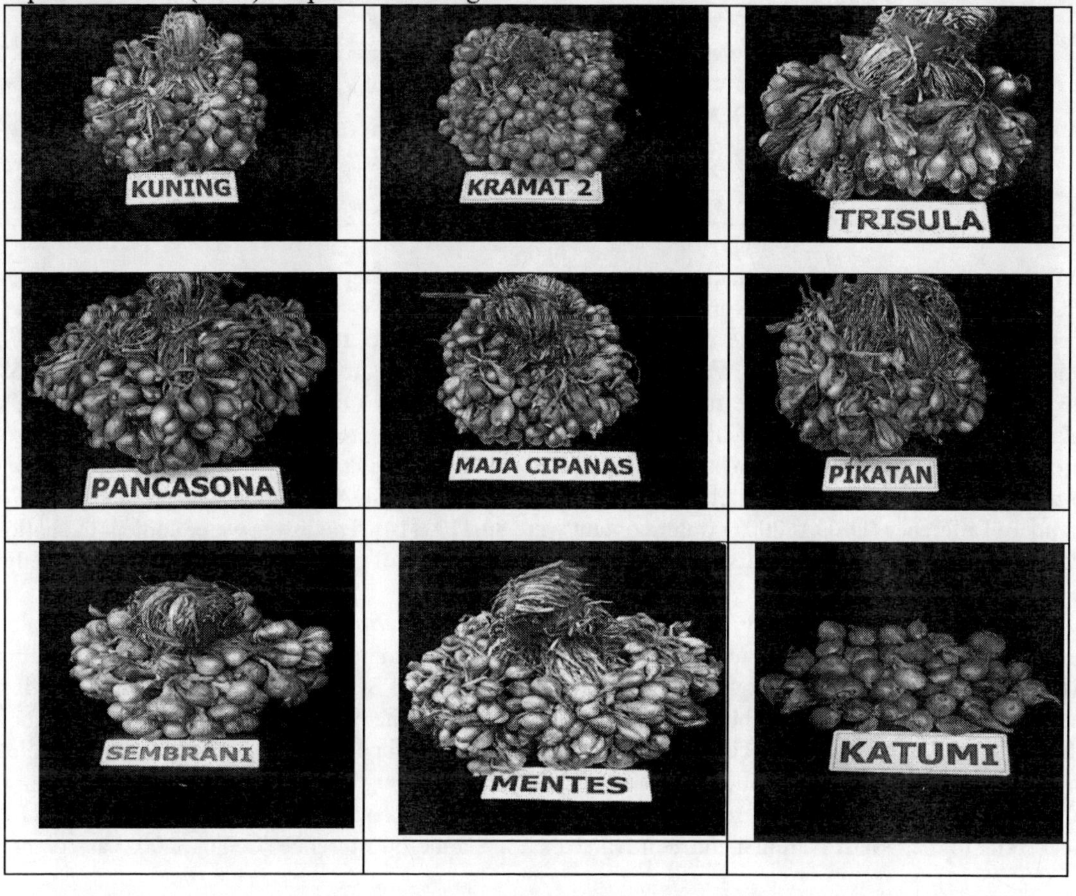

Figure 1. The red color level of various shallot NSV

3.2. Chemichal and Functional Compound Characteristics of Various Shallot New-Superior Varieties (NSV)

The results showed that varieties of shallot were significantly effect on the parameters, among others were fat total, carbohydrate, crude fiber, starch content, antioxidant capacity and quercetin. However, varieties were not significantly different on water content, ash content and protein (Table 2). Onion has different nutritional compositions depending on the variety and stage of maturity among others [11]. The drying conditions such as flavor and nutritional composition during processing or storage [12].

Table 2. Chemichal and functional compound characteristics of various shallot NSV

No	Varieties	Water content (%)	Ash content (%)	Fat (%)	Protein (%)	Carbohydrate (%)	Crude fiber (%)	Starch (%)	Quersetin (ppm)	DPPH (ppm)
1	Pikatan	87.17 ±0.3a	0.69±0.01a	0.05±0.01a	3.13±0.1b	8.96±0.5c	1.90±0.2a	4.15±0.4bc	1490.55±67.3f	19.85±0.88c
2	Pancasona	86.45±0.2a	0.65±0.01a	0.59±0.01d	4.23±0.2b	8.08±0.3a	1.99±0.1a	4.23±0.3bc	1612,22±54,3g	19.88±0.58c
3	Katumi	84.93±0.2ab	0.65±0.01a	0.19±0.02b	2.36±0.1ab	11.87±0.4g	2.25±0.2bc	4.73±0.3d	218.91±34.8a	9.72±0.32a
4	Trisula	87.10±0.2a	0.73±0.02ab	0.34±0.02c	3.41±0.2b	8.41±0.5b	2.03±0.3ab	4.01±0.4b	304.89±21.6b	24.45±0.63d
5	Majalok Lembang	87.54±0.1a	0.80±0.01b	0.38±0.01c	2.26±0.2ab	9.01±0.2d	2.10±0.2b	4.40±0.2c	199.68±31.8a	53.99±0.47h
6	Kuning	80.03±0.2b	0.75±0.01ab	0.07±0.02a	3.16±0.1b	15.99±0.3h	2.61±0.2d	3.44±0.3a	705.85±78.2d	11.86±0.19b
7	Kramat-2	81.83±0.2b	0.73±0.01ab	0.20±0.02b	0.90±0.1a	16.33±0.3i	2.00±0.2a	4.40±0.3c	704.07±97.7d	30.77±0.53f
8	Mentes	84.43±0.1ab	0.73±0.02ab	0.18±0.02b	2.75±0.2b	11.92±0.2g	2.29±0.2c	4.20±0.4bc	1766.4±134h	33.28±0.20g
9	Sembrani	85.38±0.1ab	0.63±0.01a	0.03±0.02a	2.47±0.1ab	11.49±0.3f	2.43±0.3c	4.38±0.2bc	1151.70±100.01e	19.93±0.99c
10	Bima	84.21±0.2ab	0.55±0.1a	0.07±0.02a	3.89±0.2b	10.93±0.7e	2.23±0.2bc	4.04±0.1b	443.74±20.34c	28.58±0.77e

Water content is an important factor that can affect the quality of a product. The lowest water content was Cv. Kuning that was 80.03% and the highest was Cv. Majalok variety of 87.54%. Water content in fresh onion commodities must be in accordance with the standard, the water content is too low will reduce the weight of the product. Water content of shallot according to SNI 01-3159-1992 is 80% -85% [13]. Some shallot varieties were not comply with the standard for moisture content. They were Cv. Pikatan, Cv. Pancasona, Cv. Trisula, and Cv. Majalok, however moisture content can be decreased through the drying process. Onion has a moisture content of 84.55 to 86.07% [8], this value is equivalent to the water content of VUB variety shallot. According to the USDA National Nutrient Database for Standard Reference (USDA, 2007) water content were 89.11% [14]. The lower water content of shallots will have better quality. Low water content can reduce the growth of microbes that can reduce the quality of the product [15].

The lowest ash content was Cv. Bima of 0.55% and the highest was Cv. Majalok of 0.8%. Ash content describes the mineral content contained in the material after the material is burned to be free of carbon [16]. This is not a parameter required in the Indonesian National Standard, but the determination it in the food is important to do. It is because ash content determines the mineral content present in the food. The higher the mineral content contained in food, the higher ash content present in the food. Shallots contain minerals include potassium of 334 mg / 100g, 0.80 g / 100g of iron and phosphorus of 40 mg / 100g. In contrast to onion, the most important minerals are potassium, calcium and selenium [11]. According to the USDA National Nutrient Database for Standard Reference (USDA, 2007) ash content are 0.35%[14].

The lowest fat content is Cv. Sembrani that is 0,03% and highest is Cv. Pancasona varieties equal to 0,59%. Fat is almost found in all food with different content. Shallot contain low fat, usually in the form of essential oils, but these fats can cause problems in processing such as shallot flour, this product will be clumping after being stored. According to the USDA National Nutrient Database for Standard Reference (USDA, 2007) total fat content are 0.1%[14].

The lowest protein content was Cv.Kuning of 0.9% and the highest was Cv.Pancasona of 4.23%. Proteins serve as anthocyanin precursors that contribute to red color. The anthocyanin comes from one of the amino acids, phenylalanine. Phenylalanine is one of the protein amino acids. Anthocyanin in form of cyanidin 3-glucoside [17]. Protein content is proportional to the concentration of anthocyanin in the shallot flour. Protein in the onion ranges from 0.96% to 1.04% [8]. This value tends to be lower than some varieties of NSV varieties. According to the USDA National Nutrient Database for Standard Reference (USDA, 2007) the total protein content are 1.1%[14].

The lowest carbohydrate content was Pancasona variety which was 8.08% and the highest was Cv. Kramat-2 of 16.33%. According to the USDA National Nutrient Database for Standard Reference (USDA, 2007) total carbohydrate content are 9.34% [14]. Shallots with high carbohydrate were suitable as raw material for producing both of dried slices and flour. In the processing of both products will be produced 10-12% of yield. In contrast to instant shallot flour made with a drum dryer with the addition of tapioca as filler will result more than 70% of yield [18].

The lowest crude fiber content was Cv. Pikatan of 1.90% and the highest was Cv. Sembrani of 2.43%. Crude fiber is the amount of fiber mass in foods, they were resistant to strong acids and strong alkalines. In food products, crude fiber content is always lower than dietary fiber. This is because strong acids such as sulfuric acid and strong alkalines such as sodium hydroxide have a greater ability to hydrolyze food components if they were compared to digestive enzymes [19]. The crude fiber contributes to the crispness of the shallot. Shallots with have a high coarse fiber are suitable for preserved in wholes and slices in brine products. According to the USDA National Nutrient Database for Standard Reference (USDA, 2007) dietary fiber content are 1.7% [14].

The lowest level of starch was Cv. Kuning was 3.44% and the highest was Cv. Katumi of 4.73%. Nutrients component such as fats, proteins, carbohydrates, crude fiber and starch were not included in the parameters required in fresh shallot Indonesian National standard.

The lowest antioxidant capacity was Cv. Katumi varieties that was 9.72 ppm and the highest was Cv. Majalok of 53.99 ppm. Previous studies were showed that the antioxidants present in the shallot are quercetin 3,4-digecoside and quersetin 4-monoglucoside [20]. Antioxidants in shallot contribute to inhibit of browning reaction. This is important when the shallot will be processed into another product. If in the process we will require of slicing or destructing, then it will cause injured shallot cells so will accelerate the oxidation processed and the color will turn into brownish.

Antioxidants may be enzymes (SOD, catalase, and glutathione peroxidase), vitamins (A, C, E, and B-carotene), and other compounds (flavonoids, etc.). Enzymatic antioxidants are a major defence system against oxidative stress conditions. Non-enzyme antioxidants can be nutritional compounds and nutrients. Vitamin C (L-ascorbic acid) is a water-soluble antioxidant [21].

The quantitative determination of antioxidant activity can be determined by DPPH (Diphenylpicrylhydrazyl). DPPH is a free radical compound that reacts with antioxidants that donate one electron to form Diphenylpicrylhydrazine compound. The DPPH method measures the ability of an antioxidant compound in capturing free radicals. Previous research on the color of shallot flour has a positive correlation to the antioxidant activity. The antioxidant activity is related to the amino acid product which is the result of non-enzyme browning reaction during storage. If the amino acid decreases then the antioxidant activity to inhibit the peroxyl radicals in shallot flour will increase [22].

Shallots new-superior varieties (NSV) have wide range of quercetin content. The lowest level of quercetin was Cv. Majalok that was 199.68 ppm and the highest was Cv. Mentes 1766.4 ppm. Quercetin is one of the flavonoids found in shallot. The form of quercetin in onion flour is quercetin 7,4-diglukosida [17]. Quercetin may be used as an effective lipid peroxidation inhibitor after mirisetin and kaempferol. The quercetin content in the outer and inner sections of the shallot was 83477 ± 2879 mg / kg and 1926 ± 266 mg / kg [23].

Shallot contains phytochemical compounds, allisin, alliin, allil propel disulfide, phytosterol, flavonal, flavonoid, kaemfenol, quersetin, quercetin glycoside, pectin, saponin, and others [24]. Shallot nutrient composition is very complex, where it becomes one of the main sources of flavonoids in many countries. In particular, shallot contains quercetin flavonols and quercetin derivates. In addition, it is rich in other bioactive compounds such as frutooligosaccharida and sulfur compounds [2008]. The functional nature of shallot (antioxidant capacity and DPPH), is important to know if the shallot will be use as functional food or raw materials for the pharmaceutical industry. Allium species are rich in flavonols among which quercetin $3,4^1$-o- diglucosidase represent the major components. Quercetin is known for its antioxidant and free radical scavenging power and its capability to protect agains cardiovascular deseases [26,27,28].

4. Conclusion

Shallots new superior varieties (NSV) were significantly effect on the physico-chemical parameters, such as diameter length, weight of both in main bulb and tiller bulb, fat total, carbohydrate, crude fiber, starch content, antioxidant capacity and quercetin.

Cv. Sembrani was one of variety of shallot which can be categorized in quality I in SNI for bulb diameter, meanwhile some varieties such as Cv. Bima Cv. Kuning and Cv. Trisula were categorized in quality II. Some varieties such as Cv. Katumi, Cv. Kuning, Cv. Kramat-2, Cv. Mentes, Cv. Sembrani and Cv. Bima have also within SNI based on their moisture content parameters.

Recommendations from this research:varieties such as Cv. Sembrani, Cv. Bima, Cv. Kuning and Cv. Trisula suitable for use as fresh product. Shallot varieties such as Cv. Pikatan, Cv. Pancasona, Cv. Katumi and Cv. Kramat-2 are suitable as raw materials for processed products. Cv. Mentes and Cv. Majalok were potential for raw materials of functional food and pharmaceutical industries

References

[1] Setiadi 2006 *Chili Rawit: Type and Cultivation* [In Bahasa Indonesia]. Jakarta : Penebar Swadaya.

[2] Anonymous 2014 Shallot *Superior Variety to Overcome of Scarcity.* [In Bahasa Indonesia] *Indonesia Science Magazine Edition 36.*

[3] AOAC (Association of Official Analytical Chemist) 2005 *Official Method of Analysis of the Association Analytical of Chemist*. The Association of Official Analysis Chemist, Inc., Arlington.

[4] Takaya Y, Kondo Y, Furukawa T and Niwa M 2003 *Journal of Agriculture Food Chemistry* **51** 8061-8066.

[5] Phani C R S, Viaykumar C, Umamaheswara K and Sindhuja G 2010. *International Journal of Research in Pharmaceutical and Biomedical Sciences* **1(1)**: 19-22.

[6] Putrasamedja S and Soedomo P 2007. *Journal of Rural Development* **7(3)**:133-146. (Please translate this.

[7] Azmi C, Hidayat I M, Wiguna G 2011. *Journal of Horticulture* **21(3)**:206-213.

[8] Kumar S,Imtiyaz M and Kumar A 2007. *Scientia Horticultural* **112**121-129.

[9] Rodrigo N C, Santos D T and Meireles M AA 2011. *Food Research International* **44(2)**: 499-509.

[10] Zang J, Wang D and Zhao G 2013. *Trends in Food Science and Technology* **30(2)**: 162-173.

[11] Mota C L, Luciano C, Dias A, Barroca M J and Guine R P F 2009. *Food and Bioproducts Processing* **88** 115-123.

[12] Kaymak-Ertekin F and Gedik A 2005. *Journal of Food engineering* **68** 443-453.

[13] Anonymous 1992 *Indonesian National Standard 01-3159-1992: Bawang Merah*

[14] Anonymous. 2007 *U.S Departement of Agriculture, Agricultural Research Service*, (http://www.ars.usda.gov/ba/bhnrc/ndl).

[15] Triyono A 2010 *Study the effect of the addition of several acids on the protein isolate process to the green bean protein isolate (Phaseolus radiatus L.)* (Symposium: Diponegoro University).

[16] Husien H S 2009 *Stabilized bran drying using drum dryer dryers and quality changes during storage (Thesis:* Bogor Agricultural University)

[17] Rodrigues A S, Perez-Gregorio M R, García Falcón M S, SimalGándara J and Almeida D P F 2011. *Food Chemistry* **124(1)**: 303-308.

[18] Setyadjit and Sukasih E 2015. *Procedia Food Science* **3** 396-408.

[19] Sinaga F 2009 *Study of making dietary fiber from corn kernel.* (Thesis Sumatera Utara University)

[20] Gregorio R, Jorge R, Barreiro C G and Gandara J S 2011. *Journal of Food Control.* **22(7)**: 1108-1113.

[21] Winarsi H 2007 *Natural Antioxidants and Free Radicals* [In Bahasa Indonesia]. Yogyakarta: Kanisius.
[22] Moreno F J, Martinez M C, Castillo M D and Villamiel M 2006. *Journal of Food Research International.* **39(8)**: 891-897.
[23] Nuutila A M, Puuponen R, Aarni M and Oksman-Caldentey K 2003. *Food Chemistry* **81(4)**: 485-493.
[24] Jaelani 2007 *Benefits of Shallot* [In Bahasa Indonesia]. Yogyakarta: Kanisius.
[25] Roldan E, Concepcion S M, Begona de A M and Pilar C 2008. *Food Chemistry* **(108)**: 907-916.
[26] Fossen T, Pedersen A T and Andersen Q M 1998. *Phytochemistry* **47** 281-285.
[27] Ibbara M, Perez-Vizcaino F, Cogolludo A, Duarte J, Zaragoza-Arnaez, F and Lopez-Lopez J G 2002 . *Planta Medica* **68** 307-310.
[28] Clitton P M 2004. *Journal of Biomedicine and Biotechnology* **5** 272-278

Specific primer design of mitochondrial 12S rRNA for species identification in raw meats

[1]M Cahyadi, Puruhita, F H Barido and B S Hertanto

Department of Animal Science, Faculty of Agriculture, Universitas Sebelas Maret, Surakarta, Indonesia 57126

Email : mcahyadi@uns.ac.id

Abstract. Polymerase chain reaction (PCR) is a molecular technique that widely used in agriculture area including species identification in animal-based products for halalness and food safety reasons. Amplification of DNA using PCR needs a primer pair (forward and reverse primers) to isolate specific DNA fragment in the genome. This objective of this study was to design specific primer from mitochondrial 12S rRNA region for species identification in raw beef, pork and chicken meat. Three published sequences, HQ184045, JN601075, and KT626857, were downloaded from National Center for Biotechnology Information (NCBI) website. Furthermore, those reference sequences were used to design specific primer for bovine, pig, and chicken species using primer3 v.0.4.0. A total of 15 primer pairs were picked up from primer3 software. Of these, an universal forward primer and three reverse primers which are specific for bovine, pig, and chicken species were selected to be optimized using multiplex-PCR technique. The selected primers were namely UNI_F (5'-ACC GCG GTC ATA CGA TTA AC-3'), SP_R (5'-AGT GCG TCG GCT ATT GTA GG-3'), BB_R (5'-GAA TTG GCA AGG GTT GGT AA-3'), and A_R (5'-CGG TAT GTA CGT GCC TCA GA-3'). In addition, the PCR products were visualized using 2% agarose gels under the UV light and sequenced to be aligned with reference sequences using Clustal Omega. The result showed that those primers were specifically amplified mitochondrial 12S rRNA regions from bovine, pig, and chicken using PCR. It was indicated by the existence of 155, 357, and 611 bp of DNA bands for bovine, pig, and chicken species, respectively. Moreover, sequence analysis revealed that our sequences were identically similar with reference sequences. It can be concluded that mitochondrial 12S rRNA may be used as a genetic marker for species identification in meat products.

Keywords : Design specific primer, mitochondrial RNA, species identification, raw meats

1. Introduction

The high price of beef increases the rise of adulteration of beef with other meats. The main factor is because the quarrel of price between beef and pork or chicken relatively high. Adulteration of meat products is intended to produce the final product with relatively cheaper price [1]. However, adulteration of meat products interfere the consumers, thus detection of meat contamination to know the existence of fake meat is highly necessary. An effort to detect contamination in meat can be done by using multiplex polymerase chain reaction (multiplex-PCR). Previous study proved that the multiplex-PCR method can point out the specific difference between animal species, because each animal has unique area on each of its DNA [2].

According to [3], PCR is used to amplify the specific sequence of DNA. PCR reaction components consist of template DNA, a pair of primers, deoxynucleotide triphosphates (dNTPs), buffers, and Taq DNA polymerase, which then proceed to a series of amplification cycles with a mixture of three different temperatures. Multiplex-PCR can amplify a small target DNA into millions of copies of DNA with more than one primers in a single reaction. PCR uses a pair of primers, which complement each of the two target DNA regions to be amplified. Primer candidate which is designed by the program based on the *conserved area* where the best attachment is on this area [4].

Primer is a short oligonucleotide molecule that serves as a limiting target DNA fragment which will be amplified [5]. A good primer influences the success of PCR. Primer for DNA amplification by PCR is obtained through primer design. This primer design is an important aspect for obtaining specific, effective and efficient PCR products [6]. Primer design can be complicated when a pair of primers is needed to amplify specific areas within different species [3]. Specific primers will attach on specific regions of DNA templates such as in the area of mitochondrial DNA (mt-DNA).

Mitochondrial DNA is derived maternally, so that there is limited variation within the same species while inter-species variation is very large [7]. Mitochondrial DNA has two rRNA encoding genes which are 12S and 16S rRNAs that have high sequence variations, uniqueness and high variation at the species level so that 12S rRNA areas can be used for specific detection of certain species using multiplex PCR technology [8]. Commonly, cytochrome *b* gene is used to identify species in raw, cooked and processed-meat products. However, mitochondrial 12S rRNA is widely used for forensic analyses in wide-range of species [9]. Previous studies reported that multiplex PCR of 12S rRNA, tRNA Val and 16S rRNA has been successfully identified ruminants, poultry, fish and porks [10] in feedstuffs. The use of mt DNA 12S rRNA as genetic marker in one tube PCR reaction for species identification in meat-based foods has not been reported. Hence, the objective of this study was to obtain a specific primer which used to amplify mt DNA 12S rRNA region of pig, chicken and bovine in one multiplex PCR reaction.

2. Materials and Methods

2.1. Primer design

The 12S rRNA sequences of cow, pig and chicken were obtained from the genebank database on NCBI website (National Center for Biotechnology Information). 12S rRNA sequence search was done by accessing NCBI website (http://www.ncbi.nlm.nih.gov/) which then chose nucleotide in search reference and typed keyword of section and species needed as an example (*Bos Taurus* complete mtDNA genome). Related sequences will appear on the screen and then selected the required sequence. Complete genome of the mitochondrial DNA will appear on the screen then selected the part to be taken which was 12S rRNA region and then clicked FASTA and copied.

Alignment analysis can be used to compare two or more sequences to know that the region used is truly unique and the primers obtained are not attached to other organisms. The program used for alignment analysis was Basic Local Alignment Search Tools (BLAST) program. Alignment with BLAST analysis done by accessing NCBI BLAST website (http://blast.ncbi.nlm.nih.gov/Blast.cgi) then chose the link nucleotide BLAST (nblast). After the nblast page appeared, then Allign two or more sequences columns were selected to compare 2 nucleotide sequences. Then entered the sequences which expected to compare in the form of FASTA on available column and BLAST was chosen at the bottom of the page. The results from BLAST would display a scale that shows the level of similarity of compared sequences. The BLAST results were also analyzed to show conserved regions of 12S rRNA gene sequences.

Primer candidates obtained by using PRIMER3 program which could be accessed on PRIMER3 website with 0.4.0 version (http://bioinfo.ut.ee/primer3-0.4.0/) and entered the 12S rRNA sequences of each species which will be used in the form of FASTA into provided columns without spaces and then

IOP Conf. Series: Earth and Environmental Science **102** (2017) 012038 doi:10.1088/1755-1315/102/1/012038

clicked pick primer. The results of pick primer will feature a number of primer candidates and each primer was composed by 20 nucleotide bases together with their each properties. The primer candidates that have been compiled were then analyzed on its each sequence based on the order of codons. The primer candidates obtained from the data processing of 12S rRNA region consist of forward and reverse primers.

2.2. *The DNA extraction*

The DNA of cattle, pigs and chickens were extracted according to the gsyncTM DNA extraction kit procedure for animal tissues. A total of 25 mg samples of each species of cattle, pigs and chickens were inserted into 1.5 ml microtube and added 200 µl of GST buffer then homogenized. A 20 µl Proteinase K was added to the microtube and then homogenized. Then the mixture was incubated at 60 °C overnight until the mixture becomes clear.

After overnight incubation, the supernatant was transferred to a new microtube and added 200 µl of GSB buffer, homogenized with vortex for 10 seconds. A total of 200 µl of absolute ethanol was added and homogenized. Furthermore, the mixture was moved to GS column which had been put into 2 ml collection tube. The removed mixture was centrifuged at a rate of 14,000-16,000 rpm for 1 minute. Then, 2 ml collection tube was replaced with a new one.

A total of 400 µl W1 buffer was added to the GS column and purged for 30 seconds. Then GS column was put into new collection tube and added 600 µl wash buffer, centrifugation was occurred in 3 minutes. In addition, the GS column was then transferred to 1.5 ml new microtube and added elution buffer which had previously been incubated and let stand for 3 minutes and it was centrifuged for 30 seconds for DNA to elute. The extracted DNA obtained was electrophoresis on 1% agarose gel and then it was visualized using gel document (Vilber Lourmat Infinity 1100126M, France).

2.3. *Polymerase chain reaction and DNA sequencing*

PCR was performed on a PCR machine with a total volume of 25 µl consisted by 12.5 µl 2X KAPA2G Fast Multiplex PCR Kit (KAPA Biosystems. Inc., USA), 0.5 µl (10 µM) for each primer, 1 µl template DNA of cow, pork and chicken, respectively and 8.5 µl aquabidest. The PCR reaction was performed by denaturation initiation using 95°C for 3 minutes and followed by 30 cycles of denaturation using 95°C for 15 seconds, annealing temperature at 60°C for 30 seconds, extension at 72°C for 30 seconds and final extension 72°C for 3 minutes. The result of the reaction was then electrophoresis on 2% agarose gel and visualized by using gel document.

2.4. *Data analyses*

Data obtained from the gel document visualization and sequence analysis using Chromas 2.6.2 and alignment (clustal omega) were then analyzed by using qualitative descriptive analysis. Qualitative descriptive analysis is an analysis by combining all the data obtained from the research results then compared between obtained data, sorted and then descriptively analyzed [11].

3. Results and discussion

3.1. *Primer design*

The primary candidate was obtained by using PRIMER3 program by accessing PRIMER3 website at 0.4.0 version (http://bioinfo.ut.ee/primer3-0.4.0/). The mt-DNA of 12S rRNA fragment was present in the genetic data bank NCBI with accession number HQ184045 (cow), JN601075 (pig) and KT626857 (chicken). The mt-DNA fragment inputted in the PRIMER3 program will provide several pairs of primer candidates. Forward (F) and reverse (R) primer of 12S rRNA gene from species of cattle, pigs and chickens can be seen in Table 1.

Each primer was analyzed to obtain a primer with the best criteria which was analyzed based on the primer characteristics including position, length, composition (GC content) and amplicon length of primer obtained from PRIMER3 [12]. The primer candidate has a length of 20 bp. This length was expected to be enough to bind templates at annealing temperatures and obtain a specific sequence [13].

The best pair of primers of 12S rRNA gene from cow, pig and chicken of primer design results after confirmed with the NCBI primer BLAST is presented in Table 1. The NCBI primer BLAST analysis was used to determine the area used was completely unique and the primers obtained did not stick to any other organisms [3].

Table 1. Selected primers from Mitochondrial DNA 12S rRNA region for bovine, pig and chicken

Species	Forward Primer (5' to 3')	Reverse Primer (5' to 3')	Product size (bp)
Bovine	ACCGCGGTCATACGATT AAC	AGTGCGTCGGCTATTGTAGG	155
Pig		GAATTGGCAAGGGTTGGTAA	357
Chicken		CGGTATGTACGTGCCTCAGA	611

3.2. *Optimation of PCR*

The use of 12S rRNA gene has been widely used in phylogenic studies [8], [14], [15]. 12S rRNA was one of the genes in mitochondrial DNA. 12S rRNAs which can be used as biomarkers to identify species in meat because they have a higher sequence variation between different organisms than the order within a species [8]. Mitochondrial DNA of 12S rRNA gene from species of cattle, pigs and chickens was amplified by PCR method through three stages of temperature which were denaturation, annealing and extension [16].

Primer specificity of the target species was checked by PCR using a pair of primer which were universal forward primer pair for all three species and each species-specific reverse primer. Based on a 2% agarose gel electrophoresis visualization of PCR results showed that the PCR reaction produced a single band according to the target size of one meat species and no fragments produced by non-specific amplification. The result of PCR reaction visualized with gel document using 2% of agarose gel is presented in Figure 1.

Primer designed of 12S rRNA amplifies the target DNA of the 12S rRNA gene in cattle, pigs and chickens species. Amplification of the 12S rRNA gene resulted in 155 (cattle), 357 (pigs) and 611 bp (pigs), respectively. This study has successfully detected species in meat using a DNA target of 12S rRNA gene.

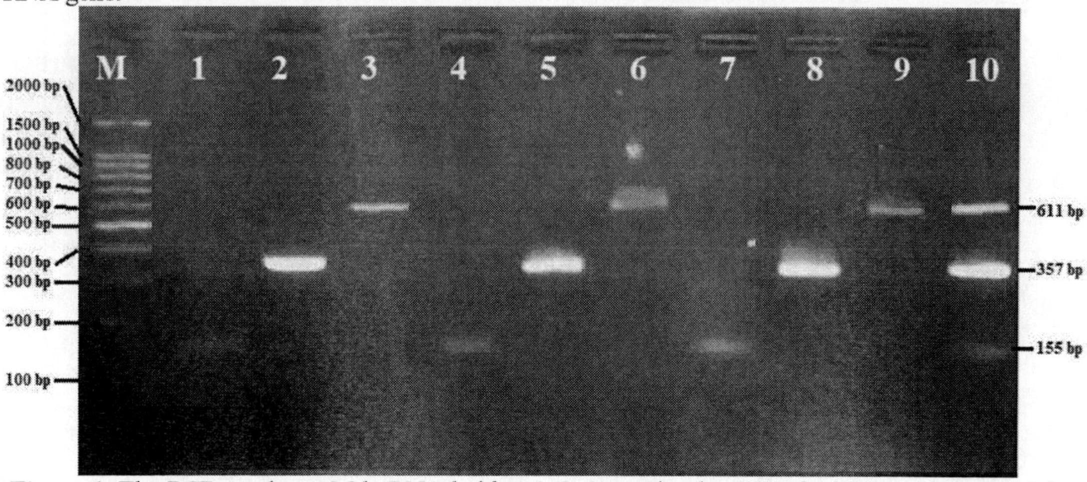

Figure 1. The PCR products. M is DNA ladder; 1, 2, 3 are simplex-PCR for bovine, pig and chicken; 4, 5, 6 are multiplex-PCR containing one DNA template (bovine, pig, and chicken, respectively); 7, 8, 9 are simplex-PCR containing three DNA templates (mixture of bovine, pig, and chicken DNA genomes); 10 is multiplex-PCR containing DNA mixture of bovine, pig, and chicken.

The results showed that the optimum annealing temperature in this study was 60°C. One of important parameter for successful PCR was PCR program optimization. One of them was annealing temperature optimization that aims to avoid miss-priming. Annealing temperatures which are too high can produce a low PCR product and annealing temperatures which are too low tent to stick elsewhere and produce non-specific products [13]. Annealing temperature can be determined by calculating Tm and 5°C annealing temperature under the actual primary Tm. However, sometimes annealing temperatures with Tm calculations did not show optimal results, resulting in PCR annealing temperature gradients [17].

3.3. Sequence analysis

Based on the results of 12S rRNA primer sequence analysis from cows, pigs and chickens that have been processed into alignment, it was found that the primers were attached in accordance region with the targeted primer which designed on 12S rRNA. Sequencing is the method used to determine the sequence of nucleotide bases (adenine (A), guanine (G), cytosine (C) and thymine (T)). DNA sequencing can be used to determine the identity and function of genes or other DNA fragments by comparing target sequences with other known DNA sequences [18]. The PCR DNA sequencing results were exposed using Chromas 2.6.2 program copied in FASTA as shown in Figure 2.

```
CLUSTAL O(1.2.4) multiple sequence alignment

UNI_F       -ACCGCGGTCATACGATTAAC
AYAM        ACCGCGGTTCATACGATTAACCCAAATCAATAGCTACCCGGCGTAAAGAGTGGCCACATGTTATCTGCA
BABI        -ACCGCGGTCATACGATTAACCCAAATTAATAGATCCACGGCGTAAAGAGTGTTTAAGAAAAAAA-AAC
SAPI        -ACCGCGGTCATACGATTAACCCAAGCTAACAGGAGTACGGCGTAAAACGTGTTAAAGCACC----ACA
              *   * *************

AYAM        CCAGCTAAGATTAAAATGCAACCAAGCTGTCATAAGCCTAAGATCCACCTAAACCCAACCCAAATCCAT
BABI        CACAATAGAGTTAAATTATAACTAAGCTGTAAAAAGCCCTAGTTAAAATAAA--------ATAAC-CCA
SAPI        CTAAATAGGGTTAAATTCTAATTAAGCTGTAAAAAGCCATAATTAAAATAAA--------AATAAATGA

AYAM        CTTAGCCTAAACGATTAATTTTAACCCACGAAAGCTAGGACCCAAACTGGGATTAGATACCCCACTATG
BABI        CGAAAGTGACTCTAATAATCCTGACACACGATAGCTAGGACCCAAACTGGGATTAGATACCCCACTATG
SAPI        CGAAAGTGACCCTACAATAGCCGACGCACT---------------------------------------

AYAM        CCTAGCCCTAAATCTAGATACCTCCCATCACACATGTATCCGCCTGAGAACTACGAGCACAAACGCTTA
BABI        CCTAGCCCTAAACCCAAATAGTTACATAACA-AAACTATTCGCCAGAG--TACTACTCGCAACTGCCTA
SAPI        ---------------------------------------------------------------------

AYAM        AAACTCTAAGGACTTGGCGGTGCCCCAAACCCACCTAGAGGAGCCTGTTCTATAATCGATAATCCACGA
BABI        AAACTCAAAGGACTTGGCGGTGCTTCACATCCACCTAGAGGAGCCTGTTCTATAATCGATAAACCCCGA
SAPI        ---------------------------------------------------------------------

AYAM        TTCACCCAACCACCCCTTGCCAGCACAGCCTACATACCGCCGTCGCCAGCCCACCTCTAATGAAAGAAC
BABI        TAGACCTTACCAACCCTTGCCAATTC-------------------------------------------
SAPI        ---------------------------------------------------------------------

AYAM        AACAGTGAGCTCAATAGCCCCTCGCTAATAAGACAGGTCAAGGTATAGCCTATGGGGTGGGAGAAATGG
BABI        ---------------------------------------------------------------------
SAPI        ---------------------------------------------------------------------

AYAM        GCTACATTTTCTAACATAGAACAAACGAAAAAGGACGTGAAACCCGCCCTTAGAAGGAGGATTTAGCAG
BABI        ---------------------------------------------------------------------
SAPI        ---------------------------------------------------------------------

AYAM        TAAAGTGAGATCATACCCCCTAAGCTCACTTTAAGACGGCTCTGAGGCACGTACATACCG
BABI        ------------------------------------------------------------
SAPI        ------------------------------------------------------------
```

Figure 2. Alignment of mt-DNA 12S rRNA sequences from three different species (Bovine, Pig, and Chicken). AYAM is sequence for chicken; BABI is sequence for Pig; SAPI is sequence for bovine. Red and blue boxes are primer regions for each species.

The sequence analysis results were processed in the form of alignment (Figure 2.) which shows the primer as the boundary of the amplified target DNA area. Primary forwards marked with red and primary reverse boxes are marked with a complementary blue box against the target DNA sequence.

The forward and reverse primers were complementary DNA sequences against the amplified sequences and attached to both ends of the target DNA in opposite directions [13].

The successful of DNA amplification depends on the primer accuracy used [19]. The primers used in the PCR process should be able to limit the area to be amplified and to limit the target DNA fragments to be amplified [5]. The primer that resides before the target area is called the forward primer and which resides after the target area is called the reverse primer.

Sequence alignment analysis results show that primer of cattle, pigs and chicken species were well known. The length of the product produced corresponds to the length of the product that has been designed. The length of PCR product by using the primary design result of mt-DNA 12S rRNA was 155 bp for beef, 357 bp for pork and 611 bp for chicken meat, respectively. The primer used as limiter of the amplified region called UNI_F was the universal forward primer that used for all three species and each species-specific primary reverse. PCR reactions produced specific PCR products because they use specific primers of species of cattle, pigs and chickens. Specific primers are primers that do not amplify other target DNA [20].

4. Conclusions

The new primers designed from mt DNA 12S rRNA of bovine, pig and chicken have been successfully obtained and applied to amplify DNA target in vitro by using multiplex PCR technique.

Acknowledgments

This research was supported and funded by the Research Scheme of the "Pusat Keunggulan–Universitas Sebelas Maret (PK-UNS)" and Mandatory research group scheme (MRG-UNS) with contract number: 623/UN27.21/PP/2017.

References

[1] Nakyinsige K, Man Y B C, Sazili A Q 2012 *Meat Sci.* **91** 207–214
[2] Matsunaga T, Chikuni K, Tanabe R, Muroya S, Shibata K, Yamada J and Shinmura Y 1999 *Meat Sci.* **51** 143–8
[3] Lima, A O S and Garces S P S 2006 *Biochem. Mol. Biol. Educ.* **34** 332–7
[4] Corpet F 1988 *Nucleic Acids Res.* **16** 10881–90
[5] Handoyo D dan Rudiretna A 2000 *Unitas* **9** 17–29
[6] Sasmito D E K, Kurniawan R dan Muhimmah I 2014 Seminar Nasional Informatika Medis (SNIMed) V (Yogyakarta: Magister Teknik Informatika/Fakultas Teknologi Industri/Universitas Islam Indonesia) p 93-102
[7] Melton T and Nelson K 2001 *Croat. Med. J.* **42** 298–303
[8] Springer M S and Douzery E 1996 *J. Mol. Evol.* **43** 357–73
[9] Dalmasso A, Fontanella E, Piatti P, Civera T, Rosati S and Bottero M T 2004 *Mol. Cell. Probes.* **18** 81–87
[10] Girish P S, Anjaneyulu A S R, Viswas K N, Anand M, Rajkumar N, Shivakumar B M and Bhaskar S 2004 *Meat Sci.* **66** 551–56
[11] Sugiyono 2009 *Metode Penelitian Kuantitatif, Kualitatif dan R&D* (Bandung: Alfabeta)
[12] Gingrich J R and Kenneth A P 2004 *Proc. Natl. Acad. Sci.* **101** 6237–42
[13] Borah P 2011 *Sci. Vis.* **11** 134–6
[14] Gatesy J, Amato G, Vrba E, Schaler G and Desalle R 1997 *Mol. Phylogenet. Evol.* **7** 303–19
[15] Wang H Y and Lee S C 2002 *Mol. Biol. Evol.* **19** 138–48
[16] Nollet L M L and Toldra F 2011 *Safety Analysis of Foods of Animal Origin* (New York: CRC Press)
[17] Prezioso V and Jahns A 2013 *Using gradient PCR to determine the Optimum Annealing Temperature* (USA: Eppendorf Scientific Inc.)
[18] Snustad D P and M J Simmons 2003 *Principles of Genetics* 3rd ed (New York John: Wiley and Sons, Inc.)

[19] Diss T 2003 The polymerase chain reaction *Molecular Biology in Cellular Pathology*, Ed J Crocker (United Kingdom: G M editors) pp 193–210
[20] Langen M, Peters U, Korner U, Gissel C, Stanislawaki D and Klein G 2010 *Meat Sci.* **86** 821–4

IOP Conf. Series: Earth and Environmental Science **102** (2017) 012039 doi:10.1088/1755-1315/102/1/012039

Effects of different extraction methods on total phenolic content and antioxidant activity in soybean cultivars

E Yusnawan[1]

[1]Indonesian Legumes and Tuber Crops Research Institute. Jalan Raya Kendalpayak Km 8 PO Box 66 Malang, East Java, Indonesia. Tel./Fax. +62-341-801468/+62-341-801496

Email : eyusnawan@litbang.pertanian.go.id

Abstract. Soybean secondary metabolites particularly phenolic compounds act as chemical defence against biotic stress such as pathogen infection. Functional properties of these compounds have also been investigated. This study aimed to determine the effects of particle size and extraction methods on total flavonoid, phenolic contents as well as antioxidant activity in soybean seeds. This study also investigated the total phenolic contents and antioxidant activity of Indonesian soybean cultivars using the optimized extraction method. Soybean flour of ≤ 177 μm as many as 0.5 g was selected for extraction with 50% acetone for estimation of total phenolic and flavonoid contents and with 80% ethanol for antioxidant activity. Treatments of twice extraction either shaking followed by maceration or ultrasound-assisted extraction followed by maceration could be used to extract the secondary metabolite contents in soybean seeds. Flavonoid, phenolic contents and antioxidant activity of twenty soybean cultivars ranged from 0.23 to 0.44 mg CE/g, from 3.70 to 5.22 mg GAE/g, and from 4.97 to 9.04 μmol TE/g, respectively. A simple extraction with small amount of soybean flour such as investigated in this present study is effective to extract secondary metabolites especially when the availability of samples is limited such as breeding materials or soybean germplasm.

Keywords : Extraction methods, phenolic content, antioxidant activity, soybean

1. Introduction

Plants naturally produce phenolic compounds during their normal growth. These chemicals are also synthesized when the plants are stressed by biotic as well as abiotic factors such as pathogen infection and radiation of ultra violet. Phenolic compounds also serve as molecular signals in symbiotic process between plant and microorganisms [1-6]. Legumes contain excess amount of phenolics, for example phenolic contents in soybean seeds (*Glycine max* (L.) Merr.) are estimated between 3.04 and 5.71 mg catechin equivalent/g [7-8]. Leguminous seeds are rich in flavonoids, phenolic acids, and procyanidins [9]. Saponins and triterpenoids are also constituent of soybean phytochemical compounds [10-11].

Production of phenolic compounds by seeds and roots at early stage of growth such as during germination has significant contribution to combat infection from several soil borne pathogen and root eating insects [12]. Soil borne pathogens such as *Rhizoctonia solani, Macrophomina phaseolina, Sclerotinia sclerotiorum* and *Phytophthora sojae* are inhibited by phenolic compounds. Foliar diseases such as *Phakopsora pachyrhizi* are also inhibited by these compounds [13-15]. These chemicals also function to protect leaves, fruits and seeds from pathogen infection at the later stage of growth [13, 16-18].

The structures of phenolics influence their activity. Phenolic compounds in leguminous plants such as cajanin, medicarpin, glyceolin, rotenone, coumestrol, phaseolin, isoflavonoid, and flavonoid have properties as phytoanticipins (constitutively formed) and phytoalexins (induced) as well as nematicidal against soil borne pathogens and insect pests [12]. Interaction between crops and plant pathogens is related to phenylpropanoids. These organic compounds also have toxic properties or inhibition to plant pathogens either as phytoanticipins or phytoalexins [14].

Extraction of phenolic compounds from plant materials are one of the crucial steps which influences the yield of the whole process before estimation of its contents either spectrophotometric or chromatographic determination was conducted [19-25]. Solid-liquid extraction (SLE) techniques which can be divided into traditional and recent methods have been widely performed in the laboratories for specific purposes [26]. Traditional methods including maceration, percolation, soxhlet (reflux) extraction, sonication, and turbo-extraction (high speed mixing) [26-27] are still used in many laboratories where modern equipments to perform recent methods have not been available yet. Information of different extraction methods to estimate secondary metabolite contents in soybean seed is still limited. The objective of this study was to investigate the effect of particle size and different traditional extraction methods on the extraction efficiency of total flavonoid, phenolic contents as well as antioxidant activity in soybean seeds. This study also investigated the contents of those secondary metabolites in selected Indonesian soybean cultivars.

2. Materials and methods

2.1. Sample preparation

Soybean seeds were ground with a grinder to obtain soybean flour. The flour was gradually sieved and grouped into three different particle sizes, i.e. 211 – 297 μm, 178 – 210 μm, and ≤ 177 μm. The flour was stored at 2-8 °C in vacuum sealed plastic bags prior to be analyzed. Moisture content was measured to convert each sample into dry weight basis.

2.2. Sample extraction

The flour (0.1, 0.5, 1.0, and 2.5 g) grouped into three particle sizes (210 – 297 μm, 177 – 209 μm, and ≤ 177 μm), in total of 12 combinations, was extracted with suitable solvents. Phenolic and flavonoid compounds were extracted with 50% acetone and antioxidant activity was extracted with 80% ethanol (Table 1). Extraction solvent (1:10 w/v) was poured into capped glass tubes containing soybean flour and the tubes were placed on a shaker. The sample was macerated for 18 h after shaking. The supernatant was collected after centrifugation. The second extraction was conducted with the same procedure. The supernatant obtained from the second extraction was combined with the first extraction..

Different extractions were carried out as follows: (1) proportion of sample and organic solvent was 1:20 (w/v). The sample was placed on an orbital shaker at 150 rpm for 2 h followed by maceration for 18 h. Supernatant was collected in an amber vial after centrifugation, (2) proportion of sample and organic solvent was 1:10 (w/v). The sample was placed on an orbital shaker at 150 rpm for 2 h prior to maceration for 18 h. The supernatant was collected. The second extraction was performed with the same procedure. Combined supernatant was pooled in an amber vial, (3) proportion of sample and organic solvent was 1:20 (w/v). The sample was macerated for 20 h. The supernatant was collected, (4) proportion of sample and organic solvent was 1:10 (w/v). The sample was macerated for 20 h and the supernatant was collected. The second extraction with the same procedure was conducted. The combined supernatant was pooled, (5) proportion of sample and organic solvent was 1:20 (w/v). Ultrasound-assisted extraction for 5 min was conducted and the sample was macerated for 20 h. The supernatant was collected, and (6) proportion of sample and organic solvent was 1:10 (w/v). Ultrasound-assisted extraction for 10 min was performed and the sample was macerated for 20 h. The second extraction with the same procedure was conducted. The combined supernatant was pooled. In short, all treatments were summarized in Table 2.

2.3. Determination of total flavonoid content
The extract for flavonoid determination was added in distilled water (1:5 v/v) in a glass tube. The extract was mixed thoroughly. As many as 150 µL of 5% $NaNO_2$ was added. The mixed solution was vortexed. Incubation was performed for 6 min. Additional of 5 min incubation was conducted after 300 µL of aluminium chloride was reacted. Then, sodium hydroxide (1M, 1000 µL) was added in the solution. The final volume was brought to 5000 µL with distilled water. The solution was mixed thoroughly. Absorbance values of the solution were recorded using a spectrophotometer at 510 nm. Catechin equivalents per gram of the sample (mg CE/g) were used to express total flavonoid contents [28-29].

2.4. Determination of total phenolic content
Folin Ciocalteu's reagent was utilized to estimate total phenolic content in soybean extract [29-31]. The soybean extract for phenolic determination in distilled water (1:60 v/v) was reacted with Folin-Ciocalteu's reagent (250 µL). Then, sodium carbonate (750 µL) was added. Incubation for 8 min was conducted after the solution was thoroughly mixed. Distilled water (950 µL) was added and the mixture was incubated in the dark room for 2 h. The values of absorbance were recorded at 765 nm using a spectrophotometer. Contents of total phenolic were expressed as gallic acid equivalents per gram of sample (mg GAE/g).

2.5. DPPH free radical scavenging activity
A solution of 1 mM methanolic DPPH was used to estimate antioxidant activity [32-33]. In DPPH solution, soybean extract (1:19 v/v) was reacted. To allow complete reaction, incubation for 30 min in the dark room was conducted after mixing. Absorbance values of each sample (A_{sample}) and control ($A_{control}$) were read using a spectrophotometer at 515 nm. Antioxidant activity of the samples as represented by percent discoloration was calculated using the following formula: percent discoloration = $[1-(A_{sample}/A_{control})] \times 100\%$. Antioxidant activity of each sample was expressed as micromoles of Trolox equivalent per gram of sample (µmol TE/g).

2.6. Total flavonoid, phenolic contents, and antioxidant activity in soybean cultivars
Soybean seeds of twenty cultivars were finely ground separately as described in the sample preparation and the flour was extracted for quantification of total flavonoids, phenolics as well as antioxidant activity as described above. A completely randomized design repeated three times was used in this study. The difference among treatments was determined with the least significant different.

3. Results and discussion

3.1. Sample quantity and particle size
The proportion of sample to extraction solvent of all treatments is 1:10 (w/v) with twice extraction. The only difference is the quantity and particle size of the samples. The purpose of this study was to determine the minimum amount of sample in combination with the particle size could be employed to extract secondary metabolites in soybean flour for practical reasons with respect to environmental and economic considerations. The use of limited amount of working samples and solvents for extraction is an effective approach to reduce chemical disposal, specifically when the chemical waste is strictly regulated. Minimal amount of solvents is one of the alternatives in which more environmental friendly procedure can be conducted [34]. The use of 0.1, 0.5, 1.0, and 2.5 g of samples with different particle sizes showed no significant effect on total flavonoid content (0.52 to 0.59 mg CE/g) (Table 1). However, there was a slight difference in total phenolic contents (2.30 to 2.82 mg GAE/g). Small

quantity of sample with bigger particle size showed least amount of phenolic content as shown in treatment 1.

The same amount of phenolic content was also observed in treatment 7 and 10 in which bigger particle size was used. The use of different quantity and particle size of samples showed different antioxidant activity. The same results as observed in total phenolic content, the use of bigger particle size of the sample resulted lesser DPPH scavenging activity. Similar to this finding, extract obtained from smaller particle size of wheat brand showed higher secondary metabolites particularly phenolic acid, flavonoid, anthocyanin, as well as carotenoid contents compared to those from coarse particle samples because of the increase of accessible surface area [35]. Ultra-fine grinding of wheat brand also resulted the increase of antioxidant capacity [36]. Considering contents of the secondary metabolites in soybean extract and the value of standard deviation, treatment of 6 was selected for further optimization.

Table 1. Total flavonoid and phenolic contents as well as antioxidant activity derived from different particle size of soybean flour

Treatment	Combination	TFC (mg CE/g)	TPC (mg GAE/g)	AA (μmol TE/g)
1	0.1 g, 210 – 297 μm	0.52 ± 0.01 a	2.43 ± 0.08 c	3.423 ± 0.10 f
2	0.1 g, 177 – 209 μm	0.52 ± 0.02 a	2.81 ± 0.03 a	4.37 ± 0.04 a
3	0.1 g, ≤ 177 μm	0.59 ± 0.04 a	2.82 ± 0.19 a	4.18 ± 0.11 bc
4	0.5 g, 210 – 297 μm	0.53 ± 0.01 a	2.35 ± 0.04 c	3.32 ± 0.14 f
5	0.5 g, 177 – 209 μm	0.53 ± 0.01 a	2.61 ± 0.02 b	4.14 ± 0.08 bc
6	0.5 g, ≤ 177 μm	0.55 ± 0.01 a	2.72 ± 0.06 ab	4.13 ± 0.02 bc
7	1.0 g, 210 – 297 μm	0.53 ± 0.02 a	2.30 ± 0.03 c	3.64 ± 0.26 e
8	1.0 g, 177 – 209 μm	0.53 ± 0.01 a	2.64 ± 0.07 b	4.20 ± 0.02 abc
9	1.0 g, ≤ 177 μm	0.54 ± 0.01 a	2.63 ± 0.04 b	4.17 ± 0.07 bc
10	2.5 g, 210 – 297 μm	0.54 ± 0.04 a	2.37 ± 0.09 c	3.83 ± 0.08 d
11	2.5 g, 177 – 209 μm	0.52 ± 0.03 a	2.61 ± 0.01 b	4.32 ± 0.03 ab
12	2.5 g, ≤ 177 μm	0.53 ± 0.05 a	2.61 ± 0.08 b	4.29 ± 0.06 ab

The numbers followed by the same letter in the same column were not significantly different based on the LSD test ($p < 0.05$)
TFC = total flavonoid content, TPC = total phenolic content, AA = DPPH scavenging activity

3.2. Traditional extraction methods

Extraction of natural products is a crucial step prior to isolation, identification and the use of secondary metabolites for further purposes and there is no specific and standard method has been established to perform the extraction [37]. A significant variation of total flavonoid content was observed in different extraction methods (Table 2). Total flavonoid content of maceration treatments (C and D treatments, 0.32 and 0.41 mg CE/g) showed lesser amount compared to those of other treatments. Unlike total flavonoid content, maceration only was not significantly influenced the total phenolic content and antioxidant activity.

Number of extraction influenced the secondary metabolite contents. Twice extraction provided more secondary metabolite contents than those in single extraction. With twice extraction, total phenolic content and DPPH scavenging activity were in the range from 2.82 to 2.83 mg GAE/g and 3.87 to 4.01 μmol TE/g, respectively. The treatments of twice extraction either shaking followed by maceration or ultrasound-assisted extraction followed by maceration (B and F treatments) could be chosen to extract secondary metabolite contents in soybean seeds. The use of ultrasonication influenced the amount of plant secondary metabolites was also reported by Rostagno *et al.*[38]. They found that ultrasonication extracted the highest amount of isoflavones when compared to soxhlet, supercritical-CO_2 and ultrasonication extractions. The advantage of ultrasonication to extract the secondary metabolites is to enhance plant cell disruption, to facilitate solvent penetration as well as mass transfer. This treatment therefore, provides more efficiency in the extraction [39-40].

Table 2. Effect of different extractions on total flavonoid, phenolic contents and antioxidant activity of soybean flour

Treatment	Extraction	TFC (mg CE/g)	TPC (mg GAE/g)	AA (µmol TE/g)
A	1x extraction, shaking 2 h + maceration 18 h	0.46 ± 0.02 [b]	2.57 ± 0.06 [b]	3.67 ± 0.05 [b]
B	2x extraction, shaking 2 h + maceration 18 h	0.54 ± 0.02 [a]	2.82 ± 0.06 [a]	4.01 ± 0.17 [a]
C	1x extraction, maceration 20 h	0.32 ± 0.01 [d]	2.58 ± 0.01 [b]	3.56 ± 0.06 [b]
D	2x extraction, maceration 20 h	0.41 ± 0.02 [c]	2.83 ± 0.08 [a]	3.87 ± 0.05 [a]
E	1x extraction, UAE 5 min + maceration 20 h	0.51 ± 0.02 [a]	2.52 ± 0.02 [b]	3.63 ± 0.05 [b]
F	2x extraction, UAE 5 min + maceration 20 h	0.52 ± 0.04 [a]	2.83 ± 0.08 [a]	3.94 ± 0.11 [a]

The numbers followed by the same letter in the same column were not significantly different based on the LSD test ($p < 0.05$)
TFC = total flavonoid content, TPC = total phenolic content, AA = DPPH scavenging activity, UAE = ultrasound-assisted extraction

3.3. Secondary metabolites in soybean cultivars

3.3.1. Total flavonoid content in soybean cultivars. There were variations of flavonoid contents in all yellow soybean seeds. The contents of total flavonoid were from 0.23 to 0.43 mg CE/g (Figure 1). Dering and Gema had the highest flavonoid contents among other soybean cultivars. The range of total flavonoid contens from 0.22 to 0.45 mg CE/g in yellow soybeans obtained from germplasm collection was also reported in previous study [31]. Nevertheless, the contents of flavonoids were slightly lower than flavonoid contents in yellow soybeans (0.25 to 0.50 mg CE/g) as investigated by previous researchers [29].

The difference in extraction methods can be one of the reasons in the difference of total flavonoid contents. Yellow soybeans contain high isoflavones which contribute to major compounds of flavonoids [41-42]. Isoflavone variation of all genotypes in this present study was of interest should be investigated further to determine its contribution to the total flavonoid content. Besides isoflavones, other compounds especially carotenoids and luteins contribute to total flavonoids in yellow soybeans [43-44]. It has been comprehensively reported that other factors also influence the concentration and distribution of secondary metabolites particularly environmental factors [11, 45].

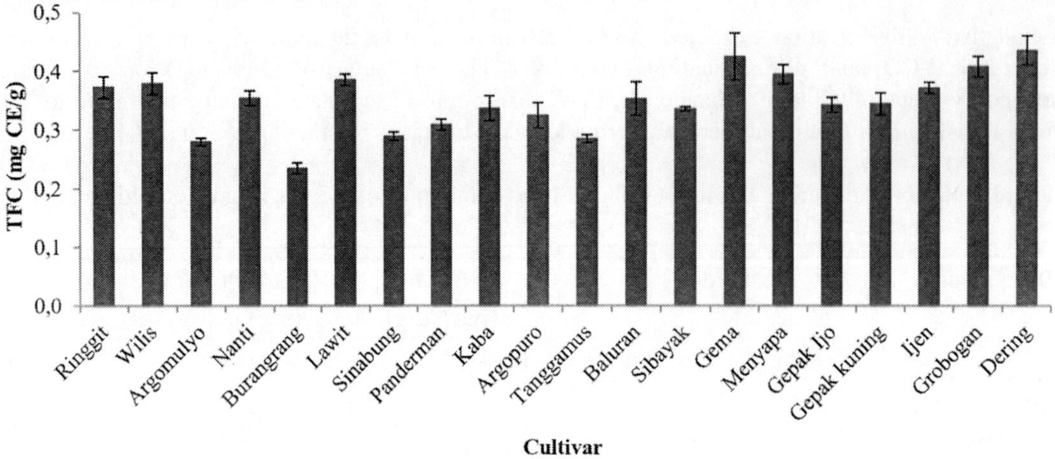

Figure 1. Total flavonoid content of twenty soybean cultivars. Error bars represent standard deviation from measurements in triplicate.

3.3.2. Total phenolic content in soybean cultivars. Similar as observed in total flavonoid contents, total phenolic contents of all soybean cultivars also showed variations (Figure 2). Interestingly, Dering cultivar which had the highest flavonoid contents also expressed the highest phenolic contents (5.22 mg GAE/g), however, similar high content of phenolic was not found in Gema cultivar (4.53 mg GAE/g). The range of phenolic contents of all soybean cultivars was from 3.70 to 5.22 mg GAE/g. This range was similar range of total phenolic contents in yellow soybean germplasm (3.49 to 5.42 mg GAE/g) as stated in a previous report [31]. Nevertheless, phenolic contents of yellow soybeans extracted from Indian genotypes (1.06 – 1.54 mg GAE/g) were lower than phenolics in Indonesian genotypes [44]. Again, different extraction as well as quantification methods used may influence quantity of phenolic contents besides genotypic variation as reported by previous researchers [42].

Seed coat, embryogenic axis and cotyledon contain phenolic compounds, however, the proportion of phenolics in the seed coat was dominant [31, 46]. Estimation of phenolic contents among different seed coat color, i.e. light yellow, yellow, green, and black soybeans has been studied and black seed coat color soybean expressed the highest phenolic contents [29, 31, 44, 46-47]. Approximately three times higher in phenolic contents is observed in black soybean than in yellow soybeans [31].

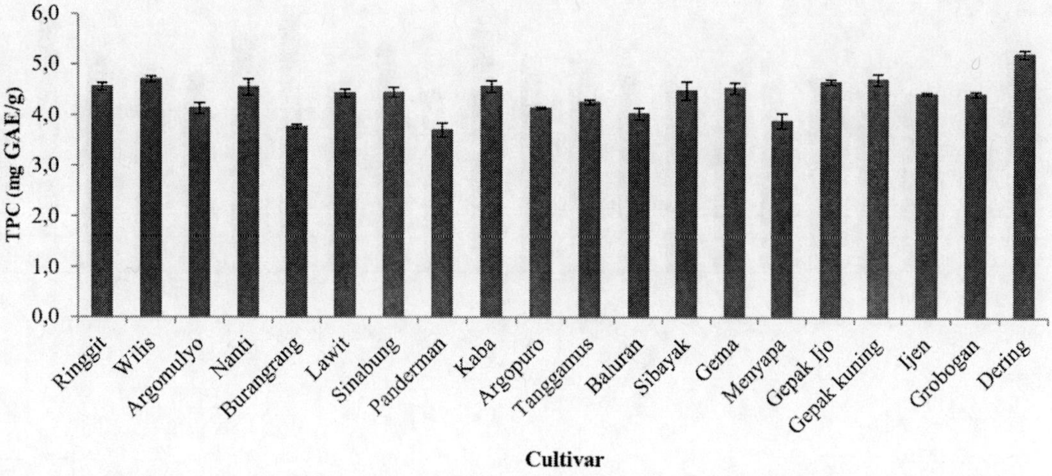

Figure 2. Total phenolic content of twenty soybean cultivars. Error bars represent standard deviation from measurements in triplicate.

3.3.3. DPPH scavenging activity of soybean cultivars. Variations among soybean cultivars were also observed in antioxidant activity (Figure 3). There was approximately one and a half-fold variation of antioxidant activity from the lowest to the highest. Grobogan which had the second highest of flavonoid content had the highest antioxidant activity. Even though total phenolics were similar contents to those in Xu and Chang study [29], DPPH scavenging activity was about 3-fold higher. This finding supports a previous study [31], in which antioxidant activity was observed approximately 3.5 times higher.

In yellow soybeans, most of antioxidant activity may be predominantly supported by isoflavones and ferulic acid in cotyledons and embryo [42]. As comparison with black seed coat color soybeans, antioxidant activity of these soybeans is mostly concentrated in its seed coat. Anthocyanins are compounds contribute to high antioxidant activity as previously observed [41]. Nevertheless, isoflavones in yellow soybeans have weaker antioxidant activity than that of procyanidins and anthocyanins in dark seed coat color soybean [46, 48].

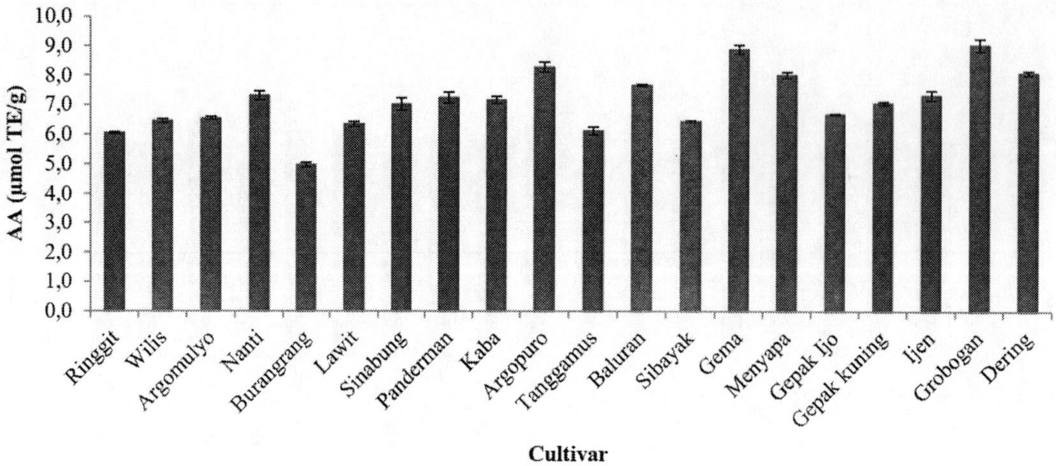

Figure 3. Antioxidant activity (AA) of twenty soybean cultivars. Error bars represent standard deviation from measurements in triplicate.

A linear correlation ($P < 0.05$) of total flavonoid and phenolic contents ($r = 0.56$) as well as total flavonoid content and antioxidant activity ($r = 0.65$) was investigated in this study. However, no significant correlation ($P > 0.05$, $r = 0.16$) was found in total phenolic content and antioxidant activity. A positive correlation among total flavonoid and phenolic contents confirmed the previous finding [31] when observing yellow soybeans. Flavonoids are one of the large groups of phenolic compounds, therefore, this significant correlation is not surprising [46]. In addition, isoflavones, one of the sub classes of flavonoids also contribute to high antioxidant activity [49]. Nevertheless, this recent result was not in line with a study conducted by previous researchers [29] in which the three parameters were correlated. Yusnawan [31] suggested to conduct other antioxidant capacity assays to obtain better understanding of correlations of the three parameters.

Phenolics in plants particularly simple phenols, lignins, and flavonoids are chemical compounds effective to inhibit fungal infection. Inhibition of hydrolytic enzymes produced by fungal pathogens triggers plant resistant to the fungal infection [15]. However, in healthy seeds for instance, total isoflavonoids are higher than seeds infected by *Phomopsis longicolla*. The amount is 5-fold higher than the diseased seeds. The same compounds are also detected twice higher in healthy seeds than in diseased seeds by *Cercospora kikuchii* [50]. Even though there is different content of flavonoids in healthy and diseased seeds, the authors suggested that the changes may be not only necessarily caused by the pathogen infections. In fact, other factors both environmental and genotipic factors such as sample condition and storage period prior to analysis may influence the chemical contents. In addition, agronomic condition, moisture, region, light, temperature and cultivars also influenced the compounds [45, 50-52]. However, the estimation of secondary metabolite contents in soybean seeds as investigated in the present study could be of value as initial information in terms of variations in chemical defenses against seed diseases among cultivars. Higher amount of secondary metabolites in certain soybean cultivars compared to the other cultivars whether may be related to more resistant to plant pathogen infection needs further studies.

In conclusion, this study revealed that extraction of fine particle sizes of soybean flour produced higher amount of flavonoid and phenolic contents as well as antioxidant activity compared to the use of coarse particles. Small amount of the sample and simple extractions either shaking followed by maceration or ultrasound-assisted extraction followed by maceration repeated twice in each treatment could be conducted to extract secondary metabolite contents. This approach was valuable to determine secondary metabolite variations particularly when a large number of samples will be analyzed with limited quantity of the samples such as breeding materials or soybean germplasm.

References

[1] Arfaoui A, El Hadrami A, Mabrouk Y, Sifi B, Boudabous A, El Hadrami I, Daayf F and Chérif M 2007 *Plant Physiol. Biochem.* **45** 470
[2] Boue S M, Cleveland T E, Carter-Wientjes C, Shih B Y, Bhatnagar D, McLachlan J M and Burow M E 2009 *J. Agric. Food Chem.***57** 2614
[3] Mandal S M, Chakraborty D and Dey S 2010 *Plant Signal. Behav.* **5** 359
[4] Couto C, Silva L R, Valentão P, Velázquez E, Peix A and Andrade P B 2011 *Food Chem.* **127** 1487
[5] Dardanelli MS et al 2012 *Appl. Soil Ecol.* **57** 31
[6] Silva L R, Pereira M J, Azevedo J, Mulas R, Velazquez E, González-Andrés F, Valentão P and Andrade P B. 2013. *Food Chem.* **141** 3636
[7] Malencic D, Popovic M and Miladinovic J 2007 *Molecules***12** 576
[8] Riedl K M, Lee J H, Renita M, St Martin S K, Schwartz S J and Vodovotz Y 2007 *J. Sci Food Agric.* **87** 1197
[9] Amarowicz R and Pegg RB 2008 Eur. *J. Lipid* Sci. *Technol.* **110** 865
[10] Lee J H, Jeon J K, Kim S G, Kim S H, Chun T and Imm J 2011 *Int. J. Food Sci. Tech.* **46** 2513
[11] Lee J H and Cho K M 2012 *Food Chem.* **131** 161
[12] Makoi J H and Ndakidemi P A 2007 *Afr. J. Biotechnol.* **6** 1358
[13] Lygin A V, Li S, Vittal R, Widholm J M, Hartman G L, and Lozovaya V V 2009 *Phytopathology***99** 1412
[14] Lygin A V, Zernova O V, Hill C B, Kholina N A, Widholm J M, Hartman G L and Lozovaya V V 2013 *Phytopathology***103** 984
[15] Bellaloui N, Mengistu A, Zobiole L H S and Shier W T 2012 *Toxin Rev.* **31** 16
[16] Oh S, Kim J A, Jeon H, Park J C, Koh Y J, Hur H and Hur J S 2008 *Plant Pathol. J.* **24** 322
[17] Mikulic-Petkovsek M, Slatnar A, Veberic R, Stampar F and Solar A 2011 *Physiol. Mol. Plant Pathol.* **76** 159
[18] Ruiz-García Y and Gómez-Plaza E 2013 *Agriculture***3** 33
[19] Cacace J and Mazza G 2003 *J. Food Sci.* **68** 240
[20] Naczk M and Shahidi F 2004 *J. Chromatogr. A***1054** 95
[21] Ma Y and Cheung P C 2007 *J. Agric. Food Chem.* **55** 4222
[22] Du G, Zhao H Y, Zhang Q W, Li G H, Yang F Q, Wang Y, Li Y C and Wang Y T 2010 *J. Chromatogr. A***1217**: 705
[23] Liu P, Yang B and Kallio H 2010 *Food Chem.* **121** 1188
[24] Jun-ming S, Bao-li S, Fen-xia H, Shu-rong Y, Hua Y and Kikuchi A 2011 *Agr. Sci. China***10** 70
[25] Sánchez-Rangel J C, Benavides J, Heredia J B, Cisneros-Zevallos L and Jacobo-Velázquez D A 2013 *Anal. Methods***5** 5990
[26] Kaufmann B and Christen P 2002 *Phytochem. Anal.* **13** 105
[27] Chukwumah Y, Walker L T and Verghese M 2009 *Int. J. Mol. Sci.* **10** 4941
[28] Heimler D, Vignolini P, Dini MG and Romani A 2005 *J Agric. Food Chem.* **53** 3053
[29] Xu B and Chang S K C 2007 *J Food Sci.* **72** S159
[30] Singleton V L, Orthofer R and Lamuela-Raventós R M 1999 *Methods in Enzymol.* vol 299, ed P Lester (Academic Press) p 152
[31] Yusnawan E 2016 *Biodiversitas***17** 704
[32] Xu B and Chang S K C 2008 *J Agric Food Chem.* **56** 7165
[33] Xu B and Chang S K C 2008 *J Agric Food Chem.* **56** 8365
[34] Magalhães L M, Santos F, Segundo M A, Reis S and Lima J L 2010 *Talanta***83** 441
[35] Brewer L R, Kubola J, Siriamornpun S, Herald T J and Shi Y 2014 *Food Chem.* **152** 483
[36] Rosa N N, Barron C, Gaiani C, Dufour C and Micard V *2013 J. Cereal Sci.* **57** 84

[37] Bucić-Kojić A, Planinić M, Tomas S, Bilić M and Velić D 2007 *J. Food Eng.* **81** 236
[38] Rostagno McA, Araújo J M and Sandi D 2002 *Food Chem.* **78** 111
[39] Toma M, Vinatoru M, Paniwnyk L, Mason T 2001 *Ultrason. Sonochem.* **8** 137
[40] Servant G, Laborde J, Hita A, Caltagirone J and Gerard A 2003 *Ultrason. Sonochem***10** 347
[41] Cho K M, Ha T J, Lee Y B, Seo W D, Kim J Y, Ryu H W, Jeong S H, Kang Y M and Lee J H 2013 *J. Funct. Foods***5** 1065
[42] Žilić S, Akıllıoğlu H G, Serpen A, Perić V and Gökmen V 2013 *Eur. Food Res. Technol.* **237** 409
[43] Monma M, Ito M, Saito M and Chikuni K 1994 *Biosci. Biotechnol. Biochem.* **58** 926
[44] Kumar V, Rani A, Dixit A K, Pratap D and Bhatnagar D 2010 *Food Res. Internat.* **43** 323
[45] Lee S J, Yan W, Ahn J K and Chung I M 2003 *Field Crops Res.* **81** 181
[46] Jeng T L, Shih Y J, Wu M T and Sung J M 2010 *Food Chem.* **123** 1112
[47] Xu B and Chang S K C 2008 *J Agric Food Chem.***56** 9102
[48] Takahata Y, Ohnishi-Kameyama M, Furuta S, Takahashi M and Suda I 2001. *J. Agric. Food Chem.* **49** 5843
[49] Lee C H, Yang L, Xu J Z, Yeung S Y V, Huang Y and Chen Z 2005 *Food Chem.* **90** 735
[50] Lee J H, Hwang S, Lee Y, Kim K, Cho K M and Lee Y B 2015 *Food Chem.* **185** 205
[51] Hoeck J A, Fehr W R, Murphy P A and Welke G A 2000 *Crop Sci.* **40** 48
[52] Lee S J, Ahn J K, Kim S H, Kim J T, Han S J, Jung M Y and Chung I M 2003 *J. Agric Food Chem.* **51** 3382

Sensory quality of mayonnaise formulated with eggs produced by laying hens fed diet enriched with purslane meal (*Portulaca oleracea*)

L R Kartikasari*, B S Hertanto, A M P Nuhriawangsa

Department of Animal Science, Faculty of Agriculture, Universitas Sebelas Maret, Jl. Ir. Sutami no 36A, Surakarta, Central Java, 57126 Indonesia

Email : lilikretna@staff.uns.ac.id

Abstract. The objective of the study was to assess the sensory quality of mayonnaise formulated by egg yolks produced by hens fed purslane meal as a source of omega-3 (n-3) fats, *alpha-linolenic acid*. Eggs used were produced by Hy-Line Brown hens (38 weeks old) fed five diets suplemented with 0, 2, 4, 6 and 8% purslane meal. A total of 40 eggs of day 29 (n = 8 eggs for each treatment) were collected to formulate mayonnaise. The ingredient composition of mayonnaise was 14% egg yolks, 9% vinegar, 74% corn oil, 1% salt, 1% sugar, and 1% mustard. The parameters included colour, emulsion stability, texture, homogeneity, aroma, taste and flavour of mayonnaise. The findings of sensory evaluation by quantitative descriptive analysis showed that the use of egg yolk from hens fed dietary purslane meal up to 8% resulted in a significant effect on the yellow color of mayonnaise, but did not significantly affect emulsion stability, texture, homogeneity, aroma, taste, and flavour. Based on consumer preference, consumers did not detect any differences in the sensory characteristics between mayonnaise prepared from control eggs and those formulated with eggs produced from dietary treatment. Apparently, there was an increase in colour liking by inclusion of purslane meal. In conclusion, eggs from laying hens fed diets supplemented with purslane meal improved the colour intensity of mayonnaise. Mayonnaise formulation using eggs from diets added with *Portulaca oleraceae* meal up to a level of 8% can be applied without influencing mayonnaise sensory characteristics and consumer acceptance.

Keywords: Egg yolks, laying hens, mayonnaise, purslane meal, sensory characteristics

1. Introduction

As animal food products, eggs are considered a high-profile ingredient. Beside eggs provide high nutritional value, eggs have multifunctional properties, including foaming, coagulation, mulsification, and flavor.[1]. Feeding laying hens is one of efforts to improve the nutrition and quality of chicken eggs. A number of nvesigators have conducted research to investigate the influence of diets enriched with omega-3 polyunsaturated fatty acids (n-3 PUFA) sources on the sensory characteristic of boiled eggs. The n-3 PUFA sources from plants include chia, camelina, canola or flax [2, 3, 4]. A previous study found that the egg n-3 fatty acid levels resulted in from hens fed dietary inclusion of flaxseed rich in n-3 PUFA (*alpha-linolenic acid*, ALA) increased egg n-3 PUFA without adversely influencing the sensory characteristics of the final products.

Portulaca oleraceae (purslane) contains alpha-tocopherol, β-catotene, and ascorbic acid, which are biologically active compounds, and is a source of many nutrients. Purslane also contains high n-3 PUFA level, ALA which is important for disease prevention, health promotion, and normal human growth.[6]. Providing diets added with purslane meal to laying hens enhanced the colour intensity of egg yolks [7]. Yolk of eggs is an excellent emulsifier for food processing. Therefore, in the production of many food emulsions, such as mayonnaise, egg yolks are key ingredients. Feedeing laying hens with purslane meal will provide higher egg yolk quality and increase the sensory quality of food products including mayonnaise. The aim of this research was to evaluate sensory quality of mayonnaise formulated with egg yolks produced by laying hens fed diets enriched with purslane meal as asource of alpha-linolenic acid up to a level of 8%.

2. Materials and Methods

2.1. Research Materials

A total of 25 eggs obtained from *Hy-Line Brown hens* aged 54 weeks fed diets supplemented with purslane meal were used in this research. Other ingredients used to produce the mayonnaise, such as corn oil, vinegar, sugar, mustard, and salt were purchased from a local grocery store.

2.2. Research Methods

One-way classification design was used for this study. The variable factor was mayonnaise formulated with eggs from laying hens fed diets containing 0 (T0), 2 (T1), 4 (T2), 6 (T3) and 8% (T4) purslane meal (w/w). Fat content of the diets was kept constant at ~ 3.5%. Birds were housed at individual cages and assigned to five experimental diets. The corn-soybean based diet, which contains energy 2,925 kkal/kg and crude protein 17% was used to prepare the diets. The composition of ingredient and nutrient value of experimental diets were shown in Table 1. At day 28, 25 egg samples in total, n = 5 egg for each treatment, were collected and processed for mayonnaise. The ingredients used in all mayonnaise formulation consisted of 14% egg yolks, 9% vinegar, 74% corn oil, 1% salt, 1% sugar, and 1% mustard.

Panellists who were willing to consume mayonnaise and had no allergies to eggs or products containing egg were recruited from students at the Faculty of Agriculture, Universitas Sebelas Maret, Surakarta. Thirty (n = 30) panellists participated in the evaluation. After investigators explained the details of the experiment, each panellist was asked to return a signed written consent form. The panellists were blinded to the experimental diets. On the days of final descriptive analysis (DA), samples of mayonnaise were prepared and each panellist assessed aroma, colour, taste, and flavour of mayonnaise samples from five different treatment groups. To evaluate the sensory characteristics of mayonnaise samples, unstructured 15-cm line scales, which were anchored at both ends (at 10 and 90%) with intensity extremes of each attributes, were used and panellists were asked to score their perception of each attribute tested. [8]. Consumer acceptance of mayonnaise was also evaluated by hedonic test using 9-point hedonic scale. Mayonnaise samples were placed in closed plastic containers, which were coded with a randomly generated three-digit number, and were presented to each panellist, along with filtered water and unsalted crackers as palate cleansers. Between every two samples, panellists were instructed to rinse their palate and had a forced 1 minute break. Samples were presented to the panellists in groups of five samples.

IOP Conf. Series: Earth and Environmental Science **102** (2017) 012040 doi:10.1088/1755-1315/102/1/012040

2.3. Data Analysis

The data in this research were sensory attributes of mayonnaise. The collected data were analysed using analysis of variance (ANOVA). When there were significant differences between treatment means, the analysis was followed by Tukey test, with significance level of P<0.05.

Table 1. Ingredient composition and nutrient value of dietary treatments

DietaryTreatments	T0	T1	T2	T3	T4
Ingredient			%		
Soybean meal	24.90	24.50	23.71	23.30	23.00
Yellow corn	53.00	53.00	53.00	53.00	53.00
Rice polish	7.53	6.57	6.15	4.85	2.20
Di-calcium Phosphate	1.30	1.34	1.34	1.34	1.34
$CaCO_3$	2.15	1.90	1.70	1.46	1.21
DL-Metionin	0.15	0.15	0.15	0.15	0.15
L-Lysin	0.10	0.10	0.10	0.10	0.10
Copra meal	1.00	1.00	1.00	1.00	1.00
Purslane meal	0.00	2.00	4.00	6.00	8.00
Limestone	4.90	4.90	4.88	4.88	4.88
Palm oil	3.00	3.07	3.00	3.20	4.65
Vitamin E	0.02	0.02	0.02	0.02	0.02
Filler	1.50	1.00	0.50	0.25	0.00
Salt	0.15	0.15	0.15	0.15	0.15
Premix	0.30	0.30	0.30	0.30	0.30
Total	100.00	100.00	100.00	100.00	100.00
Nutrient value					
DM	79.60	80.05	80.64	80.96	79.73
Ash	2.99	2.93	2.90	2.83	2.64
Ether extract	3.37	3.36	3.42	3.37	3.13
Crude protein	17.04	17.11	17.06	17.13	17.02
Crude fiber	3.34	3.66	4.00	4.33	4.48
Nitrogen free extract	59.96	59.22	58.64	57.86	55.99
ME	2925.40	2925.09	2919.68	2922.87	2852.17
Phosphorus	0.65	0.65	0.65	0.63	0.60
Available phosphorus	0.37	0.37	0.37	0.37	0.35
Calcium	3.10	3.10	3.10	3.10	3.10
Methionine	0.43	0.42	0.42	0.41	0.40
Lysine	0.96	0.94	0.92	0.90	0.88

3. Results and Discussion

The descriptive analysis of mayonnaise samples showed that there were no significant differences in the sensory characteristics including emulsion stability, texture, homogeneity, aroma, taste, flavour, or off-odour of the mayonnaise formulated with eggs from birds fed experimental diets. The average attribute scores are presented in Table 2. This indicates that there were no negative effects of eggs produced by supplementing purslane meal rich in ALA on the sensory quality of the mayonnaise. The emulsion stability of mayonnaise was not different between control mayonnaise and mayonnaise formulated with eggs from dietary treatments. This could be due to the percentage of egg yolks used to prepare mayonnaise was same. Egg yolk contains lecithin which is an emulsifier and the increase in the concentration of egg yolk improved emulsifying activity [9]. Interestingly, it appears that mayonnaise formulated with eggs from laying hens fed diets enriched with purslane meal was generally considered to elicit a higher perception of colour compared to mayonnaise control. The highest intensity of mayonnaise colour (average of intensity score of 10.98 on a scale from 0 to 15; $P < 0.05$) was reached when eggs from hens were fed diets added with the highest purslane meal level. This could be due to eggs produced from diets containing high levels of purslane meal have higher xanthophyll and carotene levels than eggs from control diet. Portulaca oleracea is an omega-3 source plant containing many nutrients, including β-catotene and xanthophyll [10]. A previous study carried out Krawczyk *et al.* [11] found that xanthophyll and carotene content of the diets influenced yolk colour of eggs. In addition, as reported by Kartikasari et al. [7], diets enriched with purslane meal up to 8% increased the colour quality of yolks. A higher yolk colour intensity of eggs from hens fed diets containing purslane meal probably resulted from the fact that purslane consumed by the laying hens was a source of xanthophylls and β-carotene [10] for yolks. Therefore, the yolks produced by hens fed dietary purslane meal have a higher amount of xanthophylls and vitamin A and this resulted in the increase in the yellow colour intensity of mayonnaise.

Table 2. Descriptive analysis of mayonnaise formulated with eggs from laying hens fed dietary treatments enriched with *Portulaca oleracea* meal.

Parameters	Dietary treatments[1]					P value	Significance
	T0	T1	T2	T3	T4		
Colour	7.35^b	6.64^b	5.90^b	7.01^b	10.98^a	0.00	**
Emulsion stability	9.47	9.41	9.09	9.20	9.60	0.97	NS
Texture	8.41	7.83	8.30	8.60	9.21	0.26	NS
Homogeneity	7.92	8.02	7.85	8.08	9.30	0.20	NS
Aroma	8.13	7.87	7.95	7.62	8.47	0.75	NS
Off-odour	4.79	4.19	4.64	4.00	3.74	0.69	NS
Taste	7.25	7.18	7.24	7.08	7.96	0.64	NS
After Taste	6.31	5.53	7.01	6.84	6.50	0.33	NS
Flavour	7.32	6.63	6.95	7.08	8.09	0.16	NS
Overall acceptance	7.32	6.94	6.95	7.11	7.93	0.44	NS

[1]Mayonnaise formulated with eggs from hens fed 0 (T0), 2 (T1), 4 (T2), 6 (T3), and 8% (T4) purslane meal diet, ** ($P<0.01$); NS= not significant

Importantly, the consumer acceptance test conducted on mayonnaise showed that eggs produced from diets containing purslane meal did not change the consumer liking of the mayonnaise including aroma and creamy compared with mayonnaise from control eggs (Table 3). In addition, consumers did not have different liking on taste and flavour between mayonnaise prepared from control eggs and those formulated with eggs produced from dietary treatments. However, mayonnaise texture prepared with eggs from hens fed up to 6% purslane meal tended to be liked more than all other mayonnaise (average of liking score of 6.07; a 9-point hedonic scale). Based on the data, a significant difference was observed in sensory panel scores for the degree of liking of mayonnaise colour, with the highest score achieved for mayonnaise formulated with eggs from 8% dietary purslane meal (average of liking score of 6.03; a 9-point hedonic scale).

Table 3. Consumer acceptance of mayonnaise formulated with eggs from laying hens fed dietary treatments enriched with *Portulaca oleracea* meal.

Parameters	Dietary treatments[1]					P Value	Significance
	T0	T1	T2	T3	T4		
Colour	5.633[ab]	5.300[ab]	4.967[b]	5.467[ab]	6.033[a]	0.05	*
Aroma	5.300	4.967	5.100	5.467	5.667	0.19	NS
Texture	5.367	5.300	5.200	5.433	6.067	0.11	NS
Creamy	5.300	5.467	5.600	5.767	6.100	0.16	NS
Taste	5.300	5.300	5.200	6.067	5.733	0.09	NS
Flavour	5.400	5.367	5.467	5.900	5.900	0.32	NS
Overall acceptance	5.600	5.400	5.567	5.867	6.100	0.23	NS

[1]Mayonnaise formulated with eggs from hens fed 0 (T0), 2 (T1), 4 (T2), 6 (T3), and 8% (T4) purslane meal diet, * significant (P<0.05); NS= not significant

4. Conclusions

It was concluded that eggs from laying hens fed experimental diets containing *Portulaca oleracea* (purslane) meal increased the colour intensity of mayonnaise and the consumer liking for the mayonnaise colour. The use of eggs from laying hens fed diets added with *Portulaca oleracea* meal up to a level of 8% can be applied to formulate mayonnaise, without influencing the sensory characteristics and consumer acceptance of the mayonnaise.

Acknowledgement
This research was funded by the research grant "Pusat Keunggulan-Universitas Sebelas Maret (PKUNS)", Project Number: 632/UN27.21/LT/2016.

References

[1] Herald TJ, Abugoush, M and Aramouni, F 2009 Physical and sensory properties of egg yolk and egg yolk substitutes in a model mayonnaise system *J. Texture Stud.* **40** 692–709
[2] Ayersa R and Coates W 2002 Dietary levels of chia: influence on hen weight, egg production and sensory quality, for two strains of hens *Br. Poult. Sci.* **43** 283-90
[3] Cherian G and Quezada N 2016 Egg quality, fatty acid composition and immunoglobulin Y content in eggs from laying hens fed full fat camelina or flax seed *J. Anim. Sci. Biotechnol.* **7** 1-8
[4] Hayat Z, Cherian G, Pasha TN, Khattak FM and Jabbar MA 2009 Effect of feeding flax and two types of antioxidants on egg production, egg quality, and lipid composition of eggs *J. Appl. Poult. Res.* **18** 541-51
[5] Kartikasari LR, Hughes RJ, Geier MS, Bastian SEP, Makrides M and Gibson RA 2014 Optimizing the n-3 fatty acid content of eggs produced by layer hens fed with alpha-linolenic acid enriched diets while maintaining sensory qualities *Proc. of the 16th Asian-Australasian Assoc. of Anim. Prod. Societies Congress* (Yogyakarta: Gadjah Mada University Press) pp 342-45
[6] Uddin MK, Juraimi AS, Hossain MS, Nahar MAU, Ali ME and Rahman MM 2014 Purslane weed (Portulaca oleracea): a prospective plant source of nutrition, omega 3 fatty acid, and antioxidant attributes *Scientic World J.* 2014 1-6
[7] Kartikasari LR, Hertanto HS, Pranoto D, Salim WN and Nuhriawangsa AMP 2017 Exterior and interior physical quality of egg of laying hens fed diets containing different dietary purslane levels *International Conference On Food Science and Engineering 2016* IOP Conf.

Series: Materials Science and Engineering 193 (2017) 012027 doi:10.1088/1757-899X/193/1/012027

[8] Bastian SEP, Collins C, Johnson TE 2010 Understanding consumer preferences for Shiraz wine and Cheddar cheese pairings. *Food Qual. Prefer.* **21** 668-678

[9] Souza CJF, Rojas EEG 2012 Emulsion of systems containing egg yolk, polysaccharides and vegetable oil *Food Sci. Tech.* **36** 543-550

[10] Raju M, Varakumar S. Lakshminarayana R, Krishnakantha TP and Baskaran V 2007 Carotenoid composition and vitamin A activity of medicinally important green leafy vegetables *Food Chem.* **101** 1598-05

[11] Krawczyk J, Sokołowicz Z and Szymczyk B 2011 Effect of housing system on cholesterol, vitamin and fatty acid content of yolk and physical characteristics of eggs from Polish native hens *Arch. Geflügelk* **75** S151-57

Performance and lipid profiles of native chickens fed diet containing skipjack fish oil as by-product of fish canning factory

J R Leke*, J S Mandey, J T Laihad, R M Tinangon, L Tangkau and C Junus

Faculty of Animal Husbandry, Sam Ratulangi University, Manado, North Sulawesi, Indonesia

*E-mail: rinileke@yahoo.com

Abstract: The study was conducted to determine the use of fish oil as by-product of fish canning factory in diet on the performance and lipid profiles of native chickens. The experiment used 100 native chicken with an average initial body weight of 48,9 gram (sd $\pm$ 9.9), was used in this study for 8 weeks experiment.These were arranged by a completely randomized design with 5 treatments, 5 replications and 4 hens in replication each. The diets were: R0 = 100% Based Diet (BD) + 0% Fish Oil (FO); Rl = 98.5% BD + 1.5% FO; R2 = 98% BD + 2% FO; R3 = 97.5% BD + 2.5% FO; R4 = 97 % BD + 3% FO. Feed and water were provided *ad libitum*. Variables were performance parameters and lipid profiles. Results showed that fish oil inclusion in diets were significantly increased feed intake, body weight gain, carcass percentage, liver, breast and thigh weight, and decreased blood cholesterol, carbohydrate and meat cholesterol, and also tended to decrease abdominal fat. However, there were no affected on feed conversion, water, protein, fat and ash of breast meat. It can be concluded that the use of fish oil in diet up to 3% could improved performance parameters of native chickens.

Keywords: *Fish oil, Lipid profiles, Native chicken, Performance*

1. Introduction

Chicken meat is widely consumed by people because of the low and medium of fat content. Unlike other animal fats, around two third of poultry fat are unsaturated fatty acids, and they are belonged to omega-3 (n-3) and omega-6 (n-6) fatty acids. Therefore, to develop the meat products with lower fat content and/or higher levels of unsaturated fatty acids, especially n-3 fatty acids, would be desirable as a strategy to help defer cardiovascular and inflammatory pathologies [1-3].

Dietary supplements such as n-3 PUFA have been tested in an attempt to decrease fat and cholesterol contents of poultry meat [4]. In poultry, it is common that dietary fat has a great influence on fatty acid profiles of poultry meat [5,6]. Polyunsaturated fatty acids (PUFA) intake including n-3 and n-6 fatty acids have beneficial effects on both animal and consumer health. However, as [7] reported that excessive intake of n-6 relative to n-3 fatty acids causes an increase in pathogenesis. Dietary fatty acids would be absorbed by monogastric animals and deposited in their tissues without significant modification [8]. It has been recorded that chickens could modify their lipid profile within a week after replacement of the dietary lipid source [9].

Fish lipids in the diet of animals improves health. Both fish oil and fish meal supply lipids and provide source of EPA and DHA for animals. Fish oil that contains unsaturated fatty acids with long omega-3 chains (LC-n-3 PUFA), ecosapentanoic acid (EPA20:5n-3) and decosahexanoic acid (DHA22:6n-3) improves health-related factors in humans and animals [5]. They are used because of their growth-promoting effects, their cheap source of energy, and their vitamin A and D content. The deposition of EPA and DHA in animal products such as meat and eggs provides a valuable source in the human diet [10].

The positive effect of fish oil in broiler diets on the performance has been already demonstrated. Fish oil is one of the available sources used in the diet to provide energy for fowl's while it's supplementation has been shown to improve body weight gain, feed conversation ratio and safety in the poultry [11-12]. Numerous studies indicated that dietary fish oil declines the level of total cholesterol, low density lipoprotein and triglycerides [13]. Also, it has been shown that fish oil can modify abdominal fat deposition in broiler chickens [14] stated that fish oil improved lipid profiles, reduced proinflammatory eicosanoid contents by inhibiting phospholipase A2 production in broilers; while zinc reduced the contents of plasma malondialdehyde, phospholipase A2, and then decreased the concentration of prostaglandin E2 of broiler. Laying hens can be fed fish oil because of the beneficial effect on egg production and hatchability.

Skipjack fish industrial waste has potential as an alternative feed source in Indonesia. That is a by-product of making wooden fish and canning fish. North Sulawesi Utara is known as one of the region's potential skipjack fishing ground in Indonesia. The production of skipjack in Sulawesi Utara in 2010 amounted to 60.168.1 tons/per year, where one ton of wet skipjack waste can produce 250 kg of dry waste with 10 kg moisture content [15], and also it can produce fish oil.

To the rural farmers, indigenous chickens are the most preferred for they are not capital intensive [16]. The chickens have little/no religious or social constraints affecting their production [17]. However, indigenous chickens have poor growth rate, low feed conversion efficiency, and their weight is 2.5 times less when compared with exotic breeds of chicken broilers [18].

Native chicken in Indonesia is believed to be the most qualified sources for healthy eggs and chicken meat among Indonesians [19]. Indonesian peoples prefer to consume products of native chicken rather than those of broiler because of its egg and meat's taste are better. And, they are believed to contain higher nutritional values which bringing about the selling price is relatively more expensive compared to that of broiler chicken products [20]. [21] Stated that productivity of indigenous chicken breeds may be doubled with improved diets and management conditions. The indigenous chickens have not attained their full production potential due to exposure to risks that influence against their survival and productivity under extensive management conditions.

Improved nutrition is critical for increasing egg and meat production in native chickens in Indonesia [22]. However, there was no information about effect of fish oil of native chickens. The objective of this research was to determine the performance and lipid profiles of native chicken fed fish oil as fish canning by-product in diets.

2. Materials and Methods

2.1. Preparation of Samples

Skipjack fish canning waste is liquid fraction mixed with meat residue of fish canning factory. The wastes were collected and then were heated 10-15 minutes at 95 ^{0}C, and filtered for separating the solid fraction (meat residue) and liquid fraction (fish oil). The fish oil which have brown color then used as treatments.

Table 1. Fatty Acid Profiles of Skipjack Fish Oil.

Fatty Acid	(mg/100 g)

Caprilic acid (C8:0)	Ud
Capric acid (C10 :0)	Ud
Lauric acid (C12 : 0)	1
Miristic acid (C14 : 0)	56
Pentadekanoic acid (C15 :0)	Ud
Palmitic acid (C16 : 0)	518
Stearic acid (C18 :0)	250
Arakhidic acid (C20 : 0)	11
Total Saturated Fatty Acid	**836**
Miristoleic acid (C14 :1)	Ud
Palmitoleic acid (C16 : 1, n-7)	63
Oleic acid (C18 : 1, n-9)	170
Linoleic acid (C18:2, n-6)	26
α-Linolenic acid (C18:3, n-3)	13
11-Eicocanoic acid (C20:1, n-9)	14
Arakhidonic acid (C20:4, n-6)	Ud
EPA (C20:5, n-3)	84
DHA (C22:6, n-3)	601
Total Unsaturated Fatty Acid	**971**
Omega-3 Fatty Acid	0.69 %
Omega-6 Fatty Acid	0.03 %
Omega-9 Fatty Acid	0.18 %

Ud = Undetected
Results of laboratory analysis of nutrition science department and integrated laboratory technology.IPB.

2.2. Birds, Diets and Experimental Design

The research was subject to one-way completely randomized design, rationing It were used One hundred native chickens animal experiment. The experiment was applied five treatments and four replications. The birds were divided into five experimental diets and each was divided into four replicate groups of five birds per replication using completely randomized design, thus there were 20 unit experiments. The dietary treatment were formulated using similar feed with 45% yellow corn, 9% rice bran, 12% fish meal, 9% soybean meal, and 25% commercial diet. Fish oil was included in five experimental diets at levels of 0%, 1.5%, 2%, 2.5%, and 3%, respectively, to substitute based diet. Fatty acid profiles of skipjack fish oil were shown in Table 1 and the composition and nutrients of diets were shown in Table 2. Feed and water were provided *ad libitum* and diets were presented in mash form. The study was conducted over a period of 8 weeks. The parameters observed were performance: carcass and abdominal fat percentage, heart, liver, breast, thigh, back and wings weight; blood cholesterol, HDL-cholesterol and LDL-cholesterol; meat cholesterol and meat chemical profiles of chickens.

Table 2. Composition and Nutrients of the Diets

	R0	R1	R2	R3	R4
Ingredients, %					
Based Diet	100	98.5	98	97.5	97
Fish Oil	0	1.5	2	2.5	3

Total	100	100	100	100	100
Calculated composition:					
Protein (%)	20.8	20.6	20.6	20.5	20.5
Fat (%)	6.83	7.1	7.48	7.85	8.23
Crude Fiber (%)	4.83	4.75	4.73	4.70	4.68
Ca (%)	0.82	0.82	0.82	0.82	0.82
P (%)	0.81	0.81	0.81	0.81	0.81
ME (Kcal/kg)	2905	2959	2969	2978	2988

One representative bird from each pen was conventionaly sacrified by cervical dislocation technique, as described in the Report of the AVMA Panel on Euthanasia [23] .This technique consists basically on cutting the carotid artery and the jugular vein along the neck (unilateral cut) [24] and its carcass parameters (ready to cook) including dressing percentage and abdominal fat were determined.

Carcass characteristics and internal organs were weighted and abdominal fat pad (including fat surrounding gizzard, bursa of fabricius, cloaca & adjacent muscles) was removed and weighed individually for 4 chicks per treatment. Breast muscles of chickens were removed and were weighed individually. Samples were stored at -20 ^{0}C until analysis. Four samples of muscle per treatment were taken from the breast and were used to cholesterol, crude fat, protein and carbohydrate, water and ash determination.

2.3. Chemical Analysis

Dry matter was determined by drying samples at 100 ^{0}C for 16 h in a pre-weighed dried crucible in a convection oven. Fat was determined by Soxhlet extraction [25]. Starch was measured using alpha-amylase method [25].

2.4. Statistical Analysis

Data were subjected with one-way analysis of variance, and it was continued to Duncan's multiple range test [26] at a probably level of 5% when the treatment indicated significant effect. All data are expressed in the form of standard error of mean. The IBM SPSS Statistics 22 software was used for the statistical processing of data.

3. Results and Discussion

The effects of dietary treatments on feed intake, body weight gain, carcass percentage and organ weight data obtained are presented in Table 3. Results showed that feed intake, body weight gain, carcass percentage and organ weight were affected by the treatment diets. Fish oil inclusion in diets were significantly increased the feed intake, body weight gain, carcass percentage, liver, breast and thigh weight, and decreased blood cholesterol, carbohydrate and meat cholesterol, and also tended to decrease abdominal fat. However, there were no significant difference between treatments on feed conversion, water, protein, fat and ash of breast meat. This result is not consistent with the findings [27] who reported that the inclusion of FO in poultry diets had no effect on feed intake.

Table 3. Effect of Fish Oil in Diet on Performance and Meat Chemical Profiles of Native Chickens

Variables	Treatments					SEM [1]	P value[2]
	0% FO	1.5% FO	2% FO	2.5% FO	3% FO		
Performances:							

Fed Intake (g/bird)	438.4^{a}	480.4^{ab}	457.0^{ab}	502.4^{b}	514.4^{b}	9.51	*
Weight Gain (g/bird)	1115^{a}	1264^{b}	1202^{a}	1257^{b}	1266^{b}	19.81	*
Feed Conversion	2.54	2.64	2.63	2.52	2.46	.04	NS
Carcass (%)	72.41^{a}	73.12^{a}	74.74^{a}	76.47^{b}	76.47^{b}	0.52	*
Abdominal Fat (%)	1.46	1.02	1.32	1.09	1.02	0.67	NS
Liver Weight (g)	28.0^{a}	29.2^{a}	30.2^{ab}	32.0^{b}	34.6^{c}	0.90	**
Heart Weight (g)	7.6	8.0	8.8	8.8	8.8	0.24	NS
Breast Weight (g)	216.60^{a}	218.48^{a}	221.40^{a}	242.24^{b}	246.40^{b}	3.49	**
Thigh Weight (g)	237.80^{a}	335.52^{b}	338.84^{b}	341.36^{b}	350.00^{c}	10.7	**
Back Weight (g)	233.50	227.40	232.50	243.00	245.50	7.72	NS
Wings Weight (g)	135.00	136.32	131.86	139.00	135.50	3.27	NS
Blood:							
Cholesterol (mg/100 g)	139.4^{b}	139.4^{b}	122.4^{a}	123.4^{a}	120.2^{a}	2.11	*
HDL-Cholesterol	102.8^{b}	99.4^{b}	98.0^{b}	98.2^{b}	93.8^{a}	3.38	*
LDL-Cholesterol	55.4^{c}	57.8^{d}	46.8^{a}	49.0^{b}	43.2^{a}	2.76	*
Meat:							
Cholesterol (mg/100g)	39.5^{c}	26.2^{b}	23.1^{ab}	22.4^{a}	25.5^{b}	1.99	**
Water (%)	73.6	73.2	74. 0	73.9	74.0	0.20	NS
Protein (%)	22.6	21.2	21.5	21.5	21.2	0.33	NS
Fat (%)	2.23	1.84	1.83	1.85	1.68	0.15	NS
Ash (%)	1.06	1.15	1.10	1.10	1.15	0.02	NS
Carbohydrate (%)	0.54^{b}	0.50^{b}	0.34^{a}	0.34^{a}	0.33^{a}	0.07	*

$^{a-c}$Means in a row with different superscripts are significantly different at the *P*-value shown
[1]SEM = pooled standard error of mean (n=4)
[2]Significance level: *$P<0.05$, **$P<0.01$, NS = not significant $P>0.05$
FO = fish oil

The results indicated that the diet with 3% FO had the best values for carcass, abdominal fat and cholesterol. A significant reduction in meat and blood cholesterol were observed when the diet was supplemented with the highest level of fish oil (3%). The fish oil used in this study was a rich source of DHA (601 mg/100 g) and contained n-3 fatty acid (0.69%) and EPA (84 mg/100 g (Table 1). That was shown the total n-3 PUFA and monounsaturated fatty acid increased with increasing levels of FO in the diet. Several trials have shown that an increase in the oil content of long chain n-3 PUFA in chicken broiler meat may be achieved by including linseed oil as a source of precursor, α-linolenic acid [28]

In recent year, high growth performance, properties of carcass and meat quality have been beneficial to the broiler production. Especially is on better body composition with higher breast meat yield and lower abdominal fat. According to [29] that the use of fish oil in native chickens up to 7.5% could reduce lipid consumption and increase carcass yield but the result as relatively similar effect to blood profile, feed and energy consumption. Also [30] reported that Thai native chicken has an advantage over broiler in terms of carcass composition and meat quality traits such as high WHC, firm meat texture and high protein as well as low fat content.

The internal organs of Thai native chickens were liver 2.17%, heart 0.44% and dressing percentage 64.54%, that can be the alternative "healthy" meat. [31] reported that chicks fed diets containing 4% fish oil had a significantly lower ether extract in breast, but a higher ether extract concentration in thigh. The present study showed that deposition of broiler abdominal fat could be tend to reduced by supplementing fish oil in the diet. As [32-33] reported that dietary fatty acid could affect abdominal fat deposition and there are several reports on the effects of polyunsaturated fatty acids on abdominal fat pad reduction in broiler chickens. Compared to fast-growing broilers, native chicken breeds and their hybrids show lower weight gain and a smaller proportion of breast muscle in the carcass, but their meat has many quality characteristics valued by modern consumers [34].

These results are in agreement with those reported previously by [35] which showed that the presence of saturated fatty acid in bird meat was dependent on the presence of these fats in the diet, their oxidation rate, and also their synthesis in the liver. [36] showed that the supplementation of feeds with fish oil at 1.5% improved performance of broiler chickens. The supplementation of 3% fish oil in the diet improved the level of n-3 fatty acid and the immune system, and decreased the levels of cholesterol and triglyceride in the plasma of broilers.[37] showed that the lipid content of serum reduced as the levels of dietary PUFA increased. However, in opposite results to our findings, [38] showed that the level of serum cholesterol was not affected noticeably by dietary rich fish oil. The differences between the studies on the lipid content of serum may be attributed to the genetic, sex and dietary factors.

4. Conclusions

Based on these results it can be concluded that the use of fish oil in diet up to 3% could improved performance parameters in native chickens

5. References

[1] Kris-Etherton P M, Harris W S and Appel L J 2002 Fish consumption, fish oil, omega-3 fatty acids, and cardiovascular disease *Circulation* **106** 2747–2757

[2] Din J N, Newby D E and Flapan A D 2004 Science, medicine, and the future – Omega 3 fatty acids and cardiovascular disease - Fishing for a natural treatment *British Med. J.* **328** 30–

[3] Ruxton *et al.*, 2007 Ruxton C H S, Reed S C, Simpson M J A and Millington K J 2007 The health benefits of omega-3 polyunsaturated fatty acids: a review of the evidence *J. Hum. Nutr. Diet* **20** 275–285

[4] Ayerza *et al.*, 2002 Ayerza R, Coates W and Lauria M 2002 Chia seed (*Salvia hispanica* L.) as an omega-3 fatty acid source for broilers: influence on fatty acid composition, cholesterol and fat content of white and dark meats, growth performance, and sensory characteristics *Poult. Sci.* **81** 826–837

[5] Crespo N and Esteve-Garcia E 2001 Dietary fatty acid profile modifies abdominal fat deposition in broiler chickens *Poult. Sci.* **80** 71-78

[6] Krejcí-Treu *et al.*, 2010 Krejcí-Treu T, Straková E, Suchý P and Herzig I 2010 Effect of vegetable oil fortified feeds on the content of fatty acids in breast and thigh muscles in broiler chickens. *Acta Vet. Brno.* **79** S21-S28

[7] Simopoulos A P 2003 Importance of the ratio of omega-6/omega-3 essential fatty acids: Evolutionary aspects *World Rev. Nutr. Diet* **92** 171-174

[8] Rhee K S, Davidson T L, Cross H R and Ziprin Y A 1990 Characteristics of pork products from swine fed a high monounsaturated fat diet: Part 1. Whole muscle products *Meat Sci.* **27** 329-341

[9] Lopez-Ferrer S, Baucells M D, Barroeta A and Grashorn M A 2000 n-3 Enrichment of chicken meat. 1. Use of very long-chain fatty acids in chicken diets and their influence on meat quality: Fish oil *Poult. Sci.* **80** 741-752

[10] Pike I H 1999 Health benefits from feeding fish oil and fish meal the role of long chain omega-3 polyunsaturated fatty acids in animal feeding *IFOMA* No 28

[11] Alparslan G and Özdogan M 2006 The effects of diet ontaining fish oil on some blood parameters and the performance values of broilers and cost efficiency *Int. J. of Poult. Sci.* **5** (**5**) 415-419

[12] Farhoomand P and Checaniazer† S 2009 Effects of graded levels of dietary fish oil on the yield and fatty acid composition of breast meat in broiler chickens *J. App. Poult. Res.* **18** 508–513

[13] Crespo N and Esteve-Garcia† E 2003 Polyunsaturated fatty acids reduce insulin and very low density lipoprotein levels in broiler chickens *Poult. Sci.* **82** 1134–1139

[14] Liu G, An S, Yuan J, Guo Y, Liu D, Chen H and Huang R 2014 Dietary fish oil and zinc reduced plasma prostaglandin E2 content by inhibiting phospholipase A2 production in broilers *J. Poult. Sci.* **51** 66-70

[15] Leke J R, Achmanu, Sjofjan O and Najoan M 2013 Egg internal quality and n-3 fatty acids of native chicken fed on skipjack fish (*Katsuwonus pelamis.* L.) industrial waste containing feed *Int. J. of Poult. Sci.* **12** (**8**) 484-488

[16] Food and Agriculture Organisation of the United Nations 2003 *Poultry keeping* www.fao.org/english/newsroom/new/2003/13200-en.html.

[17] Swatson H K, Tshovhote J, Nesamvumi E, Ranwedzi N E and Fourie C 2001 *Characterization of village free ranging poultry production systems under traditional management conditions in the Vhembe district of the Limpopo Province* pp. 1-5

[18] Jaturasitha S, Kayan A and Wicke M 2008 Carcass and meat characteristics between Thai village chickens compared with improved layer breeds and their crossbred Poult. Sci. **51** (**3**) 283-294

[19] Iriyanti N, Santosa R S and Rachmawati W S 2014 Blood profile and performance of native chicken with functional feed *Int. J. of Poult. Sci.* **13** (**11**) 645-651

[20] Ma'rifah B, Atmomarsono U and Suthama N 2013 Nitrogen retention and productive performance of crossbred native chicken due to feeding effect of kayambang (*Salvinia molesta*) *Int. J. of Sci. and Engin.* **5** (**1**) 19-24

[21] Chowdhury S D, Ahmed S and Hamid M A 2006 Improved feeding of desi chicken reared in confinement. *The Bangladesh Vet.* **23** 29-35

[22] Henuk Y L and Bailey C A 2014 Husbandry systems for native chickens in Indonesia *Proc. of the16th AAAP Anim. Sci. Congress* pp 759-762

[23] AVMA 2000 Report of the AVMA Panel of Euthanasia *J. of the American Vet. Med. Assoc.* **218** (**5**) 669-696

[24] Craig, E.W. and D.L Fletcher and P.A.Papinaho 1999 The effects of antemortem electrical stunning and postmortem electrical stimulation on biochemical and textural properties of broiler breast meat. Poult.Sci., **78**;480-494.

[25] AOAC 1990 *Official Methods of Analysis* (Association of Official Analytical Chemists Ed.) Washington DC

[26] Steel R G D and Torrie J H 1994 *Principles and Procedures of Statistics* McGraw-Hill Book Co. Inc. Pub. Ltd. London

[27] Huang Z B, Ackman R G, Ratnayake W M N and Proudfoot F G 1990 Effect of dietary fish oil on n-3 fatty acid levels in chicken eggs and thigh flesh *J. Agric. Food Chem.* **38** 743–747

[28] Zelenka ,J.,A.Jarosovo, D. Schneiderova 2008 Influence of n-3 and n-6 polyunsaturated fatty acids on sensory characteristics of chicken meat. Czech.J.Anim.Sci. **53**,299 -305.

[29] Iriyanti N, Santosa R S and Rachmawati W S 2014 Blood profile and performance of native chicken with functional feed *Int. J. of Poult. Sci.* **13** (**11**) 645-651

[30] Jaturasitha S, Kayan A and Wicke M 2008 Carcass and meat characteristics between Thai village chickens compared with improved layer breeds and their crossbred Poult. Sci. **51** (**3**) 283-294

[31] Navidshad B 2009 Effects of fish oil on growth performance and carcass characteristics of broiler chicks fed a low-protein diet *Int. J. Agric. Biol.* **11**(**5**) 635-638

[32] Vila B and Esteve-Garcia E 1996 Studies on acids oil and fatty acids for chickens. I. Influence of age, rate of inclusion and degree of saturation on fat digestibility and metabolisable energy of acid oils *Braz. Poult. Sci.* **37** 105 - 117

[33] Sanz M, Flores A, Pe´rez De Ayala P and Lo´Pez-Bote C J 1999 Higher lipid accumulation in broilers fed on saturated fats than in those fed on unsaturated fats *Braz. Poult. Sci.* **40** 95-101

[34] Sokołowicz Z, Krawczyk J and Świątkiewicz S 2016 Quality of poultry meat from native chicken breeds - a review. *Annals of Anim. Sci.*

[35] Lopez-Ferrer S, Baucells M D, Barroeta A C and Grashorn M A 1999 Omega-3 enrichment of chicken meat using fish oil: Alternative substitution with rapeseed and linseed oils *Poult. Sci.* **78** 356–365

[36] Saleh H, Rahimi S and Karimi Torshizi M A 2009 The effect of diet that contained fish oil on performance, serum parameters, the immune system and the fatty acid composition of meat in broilers *Int. J.Vet. Res.* 3 (**2**) 69-75

[37] Celebi and Utlu 2006 Influence of animal and vegetable oil in layer diets on performance and serum lipid profile. Int J Poult Sci. **5**-370-373

[38] Alparslan G and Özdogan M 2006 The effects of diet ontaining fish oil on some blood parameters and the performance values of broilers and cost efficiency *Int. J. of Poult. Sci.* **5** (**5**) 415-419

IOP Conf. Series: Earth and Environmental Science **102** (2017) 012042 doi:10.1088/1755-1315/102/1/012042

Performance and carcass quality of broiler chickens fed diet containing pineapple waste meal fermented by "ragi tape"

J S Mandey*, B Tulung, J R Leke, B F J Sondakh

Animal Husbandry Faculty, Sam Ratulangi University
Manado, North Sulawesi, Indonesia

*E-mail: jetsm_fapet@yahoo.co.id

Abstract. The study was conducted to determine the effect of pineapple waste meal fermented by "ragi tape" (FPW) in diets on the performance and carcass quality of broilers. The "ragi tape" was a traditional commercial product of yeast. Five dietary treatments containing 0, 5, 10, 15 and 20% levels of FPW with four replicates were fed to 250 broiler chickens for 42 days in a completely randomized design. Feed and water were provided *ad libitum.* The variables were performance parameters and carcass quality. Results showed that the performance in finisher, carcass percentage and abdominal fat percentage were significantly affected by dietary treatments. Carcass percentage and abdominal fat percentage were significantly decrease in the proportion of 20% of FPW. However, the carcass percentage in treatments R0 – R4 were still in a good category. The higher the levels of FPW the lower the abdominal fat percentage signed that FPW treatments up to 20% resulted good category of broiler carcass. Income over feed cost and broiler cost analysis in 20% FPW treatment obtained the highest income and the lowest was 0% FPW treatment. It can be concluded that FPW can be fed to broiler chickens at up to 20% level.

Keywords: Broiler, Carcass, Fermentation, Performance, Pineapple

1. Introduction

Nutritionist in recent time have been focused in research of agro-industrial wastes in animal nutrition especially for monogastric animals. In fact, many feeds that can be fed alternatively at cheaper cost to monogastric livestock are based on the use of agro-industrial waste that are of no food value to humans [1]. Onwuka, *et al.* [2] stated that a major strategy to develop the livestock industry in developing countries could be the use of agricultural by-products like pineapple waste, corn cobs and brewers dry grain. Pineapple waste is agro by-products from pineapple fruit.

Pineapple waste occurs as pineapple peels and core, making about 40-50% of the fresh fruit [3]. It contains mainly the sugars of sucrose, fructose and glucose [4]. Also, it contains low amounts of protein, fat, ash [5]. Pineapple peel is rich in cellulose, hemicellulose and other carbohydrates. Raw pineapple waste (on DM basis) contains about 4% crude protein, 60–72% NDF, 40–75% soluble sugars (70% sucrose, 20% glucose and 10% fructose) as well as pectin, but it is poor in minerals [6, 7], except Ca. It is locally available.

Traditional market wastes and restaurant wastes of agro-product like pineapple fruit represent a serious problem since it is usually causes major environmental problems. The process of pineapple to juice and dried pineapple fruits release in significant quantities of by-products, such as: peels, crowns and hearts. Nurhayati [8] reported that pineapple peel is produced from pineapple fruit processing. Approximately 27% of pineapple fruit is pineapple peel. The accumulation of pineapple waste in the neighbourhoods constitutes a source of environmental pollution. If fresh pineapple peels are not consumed, it often gets mouldy and sour, and therefore unlike heptonly to be used as an animal feedstuff. Therefore, some studies were conducted to develop a procedure for converting pineapple waste into animal feed [9, 10].

Problems related with the fresh form, were overcome by the sun drying technique of pineapple peels developed by Aboh, *et al.* [11]. It gave dried peels of good quality, however, the dried peels are too compact and hard for its ingestion by animals. Researchers reported that pineapple wastes have been described as equivalent to cereal grains for ruminants [6] or aș a low-nutrient feed [12].

Pineapple wastes are recommended as tremendous sources of organic raw materials and are potentially available for conversion into useful products such as animal feeds [13]. However there is a constraint to use it as poultry feedstuff due to it contains 19.8% of cellulose and 11.7% of hemicelluloses [14]. Therefore, to overcome this constraint, it is significant to explore some treatments to be applied to the peels such as crushing, to improve their ingestion without degrading the feedstuff value [11]. In any case, the high amount of fibre makes pineapple wastes more suitable to ruminants than to pigs and poultry. The bulkiness of the fresh products limits intake. Inclusion of 15 percent pineapple bran in chick diets depressed the feed conversion ratio and 20 percent inclusion decreased weight [15]. The final average live weight gains and feed conversion rate of growing rabbits significantly reduced as inclusion level of pineapple waste increased [16, 17]. However, Lamidi, *et al.* [18] found that broiler chickens could tolerate up to 10% pineapple waste in their diets without any deleterious effect. Olosunde [19] reported that sheep could tolerate up to 45% pineapple waste but 30% was superior when substituted for corn bran.Therefore, to overcome this constraint, it is significant to explore some treatments to be applied to the peels such as crushing, to improve their ingestion without degrading the feedstuff value.

It has been reported that pineapple waste contains high amounts of crude fiber and suitable sugars for growth of microorganisms. Similar to rice husk, pineapple peel also need pretreatment before offering to the poultry [13]. *Trichoderma harzianum* might be used to ferment agricultural by products those rich in fiber content to increase their quality by increasing crude protein content and reducing crude fiber content. Rice husk and pineapple peel had higher quality after fermenting with 12% of *Trichoderma harzianum*. Banana peel was not suggested to ferment with *Trichoderma harzianum* [20].

The use of microorganisms through fermentation to improve nutritional value of agro-industrial wastes, thereby offering the potential to make dramatic contributions to sustainable livestock production has been well documented [1, 21, 22]. The utilization of fungi for nutrient enhancement in agro-industrial waste by fermentation has been studied for years and their efficiency shown in substrates such as lignin, cellulose and hemicellulose polymers found in agro-industrial waste [23]. *Aspergillus niger* and *Trichoderma viride* have been successfully used in a number of fermentation studies towards solid waste management, biomass energy conservation and production of secondary metabolites in various agro-industrial wastes [24, 25, 26].

Some studies were conducted to develop a procedure for converting pineapple waste into animal feed [9, 10]. In studies by Correia, *et al.* [27], *Rhizopus oligosporous* was used to produce enhanced levels of free phenolics from pineapple residue in combination with soy flour as potential nitrogen source.

"Ragi tape" was a traditional commercial product contained *Candida parapsilosis, C. melinii, C. lactosa. C. solani, Hansenula subpelliculosa, Rhizopus oligosporus, Aspergillus flavus, A. oryzae* and *Hansenula malanga* [28], that was could used to develop a procedure for converting pineapple waste into animal feed. However, there are no reported studies of the dietary of fermented pineapple waste in diet of broiler. This study aimed to investigate the effect of the dietary level of pineapple waste fermented by "ragi tape" on carcass percentage and abdominal fat percentage of broiler chickens.

2. Materials and Methods

2.1. Birds, Diets and Experimental Design

Fresh pineapple peels collected were washed and steamed for twenty five minutes. Then cooled and mixed with 30 g "ragi tape"/kg pineapple peels, incubated on three days at room temperature. "Ragi tape' was a traditional commercial product contained *Candida parapsicosis*, *Candida melinis*, *Hansenula supbeliculosa*, *Hansenula malanga*, *Aspergillus niger*, *A. oryzae* and *Saccharomyces cerevisiae*. Part of the fermented pineapple peels then were dried and ground to fine powder using mortar and pestle.

The PWF then incorporated into the experimental diets at five levels of 0, 5, 10 and 15 and 20%. Based diet and PWF were crushed to obtain diets R0, R1, R2, R3 and R4, respectively. The proximate analysis of fermented pineapple waste (PWF) was shown in Table 1. Based diet contain ingredients: yellow corn 55%, fish meal 12%, soybean cake 15%, rice bran 7%, coconut cake 10.5% and top mix 0.5%, and the nutrients composition is shown in Table 2.

A total of 250 unsexed broiler finisher (Cobb CP 707) have been used in this experiment. Birds were weighed and maintained under standar managemental practices. Feed and water were provided *ad libitum* throughout the experimental period, and these treatments were administrated for a 42 days period. Vaccination programme against Gumboro and ND were carried out as per schedule.

Parameters were evaluated: feed intake, weight gain, feed conversion ratio, carcass percentage and abdominal fat percentage. At 42 days the experiment, one representative bird from each pen was conventionaly slaughtered by cervical dislocation, bled and eviscerated. Then, carcass parameters (ready to cook) including dressing percentage and abdominal fat were determined.

Table 1. Chemical analysis of unfermented and Fermented pineapple waste

Nutrients	Unfermented PW	Fermented PW
Crude Protein (%)	0.92	7.87
Crude Fiber (%)	18,25	17.42
Fat (%)	0.80	1.53
Ca (%)	0.58	12.73
P (%)	0.4	0.82
GE (Kcal/kg)	2782	3830

Notes: PW = pineapple waste

Table 2. Diets and calculated analysis

Treatments	R0	R1	R2	R3	R4
Based Diet	100	95	90	85	80
Fermented Pineapple Waste	0	5	10	15	20
Calculated Analysis:					
Protein	21.85	21.15	20.45	19.75	19.05
Crude Fiber	4.86	5.48	6.11	6.74	7.37
Fat	6.32	6.08	5.84	5.60	5.37
Ca	2.31	2.83	3.35	3.87	4.40
P	1.27	1.25	1.22	1.20	1.18
GE (Kcal/kg)	3883	3880	3878	3875	3872

2.2. Statistical analysis

The data were subjected to analyze for a variance technique using completely randomized design that was employed in one-way analysis of variance, and significant differences compared by Duncan's multiple range test [29]. All of statement of differences were performed at significance levels of 1% and 5%. The IBM SPSS Statistics 22 software was used for the statistical processing of data.

3. Results and Discussion

Data on performance and carcass quality of broilers affected by fermented pineapple waste (PWF) in diet is shown in Table 2. The results showed that carcass percentage and abdominal fat percentage were highly significant (P<0.01) affected by dietary treatments. Carcass percentage was significantly decrease in the proportion of 15% and 20% of PWF. However, the carcass percentage in treatments R0 – R4 were still in a good category. Abdominal fat percentage also was significant decrease in the proportion of 15% and 20% of PWF.

Table 3. Effect of Fermented Pineapple Waste in Diet on Carcass Percentage and Abdominal Fat Percentage of Broilers

Variable	Treatments					SEM	P value
	0% PWF	5% PWF	10% PWF	15% PWF	20% PWF		
Feed Intake (g/h/d)							
Starter	82.00	81.74	80.60	80.30	80.65	.206	.021
Finisher	109.21	109.66	104.05	98.10	98.59	.876	.000
Weight Gain							
Starter	26.85	26.89	26.93	26.76	26.66	.088	.892
Finisher	40.84	44.86	41.57	39.33	39.88	.361	.000
Feed Conversion Ratio							
Starter	3.02^{ab}	3.04^{c}	3.00^{ab}	2.99^{a}	3.02^{ab}	.007	.142
Finisher	2.69^{b}	2.45^{a}	2.50^{a}	2.49^{a}	2.50^{a}	.016	.000
Carcass (%)	74.89^{b}	73.28^{ab}	$73.27a^{b}$	72.27^{ab}	71.26^{a}	.411	.063
Abdominal Fat (%)	2.34^{e}	1.76^{d}	1.50^{c}	1.30^{a}	1.19^{b}	.060	.000
IOFCC (Rp)	$8{,}554^{a}$	$10{,}509^{b}$	$10{,}556^{b}$	$10{,}548^{b}$	$10{,}687^{b}$	163	.022

Notes: g/h/d = gram/head/day; ^{abcd} means with the same letters on the same row are in not significantly difference (P>0.05)

In the case of unfermented pineapple waste, Aboh, *et al.* [11] reported that final average live weight gains and feed conversion rate of growing rabbits significantly reduced as inclusion level of pineapple peels increased. Moreover, Adeyemi, *et al.* [17] and Fapohunda, *et al.* [16] observed the same trend when using pineapple peels in the rabbits feed. The difference may due to the nutrient unbalance, mainly the protein contents of diets, and based on unfermented pineapple peel. In the present study, crude protein value in the fermented pineapple peel obtained was higher than the value recorded (5.11%) by Adeyemi et al. [17], resulted good category of broiler carcass until 20% FPW treatment. Omwango et al [30] found higher crude protein content in pineapple waste fermented using the fungi *A. niger* and *T. viride* than in the unfermented pineapple waste for the 48, 72 and 96 h fermentation periods. The post fermentation increase in crude protein content could be attributed to the possible secretion of some extra cellular enzymes (protein) such as amylases, xylanases and cellulases into the pineapple waste mash by the fermenting fungi in an attempt to make use of the carbohydrates in the mash as a carbon and energy source [31]). *A. niger* has been reported to have high specific activity for cellulases and hemi-cellulases [23]. Additionally, *T. viride* and *A. niger* have found use in the production of extra cellular enzymes including cellulase, amylase and xylanase [32]. Mandey et al. [33] reported that carcass yield was significantly affected by the treatments, in which the carcass yield was highest from the birds fed diet containing 10% banana leaves fermented by *Trichoderma viride* for 10 days (74.58%).

Fungi colonize substrates for utilization of available nutrients. They synthesize and excrete high quantities of hydrolytic extra cellular enzymes, which catalyze the breakdown of nutrients to products that enter the fungal mycelia across cell membrane to promote biosynthesis and fungal metabolic activities leading to growth [31]. Therefore, increase in the growth and proliferation of fungal biomass in the form of single cell protein (SCP) or microbial protein accounts for part of the increase in the protein content after fermentation [31].

The decreasing of abdominal fat in the present study may due to th Ca content and Ca : P balance in diet. Some reports suggest that high dietary Ca can adversely affect the utilization of fat [34], nitrogen and metabolisable energy [35] in broilers. Dietary Ca concentration had a significant effect on apparent fat digestibility. Fat digestibility was reduced by increasing Ca concentrations in all intestinal segments [36]. Aboh, *et al.* [11] reported that ash, calcium and magnesium in sun dried pineapple peel were a useful mineral source for rabbits. In this experiment, the higher the levels of PWF the lower the abdominal fat percentage signed that PWF treatments up to 20% resulted good category of broiler carcass.

The significant increase in protein content of the pineapple waste after fermentation with "ragi tape" and the decrease in crude fiber concludes that fermentation of pineapple waste by "ragi tape" enriches the nutrient content of the waste and this by product can be good supplement in compounding animal feed provided that it is acceptable and highly digestible.

4. Conclusion

It can be concluded that fermented pineapple waste can be fed to broiler chickens at up to 20% level with a promising good category of broiler carcass and abdominal fat, although IOFCC was more higher.

5. References

[1] Iyayi E A and Fayoyin F K 2005. Effect of feeding cassava fruit coat meal on the nutrient digestibility and performance of broilers *Livest. Res. for Rural Develop.* **17 (19)**

[2] Onwuka C F I, Adetiloye P O and Afolami C A 1997 Use of household wastes and crop residues in small ruminant feeding in Nigeria. *Small Rum. Res.* **24 (3)** 233-37

[3] Buckle K A 1989 Biotechnology opportunities in waste treatment and utilization for the food industries In: Rogers P. L. (Ed.) Biotechnology and the Food Industry. Breach Publishers, New York. p. 261-77

[4] Krueger D A, Krueger R G and Maciel J 1992 Composition of pineapple juice. *J. of AOAC Int.* **75** 280-82

[5] Hebbar H U, Sumana B and Raghavarao K S M S 2008 Use of Reverse Micellar Systems for the Extraction and Purification of Bromelain from Pineapple Wastes *J. of Bioresources Tech.* **99 (11)** 4896-4902 doi:10.1016/j.biortech.2007.09.038.

[6] Müller Z O 1978 Feeding potential of pineapple waste for cattle *Revue Mondiale de Zootechnie* **25** 25-9

[7] Pereira E S, Regadas Filho J G L, Freitas E R, Neiva J N M and Candid M J D 2009 Energetic value from by-product of the Brazil agroindustria *Archivos de Zootecnia* **58** 455–58

[8] Nurhayati 2013 Penampilan Ayam Pedaging yang Mengkonsumsi Pakan Mengandung Tepung Kulit Nanas Disuplementasi dengan Yoghurt *Agripet* **13** (**2**) 15-20

[9] Makinde O A, Odeyinka S M and Ayandiran S K 2011 Simple and quick method for recycling pineapple waste into animal feed *Livest. Res. for Rural Develop*. **23** (**9**)

[10] Sruamsiri S, Silman P and Srinuch W 2007 Agro-industrial by-products as roughage source for beef cattle: Chemical composition, nutrient digestibility and energy values of of Ipil–Ipil leaves. *Mj. Int. J. Sci. Tech.* **9** 88-94

[11] Aboh A B, Zoffoun G A, Djenontin A J P, Babatounde S and Mensah G A 2013 Effect of graded levels of dry pineapple peel on digestibility and growth performance of rabbit *J. of Appl. Biosci*. **67** 5271-276

[12] Hepton A and Hodgson A S 2003 *Processing*. In: *The Pineapple: Botany, Production and Uses*. Bartholomew, D. P., R. E. Paul and K. G. Rohrbach (Eds), CABI Publishing.

[13] Hemalatha R and Anbuselvi S 2013 Physiochemical constituents of pineapple pulp and waste *J. Chem. Pharm. Res.* **5** (**2**) 240-42

[14] Bardiya N, Somayaji D and Khanna S 1996 Biomethanation of banana peel and pineapple waste *Bioresource Tech*. **58** (**1**) 73-76

[15] Hutagalung R I, Webb B H and Jalaludin S 1973 Evaluation of agricultural products and by-products as animal feeds 1. The nutritive value of pineapple bran for chicks *Malaysian Agric. Res.* **2** 39–47

[16] Fapohunda J B, Iji O T, Makanjuola B A and Omole A J 2008 Effect of different levels of dry pineapple waste in the diet of growing rabbits *Proc. 33rd Annual Conf. Nig. Soc. Anim. Prod.* p. 195–98

[17] Adeyemi O A, Ajado A O, Okubanjo A O and Eniolorunda O O 2010 Response of growing rabbits to graded levels of fermented and unfermented pineapple peel *Ejeafche*. **20** 898-909

[18] Lamidi A W, Fanimo A O, Eruvbetine D and Biobaku W O 2008 Effect of graded levels of pineapple (*Ananas comosus* L. Meer) crush waste on the performance, carcass yield and blood parameters of broiler chicken *Nig. J. Anim. Prod.* **35** 40-7

[19] Olosunde A O 2010 *Utilization of pineapple waste as feed for West African Dwarf (WAD) sheep*. M.Phil Thesis. Obafemi Awolowo University, Ile-Ife, Nigeria.

[20] Nurhayati and Nelwida 2014 Quality of Agricultural by Products Fermented by Trichoderma harzianum *Agripet* **14** (**2**)

[21] Iyayi E A and Aderolu Z A 2004 Enhancement of the feeding value of some agro-industrial by-products for laying hens after their solid state fermentation with *Trichodema viride Afr. J. Biotechnol.* **3** (**3**) 182-85

[22] Fasuyi A O 2005 Maize-sorghum based brewery by product as an energy substitute in broiler starter. Effect on performance, carcass, characteristics organ and muscle growth *Int. J. Poult. Sci.* **4** (**5**) 334-38

[23] Howard R L, Abotsi E, Jansen van Rensenburg E E L and Howard S 2003 Lignocellulose biotechnology: issues of bioconversion and enzyme production *Afr. J. Biotechnol.* **2**(**12**) 602-19

[24] Omojasola P F, Jilani O P, Ibiyemi S A 2008 Cellulase production by some fungi cultured on pineapple waste *Nature and Sci.* **6** (**2**) 64-79

[25] Femi-Ola T O, Oluyege J O and Gbadebo A O 2009 Citric acid production from pineapple waste *Continental J. Microbiol*. **3** 1-5

[26] Kareem S O, Akpan I and Alebiowu O O 2010 Production of citric acid by *Aspergillus niger* using pineapple waste *Malaysian J. Microbiol*. **6**(**2**) 161-65

[27] Correia R T P, McCue P, Magalhaes M M A, Macedo G R and Shetty K 2004 Production of phenolic antioxidants by the solid-state bioconversion of pineapple waste mixed with soy flour using Rhizopus oligosporus *Process Biochem*. J. **39** 2167-4902.

doi:10.1016/j.procbio.2003.11.034

[28] Dwidjoseputro D 1981 *Dasar-dasar Mikrobiologi* Cetakan ke-5 Penerbit Jambatan

[29] Snedecor G W and Cochran W G 1967 *Statistical Methods* 6th Ed. Iowa State Univ. Press, Ames, IA.

[30] Omwango E O, Njagi E N M, Orinda G O and Wanjau R N 2013 Nutrient enrichment of pineapple waste using Aspergillus niger and Trichoderma viride by solid state fermentation. *African J. of Biotech.* **12 (43)** 6193-196

[31] Raimbault M J 1998 General and microbiological aspects of solid substrate fermentation *Electronic J. Biotechnol.* **1** (**3**) 395-99

[32] Nair S C, Sindhu R, Shashidhar S 2008 Fungal xylanase production under solid state fermentation and submerged fermentation conditions *Afr. J. Microbiol. Res.* **2** 082-86

[33] Mandey J S, Leke J R, Kaunang W B and Kowel Y H S 2015 Carcass yield of broiler chickens fed banana (*Musa paradisiaca*) leaves fermented with *Trichoderma viride J. of the Indonesian Trop. Anim. Agric.* **40(4)** 229-33

[34] Sibbald I and Price K 1977 The effects of level of dietary inclusion and of calcium on the true metabolizable energy values of fats *Poult. Sci.* **56** 2070-078

[35] Shafey T M and McDonald M W 1991 The effects of dietary calcium, phosphorus, and protein on the performance and nutrient utilization of broiler chickens. *Poult. Sci.* **70** 548-53

[36] Mutucumarana R K, Ravindran V, Ravindran G and Cowieson A J 2014 Influence of dietary calcium concentration on the digestion of nutrients along the intestinal tract of broiler chickens. *J. Poult. Sci.* **51** 392-401

Motivation of farmers to cultivate organic rice in Central Java

T Dalmiyatun, B T Eddy, W Sumekar, D Mardiningsih
Department of Agriculture, Faculty of Animal and Agricultural Sciences, Diponegoro University

Corresponding Author : tutik.dalmiyatun@undip.ac.id

Abstract. The consumer's need for organic agricultural products increases sharply along with awareness of health, lifestyle and environmental concern. This research was intended to determine the relationship between social factors and the motivation of farmers for cultivating organic rice in Central Java. The research has been done by survey to farmers groups at three regions i.e. Semarang, Sragen and Demak. The determination of the location was carried out by means of purposive i.e. farmer groups that conduct organic rice cultivation (not semi organic). The determination of the sample was conducted purposively for a number of 50 people each regencies. Data were analyzed descriptive analysis and rank Spearman correlation analysis. The results showed that social factors include age, cultivated area, education, farming experience have correlation with motivation. education and cultivated area of land has a fairly close relation with correlation value 0,463% and 0,242%. Based on the motivation level, 33% of farmers have high motivation, motivation of farmers varied but most of them, 54% of total farmers stated that the motivation to cultivate organic rice is the quality of organic rice products and high income.

Keywords: Motivation, social factor, farmer, cultivate organic rice

1. Introduction

Indonesia is an agrarian country and agricultural development is a top priority in regional economic development. Indonesia has a great opportunity to become a world organic food producer. Indonesia has tropical agricultural land with diverse commodity, and the availability of abundant organic material [5] [14]. The development of organic agriculture in Indonesia began in the early 1980s marked by increasing organic agricultural land, and the number of organic producers. Organic farming is a healthy farming system by avoiding chemical fertilizers and chemical pesticides to avoid soil and water contamination and agricultural production in particular. [7] Moreover, organic farming also keeps the balance of ecosystems and natural resources directly involved in the production process. People are beginning to realize the importance of food quality, not just food in terms of quantity.

Public awareness has become a great opportunity for the development of organic farming. The main areas of organic rice producers in Indonesia are Central Java, East Java, West Java, and Yogyakarta. Rice is one of the agricultural commodities that become the basic needs of people in Indonesia, especially in Java. Central Java province has an important role because until now the province of Central Java is still a province of national food buffer. Organic farming has benefits in improving soil fertility, preventing soil erosion by wind and water, improving infiltration and soil retention capacity, reducing surface and ground water consumption and continuous soil salinization, and reducing water contamination surface and ground water [3]. The selling price of organic agricultural products is higher compared to conventional agricultural products besides the lower production cost of organic farming, so farmers get more profit.[8]. Some advantages of organically cultivating rice are consumer health; the use of organic fertilizers that restore soil fertility and environmental sustainability; and increasing farmer's income, because the selling price is higher than conventional rice [7]. Therefore, with the attractive price level, farmers will be moved and motivated to develop organic farming. The productivity and price of organic rice is higher than conventional rice.

Organic farming business in Indonesia still has a great opportunity. Indonesia has big population to be a great potential as a consumer of organic products. Although not all Indonesian can afford to buy organic agricultural products, because the price of organic agricultural products is usually quite expensive. Based on Indonesian Organic Agricultural Statistics showed the opportunity of product organic has started to be widely utilized, it proved there is an increase of organic farming land of Indonesia [1].

Organic rice has been cultivated by several farmer groups. The development of producers and organic commodities are caused by the influence of people's lifestyles consumers are attention to the importance of health and the environment. Although organic farming has many benefits but there are many farmers still cultivate conventional farming systems. The consideration in organic rice farming generally the initial productivity will decrease first before soil conditions become stable and fertile again. The stability process of each plant is different. For rice crops, production fell 3-4 times the growing season. However, ideally the land recovery period can be up to 2 years. After reaching the critical point of land recovery, rice productivity will increase, and may even surpass productivity over conventional rice. Based on the existing problems, this study aims to determine the variety and level of motivation of farmers in organic rice cultivation and to analyze social relationships include age, land area, education, farming experience, number of families with the motivation of farmers in the application of organic rice farming system.

2. Research methods

This research was conducted from May to June 2017 in Semarang, Sragen and Demak regencies. The sampling method used simple random sampling [9] [12]. Population in this research were 150 farmers who have cultivated organic rice. Research variables include age, land area, education, farming experience, and number of families and farmer motivation in the application of organic rice farming system. This research also describe diverse of farmers and the level of motivation. The type of research used descriptive analysis then the data was analyzed using rank spearman correlation analysis. The formula that can be used to calculate the correlation between two variables was as follows :

$$r_{xy} = \frac{\Sigma xy}{\sqrt{\sum x^2 \sum y^2}} \quad \text{..........[15]}$$

r is the correlation coefficient, X is a social factor, and Y is motivation

The hypothesis used is:

H0 = There is no real relationship between variables

H1 = There is a real relationship between variables

Basic decision-making test is as follows:

Accept Ho, if P value> 0,05; meaning there is no significant relationship between variables. Reject Ho, if P value <0.05; meaning there is a significant relationship between variables.

3. Results and Discussion

Motivation is the set of forces that cause people to behave in certain ways. Farmers have choosen cultivate organic farming must be have a reason, some considerations may be related to social factors of farmers, i.e. age, land area, education, farming experience, number of families. The classification of respondents by age, education, and the number of families, land area, farming experience, can be seen in Table 1. Then diverse of motivation farmers can be seen in Table 2.

Table 1. Number and percentage of respondents based on several social characteristics

Categories	Information	Number of people	Percentage (%)
Age	<30	4	2,7
	31 – 40	29	19,3
	41 – 50	54	36
	51 – 60	41	27,3
	>60	22	14,7
Education	Elementary school (not graduated)	34	22,7
	Elementary school	54	36
	Junior high school	25	16,7
	Senior high school	26	17,3
	Bachelor degree	11	7,3
Number of Family	0-1	30	20
	2-3	86	57,3
	4-5	20	13,3
	5-6	10	6,7
	>6	4	2,7
Land area	<1000m^2	4	2,7
	1000-1499 m^2	10	6,7
	1500-1999 m^2	20	13,3
	2000-2499 m^2	86	57,3
	>2500 m^2	30	20
Farming experience	<5 years	21	14
	6-10 years	21	14
	11-15 years	14	9,3
	16-20 years	32	21,3
	>20years	62	41,3

Table 1 shows that, 36 % of farmers have 41-50 years old. Based on age criteria shows that organic rice farmers in Central Java are dominated by farmers with productive age. Farmers with productive age allows farmers to accept both science and technology. The ability to receive knowledge and technology expectations can be aligned with increasing awareness and motivation of farmers to be able to develop organic rice farming business.

36% of the respondents (54 farmers) have only primary school education. Based on education criteria can be seen that the farmer's education are still low. 62 farmers or 41.3% respondents have a long experience of farming (> 20 years). Farmers are a hereditary work of the family, many of whom learn farming from helping their parents. Some farmers get knowledge from learning by doing and rely on farming experience as a teacher in running an agricultural business. Farmers have higher income levels when cultivate organic rice than conventional rice farming. This is consistent with the opinion that organic farming provides greater benefits and has a significant effect on farmer's income [2][11]. The results of previous research in several regions in Indonesia on organic rice showed farmers who cultivate organic rice have higher income than conventional rice [6] [4]. In Sragen Regency R/C ration for organic rice farming was 2.83 and for non-organic rice farming was 1.81 [8].

Farmers average have 2-3 people in the family (57.3%) or 86 respondents. Number families are family members who live in one house and rely on daily needs. Based on the land area, 86 farmers or 57.3% of respondents have own land about 2000-2499 m2. This indicates that the land area owned by farmers are still large and it is enough to be able to produce food to meet the daily needs and can be sold to cooperatives or consumers.

Table 2. Percentage of motivation farmers to cultivate organic rice.

Categories	Frequency	Percentage
Quality of product	42	28
High income	39	26
Ease to cultivate	19	12.66667
Ease Access to Marketing	22	14.66667
Group support	19	12.66667
Government support	9	6
	150	100

Motivation of farmers are diverse. 42 farmers stated that the reason farmers choose to cultivate organic rice because the quality of organic rice products is better than conventional rice. Farmers also stated that organic rice taste better, *pulen* and not easily stale. In addition, organic rice is also healthier because it is free from chemical fertilizers and pesticides. Crops yields will self consumption by the farmers and will sold to distributor and customer. Motivation of Farmers to cultivate organic rice because the price of products is higher than conventional rice. It is expected that the income will also be higher than conventional rice. Organic farming give benefit to farmers. They get benefit from organic farming because it can increase revenue by 20-30% [8]. Farmers stated that the constraints experienced during the transition from conventional to organic rice due to adjustments of about 2-3 years, at which time production will decline slightly. After reaching the critical point of land recovery, rice productivity will increase, and productivity higher than conventional rice.

Organic rice cultivation is technically different from conventional rice. 19 farmers stated the reason for choose organic rice cultivation because of the ease of cultivation, organic rice does not use chemical fertilizers and chemical pesticides. Farmers do not need to buy fertilizer because organic fertilizer produced by farmers themselves. This can reduce production costs so farmers prefer organic rice. The benefits of organic rice cultivation include consumer health, the use of organic fertilizers that restore soil fertility, and environmental sustainability, and increase farmers' income because the selling price is higher than non-organic farming. Some research suggests that organic farming provides greater benefits and it has a significant effect on farmers income [2][11].

Based on organic rice marketing, farmers stated that they have no difficulty in access to market organic rice products, it can be seen Table 2. 14,67% of farmers stated there is convenience in marketing. There are many retailers to distribute farmers' products to consumers. The most organic product' consumers were found in Semarang regency.
Group support and government support were sufficient. Groups in Semarang and Kabupaten Demak are institutionally managed well than in Sragen. Many activities organized by the group to support the development of farming business, for example training or joint discussion and other activities. Government support is also perceived as a sufficient category, including counseling, participating in

food exhibitions and some similar activities. Group support and government support are important to developing the farmers [13].

Table 3. Percentage of respondents by level of motivation

Categories	Interval skore	Frequency	Percentage
Very high	17-20	32	21,4
High	13-16	47	31,3
Medium	9-12	36	24
Lower	5-8	26	17,3
Very low	0-4	9	6
Total		150	100

Table 3 shows that 47 farmers (31.3%) have high motivation, and only 6% have very low motivation level. Motivation is seen as everything that inspires human actions including aspirations or intentions in behavior [10]. There were various levels of organic rice farmers motivation in Table 3. Motivation in carrying out organic farming is very diverse. Farmers were motivated to run organic farming because they already felt the benefits from financial factors and group support.

Table 4. Results of correlation between socio-economic factors with farmer motivation

No	Variabel	Korelasi	Signifikansi	Result
1	Age	-0,175*	0.032	Have correlation
2	Land area	0,242**	0.003	Have correlation
3	education	0,463**	0,000	Have correlation
4	Farming experience	-0,187*	0,022	Have correlation
5	Number of families' member	-0,072	0,380	No correlation

Table 4 is the result of the correlation analysis. Table 4 shows that age, education, farming experience, land area have correlation with farmer's motivation level for cultivating organic rice. However the level of the relationship were still weak. It can be seen from the correlation value. Number of families have no significant relationship, this means the number of family does not become the basis or the reason of farmers to cultivate organic rice.

4. Conclusion

Based on the results, it can be concluded that age, cultivated area, education, farming experience, have significant correlation with farmer's motivation. Education and cultivated area of land which has strongest relation. Based on the motivation level, 33% of farmers have high motivation, motivation of farmers varied. Most of the farmers stated that the motivation to cultivate organic rice are the quality of organic rice products and high income.

ACKNOWLEDGMENT

This research was supported and funded by the Research Scheme of "Penelitian unggulan" the Faculty of Animal and Agricultural Sciences (PNBP 2017).

IOP Conf. Series: Earth and Environmental Science **102** (2017) 012043 doi:10.1088/1755-1315/102/1/012043

References

[1] Ariesusanty, L., S. Nuryanti, R. Wangsa. 2010. Statistik Pertanian Organik Indonesia. AOI. Bogor.

[2] Da Costa, Anna. 2012. Can Organic Farming Enhance Livelihoods for India's Rural Poor? guardian.co.uk.http://www.guardian.co.uk/global-development/poverty-matters/ 2012/mar/15/ organic-farming-india-rural poor 15 March 2017

[3] IFOAM, 2008. Statistics of Organic Agriculture Indonesia. www.ifoam.org. access on 25 april 2017

[4] Inawati, L. 2011. Quality and Market Manager of the Indonesian Organizational Alliance (AOI), semiloka "Promoting Organic Farming in Indonesia: Opportunities and Challenges in the future". Yayasan Bina Sarana Bhakti in Cisarua, Bogor, West Java.

[5] Lenny Siahaan. 2009. Development Strategy of Organic Rice in Sisandi Farmer Group, Baru Baru Village, Toba Samosir Regency, North Sumatera. tessis. IPB. Bogor.

[6] Mansor, N. dan Azman Che Mat. 2010. The Significance of Psycholody and Environment Dimension for Malaysian Muslim Woman Entrepreneurship Venturing, International *Journal of Human Science*. Vol **7** (1), 253-269

[7] Mayrowani, H. 2012. Development of Organic Agriculture in Indonesia. Agro Economic Research Forum, Volume 30 (2), 91-108

[8] Mulyaningsih, A. 2010. Rice Production Analysis of Organic Rice Farming Method SRI (system of rice intensification); Caseuyeum Village Case Study, Haurwangi District, Cianjur District, West Java Province. Tessis. IPB.

[9] Nasution, S. 2004. *Research Methods*. Bumi Aksara. Jakarta.

[10] Rachmawati, N, Triyono and Sriyadi. 2015. Entrepreneurship Motivation Organic Rice Farmer In Bantul Regency.. Social economi Agricultural (SEPA) Journal. Vol 12 no 1. Hal 19-28

[11] Rahmawati, D. Awalia, M. M. Mustadjab, Fahriyah. 2012. Efforts to Increase Farmers' Income through the Use of Organic Fertilizer. Case Study on Maize Farmers in Surabayan Village, Sukodadi District, Lamongan District. Brawijaya University. Malang.

[12] Silalahi, Ulber. 2009. *Methods of Social Researches*. Bandung: PT. Refika Aditama.

[13] Sucihatiningsih, DWP dan Waridin.2010. Strengthening institutional capacity Agricultural extension models in improving performance through agricultural transaction costs Journal of Development Economics (JEP), Vol. 11 No. 1, pp. 13-29.

[14] Sutanto, S. 2002. Organic agriculture. Toward Alternative and Sustainable Agriculture. Penerbit Kanisius, Yogyakarta.

[15] Usman, H. dan R.P.S. Akbar. 2003. Introduction to Statistics. PT Bumi Aksara, Jakarta

Yield stability of soybean promising lines across environments

A Krisnawati and M M Adie

Indonesian Legumes and Tuber Crops Research Institute, Jl. Raya Kendalpayak KM 8 Malang 65101, East Java, Indonesia

Email: my_ayda@yahoo.com

Abstract. Soybean is grown in the wide range of environments in Indonesia, hence the genotype and environment interaction is become one of major complication in the breeding programs. The objective of this study was to evaluate the stability performance of soybean promising lines in eight locations during the dry season 2015. A completely randomized block design with four replicates was used to arrange the experiment. An additive main effects and multiplicative interaction (AMMI) model was used to examine the performance of yield stability. Soybean seed yield was affected by genotype (9.23%), environment (45.23%) and their interaction (45.62%). The first two principal component axes (PCA 1 and PCA 2) were significant ($p < 0.01$) and cumulatively contributed to 77.41% of the total variation. Based on AMMI multivariate method, two genotypes (G511H/Anj-1-6 and G511H/Anj-1-4) with seed yield of 2.67 and 2.74 t/ha, respectively, were identified as high yielding and stable promising lines across environments.

Keywords: environment, genotype, soybean, yield stability

1. Introduction

Soybeans [(*Glycine max* (L.) Merr.] in Indonesia is grown over wide environments with different environmental conditions, such as soil characteristics, mean temperature, and rainfall. The large variation in soybean yield that appears as a result the genotype by environment interaction (GEI) [1,2]. The presence of significant GEI for quantitative traits such as yield will lead to the lack of consistency in genotype performance across environments, hence may limit the progress of selecting superior genotypes [3]. It is therefore important to study the GEI to identify the yield contributing traits and yield stability [4,5].

Various methods have been proposed to explore the performance of soybean genotypes and their interaction with the environment. Currently, the additive main effects and multiplicative interaction analysis (AMMI) have been widely used to quantify the genotypic effects of the GEI [6,7,8,9,10]. AMMI biplot analysis is defined as a useful tool to explore the GEI patterns [11] with simple graphic representation of genotypes and environments in a multivariate biplot [12]. Furthermore, it captures a large portion of the GEI sum of squares and cleanly separates main and interaction effects, and provides valuable interpretation of the data [13].

There are two basic AMMI biplots, i.e. AMMI 1 and AMM2 biplots. The AMMI1 biplot plots the main effects (genotype mean and environment mean) against IPCA1 scores for both genotypes and environments. The AMMI 2 biplot, e the scores of IPCA1 are plotted against IPCA2 scores [14]. The application of AMMI biplot has been widely used in soybean crops. Evaluation of soybean lines in five environments during three seasons in Ugandarevealed the presence of a scale GEI for soybean seed yield [15]. Similarly, a significant GEI also obtained for seed yield and stability of six soybean genotypes at ten locations in Nigeria [6]. Furthermore, an evaluation on combined yield of the 24 soybean genotypes across 15 environments in Ugandan mega-environmentshad found a genotype (Maksoy 3N) with the highest mean yield [16].

Multi-environmental trials (METs) are usually used to evaluate the varietal performance [17] and to identify the most superior genotype for wide or specific cultivated zones based on yield and stability [5]. A differential response of GEI is frequent in METs [18] and must be estimated and considered when indicating cultivars for breeding program [5]. Therefore, the study on GEI and identification of yield stability is important for breeding new cultivars. The objective of this study was to evaluate the stability performance of soybean promising lines.

2. Materials and Methods

2.1. Genetic materials

The genetic materials consisted of seven soybean promising lines and two check varieties (Table 1). The check varieties (Anjasmoro and Grobogan) are high yielding and large seeded size.

Table 1. Genotype and genotype code in soybean multi location trials at eight locations. 2015

Code	Genotype	Pedigree
G1	MYT-11-2	G511H /Anjasmoro//Anjasmoro ///Anjasmoro
G2	MYT-1-6	G511H /Anjasmoro
G3	MYT-7-1	G511H /Anjasmoro//Anjasmoro ///Anjasmoro
G4	MYT-5-1	G511H /Anjasmoro//Anjasmoro
G5	MYT-2-1	G511H /Argomulyo//Argomulyo
G6	MYT-1-4	G511H /Anjasmoro
G7	MYT-1-2	G511H /Anjasmoro
G8	Anjasmoro	-
G9	Grobogan	-

2.2. Field trials

The field trials were conducted at eight locations of soybean production centers with different soil types, mean seasonal rainfall, and altitude. The detail description of test locations is presented at Table 2. The study was conducted during the dry season 2015. The first dry season was started from February to May 2015, and the second dry season was started from June to September 2015.

Table 2. Location, code and characteristics of the multi-environment trials. 2015

No.	Location	Code	Soil Type	Land Type	Climate[a]	Altitude (masl[b])
1.	Jati Kampir, Bagor, Nganjuk, East Java	E2	Regosol	Lowland	E	58
2	Kedunguneng, Bangsal, Mojokerto, East Java	E3	Gray Grumosol	Lowland	C3	72
3	Binangun, Binangun, Blitar, East Java	E4	Alluvial	Upland	C3	355
4	Sumber Banteng, Kejayan, Pasuruan[c], East Java	E5	Brown Mediteran	Upland	E	124
5	Sumber Banteng, Kejayan, Pasuruan[d], East Java	E6	Brown Mediteran	Lowland	E	124
6	Budeng, Jembrana, Jembrana, Bali[d]	E8	Alluvial	Lowland	D2	174
7	Berabah, Kediri, Tabanan, Bali	E9	Latosol	Lowland	D3	20
8	Segara Anyar, Pajut, Central Lombok, West Nusa Tenggara	E10	Regosol	Lowland	C3	24

[a]climate type based on Oldeman (1974) climate classification system: C3 = 5-6 wet months and 4-6 dry months, D2 = 3-4 wet months and 2-3 dry months, D3 = 3-4 wet months and 4-6 dry months, E = < 3 wet months
[b]masl = meter above sea level.
[c]planted during the first dry season
[d]planted during the second dry season

2.3. Experimental design and plant cultivation

The experiment was arranged in a completely randomized block design with four replicates in each location [2]. Each line was planted using 2.0 m × 4.5 m plot size, 40 cm × 15 cm plant spacing, two seeds per hill. Plants were fertilized with 50 kg ha^{-1} Urea, 100 kg ha^{-1} SP36 and 75 kg ha^{-1} KCl which applied prior sowing the seeds. Weeds, pests and diseases were intensively controlled. Irrigation was made to keep the optimum soil moisture.

2.4. Observation and Data Analysis

Parameters observed included plant height, days to flowering, days to maturity, number of branch per plant, number of node per plant, number of filled pod, 100 seed weight, and seed yield (t ha^{-1}) [14]. The data were subjected to combined analysis of variance (ANOVA). SAS ver. 9.1.3 [19] was used to perform stability analysis on seed yield by using AMMI method.

3. Result and Discussion

The analysis of variance for seed yield and yield components of the seven soybean promising lines and two check varieties tested in eight environments revealed that environment, genotype, and genotype × environment interactions (GEI) were highly significant for most all of the observed traits, except for the number of branches per plant and number of empty and filled pods (Table 3). The seed yield of nine genotypes were significantly affected by environment, genotype, and GEI. The coefficient of variation (CV) for seed yield was 12.34%, indicating the precision of experimental research and control of environmental factors [8]. The highly significant genotype indicates the different performance of genotypes across environments [5]. A research which evaluated for seed yield and stability of six soybean genotypes at ten sites also obtained significant GEI [6].

Table 3. Analysis of variance for yield and yield components of seven soybean promising lines and two check varieties tested in eight environments. 2015

Parameter	Mean Square				CV[a] (%)
	Replication|R	Environment (E)	Genotype (G)	G × E	
Days to flowering (day)	0.68[ns]	4.01**	129.88**	0.01**	2.18
Days to maturity (day)	1.67*	81.26**	126.27**	3.39**	1.21
Plant height (cm)	68.01*	2589.54**	2352.82**	127.32**	10.72
Number of branches/plant	0.09*	0.58**	0.52**	0.06[ns]	16.78
Number of node/plant	7.869[ns]	54.88**	72.36**	12.75*	22.68
Number of filled pod/plant	90.12[ns]	1029.63**	632.07**	134.06[ns]	21.25
Number of empty pod/plant	0.28**	2.18**	0.20[ns]	0.15[ns]	25.57
100 seed weight (g)	4.28**	51.63**	98.05**	4.65**	7.71
Seed yield (t/ha)	0.19**	2.32**	0.54**	0.37**	12.34

[a]CV = coefficient of variation
* = significant at 5 % probability level ($p < 0.05$) ** = significant at 1 % probability level ($p < 0.01$), ns = not significant

The AMMI analysis of variance of seed yield showed that 45.23% of the total sum of squares was due to environmental effects, 9.23% to genotypic effects, and 45.62% to GEI effects (Table 4). This result revealed that the environment and GEI were the most important source of yield variation, due to their large contribution to the total sum of squares. Similar findings were also obtained in previous studies [20,21], which showed environments has greater proportion of total yield variation followed by GEI and genotypes. However, another research by [22] on twenty soybean genotypes tested at six environments found the GEI as the most important source of variation.

Through AMMI model, the GEI had partitioned into the first two significant IPCAs with contributions of IPCA1 (58.29%) and IPCA2 (19.12%). The remaining residuals were not significant (Table 4). The IPCA1 and IPCA2 cumulatively captured 77.41% of the total GEI sum of square, thus the interaction of the nine genotypes in eight environments can be predicted by the first two principal components of genotypes and environments. This result is in agreement with previous study [23] which stated that the most accurate model for AMMI can be predicted using the first two IPCAs. Furthermore, the overall pattern of genotypes interaction with environments was interpreted using AMMI1 and AMMI2 biplot.

Table 4. Additive main effects and multiplicative interactions analysis of variance for seed yield of the soybean genotypes across environments. 2015

Source of variation	Degree of freedom	Sum of Squares	Mean Square	Explained (%)
Total	287	77.18		
Treatment	95	46.46	0.48**	
Genotype (G)	8	4.25	0.53**	9.15
Environment (E)	7	21.01	3.00**	45.23
G × E	56	21.19	0.37**	45.62
[a]IPCA1	14	12.35	0.88**	58.29
IPCA2	12	4.05	0.33*	19.12
IPCA3	10	2.78	0.27[ns]	
Residual	20	1.99	0.33[ns]	
Pooled error	192	0.15	0.15	

[a]IPCA = Interaction Principal Component;
* = significant at 5 % probability level ($p < 0.05$) ** = significant at 1 % probability level ($p < 0.01$), ns = not significant

AMMI1 biplot shows the mean seed yield performance and adaptability of nine soybean genotypes across environments (Fig.1). In this biplot, the abscissa represents the main effects (mean) and ordinate represents the effects of the interaction (IPCA1). In this figure, the vertical line in the middle of biplot represents the grand mean ($\bar{X}$ = 2.64 t ha^{-1}). Right side of the line showed genotypes with higher yield than the grand mean, and the left side were genotypes with lower yield than the grand mean. Based on those description, thus G7, G6, G5, and G2 were generally high yielding (2.70 t ha^{-1}, 2.74 t ha^{-1}, 2.68 t ha^{-1}, 2.67 t ha^{-1}, respectively) with G6 being the overall best yield of 2.74 t ha^{-1}. In contrast, G3 G4, and two check varieties (G8 and G9) were generally have low yield.

The genotypes with IPCA1 scores close to zero exhibit general adaptation, meanwhile a more specific adaptation was represented by a larger scores IPCA1 [24]. Furthermore, the relative magnitude and direction of genotypes along the abscissa and ordinate axis in biplot is useful to describe the response pattern of genotypes across environments and also to differentiate high yielding and adaptable genotypes [4,25]. Accordingly, G3 and G5 were the most unstable and contributed the most toward GEI. Furthermore, the genotypes which are characterized by means greater than grand mean and the IPCA score nearly zero are considered as generally adaptable to all environment [26]. In this study, G9, G7, G2, and G1 were over the grand mean and located near the zero point of IPCA1, indicate those four genotypes have good adaptation across environments.

The AMMI1 biplot also describe the presence of two mega-locations [27,28]. Accordingly, three environments (E3, E6, and E7) with similar IPCA1 scores had constructed one mega-environment, meanwhile another five environments had constructed another mega-environment. However, the mega-environment classification can be explored in detail through AMMI2 biplot [28].

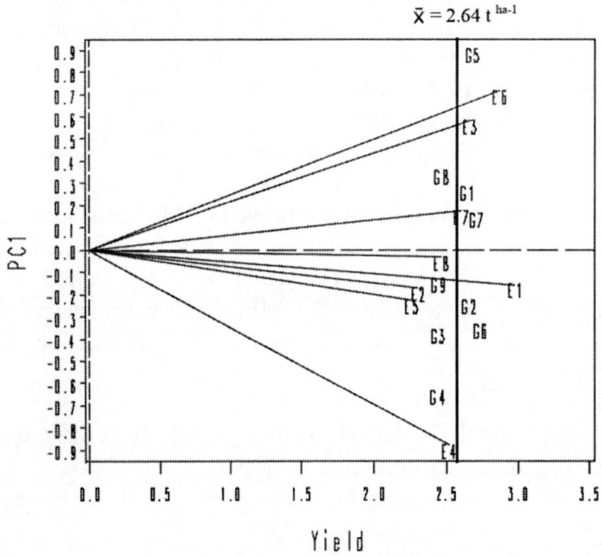

Figure1. Biplot AMMI1 (means yield vs PC1) for seed yield of seven soybean promising lines and two check varieties tested in eight environments. 2015

In AMMI2 biplot, which created using scores of genotypec and environment of the first two AMMI components (IPCA1 and IPCA2), explain the magnitude of interaction of each genotype and environment [4,26,29]. Genotypes which located near biplot origin were more stable (less interaction) compared to genotypes positioned further away (30]. In Figure 2, based on distribution of genotype points in the AMMI2 biplot, it revealed that soybean promising lines G2 and G6 scattered close to the origin, indicating less interaction (showed high stability) of these genotypes with environments. The remaining seven genotypes scattered away from the origin of the biplot showing that the genotypes were more responsive to environmental interactive forces (lower stability or large interaction).

Another research on Indonesian soybean promising lines had found four stable genotypes [2]. Similarly, a study by using AMMI had found a stable and high yielding promising line which recommended to be released as new improved soybean variety [31].

Interaction of genotypes with specific environmental conditions can be reflected by projection of each genotype to environment spokes [14,32]. For instances, the G8 had positive interaction with environments E7 and specifically adapted to this environment, whereas G3 and G9 indicated specific adaptability with E5 and E8 (Fig. 2).

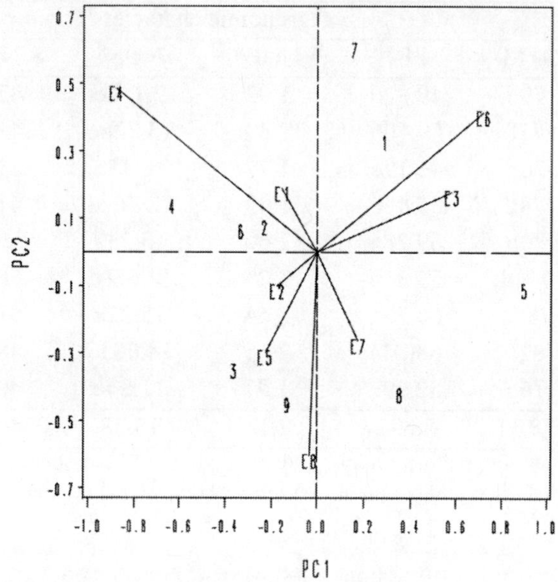

Figure2. Biplot AMMI2 (PC1 vs PC2) for seed yield of seven soybean promising lines and two check varieties tested in eight environments. 2015

The mean yield of the nine genotypes grown in eight environments was presented in Table 4. Mean yields ranged from 2.44 t ha^{-1} (G4) to 2.74 t ha^{-1} (G6). Five of the genotypes showed higher yield than the mean average yields. Among the environments, four environments (E2, E4, E5, and E8) below the mean average yields. The highest yields were recorded in environments E1 followed by E6 with mean yields of 3.03 t ha^{-1} and 2.87 t ha^{-1}, respectively.

Table 5. Seed yield of seven soybean promising lines and two check varieties tested in eight environments. 2015

Code[a]	Seed yield (t/ha)								Mean
	E1	E2	E3	E4	E5	E6	E7	E8	
G1	2.88	2.17	3.40	2.60	2.26	3.01	2.62	2.26	2.65a
G2	2.89	2.44	2.79	2.90	2.46	2.75	2.78	2.34	2.67a
G3	2.74	2.35	2.27	2.60	2.12	2.31	2.64	2.55	2.45b
G4	3.23	2.20	2.09	2.96	2.01	2.35	2.25	2.41	2.44b
G5	3.12	2.29	3.18	1.73	2.02	3.57	2.92	2.65	2.68a
G6	3.39	2.40	2.54	2.94	2.55	2.89	2.59	2.63	2.74a
G7	3.14	2.36	2.77	2.78	2.14	3.58	2.59	2.21	2.70a
G8	2.75	2.07	2.52	1.91	2.20	2.98	2.63	2.68	2.47b
G9	2.65	2.33	2.48	2.26	2.52	2.43	2.43	2.47	2.45b
Mean	3.03	2.35	2.70	2.51	2.25	2.87	2.72	2.64	2.64

[a]Genotypes and environments code refer to Table 1 and Table 2, respectively.

The performances of agronomic characters of seven soybean promising lines and two check varieties tested in eight environments were presented in Table 6. The average of days to flowering was 33 days. Seven tested genotypes have early days to maturity (< 80 days), and the rests have medium maturity (G5 and G7). The seed size of nine genotypes was categorized as large, ranged from 15.36 to 21.11 g.

Table 6. Agronomic characters of seven soybean promising lines and two check varieties tested in eight environments. 2015

Code[i]	Agronomic characters							
	DTF[a]	DTM[b]	PH[c]	NOB[d]	NON[e]	NFP[f]	NEP[g]	SW[h]
G1	34b	79c	49.82d	1.92	12.79bcd	47.85	1.85	15.63d
G2	33c	77f	59.50b	2.39	13.72bc	45.71	1.67	17.02b
G3	32d	76g	46.19e	1.77	10.73e	42.12	1.64	16.35c
G4	32d	78e	56.48b	1.49	12.36cd	41.36	1.95	17.15b
G5	35a	80b	69.28a	2.60	15.23a	50.89	2.12	15.74cd
G6	34b	79cd	53.31b	2.25	13.65bc	50.93	1.55	16.33c
G7	35a	78de	68.70a	2.64	15.22a	51.57	2.04	15.36d
G8	35a	82a	68.84a	2.15	14.00ab	48.53	1.91	15.92cd
G9	29e	76g	52.99c	1.97	11.84e	40.08	1.30	21.11a
Mean	33.13	78.31	58.91	2.13	13.28	46.56	1.78	16.73

[a]Days to flowering (days), [b]Days to maturity (days), [c]Plant height (cm), [d]Number of branches per plant, [e]Number of node per plant, [f]Number of filled pod per plant, [g]Number of empty pod per plant, [h]100 seed weight (g)
[i]Genotypes code refer to Table 1

In this study, G2 and G6 exhibit higher stability than other genotypes. The agronomic characters both of G2 and G6 were having early days to maturity (77 and 79 days, respectively), and large seed size (17.02 g and 16.33 g, respectively). Considering the mean yield, the stable promising line G6 had the highest yield (2.74 ha^{-1}). Similarly, G2 also showed a high yield (2.67 ha^{-1}) which was not significantly different to G6. Thus, those two genotypes may be recommended to be released as new improved soybean variety, or may be used as parents in crosses on breeding program.

4. Conclusion

AMMI biplot is a useful technique that was able to effectively reveal the existence of a significant GEI between nine soybean genotypes across eight environments. The stability analysis using AMMI method had successfully identified the G2 and G6 as high yielding and stable soybean promising lines across environments.

Acknowledgment
This research was supported by the Indonesian Agricultural Agency for Research and Development (IAARD), Ministry of Agriculture. We thank Arifin for his assistance during the field research.

References

[1] Susanto GWA and Adie MM 2010 Adaptability of promising soybean lines at different environmental conditions. *Penelitian Pertanian Tanaman Pangan* **29** (3): 166-170.
[2] Adie MM and Krisnawati A 2015 Soybean yield stability in eight locations and its potential for seed oil source in Indonesia. *Energy Procedia* **65**: 223–229.
[3] Fan XM, Kang M, Chen H, Zhang Y, Tan J and Xu C 2007 Yield stability of maize hybrids

evaluated in multi-environment trials in Yunnan, China. *Agron J* **99**:220–228.

[4] Sanni KA, Ariyo OJ, Ojo DK, Gregorio G, Somado EA, Sanchez I et al 2009 Additive Main Effect and Multiplicative Interaction Analysis of grain yield performances in rice genotypes across environments. *Asian J of Plant Sci* **8** (1): 48-53.

[5] Roshandel M, Pourmohammad A, Babaei HR and Shekari F 2016 Grain yield stability analysis of soybean genotypes by AMMI method. *Azarian J of Agric* **3**(6): 119-128.

[6] Amira JO, Ojo DK, Ariyo OJ, Oduwaye OA et al 2013 Relative discriminating powers of GGE and AMMI models in the selection of tropical soybean (*Glycine max* L. Merr.) genotypes. *Afr Crop Sci J* **21**: 67-73.

[7] Bose L, Jambhulkar NN and Singh ON 2014 Additive Main Effect and Multiplicative Interaction (AMMI) analysis of grain yield stability in early duration rice. *J of Animal and Plant Sci.* **24**: 1885-1897.

[8] Sousa L.B. Hamawaki O.T. Nogueira APO, Batista RO, Oliveira VM and Hamawaki RL 2015 Evaluation of soybean lines and environmental stratification using the AMMI, GGE biplot, and factor analysis methods. *Gen and Mol Res* **14**: 12660-12674.

[9] Satoto, Rumanti IA and Widyastuti Y 2016 Yield stability of new hybrid rice across locations. *Agrivita* **38** (1): 33-39.

[10] Bhartiya A, Aditya JP, Singh K, Pushpendra, Purwar JP and Agarwal A 2017 AMMI & GGE biplot analysis of multi environment yield trial of soybean in North Western Himalayan state Uttarakhand of India. *Legume Research* **40** (2): 306-312.

[11] Kaya Y, Palta C and Taner S 2002 Additive Main Effect and Multiplicative Interaction Analysis of yield performances in bread wheat genotypes across environments. *Turk J Agric For* **26**: 275-279.

[12] Yokomizo GK, Duarte JB, Vello NA and Unfried JR 2013 AMMI analysis of grain yield in soybean lines selected for resistance to Asian rust. *Pesq Agropec Bras* **48**: 1376-1384.

[13] Gauch HG 1992 *Statistical analysis of regional yield trials: AMMI analysis of factorial designs*. Elsevier Science Pub. Amsterdam, Netherland.

[14] Akter A, Hassan MJ, Kulsum MU, Islam MR, Hossain K and Rahman MM 2014 AMMI biplot analysis for stability of grain yield in hybrid rice (*Oryza sativa* L.). *J of Rice Res* **2** (2): 126- 129.

[15] Tukamuhabwa P, Asiimwe M, Nabasirye M, Kabayi P and Aphosa M 2012 Genotype by environment interaction of advanced generation soybean lines for grain yield in Uganda. *African Crop Sci J* **20**: 107-115.

[16] Obua T, Maphosa M, Tusiime G and Tukamuhabwa P 2013 Genotype main effects and genotype by-environment biplot analysis for soybean grain yield in Uganda. *African Crop Sci Conference Proc* **11**: 537 – 541.

[17] Li T, Ali J, Marcaida M III, Angeles O, Franje NJ, Revilleza JE et al 2016 Combining limited multiple environment trials data with crop modeling to identify widely adaptable rice varieties. *PLoS ONE* **11**(10)): e0164456. https://doi.org/10.1371/journal.pone.0164456

[18] Rodrigues PC, Malosetti M, Gauch HG and Van Eeuwijk FA 2014 A weighted AMMI algorithm to study genotype-by-environment interaction and QTL by-environment interaction. *Crop Sci* **54** (4): 1555-1570.

[19] SAS Institute 2011 *SAS/STAT 9.3 User's Guide*. SAS Instutue Inc: Cary, NC, USA.

[20] Bertoia LM and Aulicino MB 2014 Maize forage aptitude: Combining ability of inbred lines and stability of hybrids. *The Crop J* **2** (6): 407-418.

[21] Asfaw A, Alemayehu F, Gurum F and Atnaf M 2009 AMMI and SREG GGE biplot analysis for matching varieties onto soybean production environments in Ethiopia. *Sci Res Essays* **4**: 1322-1330.

[22] Gurmu F, Hussein M and Alemaw G 2009 Genotype × environment interactions and stability of soybean for grain iyeld and nutrition quality. *African Crop Sc J* **17** (2): 87 – 99.

[23] Gauch HG and Zobel RW 1996 *AMMI analysis of yield trials*. In Kang MS, Gauch HG, editors. Genotype-by-Environment Interaction. CRC Press, Boca Raton. New York, United States of

America.

[24] Ebdon JS and Gauch HG 2002 Additive Main Effect and Multiplicative Interaction Analysis of national turfgrass performance trials: I. Interpretation of genotype × environment interaction. *Crop Sci* **42**: 489-496.

[25] Samonte SO, Wilson LT, McClung AM and Medley JC 2005 Targeting cultivars onto rice growing environments using AMMI and SREG GGE biplot analyses. *Crop Sci* **45**: 2414–2424.

[26] Rashidi M, Farshadfar E and Jowkar MM 2013 AMMI analysis of phenotypic stability in chickpea genotypes over stress and non-stress environments. *Intl J of Agric and Crop Sci* **5**(3): 253-260.

[27] Oliviera RL, Pinho RGV, Balestre M and Ferreira DV 2010 Evaluation of maize hybrids and environmental stratification by the methods AMMI and GGE biplot. *Crop Breeding and App Biotech* **10**: 247-253.

[28] Zeleke AA and Berhanu FA 2016 AMMI and GGE models analysis of stability and GEI of common bean (*Phaseolus vulgaris* L.) lines in Ethiopia. *J of Biol Agric and Helathcare* **6**(9): 127-135.

[29] Vargas M and Crossa J 2000 *The AMMI analysis and graphing the biplot.* Biometrics and Statistics Unit, CIMMYT.

[30] Pržulj N, Mirosavijevic M, Canak P, Zoric M and Bocanski J 2015 Evaluation of spring barley performance by biplot analysis. *Cereal Res Communication* **43**(4): 692-703.

[31] Krisnawati A, Basunanda P, Nasrullah and Adie MM 2016 Genotype stability analysis of soybean using Additive Main Effect and Multiplicative Interaction (AMMI) methods. *Informatika Pertanian* **25**(1): 41-50.

[32] Osiru MO, Olanya OM, Adipala E, Kapinga R and Lemaga B 2009 Yield stability analysis of *Ipomoea batatus* L. cultivars in diverse environments. *Aust J Crop Sci* **3**(4):213-220.

IOP Conf. Series: Earth and Environmental Science **102** (2017) 012045 doi:10.1088/1755-1315/102/1/012045

Identification of soybean genotypes adaptive to tropical area and suitable for industry

M M Adie and A Krisnawati

Indonesian Legumes and Tuber Crops Research Institute, Jl Raya Kendalpayak Km 8 Malang 65101

mm_adie@yahoo.com

Abstract. Soybeans in Indonesia are mostly used for raw material of tempeh industry. This study aims to identify 150 soybean genotypes for their suitability for raw materials of tempeh and adaptability to be developed in tropical area of Indonesia. The research material consisted of 150 soybean genotypes. The field research was conducted in Malang from February to May 2016, using a randomized block design with two replicates. The identification of 150 soybean genotypes showed 30.67% of super early maturity (<75 days), 50% of early maturity (76 - 79 days), and 19.33% were medium maturity (80 - 90 days). In the group of super early maturity, 11 genotypes were yielded between 3.01 - 3.69 t/ha and the 100 seed weight ranged from 15.27 - 20.18 g. In the early maturity group, there were 23 genotypes with seed yields between 3.01 - 3.66 t/ha, and the 100 seed weight ranged from 13.90 - 20.23 g. In Indonesia, tempeh industry requires soybeans with large seed size. In this research, G511H/Anj//Anj////Anjs-8-5 was suitable to be developed in Indonesia's tropical climate and also preferred by industry for tempeh raw material due to its high yield, super early days to maturity, and large seed size.

Keywords: early days to maturity, large seed size, tempeh, tropical soybean

1. Introduction

Soybean cultivation in Indonesia is unique because it can be planted throughout the season, which starting from the beginning of the rainy season, the beginning of the dry season, and the end of the dry season. Soybean consumption in Indonesia reached 7.62 kg/ capita/year [1]. Processed soy-based food in Indonesia has large variation, for example tempeh tofu, soy sauce, milk, tauco etc, and of a wide range of soybean food products in Indonesia, tempeh and tofu are the most popular foods. Based on Susenas data [2], the level of tempeh consumption in Indonesia reached 6.95 kg/capita/year. So far, Indonesia still import soybean about 67.28% or as many as 1.96 million tons to meet the domestic demand.

Soybean characteristics are vary in each type of processed food products. Soybean for tofu raw material is more determined by characteristic of high starch content, whereas the industry prefers soybean with large seed size for the tempeh raw material. Soybean seed differences are vary between soybean production centers in the world. In Indonesia, the size of the soybean seeds is divided into three groups, i.e. small seeds (<10 g/100 seeds), medium (10 - 14 g/100 seeds), and large seeded (> 14 g/100 seeds). A research by Krisdiana [3] showed that about 93% of tempeh producers prefer soybeans with yellow seed color and large seed size, because it produced a brighter colored tempeh and larger volume. According to Ginting [4], several superior varieties of soybeans in Indonesia can deliver tempeh which have similar quality with imported ones, and also frequently showed a higher protein content. Moreover, Antarlina et al. [5] reported that soybean seed size could be a determinant factor of tempeh quality because it showed positive correlation with tempeh weight (r = 0.86**) and tempeh volume (r = 0.95**). The relationship between seed size and tempeh appearance was also reported by Astawan et al. [6] that the tempeh produced from Grobogan soybean varieties (large seed size) has the similar water, protein, and oil content with those of produced from imported soybeans which have seed size of 16-18 g/100 seeds. Research on four Brazilian soybean cultivars conducted by Bavia et al. [7] showed that the protein in tempeh increase from 36.16% of the seed protein to 43.16% in soybean protein. Thus, the processed soybean for tempeh become a potential source of protein for humans.

Soybean seed size as a determinant of quality of tempeh, was genetically controlled. Thus, an effort to increase seed size through genetic modification will provide a high chance of success. Liang et al. [8] reported that seed morphological traits, such as seed length, seed width, and seed height, were associated with seed weight and also affect the morphological quality of the seeds. The development of soybean for the provision of raw material of tempeh industry for tropical condition of Indonesia, is recommended not only based on seed size but also pay attention on the character of early days to maturity. This is due to soybeans are planted three times a year in Indonesia. Krisnawati & Adie [9] did simultaneously selection on both of large seed size (>15.8 g/100 seeds) and high seed yield (>2.53 t/ha) and obtained eight selected genotypes. Vollman et al. [10] stated that selection of early maturing soybean genotypes with improved seed protein content appears to be feasible and was limited by the moderately negative correlation between protein content and seed yield. Various studies on plant age have been widely reported. It was also reported that an increase in the number of flowers at the beginning of the flowering phase will potentially increase the number of pods and seeds at maturity phase [11].

The availability of soybean cultivars which suitable for raw material of tempeh and also adaptive in tropical area are important to meet the users' preference. The objective of the research was to identify 150 soybean genotypes for their suitability for raw materials of tempeh and adaptability to be developed in tropical area of Indonesia.

2. Methodology

2.1. Research material

The research materials were 150 soybean genotypes which consisted of 145 genotypes as a result of crosses, and five soybean varieties (Grobogan, Burangrang, Anjasmoro, Argomulyo and Dega 1).

2.2. Field study

The research study was conducted in Kendalpayak Research Station (Malang, East Java, Indonesia) which located at 8° 049645" South Latitude until 112° 625592" East Longitude, 445 m above sea level, C2 climate type of Oldeman system, and the 2191 mm/year of mean annual rainfall intensity. The study was performed from February to May 2016 in the lowland (paddy field) after rice planting. The experiment was arranged in randomized complete block design with two replicates. The plot size was 1.2 × 4.5 m, 40 cm × 15 cm plant spacing, two seeds per hill. Plants were fertilized by 250 kg Phonska/ha + 100 kg SP36, and 1 t/ha organic fertilizer which applied entirely during planting. The land used in the study was wetland from former rice paddy field. The drainage channel was made prior planting to maintain the optimum soil moisture. The weed, pest, and disease were controlled optimally. The plant harvest was done when 90% pods reach maturity.

2.3. Observation

Parameter observed was included seed yield, days to maturity, 100 seed weight, and the supporting characters which consisted of plant height, number of branches, number of nodes, number of filled pods, and number of empty pods [9].

3. Result

The performance of main characters (seed yield, days to maturity, and 100 seed weight) and the supporting characters (plant height, number of branches, number of nodes, number of filled pods, and number of empty pods) were significantly influenced by the differences of genotypes used (Table 1). It indicates that each soybean genotype produce different value of main and supporting characters.

Table 1. Analysis of variance for main and supporting characters of 150 soybean genotypes. Malang. 2016.

Character	Mean Square		CV (%)
	Replication	Genotype	
Main characters:			
Seed yield (t/ha)	28.5825 **	0.3447 **	18.46
Days to maturity (day)	88.5633 **	11.8503 **	2.50
100 seed weight (g)	10.3788 **	5.5558 **	7.52
Supporting characters:			
Plant height (cm)	1992.8021 **	112.7910 *	16.25
Number of branches/plant	1.7710 ns	1.0804 **	29.09
Number of node/plant	1.6133 ns	3.1714 *	14.51
Number of filled pod/plant	1.4145 ns	193.4303 **	23.29
Number of empty pod/plant	0.0385 ns	2.5096 **	37.47

** = significant at 1% probability level ($p < 0.01$), * = significant at 5% probability level ($p < 0.05$), ns = not significant, CV = coefficient of variation.

The descriptive statistic showed that the seed yield of 150 genotypes was ranged from 0.91 – 3.08 t/ha (average of 2.31 t/ha). The days to maturity was vary from 73 – 84 days (average of 77 days), whereas for 100 seed weight was between 13.07 to 22.65 g (average of 16.60 g/100 seeds) (Table 2). Those three characters are classified as main characters to support soybean development in the tropical area of Indonesia for tempeh raw material.

The supporting characters, i.e. plant height was ranged from 33.70 – 74.00 cm (average of 55.45 cm), the range of number of branches was 0.70 – 5.30 branches (average of 2.88 branches/plant), the range of number of nodes was 7.50 – 15.50 nodes/plant (average of 10.19 nodes/plant). Character of number of pod/plant was divided into number of filled pod (which ranged from 22.70 – 81.80 filled pods and with an average of 39.89 pods/plant) and number of empty pods which ranged from 0.10 to 8.70 empty pods (average of 1.39 empty pods/plant) (Table 2). The performance of main and supporting characters in this study showed that the plant growth was optimal and the variability of morphological characters were vary between genotypes.

Table 2. Descriptive statistic of main and supporting characters of 150 soybean genotypes. Malang. 2016.

Character	Mean	Minimal value	Maximal value	Standard deviation
Main characters:				
Seed yield (t/ha)	2.31	0.91	3.08	0.41
Days to maturity (day)	77	73	84	2.43
100 seed weight (g)	16.60	13.07	22.65	1.66
Supported characters :				
Plant height (cm)	55.45	33.70	74.00	7.48
Number of branches/plant	2.88	0.70	5.30	0.73

Number of node/plant	10.19	7.50	15.50	1.26
Number of filled pod/plant	39.89	22.70	81.80	9.80
Number of empty pod/plant	1.39	0.10	8.70	1.12

The preference of tempeh industry is large seeded soybean variety. The grouping of 150 soybean genotypes based on their seed size was classified into three groups, i.e. seed size less than 14 g/100 seeds, size between 14 – 16 g/100 seeds, and seed size over 16 g/100 seeds (Table 3). Based on those grouping, this study has obtained five soybeans with seed size less than 14 g/100 seeds, 51 genotypes of sees size between 14 – 16 g/100 seeds, and 94 genotypes with seed size over 16 g/100 seeds. Those seed size are suitable for raw material of tempeh, and also have a similar size with imported variety.

The highest yield on the group less than 14 g/100 seeds was 2.52 t/ha. All genotypes in this group were have early days to maturity. In the group of 14 – 16 g/100 seeds, the highest yield was 2.93 t/ha, and the seed-size group over > 16 g/100 seeds produced the highest yield of 3.08 t/ha. On those three seed-size groups, the lowest yield was in the group with seed size less than 14 g/100 seeds. If we compared it with the performance of the supporting characters, it clearly seen that the larger seed size will resulted in more optimal supporting characters compared to those of in the group of less than 14 g/100 seeds.

G511H/Argomulyo-2 produced the highest yield in the seed-size group less than < 14 g with the days to maturity reached 76 days. The best yielded genotype in the group 14 – 16 g was G511H/Arg//Arg///Arg///Arg-30-3 (2.91 t/ha) with the days to maturity 76 days. In the seed-size group over 16 g/100 seeds, G511H/Anj//Anj///Anj////Anjs-8-5 showed the highest yield (3.08 t/ha) with 75 days of maturity (Table 4). Thus, G511H/Anj//Anj///Anj////Anjs-8-5, was identified as the best genotype, and suitable to be developed in Indonesia for raw material of tempeh.

Table 3. Grouping of 150 soybean genotypes based on seed size. Malang. 2016.

Seed weight (g/100 seed)	Descriptive statistic	SY[a]	DM[b]	PH[c]	NB[d]	NN[e]	FP[f]	EP[g]
< 14 g	Mean	2.19	76.20	55.32	3.50	10.00	39.44	0.84
(5 genotypes)	Minimal	1.70	74.00	46.80	2.20	9.00	34.20	0.40
	Maximal	2.52	79.50	68.05	5.30	11.00	51.80	1.20
	Std	0.28	2.11	7.59	1.02	0.71	6.41	0.32
14 – 16 g	Mean	2.32	77.54	57.13	3.07	10.53	43.78	1.45
(51 genotypes)	Minimal	1.62	73.50	33.70	1.60	8.00	26.80	0.40
	Maximal	2.93	84.00	74.00	4.50	13.50	77.20	4.60
	Std	0.35	2.54	8.34	0.62	1.13	10.67	0.88
>16 g	Mean	2.31	76.60	54.54	2.74	10.01	37.80	1.38
(94 genotypes)	Minimal	0.91	73.00	39.40	0.70	7.50	22.70	0.10
	Maximal	3.08	82.50	69.40	5.00	15.50	81.80	8.70
	Std	0.45	2.31	6.80	0.73	1.30	8.76	1.24

[a] Seed yield (t/ha)
[b] Days to maturity (day)
[c] Plant height (cm)
[d] Number of branches/plant
[e] Number of node/plant
[f] Number of filled pod/plant
[g] Number of empty pod/plant

Table 4. The best genotype on each seed-size group. Malang. 2016.

Grouping	Genotype	SS[a]	SY[b]	DM[c]	PH[d]	NB[e]	NN[f]	FP[g]	EP[h]
< 14 g	G511H/Argomulyo-2	13.90	2.52	76	59.50	3.40	10.50	35.90	0.80
14 – 16 g	G511H/Arg//Arg///Arg///Arg-30-3	15.84	2.91	76	66.90	4.50	11.00	48.50	0.70
>16 g	G511H/Anj//Anj///Anj////Anjs-8-5	16.92	3.08	75	56.40	2.50	10.00	32.40	0.80

[a] Seed size(g)
[b] Seed yield (t/ha)
[c] Days to maturity (days)
[d] Plant height (cm)
[e] Number of branches/plant
[f] Number of node/plant
[g] Number of filled pod/plant
[h] Number of empty pod/plant

4. Discussion

The domestic soybean demand continues to increase as a result of the increasing population of Indonesia. Tempeh is one of the processed soybean products consumed by most of the Indonesian people. The fulfilment of soybean raw materials for industrial purposes is partly imported from other countries. The physical characteristic of imported soybeans is large seed size (about 16 g/100 seeds), and the size is in accordance with the preference of tempeh industry in Indonesia. The development of soybean to meet the needs of tempeh industry in Indonesia must have the requirements: (1) adaptive and productive in the tropical environment of Indonesia, (2) large seed size (> 14 g/100 seeds), and early days to maturity (<80 days).

Identification on 150 soybean genotypes showed that the performance of main characters (yield, 100 seed weight, and days to maturity) as well as supporting characters were differ between genotypes. Soybean seed size become the determinant of tempeh quality. The grouping of 150 soybean genotypes was based on seed size which measured through 100 seeds weight. It has classified into three groups, i.e. less than 14 g, between 14 – 16 g, and over 16 g. The result showed that most of soybean genotypes (94 genotypes) were in the seed-size group over 16 g, followed by 51 genotypes in the group of 14 – 16 g, and the rest (5 genotypes) were in the group of less than 14 g. Soybeans used in this study were derived from selection result of soybean germplasm crossing which aimed to increase the seed size and early maturing characters. Seed size is essential character to optimize the yield and also play important role in adaptation of a cultivar and influence to seed vigor [12]. Hu et al. [13] studied about seed length, seed width, and seed height as seed-size traits, and their ratios of these values as seed-shape traits of 184 recombinant inbred lines and 219 cultivated soybean accessions. They reported that seed size traits, which are controlled by multiple genes in soybean, play an important role in determining seed yield, quality and appearance. Another research reported that the matured soybean seed size was simultaneously controlled by embryo, cytoplasm, and maternal effects [14]. Related to the soybean seed character, Sudaric & Vrataric [15] reported that the direct selection on higher number of seeds/plant and higher seed yield/plant would result in greater genetic advance in grain yield. The nutrient composition on soybean is determined by variety, location, climate, and the cultivation management.

The opportunity to obtain soybean cultivar adapted with tropical condition, especially in Indonesia, is widely open. In the group of 14 – 16 g/100 seeds, the range of seed yield was between 1.62 – 2.93 t/ha with days to maturity from 73 – 84 days; whereas in the seed-size group of over 16 g/100 seeds have range of seed yield between 0.91 – 3.08 t/ha and range of 73 – 82 days to maturity. If the limit of soybean productivity is determined by 3.0 t/ha, seed size > 14 g/100 seeds, and days to maturity < 80 days, then in the seed-size group of 14 – 16 g, G511H/Arg//Arg///Arg///Arg-30-3 was able to produce 2.91 t/ha with the days to maturity 76 days. In the seed-size group over 16 g/100 seeds, G511H/Anj//Anj///Anj////Anjs-8-5 was produce up to 3.08 t/ha with days to maturity around 75 days.

The requirement of agronomic character and soybean agro-ecosystem in the tropical area of Indonesia is differ with other soybean production center countries. This is related to the existing growing season performed three times a year, i.e. in the beginning of rainy season (November/December), beginning of dry season (February/March), and dry season (June/July). A study on soybean in Argentina by Santachiara et al. [16] revealed that maturity groups (MG) III had a longer seed set period, received more incident radiation during that period, and captured more N compared to the MG V. A study in Thailand showed that the extension of days to flowering resulted in an increase of the yielding potential of early maturing [17].

The soybean variety development in di Indonesia in the last five years has aimed to develop soybean early maturing varieties (< 80 days). The newest soybean variety in Indonesia, namely Detap 1 which released in 2017, is resistant to pod shattering, the average yield 2.70 t/ha, the 100 seed weight 15.37 g, and the days to maturity 78 days. Liu et al. [18] reported that the soybean breeding program during 82 years in China for character of days to maturity has successfully decrease the days to maturity from 140.3 days to 125.6 days, whereas the average yield increase from 1283.3 kg to 2310.7 kg. The identification of soybean genotypes that are not only adaptive to Indonesia's tropical climate, but also have early days to maturity and large seed size is important for the provision of soybean raw materials for tempeh industry.

5. Conclusion

The soybean genetic approach to produce variety with large seed size and maturity of early days to maturity have a relatively high chance of success. G511H/Anj//Anj////Anjs-8-5 with yield of 3.08 t/ha, early days to maturity (75 days), and large seed size (16.92 g/100 seeds) is potential to be developed in Indonesia's tropical climate to support the provision of raw materials of tempeh industry.

References

[1] Ministry of Agriculture 2015 *The Outlook of Food Crops Commodity: Soybean (Outlook Komoditas Tanaman Pangan: Kedelai)*. Data center and Agriculture Information System. Ministry of Agriculture p 73 [in Bahasa].

[2] BPS 2014 National Survey of Social and Economy. Statistics Indonesia.

[3] Krisdiana R 2005 The preference of tofu and tempeh industry using soybean raw material in East Java (Preferensi industri tahu dan tempe dalam menggunakan bahan baku kedelai di Jawa Timur). *Research Performance in Support on Legume and Tuber Agribusiness*. Indonesian Center for Food Crops Research and Development. Bogor. [in Bahasa].

[4] Ginting E, Antarlina SS and Widowati S 2009 The suitability of improved soybean varieties for food industry ingredient (Varietas unggul kedelai untuk bahan baku industri pangan). *J. Litbang Pertanian* **28**: 79-87. [in Bahasa].

[5] Antarlina SS, Utomo JS, Ginting E and Nikkuni S 2002 Evaluation of Indonesian soybean varieties for food processing. p. 58–68. In A.A. Rahmianna and S. Nikkuni (Eds.). Soybean Production and Postharvest Technology for Innovation in Indonesia. *Proceedings of RILET-JIRCAS Workshop on Soybean Research*. Malang, 28 September 2000.

[6] Astawana M, Wresdiyati T, Widowati S, Bintari SH and Ichsania N 2013 Psychochemical characteristic and functional character of tempeh derived from various soybean varieties (Karakteristik fisikokimia dan sifat fungsional tempe yang dihasilkan dari berbagai varietas kedelai). *Pangan* **22**: 241-52. [in Bahasa].

[7] Bavia ACF, da Silva CE, Ferreira MP, Leite RS, Mandarino JMG and Carrao-Panizzi MC 2012 Chemical composition of tempeh from soybean cultivars specially developed for human consumption. *Ciência e Tecnologia de Alimentos* **32**: 613-20.

[8] Liang HZ, Li WD, Wang H and Fang XJ 2005 Genetic effects on seed traits in soybean. *Acta Genet. Sin.* **32**: 1199–204.

[9] Krisnawati A and Adie MM 2015 Selection of soybean genotypes by seed size and its prospects for industrial raw material in Indonesia. *Procedia Food Science* **3**: 355 – 63.

[10] Vollmann J, Fritz CN, Wagentristl H and Ruckenbauer P 2000 Environmental and genetic variation of soybean seed protein content under Central European growing conditions. *J. Sci. Food Agric.* **80**: 1300-06.

[11] Egli DB and Bruening WP 2002 Synchronous flowering and fruit set at phloem-isolated nodes in soybean. *Crop Sci.* **42**: 1535–40.

[12] Morrison MJ and Xue AG 2007 The influence of seed size on soybean yield in short-season regions. *Can. J. Plant Sci.* **87**: 89-91.

[13] Hu Z, Zhang H, Kan G, Ma D, Zhang D, Shi G, Hong D, Zhang G and Yu D 2013 Determination of the genetic architecture of seed size and shape via linkage and association analysis in soybean (*Glycine max* L. Merr.). *Genetica* **141**: 247-54.

[14] Ning H, Wen-Bin L, Wen-Xia L, Ji-An W, Da-Yong Z, Shu-Hong W, Xian-Xin M and Ya-Zhi Z 2005 Genetic effects analysis of seed weight in soybean (*G. max*). *Chinese Journal of Oil Crop Sci.* **27**: 69-71.

[15] Sudaric A and Vrataric M 2002 Variability and interrelationships of grain quantity and quality characteristics in soybean. *Die Bodenkultur* **53**:137-42.

[16] Santachiara G, Borras L and Rotundo JR 2017 Physiological processes leading to similar yield in contrasting soybean maturity groups. *Agron. J.* **109**: 1–10.

[17] Machikowa T and Laosuwan P 2009 Effects of extension of days to flowering on yield and other characters of early maturing soybean. *Suranaree J. Sci. Technol.* **16**: 169-74.

[18] Liu G, Yang C, Xu K, Zhang Z, Li D, Wu Z and Chen Z 2012 Development of yield and some photosynthetic characteristics during 82 years of genetic improvement of soybean genotypes in northeast China. *AJCS* **6**: 1416-22.

The effect of cassava and corn flour utilization on the physicochemical characteristics of cassava leaves snack

I Ambarsari, R Endrasari and G N Oktaningrum

Assessment Institute for Agricultural Technology Central Java, Indonesian Agency for Agricultural Research and Development.

E-mail: indrie_ambarsari@yahoo.com

Abstract. Cassava leaves are nutritious vegetable, but often regarded as an inferior commodity. One of the efforts increasing in the benefit of cassava leaves is through processing it into snack. In order to support the food diversification program and to reduce the dependence on imported commodities, the development of cassava leaves snack could be accompanied by optimizing the use of local materials to minimize the use of wheat flour. The aim of this assessment was to learn the effects of cassava and corn flour substitution on the physicochemical characteristics of cassava-leaves snack. The substitution of local flour (cassava and corn) on the snack production was carried on three levels at 15, 30, and 45%. A control treatment was using 100% wheat flour. The results showed that cassava and corn flour were potential to substitute wheat flour for making cassava-leaves snack. The substitution of cassava and corn flour as much as 45% was able to produce crispy products with a brighter color. The substitution of corn flour was resulting in snacks with the lower content of lipid than the other substitution snacks.

Keywords: cassava leaves, substitution, local flour.

1. Introduction

Cassava leaves, like other vegetables, are rich in vitamin and mineral. Cassava leaves are containing highly nutrient, such as vitamin A (1970 μg RE), vitamin C (311 mg/100 g), protein and carbohydrates [1]. During this time, cassava leaves were often regarded as inferior vegetables, which have low market price. In relation with this condition, various processing technologies have been developed not only to increase the added value of cassava leaves but also the consumption level of vegetables. One of technology that simple and suitable to be developed in village is processing the cassava leaves into snack.

Snack is one of processed food products that very popular in Indonesia. Nowadays, consuming snack has become a part of lifestyle and its production is increasing more from year to year. Based on that condition, product development such as cassava-leaves snack is one of the attempts in supporting the food diversification program and increasing vegetable consumption that aimed to improve the nutritional status of Indonesian community.

In Central Java, the production of snack from cassava leaves has been developed in such way to produce a unique flavor that favorable by customers. In its production, the producers still rely on wheat flour as the main ingredient. Dependency of local products on wheat flour as a raw material is quite apprehensive considering that wheat is an imported commodity. To overcome this condition, the use of flour from local commodities as an alternative of wheat substitution should be developed.

Corn and cassava are local food crops with a great potency to be developed as an alternative material in wheat substitution. Both of these crops are common cultivated by Indonesian farmer because its cultivation is relatively simple. Although study on the utilization of corn flour and cassava flour has been widely developed especially for bread and noodles, but its utilization for snack products was still not considered. Each type of flours has different property, so that study would evaluate the effects of corn and cassava flour utilization on the physicochemical characteristics of cassava-leaves snack.

2. Material and method

The production of cassava leaves snack refers to a method developed by women-farmers in Seboto village, district of Boyolali, Central Java. The ingredients for preparing snack are cassava-leaves (mix between green and dark leaves with ratio 1:3), wheat flour, tapioca, eggs, butter, and some condiments. The cassava and yellow corn flour substituting wheat flour in the production of snack were obtained from Boyolali farmer. The substitution level variations of flour (cassava and corn) in the production of snack were 15, 30, and 45% of the wheat flour weight. Snack made of 100% wheat flour were used as a control treatment. The instruments used in this study were grinder, electric single fryer (Fomac FRY.E2LI), slicer, digital scales, and other supporting tools.

The processes of snack production firstly began to boil the cassava leaves for 30 minutes. Afterward the boiled leaves were squeezed to reduce the water content. The leaves then finely grounded and mixed with flour and other ingredients to form sturdy dough. Next, the dough was flattened into a thin sheet before it was deep-fried in the hot oil (190 ± 5°C). The snacks were fried twice to obtain a crunchy texture. The first frying was carried out for 5 minutes, while the second was done for 1 minute. At the end of frying, snack were drained to reduce the excess oil. Products allowed to cool at room temperature, then packed in 0.08 mm polyethylene (PE) bag and stored in airtight container before further analysis. All treatments were replicated three times.

Analysis of snack quality attributes comprised its physical and chemical properties. The physical analysis included the hardness level (Instron Universal Testing Machine Instron Corporation, Canton, MA) and the color of product (Hunter Lab Scan Spectrocolorimeter). Chemical analysis consisted of moisture content [2], protein [3], and total lipid [2].

Experimental data were statistically analyzed using one-way analysis of variance (One Way ANOVA). Detailed examinations for significant differences were done using a Duncan Multiple Range Test at $p \leq 0.05$. SPSS 17.0 statistical software (SPSS Inc., Chicago, IL, USA) was used in all statistical analysis.

3. Result and Discussion

3.1. *Physical characteristics of cassava leaves snack*

Hardness level is one of the important quality parameters on the snacks because it related to the crispness, which is one of the main consumer considerations in buying snacks. Crispness may associate with some characteristics of products such as firmness, brittle, and a specific sound when it is broken [4]. In snacks, a decrease of crispness can become the main reason of product rejection [5].

The effect of local flour (cassava and corn) substitution on the hardness level of cassava-leaves snack was presented in Figure 1. The results showed that the more amount of cassava or corn flour as the substitution material then the hardness level of snack were tend to decreased, in other words the products became crunchier. The degree of hardness was also determined by the protein content of the constituent materials [6].

Protein is long peptide chains, which requires a high energy to break it. Wheat flour has higher protein content than cassava and corn flour. Wheat flour used in this study was included in the moderate protein content group ranging from 12 to 13%. In the other hand, the protein content of cassava and corn flour in this study were 2.0 and 4.0% respectively (data not shown). The lower protein content on this substitution material was alleged to be the cause of the hardness decline.

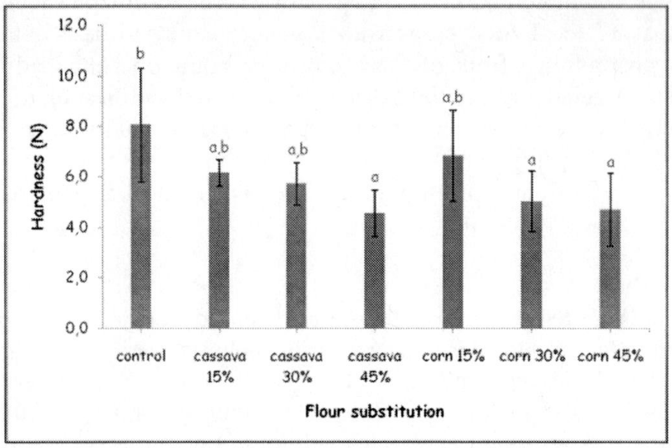

Figure 1. The effect of cassava and corn flour substitution on the hardness level of the snacks

Besides crispness, the color of product is also one of main factors that affected costumer acceptance. Color would visually appear first and often determine a product value [7]. Table 1 shows the effect of cassava and corn flour substitution on the color of cassava-leaves snack. The use of cassava and corn flour as the substitution material of wheat flour on the production of cassava-leaves snack has significant effect on the lightness (L*), index of redness/greenness (a*), and index of yellowness/blueness (b*).

Table 1. The effect of cassava and corn flour substitution on the color intensity of the snacks

Treatments	Color index		
	L*	a*	b*
Control (wheat 100%)	$33{,}69 \pm 1{,}04^{a,b}$	$0{,}77 \pm 0{,}04^{c}$	$21{,}82 \pm 1{,}26^{b,c}$
Cassava 15%	$36{,}61 \pm 1{,}08^{b,c}$	$0{,}53 \pm 0{,}56^{b,c}$	$22{,}93 \pm 0{,}20^{b,c}$
Cassava 30%	$35{,}89 \pm 3{,}49^{a,b,c}$	$0{,}12 \pm 0{,}42^{a,b,c}$	$23{,}38 \pm 3{,}17^{c}$
Cassava 45%	$37{,}84 \pm 1{,}15^{c}$	$0{,}24 \pm 0{,}46^{a,b,c}$	$21{,}56 \pm 2{,}07^{b,c}$
Corn 15%	$34{,}22 \pm 0{,}50^{a,b}$	$-0{,}59 \pm 0{,}63^{a}$	$20{,}79 \pm 1{,}29^{a,b,c}$
Corn 30%	$34{,}38 \pm 2{,}83^{a,b,c}$	$-0{,}20 \pm 0{,}44^{a,b}$	$19{,}58 \pm 2{,}01^{a,b}$
Corn 45%	$32{,}44 \pm 0{,}57^{a}$	$-0{,}09 \pm 0{,}55^{a,b,c}$	$17{,}87 \pm 1{,}14^{a}$

Note: Means in the same column with different letters (a, b, c) differ significantly ($p \leq 0.05$) by Duncan's multiple range test

Snack that was substituted with 45% cassava flour has a higher L value than control snack. Nevertheless, the utilization of cassava flour as a substitution material has no effect neither on a* index nor on b* index. The opposite condition occurred in corn flour. The utilization of corn flour as a substituted material produced snack with higher intensity level in green than control treatment. Snack that was substituted with 45% corn flour has the lowest b* index (more yellow than other treatments). This condition probably influenced by carotene content in the corn flour. The color change in fried snacks may also related with the Maillard reaction, a non-oxidative browning process which caused by an interaction between the amine group with reducing sugar [7]. The high protein content in the wheat flour might promote the brown color through the Maillard reaction during the frying process. Therefore, the higher level of wheat flour substitution will produce snacks with a brighter color.

3.2. Chemical characteristics of cassava leaves snack

Moisture content of snack that was substituted with cassava flour and corn flour were about 1.75 – 2.21% and 2.79 – 2.99%, respectively (Figure 2). The utilization of cassava flour as a substitution flour produced snack with lower moisture content. This condition probably had a connection with the initial moisture content of the flour itself and the moisture content of cassava-leaves as the main raw material.

Beside the initial moisture content of raw materials, water holding capacity of the flour was also could be affected the moisture content of the final product [8]. In this study, we presumed that the water-holding capacity of corn flour were relatively higher than cassava flour. Based on the data of several researches, the water holding capacity of cassava flour is ranged from 101 to 202% [9, 10], while corn flour and wheat flour are 617 and 778%, respectively [11]. The higher water-holding capacity allowed retaining more water molecules in snacks, and this resulted in higher moisture content [9].

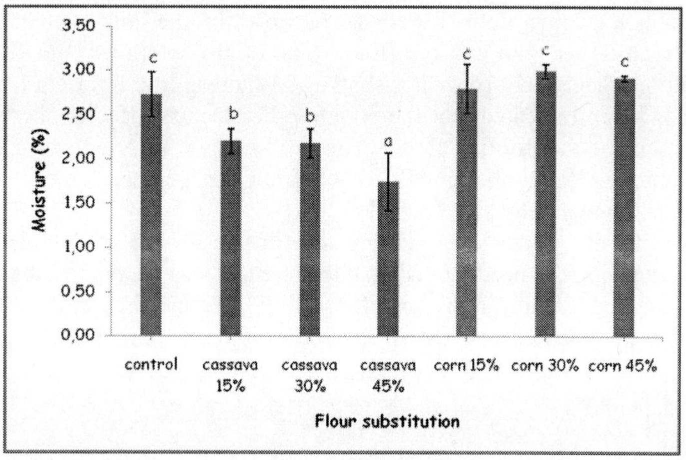

Figure 2. The effect of cassava and corn flour substitution on the moisture content of the snacks

Effect of the utilization local flour (cassava and/or corn) on the protein content of snack was shown in Figure 3. Statistical analysis showed that snacks that substituted with cassava or corn flour at 15% have equivalent protein content with the control snack. This condition indicated that cassava and/or corn flour are potential to substitute the wheat flour without changing the nutrition of the final product.

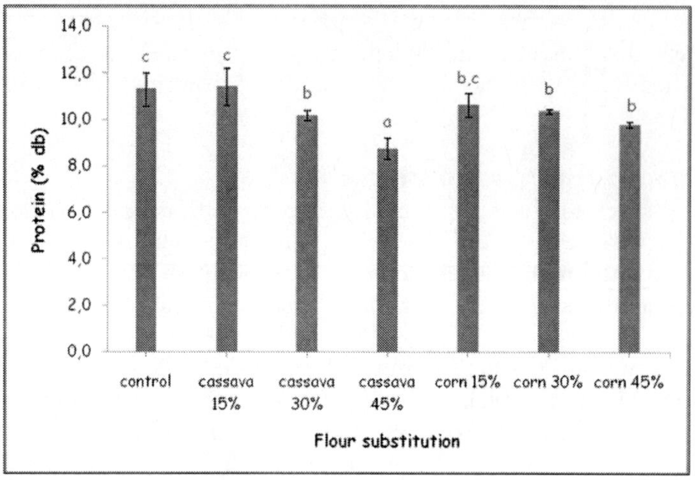

Figure 3. The effect of cassava and corn flour substitution on the protein content of the snacks

In this study, the lipid content of snack that was substituted with corn flour was lower than snack that was substituted with cassava flour (Figure 4). Meanwhile, the lipid content of corn flour that was used in this study was higher than cassava flour. The lipid content of corn flour and cassava flour were 2.04 and 1.50%, respectively (data not shown). Whereas, the lipid content of wheat flour was 8.9% [12]. This condition indicated that there were other factors affected the lipid content of snack aside from the lipid of the substitution flour. The lipid content of fried snack can be influenced by several factors, including type and condition of fried ingredients, types of cooking oil, time, temperature, frying equipment, etc. [13, 14, 15].

Figure 4 showed that the higher level of flour substitution for the production of snack induced the increase of lipid content of the product. This phenomenon was related to the degree of expansion, which followed by the degradation in the specific surface area and the number of pores in the product during frying that was not enabled a sufficient space for oil absorption [16, 17].

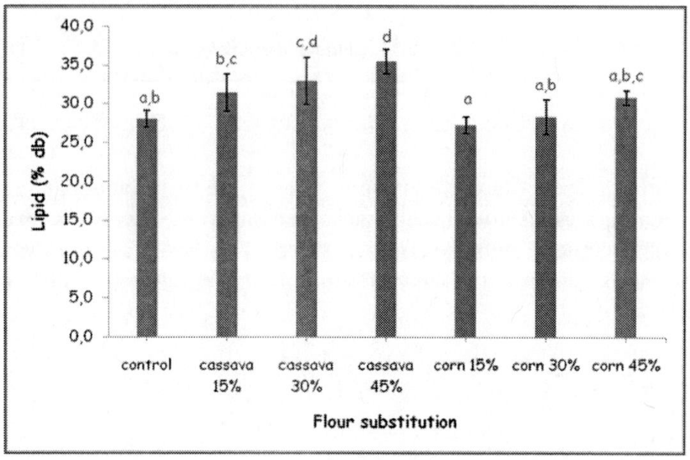

Figure 4. The effect of cassava and corn flour substitution on the lipid content of the snacks

Some researchers [18, 19, 20] stated that the oil absorption in the product was affected by the initial moisture content of fried products. During the frying process, water diffuses inside the matrix and forms a path called capillary hole that would be filled by oil. As more moisture losses during the frying process, the oil (lipid) of the product will increase [20].

4. Conclusion

The flour from local commodities such as cassava and corn are potential to substitute wheat flour on the production of vegetable snack. Different types of substitution flour in the production of cassava-leaves snack have a significant effect on physicochemical characteristics of products, specifically hardness, color, moisture content, protein, and total lipid. The utilization of cassava and/or corn flour was potential to improve the physical characteristics of the snacks such as crispness and color. However, the overmuch substitution would produce undesired snacks, which have lower protein content and higher fat content. Therefore, the suggested level of substitution in producing cassava-leaves snack was 15%.

References

[1] Uusiku N P, Oelofse A, Duodu K G, Bester M J and Faber M 2010 *Journal of Food Composition and Analysis* **23** 499-509
[2] AOAC 2000 *Official Methods of Analysis* 17th Ed (Gaithersburg USA: AOAC International)
[3] AOAC 2001 *Protein (crude) in Animal Feed, Forage (Plant Tissue), Grain, and Oilseed* (Washington: AOAC International)
[4] Luyten H, Plijter J J and Vliet T V 2004 *Journal of Texture Studies* **35** (5) 445-92
[5] Piazza L, Gigli J and Ballabio D 2007 *Chemometrics and Intelligent Laboratory Systems* **86** (1) 52-59
[6] Saelaw M and Schleining G 2010 *Journal of Food Engineering* **100** 12-24
[7] Oginni O C, Sobukola O P, Henshaw F O, Afolabi W A O and Munoz L 2015 *Food Structure* **3** 12-20
[8] Anam C and Handajani S 2010 *Caraka Tani* **25** (1) 72-78
[9] Aryee F N A, Oduro I, Ellis W O and Afuakwa J J 2006 *Food Control* **17** 916-22
[10] Doporto M C, Dini C, Mugridge A, Vina S Z and Garcia M A 2012 *Journal of Food Engineering* **113** 569-76
[11] Nusantoro B P, Haryadi, Bintoro N and Darmadji P 2005 *Agritech* **25** (3) 148-53
[12] Wulandari M and Handarsari E 2010 *Jurnal Pangan dan Gizi* **1** (2) 55-62
[13] Chatzilazarou A, Gortzi O, Lalas S, Zoidis E and Tsaknis J 2006 *Journal of Food Lipids* **13** 27-35
[14] Kalogianni E P, Karastogiannidou C and Karapantsios T D 2010 *International Journal of Food Science and Technology* **45** 765-75
[15] Ketaren S 2008 *Oil and Fat* (in Indonesia) (Jakarta: UI Press)
[16] Maneerote J, Noomhorm A and Takhar P S 2009 *LWT – Food Science and Technology* **42** 805-12
[17] Wang Y, Zhang M and Mujumdar A S 2012 *LWT – Food Science and Technology* **47** 175-82
[18] Gamble M H and Rice P 1987 *International Journal of Food Science and Technology* **22** 535-48
[19] Kassama L S and Ngadi M O 2004 *LWT* **37** 841-47
[20] Ngadi M, Li Y and Oluka S 2007 *LWT* **40** 1784-91

IOP Conf. Series: Earth and Environmental Science **102** (2017) 012047 doi:10.1088/1755-1315/102/1/012047

Physicochemical and sensory characteristics of soy sauce substituted with pigeon pea (*Cajanus cajan* (Linn.)

C Retnaningsih[1], Sumardi[1], Meiliana[1] and A Surya[1]

[1]Food Technology Department, Soegijapranata Catholic University, Jl. Pawiyatan Luhur IV/1, Bendan Duwur, Semarang 50234

nik@unika.ac.id

Abstract. The objective of this study wasto investigate the physicochemical and sensory properties of the soy sauce substituted with pigeon pea. Soybean was substituted by 20%, 50%, 75%, and 100% of pigeon pea. The observation included viscosity, total solids, protein levels, antioxidant activity, and sensory characteristics. The results showed that the more substitution of pigeon pea, the less the protein content of soy sauce and the more the antioxidant activity as well as total solids. The most favored group was 25% pigeon pea substitution. It is suggested that soy sauce could be prepared using 25% to 75% pigeon pea substitution.

Keywords: soy sauce, pigeon pea

1. Introduction

Pigeon pea *(Cajanuscajan* (Linn.))is a group of nuts with 20.7% of proteinand 1.4% of fat contents and it could be used as black soybean substitution in making soy sauce due to their similar properties in shape, color, and seed size [1]. Pigeon pea also contains antioxidants such as polyphenols, isoflavones, phenols, phenolic acids and derivatives, flavonoids, tocols, phospholipids, amino acids, peptides, phytic acid, ascorbic acid, pigments, and sterols [1]. However, pigeon pea containsa low level of antinutrition compounds such as antitrypsin, hemagglutinin, and phytic acid. Phytic acid is not desirable because it can bind minerals and proteins and decrease the bioavailability of iron in the body. The content of phytic acid in pigeon pea flour is 0.77% and can be eliminated. Washing, soaking, boiling, steaming, and fermentation process can decrease phytic acid by 91-97% [2].

Cikuray and Merapi black soybean are two best black soybean varieties and suitable for producing soy sauce due to their high protein content at around 37% [3]. Soybean has globular protein which is soluble in water or salt solution at a pH value below or above the isoelectric point [4]. Amino acids found in black soybean are leucine and lysine, essential amino acids needed by soy-breaking enzymes to produce soy sauce with tasty and distinctive tastes.

In Indonesia, soy sauce is *kecap manis*, literary means sweetened soy sauce. Soy sauce is produced by fermentation process which activates the growth of the required microorganisms. Fermentation process breaks down the long molecular chains into simpler ones. Fermented foods have higher nutritional values than their original ingredients because the microorganisms convert the complex proteins, fats, and polysaccharides into simpler and higher-digesting agents. Fermentation process alsotraditionally improvesflavor and the food quality as the products are easier to digest and lower in anti-nutritive substances [5]. This research aims to investigate the physicochemical and sensory properties of the soy sauce substituted with pigeon pea.

2. Materials and Methods

2.1. Materials and Tools

Materials used in this study includedblack soybeans, pigeon peas obtained from Gita Pertiwi Foundation in Surakarta, inoculum *tempe*Raprima, brown sugar, salt, galangal, bay leaf, lemongrass leaves, orange leaves, deaf, K_2SO_4, HgO, concentrated H_2SO_4, $Na_2S_2O_3$-NaOH , 4% boric acid, methyl red and methylene blue indicator, 0.1 N HCl, methanol, DPPH (2,2-diphenyl1-1-picrylhydrazyl), and aquades. The tools used were stove, pot, *tampah*, filter cloth, bottle, plastic bucket, strainer, wood stirrer, analytical scales, cormorant, pounder, *cobek*, viskotester, destructive, destilator, refractometer, burette, statif, erlenmeyer, bekker glass, pumpkin powder, porcelain cup, spectrophotometer, waterbath, and hotplate.

2.2. *Materials Preparation.*

Black soybeans and pigeon peas were sorted and cleaned through running water. Black soybeans were soaked in 3 liters of water for 24 hours while pigeon peas were soaked for 48 hours in water which was changed every 6 hours. The black soybeans and pigeon peas were boiled for 60 minutes to soften the texture, drained, and fully peeled. Then, they were steamed for 30 minutes, drained, and chilled at 30^0C. Last, the materials were prepared into 4 groups which contained 1 kg of materials for each group. The groups respectively contained 25%, 50%, 75%, and 100% of pigeon peas.

2.2.1. Koji Fermentation. The yeast required in koji fermentation was 1% of the soybean net weight of each group, sprinkled over the materials on a *tampah* (clean tray). The materials were incubated at room temperature for 3-4 days in a rather dark place. After the hyphae were formed and the structure of the materials became compact, the koji then was dried [2].

2.2.2. Moromi Fermentation. The dried koji was soaked into a 20% salt solution for up to 2 months. One kg of materials requires 2 liters of 20% salt solution. During this incubation process, the mixture needed to be sunbathed and stirredevery morning and closed and stored every afternoon [2].

2.2.3. Cooking Moromi Filtrate with Seasoning Spices. After 2 months, the soy sauce filtratewas filtered and cooked with Javanese sugar and mashed-roasted spices until it boiled, caramelized, and thickened, forming the desired texture. During the cooking process, the filtrate should be continuously stirred so that the heat evenly spread and all the ingredients well-blended. Next, the boiled soy sauce was filtered using a filter cloth to remove the solids [6]. The pigeon pea soy sauce formulasare modified was presented in Table 1.

Table 1. The pigeon pea soy sauce formula.

Materials	Pigeon pea 25%	Pigeon pea 50%	Pigeon pea 75%	Pigeon pea 100%
Pigeon pea (g)	250	500	750	1000
Black soybean (g)	750	500	250	-
Lemongrass (g)	10	10	10	10
Citrus leaf (g)	10	10	10	10
Bay leaf (g)	10	10	10	10
Salt (g)	200	200	200	200
Brown sugar (g)	2000	2000	2000	2000
Star anise (g)	10	10	10	10
Water (ml)	3000	3000	3000	3000

2.3. Physicochemical and Sensory Analysis

*2.3.1. Viscosity.*A 100 ml of the samples from each group was measured using rotor viscotester three [7].

2.3.2. Total Solids. The process included putting the porcelain cup into the kiln for 1 hour with a temperature of 100^{0}C, weighing it with a desiccator (a gram), putting 5 ml of soy sauce on the porcelain cup, weighing it (b gram), inserting the porcelain cup into an oven at 100^{0}C for 24 hours, cooling the cup in desiccator for 10 minutes, and weighing the cup (c gram). The total dissolved solid (TDS) is calculated by the formula [8]:

$$TDS = \frac{c - a}{b} \times 100\,\%$$

Information : a : weighing the cup
b : weighing the cup + sample before dried
c : weighing the cup + sample after dried
(Apriyantono *et al.*, 1989).

*2.3.3. Sugar Level and Salinity.*The process included diluting 1 g of sample in 10 ml of aquades, cleaning the refractometer prism and the lid with alcohol, dripping the solution on the refractometer prism, and directing light through the refractometer prism until the focal point between bright regions and dark areas was clear. The values appearing on the brix refractometer are calculated as thebrix value in soy sauce [9].

*2.3.4. Protein Level.*The process included putting 1 g of sample (1 g of aquades for blanko) into the destruction flask, adding 7 grams of K_2SO_4, 0.35 gram HgO, and 15 ml of concentrated H_2SO_4 to the flask, destructing the solution for 20 minutes at 420^{0}C or 60 minutes at 370^{0}C, cooling the solution, transferring the solution into a distillation flask while rinsing it with 50 ml of aquades, adding 400 grams of $Na_2S_2O_3$ and 25 grams of NaOH, and diluting the solution in the distillation flask until it reached a volume of 1000 ml. Next, 25 ml of 4% boric acid was prepared in a distillate erlenmeyer. The distillation was carried out until 100 ml distillate was obtained. Then, the distillate was added by 2 drops of methyl red and methylene blue mixture and titrated with 0.1 N HCl until the distillate became purple[6]. The formulas to calculate % N and% proteins are:

$$\% \text{ Nitrogen} = \frac{\text{HCl volume (sample-blanco) x N HCl x 14,008}}{\text{Sample weighing (gram) x 100}} \text{ x } 100\%$$

% protein =% N x conversion factor (6.25)

2.3.5. Antioxidant Activity. The process included measuring 0.5 ml of the sample (methanol as blanko) and extracting it using 5 ml methanol for 2 hours, taking 0.1 ml of the extraction and mixing itwith 3.9 ml of DPPH solution (2,2-diphenyl1-1-picrylhydrazyl) for 30 minutes of chemical reaction, and measuring the absorbance at t = 0 for control and t = 30 for sample using a 517 nm wavelength [10]. The antioxidant activity was measured as% discoloration calculated using formula:

% discoloration = [1- (At30 / At0)] x 100

Information

At 30 = sample, At 0 = control

2.3.6. Sensory Characteristics. The 4 groups of pigeon pea soy sauce and commercial soy sauce as control were evaluated for their color, odor, taste, and overall appearance using hedonic rating methodand 30 panelists. The attributes was expressed by a score of 1 to 5 written in a questionnaire sheet, with 1 as least preferredand 5 as the most preferred. The highest average of the resultsdetermined the most preferred group [11].

2.4. *Statistical Analysis*
Data were subjected to ANOVA (and Duncan test for any significant results with p <0.05) to determine significant differences between groups using SPSS 13 for Windows statistical application program. The Mann-Whitney test was used to determine the real difference between variables. Sensory characteristics data were analyzed using Kruskal-Wallis and Mann Whitney test [12].

3. Results and Discussion
In this study, the characteristics of the pigeon pea soy sauce were compared to the characteristics of the commercial soy sauceas control. Each group of the pigeon pea soy sauce yielded about 2,125 ml. The data of viscosity, sugar level, and salt level are presented in Table 2.

Table 2. Viscosity, sugar, and salt level of the commercial soy sauce and pigeon pea soy sauce.

Soy Sauce	Viscosity (dPaS)	Sugar level (%)	Salt level (%)
Commercial	8.5	74	7
Pigeon pea 25%	8.5	74	6
Pigeon pea 50%	8.5	74	6
Pigeon pea 75%	8.5	74	6
Pigeon pea 100%	8.5	74	6

IOP Conf. Series: Earth and Environmental Science **102** (2017) 012047 doi:10.1088/1755-1315/102/1/012047

The result showed that there were no differencesof viscosity and sugar level between pigeon pea soy sauce and commercial soy sauce. Commercial soy sauce had a higher salt level compare to pigeon pea soy sauce. The salt level might be influenced by the different formulas and moromi fermentation process. The immersion of koji in salt solution increased the soy sauce saltiness. This process also meant to prevent thegrowth of harmful microbeswhile still allowing the growth of bacteria and taste-shaping yeast. The salt solution serves to extract the dissolved nitrogen compounds resulting from the hydrolysis of the mold enzymes in soybeans. Without salt, an undesirable anaerobic fermentation process occurs [4,13]. During the salting process, every morning the koji needed to be sunbathed and stirred, while every afternoon it needed to be closed and stored again. During this incubation process, the koji must be stirred so that the salt solution can homogeneously touch the surface of the substrate and provide air to stimulate the growth of yeast and flavor-forming bacteria [13,14].

Viscosity is one of the physical attributes that determines the quality of soy sauce. Viscosity equalization is needed so that the quality of the pigeon pea soy sauce and commercial soy sauce is similar with standard viscosity of 8.5 dPaS. Therefore, the cooking process of all groups stopped when they reached the same viscosity as the commercial soy sauce. The total content of solids, protein level, and antioxidant activity of the soy sauce groups are presented in Table 3.

Table 3. Total dissolved solids, protein, and antioxidants activityof the commercial soy sauce and pigeon pea soy sauce[AB].

Soy Sauce	Total solids (%)	Protein level (%)	Antioxidant activity (%)
Commercial	15.67 ± 0.26^{d}	3.51 ± 0.02^{b}	90.31 ± 0.84^{a}
Pigeon pea 25%	13.21 ± 0.25^{a}	3.70 ± 0.04^{e}	92.59 ± 0.29^{b}
Pigeon pea 50%	13.72 ± 0.12^{b}	3.63 ± 0.02^{d}	93.15 ± 0.20^{c}
Pigeon pea 75%	14.38 ± 0.32^{c}	3.54 ± 0.02^{c}	93.55 ± 0.21^{cd}
Pigeon pea 100%	15.41 ± 0.36^{d}	3.46 ± 0.02^{a}	$93.91\pm0.14_{d}$

[A]The result is the mean value ± standard deviation.
[B]Different letters show significant difference values ($p<0.05$).

The total solids of commercial soy sauce had the same value as 100% pigeon pea soy sauce. The protein levels in all groups of soy sauce were significantly different due to the different amount of pigeon pea substitution. The more amount of pigeon pea used for substitution in soy sauce, the fewer the protein level of soy sauce because the protein level of pigeon pea is lower than black soybean. However, the higher concentration of pigeon pea added, the more the antioxidant activity of soy sauce. The carbohydrate level of pigeon pea is higher than black soybean, resulting in increasing the total solids of soy sauce. In food material, some solids are in soluble form and some are in non-solvent form. Soluble solids usually consist of organic substances, inorganic salts and, dissolved gases [2].

Protein level in pigeon pea soy sauce and commercial soy sauce are between 3-4%. According to *Standar Nasional Indonesia* (SNI) 3543-01-2013, the minimum protein levelof soy sauce is 2%. The pigeon pea substitution decreased the protein level of soy sauce [15]. Soy sauce has more carbohydrate than protein level due to the use of brown sugar as one of the seasoning spices. Moromi extract has complete nutrients with amino acids. Almost all sorts of amino acids are present in soy sauce, including hydrophobic amino acids (valine, leucine, and isoleucine), hydrophilic amino acids (glutamic, glycine, and aspartate) and aromatic amino acids (phenylalanine) [16].

Sensory analysis on pigeon pea and commercial soy sauce (Table 4) showed that 25% pigeon pea soy sauce had the highest score on the overall parameter. It showed that pigeon pea can substitute black soybean for producing soy sauce.

Table 4. Sensory Characteristics of the commercial soy sauce and pigeon pea soy sauce.

Soy Sauce	Color	Aroma	Taste	Mouth feel	Overall
Commercial	$3{,}60^{b}$	$3{,}20^{b}$	$3{,}73^{b}$	$3{,}47^{ab}$	$3{,}67^{ab}$
Pigeon pea 25%	$4{,}27^{bc}$	$3{,}83^{b}$	$3{,}60^{ab}$	$3{,}77^{b}$	$4{,}20^{b}$
Pigeon pea 50%	$3{,}87^{bc}$	$3{,}63^{b}$	$3{,}27^{ab}$	$3{,}13^{a}$	$3{,}37^{a}$
Pigeon pea 75%	$3{,}00^{a}$	$3{,}43^{ab}$	$3{,}60^{ab}$	$3{,}17^{a}$	$3{,}40^{a}$
Pigeon pea 100%	$2{,}63^{a}$	$2{,}97^{a}$	$3{,}10^{a}$	$2{,}90^{a}$	$3{,}33^{a}$

[A] The result is the mean value ± standard deviation.
[B] Different letters show significant difference values ($p<0.05$).
[C] The score interpretation are 1 = very unacceptable, 2 = unacceptable, 3 = quite acceptable, 4 = acceptable, and 5 = very acceptable.

The mouthfeel of pigeon pea 25% had the best score among other concentrations. It is related to the total solids presented in soy sauce. Pigeon pea 100% had a high total solid content which gave a specific taste in mouth feel attribute. Besides, pigeon pea 100% had a stronger flavor of pigeon pea which made it less favored by the panelists.

4. Conclusion

Based on the sensory analysis and protein level results, the best formula of pigeon pea soy sauce waspigeon pea 25%. The more amount of pigeon pea used for substitution in soy sauce, the fewer the protein level of soy sauce and the more the total solids and antioxidant activity of soy sauce. Higher amount of pigeon pea decreased the panelists' acceptance of soy sauce.

References

[1] Fitri V 2007 Antinutrition compounds in pig nuts (*Vicia faba*) and pigeon pea (*Cajanus cajan*) and inactivation methods *Bogor Agricultural Institute*
[2] Rukmana R 1999 *Pigeon pea cultivation and postharvest handling* (Yogyakarta: Kanisius)
[3] Widowati S, Erliana G and Sri S A 2009 Soybean superior variety for soybean industry raw material*Journal of Agricultural Research* **28**
[4] Pangastuti A, Yona S and Purwoko T 2004 Content carbohydrate, fat and protein in soy sauce from tempe *Surakarta*
[5] Isnawan H 2010 Changes in quality of soy sauce household scale production during the three months storage*Jurnal Teknologi dan Industri Pangan* **14**
[6] Purwoko T and Noor S H 2007 Protein content of sweet soy sauce without fermentation moromi as fermentation results of *Rhizopus oryzae* dan *R. oligosporusBiodiversitas* **8** 223-27.
[7] AOAC 1995*Official methods of analysis of the association of official agriulture chemist*(Washington DC: Association of Official Analytical Chemist)
[8] Apriyantono A and Yulianawati G D 2004Volatile component changes during fermentation ketchup*Jurnal Teknol dan Industri Pangan* **15** 100-12
[9] AOAC 2000 *Officialmethod of analysis of the AOAC 17th ed*(Washington DC: Association of Official Analytical Chemist)
[10] Williams W B, Cuvelier M E and Berset C 1995 Use of afree radical method to evaluate antioxidant activity*Technol* **28** 25-30
[11] Meilgaard M, Civille G V and Carr B T 1999 *Sensory evaluation techniques 3rd edition*(Florida: CRC Press LLC)
[12] Trihendradi C 2005 *Step by Step SPSS 13 Statistical Data Analysis* (Yogyakarta: Andi Publisher Yogyakarta)

[13] Rahayu E, Indrati R, Utami T, Harmayani E and Cahyanto M N 1993 Fermented food ingredients *Food & NutitionCollection*

[14] Rahayu A, Suranto and Tjahjadi P 2005 Analysis of carbohydrates, proteins, and fats in the making of lamtoro gung (*Leucaena leucocephala*) sauce fermented by *Aspergillus oryzaeSurakarta*

[15] Standar Nasional Indonesia (SNI) 2013 *Sweet soy sauce* (Jakarta: Center for Industrial Standardization Ministry of Industry)

[16] Andarwulan N, Fardiaz D, Anton A and Dedin F R 2007 Isolation and sweet ketchup melanoidin characterization and its role as antioxidant *Bogor Agricultural Institute*

Functional-drink rich in antioxidant cardamom-rhizome (*Amomum cardamomum* willd) suppresses inflammation and improves lipid profile

H Winarsi[1*] and S S Susilowati[2]

[1]Faculty of Health Sciences, Department of Nutrition Science, University of Jenderal Soedirman;

[2]Faculty of Health Sciences, Department of Pharmacy, University of Jenderal Soedirman, Purwokerto, Central Java, Indonesia;

*Email: winarsi12@gmail.com

Abstract. The aim of this research was to know the effect of functional drink rich in antioxidant cardamom rhizome (Fd-Carrhi) on level of IL-6, C-RP, and lipid profile of atherosclerotic. A total of 30 women with atherosclerosis, age 40-65 years old, hypertension, hypercholesterolemia, hypertriglyceridemia, lived in Purwokerto, Banyumas, Central Java, Indonesia, and were willing to sign informed consent, recruited as research subjects. They consumed simvastatin from doctors, divided by 3 groups of 10 people each. Group I, given Fd-Carrhi; II, placebo; and III, only simvastatin, for 2 months. As many as 100 ml of Fd-Carrhi or placebo were given every morning. Blood samples were taken 3 times, 1 ml, at baseline, 1 and 2 months after intervention. Blood plasma was determined levels of IL-6, C-RP, as well as total cholesterol (total-c), triglycerides (TG), LDL-c, and HDL-c. Result showed Fd-Carrhi versus placebo significantly decreased plasma level of IL-6, C-RP, total-c, and LDL-c, and otherwise increased HDL-c, but no differences were seen in TG. The findings clearly support Fd-Carrhi inhibit the development of atherosclerosis towards cardiovascular heart diseases (CHD) by suppressing IL-6 and CRP levels, and improving lipid profile.

Key words: *Cardamom rhizome,* functional drink, IL-6, CRP, lipid profile

1. Introduction

Inflammation is one of the factors that induce atherosclerosis. Clinical studies pointed that inflammation status is reflected from the high level of hs-CRP and IL-6 in plasma [1, 2]. It is stated that high level of cytokine contributes to the early stage of coronary artery disease [3]. hs-CRP is the result of hepatic inflammation whose performance is managed by cytokine IL-6, while messenger proinflammatory cytokines (IL-6) is secreted by macrophag and smooth muscle cells in atherosclerotic lesion. Therefore, hs-CRP and IL-6 can be used to predict the development of atherosclerosis toward cardiovascular disease (CVD) [4]. Kumar *et al.* [5] supported that CVD is more prevalent in women than men. High markers of inflammation is related to the increasing risk of atherosclerosis. Accordingly, Frostegård [6] ensured that antiiflammatory treatment is considered the new medication for CVD.

Hypercholesterolemia is also related to atherosclerosis. LDL-c transports cholesterol from liver into the tissues for body performance, while HDL-c transports cholesterol from the tissues to liver. However, when LDL-c level more than normal, it can accumulate in the artery and increase the risk of

plaque that clogs the artery, called atherosclerosis. In atherosclerotic lesion is found a more electronegative LDL-c than native LDL-c [7]. It is certain that the contributing factor of atherogenesis is not LDL native but oxidized LDL (LDL-ox), while the formation of LDL-ox is the effect of high level of LDL-c. Thus, it is crucial to minimize the level of either LDL-ox or LDL-c.
Hypercholesterolemia is not a disease but a metabolic disorder that can contribute to the occurrence of various degenerative diseases, especially cardiovascular disease. Several studies have reported that the emergence of degenerative diseases is due to accumulation of free radicals in the body [8,9]. One substance that is capable of controlling the work of free radicals are antioxidants. Numerous beverage products in markets are labelled rhizome-based and rich in antioxidant such as saffron-acid drinks, ginger ale, ginger drink, and many others but not supported by scientific evidence.
The cardamom rhizome, recently reported Winarsi et al. [10] contains flavonoid antioxidants of 324.51 mg / g, nearly 3 times the levels in cardamom leaves, and 15 times their levels in cardamom rods. Extracts of cardamom rhizomes given to atherosclerotic rats for 2 weeks, proven to lower total cholesterol, LDL-c, LDL-ox [10], Triglycerides, MDA, IL-6, and CRP [11], and otherwise increase HDL levels [10]. The cardamom rhizome has also been formulated into functional drinks by Winarsi and Hernayanti [12]. The drink is known to be rich in phenolic antioxidants (498.8 $\pm$ 0.01 ppm), greater than the ginger rhizome drink (447.93 ppm) found by Ibrahim et al. [13]. In addition to phenolic antioxidants, beverages based on the cardamom rhizome also contain vitamin C (36 mg/100 g), higher than sour turmeric drink (0.688 mg/100 g) [14]. The drink tastes sweet, semriwing, warms the body, and there is a preferred aftertaste [12].
The antioxidant compounds are capable of suppressing inflammation, protecting endothelial cell membranes by modulating lipid profiles [15], and prevent the development of atherosclerosis [10]. But, no data revealed the potential of Fd-Carrhi in suppressing levels of IL-6 and C-RP, and improving lipid profile. The aim of this study was to investigate the effects of Fd-Carrhi on levels of IL-6, C-RP, total cholesterol, triglycerides, LDL-c, and HDL-c in plasma of women with atherosclerosis.

2. Materials and methods

This randomized double-blind placebo-controlled trial was a continuation of the research Winarsi and Hernayanti [12] conducted through several stages. The research was approved by the ethics committee of the Medical Faculty of Diponegoro University and Central General Hospital Dr. Kariadi, Semarang, Indonesia.

2.1 Production functional drink rich in antioxidant rhizome cardamom (Fd-Carrhi)

The cardamom rhizome was obtained from cardamom farmers in the village of Sumbang, Banyumas, Central Java, Indonesia. After washing, cardamom rhizomes were thinly sliced then dried in cabinet dryer at 57°C, so the water content was 4-6%, called cardamom chip rhizome (CCR). The drinks was composed of a variety of CCR coupled with spices such as wooden cup, cinnamon, cloves, star anise, ginger, lemongrass, and lime leaves, as well as low calorie sweetener and a little salt [12].

2.2 Selection, classification, and intervention of research subjects

Research subjects were 30 atherosclerotic women, aged 40-65 years and suffered from hypertension (>140/90 mmHg) [16], hypercholesterolemic (>200 mg/dl), and obesity (BMI > 25 kg/m2) [17]. Subjects lived in Purwokerto, Banyumas Regency, Central Java, Indonesia and were willing to sign the informed consent. Subjects were randomly assigned into three groups of 10 and took simvastatin from the doctor. Group I was given Fd-Carrhi, II placebo (=Fd-Carrhi without CCR), and III only simvastatin. Intervention was performed in two months straight. Fd-Carrhi and placebo were given 100 mL/day one time a day, and delivered to the subject's house every morning between 06.00-08.00 to directly taken.

2.3 Blood sampling and test

Blood sample was taken three times 3 ml each on the baseline, then 1 and 2 month after intervention. Blood sample was taken using venoject with 10% EDTA. Blood was sentrifuged at 3,000 rpm for 10 minutes, separated the plasma [18] then evaluated the IL-6, C-RP, total cholesterol, trigliseride, LDL-c, and HDL-c.

2.4 Data Analysis
The result was expressed as mean ± SE. The obtained data were subject to ANOVA followed by Duncan test in case of significance with 5% error.

3. Result and Discussion

3.1 Effect Fd-Carrhi on IL-6 levels
In this research, the CCR-scented was semriwing, less savory aroma, and chelate. Preparation of Fd-Carrhi refers to Winarsi and Hernayanti [12]. Level of IL-6 was determined using Human IL-6 Elisa Kit, SunRed-Bio. On the baseline, the three groups were homogenous (P=0.84), but after 2-month intervention IL-6 level decreased from 82.67 to 54.69 ng/L (P=0.02) in Fd-Carrhi group. Opposite result occurred in placebo and statin group (P=0.53 and P=0.13) (Figure 1).

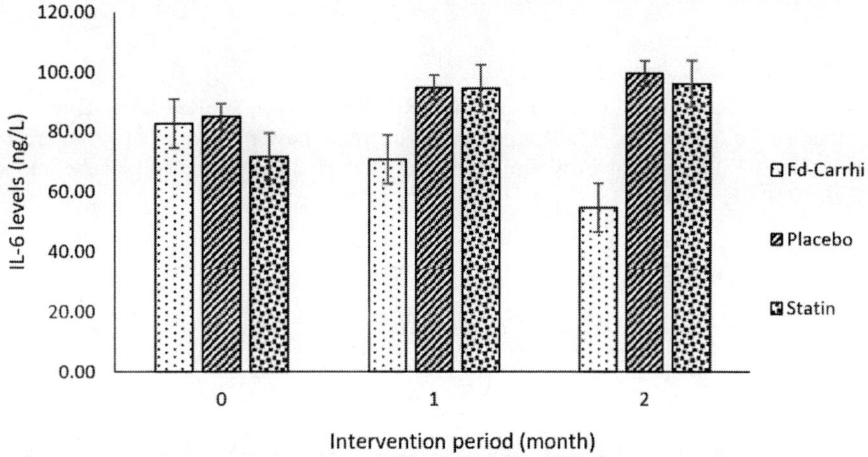

Figure 1. Effect of Fd-Carrhi on IL-6 levels in blood plasma atherosclerotic women.

Notes: Fd-Carrhi, group consuming statin + functional drink based on cardamom rhizome; Placebo, group consuming statin + placebo; Statin, group consuming statin only; n=30; P<0.05.

The decrease of IL-6 level in Fd-Carrhi group may be due to the phenpolic antioxidant in the drink. Hopkins *et al.* [19] stated that supplementing antioxidant can lower IL-6 level in blood. Phenolics are heterogenous compound derived from plant secondary metabolism. According to Bravo [20] phenolic compounds can be classified inti flavonoid and non-flavonoid. Epidemiologic data showed that high phenol diets reduced the occurrence of chronic diseases such as diabetes, cardiovascular disease, alzheimer, parkinson, and inflammation [21]. Phenolic compounds are assumed to responsible of beneficial effects on the diseases. Chronic acute inflammation contributes to the development of chronic disease such as atherosclerosis. Intervention that can modify inflammation in chronic diseases is assumed to be potential for prevention [22].
Several phenolic compounds show anti-inflammation properties. Although the antiinflamation mechanism is not thoroughly known, correlation between high intake of phenol-enriched diet and inflammation downregulation response [23]. Nieman *et al.* [24] supported that as anti inflammation, phenolic compound lowered TNF-α level through inhibiting Nuclear Factor kappa-B (NF-kB). NF-kB controls gene expression related to cytokine and chemokine proinflammation TNF-α, IL-1β and IL-6. As NF-kB is inhibited, IL-6 expression and the content decrease. Phenolic compounds work the same way as non-steroid anti-inflammatory drugs (NSAID), such as inhibiting pro inflammation mediator besides COX by limiting the activity or expression of the genes. Moreover, some phenolic compounds

increase or decrease transcription factor such as NF-kB on the tract of inflammation and antioxidant [25,26].
The structure of phenolic compounds significantly affects anti inflammation mechanism. Lättig *et al.* [27] reported that unsaturation C ring stabilized intermediate radicals through resonance. Besides, the double bond in C2 and C3 induced coplanarity between ring A and C and stimulated flavonoid interaction with the active enzyme. Eventually, lignan of phenolic compounds participates in the formation of covalent bond between flavonoid and macromolecule [28]. Meanwhile, Chuang and McIntosh [29] stated that phenolic compounds showed antiinflamation activity by inhibiting the synthesis of pro inflammation mediator, modifying eicosanoid synthesis, inhibiting activated immune cells and NOS (nitric oxide synthase) and xyclooxygenase-2 by the inhibitory effect of NF-κβ.
Some flavonoid diets are shown to modulate inflammation mediator like IL-6, such as flavonol (a type of flavonoid) that affects the concentration of blood plasma IL-6 [30]. Epidemiology data also proved the decrease in chronic disease incidence in people taking phenolic-enriched diet [31]. Every type of phenolic in extracts has different effect of pro inflammation mediator. For example, one type can inhibit pro inflammation mediator while the other inhibits the expression. Therefore, Ambriz-Pérez *et al.* [32] stated that phenolic compounds, either mono or combined, can serve as the alternative therapy for inflammation because of the ability to perform under diverse mechanisms, while medicine has specific work on the body. Fd-Carrhi is viable as an alternative natural diet for inflammation therapy to atherosclerotic patients.

3.2 Effect of Fd-Carrhi on CRP levels
C-RP level was determined using Human C-RP Elisa Kit, SunRed-Bio. Three groups were homogenous on the baseline (P=0.88); however, it decreased after 2-month intervention from 2.92 to 1.34 mg/L (P=0.047) in Fd-Carrhi group, while opposite result occurred in placebo and statin group (P=0.67 and P=0.93) (Figure 2).

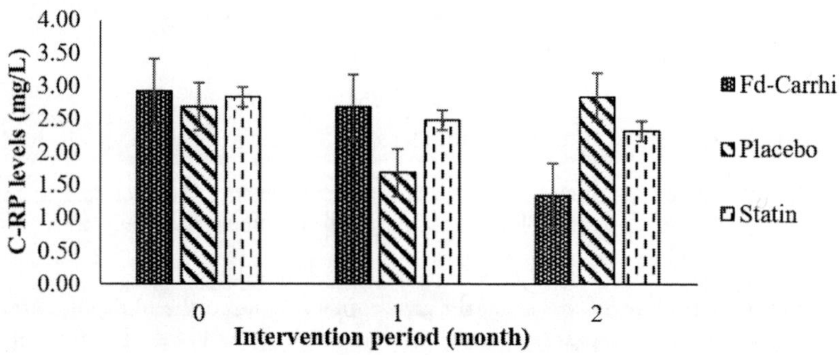

Figure 2. Effect of Fd-Carrhi on C-RP levels in blood plasma atherosclerotic women.
Notes: Fd-Carrhi, group consuming statin + functional drink based on cardamom rhizome; Placebo, group consuming statin + placebo; Statin, group consuming statin only; n=30; P<0.05.

C-RP is acute-phase protein whose level increases during chronic inflammation [33]. Epidemiology study stated that CRP is both marker and agent of atherosclerosis development [34]. CRP is also the clinical marker of the heightened risk of cardiovascular [35]. In vitro and in vivo research showed that CRP served as pro atherogenic factor and promoted atherothrombosis [36], promoted activation and dysfunction of endhotel cell [37], affected migration and proliferation of vascular smooth muscle [38], induced changes in biology matrix [39], and promoted coagulation [40].
In early study, CRP level of atherosclerotic patients was 2.72$\pm$2.23 mg/L. however, the level significantly decreased in Fd-Carrhi group compared to the placebo and statin groups, assumedly due

to phenolic compounds in Fd-Carrhi. Similar findings by Fitó *et al.* [41] stated that IL-6 and C-RP decreased after 28 coronary heart disease (CHD) patients consumed 161 mg/kg BW phenol rich olive oil. Phenolic protective effect against heart disease is derived from the ability to reduce thrombocyte aggregation [42], to improve vasorelaxant [43], to reduce lipid peroxidation [44], and to suppress procoagulant [45]. Ambriz-Pérez *et al.* [32] has traced the potential of grape phenolic that provides protection against atherosclerosis by suppressing CRP expression, the protein compounds that significantly performs in atherosclerotic pathogenesis [36]. Decreasing CRP level also occurred in the plasma of lactating mothers after consuming soy bean milk rich in isoflavone [18].

CRP serves in atherosclerosis pathogenesis because the decreased level of CRP is effective to inhibit atherosclerosis. A number of studies have proven that CRP predict the risk of CVD such as myocard infarction, coronary artery disease (CAD), stroke, peripheral artery disease, even sudden death [46]. CRP is the additional marker for 10-year risk score in predicting future CVD in healthy American women [47]. Based on the findings above, Fd-carrhi brings the atherosclerotic patients to the middle risk of CVD

3.3 Effect of Fd-Carrhi on lipid profile

Total cholesterol was determined using CHOD-PAP method (Cholesterol Oxidase-Peroxidase Aminoantipyrine Phenol). The initial total cholesterol was not different across the groups (P=0.60), but after 2-month of intervention the level decreased in Fd-Carrhi group from 291 to 186.3 mg/dl (P=3.71E-07), and in placebo group from 294.3 to 209.7 mg/dl (P=7.84E-05). Despite significant decrease in total cholesterol in placebo, the level was still more than normal. Cholesterol level in statin group remained (P=0.72) (Figure 3).

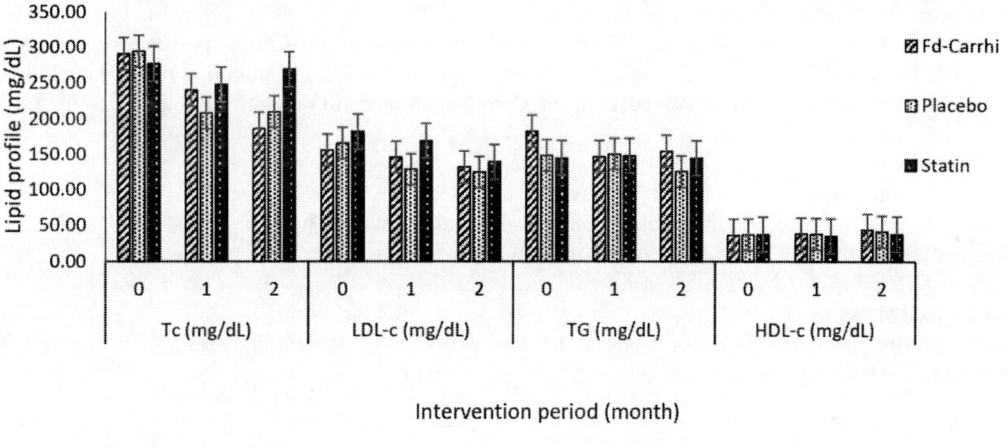

Figure 3. Effect of Fd-Carrhi on lipid profile in blood plasma atherosclerotic women.
Notes: Fd-Carrhi, group consuming statin + functional drink based on cardamom rhizome; Placebo, group consuming statin + placebo; Statin, group consuming statin only; Tc, total cholesterol; LDL-c, low density lipoprotein cholesterol; TG, triglyceride; HDL-c, high density lipoprotein cholesterol; n=30; P<0.05.

Initial triglyceride level (TG) was homogenous across the groups (P=0.29). After 2-month intervention, the level decreased but not significantly in all groups, and still within normal range (30-200 mg/dl). LDL-c level on the baseline was not different across the groups (P=0.23). After 2-month intervention, the level significantly decreased in Fd-Carrhi, placebo and statin group, that is P=0.0009; P=0.029705; and P= 0.006279, respectively (Figure 3).

High total cholesterol (T-c) and LDL-c level is the risk factor of atherosclerosis and the primary cause of CVD, but high level of high density lipoprotein cholesterol (HDL-c) is considered the protector and anti inflammation [48]. Subjects in the present research suffered from hypercholesterolemia (287.4 mg/dl), with high LDL-c (167.76 mg/dl), and low HDL-c (37.37 mg/dl). This condition is prone to CVD. However, in Fd-Carrhi group, the T-c level and LDL-c decreased but HDL-c level increased, assumedly due to the phenolic compounds in Fd-Carrhi that decreased T-c. This result supported Winarsi *et al.* [10] that flavonoid (one type of phenolic) of cardamom rhizome decreased T-c level of atherosclerotic rats induced with epinephrine. Phenolic compounds belong to flavonoid that can inhibit the activity of HMG-CoA reductase enzyme [49], thereby decreasing cholesterol synthesis and thus cholesterol content. Flavonoids inhibit cholesterol absorption by limiting the formation of micelles. The inhibited cholesterol absorption in intestines is due to insoluble flavonoid-cholesterol complex, then bound with bile acid and excreted in feces. Cholestyramine, sequestrant of bile acid distracts the circulation of enterohepatic bile acid by releasing it and preventing the reabsorption in the intestine. Consequently, bile acid pool is reduced. The more cholesterol converted into bile acid (to maintain steady level in blood circulation), the lower plasma cholesterol level [50]. Cicerale *et al.* [51] stated that in the body of 200 healthy men consuming high phenolic diets, T-c/HDL-c decreased. The increasing HDL-c is noted linear to the increasing phenolic [52]. Olive oil phenolic intake has improved HDL-c circulation around 5.1-6.7% in human [53]. Gimeno *et al.* [54] also reported LDL-c decrease after one week consumption of phenol-enriched olive oil.

Upon FD-Carrhi consumption, HDL-c level increased from 36.69 to 44.26 mg/dl (P=0.014). Similarly, placebo and statin groups experienced HDL-c increased but not significant (P=0.09) and (P=0.71), respectively (Figure 3).

The increasing HDL-c level also occurred in Fd-Carrhi subjects during 2 months which may due to phenolic content. Mechanism of HDL-c level increased by phenolic FD-carrhi was unclear, but Lamon-Fava *et al.* [55] confirms that flavonoid increases the production of apolipoprotein A1 (Apo-A1) and regulation of its expression through signaling pathways of mitogen activated protein kinase. Apolipoprotein-A1 is a compound which contributes to the formation of prebeta-HDL-c, which will then be converted to alpha-HDL-c, then mature through the process of esterification of free cholesterol to cholesterol ester by lecithin-cholesterol acyl transferase enzyme. HDLc is an antiatherogenic, antioxidant, anti inflammatory [56]. Increased concentrations of HDL-c can reduce the progression of atherosclerotic lesions.

In this study there was no change in TG levels in the group taking Fd-Carrhi, placebo, or statins, but the levels were within the normal range (167.76 mg / dl). According to Sudhop *et al.* [57] such levels belong to high normal limit criteria (borderline), therefore it needs to be controlled with Fd-Carrhi.

4. Conclusion

Fd-Carrhi can reduce of IL-6 by 33.85%, C-RP 54.24%, T-c 35.98%, and LDL-c 15.22%, while increasing HDL-c 20.63% and maintaining triglycerides level. Therefore, Fd-Carrhi can inhibit atherosclerosis toward CHD by suppresses inflammation and improves lipid profile.

Acknowledgement

We gratefully thank the Director General of The Ministry of Research, Technology and Higher Education for funding this Competence Grant Research 2017.

References

[1] Larsen CM, Faulenbach M, Vaag A, Volund A, Ehses JA, Seifert B, Mandrup-Poulsen T and Donath MY 2007 Interleukin-1-receptor antagonist in type 2 diabetes mellitus. *N Engl J Med.* **356**(15), 1517–26.

[2] Winarsi H and Purwanto A 2010 The effect of germinated-soy protein supplementation on IL-1 Beta levels in type 2 diabetics. *Jurnaltip.* **21**(1), 6–10.

[3] Dandona P, Aljada A and Chaudhuri A 2005 Metabolic syndrome: a comprehensive perspective based on interactions between obesity, diabetes, and inflammation. *Circulation.* **111**, 1448–54.

[4] Lind L 2003 Circulating markers of inflammation and atherosclerosis. *Atherosclerosis.* **169**, 203–14.

[5] Kumar EP, Mukherjee R, Senthil R, Parasuraman S and Suresh B 2012 Evaluation of oxidative stress and antioxidant status in patients with cardiovascular disease in rural populations of the Nilgiris, South India. *Int Sch Res Notices* (2012), Article ID 941068, 6p.

[6] Frostegård J 2013. Immunity, atherosclerosis and cardiovascular disease. *BMC Med.* **11**, 117–129.

[7] Mitra S, Goyal T and Mehta JL 2011 Oxidized LDL, LOX-1 and atherosclerosis. *Cardiovasc Drugs Ther.* **25**(5), 419–29.

[8] Valko M, Rhodes CJ, Moncol J, Izakovic M and Mazur M 2006 Free radicals, metals and antioxidants in oxidative stress-induced cancer. *Chem. Biol. Interact.* **160**(1), 1–40.

[9] Chatterjee S, Zareena N, Gautam S, Soumyakanti A, Prasad SV and Arun S 2007 Antioxidant activity of some phenolic constituents from green pepper (*Piper nigrum* L.) and fresh nutmeg mace (*Myristica fragrans*). *Food Chem.* **101**(2), 515–523.

[10] Winarsi H, Yuniaty A and Nuraeni I 2016a Hypocholesterolemic and attenuated oxidized-LDL of epinephrine-induced atherosclerosis rats using cardamom rhizome ethanolic extract: Study of functional-food components. *Int Food Res J.* **23**(5), 2103–2111.

[11] Winarsi H, Yuniaty A and Nuraeni I 2016b Improvement of antioxidant and immune status of atherosclerotic rats adrenaline and egg-yolks -induced using cardamom-rhizome-ethanolic-extract: An initial study of functional food. *Agriculture and Agricultural Science Procedia* **9**, 264–270.

[12] Winarsi H and Hernayanti 2016 Functional drinks rich in antioxidant cardamom rhizome. *International Food Conference 2016. Surabaya, 20-21 October 2016.*

[13] Ibrahim AM, Yunianta and Srihertyna FH 2015 The influence of the temperature and the length of time the extraction of the chemical and physical properties in the manufacture of juice drinks red ginger (*Zingiber officinale* var. Rubrum) in combination with the addition of honey as a sweetener. *Journal of Food and Agro-Industry,* **2**(3), 530–541.

[14] Mulyani S, Harsojuwono BA and Puspawati GAKD 2014 Potential acid drinks turmeric (*Curcuma domestica* Val. -*Tamarindus Indica* L.) as an antioxidant-rich beverage. *Agritech* **34**(1), 65–71.

[15] Squadrito F, Altavilla D, Crisafulli A, Saitta A, Cucinotta D, Morabito N, D'Anna R, Corrado F, Ruggeri P, Frisina N and Squadrito G 2003 Effect of genistein on endothelial function in postmenopausal women: a randomized, double-blind, controlled study. *Am J Med* **114**, 470–476.

[16] Humar R, Zimmerli L and Battegay E 2009 Angiogenesis and hypertension: an update. *J Hum Hypertens.* **23**, 773–782.

[17] Nishida C 2004 Appropriate body-mass index for Asian populations and its implications for policy and intervention strategies. *Lancet* **363**, 157–63.

[18] Winarsi H, Sasongko ND and Purwanto A 2016c Germinated-soy milk in supressing inflammation and oxidative stress in blood plasma and breast milk of lactating mothers. *Int Food Res J.* **23**(2), 646–652.

[19] Hopkins MH, Fedirko V, Jones DP, Terry PD and Bostick RM 2010 Antioxidant micronutrients and biomarkers of oxidative stress and inflammation in colorectal adenoma patients: results from a randomized, controlled clinical trial. *Cancer Epidemiol Biomarkers Prev.* **19**(3), 850–8.

[20] Bravo L 1998 Polyphenols: chemistry, dietary sources, metabolism, and nutritional significance. *Nutr. Rev.* **56**(11), 317–33.

[21] Mohammed MS, Osman WJA, Garelnabi EAE, Osman Z, Osman B, Khalid HS and Mohamed MA 2014 Secondary metabolites as anti-inflammatory agents. *J Pharmacol.* **3**(4), 275–285

[22] Allgrove J and Davison G 2014 Dark chocolate/cocoa polyphenols and oxidative stress. Polyphenols in human health and disease. (Vol 1). Elsevier. ISBN 978-0-12-398456-2. (Chapter 19).

[23] de la Lastra CA and Villegas I 2005 Resveratrol as an anti-inflammatory and anti-aging agent: Mechanisms and clinical implications. *Mol Nutr Food Res.* **49**, 405–430.

[24] Nieman DC, Henson DA, Davis JM, Murphy EA, Jenkins DP, Gross SJ, Carmichael MD, Quindry JC, Dumke CL, Utter AC, McAnulty SR, McAnulty LS and Mayer EP 2007 Quercetin's influence on exercise-induced changes in plasma cytokines and muscle and leukocyte cytokine mRNA. *J Appl Physiol.* **103**, 1728–1735.

[25] Maroon JC, Bost JW and Maroon A 2010 Natural antiinflammatory agents for pain relief. *Surg Neurol Int.* **1**, 1–80.

[26] Sergent T, Piront N, Meurice J, Toussaint O and Schneider YJ 2010 Anti-inflammatory effects of dietary phenolic compounds in an in vitro model of inflamed human intestinal epithelium. *Chem. Biol. Interact.* **188**, 659–667.

[27] Lättig J, Böhl M, Fischer P, Tischer S, Tietböhl C, Menschikowski M and Pisabarro MT 2007 Mechanism of inhibition of human secretory phospholipase A2 by flavonoids: Rationale for lead design. *J Comput Aided Mol Des.* **21**, 473–483.

[28] Lago JH, Toledo-Arruda AC, Mernak M, Barrosa KH, Martins MA, Tibério IF and Prado CM 2014 Structure-activity association of flavonoids in lung diseases. *Molecules,* **19**, 3570–3595.

[29] Chuang C-C and McIntosh MK 2011 Potential mechanisms by which polyphenol-rich grapes prevent obesitymediated inflammation and metabolic diseases. *Annu. Rev. Nutr.* **31**, 155–176.

[30] Stote K, Clevidence B, Novotny J, Henderson T, Radecki S and Baer D 2012 Effect of cocoa and green tea on biomarkers of glucose regulation, oxidative stress, inflammation and hemostasis in obese adults at risk for insulin resistance. *Eur J Clin Nutr.* **66**, 1153–1159.

[31] Rodrigo R, Miranda A and Vergara L 2011 Modulation of endogenous antioxidant system by wine polyphenols in human disease. *Clin. Chim. Acta.* **412**, 410–424.

[32] Ambriz-Pérez DL, Leyva-López N, Gutierrez-Grijalva EP and Heredia JB 2016 Phenolic compounds: Natural alternative in inflammation treatment. A Review. *Cogent Food Agric.* (2016), **2**, 1131412.

[33] Black S, Kushner I and Samols D 2004 C-reactive Protein. *J. Biol. Chem.* **279**(47): 48487–90.

[34] Verma S, Kuliszewski MA, Li SH, Szmitko PE, Zucco L, Wang CH, Badiwala MV, Mickle DA, Weisel RD, Fedak PW, Stewart DJ and Kutryk MJ 2004 C-reactive protein attenuates endothelial progenitor cell survival, differentiation, and function: further evidence of a mechanistic link between C-reactive protein and cardiovascular disease. *Circulation,* **109**(17), 2058–67.

[35] Scirica BM and Morrow DA 2006 Is C-reactive protein an innocent bystander or proatherogenic culprit? The verdict is still out. *Circulation,* **113**(17), 2128–34.

[36] Venugopal SK, Devaraj S and Jialal I 2005 Effect of C-Reactive Protein on vascular cells: evidence for a proinflammatory, proatherogenic role. *Curr Opin Nephrol Hypertens.* **14**(1), 33–37.

[37] Fichtlscherer S, Rosenberger G, Walter DH, Breuer S, Dimmeler S and Zeiher AM 2000 Elevated C-reactive protein levels and impaired endothelial vasoreactivity in patients with coronary artery disease. *Circulation,* **102**(9), 1000–6.

[38] Blaschke F, Bruemmer D, Yin F, Takata Y, Wang W, Fishbein MC, Okura T, Higaki J, Graf K, Fleck E, Hsueh WA and Law RE 2004 C-Reactive Protein induces apoptosis in human coronary vascular smooth muscle cells. *Circulation* **110**, 579–587.

[39] Montero I, Orbe J, Varo N, Beloqui O, Monreal JI, Rodríguez JA, Díez J, Libby P and Páramo JA 2006 C-Reactive Protein stimulates MMP-1 expression in U937 histiocytes through FcγRII and extracellular signal-regulated kinase pathway: An implication of CRP involvement in plaque destabilization. *J. Am. Coll. Cardiol.* **47** (7), 1369 –78.

[40] Cermak J, Key NS, Bach RR, Balla J, Jacob HS and Vercellotti GM 1993 C-reactive protein induces human peripheral blood monocytes to synthesize tissue factor. *Blood,* **82**, 513–520.

[41] Fitó M, Cladellas M, de la Torre R, Martí J, Muñoz D, Schröder H, Alcántara M, Pujadas-Bastardes M, Marrugat J, López-Sabater MC, Bruguera J and Covas MI 2008 Anti-inflammatory effect of virgin olive oil in stable coronary disease patients: a randomized, crossover, controlled trial. *Eur J Clin Nutr .* **62**, 570–574.

[42] Moschona A, Kyriakidis KD, Kleontas AD and Kyriakides ML 2017 Comparative study of natural phenolic acids and flavonols as antiplatelet and anti-inflammatory agents. *Grant Med J* **2**(4), 057–066.

[43] Lekakis J, Rallidis LS, Andreadou I, Vamvakou G, Kazantzoglou G, Magiatis P, Skaltsounis AL and Kremastinos DT 2005 Polyphenolic compounds from red grapes acutely improve endothelial function in patients with coronary heart disease. *Eur J Cardiovasc Prev Rehabil.* **12**, 596–600.

[44] Aviram M and Fuhrman B 2002 Wine fl avonoids protect against LDL oxidation and atherosclerosis. *Ann. N. Y. Acad. Sci.* **957**, 146–161.

[45] Di Santo A, Mezzetti A, Napoleone E, Di Tommaso R, Donati MB, De Gaetano G and Lorenzet R 2003 Resveratrol and quercetin down-regulate tissue factor expression by human stimulated vascular cells. *J Thromb Haemost.* **1**(5), 1089–95.

IOP Conf. Series: Earth and Environmental Science **102** (2017) 012048 doi:10.1088/1755-1315/102/1/012048

[46] Ridker PM 2003 Clinical application of C-Reactive Protein for cardiovascular disease detection and prevention. *Circulation,* **107**, 363–369.

[47] Ridker PM, Hennekens CH, Buring JE and Rifai N 2000 C-reactive protein and other markers of inflammation in the prediction of cardiovascular disease in women. *N Engl J Med* **342**, 836–843.

[48] Chrysohoou Ch, Pitsavos Ch, Skoumas J, Masoura C, Katinioti A, Panagiotakos D and Stefanadis Ch 2006 The emerging anti-inflammatory role of HDL-cholesterol, illustrated in cardiovascular disease free population; The Attica Study. *Int. J. Cardiol.* **122** (1), 29–33.

[49] Prahastuti S, Tjahjani S and Hartini E 2011 The effect of bay leaf infusion (*Syzygium polyanthun* (Wight) Walp) to decrease blood total choleserol level in dyslipidemia model wistar rats. *Planta* Med. **1**(4), 29–30.

[50] Ngamukote S, Mäkynen K, Thilawech T and Adisakwattana S 2011 Cholesterol-lowering activity of the major polyphenols in grape seed. *Molecules* 16, 5054-5061.

[51] Cicerale S, Lucas L and Keast R 2010 Biological activities of phenolic compounds present in virgin olive oil. *Int. J. Mol. Sci.* **11**(2), 458–479.

[52] Covas MI, Nyyssönen K, Poulsen HE, Kaikkonen J, Zunft HJF, Kiesewetter H, Gaddi, A, de la Torre R, Mursu J, Bäumler H, Nascetti S, Salonen JK, Fitó M, Virtanen J and Marrugat J 2006 The effect of polyphenols in olive oil on heart disease risk factors: A randomized trial. *Ann. Intern. Med.* **145** (5), 333–341.

[53] Marrugat J, Covas MI, Fitó M, Schröder H, Miró-Casas E, Gimeno E, López-Sabater MC, de la Torre R and Farré M 2000 Effects of differing phenolic content in dietary olive oils on lipids and LDL oxidation a randomized controlled trial. *Eur J Nutr.* **43**(3), 140–7.

[54] Gimeno E, Castellote AI, Lamuela-Raventós RM, de la Torre MC and López-Sabater, M. C. (2002). The effect of harvest and extraction methods on the antioxidant content (phenolics, α-tocopherol and β-carotene) in virgin olive oil. *Food Chem.* **78**, 207–211.

[55] Lamon-Fava S 2000 Genistein activates apolipoprotein A-I gene expression in the human hepatoma cell line Hep G2. *J Nutr.* **130**, 2489–92.

[56] Soran H, Schofield JD and Durrington PN 2015 Antioxidant properties of HDL. *Front Pharmacol.* **6**(222), 1-6.

[57] Sudhop T, Reber M, Tribble D, Sapre A, Taggart W, Gibbons P, Musliner T, von Bergmann K and Lütjohann D 2009 Changes in cholesterol absorption and cholesterol synthesis caused by ezetimibe and/or simvastatin in men. *J. Lipid Res.* **50**, 2117–2123.

IOP Conf. Series: Earth and Environmental Science **102** (2017) 012049 doi:10.1088/1755-1315/102/1/012049

Quality evaluation of polypropylene packaged corn yogurt during storage

Nur Aini[1,2], V Prihananto[1], B Sustriawan[1], Y Astuti[1] and M R Maulina[1]

Department of Food Science and Technology, Jenderal Soedirman University, Purwokerto, Indonesia

Corresponding author : nur.aini@unsoed.ac.id

Abstract. Packaging is an important factor to control the process of quality decrease of any food product, including to determine the shelf life. The objective of this study was to determine changes quality of corn yogurt packaged using polypropylene. The method were using was package yogurt polypropylene, then it was stored in a refrigerator at 5, 10, or 15°C during 21 days. The yogurt was analysed every 7 days over a 21-day period. The results indicate that protein content decreased during storage, while the lactic acid bacteria, total acid, pH, viscosity, and total solids were increased. At the end of storage, the amount of lactic acid bacteria still fulfil the minimum requirements of a probiotic food, with a count of 6.407 log CFU/g. Overal scoring by panelist (scores ranged from 0 to 5) have a 4.78 at the beginning of storage. By the 21st day of storage, yogurt was packaging using transparent polypropylene having a score of 3.85, and that stored in opaque white packaging having a value of 3.95.

1. Introduction

Yoghurt is one of dairy-based probiotic products which is much sold on the market. People who are allergic to milk protein require an alternative ingredient of yoghurt. For example, Yoghurt with vegetable as the raw material such as peanuts and soybeans [1], [2]. [3] developed probiotic corn extract which has similar characteristic to yogurt in which it has 8.74 log CFU/g of total lactic acid bacteria (BAL). This shows that sweet corn extract is qualified as a probiotic food which requires lactic acid bacteria of 6 log CFU/g [4].

Changes during the storage will affect the food quality. The stability of food products is related to the speed of physical damage due to the changes of chemical substance, physic and microbiology. Temperature is a factor which affects the changes in food product quality. When food is stored, the storage room temperature is constant. For fermentation products' storage, the temperature is very much affect the storage time. For fermentation products whose bacteria is alive, its storage time is short and it should be stored at 4° C. According to [5], the storage of yogurt at 4°C will decrease the viability of lactic acid bacteria.

Pasteurized or sterilized fermentation products have a longer storage time, but its condition also depends on the type of packaging [6]. The packaging commonly used for yoghurt is a polypropylene cup due to its small surface area so it can reduce the contact of the ingredients with oxygen. Generally, the polypropylene packaging used is transparent and opaque. The opaque polypropylene packaging is not transparent so it can provide different qualities during the storage.

The aim of this research is to know the quality change of corn yoghurt which is packed with transparent and opaque polypropylene at 5, 10 and 15°C.

2. Experimental details

2.1. The making of corn yogurt

The method used to produce corn yogurt was that reported by [3]. A corn extract was prepared and combined with 15% sweet potato, 10% mung bean extract, 15% sugar, and 10% skimmed milk powder. The mixture was pasteurized at 70°C for 15 minutes, then cooled to 40–43°C. A mixed culture of *Lactobacillus bulgaricus* and *Streptococcus thermophilus* were then added. The final mixture was incubated for 8 hours at 37°C.

2.2. Package and storage of corn yogurt

Corn yogurt was then packed with transparent and white cups. Storage was done on refrigerator at 5°C, 10°C and 15°C for 21 days. The data were analyzed every 7 days for 21 days.

2.3. Analysis of samples

The research variables observed was total lactic acid bacteria using plate count method and M.R.S Agar media [7]; the pH measurement was carried out by potentiometric method, by using pH meters in all experimental units, total soluble solids (using refractometer), lactic acid levels determined by titration using alkaline solution (Mann,s Acid Test), total protein level (micro Kejldahl method). Sensory analysis including color, acid flavor, aroma and preference was done by using scoring test. Panelists used were trained panelists consisting of 20 people. The scoring is on a scale of 1 to 5.

2.4. Analysis of data

Data analysis was done by using simple linier regression method and descriptive analysis.

3. Result and discussion

3.1. Viability of lactic acid bacteria

According to [8], Lactobacillus is a potential lactic acid bacteria (BAL) to make food to be functional. At first, the amount of BAL was 7,573 log CFU/ml, which fulfilled the minimum requirement of probiotic food. According to [9], the total minimum of BAL in probiotic food is 6 log CFU/g at expiration time.

During the storage, BAL decreased slightly from 0.67 to 0.86 log cycle. Until the 21st day, the amount of BAL still met the minimum requirement of probiotic food, in which the lowest was 6.407 log CFU/ml (Figure 1). The decrease of BAL during this storage was better than [5] result in which BAL of yoghurt decreased as much as 2.34 log cycles during the 28 days storage time. The decrease viability of lactic acid bacteria is caused by the excessive production of acids resulting in the lactic acid bacteria death. Meanwhile, according to [10] the more total BAL exists in food, the tighter the competition among BAL will be. It means that the more total BAL in food will decrease the availability of nutrients and the BAL survival. So, the longer the storage time, the number of BAL will decrease.

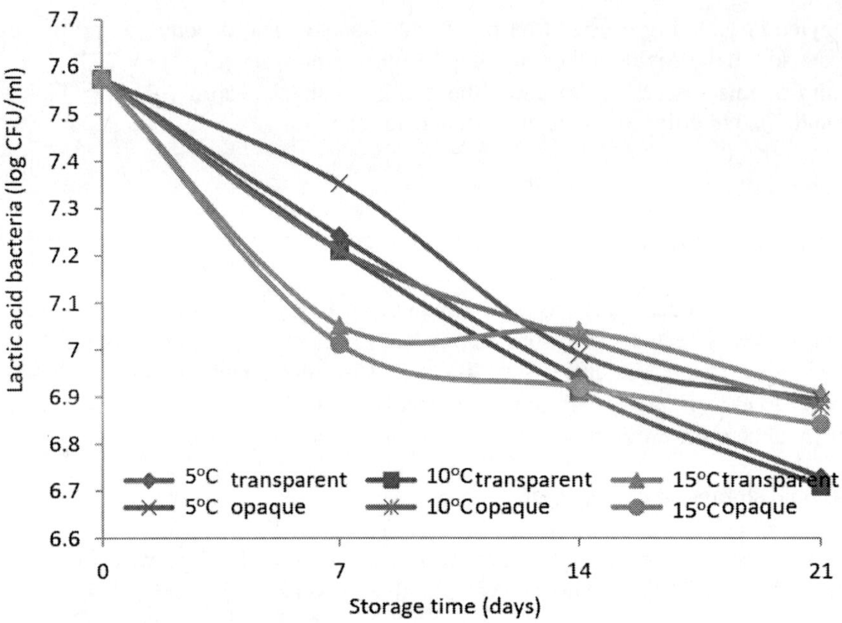

Figure 1. The number of lactic acid bacteria in corn yogurt using transparent and opaque packaging

The amount of lactic acid bacteria in yogurt using transparent and opaque packaging tended to be the same during the storage. According to [4], the ability of bacteria to grow and develop is influenced by the nutrition and environmental conditions. Nutrition that can be used by BAL to survive is proteins, fats and minerals (calcium, phosphorus and magnesium), of which have the same amount.

3.2. Protein content

During the storage, the protein content of corn yoghurt tended to decrease until the 7th day, after that, it tended to increase until the 21st day. The protein levels decreased until the 7th day because Lactobacillus was proteolytic, breaking proteins into simple peptides which was decreasing the protein levels.

Protein levels which tended to increase after the 7th day were likely because the protease enzymes degraded the proteins in microbes. Microbes are composed by various components, including 60-70% of protein. According to [11], microbial cells lysis during the storage. The protease and peptidase enzymes will break the protein down in the BAL, so the more cells lyse, the higher the protein level in yoghurt. This protein changes pattern is almost the same for transparent and opaque polypropilene packaging yogurt during the storage, but for yoghurt which was stored at the highest temperature (5°C), the protein level tends to be higher.

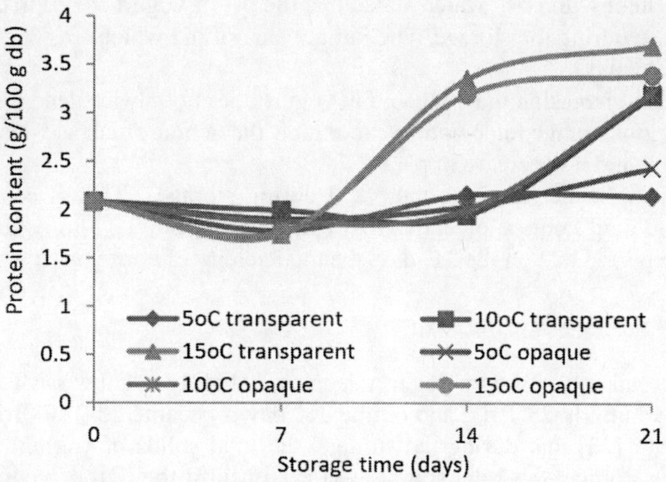

Figure 2. The protein content of corn yogurt was packaged by transparent and opaque polypropylene during the storage

3.3. Total acid and pH

The most produced organic acids in yogurt and as the main product is lactic acid [12]. During the storage, total acids in each packaging and temperature tended to increase, initially from 0.765 to 0.819-1.026 (Table 1). This shows that the production of lactic acid in corn yoghurt had increased during the storage. This is similar to [5] result in which during the 21 days storage, lactic acid levels increased to 1.11 until 1.42%. [13] also almost the same, in which during the 21 days storage, the total acidity of yogurt is 1.

According to [11], lactic acid bacteria has the optimum temperature to produce acids at 10-40°C. At 10°C and 15°C, the production of lactic acid approaches the optimal temperature of lactic acid bacteria activity so that the production of lactic acid increases. This corn yogurt also fulfill the standard of yoghurt SNI no 01-2981-1992 which has total acid of 0.5-2.

Table 1. The pH and total acid of corn yogurt was packaged by polypropylene during storage

Packaging	Storage temperature	pH on the day				Total acid on the day (%)			
		0	7	14	21	0	7	14	21
Transparent	5	4.4	4.5	4.5	4.7	0.765	0.774	0.882	0.882
	10	4.4	4.4	4.4	4.6	0.765	0.819	1.053	0.999
	15	4.4	4.5	4.5	4.6	0.765	0.81	0.882	0.918
Opaque white	5	4.4	4.5	4.5	4.6	0.765	0.783	0.864	0.819
	10	4.4	4.4	4.4	4.6	0.765	0.837	0.927	1.026
	15	4.4	4.5	4.5	4.6	0.765	0.747	0.963	0.945

Lactic acid produced during the fermentation can increase the acidity or decrease the pH. In this research, there was no decrease in pH, instead the pH increased (Table 1). On the 21st day, the pH which was originally 4.4 became 4.6-4.7. This was due to the delayed growth of lactic acid bacteria and the growth of proteolytic and other lipolytic organisms which have the ability to perform lactic acid metabolism and therefore produced base compounds. The increase of pH of yoghurt during the storage is also consistent with [14] which stated that the pH of yogurt stored in cold temperatures (4 ± 1°C) tends to increase during the storage. The initial value of pH which was 3.2 rose to 3.5 on the 10th day and 3.8 on the 35th day.

[15] reported that increasing the total acid in yogurt does not always lead to a decrease in pH. [4] also reported that a low storage temperature can inhibit the action of lactase, leading to a reduction in lactic acid production and no decrease in pH.

The production of lactic acid was increased during storage. This is consistent with [5], who found that the lactic acid content of corn yogurt increased from 1.11 to 1.42% during 21 days of storage. [13] also reported that on the 21st day, the total acidity of corn yogurt was 1%.

3.4. *Total solid and viscosity*

During the storage, total solids of corn yoghurt decreased slightly as shown in Table 2. The total solids of corn yoghurt was initially 23°Brix, and on the 21st day it became 22-22.7°Brix. This total decrease of solids is similar to [13] that during the storage, the total solids of yoghurt decrease. Despite the decreases during the storage, this total soluble solids still fulfil the SNI of yogurt standard of at least 8.2°Brix.

Table 2. Total solid and viscosity of corn yogurt was packaged by polypropylene during storage

Packaging	Storage temperature	Total solid at day (°Brix)				Viscosity at day (cP)			
		0	7	14	21	0	7	14	21
Transparent	5	23	23	23	23	160	190	170	160
	10	23	23	23	22,7	160	160	180	140
	15	23	23	23	22,7	160	170	180	140
Opaque white	5	23	23	23	22,5	160	180	180	200
	10	23	23	23	22	160	150	180	190
	15	23	23	23	22,6	160	150	180	160

The desired viscosity of yogurt is viscous liquids and homogeneous consistency. According to [16], the viscosity of yogurt is also influenced also by milk protein, which is casein. At pH close to 4.6, the solubility of casein is lost therefore the hydrophobic interaction occurs between the casein forming the structure and the main consistency of yogurt, and the process may affect the physic of yogurt, such as texture, viscosity, water holding capacity, and syneresis.

Casein is coagulated as the degree of acidity decreases due to the production of lactic acid. Lactic acid can agglomerate milk protein, sweet corn protein or mungbean protein. [16] explained that lactic acid leads to the change of casein/phosphate complexes to be soluble fraction of calcium and phosphate so that the casein destabilizes.

During the storage, there is increased yoghurt viscosity due to the increase of lactic acid. This can be seen from yogurt which was stored in opaque packaging until the 21st day and transparent packaging until the 14th day. It is caused by syneresis or separation of whey causing unstable curd. Syneresis can occur due to high storage temperatures, low total solids in milk, and no vibration during transport or during storage. Syneresis occurred because the transparent packaging can be penetrated by light that disrupts the stability of yogurt.

At the end of storage, the opaque packaging corn yoghurt has a higher viscosity than the transparent packaging. Storage was done in the refrigerator which has light bulb which directly highlights the product. During the storage, there is increased yoghurt viscosity due to the increase of lactic acid. This can be seen from yogurt which was stored in opaque packaging until the 21st day and transparent packaging until the 14th day. It is caused by syneresis or separation of whey causing unstable curd. Syneresis can occur due to high storage temperatures, low total solids in milk, and no vibration during transport or during storage. Syneresis occurred because the transparent packaging can be penetrated by light that disrupts the stability of yogurt.

Yogurt is required to have the form of a viscous fluid, with a homogenous consistency. According to [17], the viscosity of yogurt is also influenced by milk protein, i.e., casein. At a pH of approximately 4.6, casein becomes insoluble, resulting in hydrophobic interactions between casein micelles that form the main structure and consistency of yogurt. This process can influence the physical properties of yogurt, including its viscosity, water-holding capacity, and syneresis.

Casein is coagulated when acidity decreases due to lactic acid production. Lactic acid can agglomerate protein in milk, corn, or mungbean. [17] explain that the production of lactic acid resulted in a complex change whereby casein–phosphate was converted into soluble calcium and phosphate, with the casein being destabilized.

3.5. Sensory properties

The yoghurt quality is determined by texture or viscosity, the degree of acidity and flavor which is highly influential on sensory quality and consumer acceptance. Sensory analysis done included color, flavor, aroma and overall acceptance.

Corn yoghurt is yellow due to the presence of carotenoids. Color changes also occur because of oxidation reactions which can cause changes in flavor, color, and microbial growth in large quantities which also can damage the color or appearance of food.

Color is an important food attribute and becomes the first characteristic seen by consumers. The color of yoghurt according to the panelist assessment is 4.97 (yellow). Yellow color in yoghurt is caused by the presence of carotenoids in corn and sweet potatoes. The longer the storage, the more decrease the carotenoid color pigment will be. This is due to the degradation of pigments carried out by lactic acid bacteria. According to [16] the color changes of the fermentation medium which is getting faded after the fermentation can be caused by anthocyanin pigment degradation due to the decrease of acidity (pH) during the fermentation process. Yogurt which packaged by white opaque polypropylene was more yellow (4,2-4,35) than yogurt which packaged by transparent polypropylene (4.15) as shown in Table 3.

The formation of lactic acid in yogurt resulted in sour taste. The panelist rate on the yoghurt sour taste at the beginning of storage was 3.98 (acid), and at the end of the storage range at 3.95 (sour) to 4.4 (very sour). According to [18] the sour taste characteristic of yogurt is due to their lactic acid. The increased acid taste during storage in accordance with [13] that sour taste during the storage increased because the amount of lactic acid increased.

The panelist assessment of yogurt aroma indicated that yogurt aroma decreased during the storage. At first, the aroma value was 4.85 and at the end of the storage (the 21st day) it decreased to 3.5 - 4.35. This is caused by alcohol formed as the result of further degradation of lactic acid during the storage. According to [5], this is considered to be due to a contamination of fermentative yeast that will form alcohol during the storage.

According to [18], flavor yogurt compounds consists of 1) non-volatile acids (lactic acid, pyruvate or succinate); 2) volatile acids (formic acid, acetate, propionate or butyrate); 3) carbonyl compounds (acetaldehyde, acetone, or diacetyl), and 4) other compounds (amino acids and compounds formed by thermal degradation of proteins, fats or lactose). Among these compounds, acetaldehyde and diacetyl is the most predominant in determining the scent of yoghurt. According to [5], acetaldehyde and other carbonyl compounds will decrease during the storage of yoghurt products. In addition, during the storage, there may be oxidation so that the volatile compounds evaporate. This results in the decrease of yoghurt aroma during the storage.

Table 3. Sensory properties of corn yogurt was packaged by polypropylene during storage

Sensory attributes	Packaging		Day			
			0	7	14	21
Color	Transparent	5	4.97	4.4	4.35	4.15
		10	4.97	4.4	4.1	4.1
		15	4.97	4.35	4.1	4.1
	Opaque white	5	4.97	4.45	4.35	4.3
		10	4.97	4.4	4.4	4.35
		15	4.97	4.25	4.25	4.2
Flavour	Transparent	5	4.85	4.4	4.5	4.35
		10	4.85	3.85	3.75	3.75
		15	4.85	4.05	4	3.95
	Opaque white	5	4.85	4.3	4.25	4.15
		10	4.85	4.45	3.55	3.5
		15	4.85	4.35	3.85	3.7
Taste	Transparent	5	3.98	3.85	4.15	4.2
		10	3.98	3.95	4	4.15
		15	3.98	3.8	3.9	3.95
	Opaque white	5	3.98	3.8	4.05	3.95
		10	3.98	3.7	4.05	4.25
		15	3.98	3.75	4.2	4.4
Overall acceptance	Transparent	5	4.78	4.5	4.1	3.85
		10	4.78	4.4	3.3	3.15
		15	4.78	4.35	3.75	2.7
	Opaque white	5	4.78	4.55	4.15	3.85
		10	4.78	4.65	4	3.3
		15	4.78	4.5	3.2	2.85

Panelists' preference on yogurt is affected by several factors, including flavor, aroma, and consistency. According to [18] the preference on yoghurt is much related to the fermentation by lactic acid bacteria which resulted in a distinctive flavor of yoghurt. The stability of yoghurt also affects the preference, which can be seen from no occurrence of yoghurt damage in the form of wheying off or syneresis. According to [5] moderate viscosity and subtle texture are preferred by consumers.

The longer the storage time, the more decrease the panelists' preference judgment on yogurt is. At first, panelist's preference on yogurt was 4.78 and after storage for 21 days, panelists' preference on yogurt was 2.7-3.85 (Table 3). In metabolism reactions, the acid will turn into alcohol so that other aromas which arise are the result of the process. This is also reason of the decrease preference of panelists' preference on yoghurt.

4. Conclusion

During corn yogurt storage, the protein content decreased, while the total acid, pH, viscosity, and total solids remained consistent for both types of packaging at all three temperatures tested. At the end of storage, the amount of lactic acid bacteria still fulfil the minimum requirements of a probiotic food, with a count of 6.407 log CFU/g. Panel ratings of sensory variables tended to decrease as the length of storage increased. Corn yogurt was rated by a sensory assessment panel, and scores ranged from 0 to 5, with a mean rating of 4.78 at the beginning of storage. By the 21st day of storage, these ratings had decreased, with yogurt stored in transparent packaging having a rating of 3.85, and that stored in opaque white packaging having a value of 3.95.

Acknowledgement

This study was supported by the Ministry of Research, Technology, and Higher Education of the Republic of Indonesia under the National Strategic Research program in 2016 grant number 2045/UN23.14/PN/2016.

Reference

[1] J. Isanga and G. Zhang, "Production and evaluation of some physicochemical parameters of peanut milk yoghurt," *LWT-Food Sci. Technol.*, vol. 42, no. 6, pp. 1132–1138, 2009.

[2] M. Sveučilište u Zagrebu. Prehrambeno-biotehnološki fakultet., L. Hrvatsko društvo za biotehnologiju., Saveza hrvatskih društava za unapređenje prehrane., and Slovenskog mikrobiološkog društva., "Physicochemical, Textural and Sensory Characteristics of Probiotic Soy Yogurt Prepared from Germinated Soybean," *Food Technol. Biotechnol.*, vol. 48, no. 4, pp. 490–496, Dec. 2010.

[3] N. Aini, V. Prihananto, G. Wijonarko, A. Arimah, and M. Syaifudin, "Effect of Culture Concentration and Sweet Potato Prebiotic to the Properties of Sweet Corn Juice Probiotic," *Agritech*, vol. 37, no. 2, pp. 165–172, 2017.

[4] D. A. . Saccaro, A. Y. Tamime, S. Pilleggi, Ana Lu Cia O.P, and M. N. Oliveira, "The viability of three probiotic organisms grown with yoghurt starter cultures during storage for 21 days at 4 C," *Int. J. Dairy Technol.*, vol. 62, no. 3, pp. 397–404, 2009.

[5] M. R. Damin, E. Minowa, M. R. Alcantara, Oliveira, and Marice Nogueira, "Effect of Cold Storage on Culture Viability and Some Rheological Properties of Fermented Milk Prepared with Yogurt and Probiotic Bacteria," *J. Texture Stud.*, vol. 39, no. 1, pp. 40–55, Feb. 2008.

[6] D. . Johnson, S. Duncan, L. . Bianchi, and H. . Chang, "Packaging modifications for protecting flavor of extended-shelf-life milk from light," *J. Dairy Sci.*, vol. 98, pp. 1–10, 2015.

[7] P. Romano, C. Fiore, M. Paraggio, M. Caruso, and A. Capece, "Function of yeast species and strains in wine flavour," vol. 86, pp. 169–180, 2003.

[8] W. Gustaw, M. Kordowska-Wiater, and J. Kozio, "The influence of selected prebiotics on the growth of lactic acid bacteria for bio-yoghurt production," *Acta Sci. Polym. Technol.*, vol. 10,

no. 4, pp. 455–466, 2011.

[9] A. Y. Tamime, M. Saarela, A. K. Sondergaard, V. V. Mistry, and Shah. N.P, "Production and maintenance of viability of probiotic microorganisms in dairy product," *Probiotic Dairy Prod.*, vol. 3, pp. 39–63, 2005.

[10] M. Gurses, B. Cetin, and M. Sengul, "Change in Probiotic Microflora and Physico-chemical Characteristics of Rose Hip Marmalade Bio-yoghurts During Refrigerated Storage," *Asian J. Chem.*, vol. 21, no. 5, pp. 4097–4103, 2009.

[11] M. Granito and G. Álvarez, "Lactic acid fermentation of black beans (Phaseolus vulgaris): microbiological and chemical characterization," *J. Sci. Food Agric.*, vol. 86, no. 8, pp. 1164–1171, Jun. 2006.

[12] E. Mani-López, E. Palou, and A. López-Malo, "Probiotic viability and storage stability of yogurts and fermented milks prepared with several mixtures of lactic acid bacteria," *J. Dairy Sci.*, vol. 97, no. 5, pp. 2578–2590, May 2014.

[13] T. E. Obi, F. . Henshaw, and O. . Atanda, "Quality Evaluation of Plain-Stirred Probiotic Yoghurt Produced From Skim and Whole Milk Powder during Refrigerated Storage," *Electron. J. Environ. Agric. Food Chem.*, vol. 9, no. 7, pp. 1203–1213, 2010.

[14] A. G. Evrendilek, "Survival of Escherichia coli O157:H7 in yogurt drink, plain yogurt and salted (tuzlu) yogurt: Effects of storage time, temperature, background flora and product characteristics," *Int. J. Dairy Technol.*, vol. 60, no. 2, pp. 118–122, May 2007.

[15] A. B. Dantas *et al.*, "Manufacture of probiotic Minas Frescal cheese with Lactobacillus casei Zhang," *J. Dairy Sci.*, vol. 99, no. 1, pp. 18–30, Jan. 2016.

[16] A. P. Marafon, A. Sumi, M. R. Alcântara, A. Y. Tamime, and M. Nogueira de Oliveira, "Optimization of the rheological properties of probiotic yoghurts supplemented with milk proteins," *LWT - Food Sci. Technol.*, vol. 44, no. 2, pp. 511–519, 2011.

[17] A. P. Marafon, A. Sumi, M. R. Alcântara, A. Y. Tamime, and M. Nogueira de Oliveira, "Optimization of the rheological properties of probiotic yoghurts supplemented with milk proteins," *LWT - Food Sci. Technol.*, vol. 44, no. 2, pp. 511–519, 2011.

[18] L. Allgeyer, M. Miller, and S. Lee, "Sensory and microbiological quality of yogurt drinks with prebiotics and probiotics," *J. Dairy Sci.*, 2010.

Food safety of milk and dairy product of dairy cattle from heavy metal contamination

E Harlia, KN Rahmah and D Suryanto
Faculty of Animal Husbandry, Universitas Padjadjaran, 45363 Sumedang, Indonesia
Corresponding E-mail: harliaelin@yahoo.co.id

Abstract. Food safety of milk and dairy products is a prerequisite for consumption, which must be free from physical, biological and chemical contamination. Chemical contamination of heavy metals Pb (Plumbum/Lead) and Cd (Cadmium) is generally derived from the environment such as from water, grass, feed additives, medicines and farm equipment. The contamination of milk and dairy products can affect quality and food safety for human consumption. The aim of this research is to investigate contamination of heavy metals Pb and Cd on fresh milk, pasteurized milk, and dodol milk compared with the Maximum Residue Limits (MRL). The methods of this researched was through case study and data obtained analyzed descriptively. Milk samples were obtained from Bandung and surrounding areas. The number of samples used was 30 samples for each product: 30 samples of fresh milk directly obtained from dairy farm, 30 samples of pasteurized milk obtained from street vendors and 30 samples of dodol milk obtained from home industry. Parameters observed were heavy metal residues of Pb and Cd. The results showed that: 1) approximately 83% of fresh milk samples were contaminated by Pb which 57% samples were above MRL and 90% samples were contaminated by Cd above MRL; 2) 67% of pasteurized milk samples were contaminated by Pb below MRL; 3) 60% of dodol milk samples were contaminated by Pb and Cd above MRL.

Key words: food safety, Pb, Cd, residue

1. Introduction

The quality of cow's milk is influenced by many things, including genetic, feed and management. The feed and drinking water used should be free from heavy metal contamination. Livestock farming management such as sanitation of cages, workers and equipment should be guaranteed hygiene and meet food safety requirements. Dairy cattle generally feed grass and concentrate. The main source of contamination of Pb and Cd derives from air and water pollution that contaminates soil. Furthermore, all plants grown on contaminated soil will accumulate the metals in all parts (roots, stems, leaves and fruit) [11] Pb and Cd can accumulate in enormous amounts in plants such as rice, grass, some leguminous species for feed, and vegetable [9]. Most of Cd accumulates on crops, grains and grains products [10]. Milk is an ideal source of macroelement (Ca, K and P) and microelement (Cu, Fe, Zn, Se), however addition of heavy metals might contaminated milk and dairy products reachs levels that are harmful to humans [7]. Heavy metals such as zinc (Zn), copper (Cu), chromium (Cr), arsenic (As), cadmium (Cd), and lead (Pb) are potential bioaccumulative toxins of the dairy production system as soils tend to act as long term sinks or these metals through absorption onto metal oxides, particularly iron and manganese oxides, clay minerals, soil organic material, and other forms of humified natural organic material [4]. Residue of Pb and Cd contained in the food can accumulate in the livestock and human organs, in the long run can disrupt livestock and humans health. In order to prevent the accumulation of Pb and Cd on the body is by maintaining the balance of mineral intake. Government establishes Maximum Residual Limits (MRLs) to prevent contamination of heavy metals in foods. MRLs in milk and dairy products are regulated on INS (1998) while Cd is regulated on EPA (1985). Pb standards for milk and dairy products are below 1.00 ppm (INS) and Cd maximum limit is below 0.01 ppm (EPA). Food safety on milk and dairy products is an absolute requirement that must be fulfilled by farmers. In addition on affecting human health, it also impacts on domestic and global trade and the economies of countries involved in such trade [5]. Milk and dairy products must be of good quality and free from heavy metal contamination. Therefore, this research is aimed to detect food safety on milk and dairy products based on heavy metals (Pb and Cd).

2. Materials and Methods

Materials used in this research were fresh milk and dairy products. Fresh milk was obtained from 30 dairy farmers in Pemulihan Subdistrict Sumedang Regency, pasteurized milk was obtained from 3 street vendors on Bandung, dodol milk was obtained form 3 home industries in Pangalengan, with 10 repetitions for each sample. This research was an explorative study preceded by survey methods to determine sampling of milk and dairy products from dairy cow farmers, street vendors, and dodol milk industry. The parameter observed were residues of Pb and Cd in fresh milk, pasteurized milk and dodol milk. Multi Stage Sampling data obtained were analyzed descriptively, compared to the maximum limit of Pb residue regulated in INS (1998), while Cd residue compared to MRLs recommended by EPA (1985). Measurement of residual content in milk was utilizing Atomic Absorption Spectrophotometry (AAS).

3. Results and Discussion

The content of Pb residue in fresh milk, pasteurized milk and dodol milk was shown in Figure 1. It appeared that most of fresh milk samples were contaminated by Pb (83%), and as much as 53% of samples exceeded MRL standards. The average of Pb content in fresh milk sample 0.5578±0,4308 with coefficient of variation 77.23%, the average of Pb residual content in pasteurized milk sample 0.0047±0.0027 with coefficient of variation 57.45 % while the average residual content of Pb in dodol milk 0.3173±0.2990 with a coefficient of variation of 94.23%. This condition indicated that samples of fresh dairy milk, pasteurized milk and dodol milk varied greatly from each other. High in coefficient of variation showed the difference ways in handling milk by farmers, traders, and dodol milk industry. Standard deviation value indicated that there was no uniformity on dairy cattle farming management. Feed, water source, dairy cow housing hygiene, cleanliness, worker hygiene and type of milk containers were vary. In farmers who use plastic buckets as a milk container showed higher Pb content than in fresh milk on milkcan. The cleanliness of the equipment, the way of washing also affected the residual content of Pb in fresh dairy milk produced [3].

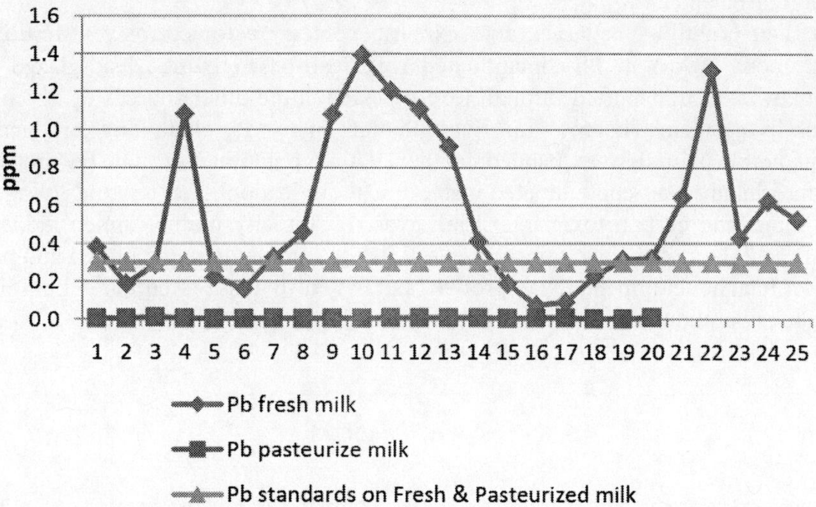

(a)

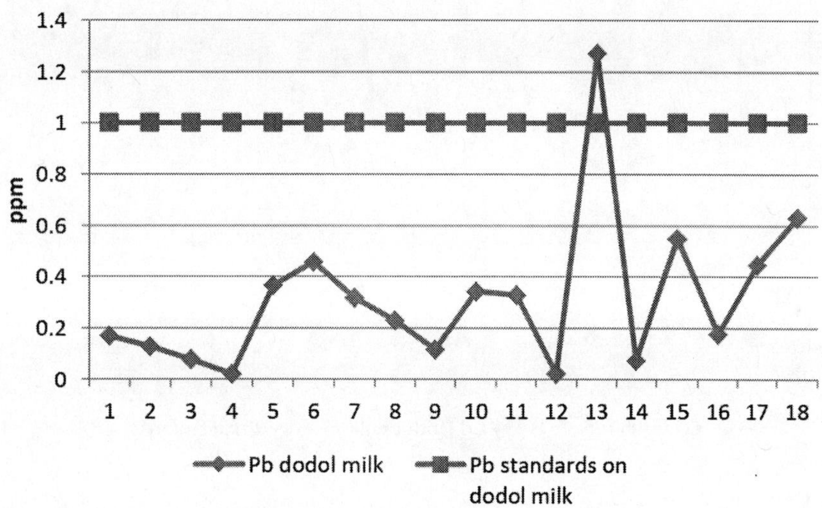

(b)

Figure 1. The content of Pb residue in milk and dairy products compared with Pb standards regulated by INS: (a) fresh milk and pasteurized milk; (b) dodol milk

Figure 1 showed showed most of fresh milk and dodol milk samples contained Pb above the MRLs. This condition was not meet food safety requirement. Accumulation of heavy metals residue on body gradually might harm human health in long term. Pb contamination on fresh milk and dodol milk were allegedly due to the presence of several sources of Pb residue from feed in the form of grass, straw, concentrate and water. In accordance with the opinion of [6], feed especially grass species grown on location near the road and industrial areas potentially contaminated Pb. Milk and dairy products became contaminated with heavy metals through materials and packaging processes [1, 2]. The source of dodol contamination derived from raw materials used which was milk. This was in accordance with the opinion of [12], that milk was one type of food that is susceptible to contamination of Pb and Cd.

Moreover, air pollution, batteries, auto exhaust, containers for corrosives, fertilizers, pesticides, insecticides, foods grown in Pb-contaminated soil, lead-based paints, lead-glazed pottery, solder, tobacco smoke, water transported through lead pipes could be other sources of Pb contamination. Pb was a naturally-occurring neurotoxin. Even though many Pb-containing products (for example gasoline and house paints) were banned in the 1970s, contamination can be found in the present majorly from drinking lead-contaminated water, breathing leadpolluted air, and living in or near older painted buildings and certain toxic industrial areas. Pb toxicity mainly aimed kidneys, the nervous system, heart and blood, bones and poses greatest risk to infants, young children and pregnant women. It could affect fetal development, delay growth, and may also cause attention deficit disorder, learning disabilities, behavioral defects, and other developmental problems [8].

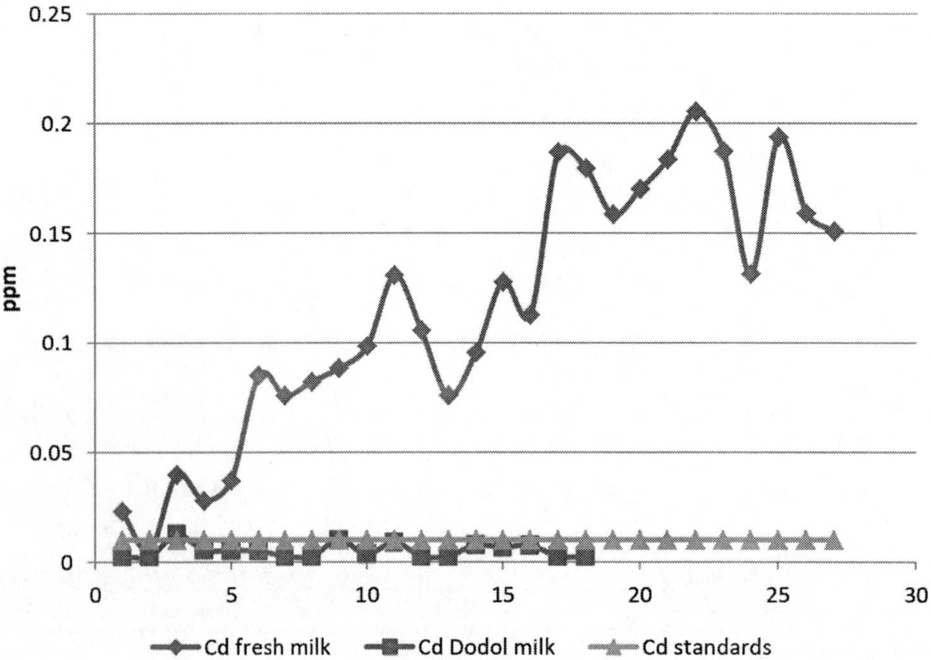

Figure 2. The content of Cd residue in milk and dairy products

Figure 2 illustrated the content of Cd in fresh milk and dodol milk. Fresh milk samples were mostly contained Cd above Cd standards by EPA, while dodol milk samples in contrary contained Cd below MRLs. Most of Cd contamination were derived from air pollution, batteries, ceramic glazes/enamels, cigarette smoke, tap and well water, food (grown in cadmium-contaminated soil), fungicides, mines, paints, power and smelting plants and seafood. Exposure to cadmium can occur through inhalation or ingestion in places or situations where cadmium products are used, manufactured, or ingested. Cigarette smoke is the biggest source of cadmium toxicity, which seems to mainly affect the lungs, kidneys, bones, and immune system. It may lead to lung cancer, prostate cancer and heart disease, and also causes yellow teeth and anemia. Cadmium also seems to contribute to autoimmune thyroid disease [8].

4. Conclusion

The heavy metal content of Pb in milk and dodol milk mostly exceeded MRLs, while pasteurized milk had met the MRLs requirement. The content of heavy metals Cd in milk largely exceeded MRLs, while dodol milk had met eligible MRLs. Fresh milk was not meet food safety standards however processed it into products can be decrease heavy metals residue.

5. References

[1] Anastasio, A., Caggiano, R., Macciato M., Paolo, C., Ragosta, M., Paino, S. and Cortesi, M. L. 2006. Heavy metal concentrations in dairy products from sheep milk collected in two regions of southern Italy, *Acta Veterinaria Scandinavica* **47**: 69-74.

[2] Ayar, A, Sert, D. and Akin, N. 2009. The trace metal levels in milk and dairy products consumed in middle AnatoliaTurkey, *Environmental Monitoring Assessment* **152**: 1-12.

[3] Birghila, S., S. Dohrinas, G. Stanciu, amd A. Spceanu. 2008. Detemination of Major and Minor Elements in Milk through ICP-AES. *Environmental Engineering and Management Journal. November/Desember 2008, Vol.* **7**, No.6, pp 805-808.

[4] Brown, G. E., and G. A. Parks. 2001. Sorption of trace elements on mineral surfaces: Modern perspectives from spectroscopic studies, and comments on sorption in the marine environment. *Int. Geol. Rev.* **43**:963–1073.

[5] Buzby, J.C. 2003 International Trade and Food Safety: Economic Theory and Case Studies. United States Department of Agriculture. *Agricultural Economic Report* Number **828**.

[6] Onder.,S. Dursun.,S. Gezgin.,A. Demirbas. 2007. Determination of Heavy Metal Pollution in Gras and Soil of City Centre Green Areas (Konya, Turkey). *Polish J.of. Envinron. Stud. Vol* **16** *No.1. P.145-154.*

[7] Qin, L. Q., Wang, X. P., Li, W., Tong, X. and Tong, W. J. 2009. The minerals and heavy metals in cow's milk from China and Japan. *Journal Health Science* **55(2)**: 300-305.

[8] Tibebu Kocharea, BerhanTamirb. 2015. Assessment of Dairy Feeds for Heavy Metals. *American Scientific Research Journal for Engineering, Technology, and Sciences (ASRJETS)* Volume **11**, No 1, pp 20-31

[9] Valérie Page and Urs Feller. 2015. Review Heavy Metals in Crop Plants: Transport and Redistribution Processes on the Whole Plant Leve. Agronomy, 5, 447-463; doi:10.3390/agronomy5030447 agronomy ISSN 2073-4395

[10] Wieczorek, J. Z. Wieczorek, T. Bieniaszewski. 2005.Cadmium and Lead Content in Cereal Grains and Soil from Cropland Adjacent to Roadways. *Polish Journal of Environmental Studies* Vol. **14**, No 4,P 535-540

[11] Yintao Lu, Hong Yao, Dan Shan, Yichen Jiang, Shichao Zhang, and Jun Yang. 2015. Heavy Metal Residues in Soil and Accumulation in Maize at Long-Term Wastewater Irrigation Area in Tongliao, China. Research Article.*Journal of Chemistry.*Volume, Article ID 628280, 9 pages http://dx.doi.org/10.1155/2015/628280

[12] Simsek, O., R. Gultekin, O. Oksuz, and S. Kurultay. 2000. The effect of environmental pollution on the heavy metal content of raw milk. *Molecular Nutrition Food Research* Vol. **44** Issue 5, October 2000, page 360–363.

IOP Conf. Series: Earth and Environmental Science **102** (2017) 012051 doi:10.1088/1755-1315/102/1/012051

The physical and microbiological quality of chicken meat in the different type of enterprise poultry slaughterhouse: a case study in Karanganyar District

B S Hertanto[1]*, C D A Nurmalasari[2], A M P Nuhriawangsa[1], M Cahyadi[1] and L R Kartikasari[1]

[1]Laboratory of Industry and Animal Product Processing, Departement of Animal Science, Universitas Sebelas Maret*, Jl. Ir. Sutami 36A Kentingan, Jebres, Surakarta, Central Java, Indonesia +6257126, Phone/Fax: +62271637457

[2]Graduated Student of Department of Animal Science, Faculty of Agriculture, Universitas Sebelas Maret*, Jl. Ir. Sutami 36A Kentingan, Jebres, Surakarta, Central Java, Indonesia +6257126, Phone/Fax: +62271637457

*E-mail: bayuhertanto@uns.ac.id

Abstract. The aim of this study was to determine the physical and microbiological quality of chicken meat produced by the different type of enterprise slaughterhouse in Karanganyar District. The number of 20 poultry slaughterhouses was determined by convenience sampling method. The samples of chicken meat were randomly collected from medium enterprise poultry slaughterhouses (n=12) and small enterprise poultry slaughterhouses (n=8). A survey was carried out among poultry slaughterhouses in Karanganyar District. All the samples were subjected to physical quality consisted of pH test, texture, and color, while microbiological quality consisted of total plate count, microbial detection of *Escherichia coli* and *Salmonella*. The data were analyzed using descriptive quantitative analysis. The study showed that chicken meat in 6 small enterprise slaughterhouses and 11 medium enterprise slaughterhouses had normal pH of 5.81 - 6.3. Color and texture of chicken meats had relatively normal in both small and medium enterprise slaughterhouses. The total plate count of chicken meat showed in both small and medium enterprise slaughterhouses was $<1x10^6$ CFU/gr. The test of bacterial contamination showed that 3 of small and medium enterprise slaughterhouses were positively contaminated by *Escherichia coli* of $>1x10^1$ CFU/gr, and *Salmonella* was detected in 1 medium enterprise slaughterhouse. The overall results of the study suggest that the potential risk of chicken meat contamination depends on the processing of chicken meat in poultry slaughterhouses.

***Keywords**: slaughterhouses, physical, microbiological, chicken meat*

1. Introduction

Ensuring supply of safe food has been one of the major challenges and concerns for producers, consumers and public health officials in the world. This is because foods excessively contaminated with pathogenic and spoilage micro-organism are undesirable and can cause food borne illnesses [1–2]. Chicken meat, a food source of animal protein, is the most widely consumed by the community. Production of chicken meat in Karanganyar District in 2009 was 2,276,795 kg and increased by 30% in 2013 around 2,963,043 kg [3]. The chicken meat contributes about 20.22% to world meat production and Indonesia is the largest contributor to ASEAN with an average of 1.56 million tons.

The growth is expected more in near future [4]. This might be due to popularity, easy availability, price, no religious taboos, highly digestible, tasty and low-calorie food [5].

Raw or undercooked chicken meat is particularly prone to contamination. The microbiological and physical quality of chicken meat as purchased by consumers depends mostly on; the slaughter process, sanitation during processing and packaging, maintenance of adequate cold chain storage from the processing to the retail level and to the consumer and finally sanitation during handling at the retail end [6–7]. Microorganisms from the environment, equipment and operators hands can contaminate meat [8]. Chicken meat is responsible for many zoonotic foodborne infections in the World. Foodborne diseases have a serious impact on the public health and the economic losses due to food poisoning are in billions [9]. Poultry and poultry meat are often found contaminated with potentially pathogenic microorganisms such as *Salmonella*, *Campylobacter*, *S. aureus*, *E. coli* and *Listeria* [5]. Previous study conducted in slaughterhouses spread in middle java (Ditrict of Sukohajo and City of Surakarta) and Province of Yogyakarta (City of Yogyakarta, District of Sleman and Bantul) showed all samples of chicken meat were contaminated by *E. coli*, *Staphylococcus sp.* and *Salmonella sp.* despite the chicken carcasses had good physical appearances [10].

According to meat contamination, poultry slaughterhouse has a role of chicken meat circulating in supply chain. It is one of the major critical points with potential effects on the hygiene of poultry meats. During slaughter operations, inter-contamination phenomena occur, which induces proliferation of bacterial pathogens on carcasses initially healthy. Special attention must be observed in chicken meat production because of possible contamination from alimentary tracts, water, packaging, utensils and handlers [11-13]. Therefore, good hygiene practices in food establishments are essential for consumer protection and the control of public health risks. The way of handling and the equipment used is very influential on the quality of meat produced, because chicken meat is one of the animal food material that is easily damaged.

This paper discusses the results of a study on assessment the hygienic conditions, conducted at a poultry slaughterhouse, which aimed to determine the physical and microbiological quality of chicken meat produced by the different type of enterprise slaughterhouse in Karanganyar District.

2. Materials and methods

2.1. Sampling size

Sampling of poultry slaughterhouses was conducted in medium enterprise poultry slaughterhouses (MEPS) (n=12) and small enterprise poultry slaughterhouses (SEPS) (n=8). The method of selecting poultry slaughterhouses in Karanganyar District used convenience sampling method. Convenience sampling method can be used with the provision if there is no information about the sampling population [14]. Sampling of broiler chicken meats was done randomly at poultry slaughterhouses in Karanganyar District.

2.2. Sample collection

One carcass of broiler chicken was randomly taken from enterprise poultry slaughterhouses after slaughtering process using a clean pair of latex gloves, a sterile bag, and cool box. Within 1 hours after sampling, samples were sent in an insulated box to Laboratory of Veterinary and Animal Health Center of Boyolali District where microbiology quality test were performed, and Laboratory of Livestock Processing Industry (Department of Animal Science, Faculty of Agriculture, Universitas Sebelas Maret*, Surakarta) for testing the physical quality of fresh chicken meat.

2.3. Microbiological test

Total Microbial Count was measured by Plate Count Agar (PCA) [15], the marking of the presence of *Escherichia coli* (*E. coli*) by [16], the marking of the presence of *Salmonella* [17].

2.4. Physical quality of chicken meat

Color test according to [18], texture test using a scale of one to five refers to [19], value of pH by [20].

3. Results and discussion

3.1. Condition of poultry slaughterhouses in Karanganyar

Results of research conducted in 12 MEPS and 8 SEPS spread in Karanganyar District stated there are 11 MEPS located in densely populated location (Table 1 and 2). All poultry slaughterhouses already has dirty and clean areas, although the use of the area is still less effective. [21] The dirty area is the area consisting of the place where the place of live poultry up to evisceration and the clean area consists of washing carcass up to the packaging stage. However, in most poultry slaughterhouses the dirty areas are used until the weighing stage of the carcass and the clean area is only used for carcass processing and packaging. Building of MEPS meets the SNI 01-6160-1999 standard on poultry slaughterhouse. The walls throughout the MEPS are already made of cement walls and are waterproof. Water used to clean the carcasses already used running water. The handling process of chicken was began with rest for a while before the chicken was cut.

Table 1. Conditions of medium enterprise slaughterhouses in Karaganyar District

Slaughterhouses tools of Indonesian National Standard	Conditions of medium enterprise slaughterhouses											
	1	2	3	4	5	6	7	8	9	10	11	12
1. The location is not in a crowded part of town	√	x	x	x	x	x	x	x	x	x	x	x
2. The buildings have dirty and clean areas	√	√	√	√	√	√	√	√	√	√	√	√
3. Buildings should be constructed of impact resistant walls, easy to clean and not flaking. The floor should be flat, waterproof material and easy to clea, and also good ventilation	√	√	√	√	√	√	√	√	√	√	√	√
4. Unloading of chicken is temporarily, rested and postmortem examination is done	√	x	√	x	√	√	x	x	√	√	√	√
5. The slaughter is done using an anti corrosive knife and is not easily damaged	√	√	√	√	√	√	√	√	√	√	√	√
6. Bleeding takes approximately 5 minutes	√	√	√	√	√	√	√	√	√	√	√	√
7. Scalding is done using anti-corrosive tools and periodically cleaned	√	√	√	√	√	√	√	√	√	√	√	√
8. Evisceration was performed using personal protective equipment (gloves, apron, headgear and mask) and anti-corrosive knives and antemortem examination	x	√	x	x	√	√	√	x	√	√	√	√
9. Carcass washing done in clean area using water flowing at least 25-35 liter/ tail	√	√	√	√	√	√	√	√	√	√	√	√
10. Carcass and parting process of chicken using a knife and anti-corrosive cutting board, and then do the packaging.	√	x	√	√	√	√	x	x	√	√	√	√

Caption: √ = in accordance with The Indonesian National Standard; x = not in accordance with The Indonesian National Standard

In other slaughterhouses, all SEPS buildings do not have dirty areas and clean areas. Poultry handling is done in the same room. The SEPS buildings are made of waterproof walls, flat floors and good ventilation. Bleeding process is done directly, because there was no resting room. The equipment used anti-corrosive tools and is cleaned regularly. This dirty equipment is a medium that can cause cross contamination from one carcass to another carcass [22]. In this study, some SEPS still use water that does not flow for the meat-cleaning process. Workers also have not used completely personal protective equipment. [15] Microbial contamination occurs due to a lot of equipment that is less clean, contamination from the hands of workers and contamination of water that has been contaminated. The process of slaughtering until the packaging is done as quickly as possible so as to minimize the occurrence of direct contact with the outside. [5] High levels of bacterial cross-contamination may occur especially during defeathering and water chilling, with intestinal contamination apparently being the only source.

Table 2. Conditions of small enterprise poultry slaughterhouses in Karaganyar District

Slaughterhouses tools of Indonesian National Standard	Conditions of small enterprise poultry slaughterhouses							
	1	2	3	4	5	6	7	8
1. The location is not in a crowded part of town	X	X	X	X	√	√	√	√
2. The buildings have dirty areas and clean areas	X	X	X	X	X	X	X	X
3. Buildings should be constructed of impact resistant walls, easy to clean and not flaking. The floor should be flat, waterproof material and easy to clean, and also good ventilation	√	√	√	√	√	√	√	√
4. Unloading of chicken is temporarily, rested and postmortem examination is done	√	√	√	√	√	x	√	√
5. The slaughter is done using an anti corrosive knife and is not easily damaged	√	√	√	√	√	√	√	√
6. Bleeding takes approximately 5 minutes	√	√	√	√	√	√	√	√
7. Scalding is done using anti-corrosive tools and periodically cleaned	√	√	√	√	√	√	√	√
8. Evisceration was performed using personal protective equipment (gloves, apron, headgear and mask) and anti-corrosive knives and antemortem examination	x	x	x	√	√	√	x	x
9. Carcass washing done in clean area using water flowing at least 25-35 liter/ tail	x	√	√	√	√	√	√	√
10.Carcass and parting process of chicken using a knife and anti-corrosive cutting board, and then do the packaging.	√	√	x	√	√	x	x	√

Caption: √ = in accordance with The Indonesian National Standard; X = not in accordance with The Indonesian National Standard

3.2. Physical quality of chicken meat

The physical properties is very influential on the appearance of meat quality. Chicken meat has pH value of 6.26 - 6.30 that meet the normal pH of chicken meat, but the pH value still above the optimum pH of chicken meat.

Pre-postmortem treatment may affect the pH quality of chicken meat as it may affect the acidity or development of rigor time [23]. Based on the research, the treatment given by workers in some MEPS and SEPS on chicken before slaughtering was not rested so that the pH produced is different from previous research. Optimal pH of broiler chicken meat without treatment ranged from 6.11 to 6.25 [24]. Under normal conditions, fresh chicken meat has a pH range of about 5.3-6.5 post slaughtering [25]. The research refers to pH standard according to 6 - 7 [26].

The chicken meat color both MEPS and SEPS showed the normal color. Color scores 2 and 3 are good color scores for chicken meat because the pH value is about 5.70 - 6.50 [27]. Chicken meat that has not been exposed to contamination will be colored bright yellowish white has a pH of 5.7 to 6.50 [28]. The color of chicken meat influenced by several factors such as pH value, the lower the pH value so the color of the meat will be more pale [29].

The texture of chicken meat showed a variety of texture scores ranging from 3 to 5. This result was caused by a variety of pH value. The normal texture of chicken meat is related to pH and color values. Texture chicken meat has fine fibers and not mushy that shows a pH ranging from 5 - 6 and has a yellowish white color. However, the textures that tend to hard have a pH> 6 and a darker color [26].

The high value of final pH will increase the ability to bind water, because the meat fluid is bound by the protein. This causes the meat to have a harder texture that causes the texture of the meat becomes harder [30].

Table 3. The physical quality of broiler chicken meats in medium and small poultry slaughterhouses of Karanganyar District

Type of poultry slaughterhouses	No.	Physical quality		
		pH	Color**	Texture***
MEPS	1	6.12	3	5
	2	6.30	3	3
	3	6.24	3	5
	4	6.26	3	3
	5	6.11	3	5
	6	6.12	3	5
	7	6.27	3	3
	8	6.27	3	3
	9	5.99	2	4
	10	6.20	3	5
	11	5.96	2	4
	12	5.81	2	4
Mean±Stadev		6.14±0.15	2.75±0.43	4.08±0.86
SEPS	1	6.10	2	4
	2	5.90	2	4
	3	5,90	2	4
	4	6.26	3	3
	5	6.18	3	5
	6	5.98	2	4
	7	6.10	2	4
	8	6.15	3	5
Mean±Stadev		6.07±0.12	2.38±0.48	4.13±0.60

Caption: **: 2 = pale yellowish white, 3 = yellowish white.; ***: 3 = meat fibers rather smooth, not elastic and hard if pressed, 4 = fine meat fibers, elastic but feels moist when pressed, 5 = fine meat fibers, elastic and feels dry if pressed

3.3. Microbiological quality of chicken meat

Microbial contamination of meat begins since bleeding of slaughter steps, especially if the tools used for blood removal are not sterile. Therefore, it is necessary to test the quality of microbiology for contamination that occurs can be controlled [25].

The results showed that total bacteria of chicken meat samples was normal. The total bacteria in both MEPS and SEPS respectively were 11.8 x 10^4 CFU/g and 8.8 x 10^4 CFU/g. [26] Regarding to carcass and chicken quality, the maximum number of microbials in chicken meat was $1x10^6$ CFU/g. Based on research, carcass washes at 12 MEPS already using running water and some equipment that was directly related to meat already using anti-corosive materials.

The results showed that there were 14 samples of meat that were declared negative contaminated bacteria of *E. coli* and 6 samples of meat expressed positively polluted. The source of *E.coli* could be caused by equipment used for the process of slaughtering, evisceration and the division of meat. The bacteria lives normally in the intestinal tract of warm-blooded animals and can contaminate food in various ways, including hand pollution, during the evolutionary process, indirect pollution through water pollution, and product packaging [31]. *E. coli* probably comes from contamination with the environment especially water when processing [32]. This is consistent with the statement of [33], Contamination of *E. coli* in meat and its derivatives usually comes from water used at the time of washing of carcass. The maximum limit of microbe contamination of *E. coli* in food is $1x10^1$ CFU/g [20].

The results of *Salmonella* test showed that there was 1 MEPS positive bacteria contaminated. Previous research by [34] states that *Salmonella* bacteria are very easy to grow in a dirty environment of poultry slaughterhouse. The standard of food requires no *Salmonella* in 25 gram of food samples. The process of evisceration is not done quickly, whereas the process of evisceration is important in the process of handling chicken meat. The highest *Salmonella* contamination in poultry slaughterhouses occurred in the process of evisceration done manually so as to potentially cause contamination of carcasses. The process of manual evisceration can be done by considering the time of evisceration, the process must be done quickly and use the water flowing so that *Salmonella* contamination can be suppressed [35].

Tabel 4. Microbiological quality of broiler chicken meats in medium and small poultry slaughterhouses of Karanganyar District

Type of poultry Slaughterhouses	No.	Total plate count (TPC)	*E. coli* test	*Salmonella* test
	1	√	+	+
	2	√	+	-
	3	√	-	-
	4	√	-	-
	5	√	-	-
MEPS	6	√	-	-
	7	√	+	-
	8	√	+	-
	9	√	-	-
	10	√	-	-
	11	√	-	-
	12	√	-	-
	1	√	+	-
	2	√	-	-
	3	√	-	-
SEPS	4	√	+	-
	5	√	+	-
	6	√	-	-
	7	√	-	-
	8	√	-	-

Caption: √ = TPC in accordance with The Indonesian National Standard; x = TPC is not in accordance with The Indonesian National Standard; + = positive *Escherichia coli, Salmonella;* - = negative *Escherichia coli, Salmonella*

4. Conclusions

It can be concluded both MEPS and SEPS have the good physical quality of chicken meat (pH, color, and texture), and the total bacterial count of all fresh chicken meat did not exceed the maximum limit of microbial contamination established under SNI 3924-2009. However, there were meat contamination of *E. coli* in both MEPS and SEPS, and *Salmonella* in MEPS of Karanganyar District.

5. Acknowledgment

The authors gratefully acknowledge to Universitas Sebelas Maret* for funding the research through PNBP 2016 (No. 632/UN27.21/LT/2016).

6. References

[1] Lindqvist R, Anderson Y, De Jong B and Norberg P 2000 *J. Food Protect.* **63** 1315
[2] Su H P, Chu SI, Tsai J L, Lee C L, Pan T M 2005 *J. Infect. Chem.* **11**146–151
[3] Department of Animal Husbandry and Animal Health of Central Java *Statistik Peternakan 2014* (Ungaran: Disnakkeshew)
[4] Agriculture Ministry of Republic of Indonesia 2015 Outlook Komoditas Pertanian Subsektor Peternakan Daging Ayam (Jakarta: Source of data and agriculture information, Secretariat General of The Ministry of Agriculture)
[5] Bhaisare D B, Thyagarajan D, Churchil R R and Punniamurthy N *Int. J. Life Sci. Res.* **2** 1
[6] Borche E and Arinder P 2002 *Meat Sci.* **62** 381
[7] Selvan P R, Narendra B, Sureshkumar S and Venkatamanujam V 2007 *Am. J. Food. Technol.* **2** 55
[8] Kumar H S, Otta S and Karunasagar I." *Lett. Appl. Microbiol* **32** 334
[9] Álvarez-Astorga M, Capita R, Aonso-Calleja C, Moreno B and Del Camoni García-Fernández M
[10] Supartono S, Raharjo S, Iskandar S 2003 *Agritech* **23** 189
[11] Ramasastry P, Ramkrishna M and Mrunalini N 1999 *meat Ind. Vet. J.* **76** 409
[12] Bryan F L 2001 *World's Poul. Sci. J.* **27** 223
[13] Nwachukwu N and Nnamani O H 2013 *Int. J. Current Microbiol. Appl. Sci.* **2** 224
[14] Sugiyono *Metodologi Penelitian Kuantitatif Kualitatif dan R &D* 2015 (Bandung: Alfabetha)
[15] Hadioetomo R *Mikrobiologi Dasar-Dasar Dalam Praktek* 1996 (Jakarta: Gramedia)
[16] Dewantoro G I, Adiningsih M W, Purnawarman T, Sunartatie T and Afif U 2009 *Jurnal Ilmu Pertanian Indonesia* **30** 211
[17] Adiningsih M W 2009 Microbiological Aspect of Frozen Chicken Passed through Merak's Crossing Port. Thesis. Graduated School. Institut Pertanian Bogor. Bogor.
[18] Indonesian National Standard SNI 3932-2008: Metode pengujian cemaran mikroba dalam daging, telur dan susu serta hasil olahannya (Jakarta: National Standards Agency)
[19] Widyati R dan Yuliahsih *Higiene dan Sanitasi Umum 2002* (Jakarta: PT Gramedia Widiarsana)
[20] AOAC 1995 *Official Methods of Analysis* ed 16th (Washington D.C: Association of Official Analytical Chemist)
[21] Indonesian National Standard *SNI 01-5150-1999: Rumah Potong Unggas* (Jakarta: Nasional Standards Agency)
[22] Meggitt C. 2003. *Food Hygiene and Safety*. Oxford: Heinemann Educational Pub.
[23] Anggraeni Y. 2005. *Sifat Fisik Daging Dada Ayam Broiler pada Berbagai Lama Postmortem di Suhu Ruang* (Bogor: Bogor Agriculture Institute)
[24] Prayitno, A H and Suryanto E 2012 *Buletin Peternakan* **34** 55
[25] Soeparno 2009 *Ilmu dan Teknologi Daging* ed 2th (Yogyakarta: Gadjah Mada University Press)
[26] Indonesian National Standard *3924-2009: Daging Ayam* (Jakarta: National Standards Agency)
[27] Paiao F G, Ferracin L M, Pedrao M, Kato T and Shimokomaki M 2013 *Skeletal Muscle Calcium Channel Ryanodine And The Development Of Pale, Soft and Exudative Meat In Poultry* (Brazil: Estadual de Londrina University)
[28] Cassens R G 1966 *General aspect of postmortem changes* (Canada: The University of Manitoba)
[29] Resnawati H 2008 *Pros. Seminar Nasional Teknologi Peternakan dan Veteriner* (Bogor: Balai Penelitian Ternak)
[30] Price J F and Schweigert B S 1971 *The Science of Meat and Meat Products* ed 3rd (USA: Michigan University)
[31] Forsythe S J 2000 *The Microbiology of Safe Food* (London: Blackwell Science)
[32] Hafid H, Nursanti dan Irawan F Y 2014 *Pros. Seminar Nasional Teknologi Peternakan dan Veteriner* (Jawa Barat: Pusat Penelitian dan Pengembangan Peternakan)
[33] Buckle K A, Edwards R A, Fleet G H and Wooton M 1987 *Ilmu Pangan ed H. Purnomo and Adiono* (Jakarta: University of Indonesia Press)

[34] Lawrie and Leward 2006 *Ilmu Daging ed A. Parakkasi dan A. Yudha* (Jakarta: University of Indonesia Press)

[35] Kusuma A S, Padaga M C and Wuragil D K 2012 *Isolasi dan Karakteririsasi Salmonella spp. di Lingkungan Rumah Potong Ayam di Kota Malang* (Malang: Brawijaya University)

The effects of sodium azide on seed germination and seedling growth of chili pepper (*Capsicum annum* L. cv. Landung)

Yafizham[1] and B Herwibawa[2]

[1] Laboratory of Ecology and Crop Production, Department of Agriculture, Faculty of Animal and Agricultural Sciences, Diponegoro University, Semarang, Indonesia
[2] Laboratory of Physiology and Crop Breeding, Department of Agriculture, Faculty of Animal and Agricultural Sciences, Diponegoro University, Semarang, Indonesia

E-mail: yafizham@live.undip.ac.id

Abstract. This study was aimed to determine the effects of sodium azide on the performance of chili pepper. Dry seeds from pure lines of chili pepper cv. Landung were used in this experiment. Each set containing 50 seeds was placed in nylon fishner bags, washed in flowing water and soaked in distilled water for four hours. After soaking, four sets of seeds were dabbed dry with tissue paper and were treated for two hours at 30 ^{0}C in appropriate solution of 0, 0.10, 0.20, 0.40, 0.80, 1.60 mM sodium azide with phosphate buffer at pH 3. After treatment, the seeds were germinated in plastic boxes containing sterilized sand and kept under laboratory condition by supplied with water everyday. The seeds for M_1 germination percentage, seedling height, plant height and number of leaves were recorded 30 days and 60 days after treatment, respectively. Seeds which radicle emerged were considered germinated, the seedling and plant height were measured from the tip of primary root to the base of the first leaf pair, and the number of leaves were counted for only fully expanded leaves. A completely randomized block design in four replicates was used throughout the experiment. Data obtained were analysed for range, mean, standard of deviation, and percent of control using Microsoft Office Excel 2007 software. It was concluded that different doses of sodium azide influenced the performances of chili pepper cv. Landung. Very low doses of sodium azide (0-1.60 mM) might be used to study the improvement of chili pepper diversity.

Keywords : sodium azide, chili pepper, germination

1. Introduction

The mutagenicity of sodium azide has been demonstrated in several plants, viz. maize [1], rice [2], wheat [3], chickpea [4], fenugreek [5], sunflower [6], rapeseed [7], tomato [8], etc. Inconsistent effects of sodium azide have been obtained in chili pepper, and it is well known that there are many factors such as varieties, ages and environmental factors [9,10,11]. There are varying opinions as to mutagenic properties of sodium azide, however, it has been reported that hydrogen ion concentration of its treatment solution as a key to its high mutagenic potency [12]. Mutagenic efficiency of sodium azide is also possiby due to the low physiologic damage azide induced in treated plants [13].

The mutagenicity of the sodium azide to an indirect action, through its inhibition of catalase and peroxidase were caused by the accumulation of hydrogen peroxidase [14]. Many studies demonstrated the sodium azide must be metabolized by plant cells to the mutagenic agent, identified as an amino acid analogue L-azidoalanine (N_3-CH_2-$CH(NH)_2$-COOH), and the production of this metabolite was

IOP Conf. Series: Earth and Environmental Science **102** (2017) 012052 doi:10.1088/1755-1315/102/1/012052

found on the O-acetylserinesulfhydrylase, in which the enzyme catalyses the condensation of azide or sulfide with O-acetylserine to produce azidoalanine or L-cysteine respectively [15]. The pH of the solution influences the mutagenic efficiency of sodium azide, and it has been shown that azide is most effective in inducing mutations at pH 3 [13].

The increasing work on the induction of mutations to increase genetic diversity using mutagen reflects the growing interest in mutation breeding [16]. It is also indicative of the mutagenic efficiency of sodium azide. However, there is no work done in the mutagenicity of sodium azide on chili pepper cv. Landung. Its use in mutation breeding may help improve the existing varieties of the capsicums. Mutant selection would be done since germination phase [17]. The characters of germination, as measured in the laboratory, is able to reflect the ability of plants to live under field conditions [18]. This study was aimed to determine the effects of sodium azide on the performance of chili pepper. The M_1 germination percentage, seedling height, plant height and number of leaves were recorded 30 days and 60 days after treatment, respectively. This study is expected to provide information that can be used in chili pepper breeding programs.

2. Materials and Methods

Dry seeds from pure lines of chili pepper cv. Landung were used in this experiment. A modified methods by [11] were employed. Each set containing 50 seeds was placed in nylon fishner bags, whased in flowing water and soaked in distilled water for four hours. After soaking in water, four sets of seeds were dabbed dry with tissue paper and were treated for two hours at 30 ^{0}C in appropriate solution of 0, 0.10, 0.20, 0.40, 0.80, and 1.60 mM sodium azide with phosphate buffer at pH 3. After treatment, the seeds were washed for 2 hours in running water and then germinated in sand sterilized for one hour at 16 p.s.i in plastic boxes and kept under laboratory condition by supplied with water everyday.

The seeds were scored for M_1 germination percentage and seedling height 30 days later, and plant height and number of leaves 60 days after treatment. Seeds which radicle emerged were considered germinated, the seedling and plant height were measured from the tip of the primary root to the base of the first leaf pair, and the number of leaves were counted for only fully expanded leaves. The entire experiment was arranged in a completely randomized block design in four replicates. Data obtained were analyzed for range, mean, standar of deviation, and percent of control using Microsoft Office Excel 2007 software.

3. Results and Discussions

Sodium azide decreased germination percentage of chili pepper (Table 1). It was found that the germination percentage ranged from 28 to 100 % and decreased 1.01 to 2.96 times. Similar experiments have been carried out in chili pepper by [10] and [11]. The germination percentage of chili pepper cv. pusa jwala and cv. hyderabad in different doses were detected from 40 to 90 %. It was explained the sensitivity to sodium azide was doses dependent, and evident that sodium azide produced physiological effects. The effect of sodium azide was also noticeable in growth inhibition as a very common effect in chili pepper [9].

Table 1. Germination percentage of seeds of chili pepper cv. Landung after sodium azide treatments

Sodium Azide (mM)	Germination Percentage (%)		Percent of control
	Range	Mean/SE	
0	99 – 100	99.67 ± 0.58	100.00
0.10	93 – 94	93.33 ± 0.58	93.65
0.20	90 – 93	91.33 ± 1.53	91.64
0.40	83 – 92	87.00 ± 4.58	87.29
0.80	67 – 80	73.67 ± 6.51	73.91
1.60	28 – 42	33.67 ± 7.37	33.78

Table 2. Seedling height of chili pepper cv. Landung after sodium azide treatments

Sodium Azide (mM)	Seeding Height (cm)		Percent of control
	Range	Mean/SE	
0	3.27 – 3.90	3.59 ± 0.32	100.00
0.10	2.30 – 3.70	3.13 ± 0.74	87.28
0.20	2.10 – 3.50	2.84 ± 0.70	79.20
0.40	2.00 – 3.20	2.67 ± 0.61	74.28
0.80	1.80 – 2.90	2.39 ± 0.55	66.57
1.60	0.59 – 1.84	1.23 ± 0.63	34.26

There is little information on the mechanism of azide mutagenesis in chili pepper. However, it has been reported that sodium azide causes the reduction of plant height of rice [19] and wheat [3]. The reports support the findings of the experiment, which is sodium azide reduced seedling height (Table 2) and plant height (Table 3). Different results shown in sunflower [20], grapevine [21], and lentil [22] where the plant height reduction is not consistent with increase of sodium azide doses. It is common knowledge that mutations whether they are point or gross chromosomal aberrations involve alterations in the DNA molecule. Alterations in the DNA structure and constituent nucleotides brought about by mutagens may change the genetic information, and possible to give different expressions.

Table 3. Plant height after 60 days of chili pepper cv. Landung after sodium azide treatments

Sodium Azide (mM)	Plant Height (cm)		Percent of control
	Range	Mean/SE	
0	14.30 – 15.70	14.93 ± 0.71	100.00
0.10	13.40 – 14.70	14.02 ± 0.65	93.91
0.20	12.80 – 13.40	13.11 ± 0.30	87.79
0.40	12.30 – 13.30	12.77 ± 0.50	85.49
0.80	11.60 – 12.40	11.98 ± 0.40	80.20
1.60	4.15 – 5.81	5.16 ± 0.89	34.58

Table 4. Number of leaves after 60 days of chili pepper cv. Landung after sodium azide treatments

Sodium Azide (mM)	Number of Leaves		Percent of control
	Range	Mean/SE	
0	9.20 – 13.10	11.14 ± 1.95	100.00
0.10	7.80 – 10.20	8.99 ± 1.20	80.68
0.20	8.70 – 10.00	9.39 ± 0.65	84.27
0.40	8.20 – 10.30	9.30 ± 1.05	83.46
0.80	8.00 – 11.60	9.27 ± 2.21	83.16
1.60	3.75 – 5.43	4.66 ± 0.85	41.82

The highest number of leaves (11.14) was observed for control (0 mM) and the lowest (4.66) with 1.60 mM (Table 4). The reduction of number of leaves was similar to research [23] and [24] of black rice and tomato, repectively, where the number of leaves decreased as the higher of sodium azide doses. However, these results were different with those obtained by others, the reduction of number of leaves was not consistent with increased in sodium azide doses of rice [25] and groundnut [26]. The inhibitory effects of sodium azide on the different biological parameters of chili pepper may probably be explained by the inhibition of mitosis, disruption of the enzymatic process or direct changes on the genes involved [27]. There are different mechanism among the given mutagen doses, so it caused changes of each individual [28]. It was evident from the results obtained that there were differences response to sodium azide were doses and individual chili pepper plant dependent. These may be attributed to difference in the uptake or metabolism of the mutagen, cellular repair mechanisms or differences in the individual metabolic activities.

4. Conclusion

It was concluded that different doses of sodium azide influenced the performances of chili pepper cv. Landung. Very low doses of sodium azide (0-1.60 mM) might be used to study the improvement of chili pepper diversity.

References

[1] Eze, J.J., Dambo, A. 2015. Mutagenic effects of sodium azide on the quality of maize seeds. *Journal of Advanced Laboratory Research in Biology* **6** (III): 77-82
[2] Herwibawa, B., Kusmiyati, F. 2017. Mutagenic effects of sodium azide on the germination in rice (*Oryza sativa* L. cv. Inpago Unsoed 1). *Jurnal Agroteknologi* **7** (2): 9-14
[3] Srivastava, P., Marker, S., Pandey, P., Tiwari, D.K. 2011. Mutagenic effects of sodium azide on the growth and yield characteristics in wheat (*Triticum aestivum* L. em. Thell.). *Asian Journal of Plant Science* **10** (3): 190-201 doi: 10.3923/ajps.2011.190.201
[4] Kulthe, M.P., Kothekar, V.S. 2011. Effects of sodium azide on yield parameters of chickpea (*Cicer arietinum* L.). *Journal of Phytology* **3** (1): 39-42
[5] Sidduqui, S., Meghvansi, M.K., Hasan, Z. 2007. Cytogenetic changes induced by sodium azide (NaN_3) on *Trigonella foenum-graecum* L. seeds. *South African Journal of Botany* **73** (4): 632-635 doi: 10.1016/j.sajb.2007.06.005
[6] Elfeky, S., Abo-Hamad, S., Saad-Allah, K.M. 2014. Physiological impact of sodium azide on *Helianthus annuus* seedlings. *International Journal of Agronomy and Agricultural Research* **4** (5): 102-109
[7] Hussain, S., Khan, W.M., Khan, M.S., Akhtar, N., Umar, N., Ali, S., Ahmed, S., Shah, S.S. 2017. Mutagenic effect of sodium azide (NaN_3) on M_2 generation of *Brassica napus* L. (variety Dunkled). *Pure and Applied Biology* **6** (1): 226-236 doi: 10.19045/bspab.2017.60018
[8] El-Kaaby, E.J.S., Al-Ajeel, S.A., Al-Anny, J.A., Al-Aubaidy, A.A., Ammar, K. 2015. Effect of the chemical mutagens sodium azide on plant regeneration of two tomato cultivars under salinity stress condition in vitro. *Journal of Life Sciences* **9**: 27-31 doi: 10.17265/1934-7391/2015.01.004
[9] Sari, N.K.Y., Pharmawati, M., Junitha, I.K. 2012. Effects of sodium azide on morphological traits of *Capsicum annuum* L. *Jurnal Metamorfosa* **1** (1): 25-28 (In Indonesian)
[10] Padmavathi, S. 2015. Compatibility study of sodium azide induced macromutants of *Capsicum annuum* L. var. pusa jwala. *International Journal of researches in biosiences, agriculture and technology* Special Issue (6): 15-18
[11] Dahot, M.U., Rafiq, M., Arif, A.M., Naqvi, S.H.A. 2012. Effect of sodium azide on the growth of *Capsicum annuum*. *Pakistan Journal of Biotechnology* **9** (1): 13-20
[12] Singh, C., Olejniczak, J. 1983. Modification of mutagenic efficiency of sodium azide. *Cytologia* **48** (3): 437-444 doi: 10.1508/cytologia.48.437
[13] Gruszka, D., Szarejko, I., Maluszynski, M. Sodium azide as a mutagen. 2012. *In*: Plant Mutation Breeeding and Biotechnology. CABI International, Wallingford, UK. pp. 159-166
[14] Kleinhofs, A., Kleinschmidt, M., Sciaky, D., Broembsen, S.V. 1975. Azide mutagenesis: in vitro studies. *Mutation Research/Fundamental and Molecular Mechanisms of Mutagenesis* **29** (3): 497-499 doi: 10.1016/0027-5107(75)90069-X
[15] Khan, S., Al-Qurainy, F., Anwar, F. 2009. Sodium azide: a chemical mutagen for enhancement of agronomic traits of crop plants. *Environment and We: An International Journal of Science and Technology* **4**: 1-21

[16] Herwibawa, B., Haryanto, T.A.D., Sakhidin. 2014. The effect of gamma irradiation and sodium azide on germination of some rice cultivars. *Agrivita Journal of Agricultural Science* **36** (1): 26-32 doi: 10.17503/Agrivita-2014-36-1-p026-032

[17] Harding, S.S., Johnson, S.D., Taylor, D.R., Dixon, C.A., Turay, M.Y. 2012. Effect of gamma rays on seed germination, seedling height, survival percentage and tiller production in some rice varieties cultivated in Sierra Leone. *American Journal of Experimental Agriculture* **2** (2): 247-255

[18] Wang, Z., Wang, J., Bao, Y., Wang, F., Zhang, H. 2010. Quantitative trait loci analysis for rice seed vigor during the germination stage. *Biomedicine and Biotechnology* **11** (12): 958-964

[19] Rao, D.R.M., Reddi, T.V.V.S. 1986. Azide mutagenesis in rice. *Proceedings of the Indian Academy of Sciences (Plant Science)* **96** (3): 205-205

[20] Mostafa, G.G. 2011. Effect of sodium azide on the growth and variability induction in *Helianthus annuus* L. *International of Plant Breeding and Genetics* **5**: 76-85 doi: 10.3923/ijpbg.2011.76.85

[21] Munir, N., Safdar, I., Naz, S. 2015. Effect of induced mutation for varietal improvement in some local grapevine cultivars. *The Journal of Animal and Plant Sciences* **25** (1): 234-242

[22] Ali, A., Yubey, K., Deka, U.K., Tomar, S.M.S. 2014. Effect of sodium azide on seed germination and related agro-metrical traits in M_1 lentil (*Lens culinaris* Medik.) generation. *World Journal of Agricultural Sciences* **10** (3): 95-102 doi: 10.5829/idosi.wjas.2014.10.3.8582

[23] Dewi, K., Meidana, G., Sudjino, Suharyanto. 2016. Effects of sodium azide (NaN_3) and cytokininon vegetative growth and yield of black rice plant (*Oryza sativa* L. 'Cempo Ireng'). *AIP Conference Proceedings* **1755**, 130005 (2016) doi: 10.1063/1.4958549

[24] Adebola, M.O. 2013. Mutagenic effects of sodium azide (NaN_3) on morphological characteristics of tomato (*Lycopersicum esculentum*). *Research Journal of Science and IT Management* **2** (4): 1-5

[25] Ali, A., Bordoloi, N., Chand, S. 2014. Effect of sodium azide on indigenous rice (*Oryza sativa* L.) varieties of Nagaland. *International Journal of Recent Trends in Science and Technology* **12** (1): 123-130

[26] Animasaun, D.A., Oyedeji, S., Azeez, M.A., Onasanya, A.O. 2014. Evaluation of the vegetative and yield performances of groundnut (*Arachis hypogaea*) varieties samnut 10 and samnut 20 treated with sodium azide. *International Journal of Scientific and Research Publications* **4** (3): 1-10

[27] Liamngee, S.M., Ogah, J.J., Amagu, K.T., Kwon-Ndung, E.H., Iorkor, D., Tervershima, J.E. 2017. Mutagenic action of sodium azide on germination and emergence in landraces of *Phaseolus vulgaris* L. on the Jos Plateau Agro-Ecological Zone. *IOSR Journal of Agriculture and Veterinary Science* **10** (2): 64-70 doi: 10.9790/2380-1002016470

[28] Herwibawa, B., Haryanto, T.A.D., Sakhidin. 2014. Peroxidase isozyme identification of some rice genotypes in M_1 generation under drought stress level of -0.03 MPa. *Agrivita Journal of Agricultural Science* **36** (3): 210-215 doi:10.17503/Agrivita-2014-36-3-210-216

IOP Conf. Series: Earth and Environmental Science **102** (2017) 012053 doi:10.1088/1755-1315/102/1/012053

Model test on the relationship feed energy and protein ratio to the production and quality of milk protein

R Hartanto*, M A C Jantra, S A B Santosa and A Purnomoadi

Faculty of Animal and Agricultural Sciences, Diponegoro University, Semarang 50275, Indonesia

*Corresponding author's e-mail: rudyharta@gmail.com

Abstract. The purpose of this research was to find an appropriate relationship model between the feed energy and protein ratio with the amount of production and quality of milk proteins. This research was conducted at Getasan Sub-district, Semarang Regency, Central Java Province, Indonesia using 40 samples (Holstein Friesian cattle, lactation period II-III and lactation month 3-4). Data were analyzed using linear and quadratic regressions, to predict the production and quality of milk protein from feed energy and protein ratio that describe the diet. The significance of model was tested using analysis of variance. Coefficient of determination (R^2), residual variance (RV) and root mean square prediction error (RMSPE) were reported for the developed equations as an indicator of the goodness of model fit. The results showed no relationship in milk protein (kg), milk casein (%), milk casein (kg) and milk urea N (mg/dl) as function of CP/TDN. The significant relationship was observed in milk production (L or kg) and milk protein (%) as function of CP/TDN, both in linear and quadratic models. In addition, a quadratic change in milk production (L) ($P = 0.003$), milk production (kg) ($P = 0.003$) and milk protein concentration (%) ($P = 0.026$) were observed with increase of CP/TDN. It can be concluded that quadratic equation was the good fitting model for this research, because quadratic equation has larger R^2, smaller RV and smaller RMSPE than those of linear equation.
Keywords: feed energy and protein ratio, milk production, milk protein, regression

1. Introduction

Milk is a nutritious food that has high nutritious value, because it contains a plenty of nutritional substances and fulfills the needs of the human body. It is important for breeders to grow their dairy cows producing high quality and large amount of milk. The quality of milk is a reflection of the quality of feed given. The composition of nutrients present in animal feed will affect the production and composition of milk. High-performance lactation dairy cattle require adequate energy and protein intake to meet production needs in terms of both quality and quantity [1, 2, 3]. Protein feed has a very important role in the synthesis of milk components (protein, lactose and milk fat) [4]. Protein content in milk is one of important measurements of milk quality. INS stated that good fresh milk contains at least 2.8% protein [5].

The consumption of protein from the diet affects the amount of production and quality of milk proteins. Milk protein production is the result of multiplication of milk protein concentration with the amount of milk production [6]. There was a significant positive linear relationship between the consumption of crude diet protein and milk protein concentration, therefore higher consumption of protein will cause higher milk protein content [4]. Dairy cows including livestock are efficient in the use of food energy into milk energy [2]. Because of the principle of dairy cattle production is energy storage in milk, the lack of energy in the ration will result in decreasing milk production [7]. Protein synthesis is also influenced by the energy content of feed, because the synthesis of milk protein cannot optimally occur unless adequate energy is provided in the body [1]. Therefore it is reasonable to suspect the relationship between the feed energy and protein ratio with the amount of production and quality of milk proteins. Until now there is no clear information about it. Based on these facts, this research will be conducted using a survey method at the farmer level to collect the data needed to find an appropriate relationship model between the feed energy and protein ratio with the amount of production and quality of milk proteins.

2. Materials and Methods

2.1. Database

This research was conducted at Getasan Sub-district, Semarang Regency, Central Java Province, Indonesia using 40 samples (Holstein Friesian cattle, lactation period II-III and lactation month 3-4). Purposive sampling was used to determine the location of farms that spread in 4 villages. The data were collected based on the amount of feeding, feed consumption and milk production for each cattle during 5 days. Feed samples were analyzed proximate to determine the content of feed ingredients which later to calculate the feed energy and protein ratio (CP/TDN) based on data of feed consumption. On the fifth day, 200 ml of milk samples from each cattle were obtained for the analysis of protein, casein and milk urea nitrogen (MUN) according to AOAC [8].

2.2. Statistical Analyses

Data were analyzed with regression analysis including linear and quadratic models using SPSS software. The feed energy and protein ratio was independent variable (X); the milk content of protein, casein and MUN were dependent variables (Y). Coefficient of determination (R^2), P-value and Residual variance (RV) values were reported for the developed equations as an indicator of the goodness of model fit [9]. Mean square prediction error (MSPE) was also reported for the developed models and the evaluated extant models, calculated as:

$$\mathbf{MSPE} = \sum_{i=1}^{n} (O_i - P_i)^2/n$$

Where Oi is the observed value, and Pi is the predicted value. An estimate of the overall prediction error prediction by square root of the MSPE (RMSPE), expressed as a proportion of the observed mean [10].

3. Results

The equations of linear and quadratic models are shown in Table 1. P-value showed that the significant relationship was observed in milk production (L or kg) and milk protein quality (%) as a function of CP/TDN, on linear and quadratic model. Other relationships were not significant ($P \geq 0,082$). Increased feed energy and protein ratio did not affect ($P < 0.05$) in a quadratic increase in milk production (L or kg) and in a quadratic decrease in milk protein quality (%).

4. Discussion

In our research, a quadratic model was the best model for prediction milk production and milk protein content as a function of CP/TDN ratio, because a quadratic model has shown larger R^2, smaller RV and smaller RMSPE than those of linear equation [9, 10], as shown in Table 1. In quadratic model for milk production, the cusp of the curve was on X = 0.24 and Y = 13.84 kg. Milk production has good correlation with consumption of the feed energy and protein ratio. The contents of energy and protein in ration have positive effect on milk production [1, 7]. The nutrient requirement of dairy cows increases with milk production, where high production of milk requires high energy and protein diets [3]. The utilization of nitrogen and energy in the rumen for milk production on dairy cows was influenced by the source of dietary crude protein (CP) and energy fed [2, 6]. In other studies, they reported that the balanced intake of energy and protein with rumen undegraded protein supplementation in early lactating dairy cows didn't impact on milk production but may increase the amount of protein in the small intestine and finally could impact on milk protein concentration and production [6].

The ratio of protein and energy availability for milk production influenced milk protein production efficiency [11, 12]. Besides, the stage of lactation period and level of milk production also influenced it [6]. The low energy and high protein content in the ration caused low protein efficiency [13]. Although most research suggested that milk production is improved by increasing CP intake, and increasing energy intake increased both concentration and production of milk protein [14], the present research showed that only concentration of milk protein was influenced by CP/TDN and production of milk protein, milk casein and MUN were not affected. This can explain that decrease of concentration of milk proteinand increase of milk production by increasing CP/TDN intake lead no no effect on production of protein, casein, MUN. In this quadratic model for concentration of milk protein, the turning point of the curve was on X = 0.23 and Y = 1.50%.

5. Conclusions

Quadratic model can explain well the protein content and milk production as a function of CP/TDN. Especially, this analysis method in local farm has great merit to improve dairy milk production.Based on the result of research, it can be concluded that quadratic equation was the good fitting model for prediction milk production and milk protein content in this research, because of larger R^2, smaller RV and smaller RMSPE than those of linear equation.

Acknowledgments

This work was supported by the "HibahPenelitian FPP UNDIP, No. 84/UN7.5.5/PP/2017" Faculty of Animal and Agricultural Sciences, Diponegoro University.

References

[1] Bath D L, Dickinson F N, Tucker H A and Appleman R D 1985 *Dairy Cattle : Principles, Practices, Problem, Profits 3th Ed* (Philadelphia: Lea Febiger)
[2] McDonald P, Edwards R A, Greenhalg J F D and Morgan C A 2011 *Animal Nutrition7th Ed* (London: Prentice Hall Inc.)
[3] NRC 2001 *Nutrient Requirements of Dairy Cattle 7th Ed* (Washington D C: National Academic Science)
[4] Syafri A, Harjanti D W and Santoso S A B 2014 *Animal Agriculture Journal* **3**(3): 450-456
[5] Indonesian National Standard 2011 *INS No. 3141.1:2011 about Milk Fresh* (Jakarta: National Standardization Agency of Indonesia)
[6] Widyobroto BP, Rochijan, Ismaya, Adiarto and Suranindyah Y Y 2016*J. Indonesian Trop. Anim. Agric.***41**(2):83-90
[7] Musnandar, E 2011 *J. Penelitian Universitas Jambi Seri Sains* **13**(2): 53-58

IOP Conf. Series: Earth and Environmental Science **102** (2017) 012053 doi:10.1088/1755-1315/102/1/012053

[8] AOAC 2006 *Methods of Analyses16th Ed* (Rockville: Publ,AOAC)

[9] Gordeyase Mas I K, Hartanto R and Prastiwi W D 2007*J. Pengembangan Peternakan Tropis.***32**(4): 285 – 292

[10] Ellis J L, *et al.*2009 *J. Anim. Sci.***87**:1334-1345

[11] Mahr-un-Nisa, Javaid A, Aasif Shahzad M and Sarwar M 2008 *Asian-Aust. J. Anim. Sci.***21**:1303-1311

[12] Santos F A P, Santos J E P, Theurer C B and Huber J T 1998 *J. Dairy Sci.***81**: 3182-3213

[13] Widyobroto B P, Budhi S P S and Agus A 2010 *J. Indonesian Trop. Anim. Agric.***35**(1):27-33

[14] DePeters E J and Cant J P 1992 *J. Dairy Sci.***75**: 2043-2070

Table 1. Summary and evaluation of linear and quadratic equations developed

Variable	Equation	P-Value	R	R^2	RV	RMSPE (%)
Milk Production (L) (Y) and CP/TDN (X)						
- Linear	$Y = 134.100 X - 15.456$	0.001	0.518	0.268	4,778	22,713
- Quadratic	$Y = 589.666 X - 1222.966 X^2 - 57.765$	0.003	0.520	0.271	4,760	22,671
Milk Production (Kg) (Y) and CP/TDN (X)						
- Linear	$Y = 139.464 X - 16.074$	0.001	0.518	0.268	5,168	22,713
- Quadratic	$Y = 613.252 X - 1271.885 X^2 - 60.076$	0.003	0.520	0.271	4,763	22,671
Milk Protein (%) (Y) and CP/TDN (X)						
- Linear	$Y = 9.038 - 35.862 X$	0.007	0.420	0.176	0,584	31,937
- Quadratic	$Y = 22.681 - 182.770 X + 394.375 X^2$	0.026	0.423	0.179	0,582	31,888
Milk Protein (Kg) (Y) and CP/TDN (X)						
- Linear	$Y = 0.305 - 0.418 X$	0.766	0.049	0.002	0,007	3,545
- Quadratic	$Y = 15.961 X - 43.971 X^2 - 1.216$	0.903	0.074	0.005	0,007	3,540
Milk Casein (%) (Y) and CP/TDN (X)						
- Linear	$Y = 1.300 + 0.137 X$	0.986	0.003	0.000	0,212	34,283
- Quadratic	$Y = 302.306 X - 811.226 X^2 - 26.765$	0.506	0.190	0.036	0,204	33,658
Milk Casein (Kg) (Y) and CP/TDN (X)						
- Linear	$Y = 1.583 X - 0.164$	0.082	0.279	0.078	0,003	4,008
- Quadratic	$Y = 30.100 X - 76.554 X^2 - 2.813$	0.144	0.315	0.099	0,003	3,961
Milk Urea N (mg/dl) (Y) and CP/TDN (X)						
- Linear	$Y = 10.009 - 15.773 X$	0.339	0.155	0.024	0,982	13,833
- Quadratic	$Y = 25.806 X - 111.617 X^2 + 6.148$	0.636	0.156	0.024	0,982	13,832

IOP Conf. Series: Earth and Environmental Science **102** (2017) 012054 doi:10.1088/1755-1315/102/1/012054

The potential of pigeon pea (Cajanus cajan) beverage as an anti-diabetic functional drink

S Ariviani[1], D R Affandi[1], E Listyaningsih[2], S Handajani[1]

[1]Department of Agricultural Technology, Faculty of Agriculture, Universitas Sebelas Maret, Jl. Ir. Sutami No. 36A, Kentingan, Surakarta 57126, Indonesia.
[2]Department of Histology, Faculty of Medicine, Universitas Sebelas Maret, Jl. Ir. Sutami No. 36A, Kentingan, Surakarta 57126, Indonesia.

setya_ariviani@yahoo.com

Abstract. The number of patients with diabetes continues to increase. Diabetes complications might induce serious diseases such as kidney, nervous, cardiovascular diseases and stroke. Diabetic complications can be prevented by keeping blood glucose and cholesterol at normal levels. This study aims to determine the potential of pigeon pea beverage for lowering glucose and total cholesterol plasma levels and increasing the antioxidant status of diabetic-hypercholesterolemia rats. The research was conducted using 18 Sprague Dawley male rats aged 3 months old with an average body weight of 154 g. The rats were divided into three groups: normal group, D-H group (diabetic-hypercholesterolemia group), and pigeon pea beverage group. The results showed that pigeon pea beverage diet showed hypoglycemic and hypocholesterolemic activities, and could improve the antioxidant status of diabetic-hypercholesterolemia rats. Plasma glucose and total cholesterol levels of diabetic-hypercholesterolemia rats decreased 33.86% and 19.78% respectively. The improvement of the plasma antioxidant status was indicated by the decrease of plasma MDA (malondialdehyde) level, reaching 37.16%. The research result provides an alternative to diabetes management by using the local bean as an anti-diabetic functional drink.

Keywords: pigeon pea beverage, hypoglicemic, hypocholesterolemic.

1. Introduction

The number of diabetes patients continues to increase, from approximately 108 million in 1980 to 415 million in 2014. This number is estimated to increase to 552 million by 2035 [1, 2]. Glucose oxidation and non-enzymatic protein glycation which occurs during diabetes leads to an excessive free radical formation and subsequently result in oxidative stress which promotes the development of diabetes complications. Intakes of antioxidants have already been proved to be prospective in the treatment of diabetes [3, 4]. According to Forbes and Cooper [5], diabetes complications cause microvascular diseases, such as kidney, nervous and eye disease, and microvascular diseases, such as cardiovascular disease and stroke. Keeping glucose and cholesterol at normal levels can prevent diabetes complications.

Consumption of legumes provides health benefits such as reduced diabetes risk, obesity and cardiovascular disease [6, 7, 8]. It was related to the components of dietary fiber [9, 10] as well as phenolic compounds [11].

Pigeon pea (Cajanus cajan) is a local commodity that has a complete nutritional composition. When compared to soybeans, this legume has higher levels of dietary fiber, total mineral, and vitamin C, as well as lower fat content [12, 13]. Compared to other legumes, pigeon pea plant has some advantages, such as being the most drought-tolerant legume, resistant to environmental stresses, having high productivity and also contribute to moisture and nutrients of the soil [14, 15]. Pigeon pea is a potential source of antioxidants. It has a higher level of total phenolic and vitamin C content, as well as higher radical scavenging capacityrather than cowpea [16]. Acevedo's study [17] reported that pigeon pea flour has a low glycemic index value (46-49), with resistant starch content reaching 30 - 33% db. Pigeon pea intake was reported to be capable of providing hypoglycemic effects in humans and diabetic rats [8], improving lipid profile of hypercholesterolemia hamster through its ability to increase HDL cholesterol level, lowering the plasma levels of triglycerides, LDL cholesterol and total cholesterol [19].

In previous research, the authors have successfully produced pigeon pea beverage as a potential functional drink in terms of its sensory and chemical characteristics, and the free radical scavenging capacity. The beverage has low-fat (2.31 g/ 100 g db) and is rich in dietary fiber (23.03 g/ 100 g db). It also has VEAC (vitamin C equivalent antioxidant capacity) of 13.72 µmol vitamin C/g db [20]. This study examines the potential of pigeon pea beverages as antidiabetic functional food through its ability to reduce plasma glucose and total cholesterol levels as well as to increase plasma antioxidant status (reducing MDA levels) of diabetic-hypercholesterolemia rats at doses equivalent to human consumption of 30 grams/day. MDA plasma has known as a key biomarker of oxidative stress and lipid damage mediated by free radicals in diabetes [21]. Some researchers have studied the hypoglycemic activity of pigeon pea, such as pigeon pea roots extract [22], pigeon pea leave extract [23, 24], pigeon pea roots and leaves extracts [25]. Amalraj and Ignacimuthu [26] studied the effect of roasting on the hypoglycemic activity of pigeon pea. The results showed that unroasted pigeon pea has a hypoglycemic effect on diabetic rats, but roasting eliminates its effect. Habib [27] reported the hypocholesterolemic and hypoglycemic effects of pigeon pea ethanol extracts in diabetic rats. Previous researches [28, 29] have studied the hypocholesterolemic effects of pigeon pea leaves extract. Pigeon pea diets in hypercholesterolemic hamsters at doses of 200-800 (g/kg feed) showed the hypocholesterolemic activity and decreased hepatic MDA levels [19]. However, research on the hypoglycemic and hypocholesterolemic effect and plasma MDA levels reducing of pigeon pea diets on diabetic-hypercholesterolemia rats has not been reported.

2. Material and Method

2.1 Materials

Materials for the preparation of beverages included pigeon pea, ginger, cinnamon, and cloves from the local market, as well as maltodextrin DE 20 (China) from the Bratachem. Assay kits for glucose (GOD-PAP) and cholesterol (CHOD-PAP) were purchased from Diasys (Diagnostic Systems GmbH, Germany). AIN-93M Mineral mix, AIN-93M Vitamin mix, and L-cystine from MP Biochemicals LLC (Santa Ana, CA, USA). TBHQ (tertbutyl hydroquinone (2-(1,1-dimethyl ethyl)-.4-benzenediol)), alloxan, cholesterol, choline bitartrate, TEP (tetra-ethoxy propane) and other chemicals were obtained from Sigma-Aldrich Co. (St. Louis, MO, USA). Casein, soybean oil, CMC (carboxy methyl cellulose), and corn starch were obtained from Food and Nutrition Laboratory (Food and Nutrition Research Center of Gadjah Mada University, Yogyakarta, Indonesia).

2.2. Preparation of pigeon pea beverage

Preparation of the pigeon pea beverage was conducted according to Patent No. P00201304596. Briefly, the legume was dehulled, soaked, boiled, dried and powdered. The powder was allowed through the 60 mesh sieve, further, maltodextrin DE 20 was added to improve the solubility.

2.3 Animals and Diets

Male Sprague Dawley rats (aged 3 months old, weighing 154 ± 16 g) from the Integrated Research and Testing Laboratory of Gadjah Mada University (Yogyakarta, Indonesia). The rats were caged individually in stainless steel cages in uncontrolled lights, at room temperature and adequate ventilation, and ad libitum feeding. The research on animal models was approved by the ethics committee of Sebelas Maret University (Surakarta, Indonesia) No.149/H27/1/17/ER/2010. The animal experiment consisted of four periods, i.e., acclimatization (seven days), hypercholesterolemia induction (five days), diabetic induction (three days) and intervention periods (two weeks), respectively. During acclimatization period, the rats were fed based on American Institute of Nutrition for maintenance (AIN-93M) standard diet [30]. After the acclimatization, the rats were randomly divided into three groups each consist of six rats, set as normal group, diabetes-hypercholesterolemia (D-H) group and pigeon pea group. The normal group rats were fed AIN-93M standard diet during the experimental periods. The pigeon pea and D-H groups were fed hypercholesteroldiets (during hypercholesterolemia induction) or standard diet (during diabetic induction). After hypercholesterolemia induction, the hypercholesterolemia rats were diabetic induced by injecting the fasting rats with alloxan (80 mg/kg body weight). The hyperglycemia was confirmed by determination of plasma glucose 3 days post alloxan injection, and the rats considered to be diabetic if the plasma glucose level was above 200 mg/dL. During intervention periods, the pigeon pea group rats were fed intervention diet and were force feeding of pigeon pea beverage dissolved in distilled water at doses of 2.7 g/kg body weight (equal to human consumption of 30 g/day), whereas the D-H group rats were fed standard diet and were force feeding of distilled water. All diets used in this study were presented in Table1. Proximate and dietary fiber analysis of pigeon pea beverage was performed according to standard AOAC methods [31, 32]. The result was presented in Table 2. The animal body weights and residual feed intake were recorded. The feed efficiency ratio (FER) was determined as the ratio of body weight gain per unit of feed consumed over a period of time.

Table 1. Diet composition used in the animal study (per 1000 g)

Ingredient	Standard diet	Hypercholesterol diet	Intervention diet
Corn starch (g)	620.7	600.7	605
Casein (g)	140	140	132.9
Sucrose (g)	100	100	100
Soybean oi (g)l	40	40	39.2
Carboxy methyl cellulose (g)	50	50	50
Mineral mix (g)	35	35	34.2
Vitamin mix (g)	10	10	10
L-cystine(g)	1.8	1.8	1.8
Choline bitartrate (g)	2.5	2.5	2.5
TBHQ (mg)	8	8	8
Cholesterol (g)	-	20	-

Table 2. Proximate and dietary fiber of pigeon pea beverage

Component	Compositions
Moisture (%)	7.82
Fat (% db)	2.31
Protein (% db)	25.18
Crude Ash (% db)	2.40
Carbohydrate (by diff) (% db)	70.11
Dietary fiber (% db)	22.25

2.4 Blood plasma collection and analysis

The blood samples were collected by retro-orbital puncture at the initial phase (after acclimatization period), after hypercholesterolemia induction and diabetic induction periods, as well as after 1 and 2 weeks intervention periods. Blood plasma isolation was conducted using EDTA (Ethylene diamine tetra acetic acid) as anticoagulants, then centrifugation at 3,000 rpm (1500 × g) for 15 min at 4°C. Plasma glucose and cholesterol levels analyses were conducted using assay kits for glucose (GOD-PAP) and cholesterol (CHOD-PAP) respectively. Plasma antioxidant status was determined by MDA measurement using thiobarbituric acid-reactive substance (TBARS-C18) method [33].

2.5 Statistical analysis

Each measurement was reported as a mean followed by the standard deviation. The IBM SPSS Statistics 22 (SPSS Inc., Chicago, USA) program were used to the research data analyses. The data were analyzed by ANOVA (analysis of variance) and the significant differences between means ($p<0.05$) were determined using DMRT (Duncan's multiple range test).

3. Results and Discussions

3.1 Effect of pigeon pea beverage diets on the body weight and feed efficiency ratio(FER) of diabetic-hypercholesterolemia rats

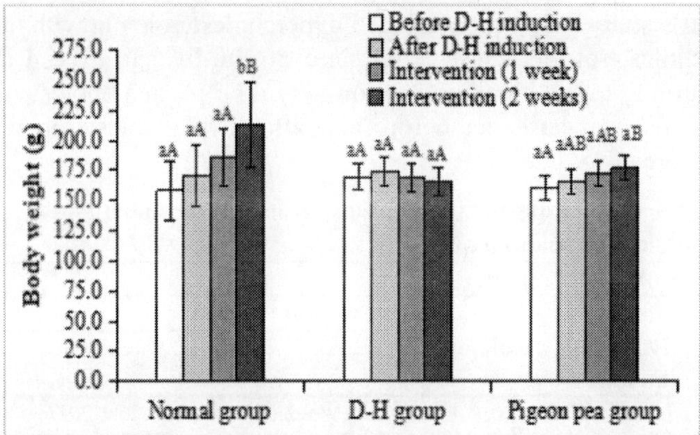

Figure 1. Effect of pigeon pea diet on the body weight of diabetic-hypercholesterolemia rats. Different lowercase letters mean significant differences ($p<0.05$) in the body weight for different groups, whereas different capital letters mean significant differences for similar groups

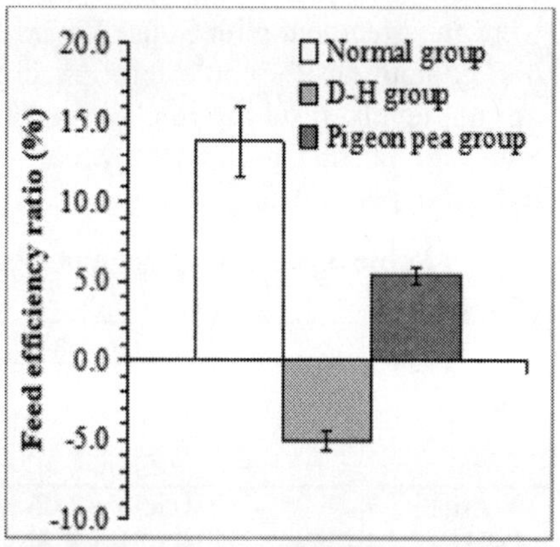

Figure 2. Effect of pigeon pea beverage diet on the feed efficiency ratio of diabetic-hypercholesterolemia rats.

Body weight and FER of all experimental rats groups were presented in figure 1 and 2 respectively. All groups of the rats showed similar body weight ($p> 0.05$) in the initial phase (before D-H induction), after D-H induction and after 1 week intervention periods (figure 1). After 2 weeks of intervention period, the normal group has significantly higher body weight than the other groups. Body weight gain was observed in both normal group rats (36.67%) and pigeon pea group rats (9.95%) after 2 weeks of intervention period, while the D-H group rats have similar body weight during experimental periods. These results suggest that D-H induction leads to the body weight loss, but it could be suppressed by the pigeon pea beverage administration during the intervention period. It was in line with the feed efficiency ratio (figure 2). The pigeon pea beverage capable to increase the feed efficiency ratio in the diabetic-hypercholesterolemia rats, although it was not as high as the normal group rats. Alloxan induces specific inhibition of glucokinase and increases the ROS (Reactive oxygen species) formation, resulting in reduced insulin secretion due to the specific inhibition of glucose-induced insulin secretion and necrosis of beta cells [34]. Therefore, the body used protein (protein washing) or fat in adipose tissue (lipolysis) became an energy source. So, the body weight was reduced. Kandulska [35] reported that alloxan treatment induced lipolysis in rats adipocyte which may cause loss of white adipose tissue, and consequently reduce rats body weight. It was attributed by the alloxan capability to form free radicals. In the previous study [20], the author observed that pigeon pea beverage demonstrates radical scavenging activity related to its phenolic compound.

The D-H group rats have negative FER value (figure 2). It was related to the rat's body weight (figure 1) which exhibits decreasing tendency at the end of the experimental period. Bang [36] and [37] reported similar results, i.e. the diabetic rats fed standard diet showed negative FER value and body weight loss.

3.2. Hypocholesterolemic effect of pigeon pea beverage diets on diabetic-hypercholesterolemia rats

Diabetic-hypocholesterolemia rats induced by alloxan and high-cholesterol diets are suitable animal models to simulate clinically diabetics complication with hypercholesterolemia [38, 39]. The effect of pigeon pea beverage supplementation on plasma total cholesterol level of diabetic-hypercholesterolemia rats is presented in table 1. The plasma total cholesterol level of all three groups rat was not significantly different at the initial phase of the experiment. The hypercholesterol diet consisting of 2% cholesterol which was used in the present study significantly elevated the plasma cholesterol level reaching 66% - 71% in both pigeon pea group and D-H group rats. These results were in agreement with Solanski and Bhat's study [40] which induced hypercholesterolemia with the diet containing 2% cholesterol. A high-cholesterol diet causes cholesterol in the liver to exceed its normal levels, resulting in cholesterol returning to the blood circulation [41]. The plasma cholesterol levels of normal group rats which were fed standard diet before and after hypercholesterolemia induction period were not significantly different.

Table 3. Effect of pigeon pea beverage diet on plasma total cholesterol level of diabetic-hypercholesterolemia rats

Group	Plasma total cholesterol levels (mg/dL)			
	initial	after diabetic-hypercholesterolemia induction	intervention period: 1 week	intervention period: 2 weeks
Normal	$105.58 \pm 3.06^{a}_{AB}$	$107.77 \pm 2.72^{ab}_{A}$	$110.19 \pm 2.63^{b}_{A}$	$106.56 \pm 2.97^{a}_{A}$
D-H	$105.31 \pm 4.32^{a}_{AB}$	$175.76 \pm 4.49^{b}_{B}$	$178.70 \pm 4.36^{bc}_{C}$	$182.68 \pm 4.90^{c}_{C}$
Pigeon pea beverage	$102.12 \pm 5.17^{a}_{A}$	$175.36 \pm 4.34^{d}_{B}$	$164.81 \pm 2.05^{c}_{B}$	$140.68 \pm 2.47^{b}_{B}$

Different subscript (capital letters) in the same column and different superscript (lowercase letters) in the same row indicated significant differences ($p<0.05$).

Table 3 indicates that supplementation of pigeon pea beverage (0.27 g/kg body weight) during the intervention period caused significantly reduced the plasma cholesterol level of diabetic-hypercholesterolemia rats, reaching 6.04% and 19.78% after 1 week and 2 week supplementation respectively. On the contrary, D-H group rats showed significantly elevated plasma cholesterol levels, reaching 1.65% and 3.92% after 1 week and 2 week intervention period. Hypocholesterolemic activity of the pigeon pea beverage may be attributed to the dietary fiber compounds and its antioxidant capacity. Ariviani [20] reported that pigeon pea beverage has soluble, insoluble and total dietary fiber of 8.45 %db, 14.58 %db, and 23.03 %db. The beverage displays radical scavenging capacity equivalent to 13.72 mmol Vitamin C/ kg dry weight, due to the vitamin C and tocopherol content as well as the phenolic compound. Dietary fiber provides a hypocholesterolemic effect through the regulation of HMG-CoA reductase expression which lowered cholesterol synthesis and enhanced cholesterol excretion in bile [42]. Asparagus fiber fraction can lower blood plasma cholesterol level of hypercholesterolemia rat [43].

The mechanism of cholesterol reduction by antioxidants is possible due to the antioxidant ability in inhibiting LDL oxidation. The oxidized LDL will be recognized by the receptor scavenger in macrophages and internalized it to form foam cells, resulting in the accumulation of cholesterol in the macrophages [44]. Because the LDL oxidation was inhibited by the antioxidant compounds, so the massive cholesterol uptake in the macrophages will be suppressed. Hypocholesterolemic potentials of the antioxidant compound in pigeon pea beverage are in line with some previous studies. Supplementation of vitamin C [45, 46, 47], tocopherols [45], and flavonoids [43] reduced total plasma cholesterol concentration in hypercholesterolemia rats. Vitamin C is necessary for the transformation of Cholesterol to bile acids (most important pathway of cholesterol catabolism) simultaneous with increased fecal and liver bile acids through activation of 7α-hydroxylase the limiting step of the Cholesterol catabolism in liver [45, 47]. Flavonoids as one of the phenolic compounds are reported to be capable of decreasing blood cholesterol content, down-regulating cholesterol synthesis and increasing the expression of LDL receptors [43].

3.3. Hypoglicemic effect of pigeon pea beverage diets on diabetic-hypercholesterolemia rats

Table 4. Effect of pigeon pea beverage diet on plasma glucose level of diabetic-hypercholesterolemia rats

Group	Plasma gluosse levels (mg/dL)			
	initial	after diabetic-hypercholesterolemia induction	intervention period 1 week	intervention period 2 weeks
Normal	$71.68 \pm 0.73^{a}{}_{A}$	$73.54 \pm 0.89^{b}{}_{A}$	$74.59 \pm 0.92^{b}{}_{A}$	$73.93 \pm 1.45^{b}{}_{A}$
D-H	$71.79 \pm 0.99^{a}{}_{A}$	$226.62 \pm 2.26^{b}{}_{B}$	$226.66 \pm 4.53^{b}{}_{C}$	$231.97 \pm 3.97^{c}{}_{C}$
Pigeon pea beverage	$71.57 \pm 0.92^{a}{}_{A}$	$227.71 \pm 2.80^{d}{}_{B}$	$188.71 \pm 3.36^{c}{}_{B}$	$150.60 \pm 1.50^{b}{}_{B}$

Different subscript in the same column and different superscript in the same row indicated significant differences ($p<0.05$)

Diabetes is a disease marked by high plasma glucose level and the presence of glucose in urine. Alloxan injection at doses of 80 mg/kg body weight proved able to induce diabetes in Sprague Dawley male rats, as indicated by elevated plasma glucose levels from normal levels (71 mg / dL) to hyperglycemia (227 mg / dL) (table 4). Diabetic action mechanism of alloxan was through its ability to induce ROS formation as well as the specific inhibition of glucokinase [34]. Table 4 indicated the hypoglicemic activity of pigeon pea beverage on diabetic-hypercholesterolemia rats. Pigeon pea group rats showed significantly lower plasma glucose level than D-H group, even though it was still higher than those of normal group rats. The results are in line with several previous studies which were reported the hypoglycemic activity of pigeon pea seed, such as [26], [19], and [18].

Capabilities of pigeon pea on suppressing severe hyperglycemia of diabetic-hypercholesterolemia rats were due to the antioxidant and dietary fiber compounds. As described previously, the ones of the diabetic mechanism of alloxan were through its ability on inducing radical (ROS) formation caused oxidative damage of pancreatic beta cells thus impairing the insulin secretion. Radical scavenging ability of pigeon pea antioxidant compound will inhibit the more severe oxidative damage caused by alloxan, thereby insulin secretion enhanced and thus lowering glucose levels. Several study reported the hypoglycemic effect of antioxidant compounds, such as [48], [49], [50]. Dietary fibers showed hypoglycemic effect through several mechanisms such as increasing the viscosity of small intestine juice and hinder diffusion of glucose, glucose binding and decrease the concentration of available glucose in the small intestine, as well as retarding alpha-amylase action through capsuling starch and the enzyme and might directly inhibit the enzyme [51].

3.4. Antioxidant status of diabetic-hypercholesterolemia rats administered with pigeon pea beverage

Oxidative stress is suggested as mechanism underlying diabetes and diabetic complications.

Oxidative stress acts as mediator of insulin resistance and its progression to glucose intolerance, subsequently support the appearance of atherosclerotic complications, and contribute to rising in many micro- and macrovascular complications [52, 21, 3]. Table 5 presents the plasma antioxidant status expressed as plasma MDA level of the normal, D-H and pigeon pea groups rats at the initial, after D-H induction, after 1 week and 2 weeks intervention periods. The Higher plasma MDA level indicates lower plasma antioxidant status. MDA was known as biomarkers of oxidative stress in diabetes mellitus related to lipids peroxidation [3].

Tabel 5. Effect of pigeon pea beverage diet on plasma MDA level of diabetic-hypercholesterolemia rats

Group	Plasma MDA levels (mmol/l)			
	initial	after diabetic-hypercholesterolemia induction	intervention period 1 week	intervention period 2 weeks
Normal	$1.45 \pm 0.17^{a}{}_{A}$	$1.72 \pm 0.19^{b}{}_{A}$	$2.11 \pm 0.23^{c}{}_{A}$	$1.89 \pm 0.19^{bc}{}_{A}$
D-H	$1.32 \pm 0.18^{a}{}_{A}$	$11.88 \pm 0.32^{b}{}_{B}$	$14.08 \pm 0.34^{c}{}_{C}$	$14.28 \pm 0.39^{c}{}_{C}$
Pigeon pea beverage	$1.56 \pm 0.24^{a}{}_{AB}$	$11.76 \pm 0.22^{c}{}_{B}$	$13.11 \pm 0.25^{d}{}_{B}$	$7.39 \pm 0.17^{b}{}_{B}$

Different subscript in the same column and different superscript in the same row indicated significant differences ($p<0.05$)

Diabetic-hypercholesterolemia induction was significantly reduced plasma antioxidant status due to the lipids peroxidation as indicated by elevated of plasma MDA levels. Supplementation with pigeon pea beverage reduced plasma MDA levels of diabetic hypercholesterolemia rats reached to 37.16% after 2 weeks intervention period. On the contrary, increasing of MDA plasma level (20.20%) was observed in D-H group rats. The plasma MDA levels (table 5) was in line with the plasma cholesterol data (table 3) and plasma glucose data (table 4). These results suggest the potential of pigeon pea beverage as a potential source of antioxidant intake to overcome oxidative stress in diabetics, thus suppressed of diabetic complication.

5. Conclusion

The pigeon pea beverages administration at doses equivalent to human consumption of 30 grams/day proven to be capable to reduce plasma glucose and total cholesterol levels, and increase plasma antioxidant status (reducing MDA levels) of diabetic-hypercholesterolemia rats. This result indicated the potential of pigeon pea beverage as an anti-diabetic functional drink. The research result provides an alternative strategy for maintaining plasma glucose and cholesterol at normal levels to prevents diabetes complication by using the local bean as an anti-diabetic functional drink.

References

[1] International Diabetes Federation (IDF). 2014. *Global Diabetes Scorecard-Tracking Progress for Action (*Brussels, Belgium). [Internet]. Available from: www.idf.org

[2] World Health Organization (WHO). 2016. *Global Report on Diabetes (*Geneva). [Internet]. Available from: http://www.who.int

[3] Ullah A, Khan A, Khan I. 2016. *Saudi Pharm J* **24**(5):547–53. Available from: http://dx.doi.org/10.1016/j.jsps.2015.03.013

[4] Khan AN, Khan RA, Ahmad M, Mushtaq N. 2015. R *J Pharmacogn Phytochem* **3**(6):217–20.

[5] Forbes JM, Cooper ME. Mechanisms of diabetic complications. 2013. *Physiol Reivew* **93**:137– 88.

[6] Kushi LH, Meyer KA, Jacobs Jr DR. 1999. *Am J Clin Nutr* **70**:451S–458S.

[7] Bazzano LA, He J, Ogden LG, Loria C, Vupputuri S, Myers L, and Whelton PK. 2001. *Arch Intern Med.* **161**(26):2573–8.
[8] Venn BJ, Mann JI. 2004. *Eur J Clin Nutr* **58**(11):1443–61. Available from: http://www.ncbi.nlm.nih.gov/pubmed/15162131
[9] Messina MJ. 1999. *Am J Clin Nutr* **70**:439S–450S.
[10] Trinidad TP, Mallillin AC, Loyola AS, Sagum RS, Encabo RR. 2010. *Br J Nutr* **103**(4):569–754. Available from: http://www.ncbi.nlm.nih.gov/pubmed/19825218
[11] Oomah BD, Tiger N, Olson M, Balasubramanian P. 2006. *Plant Foods Hum Nutr* **61**(2):91–97. Available from: http://www.ncbi.nlm.nih.gov/pubmed/16804740
[12] Torres A, Frias J, Granito M, Vidal-Valverde C. 2007. *Food Chem* **101**(1):202–11. Available from: http://linkinghub.elsevier.com/retrieve/pii/S0308814606000744
[13] Bhat R, Karim AA. 2009. *Compr Rev Food Sci Food Saf* **8**:305–331.
[14] Odeny DA. 2007. *Nat Resour Forum* **31**(4):297–305. Available from: http://doi.wiley.com/10.1111/j.1477-8947.2007.00157.x
[15] Sharma S, Agarwal N, Verma P. 2011. *J Funct Environ Bot* **1**(2):91. Available from: http://www.indianjournals.com/ijor.aspx?target=ijor:jfeb&volume=1&issue=2&article=002
[16] Oboh G. 2006. *Eur Food Res Technol* **224**(1):61–65. Available from: http://link.springer.com/10.1007/s00217-006-0289-x
[17] Acevedo BA, Avanza M V, Cháves MG, Ronda F. 2013. *J Food Eng* **119**(1):65–71. Available from: http://linkinghub.elsevier.com/retrieve/pii/S0260877413002434
[18] Kaushik G, Satya S, Khandelwal RK, Naik SN. 2010. *Diabetes Metab Syndr Clin Res Rev* **4**(1):21–40. Available from: http://linkinghub.elsevier.com/retrieve/pii/S1871402108000234
[19] Dai F-J, Hsu W-H, Huang J-J, Wu S-C. 2013. *Food Chem Toxicol* **53**:384–91. Available from: http://www.ncbi.nlm.nih.gov/pubmed/23287313
[20] Ariviani S, Handajani S, Dian RA, Listyaningsih E. 2015. *Potensi kacang gude (cajanus cajan) sebagai minuman fungsional : karakteristik kimia , sensoris dan kapasitas antioksidan in vitro (The Potential of Pigeon Pea (Cajanus cajan) as Functional Beverage : Chemical Characteristics , Sensory and In Vitro.* Seminar Nasional & Pameran Produk Pangan 2015 " Inovasi Teknologi untuk Memperkuat Peran Industri Menuju Akselerasi pemenuhan Pangan Nasional''. Semarang: PATPI (Perhimpunan Ahli Teknologi Pangan Indonesia)
[21] Tiwari BK, Pandey KB, Abidi AB, Rizvi SI. 2013. *J Biomarkers* **2013**:1–8.
[22] Nahar L, Nasrin F, Zahan R, Haque A, Haque E, Mosaddik A. 2014. *Pharmacognocy Res* **6**(2):180–187.
[23] Aja PM, Igwenyi IO, Okechukwu PCU, Orji OU, Alum EU. 2015. *Glob Vet* **14**(3):439–47.
[24] Ezike AC, Akah PA, Okoli CC, Okpala CB. Experimental evidence for the antidiabetic activity of Cajanus cajan leaves in rats. J Basic Clin Pharm [Internet]. 2010;81–4. Available from: www.jbclinpharm.com
[25] Anwar MM, Kalpana MA, Bhadra B, Rahman S, Sarker S, Chowdhury MH, et al. 2010. *Adv Nat Appl Sci* **4**(3):311–6.
[26] Amalraj T, Ignacimuthu S. 1998. *Indian J Exp Biol* **36**(10):1032–3.
[27] Habib A, Anisuzzaman AS, Khan RI, Gafur A. 2010. *J Bangladesh Pharmacol Soc* **5**:34–38.
[28] Luo Q, Sun L, Si J, Chen D. 2008. *Phytomedicine* **15**:932–9.
[29] Akinloye O., Solanke O. 2011. *J Med Plants Res* **5**(12):2521–2524.
[30] Reeves PG. 1997. *J Nutr* **127**:838–41.
[31] AOAC. 1990. *Official method of analysis of the association of official analytical chemist. 5th Edition (*Washington, DC: Association Official Analytical Chemists)
[32] AOAC. 1995. *Official method of analysis of the association of official analytical chemist. 16th edition (*Gaithersburg, Maryland, USA: Asociation of Official Analytical Chemist)
[33] Raharjo S, Sofos J., Schmidt G. 1992. *J Agric Food Chem* **40**(11):2182–5.

[34] Lenzen S. 2008. *Diabetologia* **51**:216–26.
[35] Kandulska K, Szkudelsski T, Nogowski L. 1999. *Physiol Res* **48**:113–117.
[36] Bang M-A, Kim H-A, Cho Y-J. 2009. *Nutr Res Pract* **3**(3):242–6.
[37] Lee S, Kim S, Oh S-H, Park K-Y, Lee S-D. 2008. *J Med Food* **11**(3):518–24.
[38] Xie W, Xing D, Sun H, Wang W, Ding Y, Du L. 2005. *Am J Chin Med* **33**(1):95–105.
[39] Tang L, Wei W, Chen L, Liu S. 2006. *J Ethnopharmacol* **108**:109–15.
[40] Solanki YB, Bhatt R V. 2010. *Int J Physiol Pathophysiol Pharmacol* **2**(1):57–63.
[41] Kai NS, Nee TA, Lai E, Ling C, Ping TC, Kamariah L, et al. 2015. *Asian Pac J Trop Med* **8**(1):6–13. Available from: http://dx.doi.org/10.1016/S1995-7645(14)60179-6
[42] Kaczmarczyk MM, Miller MJ, Freund GG. 2012. *Metabolism* **61**(8):1058–66. Available from: http://www.pubmedcentral.nih.gov/articlerender.fcgi?artid=3399949&tool=pmcentrez&rendert ype=abstract
[43] Vázquez-castilla S, Puerta R De, Giménez MDG, Fernandez-Arche M angeles, Guillen-Bejarano R. 2013. *Int J Mol Sci* **14**:21227–39.
[44] Ryu BH. 2000. *Biotechnol Bioprocess Enginering* **5**(5):313–4.
[45] Salem SA, Hassan DA, Mowafy AR. *Middle-East J Sci Res* **4**(4):354–60.
[46] Alkhamees OA. 2013. *J Appl Pharm Sci* **3**(4):99–105.
[47] Arafa M., Aly S., Shaheen N. 2014. *World J Pharm Res* **3**(8):1250–64.
[48] Soto C, Recoba R, Barron H, Alvarez C, Favari L. 2003. *Comp Biochem Physiol Part.* **136**:205–12.
[49] Al-azzawie HF, Alhamdani MS. 2006. *Life Sci* **78**:1371–7.
[50] Hamden K, Ali M, Masmoudi H, Ayadi FM, Jamoussi K, Elfeki A. 2009. *Biomed Pharmacother* **63**(2):95–9. Available from: http://dx.doi.org/10.1016/j.biopha.2008.02.001
[51] Ou S, Kwok K, Li Y, Fu L. 2011. *J Agric Food Chem* **49**:1026–1029.
[52] Moussa S. 2017. *Rom J Biophys* **118**(3):225–136.

IOP Conf. Series: Earth and Environmental Science 102 (2017) 012055 doi:10.1088/1755-1315/102/1/012055

Thermal degradation kinetics of phycocyanin encapsulation as an antioxidant agent

A M Nilamsari[2] , A Yunanda[2], and H Hadiyanto[1 2],

[1]Center of Biomass and Renewable Energy (CBIORE), Chemical Engineering Department, Diponegoro University
[2]Department of Chemical Engineering Diponegoro University
Jl. Prof. Soedarto SH, Tembalang, Semarang, 50275
Email: hadiyanto@live.undip.ac.id

Email: hady.hadiyanto@gmail.com

Abstract. Phycocyanin is a blue-light pigment that found in Cyanobacteria and two Eukaryotics algae such as Rhodophyta and Crytophyta. Phycocyanin is soluble in water and has a strong fluorescent properties as an antioxidant and normally used in food industry, cosmetic, biotechnology, and drug. However, Phycocyanin is easily damaged by a heating process. The aim of this study is to obtain the optimal condition of phycocyanin encapsulation with different coating materials, Chitosan and Carrageenan, by the calculation of heat resistance of antioxidant activity (D), range of temperature that increase the rate of degradation (Z), rate constant of degradation (k), and activation energy (Ea). The ratio of phycocyanin and the coating material are 2% (w/v) and 2 % (w/v).
Keyword: Phycocyanin, Antioxidant, Carrageenan, Chitosan, Degradation

1. Introduction

Antioxidants are compounds that can be used to protect food by slowing down the damage, rancidity or change of colour caused by oxidation [9]. Antioxidants are able to act as hydrogen radical contributors or free radical acceptors so as to delay the initiation phase of free radical formation. The presence of natural (such as Phycocyanin compounds) and synthetic antioxidants can inhibit lipid oxidation, prevent damage, change inorganic components in foodstuffs so as to extend shelf life [8]. T-butyl hydroxy anisol (BHA) and di-t-butyl hydroxytoluene (BHT) compounds are used as food antioxidants, but the possibility of adverse side effects is not used for therapeutic agents. The development of natural antioxidants has received considerable attention in recent years. It is intended for the purpose of preventive medicine and for the food industry. The DPPH test is an effective and rapid method for estimating antioxidant activity. The principle of this method is to measure the power of extracting an ingredient against the free radical of DPPH [3].

Development of antioxidant components is widely used in the production of functional foods. The functional properties of functional foods are determined by the bioactive components contained in them, eg dietary fiber, inulin, FOS and antioxidants [6]. Based on the source, antioxidants are divided into endogenous antioxidants and antioxidant enzymes, such as: Superoxide Dismutase (SOD), Catalase (Cat), and Glutathione Peroxidase (Gpx); as well as exogenous antioxidants, which are obtained from outside the body / food. Various natural ingredients native to Indonesia contains many antioxidants with various active ingredients, including vitamin C, E, pro vitamin A, organosulfur, α-tocopherol, flavonoid, thymoquinone, statin, niacin, Phycocyanin, and others [2].

One type of antioxidant that comes from outside the human body is Phycocyanin. Phycocyanin is the most important pigment in *Spirulina sp.* Microorganisms [4].Phycocyanin has a significant content as an antioxidant, protector of liver function, and radical compounds remover [10]. Therefore, Phycocyanin is widely used in the field of food colouring and cosmetics. Phycocyanin content in 10 grams of dried spirulina is about 1400 mg or 14%, so it has been recommended to consume 0.25-2.5 grams of Phycocyanin per day to boost the immune system and inhibit cancer growth [4]. There are several methods to protect Phycocyanin from degdradation, microencapsulation is one of them.

Microencapsulation is a process where solid materials, liquids, and even gases can be capsulated (encapsulated) with microscopic particle size, by forming thin wall covering around the material to be capsulated [1]. The techniques used for microencapsulation are spray drying, spray cooling / chilling, fluidization of bed drying, melt extrusion, melt injection, centrifugal extrusion, coacervation, liposome capture, co-crystallization, emulsification, rotational suspension rotation, and molecular inclusion [7]. The active ingredient in the microencapsulation of phycocyanin is sensitive to light, heat and oxygen so as to have a limited shelf life. Through the desired microencapsulation process can minimize the occurrence of quality degradation of phycocyanin. Therefore, thermal degradation kinetics model of the microencapsulation of phycocyanin needs to be further investigated.

2. Material and Methods

2.1. *Materials*

In this study, phycocyanin is used as raw material. In addition, side materials are also used such as: chitosan, carrageenan, DPPH, methanol, acetic acid, STTP, KCl, NaCl and aquadest. Chitosan and carrageenan were used as a coating material. Methanol, acetic acid, NaCl, and aquadest were used as solvent. STTP and KCl were used as a microencapsulation media, and DPPH as a material test. Phycocyanin is powder form of *Spirulina platensis*. While other materials obtained from laboratory in Diponegoro University.

2.2. *Variables*

The controlled variables in this research were heating time during 90 minutes with observation time for every 30 minutes. The response variable in this study was the activity of antioxidant. Independent variables in this study were temperature (40°C, 50°C, 60°C) and coating materials (chitosan 2% and carrageenan 2%).

2.3. *Making Microencapsulation of Phycocyanin*

Making of microencapsulation of Phycocyanin is devided into two methods, they are chitosan coating and carrageenan coating. Once, Coating of chitosan was done by dissolving chitosan into 1% of acetic acid solvent. One percent of acetic acid solution was prepared by dissolving 1 ml glacial acetic acid into 100 ml of aquadest. Then, 2 grams of chitosan was dissolved in 100 ml of 1% acetic acid solution, stirred with magnetic stirrer for ±4 hours until homogenous to obtain chitosan 2% w/v 100 ml. After that, dissolve Phycocyanin in aquadest with ratio of Phycocyanin and aquadest of 1:10. Then, Phycocyanin solution mixed with chitosan solution with ratio of volume of 1:1. After that, stir the solution by using magnetic stirrer until homogeneous. Then, homogeneous solution is inserted into syringe and injected into STTP (Sodium Tripolyphosphate) solution 0.5 M. Wait for about 15 minutes until the matrix perfectly formed. Then, filtered and washed the granules with 1% of NaCl. After that, dried it by freeze drying method (10°C) and wait for about 24 hours. Finally, microencapsulation with chitosan coating is formed. In the other hand, carrageenan coating was done by dissolved 2 grams of carrageenan in 100 ml of 1% NaCl solution and heated with stirring by using magnetic stirrer at temperature about 40-50 °C until gel of carrageenan are formed, and carrageenan 2% w/v 100 ml was obtained. After that, dissolved Phycocyanin in aquadest with ratio of Phycocyanin and aquadest of

1:10. Then Phycocyanin solution mixed with carrageenan solution with ratio of volume of 1:1. After that, stirred the mixture by using magnetic stirrer until homogeneous. The homogeneous solution is inserted into syringe and injected into KCl 0.5 M. Wait for about 15 minutes until the matrix perfectly formed. Then filtered and washed the granules with 1% NaCl. After that, dried by freeze drying method (10°C) and wait for about 24 hours. Finally, microencapsulation by carrageenan coating is formed.

2.4. *Antioxidant Activity Test*

The DPPH test is an effective and rapid method for estimating antioxidant activity. The principle of this method is to measure the power of extracting an ingredient against the free radical of DPPH [3]. The various variables are temperature (40°C, 50°C, and 60°C) and heating time (0, 30, 60, 90 minutes). Antioxidant activity of microencapsulation Phycocyanin analysis was conducted by dissolving 10 mg of microencapsulation into 10 ml of aquadest, then heated according to various temperature (40°C, 50°C, and 60°C). Then, took 1 ml sample for various times (0, 30, 60, 90 minutes). Make a solution of 0.1 mM DPPH by dissolving 2 mg DPPH into 50 ml of methanol. Then 1 ml of heated microencapsulated Phycocyanin, 1 ml DPPH 0.1 mM, and 5 ml of methanol are incubated for about 30 minutes. Then, take the solution that has been incubated to have an antioxidant activity test. And also, prepared blank solution in the same way without using of Phycocyanin encapsulation. Perform absorbance measurement at optimum λ in wavelength of 515 nm. Antioxidant capacity is to inhibit free radicals. To determined the equation of antioxidant activity is:

Antioxidant Activity (100%) = {1- (A sample / A blank)} x 100%.

3. Result and discussions

3.1. *Determine Thermal Degradation Kinetic of Phycocyanin Encapsulation with Carrageenan*

Thermal degradation kinetics of Phycocyanin encapsulation with carragenan are detemined by various parameters, including parameters of k (1/minute), D (minutes), Z (°C) and Ea (kJ/mol). By Arrhenius equation, the order of the reaction can be determined. After a suitable order has been determined, the rate constant of reaction (k) is known by slope of the order equation.

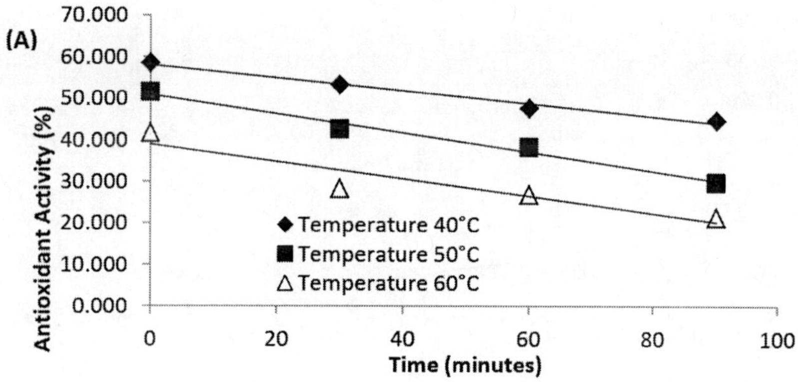

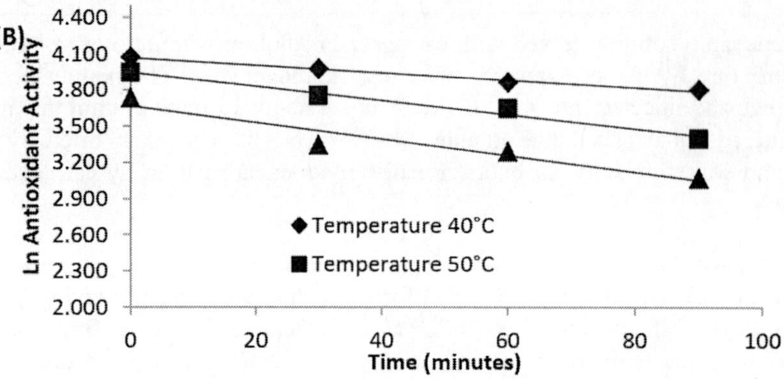

Figure 3.1. Thermal degradation kinetics of antioxidant activity (A) 0 order & (B) 1st order

Based on Figure 3.1.,maximum regression is 1st order. So, thermal degradation kinetics of antioxidant activity of carrageenan are using 1st order equation with R^2 equal to 0.9161-0.9899. The results are according to [5] that the kinetics model of antioxidant capacity by DPPH method of 1st order equation with R^2 is equal to 0.95-0.97.

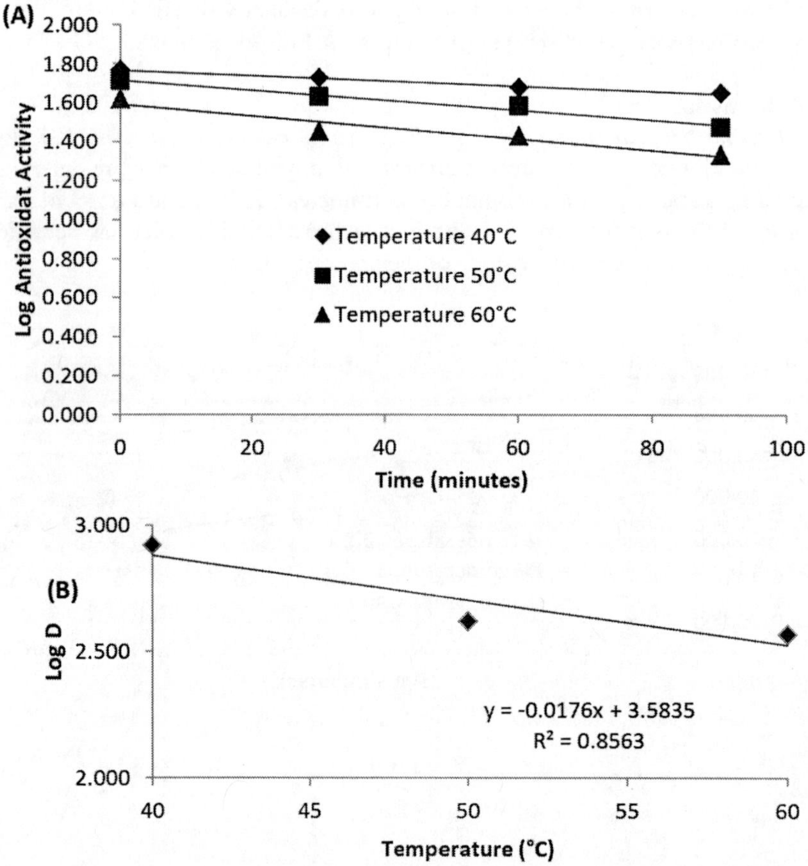

Figure 3.2. Graph (A) value D and (B) value Z

According to the data from Figure 3.2 (A) and Table 3.1, there is a degdradation process in antioxidant during heating process. The value of D, are 769 minutes for $D_{40°C}$, 400 minutes for $D_{50°C}$, and 333 minutes for $D_{60°C}$. Thus, the differences in temperature and heating time also make a different effect of degdradation kinetics of antioxidant activity in Phycocyanin encapsulation. Based on Figure 3.2 (B), it showed that regression equation is y=-0.0176x+3.5835 with R^2=0.8563.

Table 3.1 Calculation of values D, k, Z, and Ea

Temperature (°C)	D (minute)	k ($minute^{-1}$)	Z (°C)	Ea (kJ/mol)
40	769	0.0031		
50	400	0.0058	54.945	804
60	333	0.0068		

From Table 3.1., it has been known that the reaction rate constant (k) at 40°C, 50°C, and 60°C are 0.0031 min^{-1}, 0.0058 min^{-1}, 0.0068 min^{-1}.

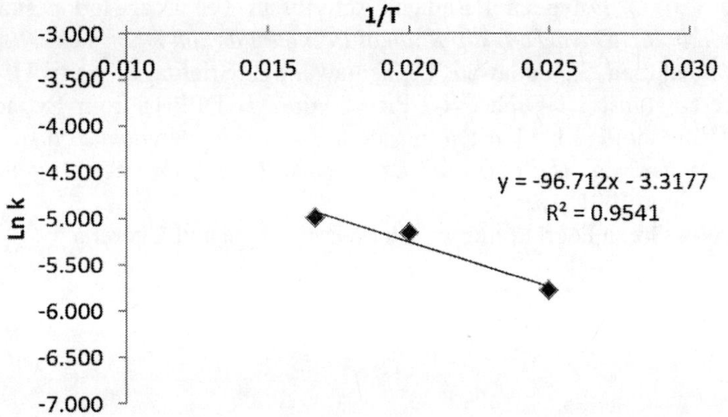

Figure 3.3. Energy activation degradation of antioxidant activity

Based on Figure 3.3., the activation energy (Ea) is determined by using linear regression between 1/T (x-axis) and Ln k (y-axis), then the slope is multiplied by R constant (8.314 J/mol K). With linear regression equation of y=-96.712x-3.3177, R^2 = 0.9541, so we obtained activation energy for 804 kJ/mol. The figure shows the minimum energy is required to initiate the degradation of antioxidant activity in Phycocyanin microencapsulation. The lower of Ea value will be easier to be decreased.

4. Conclusions

Based on the results, have been concluded that D value of carrageenan coating material for $D_{40°C}$, $D_{50°C}$, and $D_{60°C}$ are 769 minutes, 400 minutes, and 333 minutes. While D value of chitosan coating material material for $D_{40°C}$, $D_{50°C}$, and $D_{60°C}$ are 833 minutes, 417 minutes, and 370 minutes. Z value of carrageenan coating material is 54.945°C. While Z value of chitosan coating material is 56.818°C. The reaction rate constants (k) of carrageenan coating material for 40°C, 50°C, and 60°C are 0.0031 min^{-1}, 0.0058 min^{-1}, and 0.0068 min^{-1}. While the reaction rate constant (k) of chitosan coating material for 40°C, 50°C, and 60°C are 0.0027 min^{-1}, 0.0055 min^{-1}, and 0.0062 min^{-1}. Activation energy (Ea) of carrageenan coating material is 804 kJ/mol. While activation energy (Ea) of chitosan coating material is 857.3 kJ /mol.

References

[1] Ansel H C. (2007). Introduction to Pharmaceutical Dosage Forms (writen in Indonesian as *Pengantar Bentuk Sediaan Farmasi*) by Farida Ibrahim. Page: 293-294.
[2] Asri W. (2014). The Role of Antioxidant for Health (writen in Indonesian as *Peran Antioksidan bagi Kesehatan*). Center for Biomedicine and Basic Technology of Kemenkes Indonesian.
[3] Damayanti, E. (2004). Studying Antioxidant and Antibacterial Activity from Cinna-ale Cinema Spice Blend Extract (writen in Indonesia as *Mempelajari Aktivitas Antioksidan dan Antibakteri dari Ekstrak Campuran Rempah Minuman Cinna-ale*). Faculty of Agricultural Technology, Bogor Agricultural University, Bogor.Page: 52.
[4] Henrikson, R. (2009). Earth Food Spirulina: How this remarkable blue-green alga can transform your health and our planet. Ronore Enterprises, Inc., Hana.
[5] Jaiswal, A. K. and N. Abu-Ghannam. (2013). Degradation Kinetic Modelling of Colour.
[6] Marsono Y. (2007). Prospect of Functional Food Development. Paper presented at the conference of "National Food Technology Competition (NFTC)" (writen in Indonesian as *Prospek Pengembangan Makanan Fungsional*).
[7] Jan Zuidam Nicolaas and Eyal Shimoni. (2010). Overview of Microencapsulates for Use in Food. Springer Science and Business Media.
[8] Rohdiana, D. (2011). Polyphenol Radical Activity in Tea Leaf. Indonesian Journal (writen in Indonesian as *Aktivitas Daya Tangkap Radikal Polifenol dalam Daun Teh)*. 2001, **12**, 53-58.
[9] Windono, T., Soediman, S., Yudawati, U., Ermawati, E., Srielita, Erowati, TI. (2001). Free Radical Absorption Test against 1,1-Diphenyl-2-Picrylhydrazyl (DPPH) from Extract Fruit Leather and Grape Seeds (Vitis vinifera L.) (writen in Indonesian as *Uji Peredam Radikal Bebas terhadap 1,1-Diphenyl-2-Picrylhydrazyl (DPPH) dari Ekstrak Kulit Buah dan Biji Anggur)*. Probolinggo Biru and Bali. Artocarpus. 2001, **1**, 34-43.
[10] Weil, A. (2000). Green Food Spirulina, Blue-Green Agae and Chorella.

IOP Conf. Series: Earth and Environmental Science 102 (2017) 012056 doi:10.1088/1755-1315/102/1/012056

Application of iota and kappa carrageenans to traditional several food using modified cassava flour

A N Al-Baarri[1*], A M Legowo[1], H Rizqiati[1], Widayat[2], A Septianingrum[1], H N Sabrina[1], L M Arganis[1], R O Saraswati[1] and Rr C P R Mochtar[1]

[1] Food Technology Department, Faculty of Animal and Agricultural Sciences, Diponegoro University, Semarang, Indonesia
[2] Chemical Engineering Department, Faculty of Engineering, Diponegoro University, Semarang, Indonesia

Email : albari@undip.ac.id

Abstract. Carrageenan has been known well as hydrocolloids that forming viscous dispersions and gels when dispersed in water. The carrageenan has not been widely applied to traditional foods. Therefore, the aim of this research was to determine the effect of kappa and iota carrageenans in traditional food models using modified cassava flour, sugar, and coconut milk. The textural properties, i.e. hardness, cohesiveness, springiness and adhesiveness have been measured using texture analyzer. The study indicated that traditional food models added kappa carrageenan at 2% generated remarkably higher in the hardness, cohesiveness, and springiness than those added iota carrageenan. On the other hand, the reserve result were found in the adhesiveness parameter. As conclusion, kappa carrageenan scan be potentially used for producing traditional foods based on the hard-texture-oriented foods whereas iota carrageenan can be used for the traditional foods with better adhesiveness.

Key word: Iota, kappa, carrageenan, texture.

1. Introduction

Modified cassava flour is a product derived from cassava flour that uses the principle of modifying cassava cells in fermentation [3] Modified cassava flour almost similar colour and texture like wheat flour. However, modified cassava flour has no gluten, being different from wheat, rye, barley and oat[5]. It is difficult to form elastic and chewy dough by using modified cassava flour, because only wheat gluten possesses the ability to form three-dimensional network during dough formation [20]. Generally, modified cassava flour is used for making noodles, bakery, cookies with intermediate moisture food in traditional food [16].

Traditional foods modified cassava flour play important role in local identity, consumer behavior, and the transfer of cultural heritage for future generations[1] .Popularly producing of traditional food was semi-moist food (intermediate moisture food). Intermediate moisture food can reduce to a level where most microorganisms cannot grow. Intermediate moisture food has water activity 0.65-0.90 with moisture content of 15-40% [17]. The storage affects texture of intermediate moisture food, so it will get harder along its shelf life because of decreasing in moisture content by evaporation. The texture of the traditional foods are influenced by the gluten-free content resulting from modified cassava flour and water content during its shelf life. Therefore, need to stabilizer agent such as carrageenan to aid modified cassava flour dough more springy and binding water in dough.

Carrageenan has been known one of the types of hydrocolloids that form viscous dispersions or gels when dispersed in water. Carrageenan is a water soluble sulphated polysaccharide extracted from red seaweed (Rhodophyta). Generally, carrageenan consist of sulphate ester of galactose and 3,6-anhydrogalactose polysaccharides, and D-galactose residues linked alternately in a α-1,3 and β-1,4 bonds [21]. They classified based of natural polysaccharides as kappa (k-), iota (i-) and lambda (λ). Kappa (k-) and iota (i-) has been used in this experiment because of having the ability to form thermo-reversible gels in the presence of gel-promoting cations [10]. K and i-carrageenans are widely used in the food industry for improving the texture and microstructure of gels [23]. Kappa is used for thickening, water-binding, and gelling. On the other hand, the iota shows viscoelastic properties to make a model gelled food system [2] .The use of them in concrete preparation could result in a high compressive strength, flexural strength, and impact toughness [11]. Carrageenan mainly used in meat and dairy product [21].

The formulation of the research in food products can be effective in dough texture. Carrageenan have a group of natural linear sulfated polysaccharides showing the rheological properties in solid such as thickening and gelling. The use of carrageenan has been reported to be high compressive impact toughness and flexural strength [11]. The addition of carrageenan to food products containing coconut milk and sugar, showed increasing of the viscosity of the coconut milk. Therefore it can be effect on the formulation of food product. This research will be focused on the textural changes in traditional food models using modified cassava flour, sugar, and coconut milk by the addition of iota and kappa carrageenans. The advantage from this research was to learn the effect of carrageenan for traditional food texture and know the material can influence the food texture.

2. Materials and Methods

2.1 Materials

Iota and Kappa carrageenan were purchased from CV. Karagenan Indonesia. Modified cassava flour (Mocaf), brown sugar, coconut milk were bought from local market.

2.2 Dough preparation

The dough was prepared according to the traditional receipt of basic dodol ; food traditional which have main ingredient are flour, brown sugar and coconut milk. The steps of processing are preparation of dodol (mixing the dough which is consist of mocaf, brown sugar, coconut milk and adding iota also kappa carrageenan in 2% maximally) and heating the dough (± 80° C in 5 minutes) [3].

2.3 Sample formulation and preparation

The basic recipe for each preparation sample is 10 g of mocaf, 10 ml aquadest. There's no addition of iota and kappa carrageenan in T0 as a control. But in each T0 there are slight modification such as $T0_1$ with addition 4.5 g of brown sugar, $T0_2$ with addition 4 ml of coconut milk, $T0_3$ with addition both of those ingredients. In T1 2 g of Iota carrageenan was added to basic recipe with slight modifications. $T1_1$ with addition 4.5 g of brown sugar, $T1_2$ with addition 4 ml of coconut milk, $T1_3$ with addition both of those ingredients. So the next treatment In T2 2 g of kappa carrageenan was added to basic recipe and the modifications are same as T1.

2.4 Texture analysis

The texture analyses (TPA) were performed using Brookfield Texture Analyzer. Dough compressed twice up with trigger 4,5 gram, deformation 7,5 mm, and speed 2,5 mm/s. Hardness is defined as maximum force during the first compression. Cohesiveness is determined as the ratio of the positive force area during the second compression to the first one. Springiness is associated with elasticity. Adhesiveness depends on the surface properties, particularly the wetting of properties.

2.5 Statistical analysis

All measurements were carried out to assess for any significant differences between the means with one way analysis of variance (ANOVA). It was conducted by using IBM SPSS statistics 22.0 (SPSS Inc., USA). Significant differences ($p > 0.05$) were determined by Duncan's multiple range test.

3. Result and Discussion

3.1 Hardness

Table 1. Hardness of doughs of control (T0), 2 % iota (T1) and 2 % kappa (T2) carrageenans with different ingredients

Treatment	Brown Sugar	Coconut Milk	Brown Sugar + Coconut Milk
T0	97.4375	94.8125	84.4375
T1	102.4375	104.625	62.063
T2	159.31	129.4375	74.625

Textural analysis consist in compression of the sample without breaking it. hardness is the maximum force during the first cycle of texture analyser compression [12]. The changes of the hardness were recorded in Table 1 whereas increased in brown sugar dough and coconut milk dough, whereas the hardness of brown sugar and coconut milk dough was fluctuated. in Table 1. the dough with brown sugar has higher hardness than other because the sucrose could getting back hard when its turn cooling. And sample with addition coconut milk has low hardness because incorporating of coconut milk caused the increase of water content in the sample and led to the loss of hardness [9]. In table. 1 showed higher hardness in T2 (dough with 2 % kappa), but T2 containing brown sugar and coconut milk has lower hardness. So the kappa carrageenan affected the hardness of food with certain ingredient. Aside from the carrageenan and the ingredient, another treating factors could affect on hardness such as heating time and heating temperature. Hardness was influenced by the speed of water evaporating during cooking. In general, increasing of hardness could relate to ability of mastication[18].

3.2 Cohessiveness

Table 2. Cohesiveness of doughs of control (T0), 2 % iota (T1) and 2 % kappa (T2) carrageenans with different ingredients

Treatment	Brown Sugar	Coconut Milk	Brown Sugar + Coconut Milk
T0	0.375	0.55	0.6575
T1	0.3725	0.56	0.5675
T2	0.2325	0.465	1.28

The results of cohesiveness is quite differ from hardness. The changes of cohesiveness shown in Table 2. showed that cohesiveness of control (T0), iota (T1) and kappa (T2) decreased along after the addition of brown sugar, brown sugar and coconut milk. However the cohesiveness of the dough added coconut milk alone is flactuating. Cohesiveness is defined as the extent to which a material can be deformed before it ruptures in mechanics [22]. Addition of brown sugar decreased the cohesiveness of dough containing iota and kappa, and dough containing kappa is the lowest one. Different from the addition of brown sugar, the addition of coconut milk led to higher dough cohesiveness. Higher amount of coconut milk addition induced decrease of the hardness and made it into the elastic one because the high water content [6]. A combined ingredients such as brown sugar and coconut milk has a highest cohesiveness in control, but it decreased in dough containing iota-kappa carrageenan and dough containing kappa showed the lowest one. Kappa carrageenan was a better gelling agent because it had higher thickening ability than that of iota [19].

3.3 Springiness

Table 3. Springiness of doughs of control (T0), 2 % iota (T1) and 2 % kappa (T2) carrageenans with different ingredients

Treatment	Brown Sugar	Coconut Milk	Brown Sugar + Coconut Milk
T0	4.225	6.125	6.425
T1	5.175	5.925	5.625
T2	4.49	5.725	5.3

Springiness means the level of elasticity in the dough. The addition of carrageenan, coconut milk and brown sugar had an effect on the level of elasticity in the dough. A chewy and elastic dough is obtained from the gelatinization process of starch in mocaf flour and at during heating process, and gelatinized starch can absorb water and produce a thick, chewy, and elastic dough [15]. Elasticity describes material reactions when subjected to force (action). In this case, the elasticity of the sample depends on the ingredients added in the dough [4]. The addition of sugar with relatively large amounts will result in a harder texture, while with too little then the resulting texture is soft and elastic [7]. Addition In Table 3, there is a decrease in elasticity on T2 (dough with 2% kappa). T2 always the lowest in dough with different ingredients. The dough containing iota carrageenan has an elastic gel properties compared to kappa carrageenan [13]. The elastic condition of the food product is an assessment of semi-wet texture of food with glossy, dense, and non-sticky marks to the touch.

3.4 Adhesiveness

Table 4. Adhesiveness of doughs of control (T0), 2 % iota (T1) and 2 % kappa (T2) carrageenans with different ingredients

Treatment	Brown Sugar	Coconut Milk	Brown Sugar + Coconut Milk
T0	0.8575	1.6425	2.1375
T1	1.3325	2.095	1.4
T2	0.5125	2.0975	1.165

In this study, adhesiveness is featured by surface characteristic and include viscosity or viscoelastic property as well. The addition of carrageenan, brown sugar and coconut milk to modified cassava flour gave a good effect on the adhesion in dough. The higher concentration of carrageenan addition, the higher force required for a product to break [14]. The addition of brown sugar plus coconut milk to dough has the highest adhesiveness (T0). In T1 treatment (with 2% of iota), addition of brown sugar plus coconut milk to dough has the highest adhesiveness (T0). In T2 treatment (with 2% kappa), the addition of brown sugar reduced adhesiveness. These changes show that the addition of carrageenan to the dough may affect the level of adhesion in dough [8].

4. Conclusion

The results of this study showed that addition of different ingredient and carrageenan definitely affects the hardness, springiness and adhesiveness, on the other hand, the combination of ingredients did not affect the cohesiveness. Kappa carrageenan is better for the hardness of dough, while the iota carrageenan is better for cohesiveness, springiness and adhesiveness of dough. It can be suggested that the kappa enhance thickening and the iota enhance gelling.

References

[1] Albayrak M and Gunes E 2010 Traditional foods : interaction between local and global foods in Turkey *African Journal of Business Management* **4** 555-61

[2] Bosc V, Ferrari I and Michon C 2008 Adhesion to solid surfaces of gels of iota-carrageenan alone or in mixture with casein *Colloids and Surfaces A: Physicochem. Eng. Aspects* **331** 2-7

[3] Chuah T G, H H Nisah, SY T Choong, N L Chin and A H N Sheikh 2007 Effects of temperature on viscocity of dodol (concoction) *J. of Food Engineering* **80** 423-30

[4] Didier, J. K., H K Kouassi, Y D P Assoi, P A Amedee, M Vlad, M Elena, and A Emma 2016 Wheat bread dough rheological properties study dependence on the dough viscoelasticity level *J. of Recent Viotechnology* **4** 1 - 12

[5] Fiorda F A, Soares M S, Silva F A D, Grosmann M V E and Souto L R F 2013 Microestructure, texture and colour of gluten-free pasta made of amaranth flour, cassava starch and cassava bagasse *J. Food Sci. and Technol.* **54** 132-8

[6] Fuangpaiboon N and Kijroongrojana K 2015 Qualities and sensory characteristics of coconut milk ice cream containing different low glycemic index sweetener blends *International Food Research Journal.* **22** 1138-47

[7] Hasanuzzaman M, M Kamruzzaman, M Mominul, S Anjuman, M Mashiar, L Ahmed, and D Kumar 2014 A study on tomato candy prepared by dehydration technique using different sugar solutions *J. Food and Nutrition Science* **5** 1261-71

[8] Huang M, Kennedy J F, Li B, Xu X and Xie B J 2007 Characters of rice starch gel modified by gellan, carrageenan, and glucomannan : a texture profile analysis study *Carbohydrate Polymers* **69** 411-8

[9] Jakubczyk E, Gondek E and Tryzno E 2016 Application of novel acoustic measurement techniques for texture analysis of co-extruded snacks *J. Food Sci. and Technol.* **75** 582-9

[10] Lascombes C, Agoda-Tandjawa G, Boulenguer P, Garnec C L, Gilles M, Mauduit, Barey P and langendorff V 2016 Starch-carrageenan interactions in aqueous media: role of each polysaccharide chemical and macromolecular characteristics *J. Food Hydrocolloids* **30** 1-14

[11] Nakamatsu J, Kim S, Ayarza J, Ramirez E, Elgegren M and Aguilar R 2017 Eco-friendly modification of earthen construction with carrageenan: water durability and mechanical assessment *J. Construction and Building Materials* **139** 193-202

[12] Saldana E, Behrens J H, Serrano J B, Ribeiro F, Almeida M A D and Contreras-Castillo J 2015 Microstructure, texture profile and descriptive analysis of texture for traditional and light mortadella *J. Food Sci. and Technol.* **6** 13-20

[13] Serowik M, Figiel A, Nejman M, Pudlo A, Chorazyk D and Kopec W 2017 Drying characteristic and some properties of spouted bed dried semi-refined carrageenan *J. of Food Engineering* **184** 46 – 57

[14] Sidi N, Widowati E and Nursiwi A 2014 The effect of carrageenan addition on physicochemical characteristics and sensory fruit leather of pineapple and carrot (*Ananas comosus* L. Merr.) and carrot (*Daucus carota*) *J. Aplication of Food Technology* **3** 122-7

[15] Sikora M., S Kowalski, K Magdalena, R Ziobro, W Paulina, C Duska and L Alain 2010 Starch gelatinization as measured by rheological properties of the dough *J. of Food Engineering* **96** 505-9

[16] Sulistyo J and Nakahara K 2014 Cassava flour modification by microorganism *Conference Paper* Thailand 1- 9

[17] Sych J 2003 Intermediate-moisture foods *Encyclopedia of Food Microbiology (Second Edition)* (United State: Academic Press) p 372-6

[18] Valderrama-Bravo C, Lopez-Ramirez Y, Jimenez-Ambriz S, Oaxaca-Luna A, Dominiguez-Pacheco A, Hernandez-Aguilar C and Moreno-Martinez E 2014 Changes in chemical, viscoelastic, and textural properties of nixtamalization dough with nejayote *J. Food Sci. and Technol.* **30** 1-7

[19] Villanueva R D, Mendoza W G, Rodrigueza M R C, Romero J B and Montano M N E 2004 Structure and functional performance of gigartinacean kappa-iota hybrid carrageenan and solieriacean kappa-iota carrageenan blends *Food Hydrocolloids.* **18** 283-92

[20] Wang P, Jin Z and Xu X 2015 Physicochemical alterations of wheat gluten proteins upon dough formation and frozen storage – a review from gluten, glutenin and gliadin perspectives *J. Trend in Food Sci. and Technol.* **30** 1-10

[21] Webber V, Carvalho S M D, Ogliar P J, Hayashi L and Barreto L M 2012 Optimization of the extraction of carrageenan from *Kappaphycus alvarezii* using response surface methodology *Cienc. Tecnol. Aliment* **32** 812-8

[22] Yuan C, Lei D, Guangjie Z, Zhengyu J and Hui L 2016 Influence of cyclodextrins on texture behavior and freeze-thaw stability of kappa-carrageenan gel. *J. Food Chem.* **210** 600-5

[23] Zhang T, Xu X, Ji L, Li Z, Wang Y, Xue Y and Xue C 2017 Phase behaviors involved in surimi gel system : effects of phase separation on gelation of myofibrillar protein and kappa-carrageenan *J. Food of International* 1-25

IOP Conf. Series: Earth and Environmental Science **102** (2017) 012057 doi:10.1088/1755-1315/102/1/012057

Characteristics of chitinase isolated from different part of snakehead fish (*Channa striata)* digestive tract

A Baehaki[1*], S D Lestari[1], Y Wahidman[1], N Gofar[2]

[1]Fisheries Product Technology Study Program, Faculty of Agriculture, Sriwijaya University, Indralaya, South Sumatera, Indonesia
[2]Soil Sciences study Program, Faculty of Agriculture, Sriwijaya University, Indralaya, South Sumatera, Indonesia.

*E-mail: ace76_none@yahoo.com

Abstract. Naturally, snakehead fish (*Channa striata)* is a prodigious carnivore feeding mainly on live animals, including small shrimp. Based on its feeding habits, the digestrive tracts of snakehead is considered as auspicious source of various enzumes including chitinase. The purpose of this study was to partially characterize chitinase enzyme isolated from digestive tract of snakehead fish. Two parts of digestive tract, stomach and intestine were used as enzymes' source. The results showed that chitinase activity from the stomach was higher than chitinase activity from the intestine. The pH and temperature optimum of chitinase activity from digestive tract (the stomach and the intestine) were 6.0 and 70 °C, respectively.

***Keywords**: Channa striata, chitinase, digestive tract*

1. Introduction

Chitin is a polysaccharide created by β-1,4 bonding of N-acetyl glucosamine residues. Chitinases are glycosyl hydrolases, characterized for hydrolyzing β-1,4 linkage of N-acetyl glucosamine present in chitin chains [1]. Chitinases have been divided into two main categories: exochitinases and endochitinases [2]. Chitinase is used to hydrolyze chitin to produce chitin oligomers. Functional properties and physiological activities of chitin oligomers depend on molecular weight and chain length mostly [3]. The digestive tract of fish produces many enzymes and there are many reports describing the enzymes present in digestive tract, inclusing protease [4], chitinase [5], and chitosanase [6]. However, reports on chitinase activity in different part of the digestive tract (stomach and intestine) of snakehead fish (*Channa striata*) are still limited. Therefore, in this reseacrh we studied the activity of chitinase isolated from the digestive tract of snakehead fish (*Channa striata*) and partial characterization of this enzyme.

2. Materials And Methods

2.1 Chitinase activity

The chitinase activity was measured according to the Spindler method [7].

2.2 Effect of pH enzyme activity

Optimum pH was determined by assaying in buffer with pH values of 6-9 using citric acid buffer (pH 3-4), citric-phosphate buffer (pH 4-6), phosphate buffer (pH 6-8) and boric buffer (pH 8-9), in the presence of soluble chitin substrate (colloidal chitin).

IOP Conf. Series: Earth and Environmental Science **102** (2017) 012057 doi:10.1088/1755-1315/102/1/012057

2.3 Effect of temperature on enzyme activity
Optimum temperature was measured at 30, 40, 50, 60, 70 and 80 °C at pH 7, soluble chitin (colloidal chitin) was used as the substrate.

3. Result And Discussion

3.1 Chitinase Activity
The activity of chitinase isolated from digestive tract of snakehead fish (*Channa striata*) as effected by part difference can be seen in Figure 1.

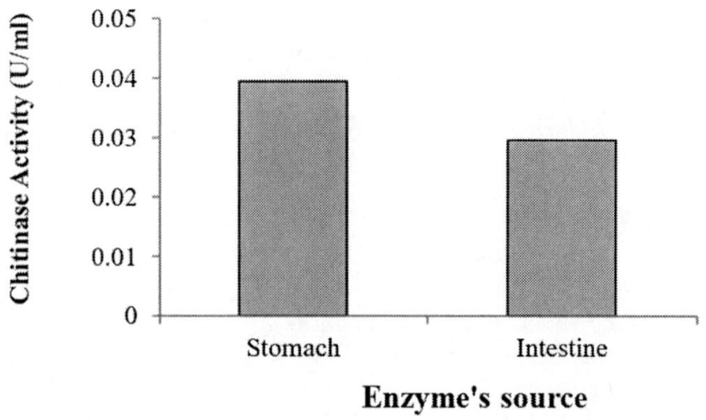

Figure 1. Chitinase activity from digestive tract of snakehead fish (*Channa striata*)

The results showed that the activity of chitinase isolated from the stomach was higher than the activity from the intestine. Chitinase activity from the stomach was 0.41 U/ml while from the intestine was 0.32 U/ml.

3.2 Effect of pH on chitinase activity
Effect of pH on the activity of chitinase isolated from digestive tract of snakehead fish (*Channa striata*) is presented in Figure 2. Optimum activity of chitinase isolated from the stomach and the intestine was achieved at pH 6. The optimum pH ranges of chitinase activity in the stomach and the intestine of Cod (*Gadus morhua*) were 4.5–5.1 and 5.1–6.5, respectively [5]. The ranges were lower than the activity of the same enzyme isolated from the intestine of discus fish (*Symphysodon aequifasciata*) which was optimum at pH 8 [4] but it were higher than the pH 2 reported for enzyme of discus fish (*Symphysodon aequifasciata*) stomach [4], and digestive proteases (acidic protease) of juvenile *Cichlasoma beani* [8].

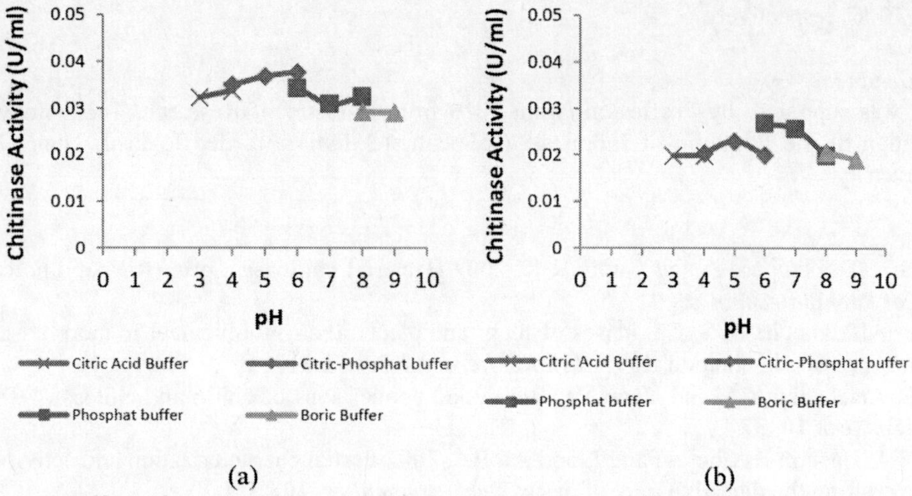

Figure 2. Effect of pH on chitinase activity in the digestive tract of *Channa striata* (a=the Stomach, b= the intestine)

3.3 Effect of temperature on chitinase activity

A temperature range between 30 °C and 80 °C were used in this study. Figure 3 showed the effect of temperature on chitinase activity from digestive tract of snakehead fish (*Channa striata*).

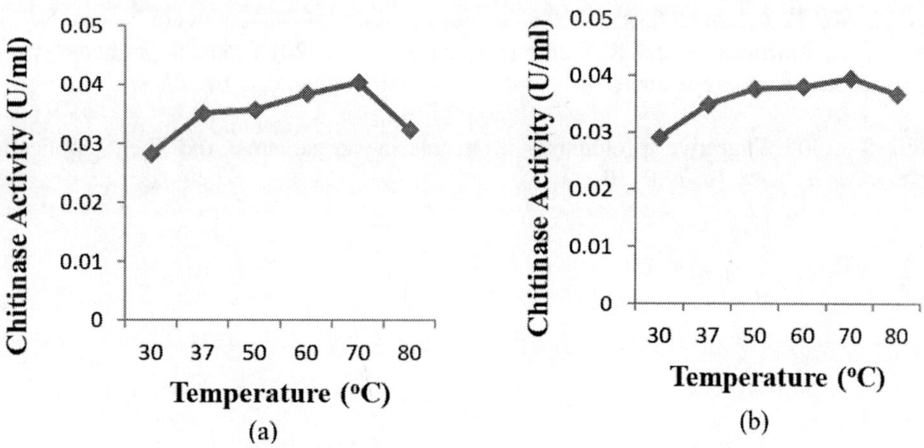

Figure 3. Effect of temperature on chitinase activity in the digestive tract of *Channa striata* (a= stomach; b=intestine)

The optimum temperature of 70°C were recorded for both chitinase (the stomach and the intestine) in this study. This were higher than optimum temperature of 55 °C reported for digestive proteases (acidic protease) of juvenile *Cichlasoma* beani [8], 65 °C for *Atractosteus tropicus* digestive proteases [9] and 50°C for purified trypsins A, B and C from skipjack tuna spleen [10].

4. Conclusion

The chitinase activity from the stomach was higher than chitinase activity from the intestine. The pH and temperature optimum of chitinase activity from digestive tract (the stomach and the intestine) were 6.0 and 70 ^{0}C respectively.

Acknowledgments
This research was supported by Proffesion Grant 2016 from Ministry of Research, Technology and Higher Education of the Republic of Indonesia (Kementristekdikti) awarded to Prof. Nuni Gofar, Sriwijaya University.

References

[1] Bhattacharya D, Nagpure A and Gupta R K, 2007 Bacterial chitinases: properties and potential, *Critical Rev. Biotechnol* **27** 21
[2] Sahai S A and Manocha M S, Chitinases of fungi and plants: their involvement in morphogenesis and host-parasite interaction, 1993 *FEMS Microbiol Rev* **11** 317
[3] Shahidi F, Arachchi J K V and Jeon YJ, 1999 Food applications of chitin and chitosans, *Trends Food Sci Tech* **10** 37
[4] Chong S C A, Hashim R, Chow-Yang L and Ali BA, 2002 Partial characterization and activities of proteases from the digestive tract of discus fish, *Aquacultre* **203** 321
[5] Danulat E and Kausch H, 1984 Chitinase activity in the digestive tract of Cod, *Gadus morhua* (L.), *J Fish Biol* **24** 125
[6] Baehaki A, Lestari S D, Gofar N and Wahidman W, 2016 Partial characterization of chitosanase from digestive tract of *Channa striata, Oriental J Chem* **32** 3231
[7] Spindler KD, Chitinase and chitosanase assays. In: Muzzarelli RAA and Peter MG (eds), 1997 Grottammare European Chitin Society, Italy 229
[8] Martínez-Cárdenas L, Álvarez-González CA, Hernández-Almeida OU , Frías-Quintana CA, Ponce-Palafox CT, and Castillo-Vargasmachuca S, 2017 Partial Characterization of Digestive Proteases in the Green Cichlid, Cichlasoma beani, *Fishes* 2
[9] Guerrero-Zarate R, Alvarez-Gonzalez CA, Perales-Garcıa MA, Olvera-Novoa N, Frıas-Quintana CA, Martınez-Garcıa R, Contreras-Sanchez WM, 2014 Partial characterization of digestive proteases in tropicalgar *Atractosteus tropicus* juveniles, *Fish Physiol Biochem* **40** 1021
[10] Klomklao S, 2008 Digestive proteinases from marine organisms and their applications, *Songklanakarin J. Sci. Technol* **30** 37

IOP Conf. Series: Earth and Environmental Science **102** (2017) 012058 doi:10.1088/1755-1315/102/1/012058

Glucose and D-Allulose contained medium to support the growth of lactic acid bacteria

A N Al-Baarri[1*], A M Legowo[1], Y B Pramono[1], D I Sari[1], W Pangestika[1]

[1] Food Technology Department, Faculty of Animal and Agricultural Sciences, Diponegoro University, Semarang, Indonesia

Email : albari@undip.ac.id

Abstract. Monosaccharide has been known as support agent for the growth of lactic acid bacteria. However the combination among monosaccharides for supporting the living of bacteria has not been understood well. This research was done for analyzing the combination glucose and D-allulose for the growth of *Lactobacillus acidophilus* and *Streptococcus thermophillus*. The NaCl medium containing glucose and D-allulose was used to analyse the growth of bacteria. The study showed that glucose and D-allulose have been detected as supportive agent to *L. acidophilus* and *S. thermophillus* specifically. As conclusion, glucose and D-allulose supported the growth of lactic acid bacteria equally. This finding might provide the beneficial information for industry to utilize D-allulose as well as glucose.
Key word: *L. achidophillus*, *S. thermophillus*, glucose, D-allulose, growth.

1. Introduction

Recently, rare sugars have attracted a great deal of attention mainly concentrated on their applications, such as potential inhibitors of various glycosidases low- or non-calorie carbohydrate sweeteners and bulking agents in the food industry. Rare sugars are monosaccharides and their derivatives that are rare in nature [1]. D-allulose is a rare sugar, is found in nature at only small amount and very difficult to synthesize chemicaly. D-allulose exits in tea plants and wheat in very small quantity and can be produced in low amount from D-fructose during heat treatment of food products [2]. Biological processes such as enzymatic reactions with various ketose epimerase, oxidoreductase and aldose isomerase as well as various microbial reactions are feasible process for the synthesis of D-allulose [3].

D-glucose (Glu) is conventional sugar which occurred widely in nature. Glu has been one of sugar which can cause browning reaction with occurs in food system containing sugar and amino acid by heat treatment. Melanoidin is formed at late stage of browning reaction. The structure of melanoidins are quite complex and are formed by reaction of heterogeneous polymers containing nitrogen [4]. Although the previous study from [5] stated that pentose sugars were more reactive than hexose to turns brown colour, Glu was still the common sugars that widely used in food industry [6].

Lactic acid bacteria (LAB) are non-sporulating, catalase-negative, Gram positive, strictly fermentative organisms, with lactic acid as a major metabolic end product of carbohydrate fermentation. Currently, this group comprises the following general: *Carnobacterium*, *Enterococcus*,

Lactobacillus, *Lactococcus*, *Leuconostoc*, *Streptococcus*, *Tetragenococcus*, *Vagococcus* and *Weissella* and can be found in soil, water, animal and mammal gastrointestinal tract, as well as in food and fermented products [7]. They cause rapid acidification of the raw material through the production of organic acids, mainly lactic acid. In addition, their production of acetic acid, ethanol, aroma compound, bacteriocins, expopoly-saccharides, and several enzymes is of importance and depend on the applied saccharides. This research was done to analyse the determine the growth of lactic acid bacteria in the presence of glucose and D-allulose.

2. Materials and Methods

2.1 Materials

Starins of lactic acid bacteria (LAB) *Lactobacillus achidophillus* (LA) and *Streptococcus thermophillus* (ST) were obtained from Integrated Laboratory, Diponegoro University, Semarang, D-allulose was obtained from Kagawa Rare Sugar Research Centre Japan. Glucose (3%). De man, Rogosa and Sharpe (MRS) broth and MRS agar were used to support the growth of bacteria.

2.2 Growth condition

The provided of LAB strain and growth condition was adopted from the methods with some modification. *L. achidophillus* and *S. thermophillus* were grown anaerobically in 9 ml MRS broth at 37°C for 24h prior to use.

2.3 Microbiological analysis

The microbiological analysis was adopted the methods of [8] with some modification. Sample (1 ml) were diluted using 9 ml sterile physiological saline (0.8% NaCl) prepared in tubes and autoclaved. The mixture was thoroughly shaken to uniform distribution, and several aliquots of diluents (10^{-1}–10^{-6}) were prepared using dilution method. For counting LAB, MRS Agar was used. Plate of MRS Agar were incubated at 37°C for 48h in the anaerobic condition. After the incubation, counts of microorganisms were determined as log CFU/mL sample.

2.4 Preparation media and inoculation

The preparation of glucose and D-allulose containing medium was done by the addition 3% of glucose and D-allulose, respectively to 0.88% NaCl. Sterile physiological saline (0.88% NaCl) was used to avoid excess electrolyte. The MRS broth was used as comparison. One millilitre of LAB strains were inserted into 14 ml solution containing MRS broth, glucose, or D-allulose.

3. Result and Discussion

The amount of LAB on glucose and D-allulose containing medium appeared non significantly different. Both of them were able to provide the population of LAB at a range 7.45–7.55 Log CFU/ml (Fig. 1). Compare to MRS containing medium, population of LAB in the sugar containing medium was lower in amount of LAB resulting the different value of 0.5-0.7 CFU/ml. It is understandable since the composition of MRS broth are more complex than sugar containing medium resulting the maximum support for the living of LAB. [10] stated that the composition of MRS broth was able to support well to the development of lactic acid bacteria. Based on these data, D-allulose might support to the growth of lactic acid bacteria as maximum as glucose. [11] stated if the medium was a simple carbohydrate, it would be more quickly hydrolysed and resulting the beneficial value for supporting the living of lactic acid bacteria.

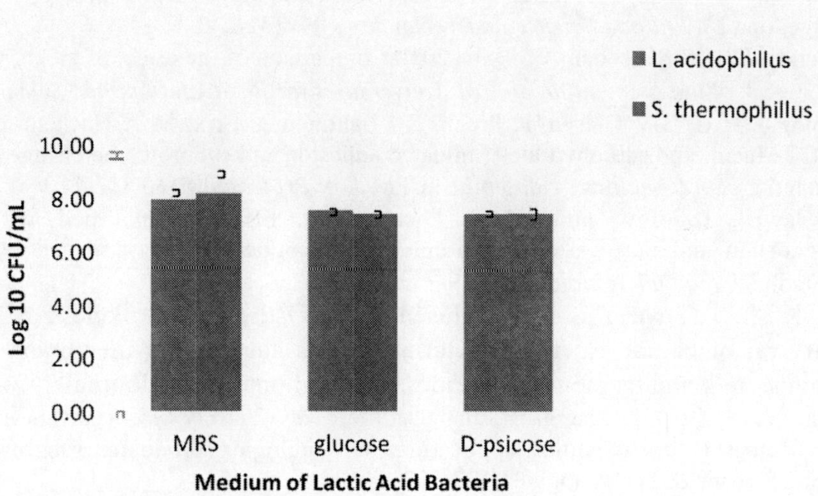

Figure 1. The detected amount of LAB in medium Containing glucose and D-allulose at concentration 3% (w/v). MRS Broth medium was used as control. Data were obtained by 12 repetitions with standard deviation.

Figure 1 showed the development of considerable amount of lactic acid bacteria in the period of 48 hours of incubation time. The number of bacteria has increased since the number of inoculated LAB was about 1 Log CFU/ml (data not shown). Based on this increase, glucose and D-allulose might equally support the living of LAB during period of 48 hours. [12] stated that the exponential growth can be observed when substrate and nutrients were present in acceptable of amount.

4. Conclusion

Based on the research it can be concluded that the glucose and D-allulose might support the living of LAB but the amount still lower than in MRS.

5. Acknowledgement

All of this research was supported by Ministry of Research, Technology, and Higher Education, Republic of Indonesia through the collaboration research scheme program.

References

[1] Sun, Y, S Hayakawa, M Ogawa and K Izumori 2007 Antioxidant properties of custard pudding dessert containing rare hexose, D-psicose *Food Control* **18** 220-227

[2] Zhang, L ., Narita Y ., L. Gao., Muhammad Ali ., M. Oshiki, and S.Okabe. 2017. Maximum specific growth rate of anammox bacteria revisited. J. Water Research **116** 296-303

[3] Dedania, S R, M J Patel, D M Patel, R C Akhani and D H Patel 2017 Immobilization on graphene oxide improves the thermal stability and bioconversion effeiciency of D-psicose 3-empimerase for rare sugar production *Enzyme and Microbial Technology* **107** 49-56

[4] Mikami, Y and Murata, M 2015 Effect of sugar and buffer types, and ph on formation of maillard pigments in the lysine model system *Food Science and Technology Research* **21**(6) 813-819

[5] Wang, H Y, Qian, H, and Yao, W R 2011 Melanoidins produced by the maillard reaction: structure and biological activity *Food Chemistry* **128** 573-584

[6] Vuyst, L D and F Leroy. 2007. Bacteriocins from lactic acid bacteria : production, purification, and food applications. *Journal of Molecular Microbiology and Biotechnology* **3** 194-199

[7] Verón, H E, H D D Risio, M I Isla and S Torres 2017 Isolation and selection of potential

probiotic lactic acid bacteria from Opuntia ficus-indica fruits that grow in Northwest Argentina *LWT- Food Science and Technology* **84** 231-240

[8] Muniandy, P, A B Shori dan A S Baba 2017 Comparison of the effect of green, white and black storage *J. of the Association of Arab Universities for Basic and Applied Science* **22** 26-30.

[9] Celebioglu, H U, S V Olesen, K Prehn, S J Lahtinen, S Brix, M A Hachem and B Svensson 2017 Mucin- and carbohydrate-stimulated adhesion and subproteome change of the probiotic bacterium *Lactobacillus acidophillus*. CFM *J. of Proteomics* **163** 102-110

[10] Svetoslav,D., Todorov., and L.M.T. Dicks. 2009. Effect of modified MRS medium on production and purification of antimicrobial peptide ST4SA produced by Enterococcus mundtii. *J. food microbiology* **15** 65-73.

[11] Moe, K.M., T. Faye, R.K. Abrahamsen, H.M. Ostlie and S. Skeie. 2012. Growth and survival of cheese ripening bacteria on milk fat globule membrane isolated from bovine milk and its monosaccharides. International Dairy Journal 259-35.

[12] Zhang, W., S Yu ., T. Zhang., B. Jiang and W. Mu. 2016. Recent advances in D-allulose: Physiological functionalities, applications, and biological production. Journal of Trends in Food Science & Technology **54** 127-137.

Mutagenic effects of gamma rays on soybean (*Glycine max* L.) germination and seedlings

F Kusmiyati[1], Sutarno[2], M G A Sas[1]and B Herwibawa[1]

[1] Laboratory of Physiology and Crop Breeding, Diponegoro University, Semarang, Central Java, Indonesia
[2] Laboratory of Ecology and Crop Production, Diponegoro University, Semarang, Central Java, Indonesia

E-mail: fkusmiyati@live.undip.ac.id

Abstract. Narrow genetic diversity is a main problem restricting the progress of soybean breeding. One way to improve genetic diversity of plant is through mutation. The purpose of this study was to investigate effect of different dose of gamma rays as induced mutagen on physiological, morphological, and anatomical markers during seed germination and seedling growth of soybean. Seeds of soybean cultivars Dering-1 were irradiated with 11 doses of gamma rays (0, 5, 10, 20, 40, 80, 160, 320, 640, 1280, and 2560 Gy [Gray]. The research design was arranged in a completely randomized block design in three replicates. Results showed that soybean seed exposed at high doses (640, 1280, and 2560 Gy) did not survive more than 20 days, the doses were then removed from anatomical evaluation. Higher doses of gamma rays siginificantly reduced germination percentage at the first count and final count, coefficient of germination velocity, germination rate index, germination index, seedling height and seedling root length, and significantly increased mean germination time, first day of germination, last day of germination, and time spread of germination. However, the effects of gamma rays were varies for density, width, and length of stomata. The LD_{50} obtained based on survival percentage was 314.78 Gy. It can be concluded that very low and low doses of gamma rays (5-320 Gy) might be used to study the improvement of soybean diversity.
Key words : seedling, stomata, morphological diversity

1. Introduction

Soybean is a major crop in many countries. Soybean can be used as food, feed, functional food and industrial products. Increasing interest in functional food and various soybean by- product is the reason for high demand of soybean [1]. However, soybean yield is decreased because of many stresses such as salinity stress. Salinity stress inhibited seed germination and seedling growth. The precise mechanisms of salt stress on seed germination and seedling growth of soybean are only partially understood [2]. Ionic and osmotic stresses are the effects of salinity stress. Ionic stress occurs when salt enters the plant reaching toxic levels, and may lead to a reduced ability to maintain water uptake, which is reffered to as the osmotic or water-deficit effect of salinity [3]. Water deficit or drought stress is one of the limiting factors which affect the numerous metabolic procces of crop [4]. Increasing salinity stress from 4 to 7 dS m^{-1} (1 dS m^{-1} = 700 mg L^{-1}) reduced soybean production by 20 – 60% [5]. Application of manure as organic fertilizer at saline soil increased growth and production of forage crops [6 – 7].

An effective way of maintaining sustainable production in salt-affected soil is through breeding of high salt tolerance soybean [8]. Narrow genetic diversity is a main problem restricting the progress of soybean breeding. One of the strategies perform crosses of parents differing in chloride accumulation tolerance, in which F_2 population segregated in a 3:1 ratio of non-necrotic to necrotic plant [9]. However, genetic improvement depends on the amount of genetic diversity present in the population, though the natural diversity can be improved through induced mutation [10]. There are various techniques to induce mutations and artificially increase variation viz. chemical mutagens such as sodium azide, methylnitrosourea, ethyl methanesulfonate and physical mutagens such as x-ray, neutron, gamma-ray etc.)[11]. Induced mutation combined with selective breeding, is highly efficient in order to screen for new traits [12].

Gamma-ray as mutagen has been used to induce changes in growth habits. Several research on the stimulative effects when using gamma rays as induced mutagens were reported. Recently, low doses of gamma-rays of seed treatment resulted significant effects on germination, shoot, and root systems of wheat [13], physic nut [14], rice[15], maize [16], pigeon pea [17], chick pea [18], lathyrus[19], castor bean[20].However, the impact of gamma rays on soybean cv. Dering 1 is still unknown. The major objective was to investigate effect of different dose of gamma rays as induced mutagen on physiological, morphological, and anatomical markers during seed germination and seedling growth of soybean. The result of such experiments could be used to study the improvement of plant diversity in soybean program for salinity tolerance.

2. Materials and Methods

2.1. Plant material

The soybean seed (*Glycine max* cv. Dering 1) were obtained from Seed Resources Management Unit, Indonesian Legumes and Tuber Crops Research Institute, Malang, Indonesia. The seeds were irradiated by gamma-rays.

2.2. Methods

2.2.1. Mutagenesis. Seeds mutagenesis were performed at the Center for Application of Isotope and Rays, National Nuclear Energy Agency of Indonesia, South Jakarta, Indonesia. The healthy seeds were irradiated by gamma-rays with doses of 0, 5, 10, 20, 40, 80, 160, 320, 640, 1280, and 2560 Gy, and arranged in a completely randomized block design in three replicates.After mutagenesis, the seeds were soaked for 24 hours in aquadest at a temperature of around 25 ^{0}C.

*2.2.2. Germination.*Three replicates of 50 seeds of each mutagen dose were germinated in plastic box containing sand, and kept under laboratory condition [21].It was supplied with water every day in order to maintain sand moisture.

2.2.3. *Physiological attributes of soybean as affected by mutagenesis*. The seeds from mutagenesis were considered germinating when the protrusion of radicle by$\geq$ 2 mm [22].

2.2.3.1. Median lethal dose (LD_{50}) was calculated at 5th day after germination, based on the number of surviving seedlings in different mutagen doses [23].

2.2.3.2. Germination percentage at the 5th day (first count) and the 8th day the final count) were calculated following the formula of Marcu *et al.*[16], in which NT = proportion of the germinating seeds of each treatment; N = number of seeds used in bioassay.

$$GP\ (\%) = \frac{NT \times 100}{N}$$

2.2.3.3. Mean germination time (MGT) was determined using the formula of Σ f.x / Σ f; f = seeds germinated on the day "x". The lower MGT values indicate the faster a population of seeds has germinated [24].

2.2.3.4. First day of germination (FDG) was the day in which the first germination occurred. The faster initiation of germination has the lower value of FDG [25].

2.2.3.5. Last day of germination (LDG) was the day in which the last germination occurred. The lower the LDG values indicate a faster ending of germination [25].

2.2.3.6. Coefficient of velocity of germination (CVG) was determined using the formula of $N_1 + N_2 + \ldots + N_x/100 \times N_1T_1 + \ldots + N_xT_x$; N = number of seeds germinated each day, T = number of days from seeding corresponding to N. The CVG gives an indication of germination rapidity. The value of CVG increases when the number of germinated seeds increases and the time required for germination decreases. The highest CVG possible is 100, if all seeds germinated on the first day [26].

2.2.3.7. Germination rate index (GRI) was determined using the formula of G1/1 + G2/2 + ... +Gx/x; G1 = germination percentage x 100 at the first day after sowing, G2 = germination percentage x 100 at the second day after sowing. Germination rate index (GRI) calculates germination percentage on each day of germination period. Higher and faster germination has higher GRI value [27].

2.2.3.8. Germination index (GI) was determined using the formula of (10 x n1) + (9 x n2) + ... + (1 x n10); n1, n2, ... n10 = number of germinated seeds on the first, second and subsequent days until the 10th day; 10, 9, ... 1 are weights given to the number of germinated seeds on the first, second and subsequent days until the 10th day, respectively. In calculation of GI, value of ten as maximum weight is given to the seeds germinated on the first day and less to those germinated later on. The value of one as lowest weight will be for seeds germinated on the 10th day. The GI emphasizes on both the germination percentage and its speed. Higher percentage and rate of germination has higher GI values [28].

2.2.3.9. Time spread of germination (TSG) was the time in days between the first and last germination occurred. Greater difference of germination speed between the fast and slow germinating sees has higher TSG values [25].

*2.2.4. Morphological attributes of soybean as affected by mutagenesis.*Seedling performance was evaluated through determination of seedling height and root length [29]. Seedling height was measured point of connection of the hypocotyl with the root, up to its point of connection of the apical bud to the stem. Seedling root length was measured from the tip of the root up to its point of connection with the hypocotyl.

*2.2.5. Anatomical attributes of soybean as affected by mutagenesis.*Stomatalsize (width and length) and density were determined using the rapid imprinting technique[30]. The abaxial leaf surface was taken firstly and washed using the tap water. Then it was dried, the transparent nail polish was applied uniformly with a brush on the surface. It was then dried at the room temperature for approximately 20 minutes. The nail polish imprints were placed on glass cover slips, and photographed under a model CX31 trinocularmicroscope (Olympus, Japan) with a mounted E330-ADU1.2X camera (Olympus, Japan).

*2.2.6. Statistical analysis.*Stomatal images were analysed to determine size using the Optilab Image Raster software. Data were analysed using CurveExpert 1.4 software for LD_{50}, and generalized linear model in the PROC-GLM procedure of SAS University Edition software for other parameters. The means were compared through Dunnett option, at probability level of 5%, in order to determine the difference in means between non-mutated and mutated seeds.

3. Results and Discusions

Soybean seed exposure to gamma rays caused serious effects on seed germination. The survival percentage of irradiated soybean cv. Dering 1 decreased as the gamma dose increased (Figure 1). At this experiment, the LD_{50} obtained based on survival percentage was 314.78 Gy. The LD_{50} is the dose corresponding to 50% decrease of the survival percentage. The desired mutation can be found arround LD_{50} [31]. Similar experiments have been carried out in soybean [32] [33],The survival percentages as LD_{50} were 457.17 Gy for cv. Argomulyo, 583 Gy for cv. CO2, and 620 Gy for cv. CO1. It is explained the radiosensitivity to gamma rays were genotypes dependent. The irrays effect was also noticeable in the delay of seed germination according to the doses, and showed growth inhibition as a very common effect.

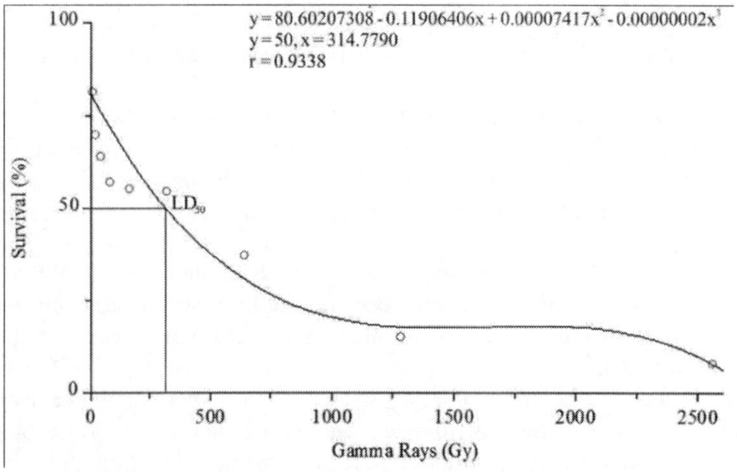

Figure 1. Dose – response curve with polynomial fit

Table 1. Effects of gamma rays on Soybean (*Glycine max* L. cv. Dering 1) germination

GR (Gy)	GP 1[1] (%)	GP 2[1] (%)	MGT[2] (day)	FDG[2] (day)	LDG[2] (day)	CVG	GRI[1] (%/day)	GI	TSG[2] (day)
0	100.00	100.00	2.08	2.00	3.00	48.08	48.67	346.00	1.00
5	92.00 **	100.00	2.47 **	2.00	6.00 **	40.47 **	45.25 *	326.33	4.00 **
10	81.33 **	91.33 **	3.36 **	2.00	6.00 **	29.73 **	29.01 **	257.33 **	4.00 **
20	70.00 **	90.67 **	3.90 **	3.00 **	6.00 **	25.72 **	25.28 **	231.33 **	3.00 **
40	64.00 **	86.67 **	4.24 **	3.00 **	6.00 **	23.61 **	22.30 **	206.33 **	3.00 **
80	57.33 **	86.00 **	4.55 **	3.00 **	6.00 **	21.99 **	20.37 **	191.33 **	3.00 **
160	55.33 **	86.00 **	4.85 **	3.00 **	6.33 **	20.64 **	18.80 **	178.67 **	3.33 **
320	54.67 **	82.67 **	5.09 **	4.00 **	6.33 **	19.64 **	16.65 **	161.67 **	2.33 **
640	37.33 **	78.67 **	5.37 **	4.00 **	6.67 **	18.64 **	14.99 **	142.67 **	2.67 **
1280	15.33 **	74.67 **	5.91 **	4.00 **	8.00 **	16.95 **	12.97 **	115.33 **	4.00 **
2560	8.00 **	62.00 **	6.01 **	4.67 **	7.67 **	16.64 **	10.42 **	92.33 **	3.00 **

GR = gamma rays; GP 1 = germination percentage at the 5th day (first count); GP 2 = germination percentage at the 8th day (final count); MGT = mean germination time; FDG = the first day of germination; LDG = last day of germination; CVG = coefficient of velocity of germination; GRI = germination rate index; GI = germination index; TSG = time spread of germination. Data were transformed by arcsine(1) and Log (X+1) (2)prior to analysis; nontransformed data are presented; * significant difference at $P \leq 0.05$; ** significant at $P \leq 0.01$

The result showed that the maximum germination percentage was observed at the control treatments (0 Gy) (Table 1). The germination percentage at the first count was significantly decreased with increasing gamma ray doses. The minimum and maximum decreased were reached at 8% and 92% for 5 and 2560 Gy, respectively. Furthermore, germination percentage at the final count was also significantly decreased with as gamma ray doses increased start from 10 Gy. The coefficient of germination velocity, germination rate index, germination index as seed germination potential showed the same decreasing pattern by increasing the rays doses. Besides, mean germination time, at the first day and the last day of germination, increased as gamma rays doses increased. However, time spread of germination increase not consistent with increasing gamma rays doses, it depends on the difference in germination speed between the fast and slow germinating . The higher gamma ray doses determined higher inhibiton of the germination process.

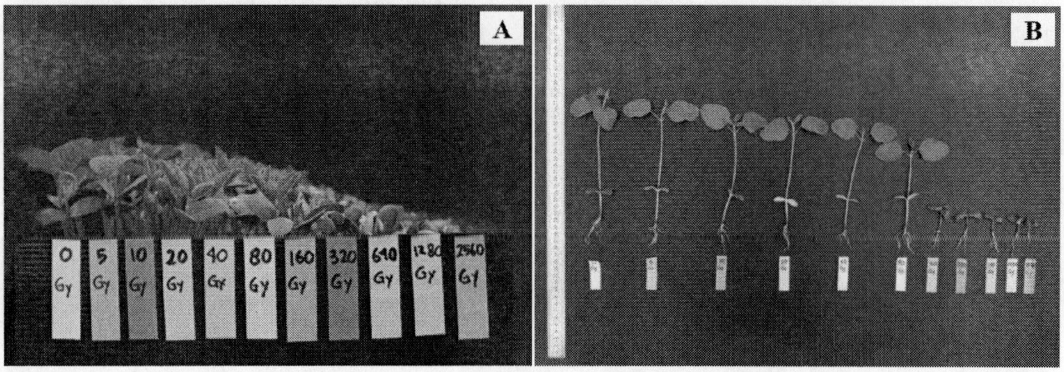

Figure 2. Effects of gamma rays on the seedling growth of Soybean (*Glycine max* L. cv. Dering 1).A = at 5th day after germination;B =at the 14th day after germination

Table 2. Effects of gamma rays on the seedling height and root length were taken at the 14th day of Soybean (*Glycine max* L. cv. Dering 1)

Gamma Rays (Gy)	Seedling Height (cm)	Seedling Root Length (cm)
0	15.07	6.97
5	15.70	5.67
10	14.20	7.57
20	14.40	5.87
40	13.47	4.30 *
80	13.20	5.77
160	8.53 **	3.20 **
320	3.77 **	3.93 *
640	2.67 **	4.20 *
1280	2.27 **	3.23 **
2560	1.33 **	2.50 **

* significant difference at $P \leq 0.05$; ** significant at $P \leq 0.01$

Gamma ray not only impact the germination potential,but also actual qualities of the germinated seedlings, such as seedling height and root length (Table 2). The biometric measurements show a significant decrease, the maximum seedling height and root length were recorded at the control (0 Gy), while the seedling height exposed to 160 – 2560 Gy decrease by 43.40 – 91.17 %, and the seedling root length exposed to 40 – 2560 Gy decreased by 38.31 – 64.13 %. Result showed that gamma ray with doses higher than 160 Gy significantly inhibited the seedling height and root length derived from irradiated seeds. These results were in compliance with the others researchers who reported that increasing the gamma ray doses were decreased the seed germination and seedling growth. Very low and low doses of gamma rays were reported cause a significant delay of germination process, decrease of survival precentage, and seedling height and root length in wheat [13], maize [16], tumeric [34], and soybean [35]. Gamma ray induced inhibition of growth through restraining cell cycle during somatic cell and damaging entire genome [36].

Table 3. Effects of gamma rays on the density, width and length of stomata were taken at 21th day of Soybean (*Glycine max* L. cv. Dering 1)

Gamma Rays(Gy)	Stomata Width(μm)	Stomata Length(μm)	Stomata Density(mm^{-1})
0	1.43	2.42	187.84
5	1.30 *	2.40	217.19
10	1.26 **	2.29	158.49
20	1.21 **	2.08 **	300.84 **
40	1.29 *	2.58	149.69 *
80	1.06 **	2.03 **	218.66
160	1.06 **	2.31	158.49
320	1.00 **	2.21 *	227.46 *

* significant difference at $P \leq 0.05$; ** significant at $P \leq 0.01$

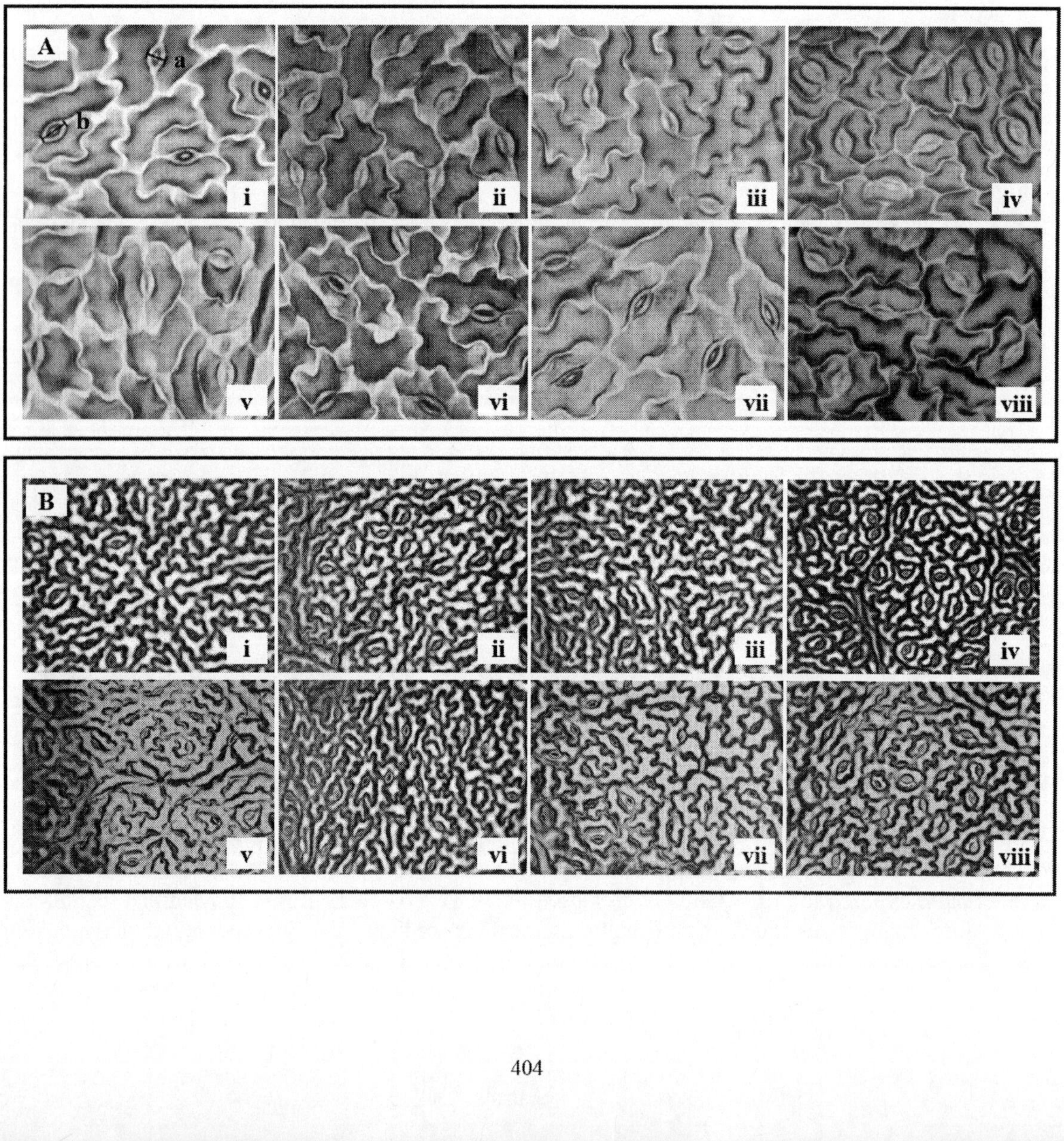

Figure 3. Panels show soybean stomatal measurements: (A) Stomata width (a) and length (b) were measured under 1.000x magnification; (B) Soybean stomata shown under 400x magnification; i = 0 Gy; ii = 5 Gy; iii = 10 Gy; iv = 20 Gy; v = 40 Gy; vi = 80 Gy; vii = 160 Gy; viii = 320 Gy

Auxin synthesis was inhibited by low doses of gamma rays, while auxin destruction was caused by higher doses of gamma rays. There were three ways of that reasons : 1) DNA is required for and is previously synthesized sequently to auxin formation, and gamma rays block occuring nucleic acid formation; 2) primary gamma rays block is in auxin synthesis, the auxin required for DNA formation; and 3) effect of gamma rays is on an undefined entity in reaction previous to and essential for both DNA and auxin synthesis [37]. The biological effect of gamma rays is also mainly due to the formation of free radicals by water hydrolisis, and results in modulation of antioxidative system, and accumulation of phenolic compounds and chlorophyll pigments [38]. However, these radicals can damage or modify important components of plant cell and affect differentially the biochemistry, physiology, anatomy, and morphology of plants [39]. The frequency of chromosomal damage cause the metabolic disorders and may be responsible for less germinability or to survive more than few days. In this study, seedling derived from seeds exposed to higher doses (640, 1280, and 2560 Gy) did not survive for more than 20 days, so it was not possible for stomata evaluations.

In figure 3, the stomatal apparatus of soybean irradiated by gamma rays were shown using nail polish imprint method. The stomatal size (width and length) and density were clearly inconsistent (decrease or increase) with increased gamma ray doses (Table 3). In this study, it was found that the stomatal width decreased 1.10 to 1.43 times, and the stomatal length decreased from 1.10 to 1.16 times. Stomatal density ranged from 149.69 to 300.84 mm^{-2} depends on gamma ray doses. It have significantly increased by 21.09 % and 60.16 % at 320 Gy and 20 Gy, respectively, and decreased 20.31 % at 40 Gy. However, Celik *et al.* [40] observed different results that soybean plant leaves developed from gamma irradiated seeds showed reductions in stomatal density compared to the control plant. The function of stomatal is correlated with process of transpiration and photosynthesis that occured in the leaves. Stomata biometric measurements have an important role in transpiration and photosynthesis process by adjusting gas exchange between the leaves and the atmosphere [41]. Stomatal size influences exchange of CO_2, hence the bigger stomatal size will increase the exchange of CO_2 [42]. It is explain the rate of photosynthesis is more efficient. Besides, the higher density of stomatal also allows the higher gas exchange, consequently, the rate of photosynthesis is higher than control [43]. The higher photosynthesis supports the plant growth. So, different stomatal size and density of each genotypes can be used as indirect selection criteria.

4. Conclusions

It was concluded that different doses of gamma rays influenced the germination and seedling growth of Soybean (*Glycine max* L.cv. Dering 1). Very low to low doses of gamma rays (5-320 Gy) might be used to study the improvement of soybean diversity.

5. Acknowledgements

Authors wish to convey gratitude to the Directorate General of Research and Community Service, the Ministry of Research, Technology and Higher Education the Republic of Indonesia for the research grant through scheme of "Superior Research of University", and Diponegoro University for the continuous support, particularly to the Institute of Research and Community Services.

References

[1] Kumawat, G., S. Gupta, M.B. Ratnaparkhe, S. Maranna, and G.K. Satpute. 2016. QTLomics in soybean: a way forward for translational genomics and breeding. *Frontiers in Plant Science* 7:1852 doi:10.3389/fpls.2016.01852

[2] Shu, K., Y. Qi, F. Chen, Y. Meng, X. Luo, H. Shuai, W. Zhou, J. Ding, J. Du, J. Liu, F. Yang,

Q. Wang, W. Liu, T. Yong, X. Wang, Y. Feng, and W. Yang. 2017. Salt stress represses soybean seed germination by negarively regulating GA biosynthesis while positively mediating ABA biosynthesis. *Frontiers in Plant Science* 8:1372 doi:10.3389/fpls.2017.01372

[3] Kusmiyati, F., Sumarsono, Karno, and E. Pangestu. 2016. Influence of rice straw mulch on saline soil: forage production, feed quality and feed intake by sheep. *ISSAAS Journal* 22 (1): 42-51

[4] Purbajanti, E.D., F. Kusmiyati, and E. Fuskhah. 2017. Growth, yield and physiological characters of three types of Indonesian rice under limited water supply. *Asian Journal of Plant Sciences* 16: 101-108 doi: 10.3923/ajps.2017.101.108

[5] Lee, J.D., J.G. Shannon, T.D. Vuong, T.D., H.T. Nguyen. 2009. Inheritance of salt tolerance in wild soybean (*Glycine soja* Sieb. And Zucc.) accession PI483463. *Journal of Heredity* 100 (6): 798-801 doi:10.1093/jhered/esp027

[6] Purbajanti, E. D., Kusmiyati F., Slamet, W and Adinurani, P. G. 2016. Chlorophyll, crop growth rate and forage yield of Brachiaria (*Brachiaria brizantha* Stapf) as the result of goat manure in various nitrogen dosage. Advances of Science and Technology for Society AIP Conf. Proc. 1755, 130013-1–130013-4; doi: 10.1063/1.4958557

[7] Kusmiyati, F., E.D. Purbajanti, Surahmanto, and Sumarsono. Forage production in saline soil treated with organic fertilizer used as feed for growing goats. *Livestock Research for Rural Development* 29 (10). http://www.lrrd.org/lrrd29/10/fkus29195.html

[8] Xu, D and D.D. Tuyen. 2012. Genetic studies on saline and sodic tolerances in soybean. *Breeding Science* 61(5):559-565 doi: 10.1270/jsbbs.61.559

[9] Guan, R., J. Chen, J. Jiang, G. Liu, Y. Liu, L. Tian, L. Yu, R. Chang, L. Qiu. 2014. Mapping and validation of a dominant salt tolerance gene in the cultivated soybean (*Glycine max*) variety Tiefeng 8. The Crop Journal 2 (6):358-365 doi:10.1016/j.cj.2014.09.001

[10] Herwibawa, B., T.A.D. Haryanto, and Sakhidin. 2014. Peroxidase isozyme identification of some rice genotypes in M_1 generation under drought stress level of -0.03 MPa. *Agrivita Journal of Agricultural Science* 36 (3): 210-215 doi:10.17503/Agrivita-2014-36-3-210-216

[11] Sikora, P., Chawade, A., Larsson, M., Olsson, J., and O. Olsson. 2011. Mutagenesis as a tool in plant genetics, functional genomics, and breeding. *International Journal of Plant Genomics* 2011, Article ID 314829 doi:10.1155/2011/314829

[12] Chen, J., C. Thammina, W. Li, H. Yu, H. Yer, R. El-Tanbouly, M. Marron, L. Katin-Grazzini, Y. Chen, J. Inguagiato, R.J. McAvoy, K. Guillard, X. Zhang, and Y. Li. 2016. Isolation of prostrate tufgrass mutants via screening of dwarf phenotype and characterization of a perennial ryegrass prostrate mutant. *Horticulture Research* 3, Article Number 16003 doi: 10.1036/hortres.2016.3

[13] Melki, M., and A. Marouani. 2010. Effects of gamma rays irrays on seed germination and growth of hard wheat. *Environmental Chemistry Letters* 8(4): 307-310 doi: 10.1007/s10311-009-0222-1

[14] Songsri, P., B. Suriharn, J. Sanitchon, S. Srisawangwong, and T. Kesmala. 2011. Effects of gamma rays on germination and growth characteristics of physic nut (*Jatropha curcas* L.). *Journal of Biological Sciences* 11(3): 268-274 doi: 10.3923/jbs.2011.268.274

[15] Harding, S.S., S.D. Johnson, D.R. Taylor, C.A. Dixon, and M.Y Turay. 2012. Effect of gamma rays on seed germination, seedling height, survival percentage and tiller production in some rice varieties cultivated in Sierra Leone. *American Journal of Experimental Agriculture* 2(2):247-255 doi: 10.9734/ajea/2012/820

[16] Marcu, D., G. Damian, C. Cosma, and V. Cristea. 2013. Gamma rays effects on seed germination, growth and pigment content, and ESR study of induced free radicals in maize (*Zea mays*). *Journal of Biological Physics* 39(4): 625-634 doi: 10.1007/s10867-013-9322-z

[17] Neelam, D., T. Tabasum, S.A. Husain, Mahmooduzaffar, and S. Subhan. 2014. Rays sensitivity of *Cajanus cajan* to gamma rayss. *Journal of Food Processing and Technology* 5:394 doi: 10.4172/2157-7110.1000394

[18] Umavathi, S., and L. Mullainathan. 2015. Physical and chemical induced mutagenesis study for identifying lethality dose in chick pea (*Cicer arietinum* L.) var. Co-4. *International Letters of Natural Sciences* 35:1-5 doi: 10.18052/www.scipress.com/ILNS.35.1

[19] Beyaz, R., C.T. Kahramanogullari, C. Yildiz, E.S. Darcin, and M. Yildiz. 2016. The effect of gamma rays on seed germination and seedling growth of *Lathyrus chrysanthus* Boiss. under in vitro conditions. *Journal of Environmental Radioactivity* 162-163: 129-133 doi: 10.1016/j.jenvrad. 2016.05.006

[20] Lopes, A.M., V.L. Bobrowski, S.D.D.A.E. Silva, S. Deuner. 2017. Morphophysiological and biochemical alterations in *Ricinus communis* L. seeds submitted to cobalt^{60} gamma rays. *Anais daAcademia Brasileira de Ciencias*89 (3): 1925-1933 doi: 10.1590/0001-3765201720170081

[21] Santorum, M., Nobrega, L.H.P., E.G. de Souza, D. dos Santos, W. Boller, and M.M. Mauli. 2013. Comparison of tests for the analysis of vigor and viability in soybean seeds and their relationship to field emergence. *Acta Scientiarum Agronomy* 35(1): 83-92 doi: 10.4025/actasciagron.v35i1.14955

[22] Moshtaghi-Kavaran, A., S. Khomari, and N. Zare. 2015. Soybean seed germination and seedling growth in response to deterioration and priming: effect of seed size. *Plant Breeding and Seed Science* 70(1): 55-67 doi:10.1515/plass-2015-0013

[23] Roslime, D.I., Herman, and I. Fiatin. 2015. Lethal dose 50 (LD_{50}) of mungbean (*Vigna radiata* L. Wilczek) cultivar Kampar. *SABRAO Journal of Breeding and Genetics* 47(4):510-516

[24] Orchard, T., 1977. Estimating the parameters of plant seedling emergence. *Seed Science and Technology* 5:61-69

[25] Al-Mudaris, M.A. 1998. Notes on various parameters recording the speed of seed germination. *Journal of Agriculture in The Tropics and Subtropics* 99 (2): 147-154

[26] Jones, K., and D. Sanders. 1987. The influence of soaking pepper seed in water or potassium salt solutions on germination at three temperatures. *Journal of Seed Technology* 11: 97-102

[27] Esechie, H. 1994. Interaction of salinity and temperature on the germination of sorghum. *Journal of Agronomy and Crop Science* 172: 194-199

[28] Arnold, R.L.B., M. Fenner, and P.J. Edwards. 1991. Changes in germinability, ABA content and ABA embryonic sensitivity in developing seeds of *Sorghum bicolor* (L.) Moench induced by water stress during grain filling. *New Phytologist* 118: 339-347 doi: 10.1111/j.1469-8137.1991.tb00986.x

[29] Filho, C.P.H., A.L.D. Goneli, T.E. Masetto, E.A.S. Martins, G.C. Oba. 2016. The effect of drying temperatures and storage of seeds on the growth of soybean seedlings. *Journal of Seed Science* 38 (4): 287-295 doi: 10.1590/2317-1545v38n4161866

[30] Inamdar, J.A., and R.C. Patel. 1969. A new technique for making plant epidermal imprints using various domestic adhesives. *Journal of Microscopy*90 (3): 269-272 doi:10.1111/j.1365-2818.1969.tb00711.x

[31] van-Harten, A.M. 1998. Mutation Breeding: Theory and Practical Application. Cambridge University Press. Cambridge, UK.

[32] Hanafiah, D.S., Trikoesoemaningtyas, S. Yahya, and D. Wirnas. 2010. Agronomic improvement of argomulyo soybean variety [*Glycine max* (L) Merr] through induced mutation by gamma irrays in M_1 and M_2 generation. *Biosfera* 27(3): 103-111

[33] Karthika, R., and B.S. Lakshmi. 2006. Effect of gamma rays and EMS on two varieties of soybean. *Asian Journal of Plant Sciences* 5 (4): 721-724 doi: 10.3923/ajps.2006.721.724

[34] Ilyas, S., and S. Naz. 2014. Effect of gamma irrays on morphological characteristics and isolation of curcuminoids and oleoresins of *Curcuma longa* L. *The Journal of Animal and Plant Sciences* 24(5): 1396-1404

[35] Pavadai, P. 2015. Studies on quantitative characters for gamma rays treatment in soybean (*Glycine max* (L.) Merr.) var. Co-1. *International Journal of Modern Cellular and Molecular Biology* 4(1): 1-10

[36] Khadimi, A.A., C. Radziah, C.M. Zain, A.N. Alhasnawi, A. Isahak, M.F. Ashraf, A. Mohamad,

F. Doni, and W.M.W. Yusoff. 2016. Impact of gamma rays exposure and growth regulators on *Oryza sativa* L. cv. MR269 callus induction. *AIP Conference Proceedings* 1784, 020006 (2016) doi: 10.1063/1.4966716

[37] Jan, S., T. Parween, T.O. Siddiqi, and X. Mahmooduzzafar. 2012. Effect of gamma (ɤ) rays on morphological, biochemical, and physiological aspects of plants and plant products. *Environmental Reviews* 20:17-39 doi: 10.1139/a11-021

[38] Akshatha, K.R. Chandrashekar, H.M. Somashekarappa, and J. Souframanien. 2013. Effect of gamma irrays on germination, growth, and biochemical parameters of *Terminalia arjuna* Roxb. *Rays Protecton and Environment* 36(1): 38-44 doi: 10.4103/0972-0464.121826

[39] Ikram, N., S. Dawar, Z. Abbas, and M.J. Zaki. 2010. Effect of (^{60}cobalt) gamma rays on growth and root rot diseases in mungbean (*Vigna radiata* L.). *Pakistan Journal of Botany* 42(3): 2165-2170

[40] Celik, O., C. Atak, and Z. Suludere. 2014. Response of soybean plants to gamma radiation: biochemical analyses and expression patterns of trichome development. *Plant Omics Journal* 7 (5): 382-391

[41] Peksen, E., A. Peksen, and C. Artik. 2006. Comparison of leaf and stomatal characteristics in faba bean (*Vicia faba* L.). *Journal of Biological Sciences* 6 (2): 360-364

[42] Qosim, W.A., R. Purwanto, G.A. Wattimena, and Witjaksono. 2011. Alteration of leaf anatomy of mangosteen (*Garcinia mangostana* L.) regenerants *in vitro* by gamma irrays. *Plant Mutation Reports* 2 (3): 4-11

[43] Widiastuti, A., Sobir, and M.R. Suhartanto. 2010. Diversity analysis of mangosteen (*Garcinia mangostana*) irradiated by gamma-ray based on morphplogical and anatomical characteristics. *Nusantara Bioscience* 2 (1): 23-33 doi: 10.13057/nusbiosci/n020104

Study of utilization liquid smoke and carrageenan as a natural antibacterial in manufacturing beef meatballs

W Widayat [1*], N Arifiani S.[1], N Yaqin[2], A. N. Al Baarri[2]

[1] Department of Chemical Engineering, Diponegoro University, Semarang
[2] Department of Food Technology, Faculty of Animal and Agricultural Sciences, Diponegoro University, Semarang
Jl. Prof. Soedarto SH, Tembalang, Semarang, 50275
[*]Email. yayat_99@yahoo.com

Abstract. This research have observed liquid smoke and carragenan ability to obstruct microbe activity. Phenol and sulfate ester in liquid smoke and carrageenan, give preservation effect by obstruct microorganisms growth on beef meatballs. Liquid smoke and carrageenan was added on a particular variation into meatballs dough. Liquid smoke variation was 0.5%-1%, and carrageenan variation was 0.5%-1.5%. Meatballs then stored up to 36 hours at room temperature and performed TPC test in every 12 hours. The results indicates that in 0-12 hours carragenan has significant effect to obstruct microorganism growth with the percentage reduction in total bacteria was 13.54% and 93.73%. In 24-36 hours liquid smoke effected to the significant effect obstruction of microorganism growth with the percentage reduction in total bacteria was 98.99% and 99.93%. The addition of liquid smoke and carrageenan did not give a significant effect on the lightness of beef meatballs produced, but provided significant effect on the storage time and the lightness of beef meatballs.

Keywords : liquid smoke, carrageenan, utilization, antibacterial, beef meatball

1. Introduction

Meatballs has a high nutrition, pH 6.0- 6.5 and high aw (> 0.9), were the maximum storage period is 1 day (12-24 hours) at room temperature [11]. Foodstuffs that have a high nutritional value in the pH range of neutral pH and high water content such as meatballs is a good medium for microbial growth. Damage caused by the emergence of microbes in foods are mucus, discoloration, mildew, the incidence of irregularities aroma, fermentation damage and decay protein ingredients. Therefore, meatballs sold in the market are usually added preservatives to extend the shelf life. However, it is still widely used preservatives are not recommended for foods such as formaldehyde and sodium borate.

Liquid smoke is derived from natural ingredients that is hemicellulose burning, cellulose and lignin from hardwoods and coconut shell to produce compounds that have antimicrobial properties, antibacterial and antioxidant such as acidic compounds and their derivatives, alcohols, phenols, aldehydes, carbonyls, ketones and pyridine [7]. Liquid smoke has several advantages over the traditional fumigation techniques in terms of ease of application, speed, uniformity of product, good characteristics desired in the final product, as well as reduce the danger of polycyclic aromatic hydrocarbons and resulting a chewy texture in processed meat products [6].

Carrageenan is a polysaccharide that soluble in water and extracted from red seaweed (Rhodophyceae). Carrageenan is a hydrocolloid consisting of potassium, sodium, magnesium and calcium sulphate esters of galactose and 3,6-anhidrogalaktosa (AG) copolymer [13]. Carrageenan has function as an emulsifiers, stabilizers, gelling, and coagulant. Carrageenan also has a water binding properties that will affect the yield and chewy texture in processed meat products. Carrageenan is used as a gelling material to replace the use of borax that harmful to health because of its ability to bind water.

The addition of liquid smoke in manufacturing meatballs have been done by [1] in which the liquid smoke has function as a water bounding agents on the meatballs in order to increase of quality and durability of the meatballs. [2] Have been doing experiments using liquid smoke in the preservation of meatballs with 0-7% concentration of liquid smoke. [3] Observed changes the quality of beef meatballs with the addition of liquid smoke which indicates that liquid smoke has significant effect on the level of preference and flavour, optimal results obtained on the addition of liquid smoke 1%. [5] Also have been studied the effect of using carrageenan in the manufacture of beef meatballs.

2.Materials and methods

2.1. *Material*

Materials in this study was beef, tapioca starch 15% of the meat weight, salt 2%, garlic 2.5%, pepper 0.8%, ice cubes, 20% MSG 1%, carrageenan, and liquid smoke [4]. Tools set for making meatballs, a set for total microbial test equipment and Digital Color Meter were used in this experiment.

2.2.Making meatballs

Making meatballs were started with cutting beef into small pieces and ground in a grinding machine. There were two stages of milling in order to obtain soft dough. Garlic 2.5%, salt 2% and pepper 0.8% which has been mashed and mixed with tapioca starch 15% and ice cubes 20% in second milling process. All compound which has been formed from mixing process, divided into 10 samples and then added liquid smoke and carrageenan with concentrations varied according to Table 1. The mixture was crushed for a half minute and then released for forming meatballs by put in warm water with a temperature of 60°C to 80°C and allowed to float. After float, meatballs were moved into the boiling water and heated until cooked, about 10 minutes. Furthermore, meatballs were stored for 36 hours at room temperature, sample were obtained for every 12 hours and total bacteria were tested by TPC. Lightness was measured using Digital Colormeter Software by Macintosh.

2.3.Total bacteria test with TPC (Total Plate Count) method

According to Indonesian National Standard (SNI) limit of total bacteria in food, especially in the processed meat product is 1x105 colonies/g (ISO 7388; 2009). To calculate total bacteria in beef meatballs used Total Plate Count (TPC) method. Storage time was 0, 12, 24, 36 hours. As much as 15 ml Plate Count Agar (PCA) was poured into the sterile petri dish and chill. One gram of meatballs sample was crushed and dissolved in distilled water until 10 mL, then homogenized the sample through shaking until the sample was well diluted in solvent. The series of dilutions were 10^{-1}, 10^{-2}, 10^{-3}, 10^{-4}, 10^{-5}. As much as 0.5 mL solution from each dilutions were put in petri dish containing PCA (Plate Count Agar) and trim using rod bent glass to form "eight" symbol. Samples were allowed to stand for 1 hour then incubated in 48 ± 2 hours at 35°C. The number of colonies was calculated by colony counter. Calculation of Total Plate Count as follows:

$$\frac{\Sigma C}{[(1\,x\,n1)+(0{,}1\,x\,n2)]\,x\,(d)} = N$$

N is total colonies product, expressed in colonies per g; ΣC is total colonies in all counted plate; n1 is total plate on the first dilution; n2 is total plate in second dilution; d is first calculated dilution (SNI 01-2332.3-2006).

3.Results and discussion

3.1 Beef meatballs total bacteria analysis

Table 1 showed that at a range of 0-12 hours, sample code V6 provided highest total bacterial concentration. This is because V6 has the smallest concentration of carrageenan compared to other samples. In this condition, carrageenan has a significant influence into sample because of the carrageenan has a hydrophilic properties so that if the number of carrageenan less than the amount of water tied too little and allowing more bacteria to grow.

Table 1. Beef meatballs total bacteria analysis

Sample Code	Liquid smoke concentration (%)	Carrageenan concentration (%)	Total bacteria/mL			
			0 hours	12 hours	24 hours	36 hours
V0	0	0	5.76×10^4	1.33×10^6	1.66×10^8	3.94×10^9
V1	0.5	0.5	3.84×10^4	4.25×10^4	5.52×10^5	3.1×10^6
V2	1	0.5	4.58×10^4	4.99×10^4	8.32×10^5	2.28×10^6
V3	0.5	1.5	3.75×10^4	3.06×10^4	5.31×10^5	1.41×10^6
V4	1	1.5	4.3×10^4	1.04×10^5	3.09×10^5	4.43×10^6
V5	0.75	1.707	5.33×10^4	1.34×10^5	6.23×10^5	1.02×10^6
V6	0.75	0.293	7.71×10^4	2.83×10^5	2.8×10^6	2.28×10^6
V7	1.103	1	6.67×10^4	6.81×10^4	5.19×10^5	9×10^5
V8	0.3965	1	3.35×10^4	4.5×10^4	4.65×10^6	7.33×10^6
V9	0.75	1	6.94×10^4	5.01×10^4	2.9×10^6	1×10^6
V10	0.75	1	3.33×10^4	2.64×10^4	3.08×10^6	2.59×10^6

At a range of 24-36 hours, V8 was highest in total bacteria value. This is because V8 had the smallest concentration of liquid smoke. The longer storage time the more yield of free water resulting in the more bacteria to grow. Carrageenan had no significant effect on the sample since the acidic conditions by more bacteria was appeared, and carrageenan would undergo the hydrolysis. Therefore, if the sample has a low concentration of liquid smoke, amount of water bound was also low thus allowing more bacteria to grow.

TPC meatballs value up to 12 hours storage still below the maximum limit specified number of microbes ($1x10^5$ CFU/g) based on SNI 01-2729-2006. The percentage reduction in total bacteria against storage time can be illustrated in Figure 1.

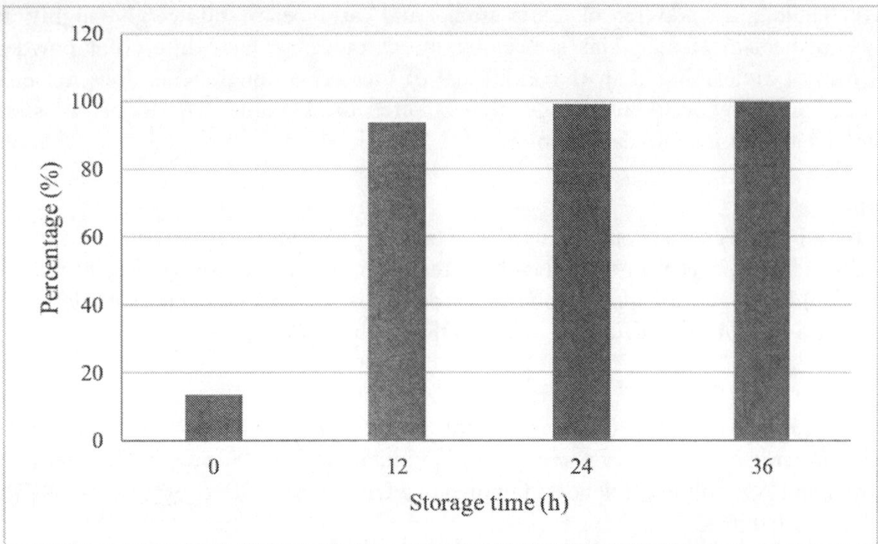

Figure 1. Percentage bacteria decrease in each storage time

Liquid smoke and carrageenan were able to inhibit bacteria growth on beef meatballs as can be seen by reduction percentage of bacteria to 13.54%, 93.73%, 98.99%, 99.93% for 0, 12, 24, and 36 hours of storage time, respectively. This is because liquid smoke contains phenols which serves to inhibit bacteria growth. As antibacterial, phenolic compound has a working mechanism by damaging bacterial cells structure and inhibits cell wall formation process that causes the lysis of a bacterial cell wall [12]. Carrageenan has sulfate ester groups that have ability to bind water. Carragenan also have a good ability in water binding and also to form three dimensional mesh that can trap water [8].

Table 2. Beef meatballs color analysis

Liquid Smoke (%)	Carragenan (%)	Lightness			
		0	12	24	36
0	0	64.81	67.77	54.67	53.13
0.5	0.5	68.42	57.53	53.08	61.88
1	0.5	64.86	61.77	56.26	57.03
0.5	1.5	67.41	63.12	54.46	57.52
1	1.5	64.93	59.77	50.01	50.60
0.75	1.707	65.23	62.98	52.01	53.98
0.75	0.293	65.10	59.55	50.87	53.75
1.103	1	64.42	65.49	50.68	56.68
0.3965	1	64.55	65.55	48.01	56.21
0.75	1	64.06	65.88	62.30	59.86
0.75	1	63.15	71.27	62.44	50.20

3.2 Beef meatballs color anaysis

Color is one of visual nature of the first seen by the consumer. Color has meaning and a very important role in food product since it can give the impression of like and dislike. Signification and role of color in food products, depends on types, signs of damage, and the manual processing [10].

An objective color measurements were performed using an Digital Colourmeter. L is a color parameter value (lightness/brightness level) stating dark and bright of sample. The greater the value of L, the more the light or bright sample. L value ranges between 0 (black) to 100 (white). L value beef meatballs can be seen in Table 2.

Based on Table 2, the addition of liquid smoke and carrageenan did not give significant effect on the lightness of beef meatballs. This is because the carrageenan has white color powder and liquid smoke in form of a clear liquid so the additional of these two components does not effect to the L value of meatballs. However, the storage time effected the L value. The longer of storage time, L value decreased and turned into darker color.

4.Conclusions

Liquid smoke and carrageenan could be used as a natural preservative in manufacturing beef meatballs because it could inhibit bacteria growth resulting the extension in the storage time of meatballs. By liwuid smoke and carrageenan application in meatballs, the decrease in bacteria could be achieved more than 90% and did not significantly effect to the lightness of meatballs.

References

[1] Abustam, E, J.C.Likadja dan A. Ma'arif. 2009. *The utilization of liquid smoke for binder in meatball made of Bali beef*. Proceeding of National Seminar "Kebangkitan Peternakan". Animal and Agricultural Sciences, Diponegoro University. ISBN : 978-979-704-746-7. (In Bahasa Indonesia)

[2] Arnim, Ferawati, and Yetti Marlida. 2012. The Effect of Liquid Smoke as Preservative for Meatballs Quality. Pakistan Journal of Nutrition 11 (11) : 1078-1080.

[3] Astati. 2013. The level of meatball quality changes made of *Pectoralis Profundus using Smoke Liquid. Jurnal Teknosains*, Vol. 7 No. 1, pp. 10-19. (In Bahasa Indonesia)

[4] Chrismanuel, Y. B. Pramono and B. E. Setyani. 2012. *The effect of utilization of Carrageenan as edible coating to pH, microbe, and H_2S of meatballs stored for 16 hour*. Animal Agriculture Journal, **1**. No. 2, p 286 -292. (In Bahasa Indonesia)

[5] Hasret Ulu. 2006. *Effects of Carrageenam and Guar* Gum *on the Cooking and Textual Properties of Low Fat Meatballs*. Food Chemistry 95 (2006) 600-605.

[6] Hattula, T., Elfving, K., Mroueh, U.M. and Luoma, T. 2001. *Use of liquid smoke flavoring as an alternative to traditional flue gas smoking of rainbow trout fillets (Oncorhyncus mykiss)*. Lebensmittel Wissenschaft und Technologie. **34**: 521-525.

[7] Koswara, Sutrisno. 2009. *Pengawet Alami Untuk Produk Dan Bahan Pangan*. Ebook pangan 2009 (23-25).

[8] Pietrasik, Z. and A. Jarmolouk., 2003. *Effect of Sodium Caseinate And K- Carrageenan on Binding and Textural Properties of Pork Muscle Gels* Enhanced *By Microbial Transglutaminase Addition*. Journal of Food Engineering. 6 (3): 285-294.

[9] SNI 01-2332.3-2006. *Plate count test for fishery food product*. Badan Standardisasi Nasional. Jakarta

[10] Soekarto, S. T. 1985. *The research of organoleptic for food industry*. Bharata Karya Aksara. Jakata. (In Bahasa Indonesia)

[11] Surjana, W. 2001. *The preservation of meatballs using chemical additives in the room storage*. Bachelor Thesis. IPB, Bogor. (In Bahasa Indonesia)

[12] Susanti, A. 2006. *Antibacterial activities of extract of Pluchea indicaless against Escherichia coli*. Veterinary Faculty. Airlangga University. Surabaya. (In Bahasa Indonenesia)

[13]Van de Velde F. *Structure and function of hybrid carrageenans*. Food Hydrocolloid. 2008;22:727–734.

IOP Conf. Series: Earth and Environmental Science **102** (2017) 012061 doi:10.1088/1755-1315/102/1/012061

Chemical and sensory characteristics of frozen wheygurt with the addition of taro and lesser yam flours as thickening agent

E Nurhartadi, R Utami, E Widowati, and B M Karunawati[1]

Department of Food Science and Technology Faculty of Agriculture
Sebelas Maret University
Jl. Ir. Sutami 36A Kentingan Jebres Surakarta 57126, INDONESIA

E-mail: edhi.nr@gmail.com

Abstract. Cheese whey is a waste product from cheese processing. It has low solid contents thus required the addition of a thickening agent. Lactic acid bacteria could utilize it in the fermented drink. This research aims to study the effect of taro and lesser yam flour addition as a thickening agent on chemical and sensory characteristics of frozen wheygurt. This research used Complete Randomized Design (CRD) with one factor that is variation ratio of taro and lesser yam flour F1 (4: 0), F2 (3: 1), F3 (2: 2), F4 (1: 3), F5 (0: 4). The number of lactic acid bacteria cell determined by using hemocytometer. The lactic acid content determined by the titrimetric method by using 0.1 N NaOH and phenolphthalein as indicator. pH value measured with pH meter. Sensory characteristics evaluated using hedonic test. The result showed that the addition of taro and lesser yam flour have a significant effect on the number of lactic acid bacteria in frozen wheygurt. The higher lesser yam flour addition, the higher lactic acid bacteria count on frozen wheygurt, due to lesser yam higher glucose and fructo-oligosaccharide content than taro. The higher lesser yam addition, the higher the lactic acid produced. The higher the total bacteria and higher levels of lactic acid, the lower the pH obtained. The conclusion of this study is addition ratio of taro and lesser yam flour effect on the chemical characteristics of frozen wheygurt. There is no difference in the level of acceptance of the panelists in sensory evaluation.

Keywords: chemical, frozen, taro, wheygurt, yam

1. Introduction

In the manufacture of cheese will be obtained byproduct cheese way. Whey has a yellowish-green liquid form. For the manufacture of cheese as much as one kilogram will produce 9 kilos of cheese way[1]. Cheese whey still contain 0.006-0.0070% fat, 0.3% -0.9% protein, 3.04-5.00% lactose, ash 0.5-0.62%[2]. It can be processed into fermented products by the activity of lactic acid bacteria because there is lactose that can be used for growth sources [3].

Frozen yogurt is now an increasing trend among other fermented beverage products. Frozen yogurt is a dessert which combines the texture of ice cream with nutritional and yogurt health benefits[4]. The quality of the texture of frozen yogurt depends on the milk fatty bean and total milk solids. The weakness of cheese way is due to the low total solids that make the derivative products have a lower viscosity than commercial fermented products[3].

To overcome the weakness is done the addition of thickening materials that can increase the viscosity of whey-based beverage products. Commonly used thickeners are hydrocolloids. One of the most common types of hydrocolloids used is starch. Starch from tubers is suitable for acids and frozen products such as frozen wheygurt[5]. Taro (*Colocasia esculenta* L.) and lesser yam (*Dioscorea esculenta* L) have the potential to be used as a thickener because of their high content (about 51.34-70.92%), and a similar ratio of amylopectin and amylose (taro 21.44: 78.56 and lesser yam 24.30: 75.7).

However, these two materials have different gelatinization profile. Taro starch has high peak gelatinization viscosity[6] thus makes ice cream made from taro starch can melt slowly[7]. Lesser yam starch has low peak gelatinization viscosity[8] thus makes ice cream can melt in an ideal time range[9]. In this study, taro and lesser yam flour are combined to find the best chemical and sensory characteristics of frozen wheygurt. Furthermore, these two tubers have prebiotic substance which can be utilized as lactic acid bacteria (LAB) growth medium in human digestion system and metabolites formation, such as oligosaccharides in taro[10] and inulin in lesser yam[11].

This study aimed to determine the effect of taro and lesser yam flour on frozen wheygurt chemical and sensory characteristics.

2. Materials and Methods

2.1. Materials

Mozzarella cheese whey obtained from cheese industry in Boyolali (Indonesia). Lactic acid bacteria used in this study were *Lactobacillus acidophilus* FNCC 0051 and *Lactobacillus plantarum* FNCC 0027 from Food and Nutrition Culture Collection Gadjah Mada University Yogyakarta (Indonesia). Taro flour bought from Naya Tepung Bogor (Indonesia) and lesser yam flour bought from Kusuka Ubiku Bantul (Indonesia). Other materials used in frozen wheygurt preparations were egg yolk, sucrose (Gulaku®), whipping cream (Anchor®, 33.33% fat).

2.2. Starter preparation

The production of frozen yogurt starter is done by using [34] method with modification. Both *L. acidophilus* FNCC 0051 and *L. plantarum* FNCC 0027 were inoculated in separate sterile deMan Rogosa Sharpe (MRS) broth medium and incubated at 37°C for 24 hours. 2% (v/v) of each culture of MRS broth were inoculated to 100 ml pasteurized skim milk and incubated at 37°C for 24 hours. The cultured skim milk contains 10^8 cells/ml of bacterial cells.

2.3. Frozen wheygurt preparation

The process of frozen wheygurt making was done by using [35] method with modification. Frozen wheygurt was prepared by mixing the materials (cheese whey, 12% (w/v) sucrose, 13% (w/v) whipping cream, 2% (w/v) egg yolk and 2,5% (w/v) thickener) for 10 minutes. The mixtures were pasteurized at 75°C for 30 min and cooled until 45°C. 2% (v/v) of each *L. acidophilus* and *L. plantarum* culture starters were inoculated into pasteurized mixtures and incubated at 37°C for 18 hours. Frozen wheygurt then refrigerated for 5 hours, frozen for 30 minutes at -18°C for 30 minutes, mixed using a mixer for 20 minutes and frozen for 24 hours.

2.4. Chemical characteristics analysis

The lactic acid bacterial cell was determined by using hemocytometer[12]. Lactic acid was determined by the titrimetric method by using NaOH 0.1 N and phenolphthalein as indicator[13]. pH value was measured with pH meter[14] Oakton PCS_Tester 35.

2.5. Sensory characteristics

Sensory characteristics were evaluated using hedonic test[5] for color, flavor, mouthfeel, viscosity and overall parameters.

3. Results and Discussion

3.1. Chemical characteristics of frozen wheygurt

3.1.1. Lactic acid bacteria count. The standard of viable lactic acid bacteria (LAB) to be consumed is 10^8 viable cells[16]. Results obtained have met the standard of LAB (Table 1). The results showed that addition ratio of taro and lesser yam flour have a significant effect on the number of LAB in frozen wheygurt. The higher lesser yam flour addition, the higher LAB count on frozen wheygurt, due to lesser yam higher glucose and FOS content than taro[17]. *Lactobacillus* does fermentation selectively, where short oligosaccharides preferred over oligosaccharides with long chains. When fermentation media are simple sugars such as sucrose and glucose in small quantities, the *Lactobacillus* bacteria group would prefer to conduct the fermentation of simple sugars[18]. LAB viability of the final product is influenced by several factors, such as bacteria strain used, the interaction between the bacteria strains, the composition of fermentation medium and inoculum size[16].

Lactobacillus acidophilus is a bacteria which grows optimally at 37-42°C. This bacteria had the highest growth rate at pH 5.5-6.0 and began to decline in pH 4.0. *L. acidophilus* is an obligate homofermentative bacteria[19]. *L. plantarum* is a facultative anaerobe heterofermentative bacteria. Mesophilic bacteria is classified as a bacteria which can grow at 15-45°C and pH 4.0-9.0[20]. *L. acidophilus* and *L. plantarum* inoculum size in their pure state have a higher number of cell density compared to the combination of both. However, combination cell density number still met the recommended standard of the viable LAB in a product[21]. The combination of *L. plantarum* and *L. acidophilus* did not affect the viability of bacteria in the final product[22]. The amount of inoculum is a key factor to ensure an adequate number of living cells in the final product. The use of high inoculum concentrations will guarantee the number of bacteria cells at the end of the incubation and maintain the viability of LAB during storage to consumption[16].

Table 1. Effect of taro and lesser yam flour as thickener agent on chemical characteristics of frozen wheygurt*

Formulation Taro: Lesser Yam	Lactic acid bacteria number (log cell/ml)	Lactic acid (%)	pH
4:0	8,916^c ± 0,207	0,405^d ± 0,023	4,48^c ± 0,023
3:1	9,284^b ± 0,086	0,470^c ± 0,020	4,64^e ± 0,013
2:2	9,366^b ± 0,049	0,542^b ± 0,016	4,55^d ± 0,024
1:3	9,536^a ± 0,064	0,566^{ab} ± 0,020	4,33^b ± 0,021
0:4	9,647^a ± 0,095	0,588^a ± 0,283	4,22^a ± 0,028

*Notation different letters in the same column indicate significantly the difference at a significance level of 5%

3.1.2. Lactic acid. The main role of lactic acid bacteria is to convert lactose into lactic acid during the fermentation of milk[16]. Total lactic acid showed concentrations of lactic acid as a result of metabolism of milk lactose into galactose and glucose[23].The results showed the addition ratio of taro and yam flours have a significant effect on the lactic acid level in frozen whey yogurt. The higher lesser yam addition, the higher the lactic acid produced. Lactic acid was found to correlate positively with total lactic acid bacteria produced where the higher the number the higher the lactic acid bacteria produced. Lesser yam addition can produce higher levels of lactic acid as lesser yam contains FOS and glucose higher than taro[17]. FOS and inulin in can also be an addition to increasing the growth of lactic acid bacteria also produce short-chain fatty acids or short chain fatty acid (SCFA). Propionic acid is one result of fermentation of prebiotics by lactic acid bacteria. The addition of flour in the manufacture of wheygurt can increase the levels of propionic acid produced[24]. The propionic acid produced is calculated as a lactic acid because the testing method used is a total titrated acid[13].

3.1.3. pH. The results showed the addition ratio of taro and lesser yam flours have a significant influence on the pH value of frozen yogurt. Due to lesser yam higher glucose and FOS contents than taro[17], thus LAB growth, metabolism, and lactic acid formation are higher in the sample higher addition of lesser yam. pH is obtained in accordance with the levels of lactic acid and total bacteria. The higher the total bacteria and higher levels of lactic acid, the lower the pH obtained.

The growth of bacterial cells followed by the formation of primary metabolites such as lactic acid derived from the change of the sugar content in the medium. The accumulation of lactic acid resulted in a decrease in pH[25]. A decrease in the degree of acidity (pH) caused by H+ ions derived from acidic compounds overhauled from lactic acid metabolism. Lactic acid is produced as the main product will dissociate producing H^+ and $CH_3CHOHCOO^-$, so the increase of lactic acid allows more H^+ to be liberated in the medium and pH can decrease further[26]. However, in this study the decrease in acidity that is not proportional to the increase in total lactic acid. This can be caused by propionic acid produced by both bacteria strain used[24] as a result of metabolism FOS and inulin from lesser yam. *L. plantarum* is a strain of bacteria which can produce metabolites other than lactic acids, such as acetic acid to lower the pH. Lactic acid is the main organic acid produced by *L. plantarum*, but acetic acid is also produced in smaller quantities[27]. In addition, the acid measured by pH meter is a concentration of H^+ ions which indicate the amount of dissociated acid, while the total titrated acid shows measurements for all the components of acid, either dissociated or not[28].

3.2. *Sensory characteristic of frozen wheygurt*

3.2.1. Color. Color is one of the first aspects which affect consumer acceptance of a product. Color is the most important quality attributes[29]. The results showed the addition ratio lesser yam and taro flours have a significant influence on the color parameters frozen wheygurt. Based on Table 2 it can be seen that in the color parameter, the sample of frozen wheygurt with the addition ratio of 4: 0 and 3: 1 is preferred by the panelist. This is due to frozen wheygurt with the addition of more taro flour ratio will have a whiter color compared to frozen wheygurt with the addition of lesser yam flour.

Table 2. Sensory characteristic of frozen wheygurt*

Formulation Taro: Lesser Yam	Color	Taste	Mouthfeel	Viscosity	Overall
4:0	2,04±0,89^a	2,92±0,86^a	2,72±0,74^a	2,92±0,86^a	2,60±0,71^a
3:1	2,28±0,68^a	2,92±1,12^a	2,72±0,79^a	2,84±0,62^a	2,96±0,68^a
2:2	2,92±0,76^b	2,76±0,83^a	2,76±0,66^a	2,80±0,71^a	2,84±0,90^a
1:3	3,08±1,15^b	2,48±0,77^a	2,68±0,80^a	2,92±0,81^a	2,72±0,79^a
0:4	2,84±1,07^b	3,04±1,10^a	3,04±0,84^a	2,88±0,78^a	3,04±0,89^a

*Notation different letters in the same column indicate significant difference at a a significance level of 5%
Score 1: extremely dislike, 2: dislike, 3: neutral, 4: like, 5: extremely like

3.2.2. Taste. Taste is a very important factor in determining consumers final decision to accept or reject a food product[30]. Taste is an important attribute in the reception of ice cream. Balance in sweetness levels needs to be maintained to produce an acceptable taste[31]. The results showed that addition ratio of taro and lesser yam flours do not affect frozen wheygurt flavors parameter. Based on the data in Table 2 on taste parameters panelists showed preference level ranged from 2.48 to 3.04 (criteria like to neutral). Sample with the highest panelist acceptance response samples with taro and lesser yam flour addition ratio 1: 3. While the sample with the lowest acceptance level is a sample with taro and lesser yam flour addition ratio 0: 4.

3.2.3. Mouthfeel. Mouthfeel parameter is a response to the texture and body of frozen yogurt in the oral cavity[32]. Body is defined as the overall quality perceived by the mouth, while the texture is defined as partial qualities that makeup as a whole[33]. The texture of the ice cream related directly to the structure. The structure depends on the size, number and arrangement of the trapped air, ice crystals, crystal lactose and fat globules[31].

The results showed that the addition ratio of taro and lesser yam flour and has no significant effect on the mouthfeel parameters of frozen wheygurt. Based on data in Table 2 on mouthfeel panelists showed preference level ranged from 2.68 to 3.04 (criteria like to neutral). Sample with the highest panelist acceptance response is a sample with taro and lesser yam addition ratio of 1: 3. While the sample with the lowest acceptance level is a sample with taro and lesser yam addition ratio of 0: 4.

3.2.4. Viscosity. The results showed taro and lesser yam ratio addition does not have a significant effect on the viscosity parameter frozen wheygurt. Table 2 on the viscosity parameter indicates the level of preference panelists ranged from 2.80 to 2.92 (criteria like to neutral). Sample with the highest panelist acceptance response is a sample with the taro and lesser yam addition ratio of 2: 2. While the samples with the lowest acceptance level are a sample with taro and lesser yam addition ratio of 4: 0 and 1: 3.

3.2.5. Overall. The overall parameter is a response which includes a general assessment results panelists that include color, aroma, texture, and taste of a sample[30]. The results showed that taro and lesser yam flours addition ratio have no significant effect on the overall parameters of frozen wheygurt. Table 2 on overall parameter indicates the level of preference panelists ranged from 2.60 to 3.04 (criteria like to neutral). Sample with the highest panelist acceptance response is a sample with taro and lesser yam flours addition ratio of 4: 0. While the sample with the lowest acceptance level is a sample with the taro and lesser yam addition ratio 0: 4.

4. Conclusions

Addition ratio of taro and lesser yam flour effect on the chemical characteristics of frozen wheygurt. Sample with higher lesser yam addition has better chemical characteristic compared to sample with higher taro addition. Addition of taro and lesser yam flour do not affect frozen wheygurt sensory evaluation.

5. References

[1] Siso M I G 1996. *Bioresource Technology* **87** 1-11.

[2] Kumar PA and Bangaraiah P 2014. *International Journal of Pharma and Bio Sciences*. **5** 1101-11

[3] Fatma, Soeparno, Nurliyani, Hidayat C, dan Taufik M 2012. *Agritech* **32** 352-61

[4] Mahrous H and Salam R 2016. *American Journal of Food and Nutrition* Vol.**6** 1-10

[5] Noraziah A Y and Theng C C 2013. *4th International Conference on Biotechnology and Food Science*

[6] Kafah F F S 2012. *Characteristics of Taro (Colocasia esculenta L. Schott) and Its Utilization for Cake Processing (Karakteristik Tepung Talas (Colocasia esculenta L. Schott) dan Pemanfaatannya dalam Pembuatan Cake)*. Skripsi Fakultas Teknologi Pertanian. Institut Pertanian Bogor

[7] Syahbania N 2012. *Study of Taro (Colocasia esculenta) Utilization as Filler Material in Ice Cream Processing (Studi Pemanfaatan Talas (Colocasia esculenta) Sebagai Bahan Pengisi dalam Pembuatan Es Krim)*. Skripsi Jurusan Teknologi Pertanian, Fakultas Pertanian. Universitas Hasanuddin

[8] Richana N and Sunarti T C 2004. Physicochemical Properties Characterization of Tuber Flour and Starch Flour of (Karakterisasi Sifat Fisikokimia Tepung Umbi dan Tepung Pati dari Umbi Ganyong, Suweg, Ubikelapa dan Gembili) *J. Pascapanen* **1** 29-37

[9] Kusumawati A 2013. Program Studi Pendidikan Biologi, Fakultas Keguruan dan Ilmu Pendidikan. Universitas Muhammadiyah Surakarta
[10] Nuraida L Palupi N S Ekasari P D Widayanti N W 2006. *Prosiding Mikrobiologi dan Bioteknologi PATPI*
[11] Prabowo A Y, Estiasih T and Purwantingrum I 2014. *Jurnal Pangan dan Agroindustri* **2** 129-35
[12] Hadioetomo R S 1993. *Basic Microbiology in Practice* (*Mikrobiologi Dasar dalam Praktek)*. Jakarta. PT Gramedia
[13] Maślanka S, Kos A, Bańczyk M, Czopek I, Adam Ł, Dorszewska J, Starczewska K 2015. *Chemik* **69** 4 241–51
[14] Robert L P 1951. *J. Chem. Educ* **28** 5, 252-53
[15] Dwi S, Anton A, Maya P S 2010. *Sensory Evaluation for Food Industry and Agro* (*Analisis Sensori untuk Industri Pangan dan Agro)*. (Bogor : IPB Press)
[16] Lourens-Hattingh A, Bennie C V 2001. *International Dairy Journal* **11**: 1-17
[17] Moongngarm A, Nathanon T, N Sirigungwan 2011. *Adv. J of Food Science and Technol* **3** 4, 269-74
[18] Seah Y N, Loo W C, Birdie S P and Fook Y C 2014. *J of Pharmacy and Nutrition Sci* **4** 92-9
[19] Matthew B, Sue P, Julian M and Eshwar M 2013. *FEMS Microbiol Letter* **349** 2, 77-87
[20] Svetoslav D T and Bernadette D G D M F 2010. *Food Reviews International* **26** 3, 205-29
[21] Syachroni 2014. Jurusan Produksi Ternak, Fakultas Peternakan, Universitas Hasanuddin. Makassar
[22] Isti H and Budi S 2012. *Jurnal Pembangunan Pedesaan* **12** 1, 56-64
[23] Antonia N C 2011. *Jurnal Teknologi Pertanian* **12** 3, 176-80
[24] Karlin R and Rahayuni A 2014. *Journal of Nutrition College* **3** 2, 16-25
[25] Kusnadi J and Afriyan T 2012. *International Confrence on Environmental and Biological Sciences*, Thailand
[26] Yudianto D, and Kusnadi J 2012. Jurusan Teknologi Hasil Pertanian. Fakultas Teknologi Pertanian. Universitas Brawijaya. Malang
[27] Tran V T, Hooi L F, Teck C L and Mohd H B 2011. *African Journal of Biotechnology* **10** 8, 1359-63
[28] Hartoto M 2003. Jurusan Teknologi Pangan dan Gizi. Fakultas Teknologi Pertanian. Institut Pertanian Bogor. Bogor
[29] Bambang K, Puji H, and Supartono W 1988. *Guidelines for Food Sensory Evaluation* (*Pedoman Uji Inderawi Bahan Pangan)*. (Yogyakarta: Gadjah Mada Press)
[30] Wasito W B 2013. Jurusan Teknologi Hasil Pertanian. Fakultas Pertanian. Universitas Sebelas Maret. Surakarta
[31] Abdullah M Rehman S Zubair H Saeed H M Kousar S and Shahid M 2003. *Pakistan Journal of Nutrition* **2** 305-11
[32] Pratama AI 2015. Fakultas Pertanian, Universitas Sebelas Maret. Surakarta
[33] Bodyfelt F, Clark S, Costello M, and Drake M (ed.) 2008. *The Sensory Evaluation of Dairy Products* 2nd Edition. (New York: Springer)
[34] Septiani A H, Kusrahayu, and Legowo A M 2013. *Animal Agriculture Journal* **2** 1: 225-31
[35] Nurwantoro, Sutaryo, Hartanti D, Sukoco H 2009 *J. Indon. Trop. Anim. Agric* **34** 1: 16-21

Acknowledgments
The publication of this paper was supported by Hibah MRG-UNS (PUT-MRG) 2017 from Universitas Sebelas Maret.

Effect of medicinal plants extracts on the incidence of mosaic disease caused by *cucumber mosaic virus* and growth of chili

H Hamidson, N Damiri* and E Angraini

Department of Plant Pests and Diseases, Faculty of Agriculture, Sriwijaya University, Jalan Raya Palembang-Prabumulih, km 32, Ogan Ilir, Indralaya 30662, South Sumatra, Indonesia

*Email: nurhayatidamiri@yahoo.co.id

Abstract. This research was conducted to study the effect of the application of several extracts of medicinal plants on the incidence of mosaic disease caused by *Cucumber Mosaic Virus* infection on the chili (*Capsicum annuum* L.) plantation. A Randomized Block Design with eight treatments including control was used throughout the experiment. Treatments consisted of *Azadiracta indica* (A), *Piper bitle* (B), *Cymbopogon citrates* (C), *Curcuma domestica* (D), *Averroa bilimbi* (E), *Datura stramonium* (F), *Annona Muricata* (G) and control (H). Each treatment consist of three replications. The parameters observed were the incidence of mosaic attack due to CMV, disease severity, plant height, wet and dry weight and production (number of fruits and the weight of total fruits) each plant. Results showed that the application of medicinal plant extracts reduced the disease severity due to CMV. Extracts of *Annona muricata* and *Datura stramonium* were most effective in suppressing disease severity caused by the virus as they significantly different from control and from a number of treatment. The plants medicinal extracts were found to have increased the plant height and total weight of the plant, fruit amount and fruit weight. Extracts of *Curcuma domestica, Piper bitle* and *Cymbopogon citrates* were the third highest in fruit amount and weight and significantly different from the control.
Key words: *Medicinal Plants, infection, CMV, Chili.*

1. Introduction

Planting chili in Indonesia often get virus attacks that cause plants to become mosaic, dwarf and disturbed growth and development so that the production also becomes low. Previous research has found that viruses that attack on lowland chili plants are *Tobacco Mosaic Virus* (TMV) and *Cucumber Mosaic Virus* (CMV). In most planting areas chili, CMV is most dominant and obtained as a single virus that attacks chili plants [1]. CMV has a large host either in the form of wild plants or cultivated plants. There are more than 100 plant families consisting of 500 genera and 1300 plant species that may host the CMV [2] and result in 50-100% yield losses [3], [4], [5]. CMV can be transmitted to healthy plants mechanically or via vector. The CMV can result in fruit being misformed and degrading fruit quality so it is not marketable [6], [7].

The control of the disease is done by controlling its vector as well as pests in chili plants, but this way is besides not effective also gives negative impact on human health and environment. Incorrect use of fungicides in both types and doses often causes problems because they can increase production costs and can leave residuals on production [8]. [9].

Utilization of non-hazardous materials such as the use of medicinal plants has begun and been investigating. It was reported by several researchers proved that medicinal plant extracts can be used as biocontrol agents to control various causes of plant diseases and are able to induce plant resistance [10,11,12,13,14].

2. Material and Methods

The study was conducted on lowland chili cultivation land which was known to be frequently attacked by *Cucumber Mosaic Virus* in Palembang South Sumatra. A Randomized Block Design with 8 treatments of medicinal plant extracts including control was used to conducted the experiment. The treatments consisted of *Azadiracta indica* (A), *Piper bitle* (B), *Cymbopogon citrates* (C), *Curcuma domestica* (D), *Averroa bilimbi* (E), *Datura stramonium* (F), *Annona Muricata* (G) and control (H). Each treatment consisted of 3 replications.

2.1 Preparation of host plant and medicinal plant extract.

Chili seeds were grown until two weeks old and from the seedlings were transplanted to the field that was prepared before.

2.2. The application of medicinal plants Extract.

The application of medicinal plant extracts was done two weeks after the chili seedlings were transferred to the field. The application was carried out by spraying the cultivated land according to the treatment. Spraying done once a week until the first harvest of chili plants. The CMV inoculation was allowed to occur naturally .

2.3 Plant Maintenance.

Plant maintenance includes: watering, weeding and fertilizing. Watering done 1-2 times, a day. Weeding conducted when necessary. Fertilization was done as recommended chili cultivation. Pest control other than CMV's vector was done mechanically. The parameters observed were the incidence of mosaic attack due to CMV, disease severity, plant height, wet and dry weight and production (number of fruits and the weight of total fruits yield) each plant. The analysis of variance (ANOVA), with the Duncan's Multiple Range Test (DMRT) comparison among means [15].

3. Result and Discussion

3.1. Result

3.1.1. Disease occurence. Symptoms of CMV attack begin to appear at week two after application but only on six treatment and control. Symptoms of leafy leave become light green and yellow but still light.

3.1.2. The severity of the disease. The results of analysis of the effect of medicinal plants extract application on the disease severity of CMV on chili plant at the end of the study were presented at Table 1.

Table 1 shows that *Annona muricata* and *Datura stramonium* extracts were most effective in suppressing disease severity caused by the *Cucumber Mosaic Virus* although not significantly different from *Piper bitle* plant extracts, *Cymbopogon citrates*, *Curcuma domestica* and but very different from *Azadiracta indica* and control.

Table 1. Effects of medicinal plants extract application on CMV attack severity on the chili plant in the field.

Treatment	Disease Severity (%)
Control	50.00 a
Azadiracta indica (A)	50.00 a
Piper bitle (B)	15,27 ab
Cymbopogon citrates (C)	12.50 ab
Curcuma domestica (D)	24.98 ab
Averroa bilimbi (E)	16.66 ab
Datura stramonium (F)	8.33 b
Annona muricata (G)	1.42 b

The figures followed by the same letter in the same column mean there is no significantly different at $p \leq 0.05$ DMRT.

3.1.3. Height and weight of plant. Result of analysis of the effect of plant extract application on height and the weight of the plants showed a noticeable difference. Further test results were presented at Table 2.

Table 2. Effects of medicinal plant extract application on height and weight of chili plant.

Treatment	Plant height (cm)	Weight plant (g)	
		Wet	dry
Piper bitle (B)	45.75 a	54.23 a	31.66 a
Averroa bilimbi (E)	48.33 a	50.23 ab	29.36 ab
Annona muricata (G)	49.71 a	47.90 abc	27.53 ab
Curcuma domestica (D)	50.29 a	45,05 abc	24.80 abc
Datura stramonium (F)	48.16 a	43.53 abc	22.60 abc
Cymbopogon citrates (C)	49.44 a	43.06 abc	22.53 abc
Azadiracta indica (A)	50.48 a	35.83 bc	18.33 bc
Control	35.66 b	33.93 bc	15.76 c

The figures followed by the same letter in the same column mean there is no different at $p \leq 0.05$ DMRT.

Table 2 shows that all medicinal plant extract treatments were capable of promoting the growth and development of chili plants. The height of chili plant which is given medicinal plant extract all different with the control. The largest wet weight and dry weight obtained in the *Piper bitle* extract treatment of 54.50 g and 31. 66 g was significantly different from *Azadiracta indica* and control treatment.

3.1.4. Production. Highest chili production was shown by *Curcuma domestica*-treated plants with average amount of 16.33 fruits/plant and average weight of 26.17 g / plant, followed by *Piper bitle* of 15 fruit / plant with weight of 23.97 g/plant. All medicinal plant extract treatments showed a tendency to increase the production of pepper plants, but that was significantly different from the controls shown only by plants extracted by *Curcuma domestica*, *Pipper bitle* and *Cymbopogon citrates* (Table 3).

Table 3. Effect of medicinal plant extract application on chili production in the field.

Treatment	Production/plant	
	Amount of fruit	Weight of fruit (g)
Curcuma domestica (D)	16.33 a	26.17 a
Piper bitle (B)	15.00 ab	23.97 ab
Cymbopogon citrates (C)	14.67 ab	23.23 ab
Datura stramonium (F)	13.00 abc	20.70 abc
Averroa bilimbi (E)	12.67 abc	20.13 abc
Azadiracta indica (A)	11.67 abc	18.03 abc
Annona muricata (G)	7.00 bc	11.23 bc
Control	5.33 c	8.63 c

The figures followed by the same letter in the same column mean there is no different at $p \leq 0.05$ DMRT.

3.2. Discussion

Phytochemicals are non-nutrient plant chemicals that have protective or disease prevention properties. Plants produce these chemicals to protect themselves but recent research proves that many phytochemicals can protect plants against disease. According to [16] and [17], plants produce secondary metabolites of plants, tannins, saponins, steroids, alkaloids and glycosides, lignin, cyanogenic, flavonoids, phenolic in their cells. It was reported that long chain chains (C_6-C_{10}) in plant extracts have been observed to have greater antifungal properties [18].
From the results of this study, there are 3 plant extracts that have the potential to be developed as a vegetable pesticide to control mosaic disease caused by *Cucumber Mosaic Virus*. *Curcuma domestica*, although only able to suppress the severity of CMV disease 50% but increase the growth of chili plant well, this is indicated by the growth and production of chili plants. Suspected phenolic content of this extract is able to suppress the development and replication of viruses in the plant. Curcumin exhibits antibacterial, antifungal and antiviral activity due to its phenolic content. Intervention of phenolic compounds by hydrogen bonding and hydrophobic interactions with microbes can make it more dangerous [19], [20].

Turmeric phamacoligional activity is closely related to curcumonoid consisting of Curcumin (CUR) and two compounds namely demothoxy curcumin (DMC) and Bisdementhoxy cucurmin. CUR interferes with the attachment of viral cells that cause stunted spread of virus in plants. Phenols in plant products are considered potentially toxic to the growth and development of pathogens and thus reduce plant diseases [21]. Phenolic metabolism serves to protect plants from biological and environmental pressures and is therefore synthesized. In response to pathogenic pathogens such as, fungi, virus, bacteria or other pathogenic agents [22].

The number of vectors that was *Aphis gossypii* found highest in control plants and plants treated by *Azadiracta indica* was an average of 40 to 100 Aphids per leaf, while in other treatment was just unde 10 Aphids per leaf. The difference of the number of Aphids was suspected due to the environment influence or location condition, According to [23] nutrient and water are an environmental factor that is crucial because most viruses require a host metabolism active for propagation purposes. Aphids beside as a pest are also a vector for the mosaic virus such as CMV. These insect are able to attack and quikly transmit the virus from infected plants to healthy plants. This vector is capable of transmitting the virus within 5 t0 10 seconds per day and translocating to other plants less than 1 minute [24]

4. Conclusion

Results of the study showed that the application of medicinal plant extracts in this study reduced the disease severity due to CMV. Extracts of *Annona muricata* and *Datura stramonium* were most effective in suppressing disease severity caused by the virus as they significantly different from control and from a number of treatment. The medicinal plants extracts were found to have increased the plant height and total weight of the plant, fruit amount and fruit weight. Extracts of *Curcuma domestica, Piper bitle* and *Cymbopogon citrates* were the third highest in fruit amount and weight and significantly different with control.

5. Acknowledgement

Authors would like to address their gratitude to those whose contributions have made possible the writing of this paper. This study was partly results of the competitive research funded by The research Institute of Sriwijaya University, Ministry of Research, Technology and Higher Education, Republic of Indonesia with contract number:1012/UN9.3.1/PP/2017

References

[1] Damiri, N., Mulawarman., Hamidson, H and Rahim, SE. 2016. Mosaic disease and chili production on different altitude in South Sumatra, Indonesia. *Proceeding of The USR International Seminar on food security (UISFS)*. Bandar Lampung, Indonesia, August 23-24. **1**:107-116.

[2] Ashfaq, M., Iqbal, S., Mukhtar, T and Shah, H. 2014. Screening for resistance to Cucumber Mosaic Virus. Cucumovirus in chili. *Journal of Animal and Plant Science*, **24**(3):791-795..

[3] Damiri, N. 2014. Mixed viral infection and growth stage on pepper (*Capsicum annum* L.) production. Pertanika. *Journal of Tropical Agricultural Science*. **37** (2): 275-283.

[4] Akbar, A., Ahmad, Z, Gegum, F, Ubairah, Raees, N. 2015. Varietal reaction of cucumber againts Cucumber Mosaic Virus. *American Journal of Plant Science*, **6**:833-838.

[5] Chen, C.C., Hsu, H.T., Chiang, F.L and Chang C.A. 2006. Serological and molecular properties of five potyvirus infecting calla lily. *Proc XIth IS on Virus disease ornamental Acta Hort*:**722**:259-264.

[6] Palukaitis, P., M. J. Roossnck., R.G. Dietzgen and R. B. Francki. 1992. Cucumber mosaic virus adv.*Virus. Res.* **41**: 281-346.

[7] Subekti, D. 2005. Evaluation of five pepper cultivars resistance towards the infection of Cucumber mosaic cucumovirus and Pepper vein mottle potyvirus: Effects of single and mix infection on pepper yield. *Tesis of Postgraduate studies of Bogor University of Agriculture.*

[8] Astuti, U.P., Wahyuni, T and Honorita, B. 2013. Technical guidelines for the manufacture of vegetable pesticides. *Agricultural Technology Assessment Center (BPTP) Bengkulu. Bengkulu.* 75 p.

[9] Ginting, C. 2006. Development of Uredospora *Hemilieia vastatrix* on ginger and turmeric extracts and cloves and betel leaves. *J. Tropical Plant Pests and Diseases* **6** (1): 52-58

[10] Shabana, Y.M., Abdalla, M.E., Shahin, A.A., El-sawy, M.M., Draz, I.S and Youssif, A.W. 2016. Efficacy of plants extracts in controlling eheat leaf rust disease caused by *Puccinia triticina*. *Egyptian journal of Basic and Applied Science*. **4**:67-73.

[11] Ayoub, M. and Niazi, A.U. 2001. Control of wheat rust by leaves extract of poisonous phanergamic plants. *J. Biol. Sci.* **1**:490-1

[12] Madhusudhan, K.N, Nalini, M.S, Prakash, H.S and Shetty, H.S. 2005. Effect of inducer against tobamovirus infection on tomato and pepper. *Int. J. Bot.* **1**:59-61.

[13] Chakraborty, D and Chakraborty, S. 2010. Bioassay guides isolation and identification of antibacterial and antifungal components from methanolic extract of green tea leaf (*Camellia sinensis*). *Res. J. Phytochem*. **4**:78-86.

[14] Srivasata, S., Sing, V.P., Kumar, R., Srivasata, M., Sinha, A., Simons, S. 2011. In vitro evaluation of carbendazim 50WP, antagonists and botanicals against *Fusarium oxysporum* f, sp. Psidii associated with rhizosphere soil of guava. *Asian J. Plant. Pathol.* **5**:46-53.

[15] Gomez, K.A. & Gomez, A.A. 1984. Statistical procedures for agricultural research (2nd edn.). *An international rice research institute book. York-Chichester-Brisbane-Toronto-Singapore. A Wiley-Intersci. Publ, John Wiley and Sons.*

[16] Harbone, J.B. 1998. Phytochemical methods. A guide to modern techniques of plant analysis. Chapman and Hall. London. Pp 182-190.

[17] Shetty, K. 1997. Biotechnology of harness the benefit of dietary phenolics: focus on Lamiaceae *Asia Pasific Journal of Clinical Nutrition*, **6**, 162-171.

[18] Holley, A.H and Patel, H. 2005. Improvement in self life and safety of perishable food by plant essential oils and smoke antimicrobial. *International Journal of Food Microbiology* **22**, 273-292.

[19] Shailendiran, D, Pawar, N, Chanchal, A, Pandey RP, Bohidar, HB., Verma, AK. 2011. Characterization and antimicrobial activity of nano curcumin and curcumin. In Nanoscience, Technology and Societal Implications (NSTSI). *International Conference on IEE*.pp.1-7.

[20] Gul, F.Z and Basheer, M. 2016. Curcumin as natural bioactive compound of medicinal plant Curcuma longa to combat against different disease. *Journal of Ayurvedic and Herbal Medicine*; **2**(5):192-199.

[21] Okwu, D.E and Okwu, M.E. 2004. Phytochemicals, vitamins and mineral contents of two Nigeria medicinal plants. *International Journal of Moleculer medicine and Advance Science* **1**, 375-381.

[22] Vattem, DA and Shetty, K. 2003. Ellagic acid production and phenolic anti oxidant activity in Cranberry lentinus (*Vaccinium macrocarpan*) mediated by *Lentinus edodes* using solid-state system. *Process Biochemistry* **34**, 367-379.

[23] Bos, L. 1994. Introduction to Plant Virology. Translated by Triharjo. Gajah Mada University Press.

[24] Dixon, G.R. 1981. Vegetable crop disease. Fist American Edition. The AVI Publishing Company.Inc. Weatport, Conecticut, Hongkong.

Exploration of *Bacillus thuringiensis* Berl. from soil and screening test its toxicity on insects of Lepidoptera order

DT Astuti[1], Y Pujiastuti[1], Suparman SHK[1] , N Damiri[1*], S Nugraha[2], ER Sembiring[2], Mulawarman[1]

[1]Program study of Crop Science, Plant Protection Department, Post graduate Program, University of Sriwijaya, Inderalaya Ogan Ilir, South Sumatra, Indonesia
[2]Plant Protection Department, Post graduate Program, University of Sriwijaya, Inderalaya Ogan Ilir, South Sumatra, Indonesia

*Email: nurhayatidamiri@yahoo.co.id

Abstract. *Bacillus thuringiensis* is a gram-positive bacterium that produces crystal proteins toxic (ᵹ-endotoxin) specific to the target insect, but is not toxic to humans and non-target organisms. This study aims to explore the origin of the soil bacterium *B. thuringiensis* sub-district Sekayu, Banyuasin, South Sumatra and toxicity to larvae of lepidoptera. Fifty soil samples were taken from Musi Banyuasin District, namely 15 from Kayuare strip 2, 20 from Kayuare and 15 from Lumpatan. Isolation, characterization, identification and screening test were conducted in the laboratorium of Pest and Disease, Agricultural Faculty, Sriwijaya University. Isolat codes were given based on the area origin of the samples. Results of the study showed that from 50 isolates of bacteria that had been isolated, there were 15 bacterial isolates, characterized by morphology and physiology the same as B. *thuringiensis*, which has round colonies, white, wrinkled edges, slippery, elevation arise, aerobic and gram-positive. Of the 15 codes that contain positive isolates of B. *thuringiensis*, we have obtained several isolates of the following codes: KJ_2D_5, KJ_2N_1, KJ_2N_4, KJ_2B_3, KJ_3R_1, KJ_3R_2, KJ_3R_3, KJ_3R_5, KJ_3J_3, KJ_3J_4, KJ_3P_1, DLM_5, $DLKK_{12}$, and $DLKK_{23}$. Results of screening tests on insects of the Lepidoptera Order showed that there were six isolates that had toxic to *Plutella xylostella* and *Spodoptera litura* insects, ie bacterial isolate codes DLM_5, KJ_3R_3, KJ_3R_5, KJ_3J_4, KJ_3P_1, and $DLKK_{23}$.

Keywords: *Exploration, Bacillus thuringiensis, Lepidoptera. isolate*

1. Introduction

Bacillus thuringiensis is a type of aerobic bacteria, spore formers, which is gram positive and is an indigenous bacteria in soil, water, plant surface, and dead insects. *B.thuringiensis* has a specific target so that it does not kill insects that are not targeted and easily decompose, and do not accumulate and pollute the environment [1]. B. thuringiensis has a protein crystal that is toxic to insects. This toxin protein is first known as parasporal crystalline inclusion but hereinafter referred to as ᵹ-endotoxin or otherwise known as Insecticidal Crystal Protein [2]. Strains of *B. thuringiensis* bacteria exhibit a wide range of specificity in various insect orders such as Lepidoptera, Diptera, Coleoptera. These bacteria produce crystalline proteins (cry proteins) during sporulation. Insects infected by *B. thuringiensis* bacteria will cause symptoms of no appetite, slow movement, and over time the insects will die, due to the crystals of proteins that dissolve in the insect's stomach.

In vegetable crops, especially plants belonging to the family brassicaceae, many types of pests that can attack the plant include *Plutella xylostella* L. (Lepidoptera: Plutellidae), *Crocidolomia pavonana* Fab. (Lepidoptera: Pyralidae), *Spodoptera litura* Fab, (Lepidoptera: Noctuidae), *Helicoverpa armigera* Hubner (Lepidoptera: Noctuidae), *Chrysodeixis orichalcea* L. (Lepidoptera: Noctuidae), *Liriomyza sp.* (Diptera: agromyzidae) and *Myzus persicae* Sulz (Homoptera: Aphidoidae) [3]. *P. xylostella* and *S. litura* is a kind of leaf-eating caterpillar pests most commonly attack plants and vegetables can cause damage to approximately 12.5% [4]. In overcoming pest attacks, farmers often use pesticides that contain active synthetic. Many of the negative impacts arising from the unwise use of chemical pesticides. Among them is the occurrence of resurgence or blasting of pest populations, resistance, death of natural enemy population and high residue level on agricultural products so it is not safe for public consumption [5]. Therefore it is necessary to find alternatives in controlling crop pests. One alternative that can be done to replace a chemical insecticide is to use the biological control of the use of living things in the form of microorganisms.

Given the superiority of *B. thuringiensis* as a pest controlling agent, the search for new strains, specific to the pathogen of a particular pest is necessary. This study aims to explore the origin of the soil bacterium Bacillus thuringiensis sub-district Sekayu, Banyuasin, South Sumatra and toxicity to larvae of Lepidoptera. Utilization of microorganisms is very helpful in exploiting microbial agents biological controller is a breakthrough in enhancing the utilization of biological resources more intensively and save the environment from pollution.

2. Material and Methods

The research was conducted in Phytopathology Laboratory of Department of Plant Protection, Faculty of Agriculture Sriwijaya University, Indralaya, Ogan Ilir, South Sumatra, Indonesia, from March to May 2017. The materials used in this research are 50 soil samples taken from Musi Banyuasin district of Sekayu sub-district. The soil samples were 15 taken from Kayuare strip 2, 20 from Kayuare strip and 15 from Lumpatan village. Accordingly, codes of isolates were given on the basis of the origin locality names of samples. Other materials were larvae of *Plutella xylostella* instar III, larvae of *Spodoptera litura,* Nutrient Agar (NA) medium, 3% KOH solution, 5% H_2O_2 solution, Crystalline Violet, Safranin, 95% Alcohol, 5% malachit green solution, Aquadest. Leaves of Caisin as larvae feed. The tools used in this research are aluminum foil, autoclave, bunsen, petri dish, erlenmeyer, measuring cup, isolation, ose needle, digital camera, label paper, laminar air flow, microphone, microscope, analytic balance, oven, tweezers, shaker, Spatel Drygalski, test tube.

2.1 Isolation of bacteria from soil.

The isolation of *B. thuringiensis* from the soil sample from Sekayu subdistrict was then performed in accordance with [6] procedures., which was diluted by weighing 1 g of soil and then put into a reaction tube containing 9 ml of aquadest then dishaker for 5 minutes, the suspension was diluted Until 10-6 is then taken as much as 0.1ml to be grown in a Petri dish filled with Nutrient Agar (NA) medium and leveled on all surfaces of the medium using Splatel Drygalski. Incubation is carried out for 48 hours at room temperature of 28 ° C, until the colony appears to have the same morphology, color, similar to *B. thuringiensis.*

2.2 Characterization of bacteria.

Characterization of Bacillus thuringiensis bacteria was performed by conducting catalyst test, gram reaction test, and microscope observation. The results of the catalase test, and the gram reaction test are presented in the form of descriptive data.

2.2.1 Catalyst Test.

The catalyst test was performed by taking a colony using needle that has the same morphological characteristics as *B.thuringiensis*, scraped on the preparation and then dropped with 5% H_2O_2 solution then stirred. In bacteria that have gram positive properties will produce air bubbles (O_2).

2.2.2 Gram reaction test. Gram reaction test is done using a needle of bacterial colony ose then put on glass preparations that had been spilled with 3% KOH solution, if the bacteria is not slimy then the bacteria is a positive bacteria that is characteristic of bacterium *B. thuringiensis.*

2.2.3 Gram staining. Gram staining is done to determine the characteristics of bacteria that are obtained gram positive bacteria or gram negative. Gram staining is done by taking 1 scratch of bacterial isolate and then placed on top of glass preparatory, flooding the bacteria with the primary dye that is crystal violet for 1 minute, then tilt the glass object and throw excess dye, then rinse with water. Next flush the odor of bacteria with iodine for 2 minutes remove excess dye, then rinse with water. Blanch with 95% alcohol then rinse with water. Then flush the bacteria with safranin for 30 seconds, remove the excess dyes and rinse with the last water to absorb the remains of water with absorbent paper, then observe under what staining microscope will be produced by the bacteria.

2.3 Selection of B. thuringiensis bacteria.
Of 10 replicated soil samples of 5 times that have been grown, which has the same characteristics as *B. thuringiensis* was selected as many as 15 isolates that have the same morphology and physiology of *B. thuringiensis* bacteria. Furthermore, in the screening test of lepidoptera larvae.

2.4 Test Screening on insects.
Screening test on exploratory bacterial strains that have been morphologically tested and microscopic observations based on spore form and spore staining. In each strain of bacteria that has demonstrated morphological properties as well as microscope observations as *B. thuringiensis*, then taken 1 ml then mixed with 20 ml aquadest. Furthermore the plant leaves are used as feed for tested insect larvae and soaked in a suspension containing 20 ml of aquadest and *B. thuringiensis* bacteria until the leaves are wet all then raised and dried. Plant leaves are put into plastic petri dish (15 cm x 15 cm) coated with paper. Furthermore, 5 larvae of insect test larvae were infected to plant leaves in plastic petri dish 15 cm x 15 cm which have been given treatment. To find isolates that are toxic to lepidoptera larvae.

3. *Result and Discussion*
3.1 Isolatation B. thuringiensis from soil samples in Musi Banyuasin Regency.
B. thuringiensis is a genus of bacteria capable of forming a dormant structure that is endospores that can be resistant to chemicals as well as physical treatments such as heat, UV and dry. Heating Treatment 800C For 10 minutes on this isolation method. It is intended that other microbes except the endospores will die. Thus only thermo-tolerant bacteria forming spores grow, and from these types of bacteria only aerobic bacteria alone can grow because the culture is incubated aerobically.

The isolation result from 50 soil samples from 3 urban villages in Sekayu subdistrict, found 15 bacterial isolates that have morphological characteristics of *B. thuringiensis*. Observations with a contrast phase microscope, these isolates show characteristics of bacteria known as biological agents (Table 1).

Table 1. Number of soil samples and *B. thuringiensis* isolate in sub-district Sekayu South Sumatra

Village	Soil sample amount	Isolat *B. Thuringeinsis amount*
1. Kayuare strip 2 *	15	5
2. Kayuare strip **	20	7
3. Lumpatan ***	15	3
Total	50	15

Legend : *= Desa kayuare Strip 2 (KJ2)
**= Desa kayuare Strip 3 (KJ3)
***=Lumpatan Village

3.2 Identification and Characterization of B. thuringiensis

From result of research which have been done result of isolation of *B. thuringiensis* from soil sample from Sekayu Subdistrict after done incubation process until emerge colony having similarity of morphology and physiology. The results showed the same morphological characteristics of white and yellow rounded colonies, wavy and slippery edges, and elevated elevations (Table 2).

Table 2. Identification of morphology and bacterial physiology of *B. Thuringiensis*

Isolate Code	Morphological test				Biochemistry test		Miscroscopic test	
	Form	Colour	Edge	Elevation	Reaction test gram	Catalyste test	P. Gram	P. Spora
KJ2D 5	Round	White	slippery	arise	positive	positive	Purple	Green
KJ2N 1	Round	White	Wavy	arise	positive	positive	Purple	Green
KJ2N 2	Round	White	Wavy	arise	positive	positive	Purple	Green
KJ2N 4	Round	White	Wavy	arise	positive	positive	Purple	Green
KJ3R 1	round	white	Wavy	arise	positive	positive	Purple	Green
KJ3R 2	round	Yell-white	slippery	arise	positive	positive	Purple	Green
KJ3R 3	round	white	slippery	arise	positive	positive	Purple	Green
KJ3R 5	round	White	slippery	arise	positive	positive	Purple	Green
KJ3J 3	round	White	Wavy	arise	positive	positive	Purple	Green
KJ3J 4	round	White	Wavy	arise	positive	positive	Purple	Green
KJ3P 1	round	White	slippery	arise	positive	positive	Purple	Green
DLM 5	round	white	Wavy	arise	positive	positive	Purple	Green
DLKK1 2	round	white	slippery	arise	positive	positive	Purple	Green
DLKK2 3	round	white	slippery	arise	positive	positive	Purple	Green

B.thuringiensis bacterial colonies that grow on the media have a round, white, slimy shape, smooth edges and raised elevation (Fig. 1). According to [8], the bacterial colonies that grow on the media on a round shape, magnitude 5-10 mm, has a white color. At the edge of this bacteria is slightly shrunken or bumpy. The bacteria have an embossed elavance and a rough surface. The bacterial colonies that grow on the first day are very small and almost flat with the media, then on the second day the bacteria will enlarge and somewhat wide but not in touch with one another. On the third day bacteria that grow on the first day will enlarge. Enlargement of bacterial elevation will be evident in solid media.

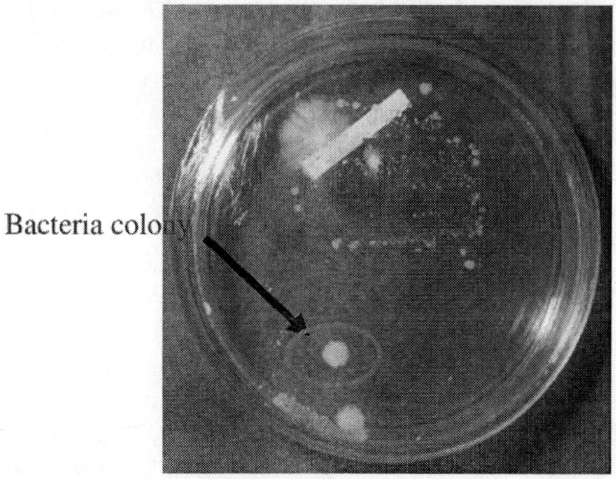

Figure 1. Morphology of colonies, colonies of bacteria growing on white medium such as bacillus colonies

IOP Conf. Series: Earth and Environmental Science **102** (2017) 012063 doi:10.1088/1755-1315/102/1/012063

The result of bacterial isolation that has been grown on Sodium Agar (NA) medium of growing colony morphologically observed shows characteristic as *B. thuringiensis* that is round, white and yellowish white, wrinkled and slippery edge, and elevation of the colony arises. Gram reaction done in the study showed a gram-positive reaction. This is because if the bacteria given KOH 3% on bacterial cultures placed on the glass preparat not participate lifted on the needle ose. If the bacteria is lifted when lifting the needle ose then the bacteria is gram negative. In this case, the untreated B.thuringiensis bacteria not raised can be said to be gram positive bacteria. The catalase test is a test where when the bacteria tested produces froth after a 5% Hydrogen Peroxide (H_2O_2) solution is given, the bacteria is gram negative. If the bacteria after given a solution of Hydrogen Peroxide (H_2O_2) 5% does not produce froth the bacteria are gram positive.

The catalase test carried out by putting a needle of bacterial culture of *B.thuringiensis* on the preparatory glass which had been dropped by 5% Hydrogen Peroxide (H2O2) solution did not produce foam then the bacterium was positive. In this bacteria after testing does not produce foam means the bacteria is a positive bacteria. This bacterium is a bacterium belonging to the form of bacil. Based on the results of gram tests that have been done on bacteria B. thuringiensis in can if this bacteria is gram positive that is visible from the color of spores produced purple, this is because bacteria B. thuringiensis has a thick peptidoglycan so it can maintain the basic color when done gram staining (Figure 2).

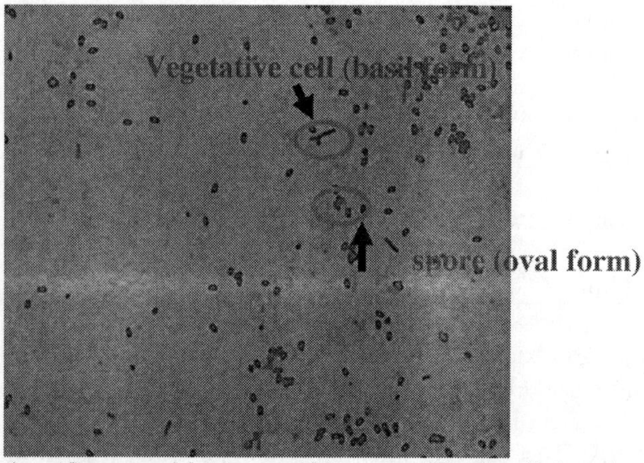

Figure 2. Gram staining (Gram positive) on a 400 x magnification light microscope

B. thuringiensis bacteria are present in all purple isolates after coloring by using a violet crystal reagent. This is evidenced by the presence of the same bacterial cells in each of the diluted isolates. Basil-shaped bacterial cells. At the beginning of the growth of stem-shaped bacteria or bacillus cells then the bacterial cells will be split and oval-shaped. According to [7], bacteria *B. thuringiensis* will form spores aerobically and at the time of sporulation these bacteria will form protein crystals. The protein crystal formed is known as delta-endotoxin.

3.3 Screening test of Lepidoptera larvae.

Of the 15 codes of isolates that have been known morphology and physiology containing *B. thuringiensis*. Furthermore, a screening test was conducted to see the toxicity of Lepidoptera insects. Test results are shown on (Table 2).

Based on the results of research that has been done found that from 15 isolates that have been tested by taking bacterial cells as much as 10^{-6} from the results of dilution, there are 6 isolates that died due to treatment. According to [9], states that with bacterial cells *B. thuringiensis* 1 x 10^6 cells / ml more virulence to targeted insects. Isolate B. thuringiensis in cells 1 x 106 cells / mL caused a high mortality in killing targeted insects. Cells owned by bacteria B. thuringiensis will enter the insect's large intestine causing the appetite to experience less appetite, the movement of the larvae becomes sluggish. Dead insects change color to black after B infected larvae thuringiensis, the average mortality of larval mortality occurs after 72 hours of application or 3 days. This is in line with the [10] which states that the symptoms of larvae infected by *B.thuringiensis* are test larvae changed the treatment to be slow and eventually stop moving and sometimes remove the green liquid from the mouth, then the dirt becomes watery (diarrhea) , And finally the larvae will die. And the color of the larva turns dark or brownish black and the body becomes soft.

Table 3. Results of screening of toxicity of bacteria Isolate against Lepidoptera larvae for 168 hours (7 days)

Isolate code	Larva Lepidoptera	
	P. xylostella	*S. litura*
KJ_2D_5	Negative	Positive
KJ_2N_1	Negative	Positive
KJ_2N_2	Negative	Positive
KJ_2N_4	Negative	Positive
KJ_2B_3	Negative	Positive
$KJ3R_1$	Negative	Positive
KJ_3R_2	Negative	Positive
KJ_3R_3	Positive	Positive
KJ_3R_5	Positive	Positive
KJ_3J_3	Negative	Positive
KJ_3J_4	Positive	Positive
$K_{J3}P_1$	Positive	Positive
DLM_5	Positive	Positive
$DLKK_{12}$	Negative	Positive
$DLKK_{23}$	Positive	Positive

Description: Negative (live insects / not toxic to treatment)
Positive (dead insects / toxic to treatment).

According to [11], the larvae infected by *B. thuringiensis* will die, the body color becomes blackish and the body is soft. When touched the skin of the larvae will break and remove the black liquid and foul-smelling. The color change in the body of the larvae that died from the B. thuringiensis fluid turned into black due to bacteria growing up to the haemokoel part so that the blood cells become poisoned [12,13].

4. Conclusion

From this study it can be concluded that from 50 isolates of bacteria which had been isolated, there were 15 bacterial isolates, characterized by morphology and physiology the same as B. *thuringiensis*, that had round colonies, white, wrinkled edges, slippery, elevation arise, aerobic and gram-positive. Of the 15 codes that contained positive isolates of B. *thuringiensis*, we obtained several isolates of the following codes: KJ_2D_5, KJ_2N_1, KJ_2N_4, KJ_2B_3, KJ_3R_1, KJ_3R_2, KJ_3R_3, KJ_3R_5, KJ_3J_3, KJ_3J_4, KJ_3P_1, DLM_5, $DLKK_{12}$, and $DLKK_{23}$. Results of screening tests on insects of the Lepidoptera order showed that there were six isolates that had toxic to *Plutella xylostella* and *Spodoptera litura* insects, ie bacterial isolate codes DLM_5, KJ_3R_3, KJ_3R_5, KJ_3J_4, KJ_3P_1, and $DLKK_{23}$.

5. Acknowledgement

The authors would like to address their gratitude to those whose contributions had made possible the writing of this paper. This study was partly results of the Profession research funded by The research Institute of Sriwijaya University, Ministry of Research, Technology and Higher Education, Republic of Indonesia with contract number:987/UN9.3.1/PP/2017.

References

[1] Hermanto, S. Jusuf, E. Hero, MS. 2013. *Exploration of Toxin Protein Bacillus thuringiensis from Land in Tangerang Regency. Valence.* **3** (**1**). Page 58-56.

[2] Jusuf, E. 2008. *Culture Collection of Potential Bacillus thuringiensis Bacterial Strains Insect Killer and the Making of a Library of Toxic Protein Coding Genes. Technical Report LIPI Biotechnology Research Center Year 2008.* pp. 18-31.

[3] Supartha NP E. Y, I Wayan S., and Ketut A. Y. 2014. *Diversity and density parasitoid associated with Plutella xylostella L. (Lepidoptera: Plutelidae) In Non-Application Cabbage Plant and Insecticide Application. J. Tropical Agroecotenology.* **3** (1): 12-21.

[4] Sriniastuti, 2005, *Effectiveness of Bacillus thuringiensis on Leaf Caterers Attacks (Plutella xylostella) on Sawi (Brassica juncea) in Sungai Selamat, Thesis, Faculty of Agriculture, Tanjungpura University, Pontianak.*

[5] Bravo, A., Likitvivatanavong, S., Gill, S., Soberón, M., 2011. *Bacillus thuringiensis: a story of a successful bioinsecticide. Insect Biochemistry and Molecular Biology.* **41**: 423-431.

[6] Rusmana, I and Hadioetomo, R.S. 1994. *Isolation of Bacillus thuringiensis Berl. from silkworm farms and their toxicity to the larvae of Crocodolomia binotalis Zell and spodoptera litura* F. *Hayati* **1** (**1**): 22-23.

[8] Rusman R. 2000. *Philosophical Science (Pps 709). Graduate / Doctoral Program. Bogor Agricultural University.*

[9] Khaswar, Rahayuningsih M, and Yulianti. 2001. *Effect of Aeration on Bioinsecticide Production by Bacillus thuringiensis Subsp. Israelensis on Bioindicator of Stirred Tank and Wave Column. Department of Industrial Technology of Agriculture, Faculty of Agricultural Technology. Bogor Agricultural University*, volum **11** (**3**), 92-100.

[10] Pujiastuti Y.2013. *Bioinsecticide Toxicity Based on Bacillus thuringiensis Origin of Ogan Ilir Ground of South Sumatra to Plutella xylostella (Lepidoptera: Plutellidae). Proceedings of the National Seminar and Annual Meeting of Dean for Agricultural Sciences BKS-PTN West Area Year 2013.*

[11] Khaeruni, A. Rahayu. *Purnamaningrum NT. 2012. Isolation of Bacillus thuringiensis Berl from Soil and its pathogenicity to Crosidolomia binotalis Zell larvae on mustard plant Brassica juncea L. Journal of Agroteknos.* **2** (**1**). Page 21-27.

[12] Poinar.G.O. And G.M. Thomas. 1982. *Diagnostic Manual for the Identification of Insect.* Pathogen. *Press. New York.*

[13] Salaki, C., Situmoran, J., Sembiring, L., and Handayani, N.S. 2010. *Isolation and characterization Indonesian indigeneous bacteria (Bacillus thuringiensis) potentially as biological pest controlling insect (Crocidolomia binotalis). Journal of Agricultural Science AGRIVITA.* **31** (**2**): 174-181.

[14] Salaki, C. 2010. *Genetic Variety Isolate Bacillus thuringiensis Berliner Endogen Indonesia As A Pest Controlling Biologist Crocidolomia binotalis Zell. (Lepidoptera; Pyralidae) On Cabbage Plant. Dissertation. Faculty of Biology Gadjah Mada University. Yogyakarta*

Characterization of *Bacillus thuringiensis* Berl. indigenous from soil and its potency as biological agents of *Spodoptera litura* (Lepidoptera: Noctuidae)

Y Pujiastuti[1*], D T Astuti[2], S R Afriyani[2], S Suparman[1], C Irsan[1], E R Sembiring[3], S Nugraha[3], Mulawarman[1] and N Damiri[1]

[1]Department of Plant Protection, Faculty of Agriculture, Sriwijaya University, Jalan Padang Selasa, Palembang, Indonesia

[2]Graduate School of Agriculture, Faculty of Agriculture, Sriwijaya University, Jalan Padang Selasa, Palembang, Indonesia

[3]Genomic and Crop Improvement Laboratory, Research Center for Biotechnology-Indonesian Institute of Sciences (LIPI), Jalan Raya Bogor Km. 46 Cibinong, Bogor, Indonesia

[*]Email: ypujiastuti@unsri.ac.id

Abstract. The objective of the study was to characterize the molecular weight of protein in order to be utilized as biological agent of *S.litura* and their cell or spores production. To investigate the molecular weight of protein was done by SDS-Page electrophoresis. Growth media used for producing *B. thuringiensis* were T3, LB broth and agricultural waste. The results showed that the molecular weight of protein ranged from 37 to 140 kDa. In DLM and DLKK23 isolates were found ranging from 37 to 40 kDa and from 110 to 130 kDa, respectively. KJ3R5 and KJ3P1 isolates were obtained having three protein bands ranging from 43 to 45, 73 to 80 and 110 to 130 kDa and 45-50, 75-80 and 130-140 kDa, respectively. It was predicted that isolates *B. thuringiensis* were belonging to Cry IA, Cry IIA, Cry IVC and Cry15c. These crystal proteins were toxic to *S. litura*. There was no protein bands found in the two last isolates (KJ3R3 and KJ3J4). Production of spore after sporulation in agricultural waste media ranged from 0.5 to 10^6 - 2.67 x 10^7 spores/ml showing medium level of toxicity to *S. litura*.

Key words : *control, electrophoresis, entomopathogene, pest, protein.*

1. Introduction

Bacillus thuringiensis is an entomopathogenic bacteria, which can produce protein and spores at the sporulation. The protein often refers to Cry toxin, because its shape resembles crystals. Cry toxins plays a role in midgut epithelial cells by inserting into the target membrane and forming pores [1]. These have a potency to control a number of insect pests belonging to Lepidoptera, Diptera and Coleoptera and are safe to the ecosystem as they do not kill any useful insect, natural enemies and human being [2 - 4]). Crystal protein was the main substance which toxic to insect pest, therefore this crystal should be isolated. The amount and molecular weight of *B. thuringiensis* crystalline proteins (Cry) was estimated by the sodium dodecyl sulphate polyacrylamide gel electrophoresis (SDS-PAGE) [5].

IOP Conf. Series: Earth and Environmental Science **102** (2017) 012064 doi:10.1088/1755-1315/102/1/012064

Recent study showed new indigenous *B. thuringiensis* strains were isolated and examined so that these alternatives were used to support the current commercial strains, such as *B. thurigiensis* subsp. Aizawai / cry proteins in use [6]. Research on *B. thuringiensis* bacterial isolate from South Sumatera has been started since 15 years ago [7]. Isolates of *B. thuringiensis* were identified, such as SMR-02 and SMR-04 [8] indicating a high level of toxicity to armyworm *Spodoptera litura*. Unfortunately, isolation of protein has never been conducted due to lack of laboratory facilities. The objective of the study was to characterize the molecular weight of proteins in order to be utilized as biological agents of *S. litura* and their cells or spores production. To investigate the molecular weight of protein was done by SDS-Page electrophoresis.

2. Materıals And Method

Soil sampling was taken from Musi Banyuasin regency of South Sumatera. Isolation of *B. thuringiensis* was conducted in Phytopathology Laboratory of Plant Protection Faculty of Agriculture Sriwijaya University and molecular weight of protein with SDS-PAGE electrophoresis was done in Genomic and Crop Improvement Laboratory, Research Center for Biotechnology- Indonesian Institute of Sciences LIPI Cibinong West Java. The study was conducted from March to July 2017.

Methods.

2.1. Isolation of Bacillus thuringiensis .

Isolation of B. thuringingiesnsi was conducted according to the method of Ozturk *et al.* [9] with some modifications. Five gram of soil samples was diluted well in 15 ml H_20 in test tube, by shaking well until perfectly diluted. One ml of upper part of dilution was taken in eppendorf tube, added by 1 µl Triton X-100, and heated in water bath 85° C for 15 minutes. With a sterile spatula, the solution was streaked on the medium NaCl Glycine Kim and Goepfert (NGKG) on petridish. Petridish was incubated at 30ºC, for 24 hours. Colonies of *B. thuringiesnsis* growed in white colour. After 24-72 hours incubation, proteinaceous parasporal inclusion bodies will be present. Identification of *B. thuringiensis* refers to Thiery and Frachon [10], by microscopic observation of bacterial cells test and gram staining.

2.2. Isolation of Bacillus thuringiensis protein

One colony of *B.thuringiensis* in the slant agar stock was transferred using a wire needle heated until red then cooled, into 1 ml T3 medium and shaked at 25 ºC overnight. This culture was then removed to three kinds of growth media, namely : 1). 10 ml of T3 medium [11], 2). 10 ml LB medium (5 g yeast extract, 10 g Tryptone and 10 g NaCl in 1000 ml volume, pH 7.2) and 3). Agricultural waste (coconut water and rice water = 1 : 1 (V/V) were shaken at 25ºC for 2 days. The cells were harvested by centrifugation at 13 K rpm for 10 min, washed in distilled water twice and collected white pellets. The pellet was suspended in distilled water. Precipitation was collected by centrifugation at 13 krpm for 10 min. It was re-suspended in water and added the same volume of 0.5 M NaCL. For collecting precipitation it was centrifuged down at 10 krpm for 15 min. This step (NaCl wash) was done at least two times and continued with distilled water wash two times. The cells were re-suspended in 5 ml of 10 mM Tris HCl pH 8 containing 10 mM EDTA and disrupted by a sonicator. This solution was kept on ice for 30 min after addition of 100 µl of mercapto-ethanol and adjustment of pH 10.5. To collect supernatant contained protein, it was centrifuged down at 10 krpm for 10 min. The last step was adjusting to pH 4.4 and kept the crystal protein on 4 ºC for long storage.

2.3. Weighing molecular mass of crystal protein by SDS-PAGE analyses

Crystal proteins isolated from *B. thuringiensis* strains, namely DLM5, KJ3R32, KJ3R53, KJ3J44, KJ3P15 and DLKK236, were analyzed by sodium dodecyl polyacrilamide gel electrophoresis (SDS-PAGE) [5]. SDS-PAGE was carried out by the method of Laemmli [12] using 10% running and 4% stacking gels. The crystal proteins was prepared as follows: 10 µl of crystal protein kept at pH 4.4 was centrifugation down at 10 k rpm for 1 min to collect precipitation. It was suspended in the equal volume of TE. After the addition of SDS sample buffer (1:4, v/v), protein sample was boiled for 3 min. The centrifugation at 7 krpm for 1 min was done to purify the protein solution. Samples were ready for electrophoresis. The 10% gel SDS-PAGE consisted of two layers, separating gel and stacking gel. The sample proteins and marker (PM1500 SMOBIO range 10-180 kDa) were loaded in the gel and run at current 60 V 30 mA 180 minutes in SDS electrophoresis buffer. After bromo phenol blue tracking dye had reached the bottom of separating gel, it needed to disconnect with power supply. The gel was removed from the glass, clean with dH20 and stained with co-massive blue overnight at room temperature. The co-massive blue was poured out and the de-staining was poured in with shaker incubation at 55 °C for 30 min. The fresh de-staining solution was added and continued de-staining until blue bands and clear background was obtained. After SDS-PAGE electrophoresis, it was analyzed the content of protein by comparing with PM1500, product of SMOBIO marker.

2.4. Cell production / spores B. thuringiensis.

Seed culture was required for the propagation of *B. thuringiensis*. The preparation was as follows: a needle isolate *B. thuringiensis* inserted into 100 ml of Nutrient Broth medium (NB) then shaken by using a shaker for 12 hours at 200 rpm. After it was taken 10 ml to put in 100 ml NB, then shaken again by using a shaker for 12 hours with a speed of 200 rpm. Furthermore, seed culture is ready for bio insecticide. Media multiplication of 50 ml coconut water + 50 ml rice water and 50 mg $CaCl_2$, 50 mg $MgSO_4$, 50 mg K_2HPO_4 and 50 mg KH_2PO_4 was prepared. The medium was inserted into a 250 ml Erlenmeyer tube. Tube was covered with aluminum foil and plastic and tied with rubbers. Further sterilized was done by insertion into the autoclave for 20 minutes with a temperature of 121°C and at a pressure of 1 atm. After cooling process, 5 ml seed culture of each Erlenmeyer was aseptically added. Furthermore, it was shaken by using a shaker for 72 hours with a speed of 200 rpm. The number of cells was produced after sporulation was calculated using a haemocytometer.

2.5. Bioassay of cell and spores of B. thuringiensis towards Spodoptera litura.

Bioassay was conducted with third instar larvae of *S. litura*. No-choice test was applied by dipping soybean leaves on 10^7 spores/ml *B. thuringiensis* medium for 5 minutes. The leaves were air-dried and placed it in a petri-dish. Third instar of Spodoptera larvae was individually inserted into the petri-dish. Each treatment was repeated 30 times or required as many as 30 larvae. Observations were made every day for 7 days, calculated the number of larvae that died or did not develop into 4th larval instar. Mortality percentage was calculated by number of insects died divided by number of insect treated x 100%.

3. Results And Discussion

3.1. Number of B. thuringiensis isolates

Among 50 isolates of *B. thuringiensis* obtained, through the process of color testing, morphology, gram positive test and spore staining was obtained 6 isolates suspected *B. thuringiensis*. Six isolates were tested for toxicity to armyworm *S. litura*. The isolates were coded as follows: DLM 5, KJ3R3, KJ3R5, KJ3J4, KJ3P1 and DLKK2 3.

3.2. Character of Molecular weight of B. thuringiensis protein

The observations on the SDS Page gel electrophoresis showed a clear band. The band indicated as a sign of protein presence in accordance with the guidelines on the PM 1500 SMOBIO marker. In Figure 1, the medium grew was T3 broth which was a suitable medium for the growth of B. thuringiensis. The results obtained that DLM5, KJ3R5 and KJ3P1 isolates possessed some clear bands. Some parts of the band looked thick showing protein content was abundance.

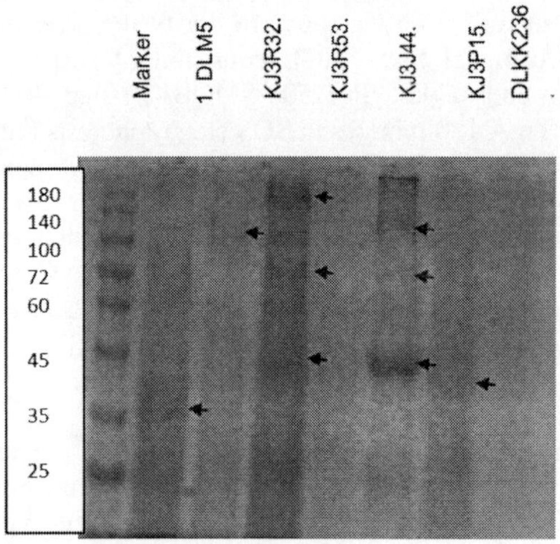

Figure 1. Profile of molecular weight of Protein of *Bacillus thuringiensis* harvested from T3 growth medium.

In Figure 2, Bt protein was produced from T3 growing medium mixed with waste materials namely coconut water and rice water. It was shown that the band was not very clearly visible. There was a relationship between isolate and growth media. Bt sporulation did not complete, leading in less production of cell/spores/crystal protein. There may be suspicion of prolong the time in the shaker. Growing with medium T3 was done for 2 days in producing perfect sporulation, whereas in waste medium waste may took longer time. It is also encountered by Palma *et al* [13] bio cidal of *B. thuringiensis* was influenced by the time in sporulation process. Furthermore it was possible that use of coconut waste and rice water was not sterilized, therefore there was a possibility for other bacteria to grow. Thus, the production of cells / spores and protein can be found in small amounts.

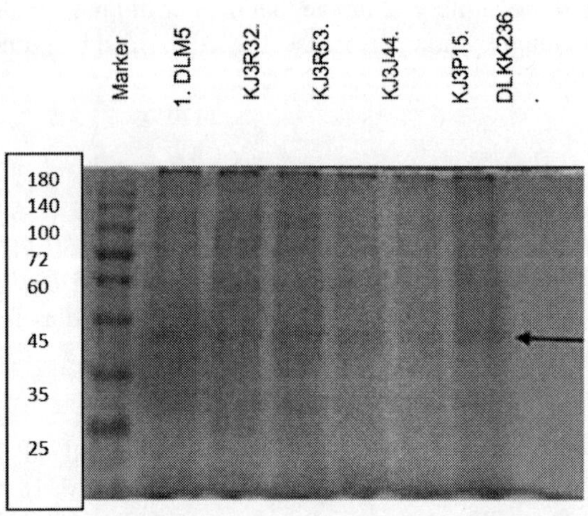

Figure 2. Profile of molecular weight of Protein of *Bacillus thuringiensis* harvested from T3 medium mixed with agricultural waste growth medium.

Figure 3 below, *B. thuringiensis* was grown in LB broth media. In the fifth column (KJ3P15) a thin band appeared at about 45-50 kDa. Bands was very light to see. The use of LB may be less suitable for growth and development of *B. thuringiensis* cells compared with T3 medium.

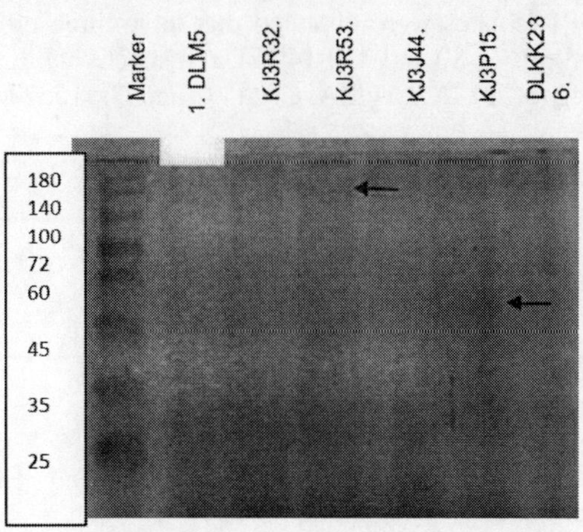

Figure 3. Profile of molecular weight of Protein of *Bacillus thuringiensis* harvested from LB broth growth medium.

3.3.Estimated weight of protein molecules

The weight of the protein molecule is one of the important indicators in determining the grouping of a strain of *B. thuringiensis*. The analysis of *B.thuringiensis* crystal composition is useful in determination of inclusion toxicity [5]. From the expected molecular weight, the targeted target insects can be determined, as stated by Hofte and Whiteley [14] and Crickmore et al. [15]) in Table 1 below.

Table 1. Molecular weight of Protein, gene type and insect target.

Gene type	Protein Type	Predicted molecular weight protein (kDa)	Host Range
cryIA(a)	CryIA(a)	133.2	Lepidoptera/Coleoptera/Diptera
cryIA(b)	CryIA(b)	131.0	Lepidoptera/Coleoptera/Diptera
cryIA(c)	CryIA(c)	133.3	Lepidoptera/Coleoptera/Diptera
cryIB	CryIB	138.0	Lepidoptera/Coleoptera/Diptera
cryIC	CryIC	134.8	Lepidoptera/Coleoptera/Diptera
cryID	CryID	132.5	Lepidoptera/Coleoptera/Diptera
cryIIA	CryIIA	70.9	Lepidoptera/Diptera
cryIIB	CryIIB	70.8	Lepidoptera
cryIIIA	CryIIIA	73.1	Coleoptera
cryIVA	CryIVA	134.4	Diptera
cryIVB	CryIVB	127.8	Diptera
cryIVC	CryIVC	77.8	Diptera
cry IVD	Cry IVD	72.4	Diptera
cyt A	Cyt A	27.4	Diptera/cyt.

Moreover, the crystalline toxins are classified into groups with amino acid similarity ([15]. Protein Cry1, Cry8, Cry9, Cry15, Cry22, Cry32, and Cry51 may cause negative effect to Lepidotera [14]. Based on band position in gel of SDS-PAGE, it was shown that the protein contained 1-8 protein with the molecular weight of protein was ranged of 37 – 140 kDa. This number of protein was similar to the result reported by Konecka et al. [5]. They found crystals contained 1–8 proteins with molecular weight of 36–155 kDa isolated during epizootics in *Cydia pomonella* L. Reyaz et al. [6] found by SDS-PAGE investigation of the spore–crystal mixture demonstrated Bt strains contained proteins of various molecular weights ranging from 150 to 28 kDa. Two protein bands at 130 kDa and 65 kDa were detected with SDS-PAGE analysis from B. thuringiensis isolated from environments of boron [15]. In the isolate of DLM and DLKK23 isolates were found one i.e. 37-40 kDa and 110-130 kDa, respectively. KJ3R5 and KJ3P1 isolates were obtained that have three protein bands namely 43-45, 73-80 and 110-130 kDa and 45-50, 75-80 and 130-140 kDa, respectively. It was predicted that isolates B. thuringiensis were belonging to Cry IA, Cry IIA, Cry IVC and Cry15c (Table 2).

Table 2. Molecular weight of Proteins of *Bacillus thuringiensis* and predicted host.

Isolate Code	Similarity of position: Sample loading in SDS-PAGE	Position	Prediction of molecular weight	Predicted Cry protein Group	Predicted host
DLM 5	Volume 20 µl running 1	Band 1 : 37-38	37-40	Cry 15Aa *	Lepidopetra
	Volume ±30 µlµl running 3	Band 1 : 38-40			
KJ3R3	No similarity				
KJ3R5	Volume 20 µl running 1	band 2 : 75	73-80	Cry IIA	Lepidoptera/ Diptera
	Volume 8 µl running 2	band 2 : 73-75			
	Volume ±30 µl running 3	band 2 : 75-80			
	Volume 8 µl running 2	band 1 : 110-130	110-130	CryIA	Lepidoptera
	Volume ±30 µlrunning 3	band 1 : 110-130			
	Volume 8 µl running 2	band 3 : 43-45	43-45	Cry 15Aa *	Lepidoptera
	Volume ±30 µl running 3	band 3 : 45			
KJ3J4	No similarity				
KJ3P1	Volume 20 µl running 1	band 1 : 130	130-140	CryIA	Lepidoptera
	Volume 8 µl running 2	band 1 : 140			
	Volume ±30 µlµl running 3	band 1 : 140			
	Volume 20 µl running 1	Band 3 : 75 (light)	75-80	CryIVc	Diptera
	Volume ±30 µlµl running 3	Band 3 : 75-80			
	Volume 20 µl running 1	Band 4 : 45	45-50	Cry 15Aa *	Lepidoptera
	Volume ±30 µlµl running 3	Band 4 : 47-50			
DLKK2 3	Volume 8 µl running 2	Band 2 : 110	110-130	CryIA	Lepidoptera
	Volume ±30 µlµl running 3	band 2 : 130			

3.4. Toxicity to target insect and spore production in growth media

In general, liquid waste media in the form of coconut water and rice water was used to grow *B. thuringiensis* cells. This technique was resulted from the multiplication of cells / spores /crystal proteins that may be done with easy-to-obtain media, useless waste and low cost, so it can be made in non-aseptic circumstances. Spore density obtained ranged from 0.50 x 10^6 to 2.52 x 10^7 spores / ml. Isolates with code KJ3R5 showed the highest result (2.63 x 10^7 spores / ml). When focused I production of crystal protein, the isolate significantly indicated the presence of three types of protein bands, ie. 43-45, 73-80 and 110-130 kDa. Similarly, KJ3P1 isolates showing spore production of 2.52 x 107 spores / ml had three bands at 45-50, 75-80 and 130-140 kDa levels. The results are presented in Table 3.

Table 3. Cell/spore production of *Bacillus thuringiensis* harvested in agricultural waste and their toxicity towards *Spodoptera litura.*

Isolate Code	Cell Production (cells/ml)	Toxicity (%)
DLM 5	2.11×10^7	53.33
KJ3R3	1.50×10^6	46.67
KJ3R5	2.63×10^7	60.00
KJ3J4	0.50×10^6	46.67
KJ3P1	2.52×10^7	53.33
DLKK2 3	2.50×10^7	46.67
Control (dH20)	-	15.00

In bioassay test of *S. litura* obtained the highest mortality result was isolate of KJ3R5. While on KJ3R3, KJ3J4 and DLKK2 3 showed similar mortality rate. Bioassay was a preliminary test which was followed by more detailed assays taking into number of the spores content possessed by the isolates. This data showed that every single isolates was toxic against lepidopteran insects. This results were supported with other finding of other researchers. Ninety percents of mortality in third instar larvae of diamond backmoth *Plutella xylostella* larvae treated for 72 hours was caused by a protein content of 11.1 g / l [16]. Research on insect bioassays with first larvae of *S. litura* showed that the purified Cry2Al1 was toxic to *S. litura* with LC_{50} 2.448 mg / ml [6]. Shah et al. found that some isolates with cry protein range 30-150 kDa were toxic against *Amsacta albistriga* (Lepidoptera:Arctiidae) [17].Therefore, it was no doubt that *B. thuringiensis* isolates indigenous South Sumatera were toxic to diamond backmoth *P. xylostella* and armyworm *S. litura.*

4. Conclusion

The molecular weight of protein ranged from 37 to 140 kDa. Isolates of DLM and DLKK23 contained two protein bands, their were 37-40 kDa and 110-130 kDa, respectively. KJ3R5 and KJ3P1 isolates had three protein bands, i.e. 43-45, 73-80 and 110-130 kDa and 45-50, 75-80 and 130-140 kDa, respectively. It was predicted that isolates *B. thuringiensis* were belonging to Cry IA, Cry IIA, Cry IVC and Cry15c. These crystal proteins were toxic to *S.litura.* Production of spore after sporulation in agricultural waste media was ranged 0.5×10^6 - 2.67×10^7 spores/ml showing medium level of toxicity to *S. litura.*

Acknowledgement

The data presented in this paper are part of the data from research funded by Sriwijaya University, Palembang Indonesia through Professional Grants Scheme in 2017.

References

[1] Bravo A, Gill S and Soberón M 2007 Mode of action of *Bacillus thuringiensis* Cry and Cyt toxins and their potential for insect control. *J. Toxicon* **49** 423-35

[2] Zhong C, Ellar J, Bishop A, Johnson C, Lin S and Hart R 2000 Characterization of *Bacillus thuringiensis* d-endotoxin which is toxic to insects in three orders *J. Invertebr Pathol* **76** 131-9

[3] Vidyarthi S, Tyagi D, Valero R and Surampalli Y 2002 Studies on the production of *Bacillus thuringiensis* based biopesticides using wastewater sludge as a raw material. *J. Water Res* **36** 4850–60

[4] Valicente H, Edgard P, Maria V, Newton C, Andréia C, Cláudia G and Ubiraci L 2010 Molecular characterization and distribution of *Bacillus thuringiensis* cry1 genes from Brazilian strains effective against the fall armyworm, *Spodoptera frugiperda*. *J. Biol Control* **53** 360–6

[5] Konecka E, Jakub B and Adam K 2014 Crystalline protein profiling and *cry* gene detection in *Bacillus thuringiensis* strains isolated during epizootics in *Cydia pomonella* L. *J.*

Biological Lett **51**(2) 83–92. Available online at:http:/www.degruyter.com/view/j/biolet DOI: 10.1515/biolet-2015-0008

[6] Reyaz A, Gunapriya L and Indra A 2017 Molecular characterization of indigenous *Bacillus thuringiensis* strains isolated from Kashmir Valley. *J. Biotech* **7** 143. DOI 10.1007/s13205-017-0756

[7] Pujiastuti Y, Jaya A and Triani A 2002 *Toxicity of Bacillus thuringiensis isolated from soil in Palembang towards Spodoptera litura* (in Bahasa Indonesia) Scientific report to Faculty of Agriculture, Sriwijaya University.

[8] Pujiastuti Y 2017 Entomopathogene : Biological Control Agents of Insect Pest. (Palembang : Noer Fikri Publisher) (in Bahasa Indonesia, ISBN 978-602-447-064-7)

[9] Ozturk, F., Acik, Abdurrahma and Suludere Z 2008 Isolation and characterization of native *Bacillus thuringiensis* strains from soil and testing the bioactivity of isolates against *Ephestia kuehniella* Zeller (Lepidoptera: Pyralidae) larvae. *Turkish J. Biochem* **33** (4) 202–8

[10] Thiery and Frachon 1997 Bacteria: Identification, Isolation, culture and preservation of entomopathogenic bacteria. *In* L Lacey (Ed) Manual of Techniques in Insect Pathology. Academic Press San Diego.USA.

[11] Travers RS, Martin PA and Reichelderfer CF 1987 Selective process for efficient isolation of *Bacillus* spp. *J. Appl. Environ. Microbiol* **53** 1263–6

[12] Laemmli K 1970 Cleavage of structural proteins during the assembly of the head of bacteriophage T4. *J. Nature London* **227** 680–5

[13] Palma L, Delia M, Colin B, Jesús M and Primitivo C 2014 *Bacillus thuringiensis* Toxins: An Overview of Their Biocidal Activity. *J. Toxins* **6**, 3296-325; DOI:10.3390/toxins6123296

[14] Hoftey H and Whitley R 1989 Insecticidal crystal proteins of *Bacillus* thuringiensis. *J. Microbiol. Rev.* **53** 242-55

[15] Crickmore N, Zeigler R, Feitelson R, Schnepf E, Van Rie J, Lereclus D, Baum J and Dean H 1998 Revision of the nomenclature for the *Bacillus thuringiensis* pesticidal crystal proteins. *J. Microbiol. Mol. Biol. Rev.* **62** 807–13

[16] Şahin B, Bekir Ç and Hatice G 2017 *Bacillus thuringiensis* Isolation from the Environments of Boron Mines and Effects of Boric Acid on Bioactivity. *GU J. Sci* **30** (1) 223-34

[17] Naimov S, Rumyana B, Rumyana K, Stefan D, Ivan M and Ruud A 2008 Solubilization, Activation, and Insecticidal Activity of *Bacillus thuringiensis* Serovar thompsoni HD542 Crystal Proteins. *J. Appl Environ Microbiol.* **74**(23) 7145–51. Published online 2008 Oct 3. doi: 10.1128/AEM.00752-08

Effect of *Pseudomonas spp* on infection of *Peronosporaparasitica* (Pers. Fr), the pathogen of downy mildew on Chinese cabbage

N Damiri, Mulawarman, A Umayah, S E Agustin and M Rahmiyah

Department of Plant Pest and Disease, Faculty of Agriculture, Sriwijaya University, Jalan Raya Palembang-Prabumulih, km 32, Ogan Ilir, Indralaya 30662, South Sumatra, Indonesia

Email: *nurhayatidamiri@gmail.com*

Abstract. This research was conducted to study the effect of the application *of Pseudomonasspp* on infection of *Peronosporaparasitica* (Pers.Fr), the pathogen of Downy mildew on Chinese cabbage. The research was conducted in the laboratory and greenhouse Department of Plant Pests and Diseases Faculty of Agriculture Sriwijaya University, Inderalaya, OganIlir South Sumatra Indonesia. The research was conducted in the laboratory and greenhouse Department of Plant Pests and Diseases Faculty of Agriculture Sriwijaya University, Inderalaya, OganIlir South Sumatra Indonesia. The research was conducted using Completely Randomized Design with ten treatments including control.ie: isolate A, Isolate B, isolate C, isolate D, isolate E, isolate F, isolate G, isolate H, isolate I and control.Each treatment consists of four replications. Results of the study showed that the application of *Pseudomonas spp.* can suppress the infection of *P. parasitica* on Chinese cabbage. The lowest disease intensity was shown by treatment C (isolate *Pseudomonas sp.*)which was significantly different from control.The best treatment in suppressing disease severity of downy mildew on chinese cabbage was isolate H which had disease severity of 37.07 percent, which was significantly different from control and other treatment.

Keywords : *Pseudomonas sp, infection, Peronosporaparasitica, Chinese cabbage.*

1. Introduction

Chinese cabbage (*Brassica juncea* L.) better known as caisin in Indonesia is a vegetable crop that is widely known to the public. This plant includes the most popular vegetables because it tastes good and contains lots of vitamins and minerals [1,2]. In the cultivation of these vegetables there are often constraints that limit production, one of which is the attack of downy mildew disease. Downy mildew caused by *Peronosporaparasaitica* (FersFr) can attack planted cabbage in any stage. Losses that can be caused by the attack of this disease can be 24 to 48% [2,3].The disease can destroy the quality of the leaves, also the quantity of the harvest isreduced due to the downy mildew infection. Young seedlings can die as a result of the infection.

Downy mildew is an obstacle in the cultivation of vegetables such as Chinese cabbage. This disease that can be found throughout a horticultural cultivation area is a member of the genus Brassica. In the development of the cotyledon, which may turn yellow, shrivel and die [3]. Downy mildew of vegetable brassica crops is widespread and can cause extensive crop damage. The first symptoms of foliar infection are dark specks that are often irregular in shape, may appear net-like, and are usually accompanied by leaf yellowing. These yellow areas enlarge and, on becoming limited by the leaf veins, take on an angular shape. If the disease is severe the yellow lesions coalesce, resulting in much of the leaf becoming brown[4].Control of disease is generally done by spraying fungicides, but the facts show that the use of chemicals continuously will result in negative impacts on the environment, increase production costs and leave residues that are harmful to the health of. According to [5], Chemical control of disease disturbs environment, subverts ecology, degrades soil productivity,mismanage water resources. The use of chemicals for fertilizer and crop protection can disturb soil equilibrium, impoverishment of soil microbial communities decreases enzyme activity, impairs nutrient absorption by root, decreases plant resistance to pathogens and lowers soil quality [6,7].

The use of microbial antagonists has been reported to have the potential to be developed as a bio-pesticide to control plant pests and diseases.Bio-control provides an alternative means of reducing these pathogens,which are otherwise difficult to control due to their survival strategies. The renewed interest inbio-control is due to its environmental friendliness, along lasting effect and safety features.Biological control is controlled by reducing the population of pathogen activity or inoculum, both active and dormant by using one or more types of microorganisms, both external and environmental manipulated, host and antagonist. Biological control by antagonistic bacteria can occur through one or more of the following mechanism: competition, antibiosis, hyper-parasite, induction of plant resistance and spurring plant growth.Both external and environmental manipulated, host and antagonist.The mechanism of antibiosis is an inhibition of pathogens by metabolic substances produced by antagonist agents such as: enzymes, volatile compounds. Competition mechanism is a process of suppressing pathogen activity by antagonists against limited sources such as organic matter, inorganic substances, space and other growth factors. The mechanism of hyper-parasite is the destruction of pathogens by substances produced antagonists such as citinase, cellulose, glicanase and other lethal agents [8,9].

Biological control through the exploration and identification of antagonistic agents to find an alternative to controlling plant pests and diseases has long been realized but not yet adopted widely. Many types of microbial origin or rhizosphere soil either bacteria or fungi are known to be used as a biological control agents such as *Bacillus spp*, *Penicilliumspp*, *Trichodermaspp*, *Gliocladium*, *Verticillium*. *Pseudomonas spp* has been shown to inhibit the development of several plant pathogens. *Pseudomanas sp.* areable to inhibit development of*Upasiasalmonicolor* the pathogen in rubber plant up to 67%[10].Bacteria that enhance plant growth and yield are referred to as plant growth promoting bacteria (PGPB) or plant growth promoting rhizobacteria (PGPR). They can produce phytohormones, improve field emergence, increase seed vigor and availability of soil nutrients, enhance plant resistance to abiotic stress and induce defense mechanisms againts the plant pathogens. The beneficial microorganisms can establish symbiotic relationships with the root system by colonizing the rhizosphere or the rhizoplane. PGPB exert positive effect on plant health and crop yield and can be used in biological control [11,12,13,14]

According to [15], *Pseudomonasflourescens*, *P. aeroginosa*, *P. malthophiliia* and *Klebsiellapnemoniae*areable to protect plants againtsFusarium disease and promote plant growth on agarwoodplant..*Pseudomonasspp* are widely used as a biological agents to control airborne contagious pathogens. These bacteria can produce antibiotic compounds such as chitinase enzymes that can hydrolyze cell walls of pathogenic fungi, siderophores and other antibiotics [16,17].*Pseudomonas flourescens*is widely used in spurring growth, resistance induction and control of plant diseases because they have antagonistic properties and can suppress disease progression by competition of Fe (III) elements and carbon elements, producing HCN, stimulating accumulation phytoalexin so that plants are more resistant and colonize the roots and stimulate the growth of plants. This research was conducted to study the effect of the application *of Pseudomonasspp* on infection of *Peronosporaparasitica* (Pers.Fr), the pathogen of Downy mildew on Chinese cabbage.

2. Material and methods

The research was conducted in the laboratory and greenhouse Department of Plant Pests and Diseases Faculty of Agriculture Sriwijaya University, Inderalaya, OganIlir South Sumatra Indonesia. The study was conducted using Completely Randomized Design with 9 treatment of *Pseudomonas spp* isolate (A, B, C, D, E, F, G, H and I)and control with 4 replications.

2.1 Isolation Pseudomonas spp

Isolation of antagonistic agents is done by taking the soil in the healthy caisinrhizosphereplant located in Palembang. Isolation is done by serial dilution method. Results of the isolation were a number of isolates that were given codes based on the locality names of their origin. The soil weighed as much as 10 g and then put into the Erlenmeyer tube which has contained 90 ml of sterile water. This mixture is homogenized by using a shaker at a rate of 150 rpm for 30 minutes. Suspension diluted up to 10^{-7}. Furthermore, as much as 0.1 ml of the suspension of 10-5-10-7 dilutions were each incorporated petridish which already contained a sterile king'B medium, flattened and then incubated for 3-4 days. The bacterial colonies that grew were subsequently purified, the characteristics of which were examined were physiological test and then ready to be used in the research.

2.2 Preparation of host plant

Chinese cabbages seeds used are local chinese cabbage. Before planting the seed sterilized surface by soaking the seeds in 1% NaOCl for 1 minute, then the seed is washed with sterile water and dried wind. Further seeds are sown on a sterile planting medium consisting of a mixture of sand, soil and manure (1: 1: 1). After one week of age , the seeds were treated with the application of isolate antagonism (A, B, C, D, E, F, G, H and I) by root immersion in antagonistic isolate suspension with spore density 10^7/ml for 4 hours. Furthermore the seedlings were transferred into polybag that has been filled with sterile media. Inoculation of *P. parasitica* was performed a week after application of *Pseudomonas spp.* isolate by spraying the suspension of P parasitica with spores density 10^4/ml to the chinese cabbage root evently.

The parameters observed in this research were disease intensity, disease severity and biomassweight of each plant. Disease intensity was calculated using leaf infected percentage, while disease severity using a formula by [18]. Biomass weight was measured using scales.The data were analyzed using the analysis of variance (ANOVA), with the Duncan's Multiple Range Test (DMRT) comparison among means [19].

3. Result and discussion

3.1 Result

3.1.1 Disease intensity. The analysis result of the effect of Pseudomonas spp. application on the disease intensity showed a significant effect. Application of *Pseudomonas sp.* Isolate A can reduce the intensity of downy mildew disease up to 59.49 percent, which is significantly different with the treatment of C, E, G, Iand control. Between the treatments of H, B, F, D and Awere not significantly different from each other, also between treatment C, E, G and I (Table 1).

Table 1. Effects of *Pseudomonas spp*ondisease intensity of downy mildew on Chinese cabbage.

Treatment	Disease intensity (%)
Control	76.35[a]
Isolate C	74.00[ab]
Isolate E	72.72[b]
Isolate G	73, 00[b]
IsolateI	70.46[b]
Isolate H	69.23[bc]
Isolate B	63.21[c]
Isolate F	60.40[c]
Isolate D	60.13[c]
Isolate A	59.49[c]

[a,b,c] The figures followed by the sameletter in the same column mean there is no significant different at p ≤ 0.05DMRT.

3.1.2 Disease severity. The application of bacteria *Psudomonasspp*as biological control of pathogen downy mildew on Chinese cabbage showed a significant effect on disease severity. All treatments of bacterial application of*P.Pseudomonasspp* showed an influence on disease severity suppression compared with control. Isolate H showed the highest suppression of *P. parasitica* infection by 37,07 percent followed by isolate C, E, G and A which was significantly different with control with 43.41%, 47.83%, 47.90% and48.77% (Table 2).

Table 2. Effects of *Pseudomonas spp* on disease severity of downy mildew on Chinese cabbage

Treatment	Disease severity		Relative percentage to the control (%)
Control	71.36	a	-
Isolate D	55.51	b	22.21
Isolate I	53.52	c	25.00
Isolate F	51.62	d	27.80
IsolaetB	51.46	d	27.88
Isolate A	48.77	de	31.65
Isolate G	47.90	ef	32.87
Isolate E	47.83	ef	32.97
Isolate C	43.41	g	39.16
Isolate H	37.07	h	48.05

[a,b,c,d,e,f,g,h] The figures followed by the same letter in the same column mean there is no significant different at p ≤ 0.05 DMRT.

3.1.3 Plant Biomass. Application of *Pseudomonasspp* as biological control agents tended to increase the weight of plants biomass compared to the control. The highest biomass weight was showed by the Chinese cabbage given F isolate treatment 1 gram/plant followed by D ,G, H and I treatment with a mean biomass weight of 0,83 gram/ plant (Figure 1).

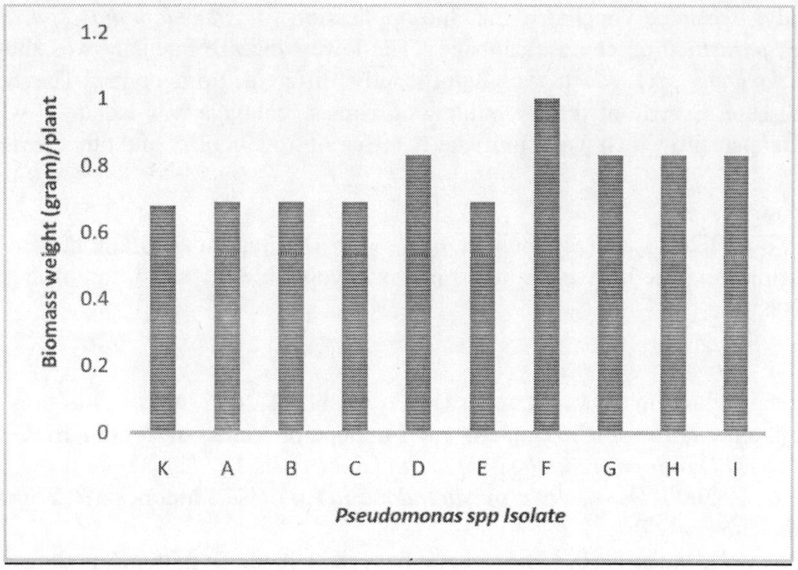

Figure 1. Effect of *Pseudomonasspp* on Chinese cabbage biomass weight.

3.2 Discussion

The suppression of the intensity of downy mildew attack may be related to the ability of bacteria to produce secondary metabolic thus protecting plants from pathogen attack[20]. According to[21], that most antagonistic bacteria have ability to adapt to their environment. The successful of biological control of plants diseases is due to some inhibitory feeding which is commonly found in biological agents such as siderophore, antibiotics, competition , enzyme, toxin and as plant growth promoting rhizo-bacteria (PGPR).[22]reported that beneficial rhizo-bacteria are known to colonize rapidly and aggressively the root system, suppresspathogenic microorganism, and enhance plant growth and development. Under certain conditionantibiotics improve the ecological fitness of these rhizo-bacteria in rhizo-sphere, which can furtherinfluence long-term biological control efficacy[23]. Some of *Pseudomonas spp* are capable to inhibiting the growth of soil borne pathogens. The inhibition mechanisms for the inhibition of pathogen proliferation are ability to colonize the root, production of antibiotics and enzymes that degrade pathogen cell wall, production of sidophores and hydrogen cyanide [24,11,25].

Natural bioactive compounds produced by antagonistic bacteria can be beneficial to plant health or agricultures. Species of bacteria can resist the plants pathogen in various ways such as by producing antibiotic compounds[26,27] It has been reported that *Pseudomonas spp*are known to be agents capable of inducing plant resistance, as antagonists and as plant growth promoting[8,28]. According to[28,30] genus Pseudomonas can posses a high potential for fixing nitrogen, solubilizing phosphate, inhibiting the development of phytophatogen, producing phytohormones and colonizing root systems. The suppression of the pathogen by antagonism bacteria occurred because the bacteria able to remove antibiotics such as pyoverdine, pyoluteorin, 2,4diacetylphloroglucinol and monoacetylphloroglucinol that could inhibitthe grow of pathogen. Some of *Pseudomonasspp* could also inhibit the development of the disease by colonized roots, nutrient competition of iron (III) and carbon element, HCN production, IAA, stimulate phytoalexin accumulation so that the plant became resistant and stimulated plant growth[30,31,32].This is in line with resultsof this study where the applied of *Pseudomonas spp*isolate are able to suppress the intensity and severity of downy mildew disease on chinese cabbage. Increased of wet weight is also occurring but not so significant.

4. Conclusion

From the study, it can be concluded that the application of *Pseudomonas spp.* can suppress the infection of *P. parasitica* on chinese cabbage. The lowest disease intensity was shown by treatment A(isolate*Pseudomonas sp.*) which was significantly different from control.The best treatment in suppressing disease severity of downy mildew on chinese cabbage was isolate H which had disease severityof 37.07 percent, which was significantly different from control and other treatment.

Acknowledgement

The authors would like to address their gratitude's to Dr. Suparman SHK, head of Plant Pest and Disease Department whose help and support had made possible the execution of the research and the writing of this paper.

References

[1] Mandiriza-Mukwirimba G, Kritzinger Q, and Aveling T 2016 A survey of brassica vegetable smallholder farmersin the Gauteng and Limpopo provinces of South Africa. *J. Agriculture and Rural Development in the Tropics and Subtropics* **117** (1) 35–44

[2] Semangun H 2000 *Disease of crops in Indonesia* (in Bahasa Indonesia) (Yogyakarta : Gajah mada University Press)

[3] Amin N, Ali B, Muzaffar S, Majeed I and Sayeed N 2013. Symptoms, pathogen biology and control of downy mildew of brassica: a minireview. *J. of Biosphere* **2** (1) 6-9

[4] Walker G, Wright P, Cameron P, Fletcher J, Walker M, Bulman S and Berry N 2016 Pests, natural enemies, diseases and disorders of vegetable brassicas in New Zealand. Vegetables NZ Inc. and The New Zealand Institute for Plant and Food Research Limited.

[5] Benuduzi A, Ambrosini A and L M P Passaglia 2012 Plant growth-promoting rhizobacteria (PGPR): Their potential as antagonists and biocontrol agents *J. Genet.Mol.Biol* **34** 1044-51

[6] Das S K and A Varma 2011 *Role of enzymes in maintaining soil healt in: Soil enzymology, soil biology* vol 22, ed G. Shukla and A. Varma (Springer-Verlag Berlin Heidelberg, USA) pp 25-42.

[7] Duke S O, Lydon J, Koskinen W C, Moorman T B, Chaney R L and R Hammerschmidt 2012 Glyphosate effects on plant mineral nutrition, crop rhizospheremicrobiota and plant disease in glyphosate-resistant crops. *J. ofAgric Food Chem.* **60** 10375-97

[8] Kloepper J W , Rodriguez-Ubana, R , Zehnder G W , Murphy J F , Sikora E and Fernandez C 1999 Plant root-bacterial interaction in biological control of soilborne disease and potential extension to systemic and foliar disease *J. Australas Plant Path* **28** (1) 21-26.

[9] Baker K F and Cook R J 1974 *Biological control of microbial plant pathogen* (San Fransisco: Freman WH)

[10] Fairuzah Z , Dalimunte C I and Aidi-Daslin 2014 The effectiveness of antagonistic bacteria *(Pseudomonas sp.)* as biological control of pink disease *(Corticiumsalmonicolor*). *Indonesian J. Net.Rubb. Res*,**32**(1) 37-44

[11] Ahemad M and M Kibret 2013 Mechanisms and application of plant growth rhizobacteria: Current perspective. *J. of King Saud University-Scince* **26** http://dx.doi.org/10.1016/j.jksus.2013.05.001.

[12] Mendes R, Kruijt M, de Bruijn I, Dekkers E, van der Voort M, Scneider J M H, Piceno Y M, DeSantis T Z, Andersen G L, Bakker P A H M and J M Raaijmakers 2012 Deciphering the rhizosphere microbiome for disease-suppressive bacteria *J. Science 332* 1097-100

[13] Figueiredo M V, Seldin L, de AraujoF F and R L R Mariano 2010 Plant growth promoting rhizobacteria: fundamentals and applications *J. Microbiology monographs***18** 21-43.

[14] Chen Y P, Rekha P D, Arun A B, Shen F T, Lai W A and C C Young 2006 Phosphate solubilizing bacteria from subtropical soil and their tricalcium phosphate solubilizing abilities *J. Appl. Soil Ecol* **34** 33-41

[15] Wahyuni T, Mulawarman and Damiri N 2015 Population dynamics of rhizobacteria and its potency as a biological control agent to control Fusarium disease in the nursery of agrawood *(Aquailariamelaccensis Lamrk,) J. Agrivita* **37**(3) 276- 83.

[16] Wang S L and Chang W T 1997 Purification and characterization of two bifungsional chitinases or lysozymes extracellularly produced by Pseudomonas aeroginosa K-187 in a shrimp and crabb shell powder medium. *J.Appl. And Enviro, Microbial* **63**(2) 380- 6

[17] Habazar T and Yaherwandi 2006 *Biological control of plants pest and diseases* (Padang : Andalas University Press)

[18] Understenhofer G 1976 The basic principle of crop protection field trial *J. Pfazenschutz-Nachricten Bayer* **29** (2) 161- 8

[19] Gomez K A and Gomez A A 1984 *Statistical procedure for agricultural research 2nd edition: An international rice research institute book* (New York-Chichester-Brisbane-Toronto-Singapore: A Wiley-intersci. Publ., John Wiley and son)

[20] Javandira C 2013 Potency of *Pseudomonas* spp. And Bacillus sp. as biological control of corn leaf blight caused by *Pantoeasp. Thesis. Brawijaya University, Malang.*

[21] Meyer K M and Leveau J H J 2012 Microbiology of the phyllosphere: a playground for testing ecological concepts *J. Ecologia* **168** 621- 9

[22] Weller D M 1988 Biological control of soil borne pathogen in the rhizosphere with bacteria *J. Annu Rev Phypathol* **26** 379-407

[23] Shanahan P, O' Sullivan D J, Simpson P, Glennon J D and O'Gara F 1992 Isolation of 2,4-diacetylphloroglucinol from a fluorescent pseudomonads and investigation of physiologicalparameters influencing its production *J. Appl Environ Microbiol* **58** 353-358.

[24] Jan A T, Azam M, Ali A and Q Haq 2011 Novel approached of beneficial Pseudomonas in mitigation of plant diseases an appraisal *J. Plant Interact* **6** 195-205.

[25] Shen X, Hu H, Peng H, Wang W and Zhang X 2013 Comparative genomic analysis of four representative plant growth-promoting rhizobacteria in Pseudomonas *BMC Genomics* **14** 271.

[26] Joseph B and R M Priya 2011 Bioactive compound from endophytes and their potential in pharmaceutical effect: a review. *Am. J. Biochem. Mol. Biol* **1**(3) 291-309

[27] Bacon C W and Hinton D M 2007 Potential for control of seedling blight of wheat caused by *Fusarium graminearum* and related species using the bacterial endophyte *Bacillus mojavensis J. Biocontrol Sci. Tech* **1** 81-94 DOI:10.1080/09583150600937006.

[28] Prezermieniecki S W, Kurowski T P and Karwowska A 2015 Plant growth promoting potential of *Pseudomonas sp.* SP0113 isolated from totable water from a closed water well *J. Arch. Biol. Sci., Belgrade* **67**(2) 663-673 DOI.10.2298/ABS14100.2029.

[29] Yu H, Yuan M, Lu W, Yang J, Dai S, Li Q, Yang Z, Dong J, Sun L, Deng Z, Zhang W, Chen M, Ping S, Han Y, Zhan Y, Yan Y, Jin Q and M Lin 2011 Complete genome sequence of the nitrogen-fixing and rhizosphere-associated bacterium Pseudomonas stutzeri strain DSM4166. *J. Bacteriol.* **193** 3422- 23

[30] Widodo, M S Sinaga, I Anas and M Mahfud 1993 The use *of Pseudomonasspp (P. flourescens)* to control the roots cudgel (*Plasmodiophorabrassicae*Wor.) on caisin (Brassica campestris L. var. chinensisRupr. Olson) *Bull. Plant Pest-disease* **6**(2) 94-105 (in Bahasa Indonesia)

[31] Blanco J M and P A H M Bakker 2007 Interaction between plants and beneficial *Pseudomonas spp*: expoiting bacterial traits for crop protection. Springer. Netherlands. *A. van Leeuw. J. Microb.* **92** (4) 367-389. DOI:10.1007/s1048-007-9167-1.

[32] Nandhini S, V Sendhivel and S Babu 2012 Endophytic bacteria from tomato and their efficacy against *Fusariumoxysporum*f.sp. lycopersici, the wilt pathogen *J. Biopest* **5** (2) 178-85

Isolation and toxicity test of *Bacillus thuringiensis* from Sekayu region soil, South Sumatra on *Spodopteralitura*

S R Afriani[1], Y Pujiastuti[1*], C Irsan[1] ,N Damiri[1] S Nugraha[2] and E R Sembiring[2]

1) Study Program of Crop Science, Post Graduate Program, Sriwijaya University, Palembang, South Sumatra, Indonesia
2) Research Center for Biotechnology, Indonesian Institute of Sciences, Jakarta, Indonesia

*Email: ypujiastuti@unsri.ac.id

Abstract. This study aimed to obtain bacterial isolates *B.thuringiensis* potential as a biological control against pests *Spodoptera litura*. The research was conducted at the Laboratory of Pest and Disease Department, Agricultural Faculty of Sriwijaya University, Campus InderalayaOgan Ilir, South Sumatera, from March to June 2017. The study was conducted with survey method and laboratory trial. The results showed that of the 50 soil samples from three villages selected through morphological observation, reaction staining, KOH test, catalase test, producing 13 bacterial isolates. Screening of the 13th toxicity of the isolates suspected *B.thuringiensis* against *S. litura* larvae was investigated. Based on the toxicity screening test the following facts were obtained: five isolates ie KJ_2M_2, KJ_3E1, KJ_3JB_1, KJ_3D_3 and KJ_3D_5 were lower toxicity than Dipel, two isolates ie KJ_3K_4 and KJ_3D_3 which had the same toxicity to Dipel. Five isolates the KJ_3E_3, KJ_3BW_5, KJ_3JB_5, KJ_3D_1 and LC_2, LC_3 known to have effectiveness until the seventh day reached 40%. There was one isolate that is KJ_3BW_5 which was more effective compared to Dipel as comparison.

Keywords : .*Bacillus thuringiensis, Toxicity, Spodoptera litura*

1. Introduction

Green mustard (*Brassica rapa*varparachinensis L.) is a vegetable that is in great demand and consumed by many people. Green mustard contains many vitamins and antioxidants [1]. The demand for green mustard is increasing, to meet the needs and demand of green mustard are needed to improve their production. Increased production of green mustard greens in Indonesia often experience some obstacles. Among these obstacles is attacked by *Spodopteralitura*larvae.

S. litura is an important pest that can cause damage to many types of vegetable plants of Bracicaceae and Solanaceaein Indonesia. Loss of results due *to Spodopteralitura* attacks in some countries such as Japan reached 80 percent, America 90 percent whereas in Indonesia the yield loss ranged from 23 to 45 percent [2]. In certain areas of vegetable plantation areas in highland of Java and Sumatra islands, the yield loss due to pests can reach 75 to 90 percent [3,4].

The use of synthetic insecticides in vegetable crops in the lowlands is quite intensive. Synthetic insecticides can rapidly decrease pest populations, thus preventing the spread of pest attacks. The use of synthetic insecticides in vegetable crops is presumably excessive in terms of type, composition, dose, time, and interval [5]. Generally farmers emphasize pest control efforts using synthetic pesticides [6,7]. Loss of yield and cost of vegetable pestscontrol globally reaches US 1 billion. It has been reported that in one season farmers use 3-4 types of insecticides with spraying frequency 11-15 times in the dry season and 6-10 times in the rainy season. These conditions resulted in the presence of high insecticide residues in the product[8,9].Therefore, new alternatives are needed to reduce pest populations and not cause negative impacts on the environment. Control of insect pests by using natural enemies from these insect pests. The natural enemiy used in this study is *Bacillus thuringiensis*, the entomopathogenic bacteria.

B. thuringiensis is a gram-positive bacterium that is rod-shaped and forms a spore. The hallmark of *B. thuringiensis*is to have protein crystals, which are toxic to insects [10]. *B. thuringiensis* is highly toxic to host insects of Lepidoptera order such as *S. litura*, thus harmless to natural enemies, as well as non-targeted living creatures such as humans and other organisms [11]. Although there have been many widely used commercial products and have identified various types of Cry proteins, but isolation and identification of *B. thuringiensis* strains are still underway. Because a large number of insect pests can not be controlled by using existing toxins. In addition, new *B. thuringiensis* strains are also required to provide alternatives when insect resistance appears to certain Bt strains [12]. This study aims to isolate *B. thuringiensis* and perform the best toxicity test of *B. thuringiensis* isolates for *Spodopteralitura* larvae on green mustard greens (*Brassica rapa varparachinensis* L.).

2. Materials and Methods

2.1 Place and Time
The research was conducted at the Laboratory of Pest and Disease Department, Agricultural Faculty of Sriwijaya University, Campus InderalayaOganIlir, South Sumatera, from March to June 2017.

2.2 Soil Sampling
Soil samples were taken from Sekayu sub-district, MusiBanyuasin, South Sumatera with 10 sampling points. *B.thuringensis*exploration was done by sampling the soil from the green mustard plant site at a depth of 10-20 cm. Soil taken as much as 250 g at each sampling point. The soil sample is then inserted into a plastic bag, tied tightly, labeled location and date of capture. The soil sample is taken to the laboratory and stored in the refrigerator to be isolated.Prior to the selection of *B.thuringensis*, bacteria were first identified.

2.3 Isolation of Bacillus thuringiensis
The soil samples taken from the field were isolated using [13] methods in the following manner. As much as 1 g of soil sample, put into a test tube containing 9 mL of sterile water. The suspension is shaken to homogeneous and heated to a temperature of about 80 °C for 10 minutes. Suspension made series dilution from 10^{-2}-10^{-6}. From aseptic dilution 10^{-6}were taken as much as 0.1 mL suspension using a micropipet flattened on a NA medium with a spreading cup method with 5 replications. The plate was isolated and placed in reverse position at 28-30 °C for 48 hours.

2.4 Identification of Bacillus thuringiensis
*2.4.1 Morphological Observations.*The bacteria were grown on NA media and were observed at 48-96 hours after inoculation. The observed parameters are colony shape, colony color, colony edge, colony surface, and colony elevation. Selection early, from many colonies that grow selected characteristics of the colony with morphological features: rod-shaped cells, motile, gram positive, circular colon, color white and yellowish colonies. Colonies of bacteria showing positive features as bacteria *B. thuringensis* were made pure cultures and then stored in a room with a temperature of 4 ° C as the stock to be used during the study.

2.4.2 *Gram staining.*Gram staining is used to distinguish bacteria that are separated generally in two major groups: gram-positive and gram-negative. Stages of gram staining of the bacteria is taken by using ose needle, then leveled in the middle of the object glass to form a thin layer and fixed. After that flood the spread of bacteria with a primary dye that is crystalline violet for 1 minute, then washed with running water, and dried wind. Flush it with iodine for two minutes, washed with water and drained. Then wash the spread with a pale solution of 95% alcohol, dropwise until the purple dye no longer appears to flow from the object glass. Then flood again with a counter-dye of safranin for 30 seconds, then wash and absorb excess water on the spread by pressingthe paper absorb carefully on it. Furthermore, microscopic observations were made.

2.4.3 *KOH Test.* KOH test is done by taking and putting 1 end of needle ose culture *B. thuringiensis* on glass preparations that had been spilled with 3% KOH solution. If bacteria are not slimy then these bacteria are gram-positive bacteria, while bacteria that produce mucus is a gram-negative bacteria. KOH test is used to distinguish gram-positive bacteria and gram-negative bacteria.

2.4.4 *Catalase Test.*The catalase test is carried out by taking and putting the end of the needle culture of*B. thuringiensis* bacteria on the preparatory glass which had previously been spilled with a solution of hydrogen peroxide (H_2O_2). Then stirred to produce bubbles, in positive catalase bacteria to produce bubbles.Whereas in the negative catalase bacteria do not produce bubbles. The catalase test was used to determine the catalase activity in the bacteria tested.

2.4.5 Breeding of Insect Bulk. Insect test used is *Spodopteralitura* obtained from farmer's land in SakoBorang, Palembang City, South Sumatera. Then the *Spodopteralitura* larvae were brought to the laboratory and maintained in a plastic container with a size of 23 cm x 50 cm covered with gauze. Then each larva is separated according to stadia. The insect test food is Chinese cabbage leaf that is replaced daily with a new one. The length of larval development is calculated until imago form. The new Imago appears separated into a plastic container that has been laid with cotton that has been previously immersed in honey to feed the imago. The test insect to be used for the efficacy test is *S.litura* instar 3 larvae from second line (F_2).

2.4.6B.thuringiensistoxicity screening test against S.litura. The treatment in the screening test was performed by administering *B.huringensis* suspension on Chinese cabbage leaf for insect larvae of *S. litura*. Prior to testing, first calculated the density of spores using haemocytometer. Each chinesecabbage leaf weighed 2.5 g. Then each chinese cabbage leaf was immersed until submerged entirely into 20 mL suspension of *B. thuringensis* and 80% tween solution for 3 minutes. Chinese cabbage leaves are dried before infestation of the larvae. Each piece of chinese cabbage leaf is inserted into a plate with a size of 12cm x12cm which has been given tissue paper. In each plate plot 10 larvae were added. As a positive control in the form of dipel and negative control of aquadest.

3 Results and Discussion

The isolation results from 10 soil samples obtained from Kayuara outlet lane 2 (3 soil samples), Kayuaralane 3 (4 soil samples), and Lumpatan (3 soil samples). The samples of soil examined were 50 isolates with traits such as the Bacillus genus. 38 isolates from the total of Bacillus colonies examined had morphological and microscopic features of *B. thuringensis*. However, only 38 isolates showed the characteristics of *B. thuringensis* such as round, white, elevated and wrinkled edges (Fig. 1). This is supported by [14] revelation that the features of *B. thuringensis* have a circular shape, rough and smooth surface of the colony, glossy and slightly glossy, yellowish white and white colonies. Microscopically, it can be seen that the form of bacterial cells suspected of *B. thuringensis* isolation has a stem cell shape and belongs to Gram positive bacteria with Gram staining (Figure 2).

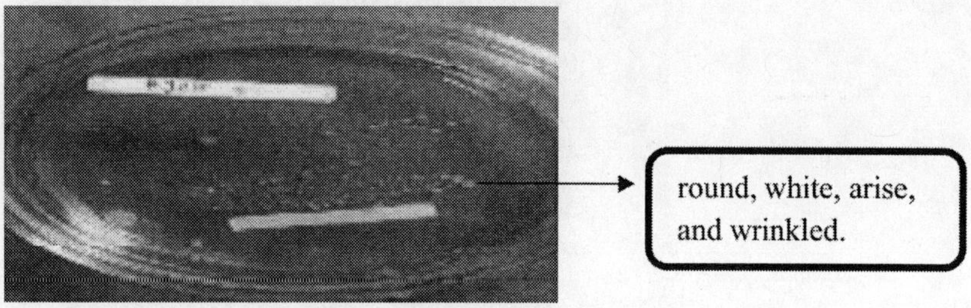

Figure 1. Morphology of bacterial colonies suspected *B thuringiensis* bacteria.

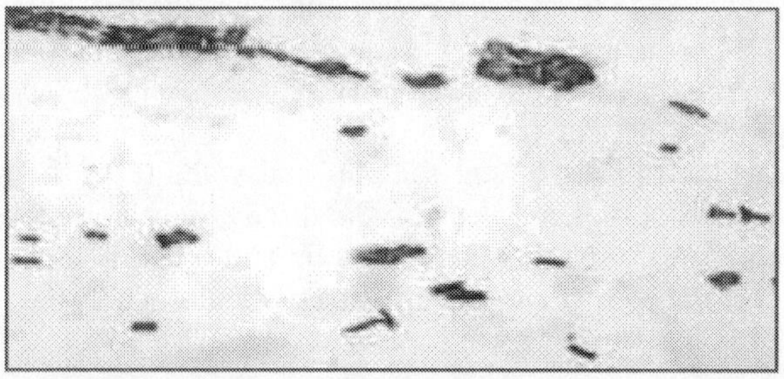

Figure 2. Bacterial vegetative cells *B. thuringiensis*have rod-shaped cells and includegram-positive bacteria, because they are purple after gram staining in 400 X magnification.

Of the 38 selected isolates were selected by KOH test and catalase test. Based on KOH test conducted, if the bacteria is lifted when lifting the needle ose then the bacteria is gram negative. In this case, the tested *B. thuringiensis* bacteria is not elevated, it can be said that this bacterium is gram positive (Figure 3). This is in line with the gram staining reaction that has been done that *B. thuringiensis* includes gram positive bacteria. While the catalase test aims to determine the reaction of catalase test in bacteria. The positive reaction of the catalase test is shown by the formation of bubbles. Conversely, if the reaction is negative, then bubbles are not formed. *B. thuringiensis*bacteria have catalase enzymes, so the catalase enzyme will work to break the H_2O_2that produces bubbles derived from the formation of oxygen gas (Figure 4). Screening of 38 selected isolates using KOH test and catalase test showed 13 isolates which will then be tested toxicity to *Spodopteralitura* larvae.

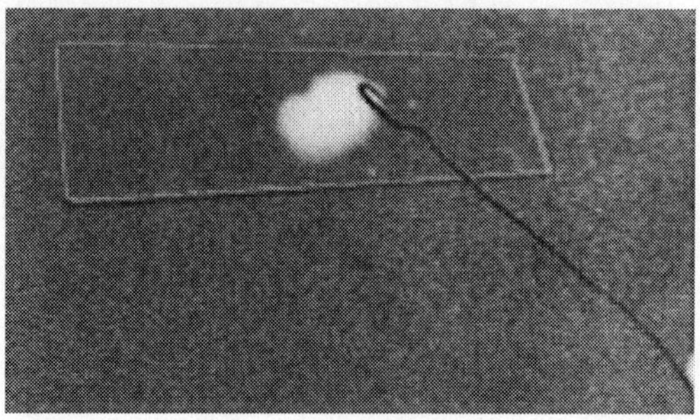

Figure 3: KOH test on *B. thuringiensis*bacteria using KOH solution that does not produce mucus (gram positive).

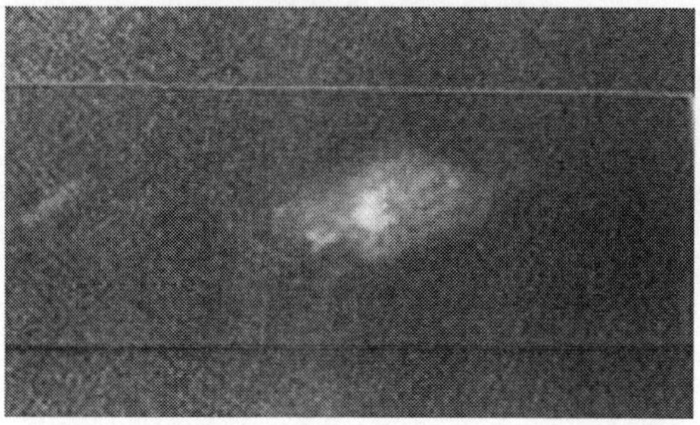

Figure 4: The catalase test on *B. thuringiensis* bacteria using H_2O_2 solution that produces bubbles.

The soil samples from Line 3 have the most isolates and line 2 has the fewest isolates. It is presumed that in the soil samples taken in line 3 have considerable sources of minerals, carbon, and nitrogen for the growth *of B. thuringiensis.*Environmental factors such as pH, oxygen solubility and temperature support bacterial growth.The low acquisition of isolated bacteria in Line 2, is most likely influenced by the content of crystal-producing bacterial spores in soil samples and viability of the spores. In relation to these habitats, multiple soil sampling is required because the discovery of entomopathogenic bacteria at any given moment is influenced by many factors including rain and other factors. There is a possibility when sampling of soil is found entomopathogenic bacteria in a particular place, but at other times can not be found again, and vice versa [14].

3.1 Screening toxicity of B.thuringiensis isolates against S.litura larvae

From selection result through morphological observation, coloring reaction, KOH test, and catalase test, 13 subsequent selection of isolates were tested for toxicity to *Spodopteralitura* instar three.

Table 1.*B.thuringiensis* toxicity screening test results against *S. litura l*arvae.

No	Isolate code	Village	Mortality (%)
1.	KJ_2M_2	KayuaraJalur 2	20
2.	KJ_3E_1	KayuaraJalur 3	13
3.	KJ_3E_3	KayuaraJalur 3	40
4.	KJ_3BW_5	KayuaraJalur 3	40
5.	KJ_3JB_1	KayuaraJalur 3	30
6.	KJ_3JB_5	KayuaraJalur 3	40
7.	KJ_3D_1	KayuaraJalur 3	40
8.	KJ_3D_3	KayuaraJalur 3	30
9.	KJ_3D_5	KayuaraJalur 3	13
10.	KJ_3K_4	KayuaraJalur 3	30
11.	LC_2	Lumpatan	40
12.	LC_3	Lumpatan	40
13.	LKB_1	Lumpatan	80
14.	Control (Aquadest)	-	7
15.	Dipel (*B.thuringiensis* commercial) as a comparison	-	30

In general, the mortality of treated *S. litura* larvae was different from that of aquades. According to the results of toxicity tests against *S.litura.B. thuringiensis* isolates obtained 5 (KJ_2M_2, KJ_3E_1, KJ_3JB_1, KJ_3D_3, KJ_3D_5) the toxicity is lower than the Dipel, a synthetic chemical insecticide, as a comparison, 2 isolates (KJ_3K_4 and KJ_3D_3) the same as the comparative toxicity. 5 isolates (KJ_3E_3, KJ_3BW_5, KJ_3JB_5, KJ_3D_1, LC_2, LC_3) that its effectiveness until the seventh day reached 40%. There is 1 isolate (KJ_3BW_5) which is more effective than comparison (Dipel). It is suspected that isolates have the same toxicity and are higher compared toDipel, since they have protein crystals that are characteristic of *B. thuringiensis*.

Factors that affect susceptibility of insect larvae are the type or type of protein crystal, alkaline pH of the alkaline insect digestive tract, and proteolytic enzymes [15]. This crystal contains a protein called δ-endotoxin, which is lethal when eaten by sensitive insects. Insect digestion can turn Bt-proteinsinto a shorter, toxic polypeptide. Toxins that have been actively interacting with epithelial cells in insect midgut. Evidence has shown that this Bt toxin causes the formation of pores (a very small hole) in the cell membrane in contribution digestion and disturbing the osmotic balance of the cells. Because the osmotic balance is disrupted, the cells become swollen and rupture and cause insect death [15].According to [16],when spores and protein crystals are eaten by sensitive insects, paralysis results in the death of the host. The bacterial crystals will dissolve in the gastrointestinal tract, in which bacteria secrete toxins that can kill insects. [17]states that insect sensitivity to destruction varies and the order of Lepidoptera has more sensitivity.Protein crystals are ingested by insects will dissolve in the alkaline environment of the insects. In target insect, the protein will be activated by an enzyme digesting protein enzyme. Protein which is activated will attach to the receptor protein located on the surface of the intestinal epithelial cells. The attachment resulted in the formation of pore or holes in the cell so that the cell undergoes lysis in the end the insect will experience indigestion and die.

Larva *S. litura*. Infected by *B. thuringiensis*shows more slow movement symptoms, decreased feeding activity, and reduced touch response. Dead larvae are blackish brown soft body. When touched the caterpillar skin will break and remove the black liquid and foul-smelling. The emergence of black color according to Steinhaus[18] caused by the body's bacteria to the haemocoel so that the blood cells become poisoned. This behavior change causes the caterpillar to become anxious and paralyzed. This paralyzed caterpillar shows that the condition of the caterpillar has been weak or the caterpillar defense system works so well that the spores.

4 Conclusion

From the results and discussions, it can be concluded that of the 50 soil samples from three villages selected through morphological observation, reaction staining, KOH test, catalase test, producing 13 bacterial isolates. Screening of the 13thtoxicity of the isolates suspected *B.thuringiensis* against *S. litura* larvae was investigated. Based on the toxicity screening test the following facts were obtained: five isolates ie KJ_2M_2, KJ_3E1, KJ_3JB_1, KJ_3D_3 and KJ_3D_5 were lower toxicity than comparison, two isolates ie KJ_3K_4 and KJ_3D_3 which had the same toxicity as Dipel. Five isolates namely KJ_3E_3, KJ_3BW_5, KJ_3JB_5, KJ_3D_1 and LC_2, LC_3 known to have effectiveness until the seventh day reached 40%. There was one isolate that is KJ_3BW_5 which was more effective than comparison.

Acknowlegment

The authors would like to address their gratitude to those whose contributions have made possible the writing of this paper. This study was partly results of the ProfessionResearch funded by The research Institute of Sriwijaya University, Ministry of Research, Technology and Higher Education, Republic of Indonesia with contract number:987/UN9.3.1/PP/2017.

References

[1] Okorogbona A O M, W Van Averbeke and T D Ramusandiwa 2011 growth and yield response of chinese cabbage *(Brassica rapa L. Subsp. Chinensis)* as affected by nutrient in air-dried and pulverized different types of animal manure using low biological activity soil. *World J. Agric. Sci.* **7** 1-12

[2] Adie M M, Krisnawati A , Mufidah A Z 2012 *The degree of resistance of soybean genotypes to grayak caterpillar pest.* National Seminar Results of research on various crops and tubers. Increased competitiveness and implementation of peanut and bulb commodities development supports the achievement of 4 agricultural developments (in Bahasa Indonesia). Puslitbangtan. Agricultural Research Agency: 29-36.

[3] Marwoto and Suharsono 2008 Strategy and component of caterpillar control technology (*Spodopteralitura fabricus)* on soy bean crop *J. Agricultural Research* **27** (4) 131-6

[4] Agency for Food Crops Research 2015 Report on the results of research and assessment (in Bahasa Indonesia). Agricultural Ministry Agricultural Research and Development Agency.

[5] Udiarto B K., E Suryaningsih and W W Hadisoeganda 1994 Base in study of identificaton and development of pest management on hingher in GarutanMajalengka(in Bahasa Indonesia). *Bul. Penel. Hort* .**27** 137- 49

[6] Weinberger K and Srinivasan R 2009 Farmer's management of cabbage and cauliflower pest in India and their approaches to crop protection. *J. Asia-Pasific Entomology,* **12** (4) 253- 4

[7] Armah F A 2011 Assesment of pesticide residues in vegetables at the farm gate: Cabbage (Brassica oleraceae) cultivation in Cape Coast Ghana. *Res. J. Env, Toxicology,***5** (3) 180-202

[8] Mazlan N and Mumford J 2005 Insecticide use in cabbage pest management in the Cameron Highland, Malaysia. *J. Crop Prot.* **24** (1) 31- 9

[9] Badii K B, Adarkwah C and Nboyine J A 2013 Insecticide use in cabbage pest management in Tamale Metroplolis of Ghana. *Greener J. Agric. Sci* **3**(5) 403- 11

[10] Khetan S K 2001 *Microbial Pest Control* (USA : Maecell Dekker, Inc) pp 3-141

[11] Bravo A, Likitvivatanavong S, Gill S and Soberón M 2011 *Bacillus thuringiensis:* a story of a successful bioinsecticide. *J. Insect Biochemistry and Molecular Biology* **41** 423-431

[12] Bahagiawati 2001 Management of insect resistance on planting the transgenic plant of Bt (in Bahasa Indonesia).*Buletin Agrobio* **4** 1-8

[13] Ohba M and Aizawa K 1978 Serological identification of Bacillus thuringiensis and related bacteria isolated in Japan *J. Invertebr. Pathol* **32** 303-9

[14] Salaki C 2010. Genetic diversity of Bacillus thuringiensis Berliner endogenous Indonesian isolates as pest controlling agent CrocidolomiabinotalisZwll. (Lepidoptera: pyralydae) on cabbage plants. Dissertation. Faculty of Biology Gajah Mada University. Yogyakarta.

[15] Hoffe H and Whiteley 1989 Insecticidal Crystal Protein of Bacillus thuringiensis *J. Microbial.***53** 245-255.

[16] Hadi M, Udi T and Rully R 2009 The entomologist's insect biology(in Bahasa Indonesia) (Yogyakarta : Graha Ilmu)

[17] Harjaka T, A Harsojo and E. Mahrub. 2012. Mushroom infection *Metarrhizium anisopliae* on *Plutellaxylostella* cabbage leaf caterpillar. *Journal of Plant Pest and Disease, UGM* 208.

[18] Steinhaus E A 1949 Principles of Insect Pathology (Newyork : McGraw Hill Book Co. Ltd)

Constant rate of paddy rice drying using air dehumidification with zeolite

F D Utari, M Djaeni*, and F Irfandy

Department of Chemical Engineering, Diponegoro University
Jl. Prof. H. Soedarto, SH, Tembalang, Semarang, Indonesia
Tel.:+62 247460058

*Email: moh.djaeni@live.undip.ac.id

Abstract. Drying using air dehumidification with zeolite has the purpose to produce paddy rice with lower moisture content and to enhance the storage life. In this study, the paddy rice was dried from initial moisture content 21% wet basis to 14% wet basis (final moisture content) using fluidized bed dryer and dehumidification with zeolite. This dryer lifted the paddy from the bottom (fluidized state) so the final moisture content was uniform. The paddy was dried in various drying temperatures (40-60°C). The moisture content was observed every 15 minutes for 120 minutes. As the control, paddy was dried without using air dehumidification. Several thin layer model was used to determine the constant rate of the drying and then the drying time can well be predicted. Result showed that Two term model was suitable to determine the constant rate of the drying. The constant rate of the drying was 19% higher in drying using air dehumidification.

Keywords : *constant rate, dehumidification, drying, paddy, zeolite*

1. Introduction

Agriculture, especially paddy *(Oryza sativa L.)* cultivation become a concern for Indonesian Government. The paddy field area in Indonesia at 2015 was 14,116,638 ha and the production was 75,397,841 ton [1]. One of the important quality of paddy is the moisture content. Inadequate drying of postharvest paddy that resulted higher moisture content can lead quality deterioration and development of insects and molds. Paddy must be dried until moisture content 14 % or less (wet basis) to get the storage life around 18 months before milling [2,3].

Paddy has been dried with two methods, using sun dryer and conventionals dryer. Sun drying is the conventional drying method that free and using renewable energy resources. But sun drying process easy to contaminated and depends on the weather condition that usually unpredicted [4]. The conventional dryer using fluidized bed dryer gives many advantages: (1) moisture content of the product is uniform because the complete mixing; (2) high drying capacity and high drying rate due to high surface area; (3) Small equipment and lower initial cost [5,6].

Fluidized bed dryer using high drying temperature (85-90 ^{0}C) was observed in previous study [6]. In higher drying temperatures, the quality of paddy like head rice yield, amylose content, and gel consistency decreased. Considering to the quality of paddy and efficient drying process, the drying process that operated in lower drying temperature and high energy efficiency must be done.

Dehumidification of the air for drying through adsorption by adsorbents (silica gel, zeolite) was used to dry the food product: cabbage, eggplant, carrot, butterbur and spinach [7]; onion [8]; carrageenan [9]; seaweed [10]. Zeolite consists of SiO_4 (neutral charge) and AlO_4 (negative charge). The ability to adsorb water comes from the negative charge that links to positive ions. There are two types of zeolite: as an engineered product and as a natural product (clay and stones). The effect of engineered zeolite is very significant at drying temperature below 80 °C [4,9].

This work aims to produce the paddy with lower moisture content using air dehumidification by zeolite. The effect of addition zeolite and drying temperature is evaluated. The experimental results are fitted to several thin layer model to describe the drying phenomenon.

2. Materials and Methods

2.1. Paddy drying

Postharvest of paddy rice provided from a local farmer in Tembalang districts, Semarang, Indonesia. Initial moisture content in paddy was 21% (wet basis), analyzed by Grain Moisture Meter, (GMK 303 RS, Gwon) Paddy rice was placed in fluidized bed dryer chamber under drying temperature (40 ^{0}C - 60 ^{0}C). The drying process took place for 120 minutes with the superficial velocity of 5.51 ms^{-1}. The moisture content was determined every 15 minutes. In the drying process, the zeolite (zeolite 3A provided by Zeochem, Switzerland) was added as the air dehumidifier. The zeolite was regenerated after each process to avoid the saturated zeolite. As a comparison, the control sample (without the addition of zeolite) was dried.

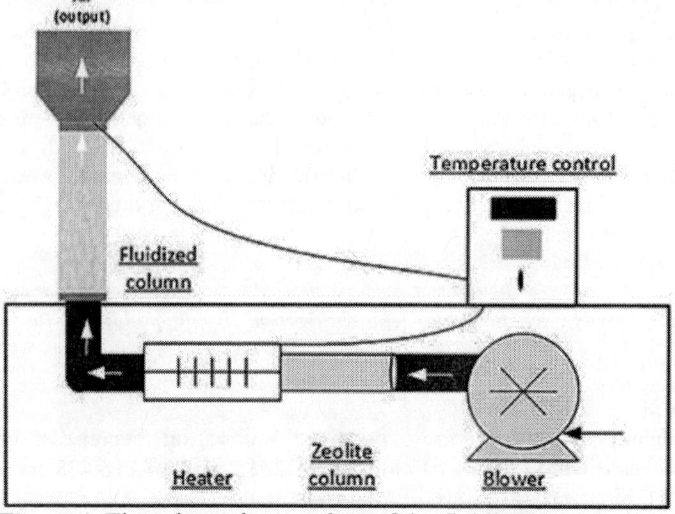

Figure 1. The schematic overview of the fluidized bed drying equipment from Djaeni et al. with modification [4].

2.2. Moisture ratio

The correlation between moisture ratios with time in paddy drying can be represented by Equation 1

$$MR = \frac{(M_t - M_e)}{(M_0 - M_e)} \quad (1)$$

where M_t was the moisture content at time t, M_0 was the initial moisture content, M_e was the equilibrium moisture content of paddy, all of them in dry basis (g g^{-1}). The equilibrium moisture content in paddy was estimated using oswin model as expressed below [11].

$$M = A[a_w/(1 - a_w)]^B \quad (2)$$

where M is the moisture content (kg kg^{-1}, dry based), a_w is the water activity, A and B are constants (dimensionless). The value of A and B can be extended as using quadratic equation (Equation 3 and 4) to get the value in higher temperatures.

$$A = 3 \times 10^{-6}T^2 - 0.0012T + 0.1418 \quad (3)$$

$$B = -0.0001T^2 + 0.0116T + 0.0876 \quad (4)$$

2.3. Model fitting

Moisture ratio of paddy drying was fitted to several thin layer model (Table 1). As an indicator, the coefficient of determination (R^2) and Root Mean Square Deviation (RSMD) were evaluated using POLYMATH Educational 6.0 software. The best model can be judged based on the highest R^2 value and the lowest RSMD value.

Table 1. Thin layer model.

Model	Model equation		References	
Newton	$MR = exp(-kt)$	(5)	O'Callaghan et al.	[12]
Handerson-Pabis	$MR = a\ exp(-kt)$	(6)	Henderson and Pabis	[13]
Page	$MR = exp(-kt^n)$	(7)	Page	[14]
Logaritmic	$MR = a\ exp(-kt) + c$	(8)	Yagcioglu et al.	[15]
Two term	$MR = a\ exp(-kt) + b\ exp\ (-ft)$	(9)	O'Callaghan et al.	[16]

3. Results and discussion

3.1. Effect of air dehumidification using zeolite

The initial moisture content of postharvest paddy was 21% (wet basis). The quality deterioration and development of insects and molds happen in paddy storage at higher moisture content. In this case, paddy must be dried until moisture content 14 % (wet basis). At the storage condition with moisture content 14%, the storage life around 18 months before milling [2,3].

Paddy was dried in fluidized bed dryer for 120 minutes. After drying process, all variable resulted the final moisture content range from 9 until 12.7 % (wet basis). The correlation between moisture content and drying time for drying with and without zeolite at drying temperature 40 ^{0}C was depicted in Figure 2. Lower moisture content was derived from the drying using zeolite as the air dehumidifier.

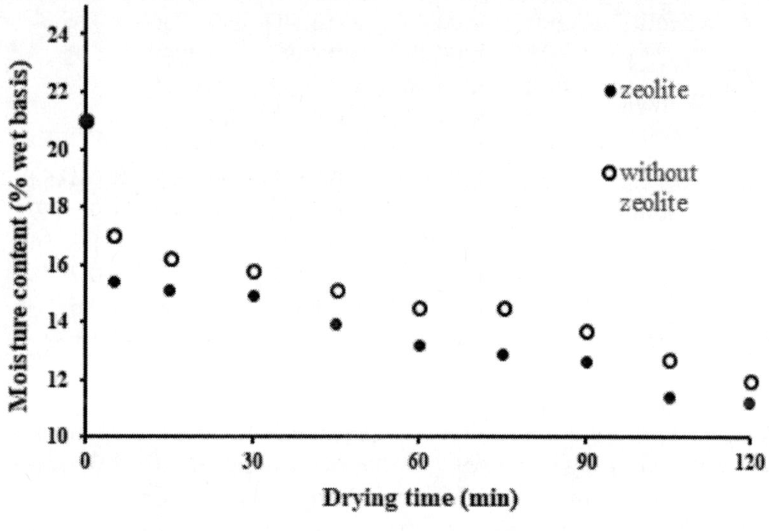

Figure 2. Effect of zeolite on moisture content.

There were two advantages of air dehumidification using zeolite, enhanced the driving force of the drying by the lower air humidity and increased the air drying temperature by release the heat of adsorption [17]. Application of zeolite as the air dehumidifier in the drying process enhanced the dryer efficiency. Energy efficiency of dryer using zeolite for food product was previously studied. The dryer that was operated with zeolite at drying temperature 50-60 ^{0}C resulted efficiency until 75% [18]

The constant rate of the drying using zeolite at drying temperature 40 ^{0}C was 19 % higher than without zeolite (Henderson-Pabis model), see Table 2. Higher constant rate of the drying resulted lower moisture content. Another study about paddy drying using mixed adsorption drying with zeolite resulted the drying time was 5-10% faster than without zeolite [4]. On the other hand, the dryer operation with zeolite in lower drying temperature give the advantage in order to retain product quality [7], [19].

Table 2. Model constant and statistical analysis on thin layer drying of paddy.

Model	Parameter	T (°C)	Model Constant				R^2	RSMD
Newton	Zeolit	40^{0}C	k:0.0064				0.2524	0.0335
		50^{0}C	k:0.0068				0.2568	0.0348
		60^{0}C	k:0.0086				0.6123	0.0319
	Without zeolit	40^{0}C	k:0.0046				0.3437	0.0258
		50^{0}C	k:0.0055				0.4738	0.0265
		60^{0}C	k:0.0070				0.7844	0.0221
Handerson-Pabis	Zeolit	40^{0}C	k:0.0040	a:0.8401			0.7390	0.0198
		50^{0}C	k:0.0042	a:0.8336			0.7399	0.0206
		60^{0}C	k:0.0062	a:0.8483			0.8493	0.0199
	Without zeolit	40^{0}C	k:0.0030	a:0.8809			0.7680	0.0153
		50^{0}C	k:0.0037	a:0.8755			0.8174	0.0156
		60^{0}C	k:0.0054	a:0.8953			0.9201	0.0135
Page	Zeolit	40^{0}C	k:0.1550	n:0.2651			0.9857	0.0046
		50^{0}C	k:0.1681	n:0.2573			0.9579	0.0083
		60^{0}C	k:0.1223	n:0.3836			0.9478	0.0117
	Without zeolit	40^{0}C	k:0.1044	n:0.2880			0.9424	0.0076
		50^{0}C	k:0.1027	n:0.3274			0.9582	0.0075
		60^{0}C	k:0.0629	n:0.4946			0.9577	0.0098
Logaritmic	Zeolit	40^{0}C	k:0.0773	a:0.3435	c:0.6126		0.8750	0.0137
		50^{0}C	k:0.0407	a:0.3340	c:0.5730		0.8111	0.0176
		60^{0}C	k:0.0183	a:0.4892	c:0.3940		0.8683	0.0186
	Without zeolit	40^{0}C	k:0.0360	a:0.2659	c:0.6695		0.8218	0.0134
		50^{0}C	k:0.0242	a:0.3243	c:0.5939		0.8554	0.0139
		60^{0}C	k:0.0127	a:0.5398	c:0.3759		0.9300	0.0126
Two term	Zeolit	40^{0}C	k:0.0025	a:0.7511	b:0.2489	f:0.4360	0.9956	0.0026
		50^{0}C	k:0.0030	a:0.7620	b:0.2378	f:9.0459	0.9917	0.0037
		60^{0}C	k:0.0048	a:0.7680	b:0.2320	f:1.3856	0.9844	0.0064
	Without zeolit	40^{0}C	k:0.0021	a:0.8252	b:0.1748	f:0.7756	0.9723	0.0053
		50^{0}C	k:0.0027	a:0.8138	b:0.1862	f:0.6506	0.9893	0.0038
		60^{0}C	k:0.0047	a:0.8513	b:0.1490	f:16.7292	0.9927	0.0041

3.2. Effect of drying temperatures

The drying was done with variation of drying temperature (40-60^{0}C). The curve of moisture content versus time was depicted in Figure 3. After the drying process, the final moisture content of paddy dried with zeolite in drying temperature 60^{0}C was 9% (wet basis), while in drying temperature 50^{0}C was 11.2% (wet basis). The result showed that the higher temperature, higher moisture that was evaporated. The constant rate of the drying using Henderson Pabis model in temperature 60^{0}C was 1.6

times higher than in temperature 50^{0}C (see Figure 2). At higher drying temperature, the air bring more sensible heat for evaporation and the relative humidity of air is low. So, the drying rate was higher in higher drying temperature [4,20,21].

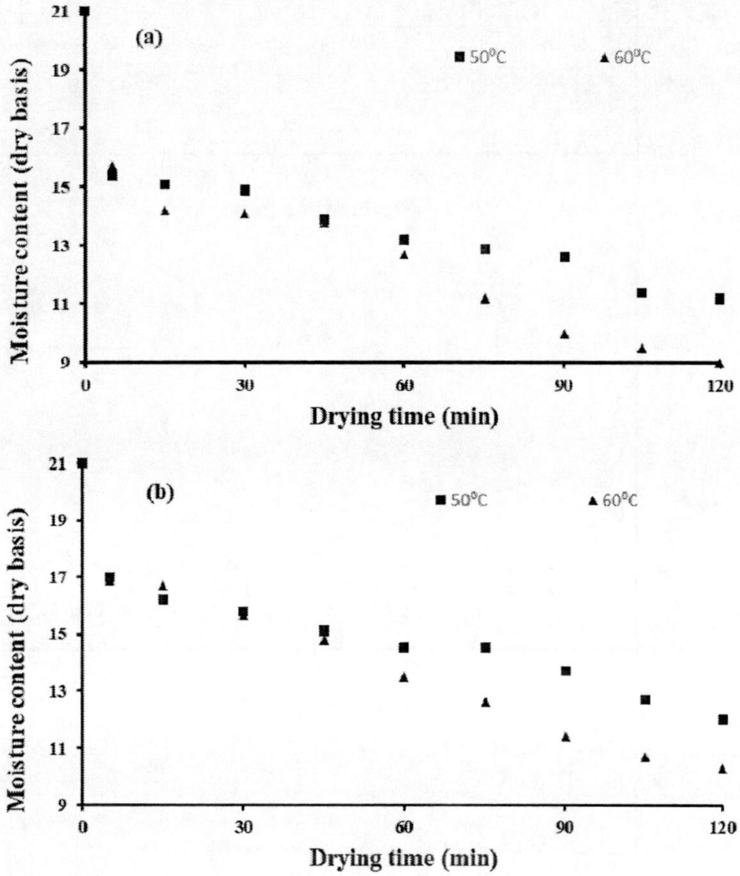

Figure 3. Effect of drying temperature on moisture content (a) zeolite (b) without zeolite.

3.3. Model fitting

The moisture ratio data of paddy drying using zeolite and without zeolite at different temperatures were fitted into several the thin layer drying model. The constant and statistical analysis are listed in Table 2. Two term gave the highest R^2 value range from 0.9723 until 0.9956 and the lowest RSMD value range from 0.0026 until 0.0064. The two term model was selected to describe the drying phenomenon in paddy drying with and without zeolite. Two term model was selected in various food drying: bael pulp drying [22], Channa fish *(Channa lucius)* [23], dika nut *(Irvingia gabonensis)* [24]. Figure 4 demonstrated the experimental curve and calculated curve using Two term model in paddy drying using zeolite in various drying temperature.

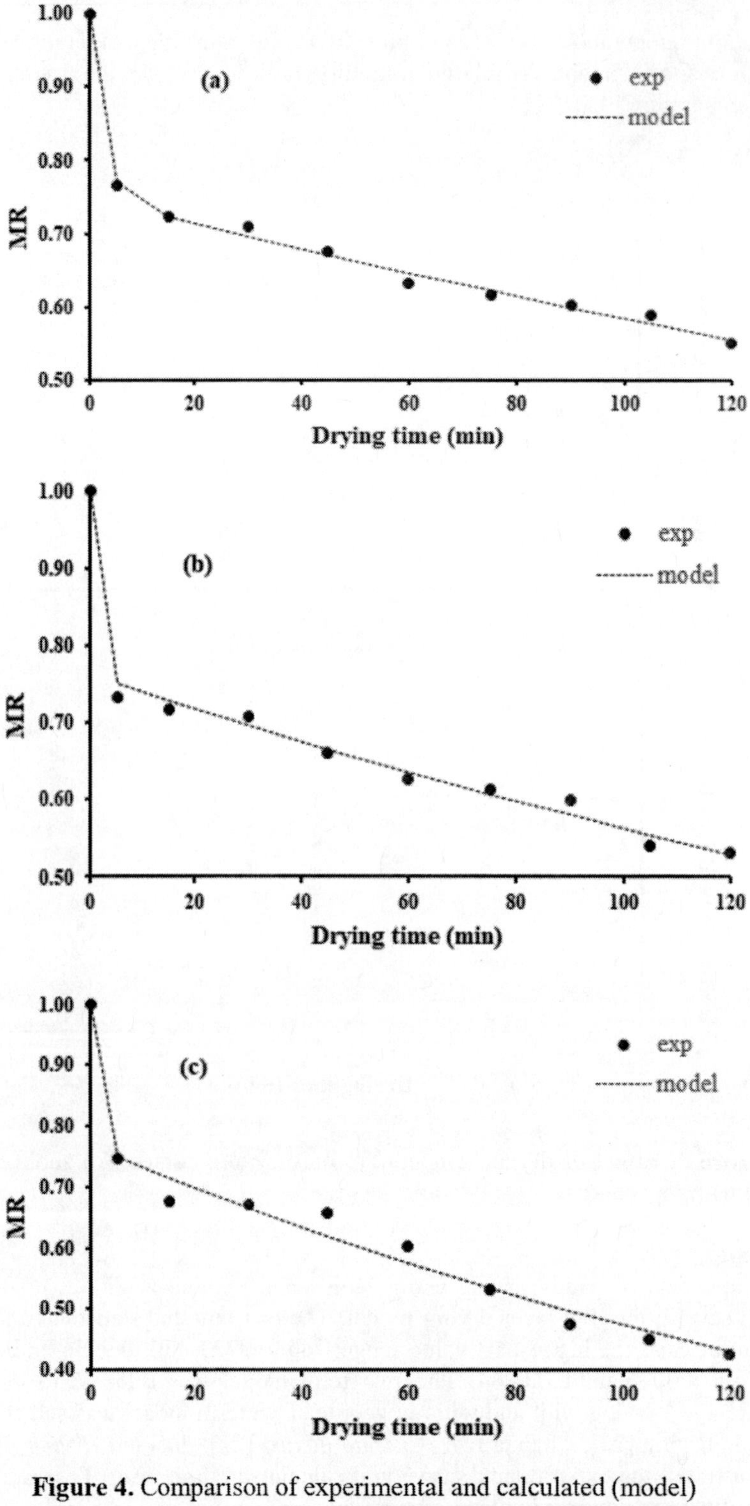

Figure 4. Comparison of experimental and calculated (model) moisture ratio curves of paddy drying using zeolite at various drying temperature (a) 40°C (b) 50°C (c) 60°C.

4. Conclusion

Paddy must be dried until moisture content 14% to avoid the quality deterioration, to avoid development of insects and molds and to enhance the storage life. The paddy drying was done with addition of zeolite under various drying temperature. Zeolite as air dehumidifier performed well. The constant rate of the drying using zeolite at drying temperature 40^0C was 19% higher than without zeolite. Zeolite enhanced the driving force of the drying by the lower air humidity and increased the air drying temperature by release the heat of adsorption. The constant rate of the drying in temperature 60^0C was 1.6 times higher than in temperature 50^0C. The drying rate was higher in higher drying temperature because the air bring more sensible heat for evaporation and the relative humidity of air is low. The two term model was selected to describe the drying phenomenon in paddy drying with and without zeolite with R^2 value range from 0.9723 until 0.9956 and the RSMD value range from 0.0026 until 0.0064

Acknowledgements

The authors would like to express their gratitude to the Indonesia's Directorate General of Higher Education and Diponegoro University for their financial supports through a number of research grants from which this paper was sourced.

References

[1] The Indonesian Central Bureau of Statistics 2015 Paddy Production based on Province (ton), 1993-2015 (in Bahasa Indonesia) Available at: https://www.bps.go.id/linkTableDinamis/view/id/865 [Accessed Sept 10, 2017]

[2] Thai C D, Allaf K. 2013 *DIC-Assisted Parboiling of Paddy Rice. Instant Controlled Pressure Drop (DIC) in Food Processing* (New York : Springer) 57–66

[3] Morrison D M, Chester L, Samuels C A N and Ledoux D R 2016 The Determination of Aflatoxins in Paddy and Milled Fractions of Rice in Guyana : Preliminary Results. Int J Biol Biomol Agric Food Biotechnol Eng. *International Journal of Biological, Biomolecular, Agricultural, Food and Biotechnological Engineering* **10** (11) 721–725

[4] Djaeni M, Ayuningtyas D, Asiah N, Hargono, Ratnawati, Jumali 2013 Paddy Drying in Mixed Adsorption Dryer with Zeolite : Drying Rate and Time *Reaktor* **14** (3) 173–178

[5] Soponronnarit S 1999 Fluidised-Bed Paddy Drying *Science Asia* **25** (1) 51–56.

[6] Wiset L, Srzednicki G, Driscoll R H, Nimmuntavin C and Plangpin S 2001 Effects of High Temperature Drying on Rice Quality *CIGR Journal of Scientific Research and Development* **III** (September 2017)

[7] Nagaya K, Li Y, Jin Z, Fukumuro M, Ando Y and Akaishi A 2006 Low-temperature desiccant-based food drying system with airflow and temperature control *J Food Eng* **75** (1) 71–77

[8] Gouda G P and Nidoni U 2014 Dehydration of Onions with Different Drying Methods *Curr Trends Technol Sci* **3**(3) 210–216.

[9] Djaeni M, Sasongko SB, Prasetyaningrum A, Jin X, van Boxtel AJ 2012 Carrageenan drying with dehumidified air: Drying characteristics and product quality *Int J Food Eng* 8(3), 1–21

[10] Djaeni M, Sari DA 2015 Low Temperature Seaweed Drying Using Dehumidified Air *Procedia Environmental Sci* 2015 **23** 2–10

[11] Mousa W, Ghazali F M, Jinap S, Ghazali H M and Radu S 2014 Sorption isotherms and isosteric heats of sorption of Malaysian Paddy *J Food Sci Technol* **51**(10), 2656–2663

[12] O'Callaghan J R., Menzies D J and Bailey P H 1971 Digital Simulation of Agricultural Drier Performance *J. Agric. Eng. Res* **16** (3) 223-244.

[13] Henderson S M and Pabis S 1961 Temperature effect on drying coefficient *J. Agric. Eng. Res* **6** 169-174.

[14] Page G 1949 MS Factors Influencing the Maximum Rates of Air Drying Shelled Corn in Thin Layers *Thesis Submitted to Purdue University* West Lafayette, IN, USA.

[15] Yagcioglu A, Degirmencioglu A and Cagatay F 1999 Drying characteristics of laurel leaves under different drying conditions *Proceedings of the 7th International Congress on Agricultural Mechanization and Energy*", (Ed.): Bascetincelik, A.. 26-27 May, Turkey Faculty of Agriculture, Çukurova University, Adana pp 565-569

[16] Henderson S M 1974 Progress in developing the thin-layer drying equation *Trans of the ASAE* **17** (6) 1167–1168

[17] Djaeni M and van Boxtel A J B 2009 Energy Efficient Multistage Zeolite Drying for Heat-Sensitive Products *Dry Technol* **27** (5) 721–722.

[18] Djaeni M, van Asselt CJ, Bartels PV, Sanders JPM, van Straten G, van Boxtel AJ 2011 Low Temperature Drying With Air Dehumidified by Zeolite for Food Products: Energy Efficiency Aspect Analysis *Int J Food Eng* **7** (6)

[19] Djaeni M, Anggoro D, Santoso GW, Agustina D, Asiah N and Hii CL 2014 Enhancing the food product drying with air dehumidified by zeolite *Adv J Food Sci Technol* **6** (7) 833–838.

[20] Ambekar SA, Gokhale SV and Lele SS 2013 Process optimization for foam mat-tray drying of passiflora edulis flavicarpa pulp and characterization of the dried powder *Int J Food Eng* **9** (4) 433–443

[21] Chinenye N M 2009 Effect of Drying Temperature and Drying Air Velocity on the Drying Rate and Drying Constant of Cocoa Bean *Agric Eng Int CIGR Ejournal* **11**

[22] Singh A, Sharma H K, Kumar N, Upadhyay A and Mishra K P 2015 Thin layer hot air drying of bael (Aegle marmelos) fruit pulp *Int Food Res J* **22** (1) 398–406.

[23] Hubackova A, Kucerova I, Chrun R, Chaloupkova P and Banout J 2014 Development of Solar Drying Model for Selected Cambodian Fish Species *Sci World J* 1–10

[24] Aregbesola OA, Ogunsina B., Sofolahan AE, Chime NN 2015 Mathematical modeling of thin layer drying characteristics of dika (*Irvingia gabonensis*) nuts and kernels *Niger Food J* **33** (1) 83–89

An analysis of the feasibility of the copra business in the village of Pendowo Harjo, sub-district of Sungsang, Banyuasin Regency

Y Z W Purba[1] and W Saleh[2]

[1] Department of Agribusiness, STIPER Sriwigama, Palembang, Indonesia
[2] Post Graduate Program, Agribusiness Department, Syakhyakirti University, Palembang Indonesia

Abstract
Copra is used as raw material of coconut oil and exported commodity. This study was conducted in the Tidal Land of Pendowo Harjo Village, Subdistrict of Sungsang, Banyuasin Regency, which aims to calculate the production costs incurred, the income earned, and to analyze the feasibility of the business of producing copra. In this research, sampling was conducted by simple random sampling method. The number of samples taken in this study was 10 individuals who were the copra business people out of 117 people of population. The results of analysis show that the productioncost incurred is Rp 1,198,076.12, and the income earned is Rp 414,598.88 per unit of the production process. Financially, the value of NPV obtained is Rp 19,668,343.86, the value of the IRR is 60.75 percent and the value of the Net B/C is 1.74. Therefore, economically, the copra business is feasible to be developed.

Keywords: *business, feasibility, copra, tidal land*

1. Introduction

Coconut (*Cocosnucifera L*) is a plantation commodity and socio-economic commodity grown from generation to generation in almost all parts of Indonesia of the total coconut planted area, 98% of is people owned coconut plants, either grown as a garden plants or as the house yard plants that concern the needs of the lives of the 1.3 million households that rely on the coconut plants [1]Coconut in Indonesia is basically allocated to meet the needs of fresh coconuts and copra. The fresh coconut in Indonesia is consumed as a foodstuff in the form of young coconut and coconut milk, raw material of homemade frying oil, raw material of dessicated coconut and coconut seeds [2]. Whereas copra is used as raw material of coconut oil and exported commodity [3].

The data obtained from the Assessment Institute of the Agricultural Technology of South Sumatra in 2013, estimated that 82.5% of the total coconut production is processed into copra. This is because in the marketing of coconut products, copra has higher price level. It can be stored for a long period of time. Besides, it is a raw material that can be further processed into consumer goods, such as cooking oil, butter, cheese and soap. Copra is the endosperm which has been dried through the drying process. Good quality copra is derived from the ripe coconut drupes whose age is around 11 to 12 months [1]. To business people, copra business still has a good opportunity in the future. With high coconut potential, the entrepreneurs are still interested and motivated to do business of coconut that is processed into copra or other derivative products. The copra farmers are very interested in dealing with the prospective foreign buyers from abroad, such Malaysia, Thailand, and other countries who want to accommodate their products.

Pendowo Harjo village is located in one of the low tide areas which has potential to be a marginal land [4]and to be made a place for managing the business of producing copra in the Subdistrict of Sungsang, Banyuasin Regency. The copra production at this location is very simple, but it is able to increase the income of the owner in fulfilling the needs of everyday life. The copra businesses in the village of PendowoHarjo is promising enough, because generally coconut palms flourish in the courtyard yard of the farmers in this village, so many farmers do copra production enterprises. Although in some other villages there are farmers who do copra production enterprises, but there many more farmers who conduct a business of copra production in the village of PendowoHarjo. How big is the business of producing copra can be developed[5].In this case there is still few research has beenconducted about feasibility analysisparticulary on copra.How ever, there is one similar researchbut on feasibility analysisof pepper farming in LahatRegency bySumantri, Sigit, Isroni[6].

Based on the aforementioned background, the problem can be formulated by calculating theproduction costs incurred and the expected profit earned in the business of producing copra and its feasibility in the Village of PendowoHarjo

Based on the existing problems, this study has the objectives and the usage as followsto calculate the amount of production costs incurred and profit earned in the business of producing copra in the village PendowoHarjo and to analyze the feasibility of copra producing business in the village PendowoHarjo.The usages of this study are as follows : the results of the study can be an information source for the researchers and it can also broaden the knowledge about the business of copra poduction business, particularly in the Village PendowoHarjo and the results of the study can be made as references for other researchers whose study is doing your math in connection with this study.

The model of approach used in this study describes that copra production is summed up from production factors while the gross income expected to be earned is claculated from total production multiplied by current price of copra obtained while the research was conducting. The net profit can be calculated from dividing gross income by production costs. The final result of this study is to assess the feasibilty of copra production business by calculating NPV, IRR and Net B/C.

2. Material and Method

This study was conducted on Low Tidal Land of PendowoHarjo Village, Subdistrict of Sungsang, Banyuasin Regency. The location determination is done intentionally (purposive) with the consideration that most of the residents of the village conduct the business of producing copra [7]

The method used in this study is survey [8]. Survey method is an approach used by means of interviewing a small number of the population, aiming to collect information about the large number of samples using questionnaires, interviews and direct observation directly in the area of the study [9, 10].

The data collected consist of the primary data and the secondary data [11]. The primary data are obtained from direct interviews with the copra business owners using questionnaires that have been prepared in advance. While the secondary data are obtained from the office, agencies and institutions that are related to this research. The sampling method used in this study is simple random sampling [12], with the number of samples taken as many as 10 copra businessmen from 117 members of the population.

The data obtained from the field are processed in tabulation, and then analyzed quantitatively and descriptively (processed, analyzed and conclusions drawn that describes the object under study). The calculation of cost of production and income were conducted according [13]. Feasibility was analyzed using Net Present Value (NPV) according to [14-16].

3. Result and Discussion

The results of the analysis reveal that it takes one week for one process of production of the copra, starting from the cutting of the coconuts, gouging and drying. For the average costs incurred and income earned, more details can be seen in the following table.

3.1 Production cost

Table 1. The Average Production Cost of Copra Producing Business in the Village PendowoHarjo, Subdistrict of Sungsang, Banyuasin Regency per unit process.

No	Description	Values (Rp)
1.	Fixed Cost	**26,727.78**
	a. Rent of Site	19,444.44
	b. Wear of Tools	7,283.34
2.	**Variable Cost**	**1,171,348.33**
	a. Means of Production	99,348.33
	b. Labor	372,000.00
3.	**Total Production Cost**	**1,198,076.11**

The average cost of production which has been incurred in the business of producing copra in Pendowo Harjo Village is Rp 1,198,076.11 consisting of fixed costs of ofRp 26,727.78 and the variable costs of Rp 1,171,348.33 per production process. The average cost of site rent is Rp 19,444.44, depreciation of tools (calculated based on the tool use) in the amount of 7,283.34 per production process (consisting of a drying floor is Rp 4,193.52, cleavers is Rp 939.00, gouging is Rp 520.01, baskets of rattan is Rp 659.17 and scales is Rp 971.64).

The average variable costs already incurred is Rp 1,171,348.33 per unit process, consisting of (means of production is Rp 799,348.33 and labor isRp 372,000.00). Means of production consists of the purchase of coconut, sacks and raffia, the average purchase of coconut is as many as 605 coconut drupes of big size at Rp 441,200.00, medium-sized coconut drupes as many as 549,40 drupes at Rp 200,125.00, and the coconut of small size as many as 582 drupes at Rp 141,283.33, with the average price of the coconut fruit of Rp 730.00 per item. The average cost of purchasing a sack is Rp 15,340.00 per process with an average price of Rp 1,360.00 per piece. The sacks are used to bag the copra produced which is ready to be weighed. The average purchase of raffia is Rp 1,400.00 per process.

The labor employed in the business of producing copra is derived from the internal and the external of the family. In this analysis, the use of labor from the internal and the external of the family is still reckoned to reveal the profit and the loss of the business done [17]. The average labor costs incurred include the cost of cutting the coconut in half is Rp 96,000.00, the cost of drying the coconut isRp48,000.00 , costs of gouging is Rp 152,000.00, the cost of drying is Rp 48,000.00, cost of sacking is USD 28,000.00 per process. The work of cutting the coconut into two requires an average of 2.4 person-days with an average wage of Rp 40,000.00 per person. The work of drying requires an average of 1.6 person-days with an average wage of Rp 30,000.00 per person. The work of gouging requires an average of 3.8 person-days with an average wage of Rp 40,000.00 per person. Drying requires the employment of an average of 1.6 person-days with an average wage of Rp 30,000.00 per person. The activity of sacking requires work of labor with an average of 0.70 HOK with an average wage of Rp 40,000.00 per person.

IOP Conf. Series: Earth and Environmental Science **102** (2017) 012068 doi:10.1088/1755-1315/102/1/012068

3.2 Revenue and Profit

Table 2. The Average Revenue And Profit of the Copra Producing Business In the village of PendowoHarjo, Subdistrict of Sungsang, Banyuasin Regency.

No	Description	Values
1.	Production (Kg)	455.56
2.	Selling Price (Rp/Kg)	3.540.00
3.	Revenue (Rp)	1,612,682.40
4.	Total Production Costs (Rp)	1,198,076.11
5.	Profit (Rp)	414,606.29

In terms of the economic sense, the production is a process of utilization of all the resources available to achieve the results that guarantee the quality and quantity, well managed, so it is worth being a traded commodity. The product which is derived from this enterprise is the copra which has been packaged in sacks. On the average in one production process (per production) the business produce as much as 455.56 kg of copra with an average selling price of Rp 3,540.00, so a revenue of Rp 1,612,682.40 is obtained and a profit of Rp 414,606.29 is obtained. The calculations show that in conducting the business of producing copra, the business owner receive an average revenue of Rp 1,612,682.40, out of production costs which is already incurred in the amount of Rp 1,198,076.11 (The revenue can cover the production costs which is already incurred, it means that the business is profitable for the business owner).

3.3 Business Feasibility

In this analysis, the financial, the ecological and the technical feasibilities are calculated. Financial feasibility analysis is conducted to determine whether or not the business is worth developing. An enterprise may results in profit, but it is not necessarily worth developing. The financial analysis involves the use of NPV, IRR, and Net B/C Ratio. Ecologically, according to the local people, the copra production enterprises have an impact on the environment where they live, because coconut water has not be used properly (coconut water is just thrown away), causing unpleasant odor. To avoid the environmental impact, product diversification should be done [18]. Technically in terms of copra production, purchase of raw material from the climbers who come directly to the owner of copra business, and the tools in use are still fairly simple, but the quantity and the quality of the copra produced is still the first priority [19, 20].

The results of financial calculations are as follows: The value of NPV is Rp 19,668,343.86 which means that the profits derived from the business for three years is Rp 19,668,343.86 at the prevailing interest rate of 18 percent per year. The IRR analysis is used to convince and evaluate the resulting level of investment. The value of IRR obtained is 60,75 percent. These results suggest that the businesses may still be continued because it is still profitable (profit of above 18 percent). The total investment equals the NPV or NPV is 0 lying between (18 % - 60.75%). The businesses earning will finish at 60.75 percent interest rate for 3 years. The analysis of Net BC ratio is the ratio between the net benefit which has been discounted positive (+) and a net benefit which has been discounted negative (-). The results of the calculation is the value of Net B/C of 1.74 which the business of copra producing is worth developing.

Table 3. The Feasibility of the Copra Production in PendowoHarjo Village, Subdistrict of Sungsang, Banyuasin Regency.

No	Description	Values
1	NPV (Rp)	19.668.343,86
2	IRR (%)	60,75
3	Net B/C	1,74

4. Conclusions

Based of the results of the study, the following conclusions are drawn :

a. In an effort to produce copra, the production cost incurred is Rp 1,198,076.11, and the profit obtained per unit of production is Rp 414,598.89.

b. Based on the 3 (three)-year experience of producing copra, it can be concluded that financially the copra production enterprise is feasible to be developed with an NPV of Rp 19,668,343.86, the value of RR of 60.75 percent, and the value of Net B/C of 1.74.

Based on the results of the study, to increase the added value and revenue from the coconut products, it is suggested that the business of copra production should not only take the benefit of the coconut flesh only, but also the coconut water which can be processed into *natadecoco*.The copra produced can be further processed into vegetable oil, although the manufacturing process is still the traditional way, it will be able to meet the needs of vegetable oil, particularly in the village of PendowoHarjo.

Acknowledgement

The authors would like to express their gratitude to the Chairman of Agricultural High School, Sriwigama Palembang, Indonesia for funding this research and to his staff for their good cooperation so that the writing of this paper gets better. Special thanks to Prof. AndiMulyana Faculty of Agriculture, Sriwijaya University, Palembang, Indonesia which provides valuable criticisms to the manuscript.

References

[1] Setyamidjaja D 2008 *Coconut Planting and Processing* (in Bahasa Indonesia) (Yogyakarta: Kanisius Publisher)

[2] Jackson E 2006 From Where Comes Coconuts. The Panama News (Volume 12, Number 16). Retrieved April 10, 2011.

[3] Suhardiman P 2001 *Planting Hybrid Coconuts* (in Bahasa Indonesia) (Jakarta : Penebar Swadaya Publisher.

[4] Sianturi D E 2010 *Alignments In Developing Tidal Land Farming*. Puslitbangtan, Bogor Agricultural Research Agency (in Bahasa Indonesia). Puslitbangtan, Badan Litbang Pertanian Bogor. http://els. bappenas.go.id/upload/other/Keberpihakan dalam.htm

[5] Marawila T, Ancev T and Odeh I 2011 The Economics of Agricultural Land Use Dynamics in Coconut Plantations of Sri Lanka. Paper presented at the 2011 Conference (55th), February 8-11, 2011, Melbourne, Australia.

[6] Sumantri, B., Sigit, BP., Isronita, M 2004. Analysis of the pepper farming financial feasibility (*Piper nigrum*) in Kunduran Village, Ulu Musi Subdistrict, Lahat Regency of South Sumatra (in Bahasa Indonesia) *J. Indonesian Agric. Sci.* **6** (1) 32-42

[7] Sriati 2004 *Methods of Social Research* (in Bahasa Indonesia) Faculty of Agriculture, Sriwijaya University.Indralaya.

[8] Beck S E and Manuel K 2008 *Practical Research Methods For Librarians And Information Professionals.* (New York: Neal-Schuman)

[9] Nasution S 2004 *Research Methods* (in Bahasa Indonesia) (Jakarta : Bumi Aksara)

[10] Bouma G D, Ling R and Wilkinson L 2009 *The Research Process Canadian edition.* (Don Mills : Oxford University Press)

[11] Odimegwu, CO. 2000. Methodological issues in the use of focus group discussion as a data collection tool. *Journal of Social Sciences*, **4**(2-3) 207-212

[12] Silalahi, Ulber. 2009. *Methods of Social Researches* (in Bahasa Indonesia) (Bandung: PT. Refika Aditama)

[13] Soekartawi 2002 *Analysis of Farming Enterprises* (in Bahasa Indonesia) (Jakarta: University of Indonesia Press)

[14] Harmoni A 2007 *The Feasibility Study* (in Bahasa Indonesia) (Jakarta : Gunadarma University Press)

[15] Nurmalina R, Sarianti T and Karyadi A 2010. *Business Feasibility Study* (in Bahasa Indonesia) (Bogor : Department of FEM Agribusiness. IPB)

[16] Kasmir J 2010 *Business Feasibility Study 2nd Edition* (in Bahasa Indonesia) (Jakarta (ID): Bumi Aksara)

[17] Pathiraja P M E K, Fernando M T N, Abeysekra A W A D R and Subasinghe S D J N 2010 An Assessment of Labour Availability in Major Coconut Growing Areas in Coconut Triangle *J. Coconut Research Institute of Sri Lanka (cocos)* **19**(1) 13-26. DOI: http://dx.doi.org/10.4038/cocos.v19i1.4747

[18] Kamaluddin 2008 *Diversified Farming of Coconut As an Effort To Increase Farmers' Income* (in Bahasa Indonesia) Riau University, Faculty of Agriculture.

[19] Fernando M T N, Zubair L M, Peiris T S G, Ranasinghe C S and Ratnasiri J 2007 *Economic value of climate variability impacts on coconut production in Sri Lanka AIACC Working Papers.* (Washington DC : The AIACC Project Office,International START Secretariat)

[20] Bourke R Michael and Harwood T 2009 *Food and Agriculture in Papua New Guenia* (Canberra: ANU Press)

Evaluation of food drying with air dehumidification system: a short review

M Djaeni*, F D Utari, S B Sasongko and A C Kumoro

Department of Chemical Engineering, Diponegoro University
Jl. Prof. H. Soedarto, SH, Tembalang, Semarang, Indonesia
Phone:+62 247460058

*Email: moh.djaeni@live.undip.ac.id

Abstract. Energy efficient drying for food and agriculture products resulting high quality products has been an important issue. Currently, about 50% of total energy for postharvest treatment was used for drying. This paper presents the evaluation of new approach namely air dehumidification system with zeolite for food drying. Zeolite is a material having affinity to water in which reduced the moisture in air. With low moisture content and relative humidity, the air can improve driving force for drying even at low temperature. Thus, the energy efficiency can be potentially enhanced and the product quality can be well retained. For proving the hypothesis, the paddy and onion have been dried using dehumidified air. As performance indicators, the drying time, product quality, and heat efficiency were evaluated. Results indicated that the drying with zeolite improved the performances significantly. At operating temperature ranging 50 – 60°C, the efficiency of drying system can reach 75% with reasonable product quality.

Keywords: *efficiency, dehumidified, humidity, quality, zeolite*

1. Introduction
The main challenge in global food demand is how to obtain high quality dry food products in efficient processing. The dry food or its extract can be a good option due to the long life storage and consumer convenience. To realize this preference, drying process offers the major role corresponding to the moisture removal from wet product. In general, the agriculture and food products with high moisture content (vegetables, herbs, starch products) are dried at low (10°C) to moderate temperatures (50-70°C) to conserve the valuable ingredients (protein, vitamins, enzymes, oil) as well as physical appearance such as color, and texture [1,2].

Currently, a huge amount of energy was used for food drying processes. In post harvest treatment, about 60 - 70% of total energy was used for drying [1,3,4]. The efficiency of energy utilization by food dryer ranged between 30 – 60%. It means that the total of energy used was about 1.5 – 3.0 times from the theoretical load.

Meanwhile, the modern drying technology has been widely developed with attractive results in the product quality aspects. However, the efficient dryer development has been scared. For example, the energy efficiency in freeze and low temperature dryers is lower than that of a conventional convective dryer. This is due to the low value of driving force for moisture transfer and higher latent heat of moisture evaporation. Recently, the energy usage has become an important issue with respect

IOP Conf. Series: Earth and Environmental Science **102** (2017) 012069 doi:10.1088/1755-1315/102/1/012069

to the limitation of fossil fuel supply. On the other hand, the availability of biomass as side product of agriculture crops has not been intensively used as renewable energy resource in drying process.

Conventional dryer with sunlight is very advantageous due to its low energy cost, and environmentally benign. The dryer has been widely used for a long time in various sectors, such as agriculture, fishery, forestry, and herbal medicine products. Unfortunately, the uses of this dryer is also restricted by main drawbacks of this dryer such as sustainability and product quality, which are rooted from its climate dependent [1].

Drying with pre-dehumidified air using zeolite as moisture sorbent has been previously discussed [5,6,7,8]. The zeolite was capable to remove moisture in the air close to 0% relative humidity [6]. Compared to the other adsorbents like silica, alumina and pillared clay, the adsorption capacity of zeolite was still higher even under low relative humidity. With the incorporation of zeolite in the dryer, the drying time was shorter, and thus save 70–75% of the total energy used by the dryer [1,9]. Furthermore, in multistage adsorption dryer design, the energy efficiency may achieve 80–90% [10]. Thus, the application of dryer with pre-dehumidified air for various food or agriculture crops drying can be an attractive option.

This research evaluates the application of air dehumidification with zeolite for food drying. Here, paddy and onion were dried using air dehumidified with zeolite [1, 11, 12]. The drying was conducted in various temperature and relative humidity. As performance indicators, the product quality and heat efficiency were evaluated. The results were also compared with a conventional drying without ari dehumidification system.

2. Materials and Methods

Based on the model and simple experimental test, the adsorption dryer with zeolite has shown attractive performances [10]. In this study, the dryer was tested in wider range application for agriculture crops drying. This work consisted of two main steps. Firstly, the two types of dryer (fluidized bed and tray dryers) were constructed in laboratory scale equipped with an air dehumidification section with zeolite. The dryers were used for different application (fluidized bed dryer for paddy drying, and tray dryer for onion drying). Secondly, the dryers performances were evaluated based on product quality, drying time, and or/ heat efficiency. The heat efficiency was estimated referring to the total heat used for evaporated water divided by the total heat introduced in the system [1,9,10]. The product quality was analyzed based on physical properties as well as important substance content.

2.1. Paddy Drying

The paddy drying system was designed as a fluidized bed dryer [11]. The dryer was equipped with a blower to deliver air for the fluidization process (see Figure 1). Initially, ambient air at a relative humidity (RH) of 70 - 80% and temperature about 30°C was flown to the heater at a velocity of 4.0 $m.s^{-1}$. The air was heated up to dryer temperature (designed as 40°C). The air was then used for paddy drying with initial moisture content of about 25% (w/w). To speed up the process, paddy was mixed with zeolite in a certain percentage (pre-determined as 20% (w/w) of 200 grams total mixture). The hot air evaporated moisture from paddy using sensible heat. At the same time, zeolite simultaneously adsorbed moisture from the air. So, relative humidity of air can be kept low from that improved the driving force for moisture transfer. In addition, the exothermic adsorption of vapour by zeolite released adsorption heat, which can increase the sensible heat of air. Hence, the heat was used for moisture evaporation from paddy surface [11].

The moisture in paddy was measured every 10 minutes using Krisbow KW06-404 Grain Moisture Meter until the moisture content in paddy close to 12%. The process was repeated for the inlet air temperature of 50 and 60°C, and the zeolite percentage in the mixture of 0 and 40%. The data were used to evaluate the drying rate and time. Meanwhile, the paddy quality was evaluated based on the swelling power [11].

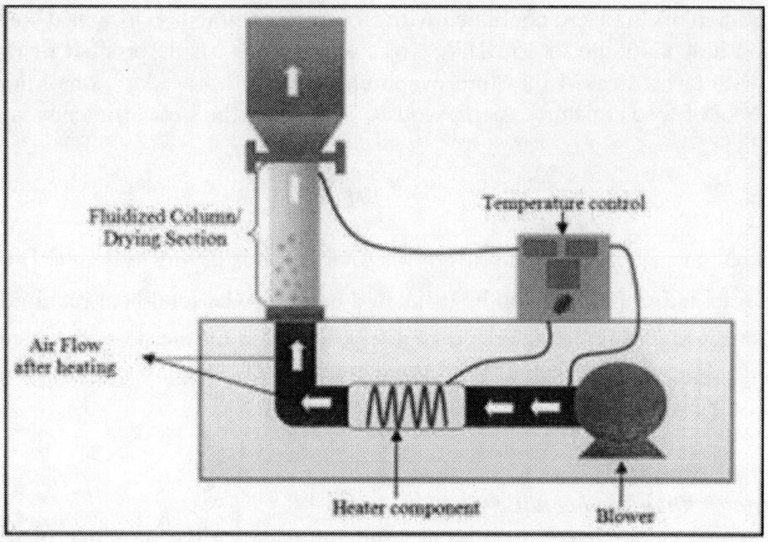

Figure 1. The schematic of fluidized bed dryer with incorporation of zeolite [1,11].

2.2. Onion Drying

Generally, onion was harvested from a farm with 88 - 92% moisture content. After drying, the average moisture content in the onion was desired to be 80 - 85% [1]. The onion drying was conducted in tray dryer completed with zeolite, as presented in Figure 2. The fresh air was heated up to 50°C by an electric heater. The air was used for onion drying with addition of a zeolite pack [1]. After evaporating the moisture from wet onion, the hot-wet air was dehumidified by zeolite. The moisture content in onion versus time was gravimetrically observed. With the data, the drying rate of moisture from onion can be developed with the model as previously published [12]. The validated model was used to find drying time for onion. The steps were repeated for different air temperature and air velocity. For comparison purpose, the onion drying without zeolite was also conducted.

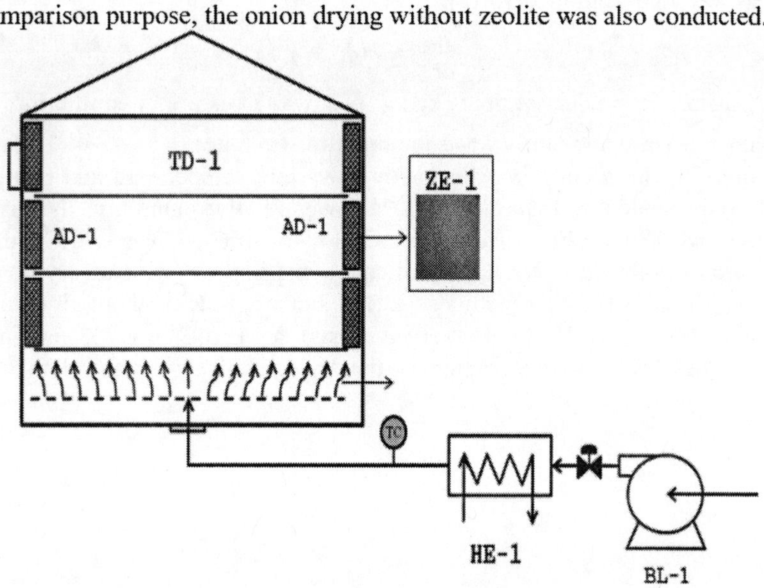

Figure 2. Onion drying with zeolite. Note: TD-1: Tray Dryer, AD-1: Adsorption Box, ZE-1: zeolite, HE-1: Air Heater, BL-1: Blower. The tray dryer dimension was 0.6x0.4x0.6 m [1]

2.3. *Heat Efficiency Estimation*

The paddy and onion drying were continued with rice husk combustion as a heat source. The rice husk with combustion heat value up to 15 MJ.kg^{-1} was obtained as a side product of rice mill industries. Theoretically, with latent heat of moisture evaporation about 2.5 MJ.kg^{-1}, one kilogram of rice husk can evaporate 6 kg of free moisture. As previously published, the heat efficiency was estimated using the equations (1) [1,9].

$$\eta = 100\%(\frac{Q_{evap}}{Q_T}) \tag{1}$$

where η was total energy efficiency (%), Q_T was total heat for drying system (kJ) and Q_{evap} was total heat for water evaporation (kJ). Q_T can be estimated based on the total heat required for heating air to dryer and air for regenerating the saturated zeolite, eliminated by the heat recycled from exhaust air [1,9]. Meanwhile, Q_{evap} was estimated based on the amount of moisture evaporated (kg) multiplied by the latent heat of water evaporation (kJ.kg^{-1}) [1,9].

3. Results and Discussion

3.1. *Paddy drying with air dehumidification*

The paddy drying was conducted at different temperatures with the presence of zeolite and without zeolite [11]. As previously discussed, zeolite was used for maintaining the moisture content of the drying air to be always low. So, the difference of this value with moisture content of paddy as driving force for drying process remains high [11]. As a result, the rate of moisture transfer from paddy to drying air was faster. The result was comparable with the dyer using the other solid adsorbents. For example, Revilla et al. showed that the drying with adsorbent such as silica, pillared clay, and zeolite speed up drying time [5]. Compared with paddy drying without zeolite conducted by Witinantakit et al., this result sounds a very positive improvement [13].

Air relative humidity and temperature determines the equilibrium moisture content in paddy as formulated with GAB model [11]. With low relative humidity and/or high temperature, the equilibrium moisture content in the product can be kept low in which speeds up drying rate proportionally as seen in equation (2) [6,11].

$$\frac{dq_{w,t}}{dt} = -k(q_{w,t} - q_e) \tag{2}$$

where $q_{w,t}$ was moisture in product at time (kg.kg^{-1}), q_e was moisture at equilibrium (kg.kg^{-1}), k was constant of drying rate (minute^{-1}) and t was sampling time (minute)

Using equation (2), the drying time for paddy at various temperature and relative humidity was then predicted, as presented in Figure 3. With a lower relative humidity, the drying time can be significantly reduced especially at temperature 40 – 70°C. Upper 70°C, the effect of air dehumidification on drying time was not significant. This improvement of drying performance was significant as compared to the other methods [13,14]. For example, without air dehumidification, the drying time of paddy from 33 to 16% (wet basis) at 150°C was 53 minutes [15]. By air dehumidification, the result was comparable with the drying at 60°C with air relative humidity of 20%.

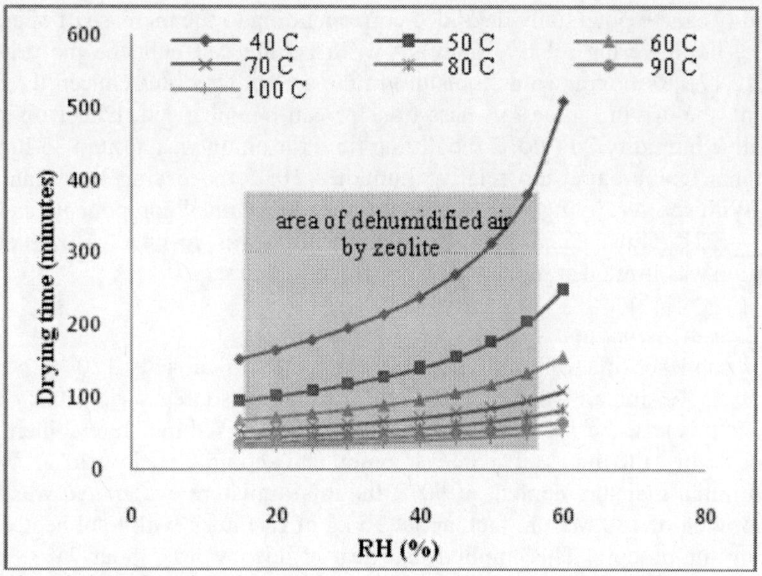

Figure 3. The drying time estimation for paddy at different temperatures and relative humidity [11].

3.2. Physical Properties of Paddy

After drying, the dried paddy was milled to remove the husk and obtain the rice. The physical properties including swelling power, percentage of head and broken rice, were observed (see Table 1). Results showed that at higher temperature the percentage of broken rice increased, for comparisons see also Sarker et al. [16]. It suggested that the paddy quality degraded. For example, at operational temperature 40 – 60°C, the percentage of head rice was about 60 - 80%. While at operational temperature 100°C, the percentage of head rice was 30%. At higher temperature, the moisture evaporation was faster. The hot vapour with large specific volume broke the structure of rice. Then, the dry paddy can be easily cracked during the milling.

Swelling power was also important to be discussed. As tabulated in Table 1, swelling power of the rice flour decreased at higher drying temperatures. Perhaps, at higher temperature the amylopectin in paddy degraded through heat moisture thermal degradation. Then, the affinity of rice starch to catch moisture decreased as indicated by the lower value of swelling power.

Table 1. Physical properties after drying and milling.

Drying Temperature °C	Physical properties			
	% (w/w) Moisture	% Head Rice	% Broken	Swelling power
40	12.60	80.40	19.60	4.20
60	12.30	78.70	21.30	3.86
80	12.20	29.87	70.13	3.83
100	11.00	13.43	86.57	3.62

3.3. Onion Drying

Onion dryings with and without the incorporation of zeolite were conducted at various temperatures. The result was presented and published in *Djaeni et al* [1]. In general, the moisture removal was highly affected by air temperature and relative humidity [7]. Higher air temperature will promote higher and faster moisture removal. The higher temperature accelerated the moisture movement or diffusivity in the tissue of onion. In addition, the available sensible heat of air for evaporating the moisture was higher. These two points increased the constant of drying rate value in which improved the water evaporation [17]. With the increase of air drying from 30 to 60°C, the drying times were found to be 60 - 70% shorter [17,18]. However, the some biochemical components, such as vitamin C, sugar and acidity can be potentially degraded corresponding to the increase of operating temperature.

Meanwhile, the lowering relative humidity with zeolite can enhance the driving force for mass transfer [2,5,12,17]. Removing some moisture in the air with zeolite reduced the relative humidity of air. As a result, the driving force for mass transfer can be enhanced. Under operational temperature 40°C and relative humidity 80 - 90%, the drying time for onion was around 35 hours. While, with the same operational temperature and relative humidity 10%, the drying time can be about 10 hours shorter [17]. With shorter drying time, the quality of biochemical components and appearance can be retained [19 – 22]. Thus, an acceptable onion quality was resulted. However, the effect of air dehumidification was limited at higher operational temperature [17].

3.4. Heat Efficiency Estimation

As previously published, the onion drying was conducted in capacity 120 kg per batch with the air temperature 50 - 55°C and air velocity of 0.7 $m.s^{-1}$. As a heat source, a rice husk combusted in furnace was used [1]. In this test, the average moisture content comprised moisture content in onion bulb layer and leaf. After drying, the moisture content in onion leaf should be below 20%. With 120 kg capacity of onion with initial moisture content of 90%, the total moisture evaporated was 72 kg needing total heat for evaporation of 180 MJ. In fact, about 16 kg of rice husk with total heat value of 240 MJ was required during the process. This implied that heat efficiency was about 75%. Meanwhile, the heat efficiency of onion drying without zeolite was around 66% [1]. With higher heat efficiency by air dehumidification, air recycle, or heat recovery, the total energy usage for moisture removal in drying can be saved [1,9]. The same story was also shown in paddy drying where the energy efficiency can be significantly improved with zeolite [1].

4. Conclusion

The air dehumidification with zeolite has been applied for tray and fluidized bed dryers. The dryers were used for paddy and onion drying. The zeolite as moisture adsorbent performed well. Therefore, the air can be significantly dehumidified. Based on product quality retention, drying time estimation, and heat utilization, the air dehumidification affected the drying performance positively. The fluidized bed and tray dryers were then operated with rice husk as a heat source. Results showed that the heat efficiency can achieve around 75%. This performance can be promising for sustainable food drying development.

Acknowledgements
The review paper was facilitated by Director General of Higher Education, Indonesia and Diponegoro University under research schema of RAPID.

References

[1] Djaeni M, Asiah N, Suherman S, Sutanto A, Nurhasanah A 2015 *Int. J. Renewable Energy Development (IJRED)* **4**(1) 20-24
[2] Djaeni M, Anggoro D, Santoso G W, Agustina D, Asiah N, Hii C L 2014 *Advance J. Food Science and Technology* **6**(7) 833-838
[3] Kudra T 2004 *Drying Technology* **22** 917-932
[4] Gilmour J E, Oliver T N, Jay S 1998 *Proc. 11th Int. Drying Simposium* volume A August 19-22 (Greece:Halkidiki) pp.573-580
[5] Osorio-Revilla G, Gallardo-Velázquez T, López-Cortés S, Arellano-Cárdenas S 2006 *Drying Technology* **24**(8) 1033-1038
[6] Djaeni M, van Asselt C J, Bartels P V, Sanders J P M, van Straten G, van Boxtel A J 2011 *Int. J. Food Engineering* **7**(6)
[7] Ratti C 2001 Journal of Food Engineering 49(4) 311-319
[8] Atuonwu J C, van Straten G, van Deventer H C, van Boxtel A J B 2011 *Chemical Engineering*

Transaction **25** 111-116
[9] Djaeni M, Bartels P, Sanders J, van Straten G, van Boxtel A J B 2008 *Drying Technology* **26**(4), 487-502
[10] Djaeni M, Bartels P V, van Asselt C J, Sanders J P M, van Straten G, van Boxtel A J B 2009 *Drying Technology* **27**(11) 1205-1216
[11] Djaeni M, Ayuningtyas D, Asiah N, Hargono H, Ratnawati R, Wiratno W, Jumali J 2013 *Reaktor* **14**(3) 173-178
[12] Djaeni M, Sasongko S B, Prasetyaningrum A, Jin X, van Boxtel A J 2012 *Int. J. Food Engineering* **8**(3)
[13] Witinantakit K, Prachayawarakorn S, Nathakaranakule A, Soponronnarit S 2006 *Drying Technology* **24**(5) 609-617
[14] Taweerarattanapanish A, Soponronnarit S, Wetchakama S, Kongseri N, Wongpiyachon S 1999 *Drying Technology* **17** 346-353
[15] Soponronnarit S 1999 *Science Asia* **25**(1) 51-56
[16] Sarker M S H, Ibrahim M N, Ab. Aziz N, Mohd. Salleh P 2014 *Energy Conversion and Management* **77** 380-395
[17] Asiah N, Djaeni M, Hii CL 2017 *Int. J. of Food Engineering* **13**(9),20160401
[18] Sarsavadia PN 2007 *Renewable Energy* **32**(15) 2529-2547
[19] Mota C L, Luciano C, Dias A, Barroca M J, Guiné R P F 2010 *Food and Bioproducts Processing* **88**(2-3) 115-123
[20] Kaymak-Ertekin F, Gedik A 2005 *J.Food Engineering* **68**(4) 443-453
[21] Arslan D, Musa Özcan M 2010 *LWT-Food Sci. and Technology* **43**(7) 1121-1127
[22] Djaeni M, Arifin UF, Sasongko S B 2016 *AIP Conf. Proc.* **1823** Article 020041

Effect of bio-organomineral fertilizer on the growth of chili (*Capsicum annum* l.)

R T Nofiyanto[1], V R Wati[1], S R Setiawati[1], W D Noviandi[1], A Kuscahyanti[2] and E Fuskhah[1*]

[1]Departement of Agriculture, Faculty of Animal and Agricultural Science, Diponegoro University, Tembalang, Semarang 50275, Indonesia
[2]Departement of Animal Science, Diponegoro University, Tembalang, Semarang 50275, Indonesia

*Email: eny_fuskhah@yahoo.com

Abstract. The productivity of chili in 2014-2015 has decreased by 29,411 tons due to conversion of agricultural land. One of the efforts to increase the production of chili is by using latosol acid as a cultivation area. The low fertility of latosol land can be overcome by fertilizing. The purpose of this research was to investigate the role of duckweed and *Methylobacterium* on chili production which cultivated at latosol soil. The design used was Completely Randomized Design (RAL) with 6 treatments and 4 replications. The treatment consisted of P0 = control, P1 = 50 kg / ha, P2 = 100 kg / ha, P3 = 150 kg / ha , P4 = 200 kg / ha, and P5 = 250 kg / ha. Assamed data were analyzed by using variance analysis (ANOVA), Duncan's Multiple Range Test (DMRT) with confidence level 95% was used for further analysis. The results showed that fertilizer application can increase the fertility of latosol soil, percentage of chili pepper plant life, height, number of productive branches, number of flower, and amount of fruit.

Keywords : *bio-organomineral, chili production, latosol soil*

1. Introduction

Indonesia as an agricultural country is a potential area for agrocomplex development. Agriculture has a great role for human life as a source of food needs. One of the potential agricultural products in Indonesia is horticultural crops such as vegetable crops. Unstable climatic changes and conversion of agricultural land in Indonesia are a major problem in horticulture cultivation. Based on statistical data about 180,000 ha of land per year turned into non-agricultural land. One attempt to overcome this is by utilizing the less productive lands in Indonesia, such as the latosol land of 84.63 million ha [1]. Latosol soils are commonly scattered in mountain slopes and mountains, and the soil is acidic. Chili is a plant that is widely cultivated in Indonesia. Domestic chili demand continues to increase as the population increases, indicating that chili is an economical commodity. However, the last three years of pepper plant productivity has decreased. One solution to overcome the low fertility of latosol land and optimization of chili productivity, among others, by fertilizing. The recommended fertilizer is organic fertilizer because it may improve the physical, chemical and biological properties of the soil. One ingredient that may be used as a mixture of organic fertilizer is the L. minor Lemna (duckweed) because it contains macro nutrients, with high N levels obtained from nitrogen fixation in waters [3]. On the other hand, Methylobacterium may increase the tolerance of stress and produce Indole Acetic Acid (IAA) hormone to accelerate plant growth [2]. The purpose of this research was to investigate the utilization of duckweed and Methylobacterium as bio-organomineral fertilizer in pellet form (Pellet

MOSS) to obtain optimal chilli production in acid souros latosol and give dosage recommendation of Pellet MOSS.

2. Materials and Methods

The research was conducted at Greenhouse of Central for Food Crops and Horticulture (BPTPHP) Ungaran, Central Java, Ecology and Plant Production Laboratory, Faculty of Animal Husbandry and Agriculture and Integrated Laboratory Diponegoro University, Semarang, from March to July 2017. Tools used were set of analytical tools, shovels, trays, scissors, sieve, bucket, hoe, trashbag, polybag size 40 cm x 35 cm, manual pelleter machine, scales, hose, stationery, camera. The materials used were Lemna minor, cow dung manure, zeolite, carrier in the form of jackfruit leaf (DN), rambutan (*Nephelium lappaceum*) leaf (DR), lettuce leaf (DS) and basil leaf (DK), water, chemicals for analysis, isolation and bacterial culturing, planting media in the form of acid sour latosol and curly chili seeds.

The design of this study used a Completely Randomized Design (RAL) monofactor 6 treatment with 4 replications. The factor is dosage of Pellet MOSS consisting of control (P0), 5 ton / ha (P1), 10 ton / ha (P2), 15 ton / ha (P3), 20 ton / ha (P4), 25 ton / ha (P5). The stages of the research are started with the preparation of chemicals for the manufacture of AMS (Ammonium Mineral Salt) media, Methylobacterium isolate from jackfruit leaf, rambutan leaf, lettuce and basil leaf. The four samples of the Methylobacterium isolate were taken to the Integrated Laboratory to be isolated and identified according to procedure [9]. Methylobacterium is then cultured in accordance with the procedure [11]. The ingredients of the fertilizer were dried and mashed, the composition of Pellet MOSS consisted of a mixture of duckweed (Lemna minor L.), cow manure, zeolite with a ratio of 1: 1: 1 and added 1: 2 tapioca starch composition ratio as pellet adhesive. Bio-organomineral fertilizers were mixed with Methylobacterium culture results then made in pellet form. Curly chili seeds are transferred in pots filled with sterile latosol soil (10 kg), each pot is planted as much as 1 chili seedlings. Pellet MOSS is applied to each pot according to the dosage treatment. Furthermore, observation of growth parameters and crop production every week for 5 weeks. Finished soil used for planting was analyzed to determine the level of fertility differences before and after given Pellet MOSS fertilizer. Parameters observed were germination, plant height, number of flowers per plant, number of productive branches, number of flowers, number of fruit and fruit production. The data obtained were analyzed by using variance analysis (ANOVA) and the data that had real effect were evaluated by Duncan's Multiple Range Test (DMRT) with 95% confidence level.

3. Result and discussions

Pellet MOSS and latosol soil media have been analyzed the nutrient content of N, P, K, C-organic and organic material (OM), the following are data obtained:

Table 1. Results of Analysis of N, P and K Pellet MOSS Content Levels.

Material	Content (%)				
	N	P	K	C-Org	OM
Pellet MOSS	4,27	0,70	1,01	4,50	45,11

The results of nutrient content analysis showed that the nutrient content contained in MOSS pellets is well characterized by high levels of N, P, K, C-organic and OM. This condition may stimulate vegetative growth and generative of chili plants. This is in accordance with the opinion of [5] that the addition of organic fertilizer can enrich the OM and C-organic as a soil enhancer. The results showed that the content of N = 1.45%, P = 0.21, K = 1.18, C-Org = 4.50, OM = 45.11. In accordance with the opinion of [1] explains that organic fertilizer with a mixture of natural minerals can enrich nutrient content and high organic matter in meeting the nutritional needs of soil and plants.

3.1. Soil fertility
The results of soil fertility showed N content on latosol soil that has been applied pellet MOSS were as follows:

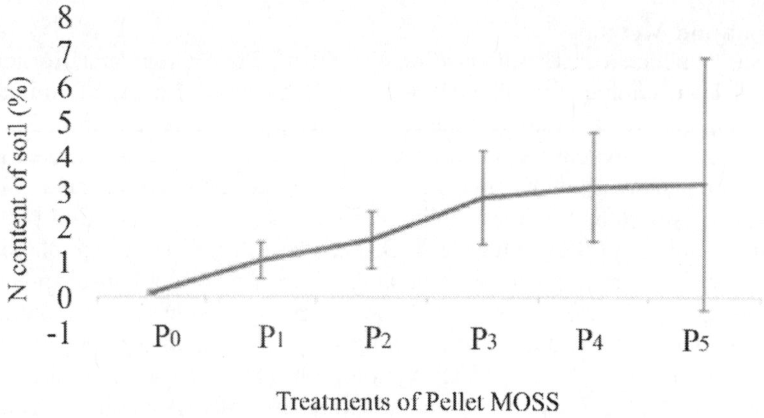

Figure 1. N content of soil after planting.

Based on the graph of NIAL analysis of MOSS pellet, it was found that the highest N at treatment P5, P4 were 3.2% and 3.1% respectively. The content indicates that the higher the dosage of the pellet MOSS given the higher the N content on the soil that pellet MOSS was applied.

3.2. Growing Power (GP)
Based on the observation of chili pepper seed germination indicated that germination process progressed faster ie age 7 Day After Seeding (DAS) from 12 - 30 HSS on Pellet MOSS treatment. The observed plant germination parameters are presented in the following observation table (Table 2)

Table 2. Calculation Results of Spice Power of Chili Plants.

Parameter	Treatment					
	P0	P1	P2	P3	P4	P5
DB (%)	74	79	80	82	86	88

It shows more and more dosage of MOSS pellet given then the higher the germination power of the plant. The amount of germination is due to the bacteria Methylobacterium spp. In a MOSS pellet that can increase the vigor or germination of a seed. This is in accordance with [2] these bacteria produce growth hormone trans-zeatin cytokines and auxin of indole acetic acid (IAA), Those two bacteria stimulate seed germination, and plant development.

Table 3. F test values on observed parameters.

Source of Varian	PH	TI	NPB	TF
Treatment	2,43^{ns}	2,36^{ns}	3,66*	1,22^{ns}

[a] PH = plant height, TI = total interest, NPB = number of productive branches, TF = Total Fruit
* = Significant; Ns = not significant

3.3. Plant Hight (PH)
Based on the high growth chart of the 75 HST chilli plant shows that the more doses of MOSS pellets were given the higher the increase in plant height (Figure 2).

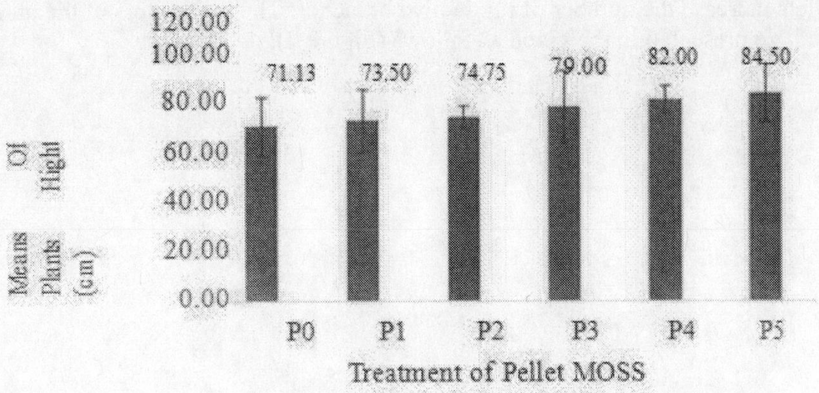

Figure 2. Charts of High Chill Plants at 75 DAP.

The best plant height was found in P5 dose, there is real difference between P0 and P4 and P5. Increased growth of pepper plants was not significant (P> 0.05). The best high growth can be obtained because of the high N content in lemna minor in MOSS pellet composition which is 6% [7] that can spur the vegetative growth of pepper plants so that the more optimal the dosage of MOSS pellet is given the higher the plant growth resulting from.

3.4. Number of Flowers (NF)

Based on the chart the number of flowers of chili plants at 75 DAP shows that the more dosage of MOSS pellets given the more the number of flowers produced. The parameters of the number of plant flowers observed are presented in the graph as follows (Figure 3).

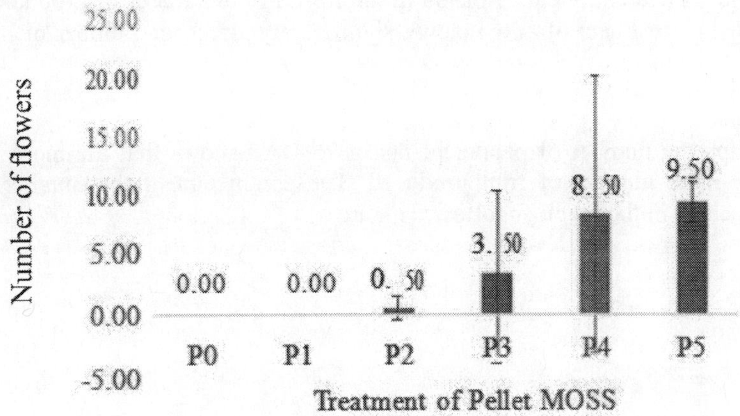

Figure 3. Graph of Chili Plant Flower Amount 75 DAP.

Based on the result of data analysis, it is known that the highest amount of flower is found in dose P5, there was significantly difference between P0, P1 and P5. The use of MOSS pellets was an impact on the growth of generative pepper plants is not significant (P> 0.05). This is supported by research of [6] organic fertilizer with doses of 30 ton / ha produced yield of crop production from eggplant family of 13 ton / ha.

3.5. Number of Productive Branches (NPB)
Based on the graph the number of productive branches of chili plants 75 DAP indicates that the more dosage of MOSS pellets increase the number of productive branches. The parameters of the number of plant flowers observed are presented in the graph as follows (Figure 4).

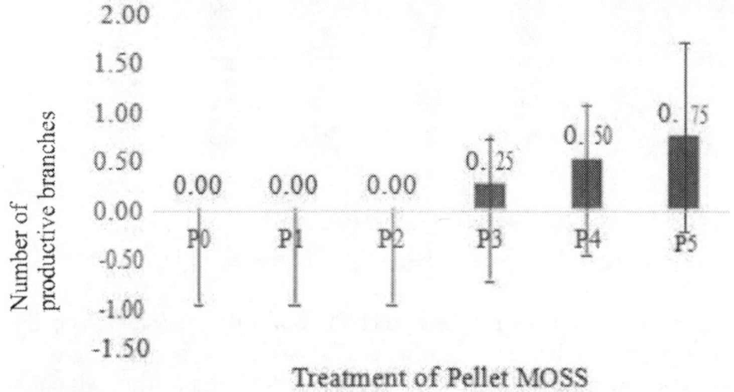

Figure 4. Number of Productive Branches of Chili Plants at 75 DAP.

The result of the data analysis revealed that the largest number of branches was found in the dosage of P5, there was a significant difference between P0, P1, P2, P3 and P5. Use of MOSS pellets Impacted significantly on the production of chili ($P < 0.05$). Indicates that the giving of organic fertilizer in the form of cow dung can give significant difference in the number of productive branches with the average number of branches as much as 82.88. Cattle manure can increase soil fertility so that it can improve the growth and yield of plants in the form of the number of productive branches. It is also supported by [4] treatment of cow manure significantly affect the number of branches at age 75 days.

3.6. Total of Fruits
Based on the graph the number of pepper plants of 75 DAP shows that the more dosage of MOSS pellets given the more number of fruit produced. The parameters of the number of plant flowers observed are presented in the graph as follows (Figure 5).

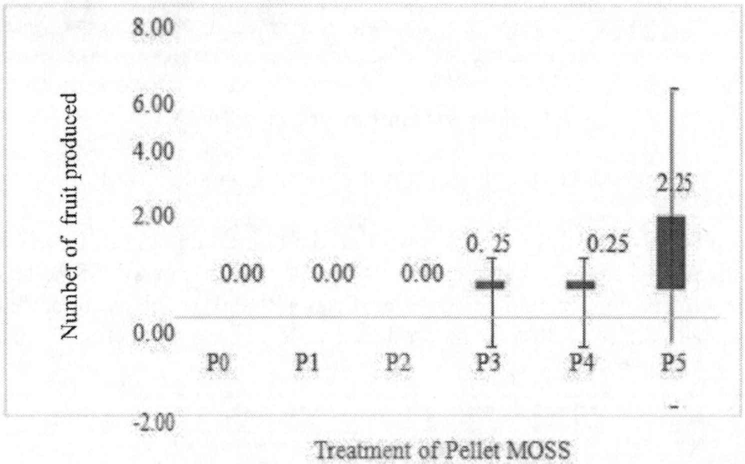

Figure 5. Chili Plant Amount 75 DAP.

The result of data analysis showed that the highest amount of fruit was found in dose P5, there was a significant difference between P0, P1 and P5. The use of MOSS pellets has resulted in a significant increase in pepper production (P <0.05). Better chili pepper production is found in P5 treatment, this is due to the high dosage of MOSS pellets providing sufficient nutrient availability for generative growth. This is supported by the opinion of [2] which states that organic fertilizer is the main source of macro nutrients such as N, P, K, Ca, Mg, and S as well as essential micro nutrients to enhance vegetative and generative growth of chili plants.

4. Conclusion

Based on the results of the research can be concluded that the application of MOSS pellet can improve soil fertility, germination, growth and production of chili. Pellet MOSS has the potential to be developed as a qualitative bio-organomineral fertilizer with a dose of fertilizer recommendation of 25 tons / ha.

Acknowledgements

The Authors would like to thank to Dr. Eny Fuskhah, M. Si as the guidance counselor who has guided us during the research, we also thanks to Mr. Ahmad as a laboran who helped during the research in the Laboratory and our friends who always support us.

References

[1] Fadila R 2015 Response of IF8 and lantern varieties of rice to the application of organo mineral fertilizer and biofertilizer to incepticol Situgede, Bogor. *J. Tropica.* **1** 55 – 74.

[2] Hayati E, T Mahmud, and R Fazil 2012 Effect of organic fertilizer and varieties on growth and yield of pepper plant (*Capciscum annum* L.). *J. Floratek.* **7** 173 – 181.

[3] Kim K, W Ying, dan P Trivedi 2010 Synergistic effects of inoculating arbuscular mycorrhizal fungi and *Methylobacterium oryzae*strains on growth and nutrient uptake of red pepper (*Capsicum annuum* L.). *J. Plant and Soil.* **327**(1) 429 – 440.

[4] Nopriani U, P D M H Karti, and I Prihantoro 2014 Productivity of duckweed (*Lemna minor*) as a forage alternative feed of livestock at different light intensities. *JITV.* **19**(4) 272 – 286.

[5] Prasetya M E 2014 Effect of pearl NPK fertilizer and cow manure on growth and yield of red curly pepper varieties of Arimbi (*Capsisum annum* L.). *J. AGRIFOR.* **13**(2) 191 – 199.

[6] Prihandini P W and T Purwanto 2007 *Technical guidance of composting made from cow dung.* Ministry of Agriculture: Center for Livestock Research and Development.

[7] Pujisiswanto H and D Pangaribuan 2007 Effect of dosage of cow manure on growth and production of tomato Lampung fruit. University of Lampung, 17 – 18 November 2008.

[8] Subhan N, Nurtika, and N Gunadi 2009 Tomato plant response to the use of compound fertilizer npk 15-15-15 on latosol soil in the dry season. *J. Horticultur.* **19**(1) 40 – 48.

[9] Sukmadi R B 2013 The activity of fitohormone indole-3-acetic acid (IAA) from several isolates of rhizosphere and endophytic bacteria. *J. Science and Technology.* **14**(3) 221 – 227.

[10] Suryanti T and Priyanto 2013 Elimination of Heavy Metal Cadmium In Wastewater Using Water Plants. *J. Environmental Enginering.* **4**(3): 143 – 147.

[11] Wibowo R S 2011 Influence of Methylobacterium spp Applications. And Dose of N, P, K Fertilizer on Rice Plant Growth (*Oryza sativa* L.). *Essay.* Faculty of Agricultur. Bogor Agricultural Institute. Bogor.

Process optimization for sensory characteristics of *serials* (*Flacourtia jangomas*) ready-to-drink (RTD) beverage

L Cimafranca[1*] and E Dizon[2]

[1]Visayas State University, Visca, Baybay City, Leyte, Philippines
[2]Institute of Food Science and Technology, College of Agriculture and Food Science, University of the Philippines Los Baños, 4031, College Laguna, Philippines

*E-mail: lynette.cimafranca@vsu.edu.ph

Abstract. *Seriales* (*Flacourtia jangomas*) is an underutilized fruit in the Philippines. The processing of the fruit into a RTD beverage was standardized by statistical methods. Plackett-Burman Design (PB) was used to determine the most significant factors that affect the sensory characteristics of the product. Response surface methodology (RSM) was applied based on the factorial Central Composite Design (CCD) to determine the optimum conditions for the maximum sensory acceptability of the *seriales* RTD beverage. Results of the PB revealed that the most significant factors were blanching time, level of *seriales* and TSS level. With different levels of blanching time (0.5, 1.0, and 1.5 min.), *seriales* level (10, 20, 30 %) and TSS value (12, 15, 18°Brix), the optimum region for sensory acceptability was perceived at 0.7 to 1.4 minutes blanching time, *seriales* level of not beyond 27 %, and TSS at any level.

Keywords: *Flacourtia jangomas, seriales*, r*eady-to-drink beverage, optimization*

1. Introduction

Flacourtia jangomas or *seriales* as commonly known in Visayan region is one among other less-known commodities in the Philippines. It belongs to the Flacourtiaceae family with varying names such as coffee plum, East Indian plum, Indian plum and Manila cherry among others [1], depending on the place where it is grown. It can be described as a large erect shrub or a small tree which may have a height of 5 to 10 m with low branches [1]. The shape of the fruits is round which are green in color when young and turns dark-maroon when ripe. The taste is characteristically acid to sweet. It has become an important fruit crop in some parts of Southeast Asia, specifically in India since it is known to the ethnic group of Gorakhpur region and North-Easternterai region for food and medicine purposes [2]. Likewise, other countries such as Malaysia and the Philippines have regarded the crop very important because it can be used as medicine to alleviate sufferings from injuries and diseases.

The chemical components of the fruit which possess bio-functional properties is the main reason for its medicinal capability. Ripe fruits are eaten to overcome nausea, diarrhea [2], liver related disorders[3] and biliousness[2,3]. Studies so far have shown that the Flacourtiaceae elaborates a diverse array of compound classes which include terpenoids, alkaloids, flavonoids and tannins, lignans and flavanolignans, glucosides, coumarins and isocoumarins, as well as xanthones, quinones, limonoids and phenazines, which are distributed in the bark, roots, leaves, and fruits of the plant [2,4].

Among other underutilized fruits, *seriales* do not receive much attention as compared to commercial fruits like guava, papaya, pineapple and mangoosteen [5]. Some reasons could be due to lack of information about the crop, its nutritional composition, physical qualities, bioactive properties and the potential benefits of consuming such. There are some research studies on *F. jangomas*, but these studies mostly are confined on the bioactive properties of leaves and roots. Studies on the exploration of the processing potential of the fruit, specifically on the production of RTD beverage have to be investigated; hence, this study was undertaken.

The objective of this paper was to optimize the processing condition for the production of RTD beverage utilizing *seriales* (*F. jangomas*) fruit. The statistical models namely Plackett-Burman design was used to determine the most significant factors that affects the sensory acceptability of the product, while Response Surface Methodology was used to determine the effect of blanching time, level of *seriales* and TSS level. The Central Composite Design (CCD) was used to optimize the process conditions that plays significant role in the sensory acceptability of *seriales* RTD beverage.

2. Materials and Methods

2.1 Experimental Materials

Fresh *seriales* fruits were obtained from the different barangays of Baybay City, Leyte, Philippines. Other ingredients such as refined sugar and food grade sodium metabisulfite were purchased in the local supermarket.

2.2 Preparation of Ready-to-Drink Seriales Beverage

The product was processed into *seriales* RTD beverage based on the generalized method of Bates et al. (2001)[6] with some modifications. The fruits were inspected and sorted. Good quality fruits were used in the processing, while fruits with inferior physical quality were discarded. Fruits were then washed thoroughly and blanched (0.5, 1.0 and 1.5 min.). Blanched fruits were added with water (1:2.5) and blended (5 min.) using a blender. Then, sodium metabisulfite (0.01%) was added, and stirred. The pH of the blend was adjusted to 3.8 and TSS to 12, 15, and 18° Brix. Pasteurization was done at 90°C for eight (8) minutes, and filled hot in glass bottles. The bottles were sealed with caps, inverted and cooled at room temperature. After cooling the products were stored in the refrigerator.

2.3 Sensory Evaluation

Sensory evaluation using quality scoring in combination with 9-point Hedonic rating scale was carried out to assess the sensory attributes of the different treatments. Samples accompanied with a sensory score sheet were presented to 60 panelists of Food Science and Technology students following an Incomplete Block Design (IBD) as laid out by Cochran and Cox (1957)[7].

2.4 Plackett-Burmann Design (PB)

Plackett-Burman design was used in the preliminary study to screen variables that affects the product formulation and product processing of ready-to-drink (RTD) beverage. Different variables that affect the product were identified. High (+) and low (-) levels were assigned to each independent variables, and these were laid out in a design matrix presented in Table 1. Frozen *seriales* represents the low level for V_1, (type of material), while fresh fruit represents the high level. V_2 refers to blanching time where no blanching represents the low, and 80°C for 3 minutes as high level. The level of *seriales* (V_3) represented 25 and 50%, while maceration time (V_4) was 2 and 5 minutes, respectively. Sodium metabisulfite (V_5)

was also assigned 0 and 0.1 %, TSS value (V_6) of 8 and 18°B, pH level (V_6) of 2.5 and 3.8, pasteurization temperature (V_8) of 93 and 95°C, polyethylene terephthalate (PET) and glass bottle as packaging material (V_{10}), and a dummy variable (V_{11}) to represent the noise factors.

Sensory evaluation followed, and results were decoded. Response variables (sensory acceptability) were then evaluated to determine the effects of the different independent variables. The principal effects of each variable on the sensory acceptability were estimated as the difference between both averages of measurements made at the higher level and at the lower level; and the significance of each variable was determined. The significant independent variables ($p<0.05$) were the ones used in the experimental design for the optimization study, while the less important variables were screened out.

Table 1. Statistical analysis using Plackett-Burman design showing the main effects of independent variables (color, aroma, taste, flavor, astringency and general acceptability) in *seriales* RTD beverage.

INDEPENDENT VARIABLES /FACTORS	RANGE OF VALUES		DEPENDENT VARIABLE/RESPONSES								
			COLOR ACCEPTABILITY[a]			AROMA ACCEPTABILITY[b]			TASTE ACCEPTABILITY[c]		
	Low (-)	High (+)	Effect	t(168)	*p* – value	Effect	t(168)	*p* - value	Effect	t(168)	*p* - value
Seriales (type)	Frozen	Fresh	0.767**	3.857	0.000	-0.011ns	-0.060	0.952	-0.156ns	-0.670	0.504
Blanching time	None	100°C/ 3 min	-0.922**	-4.640	0.000	-0.211ns	-1.141	0.255	-0.311ns	-1.341	0.182
Level of Seriales	25%	50%	-0.411*	-2.068	0.040	-0.078ns	-0.420	0.675	-0.644*	-2.777	0.006
Maceration	2 minutes	5 minutes	0.078ns	0.391	0.696	0.278ns	1.502	0.135	0.467*	2.011	0.046
Sodium benzoate level	0 %	0.1 %	0.056ns	0.280	0.780	0.100ns	0.541	0.589	0.178ns	0.766	0.445
TSS	8°B	18°B	0.500*	2.516	0.013	0.322ns	1.742	0.083	1.400**	6.033	0.000
pH	2.5	3.8	0.144ns	0.727	0.468	0.233ns	1.261	0.209	0.867**	3.735	0.000
Cooking temperature	93°C	95°C	-0.167ns	-0.839	0.403	-0.233ns	-1.261	0.209	-0.333ns	-1.437	0.153
Cooking time	8 minutes	12 minutes	0.033ns	0.167	0.867	0.033ns	0.180	0.857	-0.533*	-2.298	0.023
Packaging	PET	Glass	0.033ns	0.168	0.867	-0.233ns	-1.261	0.209	-0.200ns	-0.862	0.390
Dummy	-1	+1	0.300ns	1.509	0.133	0.033	0.180	0.857	-0.111ns	-0.479	0.633

Continuation "Table 1"...

INDEPEN-DENT VARIABLES/ FACTORS	RANGE OF VALUES		DEPENDENT VARIABLE/RESPONSES								
			FLAVOR ACCEPTABILITY			ASTRINGENCY ACCEPTABILITY			GENERAL ACCEPTABILITY		
	Low (-)	High (+)	Effect	t(168)	*p*-value	Effect	t (168)	*p*-value	Effect	t (112)	*p*–value
Seriales (type)	Frozen	Fresh	-0.122^{ns}	-0.515	0.607	0.022^{ns}	0.094	0.925	0.122^{ns}	0.548	0.584
Blanching time	None	100°C/ 3 min	-0.322^{ns}	-1.358	0.176	-0.422^{ns}	-1.782	0.077	-0.300^{ns}	-1.346	0.180
Level of Seriales	25%	50%	-0.522^{*}	-2.201	0.029	-0.333^{ns}	-1.407	0.161	-0.589^{*}	-2.642	0.009
Maceration	2 minutes	5 minutes	0.322^{ns}	1.358	0.176	0.422^{ns}	1.782	0.077	0.411^{ns}	1.845	0.067
Sodium benzoate level	0 %	0.1 %	0.189^{ns}	0.796	0.427	0.000^{ns}	0.000	1.000	0.078^{ns}	0.349	0.728
TSS	8°B	18°B	1.478^{**}	6.227	0.000	0.911^{**}	3.845	0.000	1.322^{**}	5.934	0.000
Ph	2.5	3.8	0.722^{**}	3.043	0.003	0.689^{*}	2.907	0.004	0.789^{**}	3.540	0.001
Cooking temperature	93°C	95°C	-0.389^{ns}	-1.639	0.103	-0.511^{*}	-2.157	0.032	-0.522^{*}	-2.343	0.020
Cooking time	8 minutes	12 minutes	-0.256^{ns}	-1.077	0.283	-0.378^{ns}	-1.594	0.113	-0.167^{ns}	-0.748	0.456
Packaging	PET	Glass	-0.322^{ns}	-1.358	0.176	-0.267^{ns}	-1.125	0.262	-0.233^{ns}	-1.047	0.297
Dummy	-1	+1	-0.078^{ns}	-0.328	0.744	-0.144^{ns}	-0.188	0.851	0.011^{ns}	0.050	0.960

^{ns} – no significant difference at $p > 0.05$; **- highly significant and *- significant at $p \leq 0.05$

2.5 Response Surface Methodology

Response Surface Regression (RSReg) analysis using SAS Statistical Software was used to determine the effects of independent variables on the response variables studied including ANOVA for the regression coefficients and parametric estimates. Response surface plots were made for all analysis for each run condition using STATISTICA Version 8 computer program to clarify the different effects of the factor variables on the response studied. Contour plots for the sensory acceptability were superimposed and optimum condition was determined.

3. Results and Discussion

3.1 Model Fitting

The ANOVA revealed low R^2 values (Table 2), which is expected since sensory evaluation is subjective, thereby results in parameters measuring goodness of fit is not always ideal.

Table 2. Response Variable Information and ANOVA for *seriales* RTD beverage.

Response (Acceptability)	Response Mean	Root Mse	R-Square	Coefficient Of Variation	Lack Of Fit	
					F-Value	*P*-Value
Color	6.63	1.12	0.33	16.46	5.63^{*}	0.0001
Viscosity	6.94	1.12	0.14	16.17	1.30^{ns}	0.2618
Flavor	7.04	1.11	0.08	15.86	0.47^{ns}	0.7972
Taste	7.13	1.28	0.20	17.97	2.47^{*}	0.0325
Aftertaste	6.90	1.39	0.18	20.22	2.13^{ns}	0.0619
General	6.79	1.21	0.20	17.82	4.28^{*}	0.0009

*Significant at $p<0.05$ ^{ns} not significant at $p<0.05$

In the study, the parameter estimates were calculated and are presented in Table 3. Table 2, on the other hand, summarized the response variable information and ANOVA. Response surface plots of the interaction of (a) blanching time (x-axis) and seriales level (y-axis), with TSS level constant at 15°B; (b) blanching time (x-axis) and TSS level (y-axis), with seriales level constant at 25 %); (c) seriales level (x-axis) and TSS level (y-axis), having blanching time constant at 1.0 min; and their influence on the acceptability for (A) color, (B) viscosity, (C) flavor, (D) taste, (E) aftertaste, and (F) general acceptability are illustrated in Figure1-A to F. Meanwhile, the optimum regions for the seriales RTD beverage acceptability (acceptability ≥ 7.0 in the 9-point Hedonic scale) at constant (a) 15°B TSS; (b) 25 % seriales level; (c) 1 minute blanching time are shown in Figure 1-G.

3.2 Effects of Blanching time, Seriales level, and TSS level on the response variables

3.2.1 Color Acceptability. The hypothesis test for linear, quadratic and crossproduct terms for blanching time (X_1) and *seriales* level (X_3) are significant (Table 3). The parameter estimates' table provides that the linear term for X_1 had a positive effect on color acceptability, while the rest of the terms revealed negative significant effect for every unit change in the corresponding terms, provided all other factors are held constant. The unit increase in X_1 (corresponding its linear effect) will result to 8.0754 increase in the acceptability response for color. While the linear term results to positive effect on color acceptability, the quadratic and crossproduct terms of the forenamed variable revealed to be contradictory. The square unit increase in X_1 results to decrease by 2.2429 in the color acceptability, while the crossproduct effect of X_1 and X_2 on color acceptability was observed to decrease by 0.1189.

Table 3. Parameter estimates for response surface of color, viscosity, flavor, taste, aftertaste and general acceptability for *seriales* RTD beverage.

Parameters	Responses					
	Color	**Viscosity**	**Flavor**	**Taste**	**Aftertaste**	**General Acceptability**
Intercept	-6.9882^{ns}	-0.2890^{ns}	8.5041	-5.8489	-9.1165^{ns}	-1.2407
Blanching time (X_1)	8.0754***	5.3759**	4.7656**	9.7742***	8.7090***	9.5886***
Seriales level (X_2)	0.5558^{ns}	0.4382^{ns}	0.1869	0.5677	0.8631*	0.4442^{ns}
Sugar level (X_3)	0.6533^{ns}	0.0210^{ns}	-0.7590	0.2411	0.2863^{ns}	-0.1728^{ns}
$(X_1)^2$	-2.2429***	-0.7191^{ns}	-1.8008**	-2.4503***	-1.6166*	-2.2970***
$(X_2)(X_1)$	-0.1189***	-0.0666^{ns}	0.0020	-0.1299**	-0.1018*	-0.1216**
$(X_2)^2$	-0.0120^{ns}	-0.0120^{ns}	-0.0066	-0.0231	-0.0162*	-0.0087^{ns}
$(X_3)(X_1)$	-0.0161^{ns}	-0.1376*	-0.0962	-0.0891	-0.1857*	-0.1069^{ns}
$(X_3)(X_2)$	0.0024^{ns}	0.0101^{ns}	0.0070	0.0093	-0.0043^{ns}	0.0012^{ns}
$(X_3)^2$	-0.0253^{ns}	-0.0041^{ns}	0.0241	-0.0099	0.0211^{ns}	0.0103^{ns}

*significant at p<0.05; **significant at p<0.01; ***p<0.001; ^{ns}not significant at p<0.05

Moreover, maximum color acceptability was estimated from a relatively high blanching time and low *seriales* and TSS levels as shown in Figure 1-A.

3.2.2 Viscosity Acceptability. The mean acceptability score is stipulated in Table 2. Statistically, the response mean for the acceptability of viscosity was calculated to be 6.94, which falls under the 'like slightly" category of the 9-point Hedonic Rating scale. The ANOVA displays that the linear, quadratic and crossproduct terms of the model are highly significant (Table 3). With the assumption that all other factors are constant, a unit increase in X_1 is estimated to effect an increase by 5.3759 units in the response. In contrary, an estimated decrease by 0.1376 of the response is expected with a unit increase of crossproduct term of blanching time and TSS (X_1X_3). Likewise, the quadratic term of seriales level (X_1^2) causes a significant effect on viscosity acceptability, where a square unit increase of such will relatively pull acceptability down by 0.0120 unit. The TSS level on the other hand, was revealed to be of less importance in the model. A relatively high blanching time and low *seriales* and TSS levels were suggested to derive the maximum response of the viscosity acceptability (Figure 1-B).

3.2.3 Flavor Acceptability. Fruit flavor is a complex combination of volatile compounds sourced from the fruit utilized and this affects the acceptance of consumer to the product. The volatile compounds in the *seriales* RTD beverage were derived from the material itself.

In the regression analysis, it was found that the linear and quadratic terms of the model are statistically significant, which was brought about by the factor – blanching time (Table 3). The forenamed predictor is the only factor with a significant over-all effect, which indicated the high level of importance of blanching time to the estimation of the variation in flavor acceptability.

Statistical estimation (parameter estimates) revealed a significant positive increase in the flavor acceptability of the product by 4.7656 units for every unit increase in X_1. However, a 1.8008 decrease is expected for every square unit increase on the aforesaid predictor (Table 3). This is in agreement with the study of Shamaila et al. (1996)[8] where they reported flavor attribute of carrot decreased with blanching time in the sensory evaluation conducted. Similarly, a gradual poor flavor was observed when green peas were subjected to long time/low temperature blanching[9]. It was concluded that there were correlations ($p < 0.05$) between blanching times, flavor volatiles and sensory attributes[8]. Figure 1-C illustrates that high acceptability for flavor is perceived at mid-range of blanching time.

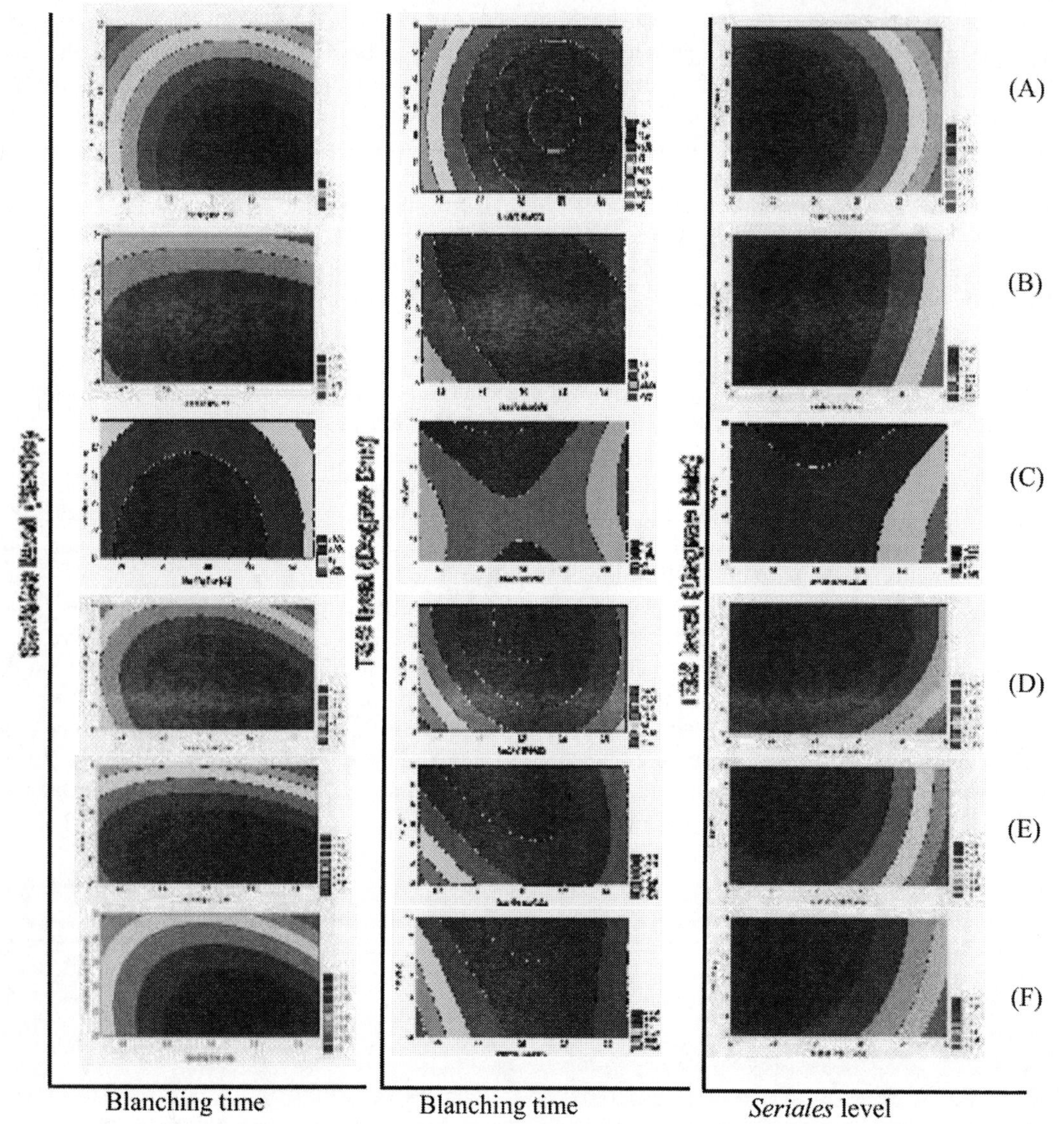

Figure 1. Response surface plots of the interaction of the various independent variables on the acceptability for (A) color, (B) viscosity, (C) flavor, (D) taste, (E) aftertaste, and (F) general acceptability for ready-to-drink *seriales* beverage (acceptability ≥ 7.0 in the 9-point Hedonic scale).

3.2.4 Taste Acceptability. The mean acceptability rating for taste was calculated at 7.13 (Table 2) which falls in the "like moderately" of the 9-point Hedonic scale. The regression analysis of the taste acceptability as affected by the independent variables (blanching time, *seriales* level and TSS level) were statistically evaluated with high significance in its linear, quadratic and crossproduct terms (Table 3).

It was made known that X_1 brought about the most significant effect on the independent factor. The linear effect of X_1 causes a positive effect on taste acceptability but a consequent negative 2.4503 units is anticipated for every unit change in the quadratic term of the aforementioned factor (X_1^2). Likewise, the unit increase in the X_1 and X_2 interaction will result in 0.1299 decrease in the response for taste acceptability.

The results of the regression analysis revealed that the response for taste acceptability was sensitive to the changes of blanching time. The contour plots in Figure 1-D predicted the maximum response surface at the middle range of blanching time, low *seriales* level and high TSS values.

3.2.5 Aftertaste Acceptability. The response mean for taste acceptability shown in Table 2 was 6.90. The acceptability rating falls within the "like slightly" of the 9-point Hedonic scale. The aftertaste acceptability was considerably high despite the possible perception of astringency which may be imparted by the tannins present in the *F. jangomas* (*seriales*) fruit[10].

The statistical analyses revealed that a unit change in the X_2 shall bring about 0.8631 increase in the acceptability score. However, a square unit (X_{22}) increase of such revealed a decrease in acceptability score by 0.0162 units in the acceptability score for aftertaste. Meanwhile, a unit change of the crossproduct term (seriales level & blanching time) was also predicted to cause 0.1002 decreased in the aftertaste acceptability. On the other hand, the quadratic terms for blanching time as well as the crossproduct term (*seriales* level & TSS level) were estimated to decrease correspondingly for every unit change of the aforesaid parameters. In terms of X_1, every unit change (linear) was estimated to increase aftertaste acceptability by 8.7090 units. The positive rating was probably due to decreased perception of tannin as an effect to blanching method induced. Many studies revealed significant reduction of tannin on fruits and vegetables when blanched [11,12,13,14]. Figure 1-E shows clear orientation of the response surface, where maximum aftertaste acceptability was determined at mid-range of blanching time and *seriales* level, while TSS level be set at high range.

3.2.6 General Acceptability. Regression coefficients for over-all acceptability showed significant difference at $p \leq 0.05$ as affected by linear and quadratic terms of blanching time, as well as the interaction of the aforesaid with the level of *seriales* (Table 3). This suggests that blanching time has a major role in the increase of over-all acceptability of the product. As shown in Figure 1-F, high over-all acceptability was oriented at mid to high range of blanching time with respect to low levels of *seriales*, and at mid-range corresponding to high level of TSS. It was noted that among all other independent variables, blanching time has affected all other dependent variables, hence, expectedly cause significant effect to the over-all acceptability as well. Results of the statistical analysis revealed that the response mean was 6.79 (Table 2) which falls along the 'like slightly' to 'like moderately' of the 9-point Hedonic scale.

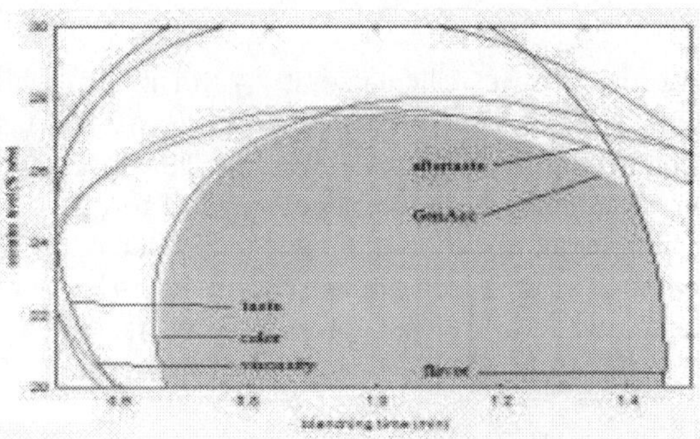

(A)

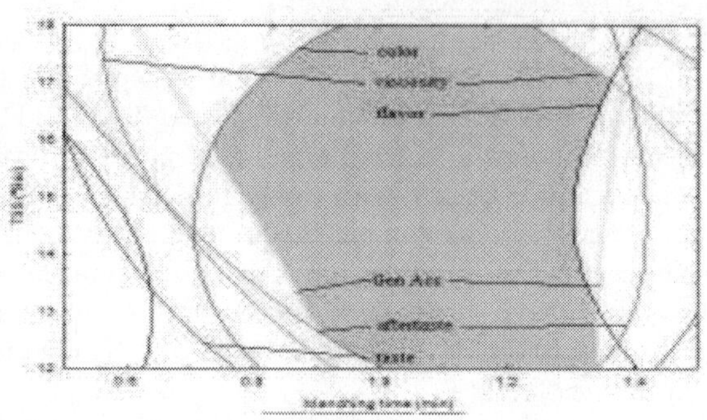

(B)

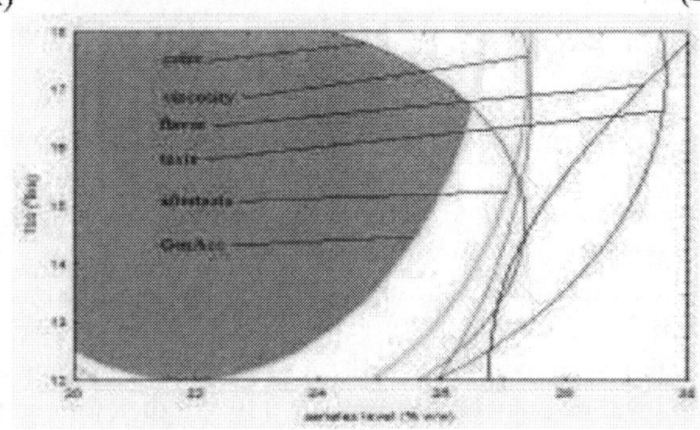

(C)

Figure 2. Optimum regions for ready-to-drink *seriales* beverage (acceptability ≥ 7.0 in the 9-point Hedonic scale) at constant (A) 15°B TSS; (B) 25 % *seriales* level; (C) 1 minute blanching time.

3.3. Attaining the Optimum Conditions for Ready-to-Drink Seriales Beverage

The shaded regions of the superimposed contour plots in Figure 2 presents the combinations of blanching time (0.7 to 1.4 minute), *seriales* level (not beyond 27 %) and the level of TSS (any level) at which the optimum acceptability of the product shall be attained.

Among the treatment combinations studied, T_{10}, T_{11}, T_{12}, and T_{15} are identified to be within the optimum region.

4. Conclusion

The results of PB analysis revealed that blanching time, level of *seriales*, and TSS level were the most important variables, and were then considered in the RSM experiment as independent factors.

Results of the response surface analysis revealed that blanching time influenced the response for the ready-to-drink *seriales* beverage the most. The linear, quadratic and its interaction with *seriales* level and TSS level have been identified to cause significant difference ($p<0.05$) in majority of the responses. The superimposed contour plots of the responses revealed optimum conditions at 0.7 to 1.4 minutes blanching time, *seriales* level of not beyond 27 %, and TSS at any level.

Recommendation

The potential of the fruit for industrial processing, specifically in the production of acceptable RTD beverage has also been studied. However, there are some areas needed to be explored. Among these are : (1) optimization study using yield as a parameter, and a study employing continuous large scale production; (2) determination of the nutritional components of the beverage; (3) packaging and shelf life study; and (4) determination of the effect of anthocyanin chemical structure on spectral characteristics and color attributes; as well as stability test of anthocyanin towards various factors such as temperature, light and pH increase, among others.

Acknowledgement

The authors would like to express gratitude to DOST-ASTHRDP, VSU and UPLB for supporting this research.

References

[1] Lim T 2013 *Flacourtia jangomas. In Edible Medicinal and Non-Medicinal Plants* (Netherland:Springer) pp 771-775

[2] Neeharika D and Pandey V 2013 Ethnobiological importance of Flacourtia jangomas (Lour.) Raeusch *Trends in Biosciences* **6** 532-534

[3] Quattrocchi U 2012. *CCC World Dictionary of Medicinal and Poisonous Plants: Common Names, Scientific Names, Eponyms, Synonyms, and Etymology.* (USA:CRC Press Taylor and Francis Group) p 1764

[4] Parvin S, Kader A, Sarkar G and Bin Hosain S 2011 In-Vitro studied of antibacterial and cytotoxic properties of Flacourtia jangomas. *International Journal of Pharmaceutical Sciences and Research* **2** 2786-2790

[5] Mohamad A 2012 Nutrient composition and antioxidant properties of 'kerekup" fruit [BS Thesis] ; Universiti Teknologi Mara

[6] Bates R, Morris J and Crandall P 2001 *Principles and practices of small- and medium scale fruit juice* (Rome:FAO Publishing and Multimedia Service Information Division) p 19

[7] Cochran W and Cox G 1957 *Experimental Design* 2nd edition (New York : John Wiley and Sons Inc)

[8] Shamaila M, Durance T and Girard B 1996 Water blanching effects on headspace volatiles and sensory attributes of carrots *Journal of Food Science* **61** 1191-1195

[9] Halpin B and Lee C 1987 Effect of blanching on enzyme activity and quality changes in green peas *Journal of Food Science* **52** 1002-1005

[10] Dubey N, Pandey V, Tewari S 2013 Antioxidant potential and phytochemical composition of unripe fruits of *Flacourtia jangomas* (Lour.) *Raeusch Medicinal Plants-International Journal of Phytomedicines and Related Industries* **5** 164-167

[11] Mosha T, Gaga H, Pace R, Laswai H and Mtebe K 1995 Effect of blanching on the content of antinutritional factors in selected vegetables *Plant Foods for Human Nutrition* **47** pp 361-367

[12] Adegunwa M, Alamu E, Bakar H and Oyeniyi C 2011 Proximate and bioactive contents of some selected vegetables in Nigeria: Processing and varietal effects *American Journal of Food and Nutrition* **1** 171-177

[13] Eze J, Okafor G and Okoyeuzu C 2014 Effects of blanching on the proximate compositions and antinutritional factors of products from African star apple (*Chrysophyllum albidium*) peels and cotyledons *International Journal of Scientific & Engineering Research* **5** 1297-1305

[14] Akande E, Adesola M and Olunlade B 2014 Effects of blanching on the qualities of Negro pepper Global *Advanced Research Journal of Food Science and Technology* **3** 99-102

Improving protein mass and cumulative body weight gain of local chicken fed ration fortified with a combination of *Lactobacillus sp.* and dahlia inulin

H I Wahyuni*, N Suthama, I Mangisah and L Krismiyanto

Faculty of Animal and Agriculture Sciences, Diponegoro University, Tembalang Campus, Semarang Indonesia

*E-mail: hihannyiw123@gmail.com

Abstract. The research aimed to evaluate meat calcium and protein content of local chicken fed diet fortified with a combination of Lactobacillus sp and Dahlia Inulin. One hundred and twenty birds of 4 months old local chicken with average body weight of 1001 g were assigned in a completely randomized design with 4 treatments and 5 replications. The treatments were the farmer formulated ration (FF) and the improved ration (IR), fortified with 1.2% inulin and 1.2 ml *Lactobacillus sp.* (FFIL and IRIL). Parameters were calcium retention, protein coefficient digestibility, meat calcium and protein mass, and cumulative body weight gain. The results showed that all parameters were significantly affected by dietary treatments. The improved ration resulted in higher calcium retention and protein coefficient digestibility than the farmer formulated ration when fed by both with and without fortification of dahlia inulin and *Lactobacillus sp.* Meat protein mass of chicken fed by both FR and IR fortified with dahlia inulin and *Lactobacillus sp.* showed higher value than chicken fed by unfortified FR and IR. Cumulative body weight gain of chicken fed by both FR and IR fortified with dahlia inulin and *Lactobacillus sp.* also showed higher value than chicken fed by without fortification. In conclusion, both FR and IR fortified with dahlia inulin and *Lactobacillus sp.* improved meat protein mass and cumulative body weight gain, especially the farmer formulated ration was pronouncedly improved by fortification of *Lactobacillus sp.* and dahlia inulin.
Keywords: *Lactobacillus sp., inulin, meat, calcium.*

1. Introduction
Local chicken are the most numerous species reared by the rural family in the villages across 34 provinces of Indonesia. As it was shown by the population in 34 provinces of Indonesia in the year 2016 was 298.672.970 birds which showed an increased population when compare to the year of 2015 that was 285.304.305 [1] But actually the population was not growing, since the growth average among those provinces was -0.6 [2]. The classical reason is that local chicken had genetically slow growing. On the other hand, this could also be attributed by the traditional rearing system implemented by the farmers such as feeding ration with nutrient under the requirement so that depressing breeding production [3]. Despite of that inferiority, consumers still choose local chicken meat rather than commercial broiler meat, since local chicken meat had low fat compare to commercial broiler meat. The high consumer preference and the higher sailing price of local chicken meat would help rural family on increasing their income, and if ever improving the production.

Increasing local chicken productivity could be achieved through feeding manipulation like adding synbiotic in the ration. Synbiotic is included in the ration as non nutritive feed additive which is a combination between probiotic and prebiotic [4]. Addition of *Lactobacillus* culture as probiotic in broiler ration had been proved to improve its performances [5,6]. On the other hand inclusion of probiotic and prebiotic at the same time in broiler ration would also influenced the productivity [7,8,9]. Broiler chickens fed probiotic had higher protein and ash in the leg and breast, and lower in fat as compared to the leg and breast meat of control chickens [10]. Infact broiler meat had higher cholesterol content as compare to the native chicken's meat i.e. 99 Vs. 71 mg/100g [11,12].

Researh had been done on the used of inulin from dahlia tuber as prebiotic on cross bred native chicken of stater period stimulated the increased of lactic acid bacteria (LAB) population and decreased the number of *E. Coli* at the small intestine hence affecting dayly weight gain at the level inclusion of 0.8% dahlia tuber meal and 1.17% of its extract form [13]. Combining dahlia inulin and *Lactobacillus sp* in crossbred local chicken's ration improved the small intestine microflora and it's performance [14], while it also improved the meat nutrient quality in term of calcium and protein content as well as the antioxidant activity [15]. The present research was conducted to evaluate protein and calcium conten of local chicken'meat fed ration fortified with a combination of *Lactobacillus sp* and Dahlia Inulin.

2. Materials and Methods

A total of 120 hens growing local chicken 4 months of aged with average body weight of 1001.35 $\pm$ 56.20 g were assigned into a completely randomized design. The treatments were the farmer's formulated ration (FF) and the improved ration (IR) fortified with synbiotic - a combination of 1.2% dahlia inulin and 1.2 ml *Lactobacillus* sp (FFIL and IRIL). Each treatment was replicated 5 times and each replicate consisted of 6 birds. Feedstuffs and nutrient composition of farmer's formulated ration and improved ration were shown at Tabel 1. The treatment rations were fed two times a day as required for 2 months, while drinking water were provided *ad libitum*.

Table 1 Feedstuffs Composition and Nutrient Content of Both Farmer's Formulated and Improved Ration (%)

Feedstuffs	Farmer's Formulated Ration	Improved Ration
Broiler Mash	19.50	-
Yellow Corn	36.00	50.00
Rice Bran	39.50	20.00
Soy Bean Meal	-	21.00
Fish Meal	-	6.00
Premix	5.00	1.00
$CaCO_3$	-	2.00
Total	100.00	100.00
Nutrient Content[a]		
Crude Protein (%)	13.04	17.44
Crude fiber (%)	8.22	4.59
Extract Ether (%)	2.30	2.30
Calcium (%)	0.71	0.84
Phosphor Total (%)	0.63	0.39
Energi Metabolism[b](Kkal/kg)	2583.11	2887.19

a). Proximate analysis;
b). Computed based on the proximate analysis value

Parameters observed were calcium retension, protein digestibility, meat calcium and protein mass, as well as body weight. The digestibility measurement was conducted using combination method of total collection and indicator [16] Meat was sampled from the breast and thigh-leg of the carcass. After removing the skin and the bone, the meat than mixed and grind until homogen. Meat sampled were taken for calcium and protein analisis using AAS and Kjeidahl method, respectively. All data gathered were analyzed using analysis of variance, and Duncan multiple range test was also used to find the different between treatment.

3. Result and Discussion

Combination inclusion of inulin dahlia and *Lactobacillus* sp. in both farmer's formulated ration (FFIL) and improved ration (IRIL) showed a significant ($P < 0.05$) increased of protein digestibility (Table 2). The lowest protein digestibility of farmer's formulated (FF) ration proved the effect of synbiotic (dahlia inulin + *Lactobacillus* sp) on the change of small intestine microbiota resulting on a healthier digestive truct. Additional of dahlia inulin as prebiotic and *Lactobacillus* sp. as probiotic depressed pathogen bacterial population like e. coli so that increasing the lactic acid bacteria population [14]. The development of beneficial microbiota at the small intestine as affected by feeding inulin as prebiotic [13,18] would also stimulating the height of villi [7,17] which in turn improving the nutrient absorption as it was shown on the increased protein digestibility of both FFIL and IRIL ration.

Tabel 2 Coefficient Digestibility of Protein, Calcium Retention, Meat Calcium and Protein Mass and Cumulative Body Weight Gain of Growing Kedu Chicken Fed Ration with Added Inulin Derived from Dahlia Tuber and *Lactobacillus* sp.

Parameters	Treatment Ration			
	Farmer's Formulated (FF)	Improved Ration (IR)	FF+Inulin+ *Lactobacillus* (FFIL)	IR+ Inulin+ *Lactobacllus* (IRIL)
Protein Coefficient Digestibility (%)	56.53^{c}	77.50^{a}	61.35^{b}	77.63^{a}
Ca Retension (g/bird)	0.76^{b}	0.90^{a}	0.66^{c}	0.87^{a}
Meat Calcium Mass (g/bird)	1.89^{c}	3.53^{a}	2.01^{c}	2.67^{b}
Meat Protein Mass (g/bird)	108.16^{c}	103.97^{c}	128.38^{a}	116.59^{b}
Cumulative Body Weight Gain (g/bird)	392.71^{b}	435.97^{b}	567.16^{a}	505.72^{ab}

^{a,b,c} in each parameter showed significantly different ($P<0.05$).

The calcium retention was not showed similar trend as the protein digestibility. Improved ration fortified with dahlia inulin and *Lactobacillus sp.* had similar calcium retention to unfortified ration. But fortification into FF showed a decreased calcium retention (Table 2). These finding were not in accordance with the previous finding which stated that prebiotic, probiotic or synbiotic would improved mineral digestion and absorbtion. Feeding prebiotic or probiotic or synbiotic would improved calcium absorbtion through increasing the calcium solubilyzation [19] as a consequences of the decreased intestinal pH which was affected by the microbial fermentation [20]

Fortification of dahlia inulin and *Lactobacillus sp.* into the improved ration (IRIL) showed a decreased meat calcium mass (Tabel 2). But fortification to the farmer formulated (FFIL) had similar meat calcium mass as compared to the unfortified. Infact the meat calcium mass measured it's deposition which actually had similar trend with calcium retention. Feeding this combination of synbiotic to cross-bred native chicken during starter period also showed no significant effect to meat calcium mass [15]. This could be due to the higher rate of calcium deposited into the bone rather than into the meat, since the bird was in the growing stage. Another reason was that decreasing the intestinal pH as influenced by fermentation of the prebiotic [20]. The condition affected calcium solubility at the small intestine [19] hence increasing it's absorption which primarily used for bone calcium deposition [21].

On the other hand, the protein showed a better deposition as compare to the calcium deposition, when the birds fed both fortified ration. Fortification of dahlia inulin and *Lactobacillus sp.* (IRIL) to both ration (FF and IR) increased meat protein mass. The increased protein deposition into the meat was a consequence of the increased protein digestibility as affected by synbiotic fortification. *Lactobacillus sp.* would utilized inulin as substrate for fermentation producing short chain fatty acid/SCFA [18]. Production of SCFA contributed on the decreased of digestive tract acidity which promoting a better microbiota [18] resulting on a healthier digestive tract so that increased nutrient absorption (protein and calcium) for meat deposition.

The low meat protein mass of the chicken fed both unfortified ration (FF dan IR) could be caused by the protein degradation process, since the meat calcium mass was higher especially of the improved ration. Calcium level in the meat could be used as an indicator for protein degradation process, since the high calcium level in the meat stimulate the enzyme activity of calcium activated neutral protease (CANP) which degradated the meat protein [22] resulting on low protein deposition. The absen of dahlia inulin and *Lactobacillus sp.* also contributed to the lowest meat protein mass, because there was no stimulation of lactic acid bacteria at the digestive tract so that less protein were absorbed producing low body weight gain.

Chicken fed farmer formulated ration fortified with dahlia inulin and *Lactobacillus sp.* (FFIL) had the highest meat protein mass but actually had the lowest protein digestibility (Table 2). The condition showed the efficiency of dietary protein utilization which was influenced by the presence of synbiotic [18,19]. The main role of synbiotic was performing healthier digestive tract that influence nutrient utilization which in turn produced the highest body weight gain.

4. Conclusion

In conclusion, both FF and IR fortified with dahlia inulin and *Lactobacillus sp.* improved meat protein mass and cumulative body weight gain of local chicken, especially the farmer formulated ration was pronouncedly improved by fortification of *Lactobacillus sp.* and dahlia inulin.

References

[1] Directorate General of Animal Husbandry and Animal Health 2016 Statistic of Animal Husbandry and Health (in Bahasa Indonesia). The Ministry of Agriculture, Republic of Indonesia

[2] BPS 2016 https://www.bps.go.id/linkTableDinamis/view/id/1088.

[3] Wahyuni H I, Suthama N Mangisah I and Teysar A S 2011 Egg quality and hatchability of in situ – reared kedu and cemani hens fed diet of farmer formulation supplemented with vitamin e (in Bahasa Indonesia) *J. Indonesian Trop. Anim. Agric.* **36** (1) 61-68

[4] Madej J P, Stefaniak T and Bednarkczyk M 2015 Effect of *in ovo*-delivered prebiotics and synbiotics on lymphoid-organs' morphology in chickens *Poult. Sci.* **94** (6) 1209-1219.

[5] Jin L Z, Ho Y W, Abdullah N and Jalaludin S 1998 Growth performance, intestinal microbial populations, and serum cholesterol of broilers fed diets containing Lactobacillus cultures *Poult Sci.* **77** 1259–1265.

[6] Mountzouris K C, Tsirtsikos P, Kalamara E, Nitsch S, Schatzmayr G and Fegeros K 2007 Evaluation of the efficacy of a probiotic containing *lactobacillus, bifidobacterium, enterococcus,* and *pediococcus* strains in promoting broiler performance and modulating cecal microflora composition and metabolic activities. *Poult Sci*. **86** 309–317.

[7] Awad W A, Ghareeb K, Abdel-Raheem S and Bohm J 2009 Effects of dietary inclusion of probiotic and synbiotic on growth performance, organ weights, and intestinal histomorphology of broiler chickens *Poult. Sci.* **88** 49-55.

[8] Nikpiran H, Taghavi M, Khodadadi A and Athari S S 2013 Influence of probiotic and prebiotic on broiler chickens performance and immune status. *J. Nov. Appl. Sci*. **2**(8)256-259.

[9] Sarangi N R, Babu L K, Kumar A, Pradhan C R, Pati P K and Mishra J P 2016 Effect of dietary supplementation of prebiotic, probiotic, and synbiotic on growth performance and carcass characteristics of broiler chickens, *Veterinary World*, **9**(3) 313-319.

[10] Khaksefidi A and S H Rahimi 2005 Effect of Probiotic Inclusion in the Diet of Broiler Chickens on Performance, Feed Efficiency and Carcass Quality *AJAS* **18**(8)1153-1156.

[11] Hasanuddin S, Yunianto V D and Tristiarti 2013 Fat and cholesterol meat of broiler chicken feed by step down protein with addition lime juice as acidifier (in Bahasa Indonesia) *Bul. Nutrisi dan Makanan* **9**(1):47-53.

[12] Fadly M, Tanwiriah W and Asmara I Y 2013 The influence of noni fruit meal (*Morinda citrifolia* l.) on ration toward abdominal fat and cholesterol of sentul chickens (*Gallus domestica*) Student e-J. **5**(4) 1-10.

[13] Krismiyanto L, Suthama N dan Wahyuni H I 2014 Feeding effect of inulin derived from dahlia variabilis tuber on intestinal microbes in starter period of crossbred native chicken. *J. Indonesian Trop. Anim. Agric.* **39** (4) 217-223.

[14] Faradila S, Suthama N and Sukamto B 2016 Combination inulin of dahlia variabilis tubers-Lactobacillus sp. Optimizing the development of intestinak microflora and growth of Pelung-Leghorn chicken crosses *J. Vet.* **17** (2) 168-175.

[15] Abdurrahman Z H, Pramono Y B and Suthama N 2016 Feeding effect of inulin derived from dahlia tuber combined with *lactobacillus* sp. on meat protein mass of crossbred kampong chicken *J. Indonesian Trop. Anim. Agric.* **41** (1) 37-44.

[16] Indreswari R, Wahyuni H I, Suthama N and Ristiana P A 2009 Calcium Utilization For Egg Shell Formation As Affected By Feeding Ratio In The Morning and Afternoon in Layer Chicken (In Bahasa Indonesia) *J. Indonesian Trop. Anim. Agric.* **34** (2) 134-138.

[17] Beski S S M and Al-Sardary S Y T 2015 Effects of Dietary Supplementation of Probiotic and Synbiotic on Broiler Chickens Hematology and Intestinal Integrity. *Int. J. Poult. Sci.* **14** 31-36.

[18] Nabizadeh A 2012 The effect of inulin on broiler chicken intestinal microflora, gut morphologi and performance *J. Anim. Sci.* **21** 725

[19] Kwiatkowska K, Winiarska-Mieczan A and Kwiecien M 2017 Feed additives regulating calcium homeostasis in the bone of poultry *Ann. Anim. Sci.,* **17** (2) 303–316 *DOI: 10.1515/aoas-2016-0031*

[20] Lopez H, C Coundray, M Levrat-Verny, C Feillet-Coudray, C Demigne dan C Remesy 2000 Fructooligosaccharides enhance mineral apparent absorption and counteract the deleterious effects of phytic acid on mineral homeostatis in rats. *J. Nutr. Biochem.* **11** 500.

[21] Chen Y C and T C Chen 2004 Mineral utilization in layers as influenced by dietary oligofructose and inulin *Int. J. Poult. Sci.***3** (7) 442-445.

[22] Suzuki K S, Ohno, Y Emori, S. Inajoh and H Kawasaki 1987 Calcium activated neutral protease (CANP) and its biological and medical implications. *Progress Clin. Biochem. J. Medical* **5** 44

Effectiveness of incorporating citric acid in cassava starch edible coatings to preserve quality of Martha tomatoes

I Ambarsari*, G N Oktaningrum and R Endrasari

Assessment Institute of Agricultural Technology Central Java, Indonesian Agency for Agricultural Research and Development, Jl. Soekarno-Hatta No.10 Km.26, Bergas, Central Java, Indonesia

*Email: indrie_ambarsari@yahoo.com

Abstract. Tomato as an agricultural product is extremely perishable. Coatings of tomatoes with edible starch extend quality and storage life of the fruits. Incorporation of citric acid as antimicrobial agent in the edible starch coatings is expected to preserve the quality of tomatoes during storage. The aim of this study was to verify the effectiveness of citric acid incorporated in cassava starch coating to preserve quality of tomatoes. The edible coatings formula consisted of cassava starch solutions (1; 2; 3%), citric acid (0.5; 1.0%) and glycerol (10%). Tomatoes were dipped to the coating solution for 10 seconds, then air-dried and stored at room temperature during 18 days. All the treatments were carried out in triplicates. Experimental data were analyzed using One Way ANOVA. The results showed that coating treatments did not affect the weight loss, moisture content, color characteristic, carotene and vitamin C content on Martha tomatoes. The low concentration of starch coating on Martha tomatoes are indicated to be the reason why there was no significant difference between coated and coated tomatoes for some parameters. However, incorporating citric acid in cassava starch-based coatings could prevent tomato fruits from firmness reduction and spoilage during storage.

Keywords: *edible coating, tomato, cassava starch.*

1. Introduction

Tomatoes are being one of the most important horticultural products in Indonesia because of their economic value and the nutrition. One medium sized tomato provides 57% of the recommended daily allotment (RDA) of vitamin C, 25% RDA of vitamin A, and 8% RDA of iron [1]. Besides consumed fresh, tomato can be baked, stewed, fried, juiced, or pickled and can be used in soups, salads, and sauces. Tomatoes are important agricultural product for food and beverage industries as well as pharmaceutical industry.

As climacteric fruits, tomato is highly perishable due to many processes affected to the loss quality after harvest and during storage including transpiration and respiration that will be followed by physiological changes such as softening tissues, reducing organic acids, color alteration, and volatile compounds losses [2, 3]. The deterioration of tomato's quality during storage and transportation is still becoming a serious post harvest problem for traders. Therefore, development of preservation technology is required to avoid postharvest losses and deterioration of tomatoes.

The use of edible coatings to preserve quality of fruits is well known. Edible coatings offer some advantages such as edibility, biocompatibility, aesthetic or glossy appearance, barrier properties, non-polluting and having low cost [4, 5]. Moreover, edible coatings act as a barrier to external elements, protect the products, extend the storage life, and prevent microbial spoilage, which is extremely important to perishable fruits [6, 7].

Polysaccharide-based coatings such as starch provide favorable functional properties that enable development of edible coatings for use in the food industry [8]. Starch edible coatings exhibit similar physical characteristics to synthetic polymers such as transparent, odorless, tasteless, semi-permeable to CO2 and resistant to O2 passage [9]. One of the important sources of starch in Indonesia is cassava. Cassava starch edible coating is cheap, tasteless, odorless, colorless, non-toxic, biodegradable, safe, and have low permeability to oxygen [10].

Based on some researches, the effectiveness of starch as an edible coating should be blend with other biopolymers, hydrophobic substances and/or antimicrobial compounds [11-14]. One of the chemical compounds that were potential to use as an antimicrobial agent in edible coating formulation is citric acid. Citric acid is generally recognized as safe chemical preservatives that can acts as antimicrobial agents. The addition of citric acid in edible coatings formula inhibited browning reaction and reduced losses because this acid acts as antimicrobial agent [15]. Compared to other organic acids, citric acid is relatively cheap and easily obtained.

Until now, the application of edible coating on Indonesian fruits is still limited. The utilization of materials that was easy to obtained and inexpensive is expected to increase the farmer interest to develop this technology on their commodities. The aim of this study was to evaluate the effect of citric acid incorporation in cassava starch edible coating on preserving the quality of Martha tomatoes during storage. The evaluation of tomato quality changes were based on their physicochemical properties such as weight loss, moisture content, vitamin C, carotene content, fruit firmness, color, as well as the percentage of fruit decay.

2. Materials and Methods

2.1. *Sample preparation*

Uniform and healthy tomatoes cv. Martha from a farmer in Boyolali, Central Java was selected as samples. The fruits were classified according to their size and maturity (based on their external color and firmness). Upon arrival at the laboratory, the tomatoes were washed in tap water and cleaned with a sterilized cloth to remove physical dirt from the surface. The tomatoes were treated within 3 hours of arrival time.

Edible coatings were prepared from cassava starch with additions of citric acid as antimicrobial agent and glycerol as plasticizer. The purpose of glycerol addition in the treatments was to enhance the coatings flexibility, since glycerol has low molecular weight that enables macromolecules interaction to increase mobility of the polymer chains and reduce the glass transition temperature of the system [16].

Experimental treatments in this study refer to Chiumarelli *et al.* [17] method, which was using cassava starch and citric acid in coating fresh-cut mango. Combination of cassava starch (1, 2, and 3% w/v solution) and citric acid (0.5 and 1.0% w/v solution) were tested as coatings. Glycerol 10% was added in all combinations as the standard treatment. All materials were mixed in distilled water and boiled at 80-90 °C for 15 minutes with a constant agitation. Edible coating solution was left until its temperature reached the room temperature (approximately 25 °C). Tomatoes were immersed in the edible coating solution for 10 seconds and air-dried using a fan. Each treatment contained 15 tomatoes and the experiment was replicated three times. Uncoated tomatoes were used as control. Tomatoes were stored at room with air conditioner at temperature 25 °C before further observation.

2.2. Data collection and analysis
The physicochemical observations were done after 18 days storage when the fruits reaching softened. The physical analysis included their weight loss (the percentage weight lost from the original weight), surface color (Hunter Lab Scan Spectrocolorimeter), textural degradation (Instron Universal Testing Machine Instron Corporation, Canton, MA), as well as the percentage of decayed fruit. The decay percentage of tomato fruit was calculated as the number of decayed fruit divided by initial number of all fruit multiplied by 100 [18]. Chemical analysis consisted of moisture content [19], vitamin C [19], and total carotene [20]. All the experiments were carried out in triplicates. Each replication was consisting 15 tomatoes.

2.3. Statistical analysis
Experimental data were analyzed statistically using one-way analysis of variance (One Way ANOVA). Detailed examinations for significant differences were done using a Duncan Multiple Range Test at $p \leq 0.05$. SPSS 17.0 statistical software (SPSS Inc., Chicago, IL, USA) was used in all statistical analysis.

3. Results and discussion

3.1. Weight loss
In this study, there were no statistically differences in weight loss among all treatments (Figure 1). The low concentration level of cassava coatings in this experiment lead to a formation of thin layer on tomatoes surface that still enable respiration activities, engendering no difference on weight loss between coated and uncoated tomatoes. Moreover, tomatoes naturally covered by a continuous wax layer that provides high resistance to water movement across the cuticle [21].

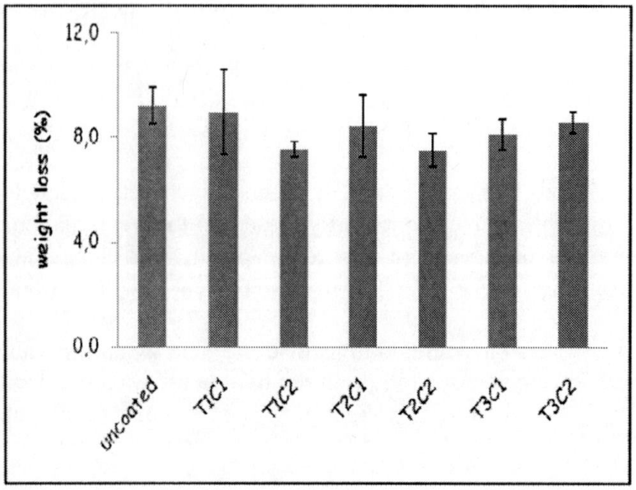

Figure 1. The effect of incorporating citric acid in cassava starch coating (T1C1 = 1% starch + 0.5% acid, T1C2 = 1% starch + 1% acid, T2C1 = 2% starch + 0.5% acid, T2C2 = 2% starch + 1% acid, T3C1 = 3% starch + 0.5% acid, T3C2 = 3% starch + 1% acid) on tomatoes weight loss after 18 days storage at room temperature.

The transpiration process as well as respiration rate triggers the weight loss on fruits during storage [22]. The basic mechanism of weight loss from fresh fruits and vegetables is by vapor pressure at different locations [23]. When the vapor pressure inside the fruit is higher than outside then the moisture will discharge resulting in the fruit weight loss. Furthermore, the respiration activity on fruits engenders fruit ripening causing reduction of fruit weight [22].

3.2. Moisture content
Moisture content is one of degradation parameter quality of tomatoes that related to weight loss. Moisture contents of coated tomatoes after 18 days storage were relatively the same (95.16 - 95.37%) (Figure 2). There were no significant differences among all treatments (uncoated and coated

tomatoes). Some researchers [24, 25] suggested that polysaccharide based coatings do not behave well as moisture barriers because of their hydrophilic nature. Other research has also reported that the use of citric acid in cassava starch formulation coating could slightly increased the dehydration of fresh-cut mango during storage, due to the cell turgidity loss [17, 26].

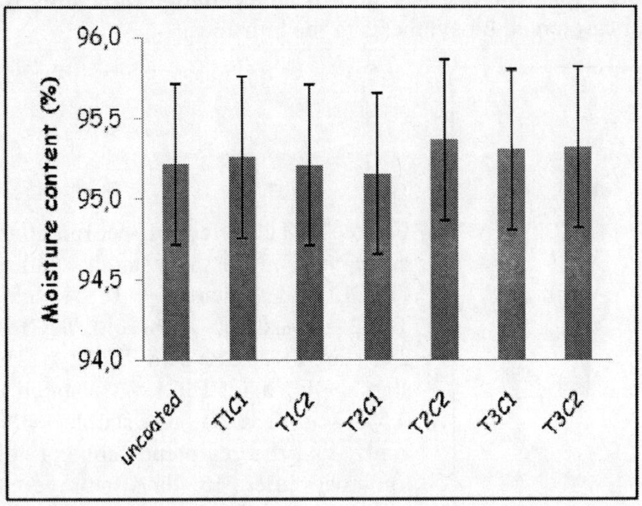

Figure 2. The effect of incorporating citric acid in cassava starch coating (T1C1 = 1% starch + 0.5% acid, T1C2 = 1% starch + 1% acid, T2C1 = 2% starch + 0.5% acid, T2C2 = 2% starch + 1% acid, T3C1 = 3% starch + 0.5% acid, T3C2 = 3% starch + 1% acid) on the moisture content of tomatoes after 18 days storage at room temperature.

3.3. Vitamin C

Tomato is rich in vitamin C. The results showed that vitamin C on coated tomatoes after 18 days storage varied between 12.24 to 15.25 mg/100 g sample (Figure 3). The low vitamin C on tomatoes samples is probably due to the decreasing of tomatoes vitamin C during storage. The vitamin C of tomato fruits were tends to decrease with the time of storage [27]. Once fruit reach the full ripe stage, vitamin C content starts to decline [28]. The loss of vitamin C can be attributed to the stress caused by the operations involved in processing, resulting in greater weight loss and juice leakage during storage [29].

In this study, there were no significant differences on vitamin C between all treatments of tomatoes during 18 days storage. This condition indicated that the coating formulation in this study is inadequate for maintaining the environment around the samples, thus enabling for the respiration activity and metabolic process. Hence, the vitamin C contained on coated and uncoated tomatoes was not significantly different.

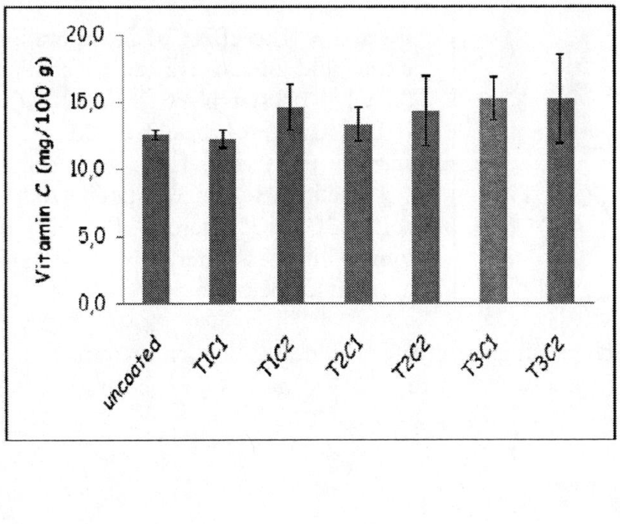

Figure 3. The effect of incorporating citric acid in cassava starch coating (T1C1 = 1% starch + 0.5% acid, T1C2 = 1% starch + 1% acid, T2C1 = 2% starch + 0.5% acid, T2C2 = 2% starch + 1% acid, T3C1 = 3% starch + 0.5% acid, T3C2 = 3% starch + 1% acid) on the vitamin C of tomatoes after 18 days storage at room temperature.

3.4. *Carotene content*

During fruit ripening, chlorophyll decomposition raises and chloroplast transform into chromoplasts, and then generates carotenoid formation [30]. The evaluation results showed that edible coating treatments have no significant effects on carotene content of tomatoes during 18 days storage (Figure 4). This result is consistent with Chiumarelli *et al.* [17], which worked on fresh-cut mango showed that cassava starch coating incorporated with citric acid has no influence on β-carotene content of fruits during storage 15 days. The possible explanation is that the low concentration of coating was not effective, thus allowing the activities of carotenoid biosynthesis in the tomatoes.

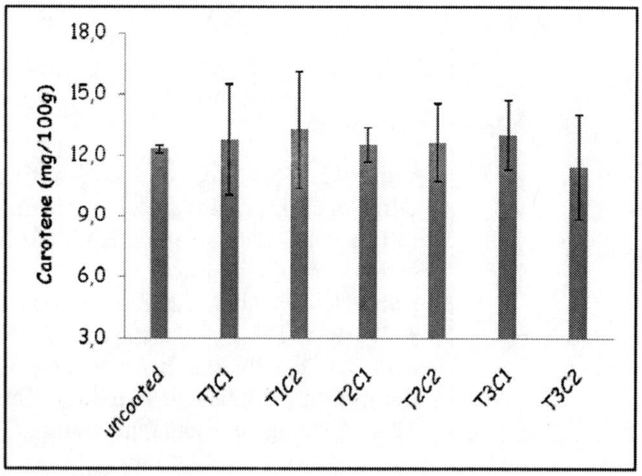

Figure 4. The effect of incorporating citric acid in cassava starch coating (T1C1 = 1% starch + 0.5% acid, T1C2 = 1% starch + 1% acid, T2C1 = 2% starch + 0.5% acid, T2C2 = 2% starch + 1% acid, T3C1 = 3% starch + 0.5% acid, T3C2 = 3% starch + 1% acid) on the carotene content of tomatoes after 18 days storage at room temperature.

3.5. *Fruit firmness*

Texture transformation is one of the obvious physical characteristic changes on tomatoes during storage. Throughout storage, a loss of mechanical properties was observed for all treatments. The observation exhibited that the coatings affected the firmness significantly. The application of 1% cassava starch coating blend with 0.5% citric acid was the most effective treatment for hindering the firmness reduction of tomatoes during 18 days of storage (Figure 5).

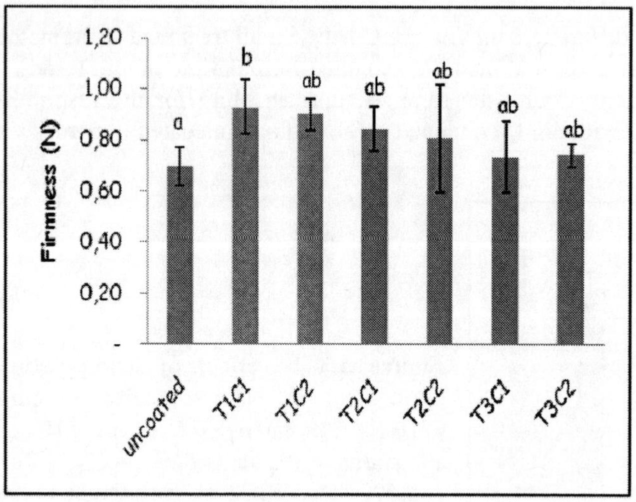

Figure 5. The effect of incorporating citric acid in cassava starch coating (T1C1 = 1% starch + 0.5% acid, T1C2 = 1% starch + 1% acid, T2C1 = 2% starch + 0.5% acid, T2C2 = 2% starch + 1% acid, T3C1 = 3% starch + 0.5% acid, T3C2 = 3% starch + 1% acid) on tomato firmness after 18 days storage at room temperature.

The higher concentration of cassava starch causing the greater firmness reduction on tomatoes. It was presumed that higher concentration of cassava starch resulted in lower O_2 concentrations and excessive CO_2 concentrations, yielding ethanol production that was triggering greater firmness reduction [31].

The fruit softening is due to deterioration in the cell structure, cell wall composition and intracellular materials [32] and a biochemical process involving the hydrolysis of pectin and starch by enzymes [28]. Edible coating enable to inhibit respiration rate thus postpone the ripening process and texture degradation during storage [33]. Low respiration rate can limit the activities of hydrolysis enzymes that triggering firmness retention on tomatoes [3, 28, 34].

3.6. Color index

The color of fresh products is one of the most significant quality factors for its acceptance. Figure 6 showed that coated tomatoes did not present significant changes on color properties when compared with control sample during 18 days storage. The similar condition was occurred on fresh-cut pineapple and strawberries where application of cassava starch edible coatings did not affect the fruit natural color characteristic [29, 35]. Other research also showed that coating application has no influences on color parameters of cherry tomatoes after 15 days of storage [21].

During ripening, the green chlorophyll pigment is degraded and there is accumulation of carotenoid, particularly lycopen giving the red color to the ripe tomato [28]. In this study, the influence of respiration rate and volatile changes on tomatoes might have been insufficient to raise a significant effect on peel color parameters of coated Martha tomatoes. It was possible that the coatings were not providing a sufficient barrier against ethylene production and gas exchange between inner and outer environment, thus there was no increasing in carotenoid synthesis. Therefore, the coated tomatoes have no different color with uncoated tomatoes. Furthermore, tomatoes that was used were selected with the uniform external color (full-developed red color), which could also explain there is no changes observed in color during storage.

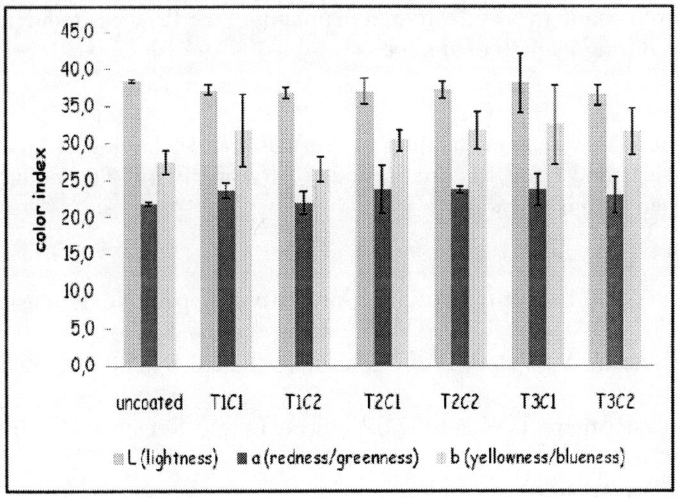

Figure 6. The effect of incorporating citric acid in cassava starch coating (T1C1 = 1% starch + 0.5% acid, T1C2 = 1% starch + 1% acid, T2C1 = 2% starch + 0.5% acid, T2C2 = 2% starch + 1% acid, T3C1 = 3% starch + 0.5% acid, T3C2 = 3% starch + 1% acid) on tomato color after 18 days storage at room temperature.

3.7. Fruit decay

Effectiveness of coating treatments on maintaining tomatoes quality can also be observed from its spoilage incidences level during storage. The results indicated that an addition of citric acid as an antimicrobial agent in coating formulation is effective on preventing decay on tomatoes. The percentage of decay level after 18 days storage on coated tomatoes was less than 5%, while on uncoated tomatoes was more than 25% (Figure 7). *Botrytis cinerea* Pers. and *Alternaria alternata* (Fr.) Keissl. are the most common fungal pathogens responsible for postharvest decay on tomato fruit, causing gray mold and black spot, respectively [36].

The decrease in decay percentage was probably due to the effect of the coating on delaying senescence, which makes the commodity more susceptible to pathogenic infection as a result of loss of cellular or tissue integrity [37]. Other research has also reported that the use of starch based coatings without antimicrobial agent in their formulation was unable to control the development of yeast and molds [38]. Starch coating material was a carbohydrate source and may have been used as a substrate for microbial growth.

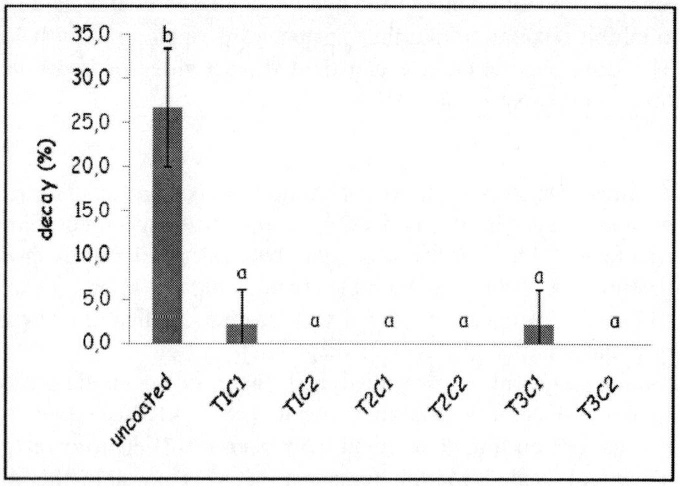

Figure 7. The effect of incorporating citric acid in cassava starch coating (T1C1 = 1% starch + 0.5% acid, T1C2 = 1% starch + 1% acid, T2C1 = 2% starch + 0.5% acid, T2C2 = 2% starch + 1% acid, T3C1 = 3% starch + 0.5% acid, T3C2 = 3% starch + 1% acid) on the percentage of tomato decay after 18 days storage at room temperature.

4. Conclusion

The coatings of cassava starch (1, 2, and 3%) with an addition of 0.5 and 1.0% citric acid did not provide sufficient barrier to transpiration, respiration activity, as well as metabolic processes of tomatoes cv. Martha. The coating treatments in this study did not affect the weight loss, fruit color, moisture, carotene, and vitamin C of Martha tomatoes as compared to uncoated tomatoes. However, an addition of citric acid on cassava starch coating was effective in maintaining the firmness reduction and microorganism growth on tomatoes during 18 days of storage

Acknowledgement

Special thanks to the post-harvest team of Assessment Institute of Agricultural Technology (AIAT) Central Java and Laboratory of Chemical and Food Material, Faculty of Agricultural Technology, Gadjah Mada University who have helped on this research.

References

[1] Lerner B R 2001 *Tomatoes* (West Lafayette: Purdue University Cooperative Extension Service)

[2] Bailen G, Guillen F, Castillo S, Serrano M, Valero D and Martinez-Remero D 2006 *J. Agric. Food Chem.* **54** 2229-35

[3] Zapata P J, Guillen F, Martinez-Romero D, Castillo S, Valero D and Serrano M 2008 *J. Sci. Food Agric.* **88** 1287-93

[4] Han J H 2000 *Food Technology* **54(3)** 56-65

[5] Vasconez M B, Flores S K, Campos C A, Alvarado J and Gerschenson L N 2009 *Food Research International* **42(7)** 762-769

[6] Dang K T H, Singh Z and Swinny E E 2008 *J. Agric. Food Chem.* **56(4)** 1361-70

[7] Han J H and Gennadios A 2005 *Innovations in food packaging* ed J H Han (Amsterdam: Elsevier) p 239-262

[8] Mehyar G F, Al-Qadiri H M, Abu-Blan H A and Swanson B G 2011 *Journal of Food Science* **76(3)** M210-17

[9] Nisperos-Carriedo M O 1994 *Edible Coatings and Films to Improve Food Quality* ed J M Krochta, E A Baldwin and M O Nisperos-Carriedo (Pennsylvania: Technomic Publ. Co.,

Inc.) p 305-335
[10] Pareta R and Edirisinghe M J 2006 *Carbohydrate Polymer* **63** 425-431
[11] Anker M, Berntsen J, Hermansson A M and Stading M 2001 *Innovative Food Science and Emerging Technologies* **3** 81-92
[12] Ayranci E and Tunc S 2003 *Food Chemistry* **80** 423-431
[13] Flores S K, Fama S L, Rojas A M, Goyanes S and Gerschenson L 2007 *Food Research International* **40** 257-265
[14] Gonzalez-Aguilar G A, Celus J, Sotelo-Mundo R R, de la Rosa L A, Garcia J R and Alvarez-Parrilla E 2008 *Int. J. Food Sci. Technol.* **43** 91-101
[15] Rocculi P, Galino F G, Mendoza F, Wadso L, Romani S, Dalla-Rosa M and Sjoholm I 2007 *Postharvest Biol. Technol.* **43** 151-157
[16] Krochta J M 2002 *Protein Based Films and Coatings* ed A Gennadios (Boca Raton: CRC Press) p 1-41
[17] Chiumarelli M, Pereira L M, Ferrari C C, Sarantopoulos C I G L and Hubinger M D 2010 *Journal of Food Science* **75(5)** E297-304
[18] El-Anany A M, Hassan G F A and Ali F M R 2009 *J. Food Technology* **7** 5-11
[19] AOAC 2005 *Official Methods of Analysis* (Washington: Association of Official Analytical Chemists)
[20] Lalel H J D, Singh Z and Tan S C 2003 *J. Hortic. Sci. Biotechnology* **78(2)** 225-233
[21] Fagundes C, Palou L, Monteiro A R and Perez-Gago M B 2014 *Postharvest Biology and Technology* **92** 1-8
[22] Mardiana K 2008 *The Utilization of Aloe Vera Gel as Edible Coating of Sweet Star Fruit* (*Averrhoa carambola* L.) (in Indonesia) (Bogor: Bogor Agricultural University)
[23] Yaman O and Bayoindirli L 2002 *Lebnsm. – Wiss. Und. Technol.* **35** 146-150
[24] Oms-Oliu G, Soliva-Fortuny R and Martin-Belloso O 2008 *LWT – Food Sci. Technol.* **41** 1862-70
[25] Yang L and Paulson A T 2000 *Food Research International* **33** 563-570
[26] Lindsay R C 1997 *Food Chemistry* ed O R Fennema (New York: Marcel Dekker) p 767-824
[27] Moneruzzaman K M, Hossain A B M S, Sani W and Saifuddin M 2008 *American Journal of Biochemistry and Biotechnology* **4(4)** 336-344
[28] Ali A, Maqbool M, Ramachandran S and Alderson P G 2010 *Postharvest Biol. Technol.* 58 42-47
[29] Bierhals V S, Chiumarelli M and Hubinger M D 2011 *J. Food Sci.* **76(1)** E62-72
[30] Vasquez-Caicedo A L, Heller A, Neidhart S and Carle R 2007 *J. Agric. Food Chem.* **54** 5769 -67
[31] Park J H, Weller C L, Vergano P J and Testin R F 1993 *J. Food Sci.* **58** 1361-64
[32] Seymour G B, Taylor J E and Tucker G A 1993 *Biochemistry of Fruit Ripening* (London, New York: Chapman and Hall)
[33] Ben-Yehoshua S 1987 *Postharvest Physiology of Vegetables* ed J Weichmann (New York: Marcel Dekker Inc) p 113-170
[34] Ahmed L, Martin-Diana A B, Rico D and Barry-Ryan C 2013 *LWT – Food Sci. Technol.* **51** 367-374
[35] Garcia L C, Pereira L M, Sarantopoulos C I G L and Hubinger M D 2010 *Food Bioprocess Technol.* **3** 834-842
[36] Wang Y, Yu T, Xia J, Yu D, Wang J and Zheng X D 2010 *Biol. Control* **53** 178-182
[37] Tanada-Palmu P S and Grosso C R F 2005 *Postharvest Biol. Technol.* **36** 199-208
[38] Riberio C, Vicente A A, Teixeira J A and Miranda C 2007 *Postharvest Biol. Technol.* **44** 63-70

Utilization of modified starch from avocado (*Persea americana* Mill.) seed in cream soup production

M Cornelia* and A Christianti

Food Technology Department, Faculty of Science and Technology, Universitas Pelita Harapan, Karawaci – Tangerang, Indonesia 15810

*Email: melanie.cornelia@uph.edu

Abstract. Avocado (*Persea americana* Mill.) seed was often seen as waste and underutilized resources, especially in the food industry. The aim of this research was to modify the structure of avocado seed starch using the cross-linking method, to improve the viscosity stability in the cream soup. In the preliminary research, starch was isolated from the seed and modified by STPP (sodium tripolyphosphate) with 2%, 4%, and 6% concentration and were reacted for 1, 2, and 3 hours. Starches were analyzed for moisture and ash content, paste clarity, gel strength, swelling power, solubility, yield, and degree of whiteness. Based on the analysis results, the best reaction time and STPP concentration was 6% at 1 hour reaction time. Native starch and the best-modified starch were applied in the cream soup and compared with commercial cream soup. Cream soups were analyzed for viscosity stability using viscometer in 0, 1, 3, and 5 hours after storage in room temperature. The result showed that cream soup using modified starch has better viscosity stability than native starch and commercial cream soup after 5 hours storage, which was 181.7 ± 4.85 cP. Sensory analysis showed that cream soup using modified starch was more acceptable than the others. Avocado seed modified starch has phosphate group that strengthen the starch chain to prevent viscosity breakdown.

Keywords: *avocado seed, cream soup, cross-linking, modified starch, viscosity stability*

1. Introduction

Consumption of avocado (*Persea americana* Mill) had increased worldwide in recent years. Avocado seed as a part of avocado fruit (25% w/w) were seen as waste and underutilized resources, especially in the food industry. Currently, the seed represents waste issue lead to environmental pollution. The way to overcome that waste problem was by processing of avocado seeds starch. Avocado seeds contain about 30% starch, so this starch has the potential to be further processed [1]. Starch could be categorized as native (natural) starch and modified starch. Natural starch could be obtained by extraction for example from seeds. Natural starch was limited usage in food industry because of stickiness, and intolerance of the acid treatment. Starch modification has been widely used to improve the functional properties of food products. Some method was used to be modified starch, i.e. chemically, physically, and enzymatically [2]. Each method will enhance the functional properties of the food product such as increasing the water binding capacity, increasing the resistance to the heating temperature, minimizing the occurrence of retrogadation or sineresis, and improving and maintaining viscosity despite stirring, etc. [3]. Modification of starch by cross linking method was one of chemical modification by cross-linking -OH within amylose or amylopectin structure, so it could form crosslink or bridge connecting one starch molecule with other starch [4]. Cream soup was a slightly different product with many soup in general, because of the texture was more viscous. The cream soup was a product that usually made from wheat flour, potato starch, or corn starch [5]. High viscosity was produced by the starch in flour, milk or cream. The problem in cream soup was the viscosity of the cream will decrease during storage, resulting the texture of the cream soup becomes more fluid. Therefore, by this research avocado seed waste was utilized by cross-linking modification of avocado seed starch, hopefully the viscosity of the cream soup was stable even it was stored at room temperature for such a long time.

2. Research Method

2.1. *Materials and equipment*

Materials used in this research were avocado seeds as main raw material and ingredients for making cream soup such as full cream milk, modified avocado starch, natural avocado starch, iodized salt, white pepper powder, garlic powder, butter, sugar and water. Chemicals were used are Sodium Tri Poly Phosphate (STPP), NaOH, and HCl.

Equipment used in this research were analytical balance, blenders, pH meters, sieve shaker, water bath, Brookfield viscometer, Minolta Chromameter CR-400, Rapid Visco Analyzer, Scanning Electron Microscope, and Fourier Transform Infra Red (FTIR).

2.2. *Methods*

2.2.1. *Research phase I.* The extraction of avocado seeds to get the natural starch using water as solvents. Avocado seed were blended with water and filtered by filter cloth. Before blending, avocado seeds were given soaking treatment using 3000 ppm sodium metabisulfite for 24 hours to prevent the browning reaction [6], [7]. The precipitated was centrifuged 3500 rpm for 10 min and dried for 24 hours at 50 °C.

The natural starch was then modified by crosslinking method with addition of 2%, 4% and 6% STPP and reacted for 1 hour, 2 hours and 3 hours. The experimental design was a Completely Randomized Design (CRD) with two factors, that were STPP concentration (2%, 4%, and 6%) and reaction time (1 hour, 2 hours, and 3 hours), with three repetitions. The process of modification were begun by mixing 50 gr of starch with 125 mL of distilled water. Furthermore, STPP was added according to the treatment and pH adjusted to 10.5 with 5% NaOH added. This solution was reacted at 45°C in a water-bath. The solution will have a pH setting back up to 5.5 with the addition of 0.1% HCl. The final solution will be centrifuged 3500 rpm for 15 minutes and the precipitate was dried for 24 hours at 50°C.

2.2.2. *Research phase II.* Application of the best natural starch and modification starch from research phase I in making cream soup. Preparing the cream soup by making chicken broth by boiling chicken in water with a ratio of 1: 2 at 90°C for 60 minutes. The ingredients used were weighed according to the formulation. The cooking process is begun by mixing the butter with the starch used. After flattening, chicken broth and full cream milk are added and cooked at 86-90°C until homogeneous. The last stage was addition of salt, white pepper powder, sugar and stirring until the cream soup thickened.The formulation of making cream soup could be seen in Table 1.

The experimental design was Completely Randomized Design (CRD) of two factors, the first factor (3 levels) were type of starch (natural avocado starch, modified avocado starch, and commercial cream soup), and the second factors (4 levels) were storage time (0 hour, 1 hour, 3 hours and 5 hours) with each three repetitions.

Table 1. Formulation of cream soup production.

Material	Composition
Chicken Broth	500 (mL)
Full cream milk	250 (mL)
Natural or Modification Starch	25 (g)
Butter	30 (g)
Sugar	8.46 (g)
Salt	4 (g)
White pepper powder	0.45 (g)

3. Results and Discussion

3.1 *Effect of % STPP addition and reaction time toward yield, moisture and ash content of modified avocado seeds starch*

The statistical test resultshowed that % STPP addition and reaction time had significant effect and interaction ($p<0.05$) on the % yield of modification starch. One hour reaction time will result the highest yield when the %STPP was increased. The longer reaction time, the yield of starch decreased. It could be caused by longer contact of starch with hot temperature will cause the components lifted during heating and drying, thus lowering yield of modification starch.The statistical test resultof water content showed that %STPP and reaction time did not give any significant effect ($p>0.05$) and no interaction. Although the addition of phosphates in different concentrations, but the same drying process so that the number of water molecules that came out of the material was not different [8].The statistical test result on ash content showed that the % STPP and reaction time had significant effect and interaction ($p<0.05$). The more bound phosphate groups, the more ash content produced because the phosphate was the ash component [9]. The longer the phosphate reacts with starch, the more time the phosphate group has to penetrate into the starch granules.

3.2 *Effect of %STPP addition and reaction time toward paste clarity, gel strength, swelling power, solubility, degree of whiteof modified avocado seeds starch*

The statistical test result on paste clarity, gel strength, swelling power, solubility and degree of white of starch showed that the addition of%STPPand reaction time had significant effect and interaction ($p<0.05$). The addition of %STPP as a cross-linking reagent will decrease the paste clarity value. The presence of phosphate compounds as cross-linking reagents will alter the structure of the starch granules as the modified starch will break the structure when heated to 95°C and decrease the paste clarity [10]. The higher %STPP added, the higher the gel strength produced. Increasing of %STPP will increase hardness and gel strength because of the higher concentration used, the stronger also the starch chains are produced. The formed gels will become more rigid and the bonds that occur between the polymers that make up the gel are stronger. The higher % of STPP added, the more decreasing the swelling power. This result was accordanceto[11]that crosslinked starch will decrease swelling power value compared to natural starch, because the modified starch granules with crosslinks will be stronger and make the starch more resistant to acid and heat medium so it is not easy to experience viscosity breakdown when done heating. Cross-modified starch will strengthen the bond between the starch chains thus retaining the starch granules for swelling. The higher %STPP added, the smaller the solubility of the starch produced. This result was according to [10]that the decreased solubility value due to crosslink modification was due to the increased density of the modified starch structure, which reduces the disintegration of starch granules. This decreased solubility value indicates a stronger starch bond than natural starch. The higher of %STPP added and the reaction time taken, the degree of white also decreased. Starch composed by amylose and amylopectin will be broken down into simple sugars. The sugars formed will produce a maillard reaction. This was because starch still has protein content that could react with simple sugars in starch causing the decrease of white degree of crosslinked starch modification [12].

3.3 *Determination of the best avocado seed modification starch*
The best choice of modification crosslinking method (STPP 6%, 1 hour reaction time) could be seen in Table 2. Based on the analysis of physical and chemical characteristics of avocado modified starch, the addition of 6% STPP with 1 hour reaction time were determined as the best crosslinking starch modification. The reaction time was determined based on best results and best effectiveness. Determination of this best results based on the value of swelling power, solubility, paste clarity, and gel strength were the best. The best crosslinked avocado seeds were analyzed for starch, amylose, amylopectin, starch gelatinization profile, starch structure, and the shape and size of starch granules.

Tabel 2. Results of analysis of natural avocado seedstarch and the best of modified avocado seedstarch.

	Analysis Result	
Parameter	Natural Starch	Modified Starch
Water (%bb)	11.33 ± 0.06	11.67 ± 0.36
Yield (%)	8.05 ± 0.27	78.79 ± 1.16
Ash(%)	0.07 ± 0.00	0.13 ± 0.00
Paste Clarity (% T_{650})	1.17 ± 0.00	0.56 ± 0.02
Gelatinization Temp (^{0}C)	83.25	80.90
Peak Viscosity (cP)	3330	4890
SetbackViscosity (cP)	2021	1709
Heat Viscosity (cP)	2900	4943
Gel Strength (g)	67.7 ± 0.44	109.73 ± 2.93
Swelling Power (g/g)	6.32 ± 0.10	4.12 ± 0.06
Solubility (%)	12.57 ± 0.06	3.00 ± 0.13
White Degree (%)	86.81 ± 0.90	69.23 ± 1.62
Starch (%)	91.48 ± 0.06	87.62 ± 0.31
Amylose (%)	42.37 ± 0.14	39.17 ± 0.07
Amylopectin (%)	49.11	48.45

3.4 *Structure of starch analysis by Fourier Transform Infra-Red (FTIR) Spectroscopy*
The natural avocado seed starch's were showed in wavelength of 1014.9357 cm^{-1}. The modified avocado seed starch's wavelength 1012.1641 cm^{-1}. The FTIR analysis showed that there are P = O groups shown at wavelengths of 1150.7464 cm^{-1} for modified avocado seed starch and 1156.287 cm^{-1} for natural avocado starch. The structure of natural and modification of avocado seed starch showed that it was not much different. This could be caused the crosslinking reactions in the starch chain did not interfere and change the structure of starch [13].Result analysis of natural and modified starch by FTIR Spectroscopy could be seen in Figure1 and Figure2.

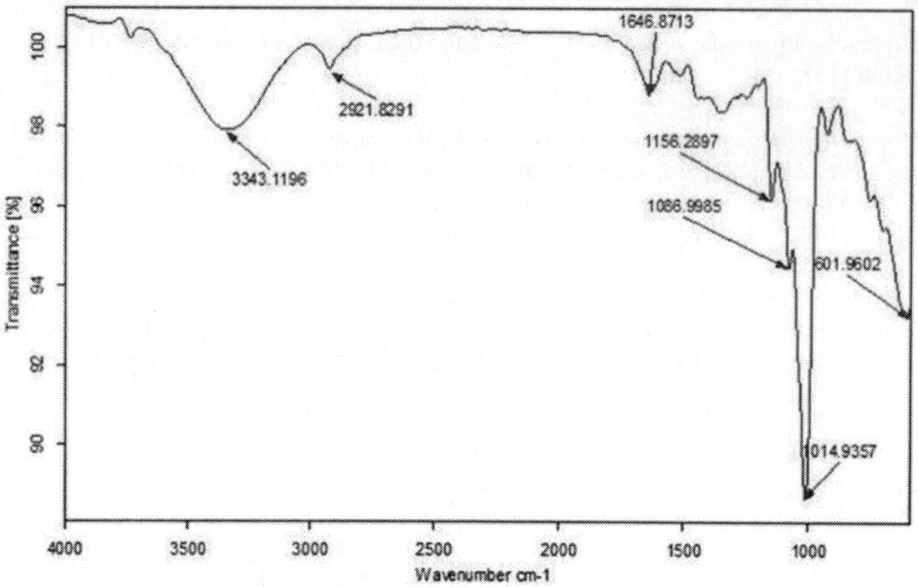

Figure1. Natural starch from avocado seed.

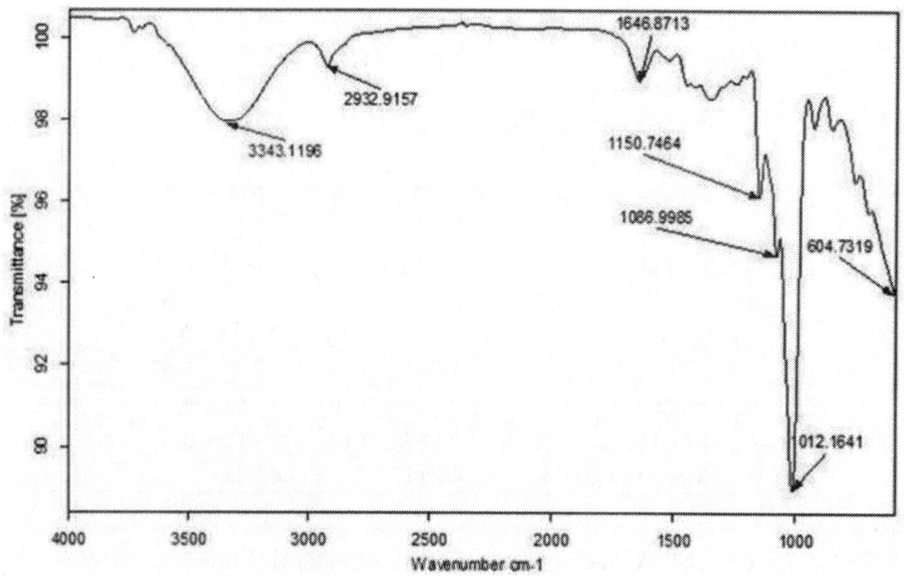

Figure2. Modified starch from avocado seed by crosslinking method.

3.5 *Starch, amylose and amylopectin content of modified avocado seed starch*
The starch content from the best modified starch was 87.62%, lower than natural avocado seed starch (91.48%). The decrease of % starch was caused by % STPP added while phosphate compound caused increasing non-starch components such as cellulose, hemicellulose, pectin and lignin. These components caused decreasing in starch content [14]. Amylose and amylopectin of modified avocado starch also decreased from natural avocado seed starch, for amylose 42.37% to 39.17% and amylopectin 48.45% from 49.11%.

3.6 *Shape and size of starch granules by Scanning Electron Microscope (SEM)*

The natural avocado starch has size between 8.82 μm to 22.94 μm with ovoid shape, oblong elliptical and circle as in Figure3. While shape and size of modified avocado starch (6%, 1 hour) as in Figure 4 showed between 5 μm to 20 μm with ovoid shape, oblong elliptical and circle. These results indicate that crosslinked starches were smaller in size than natural starches and there was no difference in shape between natural starch and modified starch. According to [15], the crosslinked starch modification has a similar shape to natural starch but has a smoother surface and did not have many pores compared to natural starch.

3.7 *The gelatinization temperature*

The gelatinization temperature of natural avocado starch was 83.25^{0}C while the modified avocado seed starch decreased to 80.90^{0}C. The decreasing may be caused by weakening of the granular structure and disintegration during the modification process [16].

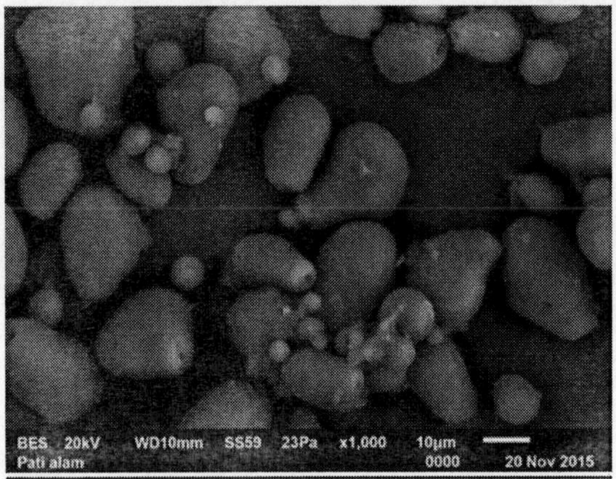

Figure 3. Natural Avocado Seed Starch with 1000x magnification (analyzed by PT. Nanotech Herbal Indonesia).

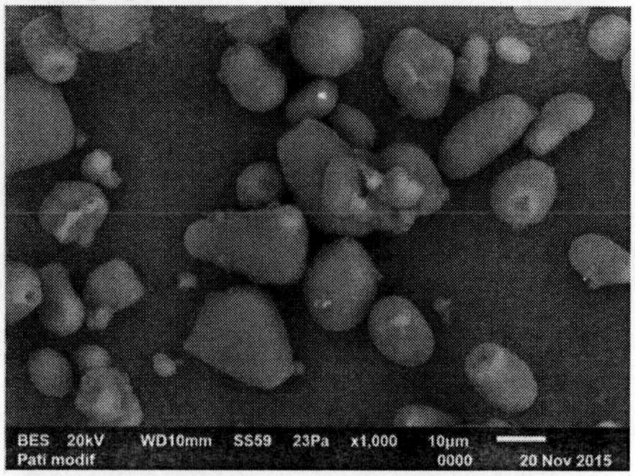

Figure 4. Modified Avocado Seed Starchwith 1000x magnification (analyzed by PT. Nanotech Herbal Indonesia).

3.8 *Peak viscosity*

The best peak viscosity of avocado seed starch yields was 4840 cP. This result according to other research result that cross-link modification process tends to increase the peak viscosity value, but only in a certain concentration of% STPP increased, the peak viscosity value will have decreased. Increased viscosity may occur due to crosslinking of amylopectin and amylose molecules so that the integrity of starch granules is stronger. Therefore, when heated, the starch hydration capacity of the starch granules is greater, and the peak viscosity increases [17].

3.9 *Heat viscosity*

Heat viscosity was a parameter used to determine the ability of starch granules in self-defense or viscosity during heating. The heating process was usually carried out at 95 °C for 20 minutes. The results of heat viscosity analysis on natural avocado seed starch and modification were 2900 cP and 4943 cP, respectively. Increasing of this heat viscosity in modified starch accordance with the theory that the higher the concentration of STPP used, and the higher the heat viscosity value and was expected in soup products because the higher the value of heat viscosity, the starch granules can more and more self-defense during the heating process. The resulting cream soup product will have a more stable viscosity compared to natural avocado seed.

3.10 *Effect of starch type and storage time on modified starch viscosity stability*
The results showed that there was a decreasing in viscosity when soup was stored longer. The decrease in viscosity occurring in natural starch could be due to the nature of the starch that was not resistant to heating and stirring so that the viscosity decreases when stored caused by the bond between starch and protein causing the decrease of viscosity [16].

3.11 Effect of type of starch toward consumer acceptance in cream soup
The scoring and hedonic tests were performed on 70 untrained panelists with 1-7 scale. For example, scale 1-7: very tasteless creamy – very creamy feel. Based on the overall hedonic test results of the cream soup products, both products were acceptable to consumers. But the overall cream soup used modification starch was preferred score 5.29 which means rather like compared to natural starch soup. Based on all sensory test parameters color, flavor, taste, aftertaste, and viscosity, consumers prefer more like the cream soup used modified starch compared to natural starch. Both types of cream soup are acceptable to the consumer, but the viscosity of the cream soup with modified starch was preferred. The viscosity affects the consumer's preference that a thicker cream soup was preferred in this research.

3.12 *Proximate analysis of selected soup cream*
The selected cream soup was a cream soup which used modified starch of avocado seed, because this cream soup had a better stability of viscosity. The proximate analysis of the selected cream soup could be seen in Table 3.

Table 3. Proximate analysis of cream soup modified avocado seed starch.

Analysis	Result (%)
Water	86.76 ± 0.23
Ash	0.97 ± 0.00
Protein	2.45 ± 0.11
Fat	1.99 ± 0.01
Carbohydrate	7.83

4. Conclusion

Extraction yield of natural starch from the avocado seed was 8.05% wet basis. The more adding of % STPP and time of reaction, ash content and gel strength increased but the paste clarity, swelling power, solubility, and white grades color decreased. The best crosslinked avocado seed starch was obtained by adding6% STPP and one hour reaction time. The modified avocado seed starch had a lower amylose content than it's natural starch. The crosslinking modification process did not affect the shape and size of the avocado starch granules by SEM analysis. In making cream soup, the modified avocado seed starch gave more stable viscosity compared to natural avocado seed starch and commercial cream soup during storage time.

References

[1] Olaeta J A, Schwartz M , Undurraga P and Contreras S 2007 Use of hass avocado (*Persea Americana* Mill.) seed as a processed product. *Proc. World Avocado Congress (Chile)* vol 6
[2] Herawati Heny 2011 Potential development of resistant starch products as functional food (in Bahasa Indonesia) *J. Litbang Pertanian* **30** (1) 31-9
[3] Abbas K A, Khalil, Sahar K and Hussin A S M 2010 Modified starches and their usages in selected food products: A review study *J. Agric. Sci.* **2** (2) 90-100.
[4] Neelam K,Vijay S and Lalit S 2012 Various techniques for the modification of starch and the applications of its derivaties. *Int. Research J. of Pharmacy* **3**(5) 25-31.

[5] Sunyoto M and Futiawati R 2012 The influence of full cream milk powder concentration on the characteristics of "Rasi" instant cream soup *J. Agricultural Sci. and Tech* **2** 1218-31

[6] Kenneth, A. K., Michael, A.O., Anthony, O.A., and Sunday, T. Isolation and physicochemical characterization of tigernut (*Cyperusesculentus*) starch as a potential industrial biomaterial *International J. Material Sci. and Appl.* **3**(2) 37-41.

[7] Subaric D, Babic J, Lalic A, Ackar D and Kopjar M 2011 Isolation and characterization of starch from different barley and oat varieties *Czech J. Food Sci.* **29** (4) 354-360

[8] Widhaswari V A and Putri W D R 2014 The effect of chemical modification with STPP onthe characteristics of purple sweet potato flour (in Bahasa Indonesia). *J. Pangandan Agroindustri* **2** (3) 121-128.

[9] Wulan S N, Widyaningsih T D and Ekasari D 2007 Modification of natural starch and starch produced from breaking the branch chain with physical/chemical treatment to increase the resistant starch content in rice starch (in Bahasa Indonesia). *J. Teknologi Pertanian* **8** (2) 80-87.

[10] Koo S H, Lee K W and Lee H G 2010 Effect of cross-linking on the physicochemical and physiological properties of corn starch *J. Food Hydrocolloids* **24** 619-25

[11] Xiao H, Lin Q, Liu G, Wu Y, Tian W, Wu W and Fu X 2011 Physicochemical properties of chemically modified starches from different botanical origin. *Scientific Research and Essays* **6** (21) 4517-25

[12] Hartanti F D, Amanto B S and Rahadian D A 2013 The study of physicochemical characteristics of breadfruit flour (*Artocarpuscommunis*) modified by variation of concentration and duration of lactic acid immersion (in Bahasa Indonesia). *J. Teknosains Pangan* **2** (4) 54-61.

[13] Li G, Li F, Zeng J and Sun J 2015 Preparation and in vitro digestibility of corn starch phospodiester *Trop. J. of Pharm. Research* **14** (3) 355-362.

[14] Susilawati, Nurdjanah S and Putri S 2008 Characteristics of physical and chemical properties of cassava (*Manihotesculenta*) based on different planting locations and harvesting age (in Bahasa Indonesia). *J. Teknologi Industri dan Hasil Pertanian* **13** (2) 59-72.

[15] Mirmoghtadaie L, Kadivar M and Shahedi M 2009 Effect of cross-linking and acetylation on oat starch properties. *Food Chemistry* **116** 709-713

[16] Aini N and Hariyadi P 2007 Waxy and non-waxy white corn pasta which are oxidised and acetyl-oxidised modified (in Bahasa Indonesian) *J. Ilmu Pertanian Indonesia* **12**(2) 108-115.

[17] Munarso S J, Muchtadi D, Fardiaz D and Syarief R 2004 Changes in physicochemical properties and functional rice flour due to the cross-modification process (in Bahasa Indonesia). *J. Pascapanen* **1** (1) 22-28.

IOP Conf. Series: Earth and Environmental Science **102** (2017) 012075 doi:10.1088/1755-1315/102/1/012075

Physical, chemical and sensory characteristics of red guava (*Psidium guajava*) velva at different fruit ripening time

D Ishartani*, F L F Rahman, R Hartanto, R Utami and L U Khasanah

Food Science and Technology Department, Faculty of Agriculture, Universitas Sebelas Maret, Jl. Ir. Sutami 36 A Surakarta, Indonesia

*Email: dishtani@yahoo.com

Abstract. This study purposed to determine the effect of red guava fruit ripening time on the physical (overrun and melting rate), chemical (vitamin C, pH, total dissolved solid) and sensory (color, taste, aroma, texture, and overall compare to control (without ripening) velva) characteristic of red guava velva. Red guava fruits were harvested at 90 days after flowering, ripened and then processed into velva. This research used Completely Randomized Design with fruit ripening time (without ripening, 4 days, and 6 days) as single factor. The research was conducted in triplicate. Chemical and physical characteristic data was analysed using One Way Analysis of Varian whether sensory characteristic data was analyzed using Independent Sample T-test. The result showed that fruit ripening time significantly affected the physical, chemical and sensory characteristic of the velva. Vitamin C, pH, and total solid of the velva were increased as the ripening time prolonged. In other hand, increasing of fruit ripening time decreased the overrun and melting rate of the velva. Red guava velva made from 6 days ripening had better sensory characteristics compared to velva made from red guava fruit without ripening or 4 day ripening. This research conclude that 6 days ripening time gives better chemical, physical and sensory characteristics of the velva compare to 4 days ripening time. Red guava fruits ripened for 6 days were recommended as raw material in velva making.

Keywords: *velva, ripening time, red guava*

1. Introduction

Indonesia is one of the tropic countries with a huge diversity of fruits. One of the popular fruits is guava. Central Bureau Statistics of Indonesia [1] reported that in 2014 guava production in Indonesia was 187406 ton while Java island produced 121.273 ton of guava, and Central Java produced 40.462 ton of guava in a year. Due to the high production of guava in every season, the processing of guava into food products is necessary to increase the farmer income. Some food products generated from red guava such as jam [2], jelly [3], and fruit leather [4]. Other promising food product developed from red guava is velva.

Velva is one of frozen dessert which was produced from fruits and contains less fat due to non milk fat utilization. Velva is also nutritious due to high content of vitamin from the fruits as ingredient. Velva production requires the high total solid compounds of fruits to produce a good texture of velva. Every fruits contained high total solid compound could be produced into velva [5]. Several fruits have been manufactured into velva such as pineapple, honeydew, soursop and guava.

Red guava contains 80.8 g water; 68 kcal energy; 2.55 g protein; 0.95 g fat, 14.32 g carbohydrate; 5.4 total dietary fiber; 8.92 g total sugar; 25 μg total carotene; 1.39 g ash; 374 μg β-carotene; 5.204

lycopene μg [6]; and 87 mg vitamin C [7].The total soluble solid content of red guava was quite high (8.5-11.4%) [6]. Therefore, red guava potentially developed into velva.

The texture of velva is also affected by stabilizer agents [8]. 0.5% gelatin has been selected as stabilizer that could produce velva with 33.91% overrun [5]. While, red guava velva produced from ingredients that contain 1% CMC as stabilizer revealed velva with 15.21% overrun and 2.68 min/g melting rate [9].

Guava is a climacteric fruits with moderate natural ethylene production. Climacteric fruits can be harvested at early physiology maturity state and storage for several days before it sold or consumed as ripened fruits. During maturity state, starch converse into sugar that revealed sweeter and softer fruits at ripening state [10]. Fruit color changes and vitamin and volatile compounds formation also occurs during maturity state [11].

Guava has a rapid maturation rate and the shelf life after harvest is relatively short ranging from 3 to 8 days [12] or 5 to 8 days [13] depending on the variety, harvest time, and environmental conditions. The guava reaches its climacteric peak between the 4th day and the 5th post-harvest day (mature-green harvested fruit) and then decreases [12] or the 4th day to the 6th day followed by rapid changing of fruit color and firmness [13]. Synthesis of ethylene increase until 8 days when fruits spoiled [14]. This indicated that difference ripening time will perform difference characteristics of fruits.

Furthermore, this study aimed to investigate the effect of ripening time of red guava on the characteristics of red guava velva. Ripening time of samples varied on 0, 4, and 6 days. The chemical (total soluble solid, vitamin C, pH), physical (overrun, melting rate) and sensory characteristics of red guava velva will determine to select the suitable ripening time of red guava to produce velva.

2. Material and Method

Red guava fruits were harvested from farmer group farms at Dusun Candi, Desa Jatirejo, Kecamatan Ngargoyoso, Kabupaten Karanganyar, Central Java at 90 days after flowering. Firm, glossy, round, not rotten and not defect fruits were selected to follow ripening process. Ripening process was carried out at ambient temperature for 0 (without ripening), 4, and 6 days in an air circulated basket (Figure 1). Ten red guava fruits was used for each ripening treatment. The basket was kept away from direct sunlight and each red guava fruit was wrapped using polyethylene plastic. Ripened red guava fruits for 0 (without ripening), 4, and 6 days are shown in Figure 2.

Figure 1. Ripening process.

IOP Conf. Series: Earth and Environmental Science **102** (2017) 012075 doi:10.1088/1755-1315/102/1/012075

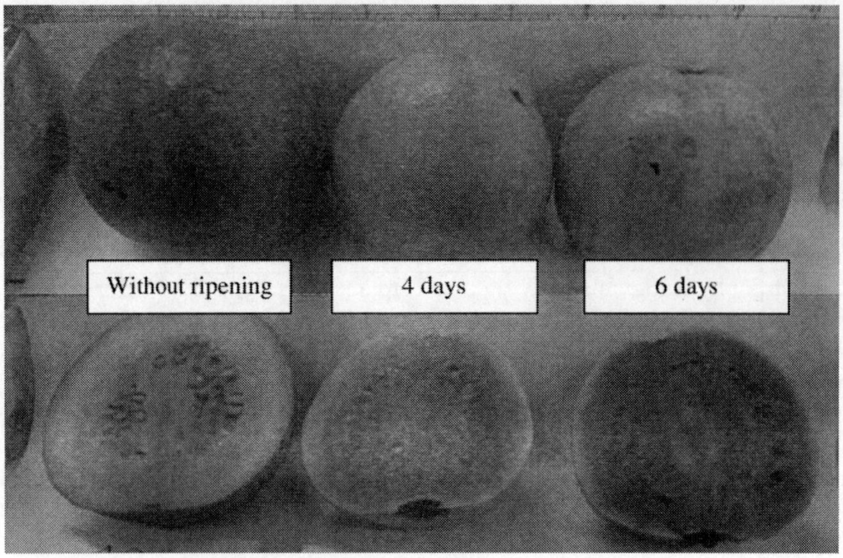

Figure 2. Ripened red guava fruits.

Ripened red guava fruits were washed, sliced, and crushed to get red guava puree. Cane sugar (25% w/w), gelatin (0,5% w/w), citric acid (0,1% w/w) and water (50% v/w) were added into red guava puree and mixed homogenously. The homogenous mix was refrigerated for 24 hours at 5-6°C followed by agitation for 30 minutes at 4°C using Ice Cream Maker (Delonghi Gelatiao ICK6000). Agitated mix was frozen at -20°C for 24 hours to obtain the red guava velva and kept in the freezer until being analyzed.

2. 1. *Determination of vitamin C.* Velva samples was diluted in aquades into 100mL solution and filtered. 25mL aliquots was transferred into Erlenmeyer, added with 1% starch solution, and then titrated using Iodin solution. Vitamin C calculated as (Iodine solution volume x vitamin C molecular weight)/samples weight [15].

2. 2. *Determination of pH.* Velva pH values was measured using digital laboratory pH meter with glass electrode.

2. 3. *Determination of total dissolved solids.* Velva total dissolved solid was measured using hand refractometer.

2. 4. *Determination of overrun.* Homogenous mix samples was weighed, frozen into velva, and then the velva samples was weighed. The weighing procedure was conducted at the same volume. Overrun calculated as [(velva weight – mix weight)/ mix weight] x 100% [16].

2. 5. *Determination of melting rate.* 100 grams velva samples was placed on the wire mesh (10 mesh) and let to melt at room temperature. The dripped portion passing the wire mesh was weighed every 10 minutes for 120 minutes. The time was plotted against the dripped portion weight to determine the slop that calculated as melting rate of the velva (grams/minutes) [16].

2. 6. *Sensory* analysis. Sensory characteristic of the velva was tested using multiple comparison test [17]. Velva made from red guava fruit without ripening process was carried out as control sample whether velva made from ripened red guava fruit for 4 days and for 6 days as test sample. 25 untrained panels were involved sensory test. Sensory parameters being observed were color, aroma, taste, texture and overall.

IOP Conf. Series: Earth and Environmental Science **102** (2017) 012075 doi:10.1088/1755-1315/102/1/012075

The analysis was conducted in triplicate. Chemical characteristic data and physical characteristic data was analysed using One Way Analysis of Varian followed by Duncan Multiple Comparison Test if there was a significant different between treatment. Sensory characteristic data was analysed using Independent Sample T-test.

3. Results and Discussion

3. 1. Vitamin C

Vitamin C of the red guava velva increased due to fruit ripening prior to velva processing and ripening time seem significantly affected the vitamin C content of the velva (Table 1). Highest vitamin C content was found in velva made from ripened red guava fruit for 6 days (76,85 mg/100 g), followed by velva made from ripened red guava fruit for 4 days (68,05 mg/100 g) and without ripening (59,25 mg/100 g). The difference of vitamin C content was suggested have correlation with vitamin C biosynthesis from simple sugar as starch degradation product during fruit ripening [18]. Ripening process was reported promote the vitamin C content up to 35% [19].

Red guava velva in this research contained higher vitamin C compare to papaya velva and tomato velva. Papaya velva contained vitamin C 46,20 mg/100 g [20] and tomato velva that contained vitamin C 27,13 mg/100g [21].

Tabel 1. Vitamin C Content of Red Guava Velva Made from Different Fruit Ripening Time.

Ripening time (days)	Vitamin C (mg/100 g)^{a,b}
Without ripening	$59{,}25^{A} \pm 5{,}18$
4	$68{,}05^{B} \pm 2{,}87$
6	$76{,}85^{C} \pm 4{,}54$

[a]Values are means ± standard deviation.
[b]Mean values followed by different superscripts are significantly different at α 0,05.

3. 2. pH

Table 2 showed that the acidity of the red guava velva was decreased as the ripening time prolonged. This phenomena probably similar to the white guava acidity changes. White guava had a certain trend of pH development during ripening process related to organic acid content. White guava at mature stage would have a pH ascending and organic acid descending [22].

The pH of the red guava velva in this research was ranged from 4,09 – 4,60. It is similar to the previous report that using 10% carboxyl methyl cellulose (CMC) as stabilizer in the red guava velva processing [9].

Tabel 2. pH of Red Guava Velva Made from Different Fruit Ripening Time.

Ripening time (days)	pH^{a,b}
Without ripening	$4{,}09^{A} \pm 0{,}77$
4	$4{,}28^{B} \pm 0{,}08$
6	$4{,}60^{C} \pm 0{,}22$

[a]Values are means ± standard deviation.
[b]Mean values followed by different superscripts are significantly different at α 0,05.

3. 3. *Total Dissolved Solids*

Total dissolved solids of red guava velva was significantly increase with ripening time increase (Table 3). Velva made from red guava fruit ripened for 6 days got highest total dissolved solid. The increasing of total dissolved solids also found in the olive sherbet as the olive maturity increase [23].

Total dissolved solids increase in the velva had correlation with maturity development of red guava fruit. Starch hydrolysis by enzymes into simple sugars during ripening caused total dissolved solids increase. This enzymatic starch hydrolysis was ascending significantly in the fourth day after harvesting [24]. Total dissolved solid of red guava velva in this research (19,80 – 21,95°Bx) was higher than tomato velva (17,67°Bx) [21] and papaya velva (19,33°Bx) [20], but lower than orange sweet potato velva (26,13°Bx) [25].

Tabel 3. Total Dissolved Solids of Red Guava Velva Made from Different Fruit Ripening Time.

Ripening time (days)	Total Dissolved Solids (°Bx)^{a,b}
Without ripening	$19{,}80^{A} \pm 0{,}79$
4	$20{,}33^{A} \pm 0{,}81$
6	$21{,}95^{B} \pm 0{,}47$

^{a}Values are means ± standard deviation.
^{b}Mean values followed by different superscripts are significantly different at α 0,05.

3. 4. *Overrun*

High overrun in frozen dessert production, including ice cream and velva, is profitable. However, consumer preference related to frozen dessert quality such as melting rate and sensory parameter should be compromised with high overrun.

Table 4 showed that red guava fruit ripening time affected the velva overrun. As the ripening time prolonged, the velva overrun was decrease significantly. This trend was contrast to the total dissolved solids shown in Table 3. The increasing of total dissolved solids would inhibit the mix to trap air so the overrun would decrease [26]. Overrun of the red guava velva in this research (46,64 – 55,64%) was higher than the red guava velva in the previous report (33,91%) [5]. It probably caused by the difference of the red guava cultivar or the processing method used.

Table 4. Overrun of Red Guava Velva Made from Different Fruit Ripening Time.

Ripening time (days)	Overrun (%)^{a,b}
Without ripening	$55{,}64^{C} \pm 1{,}29$
4	$52{,}97^{B} \pm 0{,}39$
6	$46{,}64^{A} \pm 0{,}26$

^{a}Values are means ± standard deviation.
^{b}Mean values followed by different superscripts are significantly different at α 0,05.

3. 5. *Melting Rate*

Melting rate of the red guava velva could be seen in the Table 5. Ripening time significantly affected the velva melting rate. Longer ripening time would cause lower melting rate. Lower melting rate was preferable to the consumer because the velva would not melt easily.

The melting rate of the velva was inversely proportional to the total dissolved solids (Table 3) but was proportional to the overrun (Table 4). It was similar to ice cream phenomena. Ice cream with high overrun would melt slowly because the heat transfer rate was decline due to air trapping and zigzag path which the melting fluid must flow [16, 27]. Red guava velva made from ripened fruit for 6 days got the best melting rate compare to the two others. It also got better melting rate than papaya velva (1,41 grams/minutes) [20], orange sweet potato velva (1,61 grams/minutes) [25] and tomato velva (1,63 grams/minutes) [21].

Table 5. Melting Rate of Red Guava Velva Made from Different Fruit Ripening Time.

Ripening time (days)	Melting Rate (grams/minutes)[a,b]
Without ripening	0,40C ± 0,03
4	$0,36^{B}$ ± 0,01
6	$0,33^{A}$ ± 0,02

[1a]Values are means ± standard deviation.
[b]Mean values followed by different superscripts are significantly different at α 0,05.

3. 6. *Sensory Characteristic*

Table 6 showed that either velva made from ripened fruit for 4 days or for 6 days had better score compare to control (velva made from fruit without ripening) in all sensory parameters. Ripening time did not affect the score of aroma and texture but affected the score of color, taste, and overall. Red guava velva made from ripened fruit for 4 days got better score for those color, aroma, and overall parameter than velva made from ripened fruit for 6 days. Velva made from ripened fruit for 6 days had stronger red color than the other one (Figure 3).

Table 6. Multiple Comparison Score of Red Guava Velva Made from Different Fruit Ripening Time.

Sensory Parameter	Ripening time (days)[a,b]	
	4	6
Color	$2,08^{B}$ ± 0,64	$1,40^{A}$ ± 0,50
Aroma	$3,20^{A}$ ± 1,15	$2,44^{A}$ ± 1,75
Taste	$3,44^{B}$ ± 1,44	$2,40^{A}$ ± 1,08
Texture	$3,16^{A}$ ± 1,17	$2,76^{A}$ ± 1,30
Overall	$2,72^{B}$ ± 0,79	$1,92^{A}$ ± 0,75

a. Values are means ± standard deviation. Mean values followed by different superscripts in the same row are significantly different at α 0,05.
b. 1=very much better than R; 2=much better than R; 3=better than R; 4=slightly better than R; 5=equal to R; 6=slightly worse than R; 7=worse than R; 8=much worse than R; 9=very much worse than R
R is control (without ripening)

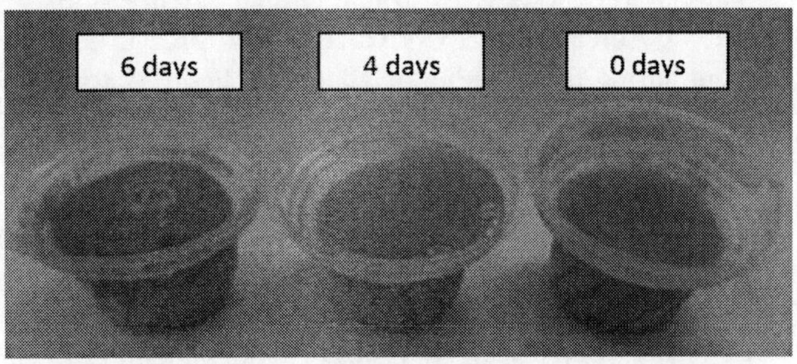

Figure 3. Velva made from ripened red guava fruits

4. Conclusion

Fruit ripening time did affect the vitamin C, pH, total dissolved solids, overrun, melting rate and sensory characteristics of the red guava velva produced. Highest vitamin C content, highest total dissolved solids, worst overrun, and best melting time was found in the velva made from fruit ripened for 6 days. Sensory characteristics of the velva made both from fruit at 4 dan 6 days ripening time were better than velva made from fruit without ripening. Therefore, making velva from fruit ripened for 6 days is recommended to be applied.

References

[1] Central Bureau Statistics of Indonesia. 2015. *Statistic of Horticulture Production in 2014* (in Bahasa Indonesia) (Jakarta: Central Bureau Statistics of Indonesia)

[2] A'yuni Q 2014 *Vitamin C Content and Sensory Characteristic of Red Guava Jam Substituted with Rosella Flower and Sour Star Fruit* (in Bahasa Indonesia). http://eprints.ums.ac.id/29549/14/NASKAH_PUBLIKASI.pdf, accessed on August, 24th 2017

[3] Widhiyandari R 2013 *The Production Process of Soft Candy Made from Red Guava (Psidium guajava L.)* (in Bahasa Indonesia) Fakultas Pertanian, UNS. Surakarta

[4] Astuti W F P, Nainggolan R J, Nurminah M 2016 Effect of Stabilizers Type on The Quality of Fruit Lether Made from Red Guava-Soursop Fruits Mix (in Bahasa Indonesia). *J. Rekayasa Pangan dan Pertanian* **4** (1) 65-71

[5] Wibowo T 1992 *Effect of Stabilizer Type and Concentration on The Quality of Red Guava Leather* (in Bahasa Indonesia) http://repository.ipb.ac.id/jspui/bitstream/123456789/38719/1/F92TWI.pdf, accessed on August, 24th 2017

[6] Chen H, Sheu M, Lin and Wu C 2007 Nutritional composition and volatile compounds in guava. *Fresh Procedure* **1**(2) 132-139

[7] Mahmud M K 2009 *Indonesia's Food Composition* (in Bahasa Indonesia) ed Zulfianto N A (Jakarta : PT Elex Media Komputindo)

[8] Sulastri T A 2008 *Effect of Arabic Gum Concentration on Pineapple Velva Quality During Cold Storage* (in Bahasa Indonesia) Fakultas Pertanian USU. Medan

[9] Maria dan Zubaidah 2014 *The Making of Probiotic (Lactobacillus acidophilus)-Guava Velva: Study on The Sucrose and CMC Percentage* (in Bahasa Indonesia) THP FTP UB. Malang

[10] Zulkarnain. 2014. *Basics of Horticulture* (in Bahasa Indonesia) (Jakarta:Bumi Aksara)

[11] Muchtadi T R, Sugiyono and Ayustaningwarno 2010 *Food Material Science* (in Bahasa Indonesia) (Bandung : CV Alfabeta)

[12] Nunez R and Flor D M 2005 *Guava (Psidium guajava L.) Fruit Phytochemicals, Antioxidant Properties And Overall Quality As Influenced By Postharvest Treatments*. Thesis University Of Florida. USA. http://ufdc.ufl.edu/UFE0011873/00001/19j, accessed on August, 24th 2017

[13] Singh and Sukhvider P 2009 *Prospective And Retrospective Approaches To Postharvest Quality*. University of Technology. Australia. Global Science Book

[14] Abreu J R, Santos C D, Abreu C M P, Correa A D and Lima L C O 2012 Sugar fractionation and pectin content during the ripening of guava cv. Pedro Sato. *Ciênc. Tecnol. Aliment.* **32** (1) 156-162

[15] AOAC. 1999. *Official Methods of Analysis of AOAC International 16th Edition*. (Maryland : AOAC)

[16] Goff HD, Hartel WR. 2013. *Ice Cream 7th Edition*. (New York : Springer)

[17] Setyaningsih D, Apriyantono A, Sari M. 2010. *Sensory Analysis for Food and Agro Industry* (in Bahasa Indonesia) (Bogor : Bogor Agricultural Institute Press)

[18] Gull J, Sultana B, Anwar F, Naseer R, Ashraf M and Ashfuzzaman M 2012 variation in antioxidant attributes at three ripening stages of guava (*Psidium guajava* L.) fruit from different geographical regions of Pakistan. *J. Molecules* **17** 3165-80

[19] Gomez M and Lajolo F 2008 Ascorbic acid metabolism in fruits: activity of enzymes involved in synthesis and degradation during ripening in mango and guava. *Jurnal of Food Science and Agriculture* **88** 756-762
[20] Rahmawati S 2014. *Physicochemical and Sensory Characteristic of Papaya (Carica papaya L.) Velva* (in Bahasa Indonesia) Ilmu dan Teknologi Pangan, Fakultas Pertanian UNS. Surakarta
[21] Sapriyanti R. 2014. Physicochemical and Sensory Characteristic of Tomato (*Lycopersicum esculentum* Mill) Velva (in Bahasa Indonesia) Prodi Ilmu dan Teknologi Pangan, Fakultas Pertanian, UNS. Surakarta
[22] Monteiro A R, Soarez F D, Pereira T and Marques M 2007 Volatile and non volatile chemical composition of the white guava fruit (*Psidium guajava*) at different stages of maturity. *J. Food Chemistry* **7** 61
[23] Sumrah and Ashraf M 2015 Effect of maturity stage of olive fruit on its sherbet quality. *J. Agric. Res.* **53** (2)
[24] Bashir A H, Abu-Bakr A and Abu-Ghoukh 2003 Compositional and enzymatic changes during guava fruit ripening. *Food Chemistry* **80** 557-563
[25] Wulandari B 2014 *Low Calorie Sweetener Application in The Orange Sweet Potatoes (Ipomea batatas L.). Velva* (in Bahasa Indonesia) Fakultas Pertanian UNS.Surakarta. Vol 3 No. 3 Juli 2014
[26] Arbuckle W S 1981 *Ice cream* (Connecticut : The AVI Pub.)
[27] Sakurai K, Kokubo S, Hakamata K, Tomita M and Yoshida S 1996 Effect of production conditions on ice cream melting resistance and hardness. *Milk Sci. Int.* **51**(8) 451-454

Isolation and identification of bacterial pathogen from mastitis milk in Central Java Indonesia

D W Harjanti[1*], R Ciptaningtyas[2], F Wahyono[1]and ET Setiatin[1]

[1]Department of Animal Science, Faculty of Animal and Agricultural Sciences, Diponegoro University, Indonesia.
[2]Department of Clinical Microbiology, Faculty of Medicine, Diponegoro University, Indonesia.

*E-mail :dianharjanti@undip.ac.id

Abstract. Mastitis is a multi-etiologic disease of the mammary gland characterized mainly by reduction in milk production and milk quality due to intramammary infection by pathogenic bacteria. Nearly 83% of lactating dairy cows in Indonesia are infected with mastitis in various inflammation degrees. This study was conducted to isolate and identify the pathogen in milk collected from mastitis-infected dairy cows. The study was carried out in ten smallholder dairy farms in Central Java Indonesia based on animal examination, California mastitis test, isolation bacterial pathogens, Gram staining, Catalase and Coagulase test, and identification of bacteria species using Vitek. Bacteriological examination of milk samples revealed 15 isolates where *Streptococcus* was predominant species (73.3%) and the coagulase negative *Staphylococcus* species was identified at the least bacteria (26.7%). The *Streptococcus* bacteria found were *Streptococcus uberis* (2 isolates), *Streptococcus sanguinis*(6 isolates), *Streptococcus dysgalactiae*ssp*dysgalactiae*(1 isolate), *Streptococcus mitis* (1 isolate) and *Streptococcus agalactiae* (1 isolate). The *Staphylococcus* isolates comprising of *Staphylococcus simulans* (1 isolate) *and Staphylococcus chromogens* (3 isolates). Contamination of raw milkwith pathogenic bacteria can cause outbreaks of human disease (milk borne disease). Thus, proper milk processing method that couldinhibit the growth or kill these pathogenic bacteria is important to ensure the safety of milk and milk products.
***Keywords** : milk, mastitis, pathogen, Central Java, Streptococcus*

1. Introduction

Indonesian dairy industry is based on smallholder farms grouped into cooperatives. Most of the farms have three to five head lactating cows. The dairy farms are based on the cut and carry system, in which the forage grasses are being gathered from the outside of the farms. Nearly 97% dairy farms are concentrated in West, Central and East Java, whereas around 3% are located in Sumatra. According to the data from the Indonesian Government of Statistic Agency, Indonesia is estimated to have more than 260 millions inhabitants in 2017, with the population growth nearly 1.67% annually. Therefore, milk and dairy product consumptions continue to increase rapidly in Indonesia. Nonetheless, the domestic milk production fails to meet demands of the population in the country, also the needs of the processing industry in terms of both quantity and quality. The Indonesian dairy farmers are facing many challenges to improve milk production and quality. One of the main challenges that impact both milk quantity and quality is a disease named Mastitis.

Mastitis is a multi-etiologic disease of the mammary gland characterized mainly by reduction in milk production and milk quality due to intramammary infection by pathogenic bacteria.The mastitis

IOP Conf. Series: Earth and Environmental Science **102** (2017) 012076 doi:10.1088/1755-1315/102/1/012076

prevalence in tropical country like Indonesia is very high as 75-83% [1]. Mastitis treatment with intramammary antibiotic has been done for many years in Indonesia. Antibiotic treatment of clinical and subclinical mastitis is a key component of mastitis control program. However, in fact, single injection may not enough to cure the inflammation. Many factors could influence to this problems. The bacteria may be resistant to the antibiotic or the commercial antibiotic used may notsusceptible to the specific bacteria species in udder. Previously, research had been conducted to identify the bacteria-causing mastitis in Indonesia. In 2002, Estuningsih et al. [2] found *Streptococcus agalactiae*from 7 farms in Java area by PCR. Furthermore, in 2005, Salasia et al. [3] did the comparative study on phenotypic and genotypic of *Staphylococcus aureus*isolates from bovine subclinical mastitis in Java and comparing them to the *Staphylococcus aureus*from Germany. Nonetheless, the bacteria causing mastitis is not only*Streptococcus agalactiae*and *Staphylococcus aureus* and the pathogenicity of the same bacteria species might be change due to environment and the specific antibiotic used. Increasing antibiotic resistance has become a serious concern worldwide. The resistance patterns of coagulate negative bacteria species were highly correlated with the antibiotic used in herd [4]. Since the species bacteria will be killed or inactivated with the specific antibiotic, hence, the antibiotic must be chosen based on the bacteria species in the infected animals. Therefore, determination of bacteria species from mastitis-infectedcow is extremely important to choose the appropriate antibiotic for treatment, also the proper milk processing method for specific dairy products. Nowadays, highly automated identification system the new VITEK gram positive and negative identification cards provide stable and decisive result of bacteria identification [5,6,7]. The result from this study will provide a data of specific mastitis-causing bacteria for further rapid detection, proper mastitis treatment and control, also precise method for milk processing to ensure food safety.

2. Material and Method

2.1.*Sample collection and bacteria isolation*

A total of twelve individual milk samples were collected from subclinical mastitis cows aseptically,after performing California Mastitis Test (CMT). The cows were obtained from 10 smallholder dairy farms in three different geographic regions in Central Java during June – July 2016. All milk samples were cultured in Microbiology Laboratory Faculty of Medicine, Diponegoro University.Refrigerated milk samples were warm at room temperature (25°C) for half an hour and then homogenized by gently shaked it in order to disperse bacteria from milk fat. One standard loop (10 μL) of milk sample was streaked on 5% sheep blood agar (Oxoid™ Blood Agar Base, CM0055) and streaked using the quadrant streaking method for each sample. The inoculated plates were then incubated at aerobic condition at 37 °C and checked after 24 and 48 hto eliminate slow growing bacteria.The plates were examined for growth, morphologic features such as colony size, shape, color and hemolytic characteristics.Growth more than one type of colonies was determined as mixed growth.Presumptive colonies were selected and sub cultured on 5% sheep blood agar (Oxoid™ Blod Agar Base CM0055) and incubated at aerobic condition at 37 °C for 24 h to get a pure culture. After incubation, pure colonies were stained using Gram stain reagent (BD Difco BBL™ ref. 212525 (Gram Crystal Violet), 212542 (Gram Iodine), 212527 (Gram Decolorizer), 212531 (Gram Safranin) and bacteria were differentiated based onGram reaction (Gram-positive or Gram-negative), cellular morphology and arrangements of the bacteria.Additional catalase test (Hydrogen Peroxide 3%) was done for Gram positive cocci, followed by coagulase test for catalase positive.

2.2.Bacteria Identification

Identification has been done by automatic method (Vitek® 2 Compact, Biomérieux, France). Gram positive and negative bacteria were identify using GP ID card and GN ID card respectively (Biomérieux, France)

3. Result and Discussion

Relatively up to date data on the etiology of bovine mastitis are available in other subtropical countries, mainly Europe America, Australia and Japan [8,9,10], but not in tropical country like Indonesia. Management conditions in Indonesia greatly differ from management systems in developed countries, in particular the feeding and milking management also the high temperature and humiditycontributing risk factorfor developing mastitis.Hence, the species and pathogenicity of pathogens in Indonesia may be differed from those in other countries.

A total of 12 mastitis milk samples were collected and streaked on the sheep blood agar. The examination was conducted on the growth, morphologic features such as colony size, shape, colour and haemolytic characteristics as presented in Table 1. There were 3 samples showed the mixed growth of two bacteria species, whereas the others only showed single bacteria species. Four isolates were positive in the catalase test, indicating that the bacteria were respire using the oxygen as aerobic or facultative anaerobic bacteria. The catalase negative bacteria may be anaerobe or they may be facultative anaerobes that only ferment and do not respire using oxygen as terminal electron acceptor. Four catalase positive bacteria that have been further tested were negative in the coagulase test. Coagulase negative bacteria have become common bovine mastitis causing pathogen in many countries and could therefore be described as emerging mastitis pathogens [11].

Identification of bacteria species from all samples were conducted by Vitek as shown in Table 2. The identified mastitis-causing bacteria were all gram positive Streptococci and Staphylococci bacteria. The *Streptococcus* bacteria found were *Streptococcus uberis* (2 isolates), *Streptococcus sanguinis* (6 isolates), *Streptococcus dysgalactiae* ssp *dysgalactiae*(1 isolate), *Streptococcus mitis* (1 isolate) and *Streptococcus agalactiae* (1 isolate), whereas *Staphylococcus* isolates comprising of *Staphylococcus simulans* (1 isolate) *and Staphylococcus chromogens* (3 isolates). Evaluation on the occurrence of mastitis and identification of causing pathogen had been conducted in many countries. In Indonesia, previous studies in 2010 atthe dairy farms in West and Central Javareported that*Staphylococcus aureus* and *Streptococcus agalactiae*were the major pathogen in mastitis milk with the prevalence rate 8.5% and 37.5% [12]. However in the present study, the *Staphylococcus aureus* is not present from all isolates. Previous study of Sugiri and Anri [12] was using a biochemical method to identify the mastitis causing bacteria, whereas we use Vitek in this study. Wallet et al [6] and Crowley et al. [7] stated that Vitek system provided rapid and accurate bacteria identification. *Staphylococci* comprises 45 species and 21 subspecies which are known to be the main etiological agents of mastitis in dairy cows in worldwide [13]. Nonetheless, the clinical relevance of *Staphylococcus aureus* when isolated and cultured from mastitis milk are remained debatable. Some consider the *Staphylococcus aureus* as the major mastitis pathogens [14] and the others regard these bacteria as minor pathogen in bovine mastitis [15,16]. Therefore, we conducted an identification of coagulase negative Staphylococcus (CoNS) at the species level using Vitekto develop the effective control strategies for CoNS in bovine mastitis. Currently, our data showed that the *Staphylococcus chromogenes* was the most frequent isolated CoNS, indicating that this bacteria probably the most predominant CoNS causing mastitis in dairy cattle in Central Java. Previously in Iran, Hosseinzadeh and Saei [13] reported that *Staphylococcus haemolyticus* (40.7%) and *Staphylococcus chromogenes* (15.7%) were the predominant CoNS in bovine mastitis cases. The catalase negative bacteria *Streptococcus uberis, Streptococcus sanguinis, Streptococcus dysgalactiae*ssp*dysgalactiae, Streptococcus mitis, Streptococcus uberis,* and *Streptococcus agalactiae* can cause persistent infections, which result in increased somatic cell count and decreased milk quality. The CMT results for those catalase negative bacteria found were varied from +++ to ++++, which can be estimated containing 1.200.000 to 5.000.000 somatic cell/ ml milk [17]. Based on current knowledge, it is difficult to know whether the coagulase negative bacteria behave as environmental or as contagious pathogens. Physiologically, lactoferrin secreted from secretory cells of mammary gland may act as antibacterial against the infection, thus, the lactoferrin concentration in milk will be higher during mastitis infection [18].Dried manure, bedding and the decreased teat sphincter patency due to the high parity or late lactation can be the main factors of intramammary bacteria infection. Biosecurity against contagious mastitis pathogens such as post milking teat dip disinfection reduce coagulase negative species in the herd and will ultimately reduce the number of somatic cells in milk [19]. The continuous monitoring of the bacteria found in current study in herds is needed. Moreover, special attention on the prevention and control of mastitis pathogens need to be placed in the top priority.

Table 1. California Mastitis Test Result, Morphologic features and characteristics of Isolates.

Sample number	CMT result	Colony types	Morphologic features	Haemolytic characteristics	Catalase test	Coagulase test
1	+++	Mixed growth	Gram positive, coccus, translucent white	Weak hemolyis	Positive	Negative
			Gram positive, coccus and twisted (like a chain), translucent white	No hemolysis	Negative	-
2	+++	Mixed growth	Gram positive, coccus, translucent white	Weak hemolysis	Negative	-
			Gram positive, coccus, yellow	β- hemolytic	Positive	Negative
3	+++	Single species	Gram positive, coccus, white	β- hemolytic	Negative	-
4	+++	Single species	Gram positive, coccus	α-hemolytic	Negative	-
5	+++	Mixed growth	Gram positive, coccus and twisted (like a chain), translucent white	No hemolysis	Negative	-
			Gram positive, coccus, yellow	β- hemolytic	Positive	Negative
6	+++	Single species	Gram positive, coccus and chained, white	α-hemolytic	Negative	-
7	++++	Single species	Gram positive, coccus and chained, white	α-hemolytic	Negative	-
8	++++	Single species	Gram positive, coccus and chained, white	α-hemolytic	Negative	-
9	++	Single species	Gram positive, coccus, yellow	β- hemolytic	Positive	Negative
10	++++	Single species	Gram positive, coccus and chained, white	α-hemolytic	Negative	-
11	++++	Single species	Gram positive, coccus and chained, white	β- hemolytic	Negative	-
12	++++	Single species	Gram positive, coccus and chained, white	α-hemolytic	Negative	-

Table 2. Identification of mastitis causing pathogens.

Identification Result	Number of isolate (s)	Prevalence
Staphylococcus simulans	1	6.7 %
Staphylococcus chromogens	3	20%
Streptococcus uberis	2	13.3%
Streptococcus sanguinis	6	40%
Streptococcus dysgalactiae ssp *dysgalactiae*	1	6.7 %
Streptococcus mitis	1	6.7 %
Streptococcus agalactiae	1	6.7 %

Conclusion

In conclusion, *Streptococcus* is predominant and the coagulase negative *Staphylococcus* species is identified at the least mastitis causing pathogens. The mastitis causing pathogens are *Streptococcus uberis*, *Streptococcus sanguinis*, *Streptococcus dysgalactiae* ssp *dysgalactiae*, *Streptococcus mitis* and *Streptococcus agalactiae*.

References

[1] Sudarwanto M H, Latif H, and Noordin M 2006The relationship of the somatic cell counting to sub- clinical mastitis to improve milk quality. *Proc. of the 1st Int. AAVS Scientific Conf. Jakarta, 12-13th July 2006*. Jakarta. Indonesia.

[2] Estuningsih S, Soedarmanto I, Fink K, Lämmler C, and Wibawan IWT 2002 Studies on *Streptococcus agalactiae* Isolated from Bovine Mastitis in Indonesia*Zoonoses and Public Health***49**(2) 185

[3] Salasia SIO, Khusnan Z, Lämmler C, and Zschöck M 2005 Comparative studies on pheno- and genotypic properties of Staphylococcus aureus isolated from bovine subclinical mastitis in central Java in Indonesia and Hesse in Germany *J. Vet. Sci.* **2** 103

[4] Schulza PJR, Smith KL, Hogan JS and Lovec BC 2004 Antimicrobial susceptibility of mastitis pathogens from first lactation and older cows*Vet.Microbiol.***102**(1) 33

[5] Funke G, and KisslingPF 2005 Performance of the new VITEK 2 card for identification of medically relevant gram-positive cocci in a routine clinical laboratory *J. Clin. Microbiol.* **43** 84

[6] Wallet F, Loïez C,Renaux E, Lemaitre N andCourcol RJ 2005 Performances of VITEK 2 Colorimetric Cards for Identification of Gram-Positive and Gram-Negative Bacteria. *J. Clin. Microbiol.***9**4402

[7] Crowley E, Bird P, Fisher K, Goetz K, Boyle M, Benzinger MJ, Juenger M, Agin J, Goins D, and Johnson R 2012 Evaluation of the VITEK 2 Gram-Negative (GN) Microbial Identification Test Card: Collaborative Study *J AOAC Int.***95**(3) 778

[8] Piepers S, Meulemeester LD, de Kruif A and Opsomer G 2007 Prevalence and distribution of mastitis pathogens in subclinically infected dairy cows in Flanders, Belgium *J. Dairy. Res.***74**(4) 478

[9] Plozza A, Lievaart JJ, Potts G. and Barkemac HW 2011 Subclinical mastitis and associated risk factors on dairy farms in New South Wales *Aust. Vet. J.* **89** 1

[10] Hayashi T, Sugita T,HataE, KatsudaK, Zhang E, Kiku Y, Sugawara K, Ozawa T, Matsubara T, Ando T, Obayashi T, Ito T,Yabusaki T, Kudo K, Yamamoto H, Koiwa M, Oshida T, Tagawa Y, and Kawai K 2013Molecular-based identification of yeasts isolated from bovine clinical mastitis in Japan*J Vet Med Sci.***75**(3) 387

[11] Pyörälä S and Taponen S 2009. Coagulase-negative *Staphylococci*-emerging mastitis pathogens *Vet. Microbiol.***134**(1) 3

[12] Sugiri, YD and AnriA 2010 Prevalence of mastitis causing pathogens (*Staphylococcus aureus* dan *Streptococcus agalactiae)* in small holder dairy farms in Java, Indonesia (in Bahasa Indonesia). Research Report of West Java Livestock Department.

[13] Hosseinzadeh S and Saei HD 2014 Staphylococcal species associated with bovine mastitis in the North West of Iran: Emerging of coagulase-negative staphylococci *Housternl J. Vet. Sci. Med.* **2** 27

[14] Zhang S and MaddoxCW. 2000 Cytotoxic activity of coagulase-negative staphylococci in bovine mastitis *Infect.Immun* **68**1102

[15] Schukken YH, Gonzalez RN, Tikofsky LL, Schulte HF, Santisteban CG, Welcome FL, Bennett GJ, Zurakowski MJ, Zadoks RN 2009CNS mastitis: nothing to worry about? *Vet. Microbiol.* **134** 9

[16] Supre K, Haesebrouck F, Zadoks RN, Vaneechoutte M, Piepers S, de Vliegher S 2011 Some coagulase-negative *Staphylococcus* species affect udder health more than others *J Dairy Sci* **94** 2329

[17] Middleton JR, HardinK, SteeverBJ and Tyler J 2004 Use of somatic cell counts and California mastitis test results from individual quarter milk samples to detect subclinical intramammary infection in dairy cattle from a herd with a high bulk tank somatic *J Am Vet Med Assoc.* **224**419

[18] Harjanti D W, Ciptaningtyas R, Al-Baarri A N and Kusumanti E 2017 Isolation and identification of lactoferrin and lactoperoxidase from the colostrum of Indonesian Ettawa crossbred goat *Advanced Science Letters* **23** 3321

[19] Mahpudin, Wahyono F, and Harjanti DW. 2017. Effectiveness of *Ageratum conyzoides* leaves extract as green antiseptic for teat dipping practices in dairy cow (in Bahasa Indonesia) *Agripet* **17**(1) 15

IOP Conf. Series: Earth and Environmental Science **102** (2017) 012077 doi:10.1088/1755-1315/102/1/012077

Understanding of social capital condition among red guava farmers in Tambahrejo Village, Pageruyung District, Kendal Regency

S Gayatri*, D Sumarjono and S Satmoko

Faculty of Animal and Agricultural Sciences, Diponegoro University

*Email: gayatri.siwi@gmail.com

Abstract. The aim of the study was to explore the potential of social capital and growing income of red guava farmers in Tambahrejo Village, Pageruyung District, Kendal Regency. Interview and observation were used for data collection. Set of Questionnaire was developed to answer research' goal. All member of farmer group I ACC (*Kelompok Tani Makmur* I ACC) were chosen as respondents in this research. Data were analyzed using multiple regressions. The result shows that there was significant relationship between social capital in community and the income of the red guava farmers. Farmer' group was found as a media to improve farmers' knowledge and networking. Farmers group facilitated farmers to market red guava product. Moreover, wife of the farmers established women group or KWT (*Kelompok Wanita Tani*). The result found that KWT contributed to improve family's income. KWT also promote activities to help product's diversification of red guava. Both farmer group and KWT provided activities such as saving and loans, it means there was trust among member of farmer group.

Keywords: *social capital, red guava, farmers*

1. Introduction

Psidium guajava or red guava has been played as important commodity for farmers and their family in Kendal Regency. Based on secondary data from Statistic Office of Kendal Regency, the production of red guava in Kendal Regency has been increased due to intervention of the local government through improvement of cultivation technique new planting area. Red guava have highly economic value for farmers, however the farmers have problems to improve productivity; plant disease control and nutrient management, for instance [1]. In the beginning, red guava had been developed as conservation crop rotation with lack of information about cultivation techniques and sufficient information on guava agribusiness. Nowadays, both farmers and local government have been thinking how to make red guava farming system become more profitable and sustainable [2]. In addition, development of agriculture sectors has been linked to improve farmers' skill and increase their participation in farmers' group. Hence, it need to integrate all the activities in agriculture sectors based on optimizing the utilization of existing local resources through farmers' group [3].

Extension services have been promoted over the years, including row spacing methods in order to improve cultivation methods. Farmers planted seeds without using exact plant spacing and it would affect productivity. The impacts of improper spacing method resulted in the growth of the guava plant. Through farmers' group activity, extension agents introduced new technology in order to improve farmers' skill [4]. It can be concluded how the farmers' group had important role in development of agriculture sector, especially in red guava farming system.

Social capital has been declared as a result of a process of social and economic organization, such as trust, reciprocity, informational and economic exchange, formal and informal group's activities. Improving social capital situation in the community will improve collective action, as well as social and economic condition [5]. Putnam [6] defined social capital as mutual trust among member of community. Social capital is designed as social institution involving network, norms, social trust and it will increase social collaboration through coordination and cooperation to reach social goals.

A research by [5] showed that agriculture sectors has been linked to social capital using indicators, such as network and membership (where the people link to) and social trust (how member organization can trust each others). In addition, a research by Bebbington in 2008, told that the success of poverty reduction program in rural farm area had been associated with improving the role of community organization through participation and social network [7]. Social capital has consequence including social access, social control, improve social economic condition by social network. Moreover, Warni and Hidir [8] had investigated several dimensions of social capital in Toraja Tribe, namely integration, linkage, organizational integrity and synergy. These all of the dimentions worked to keep member of Toraja Tribe to uphold the values of unity and togetherness. Meanwhile, there is no report on how the social capital worked to increase income among red guava farmers. Based on consideration above mentioned, this study attempted to find out the effect of social capital on improving income of red guava farmers in Tambahrejo Village, Pageruyung District, Kendal Regency.

2. Materials and methods

The study focused on of red guava farmers' group in Tambahrejo Village, Pageruyung District, Kendal Regency. Data were compiled from primary data through interview and observation. In addition, secondary data were gathered to improve data analysis process through government offices or farmers' group. It followed by data entry process and data analysis. Survey method was used in this research based on interview with red guava farmers in Kendal Regency. Set of Questionnaire was developed to answer research goal. All member of farmer group I ACC (*Kelompok Tani Makmur* I ACC) were chosen as respondents in this research. In total, there were 32 respondents in this research. Data were analyzed through editing, coding, and tabulating. Moreover, data were analyzed using descriptive analysis and multiple regression analysis. The research used the theory of social capital as independent variables. It was proposed by Putnam [6] and consisting of the elements of trust (X1), social networks (X2) and collective action (X3), meanwhile income of the farmers as dependent variables (Y).

3. Result and discussion

This study will explored three dimensions of social capital, namely trust (X1), social networks (X2) and collective action (X3), meanwhile income of the farmers was dependent variables (Y).

2.1. Trust

Based on interview with the respondents, trust was defined into three categories, trust among farmers (individual), trust in the organization and trust in the government services (organization). The respondents told that they trust their co-farmers, for example they could loan money to other farmers. The respondents responded that they had a high level of trust in the farmers' group. Respondents gave examples, the farmers believed that the farmers' group had good recording of common money or the farmers' group would help them whenever they had problems. Moreover, respondents agreed that the government gave many services to improve their condition as well as extension agents would help to solve farmers' problem. Social trust may provide opportunities for farmers to access services, information and resources that will allow them to improve their capacities in these areas [9]. Trust can create good relationship between individual or in a group. Trust is a desire to take risks in the social relationships or the feeling that others will do something as expected and will always act in a pattern of mutually supportive actions. Trust is believed that people' action will not act to harm themselves and their groups[10]. Trust is needed to maintain the integrity of a relationship. Based on data analysis, trust was one of elements to build a farmers' group. Trust arises from the existence of honesty and discipline according to mutual norms. Trust requires a process of the relationship between individual and involve in daily life [11].

2.2. Social network
Interviewed revealed that farmers will get opportunity to improve their income by joining the farmers' group. The group provided many activities that can improve income and farm management. Most of the farmers said they need to interact and communicate with co-farmers. Farmers also believed that the farmers' group will help them to get in touch with other stakeholder, government or private sectors, for instance. Moreover, based on observation and interview, wife of the farmers also established women group or KWT (*Kelompok Wanita Tani*). The result found that KWT contributed to improve family's income through women' training. KWT also promote activities to help product' diversification of red guava. Interaction and communication created mutual network [12]. One of advantage of network is to link with information or new knowledge and foster a sense of kinship among farmers. The strong social network can be seen from the cohesiveness and participation of each member whenever the group held an activity. In general, social network can be developed through the willingness of the farmers to take part and cooperate in farmers' group [6,10].

2.3. Collective actions
Collective action can be formed by activities done collectively. The respondents shared their though related to collective actions. The farmers' motivation to involve in collective action were sharing knowledge and experience, increase income, togetherness, and only two respondents answered they did not know the answer. Farmers also raised concern about the reason others farmers did not joint farmers' group, such as laziness, busy, and personal problem with other farmers. Farmers will think what kind of benefit they get when joining farmers' group. It needs the same feeling of reciprocity among group member or between farmers and the group in order maintaining and strengthening social capital. This plays an important role in the formation of groups. Mutual acceptance and mutual assistance among group members emerge from social interaction. It will create sensitive feeling to fellow group members to act according collective action[12].

2.4. Farmers' income
The definition of farmers' income in this paper based on the concept of net farm income which is all the return from farming activities after the farm' expenses have been paid [13]. The calculation of net farm income was subtraction total farm expenses from total farm income. The farmers' income of red guava farming system was IDR 6.554.698,53/year or 27.71% of all farming activities' income. This result was higher than interest rate (20%/year). It can be said that the red guava farming activities had opportunity to increase family' income. It needs government' effort to help the farmers to improve farming management, provide assistant as well as open new market of red guava products.

2.5. Multiple regression analysis
Result of regression analysis shows that trust (X1), social networks (X2) and collective action(X3), had positive influence towards farmers' income (Y).

$$Y = 1.24 + 0,24\,X_1 + 0,126\,X_2 + 1,316\,X3$$

The result of overall F test for the null hypothesis shows that all independent variables had significant influence towards farmers' income in red guava farming system. Result of t test of partial regression analysis, trust (X1) had significant influence ($P < 0,05$) towards farmers' income. Hence, if the farmers increase their trust while others factors are constant, it would increase farmers' income. Social networks (X2) had influence toward farmers' income. It can be said, if the farmers increase their social network while others factors are constant, it would increase farmers' income. Moreover, when the farmers are able to increase their collective action, it would increase farmers' income. This condition based on result of t test, collective action(X3) had significant influence ($P < 0,05$)towards farmers' income (Y).It shows that social capital is a very valuable asset. Social capital provide the basis for social cohesion. It encourages people to work with each other and not only with people they know directly will gain mutual benefits [14]. The study revealed that social capital is an important tool to explore business opportunities. All stakeholders in agribusiness system of red guava need toconnectedeach other. A farmers' group is created because of the social ties among its members on daily regular basis. Groups can be formed on the basis of cultural similarity, professional and residential religion[14]. Red guava farmers' group is a social bond between farmers who form organizations to interact in social relationships based on a common goal. The study of social capital benefit to get a better understanding of how society works to achieve common goal, increase income, for instance [15].

3. Conclusion

Based on the results, it can be concluded that social capital is very important in a group or organization. Trust, social networks and collective action have a role in maintaining relationship and increase income. It need honest attitude and discipline toward the social norms the farmers do not to deviate from the rules and agreements. Interaction can create solidarity and easy to share knowledge and information. It is important to maintain good relationships with other members. Farmers' group have to promotecollective actionto achieve common goal. Avoid conflict within the organization is also recommended.

References

[1] Sahutu S and Sjaifullah 2011 Physical and Chemical Analysis of Guava (In Bahasa Indonesia). *J. Hort.* **1**(4) 53-56.

[2] Tirtayasa M F 2009 Income Analysis of Guava Farmers of Primatani Farmers' Group in Depak City (In Bahasa Indonesia). *J. Hort.* **1**(2) 29-36.

[3] Sunda S R 2007 Morphological and Chemical Analysis of Guava in Cileungsi, Cariru and Tanjungsari Kendal Regency (Karateristik Jambu Biji (Psidium guajava L.) Berdasarkan Karakter Morpologi dan Kimia Di Kecamatan Cileungsi, Cariru, dan Tanjungsari, Kabupaten Kendal. *Jurnal Agro* **2** (2) 26-32.

[4] Tiku G V 2008 Financial Analysis of Red Guava in Majalengka Regency (In Bahasa Indonesia). [Thesis]. The Study Programme of Agribusiness Management. Faculty of Agriculture, Bogor Agricultural University.

[5] Williams K and Durrance J C 2008 Social Networks and Social Capital: Rethinking Theory in Community Informatics. *The journal of community informatics*. **4** (3) 1-22.

[6] Putnam R 1993 *Bowling Alone: The Collapse and Revival of American Community*. New York: Simon & Schuster.

[7] Bebbington A 2008 Social capital and development studies, critique and state (seen from Peru). *Progress in Development Studies* **8** (3) 271-279.

[8] Warni A S and Hidir A 2017 Social Capital in Toraja Community in Pekanbaru-Case study: Member of Sangtorayan Family Group (in Bahasa Indonesia). *JOM FISIP* **4** (1) 1-15.

[9] Cox E 2007 The Functional Value of Social Capital. *Australian Journal of Social Issues* **42** (4), 503–512.

[10] Gayatri S, Gasso-tortajada V and Vaarst M 2016 Assessing Sustainability of Smallholder Beef Cattle Farming in Indonesia: A Case Study Using the FAO SAFA Framework. *Journal of Sustainable Development* **9** (3).

IOP Conf. Series: Earth and Environmental Science **102** (2017) 012077 doi:10.1088/1755-1315/102/1/012077

[11] Fukuyama F 2005 *The Great Disruption, Humanity and New Social Structure* (in Bahasa Indonesia). (Jakarta : PT. Gramedia Pustaka Utama)
[12] Coleman J S 1988 Social Capital in the Creation of Human Capital.*The American Journal of Sociology* **94** S95-120.
[13] Luntungan A Y 2012 Income Analysis of Tomato-Apple Farm in Tompaso District, Minahasa Regency (in Bahasa Indonesia). *Jurnal Pembangunan Ekonomi dan Keuangan Daerah (PEKD)* **7** (3) 1-21.
[14] Farr J 2004 Social Capital A Conceptual History.*Political Theory* **32** (1) 6-33.
[15] Sebejo 2004 The Role of Social Capital in Economic Development. An Introduction to Study on Social Capital in Rural Indonesia (in Bahasa Indonesia) *AgroEconomic Journal* **11** (1) 77-86.

The level of organic rice farming technology at farmer group in Ketapang village, Susukan sub-district, Semarang district, Central Java Province, Indonesia

Sumarsono*, Yafizham and D W Widjajanto

Laboratory of Ecology and Crops Production, Faculty of Animal and Agricultural Sciences, Diponegoro University, Tembalang Campus, Semarang, Indonesia

*Email : marsono_53@yahoo.co.id

Abstract. The research was conducted to learn the phenomenon of the level of organic rice cultivation technology in the field at one of organic rice centers in Central Java Province, Indonesia. It was carried out using sample survey of respondents taken at 3 different of *Walisongo, Al-Barokah* and *Dewi Sri* farmer groups from organic rice farmer population in the village of Ketapang, Susukan sub-district, Semarang district. Primary data were collected quantitatively by distributing questioner to the respondent. Data were analyzed in order to evaluate the performance of inter-group observation and relationship of behavioral levels of organic rice cultivation. The results showed that the level of organic rice cultivation technology were high category (68.3%), moderate category (31.7%), and none less category, respectively. There was significant regression relationship $Y = 7.219 + 0.237 X_1 + 0,231 X_2$ ($R = 0.625$) between knowledge and attitude toward organic farming cultivation level. There was no significant difference in the application level of organic farming between *Dewi Sri* and *Al-Barokah* Farmer Groups (113.3^a vs 110.7^a), but the application level of organic farming both of *Dewi Sri* and *Al-Barokah* farmer groups were significantly ($P<0,05$) different from *Walisongo* farmer group (95.5^b).It was concluded that the level of organic technology at farmer group was moderate to high category, but still requires further extention to be more equitable among farmer groups

Keywords : *organic, rice, technology, farmer group*

1. Introduction

Rice (*Oryza sativa* L) is a strategic commodity as rice is the main source of carbohydrates and its position as source of carbohydrates for Indonesians generally cannot be replaced by other carbohydrate sources such as corn, cassava, wheat or sorghum.

Based on pre-forecast figures issued by the Ministry of Agriculture in coordination with the Central Bureau of Statistics, rice production in 2016 reached 79.141 million tons of dry milled grain (DMG) or increased by 4.96% compared to the previous year's production [1]. The increase in production occurred in Java Island and outside Java Island, respectively, of 1.22 and 2.52 million tons. One government program through the Ministry of Agriculture is the development of 1000 Organic Farming villages, while the target by the Directorate of Food Crops is 600 villages [2].

Organic agriculture was developed to address the environmental degradation problems of non-organic farming that intensively applied in the past since the green revolution. The continuous use of inorganic fertilizers leads to soil degradation characterized by decreased levels of soil organic matter [3, 4, 5]. Organic farming is a cropping production system based on biological recycling of nutrients, by means of crop and livestock wastes, and other wastes capable of improving fertility status and soil structure [6]. Organic farming strategy is to move the nutrients as soon as possible from the rest of the plant, compost, and manure into soil biomass which will be decomposed into nutrients in soil solution. The concept of organic farming is the law of return, the philosophy of feeding the soil to provide food for the plants (feeding the soil that feeds the plant) rather than feeding the plants directly.

Soil organic matter has an important role in determining the productivity of the soil, as it contains all the necessary elements of the plant, increasing the activity of soil microorganisms, providing loose soil structure, thus providing good aeration, high water storage, good root penetration, increased cation exchange capacity, lower nutrient leaching rates, in acid soils bind Al and Fe so that the availability of essential elements increases also neutralize the saline soils [7].

Organic farming is implemented by minimizing the use of synthetic fertilizers, pesticides, accelerated growth materials and other additives. To maximize the level of production possibilities, organic farming systems are implemented based on harvest rotation, residue yield, manure, green manure, waste from organic farming and pay attention to biological aspects of pest control to maintain soil productivity and support plant nutrients and control insects, and other pests [8].

Organic farming is a concern both by consumers and producers due to awareness of the dangers of using chemicals in practicing agriculture. Organic agriculture technology is available enough such as compost technology, biological fertilizer, organic pesticide, minimum or zero tillages [4]. The organic farming system can be integrated into integrated farming system, which is a system that combines the activities of agriculture, livestock, fishery, forestry and other science related to agriculture in one land, so it is expected to be one solution for increasing land productivity, development program and environmental conservation, as well as integrated village development. This agricultural system will integrate the activities of the agricultural sector and its supporting sectors both vertically and horizontally according to the potential of each region by optimizing the utilization of existing local resources [9].

Ketapang Village is one of the villages in Semarang district of Central Java Province which was very successful in implementing organic farming. The success can be proven by farmer groups that have implemented organic rice cultivation either conversion stage or certified. In addition, Ketapang village farmer group has a marketing network both domestic and abroad. The research was conducted with the aim to evaluate the phenomenon of the application of organic rice cultivation technology in one of organic rice centers in Central Java. The results are expected to be a reference for the development of organic rice farming in a wider area.

2. Research methods

The research was aimed to evaluate the phenomenon of knowledge in the field so that the explanation between the theory and practice of the application level of organic rice cultivation technology in Ketapang village,Susukansub-district of Semarang district.It was carried out using survey method. Survey was conducted on farmer population with sample of respondents taken from farmer group of organic rice farming. In addition, the survey was conducted to determine the causal relationships of the research variables as different observation groups at three different groups. The groups of respondents observed were 1) *Al-barokah* farmer group, 2) *Walisongo* farmer group and 3) *Dewi Sri*farmer group. Primary data were collected quantitatively by distributing questioner to the respondent. Meanwhile, secondary data were collected based on documents related to the research. Physical observation data were conducted on cropland and soil conditions.

The observed parameters were behavioral level of technology application including knowledge and skills of organic farming. Observations of soil conditions included organoleptic (color, texture), moisture content, solids, organic matter, ash, nitrogen and C/N ratio. The data were analyzed statistically using ANOVA to compare between observation groups and regression analysis to find out the correlation between behavior of technology application level includingknowledge, skills and attitude of farmers in organic farming cultivation.

3. Result and discussions

3.1. *Respondents characteristic*

Respondents characteristic (Table 1) were grouped into five categories, namely: 1. Farming experienced, 2. Nature of work, 3. Wetland ownership, 4. Yard ownership, and 5. Livestock ownership. Based on farming experience, the respondents were divided into farming experiences of less than 10 years, between 10-15 years and over 15 years. On the basis of the criteria of farming experienced, 51.66% of the farmers were more than 15 years experienced, 40.00% less than 10 years experienced and 8.34% between 10-15 years experienced. Meanwhile, based on the nature of the work about 68.33% and 31.67% of the farmers work as farmers and ranchers, respectively. Based on the ownership of paddy fields, 21.66%, 70% and 8.34% of the farmers were classified as farmers with paddy field less than 1000 m^2, between 1000-5000 m^2 and more than 5000 m^2, respectively. Meanwhile, based on the yard area ownership of 46.66%, 28.33% and 25.00% of the respondent farmers, respectively has a yard area with an area of less than 500 m^2, between 500-1000 m^2 and more than 1000 m^2. Based on livestock ownership about 73.33% of farmers respondents own poultry between 10-100 heads, 40.00% own goats/sheep between 2-10 heads, and 21.66% own cattle between 2-5 heads, respectively.

Table 1. Respondent characteristics Farmer Group in Ketapang village Susukansub-district

No	Identity of Respondent	Number of respondents	Percentage (%)
1.	Farming experienced		
	• <10 years	24	40.00
	• 10 – 15 years	5	8.34
	• >15 years	31	51.66
2.	Nature of job		
	• Farmers	41	68.33
	• Farmers-ranchers	19	31.67
3.	Wetland area ownership		
	• < 1.000 m^2	13	21.66
	• 1.000 – 5.000 m^2	42	70.00
	• >5.000 m^2	5	8.34
4.	Yard area ownership		
	• <500 m^2	28	46.66
	• 500-1.000 m^2	17	28.33
	• >1.000 m^2	15	25.00
5.	Livestock Ownership		
	• Cows (2-5 heads)	13	21.66
	• Goat/Sheep (2-10 heads)	24	40.00
	• Poultry (10-100 heads)	44	73.33

Table 1 shows that it is like agriculture in Indonesia, especially on the island of Java in general has problems of narrow land tenure. Increasingly reduced agricultural land and the splitting of land due to inheritance patterns make the interest of business in the field of agriculture less and less desirable. Narrow land tenure leads to lower profitability and lower farming efficiency. The problem of narrow land tenure for agricultural cultivation can be overcome by increasing productivity and business efficiency. Respondents characteristic (Table 1) requare cultivation of integrated organic agriculture into the hope of agricultural cultivation in a narrow field. The organic agriculture has improved livelihoods for smallholders while minimizing the use of external input [10]. Agricultural cultivation remains oriented to increase productivity by applying low-input technology that comes from outside, but utilizes as much as possible local resources and increases farmer's income [10].The results of research by [11]indicated that organic farming practices can positively affect the economic resilience of farmers. However, organic farming practices have more complex problems than conventional farming practices in fostering farmer groups. Large and continuous requirements lead to the use of organic fertilizers more difficult than the use of inorganicfertilizers.Table 1 shows the low level of integration of farmers-ranchers (31.67%) with livestock ownership is also low, making it less able to support large and continuous requirements leads to the use of organic fertilizers.

3.2. Level of Organic Agriculture Cultivation

The farming experience of the respondents influenced the level of organic farming, where in organic rice cultivation 68.33% in the high category, 31.66% in the moderate category and none in the low category (Tabel 2). This condition indicates that the respondent has succeeded in developing self-behaving as a group of organic rice farming farmers. The high level of cultivation of organic agriculture of respondents supported by 83.00%, 15.00% and 16.66% attitude toward organic farming in high, medium and low category, respectively. However, this condition has not been fully supported by knowledge and skills in conducting organic rice farming. Knowledge and skill were dominant in the moderate category, 55.00% and 58.33% respectively, meanwhile the remaining were in the high category, 45.00% and 41.66%, respectively.

Table 2. Level of Organic Farming Cultivation

NO	Cultivation behavior	Amount of Respondent	Proportion (%)
1.	Knowledge of organic farming		
	• High	27	45.00
	• Medium	33	55.00
	• Low	0	0.00
2.	Attitude towards Organic Farming		
	• High	50	83.00
	• Medium	9	15.00
	• Low	1	16.66
3.	Skills of Organic Farming		
	• High	25	41.66
	• Medium	35	58.33
	• Low	0	0.00
4	Level of Organic Farming Cultivation		
	• High	41	68.33
	• Medium	19	31.66
	• Low	0	0.00

The high attitudes of farmers' respondents to organic farming resulted in the decision of farmers to apply organic farming in their village. The higher the farmer's decision level for organic rice cultivation is higher the level of application of organic rice cultivation [12]. Although the level of knowledge and skills in the moderate category, but the development of organic rice farming in Ketapang village had been successful. Based on qualitative interviews with the head of farmer groups it was evident that the process and product of organic rice has been certified by the certification agency. Products and processes have been assured by the organic certification agency as organic rice so that it is marketed at a higher price than non-organic rice. Good productivity was between 6-7 tons ha^{-1} of DMG with the mainstay of local varieties of *Mentik Susu* and *Pandan Wangi*. Varied rice products are marketed in the form of premium ready-to-eat rice packaging at a good price of IDR 17,000 kg^{-1} at the farm level for both domestic and overseas marketing.

The result of soil sample analysis showed that soil fertility was good, 3.91% C-organic, 0.26% N and 14.99 C/N ratio. This indicated that the implementation of the "Standard Operating Procedure - Good Agriculture Practice" has been carried out largely by joint members of farmer groups. The level of application of organic rice cultivation was related to the availability of production facilities and the selling price of the product [12]. The application of organic rice cultivation at the same time can apply low external input sustainable agriculture (LEISA) technology because it uses input from local resources either organic fertilizer or organic pesticide [13, 4]

The application of agricultural cultivation shown by the organic rice farming skills scores was influenced by the knowledge possessed and the attitude in approving the application of organic rice farming. There weresignificant regression relationship $Y = 7{,}219 + 0{,}237\,X1 + 0{,}231\,X2$ ($R = 0{,}625$) between knowledge and attitude toward skilled level of organic farming. Regression relationship showed positive significant regression coefficient ($P<0.05$) both knowledge and attitude. This shows that organic rice farming may still be developed especially with equalization of knowledge level so that it may increase the skill in organic rice farming. Although the proportion of farmer respondents' attitudes was high in categories, attitudes often suggest a moderate relationship to behavior [14].

The level of cultivation of organic agriculture in the dominant organic rice cultivation in the high category but turned out to vary among members of the farmer group (Table 3). The farmer respondents consisted of three farmer groups, *Walisongo, Al-barokah* and *Dewi Sri*, respectively having an organic rice farming behavior score of 95.50; 110.72 and 113.28. The behavioral score of organic rice cultivation by *Walisongo* farmer group was significantly ($P <0.05$) lower than *Al-Barokah* and *Dewi Sri* farmer groups. The behavioral scores of organic rice farming among *Al-Barokah* and *Dewi Sri* farmer groups were not significantly different. Total score of behavior of organic farming in line with the score of knowledge, attitude and skill of Walisongo farmer group was significantly ($P <0{,}05$) lower than *Al-Barokah*and *Dewi Sri* farmer groups, but among *Al-Barokah* farmer group and group Dewi Sri farmers were not significantly different.
This shows that there was still a need for equitable distribution of knowledge, attitude and skills in organic rice cultivation so that all farmer groups can further develop the creation of the same organic rice farming cultivation. The development of agricultural technology is a complex process because it is influenced by internal and external factors [15]. Internal factors are derived from self-farmers themselves in receiving knowledge, encouragement attitude and willingness of application, while external factors are the availability of resources, information and technology.The future perspectives of the research is conducting an action research on increasing the level of organic farming Cultivation.

Table 3. Differences Score Behavior of Organic Farming between Farmer Group.

Farmer group	Knowledge	Attitude	Skills	Behavior
Walisongo	41.06[b]	33.68[b]	23.75[b]	95.50[b]
Al-Barokah	46.62[a]	37.00[a]	27.10[a]	110.72[a]
Dewi Sri	46.85[a]	38.57[a]	27.85[a]	113.28[a]

[a], [b] follow the values in the same column as significantly different ($P <0.05$).

3. Conclusion

It was concluded that the level of organic rice cultivation technology was generally dominant in the high category (68.33%), the rest in the moderate category (31.66%) and none in the less category. The high level of application of organic rice cultivation was closely related to knowledge and attitude in the cultivation of organic rice farming. The level of cultivation of organic farming among farmer groups was still not uniform and therefore requires development. High attitudes towards organic rice farming requires increased knowledge and skills for further development of organic paddy farming including integration support with livestockas source of organic fertilizer.

References

[1] Kurniawan A 2016 Rice Production 2016 Predicted to be Largest Throughout Indonesia's Independence. Retrieved from SindoNews.Com. https://ekbis.sindonews.com. 29 December 2016, Jakarta.

[2] Directorate of Food Crops 2016 *Technical Guidelines for Village Development of Rice Organic Agriculture* (Jakarta : Directorate General of Food Crops)

[3] Sumarsono 2008 Feed Plants At Intervention of Environmentally-Based Agricultural Systems. Professorship Inaugural Speech, Diponegoro University, Semarang.

[4] Mayrowani H 2012 Development of organic agriculture in Indonesia *Agro Economic Research Forum* **30** (2) 91-108.

[5] Dermiyati 2015 *Sustainable Organic Farming System* (in Bahasa Indonesia) (Yogyakarta : First Print. Plantaxia)

[6] Sutanto S 2002 *Organic agriculture Toward Alternative and Sustainable Agriculture* (in Bahasa Indonesia) (Yogyakarta : Kanisius Press)

[7] Sumarsono S, Anwar D, W Widjajanto and S Budianto 2010 Organic fertilizer application on performance and production of king grass in acid soil. *Proc. Int. Seminar on Tropical Animal Production (ISTAP) Community Empowerment and Tropical Animal Industry, Fac. Animal Science UGM* (Yogyakarta)

[8] Lampkin N 1994 *Organic Farming* (UK : Farming Press. Ltd)

[9] Kardiman, A 2016 *Organic Agriculture System* (in Bahasa Indonesia) First Edition (Malang : Intimedia)

[10] Bennett M and S Franzell 2013 Can organic and resource-conserving agriculture improve livelihood. *Int. J. Agric. Sustainability* **11** (2) 193-215.

[11] Widiarta A , S Adiwibowo, and Widodo 2011 Analysis of the sustainability of organic farming practices among farmers.Sodality (in Bahasa Indonesia*) J. of Transdisciplinary Sociology, Communication, and Human Ecology* **5** (1) 71-89

[12] Sriyadi, E Istiyanti, F R Fivintari 2015 Evaluation of application of standard operating procedure-good agriculture parctice (SOP-GAP) on organic rice farming in Bantul district (in Bahasa Indonesia) *J. Agraris.* **1**(2) 78-84.

[13] Purwasasmita M and A Sutaryat 2014 *Organic Rice SRI of Indonesia* (in Bahasa Indonesia) (Jakarta : Penebar Swadaya)

[14] Petrzalka P, P F Korsching and J E Malia 2010 Farmer' attitudes and behavior toward sustainable agriculture. *J. Env. Edu* **28** (1) 38-44.

[15] Meijer S S, D Catacutan, O C Ajayi, G W Sileshi and M Nieuwenhuis 2015 The role of knowledge, attitudes and perceptions in the uptake of agricultural and agroforestry innovations among smallholder farmer in sub-Saharan Africa. *Int. J. Agric. Sustainability* **13** (1) 45-54.

Evaluation of paddy quality dried with zeolite under medium temperature

F Irfandy and M Djaeni*

Department of Chemical Engineering, Faculty of Engineering, Diponegoro University Jl. Prof Soedarto, SH, Tembalang, Semarang, Indonesia 50275

*Email: moh.djaeni@live.undip.ac.id

Abstract. Drying becomes important role to obtain good quality product in post-harvest treatment for paddy. Adsorption drying process taken place on medium operation temperature can be an option to enhance paddy quality. The aim of this study was to evaluate the quality of paddy dried with zeolite as adsorbent. In case, paddy from local farmer in Semarang was dried from initial moisture content 23% (wet basis) under different temperature 40 and 60°C, both with and without zeolite. Moisture content was observed by gravimetry every 15 minutes during 60 minutes drying time. Furthermore, the quality of paddy represented by percentage of head rice and proximate was analyzed after drying process completed. Results showed that compared to conventional drying without zeolite, the adsorption dryer was more superior in term of drying time as well as product quality improvement and there are no much differences in proximate analyzed result.

Keywords: *paddy, drying, zeolite, quality, proximate*

1. Introduction

Paddy is one of the major economic crops for consuming and producing in many countries. In global, paddy is the most phytocomestible that give energysupply and the second protein source, following wheat [1].

Drying becomes important role to obtain good quality product in post-harvest treatment for paddy.The air temperature had the most superior effect both on product quality and drying rate [2]. Nowadays, there are two methods of paddy drying: direct sunlight dryer and dryer by conventional heating. Direct sunlight dryer is commonly used by farmer because it has cheaper and more simple method than dryer by conventional heating [3].Conventional dryer such fluidized bed has more advantages: shorter operational time, no dependency on the climate, low product contamination, and smaller space usage[4,5,6,7]. Fluidized bed dryer is commonly and commercially used in several country for short time food drying, sensitive food drying, and homogenous moisture content product of drying [8]

The operation temperature must be set in order to keep paddy quality and energy efficiency of drying. The temperature upper 90°C can reduce the paddy quality [9]. Food products are sensitive to the heat load during drying. For example the colorof tarragon (*Artemisia dracunculus L.*), changes to brown when drying occurs at too high temperatures [10]. Air dehumidification by adsorbent such as

IOP Conf. Series: Earth and Environmental Science **102** (2017) 012079 doi:10.1088/1755-1315/102/1/012079

silica and alumina can be used. The zeolite is an alumino-silicate compoundwith tetrahedral bound linked by oxygen that has high affinity to the water. By activation using KOH or NaOH, the adsorbing capacity can be enhanced up to 0.15-0.20 g water/g zeolite [11].

From those explanations above, paddy drying should be done at low-medium temperature. The application of zeolite in order to decrease the air dryer humidityand improve paddy quality were become an advantages we could get in this research.The aim of this study was to evaluate the quality of paddy dried with zeolite as adsorbent.The differences of final moisture content, head rice yield, and proximate composition content indried paddy with and without zeolite were also observed from this research.

2. Material and Methods

2.1. Material

Paddy from local farmer in Semarang was dried from initial moisture content 23% (wet basis) under different temperature 40 and 60°C. 220 grams of Zeolite 3A (provided by Zeochem, Switzerland) were used as adsorbent in purpose to decrease the relative humidity of air dryer.

Petroleum ether, boiling point 40–60°C was used for fat analysis. While carbohydrates analysis used 52% perchloric acid solution: 279 ml perchloric acid (specific gravity 1.70) in 100 ml distilled water; Sulphuric acid solution: 760 ml of H_2SO_4 (specific gravity 1.84) in 330 ml distilled water; Anthrone reagent. Have enough Anthrone reagent ready by preparing a 0.1% sulphuric acid solution to used the same day; Standard glucose solution. Dissolved 100 mg glucose in 100 ml water; Standard diluted glucose solution. Diluted 10 ml standard glucose solution in 100 ml distilled water (1 ml = 0.1 mg glucose).

2.2. Experimental set-up

This drying process used fluidized bed dryer from Teknik Kimia Universitas Diponegoro, Semarang.Paddy grains were fluidized in fluidized column (Figure 1). Ambient air with relative humidity 70-80% and temperature between 29-33°C was heated in the heater completed with PID controller to reach dryer temperature (40°C with zeolite at first experiment). The air dryer flow was set at 5.51 $m.s^{-1}$.It was measured with an KRISBOWInstruments AnemometerKW06-562.Moisture content was observed by grain moisture meter G-Won GMK-303RS every 15 minutes. 250 grams of paddy are used for 60 minutes of drying process. The process was repeated for the inlet air dryer temperature 60°C and operation without zeolite (at 40 and 60°C). The quality of paddy represented by percentage of head rice yield (HRY was obtained by observing the paddy with rice grader or pinset and proximate was analyzed after drying process completed. The proximate composition content analysis included ash content (ash analysis), fatcontent (soxhlet method), protein content (kjedahl method), and carbohydrate content (gravimetric method).

The ash analysiswas used to determine ash content in feedstuffs by calcination. The steps of this analysis were: place 2.5 to 5 g of dry sample in a crucible previously calcined and brought to constant weight. placed the crucible in a furnace and heated at 550°C for 12 hours; left to cool and transfered to a dryer. Then weighed the crucible again with the ash.

In the soxhlet method, the fats were extracted from the sample with petroleum ether and evaluated as a percentage of the weight before the solvent was evaporated. Steps of soxhlet method: removed extraction flasks from the kiln without touching them with the fingers, cooled in a dryer and weighed to within milligrams; Weighed 3 to 5 g of dry sample to within milligrams in an extraction thimble, handling it with tongs and place in the extraction unit; Connected the flask containing petroleum ether at $^2/_3$ of total volume to the extractor; Brought to boil and adjusted heat to obtain about 10 refluxes per hour; The length of the extraction would depend on the quantity of lipids in the sample; Very fatty materials would take 6 hours; When finished, evaporated the ether by distillation or in a rotoevaporator; Cooled the flasks in a dryer and weighed them to within milligrams. The defatted sample can be used in determining crude fibre.

Protein analysis was done by Kjeldahl's method, which evaluated the total nitrogen content of the sample after it had been digested in sulphuric acid with a mercury or selenium catalyst. To milligram precision, weighed out 1 g of sample and placed in the Kjeldahl flask; Added 10g potassium sulphate, 0.7 g mercuric oxide and 20 ml concentrated sulphuric acid;Placed the flask tilted at an angle in the digester, brought to boiling point and retain until the solution is clear; Continued to heat 30 minutes more. If foam was too abundant, added a little paraffin wax;Left to cool, gradually adding approximately 90 ml distilled, de-ionized water. When cold added 25 ml sodium sulphate solution and stir;Added one glass bead and 80 ml of 40% sodium hydroxide solution, keeping the flask tilted. Two layers will form;Quickly connected the flask to the distillation unit, heat and collect 50 ml of distillate containing ammonia in 50 ml of indicator solution;At the end of distillation, removed the receptor flask, rinsed the end of the condenser and titrated the solution with the standard chlorhydric acid solution.

Carbohydrate content was measured by Clegg-anthrone method. The steps were: Weighed out 1.0 g of dry or 2.5 g of wet sample to within 0.001 g containing approximately 60 to 300 mg of total available carbohydrates; Transfered quantitatively to a stoppered 100 ml graduated cylinder; Added 10 ml water and stired with glass rod to disperse the sample; Added 13 ml of the perchloric acid solution. Stired constantly with glass rod for 20 min; Rinsed rod with distilled water and brought volume to 100 ml. Mixed and filtered into a 250 ml volumetric flask; Rinsed the graduated cylinder with distilled water and add this to the volumetric flask; Calibrated the flask with distilled water and shake; Diluted 10 ml of the extract to 100 ml with distilled water. Used a pipette to transfer 1 ml of diluted filtrate to a test-tube; Using a pipette took two 1 mg samples of distilled water for duplicate blanks and put each one in a test-tube; Took two 1 ml duplicate blanks using the dilute glucose solution; Quickly added freshly prepared anthrone reagent to all the tubes; Caped the tubes and mixed vigorously; Placed in a water-bath and heat for 12 minutes;Cooled rapidly to room temperature; Transfered the solution to 1 cm spectrophometer cells. The green colour was stable for only 2 hours;Read the absorption rate at 630 nm against the blank.

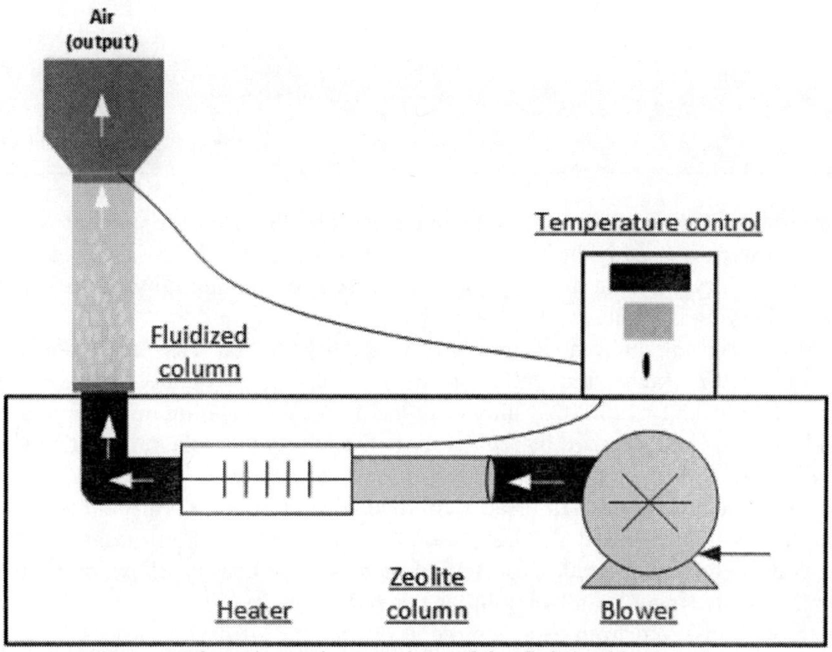

Figure 1. The schematic overview of the fluidized bed drying equipment from Djaeni et al. with modification [12].

3. Experimental Result

3.1. Moisture content and Head Rice Yield

Results showed that compared to conventional drying without zeolite in same duration of drying, the adsorption dryer was more superior in term of head rice yield as well as product quality improvement.The air temperature had effect both on moisture content and head rice yield (table1). At higher temperature, the relative humidity of air is low. Air also has more sensible heat that can be used for evaporating water [12]. The higher temperature, the higher drying rate, the lower moisture content but the lower productquality.

Table 1. Paddy drying with and without zeolite.

Treatment	Temperature (°C)	Time (min.)	Moisture Content (%)	Head Rice Yield (%)
with zeolite	40	0	23	
		15	19.5	
		30	15.2	
		45	14.6	
		60	13.8	64.7
	60	0	23	
		15	19.2	
		30	14.6	
		45	13	
		60	13	64
without zeolite	40	0	23	
		15	20.4	
		30	16.5	
		45	15.6	
		60	15.5	62.4
	60	0	23	
		15	20.1	
		30	16.2	
		45	15.3	
		60	15	60.5

In paddy drying without zeolite experiments, generally, the moisture content was higher and head rice yield was lower than paddy drying with zeolite experiments(table 1). It was caused zeolite adsorbs water from the air dryer during drying process [12]. Hence, the humidity of air can be kept low and driving force of drying is higher.

HRY was important indices in judging thequality milled rice in Indonesia; however, in somecountries, HRY was frequently used as important quality inassessing milled rice quality.According to SNI 6128:2015, Indonesia had twoclasses (premium class and medium classes) of national head rice yield standard by regulatory body. In premium class, head rice yield must be 95% of total milled rice. Medium class divided into three grades: 78%, 73%, and 60% minimum requirement of head rice yield. The head rice yield of dried paddy with and without zeolite in this experiment results had qualified the last grade (3rd grade of medium class) of SNI paddy quality 6128:2015 (min. 60%). This grade represented medium-low quality of paddy that can be found in Indonesia market (also includedin Bulog Indonesia paddy stock).

Proximate was observed from each sample at 60 minutes of drying process because they had the most propotional moisture content which is suitable for milling. The proximate analysis results showed that there were no much differences in ash, fat, protein, and carbohydrates content that contain in samples (table 2).

Table 2. Proximate analysis result.

Sample	Ash Content (%db)[a]	Fat Content (%db)[a]	Protein Content (%db)[a]	Carbohydrates Content (%db)[a]
Zeolite 40°C	11.18	0.73	6.73	66.83
Zeolite 60°C	11.10	0.73	6.73	66.86
Non-zeolite 40°C	11.24	0.71	6.75	66.80
Non-zeolite 60°C	11.32	0.70	6.76	66.32

[a]the presence of water was ignored for the purposes of the calculation

4. Conclusion

Results showed that zeolite can improve driving force for drying as indicated in lower moisture content in same temperature and duration of drying operation. The head rice yield of dried paddy with and without zeolite in this experiment results had qualified the last grade of standard by regulatory body. The proximate analyzed result showed there were no much differences in ash, fat, protein, and carbohydrates content that contained in samples.

References

[1] Pagani M, Johnson, T G and Vittuari M 2016 Energy Input in Conventional and Organic Paddy Rice Production in Missouri and Italy: A comparative case study *Journal of Environmental Management.***188** 173-182

[2] Djaeni M, Anggoro D, Santoso G W, Agustina D, Asiah N and Hii C L 2014 Enhancing the Food Product Drying with Air Dehumidified by Zeolite (in Bahasa Indonesia) *Advance Journal of Food Science and Technology* **6** (7) 833-838

[3] Oktavianty N U and Wildan 2016 Design of Measuring Instrument and Paddy Moisture Content Indicator Based Microcontroller with Photodiode Censor (in Bahasa Indonesia)*Unand Physics Journal* **5** 1

[4] Prakash B 2011 Dissertation of Doctor of Philosophy in Biological System Engineering in the Office of Graduate Studies of the University of California Davis USA *Mathematical Modelling of Moisture Movement within a Rice Kernel during Convective and Infrared Drying*

[5] Ratti C 2001 Hot Air and Freeze-drying of High-value Foods: A review*Journal of Food Engineering* **49** 311-319

[6] Djaeni M 2008 Phd Thesis *Energy efficient multistage zeolite drying for heat sensitive products* (Netherland: Wagenigen University) ISBN:978-90-8585-209-4

[7] Kiranoudis C T, Maroulis Z B and Marinos-Kouris D 1996 Drying of Solids : Selection of some Continious Operation Dryer types *Computer & Chem.Eng***20** 177-182

[8] Soponrannarit S 2003Fluidised Bed Grain Drying*Proc.3rd Asia- Pacific Drying Conference (Bangkok)* p 55-71

[9] Wiset L, Srzednicki G, Driscoll R, Nimmuntavin C and Siwapornrak P 2001Effects of High Temperature Drying on Rice Quality. Agricultural Engineering International: the CIGR *Journal of Scientific Research andDevelopment***01** 003

[10] Arabhosseini A, Huisman W, van Boxtel A and Muller J 2005 Modeling of the Equilibrium Moisture Content (EMC) of tarragon (*Artemisia Dracunculus L.*)*International Journal of Food Engineering* **1**(5) 7

[11] Djaeni M, Sasongko SB, Prasetyaningrum A, Jin X and van Boxtel A J B 2012 Paddy Drying in Mixed Adsorption Dryer with Zeolite: Drying Rate and Time Estimation*International Journal of Food Engineering* **8**(3) 32

[12] Djaeni M, Ayuningtyas D, Asiah N, Hargono, Ratnawati, Jumali and Wiratno 2013 Paddy Drying in Mixed Adsorption Dryer with Zeolite : Drying Rate and Time(in Bahasa Indonesia) *Reaktor* **14** (3) 173–178

Cross inoculation of anthracnose pathogens infecting various tropical fruits

Suparman[1*], M Rahmiyah[2], Y Pujiastuti[1], B Gunawan[1] and Arsi[1]

[1]Department of Plant Protection, Faculty of Agriculture, Sriwijaya University, Indonesia
[2]Department of Plant Science, Post Graduate Program, Sriwijaya University.

*Email: suparman@unsri.ac.id

Abstract. Anthracnose disease is very important disease of tropical fruits causing significant yield losses. The disease is caused by *Colletotrichum* spp. and infects almost all tropical fruit species, especially the succulent ones. Various species of Colletotrichum infect various tropical fruits and there are possibilities for cross inoculation to occur among tropical fruits which might cause severe infection. An experimental research was conducted to examine the effect of cross inoculation of anthracnose pathogen among papaya, eggplant, chili and common bean on the infection development and severity of the disease on each inoculated fruit species. *Colletotrichum* spp. were isolated from naturally infected papaya, eggplant, chili and common bean. Each fungal isolate was purified and identified to determine the species name. The spores of each isolate were then used to separately inoculate healthy and sterilized papaya, eggplant, chili and common bean. The results showed that cross infection developed on chili, eggplant and papaya but not on bean. Chili showed the highest susceptibility to all Colletotrichum isolates and significantly different from eggplant and papaya. The anthracnose pathogen isolated from common bean showed no pathogenicity to other hosts and might be used as cross protection inoculant to the disease in the other hosts.

Keywords: *anthracnose, cross inoculation, tropical fruits.*

1. Introduction

Papaya (*Carica papaya* L.), eggplant (*Solanum melongena* L.), chili pepper (*Capsicum annum* L.), and common bean (*Phaseolus vulgaris* L.) are highly demanded fruits in Indonesia [1]. All of the fruits are continuously available because, in the tropics, the crops are cultivated in such order that the fruits can be harvested regularly for daily needs. Like other succulent fruits, papaya, eggplant, chili pepper and common bean are subject to infection by *Colletotrichum* spp. causing anthracnose disease. The disease is very important during pre- and post-harvest phases. According to Dean *et al* [2], anthracnose disease has been considered as very important plant diseases based on its economic and scientific significance.

Colletotrichum spp. infect various fruit crops in tropical and subtropical regions, causing serious damage and significant yield losses, both quantitative and qualitatively in all growth phases of the crops. The pathogen is more threatening in humid tropics and contributes significant decrease to pre and post-harvest products [3]. Instead of causing anthracnose, the pathogens also inciting other diseases such as dieback, leaf spot, flower rot, fruit rot, stem end rot, and root rot [4, 5]. The fungus infects almost all of the above-ground parts of the host plants of all growth stages. Infection symptoms are obvious on leaves and fruit and sometimes appear on stems and flowers. Symptom on leaves is irregular or circular brown spot which eventually dries. Infection on stem causes irregular, sunken, brown spots with orange conidial masses eventually appear on the spots [6]. Symptoms appear on fruits are typical of anthracnose disease, a black round and sunken spots with black acervuli are often visible in the middle of the sot. Under favorable environmental conditions and susceptible hosts, lesions may coalesce [7]. Under natural ecosystem conditions, Colletotrichum infects several closely related host species but some species are able to infect a wider taxonomic range of hosts. It is also common to find multiple hosts infected by a single species of Colletotrichum, and single host infected by multiple species of the pathogen [8].

As a pathogen, Colletotrichum is a polyphagous fungal genus and is widely spread in both hemispheres, from tropical to subtropical countries on all continents [9]. The fungus is a hemi-biotrophic pathogen exploiting biotrophic and necrotrophic strategies to infect the host, producing appressoria, penetration peg, intracellular hyphae, and necrotrophic mycelia [10, 11]. Infection is started with the attachment of conidia to the surface of the host plant. Under humid conditions, the conidia germinate to form melanised appressoria to facilitate mechanical penetration. The penetration peg develops at the base of the appressorium to pierce the host cell. After penetration, infection vesicle and primary hyphae are formed in the epidermal cell throughout biotrophic phase and causes no harm to the host [12]. The fungus then switches to necrotrophic phase, produces secondary hyphae which then ramify inter-cellularly and kill the host cells by using hydrolytic enzymes [13].

Papaya is very susceptible to *C. gloeosporioides*, the causal pathogen of anthracnose, and losses to the disease ranges from 1 to 93% with average of 25%, depending on the postharvest handling applied [14]. In the case of eggplant, anthracnose is the most important and prevalent diseases which may cause reduction of 10 to 80% of crop production, occurring both in the field and during postharvest period [15]. Various species under genus Colletotrichum have been identified as the causal agent of anthracnose disease on eggplant, including *C.gloeosporoides* f.sp. *melongenae*, *C. nigrum*, *C.capsici*, *C. dematium* and *C. lindemuthianum* [16].

Chili pepper is another solanaceous crop severely infected by Colletotrichum causing the main problem on mature fruit. In the Colletotrichum-chili pepper pathosystem, different *Colletotrichum* species have been identified to be in association with the same host, including *C. acutatum*, *C. capsici*, and *C. gloeosporioides* [17] and cause an average yield losses up to 50% [18]. *C. capsici* is more frequently found infecting mature chili, whereas the other two Colletotrichum species infect both green and ripening chili fruits [19]. Anthracnose disease on common bean caused by *C. lindemuthianum* was reported to cause yield losses ranged from 40 to 90%, particularly in the high altitude and low temperature areas [20], since the anthracnose of common bean is favored by temperatures of 13-26°C and relative humidity above 92% [21]. *C. lindemuthianum* is the only species known as anthracnose pathogen on common bean, of which more than 25 races have been isolated and identified [22].

Colletotrichum intensely produces conidia bearing acervuli on the surface of infected host tissues. The conidia are released to become primary inoculum of anthracnose disease. The inoculum is disseminated by wind and water splashes to surrounding areas resulting in the development of secondary infections [23]. The inoculum causing secondary infection on previously healthy and younger fruit is produced by mature anthracnose spots on leaves or older fruit. Under high humidity conditions, conidia spread to immature fruit, causing quiescent infection on the fruit occurring during pre-harvest fruit development stage. The disease itself will appear and develop more apparently on harvested fruit and well known as post-harvest disease. The latent infection of anthracnose occurs due to the unsuitable conditions of the fruit to support the disease initiation. The initiation and development of anthracnose disease is also favored by the ethylene produced by the host fruit during ripening process [8]. The optimal conditions for the initiation and development of anthracnose disease in most hosts are at 25°C, even though the infection may occurs under wider range of temperature (18–33°C) and higher moisture (95–100%) [24].

Considering the continuous availability of papaya, eggplant, chili pepper and common bean under similar pre- and post-harvest conditions, and the fact that all of the four fruits species are subjected to Colletotrichum infection, it was assumed that there might be natural cross inoculations of

Colletotrichum spp. among them, as indicated by Stankova *et al* [9] that, due to the wide variability of Colletotrichum causing anthracnose, the conidia from infected host might be dangerous not only to the same host, but also to other crops. The outcomes of pathogen-host interaction vary among hosts [25] indicating host preference or host specialization [26]. The pathogen aggressiveness might be related to host susceptibility which can be determined by measuring disease or infection parameters such as incubation period, infection frequency, lesion size, and inoculum production. However, there has been not enough information of the aggressiveness of *Colletotrichum* spp. on papaya, eggplant, chili pepper and common bean on other hosts naturally grown under mixed cropping system or closed each other, and of the susceptibility of the fruits species to Colletotrichum from other hosts.

In this research, we evaluated the host specificity of the multi host pathogen Colletotrichum isolated from papaya, eggplant, chili pepper and common bean, by making cross-inoculation of the isolated pathogens. The research emphasized on the differentiation of each isolate of Colletotrichum on each inoculated host.

2. Materials and Methods

The cross-inoculation experiment was conducted in the Laboratory of Phytopathology, Department of Plant Protection, Sriwijaya University in 2016. The experiment was made using local fruits collected directly from farmers' fields in Ogan Ilir Regency and Palembang vegetable market in South Sumatra. Fruits used in this cross inoculation tests were papaya (*Carica papaya* L.), eggplant (*Solanum melongena* L.), chili pepper (*Capsicum annum* L.), and common bean (*Phaseolus vulgaris* L.). *Colletotrichum* spp. isolates used in the experiment were originated from diseased fruits obtained from the same field and market from where the healthy fruits used for cross inoculation experiment were collected.

2.1. Fungal Isolation:

Colletotrichum spp. were isolated from naturally infected papaya, eggplant, chili pepper and common bean. The pathogens were isolated from fruit without visible sporulation, by cutting fragments of 5x5 mm2 from the boundaries between anthracnose lesion and symptomless surrounding tissues. The fragments were then dipped into 1% sodium hypochlorite for 5 minutes, rinsed with sterile water, and dried with sterile tissue paper [27].

The sterilized tissue fragments were placed on potato dextrose agar in a Petri dish and incubated under room temperature until the hyphae of Colletotrichum grew and long enough to be transferred for further isolation and purification. Growing fungi showing typical colonies of *Colletotrichum* were transferred to fresh Potato Dextrose Agar medium to make pure cultures of the fungi. New colonies of *Colletotrichum* were made by single spore isolation technique. Conidia masses were streaked onto the surface of water agar 1.5% using sterilized loop and then incubated overnight under room temperature. A single germinating spore was transferred onto PDA using sterilized needle to make a pure culture [28]. Pure cultures were incubated for 14 days under daylight with temperature of 25 $\pm$ 2°C suitable for the fungal sporulation.

2.2. Pathogenicity Test

The spores used in pathogenicity test were harvested from 14 days old pure culture of *Colletotrichum* spp. isolated from the four fruit species. The spores were collected by pouring 10 mL of sterile water containing 0.1% of surfactant Tween 20. The agar surface was scrapped with paint brush to quicken the release of the spores. To make a pure spore suspension, the water containing scrapped fungal culture was filtered using double layers of cheese cloth. The spore density was adjusted to 1 x 10^6 spore mL^{-1} using haemocytometer [27, 28].

Fresh physiologically mature papaya, eggplant, chili, and common bean were graded based on size and conditions of the fruit species for further use. The selected fruits were cleaned with tap water before being sterilized by dipping into 70% ethanol for 3 minutes followed by immersing into 1% sodium hypochlorite solution for 5 minutes. The fruits were then rinsed 3 times in sterile water for 2 minutes and dried with sterile tissue paper. The sterilized fruits were placed on plastic box in which chicken wire mesh was laid two cm above the box bottom. Sterile tissue moistened with sterile water was laid on the box floor to maintain high relative humidity inside the box [29]. Each fruit species was inoculated with *Colletotrichum* spp. Isolated from all fruit species in an experiment arranged in a completely randomized design (CRD) with 4 colletorichum isolates as treatment and 5 replications. The CRD was used in the experiment because it was conducted in the laboratory under homogenous environmental conditions. Each set of treatment involving 5 fruit in each plastic box. .

Sampled fruits arranged in plastic boxes were inoculated with prepared spore suspension using micropipette. Ten inoculation points were marked on each fruit to be inoculated. Wounds were then made on the marked inoculation sites by puncturing the fruit surface using sterilized needle point to a 1 mm depth [30]. 10 µl spore suspension was dropped on each wound using micropipette. Mock inoculation was made to control fruits by dropping 10µl sterile water on wounds made on the fruits. Each set of inoculated fruits in the plastic box was incubated separately. Every plastic box containing a set of inoculated fruits was wrapped individually with transparent plastic bag to retain high humidity required for disease initiation. The boxes containing inoculated fruits were incubated under temperature of 25 $\pm$ 2°C with 12 hour light/dark cycle. The bags were removed 2 days after inoculation [27].

Inoculated fruits were visually evaluated daily to record data of infection parameters caused by *Colletotrichum* spp. cross-inoculated to papaya, eggplant, chili pepper and common bean. The parameters measures included incubation period, infection frequency, anthracnose size, number of acervuli and number of conidia per acervuli. Incubation period was measured as period (in day) when 50% of inoculation points show the first symptom of anthracnose; infection frequency was measured as the percentage of infection points developed anthracnose symptom; anthracnose was measured as the longest diameter; number of acervuli was measured is the average numbers of acervuli appeared on anthracnose symptom; and number of conidia per acervulus is the average number of conidia released from acervuli appeared on anthracnose symptom.

3. Results and Discussion

C. gloeosporioides, C. capsici and *C. lindemuthianum* were isolated from four different tropical fruits. The identification of the pathogens was based on the disease symptoms, culture characteristics, and morphological characteristics [23, 31, 32]. All Colletotrichum isolates have ability to infect other fruit species, except the one isolated from common bean. *C. gloeosporioides* isolated from papaya and eggplant and *C. capsici* isolated from chili pepper could infect all tested fruits but common bean. Inversely, *C. lindemuthianum* originated from common bean could infect only original host and failed to infect other inoculated hosts. This is in accordance with the reports by Thomazella *et al* [22] that common bean could only be infected by *C. lindmeuthianum* which is consisted of more than 25 races. The failure of *C. lindemuthianum* to infect other hosts might also be caused by environmental condition, especially the temperature of the room used to incubate the inoculated fruits which was set to 25$\pm$ 2°C which might be too high for *C. lindemuthianum* which needs optimum temperature of 13-26°C with an optimum of 17°C, and relative humidity above 92% [21]. The cross inoculation experiments demonstrated the variations in *Colletotrichum* spp. pathogenicity on their original host and on other hosts. The level of host preference among *Colletotrichum* spp. isolated from four different fruits and the susceptibility of the hosts varied significantly.

3.1. Pathogenicity of Colletotrichum spp.

All isolates of *Colltotrichum* spp. were proved as the pathogens causing anthracnose on all fruits producing typical anthracnose symptoms after being inoculated with the pathogens. This was verified by the fact that mock inoculation using sterile water had no effect on inoculated fruits. *C. gloeosporioides, C. capsici* and *C. lindemuthianum* isolated from different hosts were able to establish anthracnose not only on their original hosts but also on other hosts through cross inoculation. This demonstrate that most *Colletotrichum* spp. have wide range of host species, and many host species are subject to more than one species of Colletotrichum as reported by Barrett *et al* [33] that, under natural ecosystem, most pathogens may have wide host range, and most hosts may suffer from various pathogen infections.

3.1.1. Cross inoculation to papaya. Papaya inoculated with *Colletotrichum* spp. Isolated from papaya, eggplant and chili produced typical symptoms of anthracnose. The results of this cross inoculation revealed that Colletotrichum isolated from the four fruit species naturally infected by the fungus could initiate anthracnose on papaya at similar rate and severity (Table 1). The surprising result was the fact that the anthracnose disease developed on original hosts was not different from those developed on non-original hosts. The short incubation period (< 3 days) on all infected fruits did not automatically indicated the susceptibility of the host since other infection parameters showed the opposite. Means of incubation period (2.06 days), infection frequency (14.06%), anthracnose spot size (1.84 mm), number of acervuli per spot (0.60) and number of conidia per acervulus (1.31) showed lower values than those of natural infections, and no significant difference among hosts, except number of conidia per acervulus, in which papaya was significantly different from other three infected hosts. This significant difference is quite creditable since the conidia are very important factor in the disease epidemic. Under favorable condition, conidia produce in high rate under temperature of 25°C and relative humidity of 95-100% [24]. The size of lesson developed on papaya as original hosts when inoculated with Colletotrichum isolated from naturally infected papaya was not larger than those developed on papaya inoculated with Colletotrichum from other fruits. This result was inconsistent with those found by Lakshmi *et al* [27] that *C. gloeosporioides* isolated from a fruit species could cross-infect fruits of other species, and the fungus isolate was most aggressive when inoculated to its original host.

Table 1. Anthracnose disease produced by cross-inoculation of Colletotrichum isolated from different fruits on papaya.

Isolate	Incubation period[a]	Infection frequency[a]	Anthracnose size[a]	Number of acervuli[a]	Number of conidia ($x 10^4$)[a]
Papaya	1.75 a	13.43 a	1.32 a	1.00 a	2.22 b
Eggplant	1.81 a	15.89 a	1.49 a	0.35 a	0.77 a
Chili pepper	2.36 a	14.67 a	2.39 a	0.33 a	0.70 a
Common bean	2.33 a	12.23 a	2.15 a	0.71 a	1.55 a
Mean	2.06	14.06	1.84	0.60	1.31

[a]Means with same letters in the same column were not significantly different according to HSD 5%.

The low values of anthracnose disease parameters on papaya also indicated the low virulence of the pathogens used in the inoculation or the unfavorable environmental conditions during disease development, even though room temperature and humidity had been adjusted to suit the needs of anthracnose development. Furthermore, there has been no report of other species of Colletotrichum causing anthracnose disease on papaya except *C. gloeosporioides.* Therefore, the low values of infection parameters of anthracnose on papaya caused by Coleltorichum isolated from other fruits might not be caused by low aggressiveness of the pathogens but more likely to be caused by host specificity.

3.1.2. Cross inoculation to eggplant. Eggplant inoculated with *Colletotrichum* spp. isolated from papaya, eggplant and chili also developed anthracnose on all inoculated tested fruits with some differences in some infection parameters. As can be seen in Table 2, the results of this cross inoculation revealed that anthracnose developed almost equally in all infected tested fruits (Table 2). The surprising result was the fact that the anthracnose disease developed better on eggplant compared to those developed on papaya. All infection parameters of anthracnose developed on eggplant were higher than those on papaya, except average number of ecervuli per anthracnose spot. Means of incubation period (2.27 days), infection frequency (26.30%), anthracnose spot size (2.78) are higher than mean values of the same parameters measured on papaya. These phenomena revealed that eggplant is more susceptible to anthracnose pathogen than papaya, or the anthracnose pathogens Colltotrichum spp. are more virulent when inoculated to eggplant. Virulence of anthracnose pathogens is affected by host conditions such as availability of nutrients and enzymes required by the pathogens, or the presence of antifungal compounds [27].

The significantly higher infection frequency on papaya than that on eggplant is not quite surprising since both fruits are natural host of *C. gloeosporioides*. However, number of acervuli and number of conidia per acervulus were significantly higher on eggplant than on papaya indicating that eggplant is more suitable host for *C. gloeosporioides* isolated from eggplant. This is in accordance with previous finding that anthracnose infection was better on its natural host than on other hosts. As occurred in cross infection to papaya, cross infection to eggplant also resulted in higher number of conidia produced on natural host rather than on non-natural host and this might make difference in the next step of disease spread and development. More abundant conidia will make more effective disease spread and development. The more prolific anthracnose developed on eggplant is consistent with the report of Mathur and Kongsdal [35] that anthracnose on eggplant might be caused by *C. gloeosporoides* f. sp. *melongenae*, *C. nigrum*,*C. capsici*, *C. dematium*, *C. lindemuthianum* and *Gloeosporium melongenae*.

Table 2. Anthracnose disease produced by cross-inoculation of Colletotrichum isolated from different fruits on eggplant.

Isolate	Incubation period[a]	Infection frequency[a]	Anthracnose size[a]	Number of acervuli[a]	Number of conidia (x 10^4)[a]
Papaya	2.40 a	31.54 b	2.91a	0.41 b	0.80 b
Eggplant	2.21 a	20.82 a	2.30a	1.09b	2.35 c
Chili pepper	2.10 a	24,34 a	2.26a	0.00a	0.00 a
Common bean	2.37 a	28.51 b	3.64a	0.91b	2.29 b
Mean	2.27	26.30	2.78	0.60	1.36

[a]Means with same letters in the same column were not significantly different according to HSD 5%.

3.1.3.Cross inoculation to chili pepper. Cross inoculation of *C. capsici* isolated from chili papaya, eggplant, common bean and chili pepper itself showed that the pathogen could infect other host with significant different in infection frequency, anthracnose size, number of acervuli and number of conidia per acervulus (Table 3). The three parameters are very important infection parameters in determining the aggressiveness of the pathogen and the susceptibility of the hosts. In this case, the difference of the parameters is more likely to determine the difference of pathogens aggressiveness since the parameters were observe on the same host but of different pathogen isolates. Means of incubation period (2.13 days), infection frequency (53.63%), anthracnose spot size (3.69), number of acervuli (1.16) and number of conidia per acervulus (5.35 x 10^4) are higher than those measured on papaya and eggplant. This clearly indicates that there are different levels of host susceptibility among the tested fruit species, and chili pepper is the one showing the most susceptible to *Colletotrichum* spp. than papaya, eggplant and common bean. In general, chili pepper naturally infected by several species of Colletotrichum, including *C. acutatum*, *C. gloeosporioides*, *C. coccodes* and *C. dematium* [23] and in Indonesia, chili pepper frequently infected by *C. acutatum*, *C. capsici*, and *C. gloeosporioides* [21].

Table 3. Anthracnose disease produced by cross-inoculation of Colletotrichum isolated from different fruits on chili pepper.

Isolate	Incubation period[a]	Infection frequency[a]	Anthracnose size[a]	Number of acervuli[a]	Number of conidia (x 10^4)[a]
Papaya	2.00 a	51.16 b	3.89 b	0.91 a	6.45 a
Eggplant	1.96 a	47.02 b	3.59 b	0.53 a	3.18 a
Chili pepper	2.17 a	62.41 c	4.41 c	1.78 b	4.06 a
Common bean	2.37 a	53.91 b	2.89 a	1.40 a	7.71 a
Mean	2.13	53.63	3.69	1.16	5.35

[a]Means with same letters in the same column were not significantly different according to HSD 5%.

3.1.4.Cross inoculation to common bean. All Colletotrichum isolated from papaya, eggplant and chili pepper failed to develop anthracnose disease when inoculated to common bean. The fruit could only be infected by Colletotrichum isolated from the same fruit species which mean that the host factor is very influential to the initiation and development of the disease. Furthermore, anthracnose on common bean is favored by cool and wet weather with temperature ranges from 13 to 26°C and an optimum of 17°C [24], which could not be met during the experiment.

4. Conclusions

Cross inoculation of *Colletotrichum* spp. isolated from naturally infected papaya, eggplants and chili pepper could produce anthracnose symptoms not only on original but also on non-original hosts. Since the fruits are continuously available in the field and market and mostly cultivated under mixed cropping system, the cross inoculation potential make the vulnerability of the disease even greater and needs serious management. Colletotrichum causing anthracnose on common bean could only produce anthracnose symptoms on original host might be used to biologically control anthracnose disease of other fruits through cross protection inoculation.

References

[1] Suparman, Arsi and Cameroon R R 2016 Tropical fruit diseases in Indonesia. Survey Report (Unpublished)

[2] Dean R, Van Kan JA, Pretorius Z A, Hammnd-Kosack K E, Di Pietro A, Spanu PD, Rudd JJ, Dickman M, Kahmann R, Ellis J and Foster G D 2012 The top 10 fungal pathogens in molecular plant pathology *Mol. Plant Pathol.* **13** 414-30

[3] de Souza A, Delphino Carboni RC, Wickert E, de Macedo Lemos EG and de Goes A 2013 Lack of host specificity of *Colletotrichum* spp. isolates associated with anthracnose symptoms on mango in Brazil *Plant Pathol.* **62** 1038–47

[4] Bruce CDF, Fatima CD, Michereff SJ 2013 Biology of *Colletotrichum* spp. and epidemiology of the anthracnose in tropical fruit trees. *Revista Caatinga* **26** 130–38

[5] Nasimento R, Mizubuti ESG, Camara MPS, Ferreira MF, Maymon M, Freeman S and Michereff SJ 2010 First report of papaya fruit rot caused by *Colletotrichum magna* in Brazil. *Plant Disease* **94** 1506

[6] Debode J, van Hemelrijck W, Baeyen S, Creemers P, Heungens K A and Maes M 2009 Quantitative detection and monitoring of *Colletotrichum acutatum* in strawberry leaves using real-time PCR. *Plant Pathol.* **58** 504–14

[7] Lewis Ivey ML, Nava-Diaz C and Miller S 2004 Identification and management of *Colletotrichum acutatum* on mature bell peppers. *Plant Dis.* **88** 1198–204

[8] Shivas, RG and Tan YP 2009 A taxonomic reassessment of *Colletotrichum acutatum*, introducing *C. fioriniae* comb. et stat. nov. and *C.simmondsii* sp. nov. *Fungal Divers* **39**111–22

[9] Stankova B, Vichova J, Pokoryn R 2011 Virulence of *Colletotrichum acutatum* isolates to several host plants. *Acta Universitatis Agriculturae et Silviculturae Mendelianae Brunensis* **59** 161-69

[10] Than PP, Prohastuti H, Phoulivong S, Taylor PWJ, Hyde KD 2008 Chili anthracnose disease caused by *Colletotrichum* species . *Journal of Zhejiang University SCIENCE B* **9** *764-778*

[11] Bhadauria V, Banniza S, Vandenberg A, Selvaraj G and Wei Y 2011 EST mining identifies proteins putatively secreted by the anthracnose pathogen *Colletotrichum truncatum. BMC Genomics* **12** 327

[12] Kankanala P, Czymmek K, Valenta B 2007 Roles for rice membrane dynamics and plasmodesmata during biotrophic invasion by the blast fungus. *Plant Cell* **19** 706–24

[13] Munch S, Lingner U, Floss DS, Ludwig N, Sauer N and Deising HB 2008 The hemibiotrophic lifestyle of Colletotrichum. *J Pant Physiol.* **165** 41-51

[14] Rahman MA, Mahmud TMM, KadirJ, Abdul Rahman R and Begum MM 2008 Major post-harvest fungal diseases of papaya cv. 'Sekaki' in Selangor, Malaysia. *Pertanika J. Trop. Agric. Sci.* **31** 27-34

[15] Pedroso C, Lage DAC, Henz GP and Cafe-Filho AC 2011 Development and validation of a diagrammatic scale for estimation of anthracnose on sweet pepper fruits for epidemiological studies. *J. of Plant Pathol.* **93** 219-225

[16] Mathur BS and Kongsdal O 2001 Common Laboratory Seed Health Testing Methods for Detecting Fungi (Copenhagen: Kandrups Bogtrykkeri Publishers) p 436

[17] Voorrips RE, Finkers R, Sanjaya L and Groenwold R 2004 QTL mapping of anthracnose (*Colletotrichum* spp.) resistance in a cross between *Capsicum annuum* and *C. chinense*. *Theoretical and Applied Genetics*, **109** 1275-82

[18] Pakdeevaraporn P, Wasee S, Taylor PWJ, Mongkolporn O 2005 Inheritance of resistance to anthracnose caused by *Colletotrichum capsici* in *Capsicum Plant Breeding* **124** 206-208

[19] Kim KK, Yoon JB, Park HG, Park EW, Kim YH 2004 Structural modifications and programmed cell death of chili pepper fruits related to resistance responses to *Colletotrichum gloeosporioides* infection. *Genetics and Resistance* **94** 1295-304

[20] Mudawi HI, Idris MO, El Balla MA 2009 Anthracnose disease in common bean (*Phaseolus vulgaris* L.) in Shambat, Sudan: Disease incidence, severity and effect on yield. University of Khartoum Journal of Agricultural Sciences **17** 118-30

[21] Yesuf M and Sangchote S 2007 Survival and transmission of *Colletotrichum lindemuthianum* from naturally infected common bean seeds to the seedlings. *Trop. Sci.* **47** 96-103

[22] Thomazella C, Gonçalves-Vidigal MC, Vidigal Filho PS, Nunes WMC, Vida JB 2002 Characterization of *Colletotrichum lindemuthianum* races in Paraná state, Brazil. *Crop Breeding and Applied Biotechnology* **2** 55-60

[23] Than PP, Jeewon R, Hyde KD, Pongsupasamit S, Mongkolporn O, Taylor PWJ 2008 Characterization and pathogenicity of *Colletotrichum* species associated with anthracnose disease on chili (*Capsicum* spp.) in Thailand. *Plant Pathol.* **57** 562–72

[24] Víchova J, Stankova B, Pokorny R, Vejrazka K 2013 Evaluation of safflower varieties for resistance to the fungal pathogen, *Colletotrichum acutatum. Plant Protect. Sci.* **49** 132–36

[25] Seiwa K, Miwa Y, Sahashi N, Kanno H, Tomita M, Ueno N and Yamazaki M 2008 Pathogen attack and spatial patterns of juvenile mortality and growth in a temperate tree, Prunus grayana. *Canadian Journal of Forest Research* **38** 2445–54

[26] Sicard D, Pennings PS, Grandclement C, Acosta J, Kaltz O and Shykoff JA 2007 Specialization and local adaptation of a fungal parasite on two host plant species as revealed by two fitness traits. *Evolution* **61** 27–41

[27] Lakshmi BKM, Reddy PN and Prasad RD 2011 Cross-infection potential of *Colletotrichum gloeosporioides* Penz. isolates causing anthracnose in subtropical fruit crops. *Trop. Agric. Res.* **22** 183–193

[28] Phoulivong S, McKenzie EHC, Hyde KD 2012 Cross infection of *Colletotrichum* species; a case study with tropical fruits. *Current Research in Environmental & Applied Mycology* **2** 99– 111

[29] Than PP, Shivas RG, Jeewon R, Pongsupasamit S, Marney TS, Taylor PWJ and Hyde KD 2008 a Epitypification and phylogeny of *Colletotrichum acutatum* J.H.Simmonds. *Fungal Diversity* **28** 97–108

[30] Lin Q, Kanchana-udomkarn C, Jaunet T, Mongkolporn O 2002 Genetic analysis of resistance to pepper anthracnose caused by *Colletotrichum capsici*. *Thai J. Agric. Sci.* **35** 259–64

[31] Damn U, Cannon P, Woudenberg JHC, Johnston PR, Weir BS, Tan YP, Shivas RG and Crous PW 2012. The *Colletotrichum boninense* species complex. *Studies in Micology* **73** 1-36

[32] Bush E 2014. Anthracnose on Snap Beans. Virginia Cooperative Extension. Publication 450-719

[33] Barrett LG, Kniskern JM, Bodenhausen N, Zhang W and Bergelson J 2009 Continua of specificity and virulence in plant host–pathogen interactions: causes and consequences. *New Phytologist* **183** 513–29

[35] Mathur BS and Kongsdal O 2001 Common Laboratory Seed Health Testing Methods for Detecting Fungi (Copenhagen: Kandrups Bogtrykkeri Publishers) p 436

Application of Plackett-Burman experimental design in the development of muffin using adlay flour

J S Valmorida[1*] and K A T Castillo-Israel[2]

[1]Department of Food Science, College of Human Ecology, Central Mindanao University, 8710 Musuan, Bukidnon, Philippines
[2]Institute of Food Science and Technology, University of the Philippines - Los Baños, 4031 College, Laguna, Philippines

*Email: jsvalmorida@up.edu.ph

Abstract. The application of Plackett-Burman experimental design was made to identify significant formulation and process variables in the development of muffin using adlay flour. Out of the seven screened variables, levels of sugar, levels of butter and baking temperature had the most significant influence on the product model in terms of physicochemical and sensory acceptability. Results of the experiment further demonstrate the effectiveness of Plackett-Burman design in choosing the best adlay variety for muffin production. Hence, the statistical method used in the study permits an efficient selection of important variables needed in the development of muffin from adlay which can be optimized using response surface methodology.

Keywords: *Adlay, muffin, Plackett-Burman*

1. Introduction

Job's tears (*Coix lacryma-Jobi* L.) or coix belongs to family Poaceae. [1]. It is widely cultivated as a grain crop in Asia and has been used as food, medical or forage for a thousand of years [2]. The consumption of dehulled seed and flour of adlay have shown to improve lipid metabolism which decreased the risk of heart diseases [3].As a prebiotic, adlay provides a modifying effect on the intestinal microflora hence offers great benefit to gastrointestinal tract [4].

Bakery products are considered to be one of the most preferred food and among these products, cakes, specifically muffins are the most favored due to its organoleptic characteristics [5]. On the other hand, adlay flour is produced by milling adlay grains and it has its potential for bakeryware applications. However, being gluten-free, dough made from adlay flour will not leavened. A composite of wheat flour (70%) and Job's tear flour (30%) is considered to be a good mixture for bakery products. Studies on the development of cookies and bread using adlay flour [6,7,8]and butter cake [9] were already reported. While researches on the development of gluten-free muffin with the use of chickpea flour, waxy rice, sweet potato flour, cassava flour and composites of rice and soy flour and corn starch are already well-established [10,11], utilization of adlay flour for muffin production has not been reported. Thus, the utilization of adlay flour for its nutritive value and a source of high energy and protein with a lot of beneficial effects presents an opportunity to increase its production and consumption. Moreover, the development of gluten-free muffin from adlay also provides additional advantage over people with celiac disease. Thus, the screening of variables in the development of muffin from adlay becomes imperative, hence, this study.

2. Materials and Methods

2.1. Adlay Muffin Formulation and Preparation

The muffin recipe was a combined modifications between the standard formulation of the Central Mindanao University Bakeshop and gluten-free flour fortified muffin [12]. The formulation contained the following ingredients: adlay flour, refined sugar, refined salt, baking powder, butter, evaporated milk and fresh eggs.

Muffin was prepared by mixing dry and wet ingredients [12]. Muffin batter wastransferred into muffin pans with paper cupcake liners weighing about 50 g and were baked until done.Cooked muffins were set for five minutes, removed from the pans and allowed to cool on wire rack for one hour.

2.2.Screening of Significant Variables Using Plackett-Burman Design

Along with existing reviews [13, 12], a series of preliminary runs were conducted to identify variables that will have significant effects on adlay muffin production. These variables were used in the final screening which include the following: types of adlay flour, baking temperature, baking time, levels of butter, levels of sugar and levels of egg all of which were set at (-) low and (+) levels. An 8 run design with 7 independent variables [14] was employed during screening following the Plackett-Burman methodology.

2.3. Physicochemical Analysis

The volume of muffin was determined according to the American Association of Cereal Chemists (AACC)method 10-05.01 [15] using seed displacement method. Sample was placed in a container of known volume into which sago pearls were allowed to run until the container was full. The volume of sago pearls displaced by the sample was measured in a graduated cylinder and was considered as muffin volume.Moisture content of muffin was determined by following the method of Association of Official Analytical Chemists [15] while water activity was determined using a water activity meter (Novasina Labstart - aw). All measurements were done in triplicates.

2.4 Sensory Analysis

Product acceptability analysis was carried out by subjecting all eight treatments to sensory evaluation using a 9-point Hedonic [16]. Sensory characteristics such color, texture, gas cells, aroma, sweetness, flavor and general acceptability were assessed by 28 semi-trained panelists.

2.5 Statistical Analysis

Data on the screening of variables from the Plackett-Burman design were analyzed using the software Statistica Version 10 (Statsoft Inc., USA). Statistical *t*-values were calculated for the determination of variable significance at $p = 0.05$.

3. Results and Discussion

Placket-Burman design was used to screen formulation and process variables for adlay muffin production. It was carried out by screening seven major variables. A total of 8 runs were considered during screening set at two levels, (-) low and high (+) according to the PB matrix design. A main positive effect denotes an increase of value on the levels of variables and must be used for optimization process while a negative effect implies a decrease in response as the value of levels are increased. Hence the sign (+ or -) and the degree of the main effect were properly considered as a mechanism for the development of appropriate conditions for adlay muffin production. It shall be noted that a non-significant result for the screened variables, whether positive or negative, strongly imply that it has essentially no effect on the specific response.

3.1 Effects of processing variables on the physicochemical properties of adlay muffin

The effect of each variable on the physicochemical properties of adlay muffin is presented in Table 1. Results demostrated that the types of adlay flour showed a significant positive effects for volume and moisture content, suggesting that *gulian* variety of adlay flour is needed for optimum production of adlay muffin.

Baking temperature and time revealed a significant negative effect on the volume, moisture content and water activity of muffin samples. Previous report suggeststhat oven temperature and baking time strongly influence the development of all quality attributes of baked products[18]. Higher oven temperatures seem to produce a more humid crust which allows higher moisture loss and consequently a lower moisture content of muffins[19]. The use of higher temperature during baking allows rapid expansion of bubbles [20]and consequently the release of gas due to higher pressure, hence the lower moisture content of samples.Further, extreme baking temperatures (very high and very low)would result to undesirable crust color, volume and crumb texture [21]. Moreover, intermediate baking conditions (temp-time combination) guarantee the best microstructure and texture of muffin[22].

Low level of butter is preferred for an increased volume and moisture content of adlay muffin. It also indicate that these low levels are most likely nearer to the optimum values when subjected to RSM. Accordingly, fat plays important roles in food systems[23]. In bakery products such as muffins, fat entraps air during mixing to incorporate air bubbles into the mixture to help raise the final products.Levels of butter positively affectedthe water activity of muffin samples, suggesting that the use of higher levels would increase optimum water activity. Water constitute 20% of the whole cream where butter is made[24]which presumably might have contributed to an increased water activity of samples.

Levels of sugar positively affectedthe volume while negatively affected the moisture content and water activity of adlay muffin. The high level of sugar favored for greater muffin volume muffin from adlay. It was previously reported that optimum volume of muffin is achieved with 100% sugar in formulation while no sugar muffins exhibit higher Aw values and moisture contents [25]. The removal of sugar increases water activity of samples. A lower water activity value of cake-type muffin was observeddue to its high sugar content[26]. Sugar in bakery products has been indicated to makemajor contributions to controlling moisture retention and in improving volume and texture of baked products[27].

Table 1. Statistical analysis using Plackett-Burman design showing the main effects of the processing variables on the physicochemical properties of adlay muffin.

INDEPENDENT VARIABLES/ FACTORS	RANGE OF VALUES		EFFECT ESTIMATES		
	Low (-)	High (+)	VOLUME (cc)	MOISTURE (%)	WATER ACTIVITY (a_w)
Types of Flour	*Kiboa*	*Gulian*	+27.494**	+0.528*	-0.003^{ns}
Baking Temp. (°C)	150	200	-6.172**	-3.978**	-0.022**
Baking Time (mins.)	25	45	-7.339**	-5.817**	-0.028**
Butter (g)	10	15	-23.160**	-2.444**	+0.015**
Sugar (g)	10	30	+25.327**	-5.087**	-0.043**
Milk (g)	30	40	-24.327**	+1.962**	+0.034**
Egg (g)	10	15	+47.149**	+2.070**	-0.010*

^{ns} – no significant difference at $p > 0.05$;**– highly significant and *– significant at $p \leq 0.05$

Levels of milk showed a negative effect for volume while positive for both moisture and water activity. Milk is usually the main liquid in baked products such as muffin[28] as it allows hydration of dry ingredients and provides steam for leavening. The type of liquid with increasing water content significantly reduced the volume of gluten-free muffin[13]. Previous study has reported that gluten-free muffin generally has higher moisture contents due to higher water absorbability of gluten-free flour [29].

Based on the parameter estimates, levels of egg positively influenced the volume and moisture content while negatively affected the water activity of adlay muffin. Egg is an essential ingredient for obtaining desirable volume and texture in baked products. The protein in egg solubilize quickly in water and coagulate during baking. Foaming and emulsification properties of egg significantly contribute to batter expansion and in the stabilization of emulsions in cake and muffins [30].

3.2 Effects of processing variables on the sensory acceptability of adlay muffin

Based on Table 2the types of adlay flour gave a significant positive effect on the general acceptability of adlay muffin which imply that *gulian* variety located at the high (+) column had higher general acceptability value than *kiboa* variety. Evidently, no reviews were found in the use adlay from *gulian* and *kiboa* varieties for muffin production. Hence, the results obtained during screening experiment strongly suggest that the sensory acceptability of adlay muffin produced by using *gulian* variety is favourable considering that the levels of each variables were based on the levels in favor of *gulian.*

Among the screened variables, baking temperature, levels of butter and levels of sugar had the most significant effects on the sensory qualitites of adlay muffin. The use of higher baking temperature is preferred for an increase color, texture, aroma, sweetness, flavor and general acceptability. High temperatures and optimal baking time favors caramelization and Maillard reactions resulting to the formation of desirable aroma and flavor[19]. Browning of muffin surface improves both color and taste of the final product.

IOP Conf. Series: Earth and Environmental Science **102** (2017) 012081 doi:10.1088/1755-1315/102/1/012081

Results of the statistical analysis also revealed that the lower level of butter was highly favored for better sensory acceptability, suggesting that this low level could be near to the optimum range for adlay muffin production. Butter and fats contribute to tenderness, volume, moistness and flavor in muffin and other baked products. Fats engulf tiny air bubbles during creaming which greatly affects the textural and eating qualities of muffins. Lower fat levels are desired in some formulations as higher fat levels cause excessive tenderness of samples [28].

Another independent variable affecting adlay muffin production is the levels of sugar. Interestingly, all sensory parameters showed positive signs which mean that the use of high levels of sugar is preferred for an increase muffin acceptability. Sugar is the main ingredient in muffins and other baked products and it greatly affects the product appearance, texture and mouthfeel. Itabsorbs water and incorporates air in the shortening hence promotes tenderization and lightness in baked products. Further, it caramelized when heated adding flavor and color in baked products [31]. Moreover, the structural and sensory properties of the muffin system have been shown to be influenced by sucrose levels and is highly favored in its normal levels than its reduced counterparts [29].

Table 2. Statistical analysis using Plackett-Burman design showing the main effects of the processing variables on the sensory acceptabilityof adlay muffin.

INDEPENDENT VARIABLES/ FACTORS	RANGE OF VALUES		EFFECT ESTIMATES						
	Low (-)	High (+)	COLOR	TEXTURE	GAS CELLS	AROMA	SWEETNES S	FLAVOR	GEN. ACCEPT.
Types of Flour	*Kiboa*	*Gulian*	+0.189^{ns}	+0.039^{ns}	-0.123^{ns}	-0.018^{ns}	+0.144^{ns}	-0.028^{ns}	+0.434*
Baking Temp. (°C)	150	200	+ 1.599**	+0.932**	+0.412^{ns}	+0.642*	+0.840**	+1.171**	+1.916**
Baking Time (mins.)	25	45	+0.564**	+0.575**	+0.287^{ns}	+0.463^{ns}	+0.161^{ns}	+0.296^{ns}	+0.826**
Butter (g)	10	15	-1.010**	-0.646**	-0.483*	-0.249^{ns}	-1.054**	-0.742**	-1.326**
Sugar (g)	10	30	+1.296**	+0.503*	+0.501*	+0.731**	+1.358**	+1.046**	+1.201**
Milk (g)	30	40	-0.403^{ns}	-0.110^{ns}	+0.105^{ns}	+0.107^{ns}	-0.233^{ns}	-0.046^{ns}	-0.309^{ns}
Egg (g)	10	15	+1.120**	-0.690^{ns}	+0.344^{ns}	+0.551^{ns}	+0.896*	+0.620^{ns}	+1.534**

^{ns} – no significant difference at $p > 0.05$;**– highly significant and *– significant at $p \leq 0.05$

3.3. Selection of top three significant variables for production of adlay muffin

Evidently, all the variables screened for adlay muffin production show high levels of significance based on physicochemical properties and sensory acceptability. However, among the screened variables, levels of sugar, baking temperature and levels of butter were the top three variables that showed the highest effects. Hence, these three variables could be used in the subsequent stage of optimization through response surface methodology. Furthermore, results of the screening experiment strongly suggest on the use of adlay flour from *gulian* variety, longer baking time, lower level of milk and high level of egg for muffin production.

4. Conclusion

Results of the study clearly elucidated the importance of Plackett-Burmann as a screening design in the selection of variables (formulation and process) affecting the development of muffin from adlay. Among all the screened variables, levels of sugar, baking temperature and levels of butter were found to have great influenced on the physico-chemical and sensory acceptability of adlay muffin. Hence, optimization of these variables may be explored using response surface methodology.

5. Acknowledgement

The authors would like to express gratitude to the Department of Science and Technology – Accelerated Science and Technology Human Resource Development Program (DOST-ASTHRDP), Commission on Higher Education (CHED), Department of Agriculture – Bureau of Agricultural Research (DA-BAR) and Central Mindanao University for funding this research.

References

[1] Peñaflor L, Elepaño A and Peralta E 2014 Rice-Like Grains from Broken Rice (*Oryza sativa* L.) and Adlai (*Coix lacryma-jobi* L.). *Asian J. Agriculture and Food Science* **2** (4) 341 - 349

[2] Dechkunchon M and Thongngam M 2007 Chemical and physicochemical properties of adlay flours and starches. *45th Kasetsart University Annual Conf.* (Bangkok, Thailand)

[3] Chaisiricharoenkul J, Tongta S and Intarapichet K 2011 Structure and chemical and physicochemical properties of Job's tear *(Coix lacryma-jobi L.)* kernels and flours. *Suranaree J. Sci. Technol* **18** 109-122.

[4] Chiang W, Cheng C, Chiang M and Chung K 2000 Effects of dehulled adlay on the culture count of some microbiota and their metabolism in the gastrointestinal tract of rats. *J. Agricultural and Food Chemistry* **48** 829–832.

[5] Matsakidou A, Blekas G and Paraskevopoulou A 2010 Aroma and physical characteristics of cakes prepared by replacing margarine with extra virgin olive oil. *LWT-Food Science and Technology* **43** (6) 949-957.

[6] Chawakorn S 2006 Potential utilization of non-wheat and/or composite flour in making food products. Department of Food Technology, Thailand Institute of Applied Science, Thailand.

[7] Cheappensuk R 2006 Innovation of food products from Job's tear. Research Summary of Rajapat Pranakorn.

[8] Reungkajorn P, Wongtecha N, Kerdpiboon S and Potiset S 2007 Use of Job's tear in cookies production. Special Project, Department of Food Technology, Faculty of Natural Resources and Agro-Industry, Kasetsart University, Chalermprakiat Campus, Sakonnakorn, Thailand.

[9] Kutschera M and Krasaekoopt W 2012 The Use of Job's Tear (*Coix lacryma-jobi* L.) Flour to Substitute Cake Flour in Butter Cake *AU Journal of Technology* **15**(4).

[10] Herranz B, Canet W, Jiménez M, Fuentes R and Alvarez M 2016 Characterization of chickpea flour-based gluten-free batters and muffins with added biopolymers: rheological, physical and sensory properties. *Int. J. Food Science & Technology* **51** (5) 1087–98

[11] Johnson F 1990 Characteristics of muffins containing various levels of waxy rice flour *Cereal Chem* **67** (2) 114-118.

[12] Bhaduri S 2013 A comprehensive study on physical properties of two gluten-free flour fortified muffins. *J. Food Processing and Technology* **4**(8).

[13] Ozola L, Straumite E and Galoburda R 2012 Effect of liquid type on the quality of gluten-free muffins. *Chemical Technology* **61** (3) 27-31.

[14] Gacula M 1993 Product optimization. Design and analysis of sensory optimization, 105-235.

[15] American Association of Cereal Chemists. 2000. Approved Methods of American Association of Cereal Chemists. 10th ed. St. Paul, MN: AACC International, p 1200

[16] Mabesa L. 1986. Sensory evaluation of foods: Principles and methods. College of Agriculture, University of the Philippines Los Baños.

[17] Association of Official Analytical Chemists. 1980. *Official Method of Analysis.* (Washington, DC: Pergamon Press.) p 1080

[18] Hadiyanto H, Asselman A, Van Straten G, Boom R, Esveld DC and Van Boxtel A 2007 Quality prediction of bakery products in the initial phase of process design. *Innovative Food Science and Emerging Technologies* **8** 285–298.

[19] Ureta M, Olivera D and Salvadori V 2014 Baking of muffins: Kinetics of crust color development and optimal baking time. *Food and bioprocess technology* **7** (11) 3208-16.

[20] Mizukoshi M, Maeda H and Amano H. 1980. Model studies of cake baking. II. Expansion and heat set of cake batter during baking. *Cereal Chemistry* **57** (5) 352-355

[21] Sani N, Taip F, Kamal S and Aziz N. 2014. Effects of temperature and airflow on volume development during baking and its influence on quality of cake. *J. Engineering Science and Technology* **9** (3) 303-313.

[22] Mildner-Szkudlarz S, Bajerska J, Gornas P, Segliņa D, Pilarska A and Jesionowski T 2016 Physical and bioactive properties of muffins enriched with raspberry and cranberry pomace powder: A promising application of fruit by-products rich in biocompounds. *Plant Foods for Human Nutrition* **71** (2) 165-173.

[23] Dwyer E and Gallagher E 2001 Functional ingredients as fat replacers in cakes and pastries. Research report No 41. The National Food Centre.

[24] Lauterbach S and Albrecht J. 1994. NF94-186 functions of baking ingredients.

[25] Khouryieh H, Aramouni F and Herald T 2005 Physical and sensory characteristics of no-sugar-added/low-fat muffin. *J. Food Quality* **28** (5-6) 439-451.

[26] Sinha N 2007 Handbook of food products manufacturing, 2 volume set. (California : John Wiley & Sons)

[27] Hui Y and Nip W 2006 *"Sweeteners," in Bakery Products: Science and Technology*, ed W. Zhou and Y. H. Hui (Oxford : Wiley-Blackwell) pp 137– 160

[28] Brown A 2014 *Understanding food: principles and preparation.* (Belmont : Nelson Education)

[29] Martinez-Cervera S, Salvador A and Sanz T 2014 Comparison of different polyols as total sucrose replacers in muffins: Thermal, rheological, texture and acceptability properties. *Food Hydrocolloids* **35** 1-8.

[30] Raeker M. 1994. Functional and cake-baking properties of egg white, bovine blood plasma and their protein fractions. Iowa State University. University Microfilm International. A Bell and Howell Information Company.

[31] Gao J, Brennan M, Mason S and Brennan C 2017 Effects of sugar substitution with "Stevianna" on the sensory characteristics of muffins.Hindawi Publishing Corporation. *Journal of Food Quality.*

Evaluation of empowerment program to increase production capacity of fishery processing business in Semarang City, Indonesia

F Swastawati[1], W Roessali[2*], I Wijayanti[1], and A D Anggo[1]

[1]Faculty of Fisheries and Marine Sciences
[2]Faculty of Animal and Agricultural Sciences, Diponegoro University
Tembalang Campus, Semarang 50275 – Indonesia

*Email : wroessali@gmail.com

Abstract. This study aims to evaluate the empowerment program to increase the production capacity of fishery product processing. Empowerment program was the implementation and utilization of science and technology in the area (IPTEKDA) LIPI Indonesia for Higher Education. Activity carried out in 2016 on fish processing industry "Lumintu Group". Implementation of activities includes the transfer of technology to increase production capacity, business capital assistance in the form of production equipment, production assistance, and business management. This study uses qualitative, descriptive analysis, data collection with observation, interviews, and questionnaires. The results showed that the total number of active members was 24 people, 50% of the members specially cultivated the smoked fish that is the type of Catfish (*Arius thalassinus*) and Stingray (*Dasyatis sp*), while 45.83% of members processed boneless milkfish, and 4,17% produce salted fish. Increased average production scale of 31.82% in smoked fish business, 12.4% in boneless milkfish and 38.89% in salted fish business. Willingness to return capital in the good category, meaning that all members were able to carry out the schedule of relative payback on time. Approximately 83.3% of the group members felt that the program that followed had greatly assisted in increasing the scale of business but hoped to improve skills in terms of processing and marketing.

***Keywords**: empowerment, capital, improvement, production, fish processing*

1. Introduction

One of the regional economic resources is the fishery sector. Increased population along with increased public incomes and increased appreciation of healthy food, which encouraged increased consumption of white meat [1]. Fish is the cheapest source of animal protein [2]. Utilization of fish resources will not produce benefits and high economic value if not followed by good business processing and marketing activities. Small and medium scale enterprises (SMEs) are generally regarded as the engine of economic growth and equitable development in developing economies [3].

The high level of public consumption of fish makes the business processing opportunities also more open. Processing can be defined as the method applied to the product from harvest to consumption period [4]. Processed fish products that are known among them are smoked fish, abon, salted fish, meatballs, nuggets, and skin crackers. Fish harvesting, handling, processing, and distribution provides the livelihood for many countries [5]. Fish is a perishable food material and its flavor and texture changes rapidly during storage after death. If fish is not sold fresh, preservation methods should be applied to extend it shelf-life. These include freezing, smoking, drying and beat treatment (sterilization, pasteurization, etc). The processing and preservation of fish were very important when top quality, maximum yield and highest possible profits are to be achieved [6]. Smoke fish is one way of processing fish that serve to preserve and give the aroma with a distinctive taste [7].

Fish processing industry is generally still done traditionally and small scale. Variations of processed fish in the year 2013-2014 the amount of production is increasing. The growth of processed fishery products in Semarang City during the period of 2010-2014 (Table 1).

Table 1. Processed Fishery Product Data of Semarang City (2014) in kilogram.

PROCESSED PRODUCTS (kg)	Year				
	2010	2011	2012	2013	2014
Bone-pulls Milkfish	-	26,788	27,591	49,920	37,800
Boneless Milkfish	-	-	878,623	944,660	2,107,918
Grilled fish cake	-	21,353	22	35,424	35,940
Smoke Fish	3,898	-	-	7,063,566	6,969,156
Fish boiled spice	679	2,243	2,310	2,400	2,400
Salted	38	49,837	51,332	306,528	2,434,728
Grilled marinade	-	36,365	37,520	48,108	44,760
Shrimp paste	1,267	-	-	903,936	862,200
Peeled shrimp	97,200	-	-	1,056,780	1,046,000
Others	1112	10855	10181	308496	224138

Empowerment program aims to improve the independence of fisheries processing business actors. Associated with investment capital has been programmed during the year 2016 through empowerment program of the Application of Application and Utilization of Science and Technology In Regional (IPTEKDA) LIPI for Higher Education. The purpose of this program is to strengthen the performance of Micro Small and Medium Enterprises (MSMEs) sector in order to increase productivity and competitiveness through science and technology support scheme, capital, and entrepreneurship. Characteristics of this program are educational or not charity, science and technology oriented, and institutional strengthening of technology intermediation transfer group (KIAT) and sustainability in accordance with the local wisdom of each region.

Despite the fact that the fishery sub-sector accounts for a high enough percentage of Gross Regional Domestic Product of Central Java [8], most of the fish farmers including poor people and small-scale fish processing business operate with limited capital, traditional production methods. So, the empowerment program through technology transfer becomes important in order to improve the methods and management of production to improve business productivity. Even though, development of a universal measure of individual empowerment may not be a worthy or appropriate goal, because empowerment differs between individuals, context and time [9]. Individual empowerment measurements can give different results because empowerment can manifest in various forms of perception, behavior, competence, and action, and moreover, it may fluctuate over time [10].

One fish processing center is located in West Semarang District, especially in Krobokan Village. The fish process craftsmen in this village have formed a production group called Lumintu Group. In this group the production of preserved fish consists of smoked fish, soft-bone milkfish and salted fish. Types of fish that are usually preserved by fumigation are catfish (*Arius thalassinus*) and stingrays (*Dasyatis sp*). Milkfish especially processed to boneless milkfish and bone pulls milkfish, although sometimes produced grilled fish cake they called "*otak-otak*". Like most home industries is the scale of small businesses. Business scale is slow growing because business actors are constrained by the production equipment, such as refrigerator, high pressure cooker, and working capital. The purpose of this study is to evaluate the empowerment program to increase the production capacity of fishery product processing in Krobokan Village, Semarang Barat Sub-district, Semarang City.

2. Material and Mathod

This research was conducted from July to December 2016 in Krobokan Village, Semarang Barat Subdistrict, Semarang City. The method used in this research is the case study [11]. Census respondent collecting was done at Lumintu Group fish processing. The number of respondents was 24 people. Data Collection was obtained directly through interviews to the respondents by using questionnaire.

Methods of data analysis determine the response to the implementation of the program and perceptions of respondents to business continuity. Analyzed data was used the descriptive method. A five point Likertscale was used to elicit information on the level of responses to empowerment programs and perception of the respondents. The respondents were required to indicate the extent to which they agree or disagree with carefully constructed statements. The scale was indicated as follows: Strongly agree=5, agree=4, indifferent=3, disagree=2 and strongly disagree=1.

3. Result and Discussion

3.1. Socio-economic characteristics of respondents

Results presented in Table 1 illustrate the socio-economic characteristics of fish processors at Lumintu Group in Semarang city. The results showed that the mean age of respondents was 47.1 year. The implication is that the fish processors in the study area is constituted potential labor force[8]. As it is evident from the significantly high frequency of involvement of women in fish processing business, this implies that fish processing business is female dominated in the study area. About 66.7% of the respondents were female while only 33.3% was male. This is in agreement with [12] and [13], that woman were primarily responsible for post-harvest activities in fishery sector.

3.2. Effect of empowerment on production scale

Empowerment program whose main purpose is to increase production and productivity so that additional income can be achieved for family welfare. Table 2 shows that empowerment program has been able to increase business-scale of smoked fish business by 31.8%, 12.4% in boneless milkfish business and 38.89% in salted fishery business. Increased production is supported by technology transfer activities included in the empowerment program. Better technology makes fresh fish can be stored for the desired period without significant loss of quality. High-quality products and customer satisfaction can be achieved by fulfilling processing parameters ranging from operating to the storage and distribution of the final product. [14] and [15] have been noted that people not only adopt the technology as it is available to them. Even when technology is available and appropriate, decisions to adopt or not be influenced by personal and socio-cultural.

Table 2. Socio-economic characteristics of respondents

Variables	Frequency	Percentage	Mean
Age (year)			
Less than 40	6	25	47.1
40-49	5	20.83	
50-59	8	33.33	
>60	5	20.83	
Gender			
Male	8	33.3	
Female	16	66.7	
Educational Status			
No formal education	4	16.67	
Primary education	5	20.83	
Secondary education	4	16.67	
Tertiary education	8	33.33	
High education	3	12.50	
Processing fish experience			18.5
<10 Year	4	16.67	
10 -20 Year	8	33.33	
21-30 Year	10	41.67	
>30 Year	2	8.33	
Family Member			4.04
2-3 person	7	29.17	
4-5 person	14	58.33	
>5 person	3	12.50	

The results of the evaluation showed that 50% of the members specially cultivated the smoked fish that is the type of Catfish (*Arius thalassinus*) and Stingray (*Dasyatis sp*), while 45.83% of members processed boneless milkfish, and 4.17% produce salted fish.

Table 3. Effect of empowerment on production scale.

Producers	Production		
	Before	After	Growth
Smoke fish	52.46	69.14	31.80
Boneless milkfish	21.29	23.93	12.40
Salted fish	10.8	15	38.89

3.3. Constrain faced by fish processing producers

Of the perceived constraints to raise fish processing business (Table 3), the inadequate capital was the most severe (100%), which confirmed the findings of [3] and [16]. All of the respondents felt that high cost of fresh fish was the major constraint. Aliyah et al. [17] had noted that supply raw material was always a problem because producers were not able to afford the high cost of fresh fish.

Innovation and skills became a problem for 91.7% of respondents. The condition of fish processing in Semarang city with the level of mechanization of fish processing is still traditional and limited production. This condition is also reported by [18].[19] suggests appropriate processing technologies to enable maximum use of raw materials so as to contribute to improving economic. Research [20]on coastal community empowerment (fisherman processing) shows the condition of processed fish with the scale of micro/small and medium enterprises are still relatively powerless in obtaining economic access. In addition [21], explaining their business situation has a low level of empowerment.

Table 4. Constraint faced by fish processing producers.

Perceived Constraints	Frequency	Percentage	Percentage Rank
Inadequate capital	24	100	1
High cost of fresh fish	24	100	2
High cost innovation	22	91.7	3
Poor managerial skill	22	91.7	4
Lack of technical skill	22	91.7	5
Poor extension service	20	83.3	6
Poor transport facility	20	83.3	7
High cost of labor	18	75	8
Poor marketing structure	12	50	9
Water scarcity	6	25	10
to repay the installment of capital	4	16.7	11

Willingness to return the installment of capital in a good category, meaning that all members are able to carry out schedule of relative payback on time. Approximately 83.3% of the group members felt the program that followed had helped a large scale increase of business, but hoped to improve skills in improving the quality of processing and marketing.

4. Conclusion

Based on the findings in this study, it is concluded that the implementation and utilization of science and technology in the area (IPTEKDA) LIPI Indonesia for Higher Education for empowerment program can increase the production capacity of fishery product processing. The fish processor has the will to implement the training program and complete the installment procedure as agreed in the activity. Support for the development of resources and production technology and innovation constraints is needed to increase the production of Indonesian fishery products..

Acknowledgement
Acknowledgments for the achievement of this study thanks to financial support from the program of science and technology application and utilization in the region (IPTEKDA) from the Indonesian Institute of Sciences (LIPI) for Higher Education.

References

[1] Kusumastanto T and PKSPL-IPB Team 2007Policies and Strategies for Increasing Productivity and Competitiveness of National Fisheries Products (in Bahasa Indonesia). *Makalah.Agrinex Conference and Expo,* Jakarta 17 Maret 2007
[2] Jamu D M and Ayinla A O 2003 Potential for the Development of Aquaculture in Africa *J. Naga, Worldfish Center Quarterly* **26** (3) 7-13
[3] Agwu O and Emeti C I 2014 Issues, Challenges and prospects of small and medium scaleenterprises (SMEs) in Port-Harcourt City, Nigeria. *European Journal of Sustainable Development* **3** (1): 101-114
[4] Kallon A, Lebbie A, Sturm B, Garnett T and Wadsworth R 2017 Comparative studies of fish smoking and solar drying in the Sierra Leone artisanal fishing industry.*Journal of Stored Products and Postharvest Research.* **8** (3): 40-48.
[5] Al-Jufaili M S and Opara L U 2006 Status of fisheries postharvest industry in the Sultanate of Oman: Part1 handling and marketing system of fresh fish. *J. Fish. Int.*, **1** (2-4): 144-149.
[6] Davies R M and Davies O A 2009 Traditional and improved fish processing technology in Bayelsa State, Nigeria. *Eur. J. Sci. Res.* **26** 539-548.
[7] Lombongadil GP, Reo A R and Onibala H 2013Quality study of mackerel (Dussumieria acuta C.V) dry smoke of household industry in Tumpaan Baru Village, District of Tumpaan (in Bahasa Indonesia). *Jurnal Media Teknologi Hasil Perikanan* **1**(2): 47-53
[8] Central Bureau of Statistics of SemarangBadan Pusat Statistik 2010. City of Semarang In Figures 2010. Semarang.
[9] Zimmerman M A 2000 Empowerment Theory: Psychological, Organizational and Community Levels of Analysis. In Handbook of Community Psychology; Rappaport, J., Seidman, E., Eds. (Newyork : Kluwer Academic/Plenum Publishers) pp. 43-63
[10] Wallerstein N 2006 *What Is the Evidence of Effectiveness of Empowerment to Improve Health* WHO/EURO Copenhagen, Denmark
[11] Wirartha M 2006 *Socio-Economic Research Methods* (in Bahasa Indonesia) (Denpasar : Andi Publisher)
[12] FAO 1992 *The role of woman in small scale fisheries: Occasional paper 5.* FAO Rome, pp: 4-10
[13] Bolorunduro P I 2003 Improved fisheries technologies and approaches for their dissemination: A case study of Nigeria State. *Proc. the 16th annual Conf. of the Fisheries Society of Nigeria,* November 4-9 (Maiduguri, Nigeria) pp: 74-79
[14] Berdegue J A and Escobar G 2001 *Rural diversity, agricultural innovation policies and poverty reduction.* Agren Network papers. No. 122.ODI.
[15] Rogers E M 2003 *Diffusion of innovations (5th edition)* (New York : Free Press)
[16] Shofa S U and Navastara A M 2015Determinants of Fishery Processing Industry Development in Sidoarjo Regency through Local Economic Development (in Bahasa Indonesia). *J. Teknik ITS* **4** (2) 76-80
[17] Aliyah R, Gumilar I, and Mauline I 2015Business Development Strategies of Processing Fish Floss (Case Study of Rumah Abonin Bandung) (in Bahasa Indonesia). *JurnalPerikananKelautan*VI(2):78-84

[18] Candra A N and Setiawan A H 2013The Analysis of business prospect and development of milkfish industry in Semarang City (in Bahasa Indonesia). *Diponegoro J. of Economics* **2**(1) 1-11

[19] Davies RM 2005 Developmnet of appropriate technology of fish processing in Nigeria. *Proc. of the Workshop on Intensive Fish Farming*, February 24th (Nigeria)

[20] Susilowati I, Sudaryono A, and Agustini, T W 2004 *Development of Coastal Community Empowerment Model (Micro Small Medium Enterprises and UMKMK Cooperatives) in Supporting Food Security in Pekalongan Regency, Central Java* (in Bahasa Indonesia). RUKK Research Report Year I, LPPM Diponegoro University, Semarang

[21] Anisah R N 2007The Analysis on Empowerment ofThe Market Oriented Fishery Manufacturers (An Empirical Study in Tegal Municipality) (in Bahasa Indonesia). Master's Thesis of Coastal Resources Faculty of Fisheries and Marine Sciences Diponegoro University, Semarang

Mapping performance of the fishery industries innovation: A survey in the North Coast of Java

M Yusuf[1*], A M Legowo[2], A N Albaarri[2], Y S Darmanto[3], T W Agustini[3] and A I Setyastuti[4]

[1]Department of Food Technology, University of Muhammadiyah Semarang, Jl. Kedungmundu Raya No.18, Kedungmundu, Tembalang, Semarang City, Central Java, Indonesia
[2]Department of Food Technology, Faculty of Animal and Agricultural Sciences, Diponegoro University, Jl. Prof. H. Soedarto, S.H. Tembalang, Semarang, Central Java, Indonesia
[3]Department of Fishery Product Technology, Diponegoro University, Jl. Prof. H. Soedarto, S.H. Tembalang, Semarang, Central Java, Indonesia
[4]Fisheries Science Programe, Faculty of Fisheries and Marine Science, University of PGRI Ronggolawe Tuban, Jl. Manunggal No. 61Tuban, East Java, Indonesia

*Email: m.yusuf@unimus.ac.id

Abstract. This study aimed to put the performance indicators of industry innovation fisheries which would be used as inputs to create innovation strategies in order to win the market competition, especially in USA. Survey and in-depth interviews were conducted on 10 industries with shrimp, tuna and crab commodities representing the fishery industry in Indonesia to USA export destination. The result of mapping performance of innovation indicators of Indonesian Fishery Industry resulted the 10's strategies alternative to win the market. Based on survey result indicate that "the regulation of catch and/ or harvest of cultivation factor" is considered the weakest factor in develop innovation with a score of 3.3, while the international trade factor are considered as the strongest factor in developing innovation development with scores 5,0.Aggressive strategy by strengthening the strength owned by the internal industry was by always looking at the opportunity, so that they could take the opportunity to win the market competition at the right time.

Keywords: *Mapping, Indonesian Fishery Industry, Innovation.*

1. Introduction

Indonesia is one of the world's largest exporter of fishery commodities, the average export value of fishery products from 2012-2016 has increased every year. The total potential of fish resources in Indonesia reaches 21.05 million tons by 2015 [1]. The export destination of Indonesian fishery products is USA with percentage of export value of 9.5% / year, EU of 6.0% / year and China of 3.2% / year during

2012-2016 [2]. USA is still the largest exporting country of Indonesian fishery products with a value of 1,843,813 (US $ 1,000) by 2014 [3]. Fishery exports in 2014 reached 1.27 million tons or increased by 1.34%, while fisheries export trend in 2008-2014 increased by 6.11% and in 2014 grew by 3.69%. The main commodities of Indonesian fishery exports are shrimp, tuna (big eye, yellow fin, albacore and blue fin), tuna (little tuna), skipjack, and crabs [4].

One of the determinants of the increase in export value of Indonesian fishery products is the industry's vision that is always oriented to the market. Market orientation focuses on consumer desires, competitiveness, superiority, and integrity of all functions or performance that exist in a company to create an innovation that is able to compete in the market [5-8]. Innovation can be applied to various aspects that become indicators of the performance of a fishery industry as proposed by [9], both from ecological, economic and social aspects. Knowing or mapping performance indicators of a fishery industry innovation is very important to do because from the mapping results, it can be known about the strengths, weaknesses, opportunities, and threats of a fishery industry as input materials to create a strategy to win the market competition. This study aims to put the performance indicators of industry innovation fisheries which will be used as inputs to create innovation strategies in order to win the market competition, especially in the USA.

2. Materials and methods

Data sampling was conducted on 10 Indonesian fishery industries i.e. Shrimps (3 industry), Tuna (3 industry), and Crabs (4 industry) were spread on special areas on capital of Jakarta, west java, central java and east java .Site visits to direct observations and depth interviews were conducted on the performances indicator of fishery industries. Direct observations and depth interviews carried out with major industry actors and their stakeholders of fishery industries. The samplings were conducted during April-July 2017. Furthermore, the performance indicators were compiled by the rated a score of (1-5), where the value 1 was very weak performance / not well-performing / not effective and 5 very strong performance / well-performing / effective. The questionnaires of fishery assessment standards Performance Indicators (FPIs) adopted from [9] and has been modified.

Indonesian Fishery Industries (IFI) performance data were analyzed by descriptive and presented in the chart radar, a graphical method of displaying multivariate data in the form of two-dimensional graphs of variables measured to show the advantages and disadvantages of factors. The radar graphs with geometric projection method were to express the distribution of the weakest and the strongest factors in a multidimensional space. SWOT analysis was used as a method to map the attribute value as a comprehensive representation of the performance of the industry's competitiveness.

3. Results And Discussions

The results of the research survey and secondary data study showed that thefishery industry in Indonesia was dominated by the industries which exported shrimp and lobster (3 industries), tuna (including skipjack and little tuna) (3 industries) and crabs (4 industries). Table 1 showed the performance of the Indonesian fishery industry.

Table 1. Indicators of Indonesian fishery industry.

No	Aspects	Indicators
1	Ecological	The availability of raw materials
2		Regulation of the catch and / or the harvest of aquaculture
3	Economics	Infrastructure handling and the fish landing
4		Risk ofproductdamage
5		International trade
6		Access and market network expansion
7		Adaptation to regulations
8		Adaptation to market trends
9		Protectionagainstcompetitors' products
10		Product quality and safety
11		Product development
12		Consumer service
13		Distribution control
14		Financial capital
15	Social	Social improvement of workers' welfare
16		Training on quality improvement of workers
17		Workers' health facilities and infrastructure
18		Sanitation and hygiene facilities
19		Corporate Social Responsibility (CSR) program
20		Increasedenvironmentaleconomicvalue

The performance of the fishery industry in Indonesia could be known from the resources of strengths and weaknesses, opportunities and challenges covering all aspects from the ecological (fish stock & sustainable), harvest sector performances, and social (post-harvest sector performances). The details of the performance positions of fishing industry in Indonesia were examined from factors derived from these three aspects as adopted from [9]. For moredetails, it could be seen in Figure 1.

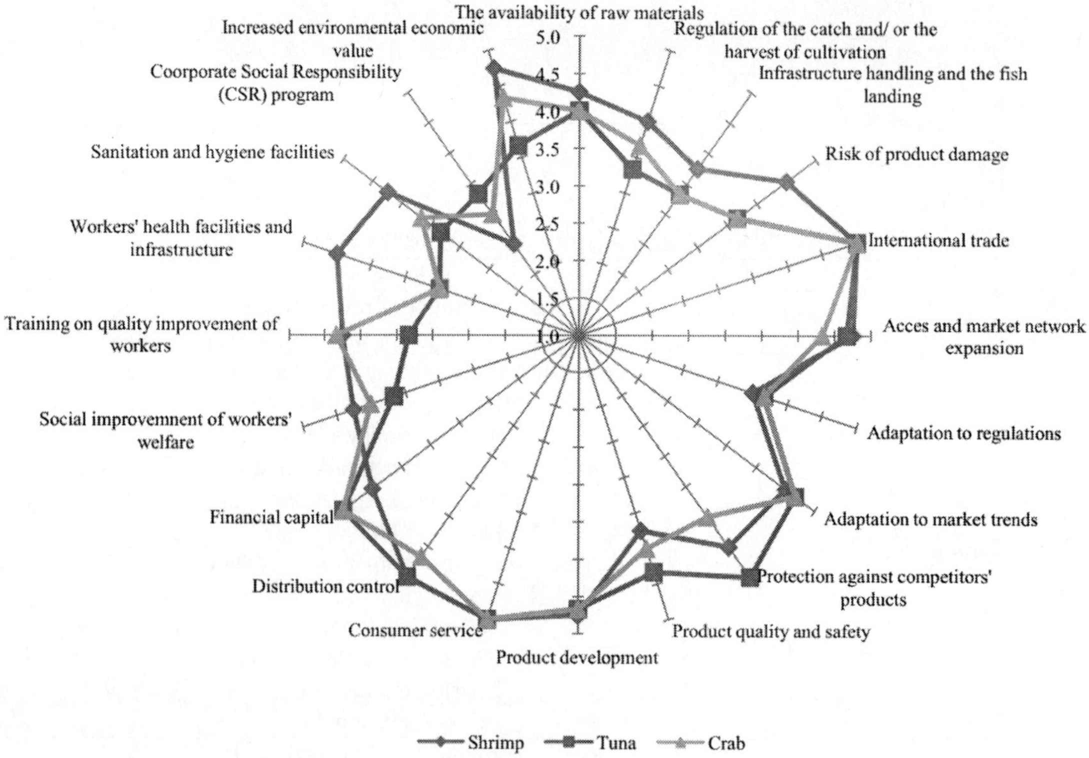

Figure 1. Mapping performances innovations of Indonesia Fishery Industry.
Value : average from 3 shrimps industries, 3 tuna industries and 4 crab industries.

3.1. Ecological factors,

Ecological factors are included the availability of raw materials and supply chains or distribution channels of fishery industry raw materials and the catch or harvest regulations. Thefisheryindustry in Indonesia was a complex system that could be described as upstream and downstream systems. Upstream system as a provider of industrial raw materials in the form of fish from the catching sector with production volume reached 6.52 million tons in 2015 and aquaculture with production volume reached 14.53 million tons in the same year[1]. The shrimp and lobster industry in Indonesia was excellent for the export market in the USA, during 2010-2014 Indonesia's shrimp exports to the USA averaged 16.92% each year with a value of 32.42% recorded in 2014 value of Indonesia's shrimp exports to USA reached 1,283 .880 (US $ 1000) [3]. The increase of aquaculture fishery could be influenced by several factors such as continuous demand factor by industry and raw material supply as well as regulatory factors related to fishing. The availability of raw materials for shrimp industry was not very influential when compared with tuna and crab industries. This was because the shrimp was one of the cultivated commodities that could be self-reproduced, so it was not so affected by the existence of ecological issues such as over fishing, Illegal Unreported Unregulated (IUU) Fishing, as well as other environmental issues. Lobsters were generally marketed in fresh form while shrimp was in the form of frozen. The types of shrimp exported in Indonesia were generally in the form of: Whole / Head On (HO), Head Less (HL), Peeled, Peeled Tail On (PTO),

Peeled and Deveined (PND), Peeled Undeveined (PUD), Butterfly, or a diversified product under the name of "Ebi Panko" (Breaded Shrimp).

Indonesia tuna export to USA during2010-2014 average per year reached 4.15%, this value was smaller when compared to the shrimp exports. The export value of tuna to USA in 2014 was 128.297 (US $ 1,000) [3]. Tuna products from Indonesia were generally marketed in the form of fresh tuna, frozen tuna or processed and preserved tuna such as whole, gilled and gutted, fillets, steaks, sticks, loin, pocket and canned variants. The ecological factors of development of the world's tuna industry faced global issues such as Illegal, Unreported and Unregulated (IUU) fishing, environmental degradation, climate change, competition with other tuna exporting countries, and trade barriers [10]. Reported by [11], about 23 stocks of tuna above 60% were fully exploited, 35% were overexploited or exhausted and the rest was exploited but very few were reported. Consideringthisfact, a Regional Fisheries Management Organizations (RFMO) was established, specifically the tuna 5 RFMO had been established, namely IOTC (Indian Ocean Tuna Commission), CCSBT (Commission for the Conservation of Southern Bluefin Tuna), WCPFC (Western and Central Pacific Ocean Highly Migratory Fisheries Commission, IATTC (Inter-American Tropical Tuna Commission (Eastern Pacific Ocean), ICCAT (International Commission for the Conservation of Atlantic Tunas) aiming to ensure tuna stocks above Maximum Sustainable Yield (MSY) [12]. As the actions to the issues, tuna fishery industry in Indonesia had made several ideas such as: 1). Adjustment of Tuna Capture Targets to be included in the RFMO category, recorded in 2011 there were 800 units (45.36%) ships measuring 100-200 GT, 500 units (28.14%) vessels measuring 60-100 GT and 295 units (16.60%) measuring 30-60 GT. 2) Utilization of registered fishing equipment that was Longline as many as 1414 units (79.57%), purseine as much as 166 units (9.34%) and others such as pole and line (1.01%), gill nets (0.11%), support vessel (3.60%) , Unknown (6.36%) [10].

Crab was one of the important fishery export commodities in Indonesia with the aim of its largest export country, USA. Crabs in Indonesia were generally exported in the form of pasteurization in cans. Supply chains and distribution channels of crab industry in Indonesia were generally managed directly by the industry by way of having a crab catching vessel leased to local fishermen, or by establishing mini plants in the crab producing areas that receive crabs from local fishermen. In the mini plant, the crab raw material was cleaned, steamed and then peeled and stored into the cooler before being sent at the plant (main plant) to be canned. The operations of capture and / or crab processing in Indonesia were spread in the North of Java Island, East Sumatra, Kalimantan and South Sulawesi. The northern territory of Java Island dominated for the crab industry, this was due to the ease of access to capture, distribution and other economic activities that supported the industry. This phenomenon made the tightening of industrial competition that led to the scarcity of raw materials, industrial crab efforts in meeting the needs of industrial raw materials by raising the purchase price at the level of fishermen or collectors but with the high purchase price of the treated industries also apply quality standards / quality products strict. It's similar type with [12], the raised price of raw material caused by the competitions of raw materials.

3.2. Economic factors

What became the serious concern to the perpetrators of the Indonesian fishery industry in the economic factors were the fishing and fish landing infrastructure, product quality and level of food security, and adaptation to market trends,but the other economic factors could not be ruled out. The fish handling and landing infrastructure were the first factor because of the good handling of crops or catches made good end product quality, so that the price of commodity of fishery industry was also higher. In the shrimp and crab industry in Indonesia, generally handling and harvesting infrastructure level was good category meaning that the company had a good understanding of the quality and standard of good product quality by applying a product quality assurance system such as Good Manufacturing Practices (GMP), Good Handling Practices (GHP), or Hazard Analysis Critical Control Point (HACCP). While in the tuna industry, the category of fish handling and landing infrastructure were still in the middle, this was because the human resources still had low understanding about the quality of the product, so that the level of application of quality assurance and food security system was still low.

From a food safety standpoint, shrimp and crab commodities faced the greater issues related to food safety products. The use of chloramphenicol (CAP) was a very strong issue for the shrimp and crab industries. Chloramphenicol (CAP) was widely used by shrimp farmers in Southeast Asia including Indonesia, and this was what caused the failure of Indonesian shrimp exports to various countries in the world [13]. As for thecrab industry during 2002 to 2013, there were 381 cases of rejection of Indonesian crab exports to the USA where 171 cases were dominated by Chloramphenicol [14]. CAP was an antibiotic used in shrimp culture that had accumulated properties in the body when we atetheshrimp or other fishery products that had been exposed to CAP. If CAP accumulated in the body for a long time, it would cause anemia and cancer [15]. The various efforts could be made to minimize the CAP content of shrimp and crabs such as the process of shrimp depurization before being processed or sent to the factory [13], but this process required greater production time and cost. During this time,the shrimp and crab industry had made efforts to minimize the CAP by way of product assessment in the laboratory either in their own factory or a competent laboratory to issue the Product Health Certificate as one of the export requirements. The trends of world trade, especially fishery products and their processed products would remain a commodity that could compete with food commodities (meat, milk, eggs, bread and others) with market share of about 31% until 2024 [16]. The price of fishery products was also influenced by demand and supply factors, including production and transportation costs, also including substitutes such as meat and flour. The changes in diet and lifestyle, the increased income, and consumer welfare also affected the demand for fishery products.

3.3 Social factors,

Consisting of improving workers' welfare, training on quality improvement of workers, health facilities and infrastructure, sanitation and hygiene facilities, Corporate Social Responsibility (CSR) programs, as well as improving the welfare of communities around the industry. Some points that needed to be improved by Indonesian fishery industry were especially in health worker facility and infrastructure, sanitation and hygiene facility and CSR program. Sanitation and hygiene facilities needed to be improved due to the basic eligibility requirements for establishing a quality assurance and product safety system (HACCP system). Based [17], the hygienic practice requirement was an essential feasibility program essential for the fishing industry and fish processing prior to the HACCP implementation. During this time, the Indonesian fishery industry to improve the sanitation and hygiene levels of production sites in cooperation with the competent authorities to assess the sanitary and hygienic levels of production sites for a certain period depending on the scale of sanitation and hygiene assessment. For instance, the level of perfect assessment could be done once a year, the mediumlevel could be done every 6 months, and once a low rate was done every 3 months. The point of Corporate Social Responsibility (CSR) of Indonesian fishery industry was still low, but indirectly for the existence of the fishery industry could improve the welfare of the community around that was with the growth of other economic activities that were able to support the existence of the fishery industry.

3.4. SWOT Analysis

The results of mapping the performance of the innovation of the Indonesian fishery industry produced several notes for Strength, Weakness, Opportunities and Threats for the development of Indonesian fishery industry strategy to be able to compete in global markets, especially USA. The following was SWOT analysis of Indonesian fishery industry. Table 2 showed the SWOT of the above industry performance indicators.

IOP Conf. Series: Earth and Environmental Science **102** (2017) 012083 doi:10.1088/1755-1315/102/1/012083

Table 2.SWOT analysis of the performance of the Indonesian fishery industry innovation.

No	Factors	PerformanceIndicators
1	Strength – 1 (S1)	International trade
2	Strength – 2 (S2)	Access and expansion of the market network
3	Strength – 3 (S3)	Adaptation to market trends
4	Strength – 4 (S4)	Product development
5	Strength – 5 (S5)	Control distribution
6	Weakness – 1 (W1)	Fish handling and landing infrastructure
7	Weakness – 2(W2)	Workers' health facility and infrastructure
8	Weakness – 3(W3)	Sanitation and Hygiene Facilities
9	Weakness – 4(W4)	Training of workers' quality improvement
10	Weakness – 5(W5)	Corporate Social Responsibility (CSR) Program
11	Opportunities – 1 (O1)	Adaptation to theregulations
12	Opportunities – 2(O2)	Increasingenvironmentaleconomicvalue
13	Opportunities – 3(O3)	Product quality and safety
14	Opportunities – 4(O4)	Consumer service
15	Opportunities – 5(O5)	Increasingworkerswelfare
16	Threats – 1 (T1)	The availability of raw materials
17	Threats – 2 (T2)	Regulation of the catch and / or the harvest of aquaculture
18	Threats – 3 (T3)	Risk ofproductdamage
19	Threats – 4 (T4)	Financial capital
20	Threats – 5 (T5)	Protectionagainstcompetitors' products

Table 3. Strategies Alternative.

Strategies alternative	Code
SO strategies	
1. Always follow the international trade regulation by enhance the resources owned.	S1O1
2. Expand the market access by prepared the quality products, product development and improving customers service.	S234O34
WO strategies	
3. Improving the infrastructure supporting industry and workers ' health, sanitation and hygiene facilities to produce the high quality product.	W12O3
4. Training of workers' quality improvement to produce the high quality and safety based on export standard.	W4O3
5. Enhancing the CSR program to increasing environmental economic value.	W5O2
ST strategies	
6. Updating the international trade with connecting the regulation of the catch and / or the harvest of aquaculture products.	S1T2
7. Controlling the distribution strat from the availability of raw materials until risk of product damage and protection against competitors' products.	S5T135
8. Enhancement of financial capital to expand the market networks	S2T5
WT strategies	
9. Improving the fish handling and landing infrastructure to ensure the availability of raw material	W1T1
10. Enhancement of financial capital to improving the fish handling and landing infrastructure, workers' health facility and infrastructure, sanitation and hygiene facilities and training of workers' quality improvement to minimize the risk of product damage.	W1234T34

4. Conclusions

The Indonesian Fishery Industry that exported to the USA faced various problems both from the internal industry itself and the external from world market trend, especially USA. The lowest performance indicators value are adaptation to regulations, social improvement of workers' welfare, workers' health facilities and infrastructure, the sanitation and hygiene facilities and CSR programs. The highest performance indicators value is international trade, adaptation to markets, the product development, and distribution control. The strategy adopted by the Indonesian Fishery Industry to maintain and improve the competitiveness in the USA market was by means of aggressive strengthening Strength which was owned by the internal industry by always looking at the Opportunities that existed, so that later on the right time could take the opportunities for winning the market competition.

Acknowledgement
This research was funded by Ministry of Research, Technology and Higher Education with Post Doctoral Scheme Project in 2017.

References

[1] Ministry of Maritime Affairs and Fisheries. 2017[a]. Accesed http://statistik.kkp.go.id/sidatik-dev/index.php?m=1
[2] Ministry of Maritime Affairs and Fisheries. 2017[b]. Accesed http://kkp.go.id/2016/09/27/nilai-ekspor-hasil-perikanan-menurut-negara-tujuan-utama-periode-januari-juni-2012-2016/
[3] Ministry of Maritime Affairs and Fisheries. 2015[b]. *Marine and Fisheries in Figures 2015*. Ministry of Maritime Affairs and Fisheries. Indonesia. Accesed http://statistik.kkp.go.id/sidatik-dev/Publikasi/src/kpda2015.pdf.
[4] Ministry of Maritime Affairs and Fisheries. 2015[a]. Ministry of Maritime Affairs and Fisheries.Accesed http://statistik.kkp.go.id/sidatik-dev/Publikasi/src/analisisdatakkp2015.pdf
[5] FAO and OECD 2011 Price Volatility in Food and Agricultural Markets: Policy Responses. Accesed http://www.amis-outlook.org/fileadmin/templates/AMIS/documents/Interagency_Report_to_the_G20_on_Food_Price_Volatility.pdf
[6] Grunert KG, Jeppesen LF, Jespersen KR, Sonne A-M, Hansen K, Trondsen T and Young JA 2005 Market orientation of value chains: A conceptual framework based on four case studies from the food industry. *European Journal of Marketing* **39** (5) 428-455.
[7] Trondsen T 2012 Value chains, business conventions, and market adaptation: A comparative analysis of Norwegian and Icelandic fish exports. *The Canadian Geographer* **56** (4) 459-473.
[8] Yusuf M and Trondsen T 2013 A market-oriented innovative quality framework for the investigation of competitive entry opportunities into new seafood markets for producers. *Int. J. of Quality and Innovation*, **2**(2) 175–192.
[9] Anderson J L, Anderson C M, Chu J, Meredith J, Asche F, Sylvia D, et al 2015 The Fishery Performances Indicators : a Management Tool for Triple Bottom Line Outcomes. *PLoS ONE,* **10** (5) : e0122809. DOI : https://doi.org/10.1371/journal.pone.0122809
[10] Sunoko R and Huang H W 2014 Indonesia tuna fisheries development and future strategy.*Marine Policy* **43** 174 – 183 DOI : https://doi.org/10.1016/j.marpol.2013.05.011
[11] Collette BB, Carpenter KE, Polidoro BA, Juan-Jorda MJ, Boustany A and Die DJ. 2011. High Value and Long Life-Double Jeopardy for Tunas and Billfishes. *Science* **333** 291.
[12] Yusuf M and Trondsen T. 2014. Competitive forces and innovation strategies : a study of the Indonesian crab industries. *J. Agribusiness in Developing and Emerging Economics* **4** (1) 78-96.

[13] Suseno H, Hudiyono P W S and Muslim S 2016 Elimination of Chloramphenicol by Back Shrimp (*Penaeusmonodon*) and White Shrimp (*Litopenaeusvannamei*) (in Bahasa Indonesia) *HAYATI J Biosci.***23** (3) 117-120 DOI : https://doi.org/10.1016/j.hjb.2016.07.001

[14] Fahmi A S, Maksum M and Suwondo E 2015 USFDA Import Refusal and Export Competitiveness of Indonesian Crab in US Market. *Agriculture and Agricultural Science Procedia* **3** 226 – 230

[15] Conti G , Copat C, Wang Z, D'Agati P, Cristaldi A and Ferrante M 2015 Determination of Illegal Antimicrobials in Aquaculture Feed and Fish : an ELISA Study *Food Control* **50** 937-41

[16] Bellman C, Tipping and Sumaila U R 2016 Global Trade in fish and fishery products : An overview *Marine Policy* **69** 181-188 DOI : https://doi.org/10.1016/j.marpol.2015.12.019

[17] Ababouch L. 2014. *Food Safety Assurance Systems : Good Practices in Fisheries and Aquaculture.* (Rome : Elsevier Inc.)

Physical, chemical and sensory properties of brownies substituted with sweet potato flour (*Ipomoea batatas* L.) with addition of black cumin oil (*Nigella sativa* L.)

I P Ligarnasari, C Anam* and A P Sanjaya
Department of Food Science and Technology, Faculty of Agriculture, Sebelas Maret University

*E-mail : choiroelanam@staff.uns.ac.id

Abstract. Effect of addition black cumin oil on the physical (hardness) characteristics, chemical (water, ash, fat, protein, carbohydrate, antioxidant IC50, total phenol and active component) characteristics and sensory (flavor, taste, texture, overall) characteristics of brownies substituted sweet potato flour were investigated. Substituted brownies was added with 0.05%, 0.10%, 0.15%, 0.20% and 0.25% of nigella sativa oil. The result showed that water content, ash, protein, fat, total phenol were increased and carbohydrate, antioxidant IC_{50} was decreased by the addition of nigella sativa oil. Due to the sensory characteristics, panelist gave the high score for substituted brownies which was added 0.05% nigella sativa oil. The result showed that the best formula of substituted brownies which was added 0.05% of nigella sativa oil had 24.89% water content, 1.19% ash content, 7.54% protein content, 37.79% fat content, 53.06% carbohydrate contain, 1043.6 ppm IC_{50} antioxidant and 0.22% total phenol. The active component on the brownies using GCMS identification were palmitic acid, oleic acid, lauric acid, theobromine and vitamin E.

Keywords: brownies, component, cumin, antioxidant

1. Introduction

Free radicals are generated through external sources by unhealthy lifestyles and environmental factors, which can lead to cell degradation. Free radicals are generated through external sources by unhealthy lifestyles and environmental factors, which can lead to cell degradation. Free radicals in the body are responsible for the rapid evolution of more than 100 kinds of diseases (cardiovascular, neurological, endocrine, respiratory, immunity, ischemia, gastrointestinal disease, tumour growth and cancer) [1]. Many efforts can be done to reduce the presence of free radicals in the body, one of them is consuming healthy foods rich in antioxidant content. One of the ingredients rich in antioxidants is purple sweet potato flour, which contains 36.8% of antioxidants. One of the applications of sweet potato flour is the cake flour substitute ingredients. Brownies are a kind of solid chocolate cake, originally a flour and hard dough made from wheat flour, eggs, fat, granulated sugar and chocolate by baking [2] . Brownies usually contain less balanced nutrients, therefore, it is regarded as food that contains a less nutritional value that can be utilized by the body. One way to improve the brownies' nutritional value is to use sweet potato flour as the cake flour substitute ingredients. Besides purple sweet potato powders, spices and medicinal plants also have great potential as a source of food that can raise the functional value of antioxidants and products [3]. One of the herbs that contain antioxidants in Indonesia is black cumin. Along with adding nutritional properties of the product, black cumin oil also utilized due to low consumption and utilization in the product. The black cumin is known by its seed which has special healing properties as known in the Quran (Al-Bukhari: 815) which states that black cumin seeds are used to treat all kinds of diseases except death itself [4]. Black cumin seed oil is also regarded for its high-level of anti-oxidant activity. Antioxidant activity of black cumin oil is caused by the fatty acid content of oleic fatty acids, tocopherol, phenol, and β-carotene. The antioxidant activity of black cumin oil is associated with the phenolic compound. It has been reported that olive oil is the only vegetable oils that contain higher phenolic content than black cumin oil [5]. The addition of black cumin oil to purple sweet potato brownies products is expected to increase the antioxidant content of the product, know the best addition of black cumin oil which was added substituate sweet potato flour brownies.

2. Materials and Methods

2.1. Materials

Black cumin oil is obtained from CV. Lansida Yogyakarta (Indonesia). Twenty five percent of Purple Sweet Potato Flour [2] used for brownie the brownies flour substitute ingredients are obtained from CV.Kusuka Ubiku, Yogyakarta (Indonesia). Supporting materials such as "Segitiga Biru" flour, dark cooking chocolate "Cholata", sugar, egg, margarine, and cooking oil "Filma" are obtained from Pasar Gede Solo (Indonesia).

2.2. Characteristics Analysis of Raw Material

The Purple sweet potato flour's water content is identified with thermogravimetry method, ash content with dry method, fat with the micro-soxhlet method, the protein with micro-Kjeldahl method [6] and antioxidant IC50 with DPPH method [7]. Black cumin oil's density is determined using pycnometer [6], IC50 antioxidant by DPPH method [7] and components identified by GCMS method [8]

2.3. Brownies production

Brownies production was divided into two steps. The first step, dark cooking chocolate was melted with 110 ml cooking oil. The second step, 3 eggs were whipped with sugar till change into white colour and expand. Then, first step and second step was mixed and was added with 56.25 wheat flour, 18.75 sweet potato flour, 10 g chocolate powder, 80 ml sweetened condensed milk, 1 g vanili, 1 g salt and nigella sativa oil (0.05%, 0.10%, 0.15%, 0.20%, and 0.25%). Then, brownies dough was poured on a baking sheet, then was steamed for 45 minutes.

2.4. Characteristics Analysis of Brownies Product

Brownies product was investigated for physical, chemical and sensory properties. Hardness was investigated as physical properties by Llyod Universe Testing Machine [9]. Chemical properties *i.e* thermogravmetry was used for water content [6], the oven was used for ash content [6], mikro-soxhlet was used for fat content [6], micro-kjedahl was used for protein content [6], by difference was used for carbohydrate contain [6], DPPH was used for antioxidant IC50 [7], GCMS Agilent Technologies Model 6890N was used for identify component product [8]. Scoring was used for sensory properties [10]. All these studys were performed in triplicate, except GCMS identify. After that, brownies product was determined the best formula by compensatory method [11]. The principle of compensatory testing technique gives a score in accordance with the contribution of a characteristic to the resulting formula. The best formula was based on the highest value of all formula.all formulas.

3. Results and Discussion

3.1. Chemical and Physical Properties of Brownies Product

The raw material of purple sweet potato flour and black cumin oil are tested before being processed into brownies. Purple sweet potato flour contains 10.37% water content, the ash content of 0.92%, the fat content of 2.83%, 4.35% protein content, and the carbohydrate content of 92.23%. The IC50 antioxidant content is 1591.5 ppm. Black cumin oil contains 412.2 ppm of antioxidants, a density of 0.922 g / ml and the identified components are oleic acid , linoleic acid (57.24%), lauric acid, palmitic acid (3.11%), limonene (2.57%), jupine (0.83%) and thymoquinone (13.80%), while the highest component carried is linoleic acid

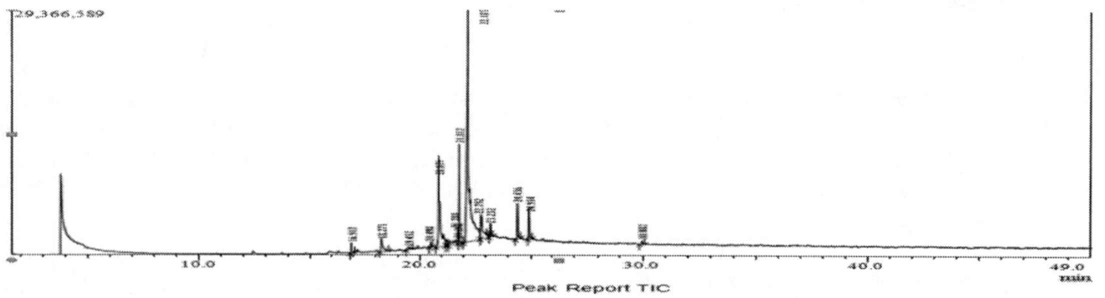

Figure 1. Chromatogram Gc-Ms of black cumin oil

The results of the physical properties of brownies' hardness (table 1). Melapa states that product hardness is influenced by water content, protein, fat, and the use of baking ingredient [12]. The decline in hardness in this study is caused by its high water content and the addition of black cumin oil. According to [13], the increase in water content in a product can decrease its hardness. The higher the water content of the brownies, the lower the hardness will be. [14] states that high water content in the material will cause the texture of the product to become softer. [15] states that the unsaturated fatty acid content and the "air binding capacity" can also cause the texture of brownies to become softer, this indicates that the higher concentration of black cumin oil will result in lower F max because the texture produced is softer. The higher the black cumin oil content, the higher the content of the water, as well as the content of ash, protein, fat, and total phenol. Meanwhile, the carbohydrate and Ic50 antioxidant content will decrease as the black cumin oil content is increased. The highest number of water content, ash, fat, protein and total phenol are recorded in the addition of 0.25% black cumin oil. It is also worth noting that the increased number of water content is caused by protein and cooking process.

The chemical properties (table 1). [16] stated that the higher the protein content the higher the water binding capacity. Along with the increasing amount of protein as the cumin oil increases, the number of hydrophilic groups will also be higher. The higher the hydrophilic group, the water absorption capacity will increase. The steaming process also has a considerable effect. [17] declared that steamed material will give the material the opportunity to absorb water vapour, resulting in an increase in the water content of the material. Increased ash content was caused by black cumin oil containing Zinc 15 mg / ml, copper 13 mg / ml, Iron 16.2 mg / ml, selenium 0.23 mg / ml, manganese 11 mg / ml, based on [18]. The increased protein content is caused by the content of 208 mg / g of protein in black cumin oil based on [18]. Increased fat content caused by the addition of black cumin oil, it is known that black cumin oil content of 70% unsaturated fatty acid and 18.5% saturated fatty acids[19] In addition, the increased fat levels are also caused by supporting materials such as dark cooking chocolate, chocolate powder, cooking oil, and condensed milk sweet.

The decrease in carbohydrate levels in brownies according to [20] is caused by the cooking process resulting in a decrease in levels of chemicals. Table 1 shows the decrease in antioxidant IC50 as black cumin oil increases. [21] states that the lower the antioxidant activity of IC50 the better the antioxidant content of the product. The amount of antioxidant IC50 in the product are also affected by beta-carotene, total phenol and tocopherol. Total tocopherol in black cumin oil ranged from 9,15-24,65 mg / 100g, β-carotene amounted to 593μg / g, phenol in black cumin seed oil 245-369 mg / kg [22]. Antioxidant activity is also influenced by purple sweet potato flour which contains anthocyanin, [23] stated that the total anthocyanin in purple sweet potato flour is 18,1-25,7 mg / 100 g. The antioxidant activity in the brownies also influenced by Theobromine extracted from dark chocolate added to brownies. In this study, the antioxidant IC50 in brownies is compared with vitamin C. Vitamin C has a value of IC50 13.39 ppm, smaller than the value of IC50 brownies because it is a pure compound. The antioxidant activity of IC50 brownies is 43.39-77.39 times lower than vitamin C. According to [25], the total increase of phenol is due to the phenolic compounds such as vanillic acid and thymoquinone found in black cumin. The vanillic acid and thymoquinone are included in Quinonic phenolic compounds. [25] states that the total phenolic content can be used as an indicator to determine the hydrophilic activity of antioxidants. The higher the total phenol, the lower IC50 antioxidant activity will be. According to [21] the lower the antioxidant activity of IC50, the better the antioxidant content of the product. It is also shown that in this study, the higher the amount of, the higher the antioxidant activity.

Table 1 Chemical and physical properties of brownies product

Black cumin oil (%)	Water content (%)	Ash (%)	Protein (%)	Fat (%)	Carbohy drate (%)	Total phenol (%)	Antioxidant Ic_{50} (ppm)	Hardness (N)
0%	24.25^{a}	1.14^{a}	7.35^{a}	36.45^{a}	$55,35^{d}$	0.19^{a}	1275.35^{f}	9.91^{a}
0.05%	24.89^{ab}	1.19^{bc}	7.54^{ab}	37.79^{b}	$53,06^{c}$	0.22^{b}	1043.60^{e}	8.47^{ab}
0.10%	25.33^{bc}	1.23^{bc}	7.66^{b}	38.52^{b}	$52.,21^{c}$	0.23^{b}	943.20^{d}	8.26^{ab}
0.15%	25.84^{cd}	1.24^{bc}	7.86^{bc}	40.10^{c}	50.33^{b}	0.25^{c}	839.25^{c}	7.74^{ab}
0.20%	26.16^{de}	1.25^{bc}	8.03^{bc}	41.08^{cd}	49.15^{a}	0.26^{d}	745.60^{b}	7.56^{ab}
0.25%	26.68^{e}	1.28^{c}	8.17^{c}	41.95^{d}	48.15^{a}	0.29^{e}	580.94^{a}	7.07^{b}

$^{a-e}$ Mean different letter within each line are significantly different ($p<0.05$)%

3.2 Identify Component of Brownies Product

Component was contained in the brownies product (table 2) were palmitic acid, oleic acid, lauric acid, theobromine, vitamin E. In this study, take the five highest of identify component. Theobromine on the brownies product was caused by ingredients such as dark cooking chocolate and chocolate powder. Haryadi reproted that theobromine has functioned as a diuretic, cardiovascular stimulant [26]. Palmitic acid on the brownies due to ingredients was used such as cooking oil, black cumin oil, dark cooking chocolate. Listyawati reported that palmitic acid as an important calori source but has lower antidote of oxidations [27]. Oleic acid on brownies was from dark cooking chocolate, black cumin oil, cooking oil. Peddywati reported that oleic acid as an antioxidant to our body, oleic acid has functioned as an antidote of oxidation which was caused by ROS. Eddison reported that oleic acid was known as omega-9 and has a function to increasing HDL and decreasing LDL [28]. Lauric acid on the brownies product was caused by ingredients additions such as cooking oil and sweetened condesened milk. Lauric acid on the brownies product has a big potential to enhance shelf life product, it was caused by lauric acid has antibacterial properties. Liberman reported that lauric acid can make a monoulaurin on the body, monolaurin as an antibacterial, antiviral was used for destroying the virus such as HIV, influenza, pathogenic bacteria, cytomegalovirus [29]. Vitamin E on the brownies was from the addition of black cumin oil. Eitenmiller reported that main of Vitamin E was tocopherol. Tocopherol was one of the strongest antioxidant as antidote free radically [30].

Table 2. Identify component identify by Gc-Ms identification

Identify component	Formula (%)				
	0.05%	0.10%	0.15%	0.20%	0.25%
Palmitic Acid	14.98	12.98	14.34	14.58	21.20
Oleic Acid	13.87	12.01	13.71	13.97	18.41
Lauric Acid	9.22	8.66	9.28	10.44	14.32
Theobromine	5.54	4.44	4.17	3.75	4.28
Vitamin E	3.35	3.57	3.68	4.07	4.33

3.3 Sensory Evaluation of Brownies Product

Brownies substituted sweet potato flour was assessed for sensory attributes using an acceptance test after production (table 3). The highest scores for brownies substituted sweet potato flour color value was found for 0.05% addition of black cumin oil. The corresponding value was 4.05. The color of brownies substituted sweet potato flour depend on ingredients was used such as dark cooking chocolate and chocolate powder, in another hand the color of brownies substituted sweet potato flour depend on the color of black cumin oil. In this study, the color value of brownies substituted sweet potato flour was not affected by the addition of black cumin oil. The highest scores for brownies substituted sweet potato flour flavor value was found for brownies substituted sweet potato flour with 0.05% addition of black cumin oil. The specific flavor of brownies substituted sweet potato flour due to the volatile component from black cumin oil, one of volatile component was tymnoquinone which can cause distinctive sting flavor [19]. The highest scores for taste value was 0.05% addition of black cumin oil. The specific taste of brownies in this study was bitter. Goerlich reported that bitter taste due to nigellin was contained on the black cumin oil [12]. In this study, the highest panelist acceptance of brownies texture was found for 0.05% addition of black cumin oil. The texture of brownies was affected by the steamed cooking process and unsaturated fatty acid on the black cumin oil.

Acceptance tests showed that no significant difference ($P > 0.05$) for the colour of the brownies substituted sweet potato flour with the addition of nigella sativa oil. In general, these results showed the good acceptance of the brownies (table 3), the brownies substituted sweet potato flour which was added 0.05% black cumin oil had the highest number for color, taste, flavor, texture, and overall acceptance than that of the other of formula brownies. The brownies substituted sweet potato flour was added 0.25% black cumin oil had the lowest score for overall acceptance. The highest panelist acceptance for an overall score on brownies substituted sweet potato flour with was added 0.05% black cumin oil.

Table 3. Sensory evaluation of brownies product

Black cumin oil	Color	Flavor	Taste	Texture	Overall
0.05%	4.05^{a}	3.88^{d}	4.07^{d}	3.60^{c}	4.05^{c}
0.10%	4.00^{a}	3.87^{cd}	3.82^{cd}	3.27^{b}	3.87^{c}
0.15%	4.00^{a}	3.4^{c}	3.55^{c}	3.22^{b}	3.55^{b}
0.20%	3.85^{a}	2.97^{b}	2.75^{b}	2.97^{ab}	2.87^{a}
0.25%	3.85^{a}	2.55^{a}	2.35^{a}	2.82^{a}	2.80^{a}

a-e Mean different letter within each line are significantly different ($p<0.05$)

3.4 Determination the Best Formula of Brownies Product

Best formula from all formulation of brownies was determained with compensatory [11] 0.05% up to 0.25% of nigella sativa oil addition have a score 0.692, 0.626, 0.504, 0.347 and 0.308, respectively.

The best formula was choosen by the high score of the formula, then brownies substituted sweet potato flour with 0.05% addition of black cumin oil as the best formula in this study.

4. Conclusions

Water content, ash content, fat content, protein content, total phenol were increased with the addition of black cumin oil. On other hand, the result of antioxidant Ic50 and carbohydrate content were decreased with the addition of black cumin oil. Identify component on the brownies were palmitic acid, oleic acid, lauric acid, theobomine and vitamin E. Due to the sensory characteristics of brownies substituate sweet potato flour with the addition of black cumin oil, the panelists provide high values for brownies which was 0.05% addition of black cumin oil. The best formula's product was addition with 0.05% black cumin oil.

References

[1] Selles AN and Garrido GG 2010 *The Challenges of Antioxidant Therapy with Natural Product. Vol 4: Antioxidants and Nutraceutical (p.2)* (Texas: Studium Press LLC)
[2] Istamar 2008 Study on Making Steamed Brownies: Substitution of Purple Sweet Potato Flour (Ipomoea batatas var Ayamurasaki) and Adding Dark Chocolate Cooking (Dark Chocolate Cooking) *Agroindustry*
[3] Winarti C and Nurdjanah N 2005 Opportunities of spices and medicinal plants as a functional food source *Journal Litbang Pertanian* **24** : 47 - 55
[4] Haron H, Linn CG, Suazana S 2014 Comparison of physicochemical analysis and antioxidant activities of Nigella sativa seeds and oils from Yemen, Iran and Malaysia *Sains Malaysiana* **43**: 535–42
[5] Salvador MD, Aranda F, Gomez-Alonso S and Fregapane G 2001 Cornicabra virgin olive oil: A study of five crop seasons, composition, quality and oxidative stability *Food Chemistry* **74**: 267-74
[6] AOAC 2005 Official Method of Analysis of the Association of Official Analitycal Chemist. 18th ed. *Maryland: AOAC International* (United States of America : William Harwitz)
[7] Amin I and Lee WY. 2005 Effect of Different Blanching Times on Antioxidant Properties in Selected Cruciferous Vegetables *Journal of the Science of Food and Agriculture* **85** (13):2314-2320
[8] Khairullah 2016 The Effects of Different Extraction Methods on Antioxidant Properties, Chemical Composition, and Thermal Behavior of Black Seed (Nigella sativa L.) Oil
[9] Bourne MC 1982 *Food Texture and Viscosity: Concept and Measurement* (New York :Academic Press)
[10] Setyaningsih D 2010 *Sensory Analysis For Food Industry and Agro* (Bogor : IPB Press)
[11] De Garmo EPWG and Sullivan JR 1984 *Engineering Economy The 7th Edition* (New York : Macmillan Publishing Comp)
[12] Melapa A, Djarkasi KJ and Ludong M 2015 Acceptance of Panellists Against Brownies Roasted Raw of Daluga Bulbs Flour (Cyrtospermamerkussi Hassk) (Schott) Food Science and Technology Studies Program
[13] Cauvain SP and Young LS 2006 *Baked Products: Science, Technology and Practice* (United Kingdoms :Blackwell Publishing)
[14] Amalia R 2011 Physicochemical and Organoleptic Characteristics of Snack Bar with Basic Material of Tempe Flour and Dried Jackfruit as CFGF Food Alternative (Casein Free Gluten Free)
[15] Ketaren S 1986 *Introduction to Oil and Fatty Food Technology* (Jakarta : UI Press)
[16] Darmono 1999 *Management of Beef Cattle Business (*Yogyakarta: Kanisius)
[17] Fellow P 1988 *Food Processing Technology Principle and Practice* (England: Chister)
[18] Kamal SS and Huat BT 2013 Extraction, isolation and characterization of anti-tumour principle, alpha-hedrin, from the seeds of Nigella sativa *Planta Med* **67** 29-22

[19] Nickavar B, Mojab F, Javidnia K and Amoli M A R 2003 Chemical composition of the fixed and volatile oils of Nigella sativa L Zeitschrift Fur *Naturforschung* **58** 629-31

[20] Sundari T 2010 Introduction of Superior Varieties and Ubikayu Cultivation Techniques of Nutrition and tubers (Research Institute: Malang)

[21] Molyneux P 2004 The use of the stable free radical diphenylpicrylhydrazyl (DPPH) for estimating antioxidant activity *Journal Science and Technology* **8** 88-92

[22] Sultan MT, Butt MS, Anjum FM, Jamil A, Akhtar S and Nasir M 2009 Nutritional profile of indigenous cultivar of black cumin seeds and antioxidant potential of its fixed and essential oil *J. Bot* **41** 1321-30

[23] Ningsih NY 2015 Effect of Length of Cooling of Resistant Starch Content of Modified Purple Sweet Potato Flour Universitas Lampung Bandar Lampung

[24] Bourgaou S, Ksouri R and Falleh H 2007 Abstract Phenollic Composition and Bioligical Activities Of Tunisia *Nigella Sativa L Elsevier Masion* **331** (1) 48-55

[25] Tubesha Z, Iqbal S and Ismail M 2011 Effects of hydrolysis conditions on recovery of antioxidants from methanolic extracts of Nigella sativa Seeds. *Journal of Medicinal Plants Research* **5** 5393-9

[26] Haryadi and Supriyanto 2012 *Chocolate Technology* (Gajah Mada University: Press Yogyakarta)

[27] Listyawati S and Setyawan AD 2008 Chemical content and anti-inflammatory test of Lantana camara L. ethanol extract on white mice (*Rattus novergicus L.*) *Bioteknologi* **5** 10-17

[28] Singh G, Marimuthu P, de Heluani CS & Catalan C 2005 Chemical constituents and antimicrobial and antioxidant potentials of essential oils and acetone extract of Nigella sativa seeds *Journal of the Science of Food and Agriculture* **85** 2297-306

[29] Liberman, H A and Kanig J L 2008 *Theory and Practice Industry Pharmatic third edition* (Jakarta : UI-Press)

[30] Eitenmiller RR, Ye L, Landeen Woo 2008 *Vitamin Analysis for the healht and food science 2nd edition* (Boca Roton: Taylor and Francis Group)

[31] Georlich Pharma International 2007 *Spesification Egyptian Black Cumin Oil* (Germany: Am Gewerbering)

Effects of microwave power and irradiation time on pectin extraction from watermelon rinds (*Citrullus lanatus*) with acetic acid using microwave assisted extraction method

A M Sari*, D Ishartani and P S Dewanty

Department of Food Science and Technology, Universitas Sebelas Maret, Ir. Sutami Street 36A, Surakarta 57126, Indonesia

*E-mail : ardheamustika@gmail.com

Abstract. The aims of this research are to study the effect of microwave power (119.7 W, 199.5 W and 279.3 W) and irradiation time (6, 9 and 12 min) on pectin extraction by using Microwave Assisted Extraction (MAE) with acetic acid and to do a preliminary characterization of pectin from watermelon rinds. A randomized factorial design with two factors was used to determine the effect of microwave power and processing time on the yield, equivalent weight, degree of methoxylation (DM), galacturonic acid content (GA) and the degree of esterification (DE) of extracted pectin. The results showed that extracted pectin from watermelon rinds using MAE method have yield ranged from 3.925% to 5.766%, with equivalent weight ranged from 1249.702 to 2007.756. Extracted pectin have a DM value ranged from 3.89% to 10.81%. Galacturonic acid content that meets with IPPA standard resulted from extraction condition of 279.3-watt microwave power for 9 min and 12 min. The degree of esterification (DE) value ranged from 56.86% to 85.76%, and this value exhibited a relatively high methoxyl pectin (>50%). The best pectin properties was obtained at a microwave power of 279.3 watts for 12 min.

Keywords: *microwave assisted extraction, pectin, watermelon*

1. Introduction

Watermelon (*Citrullus lanatus*) is a fruit usually consumed as a fresh fruit by many people in Indonesia. In 2013 Indonesia produced approximately 460,628 tons and the number was increased in 2014 to 653,995 tons [1]. The consumption of watermelon left waste rind which makes environmental problems due to watermelon rind compose approximately 30% of watermelon biomass [2]. According to the production of watermelon in 2014 at least 196,198 tons of watermelon rind was generated. Riped watermelon rinds contain 20% of cellulose, 23% of hemicellulose, 10% of lignin, and 13% of pectin [3]. Due to the pectin content, it was potentially used as raw material for pectin extraction.

Pectin is polysaccharide that composes 30% of all plant cell-wall midle lamella of the plant tissues.It is consisting of (1-4) α-D- galacturonic acid residues and some neutral sugar namely L-arabinose, D-galactose and L-rhamnosethat partially esterified at carboxylic acid with methanol or acetic acid. Every source of pectin has different composition and structure that depend on the source of pectin, the degree of ripening and the condition during extraction [4,5,10]. In food industries, pectin ussually used as food thickener, food stabilizer, water binder, emulsifier and gelling agent. The application of pectin in food product is depend on the degree of esterificaton (DE) which is represent the percentage of esterified carboxyl group with metil alcohol. Pectin with DE higher than 50% is classified as high methoxyl pectin (HMP) and pectin which has DE lower than 50% is classified as low methoxyl pectin (LMP) [10].

Various source of pectin is extracted commercially by acid sovent or diluted mineral acid at high temperature. Commercial pectin usually extracted from the waste of juice industries such as citrus peel and apple pomace. Many studies have been reported pectin extraction from other fruit waste such as pectin from orange peels [14], apple pomace [4,11], banana peels [7], watermelon rinds [2,5], and pomelo peels [13].

There are many methods use to pectin extraction that affects the yield and pectin quality. A dilute mineral acidic solvent in high temperature was used in a conventional method. Pectin extraction by using conventional methods was considered not appropriate to obtain high quantity and quality of pectin due to the length of time processing. Therefore, some novel extraction techniques have been developed to increase the process efficiency (yield per time) such as enzyme-assisted extraction (EAE), subcritical water extraction (SWE), microwave-assisted extraction (MAE), ultrasound-assisted extraction (UAE), supercritical fluid extraction (SFE), and pressurized solvent extraction (PSE) [2,12]. The comparison between conventional and novel method has been reported by [6]. They were compared conventional method, MAE, and UAE to extract grapefruit pectin.

Some previous study has been reported extraction pectin from watermelon rinds by using various method. [5] evaluated the properties of pectin from watermelon rind by using conventional extraction including rheological, emulsifying and foaming properties. Pectin was extracted with nitric acid solvent using conventional method. Extraction of pectin from watermelon rinds with destilled water using MAE exhibited maximum yield (25,79%) that obtained with microwave power of 477 W for 128 s [2]. Extraction pectin from watermelon rind with a weak solvent has not been reported. The weak acids such as citric acid are considered more interesting than strong acid from a safety application point of view. Although strong acids are more effective but weak acids solvent can minimize the environmental problems [7]. The objectives of this research were to investigate the effect of microwave power and irradiation time on pectin extraction by using Microwave Assisted Extraction (MAE) with acetic acid and to characterize the properties of pectin from watermelon rinds.

2. Experimental

2.1. Raw material and reagent

Watermelon rinds were provided by Lor In Hotel located in Karanganyar, Central Java, Indonesia. Before extraction, the fresh materials were dried and converted be a rind powder. Chemical reagen for analysis and all solvents used for extraction were purchased from Sigma-Aldrich, Singapore and Merck.

2.2. Preparation of watermelon rind powder

The watermelon rinds obtain from Lor In Hotel were cut ± 1cm in length, wide and thick after the red flesh was removed. After that, the rinds were washed and dried at 55 ^{o}C. The dried watermelon rinds were pulverized and sieved through a 60-mesh sieve. The rind powders were stored in plastic bags for pectin extraction.

2.3. Microwave assisted extraction of pectin

In this research, pectin extraction was carried out according to the method developed by Wang et al. [4]. with a some modifications. Extraction was conducted by microwave equipment (SHARP) with different microwave power (119.7-watt, 199.5-watt and 279.3-watt) and various time processing (6 min, 9 min and 12 min). A 10 g of rind powder were loaded in 1000 ml Pyrex beaker with 100 ml distilled water. An acetic acid solvent 1 N was added to adjust pH until pH of mixture was 2. After that, the mixture was placed in a microwave and heated with different microwave power and irradiation (processing) time. After the extraction process has been done, the mixture was cooled down and filtered using Whatman paper no 1. The filtrate was precipitated by adding ethanol 96% with the same volume with the filtrate (v/v) for overnight (12 h). The coagulated pectin mass was washed with 70% ethanol to remove acidity. Then, the wet pectin was dried at 45 °C in cabinet dryer until constant weight was obtained. The driedpectin was grinded in order to obtain pectin powder.

2.4. Method of analysis

The powdered pectin was analyzed for yield of pectin, equivalent weight, degree of methoxylation (DM), galacturonic acid content (GA) and the degree of esterification (DE). The yield of pectin was calculated as a ratio between the weight of dried pectin and the weight of dried watermelon rind, according to this equation:

$$PY = \frac{m_0}{m} \times 100$$

where m_0 is the weight of dried pectin and m is the weight of watermelon rind powder. Equivalent weight, DM, and GA content were determined by titration with a method described by Ranggana [8]. The DE of pectin was calculated based on DM and GA content [9]. The result of pectin yield, DM value, DE, equivalent weight and GA content were analyzed statistically with two way ANOVA using SPSS.

3. Result

3.1. Effect of process variable on pectin extracted yield and equivalent weight

3.1.1 Yield of extracted pectin. The extraction were carried out at different process condition (microwave power and duration of irradiation time) in order to study the interaction between two factor to the yield of extracted pectin. The data showed that the yield of extracted pectin was increased linearly to the enhancement of microwave power and duation of irradiation time (Fig 1a). The yield of extracted pectin from this study was ranged between 3.925%–5.766%. The interaction of microwave power and irradiation time was influenced the yield of pectin.

Some previous studies were reported the optimum condition to obtain the highest pectin using response surface methodology (RSM). [4] have been reported their study of pectin extraction using MAE from apple pomace. Their experiment data showed that the microwave power have significantly effect on the yield of extracted pectin. According to the prediction by SAS, the longer extraction time the higher yield was. The extracted pectin from watermelon rind with distilled water as a solvent that exhibited the same trend with the result in this study [2]. The optimum conditions to obtain maximum yield (25.79%) of pectin were determined to be microwave power of 477-watt, duration of time irradiation of 128 s with pH of 1,52.The enhancement of microwave power will increase the penetration of solvent into the material and lead to rapid energi transfer to the solvent and the material, generating dissolution of extracted pectin [2].

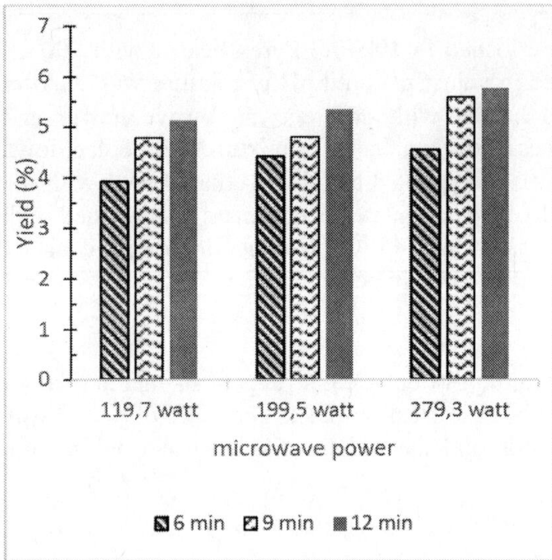

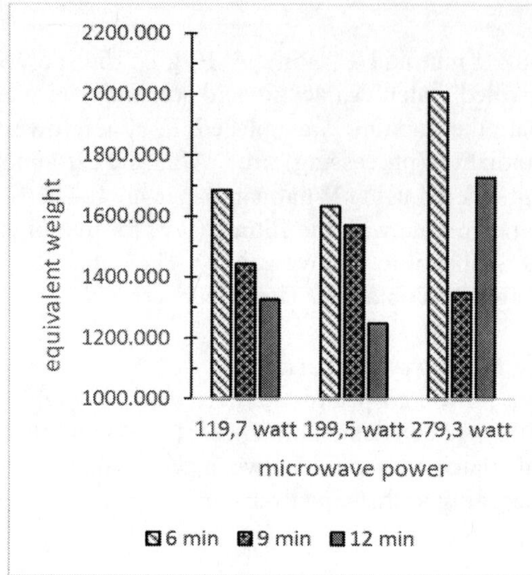

Figure 1. The effect of microwave power and irradiation time on yield (a) and equivalent weight (b)

3.1.2. Equivalent weight. The result of the effects of condition process to equivalent weight extraxted pectin has been showed in Fig 1.b. Increasing of microwave power energy lead to decrease of equivalent weight. In other hand, increasing the duration of irradiation time significantly decreased the equivalent weight. The greatest equivalent weight of pectin was obtained at 279.3-watt of microwave power and 6 min of irradiation time. According to ANOVA, interaction between microwave power and irradiation time have influenced the equivalent weight of pectin.

3.2. Effect of of process variable on qualitative charecteristics of pectin extracted

The data exhibited a significant influence of microwafe power and irradiation time on the degree of esterification of the dried pectin. The average of DE value was increased significantly with an increase in microwave power. In different irradiation time, the average of the DE value showed significantly difference.

The average DM value resulted in this study showed significant enhancement with an increasing of heating power and duration irradiation time. The highest DM value was 10.812% that obtained at 279.3-watt of microwave power and 12 min of irradiation time.

The data showed an interaction of microwave power and duration time process could effect the GA content of extracted pectin. Based on ANOVA anlysis, the average of GA content was increased with an increase in microwave power and the lenght of irradiation time. The highest GA content was exhibited in condition 279.3-watt of heating power during 12 min irradiation.

4. Conclusion

Based on the data resulted from this presen study, pectin was extracted succesfully from watermelon rind with acetic acid solvent by using MAE method. There was an influence of the microwave power and irradiation time on the yield, equivalent weight, DM value, DE value and GA content of extracted pectin from watermelon rind. The best characteristic of extracted pectin and the highest yield was obtain in extraction condition 279.3-watt of microwave power and 12 min of irradiation time.

References

[1] Central Bureau of Statistics 2015 Production data URL http://www.bps.go.id Accesed 17.10.15
[2] Maran JP, Sivakumar V, Thirugnanasambandham K and Sridhar 2014 *J. Carbohydrate Polymers* **101** 736–91
[3] Campbell M 2006 *Extraction Pectin From Watermelon Rind* Thesis Biosystems Engineering Oklahoma State University
[4] Wang S, Chen F, Wu J, Wang Z, Liau X andHu X 2007 *J. of Food Engineering* **78**: 693-700
[5] Petkowicz CLO, Vriesman LC and Williams PA 2017 *J. Food Hydrocolloids* **65**: 57-67
[6] Bagherian H, Ashtiani FZ, Fouladitajar A and Mohtashamy M 2011 *J Chem. Eng. and Processing* **50** 1237–43
[7] Oliviera TI, Rosa MF, Cavalcante FL, Pereira PHF, Moates GK, Wellner N, Mazzetto SE, Waldron KW and Azeredo H M C 2016 *J. Food Chemistry* **198** 113–8
[8] Ranganna S 1977 *Handbook of Analysis and Quality Control for Fruit and Vegetable Products Second Edition* (New Delhi: Tata McGraw-Hill Publishing Company Limited)
[9] Azad AKM, Ali MA, Akter MS, Rahman MJ and Ahmed M 2014 *J. of Food and Nutrition Sci* **2** 30–5
[10] Hosseini SS, Khodaiyan F and Yarmand MS 2016 *Int.J. of Biological Macromolecules* **82** 920–6
[11] Wikiera A, Mika M and Grabacka M 2015 *Food Hydrocolloids* **44** 156–61
[12] Adetunji LR, Adekunle A, Orsat V and Raghavann V 2017 *Food Hydrocolloids* **62** 239–50
[13] Chen Q, Zhuoyan H, Dong Yao FY and Liang H 2016 *Food Science and Tech.* **66** 538–45
[14] Hosseini SS, Khodaiyan F and Yarmand MS 2016 *Carbohydrate Polymers* **82** 59-65

Detection of *Salmonella* sp., *Vibrio* sp. and total plate count bacteria on blood cockle (*Anadara granosa*)

ER Ekawati* and S N H Yusmiati

Faculty of Health Sciences, Maarif Hasyim Latif University, Jalan Ngelom Megare 30, Sidoarjo, Indonesia

*E-mail: evysains@dosen.umaha.ac.id

Abstract. Blood cockle (*Anadara granosa*) has high level of zinc and protein, which is beneficial for therapeutic function for malnourished particularly stunting case in children. Zinc in animal foods is more absorbable than that from vegetable food. Blood cockle (*Anadara granosa*) is rich in nutrient and an excellent environment for the growth of microorganisms. This research aimed to identify the contamination of *Salmonella* sp., *Vibrio* sp. and total plate count bacteria on blood cockle (*Anadara granosa*). This was observation research with laboratory analysis. *Salmonella* sp. and Vibrio sp. were detected from blood cockle. Total plate count was determine of the total amount of the bacteria. Results detected from 20 samples of blood cockle showed that all samples were negative of *Salmonella* sp. and 1 sample positive *Vibrio* sp. The result of total plate count bacteria was $< 5 \times 10^5$ colony/g sample.

Keywords: *Salmonella sp., Vibrio sp., Anadara granosa*

1. Introduction

Food that is free from health risks cause by damage, counterfeiting and contamination by microbial and chemical compounds as well sufficient spiritual needs[1,2]. Blood cockle (*Anadara granosa*) is a seafood commodity that has commercial value and favored by consumers in Indonesia and in Asia generally. Blood cockle (*Anadara granosa*) has high level of zinc and protein, which has therapeutic function for malnourished children, particularly those with stunting case. Zinc in animal foods is more absorbable than that from vegetable food.[3,4]. Blood cockle (*Anadara granosa*) is found on muddy substrates at revier estuary with sloping beach topography to a depth of 20 m [5]. Bood cockle's (*Anadara granosa*) way of life as filter feeder causes this commodity to have high potential in accumulating pollutants substances, both microbes and heavy metals. Water in the environment is contaminated by microbial or heavy metals, which will be absorbed into the tissues of blood cockle bodies [2].

There are three general group of pathogenic bacteria in shellfish. The first is the indigenous bacteria as natural microflora, e.g *Clostridium botulinum, Vibrio* spp., *Aeromonashydrophila*. The second is the enteric (non indigenous) bacteria that are present due to fecal contamination as *Salmonella* spp., *Shigella* spp., *Escherichia coli, Staphylococcusaureus*. The third is bacterial group formed in contamination during processing, storage or preparation of cooking, such as *Bacillus cereus, Listeria monocytogenes, Staphylococcus aureus, Clostridium perfringens, Salmonella* spp.[10].

Majority of epidemic of food born-illness is due to pathogenic microorganisms that are gGenerally undetectable by human senses (usually not showing colour or flavor changes or leaving taints in the food). This bacteria are capable of growing rapidly under favourable storage conditions [6,7]. Since

1980s, the incidence of Salmonellasp. infection have increased dramatically, causing high medical costs, and decreasing the productivity of the workers [8]. *Salmonella* sp. is a major reservoirs in aquatic environment. Blood cockle and fishery product have been recognized as agent of food-borne pathogens [6,9].

Vibrio sp. is the causative agent of vibriosis that infect marine animals such as fish, shrimp and shellfish [11]. *Vibrio* sp. is mostly found in marine environment and detected in some types of species aquatic [12,13]. It is also known as a common food-borne pathogen in Asia. It is a food borne-pathogen and a major cause of gastroenteritis particularly from traditional consumption of raw or undercooked seafood [13,14]. The case of food-borne disease caused by pathogenic Vibrio sp. has been widely reported in various outbreaks [14]. *Vibrio* sp. was found to be 73% of clinical samples from hospitalized diarrhea patients within 1995-2001. Meanwhile, in Japan, within the period of 1996-1998, 20-30% cases of outbreaks occurred and in 1998 this case was reported as exceeding the number of cases caused by *Salmonella* [15]. The bacterial disease from seafood is mainly due to contaminated water [16].

This study was aimed at detection of *Salmonella* sp., *Vibrio* sp. and total plate count bacteria on blood cockle (*Anadara granosa*).

2. Materials and Methods

2.1. Sample Collection
Blood cockle samples were collected from traditional market in Sidoarjo region, Jawa Timur, Indonesia. The sample were alive in plastic containers and were transported immediately to Bacteriology Laboratory, Faculty of Heath Science, University of Maarif Hasyim Latif, Sidoarjo, Jawa Timur, Indonesia. The samples were analyzed immediately on reaching the laboratory.

2.2. Reagents and Medium
Reagents used in this study were kovac, NaCl 0,85%, Methyl Red, α-naphtol 5%, KOH 40%, and buffer phosphate pH 6,8. Bacterial medium for detection of *Salmonella* sp. used were Selenite F-Broth, MCA, SSA, KIA, and Biochemical Reactions. Bacterial medium for detection of *Vibrio* sp. used were APW, TCBS, and Biochemical reactions. Medium for total plate count bacteria were PCA.

2.3. Preparation of Blood Cockle (Anadara granosa)
The blood cockle samples was washed with sterile distilled water then mashed using a blender.

2.4. Detection of Salmonella sp.
An amount of 25 g mashed blood cockle samples were grown on selenite F-broth medium and incubated 37°C for 24 hours. Cultures on Selenite F-Broth grown on MCA and SSA were incubated for 37°C for 24 hours. Colonies taken that were not fermented lactose from MCA or small colonies and has a black node from SSA using ose then grown on KIA and incubated 37°C for 24 hours. Further grown on biochemical reactions medium and incubated 37°C for 24 hours [17].

2.5. Detection of Vibrio sp.
As much as 25 g mashed blood cockle samples were grown on 1% Alkaline Peptone Water (APW) and incubated at 37°C for 18-24 hours. Cultures on APW grown on TCBS were incubated at37°C for 18-24 hours. Suspected colonies of yellow color sized 2-3 mm were taken with ose and streak back to TCBS medium for being purified and then incubated at 37°C for 18-24 hours. Furthermore, separate bacterial colonies were taken using ose and cultured in a TSBC medium and incubated at 37°C for 18-24 hours. Furthermore, they were grown on biochemical reactions medium and incubated at 37°C for 24 hours. The positive result of *Vibrio* sp. on KIA medium was indicated by yellow base and red slope section. In SIM medium, *Vibrio* sp. venom showed a positive result, namely the growth of bacteria that spread like roots. In Simmon Citrate medium showed negative result, namely the medium did not change its color. Further

testing of *string test was* as follows:object glass with 0,5% sodium deoxycholate plus bacteria was mixed using ose, then the ose was pulled up, and positive phases took place when fibrin yarn occurred [18].

2.6. Total Plate Count Bacteria

Total Plate Count (TPC) bacteria were enumerated using pour plate technique. Smoother sample of 25 grams was mixed with 225 mL buffer phosphate to obtain dilution of 10^{-1}. The suspension samples were diluted to dilution 10^{-3}. Suspension sample of 1 mL was taken using pipette. Each dilution was put into each petri dish. PCA was poured into each petri dish and incubated for 24-48 hours at $35^0 \pm 2$ C. The number of bacterial colonies growth on PCA medium was counted by colony counter [17,18].

3. Result and Discussion

Detection of *Salmonella* sp., *Vibrio* sp. and total plate count on 20 samples blood cockle is presented in Table 1.

Tabel 1. Detection of *Salmonella* sp., *Vibrio* sp. and total plate count on blood cockle (*Anadara granosa*)

Sample Code	TPC (colony/gram)	Detection *Salmonella* sp.	Detection *Vibrio* sp.
1	1.0×10^4	Negative	Negative
2	3.8×10^3	Negative	Negative
3	5×10^3	Negative	Negative
4	6.2×10^3	Negative	Negative
5	4.7×10^3	Negative	Negative
6	1.0×10^4	Negative	Negative
7	3.4×10^4	Negative	Positive
8	5.4×10^3	Negative	Negative
9	4.5×10^4	Negative	Negative
10	1.5×10^4	Negative	Negative
11	3.7×10^3	Negative	Negative
12	2.0×10^4	Negative	Negative
13	4.0×10^3	Negative	Negative
14	1.0×10^4	Negative	Negative
15	9.3×10^3	Negative	Negative
16	1.6×10^4	Negative	Negative
17	2.7×10^4	Negative	Negative
18	1.2×10^4	Negative	Negative
19	6.1×10^3	Negative	Negative
20	7.0×10^3	Negative	Negative

The detection result from 20 samples of blood cockle (*Anadara granosa*) showed that all samples were negative *Salmonella* sp. and 1 sample positive *Vibrio* sp. The result of Total Plate Count bacteria was $< 5 \times 10^5$ colony/g sample.

Shellfish was studied in this research because shells are widely consumed and favored by the public. Food safety is a condition and effort required to prevent food from possible biological, chemical and other contaminants that may interfere, harm and endanger human health. Shellfish are said to be good and feasible for consumption by humans if they do not contain harmful bacteria, for example, *Salmonella* sp. and *Vibrio* sp.[19].

Water can be a carrier of pathogenic bacteria that is harmful to health. Health hazards or risk related to water pollution in general can be classified into two, the direct and indirect hazards. Indirect hazards may occur as the consequences of consuming fish, which is contaminated or contain harmful pollutant substances [19].

In this study, one sample was found to be positively contaminated with *Vibrio* sp. The incidence of microorganisms in shell fishes depends on the quality of water from which these animals are obtained. Periwinkle, oyster, mussels and clams are often tread in sewage polluted tidal estuaries water from which the shellfishes accumulate high levels of pathogens. Contamination of periwinkles could be as the result of processing. Food handlers could contaminate the meat during sucking. Periwinkles are filter-feeders, passing large amount of water through their bodies taking in suspended solids along with food. During this process, they pick up soil and water microorganisms, including pathogens if present. The inner tissues of healthy plants and animals are free of microorganism. However their surfaces are usually contaminated with a variety of microorganisms [20].

The result of Total Plate Count bacteria from 20 samples blood cockle was $<5x10^5$ colony/g sample. Shellfish carry a resident population of bacteria that has been found to fluctuate between 10^4 and 10^6 bacterial counts. The consumption of mollusca occasionally gives rise to some serious diseases, such as poisoning (paralytic shellfish poisoning), gastroenteritis, enteric fever, cholera, and typhoid fever. The high level of moisture, rich nutrients, including amino acids and other extracts such as nitrogenous compounds, digestible protein and psychrophilic organisms, render seafood early perishable. Most often, spoilage occurs within a short period of time even under refrigeration [20]. Handling seafood safely is important to reduce the risk of foodborne illness. Seafood must be kept in ice or in the refrigerator after being bought or collected from the river. If seafood is not to be consumed within two days after being taken from the river, it must be stored in the refrigerator or wrapped tightly in plastic, aluminum foil or moisture-proof paper and stored in the freezer [21,22].

Proper handling of fish between capture and delivery to the consumer is a crucial element in assuring final product quality. Factor of standart sanitation, handling, time and temperature of holding fish influence quality. In some cases, fish is considered free of pathogenic bacteria that have public health significance when it was firstly caught. The presence of harmful bacteria to human generally indicates poor sanitation in handling and processing and the contamination is from of human or animal. *Salmonella* have been found in fish washed with polluted water and from fish-holds washed with polluted water [23].

4. Conclusion

The detection result from 20 samples of blood cockle (*Anadara granosa*) showed that all samples was negatively contaminated with *Salmonella* sp. and one sample was positively contaminated with *Vibrio* sp. The result of Total Plate Count bacteria was $<5x10^5$ colony/g sample. Total plate count on 20 samples of blood cockle did not exceed the maximum limit required by SNI 7388:2009 about the maximum limit of contamination on food.

References

[1] Wirakartahkusumah 1994 Safety Food (Jakarta: Risalah Widya Karya Pangan dan Gizi V LIPI)

[2] Retyoadhi AY, Tri S and Erryana M. 2005 Determination of the levels of lead (Pb), total microbes and *Escherichia coli* on blood cockle marketed in the Sidoarjo Regency. *Jurnal Teknologi Pertanian* **6** 203–11

[3] Almatsier S 2004 Basic Principles Of Nutrition1st Ed. (Jakarta: Gramedia)

[4] Solang M, Bambang W and Merry A 2013 The Analysis of blood cockle (*Anadara granosa*) flour supplementation on the concentrations of Zinc, IGF-I and ephiseal plate width of femur malnourished male rats (*Rattus norvegicus*) *International Journal of Sciences and Technology.* **2** 12–24

[5] Nurjanah, Zulhamsyah and Kustiyariyah 2005 Mineral and proximal content of blood clams (*Anadara granosa*) taken from Boalemo District, Gorontalo *Buletin Teknologi Hasil Perikanan* **8** 15–24

[6] Mahmoud BSM 2012 A Dangerous Food-Borne Pathogen *Intech Open Access publisher*. http://intechopen.com .Accesed August 2017

[7] Lelieveld HLM, Unilever R and Vlaardingen D 2009 The Netherlands Sources of Contamination. Chapter 5 pp 61-72

[8] Brands DA 2006 Deadly Dideases and Epidemics *Salmonella* Chelsea House Publisher. A subsidiary of Haights Cross Communications

[9] Upadhyay BP, Utrarachkij F, Thongshoob J, Mahakunkijcharoen Y, Wongchinda N, Suthienkul O and Khusmith S 2010 Detection of *Salmonella* sp. in Freshwater and Seawater Microcosms Under Starvation. Asian Fisheries Science. **16** : 247-255

[10] Lyhs U 2009 Microbiological Methods : Fishery Products Quality, Safety and Aunthenticity. Chapter 15 pp 318-348

[11] Elmanama AA 2008 General Microbiology Laboratory Medical Technology Department Islamic university pp73-74

[12] Tepliski M, Wright AC and Lorca G 2009 Biological Approaches for controlling shellfish-associated pathogens. *Curren Opinion in Biotechnology* **20**(2) :185-190

[13] Kanjanasopa D, Benchamaporn P and Srimek C 2011 Occurrence of Vibrio parahaemolyticus in cockle (Anadaragranosa) harvested from the South Coast of Thailand, Songklanarin. *J. Sci. Technol.* **33**(3) : 295-300

[14] Cheng L, Huang J, Shi C, Thompson KD, Mackey B and Cai J 2008 Vibrio parahaemolyticusAssociatea with Mass Mortality of Post Larval Abalone, Holiotis Diversi color Supertexta in Sanya, China. *J. of the World Aquacultures Siciety*. **39**(6) : 746–57

[15] Yennie Y 2011 Isolation and Identification of Pathogenic *Vibrio parahaemolitycus* in shrimp on Pond. Thesis IPB Bogor

[16] Najjah M, Wei L, Shaharom F and Wee W 2008 Surveillance of bacterial species in fresh water and onamental fish from aquarium shop *World Applied Science J.* **3** 903–5

[17] Feng P 2007 Bacteriologycal Analytic Manual (BAM) Salmonella. http://www.fda.gou/food/Food ScienceResearch/LaboratoryMethods/ucm070149.htm Accesed August 2017

[18] Osawa 2008 Osawa Sensei's Vibrio cholera: Isolations Protocol for Environmental Samples (Seafood and River or Melted Ice Water) KOBE University, Japan

[20] Ekanem E and Edejoke G 1995 Bacteriology study of clans and periwinkles *Food Microbiology* **12** 380–5

[21] U.S. Food and Drug Administration 2013 Fresh and Frozen Seafood Selecting and serving it safely

[22] Eze VC, Edward KC and Akway JE 2014 Enumeration and identification of *E. coli*, *Salmonella*, *Vibrio cholera* from seafood in Utaewa fishing settlement, IkotAbasi Local Government Area, AkuaIbom State, Nigeria *Int. j. of Microbiologi and Application*. **1**(3) : 31–5

[23] Sciorino JA and Ravikumar R 2009 Fishery Harbour Manual on the Prevention of Polention (USA: Bay of Bengal Programme Publisher)

The effect of packaging methods (paper, active paper, and edible coating) on the characteristic of papaya MJ9 in ambient temperature storage

D Suwanti, R Utami*, Kawiji, D Praseptiangga and L U Khasanah

Department of Food Science and Technology, Agriculture Faculty, Sebelas Maret University, Jl. Ir. Sutami 36 A Kentingan 57126, Surakarta, Indonesia

*Email : rohulautami@staff.uns.ac.id

Abstract. Papaya is one of the export commodities that contain high vitamin C but having short shelf life. One method that use to extend the shelf life and maintain the quality of papaya is packaging. The aim of this study was to investigate the effect of the packaging methods (paper, active paper and edible coating) on the characteristics of papaya MJ9 (weight loss, firmness, total soluble solid (TSS), Total Titratable Acid (TTA), pH, vitamin C and total mold and yeast). The packaging methods were control (F1), wrapping paper (F2), wrapping active paper (F3), combination of edible coating and wrapping paper (F4), and combination of edible coating and wrapping active paper (F5). The result showed that paper packaging, edible coating and active paper packaging significantly affected the weight loss, firmness, total soluble solid, total titratable acids, pH, vitamin C, and total mold and yeast of papaya. The weight loss, total soluble solid and pH of packaged papaya was lower than that of control sample, however, the value of firmness and total titratable acid was higher than that of the control sample. Packaging can inhibit the increase of weight loss, total soluble solids and pH, and the decrease of firmness, total titratable acid, vitamin C and total mold and yeast. Based on the papaya characteristics, the selected packaging method was the combination of edible coating and wrapping active paper.

***Keywords:** papaya, coating, active paper, lemongrass*

1. Introduction

Papaya (*Carica papaya L*) is one of the tropical fruits commonly consumed in the form of fresh mature papaya due to vitamin C and vitamin A content [1]. According to FAO (2013), Indonesia is the 3rd largest papaya producing country after India and Brazil. One of papaya varieties cultivated in Boyolali is MJ9 papaya. The MJ9 papaya is the more sweet taste, more chewy flesh, and thicker than other varieties [2].

Fruits are perishable commodities. This is due to the high water content of fruit (70-95%). Microorganisms such as molds, bacteria, and viruses could contaminate the fruits that cause damage of papaya. Poor post-harvest handling can cause bruised and soft wounds that trigger papaya damage [3].

Food packaging can prevent and reduce damage, protect food from contamination or mechanical damage such as friction and collision. Packaging that is commonly on papaya is using baskets or paper to avoid wound on the fruit [4]. One of the food packaging being developed is an active paper packaging. Active paper packaging is the incorporation of certain additive compounds into paper packaging for the purpose of maintaining the shelf life of the product [5]. Active paper packaging with the addition of active ingredients can inhibit the growth some fungi such as *C. albicans*, *A. flavus*, *P. nalgiovense*, *P. roqueforti*, and *E. repens* [5] and prevent the damage of fresh fruit such as tomatoes [6] strawberries [7] and dragon fruit [8].

Another packaging that can be used is edible packaging, which is one type of packaging that environmentally friendly [9] and proven to inhibit oxidation [10]. The addition of an antimicrobial active component to edible coatings can increase the shelf life because the barrier properties of the coating will be reinforced with antimicrobial active components that can inhibit bacterial spoilage. Lemongrass (*Cymbopogon citratus*) is one of the natural herbal ingredients that have antimicrobial activity caused by the components of α-citral (geranial) and β-citral (neral) [11]. Edible coatings of food products have been widely practiced such as on paprika [12], tomatoes [13], melons [14].

However, the application of edible coating packaging can not protect food from mechanical hazards. Paper packaging can protect food from mechanical hazards. Paper can act as a vibration damper during distribution and simultaneously give a permeable effect on gas [15]. Therefore, this study investigated the effect of the combination of packaging methods on the characteristic of MJ9 papaya at ambient temperature storage and to determine the best treatment for preservation of papaya MJ9.

2. Materials and Methods

2.1. *Materials*

In this study, papaya obtained from farmer's group in Mojosongo, Boyolali (Indonesia) and lemongrass obtained from the local market in Surakarta (Indonesia). The materials used to produce edible coatings were citrus pectin from Cargill-Germany, glycerol, tween 80 and lemongrass essential oil. The ingredients of active paper packaging were filter paper, tapioca (Rose Brand), acetic acid, tween 80, distillation waste lemongrass oleoresin and chitosan.

2.2. *Preparation of edible coating*

The edible coating was prepared by dissolving 50 g of pectin in 1000 ml distilled water at 75 ° C for 30 minutes. Glycerol (2% w/v) was added and the solution was then allowed to cool. The essential oil (1% v/v) was added and stirred using a magnetic stirrer [16].

2.3. *Preparation of active paper packaging*

The active paper was prepared by immersion filter paper (2mmx2mm) for 24 hours. The paper was crushed, then chitosan solution in 1% acetic acid, tapioca suspension (4.5 g/50 ml of distilled water), and 2% (b/b) oleoresin emulsion were added to the pulp. The pulp was pressed and dried for 48 hours [17].

2.4. *Packaging application*

Whole papayas were dipped into the pectin-based edible coating solution then dried using a dryer. Coated papayas were repackaged with paper (F4) or active paper (F5). Uncoated papayas were packed with paper (F2) or active paper (F3). Uncoated papayas which were unpacked use as a control (F1). Each sample was prepared in duplicate and stored in baskets at ambient temperature for 18 days.

2.5. *Determined of weight loss, firmness, total soluble solids, total titratable acid, vitamin C, pH and total mold and yeast*

Weight loss was obtained by calculating the weight every 3 days. The results were shown in weight loss percentage [18]. Firmness values were measured using a fruit hardness tester (FR5105) expressed in N [19]. Total soluble solids were obtained by dripping the pulp of papayas on refractometer (Master Refractometer Manual Atago) [19]. 10 grams of papaya's pulp was dissolved in 250 ml of aquades and taken as 25 ml to be titrated with 0.1 N NaOH. Phenolphthalein was used as an indicator to determine the total percentage of titrated acids [18]. The pH measurements were performed using pH meter (pH tester 20) [20]. Vitamin C levels were analyzed by an iodometric titration method with starch indicator [21]. The growth of mold and yeast was analyzed by microbiological test using total plate count (TPC) technique [22]. All samples were analyzed on days 0, 3, 6, 9, 12, 15 and 18 during storage .

3. Result and Discussion

3.1. *Weight loss*

The results showed significantly increased ($p<0.05$) in weight loss percentage of packed papaya and unpacked (control) during storage at room temperature. The control samples had the greatest losses than packed samples (Table 1). The edible coating was a good barrier for water and oxygen so as to control the rate of respiration [23] and the paper was able to absorb the vibrations during distribution and simultaneously affected the permeability of the gas [16]. The combination packaging methods could be a good barrier so that papaya on the treatment F4 and F5 had a lower weight loss than that of the control, F2, and F3 samples. Coating of papaya MJ9 by cassava-starch based edible coating solution enriched with lemongrass oil also decreased the weight loss of fruit [24].

Table 1. Weight loss percentage of papaya MJ9 during storage

Formula	Storage (days)						
	0	3	6	9	12	15	18
F1	0.000^{Aa} ±0.000	4.367^{Db} ±0.053	8.464^{Ec} ±0.009	18.111^{Ed} ±0.059	26.86^{Ee} ±0.073	31.404^{Ef} ±0.006	38.800^{Eg} ±0.034
F2	0.000^{Aa} ±0.000	4.114^{Cb} ±0.041	7.299^{Cc} ±0.018	16.089^{Dd} ±0.029	20.053^{Ce} ±0.015	28.962^{Df} ±0.114	35.134^{Dg} ±0.039
F3	0.000^{Aa} ±0.000	4.322^{Db} ±0.071	7.623^{Dc} ±0.030	14.523^{Cd} ±0.069	20.885^{De} ±0.002	27.924^{Cf} ±0.053	34.799^{Cg} ±0.023
F4	0.000^{Aa} ±0.000	3.458^{Bb} ±0.006	6.542^{Bc} ±0.049	13.856^{Bd} ±0.083	19.061^{Be} ±0.119	27.091^{Bf} ±0.056	34.143^{Bg} ±0.073
F5	0.000^{Aa} ±0.000	3.180^{Ab} ±0.083	5.882^{Ac} ±0.028	12.158^{Ad} ±0.078	17.445^{Ae} ±0.037	25.441^{Af} ±0.082	30.801^{Ag} ±0.045

$^{A-E}$ Mean different letter within each column are significantly different ($p<0.05$)
$^{a-g}$ Mean different letter within each line are significantly different ($p<0.05$)

3.2. *Firmness*

The firmness of all samples decreased significantly ($p <0.05$) during storage. Papaya with packaging treatment could inhibit the decrease of the firmness than the control (Table 2). Papaya without packaging indicates a faster maturity rate that causes by an increased of ethylene production [18] so that caused a partial change of water-insoluble protopectin into water-soluble resulting in soft fruit [25]. Other study also stated that the firmness of papaya MJ9 treated with cassava starch-based coating could maintain the firmness better than the control [24]. The combination of edible coating and wrapping active paper (F5) can protect the fruit both mechanically and microbiologically damages so could maintain the firmness of the fruit.

Table 2. Firmness of papaya MJ9 during storage (N)

Formula	Storage (days)						
	0	3	6	9	12	15	18
F1	11.655^{Ag} ±0.025	5.381^{Af} ±0.026	3.661^{Ae} ±0.052	2.694^{Ad} ±0.018	1.921^{Ac} ±0.087	1.293^{Ab} ±0.101	0.616^{Aa} ±0.048
F2	11.573^{Ag} ±0.099	5.484^{Bf} ±0.011	4.425^{Be} ±0.075	2.661^{Ad} ±0.030	2.107^{Bc} ±0.078	1.810^{Bb} ±0.018	1.250^{Ba} ±0.014
F3	11.510^{Af} ±0.064	6.600^{Ce} ±0.024	4.553^{BCd} ±0.076	3.027^{Bc} ±0.060	3.047^{Dc} ±0.006	2.070^{Cb} ±0.053	1.444^{Ca} ±0.038
F4	11.532^{Ag} ±0.091	7.560^{Ef} ±0.042	4.607^{Ce} ±0.003	3.087^{Bd} ±0.042	2.330^{Cc} ±0.075	2.017^{Cb} ±0.021	1.608^{Da} ±0.068
F5	11.541^{Ag} ±0.076	7.445^{Df} ±0.060	5.359^{De} ±0.034	4.432^{Cd} ±0,074	3.484^{Ec} ±0.034	3.031^{Db} ±0.030	2.858^{Ea} ±0.026

$^{A-E}$ Mean different letter within each column are significantly different ($p<0.05$)
$^{a-g}$ Mean different letter within each line are significantly different ($p<0.05$)

3.3. Total soluble solids

Table 3 showed the significant increase of total soluble solids of papaya due to the breakdown of starch into simple sugars. The packaging treatment prevents oxygen contact, which inhibits respiration and transpiration [26]. The addition of the active component of lemongrass to the packaging of F3, F4, and F5 caused an inhibition of increase the total soluble solids MJ9 papaya. Other study about coating papaya MJ9 by cassava starch-based coating also state that total soluble solids were increase during storage [24].

Table 3. Total soluble solids (TSS) of papaya MJ9 during storage (°Brix)

Formula	Storage (days)						
	0	3	6	9	12	15	18
F1	7.950^{Aa} ±0.071	8.800^{Bb} ±0.000	9.650^{Cc} ±0.071	10.050^{Dd} ±0.071	11.050^{De} ±0.071	11.350^{Bf} ±0.071	11.750^{Cg} ±0.071
F2	8.350^{Ba} ±0.071	8.450^{Aa} ±0.071	9.100^{Bb} ±0.141	9.650^{Cc} ±0.071	10.800^{Cd} ±0.000	11.000^{Bd} ±0.000	11.300^{Be} ±0.141
F3	8.050^{Aa} ±0.071	8.450^{Ab} ±0.071	8.850^{Ac} ±0.071	9.150^{Ad} ±0.071	9.850^{Be} ±0.071	10.100^{Af} ±0.141	10.350^{Ag} ±0.071
F4	8.100^{Aa} ±0.000	8.500^{Ab} ±0.141	9.050^{ABc} ±0.071	9.750^{Cd} ±0.071	9.850^{Bde} ±0.071	10.100^{Aef} ±0.283	10.400^{Ag} ±0.000
F5	8.100^{Aa} +0.141	8.400^{Ab} +0.000	8.850^{Ac} ±0.071	9.400^{Bd} ±0.141	9.650^{Ae} ±0.071	10.050^{Af} ±0.071	10.300^{Ag} ±0.000

$^{A-E}$ Mean different letter within each column are significantly different ($p<0.05$)
$^{a-g}$ Mean different letter within each line are significantly different ($p<0.05$)

3.4. Total titratable acid

Table 4 showed the significant decrease of papayas total titrated acids ($p <0.05$) on all treatments during storage caused by the use of acids by the respiratory process. The active compound of the lemongrass that added to the packaging of F3, F4, and F5 can protect the papaya from mechanical and microbiological damage thus inhibiting respiration rate [27]. The controls showed the greatest decrease whereas F2 showed no significant difference compared to control on the total titratable acids parameters.

Table 4. Total titratable acid (TTA) of papaya MJ9 during storage (%)

Formula	Storage (days)						
	0	3	6	9	12	15	18
F1	0.600^{Af} ±0.056	0.480^{Ae} ±0.000	0.400^{Ad} ±0.000	0.320^{Ac} ±0.000	0.240^{Ab} ±0.000	0.260^{Ab} ±0.028	0.160^{Aa} ±0.000
F2	0.640^{Ae} ±0.000	0.520^{ABd} ±0.056	0.440^{ABc} ±0.056	0.380^{ABbc} ±0.028	0.320^{Ab} ±0.000	0.300^{ABb} ±0.028	0.160^{Aa} ±0.000
F3	0.640^{Af} ±0.000	0.560^{ABe} ±0.000	0.480^{ABd} ±0.000	0.440^{BCcd} ±0.056	0.400^{Ac} ±0.000	0.320^{Bb} ±0.000	0.240^{Ba} ±0.000
F4	0.600^{Ac} ±0.056	0.560^{ABc} ±0.000	0.480^{ABb} ±0.000	0.480^{Cb} ±0.000	0.480^{Db} ±0.000	0.340^{Ba} ±0.028	0.300^{Ca} ±0.028
F5	0.640^{Ad} ±0.000	0.600^{Bd} ±0.056	0.520^{Bc} ±0.056	0.480^{Cc} ±0.000	0.460^{Dbc} ±0.028	0.400^{Cb} ±0.000	0.320^{Ca} ±0.000

$^{A-E}$ Mean different letter within each column are significantly different ($p<0.05$)
$^{a-g}$ Mean different letter within each line are significantly different ($p<0.05$)

3.5. *Vitamin C*

Vitamin C of papaya in all treatments increased until reach the peak maturity. It is in line with other study that stated the value of vitamin C papaya increased with ripening [28]. The fruits then decayed and vitamin C decreased (Table 5). Vitamin C of papaya control decreased after 6 days storage, while papaya packed without the addition of active compounds of lemongrass decreased after 9 days storage. Papaya packed with incorporation active component of lemongrass can inhibit the maturity rate so the vitamin C decreased after 12 days storage. The decrease of vitamin C is due to the oxidation of ascorbic acid [29].

Table 5. Vitamin C of papaya MJ9 during storage (%)

Formula	Storage (days)						
	0	3	6	9	12	15	18
F1	0.090^{Ab} ±0.003	0.125^{Bc} ±0.003	0.161^{Ce} ±0.003	0.136^{Ad} ±0.006	0.116^{Ac} ±0.003	0.099^{Ab} ±0.003	0.071^{Aa} ±0.006
F2	0.092^{Aa} ±0.000	0.108^{Ab} ±0.003	0.130^{Bc} ±0.003	0.152^{ABd} ±0.003	0.134^{Ac} ±0.003	0.127^{Bc} ±0.006	0.088^{Ba} ±0.007
F3	0.092^{Aa} ±0.000	0.114^{Ab} ±0.000	0.132^{Bc} ±0.006	0.136^{Ac} ±0.006	0.158^{Bd} ±0.000	0.130^{Bc} ±0.003	0.112^{Cb} ±0.003
F4	0.094^{Aa} ±0.003	0.110^{Ab} ±0.006	0.116^{Ab} ±0.003	0.169^{Cc} ±0.009	0.213^{Cd} ±0.009	0.161^{Cc} ±0.003	0.112^{Cb} ±0.003
F5	0.090^{Aa} ±0.003	0.110^{Ab} ±0.006	0.132^{Bc} ±0.006	0.162^{BCd} ±0.006	0.281^{De} ±0.012	0.154^{Cd} ±0.006	0.119^{Cbc} ±0.000

$^{A-E}$ Mean different letter within each column are significantly different ($p<0.05$)
$^{a-g}$ Mean different letter within each line are significantly different ($p<0.05$)

3.6. *pH*

Table 6 showed an increase of all papayas pH ($p < 0.05$) during storage. An increase in pH papaya during storage was due to the conversion process of carbohydrate to sugar and the use of acids for metabolism [30]. Papaya control showed the highest pH value compared to other treatments. The packing combination (F4) and (F5) showed significantly different results compared to the control. Papaya with edible coating packaging and active paper can suppress the presence of air exchange in fruit and environmental contacts that allow the occurrence of metabolic processes due to microbial contamination. The active component in essential oils and oleoresins can interact with the cell membrane thus affecting the metabolic cycle and fruit aging [31].

Table 6. pH of papaya MJ9 during storage

Formula	Storage (days)						
	0	3	6	9	12	15	18
F1	$4,775^{Aa}$ ±0,035	$5,250^{Bb}$ ±0,000	$5,600^{Cc}$ ±0,000	$5,800^{Dd}$ ±0,000	$6,250^{Ce}$ ±0,071	$6,575^{Cf}$ ±0,035	$6,850^{Dg}$ ±0,071
F2	$4,775^{Aa}$ ±0,035	$5,250^{Bb}$ ±0,000	$5,250^{Bb}$ ±0,071	$5,600^{Cc}$ ±0,000	$5,775^{Bd}$ ±0,035	$6,175^{Be}$ ±0,035	$6,625^{Cfa}$ ±0,035
F3	$4,750^{Aa}$ ±0,071	$4,975^{Ab}$ ±0,035	$5,025^{Ab}$ ±0,035	$5,200^{Ac}$ ±0,000	$5,425^{Ad}$ ±0,035	$5,850^{Ae}$ ±0,071	$6,375^{Bf}$ ±0,035
F4	$4,850^{Aa}$ ±0,000	$5,225^{Bb}$ ±0,035	$5,300^{Bb}$ ±0,000	$5,400^{Bc}$ ±0,000	$5,550^{Ad}$ ±0,071	$5,825^{Ae}$ ±0,035	$6,350^{Bf}$ ±0,071
F5	$4,800^{Aa}$ ±0,000	$5,150^{Bb}$ ±0,071	$5,225^{Bb}$ ±0,035	$5,375^{Bc}$ ±0,035	$5,525^{Ad}$ ±0,035	$5,750^{Ae}$ ±0,071	$6,150^{Af}$ ±0,071

$^{A-E}$ Mean different letter within each column are significantly different ($p<0.05$)
$^{a-g}$ Mean different letter within each line are significantly different ($p<0.05$)

3.7. Total mold and yeast

The increase of total molds and yeast of papaya occurred in all treatments (Table 7). The controls showed a large mold and yeast growth (6,199 log CFU/g). The packaging treatment showed a significant difference compared to the control. Incorporation of essential oil and oleoresin lemongrass into treatments F3, F4 and F5 can inhibit the growth of mold and yeast of papaya. Total mold and yeast of papaya on treatment F3, F4, F5 were 3,908 log CFU/g, 5,310 log CFU/g, 3,484 log CFU/g. This is because the active component of lemongrass there are α-citral (geranial) and β-citral (neral) can be an antimicrobial agent against mold and yeast growth [32]. Other study showed that the total mold and yeast of papaya MJ9 treated with coating incorporated with 1% lemongrass oil was smaller than that of the control sample [24].

Table 7. Total mold and yeast of papaya MJ9 during storage (log CFU/g)

Formula	Storage (days)						
	0	3	6	9	12	15	18
F1	1.544^{Ba} ±0.000	2.435^{Db} ±0.078	3.108^{Ec} ±0.016	3.416^{Ed} ±0.037	4.454^{De} ±0.001	5.174^{Df} ±0.008	6.199^{Eg} ±0.016
F2	1.719^{Ca} ±0.029	2.112^{Cb} ±0.047	2.953^{Dc} ±0.016	3.047^{Dd} ±0.008	4.391^{De} ±0.016	5.203^{Df} ±0.043	5.990^{Dg} ±0.054
F3	1.238^{Aa} ±0.088	1.627^{Bb} ±0.036	2.657^{Bc} ±0.021	2.741^{Bc} ±0.030	3.219^{Bd} ±0.042	3.518^{Be} ±0.121	3.908^{Bf} ±0.047
F4	1.813^{Cb} ±0.000	1.349^{Aa} ±0.068	2.763^{Cc} ±0.011	2.901^{Cc} ±0.105	3.388^{Cd} ±0.015	4.414^{Ce} ±0.110	5.310^{Cf} ±0.109
F5	1.477^{Bb} ±0.000	1.398^{Aa} ±0.000	2.514^{Ac} ±0,033	2.498^{Ac} ±0.020	2.596^{Ad} ±0.031	2.956^{Ae} ±0.011	3.484^{Af} ±0.010

$^{A-E}$ Mean different letter within each column are significantly different ($p<0.05$)
$^{a-g}$ Mean different letter within each line are significantly different ($p<0.05$)

4. Conclusion

Packaging methods significantly affected the weight loss, firmness, total soluble solid, total titratable acids, pH, vitamin C, and total mold and yeast of papaya. The weight loss, total soluble solid and pH of packaged papaya was lower than that of the control sample, however, the value of firmness and total titratable acid was higher than that of the control sample. Addition of active components of lemongrass on the edible and paper packaging can inhibit the increase of weight loss, total soluble solids and pH, and the decrease of firmness, total titratable acid, vitamin C and total mold and yeast. Based on the papaya characteristics, the selected packaging method was the combination of edible coating and wrapping active paper with the incorporation of essential oils (1%) and oleoresin (2%).

Acknowledgements
This work was financially supported by research project of PUPT Ministry of Research, Technology, and Higher Education of Republic of Indonesia No: 041/SP2H/LT/DRPM/II/2016, Indonesia.

References

[1] Ikram E H K, Stanley R, Netzel M, Fanning K 2015 *Journal of Food Composition and Analysis* **41** 201-2011
[2] Suprapto, Anwar H, Setiapermas M N 2011 *Alternative Planting Pattern Based on Papaya in Boyolali District* (Central Java : Assessment Institute of Agricultural Technology) (in Bahasa Indonesia)
[3] Miskiyah, Winarti C, Broto W 2010 *Journal of Agricultural Research and Development* **29**(3) 79-85 (in Bahasa Indonesia)
[4] Indriyani N L P, Affandi, Sunarwati D 2008 *Healthy Papaya Garden Management* (Solok : Indonesian Tropical Fruits Research Institute) (in Bahasa Indonesia)
[5] Rodriguez A, Nerin C, Batlle R 2008 *J. Agric. Food Chem* **56** 6364-6369
[6] Rodriguez A, Nerin C, Battle R 2010 *Jurnal Agriculture Food Chemistry* **58**(11) 6780-6786
[7] Zahroh S U, Utami R, Manuhara G J 2016 *Caraka Tani Journal of Sustainable Agriculture* **31**(1) 59-70
[8] Aziz M S H, Manuhara G J, Utami R, Khasanah LU 2017 *IOP Conf. Series : Materials Science and Engineering*
[9] Utami R, Nurhartadi E, Putra A Y T 2013 *Journal of Food Technosics* **2**(2) 51-56 (in Bahasa Indonesia)
[10] Sari D K, Atmaka W, Muhammad D R A 2013 Journal of Food Technosics **2**(2) 112-120 (in Bahasa Indonesia)
[11] Shaaban H A, Ramadan M M, Amer M M, El-Sideek L E, Osman F 2013 *Journal of Applied Sciences Research* **9**(1) 5770-5779
[12] Widaningrum, Miskiyah, Winarti C 2015 *Agritech* **35**(1) 53-60 (in Bahasa Indonesia)
[13] Rudito 2005 *Journal of Agricultural Technology* **6**(1) 1-6 (in Bahasa Indonesia)
[14] Massilia R M R, Melgar J M, Belloso O M 2008 *Intl. J. Food Microbiol* **121** 313–327
[15] Rudra S G, Singh V, Jyoti S D, Shivhare U S 2013 *Food Bioscience* **3** 49-58
[16] Kawiji, Utami R, Safirani F, Praseptiangga D, Khasanah L U 2015 *Proceedings of the National Seminar PATPI* 480-490 (in Bahasa Indonesia)
[17] Manuhara G J, Khasanah L U, Utami R 2016 *IOP Conf. Series : Materials Science and Engineering* **107** 1-7
[18] Marpudi S L, Abirami L S S, Pushkala R, Srividya N 2011 *Indian Journal of Biotechnology* **10** 83-89
[19] Nasution I S, Yusmanizar, Melianda K 2012 *Journal of Indonesian Agricultural Technology and Industry* **4**(2) 21-26 (in Bahasa Indonesia)
[20] Garcia L C, Pereira L M, Sarantopoulos C I G, Hubinger M D 2011 *International Journal Packaging Technology and Science* **25** 413-425
[21] Sudarmadji S, Haryono B, Suhardi 1976 *Procedure Analysis For Food and Agriculture* (Yogyakarta : Agricultural Products Processing Department Faculty of Agricultural Technology Gajah Mada University) (in Bahasa Indonesia)
[22] Fardiaz S 1993 *Analysis of Food Microbiology* (Jakarta : Raja Grafindo Persada) (in Bahasa Indonesia)
[23] Novita D D, Sugianti C, Wulandari K P 2016 *Jurnal Teknik Pertanian Lampung* **5**(1) 49-56 (in

Bahasa Indonesia)

[24] Praseptiangga D, Utami R, Khasanah L U, Evirananda I P, Kawiji 2017 *IOP Conf. Series : Materials Science and Engineering* **176** 1-6

[25] Torres A V, Becerra-Loza A S, Sayago-Aryedi S G, Palma-Rodriguez H M, Garcia-Magana M L, Montalvo-Gonzales E 2017 *Scientia Holticulturae* **214** 221-227

[26] Winarno F G 2008 *Food and Nutrition, Technology and Consumers* (Jakarta : Gramedia Pustaka Utama) (in Bahasa Indonesia)

[27] Jannah S R, Astuti A, Utama N A 2015 *Journal of Agricultural Technology* **1**(1) 1-12 (in Bahasa Indonesia)

[28] Suryana A, Wiradinata R 2013 *Journal AGROSWAGATI* **2**(2) 54-63 (in Bahasa Indonesia)

[29] Sharmin M R, Islam M N, Alim M A 2015 *J. Bangladesh Agril. Univ.* **13**(1) 131–136

[30] Adetunji C O, Ogundare M O, Ogunkunle A T J, Kolawole O M, Adetunji J B 2014 *Asian J Agri Biol* **2**(1) 8-13

[31] Kusrini E, Usman A, Wisakanti C D, Arbianti R, Nasution D A 2015 *Makara Journal Technology* **19**(3) 148-152

[32] Supardan M D, Anisa Y, Arpi N, Satriana, Mustapha W A W 2016 *International Journal on Advanced Science Engineering Information Technology* **6**(2) 216-220

Carcass and cut yields of broiler chickens fed diet containing purslane meal rich in omega-3 fats

LR Kartikasari*, B S Hertanto and A MP Nuhriawangsa

Department of Animal Science, Faculty of Agriculture, Universitas Sebelas Maret, Jl. Ir. Sutami no 36A, Kentingan Surakarta, Central Java, Indonesia 57126

*E-mail : lilikretna@staff.uns.ac.id

Abstract. The aim of the research was to investigate the effect of diets containing *Portulaca oleraceae* (purslane) as a source of omega-3 fats on carcass and cut yields of broiler chickens. One-day old unsexed Lohmann broiler chickens (n= 180) were used and randomly allocated into 30 pens (each pen contained 6 birds). The pens were randomly assigned to five experimental diets with 6 replicates (36 birds per treatment). The diets were formulated by supplementing a basal diet with purslane meal at a level of 0, 1.5, 3.0, 4.5 and 6.0%. For a period of 42 days, water and diets were provided ad libitum. Feed intake and body weight gain were collected weekly to determine feed conversion ratio. The collected data were analysed using analysis of variance. If there were significant differences between treatment means, the analysis was continued by Duncan's New Multiple Range Test. Findings showed that diets enriched with omega-3 fats, alpha-linolenic acid did not change body weight and carcass percentage of broilers. In terms of cuts yield, there was no significant different on the percentage of breast, back and wings by feeding diets supplemented with purslane meal. However, the inclusion levels of dietary purslane meal significantly affected the percentage of thighs ($P<0.05$) with the highest weight achieved for diets supplemented with 3% purslane meal. Drumsticks tended to increase ($P= 0.056$) by feeding the experimental diets. It was concluded that the inclusion level of 6% purslane meal didn't have negative effect on carcass and cut yields of broiler chickens.

Keywords: *Broilers, Portulaca oleracea, carcass, fat*

1. Introduction

Some investigators have reported the effects of dietary inclusion of omega-3 fats to broiler diets, such as fish oils or fish meal, on broiler performance and fatty acid compositions of meat [1,2]. In order to increase n-3 LCPUFA accumulation into chicken meat, dietary supplementation of fish oils or fish meal, or in combination with n-3 PUFA (alpha-linolenic, ALA) sources, is commonly considered [2]. But, the supplementation of marine sources in the altering of fatty acid profiles of meat or meat quality may cause negative effects on the sensory characteristics of the final products due to the presence of an off-flavour [3,4], and this leads to reduce in consumer preference.

An alternative to the supplementation of fish meal or fish oil to chicken diets is to enhance n-3 LCPUFA deposition in chicken products through the inclusion of plants or seed high in the n-3 PUFA, (ALA, 18:3n-3) which is the n-3 LCPUFA precursor. [5,6] have reported an increase in chicken tissue n-3 LCPUFA including eicosapentaenoic acid (EPA), docosapentaenoic acid (DPA, 22:5n-3), and docosahexaenoic acid (DHA) by enriching diets with ALA-rich plant sources. In addition, a study conducted by [7] found that dietary inclusion of flaxseed oil into broiler diets enhanced n-3 fatty acid content of meat. Importantly, the use of plant sources rich in ALA does not negatively influence broiler performance and the sensory qualities of the meat. Therefore, feeding broiler chickens with plant seeds or oils rich in ALA can be one of dietary approach to increase n-3 fat content in chicken meat and may increase carcass quality.

Another plant that contains many biologically active compounds (β-catotene, alpha-tocopherol, and ascorbic acid) and high level of n-3 PUFA (ALA) is Portulaca oleracea (purslane) [8]. While several studies reported the effect of purslane rich in ALA on the accumulation of n-3 fat in eggs, egg quality and production performance of laying hens [9,10,11], there is little information in relation to the use of purslane in performance parameters or carcass quality. A study conducted by [12] showed that both on day 28 and 42, supplementation of purslane extract (0.4%) increased body weight gain and decreased feed conversion ratio of broilers. In addition, the dietary inclusion of 10% purslane seed did not affect the slaughter weight, carcass, and carcass cut yields of quails [13].

The growth pattern of broiler fat depots can be modified by dietary formulation and the decrease in lipid content of broiler chickens was strongly related to the dietary fatty acid composition [15]. Moreover, some investigators reported that broiler chickens fed diets containing high levels of ALA have less abdominal fat pad accumulation compared with chickens fed diets containing high levels of saturated fatty acids or monounsaturated fatty acids [15]. Therefore, the objectives of the current study were to investigate the effect of diets containing Portulaca oleraceae (purslane) as a source of omega-3 fats on carcass and cut yields of broiler chickens

2. Materials and Methods

2.1. Research Materials

A total of 180 one-day old unsexed Lohmann broiler chickens were used in this experiment. The broilers were randomly allocated into 30 pens (each pen contained 6 birds). The pens were randomly assigned to five dietary treatments with 6 replicates (36 birds per treatment). The basal diet was the corn-soybean based diets containing crude protein 23% and energy 3090.79 kkal/kg (Table 1.)

2.2. Research Methods

One-way classification design was used for this experiment. The variable factor was varying levels of purslane meal (a source of ALA) in the diet. A total of five diets were formulated by supplementing a basal diet with 0 (F0), 1.5 (F1), 3 (F2), 4.5 (F3) and 6% (F4) *Portulaca oleracea* meal (w/w). Fat level of the dietary treatments was kept constant at ~ 5.8%. The composition of ingredient composition and nutrient value of diets were shown in Table 1. The broilers were immediately weighed upon arrival and housed in cage (n= six birds per cage). Each of the experimental diet was replicated six times (n = 6 chickens for each replication). Broiler chickens were fed for 42 days and water and feed were provided *ad libitum* for the duration of experimental period. The chickens were observed at frequent intervals during the first few days to ensure that chickens got adequate feed and water and to ensure that chickens were comfortable with the environmental conditions. Performance parameter data were collected for a 42-days experimental duration. Body weight and feed intake (FI) were recorded and determined on weekly basis. Body weight gain and feed conversion ratio (FCR) were determined throughout the experimental duration (kg feed/kg body weight). A total of 30 chickens of day 35 and 42 (n = 6 chickens for each treatment) were slaughtered to evaluate carcass quality and cut yields of chickens.

2.3. Data Analysis

The obtained data in this experiment were analysed using analysis of variance (ANOVA). When there were significant differences between treatment means, further analysed using Duncan's New Multiple Range Test (DMRT) were conducted with significance level of $P<0.05$.

Table 1. Ingredient composition and nutrient content of experimental diets

Treatment	F0	F1	F2	F3	F4
			%		
Ingredient					
Yellow corn	32.00	31.52	31.04	30.56	30.08
Rice polish	3.00	2.96	2.91	2.87	2.82
Soybean meal	29.95	29.50	29.05	28.60	28.15
$CaCO_3$	0.06	0.06	0.06	0.06	0.06
Di-calcium Phosphate	2.20	2.17	2.13	2.10	2.07
L-Lysin	0.24	0.24	0.23	0.23	0.23
DL-Metionin	0.40	0.39	0.39	0.38	0.38
Premix	0.20	0.20	0.19	0.19	0.19
Salt	0.50	0.49	0.49	0.48	0.47
Limestone	1.00	0.99	0.97	0.96	0.94
Purslane meal	0.00	1.50	3.00	4.50	6.00
Coconut oil	4.50	4.43	4.37	4.30	4.23
Copra meal	14.95	14.73	14.50	14.28	14.05
DDGS	10.00	9.85	9.70	9.55	9.40
Filler	0.99	0.97	0.96	0.94	0.93
Vitamin E	0.02	0.02	0.02	0.02	0.02
Total	100.00	100.00	100.00	100.00	100.00
Nutrient value					
DM	86.97	87.07	87.17	87.27	87.37
ME	3090.79	3069.71	3048.62	3027.53	3006.45
Crude protein	23.10	23.03	22.96	22.89	22.82
Ether extract	5.77	5.78	5.78	5.78	5.78
Crude fiber	3.25	3.51	3.78	4.04	4.31
Ash	3.97	3.95	3.92	3.89	3.87
Nitrogen free extract	38.09	37.51	36.94	36.37	35.80
Calcium	1.07	1.12	1.18	1.23	1.29
Phosphorus	0.87	0.86	0.86	0.85	0.84
Available phosphorus	0.57	0.56	0.55	0.55	0.54
Lysine	1.22	1.21	1.20	1.20	1.19
Methionine	0.65	0.65	0.66	0.67	0.68

3. Results and discussion

Dietary inclusion of *Portulaca oleracea* (purslane) meal up to a level of 6% into the diets of chickens did not influence the body weight, slaughter weight, carcass weight and yields of carcass cuts (Table 2 and 3). While a study found that the use of diets containing 5.0–7.5% flaxseed resulted in significantly lower weight gain, in this current study body weight of chickens was not different between purslane meal group and control group. These findings were in agreement with a study conducted by (2014), which found that dietary inclusion of flaxseed meal 80 g/kg did not adversely affect body weight gain, carcass weight and the composition of carcass. The results of the study were also supported by [13] which reported that the supplementation of 10% purslane seed did not affect the slaughter weight, carcass, and carcass cut yields of quails [13]. The percentage of carcass in this study was approximately 72%. There were significant differences in cut yields of chickens such as thigh and drumstick. Increasing levels of dietary purslane tended to decrease thigh percentage with the lowest percentage for thigh from 6% purslane meal group (Table 3.). However, decrease in the percentage of thights indicated not much of a difference compared to the control thights and the difference in thigh percentage between control thighs and those from chickens fed 6% purslane meal was not significant. Based on the carcass characteristic data, it is indicated that there were no negative effects of including purslane meal up to a level of 6% to laying hen diets.

Importantly, the inclusion of dietary purslane meal caused a decrease in the weight and percentage of abdominal fat. It appears that feeding chickens with diets enriched with 4.5% purslane meal had a lower percentage of abdominal fat (0.53%) compared to those fed control diet (0.98%). These findings suggested that the decrease in lipid content in chickens was very associated with the composition of dietary fatty acid [15] The decreased in abdominal fat weight by feeding diets containing purslane meal is in agreement with the previous studies conducted by [15], who have reported that abdominal fat significantly decreases in chickens fed diets supplemented with linseed oil rich in n-3 PUFA at a level of 10% ($P < 0.05$).

Table 2. Body weight, carcass, cut yields and abdominal fat of broiler chickens at day 42

Parameters	F0	F1	F2	F3	F4	P value	Significance
			(g				
Body weight	1477.5	1512.5	1393.3	1445.8	1459.2	0.854	NS
Slaughter weight	1475.0	1509.2	1390.8	1444.2	1456.7	0.857	NS
Abdominal fat weight	14.36^{a}	8.90^{ab}	8.20^{b}	7.62^{b}	6.77^{b}	0.010	**
Abdominal fat (%)	0.98^{a}	0.59^{ab}	0.60^{ab}	0.53^{b}	0.46^{b}	0.010	**
Carcass weight	1053.4	1093.4	1007.2	1037.3	1050.1	0.867	NS
Breast weight	327.7	340.7	318.3	327.3	341.3	0.950	NS
Back weight	212.7	208.2	197.2	200.5	198.5	0.757	NS
Thigh and drumstick	299.7	313.2	271.7	269.7	290.5	0.345	NS
Thigh weight	151.2	163.2	138.2	146.5	139.7	0.340	NS
Drumstick weight	148.0	149.7	134.0	131.7	151.0	0.368	NS
Wing weight	100.3	105.8	105.7	116.7	102.7	0.326	NS

F0: Basal diet + 0% purslane meal, F1: basal diet + 1.5% purslane meal, F2: basal diet + 3% purslane meal, F3: basal diet + 4.5% purslane meal and F4: basal diet + 6% purslane meal
** (P<0.01); NS= not significant

Table 3. Percentage of carcass and cut yields of broiler chickens at day 42

Parameters	F0	F1	F2	F3	F4	P value	Significance
			(%)				
Carcass	71.24	72.41	72.52	71.97	72.12	0.921	NS
Breast	30.89	31.25	31.46	31.49	32.16	0.855	NS
Back	20.27	19.00	19.67	19.34	19.03	0.306	NS
Thigh and drumstick	28.40^{ab}	28.59^{a}	26.89^{ab}	26.06^{b}	27.72^{ab}	0.025	*
Thigh	14.35^{ab}	14.92^{a}	13.60^{ab}	14.12^{ab}	13.33^{b}	0.041	*
Drumstick	14.00^{ab}	13.63^{ab}	13.34^{ab}	12.7^{b}	14.4^{a}	0.028	*
Wing	9.53^{b}	9.70^{b}	10.55^{ab}	11.26^{a}	9.76^{b}	0.000	**

F0: Basal diet + 0% purslane meal, F1: basal diet + 1.5% purslane meal, F2: basal diet + 3% purslane meal, F3: basal diet + 4.5% purslane meal and F4: basal diet + 6% purslane meal
* significant (P<0.05); ** (P<0.01); NS= not significant

4. Conclusions
Based on the data, we concluded that dietary inclusion of *Portulaca oleracea* (purslane) meal up to a level of 6% didn't have negative effects on carcass quality and cut yields of broiler chickens.

Acknowledgement
This research was funded by the research grant "Pusat Unggulan Perguruan Tinggi (PUPT-DIKTI) 2017" Project Number: 873/UN27.21/PP/2017.

References
[1] Lopez-Ferrer S, Baucells MD, Barroeta AC and Grashorn MA 2001 N-3 enrichment of chicken meat. 1. Use of very long-chain fatty acids in chicken diets and their influence on meat quality: fish oil *Poult. Sci.* **80** 741–52

[2] Hulan HW, Proudfoot FG, Ackman RG and Ratnayake WM 1988 Omega-3 fatty acid levels and performance of broiler chickens fed redfish meal or redfish oil *Can. J. Anim. Sci.* **68** 533–47

[3] Bou R, Guardiola F, Barroeta AC and Codony R 2005 Effect of dietary fat sources and zinc and selenium supplements on the composition and consumer acceptability of chicken meat. *Poult. Sci.* **84** 1129–40

[4] Chekani-Azar A, Shahriar HA and Maheri-Sis N 2008 Omega-3 fatty acids enrichment and organoleptic charateristic of broiler meat. *Asian J. Anim. Vet. Adv.* **3** 62–9

[5] Febel H, Mezes M, Palfy T, Herman A, Gundel J, Lugasi A, Balogh K, Kocsis I and Blazovics A 2008 Effect of dietary fatty acid pattern on growth, body fat composition and antioxidant parameters in broilers. *J. Anim. Physiol. Anim. Nutr.* (Berl) **92** 369–76

[6] Zelenka J, Schneiderova D, Mrkvicova E and Dolezal P 2008 The effect of dietary linseed oils with different fatty acid pattern on the content of fatty acids in chicken meat. *Vet. Med.* **53** 77-85

[7] Kartikasari LR, Hughes RJ, Geier MS, Makrides M and Gibson RA 2012 Dietary alpha-linolenic acid enhances omega-3 long chain polyunsaturated fatty acid levels in chicken tissues *Prostaglandins Leukot Essent Fatty Acids* **87** 103–9

[8] Uddin MK, Juraimi AS, Hossain MS, Nahar MAU, Ali ME and Rahman MM 2014 Purslane weed (Portulaca oleracea): a prospective plant source of nutrition, omega-3 fatty acid, and antioxidant attributes *Scientific World J.* 1–6

[9] Aydin R and Dogan I 2010 Fatty acid profile and cholesterol content of egg yolk from chickens fed diets supplemented with purslane (*Portulaca oleracea* L.) *J. Sci. Food Agric.* **90** 1759–63

[10] Evaris E, Sarmiento-Franco LA, Segura-Correal JC and Capetillo CM 2015 Effect of dietary inclusion of purslane (*portulaca oleracea l.*) on yolk omega-3 fatty acids content, egg quality and productive performance of Rhode Island Red hens *Trop. Subtrop. Agroec.* **18** 33–8

[11] Kartikasari LR, Hertanto HS, Pranoto D, Salim WN and Nuhriawangsa AMP 2017 Exterior and interior physical quality of egg of laying hens fed diets containing different dietary purslane levels *International Conference On Food Science and Engineering 2016* IOP Conf. Series: Materials Science and Engineering 193 (2017) 012027 doi:10.1088/1757-899X/193/1/012027

[12] Zhao XH, He X, Yang XF and Zhong XH 2013 Effect of Portulaca oleracea extracts on growth performance and microbial populations in ceca of broilers *Poult. Sci.* **92** 1343–7

[13] Konca Y, Beyzi SB, Karabacak M and Yaylak E 2015 The effect of different dietary purslane seed (Portulaca oleracea L.) levels on carcass, blood lipid profile and antioxidant activity in quails *Tavukçuluk Araştırma Dergisi* **12** 1–6

[14] Mridula D, Kaur D, Nagra SS, Barnwal P, Gurumayum S and Singh KK 2015 Growth performance and quality characteristics of flaxseed-fed broiler chicks *J. Applied Anim. Res.* **43** 345–51

[15] Ferrini G, Baucells M D, Esteve-García E and Barroeta A C 2008 Dietary Polyunsaturated Fat Reduces Skin Fat as Well as Abdominal Fat in Broiler Chickens *Poult. Sci.* **87** 528–35

IOP Conf. Series: Earth and Environmental Science **102** (2017) 012089 doi:10.1088/1755-1315/102/1/012089

Effect of extraction temperature on characteristics of chicken legskin gelatin

M Sompie[1*] and A Triasih[2]

[1]Faculty of Animal Husbandry, Sam Ratulangi University, Jl Kampus Kleak, Manado 95115, Indonesia
[2]Faculty of Fisheries and Marine Science Sam Ratulangi University, Jl Kampus Kleak, Manado 95115, Indonesia

*E-mail: meitysompie@yahoo.com

Abstract. Gelatin is a denaturalized protein that is derived from collagen by acidic or alkaline hydrolysis and is an important functional biopolymer that has a very broad application in many industrial fields. Its functional properties depend on processing conditions as well as the raw material. The objective of the research was to study effect of extraction temperature on characteristics of native chicken legskin gelatin. This study used Completely Randomized Design (CRD) with four treatments (T1 = 50^{0}C, T2 = 55^{0}C, T3 = 60^{0}C, T4 = 65^{0}C) and five replications. Statistical analysis were carried out by one Anova and the mean difference was tested using Duncan's Multiple Range Test. The result of research indicated that, extraction temperature had significant effect ($P<0.05$) on yield, gel strength, viscocity and protein content of chicken legskin gelatin, but it had no significant effect ($P>0.05$) on water content. It was concluded that the use of extraction temperature 60^{0}C was (yields 13.75,gel strength 78.75 g bloom,viscosity 6.52 cP, protein content 84.23% and water content 6.20%).

Key words *: Chicken legskin, Extraction, Temperature*

1. Introduction

Gelatin is a protein of animal origin, that can be obtained from collagen by acidic or alkaline hydrolysis. Gelatin is a denaturalized protein that is derived from collagen and is an important functional biopolymer that has a very broad application in many industrial fields. The quality of gelatin depends on its physichochemical properties, rheological properties and manufacturing method. Scientists has been doing many research of gelatin from pigskin from goat skin and from Tuna skin [1, 2, 3]. The use of gelatin to form edible films or coatings was very well studied in the 1960's, which resulted in many patents being filed, mainly in the pharmaceutical area [4, 5]. The process of gelatin production required a curing step and temperature extraction to improve quality of gelatin [2] . The application of the curing time and concentration of acetic acid 3.5 % had significant effect on physic properties of gelatin chicken legs skin [2, 6]. However, further effects of different temperature extraction on the alkali process from chicken legs gelatin was limited information. Thus, the research was aimed to study the effect of extraction temperature on characteristics of chicken legskin gelatin.

2. Materials and Methods

Seven hundred (700) g chicken skin were used as a raw material and acetic acid (CH_3COOH) as a curing material. Gelatin was produced by the acid extraction method [2, 7] .

2.1. Procedure

Gelatine was prepared by the acid extraction method [8]. Acetic acid (CH_3COOH) concentrations of 3% (v/v) were used as a curing. The raw material were soaked for 24 hours. After soaked, samples were neutralized to pH 6, weighed and extracted. The extraction temperature process were performed on five steps (each step for 5 hours), the first step at 50 ^{0}C, second step at 55 °C, third step at 60 °C and then 65°C as a treatments. Solubilized gelatin was separated from residual skin fragments by filtration through a nylon filter. The extracted gelatin was concentrated at 70 °C for 5 hours and it was stored in the refrigerator 5-10 °C for 30 minutes, then dried at 60 °C for 24-36 hours until the gelatin sheet solid. Gelatin sheets were milled and packaged in vacuum plastic and stored in a desiccator for subsequent process. The extracted gelatin at 60 °C for 6 hours and it was stored in the refrigerator 5-10 °C for 30 minutes, then dried at 60 °C for 24-48 hours until the gelatin solid. Gelatin sheets were milled and packaged in vacuum plastic and stored in a desiccator before analysis.

2.2.Experimental design and data analysis

The experiment were determined by analysis of Completely Randomized Design (CRD) with four treatments (T1 = 50^{0}C, T2 = 55^{0}C, T3 = 60^{0}C, T4 = 65^{0}C) and five replications. Statistical analysis were carried out by one Anova and the mean difference was tested using Duncan's Multiple Range Test [11].

2.3. Parameters

The parameters of this research were yields, gel strength, viscosity, protein content and water content. The yields obtained from dry weight ratio of raw material an d the weight of the extracted native chicken legskin gelatin multiplied by 100%. Gel strength was determined with a Universal Testing Machine (Zwick/Z0.5). The value og gel strength (g Bloom) use the formula = 20 + 2.86 x 10-3D, where D = F/G x 980: F = height chart before fracture; G = constant (0.07). Viscosity was measured by gelatin powder dissolved in distilled water at temperature of 40°C with a solution concentration of 6,67%.

3. Results and Discussion

Table 1. The characteristics of native chicken legskin gelatin

Extraction (°C)	Yield (%)	Gel Strength (g Bloom)	Viscocity (cP)	Protein Content (%)	Water Content (%)
50	12.28±0.25^a	77.56±0.08^a	7.02±0.14^a	83.46±0.24^a	6.23±0.62
55	13.53±0.12^b	78.67±0.10^b	6.54±0.10^b	84.10±0.08^{ab}	6.22±0.21
60	13.75±0.05^b	78.75±0.29^b	6.52±0.16^b	84.23±0.48^b	6.20±0.23
65	14.14±0.01^{bc}	79.03±0.12^c	6.07±0.19^b	84.02±0.22^{ab}	6.20±0.32

Different letters in the same row and colum indicated the significant differences (P<0.05)
Sd = standard deviation

3.1.Yields

Statistical analysis indicated that the differences of extraction temperature had significant effect (P<0.05) on yield of native chicken legskin gelatin. The value of yield decrease with increasing level of extraction temperature. Duncan test resulted that the yield of gelatin tended to increase with increasing the level of extraction temperature. The acetic acid solution used to hydrolyze collagen making it easier solubility in hot water when the extraction of gelatin. Yield values from native chicken legskin gelatin was ranged 12.28 to 14.14 %. It values was not different with yield from chicken skin gelatin was 13.01 to 14.42 [3, 7].

3.2. Gel strength
Gel strength average of native chicken legskin gelatin was displayed in Table 1. Statistical analysis indicated that the differences of extraction temperature had significant effect (P<0.05) on gel strength of native chicken legskin gelatin. The value of gel strength tended to decrease with increasing level of extraction temperature. The presence of hydroxyproline caused the stability of the hydrogen bonds between water molecules and free hydroxyl groups of amino acids in gelatin, it is very important for gel strength.The gel formation of a stable condition that ability of a free chain to form a lot of crosslinking [9]. Gel strength of gelatin is very important on physical properties of gelatin. Gel strength values from native chicken legskin gelatin was ranged 77.56 – 79.03 g Bloom, that in line with the criteria of ISO 75-300 g Bloom [2, 3, 7, 10].

3.3. Viscocity
Statistical analysis indicated that the extraction temperature had no significant effect (P>0,05) on gelatin. The value of viscocity tended to decrease at the extraction temperature increased. In order words, the higher concentration, the viscosity was tended to inreased. This is because the viscosity of gelatin is directly proportional to the gel strength that was not significantly different between treatments. The viscosity is affected by molecular weight and amino acid chain length [10]. Increased concentrations of acetic acid in the gelatin production process can reduce the viscosity. The curing material has been breaking the peptide bonds of amino acids into short-chain molecule so that its viscosity decrease. Viscocity values from native chicken legskin was ranged 6.07 to 7.02 cP. It values is included in the ISO range 2.0 to 7.5 cP [7, 10].

3.4. Protein Content
Gelatin is the collagen protein, a group derived from the structural proteins and extracellular matrix and produced in large quantities [2, 9]. Statistical analysis indicated that the different of extraction temperature had significant effect (P<0.05) on protein content of chicken legskin gelatin. Duncan test results showed that protein content of gelatin from chicken legskin had a tended to increase with increasing temperature of extraction. This is beacause age slaughter affect the content of collagen in the skin, increasing age increased collagen protein. Protein content from chicken legskin gelatin ranged 83.46 to 84.02 % [6] That it was not different with protein content from chicken leg skin ranged 83-90 % and commercial gelatin, 91, 63% [2, 8].

3.5. Water Content
The water content average of native chicken legskin gelatin was presented in Table 1. Statistical analysis indicated that the differences in level extraction temperature had no significant effect (P>0.05) on water content of gelatin. Water content of gelatin had a tended to decreased with increasing level of temperature. This is because the denaturation in molecular changes and the amount of water that is bound to decline collagen structure. The water holding capacity will lead to volatile water during the drying gelatin and water content becomes lower [2, 7]. Water content from native chicken legskin gelatin ranged 6.20 to 6.23 %. That it was not different with water content from chicken leg skin ranged 7.12-7.44 % [3, 7, 10].

4. Conclusion
It was concluded that the use of extraction temperature 60^0C was the best characteristics of native chicken legskin gelatin (yields 13.75, gel strength 78.75 g bloom, viscosity 6.52 cP, protein content 84.23% and water content 6.20%).

References
[1] Agustin AT and Sompie M 2014 The Effect of Acetic Acid on Characteristics of Tuna Fish Skin *Proc. Int. Conf Challenges of Biotechnological Research in Food and Health* Solo pp 103-104

[2] Said MI, Triatmojo S, Erwanto Y and Fudholi A 2011 Gelatin properties of goat skin produced by calcium hydroxide as curing material *Media Peternakan* **34**(3) 184-189

[3] Sompie M, Triatmojo S, Pertiwiningrum A and Pranoto Y 2012 The effects of animal age and acetic acid concentration on pigskin gelatin characteristics *J. Indonesian Tropical Animal Agriculture* **37** (3) 176-182

[4] Antoniewski, MN, Barringer SA, Knipe CL and Zerby HN 2007 Effect of a gelatin coating on the shelf life of fresh meat *J. Food Sci.* **72** E382-387

[5] Park JW, Whiteside WS and Cho SY 2008 Mechanical and water vapor barrie roperties of extruded and heat-pressed gelatin films *LWT* **41** 692-700

[6] Swatland HJ 1984 Structure and Development of Meat Animals (New Jersey: Prentice-Hall Inc)

[7] Sompie M, Surtijono SE, Pontoh JW and Lontaan N 2015[a] Effect of acetic acid concertration and temperature extraction on physical and chemical properties of pigskin gelatin *Procedia Food Science* **3** 383-388

[8] Ockerman HW and Hansen CL 2000 Animal by Product Processing and Utilization (USA: CRC Press)

[9] Sims TJ, Bailey AJ and Field DS 1997 The chemical basis of molecular weight differences in gelatins *The Imaging Science J.* **45** 171-177

[10] Sompie M, Siswosubroto SE and Pontoh JHW 2015[b] Effect of Acetic Acid Concertration and Curing Time On the Characteristics of Native Chicken Legs skin Gelatin *Proc. The 6th ISTAP* (**2**) 714-718

The effect of alpha amylase enzyme on quality of sweet sorghum juice for chrystal sugar

T Marwati [1*], N Cahyaningrum[1], S Widodo[1], U T Astiati[2], A Budiyanto[3], Wahyudiono[3], A B Arif[3] and N Richana[3]

[1]Assessment Institute for Agricultural Technology Yogyakarta, Indonesia
[2]Faculty of Agricultural Technology, Bogor Agricultural University, Indonesia
[3]Indonesian Center for Agricultural Postharvest Research and Development Bogor

*E-mail: watipasca@yahoo.com

Abstract. Sweet sorghum juice (*Sorghum bicolor* L. Moench) has characteristics similar to sugar cane juice and potentially used for sugar substitutes that can support food security. Nevertheless the sweet sorghum juicecontain starch which impede sorghum sugar crystallization. Therefore, research on the enzymatic process is needed to convert starch into reducing sugar. The experimental design used was the Factorial Randomized Design with the first factor was alpha amylase enzyme concentration (0, 20, 40, 60, 80, 100, 120 μL/100 mL) and second factor was incubation time (0, 30, 60, 90 minute) at temperature 100°C. The experiment was conducted on fresh sweet sorghum. The results showed that the addition of the alpha amylase enzyme increased the content of reducing sugar and decreased levels of starch. Elevating concentration of alpha amylase enzyme will increase the reducing sugar content in sweet sorghum juice. The optimum alpha amylase enzyme concentration to produce the highest total sugar was 80 μL/100 mL of sweet sorghum juice with the optimum incubation time was 90 minutes. The results of this study are expected to create a new sweetener for sugar substitution.From the economic prospective aspect, sorghum is a potential crop and can be relied upon to support the success of the food diversification program which further leads to the world food security

Keywords: Sweet sorghum, enzyme, chrystal sugar

1. Introduction

Sweet sorghum (*Sorghum bicolor* (L.) Moench) juice similar with sugarcanethat may have potential as a source of sugar, such as the crystallized sugar. According to [1,2], the kind of sugar in the sorghum stalk, it can be divided into saccharin-type sweet sorghum and syrup-type sweet sorghum. Saccharin-type sweet sorghum, contains sucrose and can be used for refining crystal sugar. Syrup type sweet sorghumis used for producing syrup as it mainly contains glucose. The sugars content in sweet sorghum stalk juice mostly are sucrose and invert sugars such as glucose, fructose, maltose and xylose. Previous study reported that mannose, galactose and arabinose were not detected in sweet sorghum juice [1]. The syrup rich in vitamin C, calcium, proteins, riboflavin and nicotinic acid and was free of sulphur or any pesticide residues. The syrup being a table syrup, can be used in salad dressing, as a sweetener in baked goods or as ice-cream topping [3].

The crystallized sugar powder is easy to be used that causes the soluble sugars so practical in use, easy to be packed and distribution, and has a long shelf life because of low water content[4]. Sweet sorghum crystallized sugar (SSCS) can also be used for bread making, pancakes, salad, sweetening agent in a bakery and ice cream or pharmaceutical [5]. Sweet sorghum cultivars; Rio, CMSXS636, IESV91018LT, IESV93042SH and SPV1411 have potential in crystal raw sugar production because they have relatively higher sucrose concentration and AP greater than 75% [6].

Research on the hydrolysis process is needed to convert starch into reducing sugar. The hydrolysis process can beachieved by acid, acid-Enzyme orenzyme hydrolysis methods. However, acidhydrolysis has been replaced by enzymatic hydrolysisbecause it required the use of corrosion resistant materials,gave risecolour and salt ash and need moreenergy for processing[7]. Enzymatic hydrolysis is essential for the production of glucose syrups from starch because of the specificity of the linking bonds that characterize its physical and chemical properties. In addition the enzymatic method reduced the number of side reactions that accompany acid or alkali hydrolysis[8]. Therefore theenzyme hydrolysis isthe most advanced industrial enzyme application in the foodindustry and is one of the most successful enzymetechnology. Various grains may be used as a source ofenzyme during the hydrolysis of starch to glucose [9].

The α-amylases (1,4-α-d-glucan-4-glucanohydrolase; EC 3.2.1.1) are ubiquitous enzymes, and they play a dominant role in carbohydrate metabolism. The α-amylases from bacteriaand fungalsources have dominated applications in industrial sectors [10]. *Bacillus subtilis, B. stearothermophilus, B. licheniformis* and *B. amyloliquefaciens* are known to be good producers of α-amylase, and these have been widely used for commercial production of the enzyme for various applications.Large quantities of α-amylase were induced in sorghum when such grains were malted using traditional techniques [11]. Starch can be used to produce syrup containing fructose, glucose, or maltose, which is widely used in the food industry [12].

Sorghum starch hydrolysis using the alpha amylase enzyme has been studied by [13,14,15].The optimum conditions for the glucose production from sorghum malt using α-amylaseis gelatinization at 80°C and liquefaction of amylase at a 90°Cat pH 4[13]. The suspension should be brought to high temperatures (90-110°C) to initiate α-amylase for the starch kernel breakdown [16]. Other study found that, liquefaction under pressurized steam at 104°C was more effective compared to using a water bath at 95 °C[17]. The increase in hydrolysis of starch has the correlation with the effect of high heat or temperature on the weaker areas on the starch granule, allowing the enzyme to break the granules of starch more extensively [18].

The sweet sorghum for crystal sugar production must have a high content of sucrose in the stalks and low levels of starch which impede sorghum sugar crystallization [19,20].To achieve this goal, the starch present in the sorghum sap should be first hydrolyzed into its reducing sugars. Therefore, the aim of this research was to study the effect of alpha amylase enzyme on quality of sweet sorghum juice for crystal sugar.

2. Materials and Methods

2.1. Materials

The raw material used is sweet sorghum stalk for producing sweet sorghum juice obtained from Yogyakarta, Indonesia. The enzyme used is a thermostable α-amylase enzyme [13]derived from *Bacillus licheniformis*[11,21,22]under the trade name Termamyl[14, 24].Thereagents used were Pb-acetate, $(NH4)\ _2HPO_4$ 10% solution, Luff-Schoorl reagent, 25% H_2SO_4, 5% starch, Sodium thiosulfate 0.1N, 20% KI solution, 25% Hydrochloric acid (HCl) solution, and 50% Sodium Hydroxide solution (NaOH) solition, 0.1% phenolphthalein, 37% HCl, 3,5-Dinitrosalicyclic acid (DNSA) solution, activated carbon, glucose, and distilled water.

2.2. *Enzymatic process of sweet sorghum juice*

The sweet sorghum stalk was cleaned and then extracted by series of mills [24].The juice coming out of milling section was first screened and heated up to 100°C [25,26].A total of 100 mL of boiled sorghum sauce were taken and put into a glass bottle and various quantity of α-amylase enzyme was added. The concentration of α-amylase enzyme and incubation time were varied on the basis of statistical experimental design[14].The experimental design used was the Factorial Randomized Design with the first factor was alpha amylase enzyme concentration (0, 20, 40, 60, 80, 100, 120 µL/100 mL) and second factor was incubation time (0, 30, 60, 90 minute) at temperature 100°*C*.

2.3. *Analysis of sweet sorghum juice*

Sweet sorghum juice after the enzymatic process were analyzed for reducing sugar and total sugar by using Luff-Schoorl method[26,27],for starch using DNS method (Bintang 2010), for viscosity using a brookfield programmable DV-III rheometer [28]for brix using refractometer[29,30].and for pH using pH meter[27,31].

3. Results and Discussion

3.1. *Reducing sugar content in sweet sorghum juice*

The fresh sweet sorghum juicereducing sugaris given in Figure 1.The addition of the alpha amylase enzyme increased the content of reducing sugar and decreased levels of starch. Increased concentration of alpha amylase enzyme will increase the reducing sugar content in sweet sorghum juice. The optimum alpha amylase enzyme concentration to produce the highest total sugar was 80 µL/100 mL of sweet sorghum juice with the optimum incubation time was 90 minutes.The results of this study are in accordance with those reported by[14, 15].

The amount of α-amylase in hydrolysis process of sorghum strach gave positive effect to the response on dextrose equivalent (reducing sugar expressed as glucose). On the other hand, the liquefaction time has slightly negative effect on the amount of dextrose equivalent. As the liquefaction time increases, the amount of dextrose equivalent decreases [15].This is because the longer exposure of the enzyme to high temperatures, which are needed for gelatinization of the starch granules and for achieving a good susceptibility to enzyme action, could lead to slight enzyme deactivation [32].The best reducing sugar production by hydrolysis using α-Amylase of *manihot esculenta* was observed when temperature of 74°C, pH of 5.5 and incubation time of 90 minute[14]. Liquefaction of 25% (w/v) sorghum slurry was completed in 60minute with *B. licheniformis* a-amylase (BLA) concentration of 0.08% (v/w) of flour in the absence of CaCl2 supplementation[33]. However[17]mentioned that the optimum α-amylase enzyme concentration was 0.15% (v/w) for 60 minute of liquefaction time.

In order to get the carbon and energy stored in starch, α-amylase enzyme used to break down the polymer to smaller sugar units, which is eventually converted to the individual basic glucose units [21,22,23].The bacterial alpha-amylase randomly attacks only the alpha-1,4 bonds [34]. The enzyme mode of action depend on the starch source and enzyme system. The enzymatic hydrolysis percentages increased when the granules were incubated with alpha-amylase and that the granule susceptibility was affected by granule size and concentration of the hydrolysis products [34,35]. Reducing sugar production was strongly affected by the variation of variables on alpha amylase and glucoamylase hydrolysis of cassava starch [14].

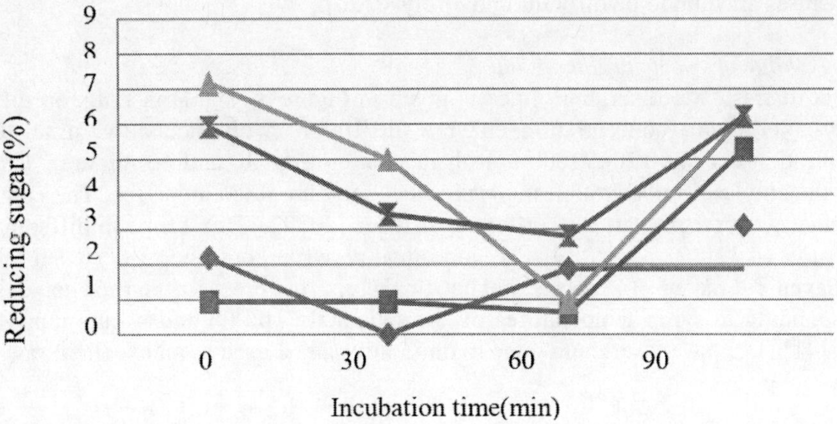

Figure 1. The reducing sugar content of fresh sweet sorghum juice at the concentration of enzyme 0(-♦-), 40(-■-), 80(-▲-), and 120(-x-) (µl/100 mL) with incubation time 0, 30, 60 and 90 minutes

3.2. Total sugar content in sweet sorghum juice

The total sugar of fresh sweet sorghum juice is given in Figure 2.The total sugar content of fresh sugar sorghum juice is in the range of 3.87 to 20.88%. The enzyme concentration of 80 µL / 100 mL of sweet sorghum juice is the optimum concentration of enzyme to produce the highest total sugar. At incubation time 90 minutes, the total sugar produced tends to be more than the other incubation time.

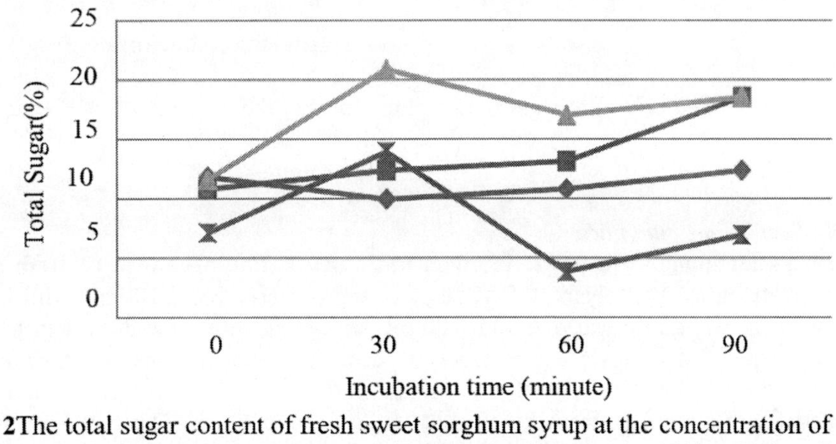

Figure 2 The total sugar content of fresh sweet sorghum syrup at the concentration of enzyme 0(-♦-), 40(-■-), 80(-▲-), and 120(-x-) (µl/100 mL) with incubation time 0, 30, 60 and 90 minutes.

The decrease in total sugar content can also be caused by the formation of other molecules such as ethanol, lactic acid, acetate, and pyruvate. Total sugar content in sweet sorghum juice may decrease after being stored for two weeks[38]. Levels of sucrose produced after storage for two weeks decreased and formed ethanol and organic acids, which formed by the presence of bacteria.

3.3 Starch content in sweet sorghum juice

Average starch content in fresh sweet sorghum tends to decrease with increasing concentration of enzyme amylase. The difference of fresh sweet sorghum starch content and after storage is caused by the hydrolysis process during storage. The results of this study fit the statement that the process of starch hydrolysis runs slowly[36]. In addition, the presence of amylopectin content in the sorghum starch causes amylum to be difficult to hydrolyze [20].

3.4. Brix value of sweet sorghum juice

The brix of fresh sweet sorghum juice is given in Figure 3. The Brix range in different treatment of hydrolysis is 16.00 (concentration enzyme 0µl/100mL with incubation time 30 minute) to 24.2 (concentration enzyme 120µl/100mL with incubation time 30 and 60 minute). This suggests that the higher the amylase concentration the brix sweet sorghum juice increases. The content of sugar in the juice of sweet sorghum varies in different varieties [40].The Brix range in different varieties of sweet sorghum is 14.32 - 22.85% ([39].The good quality syrup could be made when the raw juice was at least havethe brix of 14 degrees. The final brix of prepared syrup generally 70-75 degrees (corresponding to syrup temperatures of approximately 106^{0}C) and a minimum shelf-life of 6 to 9 months[41]. Thus sweet sorghum juice in this study can be used as an excellent raw material.

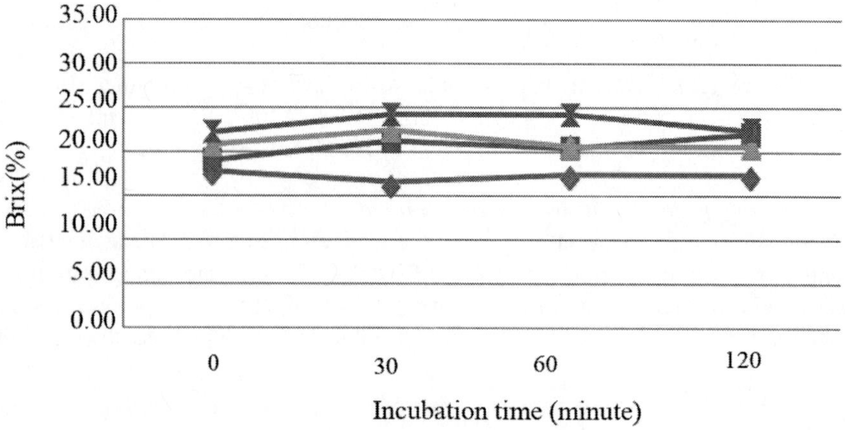

Figure 3. The brix of fresh sweet sorghum juice at the concentration of enzyme 0(-♦-), 40(-■-), 80(-▲-), and 120(-x-) (µl/100 mL) with incubation time 0, 30, 60 and 120 minutes.

3.5. pH of sweet sorghum juice

The fresh sweet sorghum juice pH is given in Figure 4. The pH range of fresh sweet sorghum in different treatment of hydrolysis is 5.85 (concentration enzyme 0µl/100mL with incubation time 90 min) to 6.67 (concentration enzyme 40µl/100 mL with incubation time 90 minute). The soluble starch hydrolysisby purified α-amylase was worked out at pH ranging from 4.0 to 9.0 at 60ºC[42]. The optimum enzyme activity was found in pH range of 6.5 to 7.5 with maximum activity at pH 7.0.The higher activity of α-amylase from *Bacillus licheniformis* mutant in pH range from 4.0 to8.0[43]. Juice can be made into crystal sugar if it has a pH between 6.00 to 6.50 [31]. In this study, fresh sugary sorghum juice was eligible for use as a raw material for making crystal sugar with concentration enzyme treatment 40µl/100mL with incubation time 90 minute.

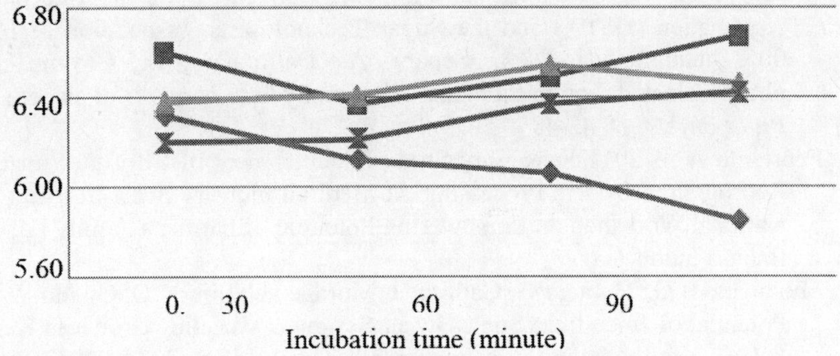

Figure 4. The pH of fresh sweet sorghum syrup at the concentration of enzyme 0(-♦-), 40(-■-), 80(-▲-), and 120(-x-) (µl/100 mL) with incubation time 0, 30, 60 and 90 minutes.

3.6. Viscosity of sweet sorghum juice

The viscosity of fresh sweet sorghum juice is given in Figure 5. The pH range of fresh sweet sorghum in different treatment of hydrolysis is 4.00 (concentration enzyme 0µl/100mL with incubation time 0 min) to 5.60 (concentration enzyme 120µl/100 mL with incubation time 60 minute).The addition of alpha amylase enzyme concentration will decrease the viscosity of hydrolyzate[44], but the viscosity of hidrolyzates depends on the kind of starch. Viscosity is influenced by the solute contained in it. The higher the viscosity of a liquid the more the dissolved substance in it [45].

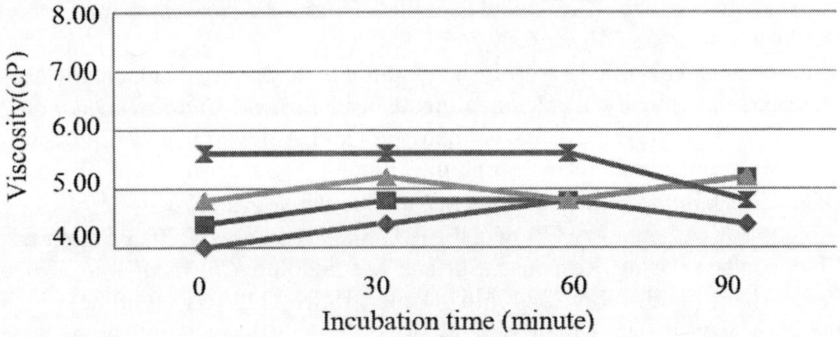

Figure 5. The viscosity of fresh sweet sorghum syrup at the concentration of enzyme 0(-♦-), 40(-■-), 80(-▲-), and 120(-x-) (µl/100 mL) with incubation time 0, 30, 60 and 90 minutes.

4. Conclussion

The addition of the alpha amylase enzyme increased the content of reducingsugar and decreased levels of starch. Increased concentration of alpha amylase enzyme will increase the reducing sugar content in sweet sorghum juice. The optimum alpha amylase enzyme concentration to produce the highest total sugar was 80 µL/100 mL of sweet sorghum juice with the optimum incubation time was 90 minutes. This study are expected to create a new sweetener for sugar substitution.

References

[1] Almodares A, Taheri R, Adeli S 2008c Stalk yield and carbohydrate composition of sweet sorghum (*Sorghum bicolor* (L.) Moench) cultivars and lines at different growth stages*J. Malesian Appl. Biol* **37** 31-36
[2] Anglani C 1998 Sorghum carbohydrates - A Rev*Plant Food Human Nutr*. **52** 77-83
[3] Rajvanshi A K., Jorapur R M, and Nimbkar N 1994 Sweet sorghum R & D for food and

fuel.Pages 29-34 To explore the possibilities of cultivation and processing of supplementary sugarcrops : proceedings of VII Joint Convention of the Deccan Sugar Technologists' Association (DSTA) and the Sugar Technologists Association of India, DSTA, Pune 411 005, IndiaAlmodares, A., Sepahi, A., Dalilitajary, H., Gavami, R. (1994a). Effect of phenological stages on biomass and carbohydrate contents of sweet sorghum cultivars. *Ann. Plant Physiol.,* **8** 42-48

[4] Febrianto A M 2011 Feasibility Study Establishment of Chrystal Sugar Processing Unit With Reprogration System Processing At Medium Industry Scale In Blitar District. Proceeding of National Workshop on Empowering Potential of Farmer's Family For Poverty Alleviation (in Bahasa Indonesia)

[5] Mukabane B G, Thiongo G, Gathitu B, Murage H, Ojijo N O, Owino W O 2014 Evaluating the Potential of Juice from Some Sweet Sorghum Varieties Grown In Kenya to Crystallize*Food Science and Quality Management***30** 31-40 ISSN 2224-6088 (Paper) ISSN 2225-0557 (Online)

[6] Silva R N Quintino F P, Monteiro N V, Asquieri E R 2010 Production of glucose and fructose syrups from cassava (*Manihot esculenta* Crantz) starch using enzymes produced by microorganisms isolated from Brazilian Cerrado soil*Tecnol.Aliment.* **30** (1) 27–31

[7] Lin Y and Tanaka S 2006 Ethanol fermentation from biomass resources: Current state and prospect*Applied Microbiol. Biotechno*l.**69** 627–642

[8] Garba M U, Mohammed A and Etim E D 2012 A kinetic study of the enzymatic hydrolysis of cassava starch *International Journal of Science and Engineering Investigations***1** (11) 65–70

[9] Pandey A, Nignam P, Soccol C R, Soccol V T, Singh D, and Mohan R 2000 Advances in microbial amylases. *Biotechnol. Appl. Biochem*., **31** 135–152

[10] Adewale I O, Agumanu E E and Otith-Okoronkwo F 2006 Comparative studies on α-amylases from malted maize (*Zea mays*), millet (*Eleusine coracana*) and sorghum (*Sorghum bicolor*)*Carbohydr. Polym*.**66**71–74

[11] Roy I and Gupta M N 2004 Hydrolysis of starch by a mixture of glucoamylase and pullulanase entrapped individually in calcium alginate beads*Enzyme Microb Technol.***34** 26–32

[12] Ayoola A A, Adeeyo O A, Efeovbokhan V C and Ajileye O A 2012 Comparative study on glucose production from sorghum bicolor and manihot esculenta species in nigeria*International Journal of Science and Technology***2** (6) 353–357

[13] Aderibigbe FA, Adejumo A L, Owolabi R U and Anozie A N 2013 Starch by α-Amylase and Glucoamylase Using Response Surface Methodology*Chemical and Process Engineering Research*. **9,** 14-23. ISSN 2224-7467 (Paper) ISSN 2225-0913 (Online)

[14] Najiah N, Maizirwan M, Ismail A K M, Rosli M Y 2014 Hydrolysis of Sorghum Starch for Ethanol Production. https://www.researchgate.net/publication/270049330 accessed date 5th August 2017

[15] Sánchez O J and Cardona C A 2008 Trends in biotechnological production of fuel ethanol from different feedstocks*Bioresour. Technol.***99** (13) 5270–5295

[16] Aggarwal N K, Nigam P, Singh D and Yadav B S 2001 Process optimization for the production of sugar for the bioethanol industry from sorghum, a non-conventional source of starch. *World J. Microbiol. Biotechnol.***17**(4) 411–415

[17] Shariffa Y N, Karim A A, Fazilah A and Zaidul I S M 2009 Enzymatic hydrolysis of granular native and mildly heat-treated tapioca and sweet potato starches at sub-gelatinization temperature *Food Hydrocolloids***23** (2) 434–440

[18] Kulp K 2000 Handbook of Cereal Science and Technology 2nd ed. Revised and Expanded (New York: CRCPress)

[19] Sirappa M P 2003 Sorgum Development Prospects in Indonesia as Alternative Commodities for Food, Feed and Industry (in Bahasa Indonesia) Jurnal Litbang Pertanian Balai Pengkajian Teknologi Pertanian Sulawesi Selatan Makassar

[20] Toby H, Xuqiu T, Gerhard F, Walter C, Mark C, David L, John M, Jay M S, Dan E R and Carl M A 2002 A novel, Hhigh performance enzyme for starch liquefaction; discovery and optimization of a low pH, thermostable –amylase *J. Biol. Chem.***277** 26501–26507

[21] Satish D S and Aniruddha B P 2007 Hydrolysis of soluble starch using Bacillus licheniformis α-amylase immobilized on superporous CELBEADS *Carbohyd. Res.***342** 997-1008

[22] Delphine P J, Marie P B, Nadine Z and Gilbert M R 2000 Kinetics of cassava starch hydrolysis with Termamyl® enzyme*Biotechnlol. Bioeng.***68** 71–77

[23] Almodares, A, Taheri R and Safavi V 2008e Sorghum, Botany, Agronomy and Biotechnology, Jahad Daneshgahi of University of Isfafan, Isfahan

[24] Quintero J A, Montoya M I, Sanchez O J, Giraldo O H and Cardona C A 2008 Fuel ethanol production from sugarcane and corn: comparative analysis for a Colombian case*Energy***33** 385–399

[25] Bintang M 2010Biochemistry: Research Techniques (in Bahasa Indonesia) (Jakarta: Erlangga)

[26] Harjadi W 1990Basic Analytical Chemistry (in Bahasa Indonesia) (Jakarta: Gramedia)

[27] Marlinah 1992Study of various treatments of keragenan separation on marine algae extraction (Eucheuma cottoni) (in Bahasa Indonesia) Skripsi. Bogor: Institut Pertanian Bogor

[28] Harisutji W 2001Quantitative Analysis Brix and Pol Nira Sugar Cane (in Bahasa Indonesia) (Pasuruan: P3GI)

[29] Khopkar S M 2007Basic Concepts of Analytical Chemistry (in Bahasa Indonesia) ed Saprorahardjo (Jakarta: UI-Press)from: Basic Concept of Analytical Chemistry

[30] Purnomo E 1997Efforts to increase the competitiveness of people's red sugar from the processing of natural sweetener plants (in Bahasa Indonesia). *Skripsi.* Bogor: Institut Pertanian Bogor

[31] Mojović L, Nikolić S, Rakin M and Vukasinović M 2006 Production of bioethanol from corn meal hydrolyzates*Fue,***85** (12-13) 1750–55

[32] Shewale S D and Pandit A B 2009 Enzymatic production of glucose from different qualities of grain sorghum and application of ultrasound to enhance the yield*Carbohydr. Res.***344** (1) 52–60

[33] Manoj A, Pradeep S, Chandraraj K and Sathyanarayana N G 2005 Hydrolysis of starch by amylase from Bacillus sp. KCA102: a statistical approach*Process Biochem.***40** 2499–2507

[34] Franco C M L, Ciacco C F, Tavares D G 1988in Aderibigbe FA, Adejumo A L, Owolabi R U and Anozie A N 2013 Starch by α-Amylase and Glucoamylase Using Response Surface Methodology*Chemical and Process Engineering Research* **9** 14-23 ISSN 2224-7467 (Paper) ISSN 2225-0913 (Online)

[35] Jayanti. (2011). Effect of pH, temperature of α-amylase enzyme hudrolysis and yeast yeast concentration for ethanol production using rice bran (in Bahasa Indonesia) Skripsi Surakarta (ID): Universitas Sebelas Maret

[36] Directorate General of Plantation 1996Sorgum sweet is a potential commodity in the eastern province of Indonesia*Proc. of the Sorgum Prospect Symposium for Agro-industry Development* 17-18 January 1995. Special Edition. Research Institute for Peanut and Tubers Crops (in Bahasa Indonesia) No **4** pp 6–12

[37] Wu X., Propheter J L, Rooney W L, Yu J and Wang D 2010 Feature of sweet sorghum juice and their performance in ethanol fermentation*J Industrial Crops and Products* **31** (1) 164–170

[38] Almodares A and Sepahi A 1996 Comparison among sweet sorghum cultivars, lines and hybrids for sugar production*Ann. Plant Physiol.* **10** 50–55

[39] Almodares A., Sepahi A., Dalilitajary H and Gavami R 1994a Effect of phenological stages on biomass and carbohydrate contents of sweet sorghum cultivars*Ann. Plant Physiol.* **8** 42–48

[40] Rajvanshi A K., Nimbkar N 2015 Sweet sorghum R&D at the Nimbkar Agricultural Research Institute (NARI)*Technical Report* http: www.nariphaltan.org/sorghum.pdf. Access date 5th

August 2017

[41] Ikram-ul-haq, Javed M M, Hameed U and Adnan F 2010 Kinetics and thermodynamic studies of alpha amylase from Bacillus licheniformis mutant *Pak. J. Bot.***42** (5) 3507–3516

[42] Lee S, Oneda H, Minoda M, Tanaka A and Inouye K 2006 Comparison of starch hydrolysis activity and thermal stability of two Bacillus Licheniformis alpha-amylases and insights into engineering alpha-amylase variants active under acidic conditions*J. Biochem.***139**(6) 997–1005

[43] Slaminska L, Wisniewska D, and Grzeskowiak A 2003 Liquefaction of starch by thermostable alpha amylase*Technologia Alimentaria* **2**(2) 17–26

[44] Marlinah 1992Study of Various Treatments of Keragenan Separation on Marine Algae extraction (*Eucheuma cottoni*) (in Bahasa Indonesia) *Skripsi* Bogor: Institut Pertanian Bogor.

Inhibitory activity of bacteriocin produced from *Lactobacillus* SCG 1223 toward *L. monocytogenes*, *S. thypimurium* and *E. coli*

T Marwati [1*], N Cahyaningrum[1], S Widodo[1], T Januarsyah[2] and Purwoko[2]

[1]Assessment Institute for Agricultural Technology Yogyakarta, Indonesia
[2]Faculty of Agricultural Technology, Bogor Agricultural University, Indonesia

*E-mail: watipasca@yahoo.com

Abstract. Bacteriocin is a protein compound which has bactericidal ability against pathogen bacteria. This research aims to study the inhibitory activity of bacteriocin produced from *Lactobacillus* SCG 1223 against *Listeria monocytogenes*, *Salmonella thypimuruim* and *Escherchia coli*. The bacteriocin produce from *Lactobacillus* SCG 1223 in the MRS broth media The experimental design used was Completely Randomized Design . The variations used in this design were percentage of inoculum (5% ,10%), medium pH (4 , 6), incubation temperature (27°C , 40°C), and incubation time (4, 10 ,14 hours). Result showed that bacteriocin from *Lactobacillus* SCG 1223 had wide spectrum toward *L. monocytogenes* ,*S. thypimuruim* and *E. coli* .The highest bacteriocin activity toward L. monocytogenes produced by Lactobacillus SCG 1223 with 10% inoculum in media with initial pH 6, incubation temperature 27°C for 14 hour, toward S. thypimurium produced by Lactobacillus SCG 1223 with in media with initial pH 6, incubation temperature 40°C for 14 hour, and toward E. coli was 1085.81 AU/ml, produced by Lactobacillus SCG 1223 in MRS broth with initial pH 4, incubation temperature 40°C for 14 hour. This study is expected to find a new food preservative that can inhibit the growth of pathogenic bacteria and extend the shelf life of food. From the economic prospective of view, bacteriocin is very promising natural alternative biopreservatives.

Keywords: Bacteriocin, Lactobacillus SCG1223,pathogenic bacteria.

1. Introduction

One of the problems that cause food to have a short shelf life and unsecured is the contamination of spoilage and pathogens bacteria during processing, transportation and storage. To overcome this is done through the use of preservatives. In recent years consumers are more interested in natural biopreservatives. Therefore, the orientation needs to be developed natural preservatives derived from plants or produced by microorganisms.

Protection of food from spoilage and pathogenic microorganisms by lactic acid bacteria (LAB) is through producing organic acids, hydrogen peroxide, diacethyl [1] antifungial compounds such as fatty acids [2] or phenullactic acid [3] and/or bacteriocins [4]. Bacteriocins constitute a heterologous subgroup of ribosomally synthesized antimicrobial peptides [4]. The suitability is due to the fact that this bacteriocin has been secured in food products, non-toxic in eukaryotic cells, becomes inactive by protease enzymes in the digestive tract so as to have no effect on microflora in the gastrointestinal tract and has a suitable characteristic in food processing [5]. The application of bacteriocin in food products can provide several advantages: extending shelf life, improving microbiological quality, reducing the intensity of physical treatment and the use of chemical preservatives [6,7].

Bacteriocin produced by lactic acid bacteria can be degraded by proteolitic enzymes in human digestion and not harmful to human health. In addition, bacteriocin also has a stability to the influence of pH and temperature [8,9,10,6] Bacteriocin has stable activity in acidic and alkaline conditions, so it is potentially exploited by industries that in the process involve both acidic and alkaline conditions . Bacteriocin has stable activity, even after treatment at temperature 20°C to 100°C so it is very good if used in industries that involve hot or cold conditions in the production process [11].

Survival of microbial during these processes strongly depends on the ability of the cells to adapt and become more tolerant to the environmental conditions [12]. The temperature and pH conditions that prevail during sourdough fermentations correspond to the range of conditions for good growth, acidification and bacteriocin production *by L. amylovorus* DCE 471 [1]. The pH factor of the media will affect the growth of bacterial cells will further affect the production of bacteriocin. The adsorptions of bacteriocin (nisin) by the producer cells are dependent on the pH of the culture broth [13]. The production of bacteriocin will increase with increasing pH to optimum pH and then decrease. Temperature has two opposing influences that increase the production of bacteriocin but also can kill bacterial lactic acid bacteria-producing bacteria. Additionally, the stability, solubility, and biological activity of nisin from *Lactobacillus* are dependent on the pH of the solution. In fermentation process, at pH < 6.0, more than 80% of nisin produced is released into the medium, at pH > 6.0, most of the nisin is associated with the cellular membrane, but not the cytoplasm. Solubility and stability increase drastically with the lowering of pH, in addition, nisin is stable at pH 2.0 and, at this pH, can be autoclaved at 121°C for 15 min without inactivation [14], on the other hand, in neutral and alkaline conditions nisin is almost insoluble. Therefore, the aim of this research was to study the inhibitory activity of bacteriocin produced from *Lactobacillus* SCG 1223 against *Listeria monocytogenes*, *Salmonella thypimuruim* and *Escherchia coli*.

2. Materials and Methods

The bacterial strains used in this study were *Lactobacillus* SCG1223 for bacteriocin production and *L. monocytogenes*, *S. thyphimurium* and *E. coli*, for assay of bacteriocin activity. Those strains were obtained from Indonesian Agency for Agricultural Research and Development.

2.1. Preparation of inoculum

Cultures are maintained at -40 °C as frozen stock cultures in equal volumes of 10% non-fat dry milk and 20% glycerol. For activation, *Lactobacillus* SCG1223 added to *de Man–Rogasa–Sharpe* (MRS) broth and incubated at 37°C for 24 hours. Bacterial activation is also performed on the *L. monocytogenes, S. thyphimurium* and *E. coli.*

2.2. *Research Design*

The experimental design used was Completely Randomized Design with four factors: percentage of inoculum, pH of medium, incubation temperature and incubation time. The variations used in this design were percentage of inoculum (5% and 10%), medium pH (4 and 6), incubation temperature (27°C and 40°C), and incubation time (4 hours, 10 hours and 14 hours). Sample code list in Table 1.

Table 1. List of sample code

Code	Percentage of inoculum (5%)			Code	Percentage of inoculum (10%)		
	pH of media	Incubation temperature (°C)	Incubation time (hour)		pH of media	Incubation temperature (°C)	Incubation time (hour)
A1	4	27	4	B1	4	27	4
A2	6	27	4	B2	6	27	4
A3	4	40	4	B3	4	40	4
A4	6	40	4	B4	6	40	4
A5	4	27	10	B5	4	27	10
A6	6	27	10	B6	6	27	10
A7	4	40	10	B7	4	40	10
A8	6	40	10	B8	6	40	10
A9	4	27	14	B9	4	27	14
A10	6	27	14	B10	6	27	14
A11	4	40	14	B11	4	40	14
A12	6	40	14	B12	6	40	14

2.3 Bacteriocin production
Fresh culture of *Lactobacillus* SCG1223 was inoculated at a rate of 5 and 10 % in MRS broth at pH 4 and 6, then incubate in shaker incubator at 27°C and 40°C for 4, 10 and 14 hours (according to Table 1). Subsequently culture was centrifuged at 10.000 rpm, 4°C for 15 minutes to produce a supernatant which was further filtered using a 0.22 µm milipore to produce a cell-free supernatant [15,16,10]. The bacteriocin activity test was performed on natural pH supernatant and neutral pH supernatant.

2.4. Bacteriocin activity test
The *Lactobacillus* SCG1223 bacteriocin activity test was performed by well agar method [17,18]. The assay plate had of Muller Hinton agar and was seeded with about 10^6 indicator bacteria (*L. monocytogenes, S. thyphimurium* and *E. coli*). Then 6 mm wells were bored out and each was filled with 50µl of the bacteriocin fractions. The agar plates were incubated at 37°C for 24 hours. The bacteriocin inhibitory activity against the indicator bacteria will be seen with the appearance of clear zones around the well. The bacteriocin activity unit is defined as the AU (Arbitrary unit), 1 AU representing the area of inhibition per sample volume of bacteriocin samples tested (mm2/ mL).

3. Results and Discussion
The bacteriocin activity toward *L. monocytogenes* pathogenic bacteria can lead to listeriosis and are very detrimental to human health. These bacteria can grow in cold conditions so it needs serious handling for the food industry that requires cold storage. Bacteriocin in natural supernatant pH produced from *Lactobacillus* SCG1223 with 5% inoculum had the highest activity at medium pH 4, incubation temperature 27°C and incubation time 14 hours (**Figure 1**).

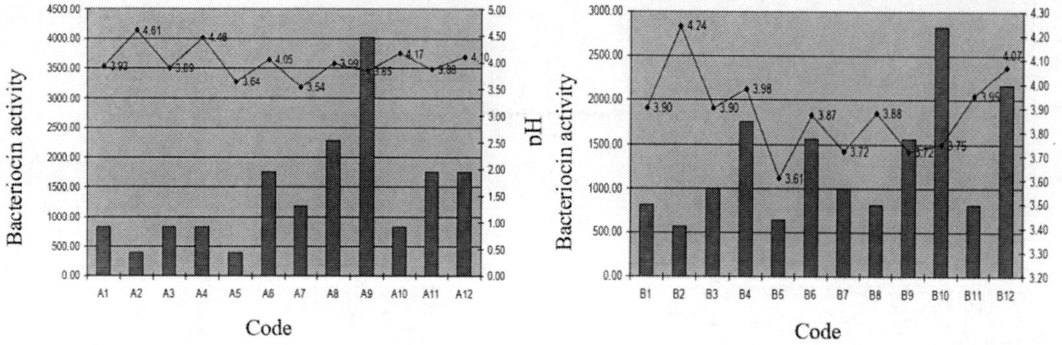

Figure1. Activity of bacteriocin in natural pH with 5% (Code A) dan 10 % (Code B) inoculum toward *L. monocytogenes*

Bacteriocin produced with 10% inoculum *Lactobacillus* SCG1223 had higher activity compared with 5% inoculum percentage (**Figure 1**). This happens because as more cells are present in the media during growth, it will further increase the production of primary metabolites. Bacteriocin is an anti-bacterial protein produced in the ribosomal pathway that occurs during the exponential phase and follows the pattern of synthesis of primary metabolites. The more cells that grow, the use of nutrients in the media will be faster, so it is suspected that with limited nutritional conditions lactic acid bacteria strain SCG 1223 get the right conditions for post-translational process that will convert prebacteriocin into active bacteriocin. Under these conditions, the highest bacteriocin activity was produced from production with a medium with an initial pH of 6, incubating at 27 ° C for 14 hours.

From **Figure 1** it is seen that, bacteriocin is detected early in the exponential phase up to the stationary phase. The best results are obtained at 14 hours incubation time where the growth phase has entered the stationary phase. This shows that *Lactobacillus* SCG 1223 produces the best active bacteriocin in the stationary phase where in this phase the rate of bacterial growth is equal to the rate of death. In the stationary phase of lactic acid production as the main product of lactic acid bacteria cell metabolism begins to decline with the decrease of growth rate. In these conditions it is suspected of post-translation process in lactic acid bacteria cells. This post-translational process will affect the activity of bacteriocin produced.

3.1. *The bacteriocin activity toward S. thypimurium*

S. typhimurium is a pathogenic bacteria that can cause salmonellosis. Bacteriocin in natural supernatant pH produced from *Lactobacillus* SCG1223 with 5% inoculum had the highest activity toward *S. Thypimurium* at medium pH 6, incubation temperature 27°C and incubation time 14 hours **(Figure 2)**.

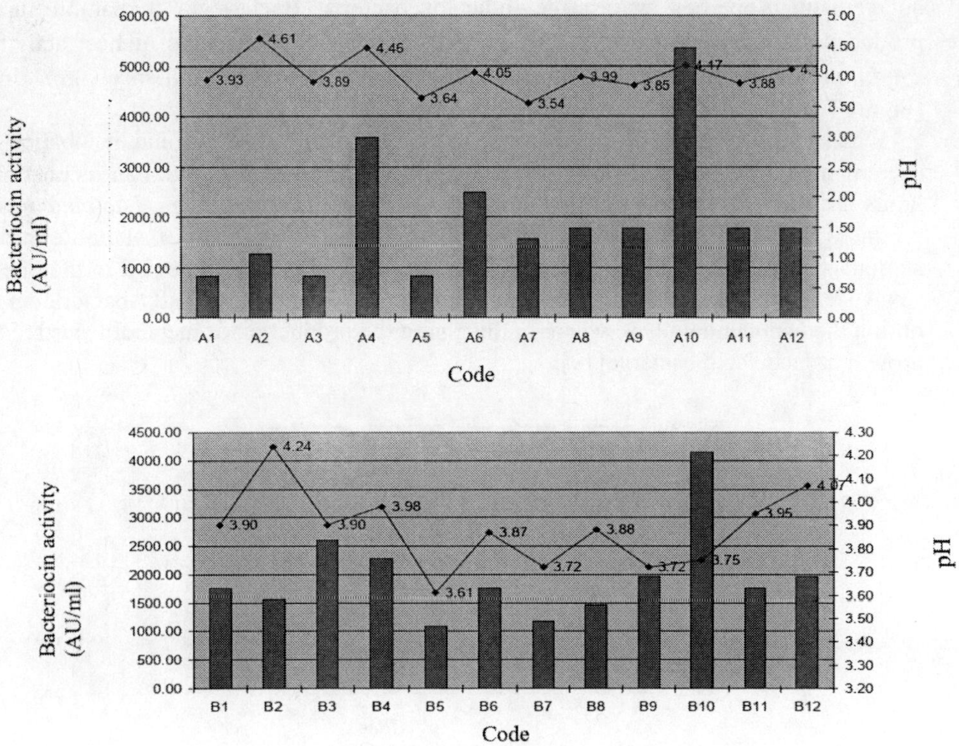

Figure2. Activity of bacteriocin in natural pH with 5% (Code A) dan 10 % (Code B) inoculum toward *S. thypimurium*

Bacteriocin produced with 10% inoculum *Lactobacillus* SCG1223 had higher activity toward *S. thypimurium* compared with 5% inoculum (Figure 5). Under these conditions, the highest bacteriocin activity was produced from production with a medium with an initial pH of 6, incubating at 27 ° C for 14 hours. These results illustrate that there is antibacterial production of compounds starting from the initial phase exponentially to the stationary phase, with the highest activity in the stationary phase. In recent years many studies have shown bacteriocin production during the stationary growth phase[19].

Lactic acid is a primary metabolite product produced in large quantities in the growth of lactic acid bacteria. Lactic acid has the ability to inhibit the growth of pathogenic bacteria. The application of lactic acid as a food preservative has been used for a long time. However, if used as a preservative in foods, lactic acid gives a sour taste and changes the texture. Bacteriocin is a natural anti-bacterial protein that also has inhibitory activity against pathogenic bacteria. Bacteriocin does not change the aroma of food and it is easily degraded by protease enzymes contained in the human body, so that its use is safe for the human body.

3.2. *The bacteriocin activity toward E. coli*

S. typhimurium is a pathogenic bacteria that can cause salmonellosis. Bacteriocin in natural supernatant pH produced from *Lactobacillus* SCG1223 with 5% inoculum had the highest activity toward *S. Thypimurium* at medium pH 6, incubation temperature 27°C and incubation time 14 hours (Figure 4).

Bacteriocin in supernatant with non-neutralized pH (pH <7) produced from *Lactobacillus* SCG1223 with all incubation treatment, provides inhibitory activity against *E.coli*. In the initial 5% inoculum use, bacteriocin produced from different treatment variations had different activities. This high activity difference is related to the resulting metabolite product which has antagonistic properties against the indicator bacteria. Bacteriocin in natural supernatant pH produced from *Lactobacillus* SCG1223 with 5% inoculum had the highest activity toward *S. Thypimurium* at medium pH 6, incubation temperature 40°C and incubation time 14 hours (Figure 7).

When compared, at the same initial pH of the same medium and incubation temperature (treatment A4, A8 and A12) it turns out that bacteriocin produced with an incubation time of 14 hours has higher inhibitory activity compared to 10 hours and 4 hours. *Lactobacillus* SCG 1223 produces another antibacterial compound which is a secondary metabolite product in the stationary phase where in this phase the bacterial growth activity is equal to the rate of death. In some lactic acid bacteria the secondary metabolite product especially bacteriocin is produced during the exponential phase where in this phase the production of metabolites is secondary to the growth of lactic acid bacteria [19].

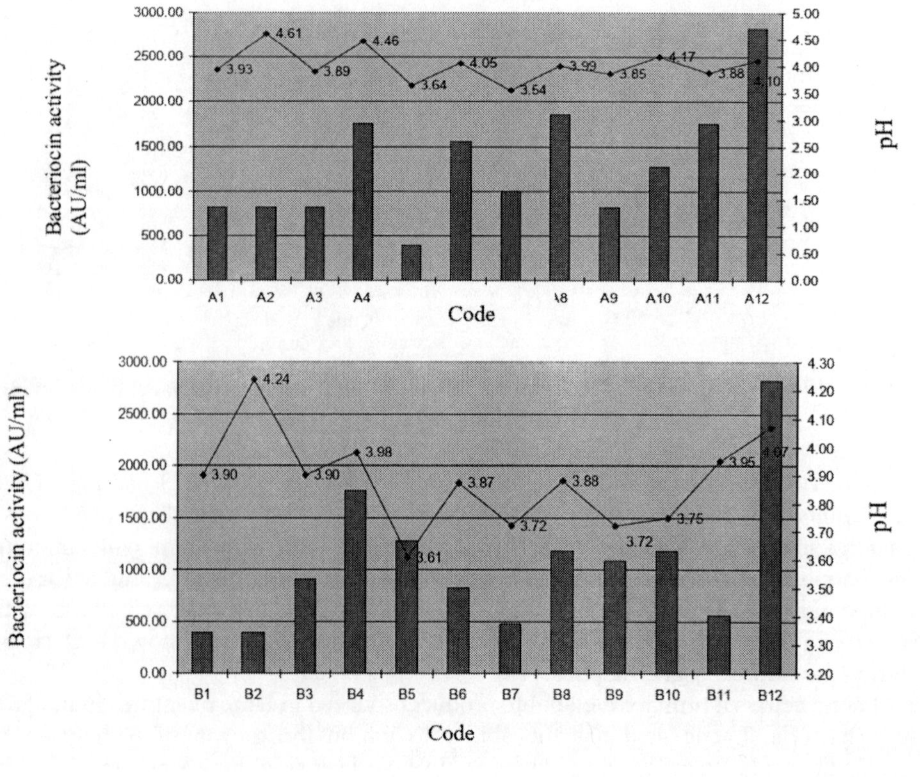

Figure 3. Activity of bacteriocin in natural pH with 5% (Code A) dan 10 % (Code B) inoculum toward *E.coli*

Bacteriocin in natural supernatant pH produced from *Lactobacillus* SCG1223 with 10% inoculum had the highest activity toward *E.coli* at medium pH 6, incubation temperature 40°C and incubation time 14 hours (**Figure 3**). The highest activity result is the same as in percentage of 5% inoculum which in stationary phase is produced higher activity. In the stationary phase a more active antibacterial metabolite product is produced than the metabolite product produced in the exponential phase. Based on the inhibitory test results, the higher the pH of the media and the incubation temperature will further increase bacterial growth and also increase the activity of the antibacterial compounds produced.

Based on these results bacteriocin production which has the highest activity against E. coli indicator bacteria occurs during exponential phase. In the stationary phase, either at the beginning of the stationary phase and at stationary stationary phase, bacteriocin activity is detected even in lower amounts than in the stationary phase. In the stationary phase of bacteriocin which is a protein begins in production where in this phase occurs the primary metabolite process. In the stationary phase it is suspected that the bacteriocin produced during the exponential phase becomes more active with the post-translational process necessary to convert prebacteriosine into active bacteriocin.

In the exponential phase, *Lactobacillus* SCG 1223 already produces bacteriocin which is a natural protein that is antibacterial. This is indicated in the neutral supernatant inhibitory assay results, with the greatest inhibitory activity against *S. thypimurium*. The synthesis of bacteriocin by producing bacteria occurs during an exponential phase trip [20] which usually follows the pattern of synthesis of primary metabolites. This production system is governed by extracromosomal plasmids [21]. However, the bacteriocin formed does not yet have a sufficiently high and widespread activity against the three indicator bacteria. This is because the process of post-translation has not been completed where at this stage activation of prebakteriosin into active bacteriocin.

Understanding of their mechanisms and inhibitory activity on different biochemical conditions that naturally exist in food is neededn for study the effectiveness of the use of bacteriocins in food preservation [22]. In general, bacterial origin of BAL has the ability to inhibit other bacteria with bactericidal effects. Mechanism of bactericidal activity several bacteriocins in general as follows: (1) molecular bacteriocins having direct contact with the cell membrane, (2) the contact process is capable of disrupting the membrane potential in the form destabilitas cytoplasmic membrane so that the cells become strong, (3) the instability of the membrane is able to make an impact The formation of a hole or a pot in the cell membrane through a process of interference with the PMF (Proton Motive Force) [23]. The leakage that occurs due to the formation of holes in the cytoplasmic membrane is indicated by the presence of inactivity of cellular molecules. The leaks that occur have an impact on the decrease in cellular pH gradient. In general, the influence of the cytoplasm hole formation as a result of their bacteriocins, leads to changes in membrane potential gradient and the release of intracellular molecule as well as the influx of extracellular substance (environment). Its effect causes stunted cell growth and results in the process of death in cells that are sensitive to bacteriocin.

As known that lactic acid bacteria have been recognized as safe. Bacteriocins produced by LAB may be a good solution to the problem of resurgence of resistant strains to antibiotics. During the last decade, a large number of bacteriocin from LAB-have been identified, and characterized biochemically and genetically. [24]. Since most of LAB strains have a GRAS status, their addition to food should not have associated legislative problems. The strain in some cases will grow and produce bacteriocin in the food. Another advantage from an economic point of view, the cultivation of these strains for inoculation purposes is in many ways relatively inexpensive. The development of resistant populations of problematic bacteria MAYBE a potential problem associated with using bacteriocins as biopreservatives in foods [25].

4. Conclusion

From the bacteriocin activity test on the natural supernatant of Lactobacillus SCG 1223, there was an effect of other antibacterial compounds. The highest bacteriocin activity toward L. monocytogenes produced by Lactobacillus SCG 1223 with 10% inoculum in media with initial pH 6, incubation temperature 27°C for 14 hour, toward S. thypimurium produced by Lactobacillus SCG 1223 with in media with initial pH 6, incubation temperature 40°C for 14 hour, and toward E. coli was 1085.81 AU/ml, produced by Lactobacillus SCG 1223 in MRS broth with initial pH 4, incubation temperature 40°C for 14 hour

References

[1] Menssens W and De Vugst L 2002 Inhibitory substances produced by *Lactobacilli* isolated from sourdougts. *Intl J. of food Microbiol* **72**: 31–43

[2] Corsetti A, Gobbetti M, and Smacchi E 1996 Antibacterial activity of sourdough lactic acid bacteria: isolation of a bacteriocin-like inhibitory substance *from Lactobacillus sanfrancisco* C57. *Food Microbiol.* **13**: 447–456

[3] Lavermicocca P, Valeria F, Evidente A, lazzaroni S, Corsetti A, and Gobbetti M 2000 Purification and characterization of novel antifungal compounds by sourdough *Lactobacillus plantarum* 21 B. *Applied and Environ Microbiol* **66**: 4084–4090

[4] De Vuyst L and Vandamme E J. 199). Lactic acid bacteria and bacteriocins: their practical importance *Bacteriocins of Lactic Acid Bacteria: Microbiology, Genetics and Applications* ed De Vuyst L and Vandamme E J (London: Blackie Academic and Professional) pp 1–11

[5] Gálvez A, Abriouel, López, R L, Gálvez, N B O, Abriouel H, López R L and Omar N B. 2007 Bacteriocin-based strategies for food biopreservation A Bacteriocin-based strategies for food biopreservation *International Journal of Food Microbiology* **120** 51–70

[6] Osmanagaoglu O, Kiran F and Nes I F 2011 A probiotic bacterium, *Pediococcus pentosaceus* OZF, isolated from human breast milk produces pediocin AcH/PA-1. *African Journal of Biotechnology* **10** (11) 2070–2079

[7] Thomas L V, Ingram R E, Bevis H E, Davies E A, Milne CF and Delves-Broughton J 2002 Effective Use of nisin to Control Bacillus and Clostridium Spoilage of Pasteurized Mashed Potato Product *Journal of Food Protection* **65** (10) 1580–1585

[8] Huang Y, Luo Y, Zhai Z, Zhang H, Yang Ch , Tian H, Zheng Li, Feng J, Liu H and Hao Y 2009 Characterization and application of an anti-Listeria bacteriocin produced by *Pediococcus pentosaceus* 05-10 isolated from Sichuan Pickle, a traditionally fermented vegetable product from China. *Food Control* **20** (11) 1030–1035

[9] Anastasiadou S, Papagianni M, Filiousis G, Ambrosiadis I. and Koidis P 2008 Pediocin SA-1, an antimicrobial peptide from *Pediococcus acidilactici* NRRL B5627: Production conditions, purification and characterization *Bioresource Technology* **99** 5384–5390

[10] Todorov S D and Dicks L M T 2009 Bacteriocin production by *Pediococcus pentosaceus* isolated from marula (*Scerocarya birrea), International Journal of Food Microbiology* doi: 10.1016/j.ijfoodmicro.2009.04.010

[11] Nurhasanah. 2004. Produksi Bakteriosin Pada Berbagai Tingkat Aerasi dan Uji Kestabilan Bakteriosin Dari Bakteri Asam Laktat Galur M6-15. *Skripsi*. Fakultas Teknologi Pertanian, Institut Pertanian Bogor

[12] Neysens P, Messens W, Gevers D, Swings and De Vuyst L 2003 Biphasic kinetics og growth and bacteriocin production with *Lactobacillus amylovorus* DCE 471 occur under stress condition. *Micobiology* **249** 1073–1082

[13] Lee S H and Kim HU 1985 Studies on lantibiotic nisin produced by *Streptococcus lactis* IFO 12007 *Korean Journal of Animal Science,* 27 476–479

[14] Penna T C V, Jozala A F, Novaes L C L, Pessoa A, Cholewa O 2005 Production of nisin by *Lactococcus lactis* in media with skimmed milk *Applied Biochemistry and Biotechnology* **122**(1-3) 619–637

[15] Yang R , Johnson M C and Ray B 1992 Novel method to extract large amount of bacteriocin from lactic acid bacteria. *Applied and Environmental Microbiology,* **58** 3355–3359

[16] Schneider R, Fernández F J, Aguilar M B , Guerrero-Legarreta I, lpuche-Solís A and Ponce-Alquicira E 2006 Partial characterization of a class IIa pediocin produced by *Pediococcus parvulus* 133 strain isolated from meat (Mexican "chorizo"). *Food Control* **17** 909–915

[17] Biswas S R, Ray P, Johnson MC and Ray B 1991 Influence of growth condition on the production of bacteriocin, pediocin AcH, by *Pediococcus acidilactici* H. *Applied and Environmental Microbiology* **57** 1265–1267

[18] Kim W J, Hong S S, Cha S K, Koo Y J 1993 Use of bacteriocinogenic *Pediococcus acidilactici* in sausage fermentation. *J. Microbial. Biotechnol*. **3** 199–203

[19] Navaro L, Zaraxaga M, Saenz J, Ruiz-Larrea F and Torres C 2000 Bacteriocin Production by Lactic Acid Bacteria Isolated From Rioja Red Wines *Journal of Applied Microbiology* **88** 44–51

[20] Samelis J, Roller S and Metaxopoulus J 1994 Sakacin B, a Bacteriocin produce by Lactobacillus sake isolated from greek dry formaterd sausage *Journal Appl Bacteriol.* **76** 475–486

[21] Ennahar S, Aoude-werner D, Sorokine D, Van Dorsselaer A, Bringel F, Hubert JC and Hesselmen C 1996 Production of Pediocin AcH by *Lactobacillus plantarum* WHE 92 *Appl. Environ Microbiol* **62** 4381–4387

[22] Atrih A, Rekhif N, Moir AJG, Lebrihi A and Lefebvre G 2001 Mode of action, purification and amino acid sequence of plantaricin C19, an anti-Listeria bacteriocin produced by *Lactobacillus plantarum* C19 *International Journal of Food Microbiology* **68** 93–104

[23] Gonzales B E, Glaasker E, Kunji ERS, Driessen AJM, Suarez JE, KOnings, WN 1996 Bactericidal mode of Action of Plantaricin S. *Appl Environ Microbiol* **62** 2701-2709

[24] Pancheniak E, de F R, Parada J L and Soccol C R 2006 Inibição de Escherichia coli hemolítica suina por Lactobacillus reuteri, produtor de reuterina. Paper presented at XX Congresso Brasileiro de Ciência e Tecnologia de Alimentos, 8-11 Oct, Curitiba, Paraná.

[25] Sullivan, L., Ross, R.P., dan Hill, C 2002 Potential of bacteriocin-producing lactic acid bacteria for improvements in food safety and quality *Biochimie* **84**593–604

Process technology of luwak coffee through bioreactor utilization

M Hadipernata* and S Nugraha

Indonesian Center for Agricultural Postharvest Research and Development, Ministry of Agriculture

Jl. Tentara Pelajar No 12 Bogor 16114, Jawa Barat Indonesia

*Email: mulya_nata@yahoo.com

Abstract. Indonesia has an advantage in producing exotic coffee that is Luwak coffee. Luwak coffee is produced from the fermentation process in digestion of civet. Luwak coffee production is still limited due to the difficulty level in the use of civet animals as the only medium of Luwak coffee making. The research was conducted by developing technology of luwak coffee production through bioreactor utilization and addition the bacteria isolate from gastric of civet. The process conditions in the bioreactor which include temperature, pH, and bacteria isolate of civet are adjusted to the process that occurs in civet digestion, including peristaltic movement on the stomach and small intestine of the civet will be replaced by the use of propellers that rotate on the bioreactor. The result of research showed that proximat analysis data of artificial/bioreactor luwak coffee did not significant different with original luwak coffee. However, the original luwak coffee has higher content of caffeine compared to bioreactor luwak coffee. Based on the cuping test the bioreactor luwak coffee has a value of 84.375, while the original luwak coffee is 84.875. As the result, bioreactor luwak coffee has excellent taste that similiar with original luwak coffee taste.

Keywords: *Luwak coffee, Bioreactor, Civet*

1. Introduction

Civet (*Paradoxurus Hermaphrodirus*) or Luwak in Indonesian language is wild animal that actively at night and tends to behave cannibals when gathered with a smaller civet. Civet eat fruits such as palm fruit, papaya, and banana. Also, civet animal eat coffee berries because the sweet taste of outer skin of coffee berries. Coffee berries undergoes a fermentation process for 12 hours in a civet digestive system containing enzymes and bacterial isolates. Non-digestible coffee beans come out with feces in the excretion process. Civet feces are collected, cleaned and dried with sunlight.

Luwak coffee has a distinctive taste and unique, so it has a high selling price in the international coffee market [1,2]. The production of luwak coffee is still limited because the civet as the only medium of luwak coffee making. The added value of luwak coffee is very large when compared to ordinary coffee. The price of luwak coffee is more than twenty time of ordinary coffee.

There are several problems from the production of Luwak coffee. The main problem is the possibility of coffee contamination with *E.coli* and *Salmonella* if luwak coffee is not properly cleaned [3]. Furthermore, there is no Halal guarantee products for Moslem people because coffee beans come out with civet feces. In addition, luwak coffee production by civet cause some cases of animal abuse.

The research related to the production of artificial luwak coffee has been done such as research on production of probiotic luwak coffee [4]. The probiotic microbial inoculum is taken from the small intestine of civet animals. During fermentation process of probiotic coffee bean, the probiotic microbial is added into coffee bean in fermentation tank. Fermentation process of probiotic luwak coffee is different compared to fermentation process in civet animal. So, the taste of probiotic luwak coffee does not have the same quality as the original luwak coffee.

The objective research is developing technology of luwak coffee production through bioreactor utilization and addition the bacteria isolate from civet animal. The process conditions in the bioreactor which include temperature, pH, and bacteria isolate of civet are adjusted to the process that occurs in civet digestion, including peristaltic movement on the stomach and small intestine of the civet will be replaced by the use of propellers that rotate on the bioreactor.

2. Materials and Methods

2.1. Materials

Coffee berries was obtained from farmer field in Pangalengan Subdistrict, Bandung, Jawa Barat. Bioreactor used was Biotron Liflus GX type from Intran Technologies Sdn Bhd, Malaysia.

2.2. Methods

Bioreactor luwak coffee was processed by artificial digestive system of civet in bioreactor with 5 lt capacity. The process of fermentation in the bioreactor can be seen in Figure 1. Coffee berries peeled then weighed 1.5 kg and put into the bioreactor. Simulated Gastric Fluid (SGF) with pH 1.2 to 2.5 were added 2.5 liter. Fermentation process in bioreactor according to the condition of civet digest, such as: pH, temperature, fermentation time, propeller stirring rate and bacterial isolates addition. The level of acidity or pH value is 1.5 to 4.5 [5]. Temperature is determined based on the temperature conditions inside the civet digest that is about 25°C to 35°C [6]. Stirring rate is set at 50 rpm.

The first stage of the in vitro digesting process is the process of acidifying the coffee beans in the SGF solution. This process lasts for 2 hours. After that pH neutralization using NaOH solution to obtain pH 6.5 to 7.5 for the second stage preparation. The third stage is the addition of lactic acid bacteria. Lactic acid bacteria used is bacterial isolate from civet animal that is 5 isolate from luwak. The amount of BAL added in the bioreactor process is 108 cfu/ml to 109 cfu/ml [6].

After fermentation process for 12 hours in bioreactor, the fermented coffee bean will be washed and then dried to a moisture content of 11-12%. Coffee beans are still wrapped in endocarp skin, then will be peeled using peeler and ready to roasted by using roaster machine, then grinded until ready for consumption. The bioreactor luwak coffee will be analyzed physically, chemically and tastefully and compared with the original luwak coffee. The original luwak coffee produced from civet animals.

The water content analysis was performed by oven method according to AOAC procedure in 1996 [7], fat content according to AOAC procedure in 2006 [8], protein content according to AOAC in 1995 [9], ash content according to AOAC 2006 [8], while carbohydrate analysis based on by difference calculation.

Determination of microbial count was done by using PCA media. A total of 10 g of sample was incorporated into 90 ml 85% NaCl and stirred until homogeneous, then diluted to dilution 10^7. The amount was then calculated by plate count method with some dilution series after incubation at 37°C. for 48 hours. Then calculated total plate count (TPC) based on standard plate count. Coliform and Salmonella analysis was performed using EMBA media and XLD media. While, yeast mold analysis was done using PDA media.

The cupping test of bioreactor luwak coffee and original luwak coffee were done in cupping test laboratory, Coffee and Cocoa Research Center in Jember. The cupping test was performed by a cupping test expert who has earned an American standard certificate and a European standard certificate.

Figure 1. Fermentation process of coffee in bioreactor

3. Results and Discussion

The fermentation process of coffee in bioreactor are describes the fermentation process of coffee that occurs in the stomach, intestines and colon civets. The addition of SGF solution is simulation of the existing solution in the civet stomach. The second stage process in bioreactor is condition in the intestine of civet animals. The third stage is the addition of lactic acid bacteria that resembles the fermentation process in the colon. The amount of BAL added in the bioreactor process is 108 cfu/ml to 109 cfu/ml according to the amount of BAL found in the civet small intestine reaches more than 107-108cfu/ml [6]. Figure 2 shows the product of artificial or bioreactor luwak coffee from coffee berries (a), coffee bean with endocarp skin (b), green bean coffee (c), and roasted coffee. Coffee bean with endocarp skin was obtain after fermentation process in bioreactor, washing treatment and drying with oven or sunlight. Green bean coffee (c) was obtain after peeling process the endocarp skin of coffee. Roasted coffee (d) was obtain after roasting treatment.

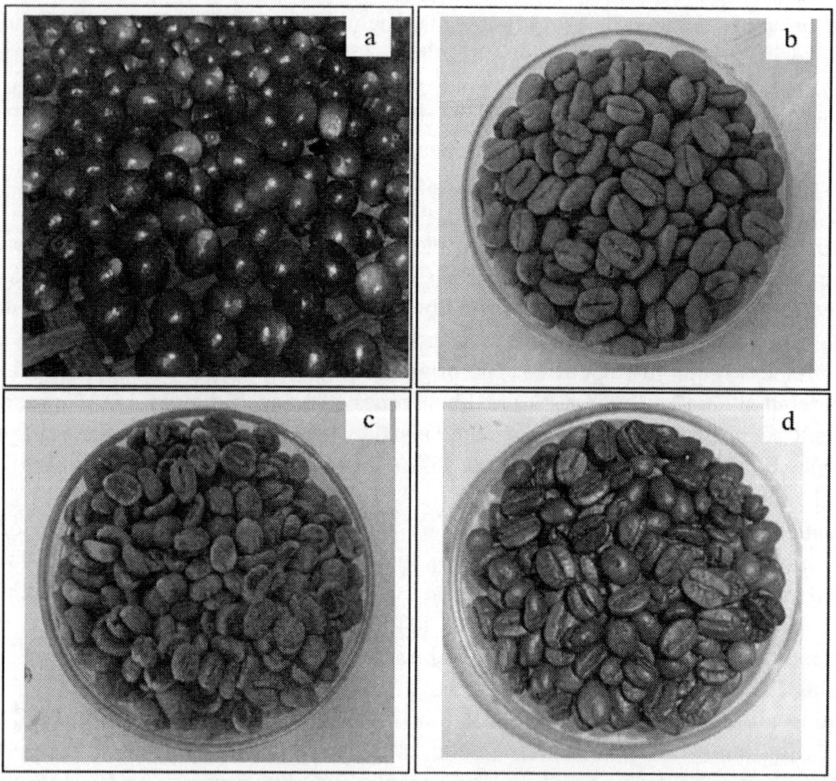

Figure 2. Coffee berries (a), coffee bean with endocarp skin (b), green bean coffee (c), and roasted coffee (d)

The research result shows the proximate analysis of coffee for water content, ash content, lipid, protein and carbohydrate in Table 1. There is no significant different for water content, ash content and lipid content of bioreactor/artificial luwak coffee compared with original luwak coffee. On the other hand, protein content of artificial luwak coffee is higher compared with original luwak coffee. Suggestion that bioreactor process appreciably suppresses the degradation of protein that it caused by fermentation process. Protein degradation of protein can be caused by the acid solution in civet digestion, the pH of gastric fluid in civet lower than pH simulated gastric fluid in the bioreactor. Addition of acid solution in this research does not cause damage to coffee beans. Martinez (2007) reported that quality of coffee bean will increase after acid and enzyme treatment [10].

Table 1. Proximate analysis of roasted coffee

Treatment	Water content (%)	Ash content (%)	Lipid (%)	Protein (%)	Carbohydrate (%)
Artificial luwak coffee	2.05 ± 0.11	5.38 ± 0.22	12.21± 0.19	13.11± 0.23[a]	67.25± 0.33
Original luwak Coffee	2.00 ± 0.15	5.34 ± 0.17	12.13 ± 0.22	11.08 ± 0.31[b]	69.44 ± 0.28

Data are presented as mean ± SD (n = 3) and values with different superscript letters (a and b) in a line show the significant difference ($P < 0.05$)

The microbiology analysis of *Salmonella*, *Coliform* and *E.coli* showed negative results which means that the undetected microbes are present in the material. Also, analysis of yeast molds in roasted coffee beans shows a negative result which means that the material does not contain yeast molds. In the calculation of total bacteria on treatment showed that the amount of bacteria in all treatments and controls was negative. Based on Indonesian's national standard for coffee powder, the maximum amount of mold allowed in maximum powdered coffee is 10^4 colonies/gram, while the maximum total plate count of bacteria is 10^6 cfu/gram [11].

Table 2. Microbiologycal analysis of roasted coffee

Treatment	*Salmonella* (cfu/g)	*Coliform* (cfu/g)	*E. Coli* (cfu/g)	Yeast Mold (cf u/g)
Artificial luwak coffee	negatif	negatif	negatif	negatif
Original luwak coffee	negatif	negatif	negatif	negatif

Salmonella, *Coliform* and *E.coli* are indicators of bacteria on the food quality and water sanitation. The existence of these three bacteria on food can cause dangerous diseases for humans. The results of microbiological analysis are still below the maximum limit of food contamination, especially for coffee beverage products in packaging [12]. For coffee beverage products, the maximum limit of *Salmonella* contamination is negative in 100 ml, while the maximum limit of *Coliform* contamination is less than 2 in 100 ml coffee. The maximum limit of *E.coli* contamination on the packaged coffee drink is negative in 100 ml. Based on microbiology analysis, artificial/bioreactor luwak coffee and original luwak coffee meets the requirements of Indonesia's National Standard.

The effect of bioreactor treatment on artificial/bioreactor coffee taste and flavor can be seen in Table 3. The quality scale grouping in cupping test are based on the Specialty Coffee Assocation of America Cupping Protocols. Cupping test is one of the important testing stages in determining the quality of a product including coffee. The organoleptic test is performed to determine the sensory differences between treatments, to describe the treatment flavor and to determine the consumer's preference for the product. One method of testing coffee flavor is cupping test. The quality parameters tested using this method include flavor, after taste, acidity, body, balance, uniformity, clean cup, sweetness, defects and overall [13].

The treatment that has a total score with a fair quality scale is the treatment with a total score below 6.0. The treatment included in the Good quality scale is treated with a total score of 6.0-6.75. The total score in the range of 7.0 - 7.75 is included in the very good quality scale, while the total score in the range of 8.0 - 8.75 is included in the excellent quality scale. The cupping test results showed that the total value of the flavor of bioreactor Luwak coffee and original luwak coffee has a value of more than 80 which means included into the class of specialty grade or special coffee group. Total score of artificial coffee and original coffee for coffee characteristic are 84.375 and 84.875. So, Luwak coffee bioreactor and original luwak coffee included into the specialty coffee, and above the premium coffee.

Yusianto *et al.* (2005) and Budryn *et al.* (2009) reported that protein will be degraded into free amino acids which is one of the typical fragrance-forming components in coffee [14,15]. During coffee roasting, maillard reaction is occurs between carbonyl groups especially from reducing sugars with amino acids, peptides and proteins [16]. The initial reaction between aldehyde or ketone groups and free amino groups is called the amino-sugar reaction [17]. The final stage of Maillard's reaction produces brown melanoidin pigments [18]. Maillard's reaction also produces a flavor in coffee [19]. During the roasting process, carbohydrates turn into water soluble polysaccharides, oligosaccharides, monomers, melanoidin, and caramels, as well as volatile components. Carbohydrates contribute to brown changes in roasted coffee, the formation of volatile components and strengthening the body. Based on American's standard coffee, the artificial luwak coffee flavor as well as the original luwak coffee belong to a very good quality scale.

Table 3. Cupping test analysis of artificial/bioreactor luwak coffee and original luwak coffee

Characteristic	Artificial luwak coffee	Original luwak coffee
Fragrance	7.75	8.00
Flavor	7.75	7.75
After Taste	7.75	7.875
Acidity	7.63	7.75
Body	7.75	7.50
Uniformity	10.00	10.00
Balance	7.88	8.00
Clean cup	10.00	10.00
Sweetness	10.00	10.00
Overall	7.88	8.00
Defect	None	None
Total score	84.375	84.875

4. Conclusion

The utilization of bioreactor technology and the addition of bacterial isolates from civet animals have produced artificial luwak coffee that has the same quality and taste as the original luwak coffee. Artificial luwak coffee technology can be developed for higher production scale compared to the production of luwak coffee using civet animal.

References

[1] Kompas 2010 Luwak Coffee, from Culture Stelseel Era to Oprah Winfrey Show 18 December 2010 p 1

[2] Hadipernata M, Nugraha S and Thahjohutomo R 2011 Improvement the Added Value of Luwak Coffee as Product Diversification at Pangalengan Subdistrict, Bandung, Jawa Barat Proceeding of National Seminar of Inovation Technology of Agricultural Postharvest 3rd

[3] Hadipernata M, Thahjohutomo R, Agustinisari I and Rahayu E 2011 Process Technology and Food Safety of Luwak Coffee *Proc. of 3rd National Seminar on Agricultural Postharvest Technology*

[4] Kompas 2010 Probiotic Luwak Coffee http://nasional.kompas.com/read/2010/05/11/18033691/Kopi.Luwak.Probiotik.Apa.Itu.

[5] Avallone S, Guyot B, Brillouet J M, Olguin E and Guiraud J P 2001 Microbiological and biochemical study of coffee fermentation *Current Microbiology Journal* **42** 252–25

[6] Hadipernata M, Nugraha S, Agustinasari I, Rahayu E, and Tjahjohutomo R 2013 Annual project report 2013: Developing Artificial Luwak Coffee. Center for Agricultural Postharvest Research and Development Not Published

[7] AOAC/Association of Official analytical Chemist 1996 Official Methods of Analysis: Official Method for Water Content. AOAC, Washington DC, USA

[8] AOAC/Association of Official analytical Chemist 1995 Official Methods of Analysis: Official Method for Protein. AOAC, Washington DC, USA.
[9] AOAC/Association of Official Analytical Chemist 2006 Official Methods of Analysis (18th edition) *Association of Official Analytical, Chemists International*, Maryland, USA
[10] Martinez LF 2008 Quality Enhancement of Coffee Beans by Acid and Enzyme Treatment. Aplication Patent No 61/067, 141
[11] SNI 01-3542 2004 Coffee powder standard. Indonesian's National Standard
[12] SNI 7388 2009 Maximum limit of microbes contamination in food. Indonesian's National Standard
[13] Cupping committee 2015 Specialty Coffee Association of America Cupping Protocols
[14] Yusianto, Halupi R, Sulistiyowati, Mawardi S and Ismayadi C 2015 Physical, chemical and flavors of some varieties of arabica coffee *Pelita Perkebunan 2005* **3** 200-222
[15] Budryn G, Nebesny E, Podsedek A, Zyzelewicz D, Materska M, Jankowski S, Janda B 2009 Effect of different extraction methods on the recovery of chlorogen Eur Food Res Technol, **228** 913–22
[16] Miller DD 1998 Food Chemistry A Laboratory Manual (New York: J Wiley & Sons Inc)
[17] Whistler R and Daniel JR 1985 Carbohydrate *Fennema OR (eds) Food Chemistry* (New York : Marcel Dekker. Inc)
[18] Bailey ME, Won Um K 1992 Maillard reaction and lipid oxidation. Di dalam: Angelo AJS. Lipid Oxidation in Food. ACS symposium series New York: August 25-30
[19] Flament I 2002 Coffee Flavor Chemistry (New York: John Wiley and Sons Ltd)

IOP Conf. Series: Earth and Environmental Science **102** (2017) 012093 doi:10.1088/1755-1315/102/1/012093

Study on *Leucaena leocochepala* seed during fermentation : sensory characteristic and changes on anti nutritional compounds and mimosine level

A Nursiwi*, D Ishartani, AM Sari and K Nisyah

Department of Food Science and Technology, Universitas Sebelas Maret, Indonesia

*E-mail: asrinursiwi@gmail.com

Abstract. Lamtoro (*Leucaena leucocephala*) seed is one of the leguminosae which have high level of protein but it contains toxic compound such as mimosine and some anti nutritional compounds such as phitic acid and tannin. The objectives of the research was to investigate the sensory characteristic and the changes onanti nutritional compounds and mimosine level in *Leucaena leucochepala* seed during fermentation.Lamtoro tempeh processing was carried out by boiling the seed, crushing to separate the hull, soaking, boiling, and fermentation. The best concentration inoculum in lamtoro tempeh processing was determined by hedonic test. Fermentation was carried out in 36 hours and every 6 hours mimosine, tannin, and phitic acid content was analyzed. From hedonic test, inoculum concentration of 1% was used in lamtoro tempeh processing. During 36 hours fermentation, phytic acid content and mimosine content was decreased significantly, from 0.0558 % to 0.0453 % and from 0.00393 % to 0.00173 % respectively. Whereas tannin content was increased signifacantly, from 0.0822 % to 0.00173 %.

Keywords *: Leucaena leucocephala, mimosine, phytic acid, tannin*

1. Introduction

Lamtoro (*Leucaena leucochepala*) plant belongs to the leguminous class which grows in several areas in Indonesia, such as the area around Gunungkidul and Trenggalek. The parts of this plant are often used, wood and twigs are used to make paper, leaves, seed, young stems and flowers are used as food sources for humans and livestock. Several areas in Indonesia use ripe seed as material for making tempeh [1].

Lamtoro seed have high level of protein and minerals. Ripe lamtoro seeds contain 30.81% protein (db), 20.45% (db) crude fiber, 8.80% (db) ash content, 1.09% (db)calcium, 0.69% (db) phosphorus, 1.50% (db) potassium, 1.11% (db) magnesium, and 0.05% sodium (db) [2]. However, lamtoro seed contain toxic compounds (mimosine) and anti-nutritional compouds such as phytic acid and tannin.

Mimosine is a non-protein amino acid structurraly having toxic properties by blocking the metabolic pathways of aromatic amino acids and tryptophan. It probably acts as tyrosine analogue because its structure resemblance to L-tyrosin that inhibits protein biosynthesis in the living body [3]. Negative impacts caused by mimosine if consumed by women and children, causes hair loss and eyebrows, causing pain in the scalp and edema occurs part of the body at a certain point [4].Feeding lamtoro tempeh in rats causing impaired thyroid function [1]. Whereas feeding lamtoro leaf with mimosine content of 1% causes hair loss, liver tissue damage, edema, and increase death percentage

[3]. Mimosine content in ripe lamtoro seed was 2-9%. It can be reduced by washing, soaking, boiling, steaming, and fermentation during tempeh processing [5].

Several groups of tannins lead to the formation of complexes of protein, metallic ions, and other macromolecules such as polysaccharides [6].It causes inhibition of iron absorption in the body and inhibition of enzymes due to the formation of the protein-tannin complex. Tannin content in 100 g lamtoro seed is 1.194 grams [7]. Phytic acid form complex bonding with Fe or other minerals such as Zn, Mg and Ca. The formation make it insoluble so that it is difficult to be absorbed by the body . Each 100 gram of lamtoro seed contains 2.325 gram of phytic acid [7].

Stages of traditional tempeh processing include boiling, cooling, removing the hull, washing, soaking, boiling, draining, cooling, inoculation, packaging and incubation [8]. Boiling, washing, and steaming reduce the antioxidant compounds in lamtoro seeds such as phytic acid and tannins. Phytic acid content in lamtoro seed reduced 37.9% by boiling andsoaking whereas tannin content reduced 92% [7]. During fermentation,complex compounds were degraded into simpler compounds through enzymes produced by inoculum (protease, amylase, lipase, and phitase) [9].

Lamtoro tempeh processing in Wonogiri use usar as inoculum. Usar contains many types of molds such as *Rhizopus sp.* and other microorganism [10]. It makes sensory quality of tempeh is not consistent [11]. Beside usar, there is a commercial tempeh inoculum 'Raprima' which can be used as an alternative in lamtoro tempeh processing. It contains *Rhizopus oligosporus* , the dominant microorganism that play a role in tempeh fermentation. The objectives of the research was to investigate the sensory characteristic and the changes on anti nutritional compounds and mimosine level in *Leucaena leucochepala* seed during fermentation.

2. Material and Methods

2.1 Materials

Lamtoro seedwas obtained from local market in Wonogiri. Comercial tempeh inoculum 'Raprima' was obtained from local market in Surakarta.

2.2 Lamtoro Tempeh Processing

Lamtoro seed was boiled with the addition of ash for about 2 hours (first boiling). Having been boiled, the lamtoro seed wasseparated with the hull. After that, the lamtoro seed was soaked for 9 hours then boiled (second boiling) for 2 hours. The boiled lamtoro seed was drained and cooled down. Afterwards, it was inoculated with comercial tempeh inoculum 'Raprima', packaged, and incubated on 27^0 C-29^0C for 36 hours.

The inoculum concentrations used were 0.2%, 0.5% and 1%. Twenty five panelists were selected for sensory evaluation using 7-points hedonic scale.

2.3 Analysis of Phytic Acid Content

The analysis of phytic acid content was determined bySpectrophotometric method [12]

2.4 Analysis of Tannin Content

The analysis of tannin content was determined by spectrophotometric method [13].

2.5 Analysis of Mimosine Content

The analysis of mimosine content was determined by using colorimetric method [14].

3. Results and Discussion

3.1 Sensory Characteristics of Lamtoro Tempeh

3.1.1 Color is one of important parameteraffecting panelists acceptance. Color parameterof lamtoro tempeh was evaluation on raw tempeh and fried tempeh. The result of color parameter evaluation was shown in Table 1.

Table 1. Sensory characteristic of lamtoro tempeh : color

Inoculum concentration	Raw tempeh	Fried tempeh
0.2 %	4.90 ± 1.39^{a}	4.50 ± 1.59^{a}
0.5 %	5.07 ± 1.14^{a}	4.33 ± 1.37^{a}
1%	5.70 ± 0.91^{b}	4.17 ± 1.36^{a}

Notes : Scoring scale 1: dislike very much, 2: dislike, 3 : dislike slightly 4: Neither like nor dislike, 5:like slightly,6: like, 7: like very much

Sensory evaluation for color parameter showed that only at inoculum concentration of 1% significantly difference with other raw tempeh, but there were no significantly difference in all fried tempeh. Raw tempeh has white color, formed as the result of the growth of mycelium on the surface of tempeh. The less inoculum concentration decrease the ability of the microorganisms to form the hypha so the mycelium less of dense and the surface of tempeh is not throughout covered by hypha.The addition of inoculum concentration properly, produce mycelium more evenly so that the color of tempe is preferably by panelist [10]. Fried tempeh has golden brown color. The color is formed due to Mailard reaction during frying process. During frying tempe occurs reaction between active side of amino acid and reducing sugar due to heatingto form melanoidin which produce brown color [15].

3.1.2. Appearance Good appearance of tempeh can be seen by the mycelium covering the tempeh surface. The dense of the mycelium determine the compactness of tempeh. Appearanceparameterof lamtoro tempeh was evaluation on raw tempeh and fried tempeh. The result of appearance parameter evaluation was shown in Table 2.

Table 2. Sensory characteristic of lamtoro tempeh : appearance

Inoculum concentrations	Raw tempeh	Fried tempeh
0.2 %	3.97 ± 1.32^{a}	4.63 ± 1.35^{a}
0.5 %	5.10 ± 0.88^{b}	4.60 ± 1.16^{a}
1 %	5.93 ± 0.58^{c}	4.73 ± 1.14^{a}

Notes : Scoring scale 1: dislike very much, 2: dislike, 3 : dislike slightly 4: neither like nor dislike, 5:like slightly, 6: like, 7: like very much

Three different inoculum concentrations in raw tempeh were significantly different. Higher inculum concentration, higher panelists acceptance. Inoculum concentration at 1% has dense and compact texture, and the appearance is similar to soybean tempeh.The addition of inoculum concentration properly, produce good mold micelium so that the arrangement of seed appears more compact [10]. There were no significantly difference in all fried tempeh.

3.1.3 Aroma The aroma of lamtoro tempeh resulted by the microorganism on lamtoro seed during fermentation.Aromaparameterof lamtoro tempeh was evaluation on raw tempeh and fried tempeh. The result of aroma parameter evaluation was shown in Table 3.

Table 3. Sensory characteristic of lamtoro tempeh : aroma

Inoculum concentrations	Raw tempeh	Fried tempeh
0.2 %	5.03 ± 1.37^{a}	5.70 ± 0.87^{a}
0.5 %	4.97 ± 1.15^{a}	5.47 ± 0.77^{a}
1 %	5.23 ± 1.25^{a}	5.73 ± 1.14^{a}

Notes : Scoring scale 1: dislike very much, 2: dislike, 3 : dislike slightly 4: neither like nor dislike, 5:like slightly, 6: like, 7: like very much

The addition of inoculum concentration properly cauces mold grow better with a lot of mycelium resulting in a distinctive tempeh aroma [10]. But in the research using different inoculum concentrations (0.2%, 0.5% and 1%) resulted no significantly difference on aroma parameter in all raw tempeh and in all fried tempeh. The aroma of lamtoro tempeh is similar to soybean tempeh, and it does not have unpleasant aroma.

3.1.4 Taste Tasteparameterof lamtoro tempeh was evaluation on fried tempeh. The result of taste parameter evaluation was shown in Table 4.

Tabel 4. Sensory characteristic of lamtoro tempeh : taste

Inoculum concentrations	Fried tempeh
0.2 %	5.13±0.90[a]
0.5 %	5.27±0.82[a]
1 %	5.33±1.24[a]

Notes : Scoring scale 1: dislike very much, 2: dislike, 3 : dislike slightly 4: neither like nor dislike, 5:like slightly, 6: like, 7: like very much

Different inoculum concentrations resulted no significantly difference on taste parameter in all fried tempeh.It has pleasant taste,formed during fermentation by *Rhizopus oligosporus* [16]. The frying process also increase the taste of lamtoro tempeh, because the frying oil is absorbed into the lamtoro tempeh. The components of frying oil and tempeh affect the characterization of tempeh [17]. The frying oil increase the tasty taste of tempeh [15].

Based on all parameter evaluated by hedonic test, the best inoculum concentration for tempeh processing was 1 %.

3.2 Anti nutritional and Mimosine Level

3.2.1 Phytic acid is one of the anti nutritional compounds found in lamtoro seeds. It is capable to bind mineral substances, so the mineral is absorbed difficulty by the body. By fermentation, phytic acid levelcan be lowered. The decrease of phytic acid level by fermentation is greater thanboiling, steaming and soaking process [18]. Phytic acid level in lamtoro seed during fermentation shown in Table 5.

Table 5. Phytic acid contents in lamtoro seed during fermentation

Fermentation time (hours)	Phytic Acid Content (%)
0	0.0558±0.0022[b]
6	0.0525±0.0029[ab]
12	0.0504±0.0035[ab]
18	0.0499±0.0019[ab]
24	0.0488±0.0058[ab]
30	0.0455±0.0070[ab]
36	0.0453±0.0101[a]

The result of the research showed that phytic acid content in lamtoro seed during fermentation was significantlydecreased from 0.0558 % before fermentation to 0.00453 % after 36 hours fermentation. The decrease in phytic acid content occurs due to the activity of the fitase enzyme produced by the *Rhizopus oligosporus* during fermentation. By phytase enzymes, phytic acid is hydrolyzed into inositol and organic phosphate compounds [19, 20]. Other study reported the similar result, phytic acid content was decreased from 1.442g/100g to 1.116g/100g after 36 hours fermentation [7].

3.2.2 Tannin Tannin are polyphenolic compounds with different molecular weight and complexity. Tannins are found in many vegetables which are used as comestibles or animal feeds. Several groups of tannins lead to the formation of complexes of protein, metallic ions, and other macromolecules such as polysaccharides [6].Tannin level in lamtoro seed during fermentation shown in Table 6.

Table 6. Tannin contents inlamtoro seed during fermentation

Fermentation time (hours)	Tannin content (%)
0	0.0822±0.0021[a]
6	0.0650±0.0114[a]
12	0.0681±0.0238[a]
18	0.0784±0.0224[a]
24	0.0904±0.0213[a]
30	0.1206±0.0321[b]
36	0.1522±0.0316[c]

The result of the research showed thattannin content on lamtoro seed during fermentation was significantly increased, from 0.0822% before fermentation to 0.1522% after 36 hours fermentation. This wassimilar with studiesthat has been done before. From the studies, reportedthat after tannins are hydrolysed by enzymes into simpler compounds, the pH increase, it weaken the bond of tannin with protein and carbohydrate, in the result, tannin will be detected. During fermentation for 48 hours tannin level was increased from 0,095 g/100 g to 0,136g/100 g [7]. Other study also had similar result, tannin content was increased after fermentation from 60 mg/100g to 122mg/100g [21].

*3.2.3 Mimosine*Mimosine is a group of aromatic amino acids with the chemical formula of (β-N-(3-hydroxy-4-pyridone)-α–amino-propenoic acid). It is a non-nutritional amino acid whose structure is nearly similar to that of tyrosine and is found in several mimosa species within the Leucaena genus. The mimosine level of lamtoro gung leaves ranged from 2% to 6% and varies according to their ripening level [22]. Mimosine level in lamtoro seed during fermentation shown in Table 7.

Table 7. Mimosine contents inlamtoro seed during fermentation

Fermentation time (hours)	Mimosine Content (%)
0	0.00393±0.00011[e]
6	0.00382±0.00022[e]
12	0.00309±0.00010[d]
18	0.00210±0.00015[b]
24	0.00230±0.00011[c]
30	0.00209±0.00010[b]
36	0.00173±0.00015[a]

The result of the research showed thatmimosine content on lamtoro seed during fermentation was significantly decreased, from 0.00393% before fermentation to 0.00173% after 36 hours fermentation. During fermentation, mimosine in lamtoro seed was metabolized by microorganism from inoculum. It caused by the mimosine structurrally resemblance to L-tyrosin [23].

4. Conclusion

From hedonic test, inoculum concentration of 1% was used in lamtoro tempeh processing. During 36 hours fermentation phytic acid content and mimosine content was decreased significantly, from 0.0558 % to 0.0453 % and from 0.00393 % to 0.00173 % respectively.Whereas tannin content was increased sigificantly from 0.0822 % to 0.00173%.

Acknowledgement
The research was financially supported by research project of PNBP UNS No : 623/UN27.21/PP/2017from Universitas Sebelas Maret, Surakarta, Indonesia.

References

[1] Listyawati S, Moeljono MPE and Handari S 2001 Histologic Picture of Thyroid Gland in Rat (*Rattus norvegicus*) after the granting of Lamtoro Gung tempeh (in Bahasa Indonesia) *Biosmart* **3**14-8
[2] Ekpenyong TE 1986 *Animal Feed Science and Technology* **15** 185
[3] Feyemi PO,Onwuka CFI, Isah OA, Jegede AV, Arigbede OM and Muchenje V 2011 *African Journal of Agricultural Research* **6** 4081-5
[4] Listyawati S 2003 Iron availability of rats (Rattus norvegicus L. given lamtoro gung tempe (in Bahasa Indonesia) *BioSMART* **5** 47-51
[5] Whiting MG 1982 *Food uses of Leucaena seeds. Tempeh lamtoro, preparation and toxicity. LeucaenaResearch Reports*3:100-01. July. (Taipei, Taiwan: Council for Agricultural Planning and Development).
[6] Mahmood SM, Ajmal K, Sarwar M, Nisa M, Lee WS, Lee S, Kim SB, Hur TY, Lee HJ and Kim HS 2007*Asian-Aust. J. Anim. Sci.* **20** 1462 – 7
[7] Almasyhuri, Yuniati H and Slamet DS 1990 Phytic acid and tannin content in Leguminous Tempeh (in Bahasa Indonesia) *PGM* **13** 65
[8] Kasmidjo RB 1990 *Tempeh : Microbiology and Processing Chemistry and It Uses*(in Bahasa Indonesia) Pusat Antar Universitas Pangan dan Gizi UGM Yogyakarta
[9] Muthmaina, Sabang SM and Supriadi 2016 Effect of fermentation time to the protein levels of fermented lamtoro gung (*Leucaena leucochepala*) fruit seed (in Bahasa Indonesia)*J. Akademik Kimia* **5** 50-4
[10] Sukardi, Wignyanto and Purwaningsih I 2008 Tempeh inoculum application test of Rhizopus oryzae with rice and cassava flour as substrate at sanan tempeh industries-Kodya Malang (in Bahasa Indonesia) *J. Teknologi Pertanian* **9** 207
[11] Dewi SR and Aziz S 2011 Isolation of Rhizopus oligosporus from tempeh inoculum in Kabupaten Banyumas (in Bahasa Indonesia) *Molekul* **6** 94
[12] Yunarti, Tri R, Zulys A, Harahap LY and Pramukti MSA 2013 *Makara Journal of Science* **17** 12-4
[13] AOAC 1975 *Official Methods of Aanalysis of the Association of Official Analytical Chemists. 12 ed.* Washington AOAC
[14] ZulI, Hanisjasymah H, Nurul AR, Norfadhilah R and Normaniza O2015.*Procedia Chemistry* **16** 164-70
[15] Winarno FG 2004 *Food Chemistry and Nutrition (*in Bahasa Indonesia) Gramedia Jakata
[16] Sayudi S, Herawati N and Akhtar A 2015 Potencial of Leucaena seed and soybean as raw material for making complementation tempeh (in Bahasa Indonesia) *JOM Faperta* **2** 1-9
[17] Aminah S 2010 Potential for mixed brown rice sprouts and soybean sprouts as functional beverage high fiber and protein (in Bahasa Indonesia) *J. Pangan dan Gizi* **01** 27-32
[18] Murtini ES, Radite AG and Sutrisno A 2011 Characteristic of chemical content and digestibilty of brown shorgum tempeh (in Bahasa Indonesia) *J. Teknol. Dan Industri Pangan* **22** 150-5
[19] Feng, Mei X, Passoth V, Jonsson CE, Alminger ML and Schnürer J 2007 *Food Microbiology* **24** 393
[20] Setiarto RHB and Widhyastuti N 2016 Reduction of Tannin and Phytic Acid on Sorghum flour by using Fermentation of *Rhizopus oligosporus*, *Lactobacillus plantarum* and *Saccharomyces cerevisiae* (in Bahasa Indonesia) *J. Ilmu-Ilmu Hayati* **15** 149-57
[21] Komari 1999 Fermentation Process of *Leucaena Seed* with Rhizopus oryzae (in Bahasa Indonesia) *J. Mikrobiologi Indonesia* **4** 19-21

IOP Conf. Series: Earth and Environmental Science **102** (2017) 012093 doi:10.1088/1755-1315/102/1/012093

[22] Laconi EB and Widiyastuti T 2010 Lamtoro Leaf *(Leucaena leucocephala)* Xanthophylls Content as the Result of Physical and Chemical Detoxification of Mimosine (in Bahasa Indonesia) *Media Peternakan* **33** 50-4

[23] Rahayu A, Suranto and Purwoko T 2005 Analysis of carbohydrates, protein, and lipids of *Leucaena* sauce fermented with *Aspergilus oryzae* (in Bahasa Indonesia*)* *Bioteknologi* **2** 14-20

Quality characteristic of liquid smoked straw mushroom (*Volvariella volvacea*) ball during storage

C W Kurniawan, W Atmaka, G J Manuhara and A P Sanjaya*

Department of Food Science and Technology, Faculty of Agriculture, Sebelas Maret University

*E-mail : adhit.sanjaya@staff.uns.ac.id

Abstract. Straw mushroom (Volvariella volvacea) ball was soaked for 15, 30, and 45 minutes with the concentration level 1%, 2%, and 3% (v/v) of the coconut shell liquid smoke. The chemical characteristics (water contains, total phenol, carbonil contains, total-N, TVB-N, and pH), microbiological characteristics (Total Plate Count), and sensory characteristics (color, flavor, taste, texture, and overalls) of the liquid smoked straw mushroom ball during 14 days storage at freezing temperature were investigated. The result showed that the water content and TVB-N were decreased after soaked and were increased after storaged. On the other hand, the result of total phenol, carbonyl content, and Total-N were increased after soaked and were decreased after storage. The level of pH and Total Plate Count of the straw mushroom ball were decreased during storage. Due to the sensory characteristics of the straw mushroom ball, the panelists provide high values for the straw mushroom ball which was soaked in 3% concentration level with 30 minutes soaked time. The best-soaked treatment was by soaked at 30 minutes with 3% concentration level liquid smoke. The straw mushroom ball has 70.95±0.10% water contains; 0.32±0.02% total phenol; 1.08±0.22% carbonyl contains; and 2.29±0.07% total-N.

Keywords: *quality characteristic, smoked straw mushroom ball*

1. Introduction

Mushroom contained high protein, low fat, and around nine types of 20 essential amino acids. Mushroom also contained various types of vitamins, minerals, pure fiber, and very low calorie [1]. Straw mushroom is one of many popular consumption mushrooms in the world but has minimally processed food. The other food ingredient that could be used as materials mix in a meatball dough with not reduced the quality criteria of beef meatballs is a straw mushroom (*Volvariella volvaceae*). Straw Mushroom (*Volvariella volvaceae*) is one type of mushroom that has high fiber content and has good flavor and texture [2]. Straw mushroom (*Volvariella volvacea*) balls could be pleasured for vegetarian and diet consumers.

The straw mushroom ball is perishable food which has a high water content and is easily damaged by oxidation or contaminated by microbiology. One of preservation method of the straw mushroom ball is using coconut shell liquid smoke as natural preservation. Liquid smoke is the product of the distillation or condensation of steam from the indirect or direct combustion of materials that contained lots of carbon and other compounds. Coconut shell liquid smoke has more than 400 components such as phenol and acid compounds that act as antibacterial and antioxidants [3]. Therefore, the utilization of liquid smoke in food product also could improve the sensory and physical characteristic, such as taste, aroma, and textural properties. In this study, the soaked time and concentration level of coconut shell liquid smoke for the straw mushroom ball were investigated to investigate the effect preservative of coconut shell liquid smoke.

2. Materials and Methods

2.1. Materials

Straw mushroom (*Volvariella volvacea*) as obtained from gedong jati, Tulung, Klaten, Central Java (Indonesia). Tapioca "Rose Brand", pepper "Ladaku", egg, salt "Segitiga", garlic, and sodium triphosphate from Gede market (Surakarta, Indonesia). Coconut shell Liquid smoke from the Laboratory Processing Engineering of Agricultural Products, Faculty of Agricultural Technology, Gadjah Mada University.

2.2. Preparation of Coconut Shell Liquid Smoke Solution

Liquid smoke that would be used as a preservative diluted by using distilled water to be 1% v/v, 2% v/v, and 3% v/v concentration level.

2.3. Straw Mushroom Ball Production

The Straw mushroom ball production procedure was described by Ruri with simple modification [4]. The straw mushroom was sorted, washed, mashed with a blender, and drained with pressure. Then, 160 g of mashed mushrooms added with other ingredients (tapioca flour 40 g, 5 g salt, 0.6 g pepper, garlic, 4 g, STPP 1 g, White egg 30 g) and mixed. Then, the straw mushroom ball dough was formed like a ball with ± 2 cm diameter. Then, boiled in heat water for 10 minutes, and soaked in ice water for 10 minutes.

2.4. Liquid Smoked Straw Mushroom Ball Production

Straw mushroom balls were soaked with coconut shell liquid smoke with concentration level (1%, 2%, and 3%) with soak time (15 minutes, 30 minutes, and 45 minutes). Then, straw mushroom balls packed with vacuum plastic with size 14x19.5 cm, stored at freezing temperature (-19°C) for 14 days. All these experiments were performed in triplicate, and the water content, total phenol, carbonyl content, Total-N, TVB-N, pH, total plate count, and sensory characteristics (color, flavor, texture, overalls) of straw mushroom balls were determined after the storage time was 0, 7 and 14 days.

3. Results and Discussion

3.1. Chemical and Physical Properties of Straw Mushroom Balls

Liquid smoke contains more than 400 components and has a function as an inhibitor to the bacterial development and is quite safe as a natural preservative such as acids, phenolics, and carbonyls [5]. Liquid smoke that used in this experiment is liquid smoke from coconut shell with chemical characteristics such as total phenol 11.81%; carbonyl contains 18.31%; TVB-N 2.7676 mg/100g; and pH level 3.

Chemical properties of straw mushroom balls are shown in **Table 1. Table 1** showed the lowest levels of water contain was achieved by straw mushroom balls which were soaked in concentration level 3% liquid smoke and soaked 45 minutes which 70.84±0.08%. SNI 01-3818-1995 reported that the water content level of meatball has maximal 70%. After soaked, the straw mushroom ball was decreased of water content. The decreased water content of the straw mushroom ball due to osmosis process. According to Purnamasari, phenol has a hypertonic characteristic; hypertonic is a high concentrated solution [6]. The hypertonic characteristic of phenol in liquid smoke caused osmosis activity on straw mushroom balls during soaking treatment. After storage, the water content of straw mushroom balls was increased. The increasing of the water content of the mushroom balls during storage might be caused by a microbial activity which can degrade protein and produce water. Microorganisms also generally produce water from respiration activity, and it can cause the increasing of the water content. According to Widyastuti, increasing the water content of the meatball also caused by syneresis process which can separate the water from the structure of starch gel and it occurred during storage at low temperature storage (cooling or freezing) [7].

Table 1 Chemical properties of straw mushroom balls

Soaked sample treatment		Water contains (%)*	Total phenol (%)*	Carbonyl contains (%)*	Total-N (%)*	TVB-N (mg N/100g)*
Untreated straw mushroom ball		72.91±0.88	0.03±0.01	0.57±0.13	2.18±0.05	4.079±3.32
1%	15 minutes	71.89±0.12	0.13±0.10	0.70±0.09	2.23±0.09	2.603±2.04
	30 minutes	71.45±0.10	0.17±0.07	0.82±0.14	2.23±0.09	2.453±1.85
	45 minutes	71.34±0.09	0.22±0.03	0.92±0.15	2.23±0.10	2.199±1.50
2%	15 minutes	71.55±0.10	0.26±0.04	0.83±0.18	2.25±0.09	1.862±1.03
	30 minutes	71.25±0.09	0.27±0.04	0.99±0.22	2.27±0.08	1.680±0.78
	45 minutes	71.14±0.08	0.28±0.03	1.01±0.21	2.28±0.07	1.146±0.03
3%	15 minutes	71.04±0.09	0.30±0.02	0.93±0.08	2.28±0.07	1.141±0.03
	30 minutes	70.95±0.10	0.32±0.02	1.08±0.22	2.29±0.07	1.127±0.04
	45 minutes	70.84±0.08	0.33±0.01	1.23±0.29	2.31±0.05	1.114±0.04

The highest total phenol was achieved by of the straw mushroom ball which was soaked in 3% concentration of liquid smoke and 45 minutes. The corresponding value was 0.33±0.01%. The presence of total phenolic compounds in untreated straw mushroom ball sample might be caused by the straw mushroom ball was contained antioxidant components from garlic and pepper as ingredients. During the storage, the decreasing of total phenol content of the straw mushroom ball was occured. The deceasing of the total phenol caused by the increasing the water contain the mushroom ball. The phenol compound could be decomposed by hydrogen of water. In another hand, microorganism has the ability to degraded phenol compound. The degraded of phenol component by microbial was influenced by several factors such as microbial type, phenol concentration and environmental conditions [8].

The carbonyl content of the untreated straw mushroom ball was 0.57±0.13%. Carbonyl content on straw mushroom balls was from the organic compound, i.e., carbonyl groups and amino acid [9], aldehydes and ketones which contain in tapioca starch. During the storage, carbonyl content in the straw mushroom ball was decreased. The decreasing of the carbonyl content is due to the occurrence of protein denaturation that resulted in the breakdown of the bonds on carbonyl groups in amino acids and syneresis process. The highest carbonyl component was achieved by the straw mushroom ball which was soaked in 3% concentration of liquid smoke and 45 min soaking time.

Table 1 shows that the total-N of the straw mushroom ball was increased caused soaked in liquid smoke. According to Buckle, proteins are divided into two parts based on their solubility, was water-soluble proteins and insoluble water [10]. The partial discharge of water results in a total water soluble solid protein being left from the straw mushroom ball. The discharge of some water also reduced the mass of the material. Thus, the protein component in the straw mushroom ball was increased and also affects to the total-N of the straw mushroom ball. Total-N value of the straw mushroom balls was decreased during storage. The decomposition of the protein caused by microbial metabolism that was consumed the nitrogen in amino acid and resulted in the decreasing of the total-N in mushroom balls. The highest total-N was achieved by the straw mushroom ball which was soaked in 3% concentration of liquid smoke and 45 min soaking time.

According to Esminingsih, the TVB-N value is an index of damage or degradation due to protein [11]. The untreated straw mushroom ball has 4.079±3.32 (mg N/100g) and has the highest value of TVB-N. During storage, the TVB-N value was increased. The degradation process of protein will produce simple nitrogen compounds. The free amino acids and nitrogenous bases are evaporated being degradation process so the TVB-N was increased.

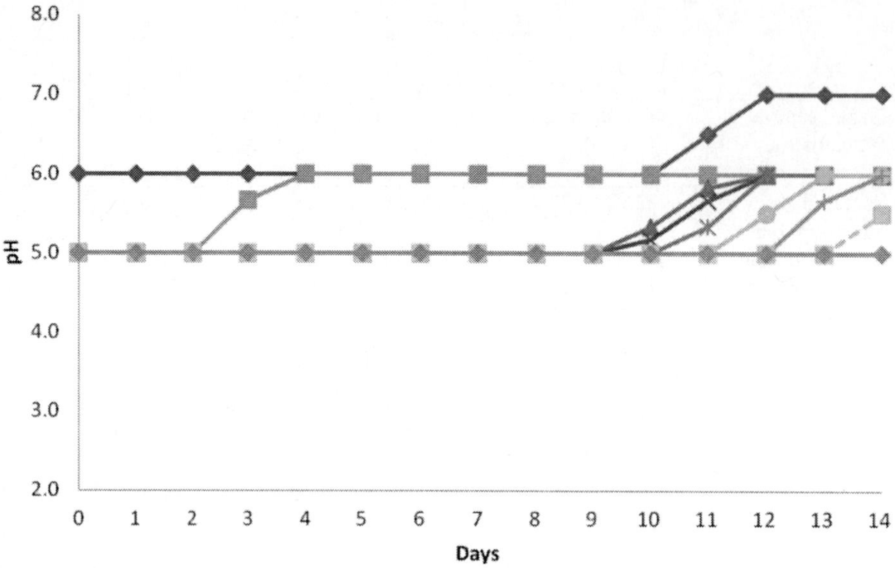

Figure 1. pH level of straw mushroom balls during storage from untreated straw mushroom ball (◆) 1%, 15 minute (■) 1%, 30 minute (▲) 1%, 45 minute (×) 2%, 15 minute (✳) 2%, 30 minute (●) 2%, 45 minute (+) 3%, 15 minute (--■--) 3%, 30 minute (--▲--) 3%, 45 minute(--◆--)

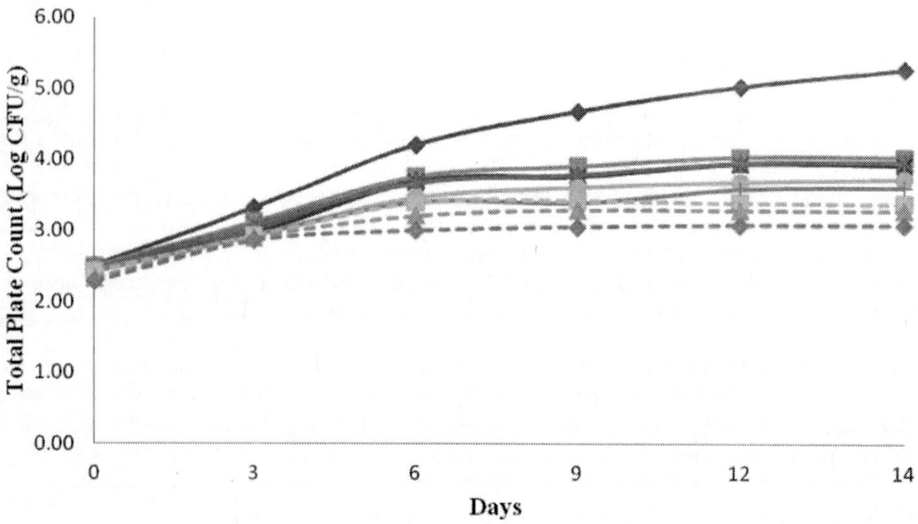

Figure 2. Total plate count of straw mushroom balls during storage from untreated straw mushroom ball (◆) 1%, 15 minute (■) 1%, 30 minute (▲) 1%, 45 minute (×) 2%, 15 minute (✳) 2%, 30 minute (●) 2%, 45 minute (+) 3%, 15 minute (--■--) 3%, 30 minute (--▲--) 3%, 45 minute (--◆--)

Straw mushroom ball has pH level 6, and it was decreased during soaked in liquid smoke. **Figure 1** shown the decreased pH level during storage. The decreasing of pH value of straw mushroom balls during storage caused by the acid component of liquid smoke. According to Zuraida, the decreasing of the pH values is due to the influence of acidity levels of coconut shell liquid smoke and the presence of acidic compounds such as 2,3-dihydroxy-benzoic acid, 3-methoxybenzoic acid methyl ester, and 4-hydroxy-benzoic acid methyl esters based on GC-MS analysis [12]. During Storage, pH level was increased. According to Goulas and Kontominas, the increasing of pH is caused by the activity of bacterial that produced proteolytic enzymes [13]. This enzyme could breakdown proteins into ammonia, trimethylamine and other volatile components that have base characteristic so that the pH level would be raised. The highest pH level was achieved by the untreated straw mushroom ball. The corresponding value was 7. The lowest pH level was achieved by a straw mushroom ball which soaked in 3% concentration of liquid smoke and 45 minutes soaking time. The corresponding value was 5.

The microorganism activity could show in **Figure 2.** Before storaged, the untreated straw mushroom ball has higher TPC value which was 2.53 (log CFU/g) and mushroom ball with 3% concentration level and 45 soaked times have the lowest TPC value which was 2.28 (log CFU/g). During storage, the presence of microorganism activity during storage at freezing temperatures can inhibit their growth. Bacteria that could slow growth at this freezing temperature are bacteria belonging to the psychrophilic group. Examples Pseudomonas, Flavobacterium, Alcaligenes, Micrococcus, Enterobacter, and Anthrobacter [10]. During storage, the higher concentration of the liquid smoke and the soaked time shown the lower the microorganism activity. According to Zuraida, liquid smoke combined with low temperatures can be either bactericidal or bactericidal depending on the concentration of liquid smoke, the temperature used, and the length of storage [12].

3.2. *Sensory Properties of Straw mushroom balls*

Straw mushroom balls were assessed for sensory attributes using an acceptance test after production (Table 2). The higher scores for straw mushroom balls flavor and taste value were found for straw mushroom balls that soaked on 1% concentration level with 15 min soaked time of liquid smoke. According to Yunus, carbonyl compounds in liquid smoke has a role to give the color and taste of the smoked product [5]. That compounds have a unique caramel-like scent. The types of carbonyl compounds present in liquid smoke include vanillin and aldehyde.

Table 2. Sensory properties of straw mushroom balls

Sample		Parameter				
		Color	Flavor	Taste	Texture	*Overall*
Untreated straw mushroom ball		3.03^{a}	2.93^{a}	3.00^{a}	3.03^{a}	3.13^{a}
1%	15 minutes	3.23^{abc}	3.93^{d}	3.97^{d}	3.13^{a}	3.33^{ab}
	30 minutes	3.40^{abc}	3.57^{bcd}	3.60^{bcd}	3.30^{a}	3.47^{ab}
	45 minutes	3.47^{abd}	3.90^{cd}	3.83^{cd}	3.40^{ab}	3.53^{abc}
2%	15 minutes	3.20^{ab}	3.53^{bcd}	3.50^{bc}	3.33^{a}	3.30^{abc}
	30 minutes	3.20^{ab}	3.23^{ab}	3.20^{ab}	3.23^{a}	3.40^{ab}
	45 minutes	3.67^{cde}	3.20^{ab}	3.23^{ab}	3.80^{bc}	3.73^{bc}
3%	15 minutes	3.50^{bcd}	3.47^{b}	3.37^{ab}	3.40^{ab}	3.57^{ab}
	30 minutes	4.03^{e}	3.50^{bc}	3.40^{ab}	3.97^{c}	3.87^{c}
	45 minutes	3.87^{de}	3.20^{ab}	3.27^{ab}	3.80^{bc}	3.83^{bc}

*Notation different letters in the same column indicate significant difference at a significance level of 5%

The higher scores for straw mushroom balls color, texture, and overall value were found on the straw mushroom balls that was soaked on 3% concentration of liquid smoke and soaked for 30 min. According to Darmadji, the colors in fumigation products especially meatballs are formed due to interactions between carbonyl compounds and amino groups [14]. Carbonyl compounds caused maillard reaction, so changed the color of meatballs to brown. According to Hadi, the protein content gave effect on the textural properties. The protein serves as a binding agent which encapsulates fat in the meatball hence the meatball emulsion become more stable, and the texture is denser [15]. The panelists of the acceptance test give a highest overall score on srawtraw mushroom balls with 3% concentration level and 30 min soaked time of liquid smoke.

4. Conclusions

The water content and TVB-N were decreased after soaked and were increased after storaged. On the other hand, the result of total phenol, carbonyl content, and Total-N were increased after soaked and were decreased after storaged. The The level of pH and Total Plate Count of the straw mushroom ball were decreased during storage. Due to the sensory characteristics of the straw mushroom ball, the panelists provide high values for the straw mushroom ball which was soaked in 3% concentration level with 30 min soaked time. The best-soaked treatment was by soaked at 30 minutes with 3% concentration level liquid smoke.

References

[1] Cahyono 2013 *The Levels of Protein and Organoleptic Test of Meatballs Based on the Composition of Different Beef and Mushroom (Volvariella volvaceae)* (Surakarta: Undergraduate Thesis)
[2] Purwanto A, Ali A dan Herawati N 2015 *Sagu* **14** 1-8
[3] Nensih S 2015 *Effect of Coconut Shell Liquid Smoke Grade I Concentrations on Shelf Life Determination of Tofu Soy Beans (Glycine max)* (in Bahasa Indonesia) (Bandung: Undergraduate Thesis)
[4] Ruri S 2014 *J. of food and agriculture processing* **1** 85-94
[5] Yunus M 2011 *J. sains dan inovasi* **7** 53-61
[6] Purnamasari E 2013 *Seminar Nasional Teknologi Peternakan dan veteriner 2013*
[7] Putri NA 2015 *Rheology Properties of Mocaf (Modified Cassava Flour) and Tapioca on the different pH condition* (in Bahasa Indonesia) (Jember: Undergraduate Thesis)
[8] Dewilda Y 2012 *J. Enviromental Engineering* **9** 59-73
[9] Dedin 2007 *J. of Food and Science Technology* **17** 3
[10] Buckle A, Edward RA, Fleet GH and Wootton M 1987 *Food Science (*Jakarta: Indonesian University Press)
[11] Esminingtyas R 2006 *Quality Changes of Smoked Dumbo Catfish (Clarias gariepinus) during Storage* (in Bahasa Indonesia)(Bogor: Undergraduate Thesis)
[12] Zuraida I, Hasbullah, Rokhani, Sukarno, Budijanto S and Prabawati S 2009 *J. Ilmu Pertanian Indonesia* 41-49
[13] Goulas AE and Kontominas MG 2005 *J. Food Chemistry* **93** 511–520
[14] Darmadji P 1996 *Agritech* **16** 19-22
[15] Hadi A 2014 *J. of Scientific Health* **7** 2

IOP Conf. Series: Earth and Environmental Science **102** (2017) 012095 doi:10.1088/1755-1315/102/1/012095

The influence of brewing water characteristic on sensory perception of pour-over local coffee

K Fibrianto*, A D Ardianti, K Pradipta and W B Sunarharum
Food Sensory Research Group, Faculty of Agricultural Technology, Brawijaya University, Indonesia

*E-mail: kiki.fibrianto@ub.ac.id

Abstract. The coffee quality can be characterized by its multisensory perceptions. The content and mineral composition and other substances of brewing water can affect the result of brewed-coffee. The water may influence in extraction capabilities and flavor clarity. The ground Dampit coffee and two commercial instant coffee with pour-over method were used in this study. Various types of commercial drinking water were used to brew the coffee. The result suggests that the different brewing water affects the intensity of sweet and chocolate aroma, as well as oily mouth-feel. Surprisingly, taste and flavour attributes were not affected by the pH of brewing water within the range of 5.5 to 9.1.

Key words: *Alkalinity, Water, Coffee, Sensory, Hardness*

1. Introduction

Coffee (*Coffea sp.)* is widely cultivated plant in the world. According to [1], coffee contains nearly a thousand volatile compounds which are developed during roasting process. As it is very rich in flavours, coffee is considered as the most popular beverage in the world. Based on the International Coffee Organization (ICO), the world coffee consumption rate in 2015 reached 152.2 million per 60 kg of coffee sachet and increased 2.0% annual average since 2011 [2]. In Indonesia, the coffee consumption in the last 4 years continues to increase by 36% from 2010-2014 by the amount of 1.03 kg/capita/year by 2014 [3].

Basically, multisensory perception on coffee could be influenced by brewing method, temperature, roasting, coffee varieties, storage condition, grinding, and brewing water [4, 5, 6]. Based on its genetic, coffee is distinguished into Arabica coffee and robusta coffee which has its own characteristic and flavor. Robusta coffee contains more dissolved solids, therefore it is widely used in commercial instant coffee to add body to the solution and increase the extraction yield [7]. In addition, based on the process, the coffee can be classified as instant coffee and ground coffee. The process of making instant coffee involves the industrial process from roasting until drying extraction. Meanwhile, ground coffee involves a more traditional and simple process of grinding.

One of the coffee-brewing methods according to SCAA is pour over. The method is relatively well known, due to its simple way of brewing. The principle is to pour boiling water into a coffee powder to extract the powdered-coffee. Pour over coffee should be waited a while until the coffee grounds settle completely before it is consumed by consumers [8].

One of the factors that influence the perception of coffee multisensory is brewing water. The content and composition of minerals and other substances can affect the result of brewing coffee. This water brewing quality can be represented by total hardness and alkalinity. Total hardness is defined as the amount of calcium and magnesium in an equivalent concentration [9, 10, 11]. According to the Specialty Coffee Association of Europe (SCAE), the total limit of hardness content or $CaCO_3$ contained in the upper limit of hardness or $CaCO_3$ is 175 ppm whereas the lower limit is 50 ppm [12]. In sensory evaluation, consumers judge water quality by its clarity, palatability and odor. Thus, the threshold test is important to ensure the water is odorless as it is required for being palatable [13].

According to [14], there is a difference of binding extraction between Ca^{2+}, Mg^{2+} and Na^{+} cations with five acid components in coffee that is lactic acid, malic acid, citric acid, quinic acid, and chlorogenic acid, then in caffeine in the form of alkaloids, and coffee flavor characteristic on eugenol. Na^{+} has the lowest binding when compared with the other two minerals. Na^{+} does not facilitate the extraction between caffeine and eugenol, because the binding is not significant as the five acids present in coffee. Binding between the minerals with the highest coffee component is on Mg^{2+}. This mineral binds more of the desired flavor of coffee. In Ca^{2+} the binding of mineral of the coffee component is not as high as the Mg^{2+} [14].

As it has been reported that some minerals in water affect the extraction of particular components of coffee, it is hypotesised that different quality of water may affect the sensory attributes of extracted coffee. Therefore, in this current study, mainly two types of water; alkaline and oxygenated water were investigated. Both waters was expected to modify the brewed coffee characteristics as they may different in mineral content and pH. According to [15], one of the factors that can affects coffee extraction is mineral content. The higher mineral content in the water may lead to higher pH of water. In addition, [16] suggests that oxygenated water may affect extraction capabilities and flavor clarity of coffee.

Different type of brewing water allegedly could affect the coffee brewing result. This study aims to provide scientific report related to influence of different types of brewing water on sensory perception of particularly Indonesian coffee. This information may be very useful for barista all over the world to specify the required water quality for optimising the used brewing techniques.

2. Materials and Methods

2.1 Preparation of sample and coffee making

There were 2 groups of coffee used in this study; namely ground and instant coffee. The ground coffees were represented by ground Robusta coffee derived from mixture of all regions in Indonesia, including Dampit-Robusta coffee. Meanwhile, the instant coffee were represented by Robusta coffee derived from mixture of all regions in Indonesia and Arabica coffee from Mandailing area of Sumatra. Pour over method was chosen as the brewing technique in this study due to its simplicity and popularity among local coffee consumers.

There were 11 different types of water based on the differences of mineral and pH content that being used in this study. They were San Benedetto, Cleo, Aqua, Fitoxy, Oxygenizer, E^{+}, Total 8^{+}, Milagros, Super O_2, tap water and refilled water.

Coffee making refers to the coffee-making standard for sensory analysis on ISO 6668:2008, as much as 7 grams of brewed coffee used in 100 ml of boiling water of 92°C-96°C with unfiltered pour over method. In this study, the coffee presented for sensory analysis was prepared by brewing 3.5 grams of coffee in 50 ml of water for 3 minutes. The coffee was served to the panels at 60±5°C for aroma evaluation first, followed by oral experience assessments.

2.2 Panellist's recruitment

The panellists were recruited from Brawijaya University students, covering both genders equally from 20 to 23 years old. The criteria of candidates are those who have basic knowledge of sensory testing, have an interest for being trained panellist, have been accustomed to the attributes of a food product and preferably have a habit to coffee consumption, having knowledge of coffee, being willing to consume coffee, having good healthy condition, having good communication skills, and able to provide a detailed description of product [17].

2.3 Panellist selection

This research uses the Quantitative Descriptive Analysis (QDA) method which is done by three stages; panellist selection, training and intensity assessment of coffee sensory attributes [18]. The candidates were required to pass two stages of selection; the aroma recognition test and basic recognition test. They needed to answer correctly at least 80% of overall aroma and basic taste in the test to get selected. Then after passing the selection, the panellists were tested their absolute threshold determine the threshold sensory capability of each panellist. The absolute threshold test was performed by 3-AFC (Alternative Forced Choice) method against five basic tastes of acid, sweet, bitter, salty, and umami [18].

Table 1. Attribute, Definition, and Reference

Attribute	Definition	Reference
	Aroma	
Sweet	Sweet *caramel-like*	*Toffieco* caramel splint (2 drops)
Chocolate	*Chocolate brewed-like*	*Van Houten* brown powder (6% b/v)
Roasting	Roasted coffee aroma	Roasted *Dampit* coffee beans (5-8 seeds)
Burnt	Burnt cassava aroma	Burnt cassava (1 slice of ±5 cm diameter)
Acid	Citrus acid aroma	*Red bell* tangerine (2 drops)
	Taste	
Bitter	Bitter taste	Pure caffeine P.A (0.03% b/v)
Vinegar acid	*Vinegar-like* taste	*Dobbel* vinegar (1% v/v)
Citric acid	Citric acid taste	Pure citric acid P.A (0.04% b/v)
Salty	Salty	*Refina* kitchen salt (0.4% b/v)
	Flavor	
Cassava	Burnt Cassava Flavour	Burnt cassava (1 slice ±5 cm diameter)
	After Taste	
Astringent	Astringent	Cranberries HBF International (1 piece)
Sweet	Sweet	Sugar (0.5% b/v)
	Mouth-feel	
Dry	Dry sensation *bean-like*	Roasted ground beans (1 seed)
Oily	Oily sensation	Butter Salt Anchor Fonterra (5 ml)

2.4 Development of attribute vocabulary and panellist training
Vocabulary development was conducted through discussions of sensory attribute in two sessions to gain aggrement of terminologies used in the test. The development of standard vocabulary was then followed by the aggreement on definitions and references of each sensory attributes [17]. The selected panellists were then trained to assess the intensity of each attribute reference that has been previously agreed using line scale [18]. References training were conducted twice a week to see the panellist's consistency on the intensity assessment of each reference [19]. The results of attribute agreed by panellists are shown in Table 1.

2.5 Descriptive analysis of sensory perceptions
Panels assessed the intensity of the sensory attributes of each sample of coffee using 15 cm unstructured line scales, anchored at 1 cm both ends. A total of 55 coffee samples were given to the panels by testing each sample of combination between each coffee and water and repeating one sample each type of coffee with each type of water randomly. Each sample was served at 60±5°C on a 120 ml paper cup containing 50 ml of coffee. Samples were presented randomly to the panellists.

2.6 Data analysis
The descriptive test data of coffee were analyzed using Analysis of Variance (ANOVA) with Generalized Linear Model (GLM) on Minitab 17, followed by Tukey test. Multivariate data analysis was also conducted by Principal Component Analysis (PCA) on Minitab 17.

2.7 Chemical and physical analysis
2.7.1 Mineral content testing with AAS (Atomic Absorption Spectroscopy). In this study, the mineral content test was done by using Atomic Absorption Spectroscopy (AAS). This test is a quantitative method of elemental analysis based on the absorption of light of a certain wavelength by metal atom in a free state [20]. If light of a certain wavelength is passed to a cell containing free atoms the some of the light will be absorbed and the absorption intensity will be proportional to the number of free metal atoms in the cell [21].

2.7.2 pH analysis. pH measurements in this study were conducted after coffee descriptive test. Coffee pH measurements were performed for each samples before and after the consuming (spit out). The pH measurements of spit out samples were done shortly after the panels consumed the coffee to avoid the pH changes during storage. pH measurement was performed using pH meter that had been calibrated using standard solution of pH 4 and 7 at room temperature. The pH meter was calibrated before taking pH measurements of the sample and calibration was done for each measurement of five samples. Each samples was measured pH twice.

3. Result and Discussion

3.1 Mineral Content of Each Water Type Based on Atomic Absorption Spectroscopy (AAS)
Each type of water tends to have different mineral content (**Table 2**). Water code 5 has the highest mineral content of Ca 6,845 mg/L, Mg 1.178 mg/L and Na 167.550 mg/L. While water code 11 has the lowest detectable mineral content of Na 0.97 mg/L. Mineral content of Ca and Mg of some types of water were not detected as they were below the detection limit.

The different mineral content in brewing water affects the pH of ground coffee type. Water which are containing lower Ca, Mg, and Na tend to induce lower pH of brewed ground coffee, compared to those of higher mineral content (**Table 3**). Similarly, the pH of the ground coffee after consumed, although the pH

IOP Conf. Series: Earth and Environmental Science **102** (2017) 012095 doi:10.1088/1755-1315/102/1/012095

increased to be higher, the brewed ground coffee that brewed with the low mineral content has a lower pH than that of other ground coffee. However, the mineral content of brewed water has slightly changed the pH on instant coffee before the consumption or after consumption. As pH of instant coffees slightly increased after consumption (**Table 3**), it suggest that interaction of saliva and coffee components may involve in perceiving coffee.

Table 2. Atomic Absorption Spectroscopy Result

Water type	Ca (mg/L)	Mg (mg/L)	Na (mg/L)	$CaCO_3$ (mg/L)
1	6.713	1.177	26.440	21.607
2	6.349	1.019	67.550	20.052
3	0.031	-	12.050	-
4	6.459	1.012	60.190	20.298
5	6.845	1.178	167.550	21.945
6	0.317	1.086	10.85	5.264
7	5.686	0.952	46.48	18.118
8	-	0.912	5.32	3.759
9	1.183	0.286	2.31	4.133
10	0.121	-	0.41	0.302
11	-	-	0.97	-

Description of water code:
1 : mineral water brand A
2 : mineral water brand B
3 : mineral water brand C
4 : mineral water refill
5 : tap water
6 : alkaline water brand A
7 : alkaline water brand B
8 : alkaline water brand C
9 : oxygenated water brand A
10 : oxygenated water brand B
11 : oxygenated water brand C

Low mineral content in drinking water can affect the pH. Low mineral content tends to have a low pH as well. Water with codes 3, 9, 10, and 11 have a low Ca, Mg, and Na minerals and have a pH range of 5.8-6.85. Low mineral content also affects the stability test results, water codes 9, 10, and 11 tend not to have stable pH. The pH of water after consumption or with saliva tends to increase toward neutral in the pH range of 6.3 to 7.

Differences in pH and mineral content of water samples varied considerably shown on Table 4. Nevertheless, the difference in water samples has only significant effect on the attributes of sweet aroma, chocolate aroma, and oily mouth-feel. The vinegar taste and citric acid taste is not affected by water differences. This is reinforced by the result of coffee pH test brewed with various types of water tend to be in the pH range 4.9-6.1. Ground type of coffee (A and B) has a pH range of 5.5-6.1 and instant coffee (C and D) has a lower pH range and a closer range of 4.9-5.1

3.2 pH Test of Coffee

Table 3. before and after pH on coffee consumption

Coffee	Water										
	1	2	3	4	5	6	7	8	9	10	11
Before Consumption											
A	5.8	5.9	5.7	5.9	6.1	5.7	5.7	5.6	5.7	5.6	5.55
B	5.8	5.9	5.7	5.9	6.1	5.9	5.85	5.55	5.6	5.55	5.5
C	4.9	5	4.9	5	5	5	5.1	5	4.95	5	4.8
D	4.9	4.9	4.8	4.9	4.9	4.9	4.9	4.8	4.9	4.8	4.8
After Consumption											
A	6.3	6.2	5.9	6.2	6.3	6.1	6.1	5.7	5.7	5.7	5.7
B	6.2	6	5.8	6.2	6.2	5.9	6.1	5.8	5.8	5.85	5.87
C	5.1	5.1	5.1	5.1	5.1	5.1	5.1	5.1	5	5	5
D	5	5	5	5	5	5	5.1	4.9	4.9	4.9	4.9
Pure Water	7.6	6.7	5.8	7	6.8	8.25	8.2	8.05	6.7	6.85	6.5

Description of water code:
1 : mineral water brand A
2 : mineral water brand B
3 : mineral water brand C
4 : mineral water refill
5 : tap water
6 : alkaline water brand A
7 : alkaline water brand B
8 : alkaline water brand C
9 : oxygenated water brand A
10 : oxygenated water brand B
11 : oxygenated water brand C

3.3 Descriptive analysis test based on coffee differences

The data (**Table 4**) showed that the difference of coffee type had a significant effect on the attributes of sweet aroma, chocolate aroma, roasting aroma, burnt aroma, citrus aroma, bitter taste, vinegar taste, salty taste, cassava flavor, astringency aftertaste, dry mouth-feel and oily mouth-feel. However, there is no significant effect on the sweet after taste attribute.

Table 4. Result table of attribute sensory

Attribute	A	B	C	D	Ground	Instant	Robusta	Arabica
				Aroma				
Sweet	3.868^{b}	3.189^{c}	3.830^{b}	4.886^{a}	3.507^{b}	4.354^{a}	3.619^{b}	4.852^{a}
Chocolate	2.976^{a}	2.784^{ab}	2.365^{bc}	2.050^{c}	2.848^{a}	2.214^{b}	2.706^{a}	2.014^{b}
Roasting	5.381^{a}	5.748^{a}	4.502^{b}	3.047^{c}	5.548^{a}	3.767^{b}	5.209^{a}	3.030^{b}
Burnt	2.260^{a}	1.954^{ab}	1.539^{b}	1.560^{b}	2.104^{a}	1.546^{b}	1.922^{a}	1.537^{b}
Citrus	0.329^{c}	0.489^{c}	1.486^{b}	2.799^{a}	0.410^{b}	2.160^{a}	0.771^{b}	2.803^{a}
				Taste				
Bitter	5.683^{a}	4.197^{b}	4.903^{ab}	4.863^{b}	4.921^{a}	4.908^{a}	4.926^{a}	4.882^{a}
Vinegar acid	1.078^{c}	0.607^{c}	3.3375^{b}	4.590^{a}	0.836^{b}	3.966^{a}	1.664^{b}	4.581^{a}
Citric acid	1.080^{c}	0.664^{c}	3.583^{b}	4.232^{a}	0.868^{b}	3.867^{a}	1.770^{b}	4.142^{a}
Salty	0.489^{c}	0.399^{c}	1.915^{b}	3.170^{a}	0.439^{b}	2.542^{a}	0.937^{b}	3.125^{a}
				Flavor				
Cassava	3.426^{a}	2.370^{b}	1.943^{bc}	1.749^{c}	2.890^{a}	1.851^{b}	2.585^{a}	1.734^{b}
				After taste				
Astringent	2.214^{b}	1.638^{c}	4.014^{a}	3.578^{a}	1.923^{b}	3.783^{a}	2.628^{b}	3.525^{a}
Sweet	0.653^{a}	0.756^{a}	0.663^{a}	0.757^{a}	0.703^{a}	0.713^{a}	0.692^{a}	0.755^{a}
				Mouthfeel				
Dry	3.345^{b}	2.146^{c}	3.615^{a}	2.772^{bc}	2.718^{b}	3.218^{a}	3.033^{a}	2.779^{a}
Oily	2.165^{b}	2.300^{b}	2.463^{ab}	2.813^{a}	2.222^{b}	2.618^{a}	2.312^{b}	2.737^{a}

Description of water code:
1 : ground coffee brand A
2 : ground local coffee B
3 : instant coffee brand A
4 : instant coffee brand B

3.4 Descriptive analysis test based on water differences

The different types of water tend to have significant effect on average attribute intensity (p-value<0.05). The attributes of sweet intensity in water code 3, while water code 1 has a high intensity on chocolate aroma and oily mouth-feel.

Table 5. Coffee sensory attributes as affected by 11 different water

Attribute	1	2	3	4	5	6	7	8	9	10	11
						Aroma					
Sweet	4.377^{abc}	4.380^{abc}	4.648^{a}	4.567^{ab}	4.467^{ab}	3.409^{bcd}	3.088^{d}	3.944^{abcd}	3.539^{abcd}	3.740^{abcd}	3.216^{cd}
Chocolate	3.478^{a}	3.347^{a}	3.056^{abc}	3.044^{abc}	3.158^{ab}	1.953^{cd}	1.795^{d}	2.019^{cd}	1.986^{cd}	2.136^{bcd}	1.912^{d}
Roasting	4.914^{a}	5.170^{a}	4.968^{a}	4.992^{a}	5.060^{a}	4.654^{a}	4.444^{a}	4.452^{a}	4.105^{a}	4.448^{a}	4.116^{a}
Burnt	2.103^{a}	2.228^{a}	1.887^{a}	2.215^{a}	2.158^{a}	1.600^{a}	1.615^{a}	1.658^{a}	1.639^{a}	1.670^{a}	1.340^{a}
Citrus	1.356^{a}	1.127^{a}	1.132^{a}	1.230^{a}	1.380^{a}	1.414^{a}	1.294^{a}	1.486^{a}	1.228^{a}	1.033^{a}	1.355^{a}
						Taste					
Bitter	4.770^{a}	4.889^{a}	4.898^{a}	4.844^{a}	5.227^{a}	5.111^{a}	4.605^{a}	5.240^{a}	4.738^{a}	4.856^{a}	4.851^{a}
Vinegar	2.541^{a}	2.503^{a}	2.679^{a}	2.872^{a}	2.439^{a}	2.090^{a}	1.994^{a}	2.365^{a}	2.465^{a}	2.437^{a}	3.047^{a}
Citric Acid	2.480^{a}	2.278^{a}	2.617^{a}	2.424^{a}	2.491^{a}	2.215^{a}	1.793^{a}	2.524^{a}	2.643^{a}	2.349^{a}	2.471^{a}
Salty	1.628^{a}	1.477^{a}	1.737^{a}	1.728^{a}	1.553^{a}	1.528^{a}	1.050^{a}	1.490^{a}	1.320^{a}	1.361^{a}	1.553^{a}
						Flavor					
Cassava	2.405^{a}	2.238^{a}	2.268^{a}	2.302^{a}	2.629^{a}	2.495^{a}	2.512^{a}	2.359^{a}	2.286^{a}	2.561^{a}	2.033^{a}
						After Taste					
Astringent	3.282^{a}	3.339^{a}	3.138^{a}	3.219^{a}	3.305^{a}	2.464^{a}	2.579^{a}	2.659^{a}	2.481^{a}	2.560^{a}	2.696^{a}
Sweet	0.734^{a}	0.603^{a}	0.764^{a}	0.896^{a}	0.857^{a}	0.706^{a}	0.526^{a}	0.712^{a}	0.636^{a}	0.537^{a}	0.810^{a}
						Mouthfeel					
Dry	3.410^{a}	3.133^{a}	3.015^{a}	2.940^{a}	3.200^{a}	2.984^{a}	2.763^{a}	2.627^{a}	2.568^{a}	2.855^{a}	3.167^{a}
Oily	2.844^{a}	2.646^{ab}	2.390^{ab}	2.696^{ab}	2.824^{a}	2.378^{ab}	1.886^{b}	2.318^{ab}	2.297^{ab}	2.288^{ab}	2.221^{ab}

3.5 Principal Component Analysis

The first two principal components (PCs) explained 80.2% of variation in the data set for 24 data variation. Sample were separated across PC1 (68%) according to samples that were scored highly for vinegar taste, citric acid taste, and salty taste. PC 2 (12.2%) separated samples with high scored in bitter taste, dry mouthfeel, and chocolate aroma.

According to **Figure 1**, there are some strong positive correlation between attributes citric acid and vinegar taste, between dry mouthfeel and bitter taste, between salty taste and citric acid taste. There also a correlation between burnt aroma and cassava flavor. The correlation between the attribute will be explained based on the table below the figure.

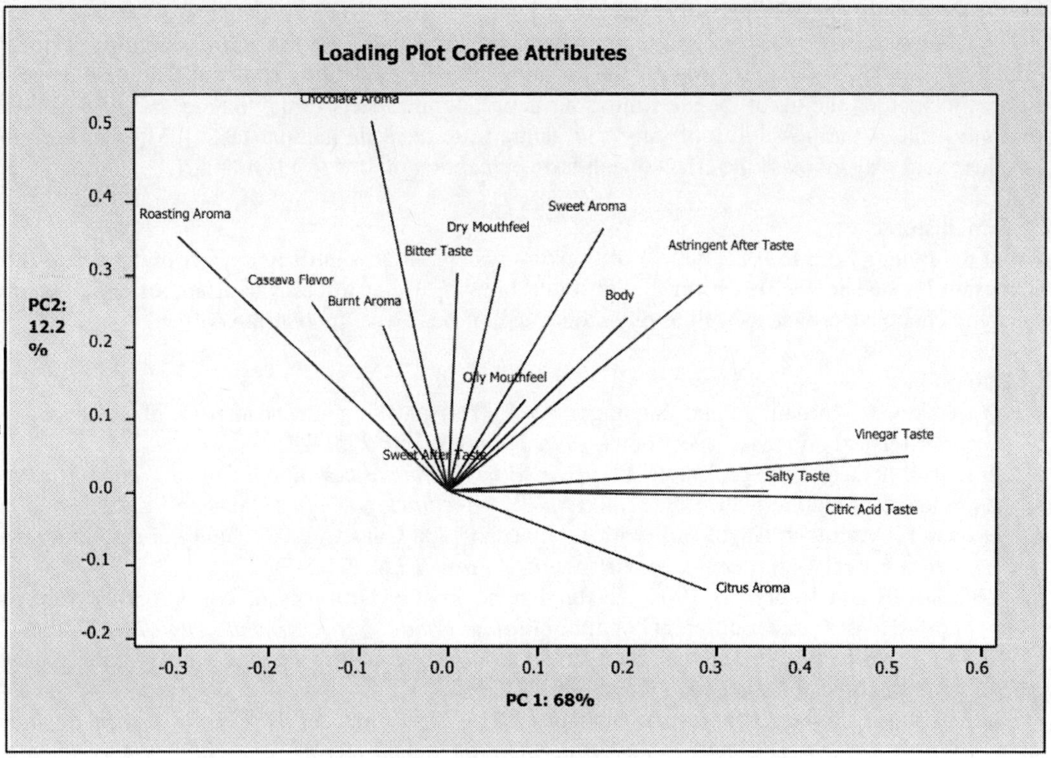

Figure 1. Principal Component Analysis

Table 6. Pearson Correlation and Paired T-test on PCA correlation

Correlation	PCC, r^2
Citric acid and vinegar taste	0.931
Dry mouth-feel and bitter taste	0.533
Salty taste and citric acid taste	0.902
Burnt aroma and cassava flavour	0.563

Critical value for 12 panels: 0.576

Based on the correlation coefficient table of pearson's product moment, the critical value limit for 12 panels is 0.576 on the P-value <0.05. While the PCC score shows the consistency of individual assessment of each panels against the intensity of the attribute, the PCC value>0.576 means that the two attributes are not significantly different.

According to the **Table 6**, the correlation between citric acid and vinegar taste has PCC>0.576. Thus it indicates that both attributes were understood somehow as the same. Vinegar is a transformation of wine or fruit a chemical process in which ethyl alcohol converted to acetic acid. Thus the correlation is between the acid in those 2 attributes is remain strong. According to [22] on the pH range between 4.5-6.5 the taste perception of citric acid and vinegar taste tend to not have a significantly differences between those two taste. In this study a correlation between the taste of vinegar and citric acid taste on coffee perception was observed within the range of pH 4.8 to 6.3.

The combination between salty taste and citric acid taste also does not show a significant difference, with the PCC>0.576. That is because the location of sub modalities sour and saltiness closed. The saltiness located on the tip of tongue while sour is on lateral aspects of the tongue. Because of this close proximity, causes the possibility of salty and acidic taste have the similar taste [23]. And according to [24] citric acid may increase the effect of salt taste perception on low levels of NaCl.

4. Conclusions

Within the range of 5.5 to 9.13, the pH of brewing water has no significant effect on taste and flavor of Indonesian brewed coffee. However, the different brewing water affects the aroma of coffee, especially sweet and chocolate aroma, as well as oily mouth-feel of pour-over Indonesian coffee.

5. References

[1] Yeretzian, C, Jordan, A and Lindinger, W 2003 Analysing the headspace of coffee by proton transfer-reaction mass spectrometry *Int. J Mass Spectr* 223-224
[2] International Coffee Organization 2016 *The Current State of the Global Coffee Trade* www.ico.org
[3] Association of Indonesian Coffee Exporters 2014 *Archipelago Coffee* (Jakarta: AEKI)
[4] Decazy F, Avelino J, Guyot B, Perriot JJ, Pineda C and Cilas C 2003 Quality of different honduran coffees in relation to several environments *J Food Sci* **68** (7) 2356-2361
[5] Nebesny E and Budryn B 2006 Evaluation of sensory attributes of coffee brews from robusta coffee roasted under different condition *Europe Food Research Technology* **224**:159-165
[6] Ross CF, Pecka K and Weller K 2006 Effect of storage conditions on the sensory quality of ground arabica coffee *J Food Qual* **29** 596-606
[7] International Coffee Organization Statistics 2011 *Breakdown of Exports of Green Arabica and Green Robusta of Countries Exporting Significant Volumes of both Types of Coffee June 2009* www.ico.org
[8] Gardjito, Murdijati and Rahadian, D 2011 *Coffee* (Yogyakarta: Kanisius)
[9] DIN 38409-6 1986 *German standard methods for the examination of water, waste water and sludge, Summary indices of actions and substances (Group H), water hardness (H 6)* (Deutsches Institut fur Normung E.V.)
[10] ASTM D 1067 2002 Standard Test Methods for Acidity or Alkalinity of *Water* (West Conshohocken: ASTM International)
[11] Environmental Protection Agency USA 1999 *Total Hardness (mg/L as CaCO 3) (Titrimetric, EDTA)* EPA Method **130.2**
[12] Lin SD 1977 *Tastes and Odors In Water Supplies: A Review* Circular-State of Illinois Department of Registration and Education **127**
[13] Wellinger M, Samo Sand Chahan Y 2015 The SCAE Water Chart Measure Aim Treat.Specialty Coffee Association of Europe (Zurich: Zurich University of Applied Sciences)
[14] Hendon CH, Colonna-Dashwood, L and Colonna-Dashwood M 2014 The role of dissolved cations in coffee *extraction J Agri and Food Chem* **62** 4947-4950
[15] Specialty Coffee Association of America (SCAA) 2009 SCAA Standard: Water for Brewing Speciality Coffee (Santa Ana, California)
[16] Wright RG and George PM 1999 Providing oxygen from source integrated in water cooler, transferring to headspace above water in bottle to maintain dissolved oxygen content at supersaturated level, dispensing oxygen enriched water *U.S. Patent No.* **5868944**
[17] Civille GV and Lawless HT 1986 The importance of language in describing perceptions *J of Sensory Studies* **1** 217–236
[18] Stone H 1992 Quantitative descriptive analysis *ASTM Manual Series* ISBN 0-8031-17566
[19] Lawless HT and Heymann H 2010 Sensory Evaluation of Food 2nd Edition (New York: Springer Science & Business Media)

[20] Skoog DA, West DM, Holler J and Stanley RC 2000 *Fundamentals of Analytical Chemistry* (Brooks Cole Publisher)
[21] Anshori JA 2005 Atomic Absorption Spectroscopy (Bandung: Padjajaran University Press)
[22] Hartwig P and McDaniel MR 1995 Flavor characteristics of lactic, malic, citric, and acetic acids at various ph levels *J Food Scie* **60** 384–388
[23] Altner H 1978 Physiology of taste Fundamentals of Sensory Physiology ed Schmidt R F (New York: Springer-Verlag)
[24] Breslin P A 1996 Interactions among salty, sour and bitter compounds *Trends in Food Scie and Tech* **7**

Chemical characteristic of salt fermented meat inoculated with *Pediococcus ssp.*

Y B Pramono[1*], E S Rahayu[2], Suparmo[2], T Utami[2], Nurwantoro[1] and V D Yunianto[1]

[1] Departement of Food Technology, Faculty of Animal and Agricultural Sciences, Diponegoro University, Semarang, Indonesia
[2] Faculty of Food Technology, Gadjah Mada University, Yogyakarta, Indonesia

*Email : yok_b_p@yahoo.com

Abstract. The research goal is knowing of the characteristict of salt fermented meat by *Pediococcus ssp.* There were microbiological, chemical, and *off-flavor* compound during fermentation. This study was conducted on research of influence of salt-meat fermentation inoculated used starter. They were included microbiological characteristics, and chemical characteristics. Microbiological characteristics observed were total bacteria, number of *coliform* groups, bacteria producing bioamine, and total lactic acid bacteria. The result showed that decreasing of *coliform* and bioamine producer bacteria, and total lactic acid bacteria decreased 3 *log cycle.* While the soluble protein increased of 7-8% and bioamine increased of 5-6 mg/100 g. And then *Off-flavour* compound, TVN and TMA increased of 36-20 mg/100g and 16-30 mg/100g, respectively. Conclusion of the research that *Pedioccoccus ssp.* influenced salt fermented meat.

Key words : Salt fermented meat, *Pediococcus ssp.*, Microbiolocal, Chemical, *off-flavour*

1. Introduction

In the meat fermentation, salt has important role due to the success of fermentation. The method of salting influenced the salt-resistant microbiota. It influenced the fermentation conditions. Some lactic acid bacteria play a role in suppressing the growth of unwanted bacteria, i.e. bacteria that cause decay and bacterial pathogens. The activity of lactic acid bacteria can also suppress deamination activity by suppressing the growth of bacteria producing bioamine. The components of this bioamine should be avoided because it can cause allergies for those who consume and is quite dangerous for people at risk of cancer because it can stimulate the development of cancer [1].

Starter *Pediococcus* ssp. has potentially to suppress bacterial decay, pathogenic bacteria, and off-flavor products. This inhibitory ability is assumed by the ability to produce metabolic products, competition with existing bacteria and the bacteriocin result. This is expected to improve the fermentation quality. [2] reported that in beef fermentation, pathogenic bacteria commonly appear were the genus Salmonella and strains of *E. colli* species. Salt suppressed the growth of microbes because it can increase the osmotic pressure that can lead plasmolisis in microbial cells. Besides that, NaCl is also hygroscopic, it can absorb water from materials that decreased *Aw* (water activity). It also caused cells dehydration that are toxic to microbes, reducing oxygen solubility in materials so that microbes can be suppressed and inhibit proteolytic activity.

The dominant high-salted fermented meat is the genus *Pediococcus* such as member of *P. damnosus* species, *P. acidilacticii*, *P. pentasaceus*. The use of lactic acid bacteria starter served to control the fermentation process. It can suppress the pathogenic bacteria and the existing proteolytic bacteria. The ability to suppress the microbes is derived from metabolic products in addition to the resulting lactic acid. These metabolic products include hydrogen peroxide, diacetyl, and bacteriocin [3]. The purpose of this research is knowing the influence of starter *Pediococcus* spp. to the characteristics of salted fermentation is done, it is expected that fermentation will take place better.

2. Material and Method

This research used lactic acid bacteria isolate from previous research result [4] is isolate of BAL which were 11 *Pediococcus* sp. YDA1, *Pediococcus* sp. YDA10, *Pediococcus* sp. YDA3, *Pediococcus* sp. YDA4, *Pediococcus* sp. YDA11, *Pediococcus* sp. YDA2, *Pediococcus* sp. YDA5, *Pediococcus* sp. YDA6, *Pediococcus* sp. YDA7, *Pediococcus* sp. YDA8, *Pediococcus* sp. YDA9. This study was conducted on research of influence of salt-meat fermentation inoculated used starter. They were included microbiological characteristics, and chemical characteristics. Microbiological characteristics observed were total bacteria, number of *coliform* groups, bacteria producing bioamine, and total lactic acid bacteria. The measured chemical changes were pH change, total titrated acid, dissolved protein, histamine amount, TVN (total volatile base nitrogen) and TMA (tri methyl amine).

The process of salted beef fermentation used 350 g minced beef plus 100 g salt and 50 g (equivalent to 50 ml water . The fermentation process is carried out without heating before being inoculated with a starter of *pediococcus* sp. Inoculation with starter culture of 50 ml, after the pre-fermentation phase was done at 4 ° C for 6 hours, then incubated 48 hours at room temperature. Sampling was done at the beginning and end of fermentation by taking 50 g at the beginning and end of fermentation, 25 g for chemical characteristics, and 25 g for microbiological characteristics.

Microbiological parameter measurement using dillution and plating method. For chemical measurements included for pH according to AOAC method [5], TVN and TMA calculations by AOAC method [6], for dissolved proteins by Lowry method [7], and acid calculation by titration. While the calculation of bioamine by Mahendrata method [8].

3. Result and Disccussion

The results of the research include the characteristics of microbiologic and chemical characteristics.

Table 1. Microbiological characteristics of salt fermented meat with starter *Pediococcus* ssp.

Rank	Isolate	Bacteria (CFU/ml)					
		Coliform		Bioamine Producer		Lactic Acid Bacteria	
		Beginning	Ending	Beginning	Ending	Beginning	Ending
1	*Pediococcus sp.* YDA1	$1{,}8x10^5$	$3{,}2x10^1$	$4{,}2x10^6$	$3{,}2x10^2$	$3{,}1x10^6$	$9{,}1x10^9$
2	*Pediococcus sp.* YDA3	$1{,}6x10^5$	$2{,}2x10^1$	$3{,}1x10^6$	$3{,}1x10^2$	$4{,}1x10^6$	$8{,}2x10^9$
3	*Pediococcus sp.* YDA4	$1{,}7x10^5$	$2{,}5x10^1$	$3{,}2x10^6$	$3{,}3x10^2$	$4{,}2x10^6$	$9{,}3x10^9$
4	*Pediococcus sp.*YDA10	$1{,}9x10^5$	$2{,}2x10^1$	$2{,}6x10^6$	$2{,}3x10^3$	$2{,}8x10^6$	$9{,}2x10^9$
5	*Pediococcus sp.*YDA11	$1{,}7x10^5$	$2{,}3x10^1$	$2{,}7x10^6$	$2{,}4x10^3$	$2{,}6x10^6$	$8{,}3x10^9$
6	*Pediococcus sp.* YDA2	$1{,}3x10^5$	$2{,}8x10^2$	$2{,}9x10^6$	$2{,}8x10^2$	$3{,}2x10^6$	$8{,}9x10^9$
7	*Pediococcus sp.* YDA5	$1{,}6x10^5$	$2{,}6x10^2$	$2{,}9x10^6$	$2{,}9x10^3$	$3{,}2x10^6$	$3{,}2x10^8$
8	*Pediococcus sp.* YDA6	$1{,}8x10^5$	$2{,}4x10^2$	$2{,}8x10^6$	$3{,}2x10^3$	$2{,}9x10^6$	$3{,}1x10^8$
9	*Pediococcus sp.* YDA7	$1{,}9x10^5$	$2{,}4x10^2$	$2{,}7x10^6$	$2{,}8x10^3$	$3{,}4x10^6$	$3{,}2x10^8$
10	*Pediococcus sp.* YDA8	$1{,}7x10^5$	$2{,}3x10^2$	$2{,}7x10^6$	$2{,}7x10^3$	$3{,}2x10^6$	$6{,}1x10^8$
11	*Pediococcus sp.* YDA9	$1{,}8x10^5$	$2{,}2x10^2$	$2{,}8x10^6$	$2{,}6x10^3$	$3{,}2x10^6$	$9{,}2x10^8$
	Control	$2{,}3x10^5$	$7{,}1x10^2$	$3{,}4x10^6$	$4{,}2x10^3$	$2{,}7x10^4$	$4{,}1x10^5$

Description: control without inoculation of lactic acid bacteria

The condition of salty meat fermentation affects the number of coliform members and bacteria producing bioamine that begins fermentation 10^5 CFU / ml to 10^2 - 10^1 CFU / ml which means starter *Pediococcus* ssp. can suppress between 4-3 log cycles during fermentation, this is presumably because in addition to producing lactic acid as a metabolic product but also produce other metabolic products that have antagonistic ability, for example pediocin as a result of *coliform* bacteria and bioamine producers will decrease in number [9].

Table 2. Chemical characteristics of fermented meat supplements with starter *pediococcus.*

Rank	Isolate	Dissolved Protein (%)		Bioamine (mg/100g)		TVN (mg/100g)		TMA (mg/100g)	
		Beginning	Ending	Beginning	Ending	Beginning	Ending	Beginning	Ending
1	*Pediococcus sp.* YDA11	1,7	10,1	2,6	6,1	8,3	44,6	4,2	21,6
2	*Pediococcus sp.* YDA1	1,7	9,9	2,6	5,8	7,8	48,3	4,1	23,9
3	*Pediococcus sp.* YDA3	1,7	9,7	2,6	6,1	7,9	46,3	3,6	22,9
4	*Pediococcus sp.* YDA4	1,8	9,8	2,6	5,9	8,9	47,6	4,2	23,6
5	*Pediococcus sp.* YDA10	1,7	9,8	2,5	6,3	8,7	45,2	4,1	22,1
6	*Pediococcus sp.* YDA2	1,7	9,8	2,3	5,1	8,1	47,2	3,9	24,1
7	*Pediococcus sp.* YDA7	1,7	8,9	2,3	8,8	8,3	63,1	4,2	31,6
8	*Pediococcus sp.* YDA6	1,7	9,0	2,6	8,9	8,3	65,6	4,3	31,8
9	*Pediococcus sp.* YDA5	1,7	9,1	2,5	8,1	8,3	66,2	4,1	34,2
10	*Pediococcus sp.* YDA9	1,7	8,9	2,4	8,9	8,3	68,2	4,2	34,1
11	*Pediococcus sp.* YDA8	1,7	8,9	2,4	8,9	8,3	68,2	4,2	34,1
	control	1,7	8,7	2,7	10,6	8,9	87,6	4,3	50,1

Measurements of TVN and TMA were performed to determine as a precursor to decay in high-protein foods. It is seen that the use of *pedioccoccus* ssp starter can reduce the formation of TVN and TMA compared to controls without the use of the starter. This proved that the starter *pediococcus* ssp. affecting the early decay products, it can happens because the starter can suppress the existing spoilage bacteria during the fermentation process [10].

Similarly, in formed bioamines, the using of *pediococcus* ssp. starter. can be suppressed when compared to control, it is suspected because of the the suppresing ability of bacteria bioamine so that the formed bioamine is lower when it is compared to the control without *pediococcus* ssp. starter.

In the dissolved protein, it is seen that the used starter has the same proteolytic ability when it is compared to the control (without starter). It is assumed because of the used starter *pediococcus* ssp. is one a type of lactic acid which has proteolytic ability [11].

4. Conclusion

Starter *pediococcus* ssp influenced the chemical characteristis of salt fermented meat throught the suppresion of the coliform group bacteria, bio amine product, TVN and TMA formation. *Pediococcus* ssp can be promising starter for ameliorating off flavor of high protein meat throught the salt fermentation process.

References

[1] Kalae P 2006 Biologically Active polyamines in beef, pork, and meat product : a review *Meat Science* **75** 1-11
[2] Phillips DJ, Summer JF, Alexanser and Dutton KM 2001[a] Microbiological quality of australian beef *J. of Food Protect* **64** (5) 692-6
[3] Netcher EW 2001 *Microbiology a Human Perspective 3rd Edition* (New York: Mc Graw Hill)
[4] Pramono YB, Rahayu ES, Suparmo, and Utami, T 2008 Isolation and identification of lactic acid bacteria on traditional meat fermentation (in Bahasa Indonesia) *J. Ind. Trop.Ani. Agri.* **33** (4)
[5] Anonim 1990 *Official Methods of Analysis of the Association of Official Analytical Chemist Association of Official Analytical Chemists* (Washington DC)
[6] Lawrie RA 1995 *Meat Science* (San Francisco: W H Freeman and Company)
[7] Lowry OH, Roseborg NJ, Farr AL, and Randall RJ 1951 Protein measurement with the folin phenol reagent *J. Biol. Chem.* **193** 265-75
[8] Mahendrata M 2003 The Change of histamin content in some fish-bashed food during storage *Indonesia Food and Nutrition Progress* **10** (1) 54-61
[9] Yin LJ, Tong YL, and Jiang ST 2005 Improvement of the functionality of minced mackerel by hydrolysis and subsequent lactic acid bacterial fermentation *J. of Food Sci.* **70** (3) 173-8
[10] Lindgren SE and Dobrogosz WJ 1990 Antagonistic activities of lactic acid bacteria in food and feed fermentation FEMS *Mic. Rev.* **87** 149-164
[11] Axelsson L 1998 Lactic Acid Bacteria : Classification and physiology *Lactic Acid Bacteria: Microbiology and Functional Aspect 2nd Edition* ed Salminen and Wright (New York: Marcel Dekker Inc)
[12] Anonim 1985 TVB and TVN Determination by Conway Method (in Bahasa Indonesia) SNI 1-4495-1985

Proteolytic activity of selected moulds in the first fermentation of black-seeded soysauce

R Yulifianti* and E Ginting

Indonesian Legumes and Tuber Crops Research Institute (ILETRI)
PO Box 66 Malang 65101, East Java, Indonesia

*Email: rahmi_stp@yahoo.com

Abstract. Black-seeded soybean is preferred as normally it has higher protein content and would give black colour to the soyfiltrate for making soysauce. Mould is usually used in the first fermentation of soysauce making to prepare *koji* with high soluble protein as a media for the subsequent bacteria fermentation in brine solution. Black-seeded soybean of Detam 1 variety was used as the soysauce ingredient. The trial was a randomized complete design with four replicates. The treatments were (1) *Rhizopus oligosporus* starter (in flour form) with two day-fermentation, (2) similar *R. oligosporus* starter with three-day fermentation, (3) *Aspergillus sojae* (pure culture) with three-day fermentation and (4) *A. sojae* in flour form with three-day fermentation. The black-seeded soybean had 100-grain weight of 11.7 g, high protein content of 42.5% dw and fat content of 14.9% dw. *Koji* prepared using *R. oligosporus* starter with two-day fermentation had the lowest protein content (48.9% dw). Both starters of *A. sojae* culture and flour gave the highest soluble protein content of *koji* (41.0-41.5% dw), followed by *R. oligosporus* starter with 3 day-fermentation (35.2% dw). Whilst the lowest value was noted in *koji* prepared using *R. oligosporus* starter with two day-fermentation (30.8% dw). This suggests that both *A. sojae* starters had similar proteolitic activity and higher than that of *R. oligosporus* starter. In terms of practical application and maintenance of the mould viability by soysauce processors, the use of *A. sojae* flour starter with three-day fermentation is suggested.

***Keywords**: fermentation, mould, protein, soybean.*

1. Introduction

Soybean consumption in Indonesia leading to increasing the total national soybean demand in 2012 amounted to 2.2 million tons. The biggest portion of the used of soybean for food or tofu and tempeh produced (83.7%), while the rest of soybeans are used in soysauce, tauco industry, and others (14.7%), seeds (1.2 %), and feed (0,4%) [1]. Soysauce is one of soybean fermentation food products that can be derived from either yellow or black-seeded soybean.

Black-seeded soybean is preferred by the industry as a raw material of soysauce because it can give a natural colour on soysauce produced [2]. In order to meet the black-seeded soybeans demand as raw material for soysauce industry, a number of soybean improved varieties with large seed sizes had been released, namely Detam-1, Detam-2, Detam-3 Prida, and Detam-4 Prida which contain high protein (> 40% db) [3] in addition to Merapi, Cikuray, and Mallika varieties, which have small seed sizes. The high of protein content of soybean seed is essential for soysauce making as it would give high protein content in the soysauce produced [4].

Fermentation on soysauce preparation consists of two phases, namely mould fermentation or *koji* fermentation (solid stage fermentation) and fermentation in a salt solution (brine fermentation), called *moromi*. Moulds that usually used in solid fermentation are *Aspergillus spp.* and *Rhizopus spp.* [5]. *Koji* fermentation takes 3-5 days, then *koji* was dried and soaked in brine solution of 20-30% for 14-28 days [6]. Microorganisms that are tolerant to high salt concentration, like *Hansenula spp., Zygosaccharomeces spp.,* and *Lactobacillus spp.*are active in the brine fermentation. Afterwards, *moromi* is blended with spices and sugar and boiled up to particular viscosity to obtain sweet soysauce. In general, the protein content of soysauce sold in the market were below the minimum level (2.5%) of national standar quality for sweet soysauce [7]. However, some studies showed that soysauce preparation without brine/*moromi* fermentation gave higher protein content relative to soysauce produced through *moromi* fermentation [5,6]. This reflects that first or mould fermentation (*koji*) play an important role in dictating the protein content of soysauce produced.

Protein content of *koji* is considerably influenced by the mould growth during fermentation [6]. *Aspergillus spp.* and *Rhizopus spp.* have different proteolytic activity in hydrolizing insoluble protein complex into peptides and amino acids [8]. The starter of *Rhizopus oligosporus* is commonly available in the market as it is widely used for tempeh fermentation. However, the starter of *Aspergillus spp.* is not available in the market and should be obtained from particular institution, like university, research institute or soysauce manufacturer. Therefore, this study was performed to identify effective level of concentration and proper length of fermentation both starters of *Rhizopus oligosporus* and *Aspergillus sojae* were used for koji fermentation through the protein content and soluble protein produced using a black-seeded soybean variety.

2. Materials and methods

The study was performed at the food chemistry and technology laboratory of Indonesian Legumes and Tuber Crops Research Insitute (Iletri), Malang in April 2013. Materials used in this study are black-seeded improved variety of soybean namely Detam-1 and moulds namely R*hizopus oligosporus* starter (flour form) produced by the Indonesian Institute of Sciences (LIPI) and A*spergillus sojae* starter (pure culture and flour form) from the food & nutrition culture collection, PAU, Gadjah Mada University.The trial was a randomized complete design with four replicates. The treatments were (1) *Rhizopus oligosporus* starter (in flour form) with two day-fermentation, (2) similar *R. oligosporus* starter with three-day fermentation, (3) *Aspergillus sojae* (pure culture) with three-day fermentation and (4) *A. sojae* in flour form with three day fermentation.

Making of koji were done by soybean seeds (each 200 g) soaked for 8 hours, then washed and boiled at 100° C for 2 hours. Soybean seed drained and cooled to a temperature of ±30°C. Furthermore, the soybean seed inoculated with moulds *R. oligosporus* for two days and three days fermentation, and inoculated with a pure culture of *A. sojae* and *A. sojae* in flour form for three days fermentation, respectively. Observations included physical (100 grain weight) and chemical characteristics of soybean seed (moisture, ash, protein, and fat content), protein and soluble protein contents of *koji* produced.

3. Results and discussions

3.1. The weight of 100 grain and chemical properties of black-seeded soybean

Improved variety of soybean namely Detam-1 used in this study belong black seed coat with a large grain weight of 11.56 g/100 grain. According to Susanto and Saneto [10], soybean with 100-grain weight around 8-10 g is grouped as small seeded, medium for 10-13 g and large seeded for >13 g Detam-1 variety is classified as black soybean variety released large seed [3].

Detam-1 had moisture content of 7.47%. All soybean seeds had moisture contents below the maximum level established by the national standard for soybean seed [11] that was <13%. Detam-1 variety had ash content of 5.48 % dw (Table 1). Ash represents the mineral content of soybean seed, particularly phosporus, calcium and iron, where the mineral content was different between varieties and growth environment [12].

Detam–1 variety had a higher protein content of 42.48 % dw (Table 1), the these values of protein content were comparable to those reported by Ginting and Adie [9] in Cikuray variety and some promising lines prepared of soysauce. Detam-1 variety used in this experiment had a lower protein content of similar variety in the description improved varieties of 45.36 % dw [3], this can be caused by differences in maturity, soil fertility, and climate, as well as fertilization and irrigation method [9,13,14]. Protein content is an essential quality trait in soybean as it positively correlated with the protein and soluble protein of koji produced [2].

Table 1. 100 grain weight and chemical properties of Detam-1 variety

Variety	100 grain weight (g)	Moisture content (%)	Ash content (% dw)	Protein content (% dw)	Fat content (% dw)
Detam-1	11.56	7.47	5.48	42.48	14.90

dw = dry weight

Fat content of black-seeded soybean, Detam-1 variety of 14.90% dw (Table 1). These value of fat content suggesting similar with the studied by Ginting and Adie [9] on the same variety, which ranged from 12.9 to 14.8% dw, but lower than Cikuray variety (black-seeded soybean), studied by Antarlina *et al.*[12] of 19.0% dw.

3.2. Characteristic of soybean koji

The main purpose of fermentation (*koji* fermentation) in the making of soysauce is hydrolyze the protein, starch, and fat in soybean seeds with enzymes produced by moulds into simpler components that can be used the next active microorganisms (bacteria and yeast) in fermentation (*moromi* fermentation). The results showed moisture content of koji was significantly different between treatments using the mould *R. oligosporus* in the flour form and *A. sojae* in the pure culture and flour form, for two days and three days fermentation. The *koji* from fermented by *A. sojae* in the pure culture showed the high of moisture content of 60.60% (Table 2), than the moisture content of *koji* using mould of *A. oryzae* for 40 hours fermentation , studied by Tanaka [14] that ranged from 20 to 35%, because there are using different mould in the fermentation process.

Table 2. Moisture, protein, dan soluble protein of *koji*

Type of mould	Moisture content (%)	Protein content (% dw)	Soluble protein (% dw)
R. oligosporus flour form (2 day fermentation)	57.29 c	48.9 b	32.5 c
R. oligosporus flour form (3 day fermentation)	57.16 c	52.2 a	35.2b
A. sojae pure culture (3 day fermentation)	60.60 a	51.3 a	41.0a
A. sojae flour form (3 day fermentation)	59.33 b	51.8 a	41.5 a
Lsd (5 %)	1.11	1.34	1.30
Cv (%)	1.19	1.64	2.16

Lsd: Least significant difference, values followed by different letters are significantly different at $P < 0.05$
Cv: Coefficient variation
dw: dry weight

Observations in this study included measuring protein content and soluble protein. Total protein is measured of nitrogen total (N) content in the sample. Whereas, soluble protein is oligopeptides easily absorbed by the digestive system [5]. Protein levels of *koji* were significantly different between treatments with the highest protein content of the *koji* mould fermentation using *R. oligosporus* mould and *A. S ojae* mould for three days fermentation, of 51.3% dw and 51.8% dw, respectively (Table 2). Protein content is one of the parameters of soysauce quality. Protein content in soysauce product is affected by the protein content of *koji* and *moromi* fermentation [9].Increasing the amount of microbial mass during the fermentation process can also increase the protein content of the analysis. Increasing the amount of mass of microbes will cause the content of fermentation products so that the protein content is a reflection of the mass amount of cells [16].

The result showed, soluble protein levels was significantly different among four treatments with the highest levels of soluble protein that was in the *koji* with *A. sojae* both of pure culture and flour form, 41.5 % and 41.0 %, respectively (Table 2). Soybean seed after undergoing a process of fermentation by moulds have increased levels of protein, because the activity of protein degradation by protease enzyme was higher than the consumption of amino acids by fungi for further growth. Rahayu [8] reported that the proteolytic activity of *A. oryzae* and *A. sojae* (observed on the amount of soluble N) from the first to the third day of fermentation is much higher (about 1.5 times) compared with *R. oligosporus* . This suggests that soy protein content is also influenced by the type of microorganisms in fermentation. This shows the proteolytic activity of *R. oligosporus* lower than *A. sojae*. The results of proteolytic activity was soluble protein and amino acids. During the first fermentation the amount of soluble nitrogen (N) ranged between 50 to 70 % of the total N and the number was increased to 72-82% during 2 months of fermentation [8].

4. Conclusion

Protein and soluble protein can be parameter of the quality of soysauce produced black-seeded soybean which can also be determined from the protein produced from *koji* fermentation. The highest protein content of *koji* fermentation using *A. sojae* starter flour form followed by *A. sojae* pure culture which values 51.8% and 51.3% dw, respectively. Similarly, the soluble protein content of values were 41.5 % and 41.0% dw, respectively.

References

1] Meryana E 2012 The best black seeded soybean in the world are in Indonesia (in Bahasa Indonesia) http://bisniskeuangan.kompas.com/read/2012/07/26/21322949/Kedelai.Hitam.Terbaik.Dunia.Ada.di.Indonesia (Accessed on March 18th 2014)

[2] Ginting E, Antarlina SS and Widowati S 2009 Improved variety for food industry raw materials (in Bahasa Indonesia) *J. Penelitian dan Pengembangan Pertanian* **28** (3) 79-87

[3] Iletri 2016 Descriptions of legumes and tubers improved varieties (in Bahasa Indionesia) Balitkabi Malang p 171

[4] Antarlina SS and Ginting E 1998 Supplementation of Some Legumes on the Manufactures of Soysauce (in Bahasa Indonesia) p 147-156 *Pros Seminar Nasional Teknologi Pangan dan Gizi* ed S. Rahardjo, Marseno DW and Supartono W Yogyakarta 15 Desember 1998 PATPI Yogyakarta-PAU Pangan dan Gizi UGM-Fakultas Teknologi Pertanian UGM.

[5] Purwoko T and Handajani NS 2007 Protein content of soysauce without moromi fermentation result *Rhizopus oryzae* and *R. Oligosporus* fermentation (in Bahasa Indonesia) *Biodiversitas* **8** (2) 223–227

[6] Septiani Y 2004 Study of carbohydrate, fat, and protein content on soysauce from tempeh (Essay) (in Bahasa Indonesia) Surakarta: FMIPA UNS

[7] Agency for National Standarization 1999 National Standart for soysauce (in Bahasa Indonesia) SNI 01-3543-1999 Badan Standarisasi Nasional Jakarta p 7

[8] Rahayu ES 1985 Hydrolysis of soy protein by *Aspergillus sojae, Aspergillus oryzae* dan *Rhizopus oligosporus*(in Bahasa Indonesia) Tesis S2 Program Studi Ilmu dan Teknologi Pangan, Fakultas Pasca Sarjana UGM Yogyakarta

[9] Ginting E and Adie MM 2007 Physicochemical properties of five black seeded soybean genotypes and the quality of soysauce produced (in Bahasa Indonesia) p 495-510 *In* Harnowo D, Rahmianna AA, Suharsono, Adie MM, Rozi F, Subandi and Makarim AK (eds) The Improvement of Nuts and Tubers Production to Support Food Self Suffiency (in Bahasa Indonesia) Puslitbang Tanaman Pangan Bogor

[10] Susanto T and Saneto B 1994 Agricultural Processing Technology (in Bahasa Indonesia) PT Bina Ilmu Surabaya

[11] Agency for National Standarization 1995National Standart for soybean seed (in Bahasa Indonesia) SNI 01-3543-1995 Badan Standarisasi Nasional Jakarta 7 p

[12] Antarlina SS, Utomo JS, Ginting E and Nikkuni S 2002 Evaluation of Indonesian soybean varieties for food processing p 58-68 *In* Rahmianna AA and Nikkuni S (eds) Soybean Production and Post Harvest Technology for Innovation *in* Indonesia. *Proc. of RILET-JIRCAS Workshop on Soybean Research* Malang 28 September 2000

[13] Kumar V, Rani A, Solanki S and Hussain SM 2006 Influence of growing environment on the biochemical composition and physical characteristics of soybean seed *J. Food Com. Anal.* **19**188–195

[14] Rotundo JL and Westgate ME 2009 Meta-analysis of environmental effects on soybean seed composition *Field Crop Res.***110** 147–156

[15] Tanaka Y, Watanabe J and Mogi Y 2012 Monitoring of the microbial communities involved in the soysauce manufacturing process by PCR-denaturing gradient gel electrophoresis *Food Microbiology* **31** 100–106

[16] Astuti AF and Wardani AK 2016 Effect of Tofu Slurry Sweet Sauce Fermentation on Its Physical, Chemical and Organoleptic Quality (in Bahasa Indonesia) *Jurnal Pangan dan Agroindustri* **47** (1) 27–83

The influence of maturity of VMAC5 (*Cocos nucifera* L. 'makapuno') on its physicochemical, proximate composition and fatty acid profile

M Leorna[1*] and K A Israel[2]

[1]Assistant Professor II and Section Head of Production and Utilization for Food Products, National Coconut Research Center-Visayas, VSU, Baybay City, Leyte, Philippines
[2]Assistant Professor, Institute of Food Science and Technology, College of Agriculture and Food Science, University of the Philippines Los Baños, Laguna, Philippines

*Email: mariselleorna@gmail.com

Abstract. "Makapuno" is one of the promising commodities of the Philippines with good nutritional quality. VMAC5 is the most recent "makapuno" variety developed by the Visayas State University, Baybay City, Leyte, Philippines. Maximum utilization of this commodity is dependent on its thorough characterization. This study aimed to determine the influence of nut maturity on its physicochemical, proximate composition and fatty acid profile. Results revealed that VMAC5 nuts were affected by maturity. An 8-month old nuts exhibited heavier weight compared to mature nuts. The total sugar decreased with maturity. TSS and pH were not affected by maturity. Proximate composition were affected by maturity with moisture content of 59.17-86.88%, crude protein of 1.58-2.62%, crude fiber of 2.24-5.43% and crude fat of 3.06-12.14% of which >50% were medium chain fatty acids (MCFA).

Keywords: coconut, makapuno, physicochemical, proximate composition, fatty acid profile

1. Introduction

Coconut is a nutritious food source being classified as a "functional food" because it provides many health benefits beyond its nutritional content [1]. Equally interesting is the coconut mutant, the "makapuno", which is reported from the Philippines as early as 1931 by [2]. The term "makapuno" comes from the Filipino word *puno* which means "full" and so "makapuno" means "almost full" [3]. The "makapuno" coconut is by far more important to food processors because of its protein content which is higher than the normal coconut [2]. Further, the crude fat fraction is more than 90% saturated fatty acids [4][5]. Most of the saturated fat in coconut oil is easily digestible and converted quickly into energy and not stored in the body as fat [6].

The National Coconut Research Center-Visayas (NCRC-V) of Visayas State University (VSU) in Baybay City, Leyte has developed 5 "makapuno" varieties (VMAC1, VMAC2, VMAC3, VMAC4, VMAC5). The physical and proximate composition of the first 4 of these VSU-developed "makapuno" varieties excluding VMAC5 which is the most recent variety, a hybrid from Tacunan x tall "makapuno" were reported by [7]. The fatty acid profile of the 5 VSU-developed "makapuno" varieties were reported by [8] and [9]. Aside from these, there is no other published work on the characteristics of VSU-developed "makapuno" varieties especially the VMAC5. This characterization of VMAC5 will provide insights in determining the nutritional content on its nut endosperm as well as the possible product formulations that can be derived from it. In addition, these profiles may also serve as guides in developing better varieties and hybrids. Moreover, this characterization is expected to help the farmers, entrepreneurs, and researchers determine the fruit maturity which is best in preparing novel products that can be derived from its endosperm.

2. Materials and Methods

2.1. Sample collection and preparation

"Makapuno" (VMAC5) nuts were procured from the experimental field of the NCRC-V, VSU, Baybay City, Leyte, Philippines. Seven (7) nuts from each of the different maturity (8, 9 and 10 months old) of VMAC5 were harvested and subjected to physical profiling. Chemical characterization followed the quartering method in preparing the sample. Three (3) replications were analyzed per parameter.

2.2. Physical properties

The weight of whole nut, dehusked nut, solid and liquid endosperm were determined using triple beam balance. The ratio of liquid and solid endosperm weight to their whole nut and dehusked nut weight was computed. Meat thickness and firmness were measured using a vernier caliper and penetrometer, respectively.

2.3. Chemical characteristics

The solid endosperm or "meat" was analyzed on its color (L*), total and reducing sugar, TSS, pH, proximate composition and fatty acid profile. The TSS, viscosity and pH were analyzed using Atago hand-held refractometer, NDJ-SS Digital LCD viscometer, and handheld pH meter, respectively. The total sugar, reducing sugar and starch content were analyzed using Phenol-sulfuric method, dinitrosalicylic assay and Anthrone method by [10][11][12], respectively. The color was determined using Lovibond LC 100. The proximate composition (moisture, crude fat, crude fiber, crude protein, and ash) was determined following the procedure of [13]. The fatty acid profile was analyzed by determining quantitatively the methyl esters of fatty acids using AOAC-IUPAC gas chromatographic method [13]. A spectrophotometric method of [14] was used to determine the lipid peroxidation.

3. Results and Discussion

The study focused on the basic characterization of the VSU-developed "makapuno" variety, VMAC5 nut. Determining the characteristics of this "makapuno" variety, helps establish its industrial matchbox significance or readiness for industrial applications and pave a way for its optimum utilization.

3.1. Physical characteristics of VMAC5 nut

Nut maturity of VMAC5 affected most of its physical characteristics. The whole nut of an 8 and 9-month old nuts were heavier (2610-3010g and 2182-2900g, respectively) compared to a 10-month old nut (1920-2300g) as shown in Figure 1. It was also observed that from 8 to 10 months old, the weight of the husk constituted around 50% of the nut total weight. The other 50% was composed of solid and liquid endosperm and shell.

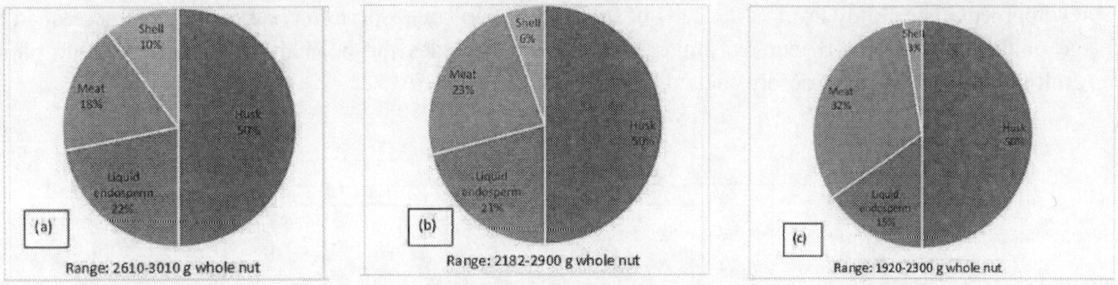

Figure 1. Composition (weight percentage) of an (a) 8-month, (b) 9-month and (c) 10-month old VMAC5 nut

Result also showed that the weight of solid endosperm or "meat" increased as the nut became mature and the meat became thicker (from 6.20 mm of an 8-month old to 13.75 and 14.00mm of a 9 and 10-month old nut, respectively as shown in Figure 2a. This is due to the higher respiratory activity of "makapuno" endosperm compared to normal coconut that resulted to tumorous nature because of the disturbance of auxin-kinin balance [15]. The quantity of the liquid endosperm decreased as it matures with the corresponding increase in weight of the solid endosperm or "meat" (Figure 2a). The percent meat based from the weight of the whole nut and dehusked nut increased with nut maturity (Figure 2b). However, the percent weight of liquid endosperm to its whole nut and dehusked nut decreased with maturity. This data is useful if there is a plan to transport the nuts to urban areas.

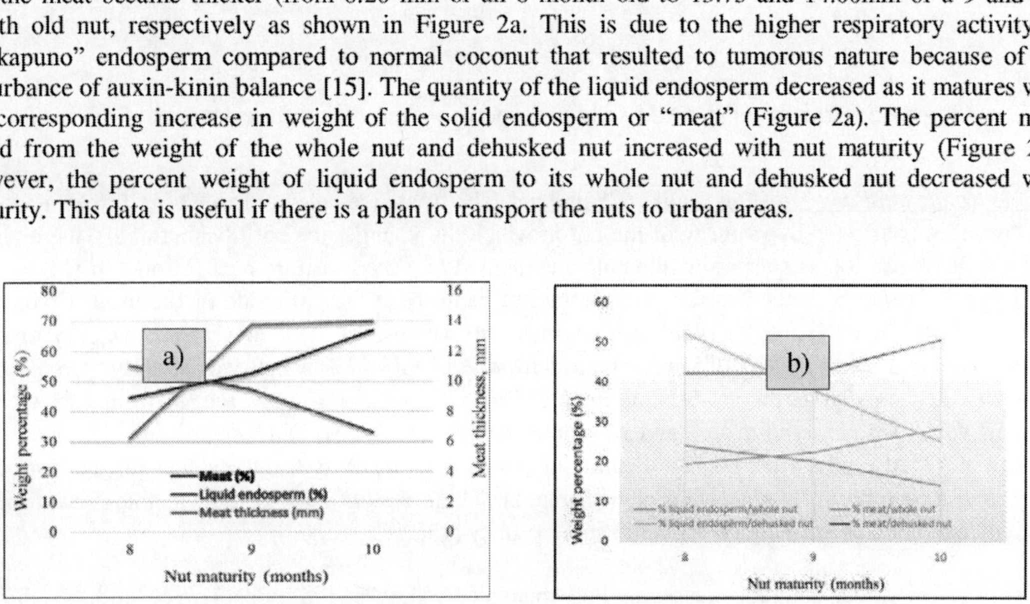

Physicochemical Characteristics of VMAC5 Solid Endosperm or "Meat"

Figure 2. a) Percent weight composition and meat thickness and b) percent weight of the liquid endosperm and meat of VMAC5 to its weight of whole nut and dehusked nut at different stages of maturity

3.1.1. Meat firmness. The firmness of the "makapuno" solid endosperm or "meat" is one of the very important attributes to consider for its potential application. Figure 3 shows that firmness of VMAC5 meat increased from 8 to 10 months in three different locations of the nut meat (eye, middle and bottom portion) where measurements were taken. A significant increase in the firmness of meat was observed from 8 months old nut (3.94 kg/cm^2 to 6.90 kg/cm^2) to 9 (7.21 -10.57 kg/cm^2) with a very minimal increased in 10 months old (7.13 to 12.17 kg/cm^2) regardless of nut locations and either front and back side of the meat. Generally, the middle (4.74 to 12.10 kg/cm^2) and bottom portion (5.31 to 12.12 kg/cm^2) of the nut meat were more firm than the eye portion (3.94 to 11.21 kg/cm^2) regardless of maturity and on either side of the meat. The very soft meat was not yet manifested up to 10 months old because according to [16] and [17], budding and fragmentation of endosperm cells were observed beyond ten months of endosperm development. The ploidy level of the cells of the "makapuno" endosperm increases with the increase in the age of the nuts. After 10 months, amitotic division supersedes the normal mitosis due to high ploidy resulting into a round and loosely packed cells.

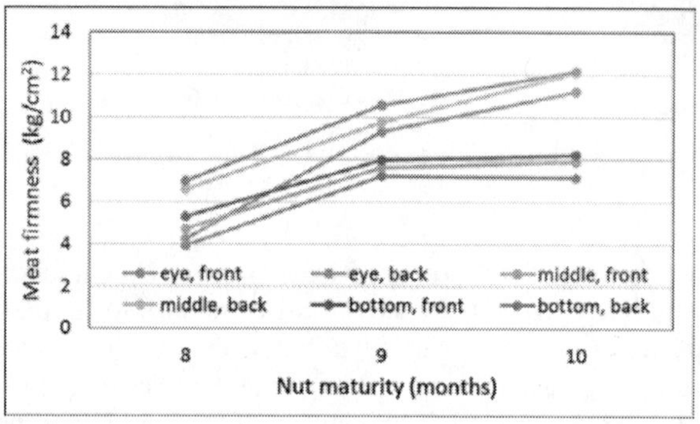

Figure 3. Firmness of VMAC5 solid endosperm or "meat" at different stages of maturity

3.1.2. Color, total and reducing sugar, TSS and pH. The front side color of the VMAC5 solid endosperm or "meat" was affected by maturity of the nut in which the younger the nut (8-month old), the higher the color value (near 100 as standard white color) compared to a more mature nuts (10-month old) as shown in Table 1. However, color was not affected by nut maturity at the back side of the meat. Color ranged from 60.05 to 89.02 (Table 1) which are all above 50 which mean that they are above grey or neutral color. The total sugar of VMAC5 meat ranged from 2.33 to 4.22% which was slightly affected by nut maturity. It decreased as the nut became mature (Table 1). Reducing sugar ranged from 1.23 to 1.43%, highest value was observed in an 8 and 10-month old nuts and lowest in a 9-month old. The TSS and pH values of VMAC5 meat were not significantly affected by nut maturity and the values ranged from 4.14 to 4.88°*Brix* and 6.86 to 7.06, respectively (Table 1). These results may be due to the slow respiratory activity exhibited by "makapuno" nuts towards maturity [15].

Table 1. Color, total and reducing sugar, TSS and pH of VMAC5 solid endosperm or "meat" at different stages of maturity

Maturity (Mos)	Meat color (L^*) Front**	Back^{ns}	Total sugar (%)	Reducing sugar (%)	TSS (°*Bx*)	pH
8	70.99a	88.40a	4.22a	1.43a	4.88a	6.86a
9	64.89b	89.02a	2.84b	1.23b	4.14a	6.98a
10	60.05c	88.23a	2.33b	1.41a	4.68a	7.06a

Means with the same letter are not significantly different at 5% using Tukey HSD.

3.1.3. Proximate composition. The proximate composition of the VMAC5 meat was generally affected by nut maturity as presented in Table 2. The meat of an 8-month old "makapuno" contained the highest moisture (86.88%) and decreased significantly as it reached 9 months old (59.17%) but increased slightly as it reached 10 months old (66.02%). This means that the dry matter composition was highest in a 9-

month old nut (38.83%) compared to 10-month (30.98%) and 8-month old nut (13.12%). These figures for moisture content were very close to the figure reported by [2] as cited by [18] which are 71.59% and 66.28%, respectively. From the dry matter component of the meat, specific components were analyzed such as crude protein, crude fat, crude fiber, ash and NFE was calculated. However, the report of these components in Table 2 was presented based on meat sample not based from dry matter composition. The crude protein content in an 8-month old nut meat was 1.58% of its meat (Table 2). This component increased significantly as the nut became mature accounting 2.62% and 2.11% in 9 and 10-month old nut, respectively. This quantity was lower compared to the values reported by [7] for the previous 4 VSU "makapuno" varieties but quite similar to the figure previously reported by [19] which is 2.4%. The report of [20] stated that mature "makapuno" endosperm has low glutamic acid and valine content. The crude fat component was also affected by nut maturity (Table 2). The meat from an 8-month old nut was lower in crude fat content (3.06%) compared with the meat from both 9 and 10-month old "makapuno" nut which contained 12.14% and 11.40%, respectively. A close figure of 17.60% was reported by [19]. The report of [7] stated that oil (crude fat) is the chief constituent of the 4 previously LSU-developed (now VSU) "makapuno" varieties (VMAC1, VMAC2, VMAC3 and VMAC4). However, it was also reported by [9] that more than 60% of the total fatty acid component of VSU-developed "makapuno" varieties are medium chain fatty acid (MCFA; chain lengths of 6 to 12 carbon atoms) [9]. It was pointed out by [21] that MCFA are more effectively absorbed and metabolized in the body as compared to saturated long-chain fatty acids (LCFA). The crude fiber content of VMAC5 meat was also affected by its maturity (Table 2). Higher crude fiber was observed at 9 (5.43%) and 10 (5.11%) months old nuts than 8 months old nuts (2.24%). This result was very close to the result of [19] which is 5.00% or 6.52 (cited by [18]) but higher compared to the result of [7] for the four (4) previous varieties evaluated that ranged only from 0.17 to 0.32%. The ash content of VMAC5 meat ranged from 0.14-0.36% and this component of the meat was not affected by its maturity. So, the NFE was higher in mature nuts (20.40% and 15.00% in 9 and 10 months old, respectively) than in younger nuts (6.10% in 8 months old). The result was close to the figure reported by [19] at 17.6% and within the values reported by [7] for the four (4) previous varieties which ranged from 4.36 to 25.34%.

Table 2. Proximate composition of VMAC5 solid endosperm or "meat" at different stages of maturity

Maturity (months)	PROXIMATE COMPOSITION (%)						
	MC	Dry matter	Crude Protein	Crude Fat	Crude Fiber	Ash	NFE
8	86.88a	13.12c	1.58b	3.06 b	2.24b	0.14a	6.10 b
9	59.17c	40.83a	2.62a	12.14a	5.43a	0.24a	20.40a
10	66.02b	33.98b	2.11a	11.40a	5.11a	0.36a	15.00a

Means with the same letter are not significantly different at 5% using Tukey HSD.

3.1.4. Fatty acid profile and lipid peroxidation. The fatty acid profile of VMAC5 meat presented in Table 3 shows that saturated fat was high in mature nuts (84.66% in 10-month and 83.97% in 9-month old nut) compared to younger nut (80.57% in 8-month old nut). However, majority of the saturated fat component were MCFA that constituted more than 50% of the total fat specifically lauric acid, a predominant fatty acid in "makapuno" which accounted more than 40% of the total fat. These results were generally lower as compared to the one reported by [9] and [22] but higher in terms of oleic acid in an 8-month old "makapuno" meat. A small amount of linolenic acid was found in the previous four (4) VSU-developed "makapuno" varieties as reported by [9] but was not detected in VMAC5. On the other hand, linoleic acid component was detected and quite high in an 8-month old nut but not reflected in the previous report. The VMAC5 meat exhibited higher lipid peroxidation for a more mature nuts (44.59% and 44.85% for a 9 and 10-month old) compared to a younger nuts (38.34% for an 8-month old).

Table 3. Fatty acid profile and lipid peroxidation of VMAC5 solid endosperm or "meat" at different stages of maturity

FATTY ACID	CONCENTRATION, %		
	8 months	9 months	10 months
Caproic (C_6)	0.20	0.32	0.28
Caprylic (C_8)	4.46	5.65	5.66
Capric (C_{10})	4.31	5.21	5.13
Lauric (C_{12})	43.41	46.91	45.85
MCFA	**52.38b**	**58.97a**	**56.92a**
Myristic (C_{14})	17.29	16.37	17.38
Palmitic (C_{16})	8.49	6.40	6.99
Stearic (C_{18})	2.33	3.04	3.30
Saturated Fat	**80.59b**	**83.97a**	**84.66a**
Oleic ($C_{18:1}$)	7.14	3.26	3.19
Linoleic ($C_{18:2}$)	1.13	0.28	0.47
Linolenic ($C_{18:3}$)	ND	ND	ND
Arachidonic (C_{20})	0.07	0.06	0.06
Behenic (C_{22})	ND	ND	ND
Lignoceric (C_{24})	0.02	ND	ND
Unsaturated Fat	**8.27a**	**3.55b**	**3.66b**
Lipid peroxidation	38.3b	44.59a	44.85a

ND – not detectable

3. Conclusion

Maturity of VMAC5 nut affected its characteristics. A thicker and more firm meat from mature nuts is ideal for processing. The superior quality of makapuno meat from a more mature nuts in terms of proximate composition and fatty acid profile also a good mark as a potential condition for food product development.

Acknowledgement

The authors would like to express gratitude to DOST-ASTHRDP, CHED, DA-BAR, VSU and UPLB for funding and supporting this research.

References

[1] Carandang E V 2004 The role and significance of the Philippine coconut research and development foundation in the advancement of the coconut industry *Philippine J. of Coconut Studie* **29**

[2] Adriano F T and Manahan M 1931 The nutritive value of green, ripe and sport coconut (buko, niyog, and macapuno) *Philippine Agriculturist* **20** (3) 195-199

[3] Lorejo R 2002 Promise of "makapuno" Industry Going Empty Inquirer News Service 2002 URL: www.inq7.net all rights reserved

[4] Jacobs L 1938 *The Chemical Analysis of Foods and Food Products. Huntington.* (New York: Robert E Krieger Publishing Company) p 87

[5] Banzon J A 1990 Coconut as Food. Diliman, Quezon City: Philippine Coconut Research and Development Foundation Inc (PCRDF)

[6] Loha-Unchit K 2005 Coconut oil – a good oil. URL: webmaster@thaifoodandtravel.com

[7] Lauzon R D 2005 Physicochemical properties and processing possibilities of macapuno cultivars developed at Leyte State University *Philippine J. of Crop Science* **30** (2) 55-60

[8] NuÑez T C and Ocoy E T 2008 Polyembryony in "makapuno" (Cocos nucifera L) hybrids *Philippine J.of Crop Science (Philippines)*

[9] Nuñez T C, Molato D D and Ocoy E T 2005 *Development of high yielding "makapuno" hybrids. Terminal Report.* National Coconut Research Center – Visayas (NCRC-V). Visayas State University, Visca, Baybay City, Leyte

[10] Dubois M, Gilles KA, Hamilton J K, Rebers P A and Smith F 1956 Colorimetric method for determination of sugars and related substances *Analytical Chemistry* **28** 350–356

[11] Miller G L 1972 Estimation of reducing sugar by dinitrosalicylic acid method *Analytical Chemistry* **31** 426-428

[12] Shallenberger RS and Birch GG 1975 *Sugar Chemistry* (Connecticut: AVI Publishing Company Inc) pp 17-46

[13] A.O.A.C. 1980 *Official Methods of Analysis. American Association of Analytical Chemists.* 13th edition (Washington DC)

[14] Leong L P and Shui G 2002 An investigation of antioxidant capacity of fruits in Singapore markets *Food chemistry* **76** (1) 69-75

[15] Argosino AS and Mendoza E M T 1993 Respiratory activity of the normal and "makapuno" endosperms *Phil J. Crop Sci.* **18**(3) 195-199

[16] Abraham A, Ninan C A and Gopinath P 1965 Cytology of development of abnormal endosperm in Philippine "makapuno" coconuts *Caryolegia* **18** 395-408

[17] Agaton O C 1983 Cytology of Callus from Coconut (*Cocos nucifera* L.) Endosperm Cultured *in vitro* [Philippines]. Thesis, Philippines Univ. Los Banos College Laguna p 89

[18] Castillo E A 1986 Chemical Characterization of Normal Coconut and "makapuno" Endosperms A dissertation in the University of the Philippines at Los Baños Laguna

[19] Food and Nutrition Research Institute (FNRI) 1997 Philippine Food Composition Table DOST-FNRI

[20] Ramirez DA 1986 Genetics of Macapuno A genetic tumor of the coconut endosperm Trans *Natl Acad. Sci. Tech.* **8** 93-110

[21] Zentek J, Buchheit-Renko S, Ferrara F, Vahjen W, Van Kessel AG and Pieper R 2011 "Nutritional and physiological role of medium-chain triglycerides and medium-chain fatty acids in piglets" *Animal Health Research Reviews* **12** (01) 83-93

[22] Rillo E P, Areza-Ubaldo M B B, Orense O D, Cueto C A, Rillo A R, Santos G A, Calub D J, Parohinog E, Nierva E, Paduano B, Arancon R N, Bachiller N S J and Sanchez A 2004 Establishment of "makapuno" embryo culture laboratories in the Philippines: A case of an effective partnership in the coconut industry *Phil. J. Coconut Studies* **29** (1&2) 13-23

Chemical structure, comparison antioxidant capacity and separation antioxidant of hen, duck and quail egg white protein hydrolysate

A Fatah[1], M Meihu[2], Q Ning[2], B E Setiani[3*] and V P Bintoro[3]

[1] Faculty of Agriculture and Animal Sciene, State Islamic University of Sultan Syarif Kasim Riau, Indonesia
[2] Live Stock Product Processing, Huazhong Agricultural University, Wuhan
[3] Faculty of Animal Agricultural Sciences, Diponegoro University, Semarang, Indonesia

***E-mail:** etz_16@yahoo.com

Abstract. Amino acid linkages as proteins are nutritional substance which important for diet intake. Purification protein procesing undergo heating procedure process followed by additional of proteolytic enzymes or acid had been resulting in protein hydrolysates. A protein hydrolysate describe as many free amino acids bound together through a complex mixture of peptides. Egg white protein hydrolysates is one of subject interested to study for human health or industry product. The objectives of the research are to determine and identification the antioxidant derived from egg white hydrolysate protein. Identification of chemical structure of albumen and albumen protein hydrolysate was examine using IR Spectrophotometry. While comparison of antioxidant capacity and antioxidant separation egg albumen was also investigate using FTIR method (Fourier Transform Infrared Spectroscopy). Hen, duck and quail albumen egg white and on hydrolisate form were used as research materials. The results were showing that different time and enzyme of hydrolysis were not influence at secondary structure of hydrolysate albumen protein. Phytochemical content such as alcohol and hydroxyl compound which have potential as functional group of antioxidant were detected in all of the samples. Their results of radical scavenging activities samples hydrolyzed by pepsin were respectively 89.40%, 50.25% and 85.13%. Whereas the radical scavenging activities of hydrolysates hydrolyzed by papain were 72.85%, 61% and 76.45% respectively.

***Keywords** : Eggs white hydrolisates, FTIR, antioxidant*

1. Introduction

Consuming eggs may help child's brain intelligence. As promised, egg consumption for adults could give physical strength. Eggs are very good protein source to consume for kids and adults as well. Egg is one of the food supply of animal protein source which has the most complete nutritional content such as protein, vitamins, lipids and other substances.

Recently many findings on egg research expose on the functional substances of eggs beyond the basic nutritional. Advance study were done to identify and characterize those functional subctances as well for example antimicrobial, anticancer, antithrombotic, hypocholesterolemic, mineral binding and antihypertensive. Many food products based on egg has high nutritive value. Besides its important reserve of highly digestible proteins, lipids, vitamins and minerals, the egg contains molecules with numerous health promoting and biotechnological properties.

Oxidative damage is one of the causes of human disease. Therefore, consumption of foods containing antioxidants or consumption of antioxidant supplements is important to prevent oxidative damage in humans [1]. Exogenous chemicals or endogenous metabolic processes will resulted in free radicals and other reactive oxygen species (ROS) in food system or human body [2]. A lot of researches reported that cancer, cardiovascular disease, hypertension, neurodegenerative disorders and aging were associate with accumulation of peroxidants (such as Alzheimer's disease).

The peptides derived from pork proteins [3], milk casein [4], soy proteins [5] and egg-yolk proteins [6] have been demonstrated to exhibit in vitro antioxidant activity. Consumers recognize these products are safe and healthy for consumption. Bioactive peptides could be produced by enzymatic hydrolysis of food proteins. Bioactive peptides resulted in antioxidant activities to against the peroxidation of lipids or fatty acids. Recently, foods development not only considers its nutritional sources but also its bioactive compounds. Thus, the research of food possessing bioactive functions has drawn attentions in recent time. The objectives of the research are to determine and identification the antioxidant derived from egg white hydrolysate protein.

2. Material and Method

2.1. *Preparation of egg white isolate.*

At the temperature and controlled pH, the preparation of egg white hydrolysate was prepared. Egg white was separate from yolk egg. The dried egg white was dissolved in distilled water at 30 mg/ml concentration and heated at 90^0C for 10 min, then hydrolyzed using Pepsin at pH 2.00, 45^0C and papain at pH 6.0, 50^0C separately. During hydrolysis, the pH was maintained in order to remain at optimal value with 1mol/L NaOH or HCl. Samples hydrolysate were taken at different time hydrolysis 2, 3, and 4h intervals and the samples were heated using boiling water for 10 minutes to inactivate the enzyme. Hydrolysates after heating treatment were centrifuged at different speed from 4,000-16000 g for 10 min on 4^0C temperature and the supernatants were collected and stored until use for the analysis. After each egg white into hidrolysate. hidrolysate each protein will be used for analysis SDS Page, antioxidant activity and other analyzes are required.

SDS PAGE was performed using 5% acrylamide in stacking gel and separating gels is 12.5% acrylamide. The first gel was prepared is 12.5% acrylamide and allowed to polymerize before with 4% stacking gel. Sample must be heated 3-5 min on boiling water. The migration process of electrophoretic was performed at 80 V in the second gel (stacking gel) and change to 120 v in first gel (separating gel, in 25 mM Tris, 192 mM Glycine and 0.1% SDS, pH 8.3. Sample was load to gel about 5-10μg. As comparison use protein marker to determine molecular size protein sample (170 kDa). After running gel were stained with comassie brilliant blue R250 for about 30 minutes, Staining solution composed of 25% ethanol, 0.1% Comassie Briliant Blue R-250 and 8% acetic acid. Destaining steps were done by soaking gel into solution (25% ethanol and 8% acetic acid) [7].

IOP Conf. Series: Earth and Environmental Science **102** (2017) 012099 doi:10.1088/1755-1315/102/1/012099

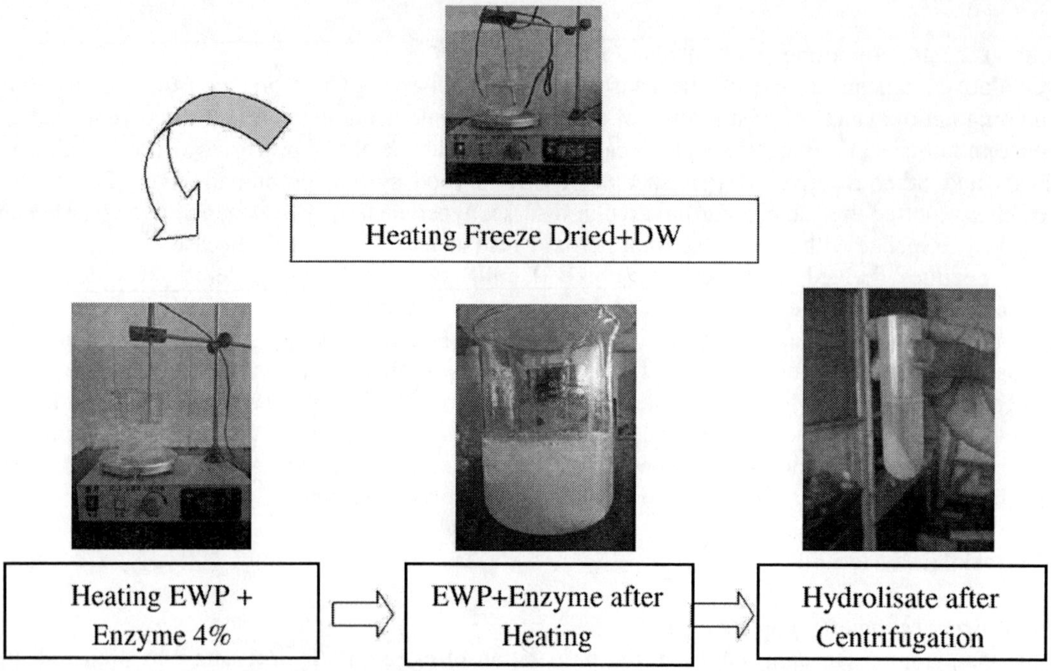

Figure 1. Preparation an egg white protein hydrolisate

2.2. Determination on antioxidant activity of hen, duck and quail egg white protein hydrolisate.
To evaluate the antioxidant activity of those kinds of egg white protein was done by DPPH method was prepare 2 mL of sample HEWPH diluted with 2 mL of DPPH solution (0.15 mM DPPH in ethanol 95%) on 15 ml Volumetric Flasks, after that the solution mixed with vortex and incubated at 37^{0}C for 30 min at dark room. After color change to yellow the absorbance of solution formed was measured at 517 nm with a spectrophotometer. BHT will use as control and standard sample respectively. Antioxidant capacity result will be analyzed using ANOVA analysis with 95% confidence level using SPSS Version 16.

2.3. Comparison antioxidant capacity of protein hydrolisate.
By selecting the highest average antioxidants from each sample consisting of different types of eggs, different kind of enzyme and different hydrolysis time, aimed to compare the antioxidant capacity of protein hydrolysate. Each sample will be selected, that have the highest antioxidant capacity and each other of samples will be compared.

2.4. The identification of the chemical structure of the protein hydrolysate using Fouries Transform Infra Red Spectrocopy(FTIR).
FTIR were used for analysis the chemical structures of different hydrolysates. Infrared spectrophotometer (Model–Nicolet Nexus-470, Thermo Nicolet Corporations, Madison, WI, USA) was used for the measurements in a transmission mode using a demountable thin layer IR-cell, with KBr windows in the frequency region 400–4000 cm^{-1}. Samples were placed at sample compartment with automatic accessory 46 recognition at the scanning speed of 2.8 mm/s. Before that sample was mix with KBr. Each spectrum was recorded in resolution of 4cm^{-1}

2.5. Separation antioxidant hydrolysates by ion exchange chromatography.
Based on antioxidant activity results, the HEWPH exhibited the highest antioxidant activity, and thus were employed for separation of antioxidant hydrolysate. The protein hydrolysate applied to a DEAE Toyopearl column (1.5×60cm) which was equilibrated with 0.1 mol/L Tris-HCl, pH 8.00.

The same buffer was used for washed out the unbound proteins. The absorb proteins were eluted stepwise with 0–30 mol/L NaCl in the Tris-HCl buffer. Fractions of 4mL were collected and the UV absorption was monitored at 280nm. Antioxidant activity was determined by each fraction. All peaks were collected analyze antioxidant activity by DPPH.

3. Result and Discussions

3.1. Preparation of egg white isolate.

The highest antioxidant activity was observed in the HEWPH 2H, HEWPH 3H and DEWPH at 3H hydrolysis, which presented 89.44%, 88.22% and 85.13%. DPPH scavenging activity of hydrolysate protein shows difference significant ($P<0.05$). High amount of peptide hydrophobicity or hydrophobic amino acids was associated with the protein hydrolisates with high level of DPPH free radical scavenging activity [8]. Therefore, the 3 kind of treatment was selected for further study. Because the protein hydrolisates have smaller molecular size that can be more beneficial for human health, so the protein hydrolisates are separated from the original protein generally.

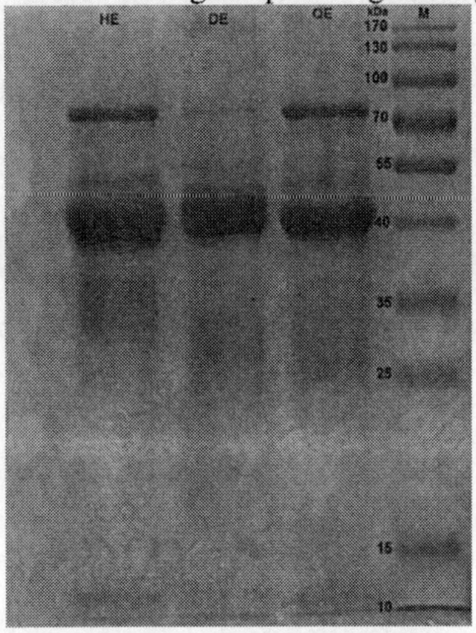

Figure 2. SDS PAGE analysis of HEPE2H, HEPE3H and DEWPH compare with kind of egg hydrolisate

Figure above tell about comparative between kinds of egg as control with sample have high antioxidant activity from all treatment. The result SDS PAGE show two major of egg white protein ovalbumin and ovotransferrin at the 45 kDa and 75 kDa respectively for three kind of egg. Ovotransferrin is an egg white glycoprotein reported has antioxidative properties [9]. The high vicinity in that figure is less than 100 kDa and the lowest vicinity from sample without treatment is 10 kDa that protein with 10 kDa is lysozyme. Different sample show same of vicinity, all samples just show 10 kDa vicinity. The protein is lysozyme. That's mean every treatment influence to protein content. This may cause by heating processing that can damage some protein in egg. But there also tell about lysozyme have antioxidant functionality. Another experiment also report about another function of lysozyme. Lysozyme also can be antimicrobial source. In addition, antioxidant activity was influenced by molecular weight of protein. In the previous research, the antioxidant activity of corn gluten meal hydrolysates was found related to molecular weight of hydrolysates and its concentration. MW 500–1500 Da peptides antioxidant activity was showed stronger than peptides above 1500 Da and peptides below 500 Da [10]. However, the overall antioxidative activity has been postulated that it has to be ascribed to the integrative effects of these actions rather than to the individual actions of peptides.

From the result it can be implied that the antioxidant activity was mainly determined influenced by the hydrolysates or peptides with low molecular wight, obtained from WPI [11]. [12] were doing a research that characterised the peptides which have antioxidant effect derived from the hydrolysates of free-lecithin egg yolk and found that the molecular wight of proteins or peptides determined their efficacy in antioxidant activity, the strongest antioxidant activity was showed by 5 kDa permeate. The DPPH radical scavenging activity of the hydrolysed WPI increased probably due to the change of protein structure, leading to radical quenching ability that become greater.

3.2. Determination on antioxidant activity of hen, duck and quail egg white protein hydrolisate.
The hydrolysates with the highest hydrolysis time and high speed produced by pepsin had higher DPPH radical scavenging activity at 100 mg/ml. This indicates that the interaction time, enzyme and speed can strongly influence the antioxidant properties of the sample. The use of different types of enzymes and different time hydrolysis show significant effect on the antioxidant capacity ($P<0.05$) for all sample HEWPH. The pepsin enzyme presented an effective effect in DPPH radical scavenging activity, higher than the papain enzyme. The Highest one on HEWPH is 89.40% on HEWPH 2H hydrolysis by pepsin.

The highest antioxidant activity is DEWPH 3H by pepsin. DEWPH 3H by pepsin have 85.13% antioxidant capacity. On this study was shown that DEWPH available to inhibition 50% of DPPH radical scavenging higher than BHT. Egg white as antioxidant source has a great potential for consumers request to provide additive-free, fresher and more natural food. All treatment and interaction on DEWPH sample show significant difference ($P<0.05$). It is possible to obtain a natural antioxidant source from Egg product. The pepsin enzyme presented an effective effect in DPPH radical scavenging activity, higher than the enzymes papain and BHT as synthetics source antioxidant.

The hydrolysates with the highest hydrolysis time and high speed produced with pepsin had higher DPPH radical scavenging at 100 mg/mL but not high than BHT. The highest on QEWPH is hydrolysate QEWPH 4H by papain 61.01%. This indicates that the time and speed can influence the antioxidant properties of the sample. The pepsin enzyme again presented an effective effect in the DPPH radical scavenging activity, larger than the enzymes papain. Different time hydrolysis, interaction treatment enzyme and hydrolysis, interaction of enzyme type and speed, interaction time and speed hydrolysis and interaction third treatment showed a significant difference ($P<0.05$).

3.3. Comparison antioxidant capacity of protein hydrolisate.
From all sample the protein hydrolysate that have antioxidant activity is hydrolysate hydrolysis by pepsin. Protein hydrolysate from hen egg (HEWPH) is the highest one. It is able inhibit scavenging radical 89.40%. Egg white proteins are widely used as functional and nutritional ingredients in food products and their hydrolysates obtained by protease treatment are water soluble and wave high nutritional value. The efficient way to recover potent bioactive peptides is by enzymatic hydrolyzing of food proteins, including antioxidant activity.

3.4. The identification of the chemical structure of the protein hydrolysate using Fouries Transform Infra Red Spectrocopy (FTIR).

3.4.1. FTIR of Egg White Protein. All spectra were baseline corrected after background correction. Figure 1 shows the result FTIR spectra for hydrolysate without treatment enzyme hydrolysis, different time hydrolysis and centrifuge speed. FTIR analysis sample without sample will be positive standard to hydrolysate hydrolysis by enzyme.

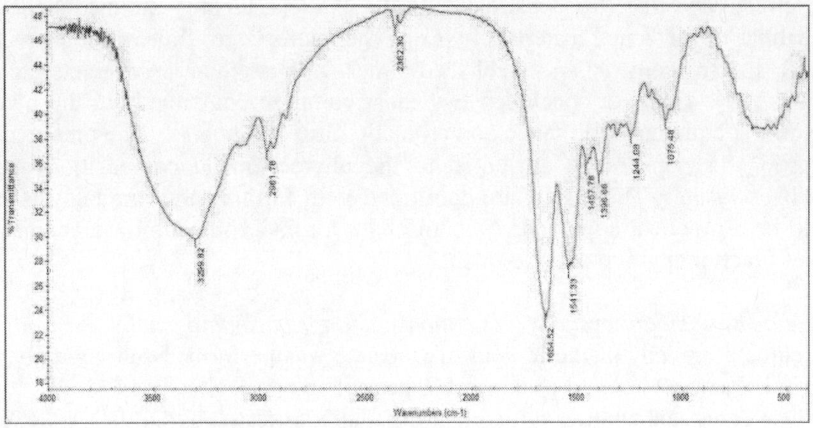

Figure 3. FTIR analysis of hen egg white protein

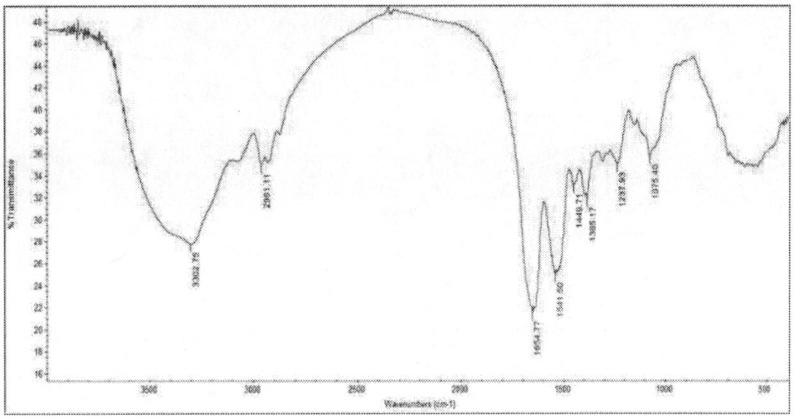

Figure 4. FTIR analysis of duck egg white protein

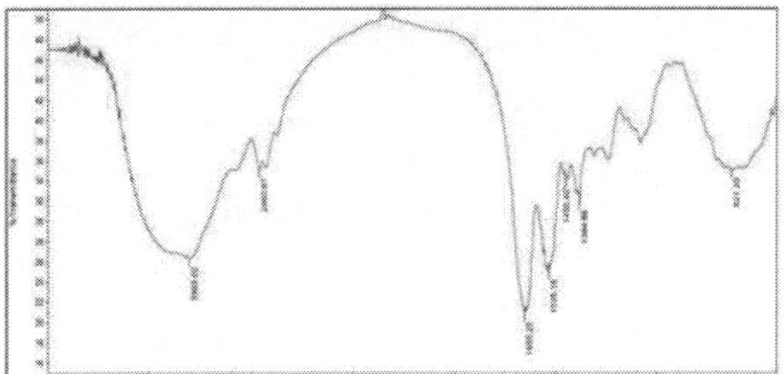

Figure 5. FTIR analysis of quail egg white protein

Figure 5 explain the range of FTIR analysis from different kind of hydrolysates. From figure 5 HEWPH range is 1396 cm^{-1} to 3299 cm^{-1}, DEWPH 1237cm^{-1} to 3302 cm^{-1} and QEWPH range 1384cm^{-1} to 3304cm^{-1}. We know that FTIR can explain chemical molecule structure in sample and chemical bonds in a molecule. Figure 5 was compared with sample which contain high antioxidant.

3.4.2. FTIR of HEWPH and DEWPH. The Fourier Transform Infrared (FTIR) spectroscopy is a valuable tool for characterization and identification compounds or functional groups (chemical

IOP Conf. Series: Earth and Environmental Science **102** (2017) 012099 doi:10.1088/1755-1315/102/1/012099

bonds). The spectroscopic analysis of polymeric molecules, including proteins, is complex due to the molecular vibrations arising from numerous atoms [13]. We can judge the origin of different extracts accurately and effectively, trace the constituents in the extracts, identify the medicinal materials and evaluate the qualities of medicinal materials by using the microscopes fingerprint characters of FT-IR spectrum,. FTIR spectroscopy is an established timesaving method to characterize and identify functional groups [14]. The data concluded that most common compounds in the hydrolysates are alcohol and hydroxyl compound, aliphatic and aromatic nitro compounds. The presence of the above mentioned functional groups might attributed to the phytochemical composition and antioxidant activity of the hydrolysates. The results are confirmed with Infrared Spectra analysis interpretation, which seems to be a practical approach. Alcohol and hydroxyl compound are chemical bonds that have potential as functional group of antioxidant.

3.5. *Separation antioxidant hydrolysates by ion exchange chromatography.*

Ion Exchange chromatography of the protein hydrolysate was performed subsequently. The result is two kind of HEWPH sample have III peak (peak I, peak II and peak III)

Every peak was separated and the antioxidant capacity was tested by DPPH. Antioxidant capacity peak of HEWPH is 58.50% 60.56% and 54.97% respectively.

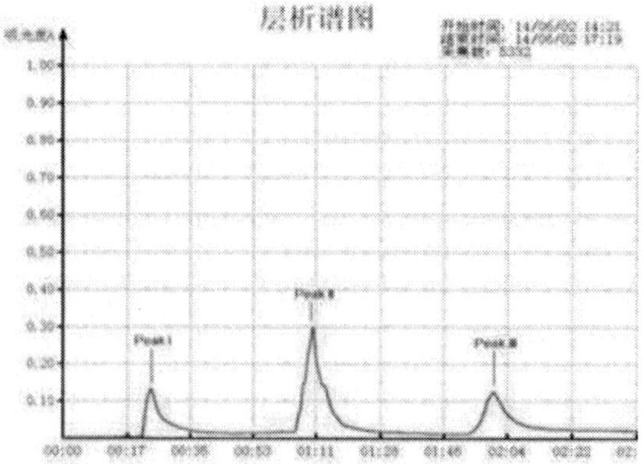

Figure 6. Antioxidant capacity and peak of HEWPH separation by ion exchage chromatography

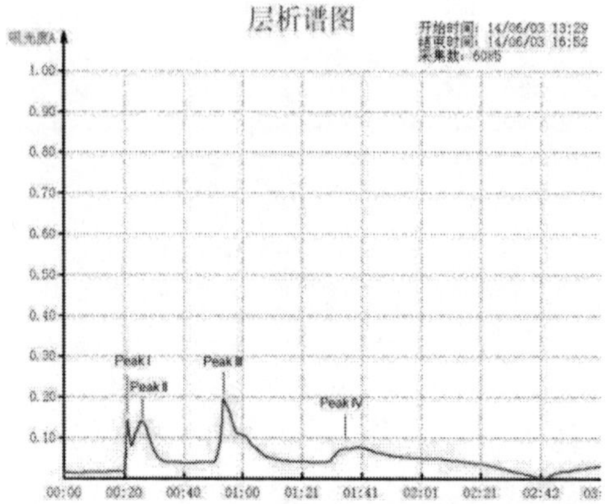

Figure 7. Antioxidant capacity and peak of DEWPH separation by ion exchange chromatography

Dissimilar with HEWPH sample, after separation, DEWPH have 4 peaks. Peak I and II bound to each other and peak III was the more larger one than others. The capacities of antioxidant that came from DEWPH was lower than HEWPH. DEWPH property of antioxidant capacity only have 32.23% on peak III and the another peak was the lowest. The antioxidant capacity of peak I, II and IV respectively is 24.87%, 22.66% and 19.28%. Antioxidant capacity of sample will be affected by different time hydrolysis of hydrolysate. Protease enzyme will affects the antioxidant activity, beside of time hydrolysis enzyme and also affects amount, size, composition of free amino acid and peptides and their amino acid sequence which in turn influences the antioxidant activity of the hydrolysates [5,15, 16]. In their research, [17] reported that different type of enzymes will produce hydrolysates from native and heated soy protein isolates. They reported that by using different type of enzymes generated in the formation of a mixture of peptides with different degrees of hydrolysis and accordingly different ranges of antioxidant activity.

Egg white proteins are found in food products and usually used as ingredients because it's functional nutrition and their hydrolysates that obtained by protease treatment are water soluble and have high nutritional value [18]. Some protein hydrolysates, such as egg white [19] potato (*Solanum tuberosum*) [4], fish muscle [20], and tuna liver [21] have been reported to take control of both ACE inhibitory capacities and antioxidant activity.

4. Conclusion

After separation of the antioxidant capacity from its hydrolysate showed different antioxidant activity and different peak. HEWPH showed unbound peak between three of peak while for DEWPH showed that two peak are bounded and one peak is unbounded. HEWPH remains the highest antioxidant capacity. HEWPH was able to inhibit free radical scavenge as 60.56% and the highest on DEWPH only 32.23%.

Acknowledgement

Department of International Cooperation and Exchange Office of Huazhong Agricultural University (Wuhan, PR China) and Master Program of Animal Science of Diponegoro University (Semarang, Indonesia) for the partial scholarship. Education Attache of the Embassy of The Republic Indonesia in Beijing to help on the Scholarship.

References

[1] Lin MY and Yen CL 1999 Antioxidative ability of lactic acid bacteria *J. Agric. Food Chem.* **47** 1460–66

[2] Wang YC,Yu RC and Chou CC 2006 Antioxidative activities of soymilk fermented with lactic acid bacteria and bifidobacteria *Food Microbiol* **23** 128–35

[3] Carlsen CU, Rasmussen KT, Kjeldsen KK, Westergaard P and Skibsted LH 2003 Pro and antioxidative activity of protein fractions from pork (longissimus dorsi) *Eur. J. Food Res. Technol.* **217** 195–200

[4] Pihlanto A, Sari A and Hannu JK 2008 ACE-inhibitory and antioxidan properties of potato (*Solanum tuberosum*) *J. Food Chemistry* **109** 104–12

[5] Chen HM, Muramoto K, Yamauchi F and Nokihara K 1996 Antioxidant activity of designed peptides based on the antioxidative peptide isolated from digests of a soybean protein *J. Agric. Food Chem.* **44** 2619–23

[6] Sakanaka S, Tachibana Y, Ishihara N and Juneja LR 2004 Antioxidant activity of egg yolk protein hydrolysates in a linoleic acid oxidation system *J. Food Chem.* **86** 99–103

[7] Desert C, Guerin-Dubiard C, Nau F and Jan G 2001 Comparison of different electrophoretic separations of hen egg white protein *J. Agric. Food Chem.* **49** No. 10

[8] Rajapakse N, Mendis E, Jung W, Je J and Kim S 2005 Purification of a radical scavenging peptide from fermented mussel sauce and its antioxidant properties *Food Research International* **38**: 175–82

[9] Shen S, Chahal B, Majumder K, You S J and Wu J 2010 Identification of novel antioxidative peptides derived from a thermolytic hydrolysate of ovotransferrin by LC–MS/MS *J. of Agricultural and Food Chemistry* **58** 7664–72

[10] Li XX, Han LJ and Chen LJ 2008 In vitro antioxidant activity of protein hydrolysates prepared from corn gluten meal *J Sci Food Agriculture* **88** 1660–6

[11] Moosman B and Behl C 2002 Secretory peptide hormones are biochemical antioxidants: structure-activity relationship *J. Molecular Pharmacology* **61** 260–8

[12] Park PJ, Jung WK, Nam KS, Shahidi F and Kim SK 2001 Purification and characterization of antioxidative peptides from protein hydrolysate of lecithin free egg yolk *J. Am. Oil Chem.*

*Soc.***78** 651–6

[13] Cooper EA and Knutson K 1995 Fourier transform infrared spectroscopy investigations of protein structure *Physical Methods to Characterize Pharmaceutical Proteins* vol 7 ed J N Herron *et al* (New York: Plenum) pp 101–43

[14] Grube M, Muter O, Strikauska S, Gavare M and Limane B 2008 Application of FTIR spectroscopy for control of the medium composition during the biodegradation of nitro aromatic compounds *J. of Indian Microbiology and Biotechnology* **35** 1545–49

[15] Jeon YJ, Byun HG and Kim S K 1999 Improvement of functional properties of cod frame protein hydrolysates using ultrafiltration membranes *J. Process Biochem* **35** 471–8

[16] Wu HC, Chen HM and Shiau CY 2003 Free amino acids and peptides as related to antioxidant properties in protein hydrolysates of mackerel (*Scomber austriasicus*) *Food Research International* **36** 949–57

[17] Penna-Ramos EA and Xiong YL 2002 Antioxidant activity of soy protein hydrolyzates in a liposomial system *J Food Sci.* **67** 2952–6

[18] Li-Chan ECY, Powrie WD and Nakai S 1995 The chemistry of eggs and egg products *Egg Science and Technology* 4th ed W J Stadelman and O J Cotterill (New York: Haworth Press)

[19] Davalos A, Miguel M, Bartolome B and Lopez–Fandino R 2004 Antioxidant activity of peptides derived from egg white proteins by enzymatic hydrolysis *J. Food Protect.***67** 1939–44

[20] Nakajima K., Yoshie-Stark Y and Ogushi M 2000 Comparison of ACE inhibitory and DPPH radical scavenging activities of fish muscle hydrolysates *J. Food Chemistry* **114** 844–51

[21] Je J, Park P and Kim S 2005 Antioxidant activity of a peptide isolated from alaska pollack (*Theragra chalcogramma*) frame protein hydrolysate *J. Food Research International* **38** 45–50

The comparison of fatty acid and cholesterol profile on fresh and mozarella cheese made by pampangan buffalo milk

H Rizqiati*, Nurwantoro and S Mulyani

Food Technology Department, Faculty of Animal and Agricultural Sciences, Diponegoro University, Semarang, Indonesia

*E-mail: heni.rizqi@gmail.com

Abstract. This research aimed to investigate the composition of fatty acid and cholesterol profile of Fresh and Mozarella Cheese from Pampangan Buffalo Milk. Material of this reseach was Pampangan buffalo milk and Mozarella cheese made from buffalo milk. Fatty acids composition were analyzed by [1] method. Result showed the major saturated fatty acid found in milk and Mozzarella cheese Pampangan buffalo milk were palmitic, stearic and miristic acid while the unsaturated fatty acid was oleic acid. The total amount of fatty acid in Mozarella cheese was lower than those in Pampangan buffalo milk.

Keyword: Buffalo milk, mozarella, fatty acids, cholesterol profile

1. Introduction

Buffalo milk was consumed by Pampangan's people, which is has a different nutrition with cow milk. [2] claimed that buffalo milk contain of 17.7% total solid, 7.3% fat, 5% protein, 4.6% lactose and 0.9% ash. [3] reported that buffalo milk has different compositions with cow's milk, because total solid, fat, protein, fat, lactose, mineral, and ash are higher than cow's milk. [4] explained that buffalo milk contained higher saturated fatty acids (palmitic and lauric) and lower mono-unsaturated fatty acid than cow milk in the same condition.

Milk fatty acid has been contributed in specific flavor, as a source of short chain-fatty acids (C6:0, C8:0, C10:0) which is made buffalo's milk become easy to absorb. Buffalo milk fatty acids produced energy and activated all of endocrine gland, organ, also body tissue without forming cholesterol and adipose tissue [5]. A higher of Medium chain-fatty acid in buffalo milk has a bacteriostatic effect.

The purpose of processing is to create or increase the flavor and nutrition, give an added value, produce a new product and extended the product's shelf-life. Buffalo's milk is a main material in mozarella cheese making because buffalo milk has a good nutrition. The making process of mozarella cheese is complex which is start with coagulation (K-casein was broke down by renet enzyme), aglomeration (produced similar protein mass) and water-released. During the process, fat maintained the membrane integrity and protein trapped in fat membrane also formed the cheese matrix [6]. [7] reported that Mozarella cheese has a 46.0% water content, 23.0% fat, 43.0% total solid fat and 1.2% salinity. In the making process of Mozarella cheese, fat degradated into fatty acids. Fatty acid in milk

IOP Conf. Series: Earth and Environmental Science **102** (2017) 012100 doi:10.1088/1755-1315/102/1/012100

in directly contributed to the aroma of cheese, however fatty acids have a role to form volatile compound through the metabolism of fatty acids [8].

This research aim to analyze total and type of fatty acid also cholesterol profile in fresh milk and mozarella cheese from Pampangan's buffalo milk.

2. Metodology

2.1. Material and Method

This research use milk and Mozzarella cheese from Pampangan's buffalo milk, obtained from Rambutan- Pampangan subdistric in Lampung. This research be held in Baranang Siang IPB Laboratory Bogor.

2.2. Experimental Procedure

This research has been analyzed the concentration and type of fatty acids, also cholestrol in milk and Mozzarella cheese from Pampangan's buffalo milk. Fatty acid analyzing is using gas chromatography through the procedure in IPB Laboratory which is following the [1].

2.3. Data Analysis

Fatty acid was identified by descriptive analyzed and compared with the previous research.

3. Result and Discussion

Total concentration of Pampangan buffalo milk's fatty acid and Mozarella cheese were recordered in Table 1. Total concentration of Fatty acid in Pampangan buffalo's milk (52.22%) is higher than Mozarella cheese (46.535%).

Table 1. Type and total unsaturated fatty acids of buffalo's milk and mozarella cheese

Fatty Acids	Concentration (%)	
	Buffalo Milk	Mozarella Cheese
Caproic. C6:0	1.31	0.86
Caprilic. C8:0	0.33	0.325
Capric. C10:0	0.65	0.65
Lauric. C12:0	0.96	0.95
tridekanoat. C13:0	0.14	0.125
miristic. C14:0	6.56	5.96
pentadekanoat. C15:0.	1.81	1.155
palmitat. C16:0	23.90	20.695
heptadekanoat. C17:0	1.03	0.93
stearat. C18:0	14.37	14.28
arakidat. C20:0	0.23	0.225
heneikosanoat. C21:0	0.09	0.12
behenat. C22:0	0.52	0.115
trikosanoat. C23:0	0.13	0.145
lignoserat. C24:0	0.15	0.86
Total	52.22	46.535

The final result shown some type of unsaturated fatty acids in fresh milk and Mozarella cheese from Pampangan's buffalo milk, such as lauric, miristic, palmitic, stearic, caproic, caprilic, capric, tridekanoat, arakidat, heneikosanoat, behenat, trikosanoat dan lignoseratrasi in different concentration. The highest concentration of saturated fatty acid in fresh milk and Mozarella cheese are palmitic, stearic, and miristic. [4] reported that buffalo's milk fatty acid contained of stearic and palmitic which is contributed to specific taste and aroma. [9] explained the highest concentration of fatty acid in buffalo's milk are palmitic, miristic and stearic acid. Saturated fatty acid's percentage in buffalo's milk is higher than cow's milk [10].

Miristic fatty acid (C14:0) has a role in increasing the *High Density Lipoprotein* (HDL) and *Low Density Lipoprotein* (LDL), antioxidant, prevent the oxidation of LDL and protect from microbe infection [11].

Buffalo milk's short chain fatty acids produced energy and activated all of endocrine gland, organ, also body tissue without forming cholesterol and adipose tissue [5]. Milk's fatty acid composition influenced by factors, such as lactation periods [12], type of buffalo [4] and feed composition [13]

In the making process of cheese involved curd forming with milk acidifocation and protease addition, so the process resulted the positive respond of fatty acid in milk and Mozarella cheese from buffalo milk. Milk fatty acid produced by lipolysis of fatty acid [8]. In this research, total fatty acid of Mozarella cheese is lower than Pampangan buffalo's milk. [5] reported that lypolisis not always increase the certain fatty acid. Lypolisis is chemical reaction which is produced the flavor of milk product. Therefore, the primer total of milk's fatty acid influenced the total fatty acid during the making process of Mozarella cheese.

Table 2. Type and Total Unsaturated Fatty Acids in Milk and Mozarella Cheese From Pampangan's Buffalo Milk

Fatty Acid	Buffalo Milk	Mozarella Cheese
miristoleinat. C14:1	0.04	0.355
palmitoleinat. C16:1	1.91	1.615
oleic. C18:1n6c	23.37	21.395
linoleat. C18:2n6c	0.76	0.725
linolenat. C18:3n3	0.33	0.02
Cis-11-Eikosenoat. C20:1	-	0.36
Cis-11-14- Eikosedienoat. C20:2	-	0.07
Cis -8.11.14 Eicosetrinoar . C20:3n6	-	0.035
arakidonat. C20:4n6	0.04	0.06
Cis-5.8.11.14.17 Eikosapentaenoat. C20:5n3.	0.04	
EPA		0.05
nervonat. C24:1	0.05	0.035
Total	26.91	24.72

Unsaturated fatty acid which is detected in fresh milk and Mozarella cheese from Pampangan's buffalo milk are miristoleic, palmitoleic, oleic, linoleic, linolenat, eikosanoat, arakidonat and eikosapentanoat shown in Table 2. Concentration of unsaturated fatty acid in Mozarella cheese is lower than Pampangan's buffalo milk. [8] explained that milk's processing did not influence the fatty acid's composition. Unsaturated fatty acid in food materials related to food's spoilage. Unsaturated fatty acid not easily oxidized.

There is a tendency of decreasing in cholesterol during processing of Pampangan's buffalo milk into Mozarella cheese. The result shown cholesterol in milk and Mozarella cheese are 1.96% and 1.04. Fat's composition which is influenced the cholesterol is lypolisis. [6,9] explained that lauric acid (C12:0), palmitic acid (C16:0) and miristic acid (C14:0) increased the serum cholesterol. Oleic acid (C18:1), linoleic acid (C18:3), and α-linoleic acid (C18:3) decreased the serum cholesterol.

4. Conclusion

Pampangan buffalo milk has a higher fatty acid than mozarella cheese. Type of fatty acid which is identified in milk and mozarella has a similiarity, however the concentration is different. There was a tendency of decreasing in cholesterol during buffalo milk processing into mozarella cheese.

Acknowledgments

Our research is funded by The Directorate of Research and Devotion to community of The Ministry of Higher Education and Culture, through the BOPTN 2014.

References

[1] AOAC 2007 Official Methods of Analysis of AOAC International 18th ed. Assoc. Off. Anal. Chem., Arlington
[2] Han XF, Lee L, Zhang L, Guo MR 2012 Chemical composition of water buffalo milk and its low-fat symbiotic yogurt development *Functional Food in Health and Disease* **2** (4) 86–106
[3] Paulina G, Nudda A. 2004. Milk Production *Dairy Sheep Nutrition* Wallingford ed Pulina G and Bencini R ((GB): CABI Publishing) pp 1-12
[4] Talpur FN, Memon N N and Bhanger M I 2007 Comparison of fatty acid and cholesterol content of Pakistani water buffalo breeds *Pak. J. Analytical and Environ. Chem.* **8** 15
[5] Murray M 2000 Mechanisms of inhibitory and regulatory effects of methylenedioxyphenyl compounds on cytochrome P450-dependent drug oxidation *Curr Drug Metab* **1** 67–84
[6] Pereira CI, Gomes AMP, Malcata XF 2009 Microstructure of cheese: Processing, technological and microbiological considerations Review *Trends Food Sci Technol* **20** 213–19
[7] McMahon DJ 2007 Product Specifications : Mozzarella Cheese Specification No. 603. McMahon Food Global Marketers.
[8] Bisig W, Eberhard P, Collomb M and Rehberger B 2007 Review - Influence of processing on the fatty acid composition and the content of conjugated linoleic acid in organic and conventional dairy products – a review *Lait* **87** 1–19
[9] Mihaylova G and Peeva T 2007 Tran's FA and conjugated linoleic acid in the buffalo milk. *Italian J. of Animal Sci.* **6** 1056–59
[10] Adnan M 1995 Food fat and problems. Gadjah Mada University, Yogyakarta. In Zulaikhah S R 2001 Characteristics of cheese from different starter concentration and long curing with abomasum sheep (Thesis) Master Programme Gadjah Mada University, Yogyakarta
[11] German JB and Dillar CJ 2004 Saturated fats: what dietary intake? *Am. J. Clin. Nutr.* **80** 550–59
[12] Qureshi MS, Mushtaq A, Khan S, Habib G and Swati ZA 2007 Variation in milk fatty acids composition with body condition in dairy buffaloes *(Bubalus bubalis. Asian Aust J Anim Sci.***23**: 340–345

[13] Patiño EM, Judis MA, Guanziroli CS, Pochon DO, Cedres JF and Doval MM 2008 Determination of milk fatty acids in buffaloes (Bubalus bubalis) from Corientes, Argentina *Revista Veterinaria.* **19** 28–32

Evaluation of resistant starch, glycemic index and fortificants content of premix rice coated with various concentrations and types of edible coating materials

W A Yulianto*, A M Susiati and H A N Adhini

Faculty of Agro-industry, University of Mercu Buana Yogyakarta, Jl. Wates Km 10, Argomulyo, Bantul, Yogyakarta 55753

*E-mail: wisnuadi@mercubuana-yogya.ac.id

Abstract. The incidence of diabetes in Indonesia has been increasing year by year. Diets with a low glycemic index and high resistant starch foods can assist diabetics in controlling their blood glucose levels. Diabetics are known to have micro-nutrient deficiencies of chromium, magnesium and vitamin D that can be overcome by consuming parboiled rice fortified by use of a coating method. The fortification of parboiled rice (premix rice) can be achieved by coating with HPMC (hydroxypropyl methyl cellulose), MC (methyl cellulose), CMC (carboxyl methyl cellulose), gum arabic and rice starch. This research aimed to evaluate the levels of resistant starch, glycemic index and fortificants of premix rice coated with different concentrations and types of edible coating materials. This research used completely randomized design, with treatments to the concentrations and the types of edible coating (HPMC, CMC, MC, gum arabic and rice starch). The concentrations of edible coating were 0.15%, 0.2% and 0.25% for cellulose derivative coatings; 25%, 30%, 35% for gum arabic and 2%, 3.5% and 5% for rice starch. This research shows that fortified premix rice coated with various concentrations and types of edible coating materials is high in resistant starch and has a low glycemic index. The coating treatment affects the levels of magnesium and vitamin D, but does not affect the levels of chromium in parboiled rice. The premix rice with a low glycemic index and high nutrient content (chromium, magnesium and vitamin D) was premix rice coated by CMC 0.25% and HPMC 0.25% with glycemic indeces of 39.34 and 38.50, respectively.

Keywords: *coating material, fortificant, glycemic, rice.*

1. Introduction

With 154.062 diabetes mellitus patients, Indonesia takes fourth place behind China (1,023,504 sufferers), India (760,429 sufferers) and the United States (223,937 patients) [1]. Based on the results of basic health research done in 2013, the prevalence of diabetics in Indonesia reached 6.9% of the population, the proportion with impaired glucose tolerance 29.9% and disturbed fasting blood sugar 36.6% [2].

Diabetics generally limit their consumption of rice based on the assumption that rice is a food that has a high glycemic response. However, not all rice is hyperglycemic [3]. Research done by [4] showed that traditional parboiled rice has a GI of 46 and can significantly lower the blood sugar profile in type 2 diabetes compared to eating non-parboiled rice (GI

55). Meanwhile, it has been reported that people with diabetes have a deficiency of chromium [5], a magnesium deficiency [6], and a vitamin D deficiency [7-8] as well. Such nutrient deficiencies can lead to an increase in blood sugar. Attempts to produce rice with a low GI and fortified with chromium have helped with success in developing the fortification process of parboiling rice with chromium [9] through the improvement of the process. Despite producing a low enough IG (36.33) and gastrointestinal starch of 11.88% (bk), the sensory test results by the panellists showed their favoured levels ranged from 'less preferred to preferred'. To improve consumer preferences a food aroma booster could be added to the rice.

Among the ingredients commonly used in food aroma boosters, pandan leaf extract is the most emphasized. The main compounds responsible for the fragrance or fragrant aroma of pandanus are: 2-acetyl-1-pyrolyine (2AP). These compounds are also found in scented rice (aromatic rice) [10-12]. Aside from being a food aroma booster, pandanus also has medical uses as a diuretic, cardio-tonic, and an anti-diabetic [13].

The addition of pandan extract and micronutrients (Cr and Mg) to rice can be done by coating (coating). Edible coating materials which can be used for coating the rice are: methyl cellulose (MC) and hydroxypropyl methyl cellulose (HPMC). Such materials, and the combination of both, have been successfully used as an edible coating material in iron fortified rice [14]. As a result, panellists were found to favour iron fortified rice which had a combination of HPMC and MC (3: 1) by 2%. Similarly, it was reported by [15], that the maximum retention of micronutrients (vitamins and minerals) in premixed rice occurs when coated with HPMC and MC combinations (3: 1) with 100% iron retention. [12] applied a rice coating with 5% rice starch and 30% sorbitol used for enriching the aroma of pandan leaf extract. Choi et al. [16] coated the rice with resistant starch type 4 to decrease the digestibility and glucose response in experiments with mice. Meanwhile, gum arabic coating at a concentration of 40% and with 20% spice extract was the best treatment for fast rice coating [17].

This study aimed to evaluate the levels of resistant starch, glycemic and fortificants (chromium, magnesium and vitamin D) and to determine the best coating type and level of premix rice fortified with chromium, magnesium and vitamin D.

2. Materials and method.

2.1. Production of Parboiled Rice

A 10 kg of Ciherang variety rice was washed and disposed. It was then soaked in 15 litres of water at a temperature of 65 $^{\circ}$C ± 2 $^{\circ}$C for 2.5 hours, after which it was drained and steamed for 25 minutes, then cooled at 0 $^{\circ}$C for 6 hours and finally dried. The drying process was done using a cabinet dryer until it had a moisture content of 13-14% (wet weight) at a temperature of 50 $^{\circ}$C. The measurement of the water was conducted using a moisture tester.

2.2. Coating of Parboiled Rice

2.2.1. Preparation of pandanus extract. Pandan leaf extracts were prepared using the modified Al-Jamal and Rasheed [18] methods. A total of 500 g of pandan extract that had been previously destroyed was soaked in 1500 ml of hot water (88 $^{\circ}$C) in a water bath for 6 hours. Furthermore, filtering, using Whatman paper no. 41, was carried out and finally a rotary vacuum concentration at a temperature of 60 $^{\circ}$C to obtain 150 ml of the herbal extract.

2.2.2. Preparation of coating materials. The parboiled rice coating was made using a variety of edible coating materials, namely HPMC (hydroxypropyl methyl cellulose), MC (methyl cellulose), CMC (carboxyl methyl cellulose): 0.15%, 0.2% and 0.25%), gum arabic (25%, 30% and 35%) and rice starch (2%, 3.5% and 5%). A composition of the coating mixture of cellulose derivatives was used as in [15] with slight modification, consisting of 5% CMC / MC / HPMC, 28.5% ethanol (95%), and 66.7% water for 5% edible film content.

2.2.3. Coating of Parboiled Rice. 500 g of parboiled rice was coated with polymer-nutrients (Cr, Mg and vitamin D). The levels of daily upper limit of intake for adults were 50 mg for vitamin D and 350 mg for magnesium [19], while for Cr the daily limit of safe intake was 50-200 mg per day for adults [20]. This quantity was used as a reference in the parboiled rice fortification. To make 500 g of parboiled rice coating requires Cr solution: 5.6 ml or 112 µg of chromium, 13 ml or 780 mg of magnesium, and 4.5 ml or 900 IU of vitamin D (multi-nutrient solution).

The use of coatings from cellulose derivatives, multi-nutrient coating on parboiled rice was worked out with polymer solution gradually. Each finished coating was dried at a temperature of 40-50 °C for 5-10 minutes to evaporate the solvent. While multi-nutrient coating with polymer solution (sorbitol 30%) in parboiled rice with gum arabic and rice starch was done in one stage, followed by a drying process.

2.3. Analysis of product and data

The premix rice obtained was analysed for glycemic starch (RS) using a modified Englyst method [21], vitamin D content by HPLC method, the Cr and Mg concentration were measured by the AAS method [22], and glycemic index testing was done using 16 healthy and non-diabetic volunteers. The design of the experiment was a complete randomized design (RAL). The results obtained were analysed with the variance (ANOVA) at 95% confidence level. If the difference was wide - each treatment was continued by Duncan Multiple Range Test. The treatment used was 15 treatments with 2 replications. After an overnight fast (10 hours), every 2 volunteers were asked to consume 1 type of cooked rice sample from the 15 products produced and afterwards, 3 volunteers were asked to consume bread as standard. The number of samples eaten is equivalent to 50 grams of carbohydrates.

3. Results and Discussion

3.1. Resistant starch

The high resistant starch (RS) content of premix rice (not significantly different), in the range of 7.85 - 10.56%, occurred in coating treatments with MC at all concentrations, 0.25% CMC coating, HPMC 0.25 %, and starch coating at concentrations of 3.5% and 5% (Table 1). Meanwhile, coating with gum arabic produced relatively low levels of RS at all concentration levels. MC coating was considered good enough to coat the premix rice when considered for its RS levels, because a low concentration level was able to produce high levels of BP-resistant starch digestibility. Previous research suggests that the resistant starch content of parboiled rice fortified with Cr increased from 9.57% to 11.99% [9]. [23] reported RS levels in Ciherang rice varieties of 1.78%. Resistant starch content of instant rice with heat moisture treatment process ranged from 7.63% to 10.18%, an increase in resistant starch content of 8.86 to 54.65% [24].

The RS that is present in the premix rice is included in type RS 3. The RS 3 is the most commonly found and is a starch fraction generally as amylose retrogradation during the cooling process of starch gelatinization [25]. The additional cooling of the *parboiling* process in this study provided opportunities for starch retrogradation. In general, the higher the level of amylose starch, the higher the content of the reistant starch and slow to digest [9]. The degree of retrogradation depends on the length of time and temperature of storage, starch sources, and other molecules contained in the system. Retrogradation that occurs in amylose is irreversible at temperatures of less than 100 ° C [26], because amylose crystals melt at temperatures above 100 ° C. It was reported by Guraya et al. [25], that a cooling process at a temperature of 1 °C for 12 hours without agitation against the non-waxy starch solution which had been cut with pullulanase ramifications brought the digestion ratio down 59%. RS has functions such as being a dietary fibre that has a hypoglycemic effect, acting as a pre-biotic, reducing the risk of gallstone formation, having a hypocholesterolemic effect, inhibiting fat accumulation acids, especially butyric acid [27-29].

Table 1. Resistant starch content of premix rice fortified with chromium, magnesium, vitamin D and pandani extract using various types and concentrations of coating materials.

Treatment		Resistant starch (%)
Coating Type	Concentration (%)	
MC	0.15	9.27 [d]
MC	0.20	10.56 [d]
MC	0.25	7.85 [bcd]
CMC	0.15	4.56 [a]
CMC	0.20	5.92 [abc]
CMC	0.25	9.86 [d]
HPMC	0.15	4.74 [a]
HPMC	0.20	5.58 [ab]
HPMC	0.25	10.51 [d]
Rice starch	2.00	3.75 [a]
Rice starch	3.50	8.43 [cd]
Rice starch	5.00	10,18 [d]
Arabic gum	25.00	5.80 [abc]
Arabic gum	30.00	5.03 [a]
Arabic gum	35.00	4.75 [a]

Note: The number followed by different letter indicates significant difference (P<0,05)

3.2. Glycemic index

The glycemic index values were calculated based on the area under the curve of each volunteer's blood sugar response in each of the premix rice treatments shown in Table 2.

Based on data analysis of blood sugar levels in Table 2, it can be seen that the IG value of rice produced premix includes rice with a low IG value, because the range of IG values of premix rice is 31-54. Food with an IG index of less than 55 includes low-IG foods. Previous research reported *parboiled* rice modified with a Cr fortification of 28.13 g per 100 g of $CrCl_3$ and Cr picolinate has GI of 25 and 26 respectively [30].

Low GI value of *parboiled* rice is affected by processing factors. Cooking starch at high temperatures for extended periods of time, and re-cooling, can cause changes in the structure of the dissolved starch granules to form degraded starch. This starch is insoluble and difficult to digest *(resistant starch)*. The amount of RS in most crude products is generally very low, but processing and storage can cause an increase [31].

Based on the value of IG, the premix rice with a coating treatment of gum arabic has a low GI, despite having relatively low levels of RS. This can be due to gum arabic being included in the 'soluble dietary fibre' type . The presence of dietary fibre can affect blood glucose levels. In general, high dietary fibre contents contribute to low IG values. Pectin and gum belong to the water soluble fibre type and determine the viscosity of dietary fibre. Therefore, the soluble function of dietary fibre is needed by diabetes mellitus patients because it can reduce the absorption of glucose in the small intestine [32].

Table 2. Glycemic index of premix rice fortified with chromium, magnesium, vitamin D and pandanus extract using various types and concentrations of coating materials.

Treatment		Glycemic Index
Coating Type	Concentration (%)	
MC	0.15	33.39 [a]
MC	0.20	31.22 [a]
MC	0.25	33.43 [a]
CMC	0.15	52.68 [b]
CMC	0.20	33.93 [a]
CMC	0.25	33.34 [a]
HPMC	0.15	54.46 [b]
HPMC	0.20	53.43 [b]
HPMC	0.25	38.50 [a]
Rice starch	2.00	53.44 [b]
Rice starch	3.50	40.17 [a]
Rice starch	5.00	37.29 [a]
Arabic gum	25.00	34,56 [a]
Arabic gum	30.00	36.60 [a]
Arabic gum	35.00	38.29 [a]

Note: The number followed by a different letter indicates significant difference ($P<0,05$)

3.3. Fortificants

3.3.1. Chromium. Fortification is the process of adding vitamins, minerals and amino acids to food products so as to improve the nutritional quality of food intake and provide health benefits with minimal risk. Table 3 shows that premix rice with various types and concentrations of coating materials did not affect the chromium content of the resulting premix rice, which ranged from 0.07 to 0.26 mg / kg. Research conducted by Yulianto et al. [9] reported that the parboiled rice obtained from a paddy soaking process with $CrCl_3$ levels of 7.47 mg / l produced rice with a 36.33 glycemic index and a total Cr content of 0.56 mg / kg. From the results of this study, it can be seen that the fortification of the coating is good enough to embed the source of Cr in parboiled rice.

Chromium is important in the control of blood sugar. The action of chromium is demonstrated by the activation of small peptides and increases to the action of insulin in the binding of insulin receptors, which causes increases in glucose or blood sugar entering the cells [19]. Adequate daily intake of chromium for men is 25-26 μg per person per day, while for women it is 19-30 μg per person per day [33]. Based on this nutritional adequacy rate, the Cr content of premix rice is much more than the adequate amount, so for this premix rice, dilution needs to be performed to meet the needs for daily chromium intake. From the analysis, data can be taken to show chromium content levels of 0.07 - 0.26 mg/1000 kg of rice, or 12.6-46.8 μg/ 180 g of rice (180 g of rice being estimated daily consumption), so the premix rice can be mixed up to 9.36 times to meet the intake requirement level of 5 μg chromium (20% of daily chromium requirement) as a fortified food.

Table 3. Chromium (mg/kg), magnesium (mg/kg) and vitamin D (mg/100 g) grade of premix rice using various types and concentrations of coating materials.

Treatment		Cr	Mg (mg/kg)	Vitamin D
Type of coating	Concentration (%)	(mg/kg)		(mg/100 g)
MC	0.15	0.21	1326.32 ^{abc}	0.19 ^{a}
MC	0.20	0.18	1348.89 ^{abc}	0.26 ^{a}
MC	0.25	0.21	1436.47 ^{c}	0.28 ^{a}
CMC	0.15	0.15	1113.94 ^{a}	2.04 ^{b}
CMC	0.20	0.12	1451.99 ^{c}	1.89 ^{b}
CMC	0.25	0.17	1322.18 ^{abc}	3.63 ^{d}
HPMC	0.15	0.14	1373.48 ^{bc}	2.01 ^{b}
HPMC	0.20	0.14	1275.09 ^{abc}	2,48 ^{c}
HPMC	0.25	0.10	1304.83 ^{abc}	3.58 ^{d}
Rice starch	2.00	0.11	1286.67 ^{abc}	0.25 ^{a}
Rice starch	3.50	0.11	1285.23 ^{abc}	0.24 ^{a}
Rice starch	5.00	0.09	1168.87 ^{ab}	0.27 ^{a}
Arabic gum	25.00	0.07	1303.93 ^{abc}	0.21 ^{a}
Arabic gum	30.00	0.11	1265.25 ^{abc}	0.20 $_{a}$
Arabic gum	35.00	0.26	1214.18 ^{abc}	0.25 ^{a}

Note: The number followed by different letter indicates significant difference (P<0,05)

3.2.2. Magnesium From Table 3 it can be seen that the use of different concentrations of the same type of coating material did not make any significant difference to the magnesium content of multi-nutrient fortified parboiled rice, except on the type of CMC coatings which showed significant differences at concentration levels of 0.15% and 0.2%. The results of this study showed that fortification coating is good enough to attach a source of magnesium to parboiled rice.

In the fortification of food, fortificant levels (Cr, Mg and vitamin D) are not allowed to surpass the limit of 20% RDA *(recommended daily dietary allowances)* per serving [34]. The intake for the mineral Mg for men = 148 mg / person / day and women 155 mg / person / day [33]. The magnesium content of parboiled rice produced, ranged from 1100-1450 mg/kg. From the analysis, data can be taken to show magnesium levels of 1100-1450 mg / kg of rice or 198-261 mg /180 g of rice, the premix rice can be mixed with ordinary rice at 1.32 to 1.74 times to meet the daily needs of 148-155 mg of magnesium, or 6.6 to 8.7 times to satisfy the requirement of 29.6-31 mg of magnesium as a fortified food.

The importance of sufficient magnesium intake, especially in individuals with diabetes mellitus, can be attributed to its role in the homeostatic maintenance of blood glucose along with the activation of factors involved in insulin sensitivity [19].

3.3.3. Vitamin D Table 3 shows that the use of different concentrations of the same type of coating material did not make any significant difference to the levels of vitamin D of the multi-nutrient fortified parboiled rice, except on the types of CMC and HPMC coatings which showed significant differences. Parboiled rice coating treatment with MC coating materials, starch and arabic gum containing vitamin D, were not significantly different. Vitamin D content in these three types of coatings ranged from 0.19 to 0.28 mg / 100 g or less than 3 mg/100g.

The requisite vitamin D intake, according to the RDA, is 5 mg / day for adults (1 mg of cholecalciferol = 40 IU of the vitamin), while the upper daily intake limit for adults is 50 mg or

200 IU for vitamin D [19]. In Indonesia, the adequacy of vitamin D in children / infants is 5 - 15 mg / person / day, men 15 - 20 mg / person / day, women 15 - 20 mg / person / day [33]. The highest vitamin D levels in premix rice were 3.6 mg / 100 g of rice or 6.48 mg / 180 g of rice (estimated daily rice intake). From these data, it can be seen that fortification coating is not good enough as an embedded source of vitamin D in parboiled rice. This can be due to several reasons, one of which is because vitamin D is insoluble in water, while the coating material used in this research is soluble in water. In addition, vitamin D tends to be sensitive to heat, while the premix - parboiled rice coating process uses heat to a temperature of 50°C.

4. Conclusion

Based on the results of this research, it was concluded that fortified premix rice coated with various concentrations and types of edible coating materials has high resistant starch and low glycemic index. Coating treatment affects the levels of magnesium and vitamin D, but does not affect the levels of chromium in parboiled rice. The premix rice with a low glycemic index and high nutrients (chromium, magnesium and vitamin D) was premix rice coated with CMC 0.25% and HPMC 0.25% producing glycemic indeces of 39.34 and 38.50, and resistant starch contents of 9.86% and 10.51%, respectively.

Acknowledgements

The author would like to thank the Directorate General of Higher Education, Ministry of Education and Culture, Indonesia, which has provided research funding through the National Strategic Research Program.

References

[1] International Diabetes Federation 2013 IDF Diabetes Atlas
[2] Health ministry of the republic of Indonesia 2013 http://www.depkes.go.id/resources/download/pusdatin/infodatin/infodatin-diabetes.pdf
[3] Widowati S 2008 *Majalah Pangan* **177** 52 51-60
[4] Larsen HN, Rasmusse OW, Rasmussen PH, Alstrup KK, Biswas SK, Tetens I, Thilsred SH and Hermansen K 2000 *EJCN* **54** 380–385
[5] Anderson RA 2008 *Proc. of Nutrition Society* (University of Ulster: Coleraine) **67** 48–53
[6] Dong J, Xun P, He K, Qin L 2011 *Diabetes Care* **34** 9 2116– 2122
[7] Driver JP, Foreman O, Mathieu C, van Etten E and Serreze D V 2008 *Clin Exp Immunol.* **151** 76-85
[8] Whiting SJ, Calvo MS 2006 *J Nutr.* **136** 1114-6
[9] Yulianto WA, Susiati AM and Slamet A 2012 Report of National Strategic Research, University of Mercu Buana Yogyakarta
[10] Paule CM and Powers JJ 1989 *J. Food Sci.* **54** 343–346
[11] Tulyathan V, Srisupattarawanich N and Suwanagul A 2010 www.thaiscience.info.
[12] Laohakunjit N and Kerdchoechuen O 2006 *Food Chem.* www.elsevier.com.locate.foodschem.
[13] Wakte KV, Thengane RJ, Jawali N and Nadaf AB 2010 *Food Chem.* **121** 595–600
[14] Mridula D and Pooja J 2011 *International Journal of Food Science and Technology* **49** 246–252
[15] Peil A, Barrett F, Rha C and Langer R 1982 *J. Food Sci.* **47** 260–262.
[16] Choi HJ, Lee CJ, Cho EJ, Choi S and Moon TW 2010 *International Journal of Food Science and Technology* **45** 2512-2621.
[17] Suryani Ch L and Slamet A 2012 *Agritech* **32** 1 44-50
[18] Al-Jamal A and Rasheed IN 2010 *African Journal of Food Science* **4** 9 615 – 617

[19] Smolin LA and Grosvenor M B 2007 Nutrition: Science and Applications (Orlando: Sounders College Publishing)
[20] Krejpcio Z 2001 *Polish Journal of Environmental Studies* **10** 6 399 -404
[21] Englyst HN, Kingman SM and Cummings J H 1992 *European Journal of Clinical Nutrition* **46** S33-S50
[22] AOAC 1995 Official Standard of Analysis of OAC International, 16th edition AOAC International, Arlington, Virginia.
[23] Widowati S, Astawan M, Muchtadi D and Wresdiyati T 2007 Prosiding Seminar Nasional PATPI Bandung 18 Juli 2007 975-987
[24] Suryani Ch L and Slamet A 2014 Prosiding Seminar Nasional Nutrisi, Keamanan Pangan dan Produk Halal Surakarta.
[25] Guraya H S, James C and Champagne ET 2001 *InterScience* **53** 2: 64-74.
[26] Fitzgerald, M 2004 Starch *Rice: Chemistry and Technology* ed Champagne E T (Minnesota: America Association of Central Chemists Inc)
[27] Matsumoto K, Yasuyoshi E, Nishi K, Honda Y, Nakaya M, Kitamura S 2016 *Journal of Func Food* **25** 556-564
[28] Srikaeo K and Janya S 2014 *Food Sci and Tech.* **50** 1129-1135.
[29] Topping DL, Balazs HB, Anthony RB, Julie MC, Lynne C, Michael AC, Matthew KM, Shusuke T 2008 *Microbial Ecology Health and Disease* **20** 103–108
[30] Yulianto WA, Susiati AM, Slamet A 2014 Report of National Strategic Research, University of Mercu Buana Yogyakarta.
[31] Marsono Y 2002 *Agritech.* **19** 124-127
[32] Sadersai VM 2003 *Introcution to Clinical Nutrition (New York:* Marcel Dekker Inc.) pp 339-354
[33] Regulation of the Minister of Health of the Republic of Indonesia No 75 of 2013 on Nutritional Sufficiency Recommended for the Indonesian Nation. 2013
[34] Muchtadi D 2006 *Food Review Indonesia* **1** 7 Agustus 48-50